RSMeans®

Plumbing Cost Data

24th Annual Edition

2001

First Printing

Foreword

R.S. Means Co., Inc. is a subsidiary of CMD Group (Construction Market Data), a leading provider of construction information, products, and services in North America and globally. CMD's flagship product line includes more than 100 regional editions, national construction project information, sales leads, and over 70 local plan room/libraries in major business centers. CMD also publishes ProFile, a directory of more than 20,000 architectural firms in the U.S. Architect's First Source (AFS) provides Publishers First Source for Products, CSI's SPEC-DATA® and CSI's MANU-SPEC® building product directories for the selection of nationally available building products. R.S. Means provides Construction cost data and training, and consulting services in print, CD-ROM and online. Associated Construction Publications (ACP) provides regional magazines covering the U.S. highways heavy construction, and heavy equipment trader. Manufacturers' Survey Associates (MSA) provides surveys, plans, and specifications. CMD Group is headquartered in Atlanta and has 1,400 employees worldwide. CMD Group is owned by Cahners Business Information (www.cahners.com), a member of the Reed Elsevier plc group (NYSE: RUK and ENL). Cahners is the leading U.S. provider of business information to 16 vertical markets, including building & construction, entertainment, manufacturing, and retail. Cahners' content portfolio encompasses 140 Web sites as well as *Variety, Publishers Weekly, Design News* and 127 other market-leading business-to-business magazines. Cahners developed the leading business-to-business Web portals e-inSITE in the electronics industry, Manufacturing.net in manufacturing as well as Buildingteam.com and HousingZone.com in construction, and maintains an active program of Internet and print launches.

Our Mission

Since 1942, R.S. Means Company, Inc. has been actively engaged in construction cost publishing and consulting throughout North America.

Today, over 50 years after the company began, our primary objective remains the same: to provide you, the construction and facilities professional, with the most current and comprehensive construction cost data possible.

Whether you are a contractor, an owner, an architect, an engineer, a facilities manager, or anyone else who needs a fast and reliable construction cost estimate, you'll find this publication to be a highly useful and necessary tool.

Today, with the constant flow of new construction methods and materials, it's difficult to find the time to look at and evaluate all the different construction cost possibilities. In addition, because labor and material costs keep changing, last year's cost information is not a reliable basis for today's estimate or budget.

That's why so many construction professionals turn to R.S. Means. We keep track of the costs for you, along with a wide range of other key information, from city cost indexes . . . to productivity rates . . . to crew composition . . . to contractor's overhead and profit rates.

R.S. Means performs these functions by collecting data from all facets of the industry, and organizing it in a format that is instantly accessible to you. From the preliminary budget to the detailed unit price estimate, you'll find the data in this book useful for all phases of construction cost determination.

The Staff, the Organization, and Our Services

When you purchase one of R.S. Means' publications, you are in effect hiring the services of a full-time staff of construction and engineering professionals.

Our thoroughly experienced and highly qualified staff works daily at collecting, analyzing, and disseminating comprehensive cost information for your needs. These staff members have years of practical construction experience and engineering training prior to joining the firm. As a result, you can count on them not only for the cost figures, but also for additional background reference information that will help you create a realistic estimate.

The Means organization is always prepared to help you solve construction problems through its five major divisions: Construction and Cost Data Publishing, Electronic Products and Services, Consulting Services, Insurance Division, and Educational Services.

Besides a full array of construction cost estimating books, Means also publishes a number of other reference works for the construction industry. Subjects include construction estimating and project and business management; special topics such as HVAC, roofing, plumbing, and hazardous waste remediation; and a library of facility management references.

In addition, you can access all of our construction cost data through your computer with Means CostWorks 2001 CD-ROM, an electronic tool that offers over 50,000 lines of Means detailed construction cost data, along with assembly and whole building cost data. You can also access Means cost information from our Web site at www.rsmeans.com

What's more, you can increase your knowledge and improve your construction estimating and management performance with a Means Construction Seminar or In-House Training Program. These two-day seminar programs offer unparalleled opportunities for everyone in your organization to get updated on a wide variety of construction-related issues.

Means also is a worldwide provider of construction cost management and analysis services for commercial and government owners and of claims and valuation services for insurers.

In short, R.S. Means can provide you with the tools and expertise for constructing accurate and dependable construction estimates and budgets in a variety of ways.

Robert Snow Means Established a Tradition of Quality That Continues Today

Robert Snow Means spent years building his company, making certain he always delivered a quality product.

Today, at R.S. Means, we do more than talk about the quality of our data and the usefulness of our books. We stand behind all of our data, from historical cost indexes... to construction materials and techniques... to current costs.

If you have any questions about our products or services, please call us toll-free at 1-800-334-3509. Our customer service representatives will be happy to assist you or visit our Web site at www.rsmeans.com

Table of Contents

How the Book Is Built: An Overview

A Powerful Construction Tool

You have in your hands one of the most powerful construction tools available today. A successful project is built on the foundation of an accurate and dependable estimate. This book will enable you to construct just such an estimate.

For the casual user the book is designed to be:

- quickly and easily understood so you can get right to your estimate
- filled with valuable information so you can understand the necessary factors that go into the cost estimate

For the regular user, the book is designed to be:

- a handy desk reference that can be quickly referred to for key costs
- a comprehensive, fully reliable source of current construction costs and productivity rates, so you'll be prepared to estimate any project
- a source book for preliminary project cost, product selections, and alternate materials and methods

To meet all of these requirements we have organized the book into the following clearly defined sections.

Quick Start

This one-page section (see following page) can quickly get you started on your estimate.

How To Use the Book: The Details

This section contains an in-depth explanation of how the book is arranged . . . and how you can use it to determine a reliable construction cost estimate. It includes information about how we develop our cost figures and how to completely prepare your estimate.

Unit Price Section

All cost data has been divided into the 16 divisions according to the MasterFormat system of classification and numbering as developed by the Construction Specifications Institute (CSI) and Construction Specifications Canada (CSC). For a listing of these divisions and an outline of their subdivisions, see the Unit Price Section Table of Contents.

Estimating tips are included at the beginning of each division.

Division 17: Quick Project Estimates: In addition to the 16 Unit Price Divisions there is a S.F. (Square Foot) and C.F. (Cubic Foot) Cost Division, Division 17. It contains costs for 58 different building types that allow you to quickly make a rough estimate for the overall cost of a project or its major components.

Assemblies Section

The cost data in this section has been organized in an "Assemblies" format. These assemblies are the functional elements of a building and are arranged according to the 12 divisions of the UniFormat classification system. For a complete explanation of a typical "Assemblies" page, see "How To Use the Assemblies Cost Tables."

Reference Section

This section includes information on Reference Numbers, Change Orders, Crew Listings, Historical Cost Indexes, City Cost Indexes, Location Factors and a listing of Abbreviations. It is visually identified by a vertical gray bar on the edge of pages.

Reference Numbers: Following selected items and major classifications throughout the book are "reference numbers" shown in bold squares. These numbers refer you to related information in other sections.

In these other sections, you'll find related tables, explanations, and estimating information. Also included are alternate pricing methods, and technical data, along with information on design and economy in construction. You'll also find helpful tips on estimating and construction.

Change Orders: This section includes information on the factors that influence the pricing of change orders.

Crew Listings: This section lists all the crews referenced. For the purposes of this book, a crew is composed of more than one trade classification and/or the addition of equipment to any trade classification. Power equipment is included in the cost of the crew. Costs are shown both with bare labor rates and with the installing contractor's overhead and profit added. For each, the total crew cost per eight-hour day and the composite cost per labor-hour are listed.

Historical Cost Indexes: These indexes provide you with data to adjust construction costs over time. If you know costs for a past project, you can use these indexes to estimate the cost to construct the same project today.

City Cost Indexes: Obviously, costs vary depending on the regional economy. You can adjust the "national average" costs in this book to 305 major cities throughout the U.S. and Canada by using the data in this section. How to use information is included.

Location Factors, to quickly adjust the data to over 930 zip code areas, are included.

Abbreviations: A listing of the abbreviations and the terms they represent, is included.

Index

A comprehensive listing of all terms and subjects in this book to help you find what you need quickly.

The Scope of This Book

This book is designed to be comprehensive and easy to use. To that end we have made certain assumptions:

1. We have established material prices based on a "national average."
2. We have computed labor costs based on a 30-city "national average" of union wage rates.
3. We have targeted the data for projects of a certain size range.

For a more detailed explanation of how the cost data is developed, see "How To Use the Book: The Details."

Project Size

This book is aimed primarily at commercial and industrial projects costing $1,000,000 and up, or large multi-family housing projects. Costs are primarily for new construction or major renovation of buildings rather than repairs or minor alterations.

With reasonable exercise of judgment the figures can be used for any building work. *However, for civil engineering structures such as bridges, dams, highways, or the like, please refer to* ***Means Heavy Construction Cost Data.***

Quick Start

If you feel you are ready to use this book and don't think you need the detailed instructions that begin on the following page, this Quick Start section is for you.

These steps will allow you to get started estimating in a matter of minutes.

1 First, decide whether you require a Unit Price or Assemblies type estimate. Unit price estimating requires a breakdown of the work to individual items. Assemblies estimates combine individual items or components into building systems.

If you need to estimate each line item separately, follow the instructions for **Unit Prices.**

If you can use an estimate for the entire assembly or system, follow the instructions for **Assemblies.**

2 Find each cost data section you need in the Table of Contents (either for Unit Prices or Assemblies).

Unit Prices: The cost data for Unit Prices has been divided into 16 divisions according to the CSI MasterFormat.

Assemblies: The cost data for Assemblies has been divided into 12 divisions according to the UniFormat.

3 Turn to the indicated section and locate the line item or assemblies table you need for your estimate. Portions of a sample page layout from both the Unit Price Listings and the Assemblies Cost Tables appear below.

Unit Prices: If there is a reference number listed at the beginning of the section, it refers to additional information you may find useful. See the referenced section for additional information.

- Note the crew code designation. You'll find full descriptions of crews in the Crew Listings including labor-hour and equipment costs.

Assemblies: The Assemblies (*not* shown in full here) are generally separated into three parts: 1) an illustration of the system to be estimated; 2) the components and related costs of a typical system; and 3) the costs for similar systems with dimensional and/or size variations. The Assemblies Section also contains reference numbers for additional useful information.

4 Determine the total number of units your job will require.

Unit Prices: Note the unit of measure for the material you're using is listed under "Unit."

- Bare Costs: These figures show unit costs for materials and installation. Labor and equipment costs are calculated according to crew costs and average daily output. Bare costs do not contain allowances for overhead, profit or taxes.
- "Labor-hours" allows you to calculate the total labor-hours to complete that task. Just multiply the quantity of work by this figure for an estimate of activity duration.

Assemblies: Note the unit of measure for the assembly or system you're estimating is listed in the Assemblies Table.

5 Then multiply the total units by . . .

Unit Prices: "Total Incl. O&P" which stands for the total cost including the installing contractor's overhead and profit. (See the "How To Use the Unit Price Pages" for a complete explanation.)

Assemblies: The "Total" in the right-hand column, which is the total cost **including the installing contractor's overhead and profit.** (See the "How To Use the Assemblies Cost Tables" section for a complete explanation.)

Material and equipment cost figures include a 10% markup. For labor markups, see the inside back cover of this book. If the work is to be subcontracted, add the general contractor's markup, typically 10%.

6 The price you calculate will be an estimate for either an individual item of work or a completed assembly or *system.*

7 Compile a list of all items or assemblies included in the total project. Summarize cost information, and add project overhead.

Localize costs by using the City Cost Indexes or Location Factors found in the Reference Section.

For a more complete explanation of the way costs are derived, please see the following sections.

Editors' Note: We urge you to spend time reading and understanding all of the supporting material and to take into consideration the reference material such as Crews Listing and the "reference numbers."

Unit Price Pages

15100 | Building Services Piping

	15110 | Valves	CREW	DAILY OUTPUT	LABOR-HOURS	UNIT	2001 BARE COSTS MAT.	LABOR	EQUIP.	TOTAL	TOTAL INCL O&P
0010	VALVES, IRON BODY R15100-090									
0020	For grooved joint, see Division 15107-690									
0100	Angle, 125 lb.									
0110	Flanged									
1020	Butterfly, wafer type, gear actuator, 200 lb.									
1030	2" size	1 Plum	14	.571	Ea.	109	19.70		128.70	149
1040	2-1/2" size	Q-1	9	1.778		112	55		167	206
1050	3" size		8	2		116	62		178	222
1060	4" size		5	3.200		145	99		244	310

Assemblies Pages

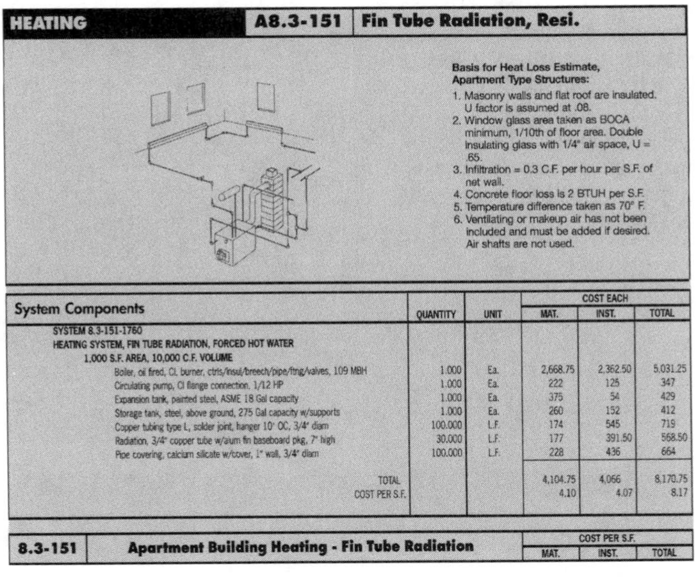

HEATING | **A8.3-151** | **Fin Tube Radiation, Resi.**

Basis for Heat Loss Estimate, Apartment Type Structures:

1. Masonry walls and flat roof are insulated. U factor is assumed at .08.
2. Window glass area taken as BOCA minimum, 1/10th of floor area. Double insulating glass with 1/4" air space, U = .65.
3. Infiltration = 0.3 C.F. per hour per S.F. of net wall.
4. Concrete floor loss is 2 BTUH per S.F.
5. Temperature difference taken as 70° F.
6. Ventilating or makeup air has not been included and must be added if desired. Air shafts are not used.

System Components	QUANTITY	UNIT	COST EACH MAT.	INST.	TOTAL
SYSTEM 8.3-151-1760					
HEATING SYSTEM, FIN TUBE RADIATION, FORCED HOT WATER					
1,000 S.F. AREA, 10,000 C.F. VOLUME					
Boiler, oil fired, CI. burner, ctrls/insul/breech/pipe/ftng/valves, 109 MBH	1.000	Ea.	2,668.75	2,362.50	5,031.25
Circulating pump, CI flange connection, 1/12 HP	1.000	Ea.	222	125	347
Expansion tank, painted steel, ASME 18 Gal capacity	1.000	Ea.	375	54	429
Storage tank, steel, above ground, 275 Gal capacity w/supports	1.000	Ea.	260	152	412
Copper tubing type L, solder joint, hanger 10' OC, 3/4" diam	100.000	L.F.	174	545	719
Radiation, 3/4" copper tube w/alum fin baseboard pkg, 7" high	30.000	L.F.	177	391.50	568.50
Pipe covering, calcium silicate w/cover, 1" wall, 3/4" diam	100.000	L.F.	228	436	664
TOTAL			4,104.75	4,066	8,170.75
COST PER S.F.			4.10	4.07	8.17

8.3-151	Apartment Building Heating - Fin Tube Radiation	COST PER S.F. MAT.	INST.	TOTAL

How to Use the Book: The Details

What's Behind the Numbers? The Development of Cost Data

The staff at R.S. Means continuously monitors developments in the construction industry in order to ensure reliable, thorough and up-to-date cost information.

While *overall* construction costs may vary relative to general economic conditions, price fluctuations within the industry are dependent upon many factors. Individual price variations may, in fact, be opposite to overall economic trends. Therefore, costs are continually monitored and complete updates are published yearly. Also, new items are frequently added in response to changes in materials and methods.

Costs—$ (U.S.)

All costs represent U.S. national averages and are given in U.S. dollars. The Means City Cost Indexes can be used to adjust costs to a particular location. The City Cost Indexes for Canada can be used to adjust U.S. national averages to local costs in Canadian dollars.

Material Costs

The R.S. Means staff contacts manufacturers, dealers, distributors, and contractors all across the U.S. and Canada to determine national average material costs. If you have access to current material costs for your specific location, you may wish to make adjustments to reflect differences from the national average. Included within material costs are fasteners for a normal installation. R.S. Means engineers use manufacturers' recommendations, written specifications and/ or standard construction practice for size and spacing of fasteners. Adjustments to material costs may be required for your specific application or location. Material costs do not include sales tax.

Labor Costs

Labor costs are based on the average of wage rates from 30 major U.S. cities. Rates are determined from labor union agreements or prevailing wages for construction trades for the current year. Rates along with overhead and profit markups are listed on the inside back cover of this book.

- If wage rates in your area vary from those used in this book, or if rate increases are expected within a given year, labor costs should be adjusted accordingly.

Labor costs reflect productivity based on actual working conditions. These figures include time spent during a normal workday on tasks other than actual installation, such as material receiving and handling, mobilization at site, site movement, breaks, and cleanup.

Productivity data is developed over an extended period so as not to be influenced by abnormal variations and reflects a typical average.

Equipment Costs

Equipment costs include not only rental, but also operating costs for equipment under normal use. The operating costs include parts and labor for routine servicing such as repair and replacement of pumps, filters and worn lines. Normal operating expendables such as fuel, lubricants, tires and electricity (where applicable) are also included. Extraordinary operating expendables with highly variable wear patterns such as diamond bits and blades are excluded. These costs are included under materials. Equipment rental rates are obtained from industry sources throughout North America—contractors, suppliers, dealers, manufacturers, and distributors.

Crew Equipment Cost/Day—The power equipment required for each crew is included in the crew cost. The daily cost for crew equipment is based on dividing the weekly bare rental rate by 5 (number of working days per week), and then adding the hourly operating cost times 8 (hours per day). This "Crew Equipment Cost/Day" is listed in Subdivision 01590.

General Conditions

Cost data in this book is presented in two ways: Bare Costs and Total Cost including O&P (Overhead and Profit). General Conditions, when applicable, should also be added to the Total Cost including O&P. The costs for General Conditions are listed in Division 1 of the Unit Price Section and the Reference Section of this book. General Conditions for the *Installing Contractor* may range from 0% to 10% of the Total Cost including O&P. For the *General* or *Prime Contractor*, costs for General Conditions may range from 5% to 15% of the Total Cost including O&P, with a figure of 10% as the most typical allowance.

Overhead and Profit

Total Cost including O&P for the *Installing Contractor* is shown in the last column on both the Unit Price and the Assemblies pages of this book. This figure is the sum of the bare material cost plus 10% for profit, the base labor cost plus total overhead and profit, and the bare equipment cost plus 10% for profit. Details for the calculation of Overhead and Profit on labor are shown on the inside back cover and in the Reference Section of this book. (See the "How to Use the Unit Price Pages" for an example of this calculation).

Factors Affecting Costs

Costs can vary depending upon a number of variables. Here's how we have handled the main factors affecting costs.

Quality—The prices for materials and the workmanship upon which productivity is based represent sound construction work. They are also in line with U.S. government specifications.

Overtime—We have made no allowance for overtime. If you anticipate premium time or work beyond normal working hours, be sure to make an appropriate adjustment to your labor costs.

Productivity—The productivity, daily output, and labor-hour figures for each line item are based on working an eight-hour day in daylight hours in moderate temperatures. For work that extends beyond normal work hours or is performed under adverse conditions, productivity may decrease. (See the section in "How To Use the Unit Price Pages" for more on productivity.)

Size of Project—The size, scope of work, and type of construction project will have a significant impact on cost. Economies of scale can reduce costs for large projects. Unit costs can often run higher for small projects. Costs in this book are intended for the size and type of project as previously described in "How the Book Is Built: An Overview." Costs for projects of a significantly different size or type should be adjusted accordingly.

Location—Material prices in this book are for metropolitan areas. However, in dense urban areas, traffic and site storage limitations may increase costs. Beyond a 20-mile radius of large cities, extra trucking or transportation charges may also increase the material costs slightly. On the other hand, lower wage rates may be in effect. Be sure to consider both these factors when preparing an estimate, particularly if the job site is located in a central city or remote rural location.

In addition, highly specialized subcontract items may require travel and per diem expenses for mechanics.

Other factors—

- season of year
- contractor management
- weather conditions
- local union restrictions
- building code requirements
- availability of:
  - adequate energy
  - skilled labor
  - building materials
- owner's special requirements/ restrictions
- safety requirements
- environmental considerations

Unpredictable Factors—General business conditions influence "in-place" costs of all items. Substitute materials and construction methods may have to be employed. These may affect the installed cost and/or life cycle costs. Such factors may be difficult to evaluate and cannot necessarily be predicted on the basis of the job's location in a particular section of the country. Thus, where these factors apply, you may find significant, but unavoidable cost variations for which you will have to apply a measure of judgment to your estimate.

Rounding of Costs

In general, all unit prices in excess of $5.00 have been rounded to make them easier to use and still maintain adequate precision of the results. The rounding rules we have chosen are in the following table.

Prices from . . .	Rounded to the nearest . . .
$.01 to $5.00	$.01
$5.01 to $20.00	$.05
$20.01 to $100.00	$.50
$100.01 to $300.00	$1.00
$300.01 to $1,000.00	$5.00
$1,000.01 to $10,000.00	$25.00
$10,000.01 to $50,000.00	$100.00
$50,000.01 and above	$500.00

Contingencies

Contingencies: The allowance for contingencies generally is to provide for unforeseen construction difficulties. On alterations or repair jobs 20% is none too much. If drawings are final and only field contingencies are being considered, 2% or 3% is probably sufficient and often nothing need be added. As far as the contract is concerned, future changes in plans will be covered by extras. The contractor should consider inflationary price trends and possible material shortages during the course of the job. If drawings are not complete or approved or a budget cost wanted, it is wise to add 5% to 10%. Contingencies, then, are a matter of judgment. Additional allowances are shown in Division 01250-200 for contingencies and job conditions and Division 01250-400 for factors to convert prices for repair and remodeling jobs.

Important Estimating Considerations

One reason for listing a job size "minimum" is to ensure that the construction craftsmen are productive for 8 hours a day on a continuous basis. Otherwise, tasks that only require 6 to 7 hours could be billed as an 8-hour day, thus providing an erroneous productivity estimate. The "productivity," or daily output of each craftsman, includes mobilization and cleanup time, break time, plan layout time, as well as an allowance to carry stock from the storage trailer or location on the job site up to 200′ into the building and on the first or second floor. If material has to be transported over greater distances or to higher floors, an additional allowance should be considered by the estimator. An allowance has also been included in the piping and fittings installation time for a leak check and minor tightening.

Equipment installation time includes all applicable items following: positioning, leveling and securing the unit in place, connecting all associated piping, ducts, vents, etc., which shall have been estimated separately, connecting to an adjacent power source, filling/bleeding, startup, adjusting the controls up and down to assure proper response, setting the integral controls/valves/ regulators/thermostats for proper operation (does not include external building type control systems, DDC systems, etc.), explaining/training owner's operator, and warranty. A reasonable breakdown of the labor costs would be as follows:

1	Movement into building, installation/ setting of equipment	35%
2.	Connection to piping/duct/power, etc.	25%
3.	Filling/flushing/cleaning/touchup, etc.	15%
4.	Startup/running adjustments	5%
5.	Training owner's rep.	5%
6.	Warranty/call back/service	15%

Note: For an explanation of "Selective Demolition" versus "Removal for Replacement" see reference number R15050-720.

Construction Section Format Relationships				
Major Sections of this Book	**CSI Format**		**Assemblies Format**	
	Division	**Description**	**Division**	**Description**
	1	General Requirements	*1	Foundations
	2	Site Work	*2	Substructures
	3	Concrete	*3	Superstructures
	*4	Masonry	*4	Exterior Closure
	5	Metals	*5	Roofing
Mechanical Related Items	6	Wood and Plastics	*6	Interior Construction
	7	Thermal and Moisture Protection	*7	Conveying Systems
	*8	Doors and Windows		
	*9	Finishes		
	10	Specialties	*10	General Conditions & Profit
Items have been rearranged to be consistent with major sections of this book.	11	Equipment	*11	Special Construction
	*12	Furnishings	12	Site Work
	13	Special Construction		
	14	Conveying Systems		
Mechanical	15	Mechanical	8	Mechanical
Electrical	16	Electrical	9	Electrical
	17	Square Foot	14	Square Foot Costs
Cost Modifications	Appendix	City Cost Index		

*Division does not appear in this book.

Note that cranes or other lifting equipment are not included on any lines in Division 15. For example, if a crane is required to lift a heavy piece of pipe into place high above a gym floor, or to put a rooftop unit on the roof of a four-story building, etc., it must be added. Due to the potential for extreme variation—from nothing additional required, to a major crane or helicopter—we feel that including a nominal amount for "lifting contingency" would be useless and detract from the accuracy of the estimate. When using equipment rental from Means do not forget to include the cost of the operator(s).

Organization of the Book

Format Division Numbers—Note that the approximate relation between the CSI format numbers and the Assemblies format are given above. Not all the divisions appear in the book.

Estimating Labor-Hours

The labor-hours expressed in this publication are based on Average Installation time, using an efficiency level of approximately 60-65% (see item 7) which has been found reasonable and acceptable by many contractors.

The book uses this national efficiency average to establish a consistent benchmark.

For bid situations, adjustments to this efficiency level should be the responsibility of the contractor bidding the project.

The unit labor-hour is divided in the following manner. A typical day for a journeyman might be:

1.	Study Plans	3%	14.4 min.
2.	Material Procurement	3%	14.4 min.
3.	Receiving and Storing	3%	14.4 min
4.	Mobilization	5%	24.0 min.
5.	Site Movement	5%	24.0 min.
6.	Layout and Marking	8%	38.4 min.
7.	Actual Installation	64%	307.2 min.
8.	Cleanup	3%	14.4 min.
9.	Breaks, Non-Productive	6%	28.8 min.
		100%	480.0 min.

If any of the percentages expressed in this breakdown do not apply to the particular work or project situation, then that percentage or a portion of it may be deducted from or added to labor hours.

Final Checklist

Estimating can be a straightforward process provided you remember the basics. Here's a checklist of some of the items you should remember to do before completing your estimate.

Did you remember to . . .

- factor in the City Cost Index for your locale
- take into consideration which items have been marked up and by how much
- mark up the entire estimate sufficiently for your purposes
- read the background information on techniques and technical matters that could impact your project time span and cost
- include all components of your project in the final estimate
- double check your figures to be sure of your accuracy
- for more information, please see "Tips for Accurate Estimating," R01100-005 in the Reference Section
- call R.S. Means if you have any questions about your estimate or the data you've found in our publications

Remember, R.S. Means stands behind its publications. If you have any questions about your estimate . . . about the costs you've used from our books . . . or even about the technical aspects of the job that may affect your estimate, feel free to call the R.S. Means editors at 1-800-334-3509.

Unit Price Section

Table of Contents

How to Use the Unit Price Pages

The following is a detailed explanation of a sample entry in the Unit Price Section. Next to each bold number below is the item being described with appropriate component of the sample entry following in parenthesis. Some prices are listed as bare costs, others as costs that include overhead and profit of the installing contractor. In most cases, if the work is to be subcontracted, the general contractor will need to add an additional markup (R.S. Means suggests using 10%) to the figures in the column "Total Incl. O&P."

1 *Division Number/Title (15100/Building Services Piping)*

Use the Unit Price Section Table of Contents to locate specific items. The sections are classified according to the CSI MasterFormat.

2 *Line Numbers (15110 200 1090)*

Each unit price line item has been assigned a unique 12-digit code based on the 5-digit CSI MasterFormat classification.

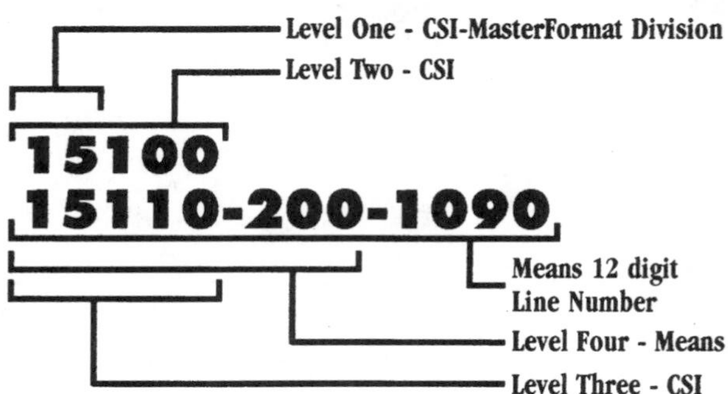

3 *Description (Valves, Iron Body)*

Each line item is described in detail. Sub-items and additional sizes are indented beneath the appropriate line items. The first line or two after the main item (in boldface) may contain descriptive information that pertains to all line items beneath this boldface listing.

4 *Reference Number Information*

R15100 -090 You'll see reference numbers shown in bold rectangles at the beginning of some sections. These refer to related items in the Reference Section, visually identified by a vertical gray bar on the edge of pages.

The relation may be: (1) an estimating procedure that should be read before estimating, (2) an alternate pricing method, or (3) technical information.

The "R" designates the Reference Section. The numbers refer to the MasterFormat classification system. An "A" denotes the Assemblies section and the numbers refer to the UniFormat classification system.

Example: The rectangle number above is directing you to refer to the reference number R15100-090. This particular reference number provides technical data on valve selection.

15100 | Building Services Piping

	15110 | Valves		CREW	DAILY OUTPUT	LABOR-HOURS	UNIT	2001 BARE COSTS MAT.	LABOR	EQUIP.	TOTAL	TOTAL INCL O&P	
200	0010	**VALVES, IRON BODY** R15100 -090										200
	0020	For grooved joint, see Division 15107-690										
	0100	Angle, 125 lb.										
	0110	Flanged										
	1020	Butterfly, wafer type, gear actuator, 200 lb.										
	1030	2″ size	1 Plum	14	.571	Ea.	109	19.70		128.70	149	
	1040	2-1/2″ size	Q-1	9	1.778		112	55		167	206	
	1050	3″ size		8	2		116	62		178	222	
	1060	4″ size		5	3.200		145	99		244	310	
	1070	5″ size	Q-2	5	4.800		175	154		329	425	
	1080	6″ size		5	4.800		198	154		352	450	
	1090	8″ size		4.50	5.333		[illegible]	171		435	550	
	1100	10″ size		4	6		[illegible]	193		523	655	
	1110	12″ size		3	8		[illegible]	257		717	895	
	1200	Wafer type, lever actuator, 200 lb.										

Crew (Q-2)

The "Crew" column designates the typical trade or crew used to install the item. If an installation can be accomplished by one trade and requires no power equipment, that trade and the number of workers are listed (for example, "2 Plum"). If an installation requires a composite crew, a crew code designation is listed (for example, "Q-2"). You'll find full details on all composite crews in the Crew Listings.

- For a complete list of all trades utilized in this book and their abbreviations, see the inside back cover.

Crews

Crew No.	Bare Costs		Incl. Subs O & P		Cost Per Labor-Hour	
Crew Q-2	Hr.	Daily	Hr.	Daily	Bare Costs	Incl. O&P
2 Plumbers	$34.45	$551.20	$52.00	$832.00	$32.15	$48.53
1 Plumber Apprentice	27.55	220.40	41.60	332.80		
24 L.H., Daily Totals		$771.60		$1164.80	$32.15	$48.53

Productivity: Daily Output (4.50)/ Labor-Hours (5.333)

The "Daily Output" represents the typical number of units the designated crew will install in a normal 8-hour day. To find out the number of days the given crew would require to complete the installation, divide your quantity by the daily output. For example:

Quantity	÷	Daily Output	=	Duration
10 Ea.	÷	4.50 Ea./ Crew Day	=	2.22 Crew Days

The "Labor-Hours" figure represents the number of labor-hours required to install one unit of work. To find out the number of labor-hours required for your particular task, multiply the quantity of the item times the number of labor-hours shown. For example:

Quantity	x	Productivity Rate	=	Duration
10 Ea.	x	5.333 Labor-Hours/Ea.	=	53.33 Labor-Hours

Unit (Ea.)

The abbreviated designation indicates the unit of measure upon which the price, production, and crew are based (Ea. = Each). For a complete listing of abbreviations refer to the Abbreviations Listing in the Reference Section of this book.

Bare Costs:

Mat. (Bare Material Cost) (264)

The unit material cost is the "bare" material cost with no overhead and profit included. *Costs shown reflect national average material prices for January of the current year and include delivery to the job site. No sales taxes are included.*

Labor (171)

The unit labor cost is derived by multiplying bare labor-hour costs for Crew Q-2 by labor-hour units. The bare labor-hour cost is found in the Crew Section under Q-2. (If a trade is listed, the hourly labor cost—the wage rate—is found on the inside back cover.)

Labor-Hour Cost Crew Q-2	x	Labor-Hour Units	=	Labor
$32.15	x	5.333	=	$171.46

Equip. (Equipment) (0)

Equipment costs for each crew are listed in the description of each crew. Tools or equipment whose value justifies purchase or ownership by a contractor are considered overhead as shown on the inside back cover. The unit equipment cost is derived by multiplying the bare equipment hourly cost by the labor-hour units.

Equipment Cost Crew Q-2	x	Labor-Hour Units	=	Equip.
$0	x	5.333	=	$0

Total (435)

The total of the bare costs is the arithmetic total of the three previous columns: mat., labor, and equip.

Material	+	Labor	+	Equip.	=	Total
$264	+	$171	+	$0	=	$435

Total Costs Including O&P

This figure is the sum of the bare material cost plus 10% for profit; the bare labor cost plus total overhead and profit (per the inside back cover or, if a crew is listed, from the crew listings); and the bare equipment cost plus 10% for profit.

Material is Bare Material Cost + 10% = $264 + $26.40	=	$290
Labor for Crew Q-2 = Labor-Hour Cost ($48.53) x Labor-Hour Units (5.333)	=	$258.81
Equip. is Bare Equip. Cost + 10% = $0 + $0	=	$ 0
Total (Rounded)	=	$550

Division 1
General Requirements

Estimating Tips

The General Requirements of any contract are very important to both the bidder and the owner. These lay the ground rules under which the contract will be executed and have a significant influence on the cost of operations. Therefore, it is extremely important to thoroughly read and understand the General Requirements both before preparing an estimate and when the estimate is complete, to ascertain that nothing in the contract is overlooked. Caution should be exercised when applying items listed in Division 1 to an estimate. Many of the items are included in the unit prices listed in the other divisions such as mark-ups on labor and company overhead.

01300 Administrative Requirements

- Before determining a final cost estimate, it is a good practice to review all the items listed in subdivision 01300 to make final adjustments for items that may need customizing to specific job conditions.

01330 Submittal Procedures

- Requirements for initial and periodic submittals can represent a significant cost to the General Requirements of a job. Thoroughly check the submittal specifications when estimating a project to determine any costs that should be included.

01400 Quality Requirements

- All projects will require some degree of Quality Control. This cost is not included in the unit cost of construction listed in each division. Depending upon the terms of the contract, the various costs of inspection and testing can be the responsibility of either the owner or the contractor. Be sure to include the required costs in your estimate.

01500 Temporary Facilities & Controls

- Barricades, access roads, safety nets, scaffolding, security and many more requirements for the execution of a safe project are elements of direct cost. These costs can easily be overlooked when preparing an estimate. When looking through the major classifications of this subdivision, determine which items apply to each division in your estimate.

01590 Equipment Rental

- This subdivision contains transportation, handling, storage, protection and product options and substitutions. Listed in this cost manual are average equipment rental rates for all types of equipment. This is useful information when estimating the time and materials requirement of any particular operation in order to establish a unit or total cost.
- A good rule of thumb is that weekly rental is 3 times daily rental and that monthly rental is 3 times weekly rental.
- The figures in the column for Crew Equipment Cost represent the rental rate used in determining the daily cost of equipment in a crew. It is calculated by dividing the weekly rate by 5 days and adding the hourly operating cost times 8 hours.

01770 Closeout Procedures

- When preparing an estimate, read the specifications to determine the requirements for Contract Closeout thoroughly. Final cleaning, record documentation, operation and maintenance data, warranties and bonds, and spare parts and maintenance materials can all be elements of cost for the completion of a contract. Do not overlook these in your estimate.

01830 Operations & Maintenance

- If maintenance and repair are included in your contract, they require special attention. To estimate the cost to remove and replace any unit usually requires a site visit to determine the accessibility and the specific difficulty at that location. Obstructions, dust control, safety, and often overtime hours must be considered when preparing your estimate.

Reference Numbers

Reference numbers are shown in bold squares at the beginning of some major classifications. These numbers refer to related items in the Reference Section. The reference information may be an estimating procedure, an alternate pricing method or technical information.

Note: Not all subdivisions listed here necessarily appear in this publication.

01100 | Summary

	01107 Professional Consultant		CREW	DAILY OUTPUT	LABOR-HOURS	UNIT	2001 BARE COSTS MAT.	LABOR	EQUIP.	TOTAL	TOTAL INCL O&P	
100	0011	**ARCHITECTURAL FEES**										100
	0020	For new construction										
	0060	Minimum				Project					4.90%	
	0090	Maximum									16%	
	0100	For alteration work, to $500,000, add to fee									50%	
	0150	Over $500,000, add to fee									25%	
200	0010	**CONSTRUCTION MANAGEMENT FEES** $1,000,000 job, minimum				Project					4.50%	200
	0050	Maximum									7.50%	
	0300	$5,000,000 job, minimum									2.50%	
	0350	Maximum									4%	
300	0010	**ENGINEERING FEES** R01107-030										300
	0020	Educational planning consultant, minimum				Project					.50%	
	0100	Maximum				″					2.50%	
	0200	Electrical, minimum				Contrct					4.10%	
	0300	Maximum									10.10%	
	0400	Elevator & conveying systems, minimum									2.50%	
	0500	Maximum									5%	
	0600	Food service & kitchen equipment, minimum									8%	
	0700	Maximum									12%	
	0800	Landscaping & site development, minimum									2.50%	
	0900	Maximum									6%	
	1000	Mechanical (plumbing & HVAC), minimum									4.10%	
	1100	Maximum									10.10%	
	1200	Structural, minimum				Project					1%	
	1300	Maximum				″					2.50%	
700	0010	**SURVEYING** Conventional, topographical, minimum	A-7	3.30	7.273	Acre	16	225		241	370	700
	0100	Maximum	A-8	.60	53.333		48	1,625		1,673	2,550	
	0300	Lot location and lines, minimum, for large quantities	A-7	2	12		25	370		395	610	
	0320	Average	″	1.25	19.200		45	595		640	975	
	0400	Maximum, for small quantities	A-8	1	32		72	970		1,042	1,575	
	0600	Monuments, 3′ long	A-7	10	2.400	Ea.	19	74.50		93.50	137	
	0800	Property lines, perimeter, cleared land	″	1,000	.024	L.F.	.03	.74		.77	1.19	
	0900	Wooded land	A-8	875	.037	″	.05	1.11		1.16	1.78	
	1500	Aerial surveying, including ground control, minimum fee, 10 acres				Total					5,500	
	1510	100 acres									9,100	
	1550	From existing photography, deduct									1,340	
	1600	2′ contours, 10 acres				Acre					440	
	1650	20 acres									300	
	1800	50 acres									90	
	1850	100 acres									80	
	2000	1000 acres									17.01	
	2050	10,000 acres									11.01	
	2150	For 1′ contours and										
	2160	dense urban areas, add to above				Acre					40%	

01250 | Contract Modification Procedures

		01250 Contract Modification Procedures	CREW	DAILY OUTPUT	LABOR-HOURS	UNIT	2001 BARE COSTS MAT.	LABOR	EQUIP.	TOTAL	TOTAL INCL O&P	
200	0010	**CONTINGENCIES** for estimate at conceptual stage				Project					20%	200
	0050	Schematic stage									15%	
	0100	Preliminary working drawing stage (Design Dev.)									10%	
	0150	Final working drawing stage				↓					3%	
300	0010	**CREWS** For building construction, see How To Use This Book										300
400	0010	**FACTORS** Cost adjustments (R01250 -010)										400
	0100	Add to construction costs for particular job requirements										
	0500	Cut & patch to match existing construction, add, minimum				Costs	2%	3%				
	0550	Maximum					5%	9%				
	0800	Dust protection, add, minimum					1%	2%				
	0850	Maximum					4%	11%				
	1100	Equipment usage curtailment, add, minimum					1%	1%				
	1150	Maximum					3%	10%				
	1400	Material handling & storage limitation, add, minimum					1%	1%				
	1450	Maximum					6%	7%				
	1700	Protection of existing work, add, minimum					2%	2%				
	1750	Maximum					5%	7%				
	2000	Shift work requirements, add, minimum						5%				
	2050	Maximum						30%				
	2300	Temporary shoring and bracing, add, minimum					2%	5%				
	2350	Maximum					5%	12%				
	2400	Work inside prisons and high security areas, add, minimum						30%				
	2450	Maximum				↓		50%				
500	0010	**JOB CONDITIONS** Modifications to total										500
	0020	project cost summaries										
	0100	Economic conditions, favorable, deduct				Project					2%	
	0200	Unfavorable, add									5%	
	0300	Hoisting conditions, favorable, deduct									2%	
	0400	Unfavorable, add									5%	
	0700	Labor availability, surplus, deduct									1%	
	0800	Shortage, add									10%	
	0900	Material storage area, available, deduct									1%	
	1000	Not available, add									2%	
	1100	Subcontractor availability, surplus, deduct									5%	
	1200	Shortage, add									12%	
	1300	Work space, available, deduct									2%	
	1400	Not available, add				↓					5%	
600	0010	**OVERTIME** For early completion of projects or where (R01100 -110)										600
	0020	labor shortages exist, add to usual labor, up to				Costs		100%				

01255 | Cost Indexes

		01255 Cost Indexes	CREW	DAILY OUTPUT	LABOR-HOURS	UNIT	MAT.	LABOR	EQUIP.	TOTAL	TOTAL INCL O&P	
200	0010	**CONSTRUCTION COST INDEX** (Reference) over 930 zip code locations in										200
	0020	The U.S. and Canada, total bldg cost, min. (Clarksdale, MS)				%					66.80%	
	0050	Average									100%	
	0100	Maximum (New York, NY)				↓					134.50%	
400	0010	**HISTORICAL COST INDEXES** (Reference) Back to 1951										400
500	0010	**LABOR INDEX** (Reference) For over 930 zip code locations in										500
	0020	the U.S. and Canada, minimum (Clarksdale, MS)				%		35%				
	0050	Average						100%				
	0100	Maximum (New York, NY)				↓		161.90%				
600	0011	**MATERIAL INDEX** For over 930 zip code locations in										600
	0020	the U.S. and Canada, minimum (Elizabethtown, KY)				%	92.20%					
	0040	Average					100%					
	0060	Maximum (Ketchikan, AK)				↓	144.20%					

01200 | Price & Payment Procedures

		01290 | Payment Procedures	CREW	DAILY OUTPUT	LABOR-HOURS	UNIT	2001 BARE COSTS MAT.	LABOR	EQUIP.	TOTAL	TOTAL INCL O&P	
800	0010	**TAXES** Sales tax, State, average R01100-090				%	4.71%					800
	0050	Maximum					7.25%					
	0200	Social Security, on first $76,200 of wages R01100-100						7.65%				
	0300	Unemployment, MA, combined Federal and State, minimum						2.10%				
	0350	Average						7%				
	0400	Maximum						8%				

01300 | Administrative Requirements

		01310 | Project Management/Coordination	CREW	DAILY OUTPUT	LABOR-HOURS	UNIT	2001 BARE COSTS MAT.	LABOR	EQUIP.	TOTAL	TOTAL INCL O&P	
150	0010	**PERMITS** Rule of thumb, most cities, minimum				Job					.50%	150
	0100	Maximum				″					2%	
200	0010	**PERFORMANCE BOND** For buildings, minimum R01100-080				Job					.60%	200
	0100	Maximum				″					2.50%	
350	0010	**INSURANCE** Builders risk, standard, minimum R01100-040				Job					.22%	350
	0050	Maximum									.59%	
	0200	All-risk type, minimum R01100-060									.25%	
	0250	Maximum									.62%	
	0400	Contractor's equipment floater, minimum				Value					.50%	
	0450	Maximum				″					1.50%	
	0600	Public liability, average				Job					1.55%	
	0800	Workers' compensation & employer's liability, average										
	0850	by trade, carpentry, general				Payroll		19.01%				
	1000	Electrical						6.79%				
	1150	Insulation						17.65%				
	1450	Plumbing						8.47%				
	1550	Sheet metal work (HVAC)						11.93%				
400	0010	**MAIN OFFICE EXPENSE** Average for General Contractors										400
	0020	As a percentage of their annual volume										
	0125	Annual volume under 1 million dollars				% Vol.				13.60%		
	0145	Up to 2.5 million dollars								8%		
	0150	Up to 4.0 million dollars								6.80%		
	0200	Up to 7.0 million dollars								5.60%		
	0250	Up to 10 million dollars								5.10%		
	0300	Over 10 million dollars								3.90%		
500	0010	**MARK-UP** For General Contractors for change R01100-070										500
	0100	of scope of job as bid										
	0200	Extra work, by subcontractors, add				%					10%	
	0250	By General Contractor, add									15%	
	0400	Omitted work, by subcontractors, deduct all but									5%	
	0450	By General Contractor, deduct all but									7.50%	
	0600	Overtime work, by subcontractors, add									15%	
	0650	By General Contractor, add									10%	
	1000	Installing contractors, on his own labor, minimum						47.80%				
	1100	Maximum						87.50%				
600	0010	**OVERHEAD** As percent of direct costs, minimum				%				5%		600
	0050	Average								13%		
	0100	Maximum								30%		

01300 | Administrative Requirements

		01310 | Project Management/Coordination	CREW	DAILY OUTPUT	LABOR-HOURS	UNIT	2001 BARE COSTS MAT.	LABOR	EQUIP.	TOTAL	TOTAL INCL O&P	
620	0010	**OVERHEAD & PROFIT** Allowance to add to items in this R01100-070										620
	0020	book that do not include Subs O&P, average				%				25%		
	0100	Allowance to add to items in this book that										
	0110	do include Subs O&P, minimum				%					5%	
	0150	Average									10%	
	0200	Maximum									15%	
	0300	Typical, by size of project, under $100,000								30%		
	0350	$500,000 project								25%		
	0400	$2,000,000 project								20%		
	0450	Over $10,000,000 project								15%		
700	0010	**FIELD PERSONNEL** Clerk average				Week		280		280	440	700
	0100	Field engineer, minimum						675		675	1,060	
	0120	Average						875		875	1,375	
	0140	Maximum						1,005		1,005	1,580	
	0160	General purpose laborer, average						925		925	1,450	
	0180	Project manager, minimum						1,265		1,265	1,985	
	0200	Average						1,420		1,420	2,230	
	0220	Maximum						1,605		1,605	2,520	
	0240	Superintendent, minimum						1,210		1,210	1,900	
	0260	Average						1,340		1,340	2,105	
	0280	Maximum						1,510		1,510	2,370	
	0290	Timekeeper, average						780		780	1,225	
		01321 | Construction Photos										
500	0010	**PHOTOGRAPHS** 8" x 10", 4 shots, 2 prints ea., std. mounting				Set	263			263	289	500
	0100	Hinged linen mounts					290			290	320	
	0200	8" x 10", 4 shots, 2 prints each, in color					310			310	340	
	0300	For I.D. slugs, add to all above					3.75			3.75	4.13	
	1500	Time lapse equipment, camera and projector, buy					3,575			3,575	3,925	
	1550	Rent per month					530			530	585	
	1700	Cameraman and film, including processing, B.&W.				Day	1,225			1,225	1,350	
	1720	Color				"	1,225			1,225	1,350	

01500 | Temporary Facilities & Controls

		01510 | Temporary Utilities	CREW	DAILY OUTPUT	LABOR-HOURS	UNIT	2001 BARE COSTS MAT.	LABOR	EQUIP.	TOTAL	TOTAL INCL O&P	
800	0010	**TEMPORARY UTILITIES**										800
	0100	Heat, incl. fuel and operation, per week, 12 hrs. per day	1 Skwk	100	.080	CSF Flr	5.10	2.39		7.49	9.40	
	0200	24 hrs. per day	"	60	.133		7.70	3.98		11.68	14.70	
	0350	Lighting, incl. service lamps, wiring & outlets, minimum	1 Elec	34	.235		2.07	8.05		10.12	14.35	
	0360	Maximum	"	17	.471		4.51	16.15		20.66	29	
	0400	Power for temp lighting only, per month, min/month 6.6 KWH								.75	1.18	
	0450	Maximum/month 23.6 KWH								2.85	3.14	
	0600	Power for job duration incl. elevator, etc., minimum								47	51.70	
	0650	Maximum								110	121	
	1000	Toilet, portable, see division 01590-400										
		01520 | Construction Facilities										
500	0010	**OFFICE** Trailer, furnished, no hookups, 20' x 8', buy	2 Skwk	1	16	Ea.	5,325	480		5,805	6,625	500
	0250	Rent per month					150			150	165	

01500 | Temporary Facilities & Controls

01520 | Construction Facilities

			Crew	Daily Output	Labor-Hours	Unit	2001 Bare Costs Mat.	Labor	Equip.	Total	Total Incl O&P	
500	0300	32′ x 8′, buy	2 Skwk	.70	22.857	Ea.	7,825	680		8,505	9,700	500
	0350	Rent per month					165			165	182	
	0400	50′ x 10′, buy	2 Skwk	.60	26.667		13,400	795		14,195	16,000	
	0450	Rent per month					295			295	325	
	0500	50′ x 12′, buy	2 Skwk	.50	32		15,800	955		16,755	18,900	
	0550	Rent per month					345			345	380	
	0700	For air conditioning, rent per month, add					35.50			35.50	39	
	0800	For delivery, add per mile				Mile	1.50			1.50	1.65	
	1000	Portable buildings, prefab, on skids, economy, 8′ x 8′	2 Carp	265	.060	S.F.	80	1.76		81.76	91	
	1100	Deluxe, 8′ x 12′	″	150	.107	″	87	3.11		90.11	100	
	1200	Storage boxes, 20′ x 8′, buy	2 Skwk	1.80	8.889	Ea.	3,300	265		3,565	4,050	
	1250	Rent per month					72.50			72.50	80	
	1300	40′ x 8′, buy	2 Skwk	1.40	11.429		3,950	340		4,290	4,875	
	1350	Rent per month					103			103	113	

01530 | Temporary Construction

			Crew	Daily Output	Labor-Hours	Unit	Mat.	Labor	Equip.	Total	Total Incl O&P	
700	0010	**PROTECTION** Stair tread, 2″ x 12″ planks, 1 use	1 Carp	75	.107	Tread	4.43	3.11		7.54	9.75	700
	0100	Exterior plywood, 1/2″ thick, 1 use		65	.123		1.48	3.59		5.07	7.25	
	0200	3/4″ thick, 1 use		60	.133		2.25	3.89		6.14	8.55	
900	0010	**WINTER PROTECTION** Reinforced plastic on wood										900
	0100	framing to close openings	2 Clab	750	.021	S.F.	.35	.49		.84	1.15	
	0200	Tarpaulins hung over scaffolding, 8 uses, not incl. scaffolding		1,500	.011		.16	.24		.40	.56	
	0250	Tarpaulin polyester reinf. w/ integral fastening system 11 mils thick		1,600	.010		.73	.23		.96	1.16	
	0300	Prefab fiberglass panels, steel frame, 8 uses		1,200	.013		.70	.30		1	1.25	

01540 | Construction Aids

			Crew	Daily Output	Labor-Hours	Unit	Mat.	Labor	Equip.	Total	Total Incl O&P	
750	0010	**SCAFFOLDING** R01540-100										750
	0015	Steel tubular, rent, 1 use/mo, no plank, erect & dismantle										
	0090	Building exterior, wall face, 1 to 5 stories	3 Carp	24	1	C.S.F.	24.50	29		53.50	72	
	0200	6 to 12 stories	4 Carp	21.20	1.509		24.50	44		68.50	95.50	
	0310	13 to 20 stories	5 Carp	20	2		24.50	58.50		83	118	
	0460	Building interior, wall face area, up to 16′ high	3 Carp	25	.960		24.50	28		52.50	70.50	
	0560	16′ to 40′ high		23	1.043		24.50	30.50		55	74	
	0800	Building interior floor area, up to 30′ high		312	.077	C.C.F.	1.87	2.24		4.11	5.55	
	0900	Over 30′ high	4 Carp	275	.116	″	1.87	3.39		5.26	7.35	
	0910	Steel tubular, heavy duty shoring, buy										
	0920	Frames 5′ high 2′ wide				Ea.	75			75	82.50	
	0925	5′ high 4′ wide					85			85	93.50	
	0930	6′ high 2′ wide					86			86	94.50	
	0935	6′ high 4′ wide					101			101	111	
	0940	Accessories										
	0945	Cross braces				Ea.	16			16	17.60	
	0950	U-head, 8″ x 8″					16			16	17.60	
	0955	J-head, 4″ x 8″					12			12	13.20	
	0960	Base plate, 8″ x 8″					13			13	14.30	
	0965	Leveling jack					30.50			30.50	33.50	
	1000	Steel tubular, regular, buy										
	1100	Frames 3′ high 5′ wide				Ea.	58			58	64	
	1150	5′ high 5′ wide					67			67	73.50	
	1200	6′-4″ high 5′ wide					84			84	92.50	
	1350	7′-6″ high 6′ wide					145			145	160	
	1500	Accessories cross braces					15			15	16.50	
	1550	Guardrail post					15			15	16.50	
	1600	Guardrail 7′ section					7.25			7.25	8	

	01540 | Construction Aids		CREW	DAILY OUTPUT	LABOR-HOURS	UNIT	2001 BARE COSTS MAT.	LABOR	EQUIP.	TOTAL	TOTAL INCL O&P	
750	1650	Screw jacks & plates R01540 -100				Ea.	24			24	26.50	750
	1700	Sidearm brackets					28			28	31	
	1750	8″ casters					33			33	36.50	
	1800	Plank 2″ x 10″ x 16′-0″					42.50			42.50	47	
	1900	Stairway section					235			235	259	
	1910	Stairway starter bar					21			21	23.50	
	1920	Stairway inside handrail					53			53	58.50	
	1930	Stairway outside handrail					73			73	80.50	
	1940	Walk-thru frame guardrail					28			28	31	
	2000	Steel tubular, regular, rent/mo.										
	2100	Frames 3′ high 5′ wide				Ea.	3.75			3.75	4.13	
	2150	5′ high 5′ wide					3.75			3.75	4.13	
	2200	6′-4″ high 5′ wide					3.75			3.75	4.13	
	2250	7′-6″ high 6′ wide					7			7	7.70	
	2500	Accessories, cross braces					.60			.60	.66	
	2550	Guardrail post					1			1	1.10	
	2600	Guardrail 7′ section					.75			.75	.83	
	2650	Screw jacks & plates					1.50			1.50	1.65	
	2700	Sidearm brackets					1.50			1.50	1.65	
	2750	8″ casters					6			6	6.60	
	2800	Outrigger for rolling tower					3			3	3.30	
	2850	Plank 2″ x 10″ x 16′-0″					5			5	5.50	
	2900	Stairway section					10			10	11	
	2910	Stairway starter bar					.10			.10	.11	
	2920	Stairway inside handrail					5			5	5.50	
	2930	Stairway outside handrail					5			5	5.50	
	2940	Walk-thru frame guardrail					2			2	2.20	
	3000	Steel tubular, heavy duty shoring, rent/mo.										
	3250	5′ high 2′ & 4′ wide				Ea.	5			5	5.50	
	3300	6′ high 2′ & 4′ wide					5			5	5.50	
	3500	Accessories, cross braces					1			1	1.10	
	3600	U - head, 8″ x 8″					1			1	1.10	
	3650	J - head, 4″ x 8″					1			1	1.10	
	3700	Base plate, 8″ x 8″					1			1	1.10	
	3750	Leveling jack					2			2	2.20	
	5700	Planks, 2″x10″x16′-0″ in place, up to 50′ high	3 Carp	144	.167			4.86		4.86	7.60	
	5800	In place over 50′ high	4 Carp	160	.200			5.85		5.85	9.10	
	6000	Heavy duty shoring for elevated slab forms to 8′-2″ high, floor area										
	6010	Set up and take down										
	6100	1 use/month	4 Carp	36	.889	C.S.F.	29.50	26		55.50	73	
	6150	2 uses/month	″	36	.889	″	14.80	26		40.80	57	
	6500	To 14′-8″ high										
	6600	1 use/month	4 Carp	18	1.778	C.S.F.	43	52		95	129	
	6650	2 uses/month	″	18	1.778	″	21.50	52		73.50	105	
755	0011	**SCAFFOLDING SPECIALTIES**										755
	1200	Sidewalk bridge, heavy duty steel posts & beams, including										
	1210	parapet protection & waterproofing										
	1220	8′ to 10′ wide, 2 posts	3 Carp	15	1.600	L.F.	42	46.50		88.50	119	
	1230	3 posts	″	10	2.400	″	63	70		133	179	
	1500	Sidewalk bridge using tubular steel										
	1510	scaffold frames, including planking	3 Carp	45	.533	L.F.	4.72	15.55		20.27	29.50	
	1600	For 2 uses per month, deduct from all above					50%					
	1700	For 1 use every 2 months, add to all above					100%					
	1900	Catwalks, 32″ wide, no guardrails, 6′ span, buy				Ea.	120			120	132	
	2000	10′ span, buy				″	190			190	209	
	2800	Hand winch-operated masons										

		01540 | Construction Aids	CREW	DAILY OUTPUT	LABOR-HOURS	UNIT	2001 BARE COSTS MAT.	LABOR	EQUIP.	TOTAL	TOTAL INCL O&P	
755	2810	scaffolding, no plank moving required										755
	2900	98′ long, 10′-6″ high, buy				Ea.	19,700			19,700	21,700	
	3000	Rent per month					790			790	865	
	3100	28′-6″ high, buy					25,700			25,700	28,200	
	3200	Rent per month					1,025			1,025	1,125	
	3400	196′ long, 28′-6″ high, buy					49,700			49,700	54,500	
	3500	Rent per month					2,000			2,000	2,175	
	3600	64′-6″ high, buy					73,000			73,000	80,000	
	3700	Rent per month					2,900			2,900	3,200	
	3720	Putlog, standard, 8′ span, with hangers, buy					61			61	67	
	3730	Rent per month					10			10	11	
	3750	12′ span, buy					92			92	101	
	3755	Rent per month					15			15	16.50	
	3760	Trussed type, 16′ span, buy					210			210	231	
	3770	Rent per month					20			20	22	
	3790	22′ span, buy					252			252	277	
	3795	Rent per month					30			30	33	
	3800	Rolling ladders with handrails, 30″ wide, buy, 2 step					144			144	158	
	4000	7 step					430			430	475	
	4050	10 step					645			645	710	
	4100	Rolling towers, buy, 5′ wide, 7′ long, 10′ high					1,150			1,150	1,250	
	4200	For 5′ high added sections, to buy, add					188			188	207	
	4300	Complete incl. wheels, railings, outriggers,										
	4350	21′ high, to buy				Ea.	1,925			1,925	2,125	
	4400	Rent/month				"	138			138	152	
800	0010	**TARPAULINS** Cotton duck, 10 oz. to 13.13 oz. per S.Y., minimum				S.F.	.46			.46	.51	800
	0050	Maximum					.55			.55	.61	
	0100	Polyvinyl coated nylon, 14 oz. to 18 oz., minimum					.45			.45	.50	
	0150	Maximum					.65			.65	.72	
	0200	Reinforced polyethylene 3 mils thick, white					.10			.10	.11	
	0300	4 mils thick, white, clear or black					.12			.12	.13	
	0400	5.5 mils thick, clear					.09			.09	.10	
	0500	White, fire retardant					.16			.16	.18	
	0600	7.5 mils, oil resistant, fire retardant					.17			.17	.19	
	0700	8.5 mils, black					.22			.22	.24	
	0710	Woven polyethylene, 6 mils thick					.45			.45	.50	
	0730	Polyester reinforced w/ integral fastening system 11 mils thick					1			1	1.10	
	0740	Mylar polyester, non-reinforced, 7 mils thick					1.10			1.10	1.21	
820	0010	**SMALL TOOLS** As % of contractor's work, minimum				Total					.50%	820
	0100	Maximum				"					2%	
		01550 | Vehicular Access & Parking										
700	0010	**ROADS AND SIDEWALKS** Temporary										700
	0050	Roads, gravel fill, no surfacing, 4″ gravel depth	B-14	715	.067	S.Y.	2.72	1.62	.25	4.59	5.80	
	0100	8″ gravel depth	"	615	.078	"	5.45	1.89	.29	7.63	9.25	
	1000	Ramp, 3/4″ plywood on 2″ x 6″ joists, 16″ O.C.	2 Carp	300	.053	S.F.	1.34	1.55		2.89	3.90	
	1100	On 2″ x 10″ joists, 16″ O.C.	"	275	.058		1.96	1.70		3.66	4.81	
	2200	Sidewalks, 2″ x 12″ planks, 2 uses	1 Carp	350	.023		.74	.67		1.41	1.85	
	2300	Exterior plywood, 2 uses, 1/2″ thick		750	.011		.25	.31		.56	.76	
	2400	5/8″ thick		650	.012		.31	.36		.67	.90	
	2500	3/4″ thick		600	.013		.37	.39		.76	1.02	

		01560 | Barriers & Enclosures	CREW	DAILY OUTPUT	LABOR-HOURS	UNIT	2001 BARE COSTS MAT.	LABOR	EQUIP.	TOTAL	TOTAL INCL O&P	
100	0010	**BARRICADES** 5′ high, 3 rail @ 2″ x 8″, fixed	2 Carp	30	.533	L.F.	10.80	15.55		26.35	36.50	100
	0150	Movable		20	.800		10.80	23.50		34.30	48.50	
	1000	Guardrail, wooden, 3′ high, 1″ x 6″, on 2″ x 4″ posts		200	.080		1.05	2.33		3.38	4.80	
	1100	2″ x 6″, on 4″ x 4″ posts		165	.097		1.65	2.83		4.48	6.25	
	1200	Portable metal with base pads, buy					15			15	16.50	
	1250	Typical installation, assume 10 reuses	2 Carp	600	.027		1.58	.78		2.36	2.96	
	1300	Barricade tape, polyethelyne, 7 mil, 3″ wide x 500′ long roll				Ea.	25			25	27.50	
250	0010	**FENCING** Chain link, 11 ga, 5′ high	2 Clab	100	.160	L.F.	3.58	3.66		7.24	9.65	250
	0100	6′ high		75	.213		3.28	4.87		8.15	11.25	
	0200	Rented chain link, 6′ high, to 500′ (up to 12 mo.)		100	.160		2.26	3.66		5.92	8.20	
	0250	Over 1000′ (up to 12 mo.)		110	.145		1.64	3.32		4.96	7	
	0350	Plywood, painted, 2″ x 4″ frame, 4′ high	A-4	135	.178		4.44	5		9.44	12.65	
	0400	4″ x 4″ frame, 8′ high	″	110	.218		7.90	6.15		14.05	18.20	
	0500	Wire mesh on 4″ x 4″ posts, 4′ high	2 Carp	100	.160		6	4.66		10.66	13.85	
	0550	8′ high	″	80	.200		9	5.85		14.85	19	
400	0010	**TEMPORARY CONSTRUCTION** See also division 01530										400
800	0010	**WATCHMAN** Service, monthly basis, uniformed person, minimum				Hr.					7.80	800
	0100	Maximum									14.15	
	0200	Person and command dog, minimum									10.30	
	0300	Maximum									15.30	
	0500	Sentry dog, leased, with job patrol (yard dog), 1 dog				Week					195	
	0600	2 dogs				″					275	
	0800	Purchase, trained sentry dog, minimum				Ea.					800	
	0900	Maximum				″					2,000	
		01580 | Project Signs										
700	0010	**SIGNS** Hi-intensity reflectorized, no posts, buy				S.F.	14			14	15.40	700

01590 | Equipment Rental

			UNIT	HOURLY OPER. COST	RENT PER DAY	RENT PER WEEK	RENT PER MONTH	CREW EQUIPMENT COST/DAY	
200	0010	**EARTHWORK EQUIPMENT RENTAL** Without operators							200
	0050	Augers for truck or trailer mounting, vertical drilling							
	0055	Fence post auger, truck mounted	Ea.	5.50	530	1,595	4,775	363	
	0060	4″ to 36″ diam., 54 H.P., gas, 10′ spindle travel		16.30	650	1,945	5,825	519.40	
	0070	14′ spindle travel		18.75	810	2,430	7,300	636	
	0080	Auger, horizontal boring machine, 12″ to 36″ diameter, 45 H.P.		6.90	335	1,000	3,000	255.20	
	0090	12″ to 48″ diameter, 65 H.P.		10.40	735	2,200	6,600	523.20	
	0100	Excavator, diesel hydraulic, crawler mounted, 1/2 C.Y. cap.		8.80	415	1,240	3,725	318.40	
	0120	5/8 C.Y. capacity		10.50	460	1,380	4,150	360	
	0140	3/4 C.Y. capacity		12.95	515	1,550	4,650	413.60	
	0150	1 C.Y. capacity		16.35	600	1,800	5,400	490.80	
	0200	1-1/2 C.Y. capacity		19.40	730	2,190	6,575	593.20	
	0300	2 C.Y. capacity		26.85	1,075	3,200	9,600	854.80	
	0320	2-1/2 C.Y. capacity		49.60	1,925	5,800	17,400	1,557	
	0340	3-1/2 C.Y. capacity		61.75	2,475	7,400	22,200	1,974	
	0341	Attachments							
	0345	Grapples		1	202	605	1,825	129	
	0350	Gradall type, truck mounted, 3 ton @ 15′ radius, 5/8 C.Y.		21.05	800	2,400	7,200	648.40	
	0370	1 C.Y. capacity		24.60	965	2,890	8,675	774.80	
	0400	Backhoe-loader, 40 to 45 H.P., 5/8 C.Y. capacity		4.60	220	660	1,975	168.80	
	0450	45 H.P. to 60 H.P., 3/4 C.Y. capacity		5.50	228	685	2,050	181	
	0460	80 H.P., 1-1/4 C.Y. capacity		6.30	297	890	2,675	228.40	
	0470	112 H.P., 1-1/2 C.Y. capacity		10.30	370	1,105	3,325	303.40	
	0480	Attachments							
	0482	Compactor, 20,000 lb		1.75	196	588	1,775	131.60	
	0485	Hydraulic hammer, 750 ft-lbs		1.30	243	729	2,175	156.20	
	0486	Hydraulic hammer, 1200 ft-lbs		1.75	294	882	2,650	190.40	
	0750	Bucket, clamshell, general purpose, 3/8 C.Y.		.75	61.50	185	555	43	
	0800	1/2 C.Y.		.75	70	210	630	48	
	0850	3/4 C.Y.		.85	80	240	720	54.80	
	0900	1 C.Y.		.90	106	319	955	71	
	0950	1-1/2 C.Y.		1.45	133	400	1,200	91.60	
	1000	2 C.Y.		1.55	167	500	1,500	112.40	
	1200	Compactor, roller, 2 drum, 2000 lb., operator walking		3.75	135	405	1,225	111	
	1250	Rammer compactor, gas, 1000 lb. blow		.75	46.50	140	420	34	
	1300	Vibratory plate, gas, 13″ plate, 1000 lb. blow		.75	43.50	130	390	32	
	1350	24″ plate, 5000 lb. blow		1.95	75	225	675	60.60	
	1950	Hammer, pavement demo., hyd., gas, self-prop., 1000 to 1250 lb.		10.45	400	1,195	3,575	322.60	
	2000	Diesel 1300 to 1500 lb.		15.60	480	1,440	4,325	412.80	
	3710	Screening plant 110 hp. w / 5′ x 10′screen		13	415	1,250	3,750	354	
	3720	5′ x 16′ screen		14.30	495	1,480	4,450	410.40	
	4891	Attachments for all skid steer loaders							
	4892	Auger	Ea.	.49	81.50	245	735	52.90	
	4893	Backhoe		.63	105	314	940	67.85	
	4894	Broom		.62	103	309	925	66.75	
	4895	Forks		.20	33.50	101	305	21.80	
	4896	Grapple		.47	78	234	700	50.55	
	4897	Concrete hammer		1.02	171	512	1,525	110.55	
	4898	Tree spade		.92	153	460	1,375	99.35	
	4899	Trencher		.73	121	364	1,100	78.65	
	4900	Trencher, chain, boom type, gas, operator walking, 12 H.P.		1.60	140	420	1,250	96.80	
	4910	Operator riding, 40 H.P.		4.55	267	800	2,400	196.40	
	5000	Wheel type, diesel, 4′ deep, 12″ wide		10.60	565	1,700	5,100	424.80	
	5100	Diesel, 6′ deep, 20″ wide		19.10	805	2,420	7,250	636.80	
	5150	Ladder type, diesel, 5′ deep, 8″ wide		9.15	425	1,275	3,825	328.20	
	5200	Diesel, 8′ deep, 16″ wide		27.05	825	2,475	7,425	711.40	
	5250	Truck, dump, tandem, 12 ton payload		18.05	375	1,130	3,400	370.40	
200	5300	Three axle dump, 16 ton payload		24.85	475	1,420	4,250	482.80	200

01590 | Equipment Rental

			UNIT	HOURLY OPER. COST	RENT PER DAY	RENT PER WEEK	RENT PER MONTH	CREW EQUIPMENT COST/DAY	
200	5350	Dump trailer only, rear dump, 16-1/2 C.Y.	Ea.	4	180	540	1,625	140	200
	5400	20 C.Y.		4.45	183	550	1,650	145.60	
	5450	Flatbed, single axle, 1-1/2 ton rating		9.75	133	400	1,200	158	
	5500	3 ton rating		12.40	142	425	1,275	184.20	
	5550	Off highway rear dump, 25 ton capacity		37.80	1,000	2,995	8,975	901.40	
	5600	35 ton capacity		38.80	1,325	3,950	11,900	1,100	
400	0010	**GENERAL EQUIPMENT RENTAL** Without operators							400
	0150	Aerial lift, scissor type, to 15' high, 1000 lb. cap., electric	Ea.	1.15	83.50	250	750	59.20	
	0160	To 25' high, 2000 lb. capacity		2	128	385	1,150	93	
	0170	Telescoping boom to 40' high, 750 lb. capacity, gas		6.50	265	795	2,375	211	
	0180	1000 lb. capacity		8.20	390	1,175	3,525	300.60	
	0190	To 60' high, 750 lb. capacity		8.30	405	1,220	3,650	310.40	
	0200	Air compressor, portable, gas engine, 60 C.F.M.		5.41	58.50	175	525	78.30	
	0300	160 C.F.M.		6.80	73.50	220	660	98.40	
	0400	Diesel engine, rotary screw, 250 C.F.M.		6.90	117	350	1,050	125.20	
	0500	365 C.F.M.		8.10	150	450	1,350	154.80	
	0600	600 C.F.M.		13.25	242	725	2,175	251	
	0700	750 C.F.M.		17.40	258	775	2,325	294.20	
	0800	For silenced models, small sizes, add		3%	5%	5%	5%		
	0900	Large sizes, add		5%	7%	7%	7%		
	0920	Air tools and accessories							
	0930	Breaker, pavement, 60 lb.	Ea.	.20	30	90	270	19.60	
	0940	80 lb.		.20	40	120	360	25.60	
	0950	Drills, hand (jackhammer) 65 lb.		.20	26.50	80	240	17.60	
	0980	Dust control per drill		1.10	16	48	144	18.40	
	1000	Hose, air with couplings, 50' long, 3/4" diameter		.02	5	15	45	3.15	
	1100	1" diameter		.03	5.65	17	51	3.65	
	1200	1-1/2" diameter		.04	9.35	28	84	5.90	
	1300	2" diameter		.11	17.65	53	159	11.50	
	1400	2-1/2" diameter		.12	20.50	62	186	13.35	
	1410	3" diameter		.16	26.50	80	240	17.30	
	1450	Drill, steel, 7/8" x 2'		.04	4.33	13	39	2.90	
	1460	7/8" x 6'		.04	5.35	16	48	3.50	
	1520	Moil points		.03	4.33	13	39	2.85	
	1560	Tamper, single, 35 lb.		.45	29.50	89	267	21.40	
	1570	Triple, 140 lb.		.90	46.50	140	420	35.20	
	1580	Wrenches, impact, air powered, up to 3/4" bolt		.10	21.50	65	195	13.80	
	1590	Up to 1-1/4" bolt		.15	40	120	360	25.20	
	1600	Barricades, barrels, reflectorized, 1 to 50 barrels		.02	2.58	7.75	23.50	1.70	
	1610	100 to 200 barrels		.01	1.93	5.80	17.40	1.25	
	1620	Barrels with flashers, 1 to 50 barrels		.02	3.25	9.75	29.50	2.10	
	1630	100 to 200 barrels		.02	2.60	7.80	23.50	1.70	
	1640	Barrels with steady burn type C lights		.03	4.33	13	39	2.85	
	1690	Butt fusion machine, electric		10.40	365	1,095	3,275	302.20	
	1695	Electro fusion machine		4.20	142	425	1,275	118.60	
	1850	Drill, rotary hammer, electric, 1-1/2" diameter		.41	25.50	76	228	18.50	
	1860	Carbide bit for above		.03	5.35	16	48	3.45	
	2100	Generator, electric, gas engine, 1.5 KW to 3 KW		1.13	40	120	360	33.05	
	2200	5 KW		1.49	58.50	175	525	46.90	
	2300	10 KW		2.40	133	400	1,200	99.20	
	2400	25 KW		6.80	153	460	1,375	146.40	
	2500	Diesel engine, 20 KW		4.35	117	350	1,050	104.80	
	2600	50 KW		6.90	133	400	1,200	135.20	
	2700	100 KW		11.88	193	580	1,750	211.05	
	2800	250 KW		29.05	335	1,000	3,000	432.40	
	2900	Heaters, space, oil or electric, 50 MBH		.59	27.50	82	246	21.10	
	3000	100 MBH		1.06	35	105	315	29.50	
	3100	300 MBH		3.41	53.50	160	480	59.30	

01590 | Equipment Rental

		01590 | Equipment Rental	UNIT	HOURLY OPER. COST	RENT PER DAY	RENT PER WEEK	RENT PER MONTH	CREW EQUIPMENT COST/DAY	
400	3150	500 MBH	Ea.	6.77	70	210	630	96.15	400
	3200	Hose, water, suction with coupling, 20' long, 2" diameter		.06	10	30	90	6.50	
	3210	3" diameter		.06	15	45	135	9.50	
	3220	4" diameter		.06	20	60	180	12.50	
	3230	6" diameter		.06	36.50	110	330	22.50	
	3240	8" diameter		.36	59.50	178	535	38.50	
	3250	Discharge hose with coupling, 50' long, 2" diameter		.05	8.35	25	75	5.40	
	3260	3" diameter		.05	10	30	90	6.40	
	3270	4" diameter		.06	13.35	40	120	8.50	
	3280	6" diameter		.07	33.50	100	300	20.55	
	3290	8" diameter		.33	54.50	163	490	35.25	
	3300	Ladders, extension type, 16' to 36' long		.14	18.35	55	165	12.10	
	3400	40' to 60' long		.18	29.50	89	267	19.25	
	3500	Light towers, towable, with diesel generator, 2000 watt		1.50	108	325	975	77	
	3600	4000 watt		1.70	130	390	1,175	91.60	
	4100	Pump, centrifugal gas pump, 1-1/2", 4 MGPH		.85	31.50	95	285	25.80	
	4200	2", 8 MGPH		1.20	33.50	100	300	29.60	
	4300	3", 15 MGPH		1.40	46.50	140	420	39.20	
	4400	6", 90 MGPH		9.30	150	450	1,350	164.40	
	4500	Submersible electric pump, 1-1/4", 55 GPM		.35	36.50	110	330	24.80	
	4600	1-1/2", 83 GPM		.41	40	120	360	27.30	
	4700	2", 120 GPM		.42	46.50	140	420	31.35	
	4800	3", 300 GPM		.75	58.50	175	525	41	
	4900	4", 560 GPM		5.30	82	246	740	91.60	
	5000	6", 1590 GPM		8	233	700	2,100	204	
	5100	Diaphragm pump, gas, single, 1-1/2" diameter		.52	29.50	88	264	21.75	
	5200	2" diameter		.80	40.50	122	365	30.80	
	5300	3" diameter		.75	47.50	143	430	34.60	
	5400	Double, 4" diameter		1.75	80	240	720	62	
	5500	Trash pump, self-priming, gas, 2" diameter		1.65	40.50	122	365	37.60	
	5600	Diesel, 4" diameter		2.65	73.50	220	660	65.20	
	5650	Diesel, 6" diameter		6.40	122	365	1,100	124.20	
	5700	Salamanders, L.P. gas fired, 100,000 B.T.U.		1.06	20	60	180	20.50	
	5800	Saw, chain, gas engine, 18" long		.15	41.50	125	375	26.20	
	5900	36" long		.15	70	210	630	43.20	
	5950	60" long		.15	83.50	250	750	51.20	
	6000	Masonry, table mounted, 14" diameter, 5 H.P.		1.30	57	171	515	44.60	
	6100	Circular, hand held, electric, 7-1/4" diameter		.16	23.50	70	210	15.30	
	6200	12" diameter		.25	33.50	100	300	22	
	6275	Shot blaster, walk behind, 20" wide		1.05	230	690	2,075	146.40	
	6350	Torch, cutting, acetylene-oxygen, 150' hose		1.50	33.50	100	300	32	
	6360	Hourly operating cost includes tips and gas		7.80				62.40	
	6410	Toilet, portable chemical		.10	16.35	49	147	10.60	
	6420	Recycle flush type		.12	20.50	61	183	13.15	
	6430	Toilet, fresh water flush, garden hose,		.14	23	69	207	14.90	
	6440	Hoisted, non-flush, for high rise		.12	20	60	180	12.95	
	6450	Toilet, trailers, minimum		.20	34	102	305	22	
	6460	Maximum		.61	102	306	920	66.10	
	6500	Trailers, platform, flush deck, 2 axle, 25 ton capacity		3.35	123	370	1,100	100.80	
	6600	40 ton capacity		4.55	223	670	2,000	170.40	
	6700	3 axle, 50 ton capacity		5	247	740	2,225	188	
	6800	75 ton capacity		6.45	325	975	2,925	246.60	
	7100	Truck, pickup, 3/4 ton, 2 wheel drive		4.75	75	225	675	83	
	7200	4 wheel drive		4.90	83.50	250	750	89.20	
	7300	Tractor, 4 x 2, 30 ton capacity, 195 H.P.		12.55	365	1,100	3,300	320.40	
	7410	250 H.P.		17.20	435	1,300	3,900	397.60	
	7620	Vacuum truck, hazardous material, 2500 gallon		7.10	305	920	2,750	240.80	
	7625	5,000 gallon	↓	7.15	410	1,225	3,675	302.20	

01590 | Equipment Rental

			UNIT	HOURLY OPER. COST	RENT PER DAY	RENT PER WEEK	RENT PER MONTH	CREW EQUIPMENT COST/DAY	
400	7700	Welder, electric, 200 amp	Ea.	4.97	43.50	130	390	65.75	400
	7800	300 amp		7.83	63.50	190	570	100.65	
	7900	Gas engine, 200 amp		4.40	55	165	495	68.20	
	8000	300 amp		5.25	70	210	630	84	
	8100	Wheelbarrow, any size		.06	9.65	29	87	6.30	
600	0010	**LIFTING AND HOISTING EQUIPMENT RENTAL** R01590 -100							600
	0100	without operators							
	0200	Crane, climbing, 106' jib, 6000 lb. capacity, 410 FPM	Ea.	38.75	1,250	3,750	11,300	1,060	
	0300	101' jib, 10,250 lb. capacity, 270 FPM	"	43.75	1,575	4,750	14,300	1,300	
	0400	Tower, static, 130' high, 106' jib,							
	0500	6200 lb. capacity at 400 FPM	Ea.	41.73	1,450	4,345	13,000	1,203	
	0600	Crawler, cable, 1/2 C.Y., 15 tons at 12' radius		4.97	530	1,590	4,775	357.75	
	0700	3/4 C.Y., 20 tons at 12' radius		18.55	540	1,625	4,875	473.40	
	0800	1 C.Y., 25 tons at 12' radius		25.90	565	1,700	5,100	547.20	
	0900	Crawler mounted, lattice boom, 1-1/2 C.Y., 40 tons at 12' radius		27	835	2,500	7,500	716	
	1000	2 C.Y., 50 tons at 12' radius		36.35	990	2,975	8,925	885.80	
	1100	3 C.Y., 75 tons at 12' radius		39.15	1,025	3,100	9,300	933.20	
	1200	100 ton capacity, standard boom		47.35	1,375	4,150	12,500	1,209	
	1300	165 ton capacity, standard boom		71.50	2,225	6,700	20,100	1,912	
	1400	200 ton capacity, 150' boom		73.25	2,575	7,700	23,100	2,126	
	1500	450' boom		105.30	3,550	10,680	32,000	2,978	
	1600	Truck mounted, lattice boom, 6 x 4, 20 tons at 10' radius		18.10	685	2,050	6,150	554.80	
	1700	25 tons at 10' radius		23.15	1,050	3,150	9,450	815.20	
	1800	8 x 4, 30 tons at 10' radius		25.95	650	1,950	5,850	597.60	
	1900	40 tons at 12' radius		31.45	835	2,500	7,500	751.60	
	2000	8 x 4, 60 tons at 15' radius		38.65	1,000	3,000	9,000	909.20	
	2050	82 tons at 15' radius		41.57	1,525	4,600	13,800	1,253	
	2100	90 tons at 15' radius		43.08	1,575	4,725	14,200	1,290	
	2200	115 tons at 15' radius		45.40	1,800	5,400	16,200	1,443	
	2300	150 tons at 18' radius		46	1,850	5,550	16,700	1,478	
	2350	165 tons at 18' radius		51.75	2,175	6,500	19,500	1,714	
	2400	Truck mounted, hydraulic, 12 ton capacity		27.35	495	1,480	4,450	514.80	
	2500	25 ton capacity		28.70	665	2,000	6,000	629.60	
	2550	33 ton capacity		31.85	915	2,750	8,250	804.80	
	2600	55 ton capacity		38.95	985	2,950	8,850	901.60	
	2700	80 ton capacity		56.80	1,450	4,355	13,100	1,325	
	2800	Self-propelled, 4 x 4, with telescoping boom, 5 ton		11.50	320	960	2,875	284	
	2900	12-1/2 ton capacity		17.15	480	1,440	4,325	425.20	
	3000	15 ton capacity		19.45	600	1,800	5,400	515.60	
	3100	25 ton capacity		22.25	705	2,120	6,350	602	
	3200	Derricks, guy, 20 ton capacity, 60' boom, 75' mast		11.44	305	920	2,750	275.50	
	3300	100' boom, 115' mast		18.63	530	1,590	4,775	467.05	
	3400	Stiffleg, 20 ton capacity, 70' boom, 37' mast		13.30	395	1,185	3,550	343.40	
	3500	100' boom, 47' mast		20.98	640	1,925	5,775	552.85	
	3550	Helicopter, small, lift to 1250 lbs. maximum, w/pilot		62.12	2,475	7,445	22,300	1,986	
	3600	Hoists, chain type, overhead, manual, 3/4 ton		.06	6.65	20	60	4.50	
	3900	10 ton		.25	28.50	85	255	19	
	4000	Hoist and tower, 5000 lb. cap., portable electric, 40' high		3.75	177	530	1,600	136	
	4100	For each added 10' section, add		.08	13.65	41	123	8.85	
	4200	Hoist and single tubular tower, 5000 lb. electric, 100' high		5.08	247	740	2,225	188.65	
	4300	For each added 6'-6" section, add		.14	23.50	71	213	15.30	
	4400	Hoist and double tubular tower, 5000 lb., 100' high		5.45	272	815	2,450	206.60	
	4500	For each added 6'-6" section, add		.15	25.50	77	231	16.60	
	4550	Hoist and tower, mast type, 6000 lb., 100' high		5.88	282	845	2,525	216.05	
	4570	For each added 10' section, add		.10	17	51	153	11	
	4600	Hoist and tower, personnel, electric, 2000 lb., 100' @ 125 FPM		12.33	750	2,245	6,725	547.65	
	4700	3000 lb., 100' @ 200 FPM		14.13	850	2,550	7,650	623.05	

01590 | Equipment Rental

		01590 Equipment Rental	UNIT	HOURLY OPER. COST	RENT PER DAY	RENT PER WEEK	RENT PER MONTH	CREW EQUIPMENT COST/DAY	
600	4800	3000 lb., 150' @ 300 FPM (R01590-100)	Ea.	15.65	950	2,855	8,575	696.20	600
	4900	4000 lb., 100' @ 300 FPM		16.20	970	2,910	8,725	711.60	
	5000	6000 lb., 100' @ 275 FPM		17.50	1,025	3,060	9,175	752	
	5100	For added heights up to 500', add	L.F.	.01	1.70	5.10	15.30	1.10	
	5200	Jacks, hydraulic, 20 ton	Ea.	.15	14.65	44	132	10	
	5500	100 ton	"	.20	40.50	122	365	26	
	6000	Jacks, hydraulic, climbing with 50' jackrods							
	6010	and control consoles, minimum 3 mo. rental							
	6100	30 ton capacity	Ea.	1.53	102	305	915	73.25	
	6150	For each added 10' jackrod section, add		.05	3.33	10	30	2.40	
	6300	50 ton capacity		2.45	163	490	1,475	117.60	
	6350	For each added 10' jackrod section, add		.06	4	12	36	2.90	
	6500	125 ton capacity		6.38	425	1,275	3,825	306.05	
	6550	For each added 10' jackrod section, add		.44	29	87	261	20.90	
	6600	Cable jack, 10 ton capacity with 200' cable		1.28	85	255	765	61.25	
	6650	For each added 50' of cable, add		.13	8.65	26	78	6.25	
700	0010	**WELLPOINT EQUIPMENT RENTAL** See also division 02240							700
	0020	Based on 2 months rental							
	0100	Combination jetting & wellpoint pump, 60 H.P. diesel	Ea.	8.29	252	755	2,275	217.30	
	0200	High pressure gas jet pump, 200 H.P., 300 psi	"	14.52	215	645	1,925	245.15	
	0300	Discharge pipe, 8" diameter	L.F.	.01	.41	1.22	3.66	.30	
	0350	12" diameter		.01	.60	1.80	5.40	.45	
	0400	Header pipe, flows up to 150 G.P.M., 4" diameter		.01	.37	1.12	3.36	.30	
	0500	400 G.P.M., 6" diameter		.01	.43	1.30	3.90	.35	
	0600	800 G.P.M., 8" diameter		.01	.60	1.80	5.40	.45	
	0700	1500 G.P.M., 10" diameter		.01	.63	1.90	5.70	.45	
	0800	2500 G.P.M., 12" diameter		.02	1.20	3.60	10.80	.90	
	0900	4500 G.P.M., 16" diameter		.02	1.53	4.60	13.80	1.10	
	0950	For quick coupling aluminum and plastic pipe, add		.02	1.58	4.75	14.25	1.10	
	1100	Wellpoint, 25' long, with fittings & riser pipe, 1-1/2" or 2" diameter	Ea.	.05	3.17	9.50	28.50	2.30	
	1200	Wellpoint pump, diesel powered, 4" diameter, 20 H.P.		4.05	145	435	1,300	119.40	
	1300	6" diameter, 30 H.P.		5.28	180	540	1,625	150.25	
	1400	8" suction, 40 H.P.		7.18	247	740	2,225	205.45	
	1500	10" suction, 75 H.P.		9.81	288	865	2,600	251.50	
	1600	12" suction, 100 H.P.		14.70	460	1,385	4,150	394.60	
	1700	12" suction, 175 H.P.		19.46	510	1,530	4,600	461.70	

01700 | Execution Requirements

01740 | Cleaning

			CREW	DAILY OUTPUT	LABOR-HOURS	UNIT	2001 BARE COSTS MAT.	LABOR	EQUIP.	TOTAL	TOTAL INCL O&P	
500	0010	**CLEANING UP** After job completion, allow, minimum				Job					.30%	500
	0040	Maximum				"					1%	
	0050	Cleanup of floor area, continuous, per day	A-5	24	.750	M.S.F.	1.65	17.20	1.64	20.49	30.50	
	0100	Final	"	11.50	1.565	"	2.63	36	3.43	42.06	62.50	
	1000	Mechanical demolition, see division 02225										

01800 | Facility Operation

01810 | Commissioning

			CREW	DAILY OUTPUT	LABOR-HOURS	UNIT	2001 BARE COSTS MAT.	LABOR	EQUIP.	TOTAL	TOTAL INCL O&P	
100	0010	**COMMISSIONING**										100
	0100	As-built drawings, punchlist, training, manuals, minimum				Project					.50%	
	0150	Maximum				"					2%	

Division Notes

	CREW	DAILY OUTPUT	LABOR-HOURS	UNIT	2001 BARE COSTS MAT.	LABOR	EQUIP.	TOTAL	TOTAL INCL O&P

Division 2
Site Construction

Estimating Tips

02200 Site Preparation

- If possible visit the site and take an inventory of the type, quantity and size of the trees. Certain trees may have a landscape resale value or firewood value. Stump disposal can be very expensive, particularly if they cannot be buried at the site. Consider using a bulldozer in lieu of hand cutting trees.
- Estimators should visit the site to determine the need for haul road, access, storage of materials, and security considerations. When estimating for access roads on unstable soil, consider using a geotextile stabilization fabric. It can greatly reduce the quantity of crushed stone or gravel. Sites of limited size and access can cause cost overruns due to lost productivity. Theft and damage is another consideration if the location is isolated. A temporary fence or security guards may be required. Investigate the site thoroughly.

02210 Subsurface Investigation

In preparing estimates on structures involving earthwork or foundations, all information concerning soil characteristics should be obtained. Look particularly for hazardous waste, evidence of prior dumping of debris, and previous stream beds.

- The costs shown for selective demolition do not include rubbish handling or disposal. These items should be estimated separately using Means data or other sources.

02300 Earthwork

- Estimating the actual cost of performing earthwork requires careful consideration of the variables involved. This includes items such as type of soil, whether or not water will be encountered, dewatering, whether or not banks need bracing, disposal of excavated earth, length of haul to fill or spoil sites, etc. If the project has large quantities of cut or fill, consider raising or lowering the site to reduce costs while paying close attention to the effect on site drainage and utilities if doing this.
- If the project has large quantities of fill, creating a borrow pit on the site can significantly lower the costs. It is very important to consider what time of year the project is scheduled for completion. Bad weather can create large cost overruns from dewatering, site repair and lost productivity from cold weather.

02700 Bases, Ballasts, Pavements/Appurtenances

- When estimating paving, keep in mind the project schedule. If an asphaltic paving project is in a colder climate and runs through to the spring, consider placing the base course in the autumn, then topping it in the spring just prior to completion. This could save considerable costs in spring repair. Keep in mind that prices for asphalt and concrete are generally higher in the cold seasons.

02500 Utility Services
02600 Drainage & Containment

- Never assume that the water, sewer and drainage lines will go in at the early stages of the project. Consider the site access needs before dividing the site in half with open trenches, loose pipe, and machinery obstructions. Always inspect the site to establish that the site drawings are complete. Check off all existing utilities on your drawing as you locate them. If you find any discrepancies, mark up the site plan for further research. Differing site conditions can be very costly if discovered later in the project.

02900 Planting

- The timing of planting and guarantee specifications often dictate the costs for establishing tree and shrub growth and a stand of grass or ground cover. Establish the work performance schedule to coincide with the local planting season. Maintenance and growth guarantees can add from 20% to 100% to the total landscaping cost. The cost to replace trees and shrubs can be as high as 5% of the total cost depending on the planting zone, soil conditions and time of year.

Reference Numbers

Reference numbers are shown in bold squares at the beginning of some major classifications. These numbers refer to related items in the Reference Section. The reference information may be an estimating procedure, an alternate pricing method or technical information.

Note: Not all subdivisions listed here necessarily appear in this publication.

	02080 | Utility Materials		CREW	DAILY OUTPUT	LABOR-HOURS	UNIT	2001 BARE COSTS MAT.	LABOR	EQUIP.	TOTAL	TOTAL INCL O&P	
400	0010	**FIRE HYDRANTS**, Mech. joints unless noted (A12.3 -922)										400
	1000	Fire hydrants, two way; excavation and backfill not incl.										
	1100	4-1/2″ valve size, depth 2′-0″	B-21	10	2.800	Ea.	660	74.50	14.20	748.70	860	
	1200	4′-6″		9	3.111		755	83	15.75	853.75	975	
	1260	6′-0″		7	4		780	106	20.50	906.50	1,050	
	1340	8′-0″		6	4.667		920	124	23.50	1,067.50	1,250	
	1420	10′-0″		5	5.600		975	149	28.50	1,152.50	1,350	
	2000	5-1/4″ valve size, depth 2′-0″		10	2.800		825	74.50	14.20	913.70	1,050	
	2080	4′-0″		9	3.111		930	83	15.75	1,028.75	1,175	
	2160	6′-0″		7	4		1,050	106	20.50	1,176.50	1,350	
	2240	8′-0″		6	4.667		1,150	124	23.50	1,297.50	1,500	
	2320	10′-0″		5	5.600		1,250	149	28.50	1,427.50	1,675	
	2350	For threeway valves, add					7%					
	2400	Lower barrel extensions with stems, 1′-0″	B-20	14	1.714		290	44.50		334.50	390	
	2440	2′-0″		13	1.846		355	47.50		402.50	465	
	2480	3′-0″		12	2		820	51.50		871.50	980	
	2520	4′-0″		10	2.400		480	62		542	625	
	5000	Indicator post										
	5020	Adjustable, valve size 4″ to 14″, 4′ bury	B-21	10	2.800	Ea.	340	74.50	14.20	428.70	505	
	5060	8′ bury		7	4		440	106	20.50	566.50	670	
	5080	10′ bury		6	4.667		485	124	23.50	632.50	755	
	5100	12′ bury		5	5.600		730	149	28.50	907.50	1,075	
	5120	14′ bury		4	7		800	186	35.50	1,021.50	1,200	
	5500	Non-adjustable, valve size 4″ to 14″, 3′ bury		10	2.800		360	74.50	14.20	448.70	525	
	5520	3′-6″ bury		10	2.800		360	74.50	14.20	448.70	525	
	5540	4′ bury		9	3.111		360	83	15.75	458.75	540	
790	0010	**UNDERGROUND MARKING TAPE**										790
	0400	Underground tape, detectable aluminum, 2″	1 Clab	150	.053	C.L.F.	3.70	1.22		4.92	6	
	0500	6″	″	140	.057	″	9.25	1.31		10.56	12.25	
800	0010	**UTILITY VAULTS** Precast concrete, 6″ thick (A12.3 -710)										800
	0050	5′ x 10′ x 6′ high, I.D.	B-13	2	28	Ea.	1,350	695	315	2,360	2,925	
	0100	6′ x 10′ x 6′ high, I.D.		2	28		1,400	695	315	2,410	2,975	
	0150	5′ x 12′ x 6′ high, I.D.		2	28		1,500	695	315	2,510	3,075	
	0200	6′ x 12′ x 6′ high, I.D.		1.80	31.111		1,675	770	350	2,795	3,400	
	0250	6′ x 13′ x 6′ high, I.D.		1.50	37.333		2,200	925	420	3,545	4,275	
	0300	8′ x 14′ x 7′ high, I.D.		1	56		2,375	1,375	630	4,380	5,450	
	0350	Hand hole, precast concrete, 1-1/2″ thick										
	0400	1′-0″ x 2′-0″ x 1′-9″, I.D., light duty	B-1	4	6	Ea.	235	141		376	480	
	0450	4′-6″ x 3′-2″ x 2′-0″, O.D., heavy duty	B-6	3	8	″	545	199	60.50	804.50	970	
900	0010	**PIPING VALVES** Water distribution, see also div. 15110										900
	3000	Butterfly valves with boxes, cast iron, mech. jt.										
	3100	4″ diameter	B-20	6	4	Ea.	228	103		331	410	
	3140	6″ diameter	″	5	4.800		262	124		386	480	
	3180	8″ diameter	B-21	4	7		315	186	35.50	536.50	680	
	3300	10″ diameter		3.50	8		465	213	40.50	718.50	890	
	3340	12″ diameter		3	9.333		610	248	47.50	905.50	1,100	
	3400	14″ diameter		2	14		1,050	375	71	1,496	1,800	
	3440	16″ diameter		2	14		1,400	375	71	1,846	2,200	
	3460	18″ diameter		1.50	18.667		1,750	495	94.50	2,339.50	2,800	
	3480	20″ diameter		1	28		2,200	745	142	3,087	3,725	
	3500	24″ diameter		.50	56		3,075	1,500	284	4,859	6,000	
	3600	With lever operator										
	3610	4″ diameter	B-20	6	4	Ea.	121	103		224	295	
	3614	6″ diameter	″	5	4.800		169	124		293	380	
	3616	8″ diameter	B-21	4	7		245	186	35.50	466.50	600	

02080 | Utility Materials

			DAILY	LABOR-		2001 BARE COSTS				TOTAL
		CREW	OUTPUT	HOURS	UNIT	MAT.	LABOR	EQUIP.	TOTAL	INCL O&P
900 3618	10" diameter	B-21	3.50	8	Ea.	395	213	40.50	648.50	805
3620	12" diameter		3	9.333		730	248	47.50	1,025.50	1,225
3622	14" diameter		2	14		1,175	375	71	1,621	1,925
3624	16" diameter		2	14		1,575	375	71	2,021	2,400
3626	18" diameter		1.50	18.667		1,825	495	94.50	2,414.50	2,875
3628	20" diameter		1	28		2,425	745	142	3,312	3,950
3630	24" diameter		.50	56		2,600	1,500	284	4,384	5,500
3700	Check valves, flanged									
3710	4" diameter	B-20	6	4	Ea.	395	103		498	595
3714	6" diameter	"	5	4.800		665	124		789	930
3716	8" diameter	B-21	4	7		1,350	186	35.50	1,571.50	1,825
3718	10" diameter		3.50	8		2,175	213	40.50	2,428.50	2,775
3720	12" diameter		3	9.333		3,325	248	47.50	3,620.50	4,100
3722	14" diameter		2	14		5,325	375	71	5,771	6,500
3724	16" diameter		2	14		7,325	375	71	7,771	8,725
3726	18" diameter		1.50	18.667		10,900	495	94.50	11,489.50	12,800
3728	20" diameter		1	28		12,600	745	142	13,487	15,200
3730	24" diameter		.50	56		18,200	1,500	284	19,984	22,600
3800	Gate valves, C.I., 250 PSI, mechanical joint, w/boxes									
3810	4" diameter	B-21A	8	5	Ea.	160	141	53	354	450
3814	6" diameter		6.80	5.882		193	166	62.50	421.50	535
3816	8" diameter		5.60	7.143		285	201	76	562	710
3818	10" diameter		4.80	8.333		460	235	88.50	783.50	970
3820	12" diameter		4.08	9.804		635	276	104	1,015	1,225
3822	14" diameter	B-21	2	14		735	375	71	1,181	1,475
3824	16" diameter		1	28		865	745	142	1,752	2,250
3826	18" diameter		.80	35		1,075	930	177	2,182	2,825
3828	20" diameter		.80	35		1,350	930	177	2,457	3,150
3830	24" diameter		.50	56		1,675	1,500	284	3,459	4,450
3831	30" diameter		.35	80		2,000	2,125	405	4,530	5,975
3832	36" diameter		.30	93.333		2,400	2,475	475	5,350	7,050
3880	Sleeve, for tapping mains, 8" x 4", add					315			315	345
3884	10" x 6", add					900			900	990
3888	12" x 6", add					900			900	990
3892	12" x 8", add					950			950	1,050
3900	Globe valves, flanged, iron body, class 125									
3910	4" diameter	B-20	10	2.400	Ea.	865	62		927	1,050
3914	6" diameter	"	9	2.667		1,625	69		1,694	1,875
3916	8" diameter	B-21	6	4.667		3,000	124	23.50	3,147.50	3,525
3918	10" diameter		5	5.600		4,725	149	28.50	4,902.50	5,475
3920	12" diameter		4	7		6,625	186	35.50	6,846.50	7,600
3922	14" diameter		3	9.333		7,100	248	47.50	7,395.50	8,225
3924	16" diameter		2	14		7,000	375	71	7,446	8,350
3926	18" diameter		.80	35		7,700	930	177	8,807	10,100
3928	20" diameter		.60	46.667		7,600	1,250	237	9,087	10,500
3930	24" diameter		.50	56		9,000	1,500	284	10,784	12,500

02110 | Excavation, Removal & Handling

			CREW	DAILY OUTPUT	LABOR-HOURS	UNIT	2001 BARE COSTS MAT.	LABOR	EQUIP.	TOTAL	TOTAL INCL O&P	
300	0010	**HAZARDOUS WASTE CLEANUP/PICKUP/DISPOSAL**										300
	0100	For contractor equipment, i.e. dozer,										
	0110	Front end loader, dump truck, etc., see div. 01590-200										
	1000	Solid pickup										
	1100	55 gal. drums				Ea.					200	
	1120	Bulk material, minimum				Ton					150	
	1130	Maximum				"					500	
	1200	Transportation to disposal site										
	1220	Truckload = 80 drums or 25 C.Y. or 18 tons										
	1260	Minimum				Mile					2.30	
	1270	Maximum				"					4	
	3000	Liquid pickup, vacuum truck, stainless steel tank										
	3100	Minimum charge, 4 hours										
	3110	1 compartment, 2200 gallon				Hr.					100	
	3120	2 compartment, 5000 gallon				"					100	
	3400	Transportation in 6900 gallon bulk truck				Mile					4.30	
	3410	In teflon lined truck				"					5	
	5000	Heavy sludge or dry vacuumable material				Hr.					100	
	6000	Dumpsite disposal charge, minimum				Ton					100	
	6020	Maximum				"					400	

02115 | Underground Tank Removal

			CREW	DAILY OUTPUT	LABOR-HOURS	UNIT	MAT.	LABOR	EQUIP.	TOTAL	TOTAL INCL O&P	
200	0010	**REMOVAL OF UNDERGROUND STORAGE TANKS** R02115-200										200
	0011	Petroleum storage tanks, non-leaking										
	0100	Excavate & load onto trailer										
	0110	3000 gal. to 5000 gal. tank	B-14	4	12	Ea.		290	45	335	500	
	0120	6000 gal to 8000 gal tank	B-3A	3	13.333			325	198	523	720	
	0130	9000 gal to 12000 gal tank	"	2	20	↓		485	297	782	1,075	
	0190	Known leaking tank add				%				100%	100%	
	0200	Remove sludge, water and remaining product from tank bottom										
	0201	of tank with vacuum truck										
	0300	3000 gal to 5000 gal tank	A-13	5	1.600	Ea.		46	104	150	185	
	0310	6000 gal to 8000 gal tank		4	2			57.50	130	187.50	230	
	0320	9000 gal to 12000 gal tank	↓	3	2.667	↓		76.50	174	250.50	305	
	0390	Dispose of sludge off-site, average				Gal.					4	
	0400	Insert inert solid CO2 "dry ice" into tank										
	0401	For cleaning/transporting tanks (1.5 lbs./100 gal. cap)	1 Clab	500	.016	Lb.	1.20	.37		1.57	1.89	
	1020	Haul tank to certified salvage dump, 100 miles round trip										
	1023	3000 gal. to 5000 gal. tank				Ea.				550	630	
	1026	6000 gal. to 8000 gal. tank								650	750	
	1029	9,000 gal. to 12,000 gal. tank				↓				875	1,000	
	1100	Disposal of contaminated soil to landfill										
	1110	Minimum				C.Y.					100	
	1111	Maximum				"					300	
	1120	Disposal of contaminated soil to										
	1121	bituminous concrete batch plant										
	1130	Minimum				C.Y.					50	
	1131	Maximum				"					100	
	2010	Decontamination of soil on site incl poly tarp on top/bottom										
	2011	Soil containment berm, and chemical treatment										
	2020	Minimum	B-11C	100	.160	C.Y.	5.60	4.26	1.81	11.67	14.70	
	2021	Maximum	"	100	.160		7.25	4.26	1.81	13.32	16.55	
	2050	Disposal of decontaminated soil, minimum									60	
	2055	Maximum				↓					125	

02210 | Subsurface Investigation

		02210 Subsurface Investigation	CREW	DAILY OUTPUT	LABOR-HOURS	UNIT	2001 BARE COSTS MAT.	LABOR	EQUIP.	TOTAL	TOTAL INCL O&P	
320	0010	**DRILLING, CORE** Reinforced concrete slab, up to 6″ thick slab										320
	0020	Including bit, layout and set up										
	0100	1″ diameter core	B-89A	28	.571	Ea.	2.30	15.05	3.62	20.97	30	
	0150	Each added inch thick, add		300	.053		.41	1.41	.34	2.16	3.02	
	0300	3″ diameter core		23	.696		5.10	18.35	4.41	27.86	39	
	0350	Each added inch thick, add		186	.086		.92	2.27	.55	3.74	5.15	
	0500	4″ diameter core		19	.842		5.10	22	5.35	32.45	46.50	
	0550	Each added inch thick, add		170	.094		1.17	2.48	.60	4.25	5.85	
	0700	6″ diameter core		14	1.143		8.45	30	7.25	45.70	64.50	
	0750	Each added inch thick, add		140	.114		1.43	3.01	.72	5.16	7.10	
	0900	8″ diameter core		11	1.455		11.55	38.50	9.20	59.25	83	
	0950	Each added inch thick, add		95	.168		1.94	4.44	1.07	7.45	10.25	
	1100	10″ diameter core		10	1.600		15.40	42	10.15	67.55	94	
	1150	Each added inch thick, add		80	.200		2.55	5.25	1.27	9.07	12.45	
	1300	12″ diameter core		9	1.778		18.45	47	11.25	76.70	106	
	1350	Each added inch thick, add		68	.235		3.06	6.20	1.49	10.75	14.70	
	1500	14″ diameter core		7	2.286		22.50	60	14.50	97	135	
	1550	Each added inch thick, add		55	.291		3.88	7.65	1.84	13.37	18.30	
	1700	18″ diameter core		4	4		29	105	25.50	159.50	225	
	1750	Each added inch thick, add		28	.571		5.10	15.05	3.62	23.77	33	
	1760	For horizontal holes, add to above								30%	30%	
	1770	Prestressed hollow core plank, 6″ thick										
	1780	1″ diameter core	B-89A	52	.308	Ea.	1.53	8.10	1.95	11.58	16.50	
	1790	Each added inch thick, add		350	.046		.26	1.20	.29	1.75	2.50	
	1800	3″ diameter core		50	.320		3.37	8.45	2.03	13.85	19.15	
	1810	Each added inch thick, add		240	.067		.56	1.76	.42	2.74	3.83	
	1820	4″ diameter core		48	.333		4.49	8.80	2.11	15.40	21	
	1830	Each added inch thick, add		216	.074		.77	1.95	.47	3.19	4.43	
	1840	6″ diameter core		44	.364		5.55	9.60	2.31	17.46	23.50	
	1850	Each added inch thick, add		175	.091		.92	2.41	.58	3.91	5.45	
	1860	8″ diameter core		32	.500		7.45	13.20	3.17	23.82	32	
	1870	Each added inch thick, add		118	.136		1.28	3.57	.86	5.71	7.95	
	1880	10″ diameter core		28	.571		10.05	15.05	3.62	28.72	38.50	
	1890	Each added inch thick, add		99	.162		1.38	4.26	1.02	6.66	9.30	
	1900	12″ diameter core		22	.727		12.25	19.15	4.61	36.01	48.50	
	1910	Each added inch thick, add		85	.188		2.04	4.96	1.19	8.19	11.30	
	1950	Minimum charge for above, 3″ diameter core		7	2.286	Total		60	14.50	74.50	110	
	2000	4″ diameter core		6.80	2.353			62	14.90	76.90	113	
	2050	6″ diameter core		6	2.667			70.50	16.90	87.40	129	
	2100	8″ diameter core		5.50	2.909			76.50	18.45	94.95	141	
	2150	10″ diameter core		4.75	3.368			89	21.50	110.50	163	
	2200	12″ diameter core		3.90	4.103			108	26	134	198	
	2250	14″ diameter core		3.38	4.734			125	30	155	229	
	2300	18″ diameter core		3.15	5.079			134	32	166	246	

02220 | Site Demolition

		02220 Site Demolition	CREW	DAILY OUTPUT	LABOR-HOURS	UNIT	MAT.	LABOR	EQUIP.	TOTAL	TOTAL INCL O&P	
875	0010	**SITE DEMOLITION** No hauling, abandon catch basin or manhole	B-6	7	3.429	Ea.		85	26	111	160	875
	0020	Remove existing catch basin or manhole, masonry		4	6			149	45	194	280	
	0030	Catch basin or manhole frames and covers, stored		13	1.846			46	13.90	59.90	86.50	
	0040	Remove and reset		7	3.429			85	26	111	160	
	0900	Hydrants, fire, remove only	B-21A	5	8			225	85	310	440	
	0950	Remove and reset	″	2	20			565	213	778	1,100	
	1710	Pavement removal, bituminous roads, 3″ thick	B-38	690	.058	S.Y.		1.50	1.36	2.86	3.81	
	1750	4″ to 6″ thick	″	420	.095	″		2.47	2.23	4.70	6.25	
	2900	Pipe removal, sewer/water, no excavation, 12″ diameter	B-6	175	.137	L.F.		3.40	1.03	4.43	6.40	
	2930	15″ diameter		150	.160			3.97	1.21	5.18	7.50	

For expanded coverage of these items see *Means Heavy Construction Cost Data 2001*

02200 | Site Preparation

02220 | Site Demolition

			Crew	Daily Output	Labor-Hours	Unit	2001 Bare Costs Mat.	Labor	Equip.	Total	Total Incl O&P	
875	2960	24" diameter	B-6	120	.200	L.F.		4.96	1.51	6.47	9.30	875
	3000	36" diameter		90	.267			6.60	2.01	8.61	12.45	
	3200	Steel, welded connections, 4" diameter		160	.150			3.72	1.13	4.85	7	
	3300	10" diameter		80	.300			7.45	2.26	9.71	14	
	4000	Sidewalk removal, bituminous, 2-1/2" thick		325	.074	S.Y.		1.83	.56	2.39	3.44	
	4100	Concrete, plain, 4"		160	.150	"		3.72	1.13	4.85	7	

02225 | Selective Demolition

			Crew	Daily Output	Labor-Hours	Unit	2001 Bare Costs Mat.	Labor	Equip.	Total	Total Incl O&P	
730	0010	**RUBBISH HANDLING** The following are to be added to the										730
	0020	demolition prices										
	0400	Chute, circular, prefabricated steel, 18" diameter	B-1	40	.600	L.F.	19.60	14.10		33.70	43.50	
	0440	30" diameter	"	30	.800	"	26.50	18.80		45.30	59	
	0725	Dumpster, weekly rental, 1 dump/week, 20C.Y. capacity (8Tons)				Week					425	
	0800	30 C.Y. capacity (10 Tons)									640	
	0840	40 C.Y. capacity (13 Tons)									775	
	1000	Dust partition, 6 mil polyethylene, 4' x 8' panels, 1" x 3" frame	2 Carp	2,000	.008	S.F.	.17	.23		.40	.54	
	1080	2" x 4" frame	"	2,000	.008	"	.33	.23		.56	.72	
	2000	Load, haul to chute & dumping into chute, 50' haul	2 Clab	24	.667	C.Y.		15.25		15.25	24	
	2040	100' haul		16.50	.970			22		22	34.50	
	2080	Over 100' haul, add per 100 L.F.		35.50	.451			10.30		10.30	16.10	
	2120	In elevators, per 10 floors, add		140	.114			2.61		2.61	4.09	
	3000	Loading & trucking, including 2 mile haul, chute loaded	B-16	45	.711			16.80	10.75	27.55	38	
	3040	Hand loading truck, 50' haul	"	48	.667			15.75	10.05	25.80	35.50	
	3080	Machine loading truck	B-17	120	.267			6.55	4.59	11.14	15.15	
	5000	Haul, per mile, up to 8 C.Y. truck	B-34B	1,165	.007			.16	.41	.57	.71	
	5100	Over 8 C.Y. truck	"	1,550	.005			.12	.31	.43	.53	
790	0010	**TORCH CUTTING** Steel, 1" thick plate	1 Clab	32	.250	L.F.		5.70		5.70	8.95	790
	0040	1" diameter bar	"	210	.038	Ea.		.87		.87	1.36	
	1000	Oxygen lance cutting, reinforced concrete walls										
	1040	12" to 16" thick walls	1 Clab	10	.800	L.F.		18.30		18.30	28.50	
	1080	24" thick walls	"	6	1.333	"		30.50		30.50	47.50	
840	0010	**WALLS AND PARTITIONS DEMOLITION**										840
	3800	Toilet partitions, slate or marble	1 Clab	5	1.600	Ea.		36.50		36.50	57	
	3820	Hollow metal	"	8	1	"		23		23	36	

02240 | Dewatering

			Crew	Daily Output	Labor-Hours	Unit	2001 Bare Costs Mat.	Labor	Equip.	Total	Total Incl O&P	
500	0010	**DEWATERING** Excavate drainage trench, 2' wide, 2' deep	B-11C	90	.178	C.Y.		4.73	2.01	6.74	9.45	500
	0100	2' wide, 3' deep, with backhoe loader	"	135	.119			3.15	1.34	4.49	6.30	
	0200	Excavate sump pits by hand, light soil	1 Clab	7.10	1.127			26		26	40.50	
	0300	Heavy soil	"	3.50	2.286			52		52	81.50	
	0500	Pumping 8 hr., attended 2 hrs. per day, including 20 L.F.										
	0550	of suction hose & 100 L.F. discharge hose										
	0600	2" diaphragm pump used for 8 hours	B-10H	4	3	Day		83.50	12.05	95.55	141	
	0650	4" diaphragm pump used for 8 hours	B-10I	4	3			83.50	23	106.50	153	
	0800	8 hrs. attended, 2" diaphragm pump	B-10H	1	12			335	48	383	565	
	0900	3" centrifugal pump	B-10J	1	12			335	61.50	396.50	580	
	1000	4" diaphragm pump	B-10I	1	12			335	91.50	426.50	610	
	1100	6" centrifugal pump	B-10K	1	12			335	228	563	760	
	1300	CMP, incl. excavation 3' deep, 12" diameter	B-6	115	.209	L.F.	6.35	5.20	1.57	13.12	16.70	
	1400	18" diameter		100	.240	"	7.90	5.95	1.81	15.66	19.90	
	1600	Sump hole construction, incl. excavation and gravel, pit		1,250	.019	C.F.	.67	.48	.14	1.29	1.63	
	1700	With 12" gravel collar, 12" pipe, corrugated, 16 ga.		70	.343	L.F.	12.10	8.50	2.59	23.19	29.50	
	1800	15" pipe, corrugated, 16 ga.		55	.436		15.45	10.85	3.29	29.59	37.50	
	1900	18" pipe, corrugated, 16 ga.		50	.480		18.15	11.90	3.62	33.67	42.50	

02200 | Site Preparation

02240 | Dewatering

		02240 Dewatering	CREW	DAILY OUTPUT	LABOR-HOURS	UNIT	2001 BARE COSTS MAT.	LABOR	EQUIP.	TOTAL	TOTAL INCL O&P	
500	2000	24″ pipe, corrugated, 14 ga.	B-6	40	.600	L.F.	22	14.90	4.52	41.42	52	500
	2200	Wood lining, up to 4′ x 4′, add	↓	300	.080	SFCA	12.30	1.99	.60	14.89	17.25	
	9950	See div. 02240-900 for wellpoints										
	9960	See div. 02240-700 for deep well systems										
700	0010	**WELLS** For dewatering 10′ to 20′ deep, 2′ diameter										700
	0020	with steel casing, minimum	B-6	165	.145	V.L.F.	2	3.61	1.10	6.71	9	
	0050	Average		98	.245		4	6.10	1.85	11.95	15.85	
	0100	Maximum	↓	49	.490	↓	10	12.15	3.69	25.84	34	
	0300	For pumps for dewatering, see division 01590-400-4100 to 4400										
	0500	For domestic water wells, see division 02520-900										
900	0010	**WELLPOINTS** For wellpoint equipment rental, see div. 01590-700 R02240-900										900
	0100	Installation and removal of single stage system										
	0110	Labor only, .75 labor-hours per L.F., minimum	1 Clab	10.70	.748	LF Hdr		17.10		17.10	26.50	
	0200	2.0 labor-hours per L.F., maximum	"	4	2	"		45.50		45.50	71.50	
	0400	Pump operation, 4 @ 6 hr. shifts										
	0410	Per 24 hour day	4 Eqlt	1.27	25.197	Day		725		725	1,100	
	0500	Per 168 hour week, 160 hr. straight, 8 hr. double time		.18	177	Week		5,100		5,100	7,750	
	0550	Per 4.3 week month	↓	.04	800	Month		23,000		23,000	34,800	
	0600	Complete installation, operation, equipment rental, fuel &										
	0610	removal of system with 2″ wellpoints 5′ O.C.										
	0700	100′ long header, 6″ diameter, first month	4 Eqlt	3.23	9.907	LF Hdr	112	285		397	555	
	0800	Thereafter, per month		4.13	7.748		89.50	223		312.50	435	
	1000	200′ long header, 8″ diameter, first month		6	5.333		100	153		253	340	
	1100	Thereafter, per month		8.39	3.814		50.50	110		160.50	222	
	1300	500′ long header, 8″ diameter, first month		10.63	3.010		39	86.50		125.50	174	
	1400	Thereafter, per month		20.91	1.530		28	44		72	97	
	1600	1,000′ long header, 10″ diameter, first month		11.62	2.754		33.50	79		112.50	157	
	1700	Thereafter, per month	↓	41.81	.765	↓	16.75	22		38.75	52	
	1900	Note: above figures include pumping 168 hrs. per week										
	1910	and include the pump operator and one stand-by pump.										

02300 | Earthwork

02315 | Excavation and Fill

		02315 Excavation and Fill	CREW	DAILY OUTPUT	LABOR-HOURS	UNIT	2001 BARE COSTS MAT.	LABOR	EQUIP.	TOTAL	TOTAL INCL O&P	
130	0010	**BEDDING** For pipe and conduit, not incl. compaction A12.3-310										130
	0050	Crushed or screened bank run gravel	B-6	150	.160	C.Y.	17	3.97	1.21	22.18	26	
	0100	Crushed stone 3/4″ to 1/2″		150	.160		18.80	3.97	1.21	23.98	28	
	0200	Sand, dead or bank	↓	150	.160		3.55	3.97	1.21	8.73	11.40	
	0500	Compacting bedding in trench	A-1	90	.089	↓		2.03	.72	2.75	3.97	
200	0010	**BORROW, LOADING AND/OR SPREADING**										200
	0020	Material only, bank run gravel				Ton	19			19	21	
	2000	Alternate pricing method										
	2020	Bank run gravel, bank measure				C.Y.	13.60			13.60	14.95	
	2100	Crushed stone, 1-1/2″ to 3/4″, loaded at pit					13.60			13.60	14.95	
	2150	3/8″ size					13.20			13.20	14.55	
	2200	Bank run sand					3.55			3.55	3.91	
	2500	Haul 2 mi.,spread, 200HP dozer, bank measure	B-15	600	.047			1.20	2.83	4.03	4.95	
	3000	Hand spread, bank run sand, bank measure	A-5	20	.900			20.50	1.97	22.47	34	
	3800	Delivery charge, minimum 12 c.y., 1 hr round trip, add	B-34B	84	.095	↓		2.28	5.75	8.03	9.80	

	02315	Excavation and Fill	CREW	DAILY OUTPUT	LABOR-HOURS	UNIT	2001 BARE COSTS MAT.	LABOR	EQUIP.	TOTAL	TOTAL INCL O&P	
200	3820	1-1/2 hr round trip	B-34B	60	.133	C.Y.		3.19	8.05	11.24	13.75	200
	3840	2 hr round trip		42	.190			4.56	11.50	16.06	19.60	
505	0010	**FILL** Spread dumped material, by dozer, no compaction	B-10B	1,000	.012	C.Y.		.33	.74	1.07	1.32	505
	0100	By hand	1 Clab	12	.667	"		15.25		15.25	24	
	0500	Gravel fill, compacted, under floor slabs, 4″ deep	B-37	10,000	.005	S.F.	.15	.12	.01	.28	.36	
	0600	6″ deep		8,600	.006		.23	.13	.02	.38	.48	
	0700	9″ deep		7,200	.007		.38	.16	.02	.56	.69	
	0800	12″ deep		6,000	.008		.53	.19	.02	.74	.90	
	1000	Alternate pricing method, 4″ deep		120	.400	C.Y.	11.35	9.65	1.11	22.11	29	
	1100	6″ deep		160	.300		11.35	7.25	.83	19.43	24.50	
	1200	9″ deep		200	.240		11.35	5.80	.67	17.82	22	
	1300	12″ deep		220	.218		11.35	5.25	.61	17.21	21.50	
	1400	Granular fill					7.60			7.60	8.35	
900	0010	**EXCAVATING, TRENCH** or continuous footing, common earth (A12.3 -110)										900
	0020	No sheeting or dewatering included										
	0050	1′ to 4′ deep, 3/8 C.Y. tractor loader/backhoe	B-11C	150	.107	C.Y.		2.84	1.21	4.05	5.70	
	0060	1/2 C.Y. tractor loader/backhoe	B-11M	200	.080			2.13	1.14	3.27	4.53	
	0090	4′ to 6′ deep, 1/2 C.Y. tractor loader/backhoe	"	200	.080			2.13	1.14	3.27	4.53	
	0100	5/8 C.Y. hydraulic backhoe	B-12Q	250	.064			1.82	1.44	3.26	4.34	
	0300	1/2 C.Y. hydraulic excavator, truck mounted	B-12J	200	.080			2.28	3.24	5.52	7	
	0500	6′ to 10′ deep, 3/4 C.Y. hydraulic backhoe, 6′ to 10′ deep	B-12F	225	.071			2.02	1.84	3.86	5.10	
	0600	1 C.Y. hydraulic excavator, truck mounted	B-12K	400	.040			1.14	1.94	3.08	3.85	
	0900	10′ to 14′ deep, 3/4 C.Y. hydraulic backhoe	B-12F	200	.080			2.28	2.07	4.35	5.75	
	1000	1-1/2 C.Y. hydraulic backhoe	B-12B	540	.030			.84	1.10	1.94	2.49	
	1300	14′ to 20′ deep, 1 C.Y. hydraulic backhoe	B-12A	320	.050			1.42	1.53	2.95	3.84	
	1400	By hand with pick and shovel 2′ to 6′ deep, light soil	1 Clab	8	1			23		23	36	
	1500	Heavy soil	"	4	2			45.50		45.50	71.50	
	1700	For tamping backfilled trenches, air tamp, add	A-1	100	.080			1.83	.65	2.48	3.57	
	1900	Vibrating plate, add	B-18	230	.104			2.45	.26	2.71	4.13	
	2100	Trim sides and bottom for concrete pours, common earth		1,500	.016	S.F.		.38	.04	.42	.63	
	2300	Hardpan		600	.040	"		.94	.10	1.04	1.58	
940	0010	**EXCAVATING, UTILITY TRENCH** Common earth (A12.3 -110)										940
	0050	Trenching with chain trencher, 12 H.P., operator walking										
	0100	4″ wide trench, 12″ deep	B-53	800	.010	L.F.		.29	.12	.41	.57	
	0150	18″ deep		750	.011			.31	.13	.44	.60	
	0200	24″ deep		700	.011			.33	.14	.47	.65	
	0300	6″ wide trench, 12″ deep		650	.012			.35	.15	.50	.70	
	0350	18″ deep		600	.013			.38	.16	.54	.76	
	0400	24″ deep		550	.015			.42	.18	.60	.82	
	0450	36″ deep		450	.018			.51	.22	.73	1.01	
	0600	8″ wide trench, 12″ deep		475	.017			.48	.20	.68	.95	
	0650	18″ deep		400	.020			.58	.24	.82	1.14	
	0700	24″ deep		350	.023			.66	.28	.94	1.30	
	0750	36″ deep		300	.027			.77	.32	1.09	1.52	
	1000	Backfill by hand including compaction, add										
	1050	4″ wide trench, 12″ deep	A-1	800	.010	L.F.		.23	.08	.31	.45	
	1100	18″ deep		530	.015			.34	.12	.46	.67	
	1150	24″ deep		400	.020			.46	.16	.62	.90	
	1300	6″ wide trench, 12″ deep		540	.015			.34	.12	.46	.66	
	1350	18″ deep		405	.020			.45	.16	.61	.89	
	1400	24″ deep		270	.030			.68	.24	.92	1.32	
	1450	36″ deep		180	.044			1.02	.36	1.38	1.99	
	1600	8″ wide trench, 12″ deep		400	.020			.46	.16	.62	.90	
	1650	18″ deep		265	.030			.69	.24	.93	1.35	
	1700	24″ deep		200	.040			.91	.32	1.23	1.79	

02300 | Earthwork

	02315 | Excavation and Fill	CREW	DAILY OUTPUT	LABOR-HOURS	UNIT	2001 BARE COSTS MAT.	LABOR	EQUIP.	TOTAL	TOTAL INCL O&P	
940	1750 36" deep (A12.3 -110)	A-1	135	.059	L.F.		1.35	.48	1.83	2.65	940
	2000 Chain trencher, 40 H.P. operator riding										
	2050 6" wide trench and backfill, 12" deep	B-54	1,200	.007	L.F.		.19	.16	.35	.47	
	2100 18" deep		1,000	.008			.23	.20	.43	.57	
	2150 24" deep		975	.008			.24	.20	.44	.58	
	2200 36" deep		900	.009			.26	.22	.48	.63	
	2250 48" deep		750	.011			.31	.26	.57	.75	
	2300 60" deep		650	.012			.35	.30	.65	.87	
	2400 8" wide trench and backfill, 12" deep		1,000	.008			.23	.20	.43	.57	
	2450 18" deep		950	.008			.24	.21	.45	.60	
	2500 24" deep		900	.009			.26	.22	.48	.63	
	2550 36" deep		800	.010			.29	.25	.54	.71	
	2600 48" deep		650	.012			.35	.30	.65	.87	
	2700 12" wide trench and backfill, 12" deep		975	.008			.24	.20	.44	.58	
	2750 18" deep		860	.009			.27	.23	.50	.66	
	2800 24" deep		800	.010			.29	.25	.54	.71	
	2850 36" deep		725	.011			.32	.27	.59	.78	
	3000 16" wide trench and backfill, 12" deep		835	.010			.28	.24	.52	.68	
	3050 18" deep		750	.011			.31	.26	.57	.75	
	3100 24" deep	↓	700	.011	↓		.33	.28	.61	.81	
	3200 Compaction with vibratory plate, add								50%	50%	
	5100 Hand excavate and trim for pipe bells after trench excavation										
	5200 8" pipe	1 Clab	155	.052	L.F.		1.18		1.18	1.85	
	5300 18" pipe	"	130	.062	"		1.41		1.41	2.20	

	02320 | Hauling	CREW	DAILY OUTPUT	LABOR-HOURS	UNIT	MAT.	LABOR	EQUIP.	TOTAL	TOTAL INCL O&P	
200	0011 **HAULING** Excavated or borrow material, loose cubic yards										200
	0015 no loading included, highway haulers										
	0020 6 C.Y. dump truck, 1/4 mile round trip, 5.0 loads/hr.	B-34A	195	.041	C.Y.		.98	1.90	2.88	3.59	
	0030 1/2 mile round trip, 4.1 loads/hr.		160	.050			1.20	2.32	3.52	4.38	
	0040 1 mile round trip, 3.3 loads/hr. (R01590 -100)		130	.062			1.47	2.85	4.32	5.40	
	0100 2 mile round trip, 2.6 loads/hr.		100	.080			1.92	3.70	5.62	7	
	0150 3 mile round trip, 2.1 loads/hr.		80	.100			2.40	4.63	7.03	8.75	
	0200 4 mile round trip, 1.8 loads/hr.	↓	70	.114			2.74	5.30	8.04	10	
	0310 12 C.Y. dump truck, 1/4 mile round trip 3.7 loads/hr.	B-34B	288	.028			.67	1.68	2.35	2.86	
	0400 2 mile round trip, 2.2 loads/hr.		180	.044			1.06	2.68	3.74	4.58	
	0450 3 mile round trip, 1.9 loads/hr.	↓	170	.047			1.13	2.84	3.97	4.84	
	1300 Hauling in medium traffic, add								20%	20%	
	1400 Heavy traffic, add								30%	30%	
	1600 Grading at dump, or embankment if required, by dozer	B-10B	1,000	.012	↓		.33	.74	1.07	1.32	
	1800 Spotter at fill or cut, if required	1 Clab	8	1	Hr.		23		23	36	

02400 | Tunneling, Boring & Jacking

	02441 | Microtunneling	CREW	DAILY OUTPUT	LABOR-HOURS	UNIT	2001 BARE COSTS MAT.	LABOR	EQUIP.	TOTAL	TOTAL INCL O&P	
400	0010 **MICROTUNNELING** Not including excavation, backfill, shoring,										400
	0020 or dewatering, average 50'/day, slurry method										
	0100 24" to 48" outside diameter, minimum				L.F.					600	
	0110 Adverse conditions, add				%					50%	

For expanded coverage of these items see *Means Heavy Construction Cost Data 2001*

02400 | Tunneling, Boring & Jacking

02441 | Microtunneling

		02441 Microtunneling	CREW	DAILY OUTPUT	LABOR-HOURS	UNIT	2001 BARE COSTS MAT.	LABOR	EQUIP.	TOTAL	TOTAL INCL O&P	
400	1000	Rent microtunneling machine, average monthly lease				Month					80,000	400
	1010	Operating technician				Day					600	
	1100	Mobilization and demobilization, minimum				Job					40,000	
	1110	Maximum				"					400,000	

02445 | Boring or Jacking Conduits

		02445 Boring or Jacking Conduits	CREW	DAILY OUTPUT	LABOR-HOURS	UNIT	2001 BARE COSTS MAT.	LABOR	EQUIP.	TOTAL	TOTAL INCL O&P	
300	0010	**HORIZONTAL BORING** Casing only, 100′ minimum,										300
	0020	not incl. jacking pits or dewatering										
	0100	Roadwork, 1/2″ thick wall, 24″ diameter casing	B-42	20	3.200	L.F.	47.50	82.50	62	192	251	
	0200	36″ diameter		16	4		75	103	77.50	255.50	330	
	0300	48″ diameter		15	4.267		110	110	82.50	302.50	385	
	0500	Railroad work, 24″ diameter		15	4.267		47.50	110	82.50	240	320	
	0600	36″ diameter		14	4.571		75	118	88.50	281.50	365	
	0700	48″ diameter	↓	12	5.333		110	137	103	350	450	
	0900	For ledge, add								145	175	
	1000	Small diameter boring, 3″, sandy soil	B-82	900	.018		13	.46	.05	13.51	15.05	
	1040	Rocky soil	"	500	.032	↓	13	.83	.09	13.92	15.65	
	1100	Prepare jacking pits, incl. mobilization & demobilization, minimum				Ea.				2,650	3,150	
	1101	Maximum				"				15,000	18,000	

02500 | Utility Services

02510 | Water Distribution

		02510 Water Distribution	CREW	DAILY OUTPUT	LABOR-HOURS	UNIT	2001 BARE COSTS MAT.	LABOR	EQUIP.	TOTAL	TOTAL INCL O&P	
350	0010	**DISTRIBUTION CONNECTION**										350
	7020	Crosses, 4″ x 4″	B-21	37	.757	Ea.	335	20	3.84	358.84	400	
	7040	6″ x 6″		25	1.120		395	30	5.70	430.70	490	
	7060	8″ x 6″		21	1.333		485	35.50	6.75	527.25	600	
	7080	8″ x 8″		21	1.333		525	35.50	6.75	567.25	645	
	7100	10″ x 6″		21	1.333		965	35.50	6.75	1,007.25	1,125	
	7120	10″ x 10″		21	1.333		1,050	35.50	6.75	1,092.25	1,225	
	7140	12″ x 6″		18	1.556		965	41.50	7.90	1,014.40	1,125	
	7160	12″ x 12″		18	1.556		1,225	41.50	7.90	1,274.40	1,425	
	7180	14″ x 6″		16	1.750		2,425	46.50	8.85	2,480.35	2,725	
	7200	14″ x 14″		16	1.750		2,550	46.50	8.85	2,605.35	2,875	
	7220	16″ x 6″		14	2		2,575	53	10.15	2,638.15	2,925	
	7240	16″ x 10″		14	2		2,625	53	10.15	2,688.15	2,975	
	7260	16″ x 16″		14	2		2,750	53	10.15	2,813.15	3,125	
	7280	18″ x 6″		10	2.800		3,850	74.50	14.20	3,938.70	4,375	
	7300	18″ x 12″		10	2.800		3,875	74.50	14.20	3,963.70	4,400	
	7320	18″ x 18″		10	2.800		4,000	74.50	14.20	4,088.70	4,525	
	7340	20″ x 6″		8	3.500		3,125	93	17.75	3,235.75	3,600	
	7360	20″ x 12″		8	3.500		3,350	93	17.75	3,460.75	3,850	
	7380	20″ x 20″		8	3.500		4,875	93	17.75	4,985.75	5,550	
	7400	24″ x 6″		6	4.667		4,025	124	23.50	4,172.50	4,650	
	7420	24″ x 12″		6	4.667		4,100	124	23.50	4,247.50	4,725	
	7440	24″ x 18″		6	4.667		6,100	124	23.50	6,247.50	6,925	
	7460	24″ x 24″		6	4.667		6,350	124	23.50	6,497.50	7,200	
	7600	Cut-in sleeves with rubber gaskets, 4″		18	1.556		111	41.50	7.90	160.40	195	
	7620	6″	↓	12	2.333	↓	141	62	11.85	214.85	265	

	02510	Water Distribution	CREW	DAILY OUTPUT	LABOR-HOURS	UNIT	2001 BARE COSTS MAT.	LABOR	EQUIP.	TOTAL	TOTAL INCL O&P	
350	7640	8″	B-21	10	2.800	Ea.	191	74.50	14.20	279.70	340	350
	7660	10″		10	2.800		265	74.50	14.20	353.70	425	
	7680	12″		9	3.111		315	83	15.75	413.75	490	
	7800	Cut-in valves with rubber gaskets, 4″		18	1.556		289	41.50	7.90	338.40	395	
	7820	6″		12	2.333		390	62	11.85	463.85	540	
	7840	8″		10	2.800		605	74.50	14.20	693.70	795	
	7860	10″		10	2.800		925	74.50	14.20	1,013.70	1,150	
	7880	12″		9	3.111		1,175	83	15.75	1,273.75	1,425	
	8000	Sleeves with rubber gaskets, 4″ x 4″		37	.757		267	20	3.84	290.84	330	
	8020	6″ x 6″		25	1.120		315	30	5.70	350.70	405	
	8040	8″ x 6″		21	1.333		395	35.50	6.75	437.25	500	
	8060	8″ x 8″		21	1.333		420	35.50	6.75	462.25	530	
	8080	10″ x 6″		21	1.333		770	35.50	6.75	812.25	915	
	8100	10″ x 10″		21	1.333		835	35.50	6.75	877.25	985	
	8120	12″ x 6″		18	1.556		770	41.50	7.90	819.40	925	
	8140	12″ x 12″		18	1.556		975	41.50	7.90	1,024.40	1,150	
	8160	14″ x 6″		16	1.750		1,925	46.50	8.85	1,980.35	2,200	
	8180	14″ x 14″		16	1.750		2,025	46.50	8.85	2,080.35	2,325	
	8200	16″ x 6″		14	2		2,050	53	10.15	2,113.15	2,375	
	8220	16″ x 10″		14	2		2,100	53	10.15	2,163.15	2,400	
	8240	16″ x 16″		14	2		2,200	53	10.15	2,263.15	2,525	
	8260	18″ x 6″		10	2.800		3,075	74.50	14.20	3,163.70	3,525	
	8280	18″ x 12″		10	2.800		3,100	74.50	14.20	3,188.70	3,550	
	8300	18″ x 18″		10	2.800		3,200	74.50	14.20	3,288.70	3,650	
	8320	20″ x 6″		8	3.500		2,500	93	17.75	2,610.75	2,925	
	8340	20″ x 12″		8	3.500		2,675	93	17.75	2,785.75	3,125	
	8360	20″ x 20″		8	3.500		3,900	93	17.75	4,010.75	4,475	
	8380	24″ x 6″		6	4.667		3,225	124	23.50	3,372.50	3,775	
	8400	24″ x 12″		6	4.667		3,275	124	23.50	3,422.50	3,825	
	8420	24″ x 18″		6	4.667		4,875	124	23.50	5,022.50	5,600	
	8440	24″ x 24″		6	4.667		5,075	124	23.50	5,222.50	5,800	
	8800	Curb box, 6′ long	B-20	20	1.200		132	31		163	194	
	8820	8′ long	″	18	1.333		133	34.50		167.50	200	
	9000	Valves, gate valve, N.R.S. post type, 4″ diameter	B-21	32	.875		233	23.50	4.44	260.94	297	
	9020	6″ diameter		20	1.400		297	37.50	7.10	341.60	390	
	9040	8″ diameter		16	1.750		475	46.50	8.85	530.35	600	
	9060	10″ diameter		16	1.750		735	46.50	8.85	790.35	890	
	9080	12″ diameter		13	2.154		935	57.50	10.90	1,003.40	1,125	
	9100	14″ diameter		11	2.545		2,200	67.50	12.90	2,280.40	2,550	
	9120	O.S.&Y., 4″ diameter		32	.875		330	23.50	4.44	357.94	400	
	9140	6″ diameter		20	1.400		425	37.50	7.10	469.60	535	
	9160	8″ diameter		16	1.750		650	46.50	8.85	705.35	795	
	9180	10″ diameter		16	1.750		935	46.50	8.85	990.35	1,100	
	9200	12″ diameter		13	2.154		1,250	57.50	10.90	1,318.40	1,450	
	9220	14″ diameter		11	2.545		2,200	67.50	12.90	2,280.40	2,550	
	9400	Check valves, rubber disc, 2-1/2″ diameter	B-20	44	.545		258	14.10		272.10	305	
	9420	3″ diameter	″	38	.632		258	16.35		274.35	310	
	9440	4″ diameter	B-21	32	.875		320	23.50	4.44	347.94	390	
	9480	6″ diameter		20	1.400		480	37.50	7.10	524.60	590	
	9500	8″ diameter		16	1.750		700	46.50	8.85	755.35	850	
	9520	10″ diameter		16	1.750		1,250	46.50	8.85	1,305.35	1,450	
	9540	12″ diameter		13	2.154		1,975	57.50	10.90	2,043.40	2,275	
	9542	14″ diameter		11	2.545		2,775	67.50	12.90	2,855.40	3,175	
	9700	Detector check valves, reducing, 4″ diameter		32	.875		640	23.50	4.44	667.94	745	
	9720	6″ diameter		20	1.400		985	37.50	7.10	1,029.60	1,150	
350	9740	8″ diameter		16	1.750		1,600	46.50	8.85	1,655.35	1,850	

02510 | Water Distribution

		02510 Water Distribution	Crew	Daily Output	Labor-Hours	Unit	2001 Bare Costs Mat.	2001 Bare Costs Labor	2001 Bare Costs Equip.	2001 Bare Costs Total	Total Incl O&P	
350	9760	10″ diameter	B-21	16	1.750	Ea.	3,175	46.50	8.85	3,230.35	3,575	350
	9800	Galvanized, 4″ diameter		32	.875		835	23.50	4.44	862.94	960	
	9820	6″ diameter		20	1.400		1,225	37.50	7.10	1,269.60	1,400	
	9840	8″ diameter		16	1.750		1,950	46.50	8.85	2,005.35	2,225	
	9860	10″ diameter		16	1.750		3,850	46.50	8.85	3,905.35	4,325	
820	0010	**PIPING, WATER DISTRIBUTION, DUCTILE IRON** cement lined										820
	0020	Not including excavation or backfill										
	2000	Pipe, class 50 water piping, 18′ lengths										
	2020	Mechanical joint, 4″ diameter	B-21A	200	.200	L.F.	8.75	5.65	2.13	16.53	20.50	
	2040	6″ diameter		160	.250		9.10	7.05	2.66	18.81	23.50	
	2060	8″ diameter		133.33	.300		11.65	8.45	3.19	23.29	29.50	
	2080	10″ diameter		114.29	.350		15.15	9.85	3.72	28.72	36	
	2100	12″ diameter		105.26	.380		19.35	10.70	4.04	34.09	42.50	
	2120	14″ diameter		100	.400		24	11.25	4.25	39.50	48	
	2140	16″ diameter		72.73	.550		26	15.50	5.85	47.35	58.50	
	2160	18″ diameter		68.97	.580		34.50	16.35	6.15	57	69.50	
	2170	20″ diameter		57.14	.700		38	19.70	7.45	65.15	80	
	2180	24″ diameter		47.06	.850		51	24	9.05	84.05	103	
	3000	Tyton, Push-on joint, 4″ diameter		400	.100		6.45	2.82	1.06	10.33	12.60	
	3020	6″ diameter		333.33	.120		7.45	3.38	1.28	12.11	14.70	
	3040	8″ diameter		200	.200		10.50	5.65	2.13	18.28	22.50	
	3060	10″ diameter		181.82	.220		15.95	6.20	2.34	24.49	29.50	
	3080	12″ diameter		160	.250		16.80	7.05	2.66	26.51	32	
	3100	14″ diameter		133.33	.300		18.50	8.45	3.19	30.14	37	
	3120	16″ diameter		114.29	.350		26	9.85	3.72	39.57	47.50	
	3140	18″ diameter		100	.400		29	11.25	4.25	44.50	54	
	3160	20″ diameter		88.89	.450		32	12.70	4.78	49.48	59.50	
	3180	24″ diameter		76.92	.520		41	14.65	5.55	61.20	73.50	
	4000	Drill and tap pressurized main (labor only)										
	4100	6″ main, 1″ to 2″ service	Q-1	3	5.333	Ea.		165		165	250	
	4150	8″ main, 1″ to 2″ service	″	2.75	5.818	″		180		180	272	
	4500	Tap and insert gate valve										
	4600	8″ main, 4″ branch	B-21	3.20	8.750	Ea.		233	44.50	277.50	415	
	4650	6″ branch		2.70	10.370			276	52.50	328.50	490	
	4700	10″ main, 4″ branch		2.70	10.370			276	52.50	328.50	490	
	4750	6″ branch		2.35	11.915			315	60.50	375.50	560	
	4800	12″ main, 6″ branch		2.35	11.915			315	60.50	375.50	560	
	4850	8″ branch		2.35	11.915			315	60.50	375.50	560	
	8000	Fittings, mechanical joint										
	8006	90° bend or elbow, 4″ diameter	B-20A	16	2	Ea.	145	55		200	244	
	8020	6″ diameter		12.80	2.500		195	68.50		263.50	320	
	8040	8″ diameter		10.67	2.999		275	82.50		357.50	430	
	8060	10″ diameter	B-21A	11.43	3.500		435	98.50	37	570.50	670	
	8080	12″ diameter		10.53	3.799		595	107	40.50	742.50	865	
	8100	14″ diameter		10	4		610	113	42.50	765.50	890	
	8120	16″ diameter		7.27	5.502		660	155	58.50	873.50	1,025	
	8140	18″ diameter		6.90	5.797		915	163	61.50	1,139.50	1,325	
	8160	20″ diameter		5.71	7.005		2,300	197	74.50	2,571.50	2,900	
	8180	24″ diameter		4.70	8.511		2,600	240	90.50	2,930.50	3,325	
	8200	Wye or tee, 4″ diameter	B-20A	10.67	2.999		213	82.50		295.50	360	
	8220	6″ diameter		8.53	3.751		160	103		263	335	
	8240	8″ diameter		7.11	4.501		234	123		357	445	
	8260	10″ diameter	B-21A	7.62	5.249		800	148	56	1,004	1,175	
	8280	12″ diameter		7.02	5.698		975	161	60.50	1,196.50	1,375	
	8300	14″ diameter		6.67	5.997		1,000	169	64	1,233	1,425	

	02510 | Water Distribution		CREW	DAILY OUTPUT	LABOR-HOURS	UNIT	2001 BARE COSTS MAT.	LABOR	EQUIP.	TOTAL	TOTAL INCL O&P
820	8320	16″ diameter	B-21A	4.85	8.247	Ea.	1,050	232	87.50	1,369.50	1,600
	8340	18″ diameter		4.60	8.696		1,975	245	92.50	2,312.50	2,650
	8360	20″ diameter		3.81	10.499		2,600	296	112	3,008	3,425
	8380	24″ diameter		3.14	12.739		3,025	360	135	3,520	4,025
830	0010	**PIPING, WATER DISTRIBUTION, POLYETHYLENE, C901**									
	0020	Not including excavation or backfill									
	1000	Piping, 160 P.S.I., 3/4″ diameter	B-20	525	.046	L.F.	.21	1.18		1.39	2.08
	1120	1″ diameter		485	.049		.35	1.28		1.63	2.39
	1140	1-1/2″ diameter		450	.053		.81	1.38		2.19	3.05
	1160	2″ diameter		365	.066		1.38	1.70		3.08	4.18
	2000	Fittings, insert type, nylon, 160 & 250 psi, cold water									
	2220	Clamp ring, stainless steel, 3/4″ diameter	B-20	345	.070	Ea.	1.10	1.80		2.90	4.03
	2240	1″ diameter		321	.075		1.10	1.93		3.03	4.24
	2260	1-1/2″ diameter		285	.084		1.14	2.18		3.32	4.66
	2280	2″ diameter		255	.094		1.14	2.43		3.57	5.05
	2300	Coupling, 3/4″ diameter		66	.364		.62	9.40		10.02	15.45
	2320	1″ diameter		57	.421		.83	10.90		11.73	17.95
	2340	1-1/2″ diameter		51	.471		1.94	12.15		14.09	21
	2360	2″ diameter		48	.500		2.33	12.95		15.28	23
	2400	Elbow, 90°, 3/4″ diameter		66	.364		1.11	9.40		10.51	15.95
	2420	1″ diameter		57	.421		1.19	10.90		12.09	18.35
	2440	1-1/2″ diameter		51	.471		2.95	12.15		15.10	22.50
	2460	2″ diameter		48	.500		4.93	12.95		17.88	26
840	0010	**PIPING, WATER DISTRIBUTION, POLYVINYL CHLORIDE**									
	0020	Not including excavation or backfill, unless specified									
	1000	AWWA C900, Class 150, SDR 18									
	1040	10″ diameter	B-20A	220	.145	L.F.	12.50	3.99		16.49	19.85
	1050	12″ diameter	"	189	.169		15.45	4.64		20.09	24
	2100	Class 160, S.D.R. 26, 1-1/2″ diameter	B-20	750	.032		.35	.83		1.18	1.69
	2120	2″ diameter		686	.035		.89	.90		1.79	2.40
	2140	2-1/2″ diameter		500	.048		1.32	1.24		2.56	3.39
	2160	3″ diameter		430	.056		1.89	1.44		3.33	4.34
	2180	4″ diameter		375	.064		3.09	1.65		4.74	6
	2200	6″ diameter		316	.076		6.65	1.96		8.61	10.45
	2210	8″ diameter		260	.092		11.30	2.39		13.69	16.15
	3010	AWWA C905, PR 100, DR 41									
	3030	14″ diameter	B-20A	213	.150	L.F.	38	4.12		42.12	48.50
	3040	16″ diameter		200	.160		39.50	4.39		43.89	50.50
	3050	18″ diameter		160	.200		40.50	5.50		46	53
	3060	20″ diameter		133	.241		50	6.60		56.60	65
	3070	24″ diameter		107	.299		72	8.20		80.20	92
	3080	30″ diameter		80	.400		102	10.95		112.95	129
	3090	36″ diameter		53	.604		130	16.55		146.55	169
	3960	Pressure pipe, class 200, SDR 21, 3/4″		1,340	.024		.26	.66		.92	1.29
	3980	1″		1,100	.029		.34	.80		1.14	1.59
	4000	1-1/2″		920	.035		.70	.95		1.65	2.23
	4010	2″		840	.038		1.08	1.05		2.13	2.79
	4020	2-1/2″		600	.053		1.60	1.46		3.06	4
	4030	3″		520	.062		2.29	1.69		3.98	5.10
	4040	4″		460	.070		3.77	1.91		5.68	7.10
	4050	6″		380	.084		8.20	2.31		10.51	12.55
	4060	8″		320	.100		13.80	2.74		16.54	19.35
	4090	Including trenching to 3′ deep, 3/4″	Q-1C	300	.080		.26	2.46	2.37	5.09	6.60
	4100	1″		280	.086		.34	2.64	2.54	5.52	7.15
	4110	1-1/2″		260	.092		.70	2.84	2.74	6.28	8.05

02510 | Water Distribution

		02510 Water Distribution	CREW	DAILY OUTPUT	LABOR-HOURS	UNIT	2001 BARE COSTS MAT.	LABOR	EQUIP.	TOTAL	TOTAL INCL O&P	
840	4120	2"	Q-1C	220	.109	L.F.	1.08	3.36	3.23	7.67	9.80	840
	4130	2-1/2"		200	.120		1.60	3.69	3.56	8.85	11.25	
	4140	3"		175	.137		2.29	4.22	4.06	10.57	13.40	
	4150	4"		150	.160		3.77	4.92	4.74	13.43	16.80	
	4160	6"	↓	125	.192	↓	8.20	5.90	5.70	19.80	24	
	4165	Fittings										
	4170	Elbow, 90°, 3/4"	B-20A	114	.281	Ea.	.39	7.70		8.09	12.25	
	4180	1"		100	.320		.78	8.80		9.58	14.30	
	4190	1-1/2"		80	.400		5.20	10.95		16.15	22.50	
	4200	2"		72	.444		6.45	12.20		18.65	26	
	4210	3"		46	.696		9.30	19.10		28.40	40	
	4220	4"		36	.889		19.60	24.50		44.10	59	
	4230	6"		24	1.333		38	36.50		74.50	98	
	4240	8"		14	2.286		64.50	62.50		127	167	
	4250	Elbow, 45°, 3/4"		114	.281		.42	7.70		8.12	12.25	
	4260	1"		100	.320		.51	8.80		9.31	14	
	4270	1-1/2"		80	.400		8.95	10.95		19.90	26.50	
	4280	2"		72	.444		8.15	12.20		20.35	27.50	
	4290	2-1/2"		54	.593		9.35	16.25		25.60	35.50	
	4300	3"		46	.696		11.80	19.10		30.90	42.50	
	4310	4"		36	.889		15.90	24.50		40.40	55	
	4320	6"		24	1.333		29	36.50		65.50	88	
	4330	8"		14	2.286		57	62.50		119.50	159	
	4340	Tee, 3/4"		76	.421		.24	11.55		11.79	17.95	
	4350	1"		66	.485		.45	13.30		13.75	21	
	4360	1-1/2"		54	.593		12.30	16.25		28.55	38.50	
	4370	2"		48	.667		13.20	18.30		31.50	42.50	
	4380	2-1/2"		36	.889		14.65	24.50		39.15	53.50	
	4390	3"		30	1.067		15.50	29.50		45	62	
	4400	4"		24	1.333		24	36.50		60.50	82.50	
	4410	6"		14.80	2.162		53.50	59.50		113	150	
	4420	8"		9	3.556		94.50	97.50		192	254	
	4430	Coupling, 3/4"		114	.281		.13	7.70		7.83	11.95	
	4440	1"		100	.320		.22	8.80		9.02	13.70	
	4450	1-1/2"		80	.400		4.43	10.95		15.38	21.50	
	4460	2"		72	.444		4.42	12.20		16.62	23.50	
	4470	2-1/2"		54	.593		4.89	16.25		21.14	30.50	
	4480	3"		46	.696		6.20	19.10		25.30	36.50	
	4490	4"		36	.889		11.25	24.50		35.75	50	
	4500	6"		24	1.333		17.15	36.50		53.65	75	
	4510	8"	↓	14	2.286	↓	37.50	62.50		100	137	
	8000	Fittings with rubber gasket										
	8003	Class 150, D.R. 18										
	8006	90° Bend , 4" diameter	B-20	100	.240	Ea.	30.50	6.20		36.70	43	
	8020	6" diameter		90	.267		54.50	6.90		61.40	71	
	8040	8" diameter		80	.300		105	7.75		112.75	128	
	8060	10" diameter		50	.480		183	12.40		195.40	221	
	8080	12" diameter		30	.800		235	20.50		255.50	292	
	8200	45° Bend, 4" diameter		100	.240		30	6.20		36.20	42.50	
	8220	6" diameter		90	.267		53	6.90		59.90	69.50	
	8240	8" diameter		50	.480		101	12.40		113.40	130	
	8260	10" diameter		50	.480		154	12.40		166.40	189	
	8280	12" diameter		30	.800		201	20.50		221.50	254	
	8300	Reducing tee 6"x4"		100	.240		92	6.20		98.20	111	
	8320	8" x 6"		90	.267		163	6.90		169.90	191	
	8330	10" x 6"	↓	90	.267	↓	210	6.90		216.90	242	

	02510 | Water Distribution		Crew	Daily Output	Labor-Hours	Unit	2001 Bare Costs Mat.	Labor	Equip.	Total	Total Incl O&P	
840	8340	10"x8"	B-20	90	.267	Ea.	218	6.90		224.90	251	840
	8350	12" x 6"		90	.267		249	6.90		255.90	285	
	8360	12" x 8"		90	.267		260	6.90		266.90	297	
	8400	Tapped service tee (threaded type) 6" x 6" x 3/4		100	.240		32.50	6.20		38.70	45.50	
	8420	6" x 6" x 3/4"		90	.267		32.50	6.90		39.40	47	
	8430	6" x 6" x 1"		90	.267		32.50	6.90		39.40	47	
	8440	6" x 6" x 1 1/2"		90	.267		32.50	6.90		39.40	47	
	8450	6" x 6" x 2"		90	.267		32.50	6.90		39.40	47	
	8460	8" x 8" x 3/4"		90	.267		138	6.90		144.90	162	
	8470	8" x 8" x 1"		90	.267		138	6.90		144.90	162	
	8480	8" x 8" x 1 1/2"		90	.267		138	6.90		144.90	162	
	8490	8" x 8" x 2"		90	.267		138	6.90		144.90	162	
	8500	Repair coupling 4"		100	.240		19.45	6.20		25.65	31	
	8520	6" diameter		90	.267		30	6.90		36.90	44	
	8540	8" diameter		50	.480		65.50	12.40		77.90	91.50	
	8560	10" diameter		50	.480		111	12.40		123.40	141	
	8580	12" diameter		50	.480		163	12.40		175.40	198	
	8600	Plug end 4"		100	.240		16.45	6.20		22.65	28	
	8620	6" diameter		90	.267		29.50	6.90		36.40	43.50	
	8640	8" diameter		50	.480		50.50	12.40		62.90	75	
	8660	10" diameter		50	.480		53	12.40		65.40	78	
	8680	12" diameter		50	.480		65.50	12.40		77.90	92	
850	0010	**PIPING, HDPE BUTT FUSION JOINTS**, SDR 21, 40′ lengths										850
	0100	4" diameter	B-22A	400	.095	L.F.	1.75	2.48	1.41	5.64	7.35	
	0200	6" diameter		380	.100		2.87	2.61	1.48	6.96	8.85	
	0300	8" diameter		320	.119		4.87	3.09	1.76	9.72	12.10	
	0400	10" diameter		300	.127		7.60	3.30	1.87	12.77	15.55	
	0500	12" diameter		260	.146		10.65	3.81	2.16	16.62	20	
	0600	14" diameter		220	.173		12.80	4.50	2.55	19.85	24	
	0700	16" diameter		180	.211		16.75	5.50	3.12	25.37	30.50	
	0800	18" diameter		140	.271		21	7.05	4.01	32.06	39	
	0900	24" diameter		100	.380		37.50	9.90	5.60	53	63	
	1000	Fittings										
	1100	Elbows, 90 degrees										
	1200	4" diameter	B-22B	32	.500	Ea.	27	13.20	3.71	43.91	54.50	
	1300	6" diameter		28	.571		72.50	15.05	4.23	91.78	108	
	1400	8" diameter		24	.667		150	17.55	4.94	172.49	198	
	1500	10" diameter		18	.889		264	23.50	6.60	294.10	335	
	1600	12" diameter		12	1.333		450	35	9.90	494.90	560	
	1700	14" diameter		9	1.778		435	47	13.15	495.15	570	
	1800	16" diameter		6	2.667		525	70.50	19.75	615.25	705	
	1900	18" diameter		4	4		705	105	29.50	839.50	980	
	2000	24" diameter		3	5.333		1,425	141	39.50	1,605.50	1,850	
	2100	Tees										
	2200	4" diameter	B-22B	30	.533	Ea.	31	14.05	3.95	49	60.50	
	2300	6" diameter		26	.615		85	16.20	4.56	105.76	124	
	2400	8" diameter		22	.727		215	19.15	5.40	239.55	273	
	2500	10" diameter		15	1.067		284	28	7.90	319.90	365	
	2600	12" diameter		10	1.600		390	42	11.85	443.85	505	
	2700	14" diameter		8	2		460	52.50	14.80	527.30	605	
	2800	16" diameter		6	2.667		540	70.50	19.75	630.25	725	
	2900	18" diameter		4	4		710	105	29.50	844.50	980	
	3000	24" diameter		2	8		1,475	211	59.50	1,745.50	2,025	
920	0010	**PIPING, WATER DISTRIBUTION, COPPER**										920
	0020	Not including excavation or backfill										

02500 | Utility Services

02510 | Water Distribution

		02510 Water Distribution	CREW	DAILY OUTPUT	LABOR-HOURS	UNIT	2001 BARE COSTS MAT.	LABOR	EQUIP.	TOTAL	TOTAL INCL O&P	
920	2000	Tubing, type K, 20′ joints, 3/4″ diameter	Q-1	400	.040	L.F.	1.32	1.24		2.56	3.32	920
	2200	1″ diameter		320	.050		1.68	1.55		3.23	4.19	
	3000	1-1/2″ diameter		265	.060		2.70	1.87		4.57	5.80	
	3020	2″ diameter		230	.070		4.31	2.16		6.47	8	
	3040	2-1/2″ diameter		146	.110		6.25	3.40		9.65	12	
	3060	3″ diameter		134	.119		8.60	3.70		12.30	15.10	
	4012	4″ diameter		95	.168		13.45	5.20		18.65	22.50	
	4016	6″ diameter	Q-2	95	.253		42.50	8.10		50.60	59	
	5000	Tubing, type L										
	5108	2″ diameter	Q-1	105	.152	L.F.	5.65	4.72		10.37	13.40	
	6010	3″ diameter		140	.114		11.20	3.54		14.74	17.65	
	6012	4″ diameter		95	.168		17.65	5.20		22.85	27.50	
	6016	6″ diameter	Q-2	100	.240		64.50	7.70		72.20	82.50	
	7020	Fittings, brass, corporation stops, 3/4″ diameter	1 Plum	19	.421	Ea.	15.50	14.50		30	39	
	7040	1″ diameter		16	.500		23.50	17.25		40.75	52	
	7060	1-1/2″ diameter		13	.615		66.50	21		87.50	106	
	7080	2″ diameter		11	.727		116	25		141	166	
	7100	Curb stops, 3/4″ diameter		19	.421		33.50	14.50		48	58.50	
	7120	1″ diameter		16	.500		46	17.25		63.25	76.50	
	7140	1-1/2″ diameter		13	.615		102	21		123	145	
	7160	2″ diameter		11	.727		168	25		193	222	

02520 | Wells

		02520 Wells	CREW	DAILY OUTPUT	LABOR-HOURS	UNIT	MAT.	LABOR	EQUIP.	TOTAL	TOTAL INCL O&P	
900	0010	**WELLS** Domestic water										900
	0100	Drilled, 4″ to 6″ diameter	B-23	120	.333	L.F.		7.75	17.10	24.85	31	
	0200	8″ diameter	″	95.20	.420	″		9.75	21.50	31.25	39	
	0400	Gravel pack well, 40′ deep, incl. gravel & casing, complete										
	0500	24″ diameter casing x 18″ diameter screen	B-23	.13	307	Total	21,000	7,150	15,800	43,950	51,500	
	0600	36″ diameter casing x 18″ diameter screen		.12	333	″	22,600	7,750	17,100	47,450	55,500	
	0800	Observation wells, 1-1/4″ riser pipe		163	.245	V.L.F.	11.55	5.70	12.60	29.85	35.50	
	0900	For flush Buffalo roadway box, add	1 Skwk	16.60	.482	Ea.	31.50	14.40		45.90	57	
	1200	Test well, 2-1/2″ diameter, up to 50′ deep (15 to 50 GPM)	B-23	1.51	26.490	″	475	615	1,350	2,440	2,975	
	1300	Over 50′ deep, add	″	121.80	.328	L.F.	12.60	7.65	16.85	37.10	44.50	
	1500	Pumps, installed in wells to 100′ deep, 4″ submersible										
	1510	1/2 H.P.	Q-1	3.22	4.969	Ea.	355	154		509	625	
	1520	3/4 H.P.		2.66	6.015		425	186		611	750	
	1600	1 H.P.		2.29	6.987		480	217		697	850	
	1700	1-1/2 H.P.	Q-22	1.60	10		620	310	320	1,250	1,500	
	1800	2 H.P.		1.33	12.030		840	375	385	1,600	1,925	
	1900	3 H.P.		1.14	14.035		1,050	435	450	1,935	2,300	
	2000	5 H.P.		1.14	14.035		1,550	435	450	2,435	2,850	
	3000	Pump, 6″ submersible, 25′ to 150′ deep, 25 H.P., 249 to 297 GPM		.89	17.978		3,900	555	580	5,035	5,775	
	3100	25′ to 500′ deep, 30 H.P., 100 to 300 GPM		.73	21.918		3,950	680	705	5,335	6,125	
	8000	Steel well casing	B-23A	3,020	.008	Lb.	.40	.21	.65	1.26	1.47	
	8110	Well screen assembly, stainless steel, 2″ diameter		273	.088	L.F.	34	2.29	7.15	43.44	49	
	8120	3″ diameter		253	.095		47.50	2.47	7.70	57.67	65	
	8130	4″ diameter		200	.120		54	3.12	9.75	66.87	75	
	8140	5″ diameter		168	.143		63.50	3.72	11.65	78.87	88.50	
	8150	6″ diameter		126	.190		76	4.96	15.50	96.46	108	
	8160	8″ diameter		98.50	.244		99.50	6.35	19.85	125.70	141	
	8170	10″ diameter		73	.329		125	8.55	27	160.55	180	
	8180	12″ diameter		62.50	.384		147	10	31.50	188.50	211	
	8190	14″ diameter		54.30	.442		166	11.50	36	213.50	240	
	8200	16″ diameter		48.30	.497		184	12.95	40.50	237.45	267	
	8210	18″ diameter		39.20	.612		228	15.95	50	293.95	330	

		02520 | Wells	CREW	DAILY OUTPUT	LABOR-HOURS	UNIT	2001 BARE COSTS MAT.	LABOR	EQUIP.	TOTAL	TOTAL INCL O&P	
900	8220	20″ diameter	B-23A	31.20	.769	L.F.	261	20	62.50	343.50	385	900
	8230	24″ diameter		23.80	1.008		325	26	82	433	485	
	8240	26″ diameter		21	1.143		355	29.50	93	477.50	540	
	8300	Slotted PVC, 1-1/4″ diameter		521	.046		1.26	1.20	3.75	6.21	7.35	
	8310	1-1/2″ diameter		488	.049		1.88	1.28	4	7.16	8.45	
	8320	2″ diameter		273	.088		2.75	2.29	7.15	12.19	14.40	
	8330	3″ diameter		253	.095		2.73	2.47	7.70	12.90	15.30	
	8340	4″ diameter		200	.120		3.39	3.12	9.75	16.26	19.30	
	8350	5″ diameter		168	.143		3.56	3.72	11.65	18.93	22.50	
	8360	6″ diameter		126	.190		4.65	4.96	15.50	25.11	30	
	8370	8″ diameter		98.50	.244		6.70	6.35	19.85	32.90	39	
	8400	Artificial gravel pack, 2″ screen, 6″ casing	B-23B	174	.138		1.89	3.59	12.15	17.63	21	
	8405	8″ casing		111	.216		2.58	5.65	19.10	27.33	32.50	
	8410	10″ casing		74.50	.322		3.25	8.40	28.50	40.15	48	
	8415	12″ casing		60	.400		4.38	10.40	35.50	50.28	60	
	8420	14″ casing		50.20	.478		5.65	12.45	42	60.10	72	
	8425	16″ casing		40.70	.590		7.80	15.35	52	75.15	89	
	8430	18″ casing		36	.667		9.10	17.35	59	85.45	102	
	8435	20″ casing		29.50	.814		10.45	21	72	103.45	123	
	8440	24″ casing		25.70	.934		11.60	24.50	82.50	118.60	141	
	8445	26″ casing		24.60	.976		13	25.50	86	124.50	148	
	8450	30″ casing		20	1.200		14.90	31	106	151.90	180	
	8455	36″ casing		16.40	1.463		16	38	129	183	219	
	8500	Develop well		8	3	Hr.	165	78	265	508	595	
	8550	Pump test well		8	3		44.50	78	265	387.50	460	
	8560	Standby well	B-23A	8	3		43.50	78	244	365.50	440	
	8570	Standby, drill rig		8	3			78	244	322	390	
	8580	Surface seal well, concrete filled		1	24	Ea.	410	625	1,950	2,985	3,575	
	8590	Well test pump, install & remove	B-23	1	40			930	2,050	2,980	3,700	
	8600	Well sterilization, chlorine	2 Clab	1	16		345	365		710	950	
	9950	See div. 02240-900 for wellpoints										
	9960	See div. 02240-700 for drainage wells										
910	0010	**PUMPS, WELL** Water system, with pressure control										910
	1000	Deep well, jet, 42 gal. galvanized tank										
	1040	3/4 HP	1 Plum	.80	10	Ea.	545	345		890	1,125	
	3000	Shallow well, jet, 30 gal. galvanized tank										
	3040	1/2 HP	1 Plum	2	4	Ea.	550	138		688	815	

		02530 | Sanitary Sewerage	CREW	DAILY OUTPUT	LABOR-HOURS	UNIT	MAT.	LABOR	EQUIP.	TOTAL	TOTAL INCL O&P	
100	0010	**SEWAGE TREATMENT** Plant, not incl. fencing or external piping										100
	0020	Steel packaged, blown air aeration plants										
	0100	1,000 GPD				Gal.				15	17.25	
	0200	5,000 GPD								10	11.50	
	0300	15,000 GPD								5.50	6.30	
	0400	30,000 GPD								5.20	6	
	0500	50,000 GPD								4	4.60	
	0600	100,000 GPD								3.50	4	
	0700	200,000 GPD								2.50	2.88	
	0800	500,000 GPD								2.45	2.80	
	1000	Concrete, extended aeration, primary and secondary treatment										
	1010	10,000 GPD				Gal.				11	12.65	
	1100	30,000 GPD								5.50	6.35	
	1200	50,000 GPD								4.50	5.18	
	1400	100,000 GPD								3.50	4.05	
	1500	500,000 GPD								2.50	2.90	

02530 | Sanitary Sewerage

		02530 Sanitary Sewerage	CREW	DAILY OUTPUT	LABOR-HOURS	UNIT	2001 BARE COSTS MAT.	LABOR	EQUIP.	TOTAL	TOTAL INCL O&P	
100	1700	Municipal wastewater treatment facility										100
	1720	1.0 MGD				Gal.				4.30	4.95	
	1740	1.5 MGD								4.25	4.90	
	1760	2.0 MGD								3.65	4.20	
	1780	3.0 MGD								2.85	3.30	
	1800	5.0 MGD								2.60	3	
	2000	Holding tank system, not incl. excavation or backfill										
	2010	Recirculating chemical water closet	2 Plum	4	4	Ea.	700	138		838	980	
	2100	For voltage converter, add	"	16	1		186	34.50		220.50	256	
	2200	For high level alarm, add	1 Plum	7.80	1.026		106	35.50		141.50	171	
105	0010	**VALVES**										105
	1000	Backwater sewer line valve										
	1010	Offset type, bronze swing check assy and bronze cover										
	1020	B & S connections										
	1040	2″ size	2 Skwk	14	1.143	Ea.	224	34		258	300	
	1050	3″ size		12.50	1.280		325	38		363	420	
	1060	4″ size		11	1.455		430	43.50		473.50	545	
	1070	6″ size	3 Skwk	15	1.600		620	48		668	755	
	1080	8″ size	"	7.50	3.200		870	95.50		965.50	1,100	
	1110	Backwater drainage control										
	1120	Offset type, w/bronze manually operated shear gate										
	1130	B & S connections										
	1160	4″ size	2 Skwk	11	1.455	Ea.	665	43.50		708.50	800	
	1170	6″ size	3 Skwk	15	1.600	"	1,075	48		1,123	1,250	
730	0010	**PIPING, DRAINAGE & SEWAGE, CONCRETE** A12.3 -510										730
	0020	Not including excavation or backfill										
	1000	Non-reinforced pipe, extra strength, B&S or T&G joints										
	1010	6″ diameter	B-14	265.04	.181	L.F.	3.48	4.38	.68	8.54	11.40	
	1020	8″ diameter		224	.214		3.83	5.20	.81	9.84	13.15	
	1030	10″ diameter		216	.222		4.25	5.35	.84	10.44	13.95	
	1040	12″ diameter		200	.240		5.20	5.80	.90	11.90	15.75	
	1050	15″ diameter		180	.267		6.10	6.45	1.01	13.56	17.80	
	1060	18″ diameter		144	.333		7.50	8.05	1.26	16.81	22	
	1070	21″ diameter		112	.429		9.20	10.35	1.62	21.17	28	
	1080	24″ diameter		100	.480		11.30	11.60	1.81	24.71	32.50	
	2000	Reinforced culvert, class 3, no gaskets										
	2010	12″ diameter	B-14	210	.229	L.F.	9.80	5.50	.86	16.16	20.50	
	2020	15″ diameter		175	.274		12.10	6.65	1.03	19.78	24.50	
	2030	18″ diameter		130	.369		12.20	8.90	1.39	22.49	29	
	2035	21″ diameter		120	.400		14.85	9.65	1.51	26.01	33	
	2040	24″ diameter		100	.480		17.80	11.60	1.81	31.21	39.50	
	2045	27″ diameter	B-13	92	.609		21.50	15.05	6.85	43.40	54.50	
	2050	30″ diameter		88	.636		24.50	15.75	7.15	47.40	59.50	
	2060	36″ diameter		72	.778		35	19.25	8.75	63	78	
	2070	42″ diameter	B-13B	72	.778		47.50	19.25	12.50	79.25	96.50	
	2080	48″ diameter		64	.875		59.50	21.50	14.10	95.10	115	
	2090	60″ diameter		48	1.167		95	29	18.80	142.80	170	
	2100	72″ diameter		40	1.400		135	34.50	22.50	192	227	
	2120	84″ diameter		32	1.750		245	43.50	28	316.50	365	
	2140	96″ diameter		24	2.333		294	57.50	37.50	389	455	
	2200	With gaskets, class 3, 12″ diameter	B-21	168	.167		10.80	4.44	.85	16.09	19.75	
	2220	15″ diameter		160	.175		12.95	4.66	.89	18.50	22.50	
	2230	18″ diameter		152	.184		16.20	4.90	.93	22.03	26.50	
	2240	24″ diameter		136	.206		24.50	5.50	1.04	31.04	36.50	
	2260	30″ diameter	B-13	88	.636		32.50	15.75	7.15	55.40	68	
	2270	36″ diameter	"	72	.778		48.50	19.25	8.75	76.50	93	

02500 | Utility Services

02530 | Sanitary Sewerage

			CREW	DAILY OUTPUT	LABOR-HOURS	UNIT	2001 BARE COSTS MAT.	LABOR	EQUIP.	TOTAL	TOTAL INCL O&P	
730	2290	48″ diameter (A12.3 -510)	B-13B	64	.875	L.F.	79.50	21.50	14.10	115.10	137	730
	2310	72″ diameter	″	40	1.400		205	34.50	22.50	262	305	
	2330	Flared ends, 6′-1″ long, 12″ diameter	B-21	190	.147		37	3.92	.75	41.67	47.50	
	2340	15″ diameter		155	.181		43	4.81	.92	48.73	55.50	
	2400	6′-2″ long, 18″ diameter		122	.230		45.50	6.10	1.16	52.76	61	
	2420	24″ diameter		88	.318		55	8.45	1.61	65.06	75.50	
	2440	36″ diameter	B-13	60	.933		84	23	10.50	117.50	140	
	3040	Vitrified plate lined, add to above, 30″ to 36″ diameter				SFCA	3.18			3.18	3.50	
	3050	42″ to 54″ diameter, add					3.40			3.40	3.74	
	3060	60″ to 72″ diameter, add					3.98			3.98	4.38	
	3070	Over 72″ diameter, add					4.24			4.24	4.66	
	3080	Radius pipe, add to pipe prices, 12″ to 60″ diameter				L.F.	50%					
	3090	Over 60″ diameter, add				″	20%					
	3500	Reinforced elliptical, 8′ lengths, C507 class 3										
	3520	14″ x 23″ inside, round equivalent 18″ diameter	B-21	82	.341	L.F.	22.50	9.10	1.73	33.33	40.50	
	3530	24″ x 38″ inside, round equivalent 30″ diameter	B-13	58	.966		41	24	10.85	75.85	94	
	3540	29″ x 45″ inside, round equivalent 36″ diameter		52	1.077		50.50	26.50	12.10	89.10	110	
	3550	38″ x 60″ inside, round equivalent 48″ diameter		38	1.474		77.50	36.50	16.55	130.55	160	
	3560	48″ x 76″ inside, round equivalent 60″ diameter		26	2.154		139	53.50	24	216.50	262	
	3570	58″ x 91″ inside, round equivalent 72″ diameter		22	2.545		187	63	28.50	278.50	335	
	3780	Concrete slotted pipe, class 4 mortar joint										
	3800	12″ diameter	B-21	168	.167	L.F.	11.15	4.44	.85	16.44	20	
	3840	18″ diameter	″	152	.184	″	17.25	4.90	.93	23.08	27.50	
	3900	Class 4 O-ring										
	3940	12″ diameter	B-21	168	.167	L.F.	11.75	4.44	.85	17.04	20.50	
	3960	18″ diameter	″	152	.184	″	16.45	4.90	.93	22.28	26.50	
780	0010	**PIPING, DRAINAGE & SEWAGE, POLYVINYL CHLORIDE** (A12.3 -510)										780
	0020	Not including excavation or backfill										
	2000	10′ lengths, S.D.R. 35, B&S, 4″ diameter	B-20	375	.064	L.F.	1.41	1.65		3.06	4.14	
	2040	6″ diameter	″	350	.069		2.41	1.77		4.18	5.45	
	2120	10″ diameter	B-21	330	.085		6.55	2.26	.43	9.24	11.20	
	2160	12″ diameter		320	.087		7.35	2.33	.44	10.12	12.20	
	2200	15″ diameter		190	.147		11.15	3.92	.75	15.82	19.20	
790	0010	**PIPING, DRAINAGE & SEWAGE, VITRIFIED CLAY** C700 (A12.3 -510)										790
	0020	Not including excavation or backfill,										
	4030	Extra strength, compression joints, C425										
	5000	4″ diameter x 4′ long	B-20	265	.091	L.F.	1.54	2.34		3.88	5.35	
	5020	6″ diameter x 5′ long	″	200	.120		2.61	3.10		5.71	7.75	
	5040	8″ diameter x 5′ long	B-21	200	.140		3.70	3.73	.71	8.14	10.65	
	5060	10″ diameter x 5′ long		190	.147		6.20	3.92	.75	10.87	13.70	
	5080	12″ diameter x 6′ long		150	.187		8.25	4.97	.95	14.17	17.90	
	5100	15″ diameter x 7′ long		110	.255		14.45	6.75	1.29	22.49	28	
	5120	18″ diameter x 7′ long		88	.318		21.50	8.45	1.61	31.56	38.50	
	5140	24″ diameter x 7′ long		45	.622		39	16.55	3.15	58.70	72.50	
	5160	30″ diameter x 7′ long	B-22	31	.968		72.50	26	6.85	105.35	128	
	5180	36″ diameter x 7′ long	″	20	1.500		113	40.50	10.65	164.15	200	
	6000	For 3′ lengths, add					30%	30%				
	6020	For 2′ lengths, add					40%	60%				
	6060	For plain joints, deduct					25%					
	7060	2′ lengths, add to above					40%					

02540 | Septic Tank Systems

			CREW	DAILY OUTPUT	LABOR-HOURS	UNIT	MAT.	LABOR	EQUIP.	TOTAL	TOTAL INCL O&P	
700	0010	**SEPTIC TANKS** Not incl. excav. or piping, precast, 1,000 gallon (A12.3 -900)	B-21	8	3.500	Ea.	490	93	17.75	600.75	700	700
	0100	2,000 gallon	″	5	5.600		980	149	28.50	1,157.50	1,350	

2 SITE CONSTRUCTION

02500 | Utility Services

		02540 | Septic Tank Systems	CREW	DAILY OUTPUT	LABOR-HOURS	UNIT	2001 BARE COSTS MAT.	LABOR	EQUIP.	TOTAL	TOTAL INCL O&P	
700	0200	5,000 gallon (A12.3 -900)	B-13	3.50	16	Ea.	4,925	395	180	5,500	6,250	700
	0300	15,000 gallon, 4 piece	B-13B	1.70	32.941		11,800	815	530	13,145	14,700	
	0400	25,000 gallon, 4 piece		1.10	50.909		25,900	1,250	820	27,970	31,400	
	0500	40,000 gallon, 4 piece		.80	70		31,800	1,725	1,125	34,650	38,900	
	0520	50,000 gallon, 5 piece	B-13C	.60	93.333		36,600	2,300	2,025	40,925	46,000	
	0540	75,000 gallon, cast in place	C-14C	.25	448		44,500	12,500	152	57,152	69,000	
	0560	100,000 gallon	"	.15	746		55,000	20,800	254	76,054	94,000	
	0600	High density polyethylene, 1,000 gallon	B-21	6	4.667		800	124	23.50	947.50	1,100	
	0700	1,500 gallon	"	4	7		1,000	186	35.50	1,221.50	1,425	
	1000	Distribution boxes, concrete, 7 outlets	2 Clab	16	1		89	23		112	134	
	1100	9 outlets	"	8	2		232	45.50		277.50	325	
	1150	Leaching field chambers, 13′ x 3′-7″ x 1′-4″, standard	B-13	16	3.500		665	86.50	39.50	791	910	
	1200	Heavy duty, 8′ x 4′ x 1′-6″		14	4		330	99	45	474	570	
	1300	13′ x 3′-9″ x 1′-6″		12	4.667		1,050	115	52.50	1,217.50	1,375	
	1350	20′ x 4′ x 1′-6″		5	11.200		775	277	126	1,178	1,425	
	1400	Leaching pit, precast concrete, 3′ diameter, 3′ deep	B-21	8	3.500		305	93	17.75	415.75	500	
	1500	6′ diameter, 3′ section		4.70	5.957		535	159	30	724	870	
	2000	Velocity reducing pit, precast conc., 6′ diameter, 3′ deep		4.70	5.957		232	159	30	421	535	
	2200	Excavation for septic tank, 3/4 C.Y. backhoe	B-12F	145	.110	C.Y.		3.14	2.85	5.99	7.90	
	2400	4′ trench for disposal field, 3/4 C.Y. backhoe	"	335	.048	L.F.		1.36	1.23	2.59	3.42	
	2600	Gravel fill, run of bank	B-6	150	.160	C.Y.	13.60	3.97	1.21	18.78	22.50	
	2800	Crushed stone, 3/4″	"	150	.160	"	21.50	3.97	1.21	26.68	31	

		02550 | Piped Energy Distribution	CREW	DAILY OUTPUT	LABOR-HOURS	UNIT	MAT.	LABOR	EQUIP.	TOTAL	TOTAL INCL O&P	
450	0010	**GAS STATION PRODUCT LINE**										450
	0020	Primary containment pipe, fiberglass-reinforced										
	0030	Plastic pipe 15′ & 30′ lengths										
	0040	2″ diameter	Q-6	425	.056	L.F.	2.91	1.84		4.75	5.95	
	0050	3″ diameter		400	.060		3.92	1.95		5.87	7.25	
	0060	4″ diameter		375	.064		5.05	2.08		7.13	8.70	
	0100	Fittings										
	0110	Elbows, 90° & 45°, bell-ends, 2″	Q-6	24	1	Ea.	29.50	32.50		62	81.50	
	0120	3″ diameter		22	1.091		31.50	35.50		67	88.50	
	0130	4″ diameter		20	1.200		42	39		81	105	
	0200	Tees, bell ends, 2″		21	1.143		36.50	37		73.50	96.50	
	0210	3″ diameter		18	1.333		37	43.50		80.50	106	
	0230	Flanges bell ends, 2″		24	1		8.65	32.50		41.15	58.50	
	0240	3″ diameter		22	1.091		10.90	35.50		46.40	65.50	
	0250	4″ diameter		20	1.200		14.85	39		53.85	75.50	
	0260	Sleeve couplings, 2″		21	1.143		7.45	37		44.45	64	
	0270	3″ diameter		18	1.333		10.30	43.50		53.80	77	
	0280	4″ diameter		15	1.600		14.25	52		66.25	94	
	0290	Threaded adapters 2″		21	1.143		10.15	37		47.15	67	
	0300	3″ diameter		18	1.333		14.80	43.50		58.30	82	
	0310	4″ diameter		15	1.600		21.50	52		73.50	102	
	0320	Reducers, 2″		27	.889		12.50	29		41.50	57.50	
	0330	3″ diameter		22	1.091		14.80	35.50		50.30	70	
	0340	4″ diameter		20	1.200		21.50	39		60.50	82.50	
	1010	Gas station product line for secondary containment (double wall)										
	1100	Fiberglass reinforced plastic pipe 25′ lengths										
	1120	Pipe, plain end, 3″	Q-6	375	.064	L.F.	4.04	2.08		6.12	7.60	
	1130	4″ diameter		350	.069		5.05	2.23		7.28	8.90	
	1140	5″ diameter		325	.074		7	2.40		9.40	11.35	
	1150	6″ diameter		300	.080		8.90	2.60		11.50	13.75	
	1200	Fittings										
	1230	Elbows, 90° & 45°, 3″	Q-6	18	1.333	Ea.	36	43.50		79.50	105	

02550 | Piped Energy Distribution

		02550 Piped Energy Distribution	CREW	DAILY OUTPUT	LABOR-HOURS	UNIT	2001 BARE COSTS MAT.	LABOR	EQUIP.	TOTAL	TOTAL INCL O&P	
450	1240	4″ diameter	Q-6	16	1.500	Ea.	64.50	49		113.50	145	450
	1250	5″ diameter		14	1.714		149	56		205	248	
	1260	6″ diameter		12	2		151	65		216	264	
	1270	Tees, 3″		15	1.600		54.50	52		106.50	139	
	1280	4″ diameter		12	2		80.50	65		145.50	187	
	1290	5″ diameter		9	2.667		163	87		250	310	
	1300	6″ diameter		6	4		170	130		300	385	
	1310	Couplings, 3″		18	1.333		25	43.50		68.50	93	
	1320	4″ diameter		16	1.500		67	49		116	147	
	1330	5″ diameter		14	1.714		139	56		195	237	
	1340	6″ diameter		12	2		144	65		209	257	
	1350	Cross-over nipples, 3″		18	1.333		5.90	43.50		49.40	72	
	1360	4″ diameter		16	1.500		6.95	49		55.95	81	
	1370	5″ diameter		14	1.714		10.30	56		66.30	95.50	
	1380	6″ diameter		12	2		10.80	65		75.80	110	
	1400	Telescoping, reducers, concentric 4″ x 3″		18	1.333		19.75	43.50		63.25	87	
	1410	5″ x 4″		17	1.412		51.50	46		97.50	126	
	1420	6″ x 5″	↓	16	1.500	↓	124	49		173	210	
464	0010	**PIPING, GAS SERVICE & DISTRIBUTION, POLYETHYLENE**										464
	0020	not including excavation or backfill										
	1000	60 psi coils, comp cplg @ 100′, 1/2″ diameter, SDR 9.3	B-20A	608	.053	L.F.	.38	1.44		1.82	2.63	
	1040	1-1/4″ diameter, SDR 11		544	.059		.65	1.61		2.26	3.18	
	1100	2″ diameter, SDR 11		488	.066		.80	1.80		2.60	3.63	
	1160	3″ diameter, SDR 11	↓	408	.078		1.68	2.15		3.83	5.15	
	1500	60 PSI 40′ joints with coupling, 3″ diameter, SDR 11	B-21A	408	.098		1.74	2.76	1.04	5.54	7.30	
	1540	4″ diameter, SDR 11		352	.114		3.84	3.20	1.21	8.25	10.45	
	1600	6″ diameter, SDR 11		328	.122		11.95	3.44	1.30	16.69	19.85	
	1640	8″ diameter, SDR 11	↓	272	.147	↓	16.30	4.14	1.56	22	26	
550	0010	**PIPE CONDUIT, PREFABRICATED** R15100-070										550
	0020	Does not include trenching, fittings or crane.										
	0300	For cathodic protection, add 12 to 14%										
	0310	of total built-up price (casing plus service pipe)										
	0580	Polyurethane insulated system, 250°F. max. temp.										
	0620	Black steel service pipe, standard wt., 1/2″ insulation										
	0660	3/4″ diam. pipe size	Q-17	54	.296	L.F.	16.55	9.30	1.86	27.71	34.50	
	0670	1″ diam. pipe size		50	.320		18.20	10.05	2.01	30.26	37.50	
	0680	1-1/4″ diam. pipe size		47	.340		20.50	10.70	2.14	33.34	41	
	0690	1-1/2″ diam. pipe size		45	.356		22	11.15	2.24	35.39	44	
	0700	2″ diam. pipe size		42	.381		23	11.95	2.40	37.35	45.50	
	0710	2-1/2″ diam. pipe size		34	.471		23	14.75	2.96	40.71	51.50	
	0720	3″ diam. pipe size		28	.571		27	17.95	3.59	48.54	60.50	
	0730	4″ diam. pipe size		22	.727		34	23	4.57	61.57	77	
	0740	5″ diam. pipe size	↓	18	.889		43.50	28	5.60	77.10	95.50	
	0750	6″ diam. pipe size	Q-18	23	1.043		50.50	34	4.37	88.87	112	
	0760	8″ diam. pipe size		19	1.263		73.50	41	5.30	119.80	149	
	0770	10″ diam. pipe size		16	1.500		97	49	6.30	152.30	187	
	0780	12″ diam. pipe size		13	1.846		121	60	7.75	188.75	232	
	0790	14″ diam. pipe size		11	2.182		135	71	9.15	215.15	265	
	0800	16″ diam. pipe size		10	2.400		154	78	10.05	242.05	299	
	0810	18″ diam. pipe size		8	3		178	97.50	12.55	288.05	355	
	0820	20″ diam. pipe size		7	3.429		198	112	14.35	324.35	400	
	0830	24″ diam. pipe size	↓	6	4		242	130	16.75	388.75	480	
	0900	For 1″ thick insulation, add					10%					
	0940	For 1-1/2″ thick insulation, add					13%					
	0980	For 2″ thick insulation, add				↓	20%					
	1500	Gland seal for system, 3/4″ diam. pipe size	Q-17	32	.500	Ea.	269	15.70	3.15	287.85	325	

02550 | Piped Energy Distribution

		Description	CREW	DAILY OUTPUT	LABOR-HOURS	UNIT	2001 BARE COSTS MAT.	LABOR	EQUIP.	TOTAL	TOTAL INCL O&P	
550	1510	1" diam. pipe size R15100-070	Q-17	32	.500	Ea.	269	15.70	3.15	287.85	325	550
	1540	1-1/4" diam. pipe size		30	.533		288	16.75	3.35	308.10	345	
	1550	1-1/2" diam. pipe size		30	.533		288	16.75	3.35	308.10	345	
	1560	2" diam. pipe size		28	.571		340	17.95	3.59	361.54	405	
	1570	2-1/2" diam. pipe size		26	.615		370	19.30	3.87	393.17	440	
	1580	3" diam. pipe size		24	.667		395	21	4.19	420.19	470	
	1590	4" diam. pipe size		22	.727		470	23	4.57	497.57	555	
	1600	5" diam. pipe size		19	.842		575	26.50	5.30	606.80	680	
	1610	6" diam. pipe size	Q-18	26	.923		610	30	3.87	643.87	725	
	1620	8" diam. pipe size		25	.960		710	31	4.02	745.02	830	
	1630	10" diam. pipe size		23	1.043		865	34	4.37	903.37	1,000	
	1640	12" diam. pipe size		21	1.143		955	37	4.79	996.79	1,100	
	1650	14" diam. pipe size		19	1.263		1,075	41	5.30	1,121.30	1,250	
	1660	16" diam. pipe size		18	1.333		1,250	43.50	5.60	1,299.10	1,450	
	1670	18" diam. pipe size		16	1.500		1,350	49	6.30	1,405.30	1,550	
	1680	20" diam. pipe size		14	1.714		1,525	56	7.20	1,588.20	1,775	
	1690	24" diam. pipe size		12	2		1,700	65	8.40	1,773.40	1,950	
	2000	Elbow, 45° for system										
	2020	3/4" diam. pipe size	Q-17	14	1.143	Ea.	170	36	7.20	213.20	249	
	2040	1" diam. pipe size		13	1.231		174	38.50	7.75	220.25	259	
	2050	1-1/4" diam. pipe size		11	1.455		198	45.50	9.15	252.65	297	
	2060	1-1/2" diam. pipe size		9	1.778		225	56	11.20	292.20	345	
	2070	2" diam. pipe size		6	2.667		216	83.50	16.75	316.25	380	
	2080	2-1/2" diam. pipe size		4	4		234	126	25	385	475	
	2090	3" diam. pipe size		3.50	4.571		270	143	29	442	545	
	2100	4" diam. pipe size		3	5.333		315	167	33.50	515.50	635	
	2110	5" diam. pipe size		2.80	5.714		405	179	36	620	755	
	2120	6" diam. pipe size	Q-18	4	6		460	195	25	680	830	
	2130	8" diam. pipe size		3	8		665	260	33.50	958.50	1,150	
	2140	10" diam. pipe size		2.40	10		850	325	42	1,217	1,475	
	2150	12" diam. pipe size		2	12		1,125	390	50.50	1,565.50	1,875	
	2160	14" diam. pipe size		1.80	13.333		1,375	435	56	1,866	2,250	
	2170	16" diam. pipe size		1.60	15		1,650	490	63	2,203	2,625	
	2180	18" diam. pipe size		1.30	18.462		2,075	600	77.50	2,752.50	3,275	
	2190	20" diam. pipe size		1	24		2,600	780	101	3,481	4,150	
	2200	24" diam. pipe size		.70	34.286		3,275	1,125	144	4,544	5,425	
	2260	For elbow, 90°, add					25%					
	2300	For tee, straight, add					85%	30%				
	2340	For tee, reducing, add					170%	30%				
	2380	For weldolet, straight, add					50%					
	2400	Polyurethane insulation, 1"										
	2410	FRP carrier and casing										
	2420	4"	Q-5	18	.889	L.F.	42.50	28		70.50	88.50	
	2422	6"	Q-6	23	1.043		73.50	34		107.50	133	
	2424	8"		19	1.263		120	41		161	194	
	2426	10"		16	1.500		161	49		210	251	
	2428	12"		13	1.846		210	60		270	320	
	2430	FRP carrier and PVC casing										
	2440	4"	Q-5	18	.889	L.F.	13.70	28		41.70	57	
	2444	8"	Q-6	19	1.263		28.50	41		69.50	93.50	
	2446	10"		16	1.500		41	49		90	119	
	2448	12"		13	1.846		52.50	60		112.50	149	
	2450	PVC carrier and casing										
	2460	4"	Q-1	36	.444	L.F.	6.75	13.80		20.55	28.50	
	2462	6"	"	29	.552		9.45	17.10		26.55	36.50	
	2464	8"	Q-2	36	.667		13.15	21.50		34.65	47	

02550 | Piped Energy Distribution

550			CREW	DAILY OUTPUT	LABOR-HOURS	UNIT	2001 BARE COSTS MAT.	LABOR	EQUIP.	TOTAL	TOTAL INCL O&P
	2466	10″ R15100-070	Q-2	32	.750	L.F.	18	24		42	56.50
	2468	12″	↓	31	.774	↓	21	25		46	60.50
	2800	Calcium silicate insulated system, high temp. (1200°F)									
	2840	Steel casing with protective exterior coating									
	2850	6-5/8″ diameter	Q-18	52	.462	L.F.	31	15	1.93	47.93	58.50
	2860	8-5/8″ diameter		50	.480		34	15.60	2.01	51.61	62.50
	2870	10-3/4″ diameter		47	.511		39.50	16.60	2.14	58.24	71
	2880	12-3/4″ diameter		44	.545		43	17.75	2.29	63.04	77
	2890	14″ diameter		41	.585		48.50	19.05	2.45	70	85
	2900	16″ diameter		39	.615		52	20	2.58	74.58	90.50
	2910	18″ diameter		36	.667		57.50	21.50	2.79	81.79	99.50
	2920	20″ diameter		34	.706		65	23	2.96	90.96	109
	2930	22″ diameter		32	.750		90	24.50	3.14	117.64	139
	2940	24″ diameter		29	.828		103	27	3.47	133.47	157
	2950	26″ diameter		26	.923		117	30	3.87	150.87	179
	2960	28″ diameter		23	1.043		146	34	4.37	184.37	216
	2970	30″ diameter		21	1.143		155	37	4.79	196.79	231
	2980	32″ diameter		19	1.263		174	41	5.30	220.30	260
	2990	34″ diameter		18	1.333		176	43.50	5.60	225.10	266
	3000	36″ diameter	↓	16	1.500		189	49	6.30	244.30	288
	3040	For multi-pipe casings, add					10%				
	3060	For oversize casings, add				↓	2%				
	3400	Steel casing gland seal, single pipe									
	3420	6-5/8″ diameter	Q-18	25	.960	Ea.	430	31	4.02	465.02	525
	3440	8-5/8″ diameter		23	1.043		505	34	4.37	543.37	610
	3450	10-3/4″ diameter		21	1.143		565	37	4.79	606.79	685
	3460	12-3/4″ diameter		19	1.263		675	41	5.30	721.30	810
	3470	14″ diameter		17	1.412		735	46	5.90	786.90	885
	3480	16″ diameter		16	1.500		865	49	6.30	920.30	1,025
	3490	18″ diameter		15	1.600		960	52	6.70	1,018.70	1,125
	3500	20″ diameter		13	1.846		1,050	60	7.75	1,117.75	1,275
	3510	22″ diameter		12	2		1,200	65	8.40	1,273.40	1,425
	3520	24″ diameter		11	2.182		1,350	71	9.15	1,430.15	1,600
	3530	26″ diameter		10	2.400		1,525	78	10.05	1,613.05	1,800
	3540	28″ diameter		9.50	2.526		1,725	82	10.60	1,817.60	2,025
	3550	30″ diameter		9	2.667		1,775	87	11.15	1,873.15	2,100
	3560	32″ diameter		8.50	2.824		1,975	92	11.85	2,078.85	2,325
	3570	34″ diameter		8	3		2,150	97.50	12.55	2,260.05	2,525
	3580	36″ diameter	↓	7	3.429		2,300	112	14.35	2,426.35	2,700
	3620	For multi-pipe casings, add				↓	5%				
	4000	Steel casing anchors, single pipe									
	4020	6-5/8″ diameter	Q-18	8	3	Ea.	385	97.50	12.55	495.05	585
	4040	8-5/8″ diameter		7.50	3.200		405	104	13.40	522.40	615
	4050	10-3/4″ diameter		7	3.429		530	112	14.35	656.35	770
	4060	12-3/4″ diameter		6.50	3.692		565	120	15.45	700.45	825
	4070	14″ diameter		6	4		665	130	16.75	811.75	945
	4080	16″ diameter		5.50	4.364		775	142	18.30	935.30	1,075
	4090	18″ diameter		5	4.800		870	156	20	1,046	1,225
	4100	20″ diameter		4.50	5.333		960	173	22.50	1,155.50	1,325
	4110	22″ diameter		4	6		1,075	195	25	1,295	1,500
	4120	24″ diameter		3.50	6.857		1,175	223	28.50	1,426.50	1,650
	4130	26″ diameter		3	8		1,325	260	33.50	1,618.50	1,875
	4140	28″ diameter		2.50	9.600		1,450	310	40	1,800	2,125
	4150	30″ diameter		2	12		1,550	390	50.50	1,990.50	2,350
	4160	32″ diameter		1.50	16		1,850	520	67	2,437	2,875
	4170	34″ diameter	↓	1	24	↓	2,075	780	101	2,956	3,550

02550 | Piped Energy Distribution

			CREW	DAILY OUTPUT	LABOR-HOURS	UNIT	2001 BARE COSTS MAT.	LABOR	EQUIP.	TOTAL	TOTAL INCL O&P	
550	4180	36" diameter (R15100-070)	Q-18	1	24	Ea.	2,250	780	101	3,131	3,750	550
	4220	For multi-pipe, add				↓	5%	20%				
	4800	Steel casing elbow										
	4820	6-5/8" diameter	Q-18	15	1.600	Ea.	530	52	6.70	588.70	670	
	4830	8-5/8" diameter		15	1.600		565	52	6.70	623.70	710	
	4850	10-3/4" diameter		14	1.714		685	56	7.20	748.20	840	
	4860	12-3/4" diameter		13	1.846		810	60	7.75	877.75	990	
	4870	14" diameter		12	2		865	65	8.40	938.40	1,050	
	4880	16" diameter		11	2.182		945	71	9.15	1,025.15	1,175	
	4890	18" diameter		10	2.400		1,075	78	10.05	1,163.05	1,300	
	4900	20" diameter		9	2.667		1,150	87	11.15	1,248.15	1,425	
	4910	22" diameter		8	3		1,250	97.50	12.55	1,360.05	1,525	
	4920	24" diameter		7	3.429		1,375	112	14.35	1,501.35	1,675	
	4930	26" diameter		6	4		1,500	130	16.75	1,646.75	1,875	
	4940	28" diameter		5	4.800		1,600	156	20	1,776	2,025	
	4950	30" diameter		4	6		1,625	195	25	1,845	2,100	
	4960	32" diameter		3	8		1,850	260	33.50	2,143.50	2,450	
	4970	34" diameter		2	12		2,025	390	50.50	2,465.50	2,875	
	4980	36" diameter	↓	2	12	↓	2,150	390	50.50	2,590.50	3,025	
	5500	Black steel service pipe, std. wt., 1" thick insulation										
	5510	3/4" diameter pipe size	Q-17	54	.296	L.F.	13.50	9.30	1.86	24.66	31	
	5540	1" diameter pipe size		50	.320		14.05	10.05	2.01	26.11	33	
	5550	1-1/4" diameter pipe size		47	.340		15.75	10.70	2.14	28.59	36	
	5560	1-1/2" diameter pipe size		45	.356		17.15	11.15	2.24	30.54	38	
	5570	2" diameter pipe size		42	.381		19.05	11.95	2.40	33.40	41.50	
	5580	2-1/2" diameter pipe size		34	.471		20	14.75	2.96	37.71	48	
	5590	3" diameter pipe size		28	.571		23	17.95	3.59	44.54	56	
	5600	4" diameter pipe size		22	.727		29	23	4.57	56.57	71.50	
	5610	5" diameter pipe size	↓	18	.889		39.50	28	5.60	73.10	91.50	
	5620	6" diameter pipe size	Q-18	23	1.043	↓	43	34	4.37	81.37	104	
	6000	Black steel service pipe, std. wt., 1-1/2" thick insul.										
	6010	3/4" diameter pipe size	Q-17	54	.296	L.F.	13.70	9.30	1.86	24.86	31	
	6040	1" diameter pipe size		50	.320		15.30	10.05	2.01	27.36	34	
	6050	1-1/4" diameter pipe size		47	.340		17.15	10.70	2.14	29.99	37.50	
	6060	1-1/2" diameter pipe size		45	.356		18.70	11.15	2.24	32.09	40	
	6070	2" diameter pipe size		42	.381		20.50	11.95	2.40	34.85	43	
	6080	2-1/2" diameter pipe size		34	.471		21.50	14.75	2.96	39.21	50	
	6090	3" diameter pipe size		28	.571		24.50	17.95	3.59	46.04	57.50	
	6100	4" diameter pipe size		22	.727		31	23	4.57	58.57	73.50	
	6110	5" diameter pipe size	↓	18	.889		39.50	28	5.60	73.10	91.50	
	6120	6" diameter pipe size	Q-18	23	1.043		45	34	4.37	83.37	106	
	6130	8" diameter pipe size		19	1.263		65	41	5.30	111.30	139	
	6140	10" diameter pipe size		16	1.500		84.50	49	6.30	139.80	173	
	6150	12" diameter pipe size	↓	13	1.846		101	60	7.75	168.75	210	
	6190	For 2" thick insulation, add					15%					
	6220	For 2-1/2" thick insulation, add					25%					
	6260	For 3" thick insulation, add				↓	30%					
	6800	Black steel service pipe, ex. hvy. wt., 1" thick insul.										
	6820	3/4" diameter pipe size	Q-17	50	.320	L.F.	14.20	10.05	2.01	26.26	33	
	6840	1" diameter pipe size		47	.340		15.10	10.70	2.14	27.94	35	
	6850	1-1/4" diameter pipe size		44	.364		17.45	11.40	2.29	31.14	39	
	6860	1-1/2" diameter pipe size		42	.381		18.55	11.95	2.40	32.90	41	
	6870	2" diameter pipe size		40	.400		19.60	12.55	2.52	34.67	43	
	6880	2-1/2" diameter pipe size		31	.516		24.50	16.20	3.25	43.95	54.50	
	6890	3" diameter pipe size		27	.593		27.50	18.60	3.73	49.83	62	
550	6900	4" diameter pipe size	↓	21	.762	↓	36	24	4.79	64.79	81	

02500 | Utility Services

02550 | Piped Energy Distribution

		02550 Piped Energy Distribution	CREW	DAILY OUTPUT	LABOR-HOURS	UNIT	2001 BARE COSTS MAT.	LABOR	EQUIP.	TOTAL	TOTAL INCL O&P	
550	6910	5" diameter pipe size (R15100-070)	Q-17	17	.941	L.F.	50.50	29.50	5.90	85.90	107	550
	6920	6" diameter pipe size	Q-18	22	1.091		56	35.50	4.57	96.07	120	
	7400	Black steel service pipe, ex. hvy. wt., 1-1/2" thick insul.										
	7420	3/4" diameter pipe size	Q-17	50	.320	L.F.	14.20	10.05	2.01	26.26	33	
	7440	1" diameter pipe size		47	.340		16.20	10.70	2.14	29.04	36.50	
	7450	1-1/4" diameter pipe size		44	.364		18.80	11.40	2.29	32.49	40.50	
	7460	1-1/2" diameter pipe size		42	.381		20.50	11.95	2.40	34.85	43	
	7470	2" diameter pipe size		40	.400		21	12.55	2.52	36.07	44.50	
	7480	2-1/2" diameter pipe size		31	.516		24.50	16.20	3.25	43.95	55	
	7490	3" diameter pipe size		27	.593		29	18.60	3.73	51.33	64	
	7500	4" diameter pipe size		21	.762		38	24	4.79	66.79	83	
	7510	5" diameter pipe size		17	.941		52	29.50	5.90	87.40	109	
	7520	6" diameter pipe size	Q-18	22	1.091		57.50	35.50	4.57	97.57	122	
	7530	8" diameter pipe size		18	1.333		86.50	43.50	5.60	135.60	167	
	7540	10" diameter pipe size		15	1.600		103	52	6.70	161.70	199	
	7550	12" diameter pipe size		13	1.846		126	60	7.75	193.75	237	
	7590	For 2" thick insulation, add					13%					
	7640	For 2-1/2" thick insulation, add					18%					
	7680	For 3" thick insulation, add					24%					
	9000	Combined steam pipe with condensate return										
	9010	8" and 4" in 24" case	Q-18	100	.240	L.F.	185	7.80	1.01	193.81	217	
	9020	6" and 3" in 20" case		104	.231		127	7.50	.97	135.47	151	
	9030	3" and 1-1/2" in 16" case		110	.218		95	7.10	.91	103.01	116	
	9040	2" and 1-1/4" in 12-3/4" case		114	.211		82.50	6.85	.88	90.23	102	
	9050	1-1/2" and 1-1/4" in 10-3/4" case		116	.207		72	6.75	.87	79.62	90.50	
	9100	Steam pipe only (no return)										
	9110	6" in 18" case	Q-18	108	.222	L.F.	104	7.25	.93	112.18	126	
	9120	4" in 14" case		112	.214		82.50	6.95	.90	90.35	102	
	9130	3" in 14" case		114	.211		76	6.85	.88	83.73	95.50	
	9140	2-1/2" in 12-3/4" case		118	.203		71	6.60	.85	78.45	89	
	9150	2" in 12-3/4" case		122	.197		68	6.40	.82	75.22	85.50	

02600 | Drainage & Containment

02620 | Subdrainage

		02620 Subdrainage	CREW	DAILY OUTPUT	LABOR-HOURS	UNIT	2001 BARE COSTS MAT.	LABOR	EQUIP.	TOTAL	TOTAL INCL O&P	
210	0010	**PIPING, SUBDRAINAGE, CONCRETE** (A12.3-530)										210
	0021	Not including excavation and backfill										
	3000	Porous wall concrete underdrain, std. strength, 4" diameter	B-20	335	.072	L.F.	1.78	1.85		3.63	4.86	
	3020	6" diameter	"	315	.076		2.31	1.97		4.28	5.65	
	3040	8" diameter	B-21	310	.090		2.86	2.40	.46	5.72	7.40	
	3100	18" diameter	"	165	.170		9.15	4.52	.86	14.53	18.10	
	4000	Extra strength, 6" diameter	B-20	315	.076		2.36	1.97		4.33	5.70	
	4020	8" diameter	B-21	310	.090		3.54	2.40	.46	6.40	8.15	
	4040	10" diameter		285	.098		7.10	2.61	.50	10.21	12.40	
	4060	12" diameter		230	.122		7.65	3.24	.62	11.51	14.20	
	4080	15" diameter		200	.140		8.50	3.73	.71	12.94	15.95	
	4100	18" diameter		165	.170		12.40	4.52	.86	17.78	21.50	
240	0010	**PIPING, SUBDRAINAGE, CORRUGATED METAL** (A12.3-530)										240
	0021	Not including excavation and backfill										

2 SITE CONSTRUCTION

02620 | Subdrainage

		02620 Subdrainage	Crew	Daily Output	Labor-Hours	Unit	2001 Bare Costs Mat.	Labor	Equip.	Total	Total Incl O&P
240	2010	Aluminum, perforated (A12.3 -530)									
	2020	6″ diameter, 18 ga.	B-14	380	.126	L.F.	2.62	3.05	.48	6.15	8.15
	2200	8″ diameter, 16 ga.		370	.130		3.81	3.14	.49	7.44	9.60
	2220	10″ diameter, 16 ga.		360	.133		4.76	3.22	.50	8.48	10.80
	2240	12″ diameter, 16 ga.		285	.168		5.35	4.07	.63	10.05	12.90
	2260	18″ diameter, 16 ga.		205	.234		8	5.65	.88	14.53	18.55
	3000	Uncoated galvanized, perforated									
	3020	6″ diameter, 18 ga.	B-20	380	.063	L.F.	4.08	1.63		5.71	7.05
	3200	8″ diameter, 16 ga.	″	370	.065		5.60	1.68		7.28	8.80
	3220	10″ diameter, 16 ga.	B-21	360	.078		8.40	2.07	.39	10.86	12.90
	3240	12″ diameter, 16 ga.		285	.098		8.80	2.61	.50	11.91	14.30
	3260	18″ diameter, 16 ga.		205	.137		13.45	3.63	.69	17.77	21
	4000	Steel, perforated, asphalt coated									
	4020	6″ diameter 18 ga.	B-20	380	.063	L.F.	3.26	1.63		4.89	6.15
	4030	8″ diameter 18 ga	″	370	.065		5.10	1.68		6.78	8.25
	4040	10″ diameter 16 ga	B-21	360	.078		5.85	2.07	.39	8.31	10.10
	4050	12″ diameter 16 ga		285	.098		6.75	2.61	.50	9.86	12
	4060	18″ diameter 16 ga		205	.137		9.20	3.63	.69	13.52	16.50
250	0010	**PIPING, SUBDRAINAGE, PLASTIC**									
	0020	Not including excavation and backfill									
	1110	10″				Ea.	3.61			3.61	3.97
	2100	Perforated PVC, 4″ diameter	B-14	314	.153	L.F.	1.44	3.69	.58	5.71	7.95
	2110	6″ diameter		300	.160		2.66	3.87	.60	7.13	9.60
	2120	8″ diameter		290	.166		3.02	4	.62	7.64	10.20
	2130	10″ diameter		280	.171		4.68	4.14	.65	9.47	12.30
	2140	12″ diameter		270	.178		6.50	4.30	.67	11.47	14.60
280	0010	**PIPING, SUBDRAINAGE, VITRIFIED CLAY**									
	0020	Not including excavation and backfill									
	3000	Perforated, 5′ lengths, C700, 4″ diameter	B-14	400	.120	L.F.	1.91	2.90	.45	5.26	7.10
	3020	6″ diameter		315	.152		3.25	3.68	.57	7.50	9.95
	3040	8″ diameter		290	.166		4.62	4	.62	9.24	12
	3060	12″ diameter		275	.175		9.60	4.22	.66	14.48	17.80
	4000	Channel pipe, 4″ diameter	B-20	430	.056		1.79	1.44		3.23	4.23
	4020	6″ diameter		335	.072		3.15	1.85		5	6.35
	4060	8″ diameter		295	.081		4.81	2.10		6.91	8.60
	4080	12″ diameter	B-21	280	.100		10.50	2.66	.51	13.67	16.25

02630 | Storm Drainage

		02630 Storm Drainage	Crew	Daily Output	Labor-Hours	Unit	Mat.	Labor	Equip.	Total	Total Incl O&P
100	0010	**PIPING, STORM DRAINAGE, CORRUGATED METAL** (A12.3 -510)									
	0020	Not including excavation or backfill									
	2000	Corrugated metal pipe, galvanized and coated									
	2020	Bituminous coated with paved invert, 20′ lengths									
	2040	8″ diameter, 16 ga.	B-14	330	.145	L.F.	5.85	3.52	.55	9.92	12.50
	2060	10″ diameter, 16 ga.		260	.185		7	4.46	.70	12.16	15.40
	2080	12″ diameter, 16 ga.		210	.229		8.45	5.50	.86	14.81	18.85
	2100	15″ diameter, 16 ga.		200	.240		10.25	5.80	.90	16.95	21.50
	2120	18″ diameter, 16 ga.		190	.253		13.15	6.10	.95	20.20	25
	2140	24″ diameter, 14 ga.		160	.300		16.15	7.25	1.13	24.53	30.50
	2160	30″ diameter, 14 ga.	B-13	120	.467		21	11.55	5.25	37.80	46.50
	2180	36″ diameter, 12 ga.		120	.467		31	11.55	5.25	47.80	57.50
	2200	48″ diameter, 12 ga.		100	.560		47	13.85	6.30	67.15	80.50
	2220	60″ diameter, 10 ga.	B-13B	75	.747		61	18.45	12	91.45	109
	2240	72″ diameter, 8 ga.	″	45	1.244		93	31	20	144	172
	2250	End sections, 8″ diameter, 16 ga.	B-14	20	2.400	Ea.	35.50	58	9.05	102.55	139
	2255	10″ diameter, 16 ga.		20	2.400		40.50	58	9.05	107.55	144
	2260	12″ diameter, 16 ga.		18	2.667		49	64.50	10.05	123.55	165

02630 | Storm Drainage

		02630 Storm Drainage	CREW	DAILY OUTPUT	LABOR-HOURS	UNIT	2001 BARE COSTS MAT.	LABOR	EQUIP.	TOTAL	TOTAL INCL O&P	
100	2265	15″ diameter, 16 ga. (A12.3-510)	B-14	18	2.667	Ea.	60.50	64.50	10.05	135.05	178	100
	2270	18″ diameter, 16 ga.		16	3		76.50	72.50	11.30	160.30	210	
	2275	24″ diameter, 16 ga.	B-13	16	3.500		109	86.50	39.50	235	298	
	2280	30″ diameter, 16 ga.		14	4		208	99	45	352	430	
	2285	36″ diameter, 14 ga.		14	4		274	99	45	418	505	
	2290	48″ diameter, 14 ga.		10	5.600		565	139	63	767	910	
	2292	60″ diameter, 14 ga.		6	9.333		775	231	105	1,111	1,325	
	2294	72″ diameter, 14 ga.	B-13B	5	11.200		1,350	277	180	1,807	2,125	
	2342	18″ diameter, 16 ga.	B-14	20	2.400		116	58	9.05	183.05	228	
	2344	24″ diameter, 14 ga.		16	3		162	72.50	11.30	245.80	305	
	2346	30″ diameter, 14 ga.		15	3.200		201	77.50	12.05	290.55	355	
	2348	36″ diameter, 14 ga.	B-13	15	3.733		271	92.50	42	405.50	485	
	2350	48″ diameter, 12 ga.	″	12	4.667		390	115	52.50	557.50	660	
	2352	60″ diameter, 10 ga.	B-13B	10	5.600		520	139	90	749	885	
	2354	72″ diameter, 10 ga.	″	6	9.333		645	231	150	1,026	1,225	
	2410	18″ diameter, 16 ga.	B-14	16	3		155	72.50	11.30	238.80	295	
	2412	24″ diameter, 14 ga.	″	16	3		233	72.50	11.30	316.80	380	
	2414	30″ diameter, 14 ga.	B-13	12	4.667		300	115	52.50	467.50	565	
	2416	36″ diameter, 14 ga.		11	5.091		370	126	57	553	670	
	2418	48″ diameter, 12 ga.		10	5.600		565	139	63	767	905	
	2420	60″ diameter, 10 ga.	B-13B	8	7		825	173	113	1,111	1,300	
	2422	72″ diameter, 10 ga.	″	5	11.200		990	277	180	1,447	1,700	
	2500	Galvanized, uncoated, 20′ lengths										
	2520	8″ diameter, 16 ga.	B-14	355	.135	L.F.	4.61	3.27	.51	8.39	10.70	
	2540	10″ diameter, 16 ga.		280	.171		5.10	4.14	.65	9.89	12.75	
	2560	12″ diameter, 16 ga.		220	.218		5.75	5.25	.82	11.82	15.45	
	2580	15″ diameter, 16 ga.		220	.218		7.65	5.25	.82	13.72	17.50	
	2600	18″ diameter, 16 ga.		205	.234		10.30	5.65	.88	16.83	21	
	2620	24″ diameter, 14 ga.		175	.274		14.10	6.65	1.03	21.78	27	
	2640	30″ diameter, 14 ga.	B-13	130	.431		17.90	10.65	4.84	33.39	41.50	
	2660	36″ diameter, 12 ga.		130	.431		29.50	10.65	4.84	44.99	54.50	
	2680	48″ diameter, 12 ga.		110	.509		39.50	12.60	5.70	57.80	69.50	
	2690	60″ diameter, 10 ga.	B-13B	78	.718		62	17.75	11.55	91.30	108	
	2695	72″ diameter, 10 ga.	″	60	.933		78	23	15.05	116.05	138	
	2711	Bends or elbows, 12″ diameter, 16 ga.	B-14	30	1.600	Ea.	69.50	38.50	6.05	114.05	143	
	2712	15″ diameter, 16 ga.		25.04	1.917		85.50	46.50	7.25	139.25	174	
	2714	18″ diameter, 16 ga.		20	2.400		96.50	58	9.05	163.55	206	
	2716	24″ diameter, 14 ga.		16	3		140	72.50	11.30	223.80	279	
	2718	30″ diameter, 14 ga.		15	3.200		176	77.50	12.05	265.55	325	
	2720	36″ diameter, 14 ga.	B-13	15	3.733		240	92.50	42	374.50	455	
	2722	48″ diameter, 12 ga.		12	4.667		330	115	52.50	497.50	595	
	2724	60″ diameter, 10 ga.		10	5.600		505	139	63	707	845	
	2726	72″ diameter, 10 ga.		6	9.333		625	231	105	961	1,150	
	2728	Wyes or tees, 12″ diameter, 16 ga.	B-14	22.48	2.135		92	51.50	8.05	151.55	190	
	2730	18″ diameter, 16 ga.		15	3.200		134	77.50	12.05	223.55	281	
	2732	24″ diameter, 14 ga.		15	3.200		208	77.50	12.05	297.55	360	
	2734	30″ diameter, 14 ga.		14	3.429		272	83	12.95	367.95	440	
	2736	36″ diameter, 14 ga.	B-13	14	4		355	99	45	499	595	
	2738	48″ diameter, 12 ga.		12	4.667		520	115	52.50	687.50	805	
	2740	60″ diameter, 10 ga.		10	5.600		760	139	63	962	1,125	
	2742	72″ diameter, 10 ga.		6	9.333		900	231	105	1,236	1,450	
	2780	End sections, 8″ diameter	B-14	24	2		55	48.50	7.55	111.05	144	
	2785	10″ diameter		22	2.182		56.50	52.50	8.25	117.25	153	
	2790	12″ diameter		35	1.371		42.50	33	5.15	80.65	104	
	2800	18″ diameter		30	1.600		65.50	38.50	6.05	110.05	139	
	2810	24″ diameter	B-13	25	2.240		95.50	55.50	25	176	219	

	02630 | Storm Drainage		CREW	DAILY OUTPUT	LABOR-HOURS	UNIT	2001 BARE COSTS MAT.	LABOR	EQUIP.	TOTAL	TOTAL INCL O&P	
100	2820	30″ diameter (A12.3 -510)	B-13	25	2.240	Ea.	181	55.50	25	261.50	315	100
	2825	36″ diameter		20	2.800		240	69.50	31.50	341	405	
	2830	48″ diameter		10	5.600		495	139	63	697	830	
	2835	60″ diameter	B-13B	5	11.200		650	277	180	1,107	1,350	
	2840	72″ diameter	″	4	14		1,175	345	225	1,745	2,050	
	2850	Couplings, 12″ diameter					10.65			10.65	11.70	
	2855	18″ diameter					15.55			15.55	17.15	
	2860	24″ diameter					21			21	23	
	2865	30″ diameter					26			26	28.50	
	2870	36″ diameter					36			36	39.50	
	2875	48″ diameter					49			49	54	
	2880	60″ diameter					77			77	84.50	
	2885	72″ diameter					107			107	117	
	3000	Corrugated galvanized or alum. oval arch culverts, coated & paved										
	3020	17″ x 13″, 16 ga., 15″ equivalent	B-14	200	.240	L.F.	17.35	5.80	.90	24.05	29	
	3040	21″ x 15″, 16 ga., 18″ equivalent		150	.320		22.50	7.75	1.21	31.46	38	
	3060	28″ x 20″, 14 ga., 24″ equivalent		125	.384		32	9.30	1.45	42.75	51.50	
	3080	35″ x 24″, 14 ga., 30″ equivalent		100	.480		39.50	11.60	1.81	52.91	63.50	
	3100	42″ x 29″, 12 ga., 36″ equivalent	B-13	100	.560		58.50	13.85	6.30	78.65	92.50	
	3120	49″ x 33″, 12 ga., 42″ equivalent		90	.622		70.50	15.40	7	92.90	109	
	3140	57″ x 38″, 12 ga., 48″ equivalent		75	.747		79.50	18.45	8.40	106.35	125	
	3160	Steel, plain oval arch culverts, plain										
	3180	17″ x 13″, 16 ga., 15″ equivalent	B-14	225	.213	L.F.	9.55	5.15	.80	15.50	19.40	
	3200	21″ x 15″, 16 ga., 18″ equivalent		175	.274		11.30	6.65	1.03	18.98	24	
	3220	28″ x 20″, 14 ga., 24″ equivalent		150	.320		18.15	7.75	1.21	27.11	33.50	
	3240	35″ x 24″, 14 ga., 30″ equivalent	B-13	108	.519		23	12.85	5.85	41.70	51.50	
	3260	42″ x 29″, 12 ga., 36″ equivalent		108	.519		37.50	12.85	5.85	56.20	68	
	3280	49″ x 33″, 12 ga., 42″ equivalent		92	.609		44.50	15.05	6.85	66.40	80	
	3300	57″ x 38″, 12 ga., 48″ equivalent		75	.747		39	18.45	8.40	65.85	81	
	3320	End sections, 17″ x 13″		22	2.545	Ea.	52	63	28.50	143.50	187	
	3340	42″ x 29″		17	3.294	″	214	81.50	37	332.50	405	
	3360	Multi-plate arch, steel	B-20	1,690	.014	Lb.	.60	.37		.97	1.24	
200	0010	**CATCH BASINS OR MANHOLES** not inclu. footing, excavation, (A12.3 -710)										200
	0020	backfill, frame and cover										
	0050	Brick, 4′ inside diameter, 4′ deep	D-1	1	16	Ea.	271	420		691	950	
	0100	6′ deep		.70	22.857		380	600		980	1,350	
	0150	8′ deep		.50	32		485	840		1,325	1,825	
	0200	For depths over 8′, add		4	4	V.L.F.	105	105		210	278	
	0400	Concrete blocks (radial), 4′ I.D., 4′ deep		1.50	10.667	Ea.	225	280		505	680	
	0500	6′ deep		1	16		298	420		718	980	
	0600	8′ deep		.70	22.857		370	600		970	1,350	
	0700	For depths over 8′, add		5.50	2.909	V.L.F.	38.50	76.50		115	160	
	0800	Concrete, cast in place, 4′ x 4′, 8″ thick, 4′ deep	C-14H	2	24	Ea.	345	690	18.95	1,053.95	1,500	
	0900	6′ deep		1.50	32		500	925	25.50	1,450.50	2,025	
	1000	8′ deep		1	48		715	1,375	38	2,128	3,000	
	1100	For depths over 8′, add		8	6	V.L.F.	85.50	173	4.74	263.24	375	
	1110	Precast, 4′ I.D., 4′ deep	B-22	4.10	7.317	Ea.	305	197	52	554	695	
	1120	6′ deep		3	10		425	269	71	765	965	
	1130	8′ deep		2	15		455	405	107	967	1,250	
	1140	For depths over 8′, add		16	1.875	V.L.F.	78.50	50.50	13.30	142.30	180	
	1150	5′ I.D., 4′ deep	B-6	3	8	Ea.	445	199	60.50	704.50	860	
	1160	6′ deep		2	12		605	298	90.50	993.50	1,225	
	1170	8′ deep		1.50	16		760	395	121	1,276	1,575	
	1180	For depths over 8′, add		12	2	V.L.F.	99	49.50	15.10	163.60	202	
	1190	6′ I.D., 4′ deep		2	12	Ea.	730	298	90.50	1,118.50	1,375	
	1200	6′ deep		1.50	16		950	395	121	1,466	1,800	100

02600 | Drainage & Containment

02630 | Storm Drainage

		02630 | Storm Drainage	CREW	DAILY OUTPUT	LABOR-HOURS	UNIT	2001 BARE COSTS MAT.	LABOR	EQUIP.	TOTAL	TOTAL INCL O&P	
200	1210	8′ deep (A12.3 -710)	B-6	1	24	Ea.	1,175	595	181	1,951	2,400	200
	1220	For depths over 8′, add		8	3	V.L.F.	153	74.50	22.50	250	310	
	1250	Slab tops, precast, 8″ thick										
	1300	4′ diameter manhole	B-6	8	3	Ea.	156	74.50	22.50	253	310	
	1400	5′ diameter manhole		7.50	3.200		275	79.50	24	378.50	450	
	1500	6′ diameter manhole		7	3.429		335	85	26	446	530	
	1600	Frames & covers, C.I., 24″ square, 500 lb.		7.80	3.077		214	76.50	23	313.50	380	
	1700	26″ D shape, 600 lb.		7	3.429		340	85	26	451	535	
	1800	Light traffic, 18″ diameter, 100 lb.		10	2.400		114	59.50	18.10	191.60	238	
	1900	24″ diameter, 300 lb.		8.70	2.759		138	68.50	21	227.50	280	
	2000	36″ diameter, 900 lb.		5.80	4.138		385	103	31	519	615	
	2100	Heavy traffic, 24″ diameter, 400 lb.		7.80	3.077		158	76.50	23	257.50	320	
	2200	36″ diameter, 1150 lb.		3	8		460	199	60.50	719.50	875	
	2300	Mass. State standard, 26″ diameter, 475 lb.		7	3.429		275	85	26	386	465	
	2400	30″ diameter, 620 lb.		7	3.429		300	85	26	411	490	
	2500	Watertight, 24″ diameter, 350 lb.		7.80	3.077		260	76.50	23	359.50	430	
	2600	26″ diameter, 500 lb.		7	3.429		261	85	26	372	445	
	2700	32″ diameter, 575 lb.		6	4		500	99.50	30	629.50	735	
	2800	3 piece cover & frame, 10″ deep,										
	2900	1200 lbs., for heavy equipment	B-6	3	8	Ea.	795	199	60.50	1,054.50	1,250	
	3000	Raised for paving 1-1/4″ to 2″ high,										
	3100	4 piece expansion ring										
	3200	20″ to 26″ diameter	1 Clab	3	2.667	Ea.	105	61		166	212	
	3300	30″ to 36″ diameter	″	3	2.667	″	147	61		208	258	
	3320	Frames and covers, existing, raised for paving, 2″, including										
	3340	row of brick, concrete collar, up to 12″ wide frame	B-6	18	1.333	Ea.	32	33	10.05	75.05	97.50	
	3360	20″ to 26″ wide frame		11	2.182		42	54	16.45	112.45	148	
	3380	30″ to 36″ wide frame		9	2.667		52.50	66	20	138.50	182	
	3400	Inverts, single channel brick	D-1	3	5.333		59	140		199	282	
	3500	Concrete		5	3.200		46	84		130	181	
	3600	Triple channel, brick		2	8		89.50	210		299.50	425	
	3700	Concrete		3	5.333		61.50	140		201.50	285	
	3800	Steps, heavyweight cast iron, 7″ x 9″	1 Bric	40	.200		9.45	5.90		15.35	19.55	
	3900	8″ x 9″		40	.200		14.20	5.90		20.10	25	
	3928	12″ x 10-1/2″		40	.200		13	5.90		18.90	23.50	
	4000	Standard sizes, galvanized steel		40	.200		11.75	5.90		17.65	22	
	4100	Aluminum		40	.200		15	5.90		20.90	25.50	

02700 | Bases, Ballasts, Pavements & Appurtenances

02740 | Flexible Pavement

		02740 | Flexible Pavement	CREW	DAILY OUTPUT	LABOR-HOURS	UNIT	2001 BARE COSTS MAT.	LABOR	EQUIP.	TOTAL	TOTAL INCL O&P	
300	0010	**ASPHALTIC CONCRETE PAVEMENT** for highways										300
	0020	and large paved areas										
	0080	Binder course, 1-1/2″ thick	B-25	7,725	.011	S.Y.	1.97	.29	.22	2.48	2.85	
	0120	2″ thick		6,345	.014		2.63	.35	.27	3.25	3.73	
	0160	3″ thick		4,905	.018		3.90	.45	.35	4.70	5.35	
	0200	4″ thick		4,140	.021		5.20	.53	.41	6.14	7.05	
	0300	Wearing course, 1″ thick	B-25B	10,575	.009		1.54	.23	.18	1.95	2.25	
	0340	1-1/2″ thick		7,725	.012		2.34	.32	.25	2.91	3.33	

2 SITE CONSTRUCTION

02700 | Bases, Ballasts, Pavements & Appurtenances

02740 | Flexible Pavement

		02740 Flexible Pavement	CREW	DAILY OUTPUT	LABOR-HOURS	UNIT	2001 BARE COSTS MAT.	LABOR	EQUIP.	TOTAL	TOTAL INCL O&P	
300	0380	2″ thick	B-25B	6,345	.015	S.Y.	3.14	.39	.30	3.83	4.39	300
	0420	2-1/2″ thick		5,480	.018		3.88	.45	.35	4.68	5.35	
	0460	3″ thick		4,900	.020		4.62	.50	.39	5.51	6.30	
	0800	Alternate method of figuring paving costs										
	0810	Binder course, 1-1/2″ thick	B-25	630	.140	Ton	26	3.50	2.71	32.21	37	
	0811	2″ thick		690	.128		26	3.20	2.47	31.67	36	
	0812	3″ thick		800	.110		26	2.76	2.13	30.89	35	
	0813	4″ thick		850	.104		26	2.60	2.01	30.61	34.50	
	0850	Wearing course, 1″ thick	B-25B	575	.167		28.50	4.26	3.33	36.09	42	
	0851	1-1/2″ thick		630	.152		28.50	3.89	3.04	35.43	41	
	0852	2″ thick		690	.139		28.50	3.55	2.78	34.83	40	
	0853	2-1/2″ thick		745	.129		28.50	3.29	2.57	34.36	39.50	
	0854	3″ thick		800	.120		28.50	3.06	2.40	33.96	39	
	1000	Pavement replacement over trench, 2″ thick	B-37	90	.533	S.Y.	3.20	12.90	1.48	17.58	25	
	1050	4″ thick		70	.686		6.35	16.55	1.91	24.81	35	
	1080	6″ thick		55	.873		10.05	21	2.43	33.48	47	
315	0011	**PAVING** Asphaltic concrete, parking lots & driveways										315
	0020	6″ stone base, 2″ binder course, 1″ topping	B-25C	9,000	.005	S.F.	1.13	.14	.17	1.44	1.63	
	0300	Binder course, 1-1/2″ thick		35,000	.001		.30	.04	.04	.38	.43	
	0400	2″ thick		25,000	.002		.39	.05	.06	.50	.58	
	0500	3″ thick		15,000	.003		.60	.08	.10	.78	.90	
	0600	4″ thick		10,800	.004		.79	.11	.14	1.04	1.19	
	0800	Sand finish course, 3/4″ thick		41,000	.001		.17	.03	.04	.24	.27	
	0900	1″ thick		34,000	.001		.21	.04	.04	.29	.34	
	1000	Fill pot holes, hot mix, 2″ thick	B-16	4,200	.008		.39	.18	.12	.69	.83	
	1100	4″ thick		3,500	.009		.56	.22	.14	.92	1.11	
	1120	6″ thick		3,100	.010		.76	.24	.16	1.16	1.38	
	1140	Cold patch, 2″ thick	B-51	3,000	.016		.45	.37	.05	.87	1.14	
	1160	4″ thick		2,700	.018		.86	.41	.06	1.33	1.65	
	1180	6″ thick		1,900	.025		1.33	.59	.08	2	2.48	

02800 | Site Improvements and Amenities

02810 | Irrigation System

		02810 Irrigation System	CREW	DAILY OUTPUT	LABOR-HOURS	UNIT	2001 BARE COSTS MAT.	LABOR	EQUIP.	TOTAL	TOTAL INCL O&P	
800	0010	**SPRINKLER IRRIGATION SYSTEM** For lawns (A12.7 -910)										800
	0100	Golf course with fully automatic system	C-17	.05	1,600	9 holes	79,000	48,400		127,400	162,500	
	0200	24′ diam. head at 15′ O.C incl. piping, auto oper., minimum	B-20	70	.343	Head	17.35	8.85		26.20	33	
	0300	Maximum		40	.600		40	15.50		55.50	68.50	
	0500	60′ diam. head at 40′ O.C. incl. piping, auto oper., minimum		28	.857		52.50	22		74.50	92.50	
	0600	Maximum		23	1.043		147	27		174	205	
	0800	Residential system, custom, 1″ supply		2,000	.012	S.F.	.26	.31		.57	.78	
	0900	1-1/2″ supply		1,800	.013	″	.29	.34		.63	.86	
	0990	For renovation work, add to above						50%				
	1020	Pop up spray head w/risers, hi-pop, full circle pattern, 4″	2 Skwk	76	.211	Ea.	3.24	6.30		9.54	13.40	
	1030	1/2 circle pattern		76	.211		3.31	6.30		9.61	13.50	
	1040	6″, full circle pattern		76	.211		7.75	6.30		14.05	18.40	
	1050	1/2 circle pattern		76	.211		7.75	6.30		14.05	18.40	
	1060	12″, full circle pattern		76	.211		9.65	6.30		15.95	20.50	
	1070	1/2 circle pattern		76	.211		9.70	6.30		16	20.50	
	1080	Pop up bubbler head w/risers, hi-pop bubbler head, 4″		76	.211		3.40	6.30		9.70	13.60	

Important: See the Reference Section for critical supporting data - Reference Nos., Crews, & City Cost Indexes

02810 | Irrigation System

	02810 Irrigation System		CREW	DAILY OUTPUT	LABOR-HOURS	UNIT	2001 BARE COSTS MAT.	LABOR	EQUIP.	TOTAL	TOTAL INCL O&P	
800	1090	6″ (A12.7 -910)	2 Skwk	76	.211	Ea.	7.75	6.30		14.05	18.40	800
	1100	12″		76	.211		9.65	6.30		15.95	20.50	
	1110	Impact full/part circle sprinklers, 28′-54′ 25-60 PSI		37	.432		9.15	12.90		22.05	30.50	
	1120	Spaced 37′-49′ @ 25-50 PSI		37	.432		21	12.90		33.90	44	
	1130	Spaced 43′-61′ @ 30-60 PSI		37	.432		44	12.90		56.90	69	
	1140	Spaced 54′-78′ @ 40-80 PSI		37	.432		96.50	12.90		109.40	127	
	1145	Impact rotor pop-up full/part commercial circle sprinklers										
	1150	Spaced 42′-65′ 35-80 PSI	2 Skwk	25	.640	Ea.	21.50	19.10		40.60	53.50	
	1160	Spaced 48′-76′ 45-85 PSI	″	25	.640	″	6.50	19.10		25.60	37	
	1165	Impact rotor pop-up part. circle comm., 53′-75′, 55-100 PSI, w/ acc										
	1170	Plastic case, metal cover	2 Skwk	25	.640	Ea.	139	19.10		158.10	183	
	1180	Rubber cover		25	.640		106	19.10		125.10	147	
	1190	Iron case, metal cover		22	.727		132	21.50		153.50	179	
	1200	Rubber cover		22	.727		139	21.50		160.50	187	
	1250	Plastic case, 2 nozzle, metal cover		25	.640		105	19.10		124.10	146	
	1260	Rubber cover		25	.640		111	19.10		130.10	152	
	1270	Iron case, 2 nozzle, metal cover		22	.727		149	21.50		170.50	198	
	1280	Rubber cover		22	.727		158	21.50		179.50	208	
	1282	Impact rotor pop-up full circle comm., 39′-99′, 30-100 PSI										
	1284	Plastic case, metal cover	2 Skwk	25	.640	Ea.	105	19.10		124.10	146	
	1286	Rubber cover		25	.640		115	19.10		134.10	157	
	1288	Iron case, metal cover		22	.727		155	21.50		176.50	204	
	1290	Rubber cover		22	.727		161	21.50		182.50	211	
	1292	Plastic case, 2 nozzle, metal cover		22	.727		112	21.50		133.50	157	
	1294	Rubber cover		22	.727		117	21.50		138.50	163	
	1296	Iron case, 2 nozzle, metal cover		20	.800		149	24		173	202	
	1298	Rubber cover		20	.800		158	24		182	212	
	1305	Electric remote control valve, plastic, 3/4″		18	.889		16.20	26.50		42.70	59.50	
	1310	1″		18	.889		24	26.50		50.50	68	
	1320	1-1/2″		18	.889		51.50	26.50		78	98	
	1330	2″		18	.889		72.50	26.50		99	121	
	1335	Quick coupling valves, brass, locking cover										
	1340	Inlet coupling valve, 3/4″	2 Skwk	18.75	.853	Ea.	40	25.50		65.50	84	
	1350	1″		18.75	.853		49.50	25.50		75	94.50	
	1360	Controller valve boxes, 6″ round boxes		18.75	.853		3.41	25.50		28.91	44	
	1370	10″ round boxes		14.25	1.123		7.50	33.50		41	61	
	1380	12″ square box		9.75	1.641		14.50	49		63.50	93	
	1388	Electromech. control, 14 day 3-60 min, auto start to 23/day										
	1390	4 station	2 Skwk	1.04	15.385	Ea.	167	460		627	905	
	1400	7 station		.64	25		166	745		911	1,350	
	1410	12 station		.40	40		254	1,200		1,454	2,150	
	1420	Dual programs, 18 station		.24	66.667		1,250	2,000		3,250	4,500	
	1430	23 station		.16	100		1,625	2,975		4,600	6,475	
	1435	Backflow preventer, bronze, 0-175 PSI, w/valves, test cocks										
	1440	3/4″	2 Skwk	2	8	Ea.	101	239		340	485	
	1450	1″		2	8		103	239		342	490	
	1460	1-1/2″		2	8		208	239		447	605	
	1470	2″		2	8		223	239		462	620	
	1475	Pressure vacuum breaker, brass, 15-150 PSI										
	1480	3/4″	2 Skwk	2	8	Ea.	60	239		299	440	
	1490	1″		2	8		82	239		321	465	
	1500	1-1/2″		2	8		162	239		401	555	
	1510	2″		2	8		150	239		389	540	
	2000	Pop-up spray, head & nozzle, low/medium volume, plastic	1 Skwk	30	.267		3.15	7.95		11.10	15.95	
	2200	Brass, economy		30	.267		5.80	7.95		13.75	18.85	
	2400	Heavy duty		30	.267		9.70	7.95		17.65	23	

	02810 | Irrigation System		CREW	DAILY OUTPUT	LABOR-HOURS	UNIT	2001 BARE COSTS MAT.	LABOR	EQUIP.	TOTAL	TOTAL INCL O&P	
800	3000	Riser mounted spray, head, low/medium volume, plastic (A12.7 -910)	1 Skwk	30	.267	Ea.	2.10	7.95		10.05	14.80	800
	3200	Brass		30	.267		2.42	7.95		10.37	15.15	
	4000	Pop-up impact sprinkler, body & case, plastic low/medium volume		30	.267		16.30	7.95		24.25	30.50	
	4200	Brass, high/medium volume		30	.267		25	7.95		32.95	40	
	4400	High volume		30	.267		71.50	7.95		79.45	91	
	5000	Quick coupling valve and key, brass, thread type		25	.320		31.50	9.55		41.05	49.50	
	5200	Lug type		25	.320		33.50	9.55		43.05	52	
	5400	Pop-up gear drive sprinkler, nozzle & case plastic, medium volume		30	.267		18.90	7.95		26.85	33.50	
	6000	Riser mounted gear drive sprinkler, nozzle and case										
	6200	Plastic, medium volume	1 Skwk	30	.267	Ea.	11.05	7.95		19	24.50	
	7000	Riser mounted impact sprinkler, body, part or										
	7200	Full circle plastic, low/medium volume	1 Skwk	25	.320	Ea.	11.05	9.55		20.60	27	
	7400	Brass, low/medium volume		25	.320		21	9.55		30.55	38	
	7600	Medium volume		25	.320		25	9.55		34.55	42.50	
	7800	Female thread		25	.320		50.50	9.55		60.05	70.50	
	8000	Riser mounted impact sprinkler, body, full circle only										
	8200	Plastic, low/medium volume	1 Skwk	25	.320	Ea.	6.70	9.55		16.25	22.50	
	8400	Brass, low/medium volume		25	.320		12.10	9.55		21.65	28.50	
	8600	High volume		25	.320		44	9.55		53.55	63.50	
	8800	Very high volume		25	.320		79	9.55		88.55	102	
900	0010	**SUBSURFACE DRIP IRRIGATION**, looped grid, pressure compensating										900
	0100	Preinserted emitter, line, hand bury, irregular area, small	3 Skwk	1,200	.020	L.F.	.32	.60		.92	1.29	
	0150	Medium		1,800	.013		.32	.40		.72	.97	
	0200	Large		2,520	.010		.32	.28		.60	.80	
	0250	Rectangular area, small		2,040	.012		.32	.35		.67	.90	
	0300	Medium		2,640	.009		.32	.27		.59	.78	
	0350	Large		3,600	.007		.32	.20		.52	.66	
	0400	Install in trench, irregular area, small		4,050	.006		.32	.18		.50	.63	
	0450	Medium		7,488	.003		.32	.10		.42	.50	
	0500	Large		16,560	.001		.32	.04		.36	.42	
	0550	Rectangular area, small		8,100	.003		.32	.09		.41	.49	
	0600	Medium		21,960	.001		.32	.03		.35	.40	
	0650	Large		33,264	.001		.32	.02		.34	.38	
	0700	Trenching and backfill	B-53	500	.016			.46	.19	.65	.91	
	0750	For vinyl tubing, 1/4", add to above						10%				
	0800	Vinyl tubing, 1/4", material only					.06			.06	.07	
	0850	Supply tubing, 1/2", material only, 100' coil					.11			.11	.12	
	0900	500' coil					.10			.10	.11	
	0950	Compression fittings	1 Skwk	90	.089	Ea.	.68	2.65		3.33	4.91	
	1000	Barbed fittings, 1/4"		360	.022		.16	.66		.82	1.22	
	1100	Flush risers		60	.133		.74	3.98		4.72	7.05	
	1150	Flush ends, figure eight		180	.044		.13	1.33		1.46	2.22	
	1200	Ball valve, 4-1/2"		20	.400		5.80	11.95		17.75	25	
	1250	4-3/4"		20	.400		6.85	11.95		18.80	26.50	
	1300	Auto flush, spring loaded		90	.089		1.84	2.65		4.49	6.20	
	1350	Volumetric		90	.089		5.80	2.65		8.45	10.50	
	1400	Air relief valve, inline with compensation tee, 1/2"		45	.178		6.20	5.30		11.50	15.15	
	1450	1"		30	.267		17.20	7.95		25.15	31.50	
	1500	Round box for flush ends, 6"		30	.267		3.41	7.95		11.36	16.25	
	1550	Fertilizer injector, non-proportional		4	2		3.31	59.50		62.81	97	
	1600	Screen filter, 3/4" screen		12	.667		18.40	19.90		38.30	51	
	1650	1" disk		8	1		48.50	30		78.50	101	
	1700	1-1/2" disk		4	2		116	59.50		175.50	221	
	1750	2" disk		3	2.667		231	79.50		310.50	380	
	1800	Typical installation 18" O.C., small, minimum				S.F.				.70	.80	
	1850	Maximum								1.30	1.50	

02800 | Site Improvements and Amenities

	02810 | Irrigation System		CREW	DAILY OUTPUT	LABOR-HOURS	UNIT	2001 BARE COSTS MAT.	LABOR	EQUIP.	TOTAL	TOTAL INCL O&P	
900	1900	Large, minimum				S.F.				.52	.60	900
	2000	Maximum				↓				.86	1	
	2100	For non-pressure compensating systems, deduct								10%	10%	
	2150	For supply piping see division 02510-840										

	02815 | Fountains		CREW	DAILY OUTPUT	LABOR-HOURS	UNIT	2001 BARE COSTS MAT.	LABOR	EQUIP.	TOTAL	TOTAL INCL O&P	
225	0010	FOUNTAINS/AERATORS										225
	0100	Pump w/controls										
	0200	Single phase, 100′ chord, 1/2 H.P. pump	2 Skwk	4.40	3.636	Ea.	2,550	109		2,659	3,000	
	0300	3/4 H.P. pump		4.30	3.721		2,925	111		3,036	3,375	
	0400	1 H.P. pump		4.20	3.810		2,950	114		3,064	3,425	
	0500	1-1/2 H.P. pump		4.10	3.902		3,050	116		3,166	3,525	
	0600	2 H.P. pump		4	4		3,100	119		3,219	3,575	
	0700	Three phase, 200′ chord, 5 H.P. pump		3.90	4.103		7,475	122		7,597	8,425	
	0800	7-1/2 H.P. pump		3.80	4.211		8,450	126		8,576	9,500	
	0900	10 H.P. pump		3.70	4.324		9,425	129		9,554	10,600	
	1000	15 H.P. pump		3.60	4.444		11,200	133		11,333	12,500	
	1100	Nozzles, minimum		8	2		187	59.50		246.50	300	
	1200	Maximum		8	2		246	59.50		305.50	365	
	1300	Lights w/mounting kits, 200 watt		18	.889		310	26.50		336.50	380	
	1400	300 watt		18	.889		345	26.50		371.50	415	
	1500	500 watt		18	.889		375	26.50		401.50	455	
	1600	Color blender	↓	12	1.333	↓	292	40		332	385	

02950 | Site Restoration & Rehabilitation

	02955 | Restoration of Underground Piping		CREW	DAILY OUTPUT	LABOR-HOURS	UNIT	2001 BARE COSTS MAT.	LABOR	EQUIP.	TOTAL	TOTAL INCL O&P	
100	0010	PIPE INTERNAL CLEANING & INSPECTION										100
	0100	Cleaning, pressure pipe systems										
	0120	Pig method, lengths 1000′ to 10,000′										
	0140	4″ diameter thru 24″ diameter, minimum				L.F.				2	2.30	
	0160	Maximum				″				5	6	
	6000	Sewage/sanitary systems										
	6100	Power rodder with header & cutters										
	6110	Mobilization charge, minimum				Total				300	350	
	6120	Mobilization charge, maximum				″				700	800	
	6140	4″ diameter				L.F.				1	1.15	
	6150	6″ diameter								1.25	1.45	
	6160	8″ diameter								1.50	1.75	
	6170	10″ diameter								1.75	2	
	6180	12″ diameter								1.90	2.20	
	6190	14″ diameter								2	2.30	
	6200	16″ diameter								2.25	2.60	
	6210	18″ diameter								2.50	2.85	
	6220	20″ diameter								2.60	3	
	6230	24″ diameter								2.90	3.35	
	6240	30″ diameter								3.50	4	
	6250	36″ diameter								4	4.60	
	6260	48″ diameter				↓				4.30	4.95	

	02955	Restoration of Underground Piping	CREW	DAILY OUTPUT	LABOR-HOURS	UNIT	2001 BARE COSTS MAT.	LABOR	EQUIP.	TOTAL	TOTAL INCL O&P	
100	6270	60" diameter				L.F.				5.60	6.45	100
	6280	72" diameter								6.20	7.15	
	9000	Inspection, television camera with film										
	9060	500 linear feet				Total				1,000	1,150	
680	0010	PIPE REPAIR										680
	0020	Not including excavation or backfill										
	0100	Clamp, stainless steel, lightweight, for steel pipe										
	0110	3" long, 1/2" diameter pipe	1 Plum	34	.235	Ea.	7.60	8.10		15.70	20.50	
	0120	3/4" diameter pipe		32	.250		7.80	8.60		16.40	21.50	
	0130	1" diameter pipe		30	.267		8.40	9.20		17.60	23	
	0140	1-1/4" diameter pipe		28	.286		8.90	9.85		18.75	24.50	
	0150	1-1/2" diameter pipe		26	.308		9.20	10.60		19.80	26	
	0160	2" diameter pipe		24	.333		10.05	11.50		21.55	28.50	
	0170	2-1/2" diameter pipe		23	.348		10.95	12		22.95	30	
	0180	3" diameter pipe		22	.364		12.10	12.55		24.65	32	
	0190	3-1/2" diameter pipe		21	.381		12.95	13.10		26.05	34	
	0200	4" diameter pipe	B-20	56	.429		13.50	11.10		24.60	32	
	0210	5" diameter pipe		53	.453		15.55	11.70		27.25	35.50	
	0220	6" diameter pipe		48	.500		17.25	12.95		30.20	39.50	
	0230	8" diameter pipe		30	.800		20	20.50		40.50	54.50	
	0240	10" diameter pipe		28	.857		52	22		74	91.50	
	0250	12" diameter pipe		24	1		55	26		81	101	
	0260	14" diameter pipe		22	1.091		57.50	28		85.50	108	
	0270	16" diameter pipe		20	1.200		62	31		93	117	
	0280	18" diameter pipe		18	1.333		65.50	34.50		100	126	
	0290	20" diameter pipe		16	1.500		72.50	39		111.50	141	
	0300	24" diameter pipe		14	1.714		80.50	44.50		125	158	
	0360	For 6" long, add					100%	40%				
	0370	For 9" long, add					200%	100%				
	0380	For 12" long, add					300%	150%				
	0390	For 18" long, add					500%	200%				
	1000	Clamp, stainless steel, with threaded service tap										
	1040	Full seal for iron, steel, PVC pipe										
	1100	6" long, 2" diameter pipe	1 Plum	17	.471	Ea.	98	16.20		114.20	133	
	1110	2-1/2" diameter pipe		16	.500		104	17.25		121.25	140	
	1120	3" diameter pipe		15.60	.513		110	17.65		127.65	148	
	1130	3-1/2" diameter pipe		15	.533		115	18.35		133.35	155	
	1140	4" diameter pipe	B-20	40	.600		127	15.50		142.50	164	
	1150	6" diameter pipe		34	.706		146	18.25		164.25	190	
	1160	8" diameter pipe		21	1.143		182	29.50		211.50	247	
	1170	10" diameter pipe		20	1.200		201	31		232	270	
	1180	12" diameter pipe		17	1.412		230	36.50		266.50	310	
	1205	For 9" long, add					20%	45%				
	1210	For 12" long, add					40%	80%				
	1220	For 18" long, add					70%	110%				
	1600	Clamp, stainless steel, single section										
	1640	Full seal for iron, steel, PVC pipe										
	1700	6" long, 2" diameter pipe	1 Plum	17	.471	Ea.	50.50	16.20		66.70	80	
	1710	2-1/2" diameter pipe		16	.500		55	17.25		72.25	86.50	
	1720	3" diameter pipe		15.60	.513		57.50	17.65		75.15	90	
	1730	3-1/2" diameter pipe		15	.533		57	18.35		75.35	90	
	1740	4" diameter pipe	B-20	40	.600		69	15.50		84.50	101	
	1750	6" diameter pipe		34	.706		86.50	18.25		104.75	124	
	1760	8" diameter pipe		21	1.143		115	29.50		144.50	174	
	1770	10" diameter pipe		20	1.200		132	31		163	194	
	1780	12" diameter pipe		17	1.412		144	36.50		180.50	215	

		02955 | Restoration of Underground Piping	CREW	DAILY OUTPUT	LABOR-HOURS	UNIT	2001 BARE COSTS MAT.	LABOR	EQUIP.	TOTAL	TOTAL INCL O&P	
680	1800	For 9" long, add				Ea.	40%	45%				680
	1810	For 12" long, add					60%	80%				
	1820	For 18" long, add					120%	110%				
	2000	Clamp, stainless steel, two section										
	2040	Full seal, for iron, steel, PVC pipe										
	2100	6" long, 4" diameter pipe	B-20	24	1	Ea.	98	26		124	149	
	2110	6" diameter pipe		20	1.200		115	31		146	176	
	2120	8" diameter pipe		13	1.846		127	47.50		174.50	214	
	2130	10" diameter pipe		12	2		184	51.50		235.50	283	
	2140	12" diameter pipe		10	2.400		201	62		263	320	
	2200	9" long, 4" diameter pipe		16	1.500		127	39		166	200	
	2210	6" diameter pipe		13	1.846		144	47.50		191.50	233	
	2220	8" diameter pipe		9	2.667		150	69		219	272	
	2230	10" diameter pipe		8	3		207	77.50		284.50	350	
	2240	12" diameter pipe		7	3.429		230	88.50		318.50	390	
	2250	14" diameter pipe		6.40	3.750		345	97		442	530	
	2260	16" diameter pipe		6	4		360	103		463	560	
	2270	18" diameter pipe		5	4.800		375	124		499	605	
	2280	20" diameter pipe		4.60	5.217		405	135		540	655	
	2290	24" diameter pipe		4	6		690	155		845	1,000	
	2320	For 12" long, add to 9"					15%	25%				
	2330	For 18" long, add to 9"					70%	55%				
	8000	For internal cleaning and inspection, see Div. 02955-100										
	8100	For pipe testing, see Div. 15955-700										
700	0010	**LINING PIPE** with cement, incl. bypass and cleaning										700
	0020	Less than 10,000 L.F., urban, 6" to 10"	C-17E	130	.615	L.F.	6	18.60	.57	25.17	36	
	0200	24" to 36"		90	.889		9.55	27	.82	37.37	53.50	
	0300	48" to 72"		80	1		15.25	30.50	.92	46.67	65.50	
800	0010	**CORROSION RESISTANCE** Wrap & coat, add to pipe, 4" dia.				L.F.	1.35			1.35	1.49	800
	0020	5" diameter					1.42			1.42	1.56	
	0040	6" diameter					1.42			1.42	1.56	
	0060	8" diameter					2.19			2.19	2.41	
	0080	10" diameter					3.27			3.27	3.60	
	0100	12" diameter					3.27			3.27	3.60	
	0120	14" diameter					4.79			4.79	5.25	
	0140	16" diameter					4.79			4.79	5.25	
	0160	18" diameter					4.84			4.84	5.30	
	0180	20" diameter					5.05			5.05	5.60	
	0200	24" diameter					6.20			6.20	6.80	
	0220	Small diameter pipe, 1" diameter, add					1.02			1.02	1.12	
	0240	2" diameter					1.12			1.12	1.23	
	0260	2-1/2" diameter					1.23			1.23	1.35	
	0280	3" diameter					1.23			1.23	1.35	
	0300	Fittings, field covered, add				S.F.	6.50			6.50	7.15	
	0500	Coating, bituminous, per diameter inch, 1 coat, add				L.F.	.24			.24	.26	
	0540	3 coat					.38			.38	.42	
	0560	Coal tar epoxy, per diameter inch, 1 coat, add					.15			.15	.17	
	0600	3 coat					.30			.30	.33	
	1000	Polyethylene H.D. extruded, .025" thk., 1/2" diameter add					.22			.22	.24	
	1020	3/4" diameter					.23			.23	.25	
	1040	1" diameter					.26			.26	.29	
	1060	1-1/4" diameter					.31			.31	.34	
	1080	1-1/2" diameter					.35			.35	.39	
	1100	.030" thk., 2" diameter					.38			.38	.42	
	1120	2-1/2" diameter					.45			.45	.50	
	1140	.035" thk., 3" diameter					.55			.55	.61	

	02955 | Restoration of Underground Piping	CREW	DAILY OUTPUT	LABOR-HOURS	UNIT	2001 BARE COSTS MAT.	LABOR	EQUIP.	TOTAL	TOTAL INCL O&P	
800	1160 3-1/2" diameter				L.F.	.63			.63	.69	800
	1180 4" diameter					.70			.70	.77	
	1200 .040" thk, 5" diameter					.97			.97	1.07	
	1220 6" diameter					1.01			1.01	1.11	
	1240 8" diameter					1.32			1.32	1.45	
	1260 10" diameter					1.61			1.61	1.77	
	1280 12" diameter					1.92			1.92	2.11	
	1300 .060" thk., 14" diameter					2.65			2.65	2.92	
	1320 16" diameter					3.11			3.11	3.42	
	1340 18" diameter					3.67			3.67	4.04	
	1360 20" diameter				▼	4.13			4.13	4.54	
	1380 Fittings, field wrapped, add				S.F.	4.25			4.25	4.68	

For information about Means Estimating Seminars, see yellow pages 11 and 12 in back of book

Division 3 Concrete

Estimating Tips

General

- Carefully check all the plans and specifications. Concrete often appears on drawings other than structural drawings, including mechanical and electrical drawings for equipment pads. The cost of cutting and patching is often difficult to estimate. See Subdivision 02225 for demolition costs.
- Always obtain concrete prices from suppliers near the job site. A volume discount can often be negotiated depending upon competition in the area. Remember to add for waste, particularly for slabs and footings on grade.

03100 Concrete Forms & Accessories

- A primary cost for concrete construction is forming. Most jobs today are constructed with prefabricated forms. The selection of the forms best suited for the job and the total square feet of forms required for efficient concrete forming and placing are key elements in estimating concrete construction. Enough forms must be available for erection to make efficient use of the concrete placing equipment and crew.
- Concrete accessories for forming and placing depend upon the systems used. Study the plans and specifications to assure that all special accessory requirements have been included in the cost estimate such as anchor bolts, inserts and hangers.

03200 Concrete Reinforcement

- Ascertain that the reinforcing steel supplier has included all accessories, cutting, bending and an allowance for lapping, splicing and waste. A good rule of thumb is 10% for lapping, splicing and waste. Also, 10% waste should be allowed for welded wire fabric.

03300 Cast-in-Place Concrete

- When estimating structural concrete, pay particular attention to requirements for concrete additives, curing methods and surface treatments. Special consideration for climate, hot or cold, must be included in your estimate. Be sure to include requirements for concrete placing equipment and concrete finishing.

03400 Precast Concrete
03500 Cementitious Decks & Toppings

- The cost of hauling precast concrete structural members is often an important factor. For this reason, it is important to get a quote from the nearest supplier. It may become economically feasible to set up precasting beds on the site if the hauling costs are prohibitive.

Reference Numbers

Reference numbers are shown in bold squares at the beginning of some major classifications. These numbers refer to related items in the Reference Section. The reference information may be an estimating procedure, an alternate pricing method or technical information.

Note: Not all subdivisions listed here necessarily appear in this publication.

03100 | Concrete Forms & Accessories

	03110 | Structural C.I.P. Forms		CREW	DAILY OUTPUT	LABOR-HOURS	UNIT	2001 BARE COSTS MAT.	LABOR	EQUIP.	TOTAL	TOTAL INCL O&P	
425	0010	**FORMS IN PLACE, EQUIPMENT FOUNDATIONS** 1 use	C-2	160	.300	SFCA	2.57	8.55		11.12	16.20	425
	0050	2 use	↓	190	.253	↓	1.42	7.20		8.62	12.80	
	0100	3 use		200	.240		1.03	6.80		7.83	11.85	
	0150	4 use		205	.234		.84	6.65		7.49	11.30	
620	0010	**ACCESSORIES, SLEEVES AND CHASES**										620
	0100	Plastic, 1 use, 9″ long, 2″ diameter	1 Carp	100	.080	Ea.	.53	2.33		2.86	4.23	
	0150	4″ diameter	↓	90	.089	↓	1.56	2.59		4.15	5.75	
	0200	6″ diameter		75	.107		2.75	3.11		5.86	7.90	
	0250	12″ diameter		60	.133		18.05	3.89		21.94	26	
	5000	Sheet metal, 2″ diameter		100	.080		.68	2.33		3.01	4.40	
	5100	4″ diameter		90	.089		.85	2.59		3.44	4.99	
	5150	6″ diameter		75	.107		1.23	3.11		4.34	6.20	
	5200	12″ diameter		60	.133		2.46	3.89		6.35	8.80	
	6000	Steel pipe, 2″ diameter		100	.080		1.52	2.33		3.85	5.30	
	6100	4″ diameter		90	.089		4.48	2.59		7.07	9	
	6150	6″ diameter		75	.107		7.60	3.11		10.71	13.20	
	6200	12″ diameter		60	.133		27	3.89		30.89	36	

03300 | Cast-In-Place Concrete

	03310 | Structural Concrete		CREW	DAILY OUTPUT	LABOR-HOURS	UNIT	2001 BARE COSTS MAT.	LABOR	EQUIP.	TOTAL	TOTAL INCL O&P	
240	0010	**CONCRETE IN PLACE** Including forms (4 uses), reinforcing										240
	0050	steel, including finishing unless otherwise indicated										
	3900	Footings, strip, 18″ x 9″, plain	C-14C	41.04	2.729	C.Y.	86	76	.93	162.93	216	
	3950	36″ x 12″, reinforced	↓	61.55	1.820	↓	88	51	.62	139.62	178	
	4000	Foundation mat, under 10 C.Y.		38.67	2.896		118	81	.98	199.98	259	
	4050	Over 20 C.Y.		56.40	1.986		105	55.50	.68	161.18	204	
	4650	Slab on grade, not including finish, 4″ thick	C-14E	60.75	1.449		79.50	41.50	.62	121.62	155	
	4700	6″ thick	″	92	.957		76	27.50	.41	103.91	128	
700	0010	**PLACING CONCRETE** and vibrating, including labor & equipment										700
	1900	Footings, continuous, shallow, direct chute	C-6	120	.400	C.Y.		9.60	.63	10.23	15.60	
	1950	Pumped	C-20	150	.427	↓		10.55	4.37	14.92	21	
	2000	With crane and bucket	C-7	90	.800			19.85	11.15	31	43	
	2100	Footings, continuous, deep, direct chute	C-6	140	.343			8.25	.54	8.79	13.35	
	2150	Pumped	C-20	160	.400			9.85	4.10	13.95	19.75	
	2200	With crane and bucket	C-7	110	.655			16.25	9.15	25.40	35	
	2900	Foundation mats, over 20 C.Y., direct chute	C-6	350	.137			3.30	.22	3.52	5.35	
	2950	Pumped	C-20	400	.160			3.95	1.64	5.59	7.90	
	3000	With crane and bucket	C-7	300	.240			5.95	3.35	9.30	12.90	
	03350 | Concrete Finishing											
300	0010	**FINISHING FLOORS** Monolithic, screed finish	1 Cefi	900	.009	S.F.		.25		.25	.37	300
	0100	Screed and bull float (darby) finish	↓	725	.011	↓		.31		.31	.46	
	0150	Screed, float, and broom finish		630	.013			.36		.36	.53	

For information about Means Estimating Seminars, see yellow pages 11 and 12 in back of book

Division 5 Metals

Estimating Tips

05050 Basic Metal Materials & Methods

- Nuts, bolts, washers, connection angles and plates can add a significant amount to both the tonnage of a structural steel job as well as the estimated cost. As a rule of thumb add 10% to the total weight to account for these accessories.
- Type 2 steel construction, commonly referred to as "simple construction," consists generally of field bolted connections with lateral bracing supplied by other elements of the building, such as masonry walls or x-bracing. The estimator should be aware, however, that shop connections may be accomplished by welding or bolting. The method may be particular to the fabrication shop and may have an impact on the estimated cost.

05200 Metal Joists

- In any given project the total weight of open web steel joists is determined by the loads to be supported and the design. However, economies can be realized in minimizing the amount of labor used to place the joists. This is done by maximizing the joist spacing and therefore minimizing the number of joists required to be installed on the job. Certain spacings and locations may be required by the design, but in other cases maximizing the spacing and keeping it as uniform as possible will keep the costs down.

05300 Metal Deck

- The takeoff and estimating of metal deck involves more than simply the area of the floor or roof and the type of deck specified or shown on the drawings. Many different sizes and types of openings may exist. Small openings for individual pipes or conduits may be drilled after the floor/roof is installed, but larger openings may require special deck lengths as well as reinforcing or structural support. The estimator should determine who will be supplying this reinforcing. Additionally, some deck terminations are part of the deck package, such as screed angles and pour stops, and others will be part of the steel contract, such as angles attached to structural members and cast-in-place angles and plates. The estimator must ensure that all pieces are accounted for in the complete estimate.

05500 Metal Fabrications

- The most economical steel stairs are those that use common materials, standard details and most importantly, a uniform and relatively simple method of field assembly. Commonly available A36 channels and plates are very good choices for the main stringers of the stairs, as are angles and tees for the carrier members. Risers and treads are usually made by specialty shops, and it is most economical to use a typical detail in as many places as possible. The stairs should be pre-assembled and shipped directly to the site. The field connections should be simple and straightforward to be accomplished efficiently and with a minimum of equipment and labor.

Reference Numbers

Reference numbers are shown in bold squares at the beginning of some major classifications. These numbers refer to related items in the Reference Section. The reference information may be an estimating procedure, an alternate pricing method or technical information.

Note: Not all subdivisions listed here necessarily appear in this publication.

		05090 | Metal Fastenings	CREW	DAILY OUTPUT	LABOR-HOURS	UNIT	2001 BARE COSTS MAT.	LABOR	EQUIP.	TOTAL	TOTAL INCL O&P	
880	0005	**WELD ROD**										880
	0010	Steel, type E6011 (all purpose), 1/8" dia, less than 500#				Lb.	1.34			1.34	1.47	
	0100	500# to 2,000#					1.21			1.21	1.33	
	0200	2,000# to 5,000#					1.14			1.14	1.25	
	0300	5/32" diameter, less than 500#					1.28			1.28	1.41	
	0310	500# to 2,000#					1.16			1.16	1.28	
	0320	2,000# to 5,000#					1.09			1.09	1.20	
	0400	3/16" dia, less than 500#					1.28			1.28	1.41	
	0500	500# to 2,000#					1.16			1.16	1.28	
	0600	2,000# to 5,000#					1.09			1.09	1.20	
	0620	Steel, type E6010 (pipe), 1/8" dia, less than 500#					1.47			1.47	1.62	
	0630	500# to 2,000#					1.31			1.31	1.44	
	0640	2,000# to 5,000#					1.22			1.22	1.34	
	0650	Steel, type E7018 (low hydrogen), 1/8" dia, less than 500#					1.26			1.26	1.39	
	0660	500# to 2,000#					1.14			1.14	1.25	
	0670	2,000# to 5,000#					1.08			1.08	1.19	
	0700	Steel, type E7024 (jet weld), 1/8" dia, less than 500#					1.34			1.34	1.47	
	0710	500# to 2,000#					1.21			1.21	1.33	
	0720	2,000# to 5,000#					1.14			1.14	1.25	
	1550	Aluminum, type 4043, 1/8" dia, less than 10#					5.15			5.15	5.65	
	1560	10# to 60#					4.42			4.42	4.86	
	1570	Over 60#					4.42			4.42	4.86	
	1600	Aluminum, type 5356, 1/8" dia, less than 10#					5.65			5.65	6.25	
	1610	10# to 60#					5.25			5.25	5.80	
	1620	Over 60#					5.25			5.25	5.80	
	1900	Cast iron (cold welding), 1/8" dia, less than 500#					16.25			16.25	17.90	
	1910	500# to 1,000#					14.20			14.20	15.65	
	1920	Over 1,000#					14			14	15.40	
	2000	Stainless steel, type 308-16, 1/8" dia, less than 500#					5.45			5.45	5.95	
	2100	500# to 1000#					4.92			4.92	5.40	
	2220	Over 1000#				↓	4.83			4.83	5.30	
900	0005	**WELDING STRUCTURAL** R05090-520										900
	0010	Field welding, 1/8" E6011, cost per welder, no oper. engr	E-14	8	1	Hr.	2.68	35	10.50	48.18	77.50	
	0200	With 1/2 operating engineer	E-13	8	1.500		2.68	49	10.50	62.18	99.50	
	0300	With 1 operating engineer	E-12	8	2	↓	2.68	63.50	10.50	76.68	122	
	0500	With no operating engineer, 2# weld rod per ton	E-14	8	1	Ton	2.68	35	10.50	48.18	77.50	
	0600	8# E6011 per ton	"	2	4		10.70	139	42	191.70	310	
	0800	With one operating engineer per welder, 2# E6011 per ton	E-12	8	2		2.68	63.50	10.50	76.68	122	
	0900	8# E6011 per ton	"	2	8	↓	10.70	254	42	306.70	485	
	1200	Continuous fillet, stick welding, incl. equipment										
	1300	Single pass, 1/8" thick, 0.1#/L.F.	E-14	150	.053	L.F.	.13	1.86	.56	2.55	4.14	
	1400	3/16" thick, 0.2#/L.F.		75	.107		.27	3.72	1.12	5.11	8.25	
	1500	1/4" thick, 0.3#/L.F.		50	.160		.40	5.60	1.68	7.68	12.40	
	1610	5/16" thick, 0.4#/L.F.		38	.211		.54	7.35	2.21	10.10	16.30	
	1800	3 passes, 3/8" thick, 0.5#/L.F.		30	.267		.67	9.30	2.80	12.77	20.50	
	2010	4 passes, 1/2" thick, 0.7#/L.F.		22	.364		.94	12.65	3.82	17.41	28	
	2200	5 to 6 passes, 3/4" thick, 1.3#/L.F.		12	.667		1.74	23	7	31.74	51.50	
	2400	8 to 11 passes, 1" thick, 2.4#/L.F.	↓	6	1.333		3.22	46.50	14	63.72	103	
	2600	For all position welding, add, minimum						20%				
	2700	Maximum						300%				
	2900	For semi-automatic welding, deduct, minimum						5%				
	3000	Maximum				↓		15%				
	4000	Cleaning and welding plates, bars, or rods										
	4010	to existing beams, columns, or trusses	E-14	12	.667	L.F.	.67	23	7	30.67	50.50	

05100 | Structural Metal Framing

05120 | Structural Steel

			CREW	DAILY OUTPUT	LABOR-HOURS	UNIT	2001 BARE COSTS MAT.	LABOR	EQUIP.	TOTAL	TOTAL INCL O&P	
440	0010	**LIGHTWEIGHT FRAMING**										440
	0400	Angle framing, field fabricated, 4" and larger	E-3	440	.055	Lb.	.36	1.83	.19	2.38	3.92	
	0450	Less than 4" angles		265	.091		.37	3.04	.32	3.73	6.25	
	0600	Channel framing, field fabricated, 8" and larger		500	.048		.37	1.61	.17	2.15	3.50	
	0650	Less than 8" channels		335	.072		.37	2.40	.25	3.02	5.05	
	1000	Continuous slotted channel framing system, minimum	2 Sswk	220	.073		1.90	2.39		4.29	6.40	
	1200	Maximum	"	135	.119		2.14	3.89		6.03	9.40	
	1250	Plate & bar stock for reinforcing beams and trusses					.67			.67	.74	
	1300	Cross bracing, rods, 3/4" diameter	E-3	700	.034		.74	1.15	.12	2.01	3.02	
	1310	7/8" diameter		850	.028		.74	.95	.10	1.79	2.64	
	1320	1" diameter		1,000	.024		.74	.80	.08	1.62	2.36	
	1330	Angle, 5" x 5" x 3/8"		440	.055		.74	1.83	.19	2.76	4.34	
	1350	Hanging lintels, average		350	.069		.74	2.30	.24	3.28	5.25	
	1380	Roof frames, 3'-0" square, 5' span	E-2	4,200	.013		.74	.43	.31	1.48	1.89	
	1400	Tie rod, not upset, 1-1/2" to 4" diameter, with turnbuckle	2 Sswk	800	.020		.80	.66		1.46	2.07	
	1420	No turnbuckle		700	.023		.77	.75		1.52	2.20	
	1500	Upset, 1-3/4" to 4" diameter, with turnbuckle		800	.020		.80	.66		1.46	2.07	
	1520	No turnbuckle		700	.023		.77	.75		1.52	2.20	
520	0005	**PIPE SUPPORT FRAMING**										520
	0010	Under 10#/L.F.	E-4	3,900	.008	Lb.	.82	.27	.02	1.11	1.42	
	0200	10.1 to 15#/L.F.		4,300	.007		.81	.25	.02	1.08	1.36	
	0400	15.1 to 20#/L.F.		4,800	.007		.80	.22	.02	1.04	1.30	
	0600	Over 20#/L.F.		5,400	.006		.78	.20	.02	1	1.24	
640	0010	**STRUCTURAL STEEL MEMBERS**										640
	0020	Shop fabricated for 1-2 story bldg., bolted conn's., 100 tons										
	0100	W 6 x 9	E-2	600	.093	L.F.	5.60	2.98	2.15	10.73	13.70	
	0300	W 8 x 10	"	600	.093	"	6.25	2.98	2.15	11.38	14.35	
	8490	For jobs less than 100 tons, add				Ton	10%					

05500 | Metal Fabrications

05540 | Floor Plates

			CREW	DAILY OUTPUT	LABOR-HOURS	UNIT	2001 BARE COSTS MAT.	LABOR	EQUIP.	TOTAL	TOTAL INCL O&P	
200	0005	**CHECKERED PLATE**										200
	0010	1/4" & 3/8", 2000 to 5000 S.F., bolted	E-4	2,900	.011	Lb.	.62	.37	.03	1.02	1.38	
	0100	Welded		4,400	.007	"	.58	.24	.02	.84	1.10	
	0300	Pit or trench cover and frame, 1/4" plate, 2' to 3' wide		100	.320	S.F.	16.25	10.65	.84	27.74	38	
	0400	For galvanizing, add				Lb.	.36			.36	.40	
	0500	Platforms, 1/4" plate, no handrails included, rectangular	E-4	4,200	.008		.98	.25	.02	1.25	1.56	
	0600	Circular	"	2,500	.013		1.37	.43	.03	1.83	2.32	

Division Notes

		CREW	DAILY OUTPUT	LABOR-HOURS	UNIT	2001 BARE COSTS MAT.	LABOR	EQUIP.	TOTAL	TOTAL INCL O&P

Division 6 Wood & Plastics

Estimating Tips

06050 Basic Wood & Plastic Materials & Methods

- Common to any wood framed structure are the accessory connector items such as screws, nails, adhesives, hangers, connector plates, straps, angles and holdowns. For typical wood framed buildings, such as residential projects, the aggregate total for these items can be significant, especially in areas where seismic loading is a concern. For floor and wall framing, nail quantities can be figured on a "pounds per thousand board feet basis", with 10 to 25 lbs. per MBF the range. Holdowns, hangers and other connectors should be taken off by the piece.

06100 Rough Carpentry

- Lumber is a traded commodity and therefore sensitive to supply and demand in the marketplace. Even in "budgetary" estimating of wood framed projects, it is advisable to call local suppliers for the latest market pricing.
- Common quantity units for wood framed projects are "thousand board feet" (MBF). A board foot is a volume of wood, 1″ x 1′ x 1′, or 144 cubic inches. Board foot quantities are generally calculated using nominal material dimensions—dressed sizes are ignored. Board foot per lineal foot of any stick of lumber can be calculated by dividing the nominal cross sectional area by 12. As an example, 2,000 lineal feet of 2 x 12 equates to 4 MBF by dividing the nominal area, 2 x 12, by 12, which equals 2, and multiplying by 2,000 to give 4,000 board feet. This simple rule applies to all nominal dimensioned lumber.
- Waste is an issue of concern at the quantity takeoff for any area of construction. Framing lumber is sold in even foot lengths, i.e., 10′, 12′, 14′, 16′, and depending on spans, wall heights and the grade of lumber, waste is inevitable. A rule of thumb for lumber waste is 5% to 10% depending on material quality and the complexity of the framing.
- Wood in various forms and shapes is used in many projects, even where the main structural framing is steel, concrete or masonry. Plywood as a back-up partition material and 2x boards used as blocking and cant strips around roof edges are two common examples. The estimator should ensure that the costs of all wood materials are included in the final estimate.

06200 Finish Carpentry

- It is necessary to consider the grade of workmanship when estimating labor costs for erecting millwork and interior finish. In practice, there are three grades: premium, custom and economy. The Means daily output for base and case moldings is in the range of 200 to 250 L.F. per carpenter per day. This is appropriate for most average custom grade projects. For premium projects an adjustment to productivity of 25% to 50% should be made depending on the complexity of the job.

Reference Numbers

Reference numbers are shown in bold squares at the beginning of some major classifications. These numbers refer to related items in the Reference Section. The reference information may be an estimating procedure, an alternate pricing method or technical information.

Note: Not all subdivisions listed here necessarily appear in this publication.

06100 | Rough Carpentry

06110 | Wood Framing

		06110 Wood Framing	Crew	Daily Output	Labor-Hours	Unit	2001 Bare Costs Mat.	Labor	Equip.	Total	Total Incl O&P	
545	0010	**FRAMING, MISCELLANEOUS**										545
	8500	Firestops, 2″ x 4″	2 Carp	.51	31.373	M.B.F.	600	915		1,515	2,075	
	8505	Pneumatic nailed		.62	25.806		600	750		1,350	1,825	
	8520	2″ x 6″		.60	26.667		590	775		1,365	1,875	
	8525	Pneumatic nailed		.73	21.858		590	635		1,225	1,650	
	8540	2″ x 8″		.60	26.667		620	775		1,395	1,900	
	8560	2″ x 12″		.70	22.857		740	665		1,405	1,850	
	8600	Nailers, treated, wood construction, 2″ x 4″		.53	30.189		785	880		1,665	2,250	
	8605	Pneumatic nailed		.64	25.157		785	735		1,520	2,025	
	8620	2″ x 6″		.75	21.333		895	620		1,515	1,950	
	8625	Pneumatic nailed		.90	17.778		895	520		1,415	1,800	

06160 | Sheathing

		06160 Sheathing	Crew	Daily Output	Labor-Hours	Unit	2001 Bare Costs Mat.	Labor	Equip.	Total	Total Incl O&P	
800	0010	**SHEATHING** Plywood on roof, CDX										800
	0050	3/8″ thick	2 Carp	1,525	.010	S.F.	.38	.31		.69	.89	
	0055	Pneumatic nailed		1,860	.009		.38	.25		.63	.80	
	0200	5/8″ thick		1,300	.012		.62	.36		.98	1.24	
	0205	Pneumatic nailed		1,586	.010		.62	.29		.91	1.14	
	0300	3/4″ thick		1,200	.013		.75	.39		1.14	1.43	
	0305	Pneumatic nailed		1,464	.011		.75	.32		1.07	1.32	
	0500	Plywood on walls with exterior CDX, 3/8″ thick		1,200	.013		.38	.39		.77	1.02	
	0505	Pneumatic nailed		1,488	.011		.38	.31		.69	.90	
	0700	5/8″ thick		1,050	.015		.62	.44		1.06	1.37	
	0705	Pneumatic nailed		1,302	.012		.62	.36		.98	1.24	
	0800	3/4″ thick		975	.016		.75	.48		1.23	1.57	
	0805	Pneumatic nailed		1,209	.013		.75	.39		1.14	1.42	

06400 | Architectural Woodwork

06410 | Custom Cabinets

		06410 Custom Cabinets	Crew	Daily Output	Labor-Hours	Unit	2001 Bare Costs Mat.	Labor	Equip.	Total	Total Incl O&P	
100	0010	**CABINETS** Corner china cabinets, stock pine,										100
	0020	80″ high, unfinished, minimum	2 Carp	6.60	2.424	Ea.	430	70.50		500.50	580	
	0100	Maximum	"	4.40	3.636	"	950	106		1,056	1,225	
	0700	Kitchen base cabinets, hardwood, not incl. counter tops,										
	0710	24″ deep, 35″ high, prefinished										
	0800	One top drawer, one door below, 12″ wide	2 Carp	24.80	.645	Ea.	126	18.80		144.80	168	
	0840	18″ wide		23.30	.687		186	20		206	236	
	0880	24″ wide		22.30	.717		220	21		241	275	
	1000	Four drawers, 12″ wide		24.80	.645		300	18.80		318.80	360	
	1040	18″ wide		23.30	.687		325	20		345	385	
	1060	24″ wide		22.30	.717		350	21		371	420	
	1200	Two top drawers, two doors below, 27″ wide		22	.727		262	21		283	320	
	1260	36″ wide		20.30	.788		300	23		323	365	
	1300	48″ wide		18.90	.847		335	24.50		359.50	410	
	1500	Range or sink base, two doors below, 30″ wide		21.40	.748		219	22		241	275	
	1540	36″ wide		20.30	.788		246	23		269	305	
	1580	48″ wide		18.90	.847		277	24.50		301.50	345	
	1800	For sink front units, deduct					53			53	58.50	

06400 | Architectural Woodwork

	06410 | Custom Cabinets		CREW	DAILY OUTPUT	LABOR-HOURS	UNIT	2001 BARE COSTS MAT.	LABOR	EQUIP.	TOTAL	TOTAL INCL O&P	
100	9000	For deluxe models of all cabinets, add					40%					100
	9500	For custom built in place, add					25%	10%				
	9550	Rule of thumb, kitchen cabinets not including										
	9560	appliances & counter top, minimum	2 Carp	30	.533	L.F.	80.50	15.55		96.05	113	
	9600	Maximum	"	25	.640	"	225	18.65		243.65	277	
400	0010	**VANITIES**										400
	8000	Vanity bases, 2 doors, 30" high, 21" deep, 24" wide	2 Carp	20	.800	Ea.	124	23.50		147.50	173	
	8050	30" wide		16	1		130	29		159	189	
	8100	36" wide		13.33	1.200		146	35		181	215	
	8150	48" wide	↓	11.43	1.400		184	41		225	266	
	9000	For deluxe models of all vanities, add to above					40%					
	9500	For custom built in place, add to above				↓	25%	10%				

For information about Means Estimating Seminars, see yellow pages 11 and 12 in back of book

For expanded coverage of these items see *Means Interior Cost Data 2001*

Division Notes

	CREW	DAILY OUTPUT	LABOR-HOURS	UNIT	2001 BARE COSTS MAT.	LABOR	EQUIP.	TOTAL	TOTAL INCL O&P

Division 7
Thermal & Moisture Protection

Estimating Tips

07100 Dampproofing & Waterproofing

- Be sure of the job specifications before pricing this subdivision. The difference in cost between waterproofing and dampproofing can be great. Waterproofing will hold back standing water. Dampproofing prevents the transmission of water vapor. Also included in this section are vapor retarding membranes.

07200 Thermal Protection

- Insulation and fireproofing products are measured by area, thickness, volume or R value. Specifications may only give what the specific R value should be in a certain situation. The estimator may need to choose the type of insulation to meet that R value.

07300 Shingles, Roof Tiles & Roof Coverings
07400 Roofing & Siding Panels

- Many roofing and siding products are bought and sold by the square. One square is equal to an area that measures 100 square feet.

This simple change in unit of measure could create a large error if the estimator is not observant. Accessories and fasteners necessary for a complete installation must be figured into any calculations for both material and labor.

07500 Membrane Roofing
07600 Flashing & Sheet Metal
07700 Roof Specialties & Accessories

- The items in these subdivisions compose a roofing system. No one component completes the installation and all must be estimated. Built-up or single ply membrane roofing systems are made up of many products and installation trades. Wood blocking at roof perimeters or penetrations, parapet coverings, reglets, roof drains, gutters, downspouts, sheet metal flashing, skylights, smoke vents or roof hatches all need to be considered along with the roofing material. Several different installation trades will need to work together on the roofing system. Inherent difficulties in the scheduling and coordination of various trades must be accounted for when estimating labor costs.

07900 Joint Sealers

- To complete the weather-tight shell the sealants and caulkings must be estimated. Where different materials meet—at expansion joints, at flashing penetrations, and at hundreds of other locations throughout a construction project—they provide another line of defense against water penetration. Often, an entire system is based on the proper location and placement of caulking or sealants. The detail drawings that are included as part of a set of architectural plans, show typical locations for these materials. When caulking or sealants are shown at typical locations, this means the estimator must include them for all the locations where this detail is applicable. Be careful to keep different types of sealants separate, and remember to consider backer rods and primers if necessary.

Reference Numbers

Reference numbers are shown in bold squares at the beginning of some major classifications. These numbers refer to related items in the Reference Section. The reference information may be an estimating procedure, an alternate pricing method or technical information.

Note: Not all subdivisions listed here necessarily appear in this publication.

07650 | Flexible Flashing

	07650 Flexible Flashing	CREW	DAILY OUTPUT	LABOR-HOURS	UNIT	2001 BARE COSTS MAT.	LABOR	EQUIP.	TOTAL	TOTAL INCL O&P
600 0010	**FLASHING** Aluminum, mill finish, .013″ thick	1 Shee	145	.055	S.F.	.35	1.85		2.20	3.25
0030	.016″ thick		145	.055		.51	1.85		2.36	3.42
0060	.019″ thick		145	.055		.77	1.85		2.62	3.71
0100	.032″ thick		145	.055		1.04	1.85		2.89	4
0200	.040″ thick		145	.055		1.45	1.85		3.30	4.46
0300	.050″ thick		145	.055		1.84	1.85		3.69	4.88
0325	Mill finish 5″ x 7″ step flashing, .016″ thick		1,920	.004	Ea.	.10	.14		.24	.33
0350	Mill finish 12″ x 12″ step flashing, .016″ thick		1,600	.005	″	.40	.17		.57	.70
0400	Painted finish, add				S.F.	.24			.24	.26
0500	Fabric-backed 2 sides, .004″ thick	1 Shee	330	.024		.93	.81		1.74	2.28
0700	.005″ thick		330	.024		1.09	.81		1.90	2.46
0750	Mastic-backed, self adhesive		460	.017		2.26	.58		2.84	3.39
0800	Mastic-coated 2 sides, .004″ thick		330	.024		.93	.81		1.74	2.28
1000	.005″ thick		330	.024		1.09	.81		1.90	2.46
1100	.016″ thick		330	.024		1.20	.81		2.01	2.58
1300	Asphalt flashing cement, 5 gallon				Gal.	3.34			3.34	3.67
1600	Copper, 16 oz, sheets, under 1000 lbs.	1 Shee	115	.070	S.F.	2.18	2.33		4.51	6
1700	Over 4000 lbs.		155	.052		2.91	1.73		4.64	5.85
1900	20 oz sheets, under 1000 lbs.		110	.073		3.90	2.44		6.34	8.05
2000	Over 4000 lbs.		145	.055		3.62	1.85		5.47	6.85
2200	24 oz sheets, under 1000 lbs.		105	.076		4.69	2.56		7.25	9.10
2300	Over 4000 lbs.		135	.059		4.34	1.99		6.33	7.85
2500	32 oz sheets, under 1000 lbs.		100	.080		6.20	2.68		8.88	11
2600	Over 4000 lbs.		130	.062		5.80	2.06		7.86	9.60
2800	Copper, paperbacked 1 side, 2 oz		330	.024		.86	.81		1.67	2.21
2900	3 oz		330	.024		1.12	.81		1.93	2.49
3100	Paperbacked 2 sides, 2 oz		330	.024		.86	.81		1.67	2.21
3150	3 oz		330	.024		1.11	.81		1.92	2.48
3200	5 oz		330	.024		1.67	.81		2.48	3.10
3250	7 oz		330	.024		2.72	.81		3.53	4.25
3400	Mastic-backed 2 sides, copper, 2 oz		330	.024		1.02	.81		1.83	2.38
3500	3 oz		330	.024		1.25	.81		2.06	2.64
3700	5 oz		330	.024		1.85	.81		2.66	3.30
3800	Fabric-backed 2 sides, copper, 2 oz		330	.024		1.09	.81		1.90	2.46
4000	3 oz		330	.024		1.40	.81		2.21	2.80
4100	5 oz		330	.024		1.90	.81		2.71	3.35
4300	Copper-clad stainless steel, .015″ thick, under 500 lbs.		115	.070		3.16	2.33		5.49	7.10
4400	Over 2000 lbs.		155	.052		3.05	1.73		4.78	6.05
4600	.018″ thick, under 500 lbs.		100	.080		4.18	2.68		6.86	8.75
4700	Over 2000 lbs.		145	.055		3.06	1.85		4.91	6.25
5800	Lead, 2.5 lb. per SF, up to 12″ wide	1 Rofc	135	.059		2.80	1.52		4.32	5.70
5900	Over 12″ wide	″	135	.059		3.25	1.52		4.77	6.20
8500	Shower pan, bituminous membrane, 7 oz	1 Shee	155	.052		1.08	1.73		2.81	3.86
8550	3 ply copper and fabric, 3 oz		155	.052		1.60	1.73		3.33	4.43
8600	7 oz		155	.052		3.30	1.73		5.03	6.30
8650	Copper, 16 oz		100	.080		3.05	2.68		5.73	7.50
8700	Lead on copper and fabric, 5 oz		155	.052		1.62	1.73		3.35	4.45
8800	7 oz		155	.052		2.87	1.73		4.60	5.85
8850	Polyvinyl chloride, .030″ thick		160	.050		.30	1.68		1.98	2.92
8900	Stainless steel sheets, 32 ga, .010″ thick		155	.052		2.16	1.73		3.89	5.05
9000	28 ga, .015″ thick		155	.052		2.55	1.73		4.28	5.50
9100	26 ga, .018″ thick		155	.052		3.25	1.73		4.98	6.25
9200	24 ga, .025″ thick		155	.052		4.22	1.73		5.95	7.30
9400	Terne coated stainless steel, .015″ thick, 28 ga		155	.052		4.05	1.73		5.78	7.15
9500	.018″ thick, 26 ga		155	.052		4.57	1.73		6.30	7.70
9600	Zinc and copper alloy (brass), .020″ thick		155	.052		3.26	1.73		4.99	6.25

07600 | Flashing & Sheet Metal

07650 | Flexible Flashing

			CREW	DAILY OUTPUT	LABOR-HOURS	UNIT	2001 BARE COSTS MAT.	LABOR	EQUIP.	TOTAL	TOTAL INCL O&P	
600	9700	.027″ thick	1 Shee	155	.052	S.F.	4.37	1.73		6.10	7.50	600
	9800	.032″ thick		155	.052		5.10	1.73		6.83	8.25	
	9900	.040″ thick	↓	155	.052	↓	6.20	1.73		7.93	9.50	

07800 | Fire & Smoke Protection

07840 | Firestopping

			CREW	DAILY OUTPUT	LABOR-HOURS	UNIT	2001 BARE COSTS MAT.	LABOR	EQUIP.	TOTAL	TOTAL INCL O&P	
100	0010	FIRESTOPPING R07800 -030										100
	0100	Metallic piping, non insulated										
	0110	Through walls, 2″ diameter	1 Carp	16	.500	Ea.	9.60	14.60		24.20	33.50	
	0120	4″ diameter		14	.571		14.65	16.65		31.30	42	
	0130	6″ diameter		12	.667		19.70	19.45		39.15	52	
	0140	12″ diameter		10	.800		35	23.50		58.50	75	
	0150	Through floors, 2″ diameter		32	.250		5.80	7.30		13.10	17.80	
	0160	4″ diameter		28	.286		8.35	8.35		16.70	22.50	
	0170	6″ diameter		24	.333		11	9.70		20.70	27.50	
	0180	12″ diameter	↓	20	.400	↓	18.50	11.65		30.15	39	
	0190	Metallic piping, insulated										
	0200	Through walls, 2″ diameter	1 Carp	16	.500	Ea.	13.60	14.60		28.20	38	
	0210	4″ diameter		14	.571		18.65	16.65		35.30	46.50	
	0220	6″ diameter		12	.667		23.50	19.45		42.95	56.50	
	0230	12″ diameter		10	.800		39	23.50		62.50	79	
	0240	Through floors, 2″ diameter		32	.250		9.80	7.30		17.10	22	
	0250	4″ diameter		28	.286		12.35	8.35		20.70	26.50	
	0260	6″ diameter		24	.333		15	9.70		24.70	31.50	
	0270	12″ diameter	↓	20	.400	↓	18.50	11.65		30.15	39	
	0280	Non metallic piping, non insulated										
	0290	Through walls, 2″ diameter	1 Carp	12	.667	Ea.	39.50	19.45		58.95	74	
	0300	4″ diameter		10	.800		49.50	23.50		73	91	
	0310	6″ diameter		8	1		69	29		98	122	
	0330	Through floors, 2″ diameter		16	.500		31	14.60		45.60	57	
	0340	4″ diameter		6	1.333		38.50	39		77.50	103	
	0350	6″ diameter	↓	6	1.333	↓	46	39		85	112	
	0370	Ductwork, insulated & non insulated, round										
	0380	Through walls, 6″ diameter	1 Carp	12	.667	Ea.	20	19.45		39.45	52.50	
	0390	12″ diameter		10	.800		40	23.50		63.50	80.50	
	0400	18″ diameter		8	1		65	29		94	117	
	0410	Through floors, 6″ diameter		16	.500		11	14.60		25.60	35	
	0420	12″ diameter		14	.571		20	16.65		36.65	48	
	0430	18″ diameter	↓	12	.667	↓	35	19.45		54.45	69	
	0440	Ductwork, insulated & non insulated, rectangular										
	0450	With stiffener/closure angle, through walls, 6″ x 12″	1 Carp	8	1	Ea.	16.65	29		45.65	64	
	0460	12″ x 24″		6	1.333		22	39		61	85.50	
	0470	24″ x 48″		4	2		63	58.50		121.50	161	
	0480	With stiffener/closure angle, through floors, 6″ x 12″		10	.800		9	23.50		32.50	46.50	
	0490	12″ x 24″		8	1		16.20	29		45.20	63.50	
	0500	24″ x 48″	↓	6	1.333	↓	32	39		71	96	
	0510	Multi trade openings										
	0520	Through walls, 6″ x 12″ ↓	1 Carp	2	4	Ea.	35	117		152	221	

07800 | Fire & Smoke Protection

	07840 | Firestopping		CREW	DAILY OUTPUT	LABOR-HOURS	UNIT	2001 BARE COSTS MAT.	LABOR	EQUIP.	TOTAL	TOTAL INCL O&P	
100	0530	12″ x 24″ R07800-030	1 Carp	1	8	Ea.	141	233		374	520	100
	0540	24″ x 48″	2 Carp	1	16		565	465		1,030	1,350	
	0550	48″ x 96″	″	.75	21.333		2,275	620		2,895	3,475	
	0560	Through floors, 6″ x 12″	1 Carp	2	4		35	117		152	221	
	0570	12″ x 24″	″	1	8		141	233		374	520	
	0580	24″ x 48″	2 Carp	.75	21.333		565	620		1,185	1,600	
	0590	48″ x 96″	″	.50	32		2,275	935		3,210	3,950	
	0600	Structural penetrations, through walls										
	0610	Steel beams, W8 x 10	1 Carp	8	1	Ea.	22	29		51	69.50	
	0620	W12 x 14		6	1.333		35	39		74	99.50	
	0630	W21 x 44		5	1.600		70	46.50		116.50	150	
	0640	W36 x 135		3	2.667		170	77.50		247.50	310	
	0650	Bar joists, 18″ deep		6	1.333		32	39		71	96	
	0660	24″ deep		6	1.333		40	39		79	105	
	0670	36″ deep		5	1.600		60	46.50		106.50	139	
	0680	48″ deep		4	2		70	58.50		128.50	168	
	0690	Construction joints, floor slab at exterior wall										
	0700	Precast, brick, block or drywall exterior										
	0710	2″ wide joint	1 Carp	125	.064	L.F.	5	1.87		6.87	8.40	
	0720	4″ wide joint	″	75	.107	″	10	3.11		13.11	15.85	
	0730	Metal panel, glass or curtain wall exterior										
	0740	2″ wide joint	1 Carp	40	.200	L.F.	11.85	5.85		17.70	22	
	0750	4″ wide joint	″	25	.320	″	16.15	9.35		25.50	32.50	
	0760	Floor slab to drywall partition										
	0770	Flat joint	1 Carp	100	.080	L.F.	4.90	2.33		7.23	9.05	
	0780	Fluted joint		50	.160		10	4.66		14.66	18.30	
	0790	Etched fluted joint		75	.107		6.50	3.11		9.61	12	
	0800	Floor slab to concrete/masonry partition										
	0810	Flat joint	1 Carp	75	.107	L.F.	11	3.11		14.11	16.95	
	0820	Fluted joint	″	50	.160	″	13	4.66		17.66	21.50	
	0830	Concrete/CMU wall joints										
	0840	1″ wide	1 Carp	100	.080	L.F.	6	2.33		8.33	10.25	
	0850	2″ wide		75	.107		11	3.11		14.11	16.95	
	0860	4″ wide		50	.160		21	4.66		25.66	30.50	
	0870	Concrete/CMU floor joints										
	0880	1″ wide	1 Carp	200	.040	L.F.	3	1.17		4.17	5.10	
	0890	2″ wide		150	.053		5.50	1.55		7.05	8.50	
	0900	4″ wide		100	.080		10.50	2.33		12.83	15.20	

07900 | Joint Sealers

	07920 | Joint Sealants		CREW	DAILY OUTPUT	LABOR-HOURS	UNIT	2001 BARE COSTS MAT.	LABOR	EQUIP.	TOTAL	TOTAL INCL O&P	
800	0010	CAULKING AND SEALANTS										800
	0020	Acoustical sealant, elastomeric, cartridges				Ea.	2.05			2.05	2.26	
	0030	Backer rod, polyethylene, 1/4″ diameter	1 Bric	4.60	1.739	C.L.F.	1.28	51.50		52.78	81	
	0050	1/2″ diameter		4.60	1.739		2.62	51.50		54.12	82.50	
	0070	3/4″ diameter		4.60	1.739		5.05	51.50		56.55	85	
	0090	1″ diameter		4.60	1.739		8	51.50		59.50	88.50	
	0100	Acrylic latex caulk, white										
	0200	11 fl. oz cartridge				Ea.	1.80			1.80	1.98	

07920 | Joint Sealants

			CREW	DAILY OUTPUT	LABOR-HOURS	UNIT	2001 BARE COSTS MAT.	LABOR	EQUIP.	TOTAL	TOTAL INCL O&P	
800	0500	1/4" x 1/2"	1 Bric	248	.032	L.F.	.15	.95		1.10	1.64	800
	0600	1/2" x 1/2"		250	.032		.29	.95		1.24	1.78	
	0800	3/4" x 3/4"		230	.035		.66	1.03		1.69	2.32	
	0900	3/4" x 1"		200	.040		.88	1.18		2.06	2.80	
	1000	1" x 1"	↓	180	.044	↓	1.10	1.32		2.42	3.24	
	1400	Butyl based, bulk				Gal.	22			22	24	
	1500	Cartridges				"	26.50			26.50	29.50	
	1700	Bulk, in place 1/4" x 1/2", 154 L.F./gal.	1 Bric	230	.035	L.F.	.14	1.03		1.17	1.75	
	1800	1/2" x 1/2", 77 L.F./gal.	"	180	.044	"	.29	1.32		1.61	2.34	
	2000	Latex acrylic based, bulk				Gal.	23			23	25.50	
	2100	Cartridges					26			26	28.50	
	2300	Polysulfide compounds, 1 component, bulk				↓	43.50			43.50	47.50	
	2600	1 or 2 component, in place, 1/4" x 1/4", 308 L.F./gal.	1 Bric	145	.055	L.F.	.14	1.63		1.77	2.67	
	2700	1/2" x 1/4", 154 L.F./gal.		135	.059		.28	1.75		2.03	3.02	
	2900	3/4" x 3/8", 68 L.F./gal.		130	.062		.64	1.82		2.46	3.52	
	3000	1" x 1/2", 38 L.F./gal.	↓	130	.062	↓	1.14	1.82		2.96	4.07	
	3200	Polyurethane, 1 or 2 component				Gal.	47			47	51.50	
	3300	Cartridges				"	46.50			46.50	51	
	3500	Bulk, in place, 1/4" x 1/4"	1 Bric	150	.053	L.F.	.15	1.58		1.73	2.61	
	3600	1/2" x 1/4"		145	.055		.31	1.63		1.94	2.86	
	3800	3/4" x 3/8", 68 L.F./gal.		130	.062		.69	1.82		2.51	3.58	
	3900	1" x 1/2"	↓	110	.073	↓	1.22	2.15		3.37	4.67	
	4100	Silicone rubber, bulk				Gal.	34			34	37.50	
	4200	Cartridges				"	40			40	44	
	4400	Neoprene gaskets, closed cell, adhesive, 1/8" x 3/8"	1 Bric	240	.033	L.F.	.20	.99		1.19	1.74	
	4500	1/4" x 3/4"		215	.037		.48	1.10		1.58	2.23	
	4700	1/2" x 1"		200	.040		1.40	1.18		2.58	3.37	
	4800	3/4" x 1-1/2"	↓	165	.048	↓	2.91	1.44		4.35	5.40	
	5500	Resin epoxy coating, 2 component, heavy duty				Gal.	26			26	28.50	
	5800	Tapes, sealant, P.V.C. foam adhesive, 1/16" x 1/4"				C.L.F.	4.60			4.60	5.05	
	5900	1/16" x 1/2"					6.80			6.80	7.50	
	5950	1/16" x 1"					11.30			11.30	12.45	
	6000	1/8" x 1/2"				↓	7.65			7.65	8.40	
	6200	Urethane foam, 2 component, handy pack, 1 C.F.				Ea.	27.50			27.50	30.50	
	6300	50.0 C.F. pack				C.F.	14.05			14.05	15.45	

For information about Means Estimating Seminars, see yellow pages 11 and 12 in back of book

Division Notes

		CREW	DAILY OUTPUT	LABOR-HOURS	UNIT	2001 BARE COSTS MAT.	LABOR	EQUIP.	TOTAL	TOTAL INCL O&P

Division 9 Finishes

Estimating Tips

General

- Room Finish Schedule: A complete set of plans should contain a room finish schedule. If one is not available, it would be well worth the time and effort to put one together. A room finish schedule should contain the room number, room name (for clarity), floor materials, base materials, wainscot materials, wainscot height, wall materials (for each wall), ceiling materials, ceiling height and special instructions.
- Surplus Finishes: Review the specifications to determine if there is any requirement to provide certain amounts of extra materials for the owner's maintenance department. In some cases the owner may require a substantial amount of materials, especially when it is a special order item or long lead time item.

09200 Plaster & Gypsum Board

- Lath is estimated by the square yard for both gypsum and metal lath, plus usually 5% allowance for waste. Furring, channels and accessories are measured by the linear foot. An extra foot should be allowed for each accessory miter or stop.
- Plaster is also estimated by the square yard. Deductions for openings vary by preference, from zero deduction to 50% of all openings over 2 feet in width. Some estimators deduct a percentage of the total yardage for openings. The estimator should allow one extra square foot for each linear foot of horizontal interior or exterior angle located below the ceiling level. Also, double the areas of small radius work.
- Each room should be measured, perimeter times maximum wall height. Ceiling areas are equal to length times width.
- Drywall accessories, studs, track, and acoustical caulking are all measured by the linear foot. Drywall taping is figured by the square foot. Gypsum wallboard is estimated by the square foot. No material deductions should be made for door or window openings under 32 S.F. Coreboard can be obtained in a 1″ thickness for solid wall and shaft work. Additions should be made to price out the inside or outside corners.
- Different types of partition construction should be listed separately on the quantity sheets. There may be walls with studs of various widths, double studded, and similar or dissimilar surface materials. Shaft work is usually different construction from surrounding partitions requiring separate quantities and pricing of the work.

09300 Tile
09400 Terrazzo

- Tile and terrazzo areas are taken off on a square foot basis. Trim and base materials are measured by the linear foot. Accent tiles are listed per each. Two basic methods of installation are used. Mud set is approximately 30% more expensive than the thin set. In terrazzo work, be sure to include the linear footage of embedded decorative strips, grounds, machine rubbing and power cleanup.

09600 Flooring

- Wood flooring is available in strip, parquet, or block configuration. The latter two types are set in adhesives with quantities estimated by the square foot. The laying pattern will influence labor costs and material waste. In addition to the material and labor for laying wood floors, the estimator must make allowances for sanding and finishing these areas unless the flooring is prefinished.
- Most of the various types of flooring are all measured on a square foot basis. Base is measured by the linear foot. If adhesive materials are to be quantified, they are estimated at a specified coverage rate by the gallon depending upon the specified type and the manufacturer's recommendations.
- Sheet flooring is measured by the square yard. Roll widths vary, so consideration should be given to use the most economical width, as waste must be figured into the total quantity. Consider also the installation methods available, direct glue down or stretched .

09700 Wall Finishes

- Wall coverings are estimated by the square foot. The area to be covered is measured, length by height of wall above baseboards, to calculate the square footage of each wall. This figure is divided by the number of square feet in the single roll which is being used. Deduct, in full, the areas of openings such as doors and windows. Where a pattern match is required allow 25%-30% waste. One gallon of paste should be sufficient to hang 12 single rolls of light to medium weight paper.

09800 Acoustical Treatment

- Acoustical systems fall into several categories. The takeoff of these materials is by the square foot of area with a 5% allowance for waste. Do not forget about scaffolding, if applicable, when estimating these systems.

09900 Paints & Coatings

- Painting is one area where bids vary to a greater extent than almost any other section of a project. This arises from the many methods of measuring surfaces to be painted. The estimator should check the plans and specifications carefully to be sure of the required number of coats.
- Protection of adjacent surfaces is not included in painting costs. When considering the method of paint application, an important factor is the amount of protection and masking required. These must be estimated separately and may be the determining factor in choosing the method of application.

Reference Numbers

Reference numbers are shown in bold squares at the beginning of some major classifications. These numbers refer to related items in the Reference Section. The reference information may be an estimating procedure, an alternate pricing method or technical information.

Note: Not all subdivisions listed here necessarily appear in this publication.

	09910 | Paints & Coatings	CREW	DAILY OUTPUT	LABOR-HOURS	UNIT	2001 BARE COSTS MAT.	LABOR	EQUIP.	TOTAL	TOTAL INCL O&P	
630	0010 **MISCELLANEOUS, INTERIOR**										630
	3800 Grilles, per side, oil base, primer coat, brushwork	1 Pord	520	.015	S.F.	.10	.40		.50	.72	
	3850 Spray		1,140	.007		.10	.18		.28	.39	
	3880 Paint 1 coat, brushwork		520	.015		.10	.40		.50	.72	
	3900 Spray		1,140	.007		.12	.18		.30	.41	
	3920 Paint 2 coats, brushwork		325	.025		.20	.64		.84	1.20	
	3940 Spray		650	.012		.23	.32		.55	.74	
	3950 Prime & paint 1 coat		325	.025		.20	.64		.84	1.20	
	3960 Prime & paint 2 coats		270	.030		.20	.77		.97	1.40	
	4250 Paint 1 coat, brushwork	2 Pord	1,300	.012	L.F.	.11	.32		.43	.61	
	4500 Louvers, one side, primer, brushwork	1 Pord	524	.015	S.F.	.06	.40		.46	.67	
	4520 Paint one coat, brushwork		520	.015		.05	.40		.45	.67	
	4530 Spray		1,140	.007		.06	.18		.24	.35	
	4540 Paint two coats, brushwork		325	.025		.10	.64		.74	1.10	
	4550 Spray		650	.012		.12	.32		.44	.62	
	4560 Paint three coats, brushwork		270	.030		.16	.77		.93	1.35	
	4570 Spray		500	.016		.17	.42		.59	.83	
	5000 Pipe, to 4" diameter, primer or sealer coat, oil base, brushwork	2 Pord	1,250	.013	L.F.	.06	.33		.39	.57	
	5100 Spray		2,165	.007		.06	.19		.25	.35	
	5200 Paint 1 coat, brushwork		1,250	.013		.06	.33		.39	.58	
	5300 Spray		2,165	.007		.05	.19		.24	.35	
	5350 Paint 2 coats, brushwork		775	.021		.11	.54		.65	.94	
	5400 Spray		1,240	.013		.12	.34		.46	.64	
	5420 Paint 3 coats, brushwork		775	.021		.16	.54		.70	1	
	5450 To 8" diameter, primer or sealer coat, brushwork		620	.026		.12	.67		.79	1.16	
	5500 Spray		1,085	.015		.19	.39		.58	.80	
	5550 Paint 1 coat, brushwork		620	.026		.17	.67		.84	1.21	
	5600 Spray		1,085	.015		.19	.39		.58	.79	
	5650 Paint 2 coats, brushwork		385	.042		.22	1.08		1.30	1.89	
	5700 Spray		620	.026		.24	.67		.91	1.30	
	5720 Paint 3 coats, brushwork		385	.042		.33	1.08		1.41	2.01	
	5750 To 12" diameter, primer or sealer coat, brushwork		415	.039		.17	1.01		1.18	1.72	
	5800 Spray		725	.022		.20	.58		.78	1.10	
	5850 Paint 1 coat, brushwork		415	.039		.17	1.01		1.18	1.72	
	6000 Spray		725	.022		.19	.58		.77	1.09	
	6200 Paint 2 coats, brushwork		260	.062		.33	1.61		1.94	2.81	
	6250 Spray		415	.039		.36	1.01		1.37	1.93	
	6270 Paint 3 coats, brushwork		260	.062		.49	1.61		2.10	2.99	
	6300 To 16" diameter, primer or sealer coat, brushwork		310	.052		.23	1.35		1.58	2.31	
	6350 Spray		540	.030		.26	.77		1.03	1.47	
	6400 Paint 1 coat, brushwork		310	.052		.23	1.35		1.58	2.30	
	6450 Spray		540	.030		.25	.77		1.02	1.46	
	6500 Paint 2 coats, brushwork		195	.082		.44	2.14		2.58	3.74	
	6550 Spray		310	.052		.49	1.35		1.84	2.59	
	6600 Radiators, per side, primer, brushwork	1 Pord	520	.015	S.F.	.06	.40		.46	.67	
	6620 Paint one coat, brushwork		520	.015		.05	.40		.45	.67	
	6640 Paint two coats, brushwork		340	.024		.10	.61		.71	1.06	
	6660 Paint three coats, brushwork		283	.028		.16	.74		.90	1.29	

For information about Means Estimating Seminars, see yellow pages 11 and 12 in back of book

Division 10 Specialties

Estimating Tips

General

- The items in this division are usually priced per square foot or each.
- Many items in Division 10 require some type of support system that is not usually furnished with the item. Examples of these systems include blocking for the attachment of grab bars and support angles for ceiling hung toilet partitions. The required blocking or supports must be added to the estimate in the appropriate division.
- Some items in Division 10, such as lockers, may require assembly before installation. Verify the amount of assembly required. Assembly can often exceed installation time.

10100 Visual Display Boards

- Toilet partitions are priced by the stall. A stall consists of a side wall, pilaster and door with hardware. Toilet tissue holders and grab bars are extra.

10600 Partitions

- The required acoustical rating of a folding partition can have a significant impact on costs. Verify the sound transmission coefficient rating of the panel priced to the specification requirements.

Reference Numbers

Reference numbers are shown in bold squares at the beginning of some major classifications. These numbers refer to related items in the Reference Section. The reference information may be an estimating procedure, an alternate pricing method or technical information.

Note: Not all subdivisions listed here necessarily appear in this publication.

10155 | Toilet Compartments			CREW	DAILY OUTPUT	LABOR-HOURS	UNIT	2001 BARE COSTS MAT.	LABOR	EQUIP.	TOTAL	TOTAL INCL O&P	
100	0010	PARTITIONS, TOILET										100
	0100	Cubicles, ceiling hung, marble	2 Marb	2	8	Ea.	1,325	234		1,559	1,800	
	0200	Painted metal	2 Carp	4	4		365	117		482	580	
	0250	Phenolic		4	4		730	117		847	980	
	0300	Plastic laminate on particle board		4	4		500	117		617	725	
	0500	Stainless steel		4	4		995	117		1,112	1,275	
	0600	For handicap units, incl. 52″ grab bars, add					293			293	320	
	0800	Floor & ceiling anchored, marble	2 Marb	2.50	6.400		1,450	187		1,637	1,875	
	1000	Painted metal	2 Carp	5	3.200		365	93.50		458.50	550	
	1050	Phenolic		5	3.200		810	93.50		903.50	1,025	
	1100	Plastic laminate on particle board		5	3.200		470	93.50		563.50	660	
	1300	Stainless steel		5	3.200		1,150	93.50		1,243.50	1,425	
	1400	For handicap units, incl. 52″ grab bars, add					253			253	278	
	1600	Floor mounted, marble	2 Marb	3	5.333		850	156		1,006	1,175	
	1700	Painted metal	2 Carp	7	2.286		340	66.50		406.50	480	
	1750	Phenolic		7	2.286		730	66.50		796.50	905	
	1800	Plastic laminate on particle board		7	2.286		465	66.50		531.50	615	
	2000	Stainless steel		7	2.286		1,175	66.50		1,241.50	1,400	
	2100	For handicap units, incl. 52″ grab bars, add					253			253	278	
	2200	For juvenile units, deduct					35			35	38.50	
	2400	Floor mounted, headrail braced, marble	2 Marb	3	5.333		820	156		976	1,150	
	2500	Painted metal	2 Carp	6	2.667		365	77.50		442.50	525	
	2550	Phenolic		6	2.667		785	77.50		862.50	980	
	2600	Plastic laminate on particle board		6	2.667		630	77.50		707.50	815	
	2800	Stainless steel		6	2.667		1,150	77.50		1,227.50	1,375	
	2900	For handicap units, incl. 52″ grab bars, add					253			253	278	
	3000	Wall hung partitions, painted metal	2 Carp	7	2.286		465	66.50		531.50	615	
	3300	Stainless steel	"	7	2.286		1,125	66.50		1,191.50	1,325	
	3400	For handicap units, incl. 52″ grab bars, add					253			253	278	
	4000	Screens, entrance, floor mounted, 58″ high, 48″ wide										
	4100	Marble	2 Marb	9	1.778	Ea.	535	52		587	665	
	4200	Painted metal	2 Carp	15	1.067		166	31		197	231	
	4300	Plastic laminate on particle board		15	1.067		294	31		325	375	
	4500	Stainless steel		15	1.067		610	31		641	720	
	4600	Urinal screen, 18″ wide, ceiling braced, marble	D-1	6	2.667		540	70		610	705	
	4700	Painted metal	2 Carp	8	2		126	58.50		184.50	230	
	4800	Plastic laminate on particle board		8	2		252	58.50		310.50	370	
	5000	Stainless steel		8	2		435	58.50		493.50	570	
	5100	Floor mounted, head rail braced										
	5200	Marble	D-1	6	2.667	Ea.	475	70		545	630	
	5300	Painted metal	2 Carp	8	2		168	58.50		226.50	276	
	5400	Plastic laminate on particle board		8	2		211	58.50		269.50	325	
	5600	Stainless steel		8	2		485	58.50		543.50	620	
	5700	Pilaster, flush, marble	D-1	9	1.778		605	46.50		651.50	735	
	5800	Painted metal	2 Carp	10	1.600		219	46.50		265.50	315	
	5900	Plastic laminate on particle board		10	1.600		288	46.50		334.50	390	
	6100	Stainless steel		10	1.600		385	46.50		431.50	495	
	6200	Post braced, marble	D-1	9	1.778		595	46.50		641.50	725	
	6300	Painted metal	2 Carp	10	1.600		229	46.50		275.50	325	
	6400	Plastic laminate on particle board		10	1.600		288	46.50		334.50	390	
	6600	Stainless steel		10	1.600		380	46.50		426.50	495	
	6700	Wall hung, bracket supported										
	6800	Painted metal	2 Carp	10	1.600	Ea.	231	46.50		277.50	325	
	6900	Plastic laminate on particle board		10	1.600		113	46.50		159.50	197	
	7100	Stainless steel		10	1.600		350	46.50		396.50	460	
	7400	Flange supported, painted metal		10	1.600		170	46.50		216.50	260	

10150 | Compartments & Cubicles

	10155 | Toilet Compartments		CREW	DAILY OUTPUT	LABOR-HOURS	UNIT	2001 BARE COSTS MAT.	LABOR	EQUIP.	TOTAL	TOTAL INCL O&P	
100	7500	Plastic laminate on particle board	2 Carp	10	1.600	Ea.	232	46.50		278.50	330	100
	7700	Stainless steel		10	1.600		405	46.50		451.50	520	
	7800	Wedge type, painted metal		10	1.600		201	46.50		247.50	294	
	8100	Stainless steel		10	1.600		415	46.50		461.50	530	

	10185 | Shower/Dressing Compartments		CREW	DAILY OUTPUT	LABOR-HOURS	UNIT	MAT.	LABOR	EQUIP.	TOTAL	TOTAL INCL O&P	
100	0010	**PARTITIONS, SHOWER** Floor mounted, no plumbing										100
	0100	Cabinet, incl. base, no door, painted steel, 1″ thick walls	2 Shee	5	3.200	Ea.	625	107		732	850	
	0300	With door, fiberglass		4.50	3.556		510	119		629	745	
	0600	Galvanized and painted steel, 1″ thick walls		5	3.200		665	107		772	895	
	0800	Stall, 1″ thick wall, no base, enameled steel		5	3.200		635	107		742	865	
	1200	Stainless steel		5	3.200		1,125	107		1,232	1,400	
	1400	For double entry type, no doors, deduct					10%					
	1500	Circular fiberglass, cabinet 36″ diameter,	2 Shee	4	4		525	134		659	780	
	1700	One piece, 36″ diameter, less door		4	4		450	134		584	700	
	1800	With door		3.50	4.571		725	153		878	1,025	
	2000	Curved shell shower, no door needed		3	5.333		665	179		844	1,000	
	2300	For fiberglass seat, add to both above					94			94	103	
	2400	Glass stalls, with doors, no receptors, chrome on brass	2 Shee	3	5.333		1,075	179		1,254	1,475	
	2700	Anodized aluminum	″	4	4		750	134		884	1,025	
	2900	Marble shower stall, stock design, with shower door	2 Marb	1.20	13.333		1,675	390		2,065	2,425	
	3000	With curtain		1.30	12.308		1,475	360		1,835	2,175	
	3200	Receptors, precast terrazzo, 32″ x 32″		14	1.143		272	33.50		305.50	350	
	3300	48″ x 34″		9.50	1.684		400	49		449	515	
	3500	Plastic, simulated terrazzo receptor, 32″ x 32″		14	1.143		89	33.50		122.50	150	
	3600	32″ x 48″		12	1.333		128	39		167	201	
	3800	Precast concrete, colors, 32″ x 32″		14	1.143		158	33.50		191.50	226	
	3900	48″ x 48″		8	2		194	58.50		252.50	305	
	4100	Shower doors, economy plastic, 24″ wide	1 Shee	9	.889		94	30		124	149	
	4200	Tempered glass door, economy		8	1		181	33.50		214.50	251	
	4400	Folding, tempered glass, aluminum frame		6	1.333		298	44.50		342.50	400	
	4500	Sliding, tempered glass, 48″ opening		6	1.333		188	44.50		232.50	275	
	4700	Deluxe, tempered glass, chrome on brass frame, minimum		8	1		229	33.50		262.50	305	
	4800	Maximum		1	8		615	268		883	1,100	
	4850	On anodized aluminum frame, minimum		2	4		107	134		241	325	
	4900	Maximum		1	8		360	268		628	815	
	5100	Shower enclosure, tempered glass, anodized alum. frame										
	5120	2 panel & door, corner unit, 32″ x 32″	1 Shee	2	4	Ea.	360	134		494	605	
	5140	Neo-angle corner unit, 16″ x 24″ x 16″	″	2	4		665	134		799	935	
	5200	Shower surround, 3 wall, polypropylene, 32″ x 32″	1 Carp	4	2		233	58.50		291.50	345	
	5220	PVC, 32″ x 32″		4	2		235	58.50		293.50	350	
	5240	Fiberglass		4	2		277	58.50		335.50	395	
	5250	2 wall, polypropylene, 32″ x 32″		4	2		194	58.50		252.50	305	
	5270	PVC		4	2		235	58.50		293.50	350	
	5290	Fiberglass		4	2		261	58.50		319.50	380	
	5300	Tub doors, tempered glass & frame, minimum	1 Shee	8	1		154	33.50		187.50	222	
	5400	Maximum		6	1.333		355	44.50		399.50	460	
	5600	Chrome plated, brass frame, minimum		8	1		201	33.50		234.50	273	
	5700	Maximum		6	1.333		400	44.50		444.50	510	
	5900	Tub/shower enclosure, temp. glass, alum. frame, minimum		2	4		275	134		409	505	
	6200	Maximum		1.50	5.333		535	179		714	865	
	6500	On chrome-plated brass frame, minimum		2	4		380	134		514	625	
	6600	Maximum		1.50	5.333		785	179		964	1,150	
	6800	Tub surround, 3 wall, polypropylene	1 Carp	4	2		154	58.50		212.50	261	
	6900	PVC		4	2		235	58.50		293.50	350	
	7000	Fiberglass, minimum		4	2		261	58.50		319.50	380	

10150 | Compartments & Cubicles

	10185 | Shower/Dressing Compartments	CREW	DAILY OUTPUT	LABOR-HOURS	UNIT	2001 BARE COSTS MAT.	LABOR	EQUIP.	TOTAL	TOTAL INCL O&P	
100	7100 Maximum	1 Carp	3	2.667	Ea.	450	77.50		527.50	615	100

10520 | Fire Protection Specialties

	10525 | Fire Prot. Specialties	CREW	DAILY OUTPUT	LABOR-HOURS	UNIT	2001 BARE COSTS MAT.	LABOR	EQUIP.	TOTAL	TOTAL INCL O&P	
200	0010 **FIRE EQUIPMENT CABINETS** Not equipped, 20 ga. steel box, (A8.2-310)										200
	0040 recessed, D.S. glass in door, box size given										
	1000 Portable extinguisher, single, 8″ x 12″ x 27″, alum. door & frame	Q-12	8	2	Ea.	74.50	62		136.50	176	
	1100 Steel door and frame		8	2		56	62		118	156	
	1200 Stainless steel door and frame		8	2		99.50	62		161.50	203	
	2000 Portable extinguisher, large, 8″ x 12″ x 36″, alum. door & frame		8	2		128	62		190	234	
	2100 Steel door and frame (A8.2-320)		8	2		82.50	62		144.50	185	
	2200 Stainless steel door and frame		8	2		131	62		193	238	
	2500 8″ x 16″ x 38″, aluminum door & frame (A8.2-320)		8	2		111	62		173	216	
	2600 Steel door and frame	↓	8	2	↓	138	62		200	246	
	3000 Hose rack assy., 1-1/2″ valve & 100′ hose, 24″ x 40″ x 5-1/2″										
	3100 Aluminum door and frame	Q-12	6	2.667	Ea.	183	82.50		265.50	325	
	3200 Steel door and frame (A8.2-390)		6	2.667		121	82.50		203.50	258	
	3300 Stainless steel door and frame	↓	6	2.667	↓	238	82.50		320.50	385	
	4000 Hose rack assy., 2-1/2″ x 1-1/2″ valve, 100′ hose, 24″ x 40″ x 8″										
	4100 Aluminum door and frame	Q-12	6	2.667	Ea.	185	82.50		267.50	330	
	4200 Steel door and frame (R10520-310)		6	2.667		125	82.50		207.50	262	
	4300 Stainless steel door and frame	↓	6	2.667	↓	240	82.50		322.50	390	
	5000 Hose rack assy., 2-1/2″ x 1-1/2″ valve, 100′ hose										
	5010 and extinguisher, 30″ x 40″ x 8″										
	5100 Aluminum door and frame	Q-12	5	3.200	Ea.	235	99		334	410	
	5200 Steel door and frame		5	3.200		156	99		255	320	
	5300 Stainless steel door and frame	↓	5	3.200	↓	278	99		377	455	
	6000 Hose rack assy., 1-1/2″ valve, 100′ hose										
	6010 and 2-1/2″ FD valve, 24″ x 44″ x 8″										
	6100 Aluminum door and frame	Q-12	5	3.200	Ea.	209	99		308	380	
	6200 Steel door and frame		5	3.200		133	99		232	297	
	6300 Stainless steel door and frame	↓	5	3.200	↓	271	99		370	450	
	7000 Hose rack assy., 1-1/2″ valve & 100′ hose, 2-1/2″ FD valve (R10520-320)										
	7010 and extinguisher, 30″ x 44″ x 8″										
	7100 Aluminum door and frame	Q-12	5	3.200	Ea.	247	99		346	420	
	7200 Steel door and frame		5	3.200		164	99		263	330	
	7300 Stainless steel door and frame	↓	5	3.200	↓	320	99		419	505	
	8000 Valve cabinet for 2-1/2″ FD angle valve, 18″ x 18″ x 8″										
	8100 Aluminum door and frame	Q-12	12	1.333	Ea.	81.50	41.50		123	153	
	8200 Steel door and frame		12	1.333		61.50	41.50		103	130	
	8300 Stainless steel door and frame	↓	12	1.333	↓	107	41.50		148.50	181	
300	0010 **FIRE EXTINGUISHERS**										300
	0120 CO2, portable with swivel horn, 5 lb.				Ea.	101			101	111	
	0140 With hose and "H" horn, 10 lb.					150			150	165	
	0160 15 lb.					172			172	189	
	0180 20 lb.					207			207	228	
	0360 Wheeled type, cart mounted, 50 lb.				↓	1,200			1,200	1,325	

10520 | Fire Protection Specialties

	10525 | Fire Prot. Specialties	CREW	DAILY OUTPUT	LABOR-HOURS	UNIT	2001 BARE COSTS MAT.	LABOR	EQUIP.	TOTAL	TOTAL INCL O&P	
300	1000 Dry chemical, pressurized										300
	1040 Standard type, portable, painted, 2-1/2 lb.				Ea.	27.50			27.50	30.50	
	1060 5 lb.					40			40	44	
	1080 10 lb.					67			67	73.50	
	1100 20 lb.					90			90	99	
	1120 30 lb.					145			145	160	
	1300 Standard type, wheeled, 150 lb.					1,400			1,400	1,550	
	2000 ABC all purpose type, portable, 2-1/2 lb.					27.50			27.50	30.50	
	2060 5 lb.					40			40	44	
	2080 9-1/2 lb.					60			60	66	
	2100 20 lb.					85			85	93.50	
	2300 Wheeled, 45 lb.					650			650	715	
	2360 150 lb.					1,450			1,450	1,600	
	3000 Dry chemical, outside cartridge to -65°F, painted, 9 lb.					200			200	220	
	3060 26 lb.					250			250	275	
	5000 Pressurized water, 2-1/2 gallon, stainless steel					75			75	82.50	
	5060 With anti-freeze					110			110	121	
	6000 Soda & acid, 2-1/2 gallon, stainless steel					75			75	82.50	
	9400 Installation of extinguishers, 12 or more, on wood	1 Carp	30	.267			7.75		7.75	12.15	
	9420 On masonry or concrete	"	15	.533	↓		15.55		15.55	24.50	

10800 | Toilet/Bath/Laundry Accessories

	10820 | Bath Accessories	CREW	DAILY OUTPUT	LABOR-HOURS	UNIT	2001 BARE COSTS MAT.	LABOR	EQUIP.	TOTAL	TOTAL INCL O&P	
100	0010 BATH ACCESSORIES										100
	0200 Curtain rod, stainless steel, 5′ long, 1″ diameter	1 Carp	13	.615	Ea.	24	17.95		41.95	54.50	
	0300 1-1/4″ diameter	"	13	.615	"	26.50	17.95		44.45	57.50	
	0500 Dispenser units, combined soap & towel dispensers,										
	0510 mirror and shelf, flush mounted	1 Carp	10	.800	Ea.	268	23.50		291.50	330	
	0600 Towel dispenser and waste receptacle,										
	0610 18 gallon capacity	1 Carp	10	.800	Ea.	310	23.50		333.50	375	
	0800 Grab bar, straight, 1-1/4″ diameter, stainless steel, 18″ long		24	.333		20	9.70		29.70	37	
	0900 24″ long		23	.348		23.50	10.15		33.65	42	
	1000 30″ long		22	.364		23.50	10.60		34.10	42.50	
	1100 36″ long		20	.400		25.50	11.65		37.15	46.50	
	1200 1-1/2″ diameter, 24″ long		23	.348		44.50	10.15		54.65	65	
	1300 36″ long		20	.400		50	11.65		61.65	73.50	
	1500 Tub bar, 1-1/4″ diameter, 24″ x 36″		14	.571		75.50	16.65		92.15	109	
	1600 Plus vertical arm		12	.667		83	19.45		102.45	122	
	1900 End tub bar, 1″ diameter, 90° angle, 16″ x 32″		12	.667		116	19.45		135.45	158	
	2300 Hand dryer, surface mounted, electric, 115 volt, 20 amp		4	2		530	58.50		588.50	675	
	2400 230 volt, 10 amp		4	2		535	58.50		593.50	675	
	2600 Hat and coat strip, stainless steel, 4 hook, 36″ long		24	.333		51	9.70		60.70	71	
	2700 6 hook, 60″ long		20	.400		79.50	11.65		91.15	106	
	3000 Mirror, with stainless steel 3/4″ square frame, 18″ x 24″		20	.400		61	11.65		72.65	85.50	
	3100 36″ x 24″		15	.533		104	15.55		119.55	139	
	3200 48″ x 24″		10	.800		141	23.50		164.50	192	
	3300 72″ x 24″		6	1.333		216	39		255	298	
	3500 With 5″ stainless steel shelf, 18″ x 24″		20	.400		94.50	11.65		106.15	122	
	3600 36″ x 24″	↓	15	.533	↓	132	15.55		147.55	170	

		10820 | Bath Accessories	CREW	DAILY OUTPUT	LABOR-HOURS	UNIT	2001 BARE COSTS MAT.	LABOR	EQUIP.	TOTAL	TOTAL INCL O&P	
100	3700	48" x 24"	1 Carp	10	.800	Ea.	153	23.50		176.50	205	100
	3800	72" x 24"		6	1.333		252	39		291	340	
	4100	Mop holder strip, stainless steel, 5 holders, 48" long		20	.400		56.50	11.65		68.15	80.50	
	4200	Napkin/tampon dispenser, recessed		15	.533		350	15.55		365.55	410	
	4300	Robe hook, single, regular		36	.222		4.84	6.50		11.34	15.45	
	4400	Heavy duty, concealed mounting		36	.222		15.60	6.50		22.10	27.50	
	4600	Soap dispenser, chrome, surface mounted, liquid		20	.400		41	11.65		52.65	63.50	
	4700	Powder		20	.400		41	11.65		52.65	63.50	
	5000	Recessed stainless steel, liquid		10	.800		141	23.50		164.50	192	
	5100	Powder		10	.800		255	23.50		278.50	320	
	5300	Soap tank, stainless steel, 1 gallon		10	.800		159	23.50		182.50	212	
	5400	5 gallon		5	1.600		219	46.50		265.50	315	
	5600	Shelf, stainless steel, 5" wide, 18 ga., 24" long		24	.333		36.50	9.70		46.20	55	
	5700	48" long		16	.500		54.50	14.60		69.10	83	
	5800	8" wide shelf, 18 ga., 24" long		22	.364		45	10.60		55.60	66	
	5900	48" long		14	.571		77.50	16.65		94.15	111	
	6000	Toilet seat cover dispenser, stainless steel, recessed		20	.400		87.50	11.65		99.15	115	
	6050	Surface mounted		15	.533		23.50	15.55		39.05	50.50	
	6100	Toilet tissue dispenser, surface mounted, SS, single roll		30	.267		11.30	7.75		19.05	24.50	
	6200	Double roll		24	.333		15.75	9.70		25.45	32.50	
	6400	Towel bar, stainless steel, 18" long		23	.348		30.50	10.15		40.65	49.50	
	6500	30" long		21	.381		49	11.10		60.10	71.50	
	6700	Towel dispenser, stainless steel, surface mounted		16	.500		36	14.60		50.60	63	
	6800	Flush mounted, recessed		10	.800		157	23.50		180.50	210	
	7000	Towel holder, hotel type, 2 guest size		20	.400		13.20	11.65		24.85	33	
	7200	Towel shelf, stainless steel, 24" long, 8" wide		20	.400		45.50	11.65		57.15	68.50	
	7400	Tumbler holder, tumbler only		30	.267		24	7.75		31.75	38	
	7500	Soap, tumbler & toothbrush		30	.267		24	7.75		31.75	38	
	7700	Wall urn ash receiver, surface mount, 11" long		12	.667		97.50	19.45		116.95	138	
	7800	7-1/2", long		18	.444		71	12.95		83.95	98.50	
	8000	Waste receptacles, stainless steel, with top, 13 gallon		10	.800		157	23.50		180.50	210	
	8100	36 gallon	↓	8	1	↓	249	29		278	320	
400	0010	**MEDICINE CABINETS** With mirror, st. st. frame, 16" x 22", unlighted	1 Carp	14	.571	Ea.	68.50	16.65		85.15	102	400
	0100	Wood frame		14	.571		95	16.65		111.65	131	
	0300	Sliding mirror doors, 20" x 16" x 4-3/4", unlighted		7	1.143		85	33.50		118.50	146	
	0400	24" x 19" x 8-1/2", lighted		5	1.600		134	46.50		180.50	220	
	0600	Triple door, 30" x 32", unlighted, plywood body		7	1.143		211	33.50		244.50	284	
	0700	Steel body		7	1.143		278	33.50		311.50	355	
	0900	Oak door, wood body, beveled mirror, single door		7	1.143		124	33.50		157.50	188	
	1000	Double door		6	1.333		315	39		354	405	
	1200	Hotel cabinets, stainless, with lower shelf, unlighted		10	.800		170	23.50		193.50	224	
	1300	Lighted	↓	5	1.600	↓	252	46.50		298.50	350	

Division 11 Equipment

Estimating Tips

General

- The items in this division are usually priced per square foot or each. Many of these items are purchased by the owner for installation by the contractor. Check the specifications for responsibilities, and include time for receiving, storage, installation and mechanical and electrical hook-ups in the appropriate divisions.
- Many items in Division 11 require some type of support system that is not usually furnished with the item. Examples of these systems include blocking for the attachment of casework and support angles for ceiling hung projection screens. The required blocking or supports must be added to the estimate in the appropriate division.
- Some items in Division 11 may require assembly or electrical hook-ups. Verify the amount of assembly required or the need for a hard electrical connection and add the appropriate costs.

Reference Numbers

Reference numbers are shown in bold squares at the beginning of some major classifications. These numbers refer to related items in the Reference Section. The reference information may be an estimating procedure, an alternate pricing method or technical information.

Note: Not all subdivisions listed here necessarily appear in this publication.

11010 | Maintenance Equipment

	11013	Floor/Wall Cleaning Equipment	CREW	DAILY OUTPUT	LABOR-HOURS	UNIT	2001 BARE COSTS MAT.	LABOR	EQUIP.	TOTAL	TOTAL INCL O&P	
800	0010	VACUUM CLEANING										800
	0020	Central, 3 inlet, residential	1 Skwk	.90	8.889	Total	555	265		820	1,025	
	0200	Commercial		.70	11.429		960	340		1,300	1,575	
	0400	5 inlet system, residential		.50	16		840	480		1,320	1,675	
	0600	7 inlet system, commercial		.40	20		1,050	595		1,645	2,075	
	0800	9 inlet system, residential	↓	.30	26.667		1,200	795		1,995	2,575	
	4010	Rule of thumb: First 1200 S.F., installed				↓					1,044	
	4020	For each additional S.F., add				S.F.					.17	

11040 | Ecclesiastical Equipment

	11041	Ecclesiastical Equipment	CREW	DAILY OUTPUT	LABOR-HOURS	UNIT	2001 BARE COSTS MAT.	LABOR	EQUIP.	TOTAL	TOTAL INCL O&P	
250	0010	CHURCH EQUIPMENT										250
	0020	Altar, wood, custom design, plain	1 Carp	1.40	5.714	Ea.	1,675	167		1,842	2,075	
	0050	Deluxe	"	.20	40	"	8,050	1,175		9,225	10,700	
	0150	Baptistry, fiberglass, 3'-6" deep, x 13'-7" long,										
	0160	steps at both ends, incl. plumbing, minimum	L-8	1	20	Ea.	2,450	605		3,055	3,625	
	0200	Maximum	"	.70	28.571		4,775	865		5,640	6,600	
	0250	Add for filter, heater and lights				↓	1,050			1,050	1,150	

11100 | Mercantile Equipment

	11102	Barber Shop Equipment	CREW	DAILY OUTPUT	LABOR-HOURS	UNIT	2001 BARE COSTS MAT.	LABOR	EQUIP.	TOTAL	TOTAL INCL O&P	
150	0010	BARBER EQUIPMENT										150
	0020	Chair, hydraulic, movable, minimum	1 Carp	24	.333	Ea.	410	9.70		419.70	465	
	0050	Maximum	"	16	.500		2,700	14.60		2,714.60	3,000	
	0500	Sink, hair washing basin, rough plumbing not incl.	1 Plum	8	1		273	34.50		307.50	350	
	1000	Sterilizer, liquid solution for tools				↓	125			125	138	

11110 | Commercial Laundry & Dry Cleaning Equipment

	11119	Laundry Cleaning	CREW	DAILY OUTPUT	LABOR-HOURS	UNIT	2001 BARE COSTS MAT.	LABOR	EQUIP.	TOTAL	TOTAL INCL O&P	
450	0010	LAUNDRY EQUIPMENT Not incl. rough-in										450
	0500	Dryers, gas fired residential, 16 lb. capacity, average	1 Plum	3	2.667	Ea.	520	92		612	710	
	1000	Commercial, 30 lb. capacity, coin operated, single		3	2.667		2,375	92		2,467	2,775	
	1100	Double stacked	↓	2	4	↓	5,100	138		5,238	5,800	

11110 | Commercial Laundry & Dry Cleaning Equipment

	11119 | Laundry Cleaning	CREW	DAILY OUTPUT	LABOR-HOURS	UNIT	2001 BARE COSTS MAT.	LABOR	EQUIP.	TOTAL	TOTAL INCL O&P	
450	1500 Industrial, 30 lb. capacity	1 Plum	2	4	Ea.	2,025	138		2,163	2,425	450
	1600 50 lb. capacity		1.70	4.706		2,650	162		2,812	3,175	
	5000 Washers, residential, 4 cycle, average		3	2.667		595	92		687	795	
	5300 Commercial, coin operated, average		3	2.667		955	92		1,047	1,200	
	6000 Combination washer/extractor, 20 lb. capacity	L-6	1.50	8		3,425	275		3,700	4,175	
	6100 30 lb. capacity		.80	15		6,600	515		7,115	8,025	
	6200 50 lb. capacity		.68	17.647		8,225	605		8,830	9,975	
	6300 75 lb. capacity		.30	40		17,000	1,375		18,375	20,800	
	6350 125 lb. capacity		.16	75		20,600	2,575		23,175	26,500	

11140 | Vehicle Service Equipment

	11141 | Service Station Equipment	CREW	DAILY OUTPUT	LABOR-HOURS	UNIT	2001 BARE COSTS MAT.	LABOR	EQUIP.	TOTAL	TOTAL INCL O&P	
150	0010 **AUTOMOTIVE**										150
	0030 Compressors, electric, 1-1/2 H.P., standard controls	L-4	1.50	16	Ea.	315	435		750	1,025	
	0550 Dual controls		1.50	16		460	435		895	1,200	
	0600 5 H.P., 115/230 volt, standard controls		1	24		1,525	655		2,180	2,725	
	0650 Dual controls		1	24		1,625	655		2,280	2,825	
	1100 Product dispenser with vapor recovery for 6 nozzles, installed, not										
	1110 including piping to storage tanks				Ea.	15,000			15,000	16,500	
	2200 Hoists, single post, 8,000# capacity, swivel arms	L-4	.40	60		3,500	1,625		5,125	6,425	
	2400 Two posts, adjustable frames, 11,000# capacity		.25	96		4,500	2,625		7,125	9,050	
	2500 24,000# capacity		.15	160		6,000	4,375		10,375	13,500	
	2700 7,500# capacity, frame supports		.50	48		5,000	1,300		6,300	7,550	
	2800 Four post, roll on ramp		.50	48		4,500	1,300		5,800	7,000	
	2810 Hydraulic lifts, above ground, 2 post, clear floor, 6000 lb cap		2.67	8.989		4,700	245		4,945	5,550	
	2815 9000 lb capacity		2.29	10.480		11,200	286		11,486	12,800	
	2820 15,000 lb capacity		2	12		12,600	325		12,925	14,400	
	2825 30,000 lb capacity		1.60	15		27,900	410		28,310	31,200	
	2830 4 post, ramp style, 25,000 lb capacity		2	12		10,900	325		11,225	12,500	
	2835 35,000 lb capacity		1	24		51,000	655		51,655	57,000	
	2840 50,000 lb capacity		1	24		57,000	655		57,655	63,500	
	2845 75,000 lb capacity		1	24		66,000	655		66,655	73,500	
	2850 For drive thru tracks, add, minimum					700			700	770	
	2855 Maximum					1,200			1,200	1,325	
	2860 Ramp extensions, 3′(set of 2)					575			575	635	
	2865 Rolling jack platform					2,000			2,000	2,200	
	2870 Elec/hyd jacking beam					5,350			5,350	5,875	
	2880 Scissor lift, portable, 6000 lb capacity					5,250			5,250	5,775	
	3000 Lube equipment, 3 reel type, with pumps, not including piping	L-4	.50	48	Set	6,000	1,300		7,300	8,650	
	3700 Pump lubrication, pneumatic, not incl. air compressor										
	3710 Oil/gear lube	Q-1	9.60	1.667	Ea.	665	51.50		716.50	815	
	3720 Grease	"	9.60	1.667		615	51.50		666.50	760	
	4000 Spray painting booth, 26′ long, complete	L-4	.40	60		12,000	1,625		13,625	15,800	

11190 | Detention Equipment

	11191 Detention Equipment		CREW	DAILY OUTPUT	LABOR-HOURS	UNIT	2001 BARE COSTS MAT.	LABOR	EQUIP.	TOTAL	TOTAL INCL O&P	
150	0010	**DETENTION EQUIPMENT**										150
	0500	Bar front, rolling, 7/8" bars, 4" O.C., 7' high, 5' wide, with hardware	E-4	2	16	Ea.	4,175	535	42	4,752	5,625	
	2000	Cells, prefab., 5' to 6' wide, 7' to 8' high, 7' to 8' deep,										
	2010	bar front, cot, not incl. plumbing	E-4	1.50	21.333	Ea.	7,400	710	56	8,166	9,500	
	3000	Toilet apparatus including wash basin, average	L-8	1.50	13.333	"	2,075	405		2,480	2,900	

11300 | Fluid Waste Treatment & Disposal Equipment

	11310 Sewage & Sludge Pumps		CREW	DAILY OUTPUT	LABOR-HOURS	UNIT	2001 BARE COSTS MAT.	LABOR	EQUIP.	TOTAL	TOTAL INCL O&P	
350	0010	**PUMPS, PNEUMATIC EJECTOR**										350
	0020	With steel receiver, level controls, inlet/outlet gate/check valves										
	0030	Cross connect. not incl. compressor, fittings or piping										
	0040	Duplex										
	0050	30 GPM	Q-2	1.70	14.118	Ea.	13,800	455		14,255	15,900	
	0060	50 GPM		1.56	15.385		19,500	495		19,995	22,100	
	0070	100 GPM		1.30	18.462		23,600	595		24,195	26,800	
	0080	150 GPM		1.10	21.818		45,400	700		46,100	51,000	
	0090	200 GPM	↓	.85	28.235		58,000	910		58,910	65,000	
	0100	250 GPM	Q-3	.91	35.165		74,500	1,150		75,650	84,000	
	0110	300 GPM	"	.57	56.140	↓	91,000	1,850		92,850	103,000	
700	0010	**SEWAGE PUMPING STATIONS** Prefabricated steel, concrete										700
	0020	or fiberglass, 200 GPM	C-17D	.17	494	Total	28,600	15,000	3,900	47,500	59,500	
	0200	1,000 GPM		.07	1,200		42,400	36,300	9,475	88,175	114,000	
	0500	Add for generator unit, 200 GPM, steel		.34	247		21,700	7,475	1,950	31,125	37,800	
	0600	Concrete		.51	164		13,800	5,000	1,300	20,100	24,500	
	1000	Add for generator unit, 1,000 GPM, steel		.30	280		23,100	8,475	2,200	33,775	41,100	
	1200	Concrete	↓	.38	221		19,500	6,700	1,750	27,950	33,900	
	1500	For drilled water well, if required, add	B-23	.50	80	↓	5,625	1,850	4,100	11,575	13,600	
	11390 Pkg Sewage Treat Plants											
900	0010	**WASTEWATER TREATMENT SYSTEM** Fiberglass, 1,000 gallon	B-21	1.29	21.705	Ea.	2,750	580	110	3,440	4,050	900
	0100	1,500 gallon	"	1.03	27.184	"	6,025	725	138	6,888	7,900	

11400 | Food Service Equipment

	11410 Food Prep Equipment		CREW	DAILY OUTPUT	LABOR-HOURS	UNIT	2001 BARE COSTS MAT.	LABOR	EQUIP.	TOTAL	TOTAL INCL O&P	
150	0010	**COMMERCIAL KITCHEN EQUIPMENT**										150
	0020	Bake oven, gas, one section	Q-1	8	2	Ea.	3,750	62		3,812	4,225	
	0300	Two sections		7	2.286		7,475	71		7,546	8,325	
	0600	Three sections	↓	6	2.667		10,700	82.50		10,782.50	11,800	
	0900	Electric convection, single deck	L-7	4	7		5,075	197		5,272	5,875	
	2350	Cooler, reach-in, beverage, 6' long	Q-1	6	2.667	↓	2,800	82.50		2,882.50	3,200	

11400 | Food Service Equipment

11410 | Food Prep Equipment

			CREW	DAILY OUTPUT	LABOR-HOURS	UNIT	2001 BARE COSTS MAT.	LABOR	EQUIP.	TOTAL	TOTAL INCL O&P	
150	2700	Dishwasher, commercial, rack type										150
	2720	10 to 12 racks per hour	Q-1	3.20	5	Ea.	2,550	155		2,705	3,050	
	2750	Semi-automatic 38 to 50 racks per hour	"	1.30	12.308		5,375	380		5,755	6,475	
	2800	Automatic, 190 to 230 racks per hour	Q-2	.70	34.286		11,000	1,100		12,100	13,800	
	2820	235 to 275 racks per hour		.50	48		22,200	1,550		23,750	26,700	
	2840	8,750 to 12,500 dishes per hour	↓	.20	120		41,700	3,850		45,550	51,500	
	4300	Freezers, reach-in, 44 C.F.	Q-1	4	4		7,400	124		7,524	8,325	
	4500	68 C.F.		3	5.333		8,725	165		8,890	9,825	
	5800	Ice cube maker, 50 pounds per day		6	2.667		1,525	82.50		1,607.50	1,800	
	6050	500 pounds per day	↓	4	4		2,825	124		2,949	3,300	
	6350	20 Gal kettle w/steam jacket, tilting w/positive lock, S/S	L-7	7	4		5,675	112		5,787	6,400	
	6600	60 gallons	"	6	4.667		6,350	131		6,481	7,175	
	7950	Hood fire protection system, minimum	Q-1	3	5.333		2,900	165		3,065	3,425	
	8050	Maximum		1	16		21,100	495		21,595	24,000	
	8300	Refrigerators, reach-in type, 44 C.F.		5	3.200		5,075	99		5,174	5,725	
	8310	With glass doors, 68 C.F.	↓	4	4	↓	6,450	124		6,574	7,275	

EQUIPMENT 11

11450 | Residential Equipment

11454 | Residential Appliances

			CREW	DAILY OUTPUT	LABOR-HOURS	UNIT	2001 BARE COSTS MAT.	LABOR	EQUIP.	TOTAL	TOTAL INCL O&P	
500	0010	RESIDENTIAL APPLIANCES										500
	0020	Cooking range, 30″ free standing, 1 oven, minimum	2 Clab	10	1.600	Ea.	249	36.50		285.50	330	
	0050	Maximum		4	4		1,200	91.50		1,291.50	1,475	
	0150	2 oven, minimum		10	1.600		1,300	36.50		1,336.50	1,475	
	0200	Maximum	↓	10	1.600		1,425	36.50		1,461.50	1,600	
	1500	Combination range, refrigerator and sink, 30″ wide, minimum	L-1	2	8		700	275		975	1,175	
	1550	Maximum		1	16		1,400	550		1,950	2,375	
	1570	60″ wide, average		1.40	11.429		2,325	395		2,720	3,150	
	1590	72″ wide, average		1.20	13.333		2,625	460		3,085	3,575	
	1600	Office model, 48″ wide		2	8		1,975	275		2,250	2,600	
	1620	Refrigerator and sink only	↓	2.40	6.667	↓	1,975	229		2,204	2,525	
	1640	Combination range, refrigerator, sink, microwave										
	1660	oven and ice maker	L-1	.80	20	Ea.	3,875	690		4,565	5,275	
	2450	Dehumidifier, portable, automatic, 15 pint					195			195	214	
	2550	40 pint					226			226	249	
	2750	Dishwasher, built-in, 2 cycles, minimum	L-1	4	4		246	138		384	475	
	2800	Maximum		2	8		271	275		546	715	
	2950	4 or more cycles, minimum		4	4		242	138		380	475	
	2960	Average		4	4		320	138		458	560	
	3000	Maximum	↓	2	8		510	275		785	975	
	3200	Dryer, automatic, minimum	L-2	3	5.333		251	137		388	490	
	3250	Maximum	"	2	8		555	205		760	930	
	3300	Garbage disposer, sink type, minimum	L-1	10	1.600		37.50	55		92.50	124	
	3350	Maximum	"	10	1.600		178	55		233	279	
	3550	Heater, electric, built-in, 1250 watt, ceiling type, minimum	1 Elec	4	2		68	68.50		136.50	177	
	3600	Maximum		3	2.667		111	91.50		202.50	259	
	3700	Wall type, minimum		4	2		96.50	68.50		165	208	
	3750	Maximum		3	2.667		128	91.50		219.50	278	
	3900	1500 watt wall type, with blower		4	2		119	68.50		187.50	233	
	3950	3000 watt	↓	3	2.667	↓	244	91.50		335.50	405	

11450 | Residential Equipment

		11454 | Residential Appliances	CREW	DAILY OUTPUT	LABOR-HOURS	UNIT	2001 BARE COSTS MAT.	LABOR	EQUIP.	TOTAL	TOTAL INCL O&P	
500	4150	Hood for range, 2 speed, vented, 30" wide, minimum	L-3	5	3.200	Ea.	35.50	101		136.50	194	500
	4200	Maximum		3	5.333		445	168		613	750	
	4300	42" wide, minimum		5	3.200		168	101		269	340	
	4330	Custom		5	3.200		600	101		701	815	
	4350	Maximum		3	5.333		730	168		898	1,050	
	4500	For ventless hood, 2 speed, add					14.05			14.05	15.50	
	4650	For vented 1 speed, deduct from maximum					27			27	30	
	4850	Humidifier, portable, 8 gallons per day					97.50			97.50	107	
	5000	15 gallons per day					157			157	173	
	5200	Icemaker, automatic, 20 lb. per day	1 Plum	7	1.143		530	39.50		569.50	645	
	5350	51 lb. per day		2	4		790	138		928	1,075	
	6400	Sump pump cellar drainer, pedestal, 1/3 H.P., molded PVC base		3	2.667		86.50	92		178.50	234	
	6450	Solid brass		2	4		179	138		317	405	
	6460	Sump pump, see also division 15440-940										
	6650	Washing machine, automatic, minimum	1 Plum	3	2.667	Ea.	325	92		417	500	
	6700	Maximum	"	1	8		640	276		916	1,125	
	6900	Water heater, electric, glass lined, 30 gallon, minimum	L-1	5	3.200		221	110		331	410	
	6950	Maximum		3	5.333		305	183		488	615	
	7100	80 gallon, minimum		2	8		420	275		695	875	
	7150	Maximum		1	16		580	550		1,130	1,475	
	7180	Water heater, gas, glass lined, 30 gallon, minimum	2 Plum	5	3.200		298	110		408	495	
	7220	Maximum		3	5.333		415	184		599	730	
	7260	50 gallon, minimum		2.50	6.400		430	220		650	810	
	7300	Maximum		1.50	10.667		600	365		965	1,225	
	7310	Water heater, see also division 15480-200										
	7350	Water softener, automatic, to 30 grains per gallon	2 Plum	5	3.200	Ea.	420	110		530	625	
	7400	To 100 grains per gallon	"	4	4		550	138		688	815	
	7450	Vent kits for dryers	1 Carp	10	.800		10.30	23.50		33.80	48	

11470 | Darkroom Equipment

		11471 | Darkroom Processing	CREW	DAILY OUTPUT	LABOR-HOURS	UNIT	2001 BARE COSTS MAT.	LABOR	EQUIP.	TOTAL	TOTAL INCL O&P	
700	0010	DARKROOM EQUIPMENT										700
	0020	Developing sink, 5" deep, 24" x 48"	Q-1	2	8	Ea.	3,950	248		4,198	4,725	
	0050	48" x 52"		1.70	9.412		4,000	292		4,292	4,850	
	0200	10" deep, 24" x 48"		1.70	9.412		4,975	292		5,267	5,925	
	0250	24" x 108"		1.50	10.667		1,800	330		2,130	2,475	
	3500	Washers, round, minimum sheet 11" x 14"		2	8		2,425	248		2,673	3,050	
	3550	Maximum sheet 20" x 24"		1	16		2,825	495		3,320	3,850	
	3800	Square, minimum sheet 20" x 24"		1	16		2,525	495		3,020	3,525	
	3900	Maximum sheet 50" x 56"		.80	20		3,950	620		4,570	5,275	
	4500	Combination tank sink, tray sink, washers, with										
	4510	dry side tables, average	Q-1	.45	35.556	Ea.	7,575	1,100		8,675	10,000	

11520 | Industrial Equipment

			CREW	DAILY OUTPUT	LABOR-HOURS	UNIT	2001 BARE COSTS MAT.	LABOR	EQUIP.	TOTAL	TOTAL INCL O&P	
250	0010	**DUST COLLECTION SYSTEMS** Commercial / industrial										250
	0120	Central vacuum units										
	0130	Includes stand, filters and motorized shaker										
	0200	500 CFM, 10" inlet, 2 HP	Q-20	2.40	8.333	Ea.	3,075	259		3,334	3,800	
	0220	1000 CFM, 10" inlet, 3 HP		2.20	9.091		3,225	282		3,507	3,975	
	0240	1500 CFM, 10" inlet, 5 HP		2	10		3,375	310		3,685	4,200	
	0260	3000 CFM, 13" inlet, 10 HP		1.50	13.333		8,625	415		9,040	10,100	
	0280	5000 CFM, 16" inlet, 2 @ 10 HP	↓	1	20	↓	9,125	620		9,745	11,000	
	1000	Vacuum tubing, galvanized										
	1100	2-1/8" OD, 16 ga.	Q-9	440	.036	L.F.	1.76	1.10		2.86	3.64	
	1110	2-1/2" OD, 16 ga.		420	.038		2.04	1.15		3.19	4.02	
	1120	3" OD, 16 ga.		400	.040		2.59	1.21		3.80	4.72	
	1130	3-1/2" OD, 16 ga.		380	.042		3.68	1.27		4.95	6	
	1140	4" OD, 16 ga.		360	.044		3.76	1.34		5.10	6.20	
	1150	5" OD, 14 ga.		320	.050		6.95	1.51		8.46	10	
	1160	6" OD, 14 ga.		280	.057		7.80	1.73		9.53	11.25	
	1170	8" OD, 14 ga.		200	.080		11.80	2.42		14.22	16.75	
	1180	10" OD, 12 ga.		160	.100		11.80	3.02		14.82	17.65	
	1190	12" OD, 12 ga.		120	.133		34	4.03		38.03	43	
	1200	14" OD, 12 ga.	↓	80	.200	↓	37.50	6.05		43.55	50.50	
	1940	Hose, flexible wire reinforced rubber										
	1956	3" dia.	Q-9	400	.040	L.F.	5.25	1.21		6.46	7.60	
	1960	4" dia.		360	.044		7.10	1.34		8.44	9.85	
	1970	5" dia.		320	.050		8.15	1.51		9.66	11.30	
	1980	6" dia.	↓	280	.057	↓	9.50	1.73		11.23	13.10	
	2000	90° Elbow, slip fit										
	2110	2-1/8" dia.	Q-9	70	.229	Ea.	7.35	6.90		14.25	18.75	
	2120	2-1/2" dia.		65	.246		10.30	7.45		17.75	23	
	2130	3" dia.		60	.267		13.95	8.05		22	28	
	2140	3-1/2" dia.		55	.291		17.25	8.80		26.05	32.50	
	2150	4" dia.		50	.320		21.50	9.65		31.15	38.50	
	2160	5" dia.		45	.356		40.50	10.75		51.25	61	
	2170	6" dia.		40	.400		56	12.10		68.10	80	
	2180	8" dia.	↓	30	.533	↓	106	16.10		122.10	142	
	2400	45° Elbow, slip fit										
	2410	2-1/8" dia.	Q-9	70	.229	Ea.	6.45	6.90		13.35	17.75	
	2420	2-1/2" dia.		65	.246		9.50	7.45		16.95	22	
	2430	3" dia.		60	.267		11.45	8.05		19.50	25	
	2440	3-1/2" dia.		55	.291		14.40	8.80		23.20	29.50	
	2450	4" dia.		50	.320		18.70	9.65		28.35	35.50	
	2460	5" dia.		45	.356		31.50	10.75		42.25	51	
	2470	6" dia.		40	.400		42.50	12.10		54.60	65.50	
	2480	8" dia.	↓	35	.457	↓	83.50	13.80		97.30	114	
	2800	90° TY, slip fit thru 6" dia										
	2810	2-1/8" dia.	Q-9	42	.381	Ea.	14.40	11.50		25.90	33.50	
	2820	2-1/2" dia.		39	.410		18.70	12.40		31.10	39.50	
	2830	3" dia.		36	.444		24.50	13.40		37.90	47.50	
	2840	3-1/2" dia.		33	.485		31.50	14.65		46.15	57	
	2850	4" dia.		30	.533		45.50	16.10		61.60	75	
	2860	5" dia.		27	.593		86	17.90		103.90	122	
	2870	6" dia.	↓	24	.667		117	20		137	160	
	2880	8" dia., butt end					249			249	274	
	2890	10" dia., butt end					450			450	495	
	2900	12" dia., butt end					450			450	495	
	2910	14" dia., butt end					815			815	895	
	2920	6 x 4" dia., butt end				↓	71.50			71.50	78.50	

EQUIPMENT 11

11500 | Industrial & Process Equipment

	11520 | Industrial Equipment		CREW	DAILY OUTPUT	LABOR-HOURS	UNIT	2001 BARE COSTS MAT.	LABOR	EQUIP.	TOTAL	TOTAL INCL O&P	
250	2930	8″ x 4″ dia., butt end				Ea.	135			135	149	250
	2940	10″ x 4″ dia., butt end					177			177	195	
	2950	12″ x 4″ dia., butt end					204			204	224	
	3100	90° Elbow, butt end segmented										
	3110	8″ dia., butt end, segmented				Ea.	161			161	177	
	3120	10″ dia., butt end, segmented					264			264	291	
	3130	12″ dia., butt end, segmented					335			335	370	
	3140	14″ dia., butt end, segmented					420			420	460	
	3200	45° Elbow, butt end segmented										
	3210	8″ dia., butt end, segmented				Ea.	137			137	151	
	3220	10″ dia., butt end, segmented					191			191	210	
	3230	12″ dia., butt end, segmented					195			195	215	
	3240	14″ dia., butt end, segmented					240			240	264	
	3400	All butt end fittings require one coupling per joint.										
	3410	Labor for fitting included with couplings.										
	3460	Compression coupling, galvanized, neoprene gasket										
	3470	2-1/8″ dia.	Q-9	44	.364	Ea.	11.20	11		22.20	29.50	
	3480	2-1/2″ dia.		44	.364		11.20	11		22.20	29.50	
	3490	3″ dia.		38	.421		14.70	12.70		27.40	36	
	3500	3-1/2″ dia.		35	.457		16.45	13.80		30.25	39.50	
	3510	4″ dia.		33	.485		17.90	14.65		32.55	42	
	3520	5″ dia.		29	.552		20.50	16.65		37.15	48	
	3530	6″ dia.		26	.615		24	18.60		42.60	54.50	
	3540	8″ dia.		22	.727		43.50	22		65.50	81.50	
	3550	10″ dia.		20	.800		63.50	24		87.50	108	
	3560	12″ dia.		18	.889		79	27		106	129	
	3570	14″ dia.		16	1		112	30		142	170	
	3800	Air gate valves, galvanized										
	3810	2-1/8″ dia.	Q-9	30	.533	Ea.	85	16.10		101.10	119	
	3820	2-1/2″ dia.		28	.571		89.50	17.25		106.75	125	
	3830	3″ dia.		26	.615		104	18.60		122.60	143	
	3840	4″ dia.		23	.696		136	21		157	183	
	3850	6″ dia.		18	.889		188	27		215	248	

11 EQUIPMENT

11600 | Laboratory Equipment

	11620 | Laboratory Equipment		CREW	DAILY OUTPUT	LABOR-HOURS	UNIT	2001 BARE COSTS MAT.	LABOR	EQUIP.	TOTAL	TOTAL INCL O&P	
350	0010	LABORATORY EQUIPMENT										350
	0020	Cabinets, base, door units, metal	2 Carp	18	.889	L.F.	127	26		153	181	
	0300	Drawer units		18	.889	"	275	26		301	345	
	1550	Counter tops, not incl. base cabinets, acidproof, minimum		82	.195	S.F.	20.50	5.70		26.20	31.50	
	1600	Maximum		70	.229		33	6.65		39.65	46.50	
	1650	Stainless steel		82	.195		66	5.70		71.70	81.50	
	2300	Service fixtures, average				Ea.	43			43	47	
	2500	For sink assembly with hot and cold water, add	1 Plum	1.40	5.714		525	197		722	870	
	2550	Glassware washer, undercounter, minimum	L-1	1.80	8.889		6,200	305		6,505	7,275	
	2600	Maximum	"	1	16		7,525	550		8,075	9,100	
	2800	Sink, one piece plastic, flask wash, hose, free standing	1 Plum	1.60	5		1,325	172		1,497	1,700	
	2850	Epoxy resin sink, 25″ x 16″ x 10″	"	2	4		147	138		285	370	

Important: See the Reference Section for critical supporting data - Reference Nos., Crews, & City Cost Indexes

11600 | Laboratory Equipment

	11620 | Laboratory Equipment		CREW	DAILY OUTPUT	LABOR-HOURS	UNIT	2001 BARE COSTS MAT.	LABOR	EQUIP.	TOTAL	TOTAL INCL O&P	
350	4200	Alternate pricing method: as percent of lab furniture										350
	4400	Installation, not incl. plumbing & duct work				% Furn.					22%	
	4800	Plumbing, final connections, simple system									10%	
	5000	Moderately complex system									15%	
	5200	Complex system									20%	
	5400	Electrical, simple system									10%	
	5600	Moderately complex system									20%	
	5800	Complex system									35%	
	6000	Safety equipment, eye wash, hand held				Ea.	210			210	231	
	6200	Deluge shower				″	500			500	550	
	6300	Rule of thumb: lab furniture including installation & connection										
	6320	High school				S.F.					29	
	6340	College									42.75	
	6360	Clinical, health care									36.75	
	6380	Industrial									59.25	

11700 | Medical Equipment

	11710 | Medical Equipment		CREW	DAILY OUTPUT	LABOR-HOURS	UNIT	2001 BARE COSTS MAT.	LABOR	EQUIP.	TOTAL	TOTAL INCL O&P	
500	0010	**MEDICAL EQUIPMENT**										500
	0015	Autopsy table, standard	1 Plum	1	8	Ea.	6,500	276		6,776	7,575	
	0020	Deluxe		.60	13.333		13,000	460		13,460	15,100	
	0700	Distiller, water, steam heated, 50 gal. capacity		1.40	5.714		13,500	197		13,697	15,200	
	1800	Heat therapy unit, humidified, 26″ x 78″ x 28″					2,475			2,475	2,725	
	2100	Hubbard tank with accessories, stainless steel,										
	2110	125 GPM at 45 psi water pressure				Ea.	19,300			19,300	21,200	
	2150	For electric overhead hoist, add					2,100			2,100	2,325	
	3600	Paraffin bath, 126°F, auto controlled					1,125			1,125	1,250	
	4200	Refrigerator, blood bank, 28.6 C.F. emergency signal					5,775			5,775	6,350	
	4210	Reach-in, 16.9 C.F.					6,175			6,175	6,775	
	4600	Station, dietary, medium, with ice					11,900			11,900	13,100	
	5000	Scrub, surgical, stainless steel, single station, minimum	1 Plum	3	2.667		970	92		1,062	1,225	
	5100	Maximum					5,825			5,825	6,400	
	5600	Sterilizers, floor loading, 26″ x 62″ x 42″, single door, steam					118,000			118,000	129,500	
	5650	Double door, steam					151,500			151,500	166,500	
	5800	General purpose, 20″ x 20″ x 38″, single door					11,400			11,400	12,500	
	6000	Portable, counter top, steam, minimum					2,825			2,825	3,100	
	6020	Maximum					4,425			4,425	4,850	
	6050	Portable, counter top, gas, 17″x15″x32-1/2″					29,200			29,200	32,100	
	6150	Manual washer/sterilizer, 16″x16″x26″	1 Plum	2	4		40,000	138		40,138	44,200	
	6200	Steam generators, electric 10 KW to 180 KW, freestanding										
	6250	Minimum	1 Elec	3	2.667	Ea.	5,950	91.50		6,041.50	6,675	
	6300	Maximum	″	.70	11.429		21,400	390		21,790	24,200	
	8100	Utensil washer-sanitizer	1 Plum	2	4		8,100	138		8,238	9,125	
	8200	Bed pan washer-sanitizer	″	2	4		5,300	138		5,438	6,025	
	8400	Whirlpool bath, mobile, sst, 18″ x 24″ x 60″					3,500			3,500	3,850	
	8450	Fixed, incl. mixing valves	1 Plum	2	4		6,850	138		6,988	7,725	

11700 | Medical Equipment

	11740 | Dental Equipment	CREW	DAILY OUTPUT	LABOR-HOURS	UNIT	2001 BARE COSTS MAT.	LABOR	EQUIP.	TOTAL	TOTAL INCL O&P	
200	0010 DENTAL EQUIPMENT										200
	0020 Central suction system, minimum	1 Plum	1.20	6.667	Ea.	2,700	230		2,930	3,325	
	0100 Maximum	"	.90	8.889		20,900	305		21,205	23,500	
	0300 Air compressor, minimum	1 Skwk	.80	10		2,975	299		3,274	3,750	
	0400 Maximum		.50	16		20,600	480		21,080	23,500	
	0600 Chair, electric or hydraulic, minimum		.50	16		3,400	480		3,880	4,500	
	0700 Maximum		.25	32		7,200	955		8,155	9,425	
	2000 Light, ceiling mounted, minimum		8	1		1,775	30		1,805	2,000	
	2100 Maximum	▼	8	1		3,575	30		3,605	3,975	
	2200 Unit light, minimum	2 Skwk	5.33	3.002		1,325	89.50		1,414.50	1,600	
	2210 Maximum		5.33	3.002		4,325	89.50		4,414.50	4,900	
	2220 Track light, minimum		3.20	5		2,350	149		2,499	2,800	
	2230 Maximum	▼	3.20	5		5,150	149		5,299	5,900	
	2300 Sterilizers, steam portable, minimum					2,975			2,975	3,275	
	2350 Maximum					10,300			10,300	11,300	
	2600 Steam, institutional					12,800			12,800	14,100	
	2650 Dry heat, electric, portable, 3 trays				▼	1,000			1,000	1,100	

For information about Means Estimating Seminars, see yellow pages 11 and 12 in back of book

11 EQUIPMENT

Division 13
Special Construction

Estimating Tips

General

- The items and systems in this division are usually estimated, purchased, supplied and installed as a unit by one or more subcontractors. The estimator must ensure that all parties are operating from the same set of specifications and assumptions and that all necessary items are estimated and will be provided. Many times the complex items and systems are covered but the more common ones such as excavation or a crane are overlooked for the very reason that everyone assumes nobody could miss them. The estimator should be the central focus and be able to ensure that all systems are complete.
- Another area where problems can develop in this division is at the interface between systems. The estimator must ensure, for instance, that anchor bolts, nuts and washers are estimated and included for the air-supported structures and pre-engineered buildings to be bolted to their foundations. Utility supply is a common area where essential items or pieces of equipment can be missed or overlooked due to the fact that each subcontractor may feel it is the others' responsibility. The estimator should also be aware of certain items which may be supplied as part of a package but installed by others, and ensure that the installing contractor's estimate includes the cost of installation. Conversely, the estimator must also ensure that items are not costed by two different subcontractors, resulting in an inflated overall estimate.

13120 Pre-Engineered Structures

- The foundations and floor slab, as well as rough mechanical and electrical, should be estimated, as this work is required for the assembly and erection of the structure. Generally, as noted in the book, the pre-engineered building comes as a shell and additional features must be included by the estimator. Here again, the estimator must have a clear understanding of the scope of each portion of the work and all the necessary interfaces.

13200 Storage Tanks

- The prices in this subdivision for above and below ground storage tanks do not include foundations or hold-down slabs. The estimator should refer to Divisions 2 and 3 for foundation system pricing. In addition to the foundations, required tank accessories such as tank gauges, leak detection devices, and additional manholes and piping must be added to the tank prices.

Reference Numbers

Reference numbers are shown in bold squares at the beginning of some major classifications. These numbers refer to related items in the Reference Section. The reference information may be an estimating procedure, an alternate pricing method or technical information.

Note: Not all subdivisions listed here necessarily appear in this publication.

13030 | Special Purpose Rooms

	13035 | Special Purpose Rooms		Crew	Daily Output	Labor-Hours	Unit	2001 Bare Costs Mat.	Labor	Equip.	Total	Total Incl O&P	
800	0010	**SAUNA** Prefabricated, incl. heater & controls, 7′ high, 6′ x 4′, C/C	L-7	2.20	12.727	Ea.	3,200	360		3,560	4,075	800
	0050	6′ x 4′, C/P		2	14		2,525	395		2,920	3,375	
	0400	6′ x 5′, C/C		2	14		3,225	395		3,620	4,125	
	0450	6′ x 5′, C/P		2	14		2,875	395		3,270	3,750	
	0600	6′ x 6′, C/C		1.80	15.556		3,800	435		4,235	4,850	
	0650	6′ x 6′, C/P		1.80	15.556		3,000	435		3,435	3,975	
	0800	6′ x 9′, C/C		1.60	17.500		4,175	490		4,665	5,350	
	0850	6′ x 9′, C/P		1.60	17.500		4,175	490		4,665	5,350	
	1000	8′ x 12′, C/C		1.10	25.455		7,350	715		8,065	9,175	
	1050	8′ x 12′, C/P		1.10	25.455		5,775	715		6,490	7,450	
	1200	8′ x 8′, C/C		1.40	20		4,925	560		5,485	6,300	
	1250	8′ x 8′, C/P		1.40	20		4,350	560		4,910	5,650	
	1400	8′ x 10′, C/C		1.20	23.333		5,400	655		6,055	6,950	
	1450	8′ x 10′, C/P		1.20	23.333		4,725	655		5,380	6,225	
	1600	10′ x 12′, C/C		1	28		7,200	785		7,985	9,125	
	1650	10′ x 12′, C/P		1	28		6,375	785		7,160	8,250	
	2500	Heaters only (incl. above), wall mounted, to 200 C.F.					505			505	555	
	2750	To 300 C.F.					530			530	580	
	3000	Floor standing, to 720 C.F., 10,000 watts, w/controls	1 Elec	3	2.667		925	91.50		1,016.50	1,150	
	3250	To 1,000 C.F., 16,000 watts	"	3	2.667		1,575	91.50		1,666.50	1,850	
940	0010	**STEAM BATH** Heater, timer & head, single, to 140 C.F.	1 Plum	1.20	6.667	Ea.	840	230		1,070	1,275	940
	0500	To 300 C.F.		1.10	7.273		970	251		1,221	1,450	
	1000	Commercial size, with blow-down assembly, to 800 C.F.		.90	8.889		3,800	305		4,105	4,625	
	1500	To 2500 C.F.		.80	10		4,000	345		4,345	4,925	
	2000	Multiple, motels, apts., 2 baths, w/ blow-down assm., 500 C.F.	Q-1	1.30	12.308		3,025	380		3,405	3,900	
	2500	4 baths	"	.70	22.857		3,675	710		4,385	5,100	
	2700	Conversion unit for residential tub, including door					1,775			1,775	1,950	

13120 | Pre-Engineered Structures

	13128 | Pre-Eng. Structures		Crew	Daily Output	Labor-Hours	Unit	2001 Bare Costs Mat.	Labor	Equip.	Total	Total Incl O&P	
200	0010	**COMFORT STATIONS** Prefab., stock, w/doors, windows & fixt.										200
	0100	Not incl. interior finish or electrical										
	0300	Mobile, on steel frame, minimum				S.F.	37.50			37.50	41	
	0350	Maximum					62.50			62.50	69	
	0400	Permanent, including concrete slab, minimum	B-12J	50	.320		138	9.10	12.95	160.05	180	
	0500	Maximum	"	43	.372		200	10.60	15.10	225.70	253	
	0600	Alternate pricing method, mobile, minimum				Fixture	1,725			1,725	1,900	
	0650	Maximum					2,575			2,575	2,825	
	0700	Permanent, minimum	B-12J	.70	22.857		9,900	650	925	11,475	12,900	
	0750	Maximum	"	.50	32		16,500	910	1,300	18,710	21,000	

13150 | Swimming Pools

	13151 | Swimming Pools	CREW	DAILY OUTPUT	LABOR-HOURS	UNIT	2001 BARE COSTS MAT.	LABOR	EQUIP.	TOTAL	TOTAL INCL O&P	
200	0010 **SWIMMING POOLS** Residential in-ground, vinyl lined, concrete sides										200
	0020 Sides including equipment, sand bottom	B-52	300	.187	SF Surf	10.40	5	1.28	16.68	20.50	
	0100 Metal or polystyrene sides	B-14	410	.117	↓	8.70	2.83	.44	11.97	14.45	
	0200 Add for vermiculite bottom				↓	.67			.67	.74	
	0500 Gunite bottom and sides, white plaster finish										
	0600 12′ x 30′ pool	B-52	145	.386	SF Surf	17.25	10.40	2.66	30.31	38	
	0720 16′ x 32′ pool	↓	155	.361	↓	15.50	9.70	2.49	27.69	35	
	0750 20′ x 40′ pool	↓	250	.224	↓	13.85	6.05	1.54	21.44	26.50	
	0810 Concrete bottom and sides, tile finish										
	0820 12′ x 30′ pool	B-52	80	.700	SF Surf	17.40	18.85	4.82	41.07	54	
	0830 16′ x 32′ pool	↓	95	.589	↓	14.40	15.85	4.06	34.31	45.50	
	0840 20′ x 40′ pool	↓	130	.431	↓	11.45	11.60	2.96	26.01	34	
	1100 Motel, gunite with plaster finish, incl. medium										
	1150 capacity filtration & chlorination	B-52	115	.487	SF Surf	21	13.10	3.35	37.45	47.50	
	1200 Municipal, gunite with plaster finish, incl. high										
	1250 capacity filtration & chlorination	B-52	100	.560	SF Surf	27.50	15.05	3.85	46.40	57.50	
	1350 Add for formed gutters				L.F.	47.50			47.50	52.50	
	1360 Add for stainless steel gutters				"	141			141	155	
	1700 Filtration and deck equipment only, as % of total				Total				20%	20%	
	1800 Deck equipment, rule of thumb, 20′ x 40′ pool				SF Pool					1.30	
	1900 5000 S.F. pool				"					1.90	
	3000 Painting pools, preparation + 3 coats, 20′ x 40′ pool, epoxy	2 Pord	.33	48.485	Total	595	1,275		1,870	2,575	
	3100 Rubber base paint, 18 gallons	"	.33	48.485	↓	450	1,275		1,725	2,425	
	3500 42′ x 82′ pool, 75 gallons, epoxy paint	3 Pord	.14	171	↓	2,500	4,475		6,975	9,575	
	3600 Rubber base paint	"	.14	171	↓	1,925	4,475		6,400	8,950	
700	0010 **SWIMMING POOL EQUIPMENT** Diving stand, stainless steel, 3 meter	2 Carp	.40	40	Ea.	4,625	1,175		5,800	6,925	700
	0300 1 meter	"	2.70	5.926	"	3,750	173		3,923	4,400	
	0900 Filter system, sand or diatomite type, incl. pump, 6,000 gal./hr.	2 Plum	1.80	8.889	Total	1,050	305		1,355	1,600	
	1020 Add for chlorination system, 800 S.F. pool	↓	3	5.333	Ea.	196	184		380	495	
	1040 5,000 S.F. pool	↓	3	5.333	"	1,150	184		1,334	1,525	
	1100 Gutter system, stainless steel, with grating, stock,										
	1110 contains supply and drainage system	E-1	20	1.200	L.F.	160	38.50	4.20	202.70	247	
	1120 Integral gutter and 5′ high wall system, stainless steel	"	10	2.400	"	240	77	8.40	325.40	405	
	2100 Lights, underwater, 12 volt, with transformer, 300 watt	1 Elec	.40	20	Ea.	116	685		801	1,150	
	2200 110 volt, 500 watt, standard	↓	.40	20	↓	124	685		809	1,150	
	2400 Low water cutoff type	↓	.40	20	↓	128	685		813	1,175	
	2800 Heaters, see division 15510-880										

13200 | Storage Tanks

	13201 | Storage Tanks	CREW	DAILY OUTPUT	LABOR-HOURS	UNIT	2001 BARE COSTS MAT.	LABOR	EQUIP.	TOTAL	TOTAL INCL O&P	
300	0010 **GROUND TANKS** Not incl. pipe or pumps, prestress conc., 250,000 gal.				Ea.					280,000	300
	3000 Steel, storage, above ground, including cradles, coating,										
	3020 fittings, not including fdn, pumps or piping										
	3040 Single wall, interior, 275 gallon	Q-5	5	3.200	Ea.	236	100		336	410	
	3060 550 gallon	"	2.70	5.926	↓	825	186		1,011	1,175	
	3080 1,000 gallon	Q-7	5	6.400	↓	1,050	213		1,263	1,475	
	3100 1,500 gallon	↓	4.75	6.737	↓	1,975	224		2,199	2,525	
	3120 2,000 gallon	↓	4.60	6.957	↓	2,000	231		2,231	2,550	

	13201 | Storage Tanks		CREW	DAILY OUTPUT	LABOR-HOURS	UNIT	2001 BARE COSTS MAT.	LABOR	EQUIP.	TOTAL	TOTAL INCL O&P
300	3320	Double wall, 500 gallon capacity	Q-5	2.40	6.667	Ea.	2,625	209		2,834	3,200
	3330	2000 gallon capacity	Q-7	4.15	7.711		4,800	256		5,056	5,650
	3340	4000 gallon capacity		3.60	8.889		8,525	295		8,820	9,825
	3350	6000 gallon capacity		2.40	13.333		10,200	445		10,645	11,900
	3360	8000 gallon capacity		2	16		11,600	530		12,130	13,600
	3370	10000 gallon capacity		1.80	17.778		12,800	590		13,390	15,000
	3380	15000 gallon capacity		1.50	21.333		19,100	710		19,810	22,100
	3390	20000 gallon capacity		1.30	24.615		21,700	820		22,520	25,100
	3400	25000 gallon capacity		1.15	27.826		26,400	925		27,325	30,400
	3410	30000 gallon capacity		1	32		28,900	1,075		29,975	33,400
800	0010	UNDERGROUND STORAGE TANKS									
	0210	Fiberglass, underground, single wall, U.L. listed, not including									
	0220	manway or hold-down strap									
	0225	550 gallon capacity	Q-5	2.67	5.993	Ea.	1,450	188		1,638	1,875
	0230	1,000 gallon capacity	"	2.46	6.504		1,825	204		2,029	2,300
	0240	2,000 gallon capacity	Q-7	4.57	7.002		2,375	233		2,608	2,950
	0250	4,000 gallon capacity		3.55	9.014		3,425	300		3,725	4,225
	0260	6,000 gallon capacity		2.67	11.985		3,825	400		4,225	4,800
	0270	8,000 gallon capacity		2.29	13.974		4,800	465		5,265	5,975
	0280	10,000 gallon capacity		2	16		5,200	530		5,730	6,525
	0282	12,000 gallon capacity		1.88	17.021		7,100	565		7,665	8,650
	0284	15,000 gallon capacity		1.68	19.048		7,575	635		8,210	9,300
	0290	20,000 gallon capacity		1.45	22.069		10,500	735		11,235	12,600
	0300	25,000 gallon capacity		1.28	25		17,400	830		18,230	20,400
	0320	30,000 gallon capacity		1.14	28.070		21,700	935		22,635	25,300
	0340	40,000 gallon capacity		.89	35.955		30,500	1,200		31,700	35,300
	0360	48,000 gallon capacity		.81	39.506		50,000	1,325		51,325	57,000
	0500	For manway, fittings and hold-downs, add					20%	15%			
	0600	For manways, add					840			840	925
	1000	For helical heating coil, add	Q-5	2.50	6.400		2,125	201		2,326	2,650
	1020	Fiberglass, underground, double wall, U.L. listed									
	1030	includes manways, not incl. hold-down straps									
	1040	600 gallon capacity	Q-5	2.42	6.612	Ea.	3,200	207		3,407	3,825
	1050	1,000 gallon capacity	"	2.25	7.111		4,250	223		4,473	5,000
	1060	2,500 gallon capacity	Q-7	4.16	7.692		6,050	256		6,306	7,025
	1070	3,000 gallon capacity		3.90	8.205		6,500	273		6,773	7,550
	1080	4,000 gallon capacity		3.64	8.791		7,100	292		7,392	8,250
	1090	6,000 gallon capacity		2.42	13.223		8,225	440		8,665	9,725
	1100	8,000 gallon capacity		2.08	15.385		10,400	510		10,910	12,200
	1110	10,000 gallon capacity		1.82	17.582		10,800	585		11,385	12,800
	1120	12,000 gallon capacity		1.70	18.824		13,200	625		13,825	15,400
	1122	15,000 gallon capacity		1.52	21.053		18,600	700		19,300	21,600
	1124	20,000 gallon capacity		1.33	24.060		22,600	800		23,400	26,100
	1126	25,000 gallon capacity		1.16	27.586		29,600	915		30,515	34,000
	1128	30,000 gallon capacity		1.03	31.068		35,400	1,025		36,425	40,600
	1140	For hold-down straps, add					2%	10%			
	1150	For hold-downs 500-4000 gal, add	Q-7	16	2	Set	162	66.50		228.50	278
	1160	For hold-downs 5000-15000 gal, add		8	4		325	133		458	555
	1170	For hold-downs 20,000 gal, add		5.33	6.004		485	200		685	835
	1180	For hold-downs 25,000 gal, add		4	8		650	266		916	1,125
	1190	For hold-downs 30,000 gal, add		2.60	12.308		975	410		1,385	1,700
	2210	Fiberglass, underground, single wall, U.L. listed, including									
	2220	hold-down straps, no manways									
	2225	550 gallon capacity	Q-5	2	8	Ea.	1,625	251		1,876	2,150
	2230	1,000 gallon capacity	"	1.88	8.511		1,975	267		2,242	2,575
	2240	2,000 gallon capacity	Q-7	3.55	9.014		2,525	300		2,825	3,225

13201 | Storage Tanks

			CREW	DAILY OUTPUT	LABOR-HOURS	UNIT	2001 BARE COSTS MAT.	LABOR	EQUIP.	TOTAL	TOTAL INCL O&P	
800	2250	4,000 gallon capacity	Q-7	2.90	11.034	Ea.	3,600	365		3,965	4,500	800
	2260	6,000 gallon capacity		2	16		4,150	530		4,680	5,350	
	2270	8,000 gallon capacity		1.78	17.978		5,125	600		5,725	6,525	
	2280	10,000 gallon capacity		1.60	20		5,525	665		6,190	7,075	
	2282	12,000 gallon capacity		1.52	21.053		7,425	700		8,125	9,200	
	2284	15,000 gallon capacity		1.39	23.022		7,900	765		8,665	9,850	
	2290	20,000 gallon capacity		1.14	28.070		11,000	935		11,935	13,500	
	2300	25,000 gallon capacity		.96	33.333		18,000	1,100		19,100	21,500	
	2320	30,000 gallon capacity		.80	40		22,700	1,325		24,025	27,000	
	3020	Fiberglass, underground, double wall, U.L. listed										
	3030	includes manways and hold-down straps										
	3040	600 gallon capacity	Q-5	1.86	8.602	Ea.	3,350	270		3,620	4,075	
	3050	1,000 gallon capacity	"	1.70	9.412		4,425	295		4,720	5,300	
	3060	2,500 gallon capacity	Q-7	3.29	9.726		6,225	325		6,550	7,325	
	3070	3,000 gallon capacity		3.13	10.224		6,675	340		7,015	7,875	
	3080	4,000 gallon capacity		2.93	10.921		7,275	365		7,640	8,550	
	3090	6,000 gallon capacity		1.86	17.204		8,550	570		9,120	10,300	
	3100	8,000 gallon capacity		1.65	19.394		10,700	645		11,345	12,800	
	3110	10,000 gallon capacity		1.48	21.622		11,100	720		11,820	13,300	
	3120	12,000 gallon capacity		1.40	22.857		13,500	760		14,260	16,100	
	3122	15,000 gallon capacity		1.28	25		18,900	830		19,730	22,100	
	3124	20,000 gallon capacity		1.06	30.189		23,100	1,000		24,100	26,900	
	3126	25,000 gallon capacity		.90	35.556		30,200	1,175		31,375	35,100	
	3128	30,000 gallon capacity		.74	43.243		36,400	1,425		37,825	42,200	
	5000	Steel underground, sti-P3, set in place, not incl. hold-down bars.										
	5500	Excavation, pad, pumps and piping not included										
	5510	Single wall, 500 gallon capacity, 7 gauge shell	Q-5	2.70	5.926	Ea.	820	186		1,006	1,175	
	5520	1,000 gallon capacity, 7 gauge shell	"	2.50	6.400		1,425	201		1,626	1,875	
	5530	2,000 gallon capacity, 1/4" thick shell	Q-7	4.60	6.957		2,300	231		2,531	2,875	
	5535	2,500 gallon capacity, 7 gauge shell	Q-5	3	5.333		2,500	167		2,667	3,000	
	5540	5,000 gallon capacity, 1/4" thick shell	Q-7	3.20	10		4,675	330		5,005	5,650	
	5560	10,000 gallon capacity, 1/4" thick shell		2	16		7,550	530		8,080	9,100	
	5580	15,000 gallon capacity, 5/16" thick shell		1.70	18.824		10,200	625		10,825	12,200	
	5600	20,000 gallon capacity, 5/16" thick shell		1.50	21.333		14,700	710		15,410	17,200	
	5610	25,000 gallon capacity, 3/8" thick shell		1.30	24.615		17,600	820		18,420	20,600	
	5620	30,000 gallon capacity, 3/8" thick shell		1.10	29.091		21,500	965		22,465	25,100	
	5630	40,000 gallon capacity, 3/8" thick shell		.90	35.556		30,700	1,175		31,875	35,500	
	5640	50,000 gallon capacity, 3/8" thick shell		.80	40		38,300	1,325		39,625	44,200	
	6200	Steel, underground, 360°, double wall, U.L. listed,										
	6210	with sti-P3 corrosion protection,										
	6220	(dielectric coating, cathodic protection, electrical										
	6230	isolation) 30 year warranty,										
	6240	not incl. manholes or hold-downs.										
	6250	500 gallon capacity	Q-5	2.40	6.667	Ea.	2,300	209		2,509	2,850	
	6260	1,000 gallon capactiy	"	2.25	7.111		3,350	223		3,573	4,000	
	6270	2,000 gallon capacity	Q-7	4.16	7.692		4,400	256		4,656	5,200	
	6280	3,000 gallon capacity		3.90	8.205		5,250	273		5,523	6,175	
	6290	4,000 gallon capacity		3.64	8.791		6,425	292		6,717	7,500	
	6300	5,000 gallon capacity		2.91	10.997		8,100	365		8,465	9,450	
	6310	6,000 gallon capacity		2.42	13.223		9,600	440		10,040	11,300	
	6320	8,000 gallon capacity		2.08	15.385		10,000	510		10,510	11,800	
	6330	10,000 gallon capacity		1.82	17.582		11,300	585		11,885	13,300	
	6340	12,000 gallon capacity		1.70	18.824		13,600	625		14,225	15,900	
	6350	15,000 gallon capacity		1.33	24.060		16,300	800		17,100	19,100	
	6360	20,000 gallon capacity		1.33	24.060		20,000	800		20,800	23,100	
	6370	25,000 gallon capacity		1.16	27.586		30,400	915		31,315	34,800	

13200 | Storage Tanks

	13201 | Storage Tanks		CREW	DAILY OUTPUT	LABOR-HOURS	UNIT	2001 BARE COSTS MAT.	LABOR	EQUIP.	TOTAL	TOTAL INCL O&P	
800	6380	30,000 gallon capacity	Q-7	1.03	31.068	Ea.	36,600	1,025		37,625	41,900	800
	6390	40,000 gallon capacity		.80	40		47,000	1,325		48,325	53,500	
	6395	50,000 gallon capacity		.73	43.836		57,000	1,450		58,450	64,500	
	6400	For hold-downs 500-2000 gal, add		16	2	Set	268	66.50		334.50	395	
	6410	For hold-downs 3000-6000 gal, add		12	2.667		350	88.50		438.50	520	
	6420	For hold-downs 8000-12,000 gal, add		11	2.909		850	96.50		946.50	1,075	
	6430	For hold-downs 15,000 gal, add		9	3.556		1,075	118		1,193	1,350	
	6440	For hold-downs 20,000 gal, add		8	4		1,175	133		1,308	1,500	
	6450	For hold-downs 20,000 gal plus, add		6	5.333		2,525	177		2,702	3,050	
	6500	For manways, add				Ea.	555			555	610	
	6600	In place with hold-downs										
	6652	550 gallon capacity	Q-5	1.84	8.696	Ea.	2,550	273		2,823	3,225	

13280 | Hazardous Material Remediation

	13281 | Hazardous Material Remediation		CREW	DAILY OUTPUT	LABOR-HOURS	UNIT	2001 BARE COSTS MAT.	LABOR	EQUIP.	TOTAL	TOTAL INCL O&P	
120	0010	**BULK ASBESTOS REMOVAL**										120
	0020	Includes disposable tools and 1 suit and respirator/day/worker										
	0200	Boiler insulation	A-9	480	.133	S.F.	.06	4.29		4.35	6.95	
	0210	With metal lath add				%				50%		
	0300	Boiler breeching or flue insulation	A-9	520	.123	S.F.		3.96		3.96	6.35	
	0310	For active boiler, add				%				100%		
	0400	Duct or AHU insulation	A-10B	440	.073	S.F.		2.35		2.35	3.76	
	0500	Duct vibration isolation joints, up to 24 Sq. In. duct	A-9	56	1.143	Ea.		37		37	59	
	0520	25 Sq. In. to 48 Sq. In. duct		48	1.333			43		43	69	
	0530	49 Sq. In. to 76 Sq. In. duct		40	1.600			51.50		51.50	82.50	
	0600	Pipe insulation, air cell type, up to 4″ diameter pipe		900	.071	L.F.		2.29		2.29	3.67	
	0610	4″ to 8″ diameter pipe		800	.080			2.58		2.58	4.13	
	0620	10″ to 12″ diameter pipe		700	.091			2.95		2.95	4.72	
	0630	14″ to 16″ diameter pipe		550	.116			3.75		3.75	6	
	0650	Over 16″ diameter pipe		650	.098	S.F.		3.17		3.17	5.10	
	0700	With glove bag up to 3″ diameter pipe		100	.640	L.F.	3.15	20.50		23.65	36.50	
	1000	Pipe fitting insulation up to 4″ diameter pipe		320	.200	Ea.		6.45		6.45	10.30	
	1100	6″ to 8″ diameter pipe		304	.211			6.80		6.80	10.85	
	1110	10″ to 12″ diameter pipe		192	.333			10.75		10.75	17.20	
	1120	14″ to 16″ diameter pipe		128	.500			16.10		16.10	26	
	1130	Over 16″ diameter pipe		176	.364	S.F.		11.70		11.70	18.75	
	1200	With glove bag, up to 8″ diameter pipe		40	1.600	Ea.	6.55	51.50		58.05	89.50	
	2000	Scrape foam fireproofing from flat surface		2,400	.027	S.F.		.86		.86	1.38	
	2100	Irregular surfaces		1,200	.053			1.72		1.72	2.75	
	3000	Remove cementitious material from flat surface		1,800	.036			1.15		1.15	1.83	
	3100	Irregular surface		1,400	.046			1.47		1.47	2.36	
	6000	Remove contaminated soil from crawl space by hand		400	.160	C.F.		5.15		5.15	8.25	
	6100	With large production vacuum loader	A-12	700	.091	″		2.95	.74	3.69	5.55	
	7000	Radiator backing, not including radiator removal	A-9	1,200	.053	S.F.		1.72		1.72	2.75	
	9000	For type C (supplied air) respirator equipment, add				%					10%	
125	0010	**ASBESTOS ABATEMENT WORK AREA** Containment and preparation.										125
	0100	Pre-cleaning, HEPA vacuum and wet wipe, flat surfaces	A-10	12,000	.005	S.F.		.17		.17	.28	
	0200	Protect carpeted area, 2 layers 6 mil poly on 3/4″ plywood	″	1,000	.064		1.50	2.06		3.56	4.95	
	0300	Separation barrier, 2″ x 4″ @ 16″, 1/2″ plywood ea. side, 8′ high	2 Carp	400	.040		1.25	1.17		2.42	3.20	

13280 | Hazardous Material Remediation

	13281 | Hazardous Material Remediation		CREW	DAILY OUTPUT	LABOR-HOURS	UNIT	2001 BARE COSTS MAT.	LABOR	EQUIP.	TOTAL	TOTAL INCL O&P	
125	0310	12′ high	2 Carp	320	.050	S.F.	1.40	1.46		2.86	3.82	125
	0320	16′ high		200	.080		1.50	2.33		3.83	5.30	
	0400	Personnel decontam. chamber, 2″ x 4″ @ 16″, 3/4″ ply ea. side		280	.057		2.50	1.67		4.17	5.35	
	0450	Waste decontam. chamber, 2″ x 4″ studs @ 16″, 3/4″ ply ea. side		360	.044		3	1.30		4.30	5.35	
	0500	Cover surfaces with polyethelene sheeting										
	0501	Including glue and tape										
	0550	Floors, each layer, 6 mil	A-10	8,000	.008	S.F.	.10	.26		.36	.52	
	0551	4 mil		9,000	.007		.06	.23		.29	.44	
	0560	Walls, each layer, 6 mil		6,000	.011		.09	.34		.43	.65	
	0561	4 mil		7,000	.009		.07	.29		.36	.55	
	0570	For heights above 12′, add						20%				
	0575	For heights above 20′, add						30%				
	0580	For fire retardant poly, add					100%					
	0590	For large open areas, deduct					10%	20%				
	0600	Seal floor penetrations with foam firestop to 36 Sq. In.	2 Carp	200	.080	Ea.	6.25	2.33		8.58	10.55	
	0610	36 Sq. In. to 72 Sq. In.		125	.128		12.50	3.73		16.23	19.60	
	0615	72 Sq. In. to 144 Sq. In.		80	.200		25	5.85		30.85	36.50	
	0620	Wall penetrations, to 36 square inches		180	.089		6.25	2.59		8.84	10.95	
	0630	36 Sq. In. to 72 Sq. In.		100	.160		12.50	4.66		17.16	21	
	0640	72 Sq. In. to 144 Sq. In.		60	.267		25	7.75		32.75	39.50	
	0800	Caulk seams with latex	1 Carp	230	.035	L.F.	.15	1.01		1.16	1.76	
	0900	Set up neg. air machine, 1-2k C.F.M. /25 M.C.F. volume	1 Asbe	4.30	1.860	Ea.		60		60	96	
130	0010	**DEMOLITION IN ASBESTOS CONTAMINATED AREA**										130
	0200	Ceiling, including suspension system, plaster and lath	A-9	2,100	.030	S.F.		.98		.98	1.57	
	0210	Finished plaster, leaving wire lath		585	.109			3.52		3.52	5.65	
	0220	Suspended acoustical tile		3,500	.018			.59		.59	.94	
	0230	Concealed tile grid system		3,000	.021			.69		.69	1.10	
	0240	Metal pan grid system		1,500	.043			1.37		1.37	2.20	
	0250	Gypsum board		2,500	.026			.82		.82	1.32	
	0260	Lighting fixtures up to 2′ x 4′		72	.889	Ea.		28.50		28.50	46	
	0400	Partitions, non load bearing										
	0410	Plaster, lath, and studs	A-9	690	.093	S.F.	.55	2.99		3.54	5.40	
	0450	Gypsum board and studs	″	1,390	.046	″		1.48		1.48	2.38	
	9000	For type C (supplied air) respirator equipment, add				%					10%	
135	0010	**ASBESTOS ABATEMENT EQUIPMENT** and supplies, buy (R13281-120)										135
	0200	Air filtration device, 2000 C.F.M.				Ea.	2,500			2,500	2,750	
	0250	Large volume air sampling pump, minimum					450			450	495	
	0260	Maximum					1,000			1,000	1,100	
	0300	Airless sprayer unit, 2 gun					2,000			2,000	2,200	
	0350	Light stand, 500 watt					250			250	275	
	0400	Personal respirators										
	0410	Negative pressure, 1/2 face, dual operation, min.				Ea.	22.50			22.50	25	
	0420	Maximum					24			24	26.50	
	0450	P.A.P.R., full face, minimum					400			400	440	
	0460	Maximum					700			700	770	
	0470	Supplied air, full face, incl. air line, minimum					450			450	495	
	0480	Maximum					600			600	660	
	0500	Personnel sampling pump, minimum					450			450	495	
	0510	Maximum					750			750	825	
	1500	Power panel, 20 unit, incl. G.F.I.					1,800			1,800	1,975	
	1600	Shower unit, including pump and filters					1,125			1,125	1,250	
	1700	Supplied air system (type C)					10,000			10,000	11,000	
	1750	Vacuum cleaner, HEPA, 16 gal., stainless steel, wet/dry					1,000			1,000	1,100	
	1760	55 gallon					2,200			2,200	2,425	
	1800	Vacuum loader, 9-18 ton/hr					90,000			90,000	99,000	
	1900	Water atomizer unit, including 55 gal. drum					230			230	253	

	13281	Hazardous Material Remediation	CREW	DAILY OUTPUT	LABOR-HOURS	UNIT	2001 BARE COSTS MAT.	LABOR	EQUIP.	TOTAL	TOTAL INCL O&P	
135	2000	Worker prot., whole body, foot, head cover & gloves, plastic R13281-120				Ea.	30			30	33	135
	2500	Respirator, single use					10.50			10.50	11.55	
	2550	Cartridge for respirator					37.50			37.50	41.50	
	2570	Glove bag, 7 mil, 50″ x 64″					8.50			8.50	9.35	
	2580	10 mil, 44″ x 60″					8.40			8.40	9.25	
	3000	HEPA vacuum for work area, minimun					1,000			1,000	1,100	
	6000	Disposable polyethelene bags, 6 mil, 3 C.F.					1.15			1.15	1.27	
	6300	Disposable fiber drums, 3 C.F.					6.50			6.50	7.15	
	6400	Pressure sensitive caution lables, 3″ x 5″					8.90			8.90	9.80	
	6450	11″ x 17″					6.50			6.50	7.15	
	6500	Negative air machine, 1800 C.F.M.					775			775	855	
140	0010	**DECONTAMINATION CONTAINMENT AREA DEMOLITION** and clean-up										140
	0100	Spray exposed substrate with surfactant (bridging)										
	0200	Flat surfaces	A-9	6,000	.011	S.F.	.35	.34		.69	.94	
	0250	Irregular surfaces		4,000	.016	″	.30	.52		.82	1.16	
	0300	Pipes, beams, and columns		2,000	.032	L.F.	.55	1.03		1.58	2.26	
	1000	Spray encapsulate polyethelene sheeting		8,000	.008	S.F.	.30	.26		.56	.74	
	1100	Roll down polyethelene sheeting		8,000	.008	″		.26		.26	.41	
	1500	Bag polyethelene sheeting		400	.160	Ea.	.75	5.15		5.90	9.10	
	2000	Fine clean exposed substrate, with nylon brush		2,400	.027	S.F.		.86		.86	1.38	
	2500	Wet wipe substrate		4,800	.013			.43		.43	.69	
	2600	Vacuum surfaces, fine brush		6,400	.010			.32		.32	.52	
	3000	Structural demolition										
	3100	Wood stud walls	A-9	2,800	.023	S.F.		.74		.74	1.18	
	3500	Window manifolds, not incl. window replacement		4,200	.015			.49		.49	.79	
	3600	Plywood carpet protection		2,000	.032			1.03		1.03	1.65	
	4000	Remove custom decontamination facility	A-10A	8	3	Ea.	15	97		112	172	
	4100	Remove portable decontamination facility	3 Asbe	12	2	″	12.50	64.50		77	117	
	5000	HEPA vacuum, shampoo carpeting	A-9	4,800	.013	S.F.	.05	.43		.48	.75	
	9000	Final cleaning of protected surfaces	A-10A	8,000	.003	″		.10		.10	.16	
145	0010	**OSHA TESTING**										145
	0100	Certified technician, minimum				Day					300	
	0110	Maximum				″					500	
	0200	Personal sampling, PCM analysis, minimum	1 Asbe	8	1	Ea.	2.75	32		34.75	54.50	
	0210	Maximum	″	4	2	″	3	64.50		67.50	106	
	0300	Industrial hygenist, minimum				Day					400	
	0310	Maximum				″					550	
	1000	Cleaned area samples	1 Asbe	8	1	Ea.	2.63	32		34.63	54.50	
	1100	PCM air sample analysis, minimum		8	1		30	32		62	84.50	
	1110	Maximum		4	2		3.09	64.50		67.59	106	
	1200	TEM air sample analysis, minimum								100	125	
	1210	Maximum								400	500	
150	0010	**ENCAPSULATION WITH SEALANTS**										150
	0100	Ceilings and walls, minimum	A-9	21,000	.003	S.F.	.25	.10		.35	.44	
	0110	Maximum		10,600	.006	″	.40	.19		.59	.75	
	0300	Pipes to 12″ diameter including minor repairs, minimum		800	.080	L.F.	.35	2.58		2.93	4.52	
	0310	Maximum		400	.160	″	1	5.15		6.15	9.35	
155	0010	**WASTE PACKAGING, HANDLING, & DISPOSAL**										155
	0100	Collect and bag bulk material, 3 C.F. bags, by hand	A-9	400	.160	Ea.	1.15	5.15		6.30	9.50	
	0200	Large production vacuum loader	A-12	880	.073		.80	2.34	.59	3.73	5.30	
	1000	Double bag and decontaminate	A-9	960	.067		2.30	2.15		4.45	5.95	
	2000	Containerize bagged material in drums, per 3 C.F. drum	″	800	.080		6.50	2.58		9.08	11.30	
	3000	Cart bags 50′ to dumpster	2 Asbe	400	.040			1.29		1.29	2.06	

13280 | Hazardous Material Remediation

	13281 | Hazardous Material Remediation	CREW	DAILY OUTPUT	LABOR-HOURS	UNIT	2001 BARE COSTS MAT.	LABOR	EQUIP.	TOTAL	TOTAL INCL O&P	
155	5000 Disposal charges, not including haul, minimum				C.Y.					50	155
	5020 Maximum				″					175	
	5100 Remove refrigerant from system	1 Plum	40	.200	Lb.		6.90		6.90	10.40	
	9000 For type C (supplied air) respirator equipment, add				%					10%	

13600 | Solar and Wind Energy Equipment

	13630 | Solar Collector Components		CREW	DAILY OUTPUT	LABOR-HOURS	UNIT	2001 BARE COSTS MAT.	LABOR	EQUIP.	TOTAL	TOTAL INCL O&P	
200	0010 SOLAR ENERGY	A8.3 -610										200
	0020 System/Package prices, not including connecting											
	0030 pipe, insulation, or special heating/plumbing fixtures	A8.3 -620										
	0150 For solar ultraviolet pipe insulation see Div. 15086											
	0500 Hot water, standard package, low temperature	A8.3 -630										
	0540 1 collector, circulator, fittings, 65 gal. tank		Q-1	.50	32	Ea.	1,350	990		2,340	3,000	
	0580 2 collectors, circulator, fittings, 120 gal. tank	A8.3 -640		.40	40		1,925	1,250		3,175	4,000	
	0620 3 collectors, circulator, fittings, 120 gal. tank			.34	47.059		2,350	1,450		3,800	4,775	
	0700 Medium temperature package	A8.3 -650										
	0720 1 collector, circulator, fittings, 80 gal. tank		Q-1	.50	32	Ea.	1,450	990		2,440	3,100	
	0740 2 collectors, circulator, fittings, 120 gal. tank	A8.3 -660		.40	40		2,075	1,250		3,325	4,175	
	0780 3 collectors, circulator, fittings, 120 gal. tank			.30	53.333		3,775	1,650		5,425	6,650	
	0980 For each additional 120 gal. tank, add	A8.3 -710					660			660	730	
	2250 Controller, liquid temperature		1 Plum	5	1.600		92	55		147	184	
	2300 Circulators, air	A8.3 -720										
	2310 Blowers											
	2330 100-300 S.F. system, 1/10 HP	A8.3 -740	Q-9	16	1	Ea.	128	30		158	188	
	2340 300-500 S.F. system, 1/5 HP			15	1.067		177	32		209	244	
	2350 Two speed, 100-300 S.F., 1/10 HP	R13600 -610		14	1.143		128	34.50		162.50	195	
	2400 Reversible fan, 20″ diameter, 2 speed			18	.889		102	27		129	154	
	2520 Space & DHW system, less duct work			.50	32		1,400	965		2,365	3,050	
	2550 Booster fan 6″ diameter, 120 CFM			16	1		33	30		63	83	
	2570 6″ diameter, 225 CFM			16	1		40.50	30		70.50	91	
	2580 8″ diameter, 150 CFM			16	1		37	30		67	87.50	
	2590 8″ diameter, 310 CFM			14	1.143		57	34.50		91.50	117	
	2600 8″ diameter, 425 CFM			14	1.143		64	34.50		98.50	124	
	2650 Rheostat			32	.500		13.95	15.10		29.05	39	
	2660 Shutter/damper			12	1.333		45	40.50		85.50	112	
	2670 Shutter motor			16	1		97	30		127	154	
	2800 Circulators, liquid, 1/25 HP, 5.3 GPM		Q-1	14	1.143		139	35.50		174.50	207	
	2820 1/20 HP, 17 GPM			12	1.333		146	41.50		187.50	224	
	2850 1/20 HP, 17 GPM, stainless steel			12	1.333		230	41.50		271.50	315	
	2870 1/12 HP, 30 GPM			10	1.600		315	49.50		364.50	420	
	3000 Collector panels, air with aluminum absorber plate											
	3010 Wall or roof mount											
	3040 Flat black, plastic glazing											
	3080 4′ x 8′		Q-9	6	2.667	Ea.	600	80.50		680.50	785	
	3100 4′ x 10′			5	3.200	″	740	96.50		836.50	965	
	3200 Flush roof mount, 10′ to 16′ x 22″ wide			96	.167	L.F.	146	5.05		151.05	169	
	3210 Manifold, by L.F. width of collectors			160	.100	″	27	3.02		30.02	34	
	3300 Collector panels, liquid with copper absorber plate											
	3320 Black chrome, tempered glass glazing											

13630 | Solar Collector Components

				CREW	DAILY OUTPUT	LABOR-HOURS	UNIT	2001 BARE COSTS MAT.	LABOR	EQUIP.	TOTAL	TOTAL INCL O&P	
200	3330	Alum. frame, 4′ x 8′, 5/32″ single glazing	A8.3 -610	Q-1	9.50	1.684	Ea.	545	52		597	680	200
	3390	Alum. frame, 4′ x 10′, 5/32″ single glazing		"	6	2.667	"	650	82.50		732.50	835	
	3440	Flat black	A8.3 -620										
	3450	Alum. frame, 3′ x 8′		Q-1	9	1.778	Ea.	430	55		485	560	
	3500	Alum. frame, 4′ x 8.5′	A8.3 -630		5.50	2.909		490	90		580	675	
	3520	Alum. frame, 4′ x 10.5′			10	1.600		595	49.50		644.50	730	
	3540	Alum. frame, 4′ x 12.5′	A8.3 -640	↓	5	3.200	↓	735	99		834	960	
	3550	Liquid with fin tube absorber plate											
	3560	Alum. frame 4′ x 8′ tempered glass	A8.3 -650	Q-1	10	1.600	Ea.	495	49.50		544.50	620	
	3580	Liquid with vacuum tubes, 4′ x 6′-10″			9	1.778		845	55		900	1,025	
	3600	Liquid, full wetted, plastic, alum. frame, 3′ x 10′	A8.3 -660		5	3.200		195	99		294	365	
	3650	Collector panel mounting, flat roof or ground rack			7	2.286	↓	64.50	71		135.50	178	
	3670	Roof clamps	A8.3 -710	↓	70	.229	Set	1.61	7.10		8.71	12.45	
	3700	Roof strap, teflon		1 Plum	205	.039	L.F.	9.50	1.34		10.84	12.50	
	3900	Differential controller with two sensors	A8.3 -720										
	3930	Thermostat, hard wired		1 Plum	8	1	Ea.	92	34.50		126.50	153	
	3950	Line cord and receptacle	A8.3 -740		12	.667		77.50	23		100.50	120	
	4050	Pool valve system			2.50	3.200		199	110		309	385	
	4070	With 12 VAC actuator	R13600 -610		2	4		256	138		394	490	
	4080	Pool pump system, 2″ pipe size			6	1.333		160	46		206	246	
	4100	Five station with digital read-out		↓	3	2.667	↓	202	92		294	360	
	4150	Sensors											
	4200	Brass plug, 1/2″ MPT		1 Plum	32	.250	Ea.	16.60	8.60		25.20	31.50	
	4210	Brass plug, reversed			32	.250		22.50	8.60		31.10	38	
	4220	Freeze prevention			32	.250		20	8.60		28.60	35	
	4240	Screw attached			32	.250		10.15	8.60		18.75	24	
	4250	Brass, immersion		↓	32	.250	↓	24.50	8.60		33.10	40	
	4300	Heat exchanger											
	4330	Fluid to air coil, up flow, 45 MBH		Q-1	4	4	Ea.	279	124		403	490	
	4380	70 MBH			3.50	4.571		315	142		457	560	
	4400	80 MBH		↓	3	5.333	↓	420	165		585	710	
	4580	Fluid to fluid package includes two circulating pumps											
	4590	expansion tank, check valve, relief valve											
	4600	controller, high temperature cutoff and sensors		Q-1	2.50	6.400	Ea.	695	198		893	1,075	
	4650	Heat transfer fluid											
	4700	Propylene glycol, inhibited anti-freeze		1 Plum	28	.286	Gal.	8.80	9.85		18.65	24.50	
	4800	Solar storage tanks, knocked down											
	4810	Air, galvanized steel clad, double wall, 4″ fiberglass											
	4820	insulation, 20 Mil PVC lining											
	4870	4′ high, 4′ x 4′, = 64 C.F./450 gallons		Q-9	2	8	Ea.	1,725	242		1,967	2,275	
	4880	4′ x 8′ = 128 C.F./900 gallons			1.50	10.667		2,250	320		2,570	2,975	
	4890	4′ x 12′ = 190 C.F./1300 gallons			1.30	12.308		3,000	370		3,370	3,875	
	4900	8′ x 8′ = 250 C.F./1700 gallons		↓	1	16		3,350	485		3,835	4,425	
	5010	6′-3″ high, 7′ x 7′ = 306 C.F./2000 gallons		Q-10	1.20	20		8,150	625		8,775	9,925	
	5020	7′ x 10′-6″ = 459 C.F./3000 gallons			.80	30		10,300	940		11,240	12,800	
	5030	7′ x 14′ = 613 C.F./4000 gallons			.60	40		12,000	1,250		13,250	15,100	
	5040	10′-6″ x 10′-6″ = 689 C.F./4500 gallons			.50	48		12,400	1,500		13,900	15,900	
	5050	10′-6″ x 14′ = 919 C.F./6000 gallons		↓	.40	60		14,600	1,875		16,475	18,900	
	5060	14′ x 14′ = 1225 C.F./8000 gallons		Q-11	.40	80		17,100	2,550		19,650	22,800	
	5070	14′ x 17′-6″ = 1531 C.F./10,000 gallons			.30	106		19,200	3,425		22,625	26,400	
	5080	17′-6″ x 17′-6″ = 1914 C.F./12,500 gallons			.25	128		22,600	4,100		26,700	31,200	
	5090	17′-6″ x 21′ = 2297 C.F./15,000 gallons			.20	160		25,200	5,125		30,325	35,700	
	5100	21′ x 21′ = 2756 C.F./18,000 gallons		↓	.18	177	↓	28,900	5,700		34,600	40,600	
	5120	30 Mil reinforced Chemflex lining,											
	5140	4′ high, 4′ x 4′ = 64 C.F./450 gallons		Q-9	2	8	Ea.	1,750	242		1,992	2,300	
	5150	4′ x 8′ = 128 C.F./900 gallons	↓	↓	1.50	10.667	↓	2,650	320		2,970	3,425	

13 SPECIAL CONSTRUCTION

	13630 | Solar Collector Components			CREW	DAILY OUTPUT	LABOR-HOURS	UNIT	2001 BARE COSTS MAT.	LABOR	EQUIP.	TOTAL	TOTAL INCL O&P
200	5160	4′ x 12′ = 190 C.F./1300 gallons	A8.3 -610	Q-9	1.30	12.308	Ea.	3,525	370		3,895	4,450
	5170	8′ x 8′ = 250 C.F./1700 gallons			1	16		3,925	485		4,410	5,075
	5190	6′-3″ high, 7′ x 7′ = 306 C.F./2000 gallons	A8.3 -620	Q-10	1.20	20		8,700	625		9,325	10,500
	5200	7′ x 10′-6″ = 459 C.F./3000 gallons			.80	30		10,600	940		11,540	13,200
	5210	7′ x 14′ = 613 C.F./4000 gallons	A8.3 -630		.60	40		12,300	1,250		13,550	15,500
	5220	10′-6″ x 10′-6″ = 689 C.F./4500 gallons			.50	48		12,700	1,500		14,200	16,300
	5230	10′-6″ x 14′ = 919 C.F./6000 gallons	A8.3 -640		.40	60		15,000	1,875		16,875	19,400
	5240	14′ x 14′ = 1225 C.F./8000 gallons		Q-11	.40	80		17,500	2,550		20,050	23,300
	5250	14′ x 17′-6″ = 1531 C.F./10,000 gallons	A8.3 -650		.30	106		19,600	3,425		23,025	26,900
	5260	17′-6″ x 17′-6″ = 1914 C.F./12,500 gallons			.25	128		23,100	4,100		27,200	31,700
	5270	17′-6″ x 21′ = 2297 C.F./15,000 gallons	A8.3 -660		.20	160		26,000	5,125		31,125	36,500
	5280	21′ x 21′ = 2756 C.F./18,000 gallons			.18	177		29,800	5,700		35,500	41,600
	5290	30 Mil reinforced Hypalon lining, add	A8.3 -710					.02%				
	7000	Solar control valves and vents										
	7050	Air purger, 1″ pipe size	A8.3 -720	1 Plum	12	.667	Ea.	41	23		64	79.50
	7070	Air eliminator, automatic 3/4″ size			32	.250		27	8.60		35.60	42.50
	7090	Air vent, automatic, 1/8″ fitting	A8.3 -740		32	.250		8.80	8.60		17.40	22.50
	7100	Manual, 1/8″ NPT			32	.250		1.94	8.60		10.54	15.15
	7120	Backflow preventer, 1/2″ pipe size	R13600 -610		16	.500		58.50	17.25		75.75	90
	7130	3/4″ pipe size			16	.500		59	17.25		76.25	91
	7150	Balancing valve, 3/4″ pipe size			20	.400		22.50	13.80		36.30	45.50
	7180	Draindown valve, 1/2″ copper tube			9	.889		187	30.50		217.50	252
	7200	Flow control valve, 1/2″ pipe size			22	.364		49.50	12.55		62.05	73.50
	7220	Expansion tank, up to 5 gal.			32	.250		58	8.60		66.60	77
	7250	Hydronic controller (aquastat)			8	1		41.50	34.50		76	97.50
	7400	Pressure gauge, 2″ dial			32	.250		21	8.60		29.60	36
	7450	Relief valve, temp. and pressure 3/4″ pipe size			30	.267		7.90	9.20		17.10	22.50
	7500	Solenoid valve, normally closed										
	7520	Brass, 3/4″ NPT, 24V		1 Plum	9	.889	Ea.	212	30.50		242.50	279
	7530	1″ NPT, 24V			9	.889		212	30.50		242.50	279
	7750	Vacuum relief valve, 3/4″ pipe size			32	.250		25	8.60		33.60	40.50
	7800	Thermometers										
	7820	Digital temperature monitoring, 4 locations		1 Plum	2.50	3.200	Ea.	135	110		245	315
	7900	Upright, 1/2″ NPT			8	1		25.50	34.50		60	80
	7970	Remote probe, 2″ dial			8	1		29	34.50		63.50	84
	7990	Stem, 2″ dial, 9″ stem			16	.500		19	17.25		36.25	47
	8250	Water storage tank with heat exchanger and electric element										
	8270	66 gal. with 2″ x 2 lb. density insulation		1 Plum	1.60	5	Ea.	650	172		822	975
	8300	80 gal. with 2″ x 2 lb. density insulation			1.60	5		745	172		917	1,075
	8380	120 gal. with 2″ x 2 lb. density insulation			1.40	5.714		840	197		1,037	1,225
	8400	120 gal. with 2″ x 2 lb. density insul., 40 S.F. heat coil			1.40	5.714		965	197		1,162	1,375
	8500	Water storage module, plastic										
	8600	Tubular, 12″ diameter, 4′ high		1 Carp	48	.167	Ea.	76	4.86		80.86	91
	8610	12″ diameter, 8′ high			40	.200		116	5.85		121.85	137
	8620	18″ diameter, 5′ high			38	.211		127	6.15		133.15	150
	8630	18″ diameter, 10′ high			32	.250		167	7.30		174.30	195
	8640	58″ diameter, 5′ high		2 Carp	32	.500		460	14.60		474.60	535
	8650	Cap, 12″ diameter						12			12	13.20
	8660	18″ diameter						15			15	16.50

13 SPECIAL CONSTRUCTION

13 SPECIAL CONSTRUCTION

	13838 | Pneumatic/Electric Controls		Crew	Daily Output	Labor-Hours	Unit	2001 Bare Costs Mat.	Labor	Equip.	Total	Total Incl O&P	
200	0010	CONTROL COMPONENTS										200
	2000	Gauges, pressure or vacuum										
	2100	2″ diameter dial	1 Stpi	32	.250	Ea.	14.20	8.70		22.90	29	
	2200	2-1/2″ diameter dial		32	.250		16.55	8.70		25.25	31.50	
	2300	3-1/2″ diameter dial		32	.250		20.50	8.70		29.20	35.50	
	2400	4-1/2″ diameter dial		32	.250		31	8.70		39.70	47	
	2700	Flanged iron case, black ring										
	2800	3-1/2″ diameter dial	1 Stpi	32	.250	Ea.	65	8.70		73.70	84.50	
	2900	4-1/2″ diameter dial		32	.250		88.50	8.70		97.20	111	
	3000	6″ diameter dial		32	.250		70.50	8.70		79.20	90.50	
	3010	Steel case, 0 - 300 psi										
	3012	2″ diam. dial	1 Stpi	16	.500	Ea.	20.50	17.45		37.95	49	
	3014	4″ diam. dial	″	16	.500	″	34	17.45		51.45	64	
	3020	Aluminum case, 0 - 300 psi										
	3022	3-1/2″ diam. dial	1 Stpi	16	.500	Ea.	47	17.45		64.45	78	
	3024	4-1/2″ diam. dial		16	.500		69.50	17.45		86.95	103	
	3026	6″ diam. dial		16	.500		116	17.45		133.45	154	
	3028	8-1/2″ diam. dial		16	.500		274	17.45		291.45	325	
	3030	Brass case, 0 - 300 psi										
	3032	2″ diam. dial	1 Stpi	16	.500	Ea.	23	17.45		40.45	51.50	
	3034	4-1/2″ diam. dial	″	16	.500	″	52.50	17.45		69.95	84	
	3040	Steel case, high pressure, 0 -10,000 psi										
	3042	4-1/2″ diam. dial	1 Stpi	16	.500	Ea.	131	17.45		148.45	171	
	3044	6-1/2″ diam. dial		16	.500		210	17.45		227.45	257	
	3046	8-1/2″ diam. dial		16	.500		380	17.45		397.45	445	
	3080	Pressure gauge, differential, magnehelic										
	3084	0 - 2″ W.C., with air filter kit	1 Stpi	6	1.333	Ea.	72	46.50		118.50	149	
	3300	For compound pressure-vacuum, add					18%					
	4000	Thermometers										
	4100	Dial type, 3-1/2″ diameter, vapor type, union connection	1 Stpi	32	.250	Ea.	95	8.70		103.70	118	
	4120	Liquid type, union connection		32	.250		133	8.70		141.70	159	
	4500	Stem type, 6-1/2″ case, 2″ stem, 1/2″ NPT		32	.250		54	8.70		62.70	72.50	
	4520	4″ stem, 1/2″ NPT		32	.250		74.50	8.70		83.20	94.50	
	4600	9″ case, 3-1/2″ stem, 3/4″ NPT		28	.286		90	9.95		99.95	114	
	4620	6″ stem, 3/4″ NPT		28	.286		96	9.95		105.95	120	
	4640	8″ stem, 3/4″ NPT		28	.286		118	9.95		127.95	145	
	4660	12″ stem, 1″ NPT		26	.308		130	10.70		140.70	159	
	4670	Bi-metal, dial type, steel case brass stem										
	4672	2″ dial, 4″ - 9″ stem	1 Stpi	16	.500	Ea.	29.50	17.45		46.95	59	
	4673	2-1/2″ dial, 4″ - 9″ stem		16	.500		32	17.45		49.45	61.50	
	4674	3-1/2″ dial, 4″ - 9″ stem		16	.500		32	17.45		49.45	61.50	
	4680	Mercury filled, industrial, union connection type										
	4682	Angle stem, 7″ scale	1 Stpi	16	.500	Ea.	74.50	17.45		91.95	109	
	4683	9″ scale		16	.500		109	17.45		126.45	147	
	4684	12″ scale		16	.500		121	17.45		138.45	160	
	4686	Straight stem, 7″ scale		16	.500		96	17.45		113.45	132	
	4687	9″ scale		16	.500		118	17.45		135.45	157	
	4688	12″ scale		16	.500		129	17.45		146.45	169	
	4690	Mercury filled, industrial, separable socket type										
	4692	Angle stem, with socket, 7″ scale	1 Stpi	16	.500	Ea.	62.50	17.45		79.95	95.50	
	4693	9″ scale		16	.500		75	17.45		92.45	109	
	4694	12″ scale		16	.500		152	17.45		169.45	194	
	4696	Straight stem, with socket, 7″ scale		16	.500		62.50	17.45		79.95	95.50	
	4697	9″ scale		16	.500		75	17.45		92.45	109	
	4698	12″ scale		16	.500		88.50	17.45		105.95	124	
	6000	Valves, motorized zone										

13800 | Building Automation & Control

13838 | Pneumatic/Electric Controls

			CREW	DAILY OUTPUT	LABOR-HOURS	UNIT	2001 BARE COSTS MAT.	LABOR	EQUIP.	TOTAL	TOTAL INCL O&P	
200	6100	Sweat connections, 1/2" C x C	1 Stpi	20	.400	Ea.	66	13.95		79.95	93.50	200
	6110	3/4" C x C		20	.400		70.50	13.95		84.45	98.50	
	6120	1" C x C		19	.421		82.50	14.65		97.15	113	
	6140	1/2" C x C, with end switch, 2 wire		20	.400		69.50	13.95		83.45	97.50	
	6150	3/4" C x C, with end switch, 2 wire		20	.400		74	13.95		87.95	103	
	6160	1" C x C, with end switch, 2 wire		19	.421		84	14.65		98.65	115	
	6180	1-1/4" C x C, w/end switch, 2 wire		15	.533		101	18.60		119.60	139	

13850 | Detection & Alarm

13851 | Detection & Alarm

			CREW	DAILY OUTPUT	LABOR-HOURS	UNIT	2001 BARE COSTS MAT.	LABOR	EQUIP.	TOTAL	TOTAL INCL O&P	
065	0010	**DETECTION SYSTEMS**, not including wires & conduits										065
	0100	Burglar alarm, battery operated, mechanical trigger	1 Elec	4	2	Ea.	249	68.50		317.50	375	
	0200	Electrical trigger		4	2		297	68.50		365.50	425	
	0400	For outside key control, add		8	1		70.50	34.50		105	129	
	0600	For remote signaling circuitry, add		8	1		112	34.50		146.50	174	
	0800	Card reader, flush type, standard		2.70	2.963		835	102		937	1,075	
	1000	Multi-code		2.70	2.963		1,075	102		1,177	1,325	
	3594	Fire, alarm control panel										
	3600	4 zone	2 Elec	2	8	Ea.	920	274		1,194	1,400	
	3800	8 zone		1	16		1,400	550		1,950	2,375	
	4000	12 zone		.67	23.988		1,825	825		2,650	3,225	
	4020	Alarm device	1 Elec	8	1		122	34.50		156.50	185	
	4050	Actuating device		8	1		292	34.50		326.50	370	
	4200	Battery and rack		4	2		690	68.50		758.50	860	
	4400	Automatic charger		8	1		445	34.50		479.50	540	
	4600	Signal bell		8	1		49.50	34.50		84	106	
	4800	Trouble buzzer or manual station		8	1		37	34.50		71.50	91.50	
	5000	Detector, rate of rise		8	1		33.50	34.50		68	88	
	5100	Fixed temperature		8	1		28	34.50		62.50	82	
	5200	Smoke detector, ceiling type		6.20	1.290		75	44.50		119.50	149	
	5400	Duct type		3.20	2.500		250	86		336	405	
	5600	Strobe and horn		5.30	1.509		95	52		147	183	
	5800	Fire alarm horn		6.70	1.194		36.50	41		77.50	101	
	6000	Door holder, electro-magnetic		4	2		77.50	68.50		146	188	
	6200	Combination holder and closer		3.20	2.500		430	86		516	605	
	6400	Code transmitter		4	2		690	68.50		758.50	860	
	6600	Drill switch		8	1		86.50	34.50		121	146	
	6800	Master box		2.70	2.963		3,100	102		3,202	3,550	
	7000	Break glass station		8	1		50	34.50		84.50	106	
	7800	Remote annunciator, 8 zone lamp		1.80	4.444		175	152		327	420	
	8000	12 zone lamp	2 Elec	2.60	6.154		300	211		511	645	
	8200	16 zone lamp	"	2.20	7.273		300	249		549	700	
350	0010	**TANK LEAK DETECTION SYSTEMS** Liquid and vapor										350
	0100	For hydrocarbons and hazardous liquids/vapors										
	0120	Controller, data acquisition, incl. printer, modem, RS232 port										
	0140	24 channel, for use with all probes				Ea.	2,050			2,050	2,250	
	0160	9 channel, for external monitoring				"	1,575			1,575	1,750	
	0200	Probes										

13850 | Detection & Alarm

	13851 | Detection & Alarm		CREW	DAILY OUTPUT	LABOR-HOURS	UNIT	2001 BARE COSTS MAT.	LABOR	EQUIP.	TOTAL	TOTAL INCL O&P	
350	0210	Well monitoring										350
	0220	Liquid phase detection				Ea.	710			710	785	
	0230	Hydrocarbon vapor, fixed position					655			655	720	
	0240	Hydrocarbon vapor, float mounted					580			580	640	
	0250	Both liquid and vapor hydrocarbon					880			880	970	
	0300	Secondary containment, liquid phase										
	0310	Pipe trench/manway sump				Ea.	320			320	350	
	0320	Double wall pipe and manual sump					310			310	340	
	0330	Double wall fiberglass annular space					286			286	315	
	0340	Double wall steel tank annular space					286			286	315	
	0500	Accessories										
	0510	Modem, non-dedicated phone line				Ea.	256			256	282	
	0600	Monitoring, internal										
	0610	Automatic tank gauge, incl. overfill				Ea.	1,025			1,025	1,125	
	0620	Product line				″	1,100			1,100	1,200	
	0700	Monitoring, special										
	0710	Cathodic protection				Ea.	610			610	670	
	0720	Annular space chemical monitor				″	830			830	915	

13900 | Fire Suppression

	13910 | Basic Fire Protection Matl/Methd		CREW	DAILY OUTPUT	LABOR-HOURS	UNIT	2001 BARE COSTS MAT.	LABOR	EQUIP.	TOTAL	TOTAL INCL O&P	
400	0010	FIRE HOSE AND EQUIPMENT A8.2-310										400
	0200	Adapters, rough brass, straight hose threads										
	0220	One piece, female to male, rocker lugs A8.2-320										
	0240	1″ x 1″				Ea.	20.50			20.50	22.50	
	0260	1-1/2″ x 1″ A8.2-390					8.80			8.80	9.70	
	0280	1-1/2″ x 1-1/2″					6.60			6.60	7.30	
	0300	2″ x 1-1/2″ R10520-310					41.50			41.50	45.50	
	0320	2″ x 2″					38			38	41.50	
	0340	2-1/2″ x 1-1/2″ R10520-320					27.50			27.50	30.50	
	0360	2-1/2″ x 2″					18.40			18.40	20	
	0380	2-1/2″ x 2-1/2″					12.85			12.85	14.15	
	0400	3″ x 2-1/2″					34.50			34.50	38	
	0420	3″ x 3″					92.50			92.50	102	
	0500	For polished brass, add					50%					
	0520	For polished chrome, add					75%					
	0700	One piece, female to male, hexagon										
	0740	1-1/2″ x 3/4″				Ea.	19.50			19.50	21.50	
	0760	2″ x 1-1/2″					30			30	33	
	0780	2-1/2″ x 1″					37			37	40.50	
	0800	2-1/2″ x 1-1/2″					28			28	30.50	
	0820	2-1/2″ x 2″					28.50			28.50	31.50	
	0840	3″ x 2-1/2″					32.50			32.50	36	
	0900	For polished chrome, add					75%					
	1100	Swivel, female to female, pin lugs										
	1120	1-1/2″ x 1-1/2″				Ea.	30			30	33	
	1200	2-1/2″ x 2-1/2″					59.50			59.50	65.50	
	1260	For polished brass, add					50%					
	1280	For polished chrome, add					75%					

13900 | Fire Suppression

	13910 | Basic Fire Protection Matl/Methd	CREW	DAILY OUTPUT	LABOR-HOURS	UNIT	2001 BARE COSTS MAT.	LABOR	EQUIP.	TOTAL	TOTAL INCL O&P
400 1400	Couplings, sngl & dbl jacket, pin lug or rocker lug, cast brass A8.2-310									
1410	1-1/2"				Ea.	25.50			25.50	28.50
1420	2-1/2" A8.2-320				"	34			34	37.50
1500	For polished brass, add					20%				
1520	For polished chrome, add A8.2-390					40%				
1580	Reducing, F x M, interior installation, cast brass									
1590	2" x 1-1/2" R10520-310				Ea.	32.50			32.50	36
1600	2-1/2" x 1-1/2"					7.35			7.35	8.10
1680	For polished brass, add R10520-320					50%				
1720	For polished chrome, add					75%				
1900	Escutcheon plate, for angle valves, polished brass, 1-1/2"					10.30			10.30	11.30
1920	2-1/2"					24.50			24.50	27
1940	3"					23			23	25
1980	For polished chrome, add					15%				
2200	Hose, less couplings									
2260	Synthetic jacket, lined, 300 lb. test, 1-1/2" diameter	Q-12	2,600	.006	L.F.	1.44	.19		1.63	1.87
2280	2-1/2" diameter		2,200	.007		2.40	.23		2.63	2.98
2360	High strength, 500 lb. test, 1-1/2" diameter		2,600	.006		1.49	.19		1.68	1.93
2380	2-1/2" diameter		2,200	.007		2.59	.23		2.82	3.19
2600	Hose rack, swinging, for 1-1/2" diameter hose,									
2620	Enameled steel, 50' & 75' lengths of hose	Q-12	20	.800	Ea.	32.50	25		57.50	73
2640	100' and 125' lengths of hose		20	.800		32.50	25		57.50	73
2680	Chrome plated, 50' and 75' lengths of hose		20	.800		50	25		75	92.50
2700	100' and 125' lengths of hose		20	.800		50	25		75	92.50
2780	For hose rack nipple, 1-1/2" polished brass, add					15.60			15.60	17.15
2820	2-1/2" polished brass, add					26			26	28.50
2840	1-1/2" polished chrome, add					16.20			16.20	17.80
2860	2-1/2" polished chrome, add					28			28	31
2990	Hose reel, swinging, for 1-1/2" polyester neoprene lined hose									
3000	50' long	Q-12	14	1.143	Ea.	67.50	35.50		103	128
3020	100' long		14	1.143		91	35.50		126.50	154
3060	For 2-1/2" cotton rubber hose, 75' long		14	1.143		108	35.50		143.50	173
3100	150' long		14	1.143		123	35.50		158.50	189
3750	Hydrants, wall, w/caps, single, flush, polished brass									
3800	2-1/2" x 2-1/2"	Q-12	5	3.200	Ea.	105	99		204	266
3840	2-1/2" x 3"		5	3.200		141	99		240	305
3860	3" x 3"		4.80	3.333		172	103		275	345
3900	For polished chrome, add					20%				
3950	Double, flush, polished brass									
4000	2-1/2" x 2-1/2" x 4"	Q-12	5	3.200	Ea.	282	99		381	460
4040	2-1/2" x 2-1/2" x 6"		4.60	3.478		405	108		513	615
4080	3" x 3" x 4"		4.90	3.265		605	101		706	820
4120	3" x 3" x 6"		4.50	3.556		625	110		735	850
4200	For polished chrome, add					10%				
4350	Double, projecting, polished brass									
4400	2-1/2" x 2-1/2" x 4"	Q-12	5	3.200	Ea.	130	99		229	293
4450	2-1/2" x 2-1/2" x 6"	"	4.60	3.478	"	251	108		359	440
4460	Valve control, dbl. flush/projecting hydrant, cap &									
4470	chain, ext. rod & cplg., escutcheon, polished brass	Q-12	8	2	Ea.	154	62		216	263
4480	Four-way square, flush, polished brass									
4540	2-1/2"(4) x 6"	Q-12	3.60	4.444	Ea.	1,525	138		1,663	1,875
5000	Nipples, straight hose to tapered iron pipe, brass									
5060	Female to female, 1-1/2" x 1-1/2"				Ea.	13.20			13.20	14.50
5080	2" x 2"					14.35			14.35	15.75
5100	2-1/2" x 2-1/2"					16.90			16.90	18.60
5200	Double male or male to female, 1" x 1"					20.50			20.50	22.50

400

13910 | Basic Fire Protection Matl/Methd

			CREW	DAILY OUTPUT	LABOR-HOURS	UNIT	2001 BARE COSTS MAT.	LABOR	EQUIP.	TOTAL	TOTAL INCL O&P	
400	5220	1-1/2" x 1"				Ea.	30.50			30.50	33.50	400
	5230	1-1/2" x 1-1/2" (A8.2 -310)					6.60			6.60	7.30	
	5260	2" x 1-1/2"					33			33	36.50	
	5270	2" x 2" (A8.2 -320)					41.50			41.50	45.50	
	5280	2-1/2" x 1-1/2"					27.50			27.50	30.50	
	5300	2-1/2" x 2" (A8.2 -390)					22.50			22.50	24.50	
	5310	2-1/2" x 2-1/2"					12.85			12.85	14.15	
	5340	For polished chrome, add (R10520 -310)					75%					
	5600	Nozzles, brass										
	5620	Adjustable fog, 3/4" booster line (R10520 -320)				Ea.	69			69	76	
	5630	1" booster line					80.50			80.50	88.50	
	5640	1-1/2" leader line					84.50			84.50	93	
	5660	2-1/2" direct connection					165			165	182	
	5680	2-1/2" playpipe nozzle					136			136	150	
	5780	For chrome plated, add					8%					
	5850	Electrical fire, adjustable fog, no shock										
	5900	1-1/2"				Ea.	263			263	289	
	5920	2-1/2"					360			360	395	
	5980	For polished chrome, add					6%					
	6200	Heavy duty, comb. adj. fog and str. stream, with handle										
	6210	1" booster line				Ea.	251			251	277	
	6240	1-1/2"					248			248	273	
	6260	2-1/2", for playpipe					284			284	315	
	6280	2-1/2" direct connection					355			355	390	
	6300	2-1/2" playpipe combination					510			510	560	
	6480	For polished chrome, add					7%					
	6500	Plain fog, polished brass, 1-1/2"					42.50			42.50	47	
	6540	Chrome plated, 1-1/2"					43.50			43.50	47.50	
	6700	Plain stream, polished brass, 1-1/2" x 10"					21.50			21.50	23.50	
	6760	2-1/2" x 15" x 7/8" or 1-1/2"					43			43	47.50	
	6860	For polished chrome, add					20%					
	7000	Underwriters playpipe, 2-1/2" x 30" with 1-1/8" tip				Ea.	179			179	197	
	7040	Less tip				"	142			142	156	
	7140	Standpipe connections, wall, w/plugs & chains										
	7160	Single, flush, brass, 2-1/2" x 2-1/2"	Q-12	5	3.200	Ea.	78.50	99		177.50	237	
	7180	2-1/2" x 3"	"	5	3.200	"	82	99		181	240	
	7240	For polished chrome, add					15%					
	7280	Double, flush, polished brass										
	7300	2-1/2" x 2-1/2" x 4"	Q-12	5	3.200	Ea.	265	99		364	440	
	7330	2-1/2" x 2-1/2" x 6"		4.60	3.478		360	108		468	560	
	7340	3" x 3" x 4"		4.90	3.265		470	101		571	675	
	7370	3" x 3" x 6"		4.50	3.556		560	110		670	785	
	7400	For polished chrome, add					15%					
	7440	For sill cock combination, add				Ea.	45			45	49.50	
	7580	Double projecting, polished brass										
	7600	2-1/2" x 2-1/2" x 4"	Q-12	5	3.200	Ea.	242	99		341	415	
	7630	2-1/2" x 2-1/2" x 6"	"	4.60	3.478	"	395	108		503	600	
	7680	For polished chrome, add					15%					
	7900	Three way, flush, polished brass										
	7920	2-1/2" (3) x 4"	Q-12	4.80	3.333	Ea.	825	103		928	1,075	
	7930	2-1/2" (3) x 6"	"	4.80	3.333		835	103		938	1,075	
	8000	For polished chrome, add					9%					
	8020	Three way, projecting, polished brass										
	8040	2-1/2"(3) x 4"	Q-12	4.80	3.333	Ea.	580	103		683	795	
	8070	2-1/2" (3) x 6"	"	4.60	3.478		585	108		693	805	
	8100	For polished chrome, add					12%					

13910 | Basic Fire Protection Matl/Methd

		13910 Basic Fire Protection Matl/Methd	Ref.	CREW	DAILY OUTPUT	LABOR-HOURS	UNIT	2001 BARE COSTS MAT.	LABOR	EQUIP.	TOTAL	TOTAL INCL O&P	
400	8200	Four way, square, flush, polished brass,	A8.2 -310										400
	8240	2-1/2″(4) x 6″		Q-12	3.60	4.444	Ea.	1,450	138		1,588	1,800	
	8300	For polished chrome, add	A8.2 -320				″	10%					
	8550	Wall, vertical, flush, cast brass											
	8600	Two way, 2-1/2″ x 2-1/2″ x 4″	A8.2 -390	Q-12	5	3.200	Ea.	585	99		684	795	
	8660	Four way, 2-1/2″(4) x 6″			3.80	4.211		1,300	131		1,431	1,625	
	8680	Six way, 2-1/2″(6) x 6″	R10520 -310		3.40	4.706		1,750	146		1,896	2,150	
	8700	For polished chrome, add						10%					
	8800	Sidewalk siamese unit, polished brass, two way	R10520 -320										
	8820	2-1/2″ x 2-1/2″ x 4″		Q-12	2.50	6.400	Ea.	310	198		508	640	
	8850	2-1/2″ x 2-1/2″ x 6″			2	8		515	248		763	940	
	8860	3″ x 3″ x 4″			2.50	6.400		445	198		643	790	
	8890	3″ x 3″ x 6″			2	8		605	248		853	1,050	
	8940	For polished chrome, add						12%					
	9100	Sidewalk siamese unit, polished brass, three way											
	9120	2-1/2″ x 2-1/2″ x 2-1/2″ x 6″		Q-12	2	8	Ea.	670	248		918	1,125	
	9160	For polished chrome, add						15%					
	9200	Storage house, hose only, primed steel						420			420	465	
	9220	Aluminum						925			925	1,025	
	9280	Hose and hydrant house, primed steel						570			570	625	
	9300	Aluminum						825			825	910	
	9340	Tools, crowbar and brackets		1 Carp	12	.667		58	19.45		77.45	94.50	
	9360	Combination hydrant wrench and spanner						15.80			15.80	17.40	
	9380	Fire axe and brackets											
	9400	6 lb.		1 Carp	12	.667	Ea.	60.50	19.45		79.95	97	
800	0010	**FIRE VALVES**	A8.2 -310										800
	0020	Angle, combination pressure adjust/restricting, rough brass											
	0030	1-1/2″	A8.2 -320	1 Spri	12	.667	Ea.	42	23		65	80.50	
	0040	2-1/2″		″	7	1.143	″	89	39.50		128.50	158	
	0042	Nonpressure adjustable/restricting, rough brass											
	0044	1-1/2″		1 Spri	12	.667	Ea.	26.50	23		49.50	63.50	
	0046	2-1/2″	A8.2 -390	″	7	1.143	″	46	39.50		85.50	110	
	0050	For polished brass, add						30%					
	0060	For polished chrome, add	R10520 -310					40%					
	0080	Wheel handle, 300 lb., 1-1/2″		1 Spri	12	.667	Ea.	26.50	23		49.50	63.50	
	0090	2-1/2″	R10520 -320	″	7	1.143	″	46	39.50		85.50	110	
	0100	For polished brass, add						35%					
	0110	For polished chrome, add						50%					
	1000	Ball drip, automatic, rough brass, 1/2″		1 Spri	20	.400	Ea.	7.70	13.80		21.50	29.50	
	1010	3/4″		″	20	.400	″	8.80	13.80		22.60	30.50	
	1100	Ball, 175 lb., sprinkler system, FM/UL, threaded, bronze											
	1120	Slow close											
	1150	1″ size		1 Spri	19	.421	Ea.	76	14.50		90.50	106	
	1160	1-1/4″ size			15	.533		82.50	18.35		100.85	119	
	1170	1-1/2″ size			13	.615		105	21		126	147	
	1180	2″ size			11	.727		132	25		157	183	
	1190	2-1/2″ size		Q-12	15	1.067		187	33		220	255	
	1230	For supervisory switch kit, all sizes											
	1240	One circuit, add		1 Spri	48	.167	Ea.	57	5.75		62.75	71.50	
	1280	Quarter turn for trim											
	1300	1/2″ size		1 Spri	22	.364	Ea.	9.80	12.55		22.35	30	
	1310	3/4″ size			20	.400		10.55	13.80		24.35	32.50	
	1320	1″ size			19	.421		11.70	14.50		26.20	35	
	1330	1-1/4″ size			15	.533		19.25	18.35		37.60	49	
	1340	1-1/2″ size			13	.615		24	21		45	58.50	

13910 | Basic Fire Protection Matl/Methd

				DAILY	LABOR-		2001 BARE COSTS				TOTAL	
			CREW	OUTPUT	HOURS	UNIT	MAT.	LABOR	EQUIP.	TOTAL	INCL O&P	
800	1350	2" size	1 Spri	11	.727	Ea.	28.50	25		53.50	69.50	800
	1400	Caps, polished brass with chain, 3/4" (A8.2 -310)					21.50			21.50	24	
	1420	1"					27.50			27.50	30.50	
	1440	1-1/2" (A8.2 -320)					7.35			7.35	8.10	
	1460	2-1/2"					11.40			11.40	12.55	
	1480	3" (A8.2 -390)					17.30			17.30	19	
	2000	Foam, control valve, 3"	1 Spri	6	1.333		1,050	46		1,096	1,225	
	2020	Supply valve, 2-1/2" (R10520 -310)		7	1.143		64	39.50		103.50	130	
	2040	Proportioner, 8"		2	4		1,200	138		1,338	1,525	
	2060	Oscillating foam monitor (R10520 -320)	Q-12	5.33	3.002		6,125	93		6,218	6,900	
	3000	Gate, hose, wheel handle, N.R.S., rough brass, 1-1/2"	1 Spri	12	.667		67	23		90	108	
	3040	2-1/2", 300 lb.	"	7	1.143		94.50	39.50		134	164	
	3080	For polished brass, add					40%					
	3090	For polished chrome, add					50%					
	3800	Hydrant, screw type, crank handle, brass										
	3840	2-1/2" size	Q-12	11	1.455	Ea.	222	45		267	310	
	3880	For chrome, same price										
	4200	Hydrolator, vent and draining, rough brass, 1-1/2"	1 Spri	12	.667	Ea.	41	23		64	79.50	
	4280	For polished brass, add					50%					
	4290	For polished chrome, add					90%					
	5000	Pressure restricting, adjustable rough brass, 1-1/2"	1 Spri	12	.667		69	23		92	111	
	5020	2-1/2"	"	7	1.143		97	39.50		136.50	167	
	5080	For polished brass, add					30%					
	5090	For polished chrome, add					45%					
	6000	Roof manifold, horiz., brass, without valves & caps										
	6040	2-1/2" x 2-1/2" x 4"	Q-12	4.80	3.333	Ea.	101	103		204	267	
	6060	2-1/2" x 2-1/2" x 6"		4.60	3.478		119	108		227	294	
	6080	2-1/2" x 2-1/2" x 2-1/2" x 4"		4.60	3.478		163	108		271	340	
	6090	2-1/2" x 2-1/2" x 2-1/2" x 6"		4.60	3.478		168	108		276	350	
	7000	Sprinkler line tester, cast brass					17.40			17.40	19.15	
	8000	Wye, leader line, ball type, swivel female x male x male										
	8040	2-1/2" x 1-1/2" x 1-1/2" polished brass				Ea.	287			287	315	
	8060	2-1/2" x 1-1/2" x 1-1/2" polished chrome				"	296			296	325	

13920 | Fire Pumps

			CREW	DAILY OUTPUT	LABOR-HOURS	UNIT	MAT.	LABOR	EQUIP.	TOTAL	TOTAL INCL O&P	
400	0010	**FIRE PUMPS** Including controller, fittings and relief valve										400
	0030	Diesel										
	0050	500 GPM, 50 psi, 27 HP, 4" pump	Q-13	.64	50	Ea.	46,600	1,650		48,250	54,000	
	0100	500 GPM, 100 psi, 62 HP, 4" pump		.60	53.333		49,300	1,750		51,050	56,500	
	0150	500 GPM, 125 psi, 78 HP, 4" pump		.56	57.143		49,900	1,875		51,775	58,000	
	0200	750 GPM, 50 psi, 44 HP, 5" pump		.60	53.333		49,700	1,750		51,450	57,000	
	0250	750 GPM, 100 psi, 80 HP, 4" pump		.56	57.143		52,000	1,875		53,875	60,000	
	0300	750 GPM, 165 psi, 203 HP, 5" pump		.52	61.538		53,500	2,025		55,525	62,000	
	0350	1000 GPM, 50 psi, 48 HP, 5" pump		.58	55.172		50,000	1,800		51,800	58,000	
	0400	1000 GPM, 100 psi, 89 HP, 4" pump		.56	57.143		53,500	1,875		55,375	61,500	
	0450	1000 GPM, 150 psi, 148 HP, 4" pump		.48	66.667		52,500	2,200		54,700	61,000	
	0470	1000 GPM, 200 psi, 280 HP, 5" pump		.40	80		66,000	2,625		68,625	76,500	
	0480	1250 GPM, 75 psi, 75 HP, 5" pump		.54	59.259		50,500	1,950		52,450	58,500	
	0500	1500 GPM, 50 psi, 66 HP, 6" pump		.50	64		54,000	2,100		56,100	62,000	
	0550	1500 GPM, 100 psi, 140 HP, 6" pump		.46	69.565		55,000	2,275		57,275	64,000	
	0600	1500 GPM, 150 psi, 228 HP, 6" pump		.42	76.190		62,000	2,500		64,500	72,500	
	0650	1500 GPM, 200 psi, 279 HP, 6" pump		.38	84.211		97,500	2,775		100,275	111,500	
	0700	2,000 GPM, 100 psi, 167 HP, 6" pump		.34	94.118		56,500	3,100		59,600	66,500	
	0750	2000 GPM, 150 psi, 284 HP, 6"pump		.30	106		76,500	3,500		80,000	89,500	
	0800	2500 GPM, 100 psi, 213 HP, 8" pump		.32	100		72,000	3,275		75,275	84,500	

13 SPECIAL CONSTRUCTION

13900 | Fire Suppression

	13920 | Fire Pumps		CREW	DAILY OUTPUT	LABOR-HOURS	UNIT	2001 BARE COSTS MAT.	LABOR	EQUIP.	TOTAL	TOTAL INCL O&P	
400	0820	2500 GPM, 150 psi, 365 HP, 8″ pump	Q-13	.26	123	Ea.	84,500	4,050		88,550	99,000	400
	0850	3000 GPM, 100 psi, 250 HP, 8″ pump		.28	114		83,000	3,750		86,750	97,000	
	0900	3000 GPM, 150 psi, 384 HP, 10″ pump		.20	160		97,000	5,250		102,250	114,500	
	0950	3500 GPM, 100 psi, 300 HP, 10″ pump		.24	133		99,500	4,375		103,875	116,000	
	1000	3500 GPM, 150 psi, 518 HP, 10″ pump	↓	.20	160	↓	140,500	5,250		145,750	162,500	
	3000	Electric										
	3100	250 GPM, 55 psi, 15 HP, 3,550 RPM, 2″ pump	Q-13	.70	45.714	Ea.	10,900	1,500		12,400	14,300	
	3200	500 GPM, 50 psi, 27 HP, 1770 RPM, 4″ pump		.68	47.059		13,500	1,550		15,050	17,300	
	3250	500 GPM, 100 psi, 47 HP, 3550 RPM, 3″ pump		.66	48.485		16,500	1,600		18,100	20,500	
	3300	500 GPM, 125 psi, 64 HP, 3550 RPM, 3″ pump		.62	51.613		18,000	1,700		19,700	22,400	
	3350	750 GPM, 50 psi, 44 HP, 1,770 RPM, 5″ pump		.64	50		19,800	1,650		21,450	24,300	
	3400	750 GPM, 100 psi, 66 HP, 3550 RPM, 4″ pump		.58	55.172		17,500	1,800		19,300	22,100	
	3450	750 GPM, 165 psi, 120 HP, 3550 RPM, 4″ pump		.56	57.143		22,900	1,875		24,775	28,000	
	3500	1000 GPM, 50 psi, 48 HP 1770 RPM, 5″ pump		.60	53.333		19,800	1,750		21,550	24,500	
	3550	1000 GPM, 100 psi, 86 HP, 3550 RPM, 5″ pump		.54	59.259		22,200	1,950		24,150	27,400	
	3600	1000 GPM, 150 psi, 142 HP, 3550 RPM, 5″ pump		.50	64		25,200	2,100		27,300	30,900	
	3650	1000 GPM, 200 psi, 245 HP, 1770 RPM, 6″ pump		.36	88.889		54,500	2,925		57,425	64,500	
	3660	1250 GPM, 75 psi, 75 HP, 1770 RPM, 5″ pump		.55	58.182		20,300	1,900		22,200	25,300	
	3700	1500 GPM, 50 psi, 66 HP, 1770 RPM, 6″ pump		.50	64		21,300	2,100		23,400	26,700	
	3750	1500 GPM, 100 psi, 139 HP, 1770 RPM, 6″ pump		.46	69.565		24,100	2,275		26,375	30,000	
	3800	1500 GPM, 150 psi, 200 HP, 1770 RPM, 6″ pump		.36	88.889		51,000	2,925		53,925	61,000	
	3850	1500 GPM, 200 psi, 279 HP, 1770 RPM, 6″ pump		.32	100		56,000	3,275		59,275	67,000	
	3900	2000 GPM, 100 psi, 167 HP, 1770 RPM, 6″ pump		.34	94.118		30,800	3,100		33,900	38,600	
	3950	2000 GPM, 150 psi, 292 HP, 1770 RPM, 6″ pump		.28	114		41,300	3,750		45,050	51,000	
	4000	2500 GPM, 100 psi, 213 HP, 1770 RPM, 8″ pump		.30	106		33,700	3,500		37,200	42,400	
	4040	2500 GPM, 135 psi, 339 HP, 1770 RPM, 8″ pump		.26	123		48,200	4,050		52,250	59,000	
	4100	3000 GPM, 100 psi, 250 HP, 1770 RPM, 8″ pump		.28	114		40,400	3,750		44,150	50,000	
	4150	3000 GPM, 140 psi, 428 HP, 1770 RPM, 10″ pump		.24	133		64,000	4,375		68,375	77,000	
	4200	3500 GPM, 100 psi, 300 HP, 1770 RPM, 10″ pump		.26	123		50,000	4,050		54,050	61,000	
	4250	3500 GPM, 140 psi, 450 HP, 1770 RPM, 10″ pump	↓	.24	133		64,000	4,375		68,375	77,000	
	5000	For jockey pump 1″, 3 HP, with control, add	Q-12	2	8	↓	2,425	248		2,673	3,050	

	13930 | Wet-Pipe Fire Supp. Sprinklers		CREW	DAILY OUTPUT	LABOR-HOURS	UNIT	MAT.	LABOR	EQUIP.	TOTAL	TOTAL INCL O&P	
400	0010	SPRINKLER SYSTEM COMPONENTS (A8.2-110)										400
	0600	Accelerator	1 Spri	8	1	Ea.	335	34.50		369.50	420	
	0800	Air compressor for dry pipe system, automatic, complete (A8.2-120)										
	0820	280 gal. system capacity, 3/4 HP	1 Spri	1.30	6.154	Ea.	700	212		912	1,100	
	0860	520 gal. system capacity, 1 HP (A8.2-130)		1.30	6.154		735	212		947	1,125	
	0910	650 gal. system capacity, 1-1/2 HP		1.30	6.154		765	212		977	1,150	
	0920	790 gal. system capacity, 2 HP (A8.2-140)		1.30	6.154		860	212		1,072	1,275	
	0960	Air pressure maintenance control		24	.333		139	11.50		150.50	170	
	1100	Alarm, electric pressure switch (circuit closer) (A8.2-150)		26	.308		131	10.60		141.60	160	
	1140	For explosion proof, max 20 PSI, contacts close or open		26	.308		261	10.60		271.60	305	
	1220	Water motor, complete with gong (R10520-110)		4	2		166	69		235	287	
	1400	Deluge system, pressured monitoring panel, 120V		18	.444		785	15.30		800.30	890	
	1600	Dehydrator package, incl. valves and nipples (R10520-120)	↓	12	.667	↓	286	23		309	350	
	1800	Firecycle system, controls, includes panel,										
	1820	batteries, solenoid valves and pressure switches (R10520-130)	Q-13	1	32	Ea.	7,775	1,050		8,825	10,200	
	1980	Detector	1 Spri	16	.500		285	17.25		302.25	340	
	2000	Release, emergency, manual, for hydraulic or pneumatic system		12	.667		87.50	23		110.50	131	
	2060	Release, thermostatic, for hydraulic or pneumatic release line		20	.400		266	13.80		279.80	315	
	2200	Sprinkler cabinets, 6 head capacity		16	.500		31	17.25		48.25	60	
	2260	12 head capacity		16	.500		33	17.25		50.25	62	
	2340	Sprinkler head escutcheons, standard, brass tone, 1″ size		40	.200		3.07	6.90		9.97	13.80	
	2360	Chrome, 1″ size	↓	40	.200	↓	1.59	6.90		8.49	12.15	

13930 | Wet-Pipe Fire Supp. Sprinklers

				CREW	DAILY OUTPUT	LABOR-HOURS	UNIT	2001 BARE COSTS MAT.	LABOR	EQUIP.	TOTAL	TOTAL INCL O&P	
400	2400	Recessed type, brass tone	A8.2-110	1 Spri	40	.200	Ea.	3.96	6.90		10.86	14.75	400
	2440	Chrome or white enamel			40	.200		1.78	6.90		8.68	12.35	
	2600	Sprinkler heads, not including supply piping	A8.2-120										
	2640	Dry, pendent, 1/2″ orifice, 3/4″ or 1″ NPT											
	2660	1/2″ to 6″ length	A8.2-130	1 Spri	14	.571	Ea.	34.50	19.70		54.20	68	
	2670	6-1/4″ to 9″ length			14	.571		37	19.70		56.70	70.50	
	2680	9-1/4″ to 12″ length	A8.2-140		14	.571		39.50	19.70		59.20	73.50	
	2690	12-1/4″ to 15″ length			14	.571		42	19.70		61.70	76	
	2700	15-1/4″ to 18″ length	A8.2-150		14	.571		44.50	19.70		64.20	79	
	2710	18-1/4″ to 21″ length			13	.615		47	21		68	83.50	
	2720	21-1/4″ to 24″ length	R10520-110		13	.615		49.50	21		70.50	86.50	
	2730	24-1/4″ to 27″ length			13	.615		52	21		73	89	
	2740	27-1/4″ to 30″ length	R10520-120		13	.615		53.50	21		74.50	91	
	2750	30-1/4″ to 33″ length			13	.615		57	21		78	94.50	
	2760	33-1/4″ to 36″ length	R10520-130		13	.615		59.50	21		80.50	97.50	
	2780	36-1/4″ to 39″ length			12	.667		62	23		85	103	
	2790	39-1/4″ to 42″ length			12	.667		64.50	23		87.50	106	
	2800	For each inch or fraction , add						.84			.84	.92	
	3600	Foam-water, pendent or upright, 1/2″ NPT		1 Spri	12	.667		50	23		73	89.50	
	3700	Standard spray, pendent or upright, brass, 135° to 286°F											
	3720	1/2″ NPT, 3/8″ orifice		1 Spri	16	.500	Ea.	6	17.25		23.25	32.50	
	3730	1/2″ NPT, 7/16″ orifice			16	.500		6	17.25		23.25	32.50	
	3740	1/2″ NPT, 1/2″ orifice			16	.500		4.13	17.25		21.38	30.50	
	3760	1/2″ NPT, 17/32″ orifice			16	.500		5.20	17.25		22.45	32	
	3780	3/4″ NPT, 17/32″ orifice			16	.500		4.95	17.25		22.20	31.50	
	3800	For open sprinklers, deduct						15%					
	3840	For chrome, add					Ea.	1.13			1.13	1.24	
	3860	For wax and lead coating, add						5.50			5.50	6.05	
	3880	For wax coating, add						2.50			2.50	2.75	
	3900	For lead coating, add						3			3	3.30	
	3920	For 360°F, same cost											
	3930	For 400°F, add					Ea.	5.50			5.50	6.05	
	3940	For 500°F, add					″	5.50			5.50	6.05	
	4200	Sidewall, vertical brass, 135 -286°F											
	4240	1/2″ NPT, 1/2″ orifice		1 Spri	16	.500	Ea.	7	17.25		24.25	33.50	
	4280	3/4″ NPT, 17/32″ orifice		″	16	.500		7.50	17.25		24.75	34.50	
	4360	For satin chrome, add						1			1	1.10	
	4400	For 360°F, same cost											
	4410	For 400°F, add					Ea.	5.50			5.50	6.05	
	4420	For 500°F, add					″	7			7	7.70	
	4500	Sidewall, horizontal, brass, 135° to 286°F											
	4520	1/2″ NPT, 1/2″ orifice		1 Spri	16	.500	Ea.	6	17.25		23.25	32.50	
	4540	For 360°F, same cost											
	4800	Recessed pendent, brass, 135° to 286°F											
	4820	1/2″ NPT, 3/8″ orifice		1 Spri	10	.800	Ea.	14.50	27.50		42	57.50	
	4830	1/2″ NPT, 7/16″ orifice			10	.800		10.25	27.50		37.75	53	
	4840	1/2″ NPT, 1/2″ orifice			10	.800		12	27.50		39.50	54.50	
	4860	1/2″ NPT, 17/32″ orifice			10	.800		9.55	27.50		37.05	52	
	4900	For satin chrome, add						.75			.75	.83	
	5000	Recessed-vertical sidewall, brass, 135°-286°F											
	5020	1/2″ NPT, 3/8″ orifice		1 Spri	10	.800	Ea.	11.25	27.50		38.75	54	
	5030	1/2″ NPT, 7/16″ orifice			10	.800		11	27.50		38.50	53.50	
	5040	1/2″ NPT, 1/2″ orifice			10	.800		10.25	27.50		37.75	53	
	5100	For bright nickel, same cost											
	5600	Concealed, complete with cover plate											
	5620	1/2″ NPT, 1/2″ orifice, 135°F to 212°F		1 Spri	9	.889	Ea.	14	30.50		44.50	62	
400													400

13930 | Wet-Pipe Fire Supp. Sprinklers

			CREW	DAILY OUTPUT	LABOR-HOURS	UNIT	2001 BARE COSTS MAT.	LABOR	EQUIP.	TOTAL	TOTAL INCL O&P	
400	5800	Window, brass, 1/2″ NPT, 1/4″ orifice (A8.2-110)	1 Spri	16	.500	Ea.	13.25	17.25		30.50	40.50	400
	5810	1/2″ NPT, 5/16″ orifice		16	.500		13.25	17.25		30.50	40.50	
	5820	1/2″ NPT, 3/8″ orifice (A8.2-120)		16	.500		13.25	17.25		30.50	40.50	
	5830	1/2″ NPT, 7/16″ orifice		16	.500		13.25	17.25		30.50	40.50	
	5840	1/2″ NPT, 1/2″ orifice (A8.2-130)		16	.500		14.05	17.25		31.30	41.50	
	5860	For polished chrome, add					1.95			1.95	2.15	
	5880	3/4″ NPT, 5/8″ orifice (A8.2-140)	1 Spri	16	.500		14.65	17.25		31.90	42	
	5890	3/4 NPT, 3/4″ orifice	″	16	.500		14.65	17.25		31.90	42	
	6000	Sprinkler head guards, bright zinc, 1/2″ NPT (A8.2-150)					2.37			2.37	2.61	
	6020	Bright zinc, 3/4″ NPT					2.37			2.37	2.61	
	6025	Residential sprinkler components, (one and two family) (R10520-110)										
	6026	Water motor alarm, with strainer	1 Spri	4	2	Ea.	166	69		235	287	
	6027	Fast response, glass bulb, 135° to 155° f. (R10520-120)										
	6028	1/2″ NPT, pendent, brass	1 Spri	16	.500	Ea.	9.05	17.25		26.30	36	
	6029	1/2″ NPT, sidewall, brass (R10520-130)		16	.500		9.45	17.25		26.70	36.50	
	6030	1/2″ NPT, pendent, brass, extended coverage		16	.500		9.45	17.25		26.70	36.50	
	6031	1/2″ NPT, sidewall, brass, extended coverage		16	.500		9	17.25		26.25	36	
	6032	3/4″ NPT sidewall, brass, extended coverage		16	.500		9.45	17.25		26.70	36.50	
	6033	For chrome, add					15%					
	6034	For polyester/teflon coating add					20%					
	6100	Sprinkler head wrenches, standard head				Ea.	14			14	15.40	
	6120	Recessed head					14.50			14.50	15.95	
	6160	Tamper switch, (valve supervisory switch)	1 Spri	16	.500		54.50	17.25		71.75	86	
	6200	Valves										
	6210	Alarm, includes										
	6220	retard chamber, trim, gauges, alarm line strainer										
	6260	3″ size	Q-12	3	5.333	Ea.	745	165		910	1,075	
	6280	4″ size	″	2	8		755	248		1,003	1,200	
	6300	6″ size	Q-13	4	8		840	263		1,103	1,325	
	6320	8″ size	″	3	10.667		995	350		1,345	1,625	
	6500	Check, swing, C.I. body, brass fittings, auto. ball drip										
	6520	4″ size	Q-12	3	5.333	Ea.	141	165		306	405	
	6540	6″ size	Q-13	4	8		261	263		524	685	
	6580	8″ size	″	3	10.667		490	350		840	1,075	
	6800	Check, wafer, butterfly type, C.I. body, bronze fittings										
	6820	4″ size	Q-12	4	4	Ea.	152	124		276	355	
	6840	6″ size	Q-13	5.50	5.818		261	191		452	575	
	6860	8″ size		5	6.400		470	210		680	835	
	6880	10″ size		4.50	7.111		710	234		944	1,125	
	7000	Deluge, assembly, incl. trim, pressure										
	7020	operated relief, emergency release, gauges										
	7040	2″ size	Q-12	2	8	Ea.	1,000	248		1,248	1,475	
	7060	3″ size		1.50	10.667		1,175	330		1,505	1,800	
	7080	4″ size		1	16		1,425	495		1,920	2,300	
	7100	6″ size	Q-13	1.80	17.778		1,750	585		2,335	2,800	
	7800	Pneumatic actuator, bronze, required on all										
	7820	pneumatic release systems, any size deluge	1 Spri	18	.444	Ea.	166	15.30		181.30	206	
	8000	Dry pipe air check valve, 3″ size	Q-12	2	8		990	248		1,238	1,475	
	8200	Dry pipe valve, incl. trim and gauges, 3″ size		2	8		1,050	248		1,298	1,525	
	8220	4″ size		1	16		1,175	495		1,670	2,050	
	8240	6″ size	Q-13	2	16		1,350	525		1,875	2,275	
	8280	For accelerator trim with gauges, add	1 Spri	8	1		360	34.50		394.50	445	
	8400	Firecycle package, includes swing check										
	8420	and flow control valves with required trim										
	8440	2″ size	Q-12	2	8	Ea.	1,225	248		1,473	1,725	
	8460	3″ size		1.50	10.667		1,900	330		2,230	2,575	

13930 | Wet-Pipe Fire Supp. Sprinklers

				CREW	DAILY OUTPUT	LABOR-HOURS	UNIT	2001 BARE COSTS MAT.	LABOR	EQUIP.	TOTAL	TOTAL INCL O&P	
400	8480	4" size	A8.2 -110	Q-12	1	16	Ea.	2,125	495		2,620	3,075	400
	8500	6" size		Q-13	1.40	22.857		2,875	750		3,625	4,275	
	8800	Flow control valve, includes trim and gauges, 2" size	A8.2 -120	Q-12	2	8		1,075	248		1,323	1,550	
	8820	3" size		"	1.50	10.667		1,275	330		1,605	1,900	
	8840	4" size	A8.2 -130	Q-13	2.80	11.429		1,500	375		1,875	2,225	
	8860	6" size		"	2	16		216	525		741	1,025	
	9200	Pressure operated relief valve, brass body	A8.2 -140	1 Spri	18	.444		119	15.30		134.30	154	
	9600	Waterflow indicator, with recycling retard and											
	9610	two single pole retard switches, 2" thru 6" pipe size	A8.2 -150	1 Spri	8	1	Ea.	109	34.50		143.50	172	

13960 | CO2 Fire Extinguishing

				CREW	DAILY OUTPUT	LABOR-HOURS	UNIT	MAT.	LABOR	EQUIP.	TOTAL	TOTAL INCL O&P	
200	0010	**AUTOMATIC FIRE SUPPRESSION SYSTEMS**	A8.2 -810										200
	0040	For detectors and control stations, see division 13850											
	0100	Control panel, single zone with batteries (2 zones det., 1 suppr.)		1 Elec	1	8	Ea.	1,275	274		1,549	1,800	
	0150	Multizone (4) with batteries (8 zones det., 4 suppr.)		"	.50	16		2,450	550		3,000	3,525	
	1000	Dispersion nozzle, CO2, 3" x 5"		1 Plum	18	.444		50	15.30		65.30	78	
	1100	FM200, 1-1/2"		"	14	.571		50	19.70		69.70	84.50	
	2000	Extinguisher, CO2 system, high pressure, 75 lb. cylinder		Q-1	6	2.667		950	82.50		1,032.50	1,175	
	2100	100 lb. cylinder		"	5	3.200		1,025	99		1,124	1,275	
	2400	FM200 system, filled, with mounting bracket											
	2460	26 lb. container		Q-1	8	2	Ea.	1,800	62		1,862	2,075	
	2480	44 lb. container			7	2.286		2,400	71		2,471	2,750	
	2500	63 lb. container			6	2.667		2,800	82.50		2,882.50	3,200	
	2520	101 lb. container			5	3.200		3,750	99		3,849	4,275	
	2540	196 lb. container			4	4		5,800	124		5,924	6,550	
	3000	Electro/mechanical release		L-1	4	4		125	138		263	345	
	3400	Manual pull station		1 Plum	6	1.333		45	46		91	119	
	4000	Pneumatic damper release		"	8	1		175	34.50		209.50	245	
	6000	Average FM200 system, minimum					C.F.	1.25			1.25	1.38	
	6020	Maximum					"	2.50			2.50	2.75	

For information about Means Estimating Seminars, see yellow pages 11 and 12 in back of book

Division 14 Conveying Systems

Estimating Tips

General

- Many products in Division 14 will require some type of support or blocking for installation not included with the item itself. Examples are supports for conveyors or tube systems, attachment points for lifts, and footings for hoists or cranes. Add these supports in the appropriate division.

14100 Dumbwaiters
14200 Elevators

- Dumbwaiters and elevators are estimated and purchased in a method similar to buying a car. The manufacturer has a base unit with standard features. Added to this base unit price will be whatever options the owner or specifications require. Increased load capacity, additional vertical travel, additional stops, higher speed, and cab finish options are items to be considered. When developing an estimate for dumbwaiters and elevators, remember that some items needed by the installers may have to be included as part of the general contract.

  Examples are:
  - shaftway
  - rail support brackets
  - machine room
  - electrical supply
  - sill angles
  - electrical connections
  - pits
  - roof penthouses
  - pit ladders

  Check the job specifications and drawings before pricing.

14300 Escalators & Moving Walks

- Escalators and moving walks are specialty items installed by specialty contractors. There are numerous options associated with these items. For specific options contact a manufacturer or contractor. In a method similar to estimating dumbwaiters and elevators, you should verify the extent of general contract work and add items as necessary.

14400 Lifts
14500 Material Handling
14600 Hoists & Cranes

- Products such as correspondence lifts, conveyors, chutes, pneumatic tube systems, material handling cranes and hoists as well as other items specified in this subdivision may require trained installers. The general contractor might not have any choice as to who will perform the installation or when it will be performed. Long lead times are often required for these products, making early decisions in scheduling necessary.

Reference Numbers

Reference numbers are shown in bold squares at the beginning of some major classifications. These numbers refer to related items in the Reference Section. The reference information may be an estimating procedure, an alternate pricing method or technical information.

Note: Not all subdivisions listed here necessarily appear in this publication.

14600 | Hoists & Cranes

	14610 | Fixed Hoists		CREW	DAILY OUTPUT	LABOR-HOURS	UNIT	2001 BARE COSTS MAT.	LABOR	EQUIP.	TOTAL	TOTAL INCL O&P	
500	0010	MATERIAL HANDLING										500
	1500	Cranes, portable hydraulic, floor type, 2,000 lb capacity				Ea.	1,800			1,800	1,975	
	1600	4,000 lb capacity					3,100			3,100	3,425	
	1800	Movable gantry type, 12′ to 15′ range, 2,000 lb capacity					2,775			2,775	3,050	
	1900	6,000 lb capacity					4,025			4,025	4,425	
	2100	Hoists, electric overhead, chain, hook hung, 15′ lift, 1 ton cap.					1,800			1,800	1,975	
	2200	3 ton capacity					2,250			2,250	2,475	
	2500	5 ton capacity				▼	5,000			5,000	5,500	
	2600	For hand-pushed trolley, add					15%					
	2700	For geared trolley, add					30%					
	2800	For motor trolley, add					75%					
	3000	For lifts over 15′, 1 ton, add				L.F.	17.10			17.10	18.80	
	3100	5 ton, add				″	40			40	43.50	
	3300	Lifts, scissor type, portable, electric, 36″ high, 2,000 lb				Ea.	3,150			3,150	3,475	
	3400	48″ high, 4,000 lb				″	3,600			3,600	3,950	

For information about Means Estimating Seminars, see yellow pages 11 and 12 in back of book

Division 15 Mechanical

Estimating Tips

15100 Building Services Piping

This subdivision is primarily basic pipe and related materials. The pipe may be used by any of the mechanical disciplines, i.e., plumbing, fire protection, heating, and air conditioning.

- The piping section lists the add to labor for elevated pipe installation. These adds apply to all elevated pipe, fittings, valves, insulation, etc., that are placed above 10′ high. CAUTION: the correct percentage may vary for the same pipe. For example, the percentage add for the basic pipe installation should be based on the maximum height that the craftsman must install for that particular section. If the pipe is to be located 14′ above the floor but it is suspended on threaded rod from beams, the bottom flange of which is 18′ high (4′ rods), then the height is actually 18′ and the add is 20%. The pipe coverer, however, does not have to go above the 14′ and so his add should be 10%.
- Most pipe is priced first as straight pipe with a joint (coupling, weld, etc.) every 10′ and a hanger usually every 10′. There are exceptions with hanger spacing such as: for cast iron pipe (5′) and plastic pipe (3 per 10′). Following each type of pipe there are several lines listing sizes and the amount to be subtracted to delete couplings and hangers. This is for pipe that is to be buried or supported together on trapeze hangers. The reason that the couplings are deleted is that these runs are usually long and frequently longer lengths of pipe are used. By deleting the couplings the estimator is expected to look up and add back the correct reduced number of couplings.
- When preparing an estimate it may be necessary to approximate the fittings. Fittings usually run between 25% and 50% of the cost of the pipe. The lower percentage is for simpler runs, and the higher number is for complex areas like mechanical rooms.

15400 Plumbing Fixtures & Equipment

- Plumbing fixture costs usually require two lines, the fixture itself and its "rough-in, supply and waste".
- Remember that gas and oil fired units need venting.

15500 Heat Generation Equipment

- When estimating the cost of an HVAC system, check to see who is responsible for providing and installing the temperature control system. It is possible to overlook controls, assuming that they would be included in the electrical estimate.
- When looking up a boiler be careful on specified capacity. Some manufacturers rate their products on output while others use input.
- Include HVAC insulation for pipe, boiler and duct (wrap and liner).
- Be careful when looking up mechanical items to get the correct pressure rating and connection type (thread, weld, flange).

15700 Heating/Ventilation/ Air Conditioning Equipment

- Combination heating and cooling units are sized by the air conditioning requirements. (See Reference No. R15710-020 for preliminary sizing guide.)
- A ton of air conditioning is nominally 400 CFM.
- Rectangular duct is taken off by the linear foot for each size, but its cost is usually estimated by the pound. Remember that SMACNA standards now base duct on internal pressure.
- Prefabricated duct is estimated and purchased like pipe: straight sections and fittings.
- Note that cranes or other lifting equipment are not included on any lines in Division 15. For example, if a crane is required to lift a heavy piece of pipe into place high above a gym floor, or to put a rooftop unit on the roof of a four-story building, etc., it must be added. Due to the potential for extreme variation—from nothing additional required, to a major crane or helicopter—we feel that including a nominal amount for "lifting contingency" would be useless and detract from the accuracy of the estimate. When using equipment rental from Means do not forget to include the cost of the operator(s).

Reference Numbers

Reference numbers are shown in bold squares at the beginning of some major classifications. These numbers refer to related items in the Reference Section. The reference information may be an estimating procedure, an alternate pricing method or technical information.

Note: Not all subdivisions listed here necessarily appear in this publication.

15 MECHANICAL

		15051 | Mechanical General	Crew	Daily Output	Labor-Hours	Unit	2001 Bare Costs Mat.	Labor	Equip.	Total	Total Incl O&P	
100	0010	**BOILERS, GENERAL** Prices do not include flue piping, elec. wiring,										100
	0020	gas or oil piping, boiler base, pad, or tankless unless noted										
	0100	Boiler H.P.: 10 kW = 34 lbs/steam/hr = 33,475 BTU/hr.										
	0120											
	0150	To convert SFR to BTU rating: Hot water, 150 x SFR;										
	0160	Forced hot water, 180 x SFR; steam, 240 x SFR										
700	0010	**PIPING** See also divisions 02500 & 02600 for site work (R15100-070)										700
	1000	Add to labor for elevated installation										
	1080	10′ to 15′ high						10%				
	1100	15′ to 20′ high						20%				
	1120	20′ to 25′ high						25%				
	1140	25′ to 30′ high						35%				
	1160	30′ to 35′ high						40%				
	1180	35′ to 40′ high						50%				
	1200	Over 40′ high						55%				

		15055 | Selective Mech Demolition	Crew	Daily Output	Labor-Hours	Unit	Mat.	Labor	Equip.	Total	Total Incl O&P	
600	0010	**PLUMBING DEMOLITION** (R15050-720)										600
	1020	Fixtures, including 10′ piping										
	1100	Bath tubs, cast iron	1 Plum	4	2	Ea.		69		69	104	
	1120	Fiberglass		6	1.333			46		46	69.50	
	1140	Steel		5	1.600			55		55	83	
	1200	Lavatory, wall hung		10	.800			27.50		27.50	41.50	
	1220	Counter top		8	1			34.50		34.50	52	
	1300	Sink, steel or cast iron, single		8	1			34.50		34.50	52	
	1320	Double		7	1.143			39.50		39.50	59.50	
	1400	Water closet, floor mounted		8	1			34.50		34.50	52	
	1420	Wall mounted		7	1.143			39.50		39.50	59.50	
	1500	Urinal, floor mounted		4	2			69		69	104	
	1520	Wall mounted		7	1.143			39.50		39.50	59.50	
	1600	Water fountains, free standing		8	1			34.50		34.50	52	
	1620	Recessed		6	1.333			46		46	69.50	
	2000	Piping, metal, to 1-1/2″ diameter		200	.040	L.F.		1.38		1.38	2.08	
	2050	2″ to 3-1/2″ diameter		150	.053			1.84		1.84	2.77	
	2100	4″ to 6″ diameter	2 Plum	100	.160			5.50		5.50	8.30	
	2150	8″ to 14″ diameter		60	.267			9.20		9.20	13.85	
	2153	16″ diameter		55	.291			10		10	15.15	
	2160	Deduct for salvage, aluminum scrap				Ton					525	
	2170	Brass scrap									420	
	2180	Copper scrap									1,025	
	2200	Lead scrap									235	
	2220	Steel scrap									65	
	2250	Water heater, 40 gal.	1 Plum	6	1.333	Ea.		46		46	69.50	
	6000	Remove and reset fixtures, minimum		6	1.333			46		46	69.50	
	6100	Maximum		4	2			69		69	104	

		15060 | Hangers & Supports	Crew	Daily Output	Labor-Hours	Unit	Mat.	Labor	Equip.	Total	Total Incl O&P	
300	0010	**PIPE HANGERS AND SUPPORTS**, TYPE numbers per MSS-SP58										300
	0050	Brackets										
	0060	Beam side or wall, malleable iron, TYPE 34										
	0070	3/8″ threaded rod size	1 Plum	48	.167	Ea.	1.73	5.75		7.48	10.55	
	0080	1/2″ threaded rod size		48	.167		3.33	5.75		9.08	12.30	
	0090	5/8″ threaded rod size		48	.167		4.43	5.75		10.18	13.50	
	0100	3/4″ threaded rod size		48	.167		4.73	5.75		10.48	13.85	
	0110	7/8″ threaded rod size		48	.167		6.10	5.75		11.85	15.35	

	15060 | Hangers & Supports	CREW	DAILY OUTPUT	LABOR-HOURS	UNIT	2001 BARE COSTS MAT.	LABOR	EQUIP.	TOTAL	TOTAL INCL O&P	
300	0120 For concrete installation, add				Ea.		30%				300
	0150 Wall, welded steel, medium, TYPE 32										
	0160 0 size, 12" wide, 18" deep	1 Plum	34	.235	Ea.	126	8.10		134.10	151	
	0170 1 size, 18" wide 24" deep		34	.235		150	8.10		158.10	177	
	0180 2 size, 24" wide, 30" deep		34	.235		198	8.10		206.10	230	
	0200 Beam attachment, welded, TYPE 22										
	0202 3/8"	Q-15	80	.200	Ea.	11.60	6.20	1.26	19.06	23.50	
	0203 1/2"		76	.211		12.35	6.55	1.32	20.22	25	
	0204 5/8"		72	.222		13.10	6.90	1.40	21.40	26.50	
	0205 3/4"		68	.235		18.20	7.30	1.48	26.98	32.50	
	0206 7/8"		64	.250		23	7.75	1.57	32.32	38.50	
	0207 1"		56	.286		24.50	8.85	1.80	35.15	42	
	0300 Clamps										
	0310 C-clamp, for mounting on steel beam flange, w/locknut, TYPE 23										
	0320 3/8" threaded rod size	1 Plum	160	.050	Ea.	2.02	1.72		3.74	4.82	
	0330 1/2" threaded rod size		160	.050		2.42	1.72		4.14	5.25	
	0340 5/8" threaded rod size		160	.050		3.46	1.72		5.18	6.40	
	0350 3/4" threaded rod size		160	.050		5	1.72		6.72	8.10	
	0352 7/8" threaded rod size		140	.057		13.10	1.97		15.07	17.35	
	0400 High temperature to 1050°F, alloy steel										
	0410 4" pipe size	Q-1	106	.151	Ea.	31	4.68		35.68	41	
	0420 6" pipe size		106	.151		52	4.68		56.68	64.50	
	0430 8" pipe size		97	.165		58.50	5.10		63.60	72	
	0440 10" pipe size		84	.190		79.50	5.90		85.40	96.50	
	0450 12" pipe size		72	.222		92.50	6.90		99.40	112	
	0460 14" pipe size		64	.250		219	7.75		226.75	253	
	0470 16" pipe size		56	.286		253	8.85		261.85	291	
	0480 Beam clamp, flange type, TYPE 25										
	0482 For 3/8" bolt	1 Plum	48	.167	Ea.	5.60	5.75		11.35	14.85	
	0483 For 1/2" bolt		44	.182		6.30	6.25		12.55	16.40	
	0484 For 5/8" bolt		40	.200		8.15	6.90		15.05	19.35	
	0485 For 3/4" bolt		36	.222		12.60	7.65		20.25	25.50	
	0486 For 1" bolt		32	.250		36.50	8.60		45.10	53	
	0500 I-beam, for mounting on bottom flange, strap iron, TYPE 21										
	0510 2" flange size	1 Plum	96	.083	Ea.	3.72	2.87		6.59	8.40	
	0520 3" flange size		95	.084		3.84	2.90		6.74	8.60	
	0530 4" flange size		93	.086		3.93	2.96		6.89	8.80	
	0540 5" flange size		92	.087		4.08	3		7.08	9	
	0550 6" flange size		90	.089		4.17	3.06		7.23	9.20	
	0560 7" flange size		88	.091		4.31	3.13		7.44	9.45	
	0570 8" flange size		86	.093		4.46	3.20		7.66	9.75	
	0600 One hole, vertical mounting, malleable iron										
	0610 1/2" pipe size	1 Plum	160	.050	Ea.	.71	1.72		2.43	3.38	
	0620 3/4" pipe size		145	.055		.88	1.90		2.78	3.84	
	0630 1" pipe size		136	.059		1.33	2.03		3.36	4.52	
	0640 1-1/4" pipe size		128	.063		1.66	2.15		3.81	5.10	
	0650 1-1/2" pipe size		120	.067		1.87	2.30		4.17	5.55	
	0660 2" pipe size		112	.071		2.82	2.46		5.28	6.80	
	0670 2-1/2" pipe size		104	.077		7.70	2.65		10.35	12.45	
	0680 3" pipe size		96	.083		10.65	2.87		13.52	16.10	
	0690 3-1/2" pipe size		90	.089		10.65	3.06		13.71	16.35	
	0700 4" pipe size		84	.095		41.50	3.28		44.78	50.50	
	0750 Riser or extension pipe, carbon steel, TYPE 8										
	0756 1/2" pipe size	1 Plum	52	.154	Ea.	2.61	5.30		7.91	10.85	
	0760 3/4" pipe size		48	.167		2.61	5.75		8.36	11.50	
	0770 1" pipe size		47	.170		2.63	5.85		8.48	11.75	

15060 | Hangers & Supports

	Line	Description	Crew	Daily Output	Labor-Hours	Unit	2001 Bare Costs Mat.	Labor	Equip.	Total	Total Incl O&P	
300	0780	1-1/4" pipe size	1 Plum	46	.174	Ea.	3.31	6		9.31	12.70	300
	0790	1-1/2" pipe size		45	.178		3.56	6.10		9.66	13.15	
	0800	2" pipe size		43	.186		3.62	6.40		10.02	13.65	
	0810	2-1/2" pipe size		41	.195		3.80	6.70		10.50	14.35	
	0820	3" pipe size		40	.200		4.05	6.90		10.95	14.85	
	0830	3-1/2" pipe size		39	.205		5	7.05		12.05	16.15	
	0840	4" pipe size		38	.211		5.10	7.25		12.35	16.55	
	0850	5" pipe size		37	.216		7.45	7.45		14.90	19.40	
	0860	6" pipe size		36	.222		8.55	7.65		16.20	21	
	0870	8" pipe size		34	.235		13.85	8.10		21.95	27.50	
	0880	10" pipe size		32	.250		22.50	8.60		31.10	38	
	0890	12" pipe size		28	.286		27	9.85		36.85	44.50	
	0900	For plastic coating 3/4" to 4", add					190%					
	0910	For copper plating 3/4" to 4", add					58%					
	0950	Two piece, complete, carbon steel, medium weight, TYPE 4										
	0960	1/2" pipe size	Q-1	137	.117	Ea.	2.10	3.62		5.72	7.75	
	0970	3/4" pipe size		134	.119		2.10	3.70		5.80	7.90	
	0980	1" pipe size		132	.121		2.22	3.76		5.98	8.10	
	0990	1-1/4" pipe size		130	.123		2.76	3.82		6.58	8.80	
	1000	1-1/2" pipe size		126	.127		2.76	3.94		6.70	9	
	1010	2" pipe size		124	.129		3.41	4		7.41	9.80	
	1020	2-1/2" pipe size		120	.133		3.67	4.13		7.80	10.30	
	1030	3" pipe size		117	.137		4.09	4.24		8.33	10.90	
	1040	3-1/2" pipe size		114	.140		5.80	4.35		10.15	12.95	
	1050	4" pipe size		110	.145		5.80	4.51		10.31	13.20	
	1060	5" pipe size		106	.151		8.40	4.68		13.08	16.30	
	1070	6" pipe size		104	.154		14.15	4.77		18.92	23	
	1080	8" pipe size		100	.160		16.90	4.96		21.86	26	
	1090	10" pipe size		96	.167		30.50	5.15		35.65	41.50	
	1100	12" pipe size		89	.180		35.50	5.55		41.05	47.50	
	1110	14" pipe size		82	.195		61.50	6.05		67.55	76.50	
	1120	16" pipe size		68	.235		66.50	7.30		73.80	84	
	1130	For galvanized, add					45%					
	1150	Insert, concrete										
	1160	Wedge type, carbon steel body, malleable iron nut										
	1170	1/4" threaded rod size	1 Plum	96	.083	Ea.	4.75	2.87		7.62	9.60	
	1180	3/8" threaded rod size		96	.083		4.99	2.87		7.86	9.85	
	1190	1/2" threaded rod size		96	.083		5.15	2.87		8.02	10.05	
	1200	5/8" threaded rod size		96	.083		5.25	2.87		8.12	10.10	
	1210	3/4" threaded rod size		96	.083		5.75	2.87		8.62	10.70	
	1220	7/8" threaded rod size		96	.083		6.70	2.87		9.57	11.70	
	1230	For galvanized, add					1.07			1.07	1.18	
	1250	Pipe guide sized for insulation										
	1260	No. 1, 1" pipe size, 1" thick insulation	1 Stpi	26	.308	Ea.	125	10.70		135.70	154	
	1270	No. 2, 1-1/4"-2" pipe size, 1" thick insulation		23	.348		148	12.10		160.10	181	
	1280	No. 3, 1-1/4"-2" pipe size, 1-1/2" thick insulation		21	.381		148	13.30		161.30	183	
	1290	No. 4, 2-1/2"-3-1/2" pipe size, 1-1/2" thick insulation		18	.444		148	15.50		163.50	187	
	1300	No. 5, 4"-5" pipe size, 1-1/2" thick insulation		16	.500		164	17.45		181.45	208	
	1310	No. 6, 5"-6" pipe size, 2" thick insulation	Q-5	21	.762		182	24		206	237	
	1320	No. 7, 8" pipe size, 2" thick insulation		16	1		320	31.50		351.50	400	
	1330	No. 8, 10" pipe size, 2" thick insulation		12	1.333		605	42		647	730	
	1340	No. 9, 12" pipe size, 2" thick insulation	Q-6	17	1.412		605	46		651	735	
	1350	No. 10, 12"-14" pipe size, 2-1/2" thick insulation		16	1.500		785	49		834	940	
	1360	No. 11, 16" pipe size, 2-1/2" thick insulation		10.50	2.286		785	74.50		859.50	975	
	1370	No. 12, 16"-18" pipe size, 3" thick insulation		9	2.667		830	87		917	1,050	
	1380	No. 13, 20" pipe size, 3" thick insulation		7.50	3.200		830	104		934	1,075	

	15060 | Hangers & Supports	CREW	DAILY OUTPUT	LABOR-HOURS	UNIT	2001 BARE COSTS MAT.	LABOR	EQUIP.	TOTAL	TOTAL INCL O&P	
300 1390	No. 14, 24" pipe size, 3" thick insulation	Q-6	7	3.429	Ea.	1,050	112		1,162	1,350	300
1400	Bands										
1410	Adjustable band, carbon steel, for non-insulated pipe, TYPE 7										
1420	1/2" pipe size	Q-1	142	.113	Ea.	.82	3.49		4.31	6.15	
1430	3/4" pipe size		140	.114		.82	3.54		4.36	6.25	
1440	1" pipe size		137	.117		.93	3.62		4.55	6.45	
1450	1-1/4" pipe size		134	.119		.97	3.70		4.67	6.65	
1460	1-1/2" pipe size		131	.122		1.09	3.79		4.88	6.90	
1470	2" pipe size		129	.124		1.50	3.84		5.34	7.45	
1480	2-1/2" pipe size		125	.128		1.58	3.97		5.55	7.75	
1490	3" pipe size		122	.131		1.76	4.07		5.83	8.10	
1500	3-1/2" pipe size		119	.134		2.73	4.17		6.90	9.30	
1510	4" pipe size		114	.140		2.73	4.35		7.08	9.55	
1520	5" pipe size		110	.145		2.98	4.51		7.49	10.10	
1530	6" pipe size		108	.148		4	4.59		8.59	11.35	
1540	8" pipe size	↓	104	.154	↓	6.25	4.77		11.02	14.05	
1550	For copper plated, add					50%					
1560	For galvanized, add					30%					
1570	For plastic coating, add					30%					
1600	Adjusting nut malleable iron, steel band, TYPE 9										
1610	1/2" pipe size, galvanized band	Q-1	137	.117	Ea.	.49	3.62		4.11	6	
1620	3/4" pipe size, galvanized band		135	.119		.49	3.67		4.16	6.10	
1630	1" pipe size, gavanized band		132	.121		.49	3.76		4.25	6.20	
1640	1-1/4" pipe size, galvanized band		129	.124		.54	3.84		4.38	6.40	
1650	1-1/2" pipe size, galvanized band		126	.127		.54	3.94		4.48	6.55	
1660	2" pipe size, galvanized band		124	.129		.57	4		4.57	6.70	
1670	2-1/2" pipe size, galvanized band		120	.133		1.08	4.13		5.21	7.45	
1680	3" pipe size, galvanized band		117	.137		1.17	4.24		5.41	7.70	
1690	3-1/2" pipe size, galvanized band		114	.140		1.45	4.35		5.80	8.15	
1700	4" pipe size, cadmium plated band		110	.145		2.17	4.51		6.68	9.20	
1710	5" pipe size, cadmium plated band		106	.151		2.94	4.68		7.62	10.30	
1720	6" pipe size, cadmium plated band		104	.154		3.29	4.77		8.06	10.80	
1730	8" pipe size, cadmium plated band	↓	100	.160		3.70	4.96		8.66	11.55	
1740	For plastic coated band, add					35%					
1750	For completely copper coated, add				↓	45%					
1800	Clevis, adjustable, carbon steel, for non-insulated pipe, TYPE 1										
1810	1/2" pipe size	Q-1	137	.117	Ea.	.81	3.62		4.43	6.35	
1820	3/4" pipe size		135	.119		.81	3.67		4.48	6.45	
1830	1" pipe size		132	.121		.86	3.76		4.62	6.60	
1840	1-1/4" pipe size		129	.124		.97	3.84		4.81	6.85	
1850	1-1/2" pipe size		126	.127		1.07	3.94		5.01	7.15	
1860	2" pipe size		124	.129		1.18	4		5.18	7.35	
1870	2-1/2" pipe size		120	.133		1.92	4.13		6.05	8.35	
1880	3" pipe size		117	.137		2.42	4.24		6.66	9.05	
1890	3-1/2" pipe size		114	.140		2.53	4.35		6.88	9.35	
1900	4" pipe size		110	.145		2.92	4.51		7.43	10	
1910	5" pipe size		106	.151		4.03	4.68		8.71	11.50	
1920	6" pipe size		104	.154		4.64	4.77		9.41	12.30	
1930	8" pipe size		100	.160		7.65	4.96		12.61	15.90	
1940	10" pipe size		96	.167		20.50	5.15		25.65	30.50	
1950	12" pipe size		89	.180		26	5.55		31.55	37	
1960	14" pipe size		82	.195		40	6.05		46.05	53	
1970	16" pipe size		68	.235		56.50	7.30		63.80	73.50	
1971	18" pipe size		54	.296		67.50	9.20		76.70	88.50	
1972	20" pipe size	↓	38	.421	↓	133	13.05		146.05	166	
1980	For galvanized, add					66%					

	15060 | Hangers & Supports		CREW	DAILY OUTPUT	LABOR-HOURS	UNIT	2001 BARE COSTS MAT.	LABOR	EQUIP.	TOTAL	TOTAL INCL O&P	
300	1990	For copper plated 1/2" to 4", add					77%					300
	2000	For light weight 1/2" to 4", deduct					13%					
	2010	Insulated pipe type, 3/4" to 12" pipe, add					180%					
	2020	Insulated pipe type, chrome-moly U-strap, add					530%					
	2250	Split ring, malleable iron, for non-insulated pipe, TYPE 11										
	2260	1/2" pipe size	Q-1	137	.117	Ea.	2.94	3.62		6.56	8.70	
	2270	3/4" pipe size		135	.119		2.96	3.67		6.63	8.80	
	2280	1" pipe size		132	.121		3.13	3.76		6.89	9.10	
	2290	1-1/4" pipe size		129	.124		3.88	3.84		7.72	10.05	
	2300	1-1/2" pipe size		126	.127		4.72	3.94		8.66	11.15	
	2310	2" pipe size		124	.129		5.35	4		9.35	11.95	
	2320	2-1/2" pipe size		120	.133		7.70	4.13		11.83	14.75	
	2330	3" pipe size		117	.137		9.65	4.24		13.89	17	
	2340	3-1/2" pipe size		114	.140		13.35	4.35		17.70	21.50	
	2350	4" pipe size		110	.145		13.35	4.51		17.86	21.50	
	2360	5" pipe size		106	.151		30	4.68		34.68	40	
	2370	6" pipe size		104	.154		30	4.77		34.77	40	
	2380	8" pipe size		100	.160		53	4.96		57.96	65.50	
	2390	For copper plated, add					8%					
	2500	Washer, flat steel										
	2502	3/8"	1 Plum	240	.033	Ea.	.04	1.15		1.19	1.77	
	2503	1/2"		220	.036		.07	1.25		1.32	1.97	
	2504	5/8"		200	.040		.15	1.38		1.53	2.25	
	2505	3/4"		180	.044		.11	1.53		1.64	2.43	
	2506	7/8"		160	.050		.15	1.72		1.87	2.77	
	2507	1"		140	.057		.16	1.97		2.13	3.15	
	2508	1-1/8"		120	.067		.20	2.30		2.50	3.69	
	2520	Nut, steel, hex										
	2522	3/8"	1 Plum	200	.040	Ea.	.22	1.38		1.60	2.32	
	2523	1/2"		180	.044		.39	1.53		1.92	2.74	
	2524	5/8"		160	.050		.82	1.72		2.54	3.50	
	2525	3/4"		140	.057		1.10	1.97		3.07	4.18	
	2526	7/8"		120	.067		1.85	2.30		4.15	5.50	
	2527	1"		100	.080		2	2.76		4.76	6.35	
	2528	1-1/8"		80	.100		2.36	3.45		5.81	7.80	
	2532	Turnbuckle, TYPE 13										
	2534	3/8"	1 Plum	80	.100	Ea.	13.50	3.45		16.95	20	
	2535	1/2"		72	.111		16.90	3.83		20.73	24.50	
	2536	5/8"		64	.125		17.55	4.31		21.86	26	
	2538	7/8"		48	.167		32.50	5.75		38.25	44	
	2539	1"		40	.200		39	6.90		45.90	53.50	
	2540	1-1/8"		32	.250		70	8.60		78.60	90	
	2544	Eye rod, welded										
	2546	3/8"	1 Plum	56	.143	Ea.	17.95	4.92		22.87	27	
	2547	1/2"		48	.167		26.50	5.75		32.25	37.50	
	2548	5/8"		40	.200		40	6.90		46.90	55	
	2549	3/4"		32	.250		51	8.60		59.60	69	
	2550	7/8"		24	.333		62.50	11.50		74	86.50	
	2551	1"		16	.500		79	17.25		96.25	113	
	2556	Eye rod, welded, linked										
	2558	3/8"	1 Plum	56	.143	Ea.	55.50	4.92		60.42	68.50	
	2559	1/2"		48	.167		75.50	5.75		81.25	91.50	
	2560	5/8"		40	.200		108	6.90		114.90	129	
	2561	3/4"		32	.250		140	8.60		148.60	167	
	2562	7/8"		24	.333		171	11.50		182.50	205	
	2563	1"		16	.500		220	17.25		237.25	267	

Important: See the Reference Section for critical supporting data - Reference Nos., Crews, & City Cost Indexes

15060 | Hangers & Supports

300		CREW	DAILY OUTPUT	LABOR-HOURS	UNIT	2001 BARE COSTS MAT.	LABOR	EQUIP.	TOTAL	TOTAL INCL O&P
2650	Rods, carbon steel									
2660	Continuous thread									
2670	1/4″ thread size	1 Plum	144	.056	L.F.	.45	1.91		2.36	3.39
2680	3/8″ thread size		144	.056		.50	1.91		2.41	3.44
2690	1/2″ thread size		144	.056		.78	1.91		2.69	3.75
2700	5/8″ thread size		144	.056		1.11	1.91		3.02	4.11
2710	3/4″ thread size		144	.056		1.97	1.91		3.88	5.05
2720	7/8″ thread size		144	.056		2.47	1.91		4.38	5.60
2721	1″ thread size	Q-1	160	.100		4.19	3.10		7.29	9.30
2722	1-1/8″ thread size	″	120	.133		5.35	4.13		9.48	12.15
2725	1/4″ thread size, bright finish	1 Plum	144	.056		.84	1.91		2.75	3.81
2726	1/2″ thread size, bright finish	″	144	.056		1.47	1.91		3.38	4.51
2730	For galvanized, add					40%				
2750	Both ends machine threaded 18″ length									
2760	3/8″ thread size	1 Plum	240	.033	Ea.	3.58	1.15		4.73	5.65
2770	1/2″ thread size		240	.033		5.75	1.15		6.90	8.05
2780	5/8″ thread size		240	.033		8.25	1.15		9.40	10.80
2790	3/4″ thread size		240	.033		12.65	1.15		13.80	15.65
2800	7/8″ thread size		240	.033		22	1.15		23.15	25.50
2810	1″ thread size		240	.033		31.50	1.15		32.65	36
2820	Rod couplings									
2821	3/8″	1 Plum	60	.133	Ea.	1.26	4.59		5.85	8.35
2822	1/2″		54	.148		1.60	5.10		6.70	9.45
2823	5/8″		48	.167		2.46	5.75		8.21	11.35
2824	3/4″		44	.182		3.59	6.25		9.84	13.40
2825	7/8″		40	.200		5.15	6.90		12.05	16.05
2826	1″		34	.235		6.90	8.10		15	19.85
2827	1-1/8″		30	.267		12.90	9.20		22.10	28
2860	Pipe hanger assy, adj. clevis, saddle, rod, clamp, insul. allowance									
2864	1/2″ pipe size	Q-5	35	.457	Ea.	21.50	14.35		35.85	45
2868	1″ pipe size		34.60	.462		27	14.50		41.50	51.50
2872	2″ pipe size		33.30	.480		30.50	15.10		45.60	56.50
2876	3″ pipe size		31.20	.513		46	16.10		62.10	75
2880	4″ pipe size		30.70	.521		48.50	16.35		64.85	78
2884	6″ pipe size		29.80	.537		60.50	16.85		77.35	92.50
2888	8″ pipe size		28	.571		96.50	17.95		114.45	133
2892	10″ pipe size		25.20	.635		132	19.90		151.90	175
2896	12″ pipe size		23.20	.690		146	21.50		167.50	194
2900	Rolls									
2910	Adjustable yoke, carbon steel with CI roll, TYPE 43									
2918	2″ pipe size	Q-1	140	.114	Ea.	11.75	3.54		15.29	18.30
2920	2-1/2″ pipe size		137	.117		11.75	3.62		15.37	18.40
2930	3″ pipe size		131	.122		13.10	3.79		16.89	20
2940	3-1/2″ pipe size		124	.129		17.05	4		21.05	25
2950	4″ pipe size		117	.137		17.65	4.24		21.89	26
2960	5″ pipe size		110	.145		20.50	4.51		25.01	29.50
2970	6″ pipe size		104	.154		25.50	4.77		30.27	35.50
2980	8″ pipe size		96	.167		38.50	5.15		43.65	50
2990	10″ pipe size		80	.200		45.50	6.20		51.70	59.50
3010	14″ pipe size		56	.286		180	8.85		188.85	211
3020	16″ pipe size		48	.333		231	10.35		241.35	270
3050	Chair, carbon steel with CI roll									
3060	2″ pipe size	1 Plum	68	.118	Ea.	15.90	4.05		19.95	23.50
3070	2-1/2″ pipe size		65	.123		17.20	4.24		21.44	25.50
3080	3″ pipe size		62	.129		18.60	4.45		23.05	27
3090	3-1/2″ pipe size		60	.133		22.50	4.59		27.09	31.50

15060 | Hangers & Supports

		15060 Hangers & Supports	CREW	DAILY OUTPUT	LABOR-HOURS	UNIT	2001 BARE COSTS MAT.	LABOR	EQUIP.	TOTAL	TOTAL INCL O&P	
300	3100	4" pipe size	1 Plum	58	.138	Ea.	25	4.75		29.75	34.50	300
	3110	5" pipe size		56	.143		27.50	4.92		32.42	37.50	
	3120	6" pipe size		53	.151		38	5.20		43.20	50	
	3130	8" pipe size		50	.160		51.50	5.50		57	65	
	3140	10" pipe size		48	.167		64.50	5.75		70.25	79.50	
	3150	12" pipe size		46	.174		94	6		100	112	
	3170	Single pipe roll, (see line 2650 for rods), TYPE 41, 1" pipe size	Q-1	137	.117		8.50	3.62		12.12	14.80	
	3180	1-1/4" pipe size		131	.122		8.50	3.79		12.29	15.05	
	3190	1-1/2" pipe size		129	.124		8.60	3.84		12.44	15.30	
	3200	2" pipe size		124	.129		9.05	4		13.05	16	
	3210	2-1/2" pipe size		118	.136		9.70	4.20		13.90	17.05	
	3220	3" pipe size		115	.139		10.25	4.31		14.56	17.80	
	3230	3-1/2" pipe size		113	.142		11.95	4.39		16.34	19.75	
	3240	4" pipe size		112	.143		13.05	4.43		17.48	21	
	3250	5" pipe size		110	.145		14.45	4.51		18.96	22.50	
	3260	6" pipe size		101	.158		18.75	4.91		23.66	28	
	3270	8" pipe size		90	.178		29	5.50		34.50	40.50	
	3280	10" pipe size		80	.200		34.50	6.20		40.70	47.50	
	3290	12" pipe size		68	.235		50	7.30		57.30	66	
	3300	Saddles (add vertical pipe riser, usually 3" diameter)										
	3310	Pipe support, complete, adjust., CI saddle, TYPE 36										
	3320	2-1/2" pipe size	1 Plum	96	.083	Ea.	127	2.87		129.87	144	
	3330	3" pipe size		88	.091		129	3.13		132.13	147	
	3340	3-1/2" pipe size		79	.101		130	3.49		133.49	148	
	3350	4" pipe size		68	.118		168	4.05		172.05	191	
	3360	5" pipe size		64	.125		174	4.31		178.31	198	
	3370	6" pipe size		59	.136		174	4.67		178.67	198	
	3380	8" pipe size		53	.151		184	5.20		189.20	210	
	3390	10" pipe size		50	.160		203	5.50		208.50	232	
	3400	12" pipe size		48	.167		222	5.75		227.75	253	
	3450	For standard pipe support, one piece, CI deduct					34%					
	3460	For stanchion support, CI with steel yoke, deduct					60%					
	3500	Multiple pipe support 2' long strut	1 Plum	30	.267	Ea.	4.40	9.20		13.60	18.70	
	3510	3' long strut		20	.400		6.60	13.80		20.40	28.50	
	3520	4' long strut		17	.471		8.80	16.20		25	34	
	3550	Insulation shield 1" thick, 1/2" pipe size, TYPE 40	1 Asbe	100	.080		5.55	2.57		8.12	10.25	
	3560	3/4" pipe size		100	.080		5.55	2.57		8.12	10.25	
	3570	1" pipe size		98	.082		6.30	2.62		8.92	11.15	
	3580	1-1/4" pipe size		98	.082		6.90	2.62		9.52	11.80	
	3590	1-1/2" pipe size		96	.083		6.95	2.68		9.63	11.95	
	3600	2" pipe size		96	.083		7.30	2.68		9.98	12.35	
	3610	2-1/2" pipe size		94	.085		7.65	2.74		10.39	12.85	
	3620	3" pipe size		94	.085		8	2.74		10.74	13.20	
	3630	2" thick, 3-1/2" pipe size		92	.087		11	2.80		13.80	16.60	
	3640	4" pipe size		92	.087		11	2.80		13.80	16.60	
	3650	5" pipe size		90	.089		16.55	2.86		19.41	23	
	3660	6" pipe size		90	.089		18.65	2.86		21.51	25	
	3670	8" pipe size		88	.091		28	2.92		30.92	35	
	3680	10" pipe size		88	.091		41.50	2.92		44.42	50	
	3690	12" pipe size		86	.093		47.50	2.99		50.49	57	
	3700	14" pipe size		86	.093		47.50	2.99		50.49	57.50	
	3710	16" pipe size		84	.095		64	3.06		67.06	75.50	
	3720	18" pipe size		84	.095		64	3.06		67.06	75.50	
	3730	20" pipe size		82	.098		71	3.14		74.14	83	
	3732	24" pipe size		80	.100		84.50	3.22		87.72	98	
	3733	30" pipe size		78	.103		92	3.30		95.30	106	

15 MECHANICAL

15060 | Hangers & Supports		CREW	DAILY OUTPUT	LABOR-HOURS	UNIT	2001 BARE COSTS MAT.	LABOR	EQUIP.	TOTAL	TOTAL INCL O&P
300 3735	36" pipe size	1 Asbe	75	.107	Ea.	101	3.43		104.43	117
3750	Covering protection saddle, TYPE 39									
3760	1" covering size									
3770	3/4" pipe size	1 Plum	68	.118	Ea.	10	4.05		14.05	17.10
3780	1" pipe size		68	.118		10	4.05		14.05	17.10
3790	1-1/4" pipe size		68	.118		10	4.05		14.05	17.10
3800	1-1/2" pipe size		66	.121		10.85	4.18		15.03	18.25
3810	2" pipe size		66	.121		10.85	4.18		15.03	18.25
3820	2-1/2" pipe size		64	.125		10.85	4.31		15.16	18.45
3830	3" pipe size		64	.125		15.65	4.31		19.96	23.50
3840	3-1/2" pipe size		62	.129		16.70	4.45		21.15	25
3850	4" pipe size		62	.129		16.70	4.45		21.15	25
3860	5" pipe size		60	.133		16.70	4.59		21.29	25.50
3870	6" pipe size	↓	60	.133	↓	19.90	4.59		24.49	29
3900	1-1/2" covering size									
3910	3/4" pipe size	1 Plum	68	.118	Ea.	12.50	4.05		16.55	19.85
3920	1" pipe size		68	.118		12.50	4.05		16.55	19.85
3930	1-1/4" pipe size		68	.118		12.50	4.05		16.55	19.85
3940	1-1/2" pipe size		66	.121		12.50	4.18		16.68	20
3950	2" pipe size		66	.121		13.55	4.18		17.73	21.50
3960	2-1/2" pipe size		64	.125		16	4.31		20.31	24
3970	3" pipe size		64	.125		16	4.31		20.31	24
3980	3-1/2" pipe size		62	.129		16.25	4.45		20.70	24.50
3990	4" pipe size		62	.129		16.25	4.45		20.70	24.50
4000	5" pipe size		60	.133		16.25	4.59		20.84	25
4010	6" pipe size		60	.133		23.50	4.59		28.09	32.50
4020	8" pipe size		58	.138		29.50	4.75		34.25	39
4022	10" pipe size		56	.143		29.50	4.92		34.42	39.50
4024	12" pipe size	↓	54	.148	↓	43	5.10		48.10	54.50
4028	2" covering size									
4029	2-1/2" pipe size	1 Plum	62	.129	Ea.	15.35	4.45		19.80	23.50
4032	3" pipe size		60	.133		18.05	4.59		22.64	27
4033	4" pipe size		58	.138		18.05	4.75		22.80	27
4034	6" pipe size		56	.143		26	4.92		30.92	36
4035	8" pipe size		54	.148		30.50	5.10		35.60	41
4080	10" pipe size		58	.138		32.50	4.75		37.25	42.50
4090	12" pipe size		56	.143		48	4.92		52.92	60.50
4100	14" pipe size		56	.143		48	4.92		52.92	60.50
4110	16" pipe size		54	.148		67.50	5.10		72.60	81.50
4120	18" pipe size		54	.148		67.50	5.10		72.60	81.50
4130	20" pipe size		52	.154		75	5.30		80.30	90.50
4150	24" pipe size		50	.160		87.50	5.50		93	104
4160	30" pipe size		48	.167		96	5.75		101.75	115
4180	36" pipe size	↓	45	.178	↓	106	6.10		112.10	126
4186	2-1/2" covering size									
4187	3" pipe size	1 Plum	58	.138	Ea.	17.45	4.75		22.20	26.50
4188	4" pipe size		56	.143		17.45	4.92		22.37	26.50
4189	6" pipe size		52	.154		29	5.30		34.30	40
4190	8" pipe size		48	.167		35.50	5.75		41.25	48
4191	10" pipe size		44	.182		39	6.25		45.25	52.50
4192	12" pipe size		40	.200		52	6.90		58.90	68
4193	14" pipe size		36	.222		52	7.65		59.65	69
4194	16" pipe size		32	.250		77	8.60		85.60	97.50
4195	18" pipe size	↓	28	.286	↓	82	9.85		91.85	105
4200	Sockets									
4210	Rod end, malleable iron, TYPE 16									300

	15060 | Hangers & Supports	CREW	DAILY OUTPUT	LABOR-HOURS	UNIT	2001 BARE COSTS MAT.	LABOR	EQUIP.	TOTAL	TOTAL INCL O&P	
300	4220 1/4" thread size	1 Plum	240	.033	Ea.	.92	1.15		2.07	2.74	300
	4230 3/8" thread size		240	.033		1.49	1.15		2.64	3.37	
	4240 1/2" thread size		230	.035		1.78	1.20		2.98	3.77	
	4250 5/8" thread size		225	.036		3.52	1.23		4.75	5.70	
	4260 3/4" thread size		220	.036		5.75	1.25		7	8.20	
	4270 7/8" thread size		210	.038		11.25	1.31		12.56	14.35	
	4290 Strap, 1/2" pipe size, TYPE 26	Q-1	142	.113		2.23	3.49		5.72	7.70	
	4300 3/4" pipe size		140	.114		2.27	3.54		5.81	7.85	
	4310 1" pipe size		137	.117		2.32	3.62		5.94	8	
	4320 1-1/4" pipe size		134	.119		2.61	3.70		6.31	8.45	
	4330 1-1/2" pipe size		131	.122		3.27	3.79		7.06	9.30	
	4340 2" pipe size		129	.124		3.37	3.84		7.21	9.50	
	4350 2-1/2" pipe size		125	.128		5.90	3.97		9.87	12.45	
	4360 3" pipe size		122	.131		6.45	4.07		10.52	13.25	
	4370 3-1/2" pipe size		119	.134		10.80	4.17		14.97	18.20	
	4380 4" pipe size		114	.140		12.10	4.35		16.45	19.85	
	4400 U-bolt, carbon steel										
	4410 Standard, with nuts, TYPE 42										
	4420 1/2" pipe size	1 Plum	160	.050	Ea.	1.04	1.72		2.76	3.74	
	4430 3/4" pipe size		158	.051		1.04	1.74		2.78	3.77	
	4450 1" pipe size		152	.053		1.14	1.81		2.95	3.99	
	4460 1-1/4" pipe size		148	.054		1.26	1.86		3.12	4.20	
	4470 1-1/2" pipe size		143	.056		1.31	1.93		3.24	4.35	
	4480 2" pipe size		139	.058		1.35	1.98		3.33	4.48	
	4490 2-1/2" pipe size		134	.060		1.87	2.06		3.93	5.15	
	4500 3" pipe size		128	.063		3.42	2.15		5.57	7	
	4510 3-1/2" pipe size		122	.066		3.54	2.26		5.80	7.30	
	4520 4" pipe size		117	.068		3.66	2.36		6.02	7.60	
	4530 5" pipe size		114	.070		3.92	2.42		6.34	7.95	
	4540 6" pipe size		111	.072		3.96	2.48		6.44	8.10	
	4550 8" pipe size		109	.073		4.74	2.53		7.27	9	
	4560 10" pipe size		107	.075		8.30	2.58		10.88	13	
	4570 12" pipe size		104	.077		10.60	2.65		13.25	15.70	
	4580 For plastic coating on 1/2" thru 6" size, add					150%					
	4700 U-hook, carbon steel, requires mounting screws or bolts										
	4710 3/4" thru 2" pipe size										
	4720 6" long	1 Plum	96	.083	Ea.	3.20	2.87		6.07	7.85	
	4730 8" long		96	.083		3.42	2.87		6.29	8.10	
	4740 10" long		96	.083		4.32	2.87		7.19	9.10	
	4750 12" long		96	.083		4.79	2.87		7.66	9.60	
	4760 For copper plated, add					50%					
	8000 Pipe clamp, plastic, 1/2" CTS	1 Plum	80	.100		.20	3.45		3.65	5.40	
	8010 3/4" CTS		73	.110		.21	3.78		3.99	5.95	
	8020 1" CTS		68	.118		.47	4.05		4.52	6.60	
	8080 Economy clamp, 1/4" CTS		175	.046		.05	1.57		1.62	2.44	
	8090 3/8" CTS		168	.048		.06	1.64		1.70	2.55	
	8100 1/2" CTS		160	.050		.06	1.72		1.78	2.67	
	8110 3/4" CTS		145	.055		.12	1.90		2.02	3	
	8200 Half clamp, 1/2" CTS		80	.100		.18	3.45		3.63	5.40	
	8210 3/4" CTS		73	.110		.19	3.78		3.97	5.90	
	8300 Suspension clamp, 1/2" CTS		80	.100		.20	3.45		3.65	5.40	
	8310 3/4" CTS		73	.110		.21	3.78		3.99	5.95	
	8320 1" CTS		68	.118		.58	4.05		4.63	6.75	
	8400 Insulator, 1/2" CTS		80	.100		.34	3.45		3.79	5.55	
	8410 3/4" CTS		73	.110		.35	3.78		4.13	6.10	
	8420 1" CTS		68	.118		.37	4.05		4.42	6.50	

	15060 | Hangers & Supports		Crew	Daily Output	Labor-Hours	Unit	2001 Bare Costs Mat.	Labor	Equip.	Total	Total Incl O&P	
300	8500	J hook clamp with nail 1/2" CTS	1 Plum	240	.033	Ea.	.10	1.15		1.25	1.84	300
	8501	3/4" CTS	↓	240	.033	↓	.10	1.15		1.25	1.84	
	15070 | Mech Sound/Vibration/Seismic Ctrl											
800	0010	VIBRATION ABSORBERS										800
	0100	Hangers, neoprene flex										
	0200	10 - 120 lb. capacity				Ea.	9.05			9.05	9.95	
	0220	75 - 550 lb. capacity					13.50			13.50	14.85	
	0240	250 - 1100 lb. capacity					25			25	27.50	
	0260	1000 - 4000 lb. capacity					56			56	61.50	
	0500	Spring flex, 60 lb. capacity					15.80			15.80	17.35	
	0520	450 lb. capacity					22.50			22.50	25	
	0540	900 lb. capacity					22.50			22.50	25	
	0560	1100-1300 lb. capacity				↓	26			26	28.50	
	0600	Rubber in shear										
	0610	45 - 340 lb., up to 1/2" rod size	1 Stpi	22	.364	Ea.	11	12.65		23.65	31.50	
	0620	130 - 700 lb., up to 3/4" rod size		20	.400		26	13.95		39.95	49.50	
	0630	50 - 1000 lb., up to 3/4" rod size	↓	18	.444		31	15.50		46.50	57.50	
	1000	Mounts, neoprene, 45 - 380 lb. capacity					7.55			7.55	8.30	
	1020	250 - 1100 lb. capacity					22.50			22.50	25	
	1040	1000 - 4000 lb. capacity					50.50			50.50	55.50	
	1100	Spring flex, 60 lb. capacity					30.50			30.50	33.50	
	1120	165 lb. capacity					30.50			30.50	33.50	
	1140	260 lb. capacity					30.50			30.50	33.50	
	1160	450 lb. capacity					40			40	44.50	
	1180	600 lb. capacity					40			40	44.50	
	1200	750 lb. capacity					40			40	44.50	
	1220	900 lb. capacity					40			40	44.50	
	1240	1100 lb. capacity					44.50			44.50	49	
	1260	1300 lb. capacity					44.50			44.50	49	
	1280	1500 lb. capacity					65			65	71.50	
	1300	1800 lb. capacity					65			65	71.50	
	1320	2200 lb. capacity					72			72	79.50	
	1340	2600 lb. capacity				↓	72			72	79.50	
	1399	Spring type										
	1400	2 Piece										
	1410	50 - 1000 lb.	1 Stpi	12	.667	Ea.	97	23		120	142	
	1420	1100 - 1600 lb.	"	12	.667	"	106	23		129	152	
	1500	Double spring open										
	1510	150 - 450 lb.	1 Stpi	24	.333	Ea.	66	11.60		77.60	90	
	1520	500 - 1000 lb.		24	.333		66	11.60		77.60	90	
	1530	1100 - 1600 lb.		24	.333		66	11.60		77.60	90	
	1540	1700 - 2400 lb.		24	.333		73	11.60		84.60	98	
	1550	2500 - 3400 lb.	↓	24	.333		121	11.60		132.60	151	
	2000	Pads, cork rib, 18" x 18" x 1", 10-50 psi					83.50			83.50	91.50	
	2020	18" x 36" x 1", 10-50 psi					167			167	183	
	2100	Shear flexible pads, 18" x 18" x 3/8", 20-70 psi					39.50			39.50	43.50	
	2120	18" x 36" x 3/8", 20-70 psi				↓	78			78	85.50	
	2150	Laminated neoprene and cork										
	2160	1" thick	1 Stpi	16	.500	S.F.	43	17.45		60.45	74	
	3000	Note overlap in capacities due to deflections										
	15075 | Mech Identification											
400	0010	PIPING SYSTEM IDENTIFICATION LABELS										400
	0100	Indicate contents and flow direction										

		15075 | Mech Identification	CREW	DAILY OUTPUT	LABOR-HOURS	UNIT	2001 BARE COSTS MAT.	LABOR	EQUIP.	TOTAL	TOTAL INCL O&P	
400	0106	Pipe markers										400
	0110	Plastic snap around										
	0114	1/2" pipe	1 Plum	80	.100	Ea.	3	3.45		6.45	8.50	
	0116	3/4" pipe		80	.100		3	3.45		6.45	8.50	
	0118	1" pipe		80	.100		3	3.45		6.45	8.50	
	0120	2" pipe		75	.107		3.95	3.67		7.62	9.90	
	0122	3" pipe		70	.114		7.40	3.94		11.34	14.10	
	0124	4" pipe		60	.133		7.40	4.59		11.99	15.10	
	0126	6" pipe		60	.133		8	4.59		12.59	15.75	
	0128	8" pipe		56	.143		11	4.92		15.92	19.55	
	0130	10" pipe		56	.143		11	4.92		15.92	19.55	
	0200	Over 10" pipe size		50	.160		13.50	5.50		19	23	
	1110	Self adhesive										
	1114	1" pipe	1 Plum	80	.100	Ea.	2.10	3.45		5.55	7.50	
	1116	2" pipe		75	.107		2.10	3.67		5.77	7.85	
	1118	3" pipe		70	.114		3.40	3.94		7.34	9.70	
	1120	4" pipe		60	.133		3.40	4.59		7.99	10.70	
	1122	6" pipe		60	.133		3.40	4.59		7.99	10.70	
	1124	8" pipe		56	.143		4.70	4.92		9.62	12.60	
	1126	10" pipe		56	.143		4.70	4.92		9.62	12.60	
	1200	Over 10" pipe size		50	.160		6.55	5.50		12.05	15.50	
	2000	Valve tags										
	2010	Numbered plus identifying legend										
	2100	Brass, 2" diameter	1 Plum	40	.200	Ea.	1.75	6.90		8.65	12.35	
	2200	Plastic, 1-1/2" Dia.	"	40	.200	"	1.60	6.90		8.50	12.15	

		15082 | Duct Insulation	CREW	DAILY OUTPUT	LABOR-HOURS	UNIT	MAT.	LABOR	EQUIP.	TOTAL	TOTAL INCL O&P	
200	0010	**INSULATION**										200
	0100	Rule of thumb, as a percentage of total mechanical costs				Job				10%		
	0110	Insulation req'd is based on the surface size/area to be covered										
	1000	Boiler, 1-1/2" calcium silicate, 1/2" cement finish	Q-14	50	.320	S.F.	2.53	9.25		11.78	17.65	
	1020	2" fiberglass	"	80	.200	"	1.49	5.80		7.29	10.90	
	2000	Breeching, 2" calcium silicate with 1/2" cement finish, no lath										
	2020	Rectangular	Q-14	42	.381	S.F.	2.86	11		13.86	21	
	2040	Round	"	38.70	.413	"	3.21	11.95		15.16	22.50	
	2300	Calcium silicate block, + 200°F to + 1200°F										
	2310	On irregular surfaces, valves and fittings										
	2340	1" thick	Q-14	30	.533	S.F.	2	15.45		17.45	26.50	
	2360	1-1/2" thick	"	25	.640	"	2.34	18.50		20.84	32	
	2410	On plane surfaces										
	2420	1" thick	Q-14	126	.127	S.F.	2	3.67		5.67	8.10	
	2430	1-1/2" thick	"	120	.133	"	2.34	3.86		6.20	8.75	
	2900	Domestic water heater wrap kit										
	2920	1-1/2" with vinyl jacket, 20-60 gal.	1 Plum	8	1	Ea.	16.25	34.50		50.75	70	
	2930	Insulated protectors, (ADA)										
	2935	For exposed piping under sinks or lavatories.										
	2940	Vinyl coated foam, velcro tabs										
	2945	P Trap, 1-1/4" or 1-1/2"	1 Plum	32	.250	Ea.	12.85	8.60		21.45	27	
	2960	Valve and supply cover										
	2965	1/2", 3/8", and 7/16" pipe size	1 Plum	32	.250	Ea.	12.85	8.60		21.45	27	
	2970	Extension drain cover										
	2975	1-1/4", or 1-1/2" pipe size	1 Plum	32	.250	Ea.	13.15	8.60		21.75	27.50	
	2980	Tailpiece offset (wheelchair)										
	2985	1-1/4" pipe size	1 Plum	32	.250	Ea.	13.50	8.60		22.10	28	
	3780	Foam, rubber										

Important: See the Reference Section for critical supporting data - Reference Nos., Crews, & City Cost Indexes

15082 | Duct Insulation

200		CREW	DAILY OUTPUT	LABOR-HOURS	UNIT	2001 BARE COSTS MAT.	LABOR	EQUIP.	TOTAL	TOTAL INCL O&P	
3782	1″ thick	1 Stpi	50	.160	S.F.	2.50	5.60		8.10	11.15	200
4000	Pipe covering (price copper tube one size less than IPS)										
4040	Air cell, asbestos free corrugated felt, with cover										
4100	3 ply, 1/2″ iron pipe size	Q-14	145	.110	L.F.	.75	3.19		3.94	5.95	
4130	3/4″ iron pipe size		145	.110		.85	3.19		4.04	6.05	
4140	1″ iron pipe size		145	.110		.90	3.19		4.09	6.10	
4150	1-1/4″ iron pipe size		140	.114		1	3.31		4.31	6.40	
4160	1-1/2″ iron pipe size		140	.114		1.10	3.31		4.41	6.50	
4170	2″ iron pipe size		135	.119		1.20	3.43		4.63	6.80	
4180	2-1/2″ iron pipe size		135	.119		1.35	3.43		4.78	7	
4190	3″ iron pipe size		130	.123		1.50	3.56		5.06	7.35	
4200	3-1/2″ iron pipe size		130	.123		1.76	3.56		5.32	7.65	
4210	4″ iron pipe size		125	.128		1.95	3.70		5.65	8.10	
4220	5″ iron pipe size		120	.133		2.25	3.86		6.11	8.70	
4230	6″ iron pipe size		115	.139		2.40	4.03		6.43	9.10	
4240	8″ iron pipe size		105	.152		3.30	4.41		7.71	10.70	
4250	10″ iron pipe size		95	.168		3.90	4.87		8.77	12.10	
4260	12″ iron pipe size		90	.178		4.50	5.15		9.65	13.20	
4280	Cellular glass, closed cell foam, all service jacket, sealant,										
4281	working temp. (-450°F to +900°F), 0 water vapor xmission										
4284	1″ wall,										
4286	1/2″ iron pipe size	Q-14	120	.133	L.F.	1.73	3.86		5.59	8.10	
4300	1-1/2″ wall,										
4301	1″ iron pipe size	Q-14	105	.152	L.F.	2.78	4.41		7.19	10.10	
4304	2-1/2″ iron pipe size		90	.178		4.68	5.15		9.83	13.40	
4306	3″ iron pipe size		85	.188		4.73	5.45		10.18	13.90	
4308	4″ iron pipe size		70	.229		6.30	6.60		12.90	17.50	
4310	5″ iron pipe size		65	.246		7.25	7.10		14.35	19.35	
4320	2″ wall,										
4322	1″ iron pipe size	Q-14	100	.160	L.F.	4.13	4.63		8.76	11.95	
4324	2-1/2″ iron pipe size		85	.188		6.15	5.45		11.60	15.45	
4326	3″ iron pipe size		80	.200		6.30	5.80		12.10	16.15	
4328	4″ iron pipe size		65	.246		7.05	7.10		14.15	19.15	
4330	5″ iron pipe size		60	.267		8.45	7.70		16.15	21.50	
4332	6″ iron pipe size		50	.320		10	9.25		19.25	26	
4336	8″ iron pipe size		40	.400		12.95	11.55		24.50	33	
4338	10″ iron pipe size		35	.457		13.35	13.25		26.60	35.50	
4350	2-1/2″ wall,										
4360	12″ iron pipe size	Q-14	32	.500	L.F.	21	14.45		35.45	46.50	
4362	14″ iron pipe size	″	28	.571	″	22.50	16.55		39.05	51.50	
4370	3″ wall,										
4378	6″ iron pipe size	Q-14	48	.333	L.F.	13.35	9.65		23	30	
4380	8″ iron pipe size		38	.421		15.60	12.20		27.80	36.50	
4382	10″ iron pipe size		33	.485		20.50	14.05		34.55	45	
4384	16″ iron pipe size		25	.640		26.50	18.50		45	59	
4386	18″ iron pipe size		22	.727		29	21		50	65.50	
4388	20″ iron pipe size		20	.800		34.50	23		57.50	74.50	
4400	3-1/2″ wall,										
4412	12″ iron pipe size	Q-14	27	.593	L.F.	30	17.15		47.15	60.50	
4414	14″ iron pipe size	″	25	.640	″	32	18.50		50.50	64.50	
4430	4″ wall,										
4446	16″ iron pipe size	Q-14	22	.727	L.F.	32.50	21		53.50	69	
4448	18″ iron pipe size		20	.800		42	23		65	83	
4450	20″ iron pipe size		18	.889		43.50	25.50		69	89	
4480	Fittings, average with fabric and mastic										
4484	1″ wall,										

MECHANICAL 15

15082 | Duct Insulation

			CREW	DAILY OUTPUT	LABOR-HOURS	UNIT	2001 BARE COSTS MAT.	LABOR	EQUIP.	TOTAL	TOTAL INCL O&P
200	4486	1/2″ iron pipe size	1 Asbe	40	.200	Ea.	3.07	6.45		9.52	13.70
	4500	1-1/2″ wall,									
	4502	1″ iron pipe size	1 Asbe	38	.211	Ea.	5.30	6.75		12.05	16.65
	4504	2-1/2″ iron pipe size		32	.250		6.60	8.05		14.65	20
	4506	3″ iron pipe size		30	.267		7.25	8.55		15.80	21.50
	4508	4″ iron pipe size		28	.286		11.15	9.20		20.35	27
	4510	5″ iron pipe size		24	.333		12.25	10.70		22.95	30.50
	4520	2″ wall,									
	4522	1″ iron pipe size	1 Asbe	36	.222	Ea.	6.15	7.15		13.30	18.20
	4524	2-1/2″ iron pipe size		30	.267		7.25	8.55		15.80	21.50
	4526	3″ iron pipe size		28	.286		8.70	9.20		17.90	24.50
	4528	4″ iron pipe size		24	.333		11.70	10.70		22.40	30
	4530	5″ iron pipe size		22	.364		14.75	11.70		26.45	35
	4532	6″ iron pipe size		20	.400		17.25	12.85		30.10	39.50
	4536	8″ iron pipe size		12	.667		26	21.50		47.50	63.50
	4538	10″ iron pipe size		8	1		32.50	32		64.50	87
	4550	2-1/2″ wall,									
	4560	12″ iron pipe size	1 Asbe	6	1.333	Ea.	55.50	43		98.50	130
	4562	14″ iron pipe size	″	4	2	″	64.50	64.50		129	174
	4570	3″ wall,									
	4578	6″ iron pipe size	1 Asbe	16	.500	Ea.	20	16.10		36.10	48
	4580	8″ iron pipe size		10	.800		29.50	25.50		55	73.50
	4582	10″ iron pipe size		6	1.333		42	43		85	115
	4584	16″ iron pipe size		2	4		100	129		229	315
	4586	18″ iron pipe size		2	4		129	129		258	350
	4588	20″ iron pipe size		2	4		153	129		282	375
	4600	3-1/2″ wall,									
	4612	12″ iron pipe size	1 Asbe	4	2	Ea.	72.50	64.50		137	183
	4614	14″ iron pipe size	″	2	4	″	80	129		209	294
	4630	4″ wall,									
	4646	16″ iron pipe size	1 Asbe	2	4	Ea.	120	129		249	340
	4648	18″ iron pipe size		2	4		147	129		276	370
	4650	20″ iron pipe size		2	4		172	129		301	395
	4900	Calcium silicate, with cover									
	5100	1″ wall, 1/2″ iron pipe size	Q-14	170	.094	L.F.	2.07	2.72		4.79	6.65
	5130	3/4″ iron pipe size		170	.094		2.08	2.72		4.80	6.65
	5140	1″ iron pipe size		170	.094		1.92	2.72		4.64	6.45
	5150	1-1/4″ iron pipe size		165	.097		1.98	2.81		4.79	6.65
	5160	1-1/2″ iron pipe size		165	.097		2.03	2.81		4.84	6.70
	5170	2″ iron pipe size		160	.100		2.41	2.89		5.30	7.30
	5180	2-1/2″ iron pipe size		160	.100		2.58	2.89		5.47	7.45
	5190	3″ iron pipe size		150	.107		2.74	3.09		5.83	7.95
	5200	4″ iron pipe size		140	.114		3.18	3.31		6.49	8.80
	5210	5″ iron pipe size		135	.119		3.51	3.43		6.94	9.35
	5220	6″ iron pipe size		130	.123		3.81	3.56		7.37	9.90
	5280	1-1/2″ wall, 1/2″ iron pipe size		150	.107		2.27	3.09		5.36	7.45
	5310	3/4″ iron pipe size		150	.107		2.28	3.09		5.37	7.45
	5320	1″ iron pipe size		150	.107		2.49	3.09		5.58	7.70
	5330	1-1/4″ iron pipe size		145	.110		2.68	3.19		5.87	8.05
	5340	1-1/2″ iron pipe size		145	.110		2.81	3.19		6	8.20
	5350	2″ iron pipe size		140	.114		3.07	3.31		6.38	8.70
	5360	2-1/2″ iron pipe size		140	.114		3.35	3.31		6.66	9
	5370	3″ iron pipe size		135	.119		3.51	3.43		6.94	9.35
	5380	4″ iron pipe size		125	.128		4.10	3.70		7.80	10.45
	5390	5″ iron pipe size		120	.133		4.60	3.86		8.46	11.25
	5400	6″ iron pipe size		110	.145		4.74	4.21		8.95	11.95

	15082 | Duct Insulation	CREW	DAILY OUTPUT	LABOR-HOURS	UNIT	2001 BARE COSTS MAT.	LABOR	EQUIP.	TOTAL	TOTAL INCL O&P	
200	5460 2" wall, 1/2" iron pipe size	Q-14	135	.119	L.F.	3.62	3.43		7.05	9.50	200
	5490 3/4" iron pipe size		135	.119		3.63	3.43		7.06	9.50	
	5500 1" iron pipe size		135	.119		3.83	3.43		7.26	9.70	
	5510 1-1/4" iron pipe size		130	.123		4.26	3.56		7.82	10.40	
	5520 1-1/2" iron pipe size		130	.123		4.30	3.56		7.86	10.45	
	5530 2" iron pipe size		125	.128		4.52	3.70		8.22	10.90	
	5540 2-1/2" iron pipe size		125	.128		5.45	3.70		9.15	11.90	
	5550 3" iron pipe size		120	.133		6	3.86		9.86	12.80	
	5560 4" iron pipe size		115	.139		6.10	4.03		10.13	13.20	
	5570 5" iron pipe size		110	.145		6.95	4.21		11.16	14.40	
	5580 6" iron pipe size	↓	105	.152	↓	7.80	4.41		12.21	15.65	
	5600 Calcium silicate, no cover										
	5720 1" wall, 1/2" iron pipe size	Q-14	180	.089	L.F.	1.93	2.57		4.50	6.25	
	5740 3/4" iron pipe size		180	.089		1.93	2.57		4.50	6.25	
	5750 1" iron pipe size		180	.089		1.80	2.57		4.37	6.10	
	5760 1-1/4" iron pipe size		175	.091		1.81	2.65		4.46	6.25	
	5770 1-1/2" iron pipe size		175	.091		1.83	2.65		4.48	6.25	
	5780 2" iron pipe size		170	.094		2.18	2.72		4.90	6.75	
	5790 2-1/2" iron pipe size		170	.094		2.34	2.72		5.06	6.95	
	5800 3" iron pipe size		160	.100		2.47	2.89		5.36	7.35	
	5810 4" iron pipe size		150	.107		2.83	3.09		5.92	8.05	
	5820 5" iron pipe size		145	.110		3.13	3.19		6.32	8.55	
	5830 6" iron pipe size		140	.114		3.38	3.31		6.69	9	
	5900 1-1/2" wall, 1/2" iron pipe size		160	.100		2.07	2.89		4.96	6.90	
	5920 3/4" iron pipe size		160	.100		2.07	2.89		4.96	6.90	
	5930 1" iron pipe size		160	.100		2.26	2.89		5.15	7.10	
	5940 1-1/4" iron pipe size		155	.103		2.54	2.99		5.53	7.55	
	5950 1-1/2" iron pipe size		155	.103		2.55	2.99		5.54	7.60	
	5960 2" iron pipe size		150	.107		2.79	3.09		5.88	8	
	5970 2-1/2" iron pipe size		150	.107		3.07	3.09		6.16	8.30	
	5980 3" iron pipe size		145	.110		3.20	3.19		6.39	8.60	
	5990 4" iron pipe size		135	.119		3.72	3.43		7.15	9.60	
	6000 5" iron pipe size		130	.123		4.17	3.56		7.73	10.30	
	6010 6" iron pipe size		120	.133		4.27	3.86		8.13	10.90	
	6020 7" iron pipe size		115	.139		5.15	4.03		9.18	12.10	
	6030 8" iron pipe size		105	.152		5.80	4.41		10.21	13.40	
	6040 9" iron pipe size		100	.160		6.95	4.63		11.58	15.05	
	6050 10" iron pipe size		95	.168		7.70	4.87		12.57	16.30	
	6060 12" iron pipe size		90	.178		9.10	5.15		14.25	18.25	
	6070 14" iron pipe size		85	.188		10.35	5.45		15.80	20	
	6080 16" iron pipe size		80	.200		11.60	5.80		17.40	22	
	6090 18" iron pipe size		75	.213		13.25	6.15		19.40	24.50	
	6120 2" wall, 1/2" iron pipe size		145	.110		3.38	3.19		6.57	8.80	
	6140 3/4" iron pipe size		145	.110		3.38	3.19		6.57	8.80	
	6150 1" iron pipe size		145	.110		3.56	3.19		6.75	9	
	6160 1-1/4" iron pipe size		140	.114		3.98	3.31		7.29	9.70	
	6170 1-1/2" iron pipe size		140	.114		4.01	3.31		7.32	9.70	
	6180 2" iron pipe size		135	.119		4.22	3.43		7.65	10.15	
	6190 2-1/2" iron pipe size		135	.119		4.59	3.43		8.02	10.55	
	6200 3" iron pipe size		130	.123		4.84	3.56		8.40	11	
	6210 4" iron pipe size		125	.128		5.70	3.70		9.40	12.20	
	6220 5" iron pipe size		120	.133		6.50	3.86		10.36	13.35	
	6230 6" iron pipe size		115	.139		7.30	4.03		11.33	14.50	
	6240 7" iron pipe size		110	.145		7.95	4.21		12.16	15.50	
	6250 8" iron pipe size		105	.152		8.70	4.41		13.11	16.65	
	6260 9" iron pipe size	↓	100	.160	↓	9.85	4.63		14.48	18.25	

15082 | Duct Insulation

			CREW	DAILY OUTPUT	LABOR-HOURS	UNIT	2001 BARE COSTS MAT.	LABOR	EQUIP.	TOTAL	TOTAL INCL O&P	
200	6270	10" iron pipe size	Q-14	95	.168	L.F.	10.90	4.87		15.77	19.80	200
	6280	12" iron pipe size		90	.178		12.20	5.15		17.35	21.50	
	6290	14" iron pipe size		85	.188		13.40	5.45		18.85	23.50	
	6300	16" iron pipe size		80	.200		15.30	5.80		21.10	26	
	6310	18" iron pipe size		75	.213		16.70	6.15		22.85	28.50	
	6320	20" iron pipe size		65	.246		19.45	7.10		26.55	33	
	6330	22" iron pipe size		60	.267		21	7.70		28.70	35.50	
	6340	24" iron pipe size		55	.291		21.50	8.40		29.90	37.50	
	6360	3" wall, 1/2" iron pipe size		115	.139		6.15	4.03		10.18	13.20	
	6380	3/4" iron pipe size		115	.139		6.15	4.03		10.18	13.20	
	6390	1" iron pipe size		115	.139		5.55	4.03		9.58	12.55	
	6400	1-1/4" iron pipe size		110	.145		5.75	4.21		9.96	13.05	
	6410	1-1/2" iron pipe size		110	.145		6.10	4.21		10.31	13.45	
	6420	2" iron pipe size		105	.152		6.50	4.41		10.91	14.20	
	6430	2-1/2" iron pipe size		105	.152		7.75	4.41		12.16	15.60	
	6440	3" iron pipe size		100	.160		8.50	4.63		13.13	16.75	
	6450	4" iron pipe size		95	.168		9.80	4.87		14.67	18.60	
	6460	5" iron pipe size		90	.178		10.65	5.15		15.80	20	
	6470	6" iron pipe size		90	.178		12.05	5.15		17.20	21.50	
	6480	7" iron pipe size		85	.188		13.75	5.45		19.20	24	
	6490	8" iron pipe size		85	.188		14.25	5.45		19.70	24.50	
	6500	9" iron pipe size		80	.200		16.10	5.80		21.90	27	
	6510	10" iron pipe size		75	.213		17.25	6.15		23.40	29	
	6520	12" iron pipe size		70	.229		19.10	6.60		25.70	31.50	
	6530	14" iron pipe size		65	.246		21.50	7.10		28.60	35	
	6540	16" iron pipe size		60	.267		23.50	7.70		31.20	38.50	
	6550	18" iron pipe size		55	.291		26	8.40		34.40	42	
	6560	20" iron pipe size		50	.320		29	9.25		38.25	47	
	6570	22" iron pipe size		45	.356		31.50	10.30		41.80	51.50	
	6580	24" iron pipe size		40	.400		34	11.55		45.55	56	
	6600	Fiberglass, with all service jacket										
	6640	1/2" wall, 1/2" iron pipe size	Q-14	250	.064	L.F.	.88	1.85		2.73	3.94	
	6660	3/4" iron pipe size		240	.067		.98	1.93		2.91	4.17	
	6670	1" iron pipe size		230	.070		1.06	2.01		3.07	4.39	
	6680	1-1/4" iron pipe size		220	.073		1.16	2.10		3.26	4.65	
	6690	1-1/2" iron pipe size		220	.073		1.24	2.10		3.34	4.73	
	6700	2" iron pipe size		210	.076		1.33	2.20		3.53	4.99	
	6710	2-1/2" iron pipe size		200	.080		1.53	2.31		3.84	5.40	
	6720	3" iron pipe size		190	.084		1.70	2.44		4.14	5.75	
	6730	3-1/2" iron pipe size		180	.089		1.83	2.57		4.40	6.15	
	6740	4" iron pipe size		160	.100		2.17	2.89		5.06	7	
	6750	5" iron pipe size		150	.107		2.51	3.09		5.60	7.70	
	6760	6" iron pipe size		120	.133		2.64	3.86		6.50	9.10	
	6840	1" wall, 1/2" iron pipe size		240	.067		1.06	1.93		2.99	4.26	
	6860	3/4" iron pipe size		230	.070		1.21	2.01		3.22	4.55	
	6870	1" iron pipe size		220	.073		1.24	2.10		3.34	4.73	
	6880	1-1/4" iron pipe size		210	.076		1.40	2.20		3.60	5.05	
	6890	1-1/2" iron pipe size		210	.076		1.51	2.20		3.71	5.20	
	6900	2" iron pipe size		200	.080		1.65	2.31		3.96	5.55	
	6910	2-1/2" iron pipe size		190	.084		1.81	2.44		4.25	5.90	
	6920	3" iron pipe size		180	.089		2.05	2.57		4.62	6.40	
	6930	3-1/2" iron pipe size		170	.094		2.26	2.72		4.98	6.85	
	6940	4" iron pipe size		150	.107		2.64	3.09		5.73	7.85	
	6950	5" iron pipe size		140	.114		3.04	3.31		6.35	8.65	
	6960	6" iron pipe size		120	.133		3.23	3.86		7.09	9.75	
	6970	7" iron pipe size		110	.145		3.98	4.21		8.19	11.15	

	15082 | Duct Insulation		CREW	DAILY OUTPUT	LABOR-HOURS	UNIT	2001 BARE COSTS MAT.	LABOR	EQUIP.	TOTAL	TOTAL INCL O&P
200	6980	8″ iron pipe size	Q-14	100	.160	L.F.	4.61	4.63		9.24	12.45
	6990	9″ iron pipe size		90	.178		5	5.15		10.15	13.75
	7000	10″ iron pipe size		90	.178		5.40	5.15		10.55	14.20
	7010	12″ iron pipe size		80	.200		6.15	5.80		11.95	16
	7020	14″ iron pipe size		80	.200		7.10	5.80		12.90	17.05
	7030	16″ iron pipe size		70	.229		8.45	6.60		15.05	19.90
	7040	18″ iron pipe size		70	.229		10.30	6.60		16.90	22
	7050	20″ iron pipe size		60	.267		11.50	7.70		19.20	25
	7060	24″ iron pipe size		60	.267		14	7.70		21.70	28
	7080	1-1/2″ wall, 1/2″ iron pipe size		230	.070		2.16	2.01		4.17	5.60
	7100	3/4″ iron pipe size		220	.073		2.26	2.10		4.36	5.85
	7110	1″ iron pipe size		210	.076		2.37	2.20		4.57	6.15
	7120	1-1/4″ iron pipe size		200	.080		2.54	2.31		4.85	6.50
	7130	1-1/2″ iron pipe size		200	.080		2.68	2.31		4.99	6.65
	7140	2″ iron pipe size		190	.084		2.93	2.44		5.37	7.10
	7150	2-1/2″ iron pipe size		180	.089		3.21	2.57		5.78	7.65
	7160	3″ iron pipe size		170	.094		3.31	2.72		6.03	8
	7170	3-1/2″ iron pipe size		160	.100		3.63	2.89		6.52	8.60
	7180	4″ iron pipe size		140	.114		3.78	3.31		7.09	9.45
	7190	5″ iron pipe size		130	.123		4.22	3.56		7.78	10.35
	7200	6″ iron pipe size		110	.145		4.34	4.21		8.55	11.50
	7210	7″ iron pipe size		100	.160		5.05	4.63		9.68	12.95
	7220	8″ iron pipe size		90	.178		5.45	5.15		10.60	14.25
	7230	9″ iron pipe size		85	.188		6.30	5.45		11.75	15.65
	7240	10″ iron pipe size		80	.200		6.85	5.80		12.65	16.80
	7250	12″ iron pipe size		75	.213		7.45	6.15		13.60	18.10
	7260	14″ iron pipe size		70	.229		8.65	6.60		15.25	20
	7270	16″ iron pipe size		65	.246		9.70	7.10		16.80	22
	7280	18″ iron pipe size		60	.267		12.55	7.70		20.25	26
	7290	20″ iron pipe size		55	.291		13.30	8.40		21.70	28
	7300	24″ iron pipe size		50	.320		15.55	9.25		24.80	32
	7320	2″ wall, 1/2″ iron pipe size		220	.073		3.49	2.10		5.59	7.20
	7340	3/4″ iron pipe size		210	.076		3.49	2.20		5.69	7.35
	7350	1″ iron pipe size		200	.080		3.73	2.31		6.04	7.80
	7360	1-1/4″ iron pipe size		190	.084		3.95	2.44		6.39	8.25
	7370	1-1/2″ iron pipe size		190	.084		4.17	2.44		6.61	8.50
	7380	2″ iron pipe size		180	.089		4.33	2.57		6.90	8.90
	7390	2-1/2″ iron pipe size		170	.094		4.65	2.72		7.37	9.45
	7400	3″ iron pipe size		160	.100		5	2.89		7.89	10.15
	7410	3-1/2″ iron pipe size		150	.107		5.30	3.09		8.39	10.75
	7420	4″ iron pipe size		130	.123		5.75	3.56		9.31	12.05
	7430	5″ iron pipe size		120	.133		5.75	3.86		9.61	12.50
	7440	6″ iron pipe size		100	.160		6.65	4.63		11.28	14.75
	7450	7″ iron pipe size		90	.178		7.65	5.15		12.80	16.65
	7460	8″ iron pipe size		80	.200		8.25	5.80		14.05	18.35
	7470	9″ iron pipe size		75	.213		9.10	6.15		15.25	19.95
	7480	10″ iron pipe size		70	.229		9.85	6.60		16.45	21.50
	7490	12″ iron pipe size		65	.246		10.85	7.10		17.95	23.50
	7500	14″ iron pipe size		60	.267		12.25	7.70		19.95	26
	7510	16″ iron pipe size		55	.291		14	8.40		22.40	29
	7520	18″ iron pipe size		50	.320		16.70	9.25		25.95	33
	7530	20″ iron pipe size		45	.356		18.50	10.30		28.80	37
	7540	24″ iron pipe size	↓	40	.400	↓	21	11.55		32.55	41.50
	7600	For fiberglass with standard canvas jacket, deduct					5%				
	7660	For fittings, add 3 L.F. for each fitting									
200	7680	plus 4 L.F. for each flange of the fitting									

15082 | Duct Insulation

							2001 BARE COSTS					
			CREW	DAILY OUTPUT	LABOR-HOURS	UNIT	MAT.	LABOR	EQUIP.	TOTAL	TOTAL INCL O&P	
200	7700	For equipment or congested areas, add						20%				200
	7720	Finishes, for .010" aluminum jacket, add	Q-14	200	.080	S.F.	.31	2.31		2.62	4.05	
	7740	For .016" aluminum jacket, add		200	.080		.40	2.31		2.71	4.15	
	7760	For .010" stainless steel, add	↓	160	.100	↓	1.20	2.89		4.09	5.95	
	7780	For single layer of felt, add					10%	10%				
	7784	Polyethylene tubing flexible closed cell foam, UV resistant										
	7828	Standard temperature (-90° to +212° F)										
	7830	3/8" wall, 1/8" iron pipe size	1 Asbe	130	.062	L.F.	.25	1.98		2.23	3.45	
	7831	1/4" iron pipe size		130	.062		.28	1.98		2.26	3.48	
	7832	3/8" iron pipe size		130	.062		.29	1.98		2.27	3.49	
	7833	1/2" iron pipe size		126	.063		.33	2.04		2.37	3.63	
	7834	3/4" iron pipe size		122	.066		.36	2.11		2.47	3.78	
	7835	1" iron pipe size		120	.067		.46	2.14		2.60	3.94	
	7836	1-1/4" iron pipe size		118	.068		.59	2.18		2.77	4.14	
	7837	1-1/2" iron pipe size		118	.068		.66	2.18		2.84	4.22	
	7838	2" iron pipe size		116	.069		.88	2.22		3.10	4.52	
	7839	2-1/2" iron pipe size		114	.070		1.20	2.26		3.46	4.93	
	7840	3" iron pipe size		112	.071		1.34	2.30		3.64	5.15	
	7842	1/2" wall, 1/8" iron pipe size		120	.067		.39	2.14		2.53	3.86	
	7843	1/4" iron pipe size		120	.067		.41	2.14		2.55	3.88	
	7844	3/8" iron pipe size		120	.067		.42	2.14		2.56	3.89	
	7845	1/2" iron pipe size		118	.068		.51	2.18		2.69	4.05	
	7846	3/4" iron pipe size		116	.069		.53	2.22		2.75	4.13	
	7847	1" iron pipe size		114	.070		.68	2.26		2.94	4.36	
	7848	1-1/4" iron pipe size		112	.071		.83	2.30		3.13	4.59	
	7849	1-1/2" iron pipe size		110	.073		1.01	2.34		3.35	4.86	
	7850	2" iron pipe size		108	.074		1.25	2.38		3.63	5.20	
	7851	2-1/2" iron pipe size		106	.075		1.62	2.43		4.05	5.65	
	7852	3" iron pipe size		104	.077		1.76	2.47		4.23	5.90	
	7853	3-1/2" iron pipe size		102	.078		2.55	2.52		5.07	6.85	
	7854	4" iron pipe size		100	.080		2.79	2.57		5.36	7.20	
	7855	3/4" wall, 1/8" iron pipe size		110	.073		.91	2.34		3.25	4.75	
	7856	1/4" iron pipe size		110	.073		.91	2.34		3.25	4.75	
	7857	3/8" iron pipe size		108	.074		1.05	2.38		3.43	4.97	
	7858	1/2" iron pipe size		106	.075		1.26	2.43		3.69	5.30	
	7859	3/4" iron pipe size		104	.077		1.50	2.47		3.97	5.60	
	7860	1" iron pipe size		102	.078		1.75	2.52		4.27	5.95	
	7861	1-1/4" iron pipe size		100	.080		2.37	2.57		4.94	6.75	
	7862	1-1/2" iron pipe size		100	.080		2.70	2.57		5.27	7.10	
	7863	2" iron pipe size		98	.082		3.22	2.62		5.84	7.75	
	7864	2-1/2" iron pipe size		96	.083		3.99	2.68		6.67	8.70	
	7865	3" iron pipe size		94	.085		5.10	2.74		7.84	10.05	
	7866	3-1/2" iron pipe size		92	.087		6.30	2.80		9.10	11.45	
	7867	4" iron pipe size		90	.089		6.70	2.86		9.56	12	
	7868	1" wall, 1/4" iron pipe size		100	.080		1.96	2.57		4.53	6.30	
	7869	3/8" iron pipe size		98	.082		1.96	2.62		4.58	6.35	
	7870	1/2" iron pipe size		96	.083		2.45	2.68		5.13	7	
	7871	3/4" iron pipe size		94	.085		2.48	2.74		5.22	7.10	
	7872	1" iron pipe size		92	.087		2.87	2.80		5.67	7.65	
	7873	1-1/4" iron pipe size		90	.089		4.02	2.86		6.88	9	
	7874	1-1/2" iron pipe size		90	.089		4.31	2.86		7.17	9.30	
	7875	2" iron pipe size		88	.091		5.65	2.92		8.57	10.95	
	7876	2-1/2" iron pipe size		86	.093		7.85	2.99		10.84	13.45	
	7877	3" iron pipe size	↓	84	.095	↓	8.20	3.06		11.26	13.95	
	7878	Contact cement, quart can				Ea.	18.55			18.55	20.50	
	7879	Rubber tubing, flexible closed cell foam										

15 MECHANICAL

Important: See the Reference Section for critical supporting data - Reference Nos., Crews, & City Cost Indexes

15082 | Duct Insulation

			CREW	DAILY OUTPUT	LABOR-HOURS	UNIT	2001 BARE COSTS MAT.	LABOR	EQUIP.	TOTAL	TOTAL INCL O&P	
200	7880	3/8" wall, 1/4" iron pipe size	1 Asbe	120	.067	L.F.	.28	2.14		2.42	3.74	200
	7900	3/8" iron pipe size		120	.067		.31	2.14		2.45	3.77	
	7910	1/2" iron pipe size		115	.070		.36	2.24		2.60	3.98	
	7920	3/4" iron pipe size		115	.070		.44	2.24		2.68	4.06	
	7930	1" iron pipe size		110	.073		.49	2.34		2.83	4.29	
	7940	1-1/4" iron pipe size		110	.073		.60	2.34		2.94	4.41	
	7950	1-1/2" iron pipe size		110	.073		.72	2.34		3.06	4.54	
	8100	1/2" wall, 1/4" iron pipe size		90	.089		.58	2.86		3.44	5.20	
	8120	3/8" iron pipe size		90	.089		.57	2.86		3.43	5.20	
	8130	1/2" iron pipe size		89	.090		.45	2.89		3.34	5.15	
	8140	3/4" iron pipe size		89	.090		.51	2.89		3.40	5.20	
	8150	1" iron pipe size		88	.091		.55	2.92		3.47	5.30	
	8160	1-1/4" iron pipe size		87	.092		.69	2.96		3.65	5.50	
	8170	1-1/2" iron pipe size		87	.092		.83	2.96		3.79	5.65	
	8180	2" iron pipe size		86	.093		1.59	2.99		4.58	6.55	
	8190	2-1/2" iron pipe size		86	.093		1.58	2.99		4.57	6.55	
	8200	3" iron pipe size		85	.094		1.66	3.03		4.69	6.70	
	8210	3-1/2" iron pipe size		85	.094		2.22	3.03		5.25	7.30	
	8220	4" iron pipe size		80	.100		2.62	3.22		5.84	8.05	
	8230	5" iron pipe size		80	.100		3.61	3.22		6.83	9.10	
	8300	3/4" wall, 1/4" iron pipe size		90	.089		.65	2.86		3.51	5.30	
	8320	3/8" iron pipe size		90	.089		.72	2.86		3.58	5.35	
	8330	1/2" iron pipe size		89	.090		.78	2.89		3.67	5.50	
	8340	3/4" iron pipe size		89	.090		.86	2.89		3.75	5.60	
	8350	1" iron pipe size		88	.091		1.14	2.92		4.06	5.95	
	8360	1-1/4" iron pipe size		87	.092		1.30	2.96		4.26	6.15	
	8370	1-1/2" iron pipe size		87	.092		1.76	2.96		4.72	6.70	
	8380	2" iron pipe size		86	.093		2.34	2.99		5.33	7.35	
	8390	2-1/2" iron pipe size		86	.093		3.12	2.99		6.11	8.20	
	8400	3" iron pipe size		85	.094		3.19	3.03		6.22	8.35	
	8410	3-1/2" iron pipe size		85	.094		3.90	3.03		6.93	9.15	
	8420	4" iron pipe size		80	.100		4.51	3.22		7.73	10.10	
	8430	5" iron pipe size		80	.100		5.45	3.22		8.67	11.15	
	8440	6" iron pipe size		80	.100		6.55	3.22		9.77	12.40	
	8444	1" wall, 1/2" iron pipe size		86	.093		1.63	2.99		4.62	6.60	
	8445	3/4" iron pipe size		84	.095		1.82	3.06		4.88	6.90	
	8446	1" iron pipe size		84	.095		2.21	3.06		5.27	7.35	
	8447	1-1/4" iron pipe size		82	.098		2.57	3.14		5.71	7.85	
	8448	1-1/2" iron pipe size		82	.098		3.38	3.14		6.52	8.70	
	8449	2" iron pipe size		80	.100		4.51	3.22		7.73	10.10	
	8450	2-1/2" iron pipe size		80	.100		5.90	3.22		9.12	11.60	
	8456	Rubber insulation tape, 1/8" x 2" x 30'				Ea.	10.90			10.90	12	
	8460	Polyolefin tubing, flexible closed cell foam, UV stabilized, work										
	8462	temp.-165° F to +210° F, 0 water vapor transmission.										
	8464	3/8" wall, 1/8" iron pipe size	1 Asbe	140	.057	L.F.	.24	1.84		2.08	3.20	
	8466	1/4" iron pipe size		140	.057		.25	1.84		2.09	3.22	
	8468	3/8" iron pipe size		140	.057		.27	1.84		2.11	3.24	
	8470	1/2" iron pipe size		136	.059		.30	1.89		2.19	3.36	
	8472	3/4" iron pipe size		132	.061		.35	1.95		2.30	3.51	
	8474	1" iron pipe size		130	.062		.38	1.98		2.36	3.59	
	8476	1-1/4" iron pipe size		128	.063		.50	2.01		2.51	3.77	
	8478	1-1/2" iron pipe size		128	.063		.62	2.01		2.63	3.90	
	8480	2" iron pipe size		126	.063		.69	2.04		2.73	4.03	
	8482	2-1/2" iron pipe size		123	.065		1	2.09		3.09	4.45	
	8484	3" iron pipe size		121	.066		1.56	2.13		3.69	5.15	
	8486	4" iron pipe size		118	.068		2.17	2.18		4.35	5.90	

15082 | Duct Insulation

		15082 Duct Insulation	CREW	DAILY OUTPUT	LABOR-HOURS	UNIT	2001 BARE COSTS MAT.	LABOR	EQUIP.	TOTAL	TOTAL INCL O&P	
200	8500	1/2" wall, 1/8" iron pipe size	1 Asbe	130	.062	L.F.	.35	1.98		2.33	3.56	200
	8502	1/4" iron pipe size		130	.062		.37	1.98		2.35	3.58	
	8504	3/8" iron pipe size		130	.062		.40	1.98		2.38	3.61	
	8506	1/2" iron pipe size		128	.063		.44	2.01		2.45	3.70	
	8508	3/4" iron pipe size		126	.063		.51	2.04		2.55	3.83	
	8510	1" iron pipe size		123	.065		.55	2.09		2.64	3.96	
	8512	1-1/4" iron pipe size		121	.066		.69	2.13		2.82	4.17	
	8514	1-1/2" iron pipe size		119	.067		.83	2.16		2.99	4.37	
	8516	2" iron pipe size		117	.068		.97	2.20		3.17	4.59	
	8518	2-1/2" iron pipe size		114	.070		1.31	2.26		3.57	5.05	
	8520	3" iron pipe size		112	.071		1.84	2.30		4.14	5.70	
	8522	4" iron pipe size		110	.073		2.28	2.34		4.62	6.25	
	8534	3/4" wall, 1/8" iron pipe size		120	.067		.53	2.14		2.67	4.01	
	8536	1/4" iron pipe size		120	.067		.56	2.14		2.70	4.05	
	8538	3/8" iron pipe size		117	.068		.65	2.20		2.85	4.24	
	8540	1/2" iron pipe size		114	.070		.72	2.26		2.98	4.40	
	8542	3/4" iron pipe size		112	.071		.94	2.30		3.24	4.71	
	8544	1" iron pipe size		110	.073		1.09	2.34		3.43	4.95	
	8546	1-1/4" iron pipe size		108	.074		1.45	2.38		3.83	5.40	
	8548	1-1/2" iron pipe size		108	.074		1.71	2.38		4.09	5.70	
	8550	2" iron pipe size		106	.075		1.99	2.43		4.42	6.10	
	8552	2-1/2" iron pipe size		104	.077		2.51	2.47		4.98	6.70	
	8554	3" iron pipe size		102	.078		3.24	2.52		5.76	7.60	
	8556	4" iron pipe size		100	.080		3.98	2.57		6.55	8.50	
	8570	1" wall, 1/8" iron pipe size		110	.073		.96	2.34		3.30	4.81	
	8572	1/4" iron pipe size		108	.074		.99	2.38		3.37	4.90	
	8574	3/8" iron pipe size		106	.075		1.02	2.43		3.45	5	
	8576	1/2" iron pipe size		104	.077		1.07	2.47		3.54	5.15	
	8578	3/4" iron pipe size		102	.078		1.40	2.52		3.92	5.60	
	8580	1" iron pipe size		100	.080		1.62	2.57		4.19	5.90	
	8582	1-1/4" iron pipe size		97	.082		1.84	2.65		4.49	6.25	
	8584	1-1/2" iron pipe size		97	.082		2.11	2.65		4.76	6.55	
	8586	2" iron pipe size		95	.084		2.76	2.71		5.47	7.40	
	8588	2-1/2" iron pipe size		93	.086		3.60	2.77		6.37	8.40	
	8590	3" iron pipe size	↓	91	.088	↓	3.95	2.83		6.78	8.90	
	8600	Seam Roller				Ea.	6.25			6.25	6.85	
	8602	Fuse Seal applicator (Butane)				"	152			152	167	
	8604	Fuse Seal 1/2" sticks				Lb.	12.20			12.20	13.45	
	8606	Leaktite adhesive				Qt.	18.30			18.30	20	
	8608	Leaktite adhesive				Gal.	51.50			51.50	56.50	
	8610	NOTE: Preslit/preglued vs unslit, same price										

15 MECHANICAL

15100 | Building Services Piping

15106 | Glass Pipe & Fittings

		15106 Glass Pipe & Fittings	CREW	DAILY OUTPUT	LABOR-HOURS	UNIT	2001 BARE COSTS MAT.	LABOR	EQUIP.	TOTAL	TOTAL INCL O&P	
120	0010	**PIPE, GLASS** Borosilicate, couplings & hangers 10' O.C. R15100-070										120
	0020	Drainage										
	1100	1-1/2" diameter	Q-1	52	.308	L.F.	8.35	9.55		17.90	23.50	
	1120	2" diameter	↓	44	.364	↓	10.95	11.25		22.20	29	

	15106 | Glass Pipe & Fittings		CREW	DAILY OUTPUT	LABOR-HOURS	UNIT	2001 BARE COSTS MAT.	LABOR	EQUIP.	TOTAL	TOTAL INCL O&P
120	1140	3" diameter (R15100-070)	Q-1	39	.410	L.F.	14.55	12.70		27.25	35
	1160	4" diameter		30	.533		26	16.55		42.55	53.50
	1180	6" diameter		26	.615		47	19.10		66.10	80.50
	1870	To delete coupling & hanger, subtract									
	1880	1-1/2" diam. to 2" diam.					19%	22%			
	1890	3" diam. to 6" diam.					20%	17%			
	2000	Process supply (pressure), beaded joints									
	2040	1/2" diameter	1 Plum	36	.222	L.F.	7.55	7.65		15.20	19.85
	2060	3/4" diameter		31	.258		9.05	8.90		17.95	23.50
	2080	1" diameter		27	.296		11.80	10.20		22	28.50
	2100	1-1/2" diameter	Q-1	47	.340		14.20	10.55		24.75	31.50
	2120	2" diameter		39	.410		18.80	12.70		31.50	39.50
	2140	3" diameter		34	.471		27	14.60		41.60	51.50
	2160	4" diameter		25	.640		37.50	19.85		57.35	71.50
	2180	6" diameter		21	.762		67.50	23.50		91	110
	2860	To delete coupling & hanger, subtract									
	2870	1/2" diam. to 1" diam.					25%	33%			
	2880	1-1/2" diam. to 3" diam.					22%	21%			
	2890	4" diam. to 6" diam.					23%	15%			
	3000	Beaded joint, armored, translucent									
	3040	1/2" diameter	1 Plum	36	.222	L.F.	17.50	7.65		25.15	31
	3060	3/4" diameter		31	.258		19.80	8.90		28.70	35.50
	3080	1" diameter		27	.296		23	10.20		33.20	41
	3100	1-1/2" diameter	Q-1	47	.340		29	10.55		39.55	48
	3120	2" diameter		39	.410		37.50	12.70		50.20	60.50
	3140	3" diameter		34	.471		62	14.60		76.60	90.50
	3160	4" diameter		25	.640		88	19.85		107.85	127
	3180	6" diameter		21	.762		157	23.50		180.50	208
	3710	To delete coupling & hanger, subtract									
	3720	1/2" diam. to 3/4" diam.					30%	27%			
	3730	1" diam. to 3" diam.					28%	22%			
	3740	4" diam. to 6" diam.					24%	15%			
	3800	Conical joint, transparent									
	3880	1" diameter	1 Plum	27	.296	L.F.	14.25	10.20		24.45	31
	3900	1-1/2" diameter	Q-1	47	.340		18.15	10.55		28.70	36
	3920	2" diameter		39	.410		24	12.70		36.70	45.50
	3940	3" diameter		34	.471		36	14.60		50.60	61.50
	3960	4" diameter		25	.640		57	19.85		76.85	92.50
	3980	6" diameter		21	.762		85.50	23.50		109	130
	4500	To delete couplings & hangers, subtract									
	4510	1" diam.					23%	27%			
	4520	1-1/2" diam. to 4" diam.					17%	23%			
	4530	6" diam.					22%	26%			
160	0010	**PIPE, GLASS, FITTINGS**									
	0020	Drainage, beaded ends									
	0040	Coupling & labor required at joints not incl. in fitting									
	0050	price. Add 1 per joint for installed price									
	0070	90° Bend or sweep, 1-1/2"				Ea.	16.85			16.85	18.50
	0090	2"					21			21	23.50
	0100	3"					35			35	38.50
	0110	4"					56			56	61.50
	0120	6" (sweep only)					171			171	188
	0200	45° Bend or sweep same as 90°									
	0350	Tee, single sanitary, 1-1/2"				Ea.	27			27	30
	0370	2"					27			27	30

	15106 | Glass Pipe & Fittings	CREW	DAILY OUTPUT	LABOR-HOURS	UNIT	2001 BARE COSTS MAT.	LABOR	EQUIP.	TOTAL	TOTAL INCL O&P
160	0380 3"				Ea.	40.50			40.50	44.50
	0390 4"					72.50			72.50	80
	0400 6"					195			195	214
	0410 Tee, straight, 1-1/2"					33.50			33.50	37
	0430 2"					33.50			33.50	37
	0440 3"					48.50			48.50	53.50
	0450 4"					85.50			85.50	94
	0460 6"					210			210	231
	0500 Coupling, stainless steel, TFE seal ring									
	0520 1-1/2"	Q-1	32	.500	Ea.	11.65	15.50		27.15	36.50
	0530 2"		30	.533		14.90	16.55		31.45	41.50
	0540 3"		25	.640		20	19.85		39.85	52
	0550 4"		23	.696		34.50	21.50		56	70.50
	0560 6"		20	.800		77	25		102	123
	0600 Coupling, stainless steel, bead to plain end									
	0610 1-1/2"	Q-1	36	.444	Ea.	15.10	13.80		28.90	37.50
	0620 2"		34	.471		18.80	14.60		33.40	42.50
	0630 3"		29	.552		32	17.10		49.10	61
	0640 4"		27	.593		48	18.35		66.35	80
	0650 6"		24	.667		162	20.50		182.50	209
	2000 Process supply (pressure), beaded ends									
	2050 90° Sweep elbow, 1/2"				Ea.	17.50			17.50	19.25
	2070 3/4"					20.50			20.50	22.50
	2080 1"					27			27	30
	2090 1-1/2"					28.50			28.50	31.50
	2100 2"					38.50			38.50	42.50
	2120 3"					61			61	67
	2130 4"					85.50			85.50	94
	2140 6"					216			216	237
	2150 45° Sweep elbow, same as 90°									
	2250 Tee, straight, 1/2"				Ea.	29.50			29.50	32.50
	2270 3/4"					29.50			29.50	32.50
	2280 1"					33			33	36.50
	2290 1-1/2"					48.50			48.50	53.50
	2300 2"					58			58	63.50
	2310 3"					70.50			70.50	77.50
	2320 4"					118			118	130
	2330 6"					200			200	220
	2350 Coupling, Viton liner, for temperatures to 400°F									
	2370 1/2"	Q-1	40	.400	Ea.	25	12.40		37.40	46
	2380 3/4"		37	.432		28.50	13.40		41.90	51.50
	2390 1"		35	.457		34	14.15		48.15	58.50
	2400 1-1/2"		32	.500		38.50	15.50		54	65.50
	2410 2"		30	.533		46	16.55		62.55	75.50
	2420 3"		25	.640		84.50	19.85		104.35	123
	2430 4"		23	.696		102	21.50		123.50	145
	2440 6"		20	.800		197	25		222	255
	2550 For beaded joint armored fittings, add					200%				
	2600 Conical ends. Flange set, gasket & labor not incl. in fitting									
	2620 price. Add 1 per joint for installed price.									
	2650 90° Sweep elbow, 1"				Ea.	37.50			37.50	41
	2670 1-1/2"					55.50			55.50	61
	2680 2"					64.50			64.50	71
	2690 3"					114			114	126
	2700 4"					151			151	166
	2710 6"					243			243	268

15106 | Glass Pipe & Fittings

			CREW	DAILY OUTPUT	LABOR-HOURS	UNIT	2001 BARE COSTS MAT.	LABOR	EQUIP.	TOTAL	TOTAL INCL O&P	
160	2750	Cross (straight), add				Ea.	55%					160
	2850	Tee, add					20%					
	2950	Lateral "Y", 1"					71.50			71.50	78.50	
	2970	1-1/2"					82.50			82.50	90.50	
	2980	2"					100			100	110	
	2990	3"					144			144	158	
	3000	4"					225			225	247	
	3010	6"					475			475	525	
	3050	Flange set and Gasket, 1"	Q-1	28	.571		29.50	17.70		47.20	59	
	3070	1-1/2"		26	.615		33	19.10		52.10	65	
	3080	2"		24	.667		37	20.50		57.50	72	
	3090	3"		22	.727		52.50	22.50		75	91.50	
	3100	4"		16	1		125	31		156	184	
	3110	6"		12	1.333		202	41.50		243.50	286	

15107 | Metal Pipe & Fittings

			CREW	DAILY OUTPUT	LABOR-HOURS	UNIT	MAT.	LABOR	EQUIP.	TOTAL	TOTAL INCL O&P	
220	0010	**PIPE, BRASS** Plain end,										220
	0900	Field threaded, coupling & clevis hanger 10' O.C.										
	0920	Regular weight										
	0980	1/8" diameter	1 Plum	62	.129	L.F.	2.55	4.45		7	9.50	
	1000	1/4" diameter		57	.140		3.18	4.84		8.02	10.80	
	1100	3/8" diameter		52	.154		3.79	5.30		9.09	12.15	
	1120	1/2" diameter		48	.167		4.51	5.75		10.26	13.60	
	1140	3/4" diameter		46	.174		6	6		12	15.65	
	1160	1" diameter		43	.186		8.50	6.40		14.90	19	
	1180	1-1/4" diameter	Q-1	72	.222		12.70	6.90		19.60	24.50	
	1200	1-1/2" diameter		65	.246		14.95	7.65		22.60	28	
	1220	2" diameter		53	.302		20	9.35		29.35	36	
	1240	2-1/2" diameter		41	.390		30.50	12.10		42.60	52	
	1260	3" diameter		31	.516		43	16		59	71.50	
	1280	3-1/2" diameter	Q-2	39	.615		71.50	19.80		91.30	109	
	1300	4" diameter	"	37	.649		71.50	21		92.50	110	
	1930	To delete coupling & hanger, subtract										
	1940	1/8" diam. to 1/2" diam.					14%	46%				
	1950	3/4" diam. to 1-1/2" diam.					8%	47%				
	1960	2" diam. to 4" diam.					10%	37%				
260	0010	**PIPE, BRASS, FITTINGS** Rough bronze, threaded										260
	1000	Standard wt., 90° Elbow										
	1040	1/8"	1 Plum	13	.615	Ea.	7	21		28	39.50	
	1060	1/4"		13	.615		7	21		28	39.50	
	1080	3/8"		13	.615		7	21		28	39.50	
	1100	1/2"		12	.667		7	23		30	42	
	1120	3/4"		11	.727		9.35	25		34.35	48.50	
	1140	1"		10	.800		15.20	27.50		42.70	58	
	1160	1-1/4"	Q-1	17	.941		24.50	29		53.50	71	
	1180	1-1/2"		16	1		30.50	31		61.50	80.50	
	1200	2"		14	1.143		49	35.50		84.50	108	
	1220	2-1/2"		11	1.455		119	45		164	199	
	1240	3"		8	2		181	62		243	293	
	1260	4"	Q-2	11	2.182		365	70		435	510	
	1280	5"		8	3		1,025	96.50		1,121.50	1,275	
	1300	6"		7	3.429		1,500	110		1,610	1,825	
	1500	45° Elbow, 1/8"	1 Plum	13	.615		8.60	21		29.60	41.50	
	1540	1/4"		13	.615		8.60	21		29.60	41.50	
	1560	3/8"		13	.615		8.60	21		29.60	41.50	
	1580	1/2"		12	.667		8.60	23		31.60	44	

15107 | Metal Pipe & Fittings

		15107 Metal Pipe & Fittings	CREW	DAILY OUTPUT	LABOR-HOURS	UNIT	2001 BARE COSTS MAT.	LABOR	EQUIP.	TOTAL	TOTAL INCL O&P	
260	1600	3/4"	1 Plum	11	.727	Ea.	12.20	25		37.20	51.50	260
	1620	1"		10	.800		21	27.50		48.50	64.50	
	1640	1-1/4"	Q-1	17	.941		33	29		62	80.50	
	1660	1-1/2"		16	1		41.50	31		72.50	93	
	1680	2"		14	1.143		67.50	35.50		103	128	
	1700	2-1/2"		11	1.455		128	45		173	209	
	1720	3"		8	2		198	62		260	310	
	1740	4"	Q-2	11	2.182		500	70		570	655	
	1760	5"		8	3		745	96.50		841.50	960	
	1780	6"		7	3.429		745	110		855	980	
	2000	Tee, 1/8"	1 Plum	9	.889		8.20	30.50		38.70	55	
	2040	1/4"		9	.889		8.20	30.50		38.70	55	
	2060	3/8"		9	.889		8.20	30.50		38.70	55	
	2080	1/2"		8	1		8.20	34.50		42.70	61	
	2100	3/4"		7	1.143		11.65	39.50		51.15	72.50	
	2120	1"		6	1.333		21	46		67	92.50	
	2140	1-1/4"	Q-1	10	1.600		36	49.50		85.50	115	
	2160	1-1/2"		9	1.778		41	55		96	128	
	2180	2"		8	2		68	62		130	168	
	2200	2-1/2"		7	2.286		161	71		232	285	
	2220	3"		5	3.200		247	99		346	420	
	2240	4"	Q-2	7	3.429		610	110		720	840	
	2260	5"		5	4.800		1,275	154		1,429	1,625	
	2280	6"		4	6		2,000	193		2,193	2,500	
	2500	Coupling, 1/8"	1 Plum	26	.308		5.85	10.60		16.45	22.50	
	2540	1/4"		22	.364		5.85	12.55		18.40	25.50	
	2560	3/8"		18	.444		5.85	15.30		21.15	29.50	
	2580	1/2"		15	.533		5.85	18.35		24.20	34	
	2600	3/4"		14	.571		8.20	19.70		27.90	38.50	
	2620	1"		13	.615		14	21		35	47.50	
	2640	1-1/4"	Q-1	22	.727		23.50	22.50		46	59.50	
	2660	1-1/2"		20	.800		30.50	25		55.50	71	
	2680	2"		18	.889		50.50	27.50		78	97	
	2700	2-1/2"		14	1.143		85.50	35.50		121	148	
	2720	3"		10	1.600		119	49.50		168.50	206	
	2740	4"	Q-2	12	2		244	64.50		308.50	365	
	2760	5"		10	2.400		460	77		537	620	
	2780	6"		9	2.667		650	85.50		735.50	845	
	3000	Union, 125 lb										
	3020	1/8"	1 Plum	12	.667	Ea.	19.90	23		42.90	56.50	
	3040	1/4"		12	.667		19.90	23		42.90	56.50	
	3060	3/8"		12	.667		19.90	23		42.90	56.50	
	3080	1/2"		11	.727		19.90	25		44.90	60	
	3100	3/4"		10	.800		27.50	27.50		55	71.50	
	3120	1"		9	.889		36	30.50		66.50	85.50	
	3140	1-1/4"	Q-1	16	1		52	31		83	105	
	3160	1-1/2"		15	1.067		62	33		95	119	
	3180	2"		13	1.231		95.50	38		133.50	163	
	3200	2-1/2"		10	1.600		267	49.50		316.50	370	
	3220	3"		7	2.286		470	71		541	625	
	3240	4"	Q-2	10	2.400		1,025	77		1,102	1,250	
320	0010	**PIPE, CAST IRON** Soil, on hangers 5' O.C. R15100-070										320
	0020	Single hub, service wt., lead & oakum joints 10' O.C.										
	2120	2" diameter	Q-1	63	.254	L.F.	4.71	7.85		12.56	17.10	
	2140	3" diameter		60	.267		6.35	8.25		14.60	19.45	

15 MECHANICAL

15107 | Metal Pipe & Fittings

		15107 Metal Pipe & Fittings	CREW	DAILY OUTPUT	LABOR-HOURS	UNIT	2001 BARE COSTS MAT.	LABOR	EQUIP.	TOTAL	TOTAL INCL O&P	
320	2160	4" diameter (R15100-070)	Q-1	55	.291	L.F.	7.95	9		16.95	22.50	320
	2180	5" diameter	Q-2	76	.316		10.60	10.15		20.75	27	
	2200	6" diameter	"	73	.329		12.90	10.55		23.45	30	
	2220	8" diameter	Q-3	59	.542		20	17.80		37.80	49	
	2240	10" diameter		54	.593		33.50	19.45		52.95	66.50	
	2260	12" diameter		48	.667		47	22		69	84.50	
	2261	15" diameter		40	.800		71.50	26.50		98	118	
	2320	For service weight, double hub, add					10%					
	2340	For extra heavy, single hub, add					48%	4%				
	2360	For extra heavy, double hub, add					71%	4%				
	2400	Lead for caulking, (1#/dia in)				Lb.	.98			.98	1.08	
	2420	Oakum for caulking, (1/8#/dia in)				"	4.66			4.66	5.15	
	2960	To delete hangers, subtract										
	2970	2" diam. to 4" diam.					16%	19%				
	2980	5" diam. to 8" diam.					14%	14%				
	2990	10" diam. to 15" diam.					13%	19%				
	3000	Single hub, service wt, push-on gasket joints 10' O.C.										
	3010	2" diameter	Q-1	66	.242	L.F.	4.97	7.50		12.47	16.80	
	3020	3" diameter		63	.254		6.70	7.85		14.55	19.25	
	3030	4" diameter		57	.281		8.35	8.70		17.05	22.50	
	3040	5" diameter	Q-2	79	.304		11.25	9.75		21	27	
	3050	6" diameter	"	75	.320		13.60	10.30		23.90	30.50	
	3060	8" diameter	Q-3	62	.516		21.50	16.95		38.45	49	
	3070	10" diameter		56	.571		36	18.75		54.75	68	
	3080	12" diameter		49	.653		50	21.50		71.50	87.50	
	3082	15" diameter		40	.800		88.50	26.50		115	137	
	3100	For service weight, double hub, add					65%					
	3110	For extra heavy, single hub, add					48%	4%				
	3120	For extra heavy, double hub, add					29%	4%				
	3130	To delete hangers, subtract										
	3140	2" diam. to 4" diam.					12%	21%				
	3150	5" diam. to 8" diam.					10%	16%				
	3160	10" diam. to 15" diam.					9%	21%				
	4000	No hub, couplings 10' O.C.										
	4100	1-1/2" diameter	Q-1	71	.225	L.F.	5.05	7		12.05	16.10	
	4120	2" diameter		67	.239		5.10	7.40		12.50	16.80	
	4140	3" diameter		64	.250		6.60	7.75		14.35	18.95	
	4160	4" diameter		58	.276		8.15	8.55		16.70	22	
	4180	5" diameter	Q-2	83	.289		11.40	9.30		20.70	26.50	
	4200	6" diameter	"	79	.304		14.20	9.75		23.95	30.50	
	4220	8" diameter	Q-3	69	.464		22.50	15.25		37.75	47.50	
	4240	10" diameter	"	61	.525		37	17.25		54.25	66.50	
	4280	To delete hangers, subtract										
	4290	1-1/2" diam. to 6" diam.					22%	47%				
	4300	8" diam. to 10" diam.					21%	44%				
360	0010	**PIPE, CAST IRON, FITTINGS** Soil										360
	0040	Hub and spigot, service weight, lead & oakum joints										
	0080	1/4 bend, 2"	Q-1	16	1	Ea.	6.10	31		37.10	53.50	
	0120	3"		14	1.143		10.60	35.50		46.10	65	
	0140	4"		13	1.231		18.60	38		56.60	78	
	0160	5"	Q-2	18	1.333		23	43		66	90	
	0180	6"	"	17	1.412		29	45.50		74.50	100	
	0200	8"	Q-3	11	2.909		93.50	95.50		189	247	
	0220	10"		10	3.200		127	105		232	299	
	0224	12"		9	3.556		172	117		289	365	

15107 | Metal Pipe & Fittings

		CREW	DAILY OUTPUT	LABOR-HOURS	UNIT	2001 BARE COSTS MAT.	LABOR	EQUIP.	TOTAL	TOTAL INCL O&P	
360	0226 15″	Q-3	7	4.571	Ea.	550	150		700	830	360
	0251 Long sweep elbow										
	0252 2″	Q-1	16	1	Ea.	10.60	31		41.60	58.50	
	0253 3″		14	1.143		16.20	35.50		51.70	71.50	
	0254 4″		13	1.231		25	38		63	85	
	0255 6″	Q-2	17	1.412		44.50	45.50		90	118	
	0256 8″	Q-3	11	2.909		115	95.50		210.50	270	
	0257 10″		10	3.200		183	105		288	360	
	0258 12″		9	3.556		297	117		414	500	
	0259 15″		7	4.571		500	150		650	775	
	0266 Closet bend, 3″ diameter with flange 10″ x 16″	Q-1	14	1.143		41.50	35.50		77	99	
	0268 16″x16″		12	1.333		59.50	41.50		101	128	
	0270 Closet bend, 4″ diameter, 1″ x 4″ ring, 6″ x 16″		13	1.231		42.50	38		80.50	104	
	0280 8″ x 16″		13	1.231		43	38		81	105	
	0290 10″ x 12″		12	1.333		40.50	41.50		82	107	
	0300 10″ x 18″		11	1.455		56	45		101	130	
	0310 12″ x 16″		11	1.455		48	45		93	121	
	0330 16″ x 16″		10	1.600		61	49.50		110.50	142	
	0340 1/8 bend, 2″		16	1		4.88	31		35.88	52.50	
	0350 3″		14	1.143		8.90	35.50		44.40	63.50	
	0360 4″		13	1.231		12.90	38		50.90	71.50	
	0380 5″	Q-2	18	1.333		17.85	43		60.85	84	
	0400 6″	″	17	1.412		22	45.50		67.50	92.50	
	0420 8″	Q-3	11	2.909		65.50	95.50		161	216	
	0440 10″		10	3.200		94	105		199	263	
	0460 12″		9	3.556		178	117		295	370	
	0461 15″		7	4.571		355	150		505	615	
	0500 Sanitary tee, 2″	Q-1	10	1.600		9.75	49.50		59.25	85.50	
	0540 3″		9	1.778		15.40	55		70.40	100	
	0620 4″		8	2		22	62		84	118	
	0700 5″	Q-2	12	2		43.50	64.50		108	145	
	0800 6″	″	11	2.182		49.50	70		119.50	161	
	0880 8″	Q-3	7	4.571		182	150		332	425	
	0881 10″		7	4.571		240	150		390	490	
	0882 12″		6	5.333		400	175		575	705	
	0883 15″		4	8		785	263		1,048	1,250	
	0900 Sanitary tee, tapped										
	0901 2″ x 2″	Q-1	9	1.778	Ea.	14.75	55		69.75	99	
	0902 3″ x 2″		8	2		16.65	62		78.65	112	
	0903 4″ x 2″		7	2.286		27	71		98	137	
	0910 Sanitary cross, tapped [double tapped sanitary tee]										
	0911 2″ x 2″	Q-1	7	2.286	Ea.	17.60	71		88.60	126	
	0912 3″ x 2″		6	2.667		19.85	82.50		102.35	147	
	0913 4″ x 2″		5	3.200		26.50	99		125.50	180	
	0920 Vent branch (tee)										
	0922 2″	Q-1	10	1.600	Ea.	20.50	49.50		70	97.50	
	0926 4″	″	8	2		38	62		100	136	
	0928 6″	Q-2	11	2.182		47	70		117	158	
	0940 Sanitary Tee, reducing										
	0942 3″ x 2″	Q-1	10	1.600	Ea.	15.40	49.50		64.90	92	
	0943 4″ x 3″		9	1.778		20	55		75	105	
	0944 4″ x 2″		9	1.778		18.60	55		73.60	104	
	0945 5″ x 3″	Q-2	9	2.667		41	85.50		126.50	174	
	0946 6″ x 4″		12	2		56	64.50		120.50	159	
	0947 6″ x 3″		8	3		38	96.50		134.50	188	
	0948 6″ x 2″		12.50	1.920		37	61.50		98.50	134	

15 MECHANICAL

15107 | Metal Pipe & Fittings

			CREW	DAILY OUTPUT	LABOR-HOURS	UNIT	2001 BARE COSTS MAT.	LABOR	EQUIP.	TOTAL	TOTAL INCL O&P	
360	0949	8″ x 6″	Q-3	9	3.556	Ea.	111	117		228	298	360
	0950	8″ x 5″		9	3.556		111	117		228	298	
	0951	8″ x 4″		10	3.200		84	105		189	251	
	0954	10″ x 6″		8	4		189	131		320	405	
	0958	12″ x 8″		7	4.571		345	150		495	600	
	0962	15″ x 10″		5	6.400		595	210		805	965	
	1000	Tee, 2″	Q-1	10	1.600		13.85	49.50		63.35	90	
	1060	3″		9	1.778		22.50	55		77.50	108	
	1120	4″		8	2		29	62		91	125	
	1200	5″	Q-2	12	2		47	64.50		111.50	149	
	1300	6″	″	11	2.182		60.50	70		130.50	173	
	1380	8″	Q-3	7	4.571		140	150		290	380	
	1400	Combination Y and 1/8 bend										
	1420	2″	Q-1	10	1.600	Ea.	13.85	49.50		63.35	90	
	1460	3″		9	1.778		21	55		76	106	
	1520	4″		8	2		29	62		91	126	
	1540	5″	Q-2	12	2		55	64.50		119.50	158	
	1560	6″		11	2.182		69.50	70		139.50	183	
	1580	8″		7	3.429		164	110		274	345	
	1582	12″	Q-3	6	5.333		390	175		565	695	
	1584	Combination Y & 1/8 bend, reducing										
	1586	3″ x 2″	Q-1	10	1.600	Ea.	15.80	49.50		65.30	92.50	
	1587	4″ x 2″		9.50	1.684		13.55	52		65.55	94	
	1588	4″ x 3″		9	1.778		15.75	55		70.75	100	
	1589	6″ x 2″	Q-2	12.50	1.920		28.50	61.50		90	124	
	1590	6″ x 3″		12	2		30	64.50		94.50	131	
	1591	6″ x 4″		11	2.182		31.50	70		101.50	141	
	1592	8″ x 2″	Q-3	11	2.909		53.50	95.50		149	203	
	1593	8″ x 4″		10	3.200		52.50	105		157.50	217	
	1594	8″ x 6″		9	3.556		72.50	117		189.50	256	
	1600	Double Y, 2″	Q-1	8	2		10.55	62		72.55	105	
	1610	3″		7	2.286		18.25	71		89.25	127	
	1620	4″		6.50	2.462		25.50	76.50		102	143	
	1630	5″	Q-2	9	2.667		41	85.50		126.50	174	
	1640	6″	″	8	3		65	96.50		161.50	218	
	1650	8″	Q-3	5.50	5.818		159	191		350	465	
	1660	10″		5	6.400		224	210		434	560	
	1670	12″		4.50	7.111		325	234		559	710	
	1676	Combination double Y & 1/8 bend										
	1678	2″	Q-1	8	2	Ea.	16.85	62		78.85	112	
	1680	3″		7	2.286		23	71		94	132	
	1682	4″		6.50	2.462		37.50	76.50		114	157	
	1684	6″	Q-2	8	3		130	96.50		226.50	289	
	1690	Combination double Y & 1/8 bend, CI, reducing										
	1692	3″ x 2″	Q-1	8	2	Ea.	18.45	62		80.45	114	
	1694	4″ x 2″		7	2.286		23	71		94	133	
	1696	4″ x 3″		7	2.286		26.50	71		97.50	136	
	1698	6″ x 4″	Q-2	9	2.667		85	85.50		170.50	223	
	1700	Double Y, CI, reducing										
	1702	3″ x 2″	Q-1	8	2	Ea.	15.55	62		77.55	111	
	1703	4″ x 2″		7	2.286		30	71		101	140	
	1704	4″ x 3″		7	2.286		33	71		104	144	
	1705	6″ x 2″	Q-2	10	2.400		72	77		149	196	
	1706	6″ x 3″		9.50	2.526		72	81		153	203	
	1707	6″ x 4″		9	2.667		71	85.50		156.50	207	
	1708	8″ x 4″	Q-3	8.50	3.765		152	124		276	355	

15107 | Metal Pipe & Fittings

			DAILY OUTPUT	LABOR-HOURS		2001 BARE COSTS				TOTAL
		CREW			UNIT	MAT.	LABOR	EQUIP.	TOTAL	INCL O&P
360 1709	8″ x 6″	Q-3	8	4	Ea.	180	131		311	395 360
1711	10″ x 6″		8	4		299	131		430	530
1712	10″ x 8″		6.50	4.923		330	162		492	610
1713	12″ x 6″		7.50	4.267		385	140		525	630
1714	12″ x 8″		6	5.333		415	175		590	720
1740	Reducer, 3″ x 2″	Q-1	15	1.067		7.70	33		40.70	58.50
1750	4″ x 2″		14.50	1.103		8.15	34		42.15	60.50
1760	4″ x 3″		14	1.143		10.10	35.50		45.60	64.50
1770	5″ x 2″		14	1.143		13.35	35.50		48.85	68
1780	5″ x 3″		13.50	1.185		12.95	36.50		49.45	70
1790	5″ x 4″		13	1.231		13	38		51	72
1800	6″ x 2″		13.50	1.185		18	36.50		54.50	75.50
1810	6″ x 3″		13	1.231		19.70	38		57.70	79
1830	6″ x 4″	Q-2	12.50	1.920		20.50	61.50		82	116
1840	6″ x 5″	Q-1	11	1.455		22	45		67	92
1860	8″ x 2″	Q-2	14	1.714		35	55		90	122
1880	8″ x 3″		13.50	1.778		35	57		92	125
1900	8″ x 4″		13	1.846		34	59.50		93.50	127
1920	8″ x 5″		12	2		35	64.50		99.50	136
1940	8″ x 6″		12	2		35	64.50		99.50	136
1942	10″ x 4″		12.50	1.920		51	61.50		112.50	149
1943	10″ x 6″		12	2		54	64.50		118.50	157
1944	10″ x 8″		9.50	2.526		57.50	81		138.50	186
1945	12″ x 4″		11.50	2.087		83.50	67		150.50	193
1946	12″ x 6″		11	2.182		89	70		159	204
1947	12″ x 8″		9	2.667		92	85.50		177.50	230
1948	12″ x 10″		8.50	2.824		93.50	91		184.50	240
1949	15″ x 6″	Q-3	12	2.667		174	87.50		261.50	325
1950	15″ x 8″		10	3.200		189	105		294	365
1951	15″ x 10″		9.50	3.368		199	111		310	385
1952	15″ x 12″		9	3.556		199	117		316	395
1960	Increaser, 2″ x 3″	Q-1	15	1.067		17.50	33		50.50	69.50
1980	2″ x 4″		14	1.143		17.50	35.50		53	73
2000	2″ x 5″		13	1.231		22.50	38		60.50	82
2020	3″ x 4″		13	1.231		19.20	38		57.20	78.50
2040	3″ x 5″		13	1.231		22.50	38		60.50	82
2060	3″ x 6″		12	1.333		27.50	41.50		69	93
2070	4″ x 5″		13	1.231		22.50	38		60.50	82
2080	4″ x 6″		12	1.333		27.50	41.50		69	93
2090	4″ x 8″	Q-2	13	1.846		56.50	59.50		116	152
2100	5″ x 6″	Q-1	11	1.455		41	45		86	113
2110	5″ x 8″	Q-2	12	2		66	64.50		130.50	170
2120	6″ x 8″		12	2		66	64.50		130.50	170
2130	6″ x 10″		8	3		113	96.50		209.50	271
2140	8″ x 10″		6.50	3.692		113	119		232	305
2150	10″ x 12″		5.50	4.364		214	140		354	450
2500	Y, 2″	Q-1	10	1.600		10.10	49.50		59.60	86
2510	3″		9	1.778		18.60	55		73.60	104
2520	4″		8	2		25	62		87	121
2530	5″	Q-2	12	2		42	64.50		106.50	144
2540	6″	″	11	2.182		57.50	70		127.50	170
2550	8″	Q-3	7	4.571		141	150		291	380
2560	10″		6	5.333		228	175		403	515
2570	12″		5	6.400		390	210		600	745
2580	15″		4	8		830	263		1,093	1,300
2581	Y, reducing									

	15107 | Metal Pipe & Fittings		CREW	DAILY OUTPUT	LABOR-HOURS	UNIT	2001 BARE COSTS MAT.	LABOR	EQUIP.	TOTAL	TOTAL INCL O&P	
360	2582	3″ x 2″	Q-1	10	1.600	Ea.	14.60	49.50		64.10	91	360
	2584	4″ x 2″		9.50	1.684		19.45	52		71.45	101	
	2586	4″ x 3″		9	1.778		21.50	55		76.50	107	
	2588	6″ x 2″	Q-2	12.50	1.920		36.50	61.50		98	133	
	2590	6″ x 3″		12	2		37.50	64.50		102	138	
	2592	6″ x 4″		12	2		39	64.50		103.50	140	
	2594	8″ x 2″	Q-3	11	2.909		70.50	95.50		166	222	
	2596	8″ x 3″		10.50	3.048		70.50	100		170.50	229	
	2598	8″ x 4″		10	3.200		70.50	105		175.50	237	
	2600	8″ x 6″		9	3.556		87.50	117		204.50	273	
	2602	10″ x 3″		10	3.200		129	105		234	300	
	2604	10″ x 4″		9.50	3.368		127	111		238	305	
	2606	10″ x 6″		8	4		138	131		269	350	
	2608	10″ x 8″		8	4		190	131		321	405	
	2610	12″ x 4″		9	3.556		214	117		331	410	
	2612	12″ x 6″		8.50	3.765		221	124		345	430	
	2614	12″ x 8″		8	4		265	131		396	490	
	2616	12″ x 10″		7.50	4.267		340	140		480	585	
	2618	15″ x 4″		7	4.571		655	150		805	945	
	2620	15″ x 6″		6.50	4.923		685	162		847	995	
	2622	15″ x 8″		6.50	4.923		695	162		857	1,000	
	2624	15″ x 10″		5	6.400		710	210		920	1,100	
	2626	15″ x 12″		4.50	7.111		725	234		959	1,150	
	3000	For extra heavy, add					44%	4%				
	3600	Hub and spigot, service weight gasket joint										
	3605	Note: gaskets and joint labor have										
	3606	been included with all listed fittings.										
	3610	1/4 bend, 2″	Q-1	20	.800	Ea.	8.70	25		33.70	47	
	3620	3″		17	.941		14	29		43	59.50	
	3630	4″		15	1.067		23	33		56	75	
	3640	5″	Q-2	21	1.143		29.50	36.50		66	88	
	3650	6″	″	19	1.263		35.50	40.50		76	101	
	3660	8″	Q-3	12	2.667		109	87.50		196.50	251	
	3670	10″		11	2.909		150	95.50		245.50	310	
	3680	12″		10	3.200		202	105		307	380	
	3690	15″		8	4		585	131		716	845	
	3700	Closet bend, 3″ diameter with ring 10″ x 16″	Q-1	17	.941		45	29		74	93.50	
	3710	16″ x 16″		15	1.067		63	33		96	119	
	3730	Closet bend, 4″ diameter, 1″ x 4″ ring, 6″ x 16″		15	1.067		46.50	33		79.50	102	
	3740	8″ x 16″		15	1.067		47	33		80	102	
	3750	10″ x 12″		14	1.143		44.50	35.50		80	103	
	3760	10″ x 18″		13	1.231		60.50	38		98.50	124	
	3770	12″ x 16″		13	1.231		52.50	38		90.50	115	
	3780	16″ x 16″		12	1.333		65	41.50		106.50	134	
	3800	1/8 bend, 2″		20	.800		7.45	25		32.45	45.50	
	3810	3″		17	.941		12.30	29		41.30	57.50	
	3820	4″		15	1.067		17.25	33		50.25	69	
	3830	5″	Q-2	21	1.143		24.50	36.50		61	82.50	
	3840	6″	″	19	1.263		29	40.50		69.50	93.50	
	3850	8″	Q-3	12	2.667		81	87.50		168.50	221	
	3860	10″		11	2.909		118	95.50		213.50	273	
	3870	12″		10	3.200		208	105		313	390	
	3880	15″		8	4		390	131		521	630	
	3900	Sanitary Tee, 2″	Q-1	12	1.333		14.90	41.50		56.40	79	
	3910	3″		10	1.600		22	49.50		71.50	99.50	
	3920	4″		9	1.778		30.50	55		85.50	117	

	15107 | Metal Pipe & Fittings	CREW	DAILY OUTPUT	LABOR-HOURS	UNIT	2001 BARE COSTS MAT.	LABOR	EQUIP.	TOTAL	TOTAL INCL O&P	
360	3930 5″	Q-2	13	1.846	Ea.	57	59.50		116.50	152	360
	3940 6″	″	11	2.182		63.50	70		133.50	176	
	3950 8″	Q-3	8	4		212	131		343	430	
	3980 Tee, 2″	Q-1	12	1.333		19	41.50		60.50	83.50	
	3990 3″		10	1.600		29.50	49.50		79	107	
	4000 4″		9	1.778		37.50	55		92.50	124	
	4010 5″	Q-2	13	1.846		60.50	59.50		120	156	
	4020 6″	″	11	2.182		74	70		144	188	
	4030 8″	Q-3	8	4		171	131		302	385	
	4060 Combination Y and 1/8 bend										
	4070 2″	Q-1	12	1.333	Ea.	19	41.50		60.50	83.50	
	4080 3″		10	1.600		28	49.50		77.50	106	
	4090 4″		9	1.778		37.50	55		92.50	125	
	4100 5″	Q-2	13	1.846		68.50	59.50		128	165	
	4110 6″	″	11	2.182		83.50	70		153.50	198	
	4120 8″	Q-3	8	4		194	131		325	410	
	4160 Double Y, 2″	Q-1	10	1.600		18.30	49.50		67.80	95	
	4170 3″		8	2		28.50	62		90.50	125	
	4180 4″		7	2.286		38	71		109	149	
	4190 5″	Q-2	10	2.400		60.50	77		137.50	183	
	4200 6″	″	9	2.667		86	85.50		171.50	224	
	4210 8″	Q-3	6	5.333		205	175		380	490	
	4220 10″		5	6.400		294	210		504	640	
	4230 12″		4.50	7.111		410	234		644	810	
	4260 Reducer, 3″ x 2″	Q-1	17	.941		13.70	29		42.70	59	
	4270 4″ x 2″		16.50	.970		15.05	30		45.05	62	
	4280 4″ x 3″		16	1		17.85	31		48.85	66.50	
	4290 5″ x 2″		16	1		22.50	31		53.50	72	
	4300 5″ x 3″		15.50	1.032		23	32		55	74	
	4310 5″ x 4″		15	1.067		24	33		57	76.50	
	4320 6″ x 2″		15.50	1.032		27.50	32		59.50	79	
	4330 6″ x 3″		15	1.067		30	33		63	83	
	4340 6″ x 5″		13	1.231		35.50	38		73.50	96.50	
	4350 8″ x 2″	Q-2	16	1.500		53	48		101	131	
	4360 8″ x 3″		15	1.600		53.50	51.50		105	137	
	4370 8″ x 4″		15	1.600		53.50	51.50		105	137	
	4380 8″ x 5″		14	1.714		57	55		112	146	
	4390 8″ x 6″		14	1.714		57	55		112	146	
	4430 Increaser, 2″ x 3″	Q-1	17	.941		21	29		50	67	
	4440 2″ x 4″		16	1		22	31		53	71	
	4450 2″ x 5″		15	1.067		29	33		62	82	
	4460 3″ x 4″		15	1.067		23.50	33		56.50	76	
	4470 3″ x 5″		15	1.067		29	33		62	82	
	4480 3″ x 6″		14	1.143		34.50	35.50		70	91.50	
	4490 4″ x 5″		15	1.067		29	33		62	82	
	4500 4″ x 6″		14	1.143		34.50	35.50		70	91.50	
	4510 4″ x 8″	Q-2	15	1.600		71.50	51.50		123	157	
	4520 5″ x 6″	Q-1	13	1.231		48	38		86	110	
	4530 5″ x 8″	Q-2	14	1.714		81.50	55		136.50	173	
	4540 6″ x 8″		14	1.714		81.50	55		136.50	173	
	4550 6″ x 10″		10	2.400		137	77		214	266	
	4560 8″ x 10″		8.50	2.824		137	91		228	287	
	4570 10″ x 12″		7.50	3.200		244	103		347	425	
	4600 Y, 2″	Q-1	12	1.333		15.30	41.50		56.80	79.50	
	4610 3″		10	1.600		25.50	49.50		75	103	
	4620 4″		9	1.778		33.50	55		88.50	120	

15107 | Metal Pipe & Fittings

360

		CREW	DAILY OUTPUT	LABOR-HOURS	UNIT	2001 BARE COSTS MAT.	LABOR	EQUIP.	TOTAL	TOTAL INCL O&P
4630	5"	Q-2	13	1.846	Ea.	55.50	59.50		115	151
4640	6"	"	11	2.182		71.50	70		141.50	185
4650	8"	Q-3	8	4		171	131		302	385
4660	10"		7	4.571		274	150		424	525
4670	12"	↓	6	5.333	↓	450	175		625	760
4900	For extra heavy, add					44%	4%			
4940	Gasket and making push-on joint									
4950	2"	Q-1	40	.400	Ea.	2.59	12.40		14.99	21.50
4960	3"		35	.457		3.41	14.15		17.56	25.50
4970	4"	↓	32	.500		4.31	15.50		19.81	28
4980	5"	Q-2	43	.558		6.65	17.95		24.60	34.50
4990	6"	"	40	.600		6.95	19.30		26.25	36.50
5000	8"	Q-3	32	1		15.30	33		48.30	66.50
5010	10"		29	1.103		23.50	36.50		60	80
5020	12"		25	1.280		30	42		72	96.50
5022	15"	↓	21	1.524	↓	35.50	50		85.50	115
5030	Note: gaskets and joint labor have									
5040	Been included with all listed fittings.									
5990	No hub									
6000	Cplg. & labor required at joints not incl. in fitting									
6010	price. Add 1 coupling per joint for installed price									
6020	1/4 Bend, 1-1/2"				Ea.	4.47			4.47	4.92
6060	2"					4.85			4.85	5.35
6080	3"					6.35			6.35	6.95
6120	4"					9.70			9.70	10.65
6140	5"					21.50			21.50	24
6160	6"					24			24	26.50
6180	8"					67.50			67.50	74
6181	10"					107			107	117
6184	1/4 Bend, long sweep, 1-1/2"					10.45			10.45	11.50
6186	2"					10.45			10.45	11.50
6188	3"					12.45			12.45	13.70
6189	4"					19.80			19.80	22
6190	5"					36.50			36.50	40
6191	6"					44.50			44.50	49
6192	8"					109			109	120
6193	10"					195			195	215
6200	1/8 Bend, 1-1/2"					3.72			3.72	4.09
6210	2"					4.13			4.13	4.54
6212	3"					5.60			5.60	6.15
6214	4"					7.10			7.10	7.80
6216	5"					15.15			15.15	16.65
6218	6"					16.45			16.45	18.10
6220	8"					47.50			47.50	52
6222	10"				↓	88			88	97
6364	Closet flange									
6366	4"	Q-1	33	.485	Ea.	6.75	15.05		21.80	30
6370	Closet bend, no hub									
6376	4" x 16"	Q-1	16.50	.970	Ea.	33.50	30		63.50	82
6380	Sanitary Tee, tapped, 1-1/2"					8.15			8.15	9
6382	2" x 1-1/2"					8.15			8.15	9
6384	2"					9			9	9.90
6386	3" x 2"					11.55			11.55	12.70
6388	3"					21.50			21.50	23.50
6390	4" x 1-1/2"					10.25			10.25	11.30
6392	4" x 2"				↓	11.55			11.55	12.70

360

15107 | Metal Pipe & Fittings

			Crew	Daily Output	Labor-Hours	Unit	2001 Bare Costs Mat.	Labor	Equip.	Total	Total Incl O&P
360	6394	6" x 1-1/2"				Ea.	24			24	26.50
	6396	6" x 2"					24.50			24.50	27
	6459	Sanitary Tee, 1-1/2"					6.15			6.15	6.80
	6460	2"					6.75			6.75	7.40
	6470	3"					8.20			8.20	9
	6472	4"					12.70			12.70	13.95
	6474	5"					36			36	39.50
	6476	6"					37			37	40.50
	6478	8"					149			149	164
	6480	10"					160			160	176
	6724	Sanitary Tee, reducing									
	6725	3" x 2"				Ea.	7.45			7.45	8.20
	6726	4" x 3"					11.65			11.65	12.85
	6727	5" x 4"					28			28	30.50
	6728	6" x 3"					27			27	30
	6729	8" x 4"					79			79	86.50
	6730	Y, 1-1/2"					6.30			6.30	6.95
	6740	2"					6.30			6.30	6.95
	6750	3"					8.95			8.95	9.85
	6760	4"					14.55			14.55	16
	6762	5"					34			34	37
	6764	6"					39			39	42.50
	6768	8"					91.50			91.50	100
	6791	Y, reducing, 3" x 2"					6.75			6.75	7.40
	6792	4" x 2"					9.30			9.30	10.25
	6793	5" x 2"					21			21	23
	6794	6" x 2"					23.50			23.50	25.50
	6795	6" x 4"					30.50			30.50	33.50
	6796	8" x 4"					52			52	57.50
	6797	8" x 6"					64			64	70.50
	6798	10" x 6"					144			144	159
	6799	10" x 8"					173			173	190
	6800	Double Y, 2"					9.70			9.70	10.65
	6920	3"					17.85			17.85	19.65
	7000	4"					36.50			36.50	40
	7100	6"					64.50			64.50	71
	7120	8"					184			184	202
	7200	Combination Y and 1/8 Bend									
	7220	1-1/2"				Ea.	6.65			6.65	7.30
	7260	2"					7.10			7.10	7.80
	7320	3"					10.80			10.80	11.90
	7400	4"					19.75			19.75	21.50
	7480	5"					43.50			43.50	48
	7500	6"					52.50			52.50	57.50
	7520	8"					136			136	150
	7800	Reducer, 3" x 2"					3.37			3.37	3.71
	7820	4" x 2"					5.25			5.25	5.80
	7840	4" x 3"					5.25			5.25	5.80
	8000	Coupling, standard (by CISPI Mfrs.)									
	8020	1-1/2"	Q-1	48	.333	Ea.	3.82	10.35		14.17	19.80
	8040	2"		44	.364		3.82	11.25		15.07	21
	8080	3"		38	.421		4.54	13.05		17.59	24.50
	8120	4"		33	.485		5.40	15.05		20.45	28.50
	8160	5"	Q-2	44	.545		13.25	17.55		30.80	41
	8180	6"	"	40	.600		13.80	19.30		33.10	44
	8200	8"	Q-3	33	.970		26	32		58	76.50

15 MECHANICAL

15107 | Metal Pipe & Fittings

			CREW	DAILY OUTPUT	LABOR-HOURS	UNIT	2001 BARE COSTS MAT.	LABOR	EQUIP.	TOTAL	TOTAL INCL O&P	
360	8220	10"	Q-3	26	1.231	Ea.	36	40.50		76.50	101	360
	8300	Coupling, cast iron clamp & neoprene gasket (by MG)										
	8310	1-1/2"	Q-1	48	.333	Ea.	7.75	10.35		18.10	24	
	8320	2"		44	.364		8	11.25		19.25	26	
	8330	3"		38	.421		8.35	13.05		21.40	29	
	8340	4"		33	.485		9.80	15.05		24.85	33.50	
	8350	5"	Q-2	44	.545		18.55	17.55		36.10	47	
	8360	6"	"	40	.600		25.50	19.30		44.80	57	
	8380	8"	Q-3	33	.970		44.50	32		76.50	96.50	
	8400	10"	"	26	1.231		68	40.50		108.50	136	
	8410	Reducing, no hub										
	8416	2" x 1-1/2"	Q-1	44	.364	Ea.	9.10	11.25		20.35	27	
	8600	Coupling, Stainless steel										
	8620	1-1/2"	Q-1	48	.333	Ea.	7.40	10.35		17.75	24	
	8630	2"		44	.364		7.70	11.25		18.95	25.50	
	8640	2" x 1-1/2"		44	.364		8.20	11.25		19.45	26	
	8650	3"		38	.421		8.30	13.05		21.35	29	
	8660	4"		33	.485		9.40	15.05		24.45	33	
	8670	4" x 3"		33	.485		13.90	15.05		28.95	38	
	8680	5"	Q-2	44	.545		20	17.55		37.55	48.50	
	8690	6"	"	40	.600		21.50	19.30		40.80	52.50	
	8700	8"	Q-3	33	.970		36	32		68	87.50	
	8710	10"	"	26	1.231		46	40.50		86.50	112	
380	0010	**PIPE, CAST IRON/STEEL, FITTINGS** Special										380
	0020	Cast iron, drainage, threaded, black										
	0030	90° Elbow, straight										
	0031	1-1/2" pipe size	Q-1	20	.800	Ea.	11.35	25		36.35	50	
	0032	2" pipe size		18	.889		17.20	27.50		44.70	60.50	
	0033	3" pipe size		10	1.600		64	49.50		113.50	146	
	0034	4" pipe size		6	2.667		102	82.50		184.50	237	
	0040	90° Long turn elbow, straight										
	0041	1-1/2" pipe size	Q-1	20	.800	Ea.	15.85	25		40.85	55	
	0042	2" pipe size		18	.889		23.50	27.50		51	67.50	
	0043	3" pipe size		10	1.600		78.50	49.50		128	161	
	0044	4" pipe size		6	2.667		119	82.50		201.50	256	
	0050	90° Street elbow, straight										
	0051	1-1/2" pipe size	Q-1	20	.800	Ea.	16.20	25		41.20	55.50	
	0052	2" pipe size	"	18	.889	"	21.50	27.50		49	65.50	
	0060	45° Elbow										
	0061	1-1/2" pipe size	Q-1	20	.800	Ea.	11.35	25		36.35	50	
	0062	2" pipe size		18	.889		16.55	27.50		44.05	59.50	
	0063	3" pipe size		10	1.600		63	49.50		112.50	144	
	0064	4" pipe size		6	2.667		98	82.50		180.50	233	
	0070	45° Street elbow										
	0071	1-1/2" pipe size	Q-1	20	.800	Ea.	15.20	25		40.20	54	
	0072	2" pipe size	"	18	.889	"	24.50	27.50		52	68.50	
	0092	Tees, straight										
	0093	1-1/2" pipe size	Q-1	13	1.231	Ea.	18.90	38		56.90	78.50	
	0094	2" pipe size	"	11	1.455	"	31	45		76	102	
	0100	TY's, straight										
	0101	1-1/2" pipe size	Q-1	13	1.231	Ea.	18.55	38		56.55	78	
	0102	2" pipe size		11	1.455		30.50	45		75.50	102	
	0103	3" pipe size		6	2.667		85.50	82.50		168	220	
	0104	4" pipe size		4	4		165	124		289	370	
	0120	45° Y branch, straight										

MECHANICAL 15

15107 | Metal Pipe & Fittings		CREW	DAILY OUTPUT	LABOR-HOURS	UNIT	2001 BARE COSTS MAT.	LABOR	EQUIP.	TOTAL	TOTAL INCL O&P
380 0121	1-1/2" pipe size	Q-1	13	1.231	Ea.	22.50	38		60.50	82.50
0122	2" pipe size		11	1.455		42	45		87	114
0123	3" pipe size		6	2.667		109	82.50		191.50	245
0124	4" pipe size		4	4		216	124		340	425
0147	Double Y branch, straight									
0148	1-1/2" pipe size	Q-1	10	1.600	Ea.	42.50	49.50		92	122
0149	2" pipe size	"	7	2.286	"	43	71		114	155
0160	P trap									
0161	1-1/2" pipe size	Q-1	15	1.067	Ea.	33.50	33		66.50	87
0162	2" pipe size		13	1.231		53.50	38		91.50	116
0163	3" pipe size		7	2.286		130	71		201	250
0164	4" pipe size		5	3.200		310	99		409	490
0180	Tucker connection									
0181	1-1/2" pipe size	Q-1	24	.667	Ea.	24.50	20.50		45	57.50
0182	2" pipe size	"	21	.762	"	30	23.50		53.50	68.50
0205	Cast iron, drainage, threaded, galvanized									
0206	90° Elbow, straight									
0207	1-1/2" pipe size	Q-1	20	.800	Ea.	15.85	25		40.85	55
0208	2" pipe size		18	.889		23	27.50		50.50	67
0209	3" pipe size		10	1.600		83	49.50		132.50	167
0210	4" pipe size		6	2.667		137	82.50		219.50	276
0216	90° Long turn elbow, straight									
0217	1-1/2" pipe size	Q-1	20	.800	Ea.	21	25		46	60.50
0218	2" pipe size		18	.889		31	27.50		58.50	75.50
0219	3" pipe size		10	1.600		98	49.50		147.50	183
0220	4" pipe size		6	2.667		162	82.50		244.50	305
0226	90° Street elbow, straight									
0227	1-1/2" pipe size	Q-1	20	.800	Ea.	20.50	25		45.50	60
0228	2" pipe size	"	18	.889	"	31.50	27.50		59	76.50
0236	45° Elbow									
0237	1-1/2" pipe size	Q-1	20	.800	Ea.	15.85	25		40.85	55
0238	2" pipe size		18	.889		22.50	27.50		50	66.50
0239	3" pipe size		10	1.600		78.50	49.50		128	161
0240	4" pipe size		6	2.667		128	82.50		210.50	265
0246	45° Street elbow									
0247	1-1/2" pipe size	Q-1	20	.800	Ea.	21	25		46	60.50
0248	2" pipe size	"	18	.889	"	32	27.50		59.50	77
0268	Tees, straight									
0269	1-1/2" pipe size	Q-1	13	1.231	Ea.	27	38		65	87
0270	2" pipe size	"	11	1.455	"	45	45		90	118
0276	TY's, straight									
0277	1-1/2" pipe size	Q-1	13	1.231	Ea.	24.50	38		62.50	84
0278	2" pipe size		11	1.455		39.50	45		84.50	112
0279	3" pipe size		6	2.667		119	82.50		201.50	256
0280	4" pipe size		4	4		214	124		338	420
0296	45° Y branch, straight									
0297	1-1/2" pipe size	Q-1	13	1.231	Ea.	35	38		73	96
0298	2" pipe size		11	1.455		54	45		99	128
0299	3" pipe size		6	2.667		144	82.50		226.50	284
0300	4" pipe size		4	4		294	124		418	510
0323	Double Y branch, straight									
0324	1-1/2" pipe size	Q-1	10	1.600	Ea.	57	49.50		106.50	138
0325	2" pipe size	"	7	2.286	"	77.50	71		148.50	193
0336	P Trap									
0337	1-1/2" pipe size	Q-1	15	1.067	Ea.	43.50	33		76.50	98
0338	2" pipe size		13	1.231		76	38		114	141

15107 | Metal Pipe & Fittings

		15107 Metal Pipe & Fittings	CREW	DAILY OUTPUT	LABOR-HOURS	UNIT	2001 BARE COSTS MAT.	LABOR	EQUIP.	TOTAL	TOTAL INCL O&P	
380	0339	3″ pipe size	Q-1	7	2.286	Ea.	182	71		253	305	380
	0340	4″ pipe size	↓	5	3.200	↓	455	99		554	650	
	0356	Tucker connection										
	0357	1-1/2″ pipe size	Q-1	24	.667	Ea.	29.50	20.50		50	63.50	
	0358	2″ pipe size	″	21	.762	″	41.50	23.50		65	81	
420	0010	**PIPE, COPPER** Solder joints R15100-050										420
	0100	Solder										
	0120	Solder, lead free, roll R15100-070				Lb.	8.70			8.70	9.55	
	1000	Type K tubing, couplings & clevis hangers 10′ O.C.										
	1100	1/4″ diameter	1 Plum	84	.095	L.F.	1.15	3.28		4.43	6.20	
	1120	3/8″ diameter		82	.098		1.27	3.36		4.63	6.45	
	1140	1/2″ diameter		78	.103		1.36	3.53		4.89	6.85	
	1160	5/8″ diameter		77	.104		1.52	3.58		5.10	7.05	
	1180	3/4″ diameter		74	.108		1.98	3.72		5.70	7.75	
	1200	1″ diameter		66	.121		2.40	4.18		6.58	8.95	
	1220	1-1/4″ diameter		56	.143		2.88	4.92		7.80	10.60	
	1240	1-1/2″ diameter		50	.160		3.53	5.50		9.03	12.20	
	1260	2″ diameter	↓	40	.200		5.25	6.90		12.15	16.15	
	1280	2-1/2″ diameter	Q-1	60	.267		7.60	8.25		15.85	21	
	1300	3″ diameter		54	.296		10.30	9.20		19.50	25	
	1320	3-1/2″ diameter		42	.381		12.85	11.80		24.65	32	
	1330	4″ diameter		38	.421		15.95	13.05		29	37.50	
	1340	5″ diameter	↓	32	.500		41.50	15.50		57	69.50	
	1360	6″ diameter	Q-2	38	.632		60	20.50		80.50	96.50	
	1380	8″ diameter	″	34	.706		107	22.50		129.50	152	
	1390	For other than full hard temper, add				↓	13%					
	1440	For silver solder, add						15%				
	1800	For medical clean, (oxygen class), add					12%					
	1950	To delete cplgs. & hngrs., 1/4″-1″ pipe, subtract					27%	60%				
	1960	1-1/4″ -3″ pipe, subtract					14%	52%				
	1970	3-1/2″-5″ pipe, subtract					10%	60%				
	1980	6″-8″ pipe, subtract					19%	53%				
	2000	Type L tubing, couplings & hangers 10′ O.C.										
	2100	1/4″ diameter	1 Plum	88	.091	L.F.	.97	3.13		4.10	5.80	
	2120	3/8″ diameter		84	.095		1.13	3.28		4.41	6.20	
	2140	1/2″ diameter		81	.099		1.25	3.40		4.65	6.55	
	2160	5/8″ diameter		79	.101		1.60	3.49		5.09	7	
	2180	3/4″ diameter		76	.105		1.59	3.63		5.22	7.20	
	2200	1″ diameter		68	.118		2.02	4.05		6.07	8.35	
	2220	1-1/4″ diameter		58	.138		2.54	4.75		7.29	9.95	
	2240	1-1/2″ diameter		52	.154		3.07	5.30		8.37	11.35	
	2260	2″ diameter	↓	42	.190		4.47	6.55		11.02	14.80	
	2280	2-1/2″ diameter	Q-1	62	.258		6.50	8		14.50	19.25	
	2300	3″ diameter		56	.286		8.65	8.85		17.50	23	
	2320	3-1/2″ diameter		43	.372		11.05	11.55		22.60	29.50	
	2340	4″ diameter		39	.410		13.55	12.70		26.25	34	
	2360	5″ diameter	↓	34	.471		38	14.60		52.60	63.50	
	2380	6″ diameter	Q-2	40	.600		47.50	19.30		66.80	81	
	2400	8″ diameter	″	36	.667		82	21.50		103.50	123	
	2410	For other than full hard temper, add				↓	21%					
	2590	For silver solder, add						15%				
	2900	For medical clean, (oxygen class), add					12%					
	2940	To delete cplgs. & hngrs., 1/4″-1″ pipe, subtract					37%	63%				
	2960	1-1/4″-3″ pipe, subtract					12%	53%				
	2970	3-1/2″-5″ pipe, subtract	↓				12%	63%				

15107 | Metal Pipe & Fittings

		Description	CREW	DAILY OUTPUT	LABOR-HOURS	UNIT	2001 BARE COSTS MAT.	LABOR	EQUIP.	TOTAL	TOTAL INCL O&P	
420	2980	6"-8" pipe, subtract R15100-050					24%	55%				420
	3000	Type M tubing, couplings & hangers 10' O.C.										
	3100	1/4" diameter R15100-070	1 Plum	90	.089	L.F.	1.01	3.06		4.07	5.75	
	3120	3/8" diameter		87	.092		1.02	3.17		4.19	5.90	
	3140	1/2" diameter		84	.095		1.05	3.28		4.33	6.10	
	3160	5/8" diameter		81	.099		1.26	3.40		4.66	6.55	
	3180	3/4" diameter		78	.103		1.34	3.53		4.87	6.80	
	3200	1" diameter		70	.114		1.70	3.94		5.64	7.80	
	3220	1-1/4" diameter		60	.133		2.17	4.59		6.76	9.35	
	3240	1-1/2" diameter		54	.148		2.78	5.10		7.88	10.75	
	3260	2" diameter		44	.182		4.17	6.25		10.42	14.05	
	3280	2-1/2" diameter	Q-1	64	.250		5.75	7.75		13.50	18.05	
	3300	3" diameter		58	.276		7.30	8.55		15.85	21	
	3320	3-1/2" diameter		45	.356		9.30	11		20.30	27	
	3340	4" diameter		40	.400		12.15	12.40		24.55	32	
	3360	5" diameter		36	.444		36.50	13.80		50.30	61	
	3370	6" diameter	Q-2	42	.571		46	18.35		64.35	78.50	
	3380	8" diameter	"	38	.632		77	20.50		97.50	115	
	3440	For silver solder, add						15%				
	3960	To delete cplgs. & hngrs., 1/4"-1" pipe, subtract					35%	65%				
	3970	1-1/4"-3" pipe, subtract					19%	56%				
	3980	3-1/2"-5" pipe, subtract					13%	65%				
	3990	6"-8" pipe, subtract					28%	58%				
	4000	Type DWV tubing, couplings & hangers 10' O.C.										
	4100	1-1/4" diameter	1 Plum	60	.133	L.F.	2.14	4.59		6.73	9.30	
	4120	1-1/2" diameter		54	.148		2.54	5.10		7.64	10.50	
	4140	2" diameter		44	.182		3.19	6.25		9.44	12.95	
	4160	3" diameter	Q-1	58	.276		5.25	8.55		13.80	18.65	
	4180	4" diameter		40	.400		8.95	12.40		21.35	28.50	
	4200	5" diameter		36	.444		26	13.80		39.80	49.50	
	4220	6" diameter	Q-2	42	.571		36.50	18.35		54.85	67.50	
	4240	8" diameter	"	38	.632		80	20.50		100.50	119	
	4730	To delete cplgs. & hngrs., 1-1/4"-2" pipe, subtract					16%	53%				
	4740	2-1/2"-4" pipe, subtract					13%	60%				
	4750	5"-8" pipe, subtract					23%	58%				
	5200	ACR tubing, type L, hard temper, cleaned and										
	5220	capped, no couplings or hangers										
	5240	3/8" OD				L.F.	.40			.40	.44	
	5250	1/2" OD					.54			.54	.59	
	5260	5/8" OD					.75			.75	.83	
	5270	3/4" OD					.99			.99	1.09	
	5280	7/8" OD					1.17			1.17	1.29	
	5290	1-1/8" OD					1.66			1.66	1.83	
	5300	1-3/8" OD					2.29			2.29	2.52	
	5310	1-5/8" OD					2.89			2.89	3.18	
	5320	2-1/8" OD					4.65			4.65	5.10	
	5330	2-5/8" OD					6.60			6.60	7.25	
	5340	3-1/8" OD					9.15			9.15	10.05	
	5350	3-5/8" OD					11.55			11.55	12.70	
	5360	4-1/8" OD					15			15	16.50	
	5380	ACR tubing, type L, annealed, cleaned and capped										
	5381	No couplings or hangers										
	5384	3/8"	1 Stpi	160	.050	L.F.	.40	1.74		2.14	3.07	
	5385	1/2"		160	.050		.54	1.74		2.28	3.22	
	5386	5/8"		160	.050		.75	1.74		2.49	3.46	
420	5387	3/4"		130	.062		.99	2.14		3.13	4.33	

15 MECHANICAL

15107 | Metal Pipe & Fittings

				CREW	DAILY OUTPUT	LABOR-HOURS	UNIT	2001 BARE COSTS MAT.	LABOR	EQUIP.	TOTAL	TOTAL INCL O&P	
420	5388	7/8"	R15100-050	1 Stpi	130	.062	L.F.	1.17	2.14		3.31	4.53	420
	5389	1-1/8"			115	.070		1.66	2.42		4.08	5.50	
	5390	1-3/8"	R15100-070		100	.080		2.29	2.79		5.08	6.75	
	5391	1-5/8"			90	.089		2.89	3.10		5.99	7.85	
	5392	2-1/8"			80	.100		4.65	3.49		8.14	10.35	
	5393	2-5/8"		Q-5	125	.128		6.60	4.02		10.62	13.30	
	5394	3-1/8"			105	.152		9.15	4.78		13.93	17.25	
	5395	4-1/8"			95	.168		15	5.30		20.30	24.50	
	5800	Refrigeration tubing, dryseal, 50′ coils											
	5840	1/8" OD					Coil	10.35			10.35	11.40	
	5850	3/16" OD						11.35			11.35	12.50	
	5860	1/4" OD						14			14	15.40	
	5870	5/16" OD						17.95			17.95	19.75	
	5880	3/8" OD						21			21	23	
	5890	1/2" OD						26.50			26.50	29.50	
	5900	5/8" OD						36			36	39.50	
	5910	3/4" OD						42			42	46	
	5920	7/8" OD						62			62	68.50	
	5930	1-1/8" OD						89			89	97.50	
	5940	1-3/8" OD						124			124	136	
	5950	1-5/8" OD						159			159	175	
460	0010	**PIPE, COPPER, FITTINGS** Wrought unless otherwise noted											460
	0040	Solder joints, copper x copper											
	0070	90° elbow, 1/4"		1 Plum	22	.364	Ea.	.60	12.55		13.15	19.55	
	0090	3/8"			22	.364		.60	12.55		13.15	19.55	
	0100	1/2"			20	.400		.22	13.80		14.02	21	
	0110	5/8"			19	.421		.48	14.50		14.98	22.50	
	0120	3/4"			19	.421		.48	14.50		14.98	22.50	
	0130	1"			16	.500		1.12	17.25		18.37	27	
	0140	1-1/4"			15	.533		1.69	18.35		20.04	29.50	
	0150	1-1/2"			13	.615		2.64	21		23.64	35	
	0160	2"			11	.727		4.80	25		29.80	43.50	
	0170	2-1/2"		Q-1	13	1.231		9.15	38		47.15	67.50	
	0180	3"			11	1.455		12.80	45		57.80	82	
	0190	3-1/2"			10	1.600		42.50	49.50		92	122	
	0200	4"			9	1.778		33	55		88	119	
	0210	5"			6	2.667		121	82.50		203.50	258	
	0220	6"		Q-2	9	2.667		169	85.50		254.50	315	
	0230	8"		"	8	3		680	96.50		776.50	895	
	0250	45° elbow, 1/4"		1 Plum	22	.364		1.20	12.55		13.75	20	
	0270	3/8"			22	.364		.98	12.55		13.53	20	
	0280	1/2"			20	.400		.98	13.80		14.78	22	
	0290	5/8"			19	.421		1.86	14.50		16.36	24	
	0300	3/4"			19	.421		.38	14.50		14.88	22.50	
	0310	1"			16	.500		.65	17.25		17.90	26.50	
	0320	1-1/4"			15	.533		1.16	18.35		19.51	29	
	0330	1-1/2"			13	.615		1.63	21		22.63	34	
	0340	2"			11	.727		2.66	25		27.66	41	
	0350	2-1/2"		Q-1	13	1.231		4.45	38		42.45	62.50	
	0360	3"			13	1.231		14.05	38		52.05	73	
	0370	3-1/2"			10	1.600		24.50	49.50		74	102	
	0380	4"			9	1.778		30	55		85	116	
	0390	5"			6	2.667		116	82.50		198.50	253	
	0400	6"		Q-2	9	2.667		174	85.50		259.50	320	
	0410	8"		"	8	3		620	96.50		716.50	830	

15107 | Metal Pipe & Fittings

			CREW	DAILY OUTPUT	LABOR-HOURS	UNIT	2001 BARE COSTS MAT.	LABOR	EQUIP.	TOTAL	TOTAL INCL O&P	
460	0450	Tee, 1/4″	1 Plum	14	.571	Ea.	1.35	19.70		21.05	31	460
	0470	3/8″		14	.571		1.03	19.70		20.73	30.50	
	0480	1/2″		13	.615		.36	21		21.36	32.50	
	0490	5/8″		12	.667		2.23	23		25.23	37	
	0500	3/4″		12	.667		.89	23		23.89	35.50	
	0510	1″		10	.800		2.59	27.50		30.09	44.50	
	0520	1-1/4″		9	.889		3.92	30.50		34.42	50.50	
	0530	1-1/2″		8	1		5.45	34.50		39.95	58	
	0540	2″		7	1.143		8.45	39.50		47.95	69	
	0550	2-1/2″	Q-1	8	2		17.10	62		79.10	112	
	0560	3″		7	2.286		26	71		97	136	
	0570	3-1/2″		6	2.667		75.50	82.50		158	208	
	0580	4″		5	3.200		63	99		162	220	
	0590	5″		4	4		197	124		321	405	
	0600	6″	Q-2	6	4		269	129		398	490	
	0610	8″	″	5	4.800		1,100	154		1,254	1,425	
	0612	Tee, reducing on the outlet, 1/4″	1 Plum	15	.533		2.17	18.35		20.52	30	
	0613	3/8″		15	.533		2.04	18.35		20.39	29.50	
	0614	1/2″		14	.571		1.78	19.70		21.48	31.50	
	0615	5/8″		13	.615		3.14	21		24.14	35.50	
	0616	3/4″		12	.667		.81	23		23.81	35.50	
	0617	1″		11	.727		2.59	25		27.59	41	
	0618	1-1/4″		10	.800		3.81	27.50		31.31	45.50	
	0619	1-1/2″		9	.889		4.04	30.50		34.54	50.50	
	0620	2″		8	1		6.40	34.50		40.90	59	
	0621	2-1/2″	Q-1	9	1.778		19.95	55		74.95	105	
	0622	3″		8	2		24	62		86	120	
	0623	4″		6	2.667		42.50	82.50		125	172	
	0624	5″		5	3.200		197	99		296	365	
	0625	6″	Q-2	7	3.429		269	110		379	460	
	0626	8″	″	6	4		1,100	129		1,229	1,400	
	0630	Tee, reducing on the run, 1/4″	1 Plum	15	.533		2.70	18.35		21.05	30.50	
	0631	3/8″		15	.533		3.62	18.35		21.97	31.50	
	0632	1/2″		14	.571		2.45	19.70		22.15	32	
	0633	5/8″		13	.615		3.62	21		24.62	36	
	0634	3/4″		12	.667		.99	23		23.99	35.50	
	0635	1″		11	.727		3.08	25		28.08	41.50	
	0636	1-1/4″		10	.800		5	27.50		32.50	47	
	0637	1-1/2″		9	.889		8.30	30.50		38.80	55	
	0638	2″		8	1		11	34.50		45.50	64	
	0639	2-1/2″	Q-1	9	1.778		27	55		82	113	
	0640	3″		8	2		36.50	62		98.50	134	
	0641	4″		6	2.667		84.50	82.50		167	218	
	0642	5″		5	3.200		197	99		296	365	
	0643	6″	Q-2	7	3.429		269	110		379	460	
	0644	8″	″	6	4		1,100	129		1,229	1,400	
	0650	Coupling, 1/4″	1 Plum	24	.333		.15	11.50		11.65	17.50	
	0670	3/8″		24	.333		.18	11.50		11.68	17.55	
	0680	1/2″		22	.364		.15	12.55		12.70	19.05	
	0690	5/8″		21	.381		.46	13.10		13.56	20.50	
	0700	3/4″		21	.381		.32	13.10		13.42	20	
	0710	1″		18	.444		.63	15.30		15.93	23.50	
	0715	1-1/4″		17	.471		1.10	16.20		17.30	25.50	
	0716	1-1/2″		15	.533		1.45	18.35		19.80	29	
	0718	2″		13	.615		2.41	21		23.41	34.50	
	0721	2-1/2″	Q-1	15	1.067		5.10	33		38.10	55.50	

15107 | Metal Pipe & Fittings

	Line	Description	Crew	Daily Output	Labor-Hours	Unit	2001 Bare Costs Mat.	Labor	Equip.	Total	Total Incl O&P	
460	0722	3"	Q-1	13	1.231	Ea.	7.85	38		45.85	66	460
	0724	3-1/2"		8	2		14.75	62		76.75	110	
	0726	4"		7	2.286		15.40	71		86.40	124	
	0728	5"		6	2.667		36	82.50		118.50	165	
	0731	6"	Q-2	8	3		56	96.50		152.50	208	
	0732	8"	"	7	3.429		193	110		303	380	
	0741	Coupling, reducing, concentric										
	0743	1/2"	1 Plum	23	.348	Ea.	.35	12		12.35	18.50	
	0745	3/4"		21.50	.372		.62	12.80		13.42	20	
	0747	1"		19.50	.410		1.13	14.15		15.28	22.50	
	0749	1-1/2"		16	.500		1.42	17.25		18.67	27.50	
	0751	2"		14	.571		3.64	19.70		23.34	33.50	
	0753	3"	Q-1	14	1.143		10.60	35.50		46.10	65	
	0755	4"	"	8	2		24.50	62		86.50	121	
	0757	5"	Q-2	7.50	3.200		78	103		181	241	
	0759	6"		7	3.429		133	110		243	310	
	0761	8"		6.50	3.692		219	119		338	420	
	0771	Cap, sweat										
	0773	1/2"	1 Plum	40	.200	Ea.	.15	6.90		7.05	10.55	
	0775	3/4"		38	.211		.26	7.25		7.51	11.25	
	0777	1"		32	.250		.62	8.60		9.22	13.70	
	0779	1-1/2"		26	.308		1.28	10.60		11.88	17.40	
	0781	2"		22	.364		2.34	12.55		14.89	21.50	
	0791	Flange, sweat										
	0793	3"	Q-1	22	.727	Ea.	26	22.50		48.50	62.50	
	0795	4"		18	.889		36.50	27.50		64	82	
	0797	5"		12	1.333		70.50	41.50		112	140	
	0799	6"	Q-2	18	1.333		74	43		117	146	
	0801	8"	"	16	1.500		193	48		241	285	
	0850	Unions, 1/4"	1 Plum	21	.381		3.83	13.10		16.93	24	
	0870	3/8"		21	.381		3.65	13.10		16.75	24	
	0880	1/2"		19	.421		2.10	14.50		16.60	24.50	
	0890	5/8"		18	.444		8.75	15.30		24.05	32.50	
	0900	3/4"		18	.444		2.63	15.30		17.93	26	
	0910	1"		15	.533		4.72	18.35		23.07	32.50	
	0920	1-1/4"		14	.571		8.10	19.70		27.80	38.50	
	0930	1-1/2"		12	.667		10.65	23		33.65	46	
	0940	2"		10	.800		18.15	27.50		45.65	61.50	
	0950	2-1/2"	Q-1	12	1.333		37.50	41.50		79	104	
	0960	3"	"	10	1.600		97.50	49.50		147	182	
	0980	Adapter, copper x male IPS, 1/4"	1 Plum	20	.400		2.11	13.80		15.91	23.50	
	0990	3/8"		20	.400		1.05	13.80		14.85	22	
	1000	1/2"		18	.444		.46	15.30		15.76	23.50	
	1010	3/4"		17	.471		.76	16.20		16.96	25.50	
	1020	1"		15	.533		1.84	18.35		20.19	29.50	
	1030	1-1/4"		13	.615		2.72	21		23.72	35	
	1040	1-1/2"		12	.667		3.12	23		26.12	38	
	1050	2"		11	.727		5.30	25		30.30	44	
	1060	2-1/2"	Q-1	10.50	1.524		16.85	47		63.85	90	
	1070	3"		10	1.600		23	49.50		72.50	101	
	1080	3-1/2"		9	1.778		38	55		93	125	
	1090	4"		8	2		38	62		100	136	
	1200	5"		6	2.667		182	82.50		264.50	325	
	1210	6"	Q-2	8.50	2.824		203	91		294	360	
	1214	Adapter, copper x female IPS										
	1216	1/2"	1 Plum	18	.444	Ea.	.67	15.30		15.97	23.50	

	15107 | Metal Pipe & Fittings	CREW	DAILY OUTPUT	LABOR-HOURS	UNIT	2001 BARE COSTS MAT.	LABOR	EQUIP.	TOTAL	TOTAL INCL O&P	
460	1218 3/4"	1 Plum	17	.471	Ea.	.93	16.20		17.13	25.50	460
	1220 1"		15	.533		2.13	18.35		20.48	30	
	1222 1-1/2"		12	.667		4.89	23		27.89	40	
	1224 2"		11	.727		6.65	25		31.65	45.50	
	1250 Cross, 1/2"		10	.800		2.78	27.50		30.28	44.50	
	1260 3/4"		9.50	.842		5.40	29		34.40	50	
	1270 1"		8	1		9.15	34.50		43.65	62	
	1280 1-1/4"		7.50	1.067		12.50	37		49.50	69.50	
	1290 1-1/2"		6.50	1.231		18.70	42.50		61.20	84.50	
	1300 2"		5.50	1.455		35.50	50		85.50	115	
	1310 2-1/2"	Q-1	6.50	2.462		81.50	76.50		158	205	
	1320 3"		5.50	2.909		107	90		197	254	
	1340 4"		4.70	3.404		150	106		256	325	
	1350 5"		3.20	5		213	155		368	470	
	1360 6"	Q-2	4.70	5.106		298	164		462	580	
	1500 Tee fitting, mechanically formed, (Type 1).										
	1520 1/2" run size, 3/8" to 1/2" branch size	1 Plum	80	.100	Ea.		3.45		3.45	5.20	
	1530 3/4" run size, 3/8" to 3/4" branch size		60	.133			4.59		4.59	6.95	
	1540 1" run size, 3/8" to 1" branch size		54	.148			5.10		5.10	7.70	
	1550 1-1/4" run size, 3/8" to 1-1/4" branch size		48	.167			5.75		5.75	8.65	
	1560 1-1/2" run size, 3/8" to 1-1/2" branch size		40	.200			6.90		6.90	10.40	
	1570 2" run size, 3/8" to 2" branch size		35	.229			7.85		7.85	11.90	
	1580 2-1/2" run size, 1/2" to 2" branch size		32	.250			8.60		8.60	13	
	1590 3" run size, 1" to 2" branch size		26	.308			10.60		10.60	16	
	1600 4" run size, 1" to 2" branch size		24	.333			11.50		11.50	17.35	
	1640 Tee fitting, mechanically formed, (Type 2)										
	1650 2-1/2" run size, 2-1/2" branch size	1 Plum	12.50	.640	Ea.		22		22	33.50	
	1660 3" run size, 2-1/2" to 3" branch size		12	.667			23		23	34.50	
	1670 3-1/2" run size, 2-1/2" to 3-1/2" branch size		11	.727			25		25	38	
	1680 4" run size, 2-1/2" to 4" branch size		10.50	.762			26.50		26.50	39.50	
	1698 5" run size, 2" to 4" branch size		9.50	.842			29		29	44	
	1700 6" run size, 2" to 4" branch size		8.50	.941			32.50		32.50	49	
	1710 8" run size, 2" to 4" branch size		7	1.143			39.50		39.50	59.50	
	1800 ACR fittings										
	1802 Tee, straight										
	1808 5/8"	1 Stpi	12	.667	Ea.	2.23	23		25.23	37.50	
	1810 3/4"		12	.667		.89	23		23.89	36	
	1812 7/8"		10	.800		.89	28		28.89	43	
	1814 1-1/8"		10	.800		5.30	28		33.30	48	
	1816 1-3/8"		9	.889		8.05	31		39.05	56	
	1818 1-5/8"		8	1		11.15	35		46.15	65	
	1820 2-1/8"		7	1.143		17.35	40		57.35	79	
	1822 2-5/8"	Q-5	8	2		35	63		98	134	
	1824 3-1/8"		7	2.286		53.50	71.50		125	167	
	1826 4-1/8"		5	3.200		129	100		229	294	
	1830 90° elbow										
	1836 5/8"	1 Stpi	19	.421	Ea.	1.40	14.65		16.05	23.50	
	1838 3/4"		19	.421		2.25	14.65		16.90	24.50	
	1840 7/8"		16	.500		2.36	17.45		19.81	29	
	1842 1-1/8"		16	.500		3.36	17.45		20.81	30	
	1844 1-3/8"		15	.533		3.47	18.60		22.07	32	
	1846 1-5/8"		13	.615		2.64	21.50		24.14	35.50	
	1848 2-1/8"		11	.727		4.80	25.50		30.30	44	
	1850 2-5/8"	Q-5	13	1.231		18.75	38.50		57.25	79	
	1852 3-1/8"		11	1.455		12.80	45.50		58.30	83	
	1854 4-1/8"		9	1.778		33	56		89	120	

Important: See the Reference Section for critical supporting data - Reference Nos., Crews, & City Cost Indexes

15107 | Metal Pipe & Fittings

460		CREW	DAILY OUTPUT	LABOR-HOURS	UNIT	2001 BARE COSTS MAT.	LABOR	EQUIP.	TOTAL	TOTAL INCL O&P
1860	Coupling									
1866	5/8"	1 Stpi	21	.381	Ea.	.32	13.30		13.62	20.50
1868	3/4"		21	.381		.46	13.30		13.76	20.50
1870	7/8"		18	.444		.32	15.50		15.82	24
1872	1-1/8"		18	.444		.63	15.50		16.13	24
1874	1-3/8"		17	.471		1.10	16.40		17.50	26
1876	1-5/8"		15	.533		1.45	18.60		20.05	29.50
1878	2-1/8"		13	.615		2.41	21.50		23.91	35
1880	2-5/8"	Q-5	15	1.067		5.10	33.50		38.60	56
1882	3-1/8"		13	1.231		7.85	38.50		46.35	67
1884	4-1/8"		7	2.286		15.40	71.50		86.90	125
2000	DWV, solder joints, copper x copper									
2030	90° Elbow, 1-1/4"	1 Plum	13	.615	Ea.	1.97	21		22.97	34
2050	1-1/2"		12	.667		2.64	23		25.64	37.50
2070	2"		10	.800		3.83	27.50		31.33	45.50
2090	3"	Q-1	10	1.600		9.45	49.50		58.95	85.50
2100	4"	"	9	1.778		44.50	55		99.50	132
2150	45° Elbow, 1-1/4"	1 Plum	13	.615		1.80	21		22.80	34
2170	1-1/2"		12	.667		1.50	23		24.50	36
2180	2"		10	.800		3.43	27.50		30.93	45.50
2190	3"	Q-1	10	1.600		7.30	49.50		56.80	83
2200	4"	"	9	1.778		31	55		86	117
2250	Tee, Sanitary, 1-1/4"	1 Plum	9	.889		3.87	30.50		34.37	50.50
2270	1-1/2"		8	1		4.82	34.50		39.32	57.50
2290	2"		7	1.143		5.60	39.50		45.10	65.50
2310	3"	Q-1	7	2.286		20	71		91	129
2330	4"	"	6	2.667		51	82.50		133.50	181
2400	Coupling, 1-1/4"	1 Plum	14	.571		.92	19.70		20.62	30.50
2420	1-1/2"		13	.615		1.14	21		22.14	33.50
2440	2"		11	.727		1.58	25		26.58	39.50
2460	3"	Q-1	11	1.455		3.06	45		48.06	71.50
2480	4"	"	10	1.600		7.35	49.50		56.85	83
2600	Traps, see division 15155									
3500	Compression joint fittings									
3510	As used for plumbing and oil burner work									
3520	Fitting price includes nuts and sleeves									
3540	Sleeve, 1/8"				Ea.	.02			.02	.02
3550	3/16"					.02			.02	.02
3560	1/4"					.02			.02	.02
3570	5/16"					.03			.03	.03
3580	3/8"					.04			.04	.04
3600	1/2"					.12			.12	.13
3620	Nut, 1/8"					.08			.08	.09
3630	3/16"					.10			.10	.11
3640	1/4"					.11			.11	.12
3650	5/16"					.12			.12	.13
3660	3/8"					.14			.14	.15
3670	1/2"					.33			.33	.36
3710	Union, 1/8"	1 Plum	26	.308		.56	10.60		11.16	16.60
3720	3/16"		24	.333		.54	11.50		12.04	17.95
3730	1/4"		24	.333		.50	11.50		12	17.90
3740	5/16"		23	.348		.64	12		12.64	18.80
3750	3/8"		22	.364		.88	12.55		13.43	19.85
3760	1/2"		22	.364		1.39	12.55		13.94	20.50
3780	5/8"		21	.381		3.83	13.10		16.93	24
3820	Union tee, 1/8"		17	.471		1.50	16.20		17.70	26

15107 | Metal Pipe & Fittings

		CREW	DAILY OUTPUT	LABOR-HOURS	UNIT	2001 BARE COSTS MAT.	LABOR	EQUIP.	TOTAL	TOTAL INCL O&P	
460	3830 3/16"	1 Plum	16	.500	Ea.	1.06	17.25		18.31	27	460
	3840 1/4"		15	.533		1.04	18.35		19.39	28.50	
	3850 5/16"		15	.533		1.29	18.35		19.64	29	
	3860 3/8"		15	.533		2.07	18.35		20.42	30	
	3870 1/2"		15	.533		3.28	18.35		21.63	31	
	3910 Union elbow, 1/4"		24	.333		.72	11.50		12.22	18.15	
	3920 5/16"		23	.348		1.06	12		13.06	19.25	
	3930 3/8"		22	.364		1.53	12.55		14.08	20.50	
	3940 1/2"		22	.364		2.49	12.55		15.04	21.50	
	3980 Female connector, 1/8"		26	.308		.46	10.60		11.06	16.50	
	4000 3/16" x 1/8"		24	.333		.40	11.50		11.90	17.80	
	4010 1/4" x 1/8"		24	.333		.41	11.50		11.91	17.80	
	4020 1/4"		24	.333		.57	11.50		12.07	18	
	4030 3/8" x 1/4"		22	.364		.70	12.55		13.25	19.65	
	4040 1/2" x 3/8"		22	.364		1.22	12.55		13.77	20	
	4050 5/8" x 1/2"		21	.381		1.76	13.10		14.86	21.50	
	4090 Male connector, 1/8"		26	.308		.36	10.60		10.96	16.40	
	4100 3/16" x 1/8"		24	.333		.39	11.50		11.89	17.80	
	4110 1/4" x 1/8"		24	.333		.39	11.50		11.89	17.80	
	4120 1/4"		24	.333		.39	11.50		11.89	17.80	
	4130 5/16" x 1/8"		23	.348		.41	12		12.41	18.55	
	4140 5/16" x 1/4"		23	.348		.45	12		12.45	18.60	
	4150 3/8" x 1/8"		22	.364		.29	12.55		12.84	19.20	
	4160 3/8" x 1/4"		22	.364		.47	12.55		13.02	19.40	
	4170 3/8"		22	.364		.77	12.55		13.32	19.75	
	4180 3/8" x 1/2"		22	.364		.82	12.55		13.37	19.80	
	4190 1/2" x 3/8"		22	.364		.75	12.55		13.30	19.75	
	4200 1/2"		22	.364		1.13	12.55		13.68	20	
	4210 5/8" x 1/2"		21	.381		1.46	13.10		14.56	21.50	
	4240 Male elbow, 1/8"		26	.308		.54	10.60		11.14	16.60	
	4250 3/16" x 1/8"		24	.333		.55	11.50		12.05	17.95	
	4260 1/4" x 1/8"		24	.333		.47	11.50		11.97	17.85	
	4270 1/4"		24	.333		.58	11.50		12.08	18	
	4280 3/8" x 1/4"		22	.364		.70	12.55		13.25	19.65	
	4290 3/8"		22	.364		.90	12.55		13.45	19.90	
	4300 1/2" x 1/4"		22	.364		1.53	12.55		14.08	20.50	
	4310 1/2" x 3/8"		22	.364		1.25	12.55		13.80	20.50	
	4340 Female elbow, 1/8"		26	.308		1.03	10.60		11.63	17.15	
	4350 1/4" x 1/8"		24	.333		.68	11.50		12.18	18.10	
	4360 1/4"		24	.333		.86	11.50		12.36	18.30	
	4370 3/8" x 1/4"		22	.364		1.04	12.55		13.59	20	
	4380 1/2" x 3/8"		22	.364		1.57	12.55		14.12	20.50	
	4390 1/2"		22	.364		1.58	12.55		14.13	20.50	
	4420 Male run tee, 1/4" x 1/8"		15	.533		.93	18.35		19.28	28.50	
	4430 5/16" x 1/8"		15	.533		1.15	18.35		19.50	29	
	4440 3/8" x 1/4"		15	.533		1.47	18.35		19.82	29	
	4480 1/4" x 1/8"		15	.533		.90	18.35		19.25	28.50	
	4490 1/4"		15	.533		1.30	18.35		19.65	29	
	4500 3/8" x 1/4"		15	.533		1.42	18.35		19.77	29	
	4510 1/2" x 3/8"		15	.533		2.39	18.35		20.74	30	
	4520 1/2"	↓	15	.533	↓	2.92	18.35		21.27	30.50	
	4800 Flare joint fittings										
	4810 Refrigeration fittings										
	4820 Flare joint nuts and labor not incl. in price. Add 1 nut per jnt.										
	4830 90° Elbow, 1/4"				Ea.	1.96			1.96	2.16	
	4840 3/8"				↓	3.03			3.03	3.33	

15 MECHANICAL

15107 | Metal Pipe & Fittings		CREW	DAILY OUTPUT	LABOR-HOURS	UNIT	2001 BARE COSTS MAT.	LABOR	EQUIP.	TOTAL	TOTAL INCL O&P
460	4850 1/2"				Ea.	4.44			4.44	4.88
	4860 5/8"					6.10			6.10	6.70
	4870 3/4"					11.30			11.30	12.45
	5030 Tee, 1/4"					3.15			3.15	3.47
	5040 5/16"					3.32			3.32	3.65
	5050 3/8"					4.27			4.27	4.70
	5060 1/2"					7.80			7.80	8.60
	5070 5/8"					8.80			8.80	9.70
	5080 3/4"					11.20			11.20	12.35
	5140 Union, 3/16"					1.97			1.97	2.17
	5150 1/4"					.60			.60	.66
	5160 5/16"					.67			.67	.74
	5170 3/8"					1.61			1.61	1.77
	5180 1/2"					2.37			2.37	2.61
	5190 5/8"					3.71			3.71	4.08
	5200 3/4"					5.20			5.20	5.70
	5260 Long flare nut, 3/16"	1 Stpi	42	.190		.42	6.65		7.07	10.45
	5270 1/4"		41	.195		.48	6.80		7.28	10.80
	5280 5/16"		40	.200		.73	6.95		7.68	11.30
	5290 3/8"		39	.205		1.13	7.15		8.28	12.05
	5300 1/2"		38	.211		1.70	7.35		9.05	12.90
	5310 5/8"		37	.216		3.42	7.55		10.97	15.10
	5320 3/4"		34	.235		5.80	8.20		14	18.75
	5380 Short flare nut, 3/16"		42	.190		.29	6.65		6.94	10.30
	5390 1/4"		41	.195		.39	6.80		7.19	10.70
	5400 5/16"		40	.200		.57	6.95		7.52	11.15
	5410 3/8"		39	.205		.47	7.15		7.62	11.30
	5420 1/2"		38	.211		.68	7.35		8.03	11.80
	5430 5/8"		36	.222		.96	7.75		8.71	12.75
	5440 3/4"		34	.235		3.36	8.20		11.56	16.10
	5500 90° Elbow flare by MIPS, 1/4"					1.38			1.38	1.52
	5510 3/8"					1.92			1.92	2.11
	5520 1/2"					3.08			3.08	3.39
	5530 5/8"					3.91			3.91	4.30
	5540 3/4"					5.60			5.60	6.15
	5600 Flare by FIPS, 1/4"					4.09			4.09	4.50
	5610 3/8"					3.57			3.57	3.93
	5620 1/2"					6.75			6.75	7.45
	5670 Flare by sweat, 1/4"					1.42			1.42	1.56
	5680 3/8"					1.38			1.38	1.52
	5690 1/2"					2.51			2.51	2.76
	5700 5/8"					3.47			3.47	3.82
	5760 Tee flare by IPS, 1/4"					2.19			2.19	2.41
	5770 3/8"					2.16			2.16	2.38
	5780 1/2"					4.45			4.45	4.90
	5790 5/8"					4.53			4.53	4.98
	5850 Connector, 1/4"					.43			.43	.47
	5860 3/8"					.80			.80	.88
	5870 1/2"					.82			.82	.90
	5880 5/8"					.94			.94	1.03
	5890 3/4"					5.30			5.30	5.85
	5950 Seal cap, 1/4"					.21			.21	.23
	5960 3/8"					.26			.26	.29
	5970 1/2"					.42			.42	.46
	5980 5/8"					1.26			1.26	1.39
	5990 3/4"					4.39			4.39	4.83

MECHANICAL 15

15107 | Metal Pipe & Fittings

		Description	CREW	DAILY OUTPUT	LABOR-HOURS	UNIT	2001 BARE COSTS MAT.	LABOR	EQUIP.	TOTAL	TOTAL INCL O&P	
460	6000	Water service fittings										460
	6010	Flare joints nut and labor are included in the fitting price.										
	6020	90° Elbow, cxc, 3/8"	1 Plum	19	.421	Ea.	5.80	14.50		20.30	28.50	
	6030	1/2"		18	.444		5.80	15.30		21.10	29.50	
	6040	3/4"		16	.500		8.20	17.25		25.45	35	
	6050	1"		15	.533		15.55	18.35		33.90	44.50	
	6060	1-1/4"		12	.667		47	23		70	86.50	
	6070	1-1/2"		12	.667		50.50	23		73.50	90	
	6080	2"		10	.800		53.50	27.50		81	101	
	6090	C x MPT, 3/8"		19	.421		9.05	14.50		23.55	32	
	6100	1/2"		18	.444		9.05	15.30		24.35	33	
	6110	3/4"		16	.500		10.80	17.25		28.05	38	
	6120	1"		15	.533		13.95	18.35		32.30	43	
	6130	1-1/4"		13	.615		30	21		51	65	
	6140	1-1/2"		12	.667		30	23		53	67.50	
	6150	2"		10	.800		77.50	27.50		105	127	
	6160	C x FPT, 3/8"		19	.421		4.49	14.50		18.99	27	
	6170	1/2"		18	.444		4.49	15.30		19.79	28	
	6180	3/4"		16	.500		6.20	17.25		23.45	33	
	6190	1"		15	.533		13.50	18.35		31.85	42.50	
	6200	1-1/4"		13	.615		42	21		63	78	
	6210	1-1/2"		12	.667		97	23		120	142	
	6220	2"		10	.800		148	27.50		175.50	205	
	6230	Tee, cxcxc, 3/8"		13	.615		10.70	21		31.70	44	
	6240	1/2"		12	.667		10.70	23		33.70	46.50	
	6250	3/4"		11	.727		11.65	25		36.65	51	
	6260	1"		10	.800		23	27.50		50.50	66.50	
	6270	1-1/4"		8	1		76	34.50		110.50	136	
	6280	1-1/2"		8	1		91	34.50		125.50	152	
	6290	2"		7	1.143		163	39.50		202.50	240	
	6300	Tee, cxcxFPT, 3/8"		13	.615		8.60	21		29.60	41.50	
	6310	1/2"		12	.667		12.70	23		35.70	48.50	
	6320	3/4"		11	.727		18	25		43	58	
	6330	Tube nut, cxnut seat, 3/8"		40	.200		1.13	6.90		8.03	11.65	
	6340	1/2"		38	.211		1.13	7.25		8.38	12.20	
	6350	3/4"		34	.235		1.91	8.10		10.01	14.35	
	6360	1"		32	.250		3.47	8.60		12.07	16.80	
	6370	1-1/4"		25	.320		20.50	11		31.50	39	
	6380	Coupling, cxc, 3/8"		19	.421		5.80	14.50		20.30	28.50	
	6390	1/2"		18	.444		5.80	15.30		21.10	29.50	
	6400	3/4"		16	.500		6.80	17.25		24.05	33.50	
	6410	1"		15	.533		13.10	18.35		31.45	42	
	6420	1-1/4"		12	.667		21	23		44	58	
	6430	1-1/2"		12	.667		33	23		56	71	
	6440	2"		10	.800		50	27.50		77.50	96.50	
	6450	Adapter, cxFPT, 3/8"		19	.421		4.23	14.50		18.73	26.50	
	6460	1/2"		18	.444		4.23	15.30		19.53	27.50	
	6470	3/4"		16	.500		5	17.25		22.25	31.50	
	6480	1"		15	.533		11.25	18.35		29.60	40	
	6490	1-1/4"		13	.615		21.50	21		42.50	55.50	
	6500	1-1/2"		12	.667		21.50	23		44.50	58	
	6510	2"		10	.800		30	27.50		57.50	74.50	
	6520	Adapter, cxMPT, 3/8"		19	.421		3.64	14.50		18.14	26	
	6530	1/2"		18	.444		3.64	15.30		18.94	27	
	6540	3/4"		16	.500		5.10	17.25		22.35	31.50	
	6550	1"		15	.533		9.05	18.35		27.40	37.50	

15107 | Metal Pipe & Fittings

			CREW	DAILY OUTPUT	LABOR-HOURS	UNIT	2001 BARE COSTS MAT.	LABOR	EQUIP.	TOTAL	TOTAL INCL O&P	
460	6560	1-1/4"	1 Plum	13	.615	Ea.	24	21		45	58	460
	6570	1-1/2"		12	.667		24	23		47	60.50	
	6580	2"		10	.800		32	27.50		59.50	77	
	6990	Polybutylene/polyethylene pipe, See 15108 for plastic ftng.										
	7000	Insert type Brass/copper, 100 psi @ 180°F, CTS										
	7010	Adapter MPT 3/8" x 3/8" CTS	1 Plum	29	.276	Ea.	.97	9.50		10.47	15.40	
	7020	1/2" x 1/2"		26	.308		.97	10.60		11.57	17.05	
	7030	3/4" x 1/2"		26	.308		1.44	10.60		12.04	17.60	
	7040	3/4" x 3/4"		25	.320		1.42	11		12.42	18.20	
	7050	Adapter CTS 1/2" x 1/2" sweat		24	.333		.36	11.50		11.86	17.75	
	7060	3/4" x 3/4" sweat		22	.364		.47	12.55		13.02	19.40	
	7070	Coupler center set 3/8" CTS		25	.320		.49	11		11.49	17.20	
	7080	1/2" CTS		23	.348		.38	12		12.38	18.50	
	7090	3/4" CTS		22	.364		.47	12.55		13.02	19.40	
	7100	Elbow 90°, copper 3/8"		25	.320		.65	11		11.65	17.35	
	7110	1/2" CTS		23	.348		.69	12		12.69	18.85	
	7120	3/4" CTS		22	.364		.84	12.55		13.39	19.80	
	7130	Tee copper 3/8" CTS		17	.471		.85	16.20		17.05	25.50	
	7140	1/2" CTS		15	.533		.66	18.35		19.01	28	
	7150	3/4" CTS		14	.571		.95	19.70		20.65	30.50	
	7160	3/8" x 3/8" x 1/2"		16	.500		1.30	17.25		18.55	27.50	
	7170	1/2" x 3/8" x 1/2"		15	.533		1.30	18.35		19.65	29	
	7180	3/4" x 1/2" x 3/4"		14	.571		1.30	19.70		21	31	
480	0010	**PIPE/TUBE, GROOVED-JOINT FOR COPPER**										480
	4000	Fittings: coupling and labor required at joints not incl.										
	4001	in fitting price. add 1 per joint for installed price.										
	4010	Coupling, rigid style										
	4018	2" diameter	1 Plum	50	.160	Ea.	8.05	5.50		13.55	17.15	
	4020	2-1/2" diameter	Q-1	80	.200		9.15	6.20		15.35	19.40	
	4022	3" diameter		67	.239		10.25	7.40		17.65	22.50	
	4024	4" diameter		50	.320		15.75	9.90		25.65	32.50	
	4026	5" diameter		40	.400		27	12.40		39.40	48	
	4028	6" diameter	Q-2	50	.480		36	15.45		51.45	63	
	4100	Elbow, 90° or 45°										
	4108	2" diameter	1 Plum	25	.320	Ea.	12.50	11		23.50	30.50	
	4110	2-1/2" diameter	Q-1	40	.400		13.60	12.40		26	33.50	
	4112	3" diameter		33	.485		19.25	15.05		34.30	43.50	
	4114	4" diameter		25	.640		45	19.85		64.85	79.50	
	4116	5" diameter		20	.800		129	25		154	180	
	4118	6" diameter	Q-2	25	.960		206	31		237	274	
	4200	Tee										
	4208	2" diameter	1 Plum	17	.471	Ea.	21	16.20		37.20	47.50	
	4210	2-1/2" diameter	Q-1	27	.593		22	18.35		40.35	52	
	4212	3" diameter		22	.727		33.50	22.50		56	70.50	
	4214	4" diameter		17	.941		74	29		103	125	
	4216	5" diameter		13	1.231		208	38		246	287	
	4218	6" diameter	Q-2	17	1.412		257	45.50		302.50	350	
	4300	Reducer, concentric										
	4310	3" x 2-1/2" diameter	Q-1	35	.457	Ea.	18.20	14.15		32.35	41.50	
	4312	4" x 2-1/2" diameter		32	.500		38	15.50		53.50	65	
	4314	4" x 3" diameter		29	.552		38	17.10		55.10	67.50	
	4316	5" x 3" diameter		25	.640		108	19.85		127.85	149	
	4318	5" x 4" diameter		22	.727		108	22.50		130.50	153	
	4320	6" x 3" diameter	Q-2	28	.857		117	27.50		144.50	170	
	4322	6" x 4" diameter		26	.923		117	29.50		146.50	173	

	15107 | Metal Pipe & Fittings		CREW	DAILY OUTPUT	LABOR-HOURS	UNIT	2001 BARE COSTS MAT.	LABOR	EQUIP.	TOTAL	TOTAL INCL O&P	
480	4324	6" x 5" diameter	Q-2	24	1	Ea.	117	32		149	177	480
	4350	Flange, w/groove gasket										
	4351	ANSI class 125 and 150										
	4355	2" diameter	1 Plum	23	.348	Ea.	65.50	12		77.50	90	
	4356	2-1/2" diameter	Q-1	37	.432		68.50	13.40		81.90	95.50	
	4358	3" diameter		31	.516		71.50	16		87.50	103	
	4360	4" diameter		23	.696		78.50	21.50		100	119	
	4362	5" diameter		19	.842		106	26		132	156	
	4364	6" diameter	Q-2	23	1.043		115	33.50		148.50	178	
500	0010	**PIPE, CORROSION RESISTANT** No couplings or hangers R15100-070										500
	0020	Iron alloy, drain, mechanical joint										
	1000	1-1/2" diameter	Q-1	70	.229	L.F.	22.50	7.10		29.60	35.50	
	1100	2" diameter		66	.242		26	7.50		33.50	40	
	1120	3" diameter		60	.267		36	8.25		44.25	52.50	
	1140	4" diameter		52	.308		46.50	9.55		56.05	65.50	
	1980	Iron alloy, drain, B&S joint										
	2000	2" diameter	Q-1	54	.296	L.F.	28	9.20		37.20	45	
	2100	3" diameter		52	.308		40.50	9.55		50.05	59	
	2120	4" diameter		48	.333		56	10.35		66.35	77	
	2140	6" diameter	Q-2	59	.407		93.50	13.10		106.60	123	
	2160	8" diameter	"	54	.444		169	14.30		183.30	208	
	2980	Plastic, epoxy, fiberglass filament wound										
	3000	2" diameter	Q-1	62	.258	L.F.	5.20	8		13.20	17.80	
	3100	3" diameter		51	.314		7.10	9.75		16.85	22.50	
	3120	4" diameter		45	.356		9.40	11		20.40	27	
	3140	6" diameter		32	.500		14.40	15.50		29.90	39.50	
	3160	8" diameter	Q-2	38	.632		19.80	20.50		40.30	52.50	
	3180	10" diameter		32	.750		34	24		58	74	
	3200	12" diameter		28	.857		43	27.50		70.50	89	
	3980	Polyester, fiberglass filament wound										
	4000	2" diameter	Q-1	62	.258	L.F.	9.85	8		17.85	23	
	4100	3" diameter		51	.314		13.20	9.75		22.95	29	
	4120	4" diameter		45	.356		14.90	11		25.90	33	
	4140	6" diameter		32	.500		23.50	15.50		39	49.50	
	4160	8" diameter	Q-2	38	.632		32	20.50		52.50	65.50	
	4180	10" diameter		32	.750		41	24		65	81.50	
	4200	12" diameter		28	.857		47	27.50		74.50	93.50	
	4980	Polypropylene, acid resistant, fire retardant, schedule 40										
	5000	1-1/2" diameter	Q-1	68	.235	L.F.	.95	7.30		8.25	12.05	
	5100	2" diameter		62	.258		1.29	8		9.29	13.50	
	5120	3" diameter		51	.314		2.70	9.75		12.45	17.65	
	5140	4" diameter		45	.356		3.83	11		14.83	21	
	5160	6" diameter		32	.500		6.45	15.50		21.95	30.50	
	5980	Proxylene, fire retardant, Schedule 40										
	6000	1-1/2" diameter	Q-1	68	.235	L.F.	3.21	7.30		10.51	14.55	
	6100	2" diameter		62	.258		4.40	8		12.40	16.95	
	6120	3" diameter		51	.314		7.95	9.75		17.70	23.50	
	6140	4" diameter		45	.356		11.30	11		22.30	29	
	6160	6" diameter		32	.500		20.50	15.50		36	46	
	6820	For Schedule 80, add					35%	2%				
	9000	For "external pipe protection" (See Div.026-652)										
560	0010	**PIPE, CORROSION RESISTANT, FITTINGS**										560
	0030	Iron alloy										
	0050	Mechanical joint										
	0060	1/4 Bend, 1-1/2"	Q-1	12	1.333	Ea.	28	41.50		69.50	93.50	

Important: See the Reference Section for critical supporting data - Reference Nos., Crews, & City Cost Indexes

15107	Metal Pipe & Fittings	CREW	DAILY OUTPUT	LABOR-HOURS	UNIT	2001 BARE COSTS MAT.	LABOR	EQUIP.	TOTAL	TOTAL INCL O&P	
560 0080	2″	Q-1	10	1.600	Ea.	36.50	49.50		86	115	560
0090	3″		9	1.778		51	55		106	139	
0100	4″		8	2		70.50	62		132.50	171	
0110	1/8 Bend, 1-1/2″		12	1.333		28	41.50		69.50	93.50	
0130	2″		10	1.600		36.50	49.50		86	115	
0140	3″		9	1.778		51	55		106	139	
0150	4″		8	2		70.50	62		132.50	171	
0160	Tee and Y, sanitary, straight										
0170	1-1/2″	Q-1	8	2	Ea.	46	62		108	144	
0180	2″		7	2.286		53.50	71		124.50	166	
0190	3″		6	2.667		70.50	82.50		153	203	
0200	4″		5	3.200		140	99		239	305	
0360	Coupling, 1-1/2″		14	1.143		26.50	35.50		62	82.50	
0380	2″		12	1.333		30.50	41.50		72	96	
0390	3″		11	1.455		32.50	45		77.50	104	
0400	4″		10	1.600		36.50	49.50		86	115	
0500	Bell & Spigot										
0510	1/4 and 1/16 bend, 2″	Q-1	16	1	Ea.	39	31		70	90	
0520	3″		14	1.143		58	35.50		93.50	117	
0530	4″		13	1.231		75	38		113	140	
0540	6″	Q-2	17	1.412		269	45.50		314.50	365	
0550	8″	″	12	2		645	64.50		709.50	805	
0620	1/8 bend, 2″	Q-1	16	1		39	31		70	90	
0640	3″		14	1.143		58.50	35.50		94	118	
0650	4″		13	1.231		75	38		113	140	
0660	6″	Q-2	17	1.412		181	45.50		226.50	268	
0680	8″	″	12	2		535	64.50		599.50	685	
0700	Tee, sanitary, 2″	Q-1	10	1.600		51	49.50		100.50	131	
0710	3″		9	1.778		81.50	55		136.50	173	
0720	4″		8	2		122	62		184	228	
0730	6″	Q-2	11	2.182		224	70		294	355	
0740	8″	″	8	3		810	96.50		906.50	1,050	
1800	Y, sanitary, 2″	Q-1	10	1.600		50	49.50		99.50	130	
1820	3″		9	1.778		85	55		140	177	
1830	4″		8	2		139	62		201	247	
1840	6″	Q-2	11	2.182		226	70		296	355	
1850	8″	″	8	3		1,050	96.50		1,146.50	1,300	
3000	Epoxy, filament wound										
3030	Quick-lock joint										
3040	90° Elbow, 2″	Q-1	28	.571	Ea.	58.50	17.70		76.20	91	
3060	3″		16	1		67.50	31		98.50	122	
3070	4″		13	1.231		92	38		130	159	
3080	6″		8	2		134	62		196	242	
3090	8″	Q-2	9	2.667		248	85.50		333.50	400	
3100	10″		7	3.429		310	110		420	505	
3110	12″		6	4		445	129		574	685	
3120	45° Elbow, 2″	Q-1	28	.571		45.50	17.70		63.20	76.50	
3130	3″		16	1		64.50	31		95.50	118	
3140	4″		13	1.231		88	38		126	155	
3150	6″		8	2		134	62		196	242	
3160	8″	Q-2	9	2.667		248	85.50		333.50	400	
3170	10″		7	3.429		310	110		420	505	
3180	12″		6	4		445	129		574	685	
3190	Tee, 2″	Q-1	19	.842		140	26		166	193	
3200	3″		11	1.455		169	45		214	254	
3210	4″		9	1.778		203	55		258	305	

15107 | Metal Pipe & Fittings

		CREW	DAILY OUTPUT	LABOR-HOURS	UNIT	2001 BARE COSTS MAT.	LABOR	EQUIP.	TOTAL	TOTAL INCL O&P
560	3220 6″	Q-1	5	3.200	Ea.	310	99		409	490
	3230 8″	Q-2	6	4		400	129		529	635
	3240 10″		5	4.800		520	154		674	805
	3250 12″		4	6		755	193		948	1,125
	4000 Polypropylene, acid resistant									
	4020 Non-pressure, electrofusion joints									
	4050 1/4 bend, 1-1/2″	1 Plum	16	.500	Ea.	12.35	17.25		29.60	39.50
	4060 2″	Q-1	28	.571		15.35	17.70		33.05	43.50
	4080 3″		17	.941		26	29		55	72.50
	4090 4″		14	1.143		41.50	35.50		77	99
	4110 6″		8	2		102	62		164	206
	4150 1/4 Bend, long sweep									
	4170 1-1/2″	1 Plum	16	.500	Ea.	12.80	17.25		30.05	40
	4180 2″	Q-1	28	.571		17.45	17.70		35.15	45.50
	4200 3″		17	.941		30	29		59	77
	4210 4″		14	1.143		43	35.50		78.50	101
	4250 1/8 bend, 1-1/2″	1 Plum	16	.500		12.15	17.25		29.40	39.50
	4260 2″	Q-1	28	.571		14.80	17.70		32.50	43
	4280 3″		17	.941		27	29		56	74
	4290 4″		14	1.143		31	35.50		66.50	87.50
	4310 6″		8	2		85.50	62		147.50	188
	4400 Tee, sanitary									
	4420 1-1/2″	1 Plum	10	.800	Ea.	15.35	27.50		42.85	58.50
	4430 2″	Q-1	17	.941		18.40	29		47.40	64
	4450 3″		11	1.455		37	45		82	109
	4460 4″		9	1.778		55	55		110	144
	4480 6″		5	3.200		93.50	99		192.50	253
	4490 Tee, sanitary reducing, 2″ x 2″ x 1-1/2″		17	.941		18.35	29		47.35	64
	4491 3″ x 3″ x 1-1/2″		11	1.455		32	45		77	103
	4492 3″ x 3″ x 2″		11	1.455		34.50	45		79.50	106
	4493 4″ x 4″ x 2″		10	1.600		49.50	49.50		99	129
	4494 4″ x 4″ x 3″		9	1.778		51.50	55		106.50	140
	4496 6″ x 6″ x 4″		5	3.200		93.50	99		192.50	253
	4500 Tee/wye, long turn									
	4520 1-1/2″	1 Plum	10	.800	Ea.	20.50	27.50		48	64
	4530 2″	Q-1	17	.941		27.50	29		56.50	74
	4550 3″		11	1.455		45	45		90	118
	4570 4″		9	1.778		62	55		117	151
	4650 Wye 45°, 1-1/2″	1 Plum	10	.800		15.65	27.50		43.15	59
	4652 2″	Q-1	17	.941		22.50	29		51.50	68.50
	4653 3″		11	1.455		39	45		84	111
	4654 4″		9	1.778		57	55		112	146
	4656 6″		5	3.200		145	99		244	310
	4660 Wye, reducing									
	4662 2″ x 2″ x 1-1/2″	Q-1	17	.941	Ea.	22	29		51	68
	4666 3″ x 3″ x 2″		11	1.455		36.50	45		81.50	108
	4668 4″ x 4″ x 2″		10	1.600		52.50	49.50		102	133
	4669 4″ x 4″ x 3″		9	1.778		55	55		110	143
	4671 6″ x 6″ x 2″		6	2.667		91.50	82.50		174	226
	4673 6″ x 6″ x 3″		5.50	2.909		93	90		183	238
	4675 6″ x 6″ x 4″		5	3.200		94	99		193	253
	4678 Combination Y & 1/8 bend									
	4681 1-1/2″	1 Plum	10	.800	Ea.	20.50	27.50		48	64
	4683 2″	Q-1	17	.941		27.50	29		56.50	74
	4684 3″		11	1.455		43.50	45		88.50	116
	4685 4″		9	1.778		62	55		117	151

15 MECHANICAL

15107 | Metal Pipe & Fittings

			Crew	Daily Output	Labor-Hours	Unit	2001 Bare Costs Mat.	Labor	Equip.	Total	Total Incl O&P	
560	4689	Combination Y & 1/8 bend, reducing										560
	4692	2″ x 2″ x 1-1/2″	Q-1	17	.941	Ea.	24.50	29		53.50	71	
	4694	3″ x 3″ x 1-1/2″		12	1.333		36	41.50		77.50	103	
	4695	3″ x 3″ x 2″		11	1.455		39	45		84	111	
	4697	4″ x 4″ x 2″		10	1.600		54.50	49.50		104	135	
	4699	4″ x 4″ x 3″		9	1.778		57	55		112	146	
	4710	Hub adapter										
	4712	1-1/2″	1 Plum	16	.500	Ea.	14.70	17.25		31.95	42	
	4713	2″	Q-1	28	.571		15.70	17.70		33.40	44	
	4714	3″		17	.941		19.75	29		48.75	65.50	
	4715	4″		14	1.143		29	35.50		64.50	85	
	4719	Mechanical joint adapter										
	4721	1-1/2″	1 Plum	16	.500	Ea.	14.75	17.25		32	42.50	
	4722	2″	Q-1	28	.571		15.55	17.70		33.25	43.50	
	4723	3″		17	.941		19.20	29		48.20	65	
	4724	4″		14	1.143		28.50	35.50		64	84.50	
	4728	Couplings										
	4731	1-1/2″	1 Plum	16	.500	Ea.	9.65	17.25		26.90	36.50	
	4732	2″	Q-1	28	.571		12.05	17.70		29.75	40	
	4733	3″		17	.941		15.45	29		44.45	61	
	4734	4″		14	1.143		22	35.50		57.50	77.50	
	4736	6″		8	2		35	62		97	132	
620	0010	**PIPE, STEEL** R15100-070										620
	0020	All pipe sizes are to Spec. A-53 unless noted otherwise										
	0030	Schedule 10, see 15107-690-0500										
	0050	Schedule 40, threaded, with couplings, and clevis type										
	0060	hangers sized for covering, 10′ O.C.										
	0540	Black, 1/4″ diameter	1 Plum	66	.121	L.F.	1.29	4.18		5.47	7.70	
	0550	3/8″ diameter		65	.123		1.40	4.24		5.64	7.95	
	0560	1/2″ diameter		63	.127		1.22	4.37		5.59	7.95	
	0570	3/4″ diameter		61	.131		1.36	4.52		5.88	8.30	
	0580	1″ diameter		53	.151		1.73	5.20		6.93	9.75	
	0590	1-1/4″ diameter	Q-1	89	.180		2.04	5.55		7.59	10.65	
	0600	1-1/2″ diameter		80	.200		2.29	6.20		8.49	11.85	
	0610	2″ diameter		64	.250		2.89	7.75		10.64	14.90	
	0620	2-1/2″ diameter		50	.320		4.59	9.90		14.49	20	
	0630	3″ diameter		43	.372		5.80	11.55		17.35	24	
	0640	3-1/2″ diameter		40	.400		7.70	12.40		20.10	27	
	0650	4″ diameter		36	.444		8.55	13.80		22.35	30.50	
	0660	5″ diameter		26	.615		12.55	19.10		31.65	43	
	0670	6″ diameter	Q-2	31	.774		16.50	25		41.50	55.50	
	0680	8″ diameter		27	.889		22.50	28.50		51	67.50	
	0690	10″ diameter		23	1.043		39	33.50		72.50	93.50	
	0700	12″ diameter		18	1.333		49.50	43		92.50	119	
	0809	A-106, gr. A/B, seamless w/cplgs. & hangers										
	0811	1/4″ diameter	1 Plum	66	.121	L.F.	2.13	4.18		6.31	8.65	
	0812	3/8″ diameter		65	.123		2.27	4.24		6.51	8.90	
	0813	1/2″ diameter		63	.127		2.65	4.37		7.02	9.50	
	0814	3/4″ diameter		61	.131		2.82	4.52		7.34	9.90	
	0815	1″ diameter		53	.151		3.33	5.20		8.53	11.50	
	0816	1-1/4″ diameter	Q-1	89	.180		3.67	5.55		9.22	12.45	
	0817	1-1/2″ diameter		80	.200		3.86	6.20		10.06	13.60	
	0819	A-53, 2″ diameter		64	.250		4.17	7.75		11.92	16.30	
	0821	2-1/2″ diameter		50	.320		7.20	9.90		17.10	23	
	0822	3″ diameter		43	.372		9	11.55		20.55	27.50	
	0823	4″ diameter		36	.444		14.10	13.80		27.90	36.50	

15107 | Metal Pipe & Fittings

			CREW	DAILY OUTPUT	LABOR-HOURS	UNIT	2001 BARE COSTS MAT.	LABOR	EQUIP.	TOTAL	TOTAL INCL O&P
620	1220	To delete coupling & hanger, subtract (R15100-070)									
	1230	1/4" diam. to 3/4" diam.					31%	56%			
	1240	1" diam. to 1-1/2" diam.					23%	51%			
	1250	2" diam. to 4" diam.					23%	41%			
	1260	5" diam. to 12" diam.					21%	45%			
	1290	Galvanized, 1/4" diameter	1 Plum	66	.121	L.F.	1.49	4.18		5.67	7.95
	1300	3/8" diameter		65	.123		1.60	4.24		5.84	8.15
	1310	1/2" diameter		63	.127		1.37	4.37		5.74	8.10
	1320	3/4" diameter		61	.131		1.53	4.52		6.05	8.50
	1330	1" diameter		53	.151		1.81	5.20		7.01	9.85
	1340	1-1/4" diameter	Q-1	89	.180		2.16	5.55		7.71	10.75
	1350	1-1/2" diameter		80	.200		2.39	6.20		8.59	12
	1360	2" diameter		64	.250		3.03	7.75		10.78	15.05
	1370	2-1/2" diameter		50	.320		4.52	9.90		14.42	19.95
	1380	3" diameter		43	.372		5.60	11.55		17.15	23.50
	1390	3-1/2" diameter		40	.400		7.10	12.40		19.50	26.50
	1400	4" diameter		36	.444		8.05	13.80		21.85	30
	1410	5" diameter		26	.615		14.90	19.10		34	45.50
	1420	6" diameter	Q-2	31	.774		20.50	25		45.50	60
	1430	8" diameter		27	.889		33	28.50		61.50	79.50
	1440	10" diameter		23	1.043		48.50	33.50		82	104
	1450	12" diameter		18	1.333		66.50	43		109.50	138
	1750	To delete coupling & hanger, subtract									
	1760	1/4" diam. to 3/4" diam.					31%	56%			
	1770	1" diam. to 1-1/2" diam.					23%	51%			
	1780	2" diam. to 4" diam.					23%	41%			
	1790	5" diam. to 12" diam.					21%	45%			
	1900	Pipe nipple std blk 2"lg 1/2"	1 Plum	19	.421	Ea.	.42	14.50		14.92	22.50
	1910	Pipe nipple std blk 2"lg 1" diameter	"	15	.533		.70	18.35		19.05	28.50
	1920	3"lg 2-1/2"	Q-1	16	1		4.67	31		35.67	52
	2000	Welded, sch. 40, on yoke & roll hangers, sized for covering,									
	2010	10' O.C. (no hangers incl. for 14" diam. and up)									
	2040	Black, 1" diameter	Q-15	93	.172	L.F.	2.60	5.35	1.08	9.03	12.10
	2050	1-1/4" diameter		84	.190		3.23	5.90	1.20	10.33	13.75
	2060	1-1/2" diameter		76	.211		3.41	6.55	1.32	11.28	15.05
	2070	2" diameter		61	.262		3.88	8.15	1.65	13.68	18.40
	2080	2-1/2" diameter		47	.340		5.05	10.55	2.14	17.74	24
	2090	3" diameter		43	.372		5.45	11.55	2.34	19.34	26
	2100	3-1/2" diameter		39	.410		6.85	12.70	2.58	22.13	29.50
	2110	4" diameter		37	.432		7.90	13.40	2.72	24.02	31.50
	2120	5" diameter		32	.500		13.10	15.50	3.15	31.75	41.50
	2130	6" diameter	Q-16	36	.667		16.80	21.50	2.79	41.09	54
	2140	8" diameter		29	.828		24.50	26.50	3.47	54.47	71
	2150	10" diameter		24	1		43	32	4.19	79.19	101
	2160	12" diameter		19	1.263		53	40.50	5.30	98.80	125
	2170	14" diameter		15	1.600		34	51.50	6.70	92.20	122
	2180	16" diameter		13	1.846		32	59.50	7.75	99.25	133
	2190	18" diameter		11	2.182		56.50	70	9.15	135.65	179
	2200	20" diameter		9	2.667		64	85.50	11.15	160.65	212
	2220	24" diameter		8	3		71.50	96.50	12.55	180.55	238
	2230	26" diameter		7.50	3.200		77.50	103	13.40	193.90	255
	2240	30" diameter		6	4		84.50	129	16.75	230.25	305
	2260	36" diameter		4.50	5.333		96.50	171	22.50	290	390
	2560	To delete hanger, subtract									
	2570	1" diam. to 1-1/2" diam.					15%	34%			
	2580	2" diam. to 3-1/2" diam.					9%	21%			

620	15107 | Metal Pipe & Fittings	CREW	DAILY OUTPUT	LABOR-HOURS	UNIT	2001 BARE COSTS MAT.	LABOR	EQUIP.	TOTAL	TOTAL INCL O&P
2590	4" diam. to 12" diam. R15100-070					5%	12%			
3250	Flanged, 150 lb. weld neck, on yoke & roll hangers									
3260	sized for covering, 10' O.C.									
3290	Black, 1" diameter	Q-15	70	.229	L.F.	6.05	7.10	1.44	14.59	18.95
3300	1-1/4" diameter		64	.250		6.70	7.75	1.57	16.02	21
3310	1-1/2" diameter		58	.276		6.85	8.55	1.74	17.14	22.50
3320	2" diameter		45	.356		7.70	11	2.24	20.94	27.50
3330	2-1/2" diameter		36	.444		9.45	13.80	2.80	26.05	34.50
3340	3" diameter		32	.500		10.30	15.50	3.15	28.95	38.50
3350	3-1/2" diameter		29	.552		12.60	17.10	3.47	33.17	43.50
3360	4" diameter		26	.615		13.55	19.10	3.87	36.52	48
3370	5" diameter		21	.762		22	23.50	4.79	50.29	65.50
3380	6" diameter	Q-16	25	.960		26	31	4.02	61.02	79.50
3390	8" diameter		19	1.263		39	40.50	5.30	84.80	110
3400	10" diameter		16	1.500		67.50	48	6.30	121.80	154
3410	12" diameter		14	1.714		92	55	7.20	154.20	192
3470	For 300 lb. flanges, add					63%				
3480	For 600 lb. flanges, add					310%				
3960	To delete flanges & hanger, subtract									
3970	1" diam. to 2" diam.					76%	65%			
3980	2-1/2" diam. to 4" diam.					62%	59%			
3990	5" diam. to 12" diam.					60%	46%			
4750	Schedule 80, threaded, with couplings, and clevis type hangers									
4760	sized for covering, 10' O.C.									
4790	Black, 1/4" diameter	1 Plum	54	.148	L.F.	1.52	5.10		6.62	9.35
4800	3/8" diameter		53	.151		1.56	5.20		6.76	9.55
4810	1/2" diameter		52	.154		1.69	5.30		6.99	9.85
4820	3/4" diameter		50	.160		1.90	5.50		7.40	10.40
4830	1" diameter		45	.178		2.30	6.10		8.40	11.80
4840	1-1/4" diameter	Q-1	75	.213		3	6.60		9.60	13.30
4850	1-1/2" diameter		69	.232		3.28	7.20		10.48	14.45
4860	2" diameter		56	.286		4.38	8.85		13.23	18.15
4870	2-1/2" diameter		44	.364		6.35	11.25		17.60	24
4880	3" diameter		38	.421		7.85	13.05		20.90	28.50
4890	3-1/2" diameter		35	.457		9.70	14.15		23.85	32
4900	4" diameter		32	.500		11.40	15.50		26.90	36
4910	5" diameter		23	.696		20	21.50		41.50	54.50
4920	6" diameter	Q-2	28	.857		32	27.50		59.50	76.50
4930	8" diameter		23	1.043		52.50	33.50		86	108
4940	10" diameter		20	1.200		80.50	38.50		119	147
4950	12" diameter		15	1.600		93	51.50		144.50	180
5430	To delete coupling & hanger, subtract									
5440	1/4" diam. to 1/2" diam.					31%	54%			
5450	3/4" diam. to 1-1/2" diam.					28%	49%			
5460	2" diam. to 4" diam.					21%	40%			
5470	5" diam. to 12" diam.					17%	43%			
5510	Galvanized, 1/4" diameter	1 Plum	54	.148	L.F.	1.72	5.10		6.82	9.60
5520	3/8" diameter		53	.151		1.76	5.20		6.96	9.80
5530	1/2" diameter		52	.154		1.89	5.30		7.19	10.10
5540	3/4" diameter		50	.160		2.15	5.50		7.65	10.65
5550	1" diameter		45	.178		2.49	6.10		8.59	12
5560	1-1/4" diameter	Q-1	75	.213		3.27	6.60		9.87	13.60
5570	1-1/2" diameter		69	.232		3.61	7.20		10.81	14.80
5580	2" diameter		56	.286		4.84	8.85		13.69	18.65
5590	2-1/2" diameter		44	.364		7.05	11.25		18.30	25
5600	3" diameter		38	.421		8.75	13.05		21.80	29.50

MECHANICAL 15

15107 | Metal Pipe & Fittings

15107	Metal Pipe & Fittings	Crew	Daily Output	Labor-Hours	Unit	2001 Bare Costs Mat.	Labor	Equip.	Total	Total Incl O&P
620 5610	3-1/2" diameter (R15100-070)	Q-1	35	.457	L.F.	10.85	14.15		25	33.50
5620	4" diameter		32	.500		12.75	15.50		28.25	37.50
5630	5" diameter		23	.696		24.50	21.50		46	59
5640	6" diameter	Q-2	28	.857		32.50	27.50		60	77
5650	8" diameter		23	1.043		51.50	33.50		85	107
5660	10" diameter		20	1.200		79.50	38.50		118	146
5670	12" diameter		15	1.600		91.50	51.50		143	178
5930	To delete coupling & hanger, subtract									
5940	1/4" diam. to 1/2" diam.					31%	54%			
5950	3/4" diam. to 1-1/2" diam.					28%	49%			
5960	2" diam. to 4" diam.					21%	40%			
5970	5" diam. to 12" diam.					17%	43%			
6000	Welded, on yoke & roller hangers									
6010	sized for covering, 10' O.C.									
6040	Black, 1" diameter	Q-15	85	.188	L.F.	2.82	5.85	1.18	9.85	13.20
6050	1-1/4" diameter		79	.203		3.58	6.30	1.27	11.15	14.85
6060	1-1/2" diameter		72	.222		3.86	6.90	1.40	12.16	16.20
6070	2" diameter		57	.281		4.52	8.70	1.77	14.99	20
6080	2-1/2" diameter		44	.364		6.05	11.25	2.29	19.59	26
6090	3" diameter		40	.400		7.15	12.40	2.52	22.07	29.50
6100	3-1/2" diameter		34	.471		8.65	14.60	2.96	26.21	35
6110	4" diameter		33	.485		9.80	15.05	3.05	27.90	36.50
6120	5" diameter		26	.615		17.85	19.10	3.87	40.82	53
6130	6" diameter	Q-16	30	.800		29.50	25.50	3.35	58.35	75
6140	8" diameter		25	.960		50.50	31	4.02	85.52	106
6150	10" diameter		20	1.200		83	38.50	5.05	126.55	155
6160	12" diameter		15	1.600		94.50	51.50	6.70	152.70	189
6540	To delete hanger, subtract									
6550	1" diam. to 1-1/2" diam.					30%	14%			
6560	2" diam. to 3" diam.					23%	9%			
6570	3-1/2" diam. to 5" diam.					12%	6%			
6580	6" diam. to 12" diam.					10%	4%			
7250	Flanged, 300 lb. weld neck, on yoke & roll hangers									
7260	sized for covering, 10' O.C.									
7290	Black, 1" diameter	Q-15	66	.242	L.F.	7.15	7.50	1.52	16.17	21
7300	1-1/4" diameter		61	.262		7.90	8.15	1.65	17.70	23
7310	1-1/2" diameter		54	.296		8.20	9.20	1.86	19.26	25
7320	2" diameter		42	.381		9.60	11.80	2.40	23.80	31
7330	2-1/2" diameter		33	.485		11.70	15.05	3.05	29.80	38.50
7340	3" diameter		29	.552		12.85	17.10	3.47	33.42	44
7350	3-1/2" diameter		24	.667		18.15	20.50	4.19	42.84	55.50
7360	4" diameter		23	.696		19.30	21.50	4.38	45.18	58.50
7370	5" diameter		19	.842		32.50	26	5.30	63.80	81
7380	6" diameter	Q-16	23	1.043		42	33.50	4.37	79.87	101
7390	8" diameter		17	1.412		72	45.50	5.90	123.40	154
7400	10" diameter		14	1.714		128	55	7.20	190.20	232
7410	12" diameter		12	2		151	64.50	8.40	223.90	272
7470	For 600 lb. flanges, add					100%				
7940	To delete flanges & hanger, subtract									
7950	1" diam. to 1-1/2" diam.					75%	66%			
7960	2" diam. to 3" diam.					62%	60%			
7970	3-1/2" diam. to 5" diam.					54%	66%			
7980	6" diam. to 12" diam.					55%	62%			
9000	Threading pipe labor, one end, all schedules through 80									
9010	1/4" through 3/4" pipe size	1 Plum	80	.100	Ea.		3.45		3.45	5.20
9020	1" through 2" pipe size		73	.110			3.78		3.78	5.70

15107 | Metal Pipe & Fittings

			CREW	DAILY OUTPUT	LABOR-HOURS	UNIT	2001 BARE COSTS MAT.	LABOR	EQUIP.	TOTAL	TOTAL INCL O&P	
620	9030	2-1/2" pipe size R15100-070	1 Plum	53	.151	Ea.		5.20		5.20	7.85	620
	9040	3" pipe size		50	.160			5.50		5.50	8.30	
	9050	3-1/2" pipe size	Q-1	89	.180			5.55		5.55	8.40	
	9060	4" pipe size		73	.219			6.80		6.80	10.25	
	9070	5" pipe size		53	.302			9.35		9.35	14.15	
	9080	6" pipe size		46	.348			10.80		10.80	16.30	
	9090	8" pipe size		29	.552			17.10		17.10	26	
	9100	10" pipe size		21	.762			23.50		23.50	35.50	
	9110	12" pipe size		13	1.231			38		38	57.50	
	9200	Welding labor per joint										
	9210	Schedule 40,										
	9230	1/2" pipe size	Q-15	32	.500	Ea.		15.50	3.15	18.65	27	
	9240	3/4" pipe size		27	.593			18.35	3.73	22.08	31.50	
	9250	1" pipe size		23	.696			21.50	4.38	25.88	37.50	
	9260	1-1/4" pipe size		20	.800			25	5.05	30.05	43	
	9270	1-1/2" pipe size		19	.842			26	5.30	31.30	45.50	
	9280	2" pipe size		16	1			31	6.30	37.30	54	
	9290	2-1/2" pipe size		13	1.231			38	7.75	45.75	66	
	9300	3" pipe size		12	1.333			41.50	8.40	49.90	72	
	9310	4" pipe size		10	1.600			49.50	10.05	59.55	86	
	9320	5" pipe size		9	1.778			55	11.20	66.20	95.50	
	9330	6" pipe size		8	2			62	12.60	74.60	107	
	9340	8" pipe size		5	3.200			99	20	119	172	
	9350	10" pipe size		4	4			124	25	149	215	
	9360	12" pipe size		3	5.333			165	33.50	198.50	287	
	9370	14" pipe size		2.60	6.154			191	38.50	229.50	330	
	9380	16" pipe size		2.20	7.273			225	46	271	390	
	9390	18" pipe size		2	8			248	50.50	298.50	430	
	9400	20" pipe size		1.80	8.889			276	56	332	475	
	9410	22" pipe size		1.70	9.412			292	59	351	505	
	9420	24" pipe size		1.50	10.667			330	67	397	575	
	9450	Schedule 80,										
	9460	1/2" pipe size	Q-15	27	.593	Ea.		18.35	3.73	22.08	31.50	
	9470	3/4" pipe size		23	.696			21.50	4.38	25.88	37.50	
	9480	1" pipe size		20	.800			25	5.05	30.05	43	
	9490	1-1/4" pipe size		19	.842			26	5.30	31.30	45.50	
	9500	1-1/2" pipe size		18	.889			27.50	5.60	33.10	47.50	
	9510	2" pipe size		15	1.067			33	6.70	39.70	57.50	
	9520	2-1/2" pipe size		12	1.333			41.50	8.40	49.90	72	
	9530	3" pipe size		11	1.455			45	9.15	54.15	78	
	9540	4" pipe size		8	2			62	12.60	74.60	107	
	9550	5" pipe size		6	2.667			82.50	16.75	99.25	143	
	9560	6" pipe size		5	3.200			99	20	119	172	
	9570	8" pipe size		4	4			124	25	149	215	
	9580	10" pipe size		3	5.333			165	33.50	198.50	287	
	9590	12" pipe size		2	8			248	50.50	298.50	430	
	9600	14" pipe size	Q-16	2.60	9.231			297	38.50	335.50	495	
	9610	16" pipe size		2.30	10.435			335	43.50	378.50	555	
	9620	18" pipe size		2	12			385	50.50	435.50	635	
	9630	20" pipe size		1.80	13.333			430	56	486	705	
	9640	22" pipe size		1.60	15			480	63	543	800	
	9650	24" pipe size		1.50	16			515	67	582	850	
640	0010	**PIPE, STEEL, FITTINGS** Threaded										640
	0020	Cast Iron										
	0040	Standard weight, black										
	0060	90° Elbow, straight										

MECHANICAL 15

15107 | Metal Pipe & Fittings

			CREW	DAILY OUTPUT	LABOR-HOURS	UNIT	2001 BARE COSTS MAT.	LABOR	EQUIP.	TOTAL	TOTAL INCL O&P	
640	0070	1/4″	1 Plum	16	.500	Ea.	3.34	17.25		20.59	29.50	640
	0080	3/8″		16	.500		4.71	17.25		21.96	31	
	0090	1/2″		15	.533		2.12	18.35		20.47	30	
	0100	3/4″		14	.571		2.21	19.70		21.91	32	
	0110	1″		13	.615		2.61	21		23.61	35	
	0120	1-1/4″	Q-1	22	.727		3.71	22.50		26.21	38	
	0130	1-1/2″		20	.800		5.15	25		30.15	43	
	0140	2″		18	.889		8	27.50		35.50	50.50	
	0150	2-1/2″		14	1.143		19.20	35.50		54.70	74.50	
	0160	3″		10	1.600		31.50	49.50		81	110	
	0170	3-1/2″		8	2		65	62		127	165	
	0180	4″		6	2.667		58.50	82.50		141	189	
	0190	5″		5	3.200		113	99		212	275	
	0200	6″	Q-2	7	3.429		144	110		254	325	
	0210	8″	″	6	4		299	129		428	525	
	0250	45° Elbow, straight										
	0260	1/4″	1 Plum	16	.500	Ea.	3.57	17.25		20.82	30	
	0270	3/8″		16	.500		3.83	17.25		21.08	30	
	0280	1/2″		15	.533		3.24	18.35		21.59	31	
	0300	3/4″		14	.571		3.26	19.70		22.96	33	
	0320	1″		13	.615		3.82	21		24.82	36	
	0330	1-1/4″	Q-1	22	.727		5.15	22.50		27.65	39.50	
	0340	1-1/2″		20	.800		8.50	25		33.50	47	
	0350	2″		18	.889		9.85	27.50		37.35	52.50	
	0360	2-1/2″		14	1.143		25	35.50		60.50	81	
	0370	3″		10	1.600		40.50	49.50		90	120	
	0380	3-1/2″		8	2		82	62		144	184	
	0400	4″		6	2.667		82	82.50		164.50	216	
	0420	5″		5	3.200		140	99		239	305	
	0440	6″	Q-2	7	3.429		180	110		290	365	
	0460	8″	″	6	4		360	129		489	590	
	0500	Tee, straight										
	0510	1/4″	1 Plum	10	.800	Ea.	4.66	27.50		32.16	46.50	
	0520	3/8″		10	.800		4.96	27.50		32.46	47	
	0530	1/2″		9	.889		3.29	30.50		33.79	49.50	
	0540	3/4″		9	.889		3.85	30.50		34.35	50	
	0550	1″		8	1		3.42	34.50		37.92	56	
	0560	1-1/4″	Q-1	14	1.143		6.25	35.50		41.75	60.50	
	0570	1-1/2″		13	1.231		8.15	38		46.15	66.50	
	0580	2″		11	1.455		11.30	45		56.30	80.50	
	0590	2-1/2″		9	1.778		29.50	55		84.50	116	
	0600	3″		6	2.667		45	82.50		127.50	175	
	0610	3-1/2″		5	3.200		88	99		187	247	
	0620	4″		4	4		88	124		212	284	
	0630	5″		3	5.333		178	165		343	445	
	0640	6″	Q-2	4	6		202	193		395	515	
	0650	8″	″	3	8		455	257		712	890	
	0660	Tee, reducing run and outlet										
	0661	1/2″	1 Plum	9	.889	Ea.	7.05	30.50		37.55	54	
	0662	3/4″		9	.889		6.10	30.50		36.60	52.50	
	0663	1″		8	1		6.30	34.50		40.80	59	
	0664	1-1/4″	Q-1	14	1.143		6.20	35.50		41.70	60.50	
	0665	1-1/2″		13	1.231		13.25	38		51.25	72	
	0666	2″		11	1.455		17.55	45		62.55	87.50	
	0667	2-1/2″		9	1.778		32	55		87	118	
	0668	3″		6	2.667		56.50	82.50		139	188	

15 MECHANICAL

	15107 | Metal Pipe & Fittings		CREW	DAILY OUTPUT	LABOR-HOURS	UNIT	2001 BARE COSTS MAT.	LABOR	EQUIP.	TOTAL	TOTAL INCL O&P	
640	0669	3-1/2″	Q-1	5	3.200	Ea.	100	99		199	260	640
	0670	4″		4	4		100	124		224	297	
	0671	5″		3	5.333		207	165		372	480	
	0672	6″	Q-2	4	6		255	193		448	570	
	0673	8″	″	3	8		289	257		546	705	
	0674	Reducer, concentric										
	0675	3/4″	1 Plum	18	.444	Ea.	5.40	15.30		20.70	29	
	0676	1″	″	15	.533		4	18.35		22.35	32	
	0677	1-1/4″	Q-1	26	.615		10.40	19.10		29.50	40.50	
	0678	1-1/2″		24	.667		14.70	20.50		35.20	47	
	0679	2″		21	.762		19.25	23.50		42.75	56.50	
	0680	2-1/2″		18	.889		27.50	27.50		55	71.50	
	0681	3″		14	1.143		37.50	35.50		73	95	
	0682	3-1/2″		12	1.333		65.50	41.50		107	135	
	0683	4″		10	1.600		65.50	49.50		115	148	
	0684	5″		6	2.667		104	82.50		186.50	240	
	0685	6″	Q-2	8	3		119	96.50		215.50	277	
	0686	8″	″	7	3.429		148	110		258	330	
	0687	Reducer, eccentric										
	0688	3/4″	1 Plum	16	.500	Ea.	12.30	17.25		29.55	39.50	
	0689	1″	″	14	.571		12.30	19.70		32	43	
	0690	1-1/4″	Q-1	25	.640		18.30	19.85		38.15	50	
	0691	1-1/2″		22	.727		26	22.50		48.50	62.50	
	0692	2″		20	.800		33	25		58	73.50	
	0693	2-1/2″		16	1		49	31		80	101	
	0694	3″		12	1.333		60.50	41.50		102	129	
	0695	3-1/2″		10	1.600		119	49.50		168.50	205	
	0696	4″		9	1.778		119	55		174	213	
	0697	5″		5	3.200		145	99		244	310	
	0698	6″	Q-2	8	3		198	96.50		294.50	365	
	0699	8″	″	7	3.429		740	110		850	980	
	0700	Standard weight, galvanized cast iron										
	0720	90° Elbow, straight										
	0730	1/4″	1 Plum	16	.500	Ea.	5.55	17.25		22.80	32	
	0740	3/8″		16	.500		5.55	17.25		22.80	32	
	0750	1/2″		15	.533		5.55	18.35		23.90	33.50	
	0760	3/4″		14	.571		6.20	19.70		25.90	36.50	
	0770	1″		13	.615		7.15	21		28.15	40	
	0780	1-1/4″	Q-1	22	.727		11.10	22.50		33.60	46	
	0790	1-1/2″		20	.800		15.15	25		40.15	54	
	0800	2″		18	.889		23	27.50		50.50	66.50	
	0810	2-1/2″		14	1.143		46.50	35.50		82	105	
	0820	3″		10	1.600		70.50	49.50		120	153	
	0830	3-1/2″		8	2		111	62		173	216	
	0840	4″		6	2.667		130	82.50		212.50	267	
	0850	5″		5	3.200		130	99		229	292	
	0860	6″	Q-2	7	3.429		289	110		399	480	
	0870	8″	″	6	4		590	129		719	845	
	0900	45° Elbow, straight										
	0910	1/4″	1 Plum	16	.500	Ea.	6.30	17.25		23.55	33	
	0920	3/8″		16	.500		6.30	17.25		23.55	33	
	0930	1/2″		15	.533		6.30	18.35		24.65	34.50	
	0940	3/4″		14	.571		7.20	19.70		26.90	37.50	
	0950	1″		13	.615		8.45	21		29.45	41.50	
	0960	1-1/4″	Q-1	22	.727		12.60	22.50		35.10	48	
	0970	1-1/2″		20	.800		18.85	25		43.85	58.50	

	15107 | Metal Pipe & Fittings	CREW	DAILY OUTPUT	LABOR-HOURS	UNIT	2001 BARE COSTS MAT.	LABOR	EQUIP.	TOTAL	TOTAL INCL O&P	
640	0980 2"	Q-1	18	.889	Ea.	26.50	27.50		54	70.50	640
	0990 2-1/2"		14	1.143		52	35.50		87.50	111	
	1000 3"		10	1.600		83	49.50		132.50	166	
	1010 3-1/2"		8	2		152	62		214	261	
	1020 4"		6	2.667		152	82.50		234.50	292	
	1030 5"		5	3.200		189	99		288	360	
	1040 6"	Q-2	7	3.429		244	110		354	435	
	1050 8"	"	6	4		485	129		614	730	
	1100 Tee, straight										
	1110 1/4"	1 Plum	10	.800	Ea.	7.05	27.50		34.55	49.50	
	1120 3/8"		10	.800		7.05	27.50		34.55	49.50	
	1130 1/2"		9	.889		6.15	30.50		36.65	53	
	1140 3/4"		9	.889		9.10	30.50		39.60	56	
	1150 1"		8	1		9.65	34.50		44.15	62.50	
	1160 1-1/4"	Q-1	14	1.143		16.85	35.50		52.35	72	
	1170 1-1/2"		13	1.231		22	38		60	82	
	1180 2"		11	1.455		28	45		73	98.50	
	1190 2-1/2"		9	1.778		60	55		115	149	
	1200 3"		6	2.667		155	82.50		237.50	296	
	1210 3-1/2"		5	3.200		159	99		258	325	
	1220 4"		4	4		178	124		302	385	
	1230 5"		3	5.333		241	165		406	515	
	1240 6"	Q-2	4	6		272	193		465	590	
	1250 8"	"	3	8		620	257		877	1,075	
	1300 Extra heavy weight, black										
	1310 Couplings, steel straight										
	1320 1/4"	1 Plum	19	.421	Ea.	2.72	14.50		17.22	25	
	1330 3/8"		19	.421		2.97	14.50		17.47	25.50	
	1340 1/2"		19	.421		4.02	14.50		18.52	26.50	
	1350 3/4"		18	.444		4.28	15.30		19.58	27.50	
	1360 1"		15	.533		5.45	18.35		23.80	33.50	
	1370 1-1/4"	Q-1	26	.615		8.75	19.10		27.85	38.50	
	1380 1-1/2"		24	.667		8.75	20.50		29.25	40.50	
	1390 2"		21	.762		13.35	23.50		36.85	50	
	1400 2-1/2"		18	.889		19.80	27.50		47.30	63.50	
	1410 3"		14	1.143		23.50	35.50		59	79.50	
	1420 3-1/2"		12	1.333		31.50	41.50		73	97.50	
	1430 4"		10	1.600		37	49.50		86.50	116	
	1440 5"		6	2.667		53	82.50		135.50	184	
	1450 6"	Q-2	8	3		65	96.50		161.50	218	
	1460 8"		7	3.429		74	110		184	248	
	1470 10"		6	4		116	129		245	320	
	1480 12"		4	6		155	193		348	460	
	1510 90° Elbow, straight										
	1520 1/2"	1 Plum	15	.533	Ea.	9.60	18.35		27.95	38	
	1530 3/4"		14	.571		10.15	19.70		29.85	40.50	
	1540 1"		13	.615		12.25	21		33.25	45.50	
	1550 1-1/4"	Q-1	22	.727		18.15	22.50		40.65	54	
	1560 1-1/2"		20	.800		22.50	25		47.50	62.50	
	1580 2"		18	.889		28	27.50		55.50	72.50	
	1590 2-1/2"		14	1.143		54.50	35.50		90	114	
	1600 3"		10	1.600		100	49.50		149.50	185	
	1610 4"		6	2.667		167	82.50		249.50	310	
	1620 6"	Q-2	7	3.429		435	110		545	640	
	1650 45° Elbow, straight										
	1660 1/2"	1 Plum	15	.533	Ea.	11.35	18.35		29.70	40	

15 MECHANICAL

Important: See the Reference Section for critical supporting data - Reference Nos., Crews, & City Cost Indexes

15107 | Metal Pipe & Fittings

640		CREW	DAILY OUTPUT	LABOR-HOURS	UNIT	2001 BARE COSTS MAT.	LABOR	EQUIP.	TOTAL	TOTAL INCL O&P
1670	3/4"	1 Plum	14	.571	Ea.	12.90	19.70		32.60	43.50
1680	1"		13	.615		15.55	21		36.55	49
1690	1-1/4"	Q-1	22	.727		22	22.50		44.50	58
1700	1-1/2"		20	.800		28	25		53	68.50
1710	2"		18	.889		40	27.50		67.50	85.50
1720	2-1/2"		14	1.143		60	35.50		95.50	120
1800	Tee, straight									
1810	1/2"	1 Plum	9	.889	Ea.	14.80	30.50		45.30	62.50
1820	3/4"		9	.889		15.10	30.50		45.60	62.50
1830	1"		8	1		18.50	34.50		53	72.50
1840	1-1/4"	Q-1	14	1.143		27.50	35.50		63	83.50
1850	1-1/2"		13	1.231		35	38		73	96
1860	2"		11	1.455		43.50	45		88.50	116
1870	2-1/2"		9	1.778		77	55		132	168
1880	3"		6	2.667		133	82.50		215.50	272
1890	4"		4	4		211	124		335	420
1900	6"	Q-2	4	6		435	193		628	770
4000	Standard weight, black									
4010	Couplings, steel straight, merchants									
4030	1/4"	1 Plum	19	.421	Ea.	.73	14.50		15.23	23
4040	3/8"		19	.421		.88	14.50		15.38	23
4050	1/2"		19	.421		.93	14.50		15.43	23
4060	3/4"		18	.444		1.19	15.30		16.49	24.50
4070	1"		15	.533		1.67	18.35		20.02	29.50
4080	1-1/4"	Q-1	26	.615		2.13	19.10		21.23	31.50
4090	1-1/2"		24	.667		2.70	20.50		23.20	34
4100	2"		21	.762		3.86	23.50		27.36	40
4110	2-1/2"		18	.889		11	27.50		38.50	53.50
4120	3"		14	1.143		15.45	35.50		50.95	70.50
4130	3-1/2"		12	1.333		27.50	41.50		69	92.50
4140	4"		10	1.600		27.50	49.50		77	105
4150	5"		6	2.667		50	82.50		132.50	180
4160	6"	Q-2	8	3		60	96.50		156.50	212
4166	Plug, 1/4"	1 Plum	38	.211		1.06	7.25		8.31	12.10
4167	3/8"		38	.211		1.06	7.25		8.31	12.10
4168	1/2"		38	.211		.84	7.25		8.09	11.85
4169	3/4"		32	.250		1.06	8.60		9.66	14.15
4170	1"		30	.267		1.06	9.20		10.26	15
4171	1-1/4"	Q-1	52	.308		1.70	9.55		11.25	16.25
4172	1-1/2"		48	.333		2.40	10.35		12.75	18.25
4173	2"		42	.381		2.80	11.80		14.60	21
4176	2-1/2"		36	.444		6.80	13.80		20.60	28.50
4200	Standard weight, galvanized									
4210	Couplings, steel straight, merchants									
4230	1/4"	1 Plum	19	.421	Ea.	.57	14.50		15.07	22.50
4240	3/8"		19	.421		.73	14.50		15.23	23
4250	1/2"		19	.421		.78	14.50		15.28	23
4260	3/4"		18	.444		.97	15.30		16.27	24
4270	1"		15	.533		1.35	18.35		19.70	29
4280	1-1/4"	Q-1	26	.615		1.73	19.10		20.83	31
4290	1-1/2"		24	.667		2.15	20.50		22.65	33.50
4300	2"		21	.762		3.22	23.50		26.72	39
4310	2-1/2"		18	.889		6.85	27.50		34.35	49
4320	3"		14	1.143		9.10	35.50		44.60	63.50
4330	3-1/2"		12	1.333		16.05	41.50		57.55	80
4340	4"		10	1.600		16.05	49.50		65.55	92.50

15107 | Metal Pipe & Fittings		CREW	DAILY OUTPUT	LABOR-HOURS	UNIT	2001 BARE COSTS MAT.	LABOR	EQUIP.	TOTAL	TOTAL INCL O&P
640 4350	5″	Q-1	6	2.667	Ea.	30	82.50		112.50	158
4360	6″	Q-2	8	3		36.50	96.50		133	187
4370	Plug, galvanized, square head									
4374	1/2″	1 Plum	38	.211	Ea.	2.15	7.25		9.40	13.30
4375	3/4″		32	.250		2.15	8.60		10.75	15.35
4376	1″		30	.267		2.15	9.20		11.35	16.20
4377	1-1/4″	Q-1	52	.308		3.59	9.55		13.14	18.35
4378	1-1/2″		48	.333		4.85	10.35		15.20	21
4379	2″		42	.381		6.05	11.80		17.85	24.50
4380	2-1/2″		36	.444		12.70	13.80		26.50	35
4381	3″		28	.571		16.30	17.70		34	44.50
4382	4″		20	.800		39.50	25		64.50	81
4384	6″		8	2		69.50	62		131.50	170
5000	Malleable iron, 150 lb.									
5020	Black									
5040	90° elbow, straight									
5060	1/4″	1 Plum	16	.500	Ea.	1.75	17.25		19	28
5070	3/8″		16	.500		1.75	17.25		19	28
5080	1/2″		15	.533		1.21	18.35		19.56	29
5090	3/4″		14	.571		1.47	19.70		21.17	31
5100	1″		13	.615		2.56	21		23.56	35
5110	1-1/4″	Q-1	22	.727		4.20	22.50		26.70	38.50
5120	1-1/2″		20	.800		5.55	25		30.55	43.50
5130	2″		18	.889		9.55	27.50		37.05	52
5140	2-1/2″		14	1.143		21	35.50		56.50	77
5150	3″		10	1.600		31	49.50		80.50	109
5160	3-1/2″		8	2		67	62		129	167
5170	4″		6	2.667		67	82.50		149.50	199
5180	5″		5	3.200		206	99		305	375
5190	6″	Q-2	7	3.429		206	110		316	390
5250	45° elbow, straight									
5270	1/4″	1 Plum	16	.500	Ea.	2.64	17.25		19.89	29
5280	3/8″		16	.500		2.64	17.25		19.89	29
5290	1/2″		15	.533		2.01	18.35		20.36	29.50
5300	3/4″		14	.571		2.47	19.70		22.17	32
5310	1″		13	.615		3.13	21		24.13	35.50
5320	1-1/4″	Q-1	22	.727		5.55	22.50		28.05	40
5330	1-1/2″		20	.800		6.85	25		31.85	45
5340	2″		18	.889		10.35	27.50		37.85	53
5350	2-1/2″		14	1.143		30	35.50		65.50	86.50
5360	3″		10	1.600		39	49.50		88.50	118
5370	3-1/2″		8	2		76.50	62		138.50	178
5380	4″		6	2.667		76.50	82.50		159	209
5390	5″		5	3.200		274	99		373	450
5400	6″	Q-2	7	3.429		274	110		384	465
5450	Tee, straight									
5470	1/4″	1 Plum	10	.800	Ea.	2.56	27.50		30.06	44.50
5480	3/8″		10	.800		2.56	27.50		30.06	44.50
5490	1/2″		9	.889		1.64	30.50		32.14	48
5500	3/4″		9	.889		2.34	30.50		32.84	48.50
5510	1″		8	1		3.99	34.50		38.49	56.50
5520	1-1/4″	Q-1	14	1.143		6.45	35.50		41.95	60.50
5530	1-1/2″		13	1.231		8.05	38		46.05	66.50
5540	2″		11	1.455		13.70	45		58.70	83
5550	2-1/2″		9	1.778		29.50	55		84.50	116
5560	3″		6	2.667		43.50	82.50		126	173

640

Important: See the Reference Section for critical supporting data - Reference Nos., Crews, & City Cost Indexes

15107 | Metal Pipe & Fittings

640

		CREW	DAILY OUTPUT	LABOR-HOURS	UNIT	2001 BARE COSTS MAT.	LABOR	EQUIP.	TOTAL	TOTAL INCL O&P
5570	3-1/2"	Q-1	5	3.200	Ea.	105	99		204	266
5580	4"		4	4		105	124		229	305
5590	5"		3	5.333		235	165		400	510
5600	6"	Q-2	4	6		305	193		498	625
5601	Tee, reducing, on outlet									
5602	1/2"	1 Plum	9	.889	Ea.	4.33	30.50		34.83	51
5603	3/4"		9	.889		3.61	30.50		34.11	50
5604	1"		8	1		4.55	34.50		39.05	57
5605	1-1/4"	Q-1	14	1.143		7.90	35.50		43.40	62
5606	1-1/2"		13	1.231		11.35	38		49.35	70
5607	2"		11	1.455		15.50	45		60.50	85
5608	2-1/2"		9	1.778		41	55		96	128
5609	3"		6	2.667		51.50	82.50		134	182
5610	3-1/2"		5	3.200		124	99		223	286
5611	4"		4	4		124	124		248	325
5612	5"		3	5.333		137	165		302	400
5613	6"	Q-2	5	4.800		144	154		298	390
5650	Coupling									
5670	1/4"	1 Plum	19	.421	Ea.	2.17	14.50		16.67	24.50
5680	3/8"		19	.421		2.17	14.50		16.67	24.50
5690	1/2"		19	.421		1.68	14.50		16.18	24
5700	3/4"		18	.444		1.97	15.30		17.27	25
5710	1"		15	.533		2.95	18.35		21.30	31
5720	1-1/4"	Q-1	26	.615		3.82	19.10		22.92	33
5730	1-1/2"		24	.667		5.15	20.50		25.65	36.50
5740	2"		21	.762		7.65	23.50		31.15	44
5750	2-1/2"		18	.889		21	27.50		48.50	64.50
5760	3"		14	1.143		28.50	35.50		64	85
5770	3-1/2"		12	1.333		57.50	41.50		99	126
5780	4"		10	1.600		57.50	49.50		107	138
5790	5"		6	2.667		125	82.50		207.50	262
5800	6"	Q-2	8	3		125	96.50		221.50	283
5840	Reducer, concentric, 1/4"	1 Plum	19	.421		2.29	14.50		16.79	24.50
5850	3/8"		19	.421		2.29	14.50		16.79	24.50
5860	1/2"		19	.421		2.32	14.50		16.82	24.50
5870	3/4"		16	.500		2.75	17.25		20	29
5880	1"		15	.533		4.34	18.35		22.69	32.50
5890	1-1/4"	Q-1	26	.615		5.20	19.10		24.30	34.50
5900	1-1/2"		24	.667		7.45	20.50		27.95	39
5910	2"		21	.762		10.75	23.50		34.25	47.50
5911	2-1/2"		18	.889		22	27.50		49.50	66
5912	3"		14	1.143		30	35.50		65.50	86.50
5913	3-1/2"		12	1.333		62	41.50		103.50	131
5914	4"		10	1.600		62	49.50		111.50	144
5915	5"		6	2.667		116	82.50		198.50	253
5916	6"	Q-2	8	3		150	96.50		246.50	310
5981	Bushing, 1/4"	1 Plum	19	.421		2.18	14.50		16.68	24.50
5982	3/8"		19	.421		2.18	14.50		16.68	24.50
5983	1/2"		19	.421		2.18	14.50		16.68	24.50
5984	3/4"		16	.500		2.18	17.25		19.43	28.50
5985	1"		15	.533		3.20	18.35		21.55	31
5986	1-1/4"	Q-1	26	.615		3.97	19.10		23.07	33.50
5987	1-1/2"		24	.667		4.81	20.50		25.31	36.50
5988	2"		21	.762		5.25	23.50		28.75	41.50
5989	Cap, 1/4"	1 Plum	38	.211		1.61	7.25		8.86	12.70
5991	3/8"		38	.211		1.36	7.25		8.61	12.45

15 MECHANICAL

15107 | Metal Pipe & Fittings		CREW	DAILY OUTPUT	LABOR-HOURS	UNIT	2001 BARE COSTS MAT.	LABOR	EQUIP.	TOTAL	TOTAL INCL O&P	
640 5992	1/2"	1 Plum	38	.211	Ea.	1.23	7.25		8.48	12.30	640
5993	3/4"		32	.250		1.67	8.60		10.27	14.85	
5994	1"		30	.267		2.01	9.20		11.21	16.05	
5995	1-1/4"	Q-1	52	.308		2.66	9.55		12.21	17.35	
5996	1-1/2"		48	.333		3.64	10.35		13.99	19.60	
5997	2"		42	.381		5.30	11.80		17.10	23.50	
6000	For galvanized elbows, tees, and couplings add					20%					
6058	For galvanized reducers, caps and plugs add					20%					
6100	90° elbow, galvanized, 150 lb., reducing										
6110	3/4" x 1/2"	1 Plum	15.40	.519	Ea.	3.24	17.90		21.14	30.50	
6112	1" x 3/4"		14	.571		4.24	19.70		23.94	34	
6114	1" x 1/2"		14.50	.552		4.50	19		23.50	33.50	
6116	1-1/4" x 1"	Q-1	24.20	.661		7.10	20.50		27.60	39	
6118	1-1/4" x 3/4"		25.40	.630		8.55	19.55		28.10	39	
6120	1-1/4" x 1/2"		26.20	.611		9.05	18.95		28	38.50	
6122	1-1/2" x 1-1/4"		21.60	.741		11.35	23		34.35	47	
6124	1-1/2" x 1"		23.50	.681		11.35	21		32.35	44.50	
6126	1-1/2" x 3/4"		24.60	.650		11.35	20		31.35	43	
6128	2" x 1-1/2"		20.50	.780		13.30	24		37.30	51	
6130	2" x 1-1/4"		21	.762		15.20	23.50		38.70	52.50	
6132	2" x 1"		22.80	.702		15.80	22		37.80	50.50	
6134	2" x 3/4"		23.90	.669		16.10	21		37.10	49	
6136	2-1/2" x 2"		12.30	1.301		45	40.50		85.50	111	
6138	2-1/2" x 1-1/2"		12.50	1.280		50	39.50		89.50	115	
6140	3" x 2-1/2"		8.60	1.860		72.50	57.50		130	167	
6142	3" x 2"		11.80	1.356		71	42		113	142	
6144	4" x 3"		8.20	1.951		140	60.50		200.50	246	
6160	90° elbow, black, 150 lb., reducing										
6170	1" x 3/4"	1 Plum	14	.571	Ea.	3.06	19.70		22.76	33	
6174	1-1/2" x 1"	Q-1	23.50	.681		7.25	21		28.25	40	
6178	1-1/2" x 3/4"		24.60	.650		8.30	20		28.30	39.50	
6182	2" x 1-1/2"		20.50	.780		11.95	24		35.95	49.50	
6186	2" x 1"		22.80	.702		11.45	22		33.45	45.50	
6190	2" x 3/4"		23.90	.669		12.70	21		33.70	45.50	
6194	2-1/2" x 2"		12.30	1.301		29	40.50		69.50	93	
7000	Union, with brass seat										
7010	1/4"	1 Plum	15	.533	Ea.	8.60	18.35		26.95	37	
7020	3/8"		15	.533		5.95	18.35		24.30	34	
7030	1/2"		14	.571		5.40	19.70		25.10	35.50	
7040	3/4"		13	.615		6.15	21		27.15	39	
7050	1"		12	.667		8.05	23		31.05	43.50	
7060	1-1/4"	Q-1	21	.762		11.55	23.50		35.05	48	
7070	1-1/2"		19	.842		14.40	26		40.40	55.50	
7080	2"		17	.941		16.75	29		45.75	62.50	
7090	2-1/2"		13	1.231		50	38		88	113	
7100	3"		9	1.778		60	55		115	149	
7120	Union, galvanized										
7124	1/2"	1 Plum	14	.571	Ea.	6.75	19.70		26.45	37	
7125	3/4"		13	.615		7.75	21		28.75	40.50	
7126	1"		12	.667		10.15	23		33.15	45.50	
7127	1-1/4"	Q-1	21	.762		14.65	23.50		38.15	51.50	
7128	1-1/2"		19	.842		17.80	26		43.80	59	
7129	2"		17	.941		20.50	29		49.50	66.50	
7130	2-1/2"		13	1.231		71	38		109	136	
7131	3"		9	1.778		99.50	55		154.50	192	
7500	Malleable iron, 300 lb										

15107 | Metal Pipe & Fittings

640		CREW	DAILY OUTPUT	LABOR-HOURS	UNIT	2001 BARE COSTS MAT.	LABOR	EQUIP.	TOTAL	TOTAL INCL O&P
7520	Black									
7540	90° Elbow, straight, 1/4"	1 Plum	16	.500	Ea.	6.15	17.25		23.40	33
7560	3/8"		16	.500		5.45	17.25		22.70	32
7570	1/2"		15	.533		7.10	18.35		25.45	35.50
7580	3/4"		14	.571		8	19.70		27.70	38.50
7590	1"		13	.615		10.25	21		31.25	43.50
7600	1-1/4"	Q-1	22	.727		14.95	22.50		37.45	50.50
7610	1-1/2"		20	.800		17.60	25		42.60	57
7620	2"		18	.889		25.50	27.50		53	69.50
7630	2-1/2"		14	1.143		60	35.50		95.50	120
7640	3"		10	1.600		75.50	49.50		125	158
7650	4"		6	2.667		157	82.50		239.50	298
7700	45° Elbow, straight, 1/4"	1 Plum	16	.500		9.15	17.25		26.40	36
7720	3/8"		16	.500		9.15	17.25		26.40	36
7730	1/2"		15	.533		10.20	18.35		28.55	38.50
7740	3/4"		14	.571		11.35	19.70		31.05	42
7750	1"		13	.615		12.55	21		33.55	46
7760	1-1/4"	Q-1	22	.727		20	22.50		42.50	56
7770	1-1/2"		20	.800		26	25		51	66.50
7780	2"		18	.889		39.50	27.50		67	85
7790	2-1/2"		14	1.143		78.50	35.50		114	140
7800	3"		10	1.600		103	49.50		152.50	189
7810	4"		6	2.667		200	82.50		282.50	345
7850	Tee, straight, 1/4"	1 Plum	10	.800		7.75	27.50		35.25	50
7870	3/8"		10	.800		8.20	27.50		35.70	50.50
7880	1/2"		9	.889		10.45	30.50		40.95	57.50
7890	3/4"		9	.889		11.30	30.50		41.80	58.50
7900	1"		8	1		13.60	34.50		48.10	67
7910	1-1/4"	Q-1	14	1.143		19.45	35.50		54.95	75
7920	1-1/2"		13	1.231		22.50	38		60.50	82.50
7930	2"		11	1.455		33.50	45		78.50	105
7940	2-1/2"		9	1.778		72.50	55		127.50	163
7950	3"		6	2.667		116	82.50		198.50	253
7960	4"		4	4		250	124		374	460
7980	6"	Q-2	4	6		296	193		489	615
7990	8"	"	3	8		465	257		722	900
8050	Couplings, straight, 1/4"	1 Plum	19	.421		5.55	14.50		20.05	28
8070	3/8"		19	.421		5.55	14.50		20.05	28
8080	1/2"		19	.421		6.20	14.50		20.70	29
8090	3/4"		18	.444		7.15	15.30		22.45	31
8100	1"		15	.533		8.15	18.35		26.50	36.50
8110	1-1/4"	Q-1	26	.615		9.70	19.10		28.80	39.50
8120	1-1/2"		24	.667		14.50	20.50		35	47
8130	2"		21	.762		21	23.50		44.50	58.50
8140	2-1/2"		18	.889		34	27.50		61.50	79
8150	3"		14	1.143		49.50	35.50		85	108
8160	4"		10	1.600		69	49.50		118.50	151
8200	Galvanized									
8220	90° Elbow, straight, 1/4"	1 Plum	16	.500	Ea.	10.45	17.25		27.70	37.50
8222	3/8"		16	.500		10.65	17.25		27.90	37.50
8224	1/2"		15	.533		12.65	18.35		31	41.50
8226	3/4"		14	.571		14.20	19.70		33.90	45
8228	1"		13	.615		18.25	21		39.25	52
8230	1-1/4"	Q-1	22	.727		28.50	22.50		51	65.50
8232	1-1/2"		20	.800		31	25		56	71.50
8234	2"		18	.889		52	27.50		79.50	99

640

MECHANICAL 15

15 MECHANICAL

15107 | Metal Pipe & Fittings

	Line	Description	Crew	Daily Output	Labor-Hours	Unit	2001 Bare Costs Mat.	Labor	Equip.	Total	Total Incl O&P	
640	8236	2-1/2"	Q-1	14	1.143	Ea.	102	35.50		137.50	166	640
	8238	3"		10	1.600		137	49.50		186.50	226	
	8240	4"		6	2.667		310	82.50		392.50	465	
	8280	45° Elbow, straight										
	8282	1/2"	1 Plum	15	.533	Ea.	18.85	18.35		37.20	48	
	8284	3/4"		14	.571		21.50	19.70		41.20	53.50	
	8286	1"		13	.615		24	21		45	58	
	8288	1-1/4"	Q-1	22	.727		37	22.50		59.50	74.50	
	8290	1-1/2"		20	.800		47.50	25		72.50	90	
	8292	2"		18	.889		66	27.50		93.50	114	
	8310	Tee, straight, 1/4"	1 Plum	10	.800		14.85	27.50		42.35	58	
	8312	3/8"		10	.800		15.10	27.50		42.60	58	
	8314	1/2"		9	.889		18.90	30.50		49.40	67	
	8316	3/4"		9	.889		21	30.50		51.50	69	
	8318	1"		8	1		25.50	34.50		60	80.50	
	8320	1-1/4"	Q-1	14	1.143		36.50	35.50		72	94	
	8322	1-1/2"		13	1.231		38	38		76	99.50	
	8324	2"		11	1.455		61	45		106	135	
	8326	2-1/2"		9	1.778		146	55		201	243	
	8328	3"		6	2.667		218	82.50		300.50	365	
	8330	4"		4	4		425	124		549	650	
	8380	Couplings, straight, 1/4"	1 Plum	19	.421		9.85	14.50		24.35	33	
	8382	3/8"		19	.421		9.85	14.50		24.35	33	
	8384	1/2"		19	.421		10.10	14.50		24.60	33	
	8386	3/4"		18	.444		11.40	15.30		26.70	35.50	
	8388	1"		15	.533		15.10	18.35		33.45	44	
	8390	1-1/4"	Q-1	26	.615		19.65	19.10		38.75	50.50	
	8392	1-1/2"		24	.667		28	20.50		48.50	61.50	
	8394	2"		21	.762		33.50	23.50		57	72.50	
	8396	2-1/2"		18	.889		67	27.50		94.50	115	
	8398	3"		14	1.143		93.50	35.50		129	157	
	8399	4"		10	1.600		100	49.50		149.50	185	
	8529	Black										
	9500	Union with brass seat, 1/4"	1 Plum	15	.533	Ea.	12.50	18.35		30.85	41.50	
	9530	3/8"		15	.533		12.75	18.35		31.10	41.50	
	9540	1/2"		14	.571		10.30	19.70		30	41	
	9550	3/4"		13	.615		11.40	21		32.40	44.50	
	9560	1"		12	.667		14.90	23		37.90	51	
	9570	1-1/4"	Q-1	21	.762		24.50	23.50		48	62	
	9580	1-1/2"		19	.842		25.50	26		51.50	67.50	
	9590	2"		17	.941		31.50	29		60.50	78.50	
	9600	2-1/2"		13	1.231		102	38		140	170	
	9610	3"		9	1.778		131	55		186	228	
	9620	4"		5	3.200		325	99		424	505	
	9630	Union, all iron, 1/4"	1 Plum	15	.533		19.50	18.35		37.85	49	
	9650	3/8"		15	.533		20.50	18.35		38.85	50	
	9660	1/2"		14	.571		20.50	19.70		40.20	52	
	9670	3/4"		13	.615		22	21		43	56	
	9680	1"		12	.667		27	23		50	64	
	9690	1-1/4"	Q-1	21	.762		41.50	23.50		65	81	
	9700	1-1/2"		19	.842		50.50	26		76.50	95	
	9710	2"		17	.941		64	29		93	115	
	9720	2-1/2"		13	1.231		216	38		254	295	
	9730	3"		9	1.778		216	55		271	320	
	9750	For galvanized unions, add					15%					

15107 | Metal Pipe & Fittings

		CREW	DAILY OUTPUT	LABOR-HOURS	UNIT	2001 BARE COSTS MAT.	LABOR	EQUIP.	TOTAL	TOTAL INCL O&P
660	0010	**PIPE, STEEL, FITTINGS** Flanged, welded and special type									660
	0020	flanged joints, C.I., standard weight, black. One gasket & bolt								
	0040	set required at each joint, not included (see line 0620)								
	0060	90° Elbow, straight, 1-1/2″ pipe size	Q-1	14	1.143	Ea.	127	35.50		162.50	194
	0080	2″ pipe size		13	1.231		100	38		138	168
	0090	2-1/2″ pipe size		12	1.333		108	41.50		149.50	182
	0100	3″ pipe size		11	1.455		129	45		174	209
	0110	4″ pipe size		8	2		111	62		173	217
	0120	5″ pipe size		7	2.286		206	71		277	335
	0130	6″ pipe size	Q-2	9	2.667		176	85.50		261.50	320
	0140	8″ pipe size		8	3		273	96.50		369.50	445
	0150	10″ pipe size		7	3.429		605	110		715	830
	0160	12″ pipe size		6	4		1,050	129		1,179	1,350
	0171	90° Elbow, reducing									
	0172	2-1/2″ by 2″ pipe size	Q-1	12	1.333	Ea.	246	41.50		287.50	335
	0173	3″ by 2-1/2″ pipe size		11	1.455		260	45		305	355
	0174	4″ by 3″ pipe size		8	2		294	62		356	420
	0175	5″ by 3″ pipe size		7	2.286		380	71		451	520
	0176	6″ by 4″ pipe size	Q-2	9	2.667		420	85.50		505.50	595
	0177	8″ by 6″ pipe size		8	3		620	96.50		716.50	830
	0178	10″ by 8″ pipe size		7	3.429		980	110		1,090	1,250
	0179	12″ by 10″ pipe size		6	4		1,700	129		1,829	2,050
	0200	45° Elbow, straight, 1-1/2″ pipe size	Q-1	14	1.143		155	35.50		190.50	225
	0220	2″ pipe size		13	1.231		152	38		190	225
	0230	2-1/2″ pipe size		12	1.333		155	41.50		196.50	234
	0240	3″ pipe size		11	1.455		163	45		208	248
	0250	4″ pipe size		8	2		169	62		231	280
	0260	5″ pipe size		7	2.286		315	71		386	450
	0270	6″ pipe size	Q-2	9	2.667		277	85.50		362.50	435
	0280	8″ pipe size		8	3		405	96.50		501.50	590
	0290	10″ pipe size		7	3.429		860	110		970	1,100
	0300	12″ pipe size		6	4		1,325	129		1,454	1,650
	0310	Cross, straight									
	0311	2-1/2″ pipe size	Q-1	6	2.667	Ea.	335	82.50		417.50	495
	0312	3″ pipe size		5	3.200		335	99		434	520
	0313	4″ pipe size		4	4		435	124		559	665
	0314	5″ pipe size		3	5.333		710	165		875	1,025
	0315	6″ pipe size	Q-2	5	4.800		710	154		864	1,025
	0316	8″ pipe size		4	6		1,150	193		1,343	1,550
	0317	10″ pipe size		3	8		1,750	257		2,007	2,325
	0318	12″ pipe size		2	12		2,650	385		3,035	3,500
	0350	Tee, straight, 1-1/2″ pipe size	Q-1	10	1.600		109	49.50		158.50	195
	0370	2″ pipe size		9	1.778		109	55		164	203
	0380	2-1/2″ pipe size		8	2		159	62		221	268
	0390	3″ pipe size		7	2.286		165	71		236	289
	0400	4″ pipe size		5	3.200		169	99		268	335
	0410	5″ pipe size		4	4		355	124		479	575
	0420	6″ pipe size	Q-2	6	4		246	129		375	465
	0430	8″ pipe size		5	4.800		385	154		539	660
	0440	10″ pipe size		4	6		880	193		1,073	1,250
	0450	12″ pipe size		3	8		1,375	257		1,632	1,925
	0459	Tee, reducing on outlet									
	0460	2-1/2″ by 2″ pipe size	Q-1	8	2	Ea.	236	62		298	355
	0461	3″ by 2-1/2″ pipe size		7	2.286		270	71		341	405
	0462	4″ by 3″ pipe size		5	3.200		660	99		759	880
	0463	5″ by 4″ pipe size		4	4		845	124		969	1,125

15107 | Metal Pipe & Fittings		CREW	DAILY OUTPUT	LABOR-HOURS	UNIT	2001 BARE COSTS MAT.	LABOR	EQUIP.	TOTAL	TOTAL INCL O&P
660 0464	6" by 4" pipe size	Q-2	6	4	Ea.	845	129		974	1,125 660
0465	8" by 6" pipe size		5	4.800		910	154		1,064	1,225
0466	10" by 8" pipe size		4	6		1,425	193		1,618	1,850
0467	12" by 10" pipe size		3	8		2,100	257		2,357	2,700
0476	Reducer, concentric									
0477	3" by 2-1/2"	Q-1	12	1.333	Ea.	250	41.50		291.50	340
0478	4" by 3"		9	1.778		267	55		322	375
0479	5" by 4"		8	2		420	62		482	555
0480	6" by 4"	Q-2	10	2.400		470	77		547	630
0481	8" by 6"		9	2.667		535	85.50		620.50	715
0482	10" by 8"		8	3		860	96.50		956.50	1,100
0483	12" by 10"		7	3.429		1,550	110		1,660	1,875
0492	Reducer, eccentric									
0493	4" by 3"	Q-1	8	2	Ea.	325	62		387	455
0494	5" by 4"	"	7	2.286		420	71		491	565
0495	6" by 4"	Q-2	9	2.667		450	85.50		535.50	625
0496	8" by 6"		8	3		590	96.50		686.50	795
0497	10" by 8"		7	3.429		1,025	110		1,135	1,300
0498	12" by 10"		6	4		1,575	129		1,704	1,925
0500	For galvanized elbows and tees, add					100%				
0520	For extra heavy weight elbows and tees, add					140%				
0620	Gasket and bolt set, 150#, 1/2" thru 1-1/2" pipe size	1 Plum	20	.400		3.05	13.80		16.85	24.50
0630	2" pipe size		13	.615		4.87	21		25.87	37.50
0640	2-1/2" pipe size		12	.667		4.64	23		27.64	39.50
0650	3" pipe size		11	.727		5.40	25		30.40	44
0660	3-1/2" pipe size		9	.889		9.80	30.50		40.30	57
0670	4" pipe size		8	1		9.80	34.50		44.30	63
0680	5" pipe size		7	1.143		14.55	39.50		54.05	75.50
0690	6" pipe size		6	1.333		15.25	46		61.25	86.50
0700	8" pipe size		5	1.600		16.25	55		71.25	101
0710	10" pipe size		4.50	1.778		33.50	61		94.50	130
0720	12" pipe size		4.20	1.905		36	65.50		101.50	139
0730	14" pipe size		4	2		51	69		120	161
0740	16" pipe size		3	2.667		67	92		159	213
0750	18" pipe size		2.70	2.963		93.50	102		195.50	257
0760	20" pipe size		2.30	3.478		125	120		245	320
0780	24" pipe size		1.90	4.211		167	145		312	405
0790	26" pipe size		1.60	5		189	172		361	470
0810	30" pipe size		1.40	5.714		234	197		431	555
0830	36" pipe size		1.10	7.273		288	251		539	695
0850	For 300 lb gasket set, add					40%				
2000	Flanged unions, 125 lb., black, 1/2" pipe size	1 Plum	17	.471	Ea.	25.50	16.20		41.70	52.50
2040	3/4" pipe size		17	.471		33.50	16.20		49.70	61.50
2050	1" pipe size		16	.500		32.50	17.25		49.75	61.50
2060	1-1/4" pipe size	Q-1	28	.571		38.50	17.70		56.20	69
2070	1-1/2" pipe size		27	.593		35.50	18.35		53.85	66.50
2080	2" pipe size		26	.615		42.50	19.10		61.60	76
2090	2-1/2" pipe size		24	.667		57.50	20.50		78	94.50
2100	3" pipe size		22	.727		65	22.50		87.50	106
2110	3-1/2" pipe size		18	.889		133	27.50		160.50	188
2120	4" pipe size		16	1		89	31		120	145
2130	5" pipe size		14	1.143		157	35.50		192.50	227
2140	6" pipe size	Q-2	19	1.263		190	40.50		230.50	271
2150	8" pipe size	"	16	1.500		325	48		373	435
2200	For galvanized unions, add					150%				
2290	Threaded flange									

15 MECHANICAL

15107 | Metal Pipe & Fittings

660		CREW	DAILY OUTPUT	LABOR-HOURS	UNIT	2001 BARE COSTS MAT.	LABOR	EQUIP.	TOTAL	TOTAL INCL O&P
2300	Cast iron									
2310	Black, 125 lb., per flange									
2320	1" pipe size	1 Plum	27	.296	Ea.	14.55	10.20		24.75	31.50
2330	1-1/4" pipe size	Q-1	44	.364		17.55	11.25		28.80	36.50
2340	1-1/2" pipe size		40	.400		16.20	12.40		28.60	36.50
2350	2" pipe size		36	.444		16.20	13.80		30	39
2360	2-1/2" pipe size		28	.571		18.90	17.70		36.60	47.50
2370	3" pipe size		20	.800		24.50	25		49.50	64
2380	3-1/2" pipe size		16	1		32.50	31		63.50	82.50
2390	4" pipe size		12	1.333		32.50	41.50		74	98.50
2400	5" pipe size		10	1.600		46	49.50		95.50	126
2410	6" pipe size	Q-2	14	1.714		52	55		107	140
2420	8" pipe size		12	2		82	64.50		146.50	188
2430	10" pipe size		10	2.400		142	77		219	273
2440	12" pipe size		8	3		237	96.50		333.50	405
2460	For galvanized flanges, add					95%				
2490	Blind flange									
2492	Cast iron									
2494	Black, 125 lb., per flange									
2496	1" pipe size	1 Plum	27	.296	Ea.	24	10.20		34.20	42
2500	1-1/2" pipe size	Q-1	40	.400		27	12.40		39.40	48
2502	2" pipe size		36	.444		30.50	13.80		44.30	54.50
2504	2-1/2" pipe size		28	.571		33.50	17.70		51.20	63
2506	3" pipe size		20	.800		41	25		66	82.50
2508	4" pipe size		12	1.333		52.50	41.50		94	120
2510	5" pipe size		10	1.600		83.50	49.50		133	167
2512	6" pipe size	Q-2	14	1.714		92.50	55		147.50	185
2514	8" pipe size		12	2		145	64.50		209.50	257
2516	10" pipe size		10	2.400		210	77		287	345
2518	12" pipe size		8	3		298	96.50		394.50	475
2520	For galvanized flanges, add					80%				
2570	Threaded flange									
2580	Forged steel,									
2590	Black 150 lb., per flange									
2600	1/2" pipe size	1 Plum	30	.267	Ea.	21.50	9.20		30.70	38
2610	3/4" pipe size		28	.286		21.50	9.85		31.35	39
2620	1" pipe size		27	.296		21.50	10.20		31.70	39.50
2630	1-1/4" pipe size	Q-1	44	.364		21.50	11.25		32.75	41
2640	1-1/2" pipe size		40	.400		21.50	12.40		33.90	42.50
2650	2" pipe size		36	.444		21.50	13.80		35.30	45
2660	2-1/2" pipe size		28	.571		26.50	17.70		44.20	55.50
2670	3" pipe size		20	.800		26.50	25		51.50	66.50
2690	4" pipe size		12	1.333		31	41.50		72.50	96.50
2700	5" pipe size		10	1.600		45	49.50		94.50	125
2710	6" pipe size	Q-2	14	1.714		51	55		106	139
2720	8" pipe size		12	2		86	64.50		150.50	192
2730	10" pipe size		10	2.400		156	77		233	288
2860	Black 300 lb., per flange									
2870	1/2" pipe size	1 Plum	30	.267	Ea.	24	9.20		33.20	40.50
2880	3/4" pipe size		28	.286		24	9.85		33.85	41.50
2890	1" pipe size		27	.296		24	10.20		34.20	42
2900	1-1/4" pipe size	Q-1	44	.364		24	11.25		35.25	43.50
2910	1-1/2" pipe size		40	.400		24	12.40		36.40	45
2920	2" pipe size		36	.444		25	13.80		38.80	48.50
2930	2-1/2" pipe size		28	.571		35	17.70		52.70	65
2940	3" pipe size		20	.800		37.50	25		62.50	79

660

MECHANICAL 15

	15107 | Metal Pipe & Fittings		CREW	DAILY OUTPUT	LABOR-HOURS	UNIT	2001 BARE COSTS MAT.	LABOR	EQUIP.	TOTAL	TOTAL INCL O&P	
660	2960	4″ pipe size	Q-1	12	1.333	Ea.	53	41.50		94.50	121	660
	2970	6″ pipe size	Q-2	14	1.714		93.50	55		148.50	186	
	3000	Weld joint, butt, carbon steel, standard weight										
	3040	90° elbow, long radius										
	3050	1/2″ pipe size	Q-15	16	1	Ea.	12.50	31	6.30	49.80	67.50	
	3060	3/4″ pipe size		16	1		12.50	31	6.30	49.80	67.50	
	3070	1″ pipe size		16	1		6.40	31	6.30	43.70	61	
	3080	1-1/4″ pipe size		14	1.143		6.40	35.50	7.20	49.10	68.50	
	3090	1-1/2″ pipe size		13	1.231		6.40	38	7.75	52.15	73	
	3100	2″ pipe size		10	1.600		6.40	49.50	10.05	65.95	93	
	3110	2-1/2″ pipe size		8	2		8.65	62	12.60	83.25	117	
	3120	3″ pipe size		7	2.286		9.90	71	14.40	95.30	134	
	3130	4″ pipe size		5	3.200		16.95	99	20	135.95	191	
	3136	5″ pipe size		4	4		35.50	124	25	184.50	254	
	3140	6″ pipe size	Q-16	5	4.800		40.50	154	20	214.50	300	
	3150	8″ pipe size		3.75	6.400		76	206	27	309	425	
	3160	10″ pipe size		3	8		139	257	33.50	429.50	580	
	3170	12″ pipe size		2.50	9.600		199	310	40	549	730	
	3180	14″ pipe size		2	12		275	385	50.50	710.50	940	
	3190	16″ pipe size		1.50	16		375	515	67	957	1,275	
	3191	18″ pipe size		1.25	19.200		485	615	80.50	1,180.50	1,550	
	3192	20″ pipe size		1.15	20.870		715	670	87.50	1,472.50	1,900	
	3194	24″ pipe size		1.02	23.529		945	755	98.50	1,798.50	2,300	
	3200	45° Elbow, long										
	3210	1/2″ pipe size	Q-15	16	1	Ea.	11.85	31	6.30	49.15	67	
	3220	3/4″ pipe size		16	1		11.85	31	6.30	49.15	67	
	3230	1″ pipe size		16	1		6.10	31	6.30	43.40	60.50	
	3240	1-1/4″ pipe size		14	1.143		6.10	35.50	7.20	48.80	68	
	3250	1-1/2″ pipe size		13	1.231		6.10	38	7.75	51.85	72.50	
	3260	2″ pipe size		10	1.600		6.10	49.50	10.05	65.65	93	
	3270	2-1/2″ pipe size		8	2		7.35	62	12.60	81.95	115	
	3280	3″ pipe size		7	2.286		8.65	71	14.40	94.05	132	
	3290	4″ pipe size		5	3.200		15.05	99	20	134.05	189	
	3296	5″ pipe size		4	4		22	124	25	171	239	
	3300	6″ pipe size	Q-16	5	4.800		32	154	20	206	290	
	3310	8″ pipe size		3.75	6.400		54	206	27	287	400	
	3320	10″ pipe size		3	8		98	257	33.50	388.50	535	
	3330	12″ pipe size		2.50	9.600		139	310	40	489	660	
	3340	14″ pipe size		2	12		189	385	50.50	624.50	845	
	3341	16″ pipe size		1.50	16		258	515	67	840	1,125	
	3342	18″ pipe size		1.25	19.200		345	615	80.50	1,040.50	1,400	
	3343	20″ pipe size		1.15	20.870		455	670	87.50	1,212.50	1,625	
	3345	24″ pipe size		1.05	22.857		660	735	96	1,491	1,925	
	3346	26″ pipe size		.85	28.235		1,325	910	118	2,353	2,975	
	3347	30″ pipe size		.45	53.333		1,475	1,725	223	3,423	4,475	
	3349	36″ pipe size		.38	63.158		2,450	2,025	265	4,740	6,075	
	3350	Tee, straight										
	3352	For reducing tees and concentrics see starting line 4600										
	3360	1/2″ pipe size	Q-15	10	1.600	Ea.	30.50	49.50	10.05	90.05	120	
	3370	3/4″ pipe size		10	1.600		16.95	49.50	10.05	76.50	105	
	3380	1″ pipe size		10	1.600		16.95	49.50	10.05	76.50	105	
	3390	1-1/4″ pipe size		9	1.778		19.50	55	11.20	85.70	117	
	3400	1-1/2″ pipe size		8	2		19.50	62	12.60	94.10	129	
	3410	2″ pipe size		6	2.667		16.95	82.50	16.75	116.20	162	
	3420	2-1/2″ pipe size		5	3.200		21.50	99	20	140.50	196	
	3430	3″ pipe size		4	4		23.50	124	25	172.50	241	

15107 | Metal Pipe & Fittings

	15107 Metal Pipe & Fittings	CREW	DAILY OUTPUT	LABOR-HOURS	UNIT	2001 BARE COSTS MAT.	LABOR	EQUIP.	TOTAL	TOTAL INCL O&P
660 3440	4" pipe size	Q-15	3	5.333	Ea.	33.50	165	33.50	232	325
3446	5" pipe size		2.50	6.400		61	198	40.50	299.50	410
3450	6" pipe size	Q-16	3	8		61	257	33.50	351.50	495
3460	8" pipe size		2.50	9.600		111	310	40	461	635
3470	10" pipe size		2	12		186	385	50.50	621.50	840
3480	12" pipe size		1.60	15		280	480	63	823	1,100
3481	14" pipe size		1.30	18.462		470	595	77.50	1,142.50	1,500
3482	16" pipe size		1	24		565	770	101	1,436	1,900
3483	18" pipe size		.80	30		895	965	126	1,986	2,575
3484	20" pipe size		.75	32		1,375	1,025	134	2,534	3,225
3486	24" pipe size		.70	34.286		1,725	1,100	144	2,969	3,725
3487	26" pipe size		.55	43.636		2,400	1,400	183	3,983	4,975
3488	30" pipe size		.30	80		2,600	2,575	335	5,510	7,100
3490	36" pipe size		.20	120		4,100	3,850	505	8,455	10,900
3491	Eccentric reducer, 1-1/2" pipe size	Q-15	14	1.143		17.30	35.50	7.20	60	80.50
3492	2" pipe size		11	1.455		17.30	45	9.15	71.45	97
3493	2-1/2" pipe size		9	1.778		14.70	55	11.20	80.90	112
3494	3" pipe size		8	2		16.65	62	12.60	91.25	126
3495	4" pipe size		6	2.667		17.90	82.50	16.75	117.15	163
3496	6" pipe size	Q-16	5	4.800		36	154	20	210	295
3497	8" pipe size		4	6		70	193	25	288	395
3498	10" pipe size		3	8		91	257	33.50	381.50	525
3499	12" pipe size		2.50	9.600		148	310	40	498	675
3501	Cap, 1-1/2" pipe size	Q-15	28	.571		7.05	17.70	3.59	28.34	38
3502	2" pipe size		22	.727		6.40	22.50	4.57	33.47	46
3503	2-1/2" pipe size		18	.889		6.40	27.50	5.60	39.50	54.50
3504	3" pipe size		16	1		6.40	31	6.30	43.70	61
3505	4" pipe size		12	1.333		8.30	41.50	8.40	58.20	81
3506	6" pipe size	Q-16	10	2.400		15.70	77	10.05	102.75	144
3507	8" pipe size		8	3		24	96.50	12.55	133.05	186
3508	10" pipe size		6	4		42	129	16.75	187.75	259
3509	12" pipe size		5	4.800		52.50	154	20	226.50	315
3511	14" pipe size		4	6		96.50	193	25	314.50	425
3512	16" pipe size		4	6		113	193	25	331	445
3513	18" pipe size		3	8		130	257	33.50	420.50	570
3517	Weld joint, butt, carbon steel, extra strong									
3519	90° elbow, long									
3520	1/2" pipe size	Q-15	13	1.231	Ea.	15.70	38	7.75	61.45	83.50
3530	3/4" pipe size		12	1.333		15.70	41.50	8.40	65.60	89
3540	1" pipe size		11	1.455		8.30	45	9.15	62.45	87
3550	1-1/4" pipe size		10	1.600		8.30	49.50	10.05	67.85	95
3560	1-1/2" pipe size		9	1.778		8.30	55	11.20	74.50	104
3570	2" pipe size		8	2		8.30	62	12.60	82.90	117
3580	2-1/2" pipe size		7	2.286		12.15	71	14.40	97.55	136
3590	3" pipe size		6	2.667		15.05	82.50	16.75	114.30	160
3600	4" pipe size		4	4		25.50	124	25	174.50	243
3606	5" pipe size		3.50	4.571		50.50	142	29	221.50	300
3610	6" pipe size	Q-16	4.50	5.333		61	171	22.50	254.50	350
3620	8" pipe size		3.50	6.857		115	220	28.50	363.50	495
3630	10" pipe size		2.50	9.600		206	310	40	556	735
3640	12" pipe size		2.25	10.667		291	345	44.50	680.50	890
3650	45° Elbow, long									
3660	1/2" pipe size	Q-15	13	1.231	Ea.	15.70	38	7.75	61.45	83.50
3670	3/4" pipe size		12	1.333		15.70	41.50	8.40	65.60	89
3680	1" pipe size		11	1.455		8.30	45	9.15	62.45	87
3690	1-1/4" pipe size		10	1.600		8.30	49.50	10.05	67.85	95

MECHANICAL 15

15107 | Metal Pipe & Fittings

			DAILY	LABOR-		2001 BARE COSTS				TOTAL	
		CREW	OUTPUT	HOURS	UNIT	MAT.	LABOR	EQUIP.	TOTAL	INCL O&P	
660	3700 1-1/2" pipe size	Q-15	9	1.778	Ea.	8.30	55	11.20	74.50	104	660
	3710 2" pipe size		8	2		7.70	62	12.60	82.30	116	
	3720 2-1/2" pipe size		7	2.286		16.30	71	14.40	101.70	141	
	3730 3" pipe size		6	2.667		18.25	82.50	16.75	117.50	163	
	3740 4" pipe size		4	4		17.90	124	25	166.90	234	
	3746 5" pipe size		3.50	4.571		41.50	142	29	212.50	292	
	3750 6" pipe size	Q-16	4.50	5.333		47.50	171	22.50	241	335	
	3760 8" pipe size		3.50	6.857		77.50	220	28.50	326	450	
	3770 10" pipe size		2.50	9.600		125	310	40	475	645	
	3780 12" pipe size		2.25	10.667		186	345	44.50	575.50	775	
	3800 Tee, straight										
	3810 1/2" pipe size	Q-15	9	1.778	Ea.	38	55	11.20	104.20	137	
	3820 3/4" pipe size		8.50	1.882		19.50	58.50	11.85	89.85	123	
	3830 1" pipe size		8	2		20	62	12.60	94.60	129	
	3840 1-1/4" pipe size		7	2.286		20	71	14.40	105.40	145	
	3850 1-1/2" pipe size		6	2.667		23.50	82.50	16.75	122.75	169	
	3860 2" pipe size		5	3.200		20	99	20	139	194	
	3870 2-1/2" pipe size		4	4		35.50	124	25	184.50	254	
	3880 3" pipe size		3.50	4.571		42.50	142	29	213.50	293	
	3890 4" pipe size		2.50	6.400		54	198	40.50	292.50	405	
	3896 5" pipe size		2.25	7.111		110	220	44.50	374.50	505	
	3900 6" pipe size	Q-16	2.25	10.667		84.50	345	44.50	474	660	
	3910 8" pipe size		2	12		172	385	50.50	607.50	825	
	3920 10" pipe size		1.75	13.714		270	440	57.50	767.50	1,025	
	3930 12" pipe size		1.50	16		405	515	67	987	1,300	
	4190 Weld fittings, reducing, standard weight										
	4200 Welding ring w/spacer pins, 2" pipe size				Ea.	2			2	2.20	
	4210 2-1/2" pipe size					2.10			2.10	2.31	
	4220 3" pipe size					2.50			2.50	2.75	
	4230 4" pipe size					2.70			2.70	2.97	
	4236 5" pipe size					2.95			2.95	3.25	
	4240 6" pipe size					3.40			3.40	3.74	
	4250 8" pipe size					5.65			5.65	6.20	
	4260 10" pipe size					6.80			6.80	7.50	
	4270 12" pipe size					7.75			7.75	8.55	
	4280 14" pipe size					8.60			8.60	9.45	
	4290 16" pipe size					9.85			9.85	10.85	
	4300 18" pipe size					10.35			10.35	11.40	
	4310 20" pipe size					12			12	13.20	
	4330 24" pipe size					13.95			13.95	15.35	
	4340 26" pipe size					14			14	15.40	
	4350 30" pipe size					17.65			17.65	19.40	
	4370 36" pipe size					24.50			24.50	27	
	4600 Tee, reducing on outlet										
	4601 2-1/2" x 2" pipe size	Q-15	5	3.200	Ea.	30.50	99	20	149.50	206	
	4602 3" x 2-1/2" pipe size		4	4		30.50	124	25	179.50	248	
	4603 3-1/2" x 3" pipe size		3.50	4.571		19.85	142	29	190.85	268	
	4604 4" x 3" pipe size		3	5.333		38.50	165	33.50	237	330	
	4605 5" x 4" pipe size		2.50	6.400		84	198	40.50	322.50	435	
	4606 6" x 5" pipe size	Q-16	3	8		76	257	33.50	366.50	510	
	4607 8" x 6" pipe size		2.50	9.600		145	310	40	495	670	
	4608 10" x 8" pipe size		2	12		245	385	50.50	680.50	905	
	4609 12" x 10" pipe size		1.60	15		253	480	63	796	1,075	
	4610 16" x 12" pipe size		1.50	16		590	515	67	1,172	1,500	
	4611 14" x 12" pipe size		1.52	15.789		505	510	66	1,081	1,400	
	4618 Reducer, concentric										

15107 | Metal Pipe & Fittings

			CREW	DAILY OUTPUT	LABOR-HOURS	UNIT	2001 BARE COSTS MAT.	LABOR	EQUIP.	TOTAL	TOTAL INCL O&P	
660	4619	2-1/2" by 2" pipe size	Q-15	10	1.600	Ea.	10.90	49.50	10.05	70.45	98	660
	4620	3" by 2-1/2" pipe size		9	1.778		8.65	55	11.20	74.85	105	
	4621	3-1/2" by 3" pipe size		8	2		20.50	62	12.60	95.10	130	
	4622	4" by 2-1/2" pipe size		7	2.286		13.45	71	14.40	98.85	138	
	4623	5" by 3" pipe size	↓	7	2.286		22	71	14.40	107.40	147	
	4624	6" by 4" pipe size	Q-16	6	4		26	129	16.75	171.75	241	
	4625	8" by 6" pipe size		5	4.800		35	154	20	209	294	
	4626	10" by 8" pipe size		4	6		52.50	193	25	270.50	375	
	4627	12" by 10" pipe size	↓	3	8	↓	82	257	33.50	372.50	515	
	4660	Reducer, eccentric										
	4662	3" x 2"	Q-15	8	2	Ea.	14.40	62	12.60	89	123	
	4664	4" x 3"		6	2.667		16.65	82.50	16.75	115.90	162	
	4666	4" x 2"	↓	6	2.667		22	82.50	16.75	121.25	167	
	4670	6" x 4"	Q-16	5	4.800		33.50	154	20	207.50	292	
	4672	6" x 3"		5	4.800		62.50	154	20	236.50	325	
	4676	8" x 6"		4	6		50.50	193	25	268.50	375	
	4678	8" x 4"		4	6		88	193	25	306	415	
	4682	10" x 8"		3	8		60	257	33.50	350.50	495	
	4684	10" x 6"		3	8		107	257	33.50	397.50	545	
	4688	12" x 10"		2.50	9.600		94.50	310	40	444.50	615	
	4690	12" x 8"		2.50	9.600		148	310	40	498	675	
	4691	14" x 12"		2.20	10.909		178	350	45.50	573.50	775	
	4693	16" x 14"		1.80	13.333		242	430	56	728	975	
	4694	16" x 12"		2	12		299	385	50.50	734.50	965	
	4696	18" x 16"	↓	1.60	15		395	480	63	938	1,225	
	5630	T-O-L, 1/4" pipe size, nozzle	Q-15	23	.696		4.31	21.50	4.38	30.19	42	
	5631	3/8" pipe size, nozzle		23	.696		4.31	21.50	4.38	30.19	42	
	5632	1/2" pipe size, nozzle		22	.727		4.31	22.50	4.57	31.38	44	
	5633	3/4" pipe size, nozzle		21	.762		5.05	23.50	4.79	33.34	46.50	
	5634	1" pipe size, nozzle		20	.800		5.70	25	5.05	35.75	49.50	
	5635	1-1/4" pipe size, nozzle		18	.889		8.05	27.50	5.60	41.15	56.50	
	5636	1-1/2" pipe size, nozzle		16	1		8.75	31	6.30	46.05	63.50	
	5637	2" pipe size, nozzle		12	1.333		9.05	41.50	8.40	58.95	81.50	
	5638	2-1/2" pipe size, nozzle		10	1.600		28.50	49.50	10.05	88.05	118	
	5639	4" pipe size, nozzle		6	2.667		64.50	82.50	16.75	163.75	214	
	5640	W-O-L, 1/4" pipe size, nozzle		23	.696		7.40	21.50	4.38	33.28	45.50	
	5641	3/8" pipe size, nozzle		23	.696		7.40	21.50	4.38	33.28	45.50	
	5642	1/2" pipe size, nozzle		22	.727		7.40	22.50	4.57	34.47	47	
	5643	3/4" pipe size, nozzle		21	.762		7.60	23.50	4.79	35.89	49	
	5644	1" pipe size, nozzle		20	.800		7.95	25	5.05	38	52	
	5645	1-1/4" pipe size, nozzle		18	.889		9.65	27.50	5.60	42.75	58.50	
	5646	1-1/2" pipe size, nozzle		16	1		9.65	31	6.30	46.95	64.50	
	5647	2" pipe size, nozzle		12	1.333		10.10	41.50	8.40	60	83	
	5648	2-1/2" pipe size, nozzle		10	1.600		23	49.50	10.05	82.55	111	
	5649	3" pipe size, nozzle		8	2		26.50	62	12.60	101.10	137	
	5650	4" pipe size, nozzle		6	2.667		34	82.50	16.75	133.25	181	
	5651	5" pipe size, nozzle		5	3.200		82.50	99	20	201.50	263	
	5652	6" pipe size, nozzle		4	4		105	124	25	254	330	
	5653	8" pipe size, nozzle		3	5.333		189	165	33.50	387.50	495	
	5654	10" pipe size, nozzle		2.60	6.154		275	191	38.50	504.50	630	
	5655	12" pipe size, nozzle		2.20	7.273		430	225	46	701	865	
	5656	14" pipe size, nozzle		2	8		800	248	50.50	1,098.50	1,300	
	5660	L-O-L, 1/4" pipe size, outlet		23	.696		47	21.50	4.38	72.88	89	
	5661	3/8" pipe size, outlet		23	.696		47	21.50	4.38	72.88	89	
	5662	1/2" pipe size, outlet		22	.727		47	22.50	4.57	74.07	90.50	
	5663	3/4" pipe size, outlet	↓	21	.762	↓	51.50	23.50	4.79	79.79	98	

MECHANICAL 15

15107 | Metal Pipe & Fittings

			CREW	DAILY OUTPUT	LABOR-HOURS	UNIT	2001 BARE COSTS MAT.	LABOR	EQUIP.	TOTAL	TOTAL INCL O&P	
660	5664	1" pipe size, outlet	Q-15	20	.800	Ea.	69.50	25	5.05	99.55	120	660
	5665	1-1/4" pipe size, outlet		18	.889		108	27.50	5.60	141.10	167	
	5666	1-1/2" pipe size, outlet		16	1		108	31	6.30	145.30	173	
	5667	2" pipe size, outlet		12	1.333		171	41.50	8.40	220.90	261	
	5674	S-O-L, 1/4" pipe size, outlet		23	.696		4.81	21.50	4.38	30.69	42.50	
	5675	3/8" pipe size, outlet		23	.696		4.81	21.50	4.38	30.69	42.50	
	5676	1/2" pipe size, outlet		22	.727		4.81	22.50	4.57	31.88	44.50	
	5677	3/4" pipe size, outlet		21	.762		4.88	23.50	4.79	33.17	46	
	5678	1" pipe size, outlet		20	.800		5.30	25	5.05	35.35	49	
	5679	1-1/4" pipe size, outlet		18	.889		8.20	27.50	5.60	41.30	56.50	
	5680	1-1/2" pipe size, outlet		16	1		8.60	31	6.30	45.90	63.50	
	5681	2" pipe size, outlet		12	1.333		9.65	41.50	8.40	59.55	82.50	
	6000	Weld-on flange, forged steel										
	6020	Slip-on, 150 lb. flange (welded front and back)										
	6050	1/2" pipe size	Q-15	18	.889	Ea.	12.15	27.50	5.60	45.25	61	
	6060	3/4" pipe size		18	.889		12.15	27.50	5.60	45.25	61	
	6070	1" pipe size		17	.941		11.20	29	5.90	46.10	63	
	6080	1-1/4" pipe size		16	1		11.20	31	6.30	48.50	66	
	6090	1-1/2" pipe size		15	1.067		11.20	33	6.70	50.90	69.50	
	6100	2" pipe size		12	1.333		11.20	41.50	8.40	61.10	84	
	6110	2-1/2" pipe size		10	1.600		13.45	49.50	10.05	73	101	
	6120	3" pipe size		9	1.778		15.70	55	11.20	81.90	113	
	6130	3-1/2" pipe size		7	2.286		19.50	71	14.40	104.90	144	
	6140	4" pipe size		6	2.667		19.50	82.50	16.75	118.75	165	
	6150	5" pipe size		5	3.200		35	99	20	154	211	
	6160	6" pipe size	Q-16	6	4		31.50	129	16.75	177.25	247	
	6170	8" pipe size		5	4.800		48	154	20	222	310	
	6180	10" pipe size		4	6		83.50	193	25	301.50	410	
	6190	12" pipe size		3	8		124	257	33.50	414.50	565	
	6191	14" pipe size		2.50	9.600		164	310	40	514	690	
	6192	16" pipe size		1.80	13.333		271	430	56	757	1,000	
	6200	300 lb. flange										
	6210	1/2" pipe size	Q-15	17	.941	Ea.	20	29	5.90	54.90	72.50	
	6220	3/4" pipe size		17	.941		20	29	5.90	54.90	72.50	
	6230	1" pipe size		16	1		20	31	6.30	57.30	76	
	6240	1-1/4" pipe size		13	1.231		20	38	7.75	65.75	88	
	6250	1-1/2" pipe size		12	1.333		20	41.50	8.40	69.90	94	
	6260	2" pipe size		11	1.455		23	45	9.15	77.15	104	
	6270	2-1/2" pipe size		9	1.778		26	55	11.20	92.20	124	
	6280	3" pipe size		7	2.286		32	71	14.40	117.40	158	
	6290	4" pipe size		6	2.667		42.50	82.50	16.75	141.75	190	
	6300	5" pipe size		4	4		54	124	25	203	274	
	6310	6" pipe size	Q-16	5	4.800		53	154	20	227	315	
	6320	8" pipe size		4	6		88.50	193	25	306.50	415	
	6330	10" pipe size		3.40	7.059		163	227	29.50	419.50	555	
	6340	12" pipe size		2.80	8.571		200	276	36	512	675	
	6400	Welding neck, 150 lb. flange										
	6410	1/2" pipe size	Q-15	40	.400	Ea.	16	12.40	2.52	30.92	39	
	6420	3/4" pipe size		36	.444		16	13.80	2.80	32.60	41.50	
	6430	1" pipe size		32	.500		16	15.50	3.15	34.65	44.50	
	6440	1-1/4" pipe size		29	.552		16	17.10	3.47	36.57	47.50	
	6450	1-1/2" pipe size		26	.615		16	19.10	3.87	38.97	51	
	6460	2" pipe size		20	.800		17.30	25	5.05	47.35	62	
	6470	2-1/2" pipe size		16	1		18.55	31	6.30	55.85	74.50	
	6480	3" pipe size		14	1.143		20.50	35.50	7.20	63.20	84	
	6500	4" pipe size		10	1.600		24.50	49.50	10.05	84.05	113	

15107 | Metal Pipe & Fittings

			CREW	DAILY OUTPUT	LABOR-HOURS	UNIT	2001 BARE COSTS MAT.	LABOR	EQUIP.	TOTAL	TOTAL INCL O&P	
660	6510	5″ pipe size	Q-15	8	2	Ea.	38.50	62	12.60	113.10	150	660
	6520	6″ pipe size	Q-16	10	2.400		37.50	77	10.05	124.55	168	
	6530	8″ pipe size		7	3.429		65.50	110	14.35	189.85	254	
	6540	10″ pipe size		6	4		105	129	16.75	250.75	330	
	6550	12″ pipe size		5	4.800		153	154	20	327	425	
	6551	14″ pipe size		4.50	5.333		221	171	22.50	414.50	525	
	6552	16″ pipe size		3	8		375	257	33.50	665.50	840	
	6553	18″ pipe size		2.50	9.600		530	310	40	880	1,100	
	6554	20″ pipe size		2.30	10.435		645	335	43.50	1,023.50	1,275	
	6556	24″ pipe size		2	12		865	385	50.50	1,300.50	1,575	
	6557	26″ pipe size		1.70	14.118		1,475	455	59	1,989	2,375	
	6558	30″ pipe size		.90	26.667		1,925	855	112	2,892	3,550	
	6559	36″ pipe size	↓	.75	32	↓	2,200	1,025	134	3,359	4,125	
	6560	300 lb. flange										
	6570	1/2″ pipe size	Q-15	36	.444	Ea.	20	13.80	2.80	36.60	46	
	6580	3/4″ pipe size		34	.471		20	14.60	2.96	37.56	47.50	
	6590	1″ pipe size		30	.533		20	16.55	3.35	39.90	50.50	
	6600	1-1/4″ pipe size		28	.571		20	17.70	3.59	41.29	52.50	
	6610	1-1/2″ pipe size		24	.667		20	20.50	4.19	44.69	57.50	
	6620	2″ pipe size		18	.889		23	27.50	5.60	56.10	73	
	6630	2-1/2″ pipe size		14	1.143		26	35.50	7.20	68.70	90	
	6640	3″ pipe size		12	1.333		26	41.50	8.40	75.90	100	
	6650	4″ pipe size		8	2		42.50	62	12.60	117.10	154	
	6660	5″ pipe size	↓	7	2.286		66	71	14.40	151.40	195	
	6670	6″ pipe size	Q-16	9	2.667		62	85.50	11.15	158.65	210	
	6680	8″ pipe size		6	4		107	129	16.75	252.75	330	
	6690	10″ pipe size		5	4.800		207	154	20	381	485	
	6700	12″ pipe size	↓	4	6	↓	265	193	25	483	610	
	7740	Plain ends for plain end pipe, mechanically coupled										
	7750	Cplg & labor required at joints not included add 1 per										
	7760	joint for installed price, see line 9180										
	7770	Malleable iron, painted, unless noted otherwise										
	7800	90° Elbow 1″				Ea.	22.50			22.50	25	
	7810	1-1/2″					22.50			22.50	25	
	7820	2″					22.50			22.50	25	
	7830	2-1/2″					22.50			22.50	25	
	7840	3″					27			27	30	
	7860	4″					39.50			39.50	43.50	
	7870	5″ welded steel					137			137	151	
	7880	6″					109			109	120	
	7890	8″ welded steel					230			230	253	
	7900	10″ welded steel					289			289	320	
	7910	12″ welded steel					370			370	405	
	7970	45° Elbow 1″					22.50			22.50	25	
	7980	1-1/2″					22.50			22.50	25	
	7990	2″					22.50			22.50	25	
	8000	2-1/2″					22.50			22.50	25	
	8010	3″					27			27	30	
	8030	4″					39.50			39.50	43.50	
	8040	5″ welded steel					137			137	151	
	8050	6″					109			109	120	
	8060	8″					126			126	139	
	8070	10″ welded steel					176			176	194	
	8080	12″ welded steel					228			228	251	
	8140	Tee, straight 1″					30.50			30.50	33.50	
	8150	1-1/2″				↓	30.50			30.50	33.50	

15107 | Metal Pipe & Fittings		CREW	DAILY OUTPUT	LABOR-HOURS	UNIT	2001 BARE COSTS MAT.	LABOR	EQUIP.	TOTAL	TOTAL INCL O&P
660	8160 2"				Ea.	30.50			30.50	33.50
	8170 2-1/2"					30.50			30.50	33.50
	8180 3"					44			44	48
	8200 4"					61			61	67
	8210 5" welded steel					134			134	147
	8220 6"					165			165	182
	8230 8" welded steel					229			229	252
	8240 10" welded steel					350			350	385
	8250 12" welded steel				↓	475			475	520
	8340 Segmentally welded steel, painted									
	8390 Wye 2"				Ea.	89			89	98
	8400 2-1/2"					89			89	98
	8410 3"					100			100	110
	8430 4"					149			149	164
	8440 5"					181			181	199
	8450 6"					249			249	274
	8460 8"					325			325	355
	8470 10"					475			475	525
	8480 12"					735			735	805
	8540 Wye, lateral 2"					95.50			95.50	105
	8550 2-1/2"					95.50			95.50	105
	8560 3"					106			106	117
	8580 4"					161			161	177
	8590 5"					209			209	230
	8600 6"					267			267	294
	8610 8"					390			390	430
	8620 10"					550			550	605
	8630 12"					745			745	820
	8690 Cross, 2"					101			101	111
	8700 2-1/2"					101			101	111
	8710 3"					120			120	132
	8730 4"					165			165	182
	8740 5"					235			235	258
	8750 6"					310			310	340
	8760 8"					395			395	435
	8770 10"					580			580	640
	8780 12"					835			835	920
	8800 Tees, reducing 2" x 1"					60			60	66
	8810 2" x 1-1/2"					60			60	66
	8820 3" x 1"					60			60	66
	8830 3" x 1-1/2"					60			60	66
	8840 3" x 2"					60			60	66
	8850 4" x 1"					87.50			87.50	96.50
	8860 4" x 1-1/2"					87.50			87.50	96.50
	8870 4" x 2"					87.50			87.50	96.50
	8880 4" x 2-1/2"					90			90	99
	8890 4" x 3"					90			90	99
	8900 6" x 2"					146			146	160
	8910 6" x 3"					154			154	169
	8920 6" x 4"					154			154	169
	8930 8" x 2"					187			187	206
	8940 8" x 3"					196			196	216
	8950 8" x 4"					200			200	220
	8960 8" x 5"					205			205	226
	8970 8" x 6"					205			205	226
	8980 10" x 4"				↓	294			294	325

15 MECHANICAL

Important: See the Reference Section for critical supporting data - Reference Nos., Crews, & City Cost Indexes

15107 | Metal Pipe & Fittings

		Description	CREW	DAILY OUTPUT	LABOR-HOURS	UNIT	2001 BARE COSTS MAT.	LABOR	EQUIP.	TOTAL	TOTAL INCL O&P	
660	8990	10" x 6"				Ea.	310			310	345	660
	9000	10" x 8"					310			310	345	
	9010	12" x 6"					455			455	500	
	9020	12" x 8"					455			455	500	
	9030	12" x 10"				↓	470			470	520	
	9080	Adapter nipples 3" long										
	9090	1"				Ea.	7.80			7.80	8.60	
	9100	1-1/2"					7.80			7.80	8.60	
	9110	2"					7.80			7.80	8.60	
	9120	2-1/2"					9.15			9.15	10.10	
	9130	3"					11.30			11.30	12.45	
	9140	4"					18.95			18.95	21	
	9150	6"				↓	51			51	56.50	
	9180	Coupling, mechanical, plain end pipe to plain end pipe or fitting										
	9190	1"	Q-1	29	.552	Ea.	24	17.10		41.10	52.50	
	9200	1-1/2"		28	.571		28.50	17.70		46.20	58	
	9210	2"		27	.593		28.50	18.35		46.85	59	
	9220	2-1/2"		26	.615		28.50	19.10		47.60	60.50	
	9230	3"		25	.640		42.50	19.85		62.35	76.50	
	9240	3-1/2"		24	.667		49	20.50		69.50	85	
	9250	4"	↓	22	.727		49	22.50		71.50	88	
	9260	5"	Q-2	28	.857		70	27.50		97.50	119	
	9270	6"		24	1		85.50	32		117.50	143	
	9280	8"		19	1.263		152	40.50		192.50	229	
	9290	10"		16	1.500		197	48		245	290	
	9300	12"	↓	12	2	↓	248	64.50		312.50	370	
	9310	Outlets for precut holes through pipe wall										
	9331	Strapless type, with gasket										
	9332	4" to 8" pipe x 1/2"	1 Plum	13	.615	Ea.	28.50	21		49.50	63.50	
	9333	4" to 8" pipe x 3/4"		13	.615		30.50	21		51.50	65.50	
	9334	10" pipe and larger x 1/2"		11	.727		28.50	25		53.50	69.50	
	9335	10" pipe and larger x 3/4"	↓	11	.727	↓	30.50	25		55.50	71.50	
	9341	Thermometer wells with gasket										
	9342	4" to 8" pipe, 6" stem	1 Plum	14	.571	Ea.	43.50	19.70		63.20	77.50	
	9343	8" pipe and larger, 6" stem	"	13	.615	"	43.50	21		64.50	80	
	9400	Mechanical joint ends for plain end pipe										
	9410	Malleable iron, black										
	9420	90° Elbows, 1-1/4"	Q-1	29	.552	Ea.	10.40	17.10		27.50	37.50	
	9430	1-1/2"		27	.593		11.40	18.35		29.75	40	
	9440	2"	↓	24	.667	↓	13.45	20.50		33.95	46	
	9490	Tee, reducing outlet										
	9510	1-1/4" x 1/2"	Q-1	18	.889	Ea.	8.30	27.50		35.80	50.50	
	9520	1-1/4" x 3/4"		18	.889		8.30	27.50		35.80	50.50	
	9530	1-1/4" x 1"		18	.889		8.30	27.50		35.80	50.50	
	9540	1-1/2" x 1/2"		17	.941		8.80	29		37.80	53.50	
	9550	1-1/2" x 3/4"		17	.941		8.80	29		37.80	53.50	
	9560	1-1/2" x 1"		17	.941		8.80	29		37.80	53.50	
	9570	2" x 1/2"		15	1.067		10.80	33		43.80	62	
	9580	2" x 3/4"		15	1.067		10.80	33		43.80	62	
	9590	2" x 1"	↓	15	1.067	↓	10.80	33		43.80	62	
	9640	Tee, reducing run and outlet										
	9660	1-1/4" x 1" x 1/2"	Q-1	18	.889	Ea.	8.90	27.50		36.40	51.50	
	9670	1-1/4" x 1" x 3/4"		18	.889		8.90	27.50		36.40	51.50	
	9680	1-1/4" x 1" x 1"		18	.889		8.90	27.50		36.40	51.50	
	9690	1-1/2" x 1-1/4" x 1/2"		17	.941		9.45	29		38.45	54.50	
	9700	1-1/2" x 1-1/4" x 3/4"	↓	17	.941	↓	9.45	29		38.45	54.50	

	15107 | Metal Pipe & Fittings		CREW	DAILY OUTPUT	LABOR-HOURS	UNIT	2001 BARE COSTS MAT.	LABOR	EQUIP.	TOTAL	TOTAL INCL O&P	
660	9710	1-1/2" x 1-1/4" x 1"	Q-1	17	.941	Ea.	9.45	29		38.45	54.50	660
	9720	2" x 1-1/2" x 1/2"		15	1.067		11.40	33		44.40	62.50	
	9730	2" x 1-1/2" x 3/4"		15	1.067		11.40	33		44.40	62.50	
	9740	2" x 1-1/2" x 1"	↓	15	1.067	↓	11.40	33		44.40	62.50	
	9790	Tee, outlet										
	9810	3" x 1-1/4"	Q-1	29	.552	Ea.	20	17.10		37.10	48	
	9820	3" x 1-1/2"		28	.571		20	17.70		37.70	48.50	
	9830	3" x 2"		26	.615		22.50	19.10		41.60	54	
	9840	4" x 1-1/4"		28	.571		25	17.70		42.70	54	
	9850	4" x 1-1/2"		26	.615		26.50	19.10		45.60	58	
	9860	4" x 2"	↓	24	.667	↓	26.50	20.50		47	60	
	9940	For galvanized fittings for plain end pipe, add					20%					
690	0010	**PIPE, GROOVED-JOINT STEEL FITTINGS & VALVES** R15100-070										690
	0020	Pipe includes coupling & clevis type hanger 10' O.C.										
	0500	Schedule 10, black										
	0550	2" diameter	1 Plum	43	.186	L.F.	3.04	6.40		9.44	13	
	0560	2-1/2" diameter	Q-1	61	.262		3.80	8.15		11.95	16.50	
	0570	3" diameter		55	.291		4.35	9		13.35	18.40	
	0580	3-1/2" diameter		53	.302		5.60	9.35		14.95	20.50	
	0590	4" diameter		49	.327		5.70	10.10		15.80	21.50	
	0600	5" diameter	↓	40	.400		9.15	12.40		21.55	29	
	0610	6" diameter	Q-2	46	.522		10.05	16.75		26.80	36.50	
	0620	8" diameter	"	41	.585	↓	19.15	18.80		37.95	49.50	
	0700	To delete couplings & hangers, subtract										
	0710	2" diam. to 5" diam.					25%	20%				
	0720	6" diam. to 8" diam.					27%	15%				
	1000	Schedule 40, black										
	1040	3/4" diameter	1 Plum	71	.113	L.F.	1.92	3.88		5.80	7.95	
	1050	1" diameter		63	.127		2.03	4.37		6.40	8.85	
	1060	1-1/4" diameter		58	.138		2.51	4.75		7.26	9.90	
	1070	1-1/2" diameter		51	.157		2.79	5.40		8.19	11.20	
	1080	2" diameter	↓	40	.200		3.21	6.90		10.11	13.95	
	1090	2-1/2" diameter	Q-1	57	.281		4.71	8.70		13.41	18.35	
	1100	3" diameter		50	.320		5.30	9.90		15.20	21	
	1110	4" diameter		45	.356		7.55	11		18.55	25	
	1120	5" diameter	↓	37	.432		12.75	13.40		26.15	34	
	1130	6" diameter	Q-2	42	.571		16.55	18.35		34.90	45.50	
	1140	8" diameter		37	.649		25	21		46	59	
	1150	10" diameter		31	.774		37.50	25		62.50	78.50	
	1160	12" diameter		27	.889		49	28.50		77.50	97	
	1170	14" diameter		20	1.200		53.50	38.50		92	117	
	1180	16" diameter		17	1.412		57.50	45.50		103	132	
	1190	18" diameter		14	1.714		86.50	55		141.50	178	
	1200	20" diameter		12	2		113	64.50		177.50	221	
	1210	24" diameter	↓	10	2.400	↓	124	77		201	252	
	1740	To delete coupling & hanger, subtract										
	1750	3/4" diam. to 2" diam.					65%	27%				
	1760	2-1/2" diam. to 5" diam.					41%	18%				
	1770	6" diam. to 12" diam.					31%	13%				
	1780	14" diam. to 24" diam.					35%	10%				
	1800	Galvanized										
	1840	3/4" diameter	1 Plum	71	.113	L.F.	2.11	3.88		5.99	8.15	
	1850	1" diameter		63	.127		2.17	4.37		6.54	9	
	1860	1-1/4" diameter		58	.138		2.69	4.75		7.44	10.10	
	1870	1-1/2" diameter		51	.157		3.01	5.40		8.41	11.45	
	1880	2" diameter	↓	40	.200	↓	3.51	6.90		10.41	14.25	

15107 | Metal Pipe & Fittings

		Description	CREW	DAILY OUTPUT	LABOR-HOURS	UNIT	2001 BARE COSTS MAT.	LABOR	EQUIP.	TOTAL	TOTAL INCL O&P	
690	1890	2-1/2" diameter (R15100-070)	Q-1	57	.281	L.F.	4.87	8.70		13.57	18.50	690
	1900	3" diameter		50	.320		5.90	9.90		15.80	21.50	
	1910	4" diameter		45	.356		8.20	11		19.20	25.50	
	1920	5" diameter		37	.432		15.65	13.40		29.05	37.50	
	1930	6" diameter	Q-2	42	.571		20.50	18.35		38.85	50	
	1940	8" diameter		37	.649		30.50	21		51.50	65.50	
	1950	10" diameter		31	.774		46	25		71	88	
	1960	12" diameter		27	.889		61	28.50		89.50	110	
	2540	To delete coupling & hanger, subtract										
	2550	3/4" diam. to 2" diam.					36%	27%				
	2560	2-1/2" diam. to 5" diam.					19%	18%				
	2570	6" diam. to 12" diam.					14%	13%				
	2600	Schedule 80, black										
	2610	3/4" diameter	1 Plum	65	.123	L.F.	2.15	4.24		6.39	8.75	
	2650	1" diameter		61	.131		2.31	4.52		6.83	9.35	
	2660	1-1/4" diameter		55	.145		2.92	5		7.92	10.75	
	2670	1-1/2" diameter		49	.163		3.30	5.60		8.90	12.10	
	2680	2" diameter		38	.211		3.96	7.25		11.21	15.30	
	2690	2-1/2" diameter	Q-1	54	.296		5.45	9.20		14.65	19.85	
	2700	3" diameter		48	.333		6.75	10.35		17.10	23	
	2710	4" diameter		44	.364		9.60	11.25		20.85	27.50	
	2720	5" diameter		35	.457		17.65	14.15		31.80	41	
	2730	6" diameter	Q-2	40	.600		27.50	19.30		46.80	59.50	
	2740	8" diameter		35	.686		49.50	22		71.50	87.50	
	2750	10" diameter		29	.828		77.50	26.50		104	125	
	2760	12" diameter		24	1		86	32		118	143	
	3240	To delete coupling & hanger, subtract										
	3250	3/4" diam. to 2" diam.					30%	25%				
	3260	2-1/2" diam. to 5" diam.					14%	17%				
	3270	6" diam. to 12" diam.					12%	12%				
	3300	Galvanized										
	3310	3/4" diameter	1 Plum	65	.123	L.F.	2.40	4.24		6.64	9.05	
	3350	1" diameter		61	.131		2.50	4.52		7.02	9.55	
	3360	1-1/4" diameter		55	.145		3.19	5		8.19	11.05	
	3370	1-1/2" diameter		46	.174		3.63	6		9.63	13.05	
	3380	2" diameter		38	.211		4.35	7.25		11.60	15.75	
	3390	2-1/2" diameter	Q-1	54	.296		6	9.20		15.20	20.50	
	3400	3" diameter		48	.333		7.60	10.35		17.95	24	
	3410	4" diameter		44	.364		10.80	11.25		22.05	29	
	3420	5" diameter		35	.457		21	14.15		35.15	44.50	
	3430	6" diameter	Q-2	40	.600		27.50	19.30		46.80	59.50	
	3440	8" diameter		35	.686		47.50	22		69.50	85.50	
	3450	10" diameter		29	.828		75	26.50		101.50	123	
	3460	12" diameter		24	1		83.50	32		115.50	140	
	3920	To delete coupling & hanger, subtract										
	3930	3/4" diam. to 2" diam.					30%	25%				
	3940	2-1/2" diam. to 5" diam.					15%	17%				
	3950	6" diam. to 12" diam.					11%	12%				
	3990	Fittings: cplg. & labor required at joints not incl. in fitting										
	3994	price. Add 1 per joint for installed price.										
	4000	Elbow, 90° or 45°, painted										
	4030	3/4" diameter	1 Plum	50	.160	Ea.	13.65	5.50		19.15	23.50	
	4040	1" diameter		50	.160		13.65	5.50		19.15	23.50	
	4050	1-1/4" diameter		40	.200		13.65	6.90		20.55	25.50	
	4060	1-1/2" diameter		33	.242		13.65	8.35		22	27.50	
	4070	2" diameter		25	.320		13.65	11		24.65	31.50	

MECHANICAL 15

15107 | Metal Pipe & Fittings

690			CREW	DAILY OUTPUT	LABOR-HOURS	UNIT	2001 BARE COSTS MAT.	LABOR	EQUIP.	TOTAL	TOTAL INCL O&P	690
	4080	2-1/2" diameter R15100-070	Q-1	40	.400	Ea.	13.65	12.40		26.05	33.50	
	4090	3" diameter		33	.485		24.50	15.05		39.55	49	
	4100	4" diameter		25	.640		26.50	19.85		46.35	59	
	4110	5" diameter	↓	20	.800		64	25		89	108	
	4120	6" diameter	Q-2	25	.960		75	31		106	129	
	4130	8" diameter		21	1.143		157	36.50		193.50	229	
	4140	10" diameter		18	1.333		286	43		329	380	
	4150	12" diameter		15	1.600		455	51.50		506.50	585	
	4170	14" diameter		12	2		820	64.50		884.50	995	
	4180	16" diameter	↓	11	2.182		1,075	70		1,145	1,275	
	4190	18" diameter	Q-3	15	2.133		1,350	70		1,420	1,575	
	4200	20" diameter		13	2.462		1,775	81		1,856	2,075	
	4210	24" diameter	↓	11	2.909		2,575	95.50		2,670.50	2,975	
	4250	For galvanized elbows, add				↓	26%					
	4690	Tee, painted										
	4700	3/4" diameter	1 Plum	38	.211	Ea.	21	7.25		28.25	34	
	4740	1" diameter		33	.242		21	8.35		29.35	35.50	
	4750	1-1/4" diameter		27	.296		21	10.20		31.20	38.50	
	4760	1-1/2" diameter		22	.364		21	12.55		33.55	42	
	4770	2" diameter	↓	17	.471		21	16.20		37.20	47.50	
	4780	2-1/2" diameter	Q-1	27	.593		21	18.35		39.35	50.50	
	4790	3" diameter		22	.727		29.50	22.50		52	66.50	
	4800	4" diameter		17	.941		45	29		74	93.50	
	4810	5" diameter	↓	13	1.231		105	38		143	174	
	4820	6" diameter	Q-2	17	1.412		122	45.50		167.50	203	
	4830	8" diameter		14	1.714		268	55		323	375	
	4840	10" diameter		12	2		555	64.50		619.50	705	
	4850	12" diameter		10	2.400		775	77		852	970	
	4851	14" diameter		9	2.667		820	85.50		905.50	1,025	
	4852	16" diameter	↓	8	3		930	96.50		1,026.50	1,175	
	4853	18" diameter	Q-3	11	2.909		1,150	95.50		1,245.50	1,425	
	4854	20" diameter		10	3.200		1,675	105		1,780	1,975	
	4855	24" diameter	↓	8	4		2,525	131		2,656	3,000	
	4900	For galvanized tees, add				↓	24%					
	4906	Couplings, rigid style, painted										
	4908	1" diameter	1 Plum	100	.080	Ea.	9.90	2.76		12.66	15.05	
	4909	1-1/4" diameter		100	.080		9.90	2.76		12.66	15.05	
	4910	1-1/2" diameter		67	.119		9.90	4.11		14.01	17.10	
	4912	2" diameter	↓	50	.160		10.15	5.50		15.65	19.50	
	4914	2-1/2" diameter	Q-1	80	.200		11.70	6.20		17.90	22	
	4916	3" diameter		67	.239		13.65	7.40		21.05	26.50	
	4918	4" diameter		50	.320		19.35	9.90		29.25	36.50	
	4920	5" diameter	↓	40	.400		25	12.40		37.40	46	
	4922	6" diameter	Q-2	50	.480		33.50	15.45		48.95	60.50	
	4924	8" diameter		42	.571		53	18.35		71.35	85.50	
	4926	10" diameter		35	.686		95	22		117	138	
	4928	12" diameter		32	.750		106	24		130	154	
	4930	14" diameter		24	1		139	32		171	202	
	4931	16" diameter		20	1.200		181	38.50		219.50	257	
	4932	18" diameter		18	1.333		209	43		252	295	
	4933	20" diameter	↓	16	1.500		249	48		297	345	
	4934	24" diameter	Q-9	13	1.231	↓	365	37		402	465	
	4940	Flexible, standard, painted										
	4950	3/4" diameter	1 Plum	100	.080	Ea.	7.10	2.76		9.86	12	
	4960	1" diameter		100	.080		7.10	2.76		9.86	12	
	4970	1-1/4" diameter	↓	80	.100	↓	9.45	3.45		12.90	15.60	

15107 | Metal Pipe & Fittings

		CREW	DAILY OUTPUT	LABOR-HOURS	UNIT	2001 BARE COSTS MAT.	LABOR	EQUIP.	TOTAL	TOTAL INCL O&P
690 4980	1-1/2" diameter R15100-070	1 Plum	67	.119	Ea.	10.35	4.11		14.46	17.55 690
4990	2" diameter		50	.160		10.90	5.50		16.40	20.50
5000	2-1/2" diameter	Q-1	80	.200		13.05	6.20		19.25	24
5010	3" diameter		67	.239		14.45	7.40		21.85	27
5020	3-1/2" diameter		57	.281		21	8.70		29.70	36
5030	4" diameter		50	.320		21	9.90		30.90	38
5040	5" diameter		40	.400		32.50	12.40		44.90	54
5050	6" diameter	Q-2	50	.480		38	15.45		53.45	65.50
5070	8" diameter		42	.571		62	18.35		80.35	96
5090	10" diameter		35	.686		103	22		125	148
5110	12" diameter		32	.750		118	24		142	166
5120	14" diameter		24	1		143	32		175	206
5130	16" diameter		20	1.200		188	38.50		226.50	264
5140	18" diameter		18	1.333		220	43		263	305
5150	20" diameter		16	1.500		345	48		393	455
5160	24" diameter		13	1.846		380	59.50		439.50	505
5176	Lightweight style, painted									
5178	1-1/2" diameter	1 Plum	67	.119	Ea.	9	4.11		13.11	16.10
5180	2" diameter	"	50	.160		9.20	5.50		14.70	18.45
5182	2-1/2" diameter	Q-1	80	.200		10.70	6.20		16.90	21
5184	3" diameter		67	.239		12.45	7.40		19.85	25
5186	3-1/2" diameter		57	.281		17.60	8.70		26.30	32.50
5188	4" diameter		50	.320		17.60	9.90		27.50	34.50
5190	5" diameter		40	.400		25.50	12.40		37.90	46.50
5192	6" diameter	Q-2	50	.480		30.50	15.45		45.95	57
5194	8" diameter		42	.571		48	18.35		66.35	80.50
5196	10" diameter		35	.686		129	22		151	176
5198	12" diameter		32	.750		144	24		168	196
5200	For galvanized couplings, add					33%				
5220	Tee, reducing, painted									
5225	2" x 1-1/2" diameter	Q-1	38	.421	Ea.	45.50	13.05		58.55	69.50
5226	2-1/2" x 2" diameter		28	.571		45.50	17.70		63.20	76.50
5227	3" x 2-1/2" diameter		23	.696		40	21.50		61.50	76.50
5228	4" x 3" diameter		18	.889		54	27.50		81.50	101
5229	5" x 4" diameter		15	1.067		116	33		149	178
5230	6" x 4" diameter	Q-2	18	1.333		128	43		171	206
5231	8" x 6" diameter		15	1.600		268	51.50		319.50	370
5232	10" x 8" diameter		13	1.846		350	59.50		409.50	475
5233	12" x 10" diameter		11	2.182		535	70		605	695
5234	14" x 12" diameter		10	2.400		600	77		677	775
5235	16" x 12" diameter		9	2.667		740	85.50		825.50	945
5236	18" x 12" diameter	Q-3	12	2.667		890	87.50		977.50	1,100
5237	18" x 16" diameter		11	2.909		1,125	95.50		1,220.50	1,375
5238	20" x 16" diameter		10	3.200		1,475	105		1,580	1,775
5239	24" x 20" diameter		9	3.556		2,325	117		2,442	2,725
5240	Reducer, concentric, painted									
5241	2-1/2" x 2" diameter	Q-1	43	.372	Ea.	15.85	11.55		27.40	35
5242	3" x 2-1/2" diameter		35	.457		19.20	14.15		33.35	42.50
5243	4" x 3" diameter		29	.552		23	17.10		40.10	51.50
5244	5" x 4" diameter		22	.727		32.50	22.50		55	69.50
5245	6" x 4" diameter	Q-2	26	.923		37.50	29.50		67	86.50
5246	8" x 6" diameter		23	1.043		98.50	33.50		132	159
5247	10" x 8" diameter		20	1.200		201	38.50		239.50	279
5248	12" x 10" diameter		16	1.500		360	48		408	470
5255	Eccentric, painted									
5256	2-1/2" x 2" diameter	Q-1	42	.381	Ea.	33.50	11.80		45.30	55

15107 | Metal Pipe & Fittings

	Line	Description	Crew	Daily Output	Labor-Hours	Unit	2001 Bare Costs Mat.	Labor	Equip.	Total	Total Incl O&P
690	5257	3" x 2-1/2" diameter (R15100-070)	Q-1	34	.471	Ea.	38.50	14.60		53.10	64.50
	5258	4" x 3" diameter		28	.571		47	17.70		64.70	78.50
	5259	5" x 4" diameter		21	.762		64	23.50		87.50	106
	5260	6" x 4" diameter	Q-2	25	.960		75	31		106	129
	5261	8" x 6" diameter		22	1.091		152	35		187	220
	5262	10" x 8" diameter		19	1.263		415	40.50		455.50	520
	5263	12" x 10" diameter		15	1.600		575	51.50		626.50	710
	5270	Coupling, reducing, painted									
	5272	2" x 1-1/2" diameter	1 Plum	52	.154	Ea.	16	5.30		21.30	25.50
	5274	2-1/2" x 2" diameter	Q-1	82	.195		21	6.05		27.05	32
	5276	3" x 2" diameter		69	.232		24	7.20		31.20	37.50
	5278	4" x 2" diameter		52	.308		38	9.55		47.55	56.50
	5280	5" x 4" diameter		42	.381		43	11.80		54.80	65.50
	5282	6" x 4" diameter	Q-2	52	.462		65.50	14.85		80.35	94.50
	5284	8" x 6" diameter	"	44	.545		98	17.55		115.55	135
	5290	Outlet coupling, painted									
	5294	1-1/2" x 1" pipe size	1 Plum	65	.123	Ea.	19.95	4.24		24.19	28.50
	5296	2" x 1" pipe size	"	48	.167		20.50	5.75		26.25	31
	5298	2-1/2" x 1" pipe size	Q-1	78	.205		32	6.35		38.35	44.50
	5300	2-1/2" x 1" pipe size	1 Plum	70	.114		36	3.94		39.94	45.50
	5302	3" x 1" pipe size	Q-1	65	.246		30.50	7.65		38.15	45
	5304	4" x 3/4" pipe size		48	.333		45	10.35		55.35	65.50
	5306	4" x 1-1/2" pipe size		46	.348		64.50	10.80		75.30	87.50
	5308	6" x 1-1/2" pipe size	Q-2	44	.545		91	17.55		108.55	127
	5750	Flange, w/groove gasket, black steel									
	5752	See 15107-660-0620 for gasket & bolt set									
	5760	ANSI class 125 and 150, painted									
	5780	2" pipe size	1 Plum	23	.348	Ea.	45.50	12		57.50	68
	5790	2-1/2" pipe size	Q-1	37	.432		56.50	13.40		69.90	82
	5800	3" pipe size		31	.516		61	16		77	91
	5820	4" pipe size		23	.696		81.50	21.50		103	123
	5830	5" pipe size		19	.842		95	26		121	144
	5840	6" pipe size	Q-2	23	1.043		103	33.50		136.50	165
	5850	8" pipe size		17	1.412		117	45.50		162.50	197
	5860	10" pipe size		14	1.714		185	55		240	287
	5870	12" pipe size		12	2		242	64.50		306.50	365
	5880	14" pipe size		10	2.400		555	77		632	725
	5890	16" pipe size		9	2.667		645	85.50		730.50	840
	5900	18" pipe size		6	4		795	129		924	1,075
	5910	20" pipe size		5	4.800		955	154		1,109	1,275
	5920	24" pipe size		4.50	5.333		1,225	171		1,396	1,600
	5940	ANSI class 350, painted									
	5946	2" pipe size	1 Plum	23	.348	Ea.	57	12		69	80.50
	5948	2-1/2" pipe size	Q-1	37	.432		66	13.40		79.40	92.50
	5950	3" pipe size		31	.516		90	16		106	123
	5952	4" pipe size		23	.696		120	21.50		141.50	165
	5954	5" pipe size		19	.842		137	26		163	190
	5956	6" pipe size	Q-2	23	1.043		159	33.50		192.50	226
	5958	8" pipe size		17	1.412		183	45.50		228.50	270
	5960	10" pipe size		14	1.714		291	55		346	405
	5962	12" pipe size	1 Plum	12	.667		310	23		333	375
	6100	Coupling, for PVC plastic pipe									
	6110	2" diameter	1 Plum	50	.160	Ea.	12.20	5.50		17.70	21.50
	6112	2-1/2" diameter	Q-1	80	.200		17.10	6.20		23.30	28
	6114	3" diameter		67	.239		21	7.40		28.40	34
	6116	4" diameter		50	.320		27	9.90		36.90	45

15 MECHANICAL

15107 | Metal Pipe & Fittings

690

	15107 Metal Pipe & Fittings	CREW	DAILY OUTPUT	LABOR-HOURS	UNIT	2001 BARE COSTS MAT.	LABOR	EQUIP.	TOTAL	TOTAL INCL O&P
6118	6" diameter (R15100-070)	Q-2	50	.480	Ea.	45	15.45		60.45	73
6120	8" diameter		42	.571		74	18.35		92.35	109
6122	10" diameter		35	.686		121	22		143	167
6124	12" diameter		32	.750		153	24		177	206
7400	Suction diffuser									
7402	Grooved end inlet x flanged outlet									
7410	3" x 3"	Q-1	50	.320	Ea.	480	9.90		489.90	540
7412	4" x 4"		38	.421		650	13.05		663.05	735
7414	5" x 5"		30	.533		760	16.55		776.55	860
7416	6" x 6"	Q-2	38	.632		955	20.50		975.50	1,075
7418	8" x 8"		27	.889		1,775	28.50		1,803.50	2,000
7420	10" x 10"		20	1.200		2,425	38.50		2,463.50	2,725
7422	12" x 12"		16	1.500		4,000	48		4,048	4,475
7424	14" x 14"		15	1.600		5,025	51.50		5,076.50	5,625
7426	16" x 14"		14	1.714		5,175	55		5,230	5,775
7500	Strainer, tee type, painted									
7506	2" pipe size	1 Plum	38	.211	Ea.	298	7.25		305.25	340
7508	2-1/2" pipe size	Q-1	62	.258		315	8		323	355
7510	3" pipe size		50	.320		350	9.90		359.90	400
7512	4" pipe size		38	.421		400	13.05		413.05	460
7514	5" pipe size		30	.533		575	16.55		591.55	660
7516	6" pipe size	Q-2	38	.632		625	20.50		645.50	720
7518	8" pipe size		27	.889		960	28.50		988.50	1,100
7520	10" pipe size		20	1.200		1,400	38.50		1,438.50	1,600
7522	12" pipe size		16	1.500		1,825	48		1,873	2,075
7524	14" pipe size		15	1.600		6,475	51.50		6,526.50	7,200
7526	16" pipe size		14	1.714		8,050	55		8,105	8,925
7570	Expansion joint, max. 3" travel									
7572	2" diameter	1 Plum	38	.211	Ea.	245	7.25		252.25	280
7574	3" diameter	Q-1	50	.320		288	9.90		297.90	330
7576	4" diameter	"	38	.421		360	13.05		373.05	415
7578	6" diameter	Q-2	38	.632		625	20.50		645.50	720
7800	Ball valve w/handle, carbon steel trim									
7810	1-1/2" pipe size	1 Plum	50	.160	Ea.	73	5.50		78.50	88.50
7812	2" pipe size	"	38	.211		100	7.25		107.25	121
7814	2-1/2" pipe size	Q-1	62	.258		207	8		215	240
7816	3" pipe size		50	.320		325	9.90		334.90	375
7818	4" pipe size		38	.421		505	13.05		518.05	575
7820	6" pipe size	Q-2	30	.800		1,475	25.50		1,500.50	1,675
7830	With gear operator									
7834	2-1/2" pipe size	Q-1	62	.258	Ea.	365	8		373	410
7836	3" pipe size		50	.320		480	9.90		489.90	545
7838	4" pipe size		38	.421		650	13.05		663.05	735
7840	6" pipe size	Q-2	30	.800		1,650	25.50		1,675.50	1,875
7870	Check valve									
7874	2-1/2" pipe size	Q-1	62	.258	Ea.	131	8		139	156
7876	3" pipe size		50	.320		154	9.90		163.90	185
7878	4" pipe size		38	.421		163	13.05		176.05	199
7880	5" pipe size		30	.533		272	16.55		288.55	325
7882	6" pipe size	Q-2	38	.632		320	20.50		340.50	385
7884	8" pipe size		27	.889		440	28.50		468.50	530
7886	10" pipe size		20	1.200		1,275	38.50		1,313.50	1,450
7888	12" pipe size		16	1.500		1,500	48		1,548	1,725
7900	Plug valve, balancing, w/lever operator									
7906	3" pipe size	Q-1	50	.320	Ea.	268	9.90		277.90	310
7908	4" pipe size	"	38	.421		295	13.05		308.05	345

15107 | Metal Pipe & Fittings

	15107 Metal Pipe & Fittings	CREW	DAILY OUTPUT	LABOR-HOURS	UNIT	2001 BARE COSTS MAT.	LABOR	EQUIP.	TOTAL	TOTAL INCL O&P	
690	7909 6″ pipe size (R15100-070)	Q-2	30	.800	Ea.	530	25.50		555.50	620	690
	7916 With gear operator										
	7920 3″ pipe size	Q-1	50	.320	Ea.	485	9.90		494.90	550	
	7922 4″ pipe size	″	38	.421		510	13.05		523.05	580	
	7926 8″ pipe size	Q-2	27	.889		995	28.50		1,023.50	1,150	
	7928 10″ pipe size		20	1.200		1,575	38.50		1,613.50	1,775	
	7930 12″ pipe size		16	1.500		2,275	48		2,323	2,600	
	8000 Butterfly valve, 2 position handle, with standard trim										
	8010 1-1/2″ pipe size	1 Plum	50	.160	Ea.	115	5.50		120.50	135	
	8020 2″ pipe size	″	38	.211		115	7.25		122.25	138	
	8030 3″ pipe size	Q-1	50	.320		166	9.90		175.90	197	
	8050 4″ pipe size	″	38	.421		182	13.05		195.05	221	
	8070 6″ pipe size	Q-2	38	.632		370	20.50		390.50	435	
	8080 8″ pipe size		27	.889		710	28.50		738.50	830	
	8090 10″ pipe size		20	1.200		1,125	38.50		1,163.50	1,275	
	8200 With stainless steel trim										
	8240 1-1/2″ pipe size	1 Plum	50	.160	Ea.	146	5.50		151.50	169	
	8250 2″ pipe size	″	38	.211		146	7.25		153.25	172	
	8270 3″ pipe size	Q-1	50	.320		197	9.90		206.90	232	
	8280 4″ pipe size	″	38	.421		214	13.05		227.05	255	
	8300 6″ pipe size	Q-2	38	.632		400	20.50		420.50	470	
	8310 8″ pipe size		27	.889		825	28.50		853.50	950	
	8320 10″ pipe size		20	1.200		1,225	38.50		1,263.50	1,400	
	8322 12″ pipe size		16	1.500		1,375	48		1,423	1,575	
	8324 14″ pipe size		15	1.600		1,800	51.50		1,851.50	2,075	
	8326 16″ pipe size		14	1.714		2,475	55		2,530	2,800	
	8328 18″ pipe size	Q-3	12	2.667		3,050	87.50		3,137.50	3,475	
	8330 20″ pipe size		11	2.909		4,125	95.50		4,220.50	4,675	
	8332 24″ pipe size		10	3.200		5,250	105		5,355	5,925	
	8336 Note: sizes 8″ up w/manual gear operator										
	9000 Cut one groove, labor										
	9010 3/4″ pipe size	Q-1	152	.105	Ea.		3.26		3.26	4.93	
	9020 1″ pipe size		140	.114			3.54		3.54	5.35	
	9030 1-1/4″ pipe size		124	.129			4		4	6.05	
	9040 1-1/2″ pipe size		114	.140			4.35		4.35	6.55	
	9050 2″ pipe size		104	.154			4.77		4.77	7.20	
	9060 2-1/2″ pipe size		96	.167			5.15		5.15	7.80	
	9070 3″ pipe size		88	.182			5.65		5.65	8.50	
	9080 3-1/2″ pipe size		83	.193			6		6	9	
	9090 4″ pipe size		78	.205			6.35		6.35	9.60	
	9100 5″ pipe size		72	.222			6.90		6.90	10.40	
	9110 6″ pipe size		70	.229			7.10		7.10	10.70	
	9120 8″ pipe size		54	.296			9.20		9.20	13.85	
	9130 10″ pipe size		38	.421			13.05		13.05	19.70	
	9140 12″ pipe size		30	.533			16.55		16.55	25	
	9150 14″ pipe size		20	.800			25		25	37.50	
	9160 16″ pipe size		19	.842			26		26	39.50	
	9170 18″ pipe size		18	.889			27.50		27.50	41.50	
	9180 20″ pipe size		17	.941			29		29	44	
	9190 24″ pipe size		15	1.067			33		33	50	
	9210 Roll one groove										
	9220 3/4″ pipe size	Q-1	266	.060	Ea.		1.86		1.86	2.82	
	9230 1″ pipe size		228	.070			2.18		2.18	3.28	
	9240 1-1/4″ pipe size		200	.080			2.48		2.48	3.74	
	9250 1-1/2″ pipe size		178	.090			2.79		2.79	4.21	
	9260 2″ pipe size		116	.138			4.28		4.28	6.45	

15 MECHANICAL

		15107 | Metal Pipe & Fittings	CREW	DAILY OUTPUT	LABOR-HOURS	UNIT	2001 BARE COSTS MAT.	LABOR	EQUIP.	TOTAL	TOTAL INCL O&P	
690	9270	2-1/2" pipe size R15100-070	Q-1	110	.145	Ea.		4.51		4.51	6.80	690
	9280	3" pipe size		100	.160			4.96		4.96	7.50	
	9290	3-1/2" pipe size		94	.170			5.30		5.30	7.95	
	9300	4" pipe size		86	.186			5.75		5.75	8.70	
	9310	5" pipe size		84	.190			5.90		5.90	8.90	
	9320	6" pipe size		80	.200			6.20		6.20	9.35	
	9330	8" pipe size		66	.242			7.50		7.50	11.35	
	9340	10" pipe size		58	.276			8.55		8.55	12.90	
	9350	12" pipe size		46	.348			10.80		10.80	16.30	
	9360	14" pipe size		30	.533			16.55		16.55	25	
	9370	16" pipe size		28	.571			17.70		17.70	26.50	
	9380	18" pipe size		27	.593			18.35		18.35	27.50	
	9390	20" pipe size		25	.640			19.85		19.85	30	
	9400	24" pipe size		23	.696			21.50		21.50	32.50	
920	0010	**PIPE, STAINLESS STEEL** R15100-070										920
	0020	Welded, with clevis type hangers 10' O.C.										
	0500	Schedule 5, type 304										
	0540	1/2" diameter	Q-15	128	.125	L.F.	4.67	3.88	.79	9.34	11.85	
	0550	3/4" diameter		116	.138		5.25	4.28	.87	10.40	13.20	
	0560	1" diameter		103	.155		6.40	4.82	.98	12.20	15.40	
	0570	1-1/4" diameter		93	.172		7.30	5.35	1.08	13.73	17.30	
	0580	1-1/2" diameter		85	.188		8.40	5.85	1.18	15.43	19.30	
	0590	2" diameter		69	.232		10.05	7.20	1.46	18.71	23.50	
	0600	2-1/2" diameter		53	.302		14.05	9.35	1.90	25.30	31.50	
	0610	3" diameter		48	.333		16.90	10.35	2.10	29.35	36.50	
	0620	4" diameter		44	.364		21.50	11.25	2.29	35.04	43	
	0630	5" diameter		36	.444		43	13.80	2.80	59.60	71.50	
	0640	6" diameter	Q-16	42	.571		40	18.35	2.39	60.74	74	
	0650	8" diameter		34	.706		61	22.50	2.96	86.46	105	
	0660	10" diameter		26	.923		85.50	29.50	3.87	118.87	143	
	0670	12" diameter		21	1.143		112	36.50	4.79	153.29	184	
	0700	To delete hangers, subtract										
	0710	1/2" diam. to 1-1/2" diam.					8%	19%				
	0720	2" diam. to 5" diam.					4%	9%				
	0730	6" diam. to 12" diam.					3%	4%				
	0750	For small quantities, add				L.F.	10%					
	1250	Schedule 5, type 316										
	1290	1/2" diameter	Q-15	128	.125	L.F.	5.85	3.88	.79	10.52	13.15	
	1300	3/4" diameter		116	.138		6.75	4.28	.87	11.90	14.80	
	1310	1" diameter		103	.155		7.65	4.82	.98	13.45	16.80	
	1320	1-1/4" diameter		93	.172		9.15	5.35	1.08	15.58	19.35	
	1330	1-1/2" diameter		85	.188		10.40	5.85	1.18	17.43	21.50	
	1340	2" diameter		69	.232		12.60	7.20	1.46	21.26	26.50	
	1350	2-1/2" diameter		53	.302		18.10	9.35	1.90	29.35	36	
	1360	3" diameter		48	.333		23	10.35	2.10	35.45	43	
	1370	4" diameter		44	.364		28.50	11.25	2.29	42.04	50.50	
	1380	5" diameter		36	.444		53.50	13.80	2.80	70.10	83	
	1390	6" diameter	Q-16	42	.571		52.50	18.35	2.39	73.24	87.50	
	1400	8" diameter		34	.706		78.50	22.50	2.96	103.96	124	
	1410	10" diameter		26	.923		113	29.50	3.87	146.37	173	
	1420	12" diameter		21	1.143		146	36.50	4.79	187.29	221	
	1490	For small quantities, add					10%					
	1940	To delete hanger, subtract										
	1950	1/2" diam. to 1-1/2" diam.					5%	19%				
	1960	2" diam. to 5" diam.					3%	9%				
	1970	6" diam. to 12" diam.					2%	4%				

MECHANICAL 15

15107 | Metal Pipe & Fittings

920		Description	CREW	DAILY OUTPUT	LABOR-HOURS	UNIT	2001 BARE COSTS MAT.	LABOR	EQUIP.	TOTAL	TOTAL INCL O&P
920	2000	Schedule 10, type 304 R15100-070									
	2040	1/4" diameter	Q-15	131	.122	L.F.	3.69	3.79	.77	8.25	10.60
	2050	3/8" diameter		128	.125		3.89	3.88	.79	8.56	11
	2060	1/2" diameter		125	.128		4.77	3.97	.81	9.55	12.15
	2070	3/4" diameter		113	.142		5.50	4.39	.89	10.78	13.70
	2080	1" diameter		100	.160		7.85	4.96	1.01	13.82	17.25
	2090	1-1/4" diameter		91	.176		9.45	5.45	1.11	16.01	19.85
	2100	1-1/2" diameter		83	.193		10.40	6	1.21	17.61	21.50
	2110	2" diameter		67	.239		12.85	7.40	1.50	21.75	27
	2120	2-1/2" diameter		51	.314		16	9.75	1.97	27.72	34.50
	2130	3" diameter		46	.348		19.60	10.80	2.19	32.59	40
	2140	4" diameter		42	.381		25.50	11.80	2.40	39.70	48.50
	2150	5" diameter		35	.457		34	14.15	2.88	51.03	62
	2160	6" diameter	Q-16	40	.600		40	19.30	2.51	61.81	76
	2170	8" diameter		33	.727		61	23.50	3.05	87.55	106
	2180	10" diameter		25	.960		85.50	31	4.02	120.52	145
	2190	12" diameter		21	1.143		107	36.50	4.79	148.29	179
	2250	For small quantities, add					10%				
	2650	To delete hanger, subtract									
	2660	1/4" diam. to 3/4" diam.					9%	22%			
	2670	1" diam. to 2" diam.					4%	15%			
	2680	2-1/2" diam. to 5" diam.					3%	8%			
	2690	6" diam. to 12" diam.					3%	4%			
	2750	Schedule 10, type 316									
	2790	1/4" diameter	Q-15	131	.122	L.F.	4.14	3.79	.77	8.70	11.10
	2800	3/8" diameter		128	.125		4.59	3.88	.79	9.26	11.75
	2810	1/2" diameter		125	.128		5.85	3.97	.81	10.63	13.30
	2820	3/4" diameter		113	.142		6.75	4.39	.89	12.03	15.10
	2830	1" diameter		100	.160		8.85	4.96	1.01	14.82	18.35
	2840	1-1/4" diameter		91	.176		11.85	5.45	1.11	18.41	22.50
	2850	1-1/2" diameter		83	.193		13.30	6	1.21	20.51	25
	2860	2" diameter		67	.239		16.35	7.40	1.50	25.25	31
	2870	2-1/2" diameter		51	.314		21	9.75	1.97	32.72	40.50
	2880	3" diameter		46	.348		27	10.80	2.19	39.99	48
	2890	4" diameter		42	.381		32.50	11.80	2.40	46.70	56.50
	2900	5" diameter		35	.457		48.50	14.15	2.88	65.53	77.50
	2910	6" diameter	Q-16	40	.600		52	19.30	2.51	73.81	89.50
	2920	8" diameter		33	.727		85.50	23.50	3.05	112.05	133
	2930	10" diameter		25	.960		113	31	4.02	148.02	176
	2940	12" diameter		21	1.143		142	36.50	4.79	183.29	217
	2990	For small quantities, add					10%				
	3430	To delete hanger, subtract									
	3440	1/4" diam. to 3/4" diam.					6%	22%			
	3450	1" diam. to 2" diam.					3%	15%			
	3460	2-1/2" diam. to 5" diam.					2%	8%			
	3470	6" diam. to 12" diam.					2%	4%			
	3500	Threaded, couplings and hangers 10' O.C.									
	3520	Schedule 40, type 304									
	3540	1/4" diameter	1 Plum	54	.148	L.F.	4.87	5.10		9.97	13.05
	3550	3/8" diameter		53	.151		5.55	5.20		10.75	13.95
	3560	1/2" diameter		52	.154		6.55	5.30		11.85	15.25
	3570	3/4" diameter		51	.157		8.50	5.40		13.90	17.50
	3580	1" diameter		45	.178		10.70	6.10		16.80	21
	3590	1-1/4" diameter	Q-1	76	.211		12.60	6.55		19.15	23.50
	3600	1-1/2" diameter		69	.232		14.70	7.20		21.90	27
	3610	2" diameter		57	.281		19.35	8.70		28.05	34.50

15107 | Metal Pipe & Fittings		CREW	DAILY OUTPUT	LABOR-HOURS	UNIT	2001 BARE COSTS MAT.	LABOR	EQUIP.	TOTAL	TOTAL INCL O&P
920 3620	2-1/2" diameter R15100-070	Q-1	44	.364	L.F.	31	11.25		42.25	51
3630	3" diameter		38	.421		40	13.05		53.05	63.50
3640	4" diameter	Q-2	51	.471		55	15.15		70.15	84
3740	For small quantities, add					10%				
4200	To delete couplings & hangers, subtract									
4210	1/4" diam. to 3/4" diam.					15%	56%			
4220	1" diam. to 2" diam.					18%	49%			
4230	2-1/2" diam. to 4" diam.					34%	40%			
4250	Schedule 40, type 316									
4290	1/4" diameter	1 Plum	54	.148	L.F.	5.80	5.10		10.90	14.05
4300	3/8" diameter		53	.151		7.50	5.20		12.70	16.10
4310	1/2" diameter		52	.154		8.25	5.30		13.55	17.05
4320	3/4" diameter		51	.157		10.35	5.40		15.75	19.50
4330	1" diameter		45	.178		13.60	6.10		19.70	24
4340	1-1/4" diameter	Q-1	76	.211		16	6.55		22.55	27.50
4350	1-1/2" diameter		69	.232		18.75	7.20		25.95	31.50
4360	2" diameter		57	.281		25	8.70		33.70	40.50
4370	2-1/2" diameter		44	.364		40.50	11.25		51.75	61.50
4380	3" diameter		38	.421		54.50	13.05		67.55	79.50
4390	4" diameter	Q-2	51	.471		73.50	15.15		88.65	104
4490	For small quantities, add					10%				
4900	To delete couplings & hangers, subtract									
4910	1/4" diam. to 3/4" diam.					12%	56%			
4920	1" diam. to 2" diam.					14%	49%			
4930	2-1/2" diam. to 4" diam.					27%	40%			
5000	Schedule 80, type 304									
5040	1/4" diameter	1 Plum	53	.151	L.F.	6.40	5.20		11.60	14.90
5050	3/8" diameter		52	.154		7.30	5.30		12.60	16
5060	1/2" diameter		51	.157		9.10	5.40		14.50	18.15
5070	3/4" diameter		48	.167		11.65	5.75		17.40	21.50
5080	1" diameter		43	.186		15.90	6.40		22.30	27
5090	1-1/4" diameter	Q-1	73	.219		22.50	6.80		29.30	35.50
5100	1-1/2" diameter		67	.239		26.50	7.40		33.90	40
5110	2" diameter		54	.296		33.50	9.20		42.70	50.50
5190	For small quantities, add					10%				
5700	To delete couplings & hangers, subtract									
5710	1/4" diam. to 3/4" diam.					10%	53%			
5720	1" diam. to 2" diam.					14%	47%			
5750	Schedule 80, type 316									
5790	1/4" diameter	1 Plum	53	.151	L.F.	7.90	5.20		13.10	16.50
5800	3/8" diameter		52	.154		8.95	5.30		14.25	17.80
5810	1/2" diameter		51	.157		12.75	5.40		18.15	22
5820	3/4" diameter		48	.167		14.45	5.75		20.20	24.50
5830	1" diameter		43	.186		19.90	6.40		26.30	31.50
5840	1-1/4" diameter	Q-1	73	.219		29	6.80		35.80	42.50
5850	1-1/2" diameter		67	.239		33.50	7.40		40.90	47.50
5860	2" diameter		54	.296		42.50	9.20		51.70	61
5950	For small quantities, add					10%				
7000	To delete couplings & hangers, subtract									
7010	1/4" diam. to 3/4" diam.					9%	53%			
7020	1" diam. to 2" diam.					14%	47%			
8000	Weld joints with clevis type hangers 10' O.C.									
8010	Schedule 40, type 304 size									
8050	1/8" pipe size	Q-15	126	.127	L.F.	3.44	3.94	.80	8.18	10.60
8060	1/4" pipe size		125	.128		4.28	3.97	.81	9.06	11.60
8070	3/8" pipe		122	.131		4.84	4.07	.82	9.73	12.35 920

15 MECHANICAL

15107 | Metal Pipe & Fittings

			CREW	DAILY OUTPUT	LABOR-HOURS	UNIT	2001 BARE COSTS MAT.	LABOR	EQUIP.	TOTAL	TOTAL INCL O&P	
920	8080	1/2" pipe size R15100-070	Q-15	118	.136	L.F.	6.10	4.20	.85	11.15	14	920
	8090	3/4" pipe size		109	.147		7.25	4.55	.92	12.72	15.85	
	8100	1" pipe size		95	.168		9.05	5.20	1.06	15.31	19	
	8110	1-1/4" pipe size		86	.186		11.15	5.75	1.17	18.07	22.50	
	8120	1-1/2" pipe size		78	.205		13	6.35	1.29	20.64	25.50	
	8130	2" pipe size		62	.258		16.65	8	1.62	26.27	32	
	8140	2-1/2" pipe size		49	.327		25	10.10	2.05	37.15	45	
	8150	3" pipe size		44	.364		32	11.25	2.29	45.54	54.50	
	8160	3-1/2" pipe size		44	.364		37.50	11.25	2.29	51.04	61	
	8170	4" pipe size		39	.410		43	12.70	2.58	58.28	69.50	
	8180	5" pipe size		32	.500		62.50	15.50	3.15	81.15	95.50	
	8190	6" pipe size	Q-16	37	.649		75	21	2.72	98.72	117	
	8200	8" pipe size		29	.828		121	26.50	3.47	150.97	177	
	8210	10" pipe size		24	1		204	32	4.19	240.19	278	
	8220	12" pipe size		20	1.200		231	38.50	5.05	274.55	320	
	8300	Schedule 40, type 316										
	8310	1/8" pipe size	Q-15	126	.127	L.F.	4.31	3.94	.80	9.05	11.55	
	8320	1/4" pipe size		125	.128		5.10	3.97	.81	9.88	12.50	
	8330	3/8" pipe size		122	.131		6.70	4.07	.82	11.59	14.40	
	8340	1/2" pipe size		118	.136		7.65	4.20	.85	12.70	15.75	
	8350	3/4" pipe size		109	.147		8.85	4.55	.92	14.32	17.60	
	8360	1" pipe size		95	.168		11.60	5.20	1.06	17.86	22	
	8370	1-1/4" pipe size		86	.186		14.30	5.75	1.17	21.22	25.50	
	8380	1-1/2" pipe size		78	.205		16.70	6.35	1.29	24.34	29.50	
	8390	2" pipe size		62	.258		21.50	8	1.62	31.12	38	
	8400	2-1/2" pipe size		49	.327		33	10.10	2.05	45.15	54	
	8410	3" pipe size		44	.364		44.50	11.25	2.29	58.04	68.50	
	8420	3-1/2" pipe size		44	.364		51.50	11.25	2.29	65.04	76.50	
	8430	4" pipe size		39	.410		58.50	12.70	2.58	73.78	86.50	
	8440	5" pipe size		32	.500		88.50	15.50	3.15	107.15	124	
	8450	6" pipe size	Q-16	37	.649		99.50	21	2.72	123.22	143	
	8460	8" pipe size		29	.828		185	26.50	3.47	214.97	248	
	8470	10" pipe size		24	1		248	32	4.19	284.19	325	
	8480	12" pipe size		20	1.200		335	38.50	5.05	378.55	435	
	8500	Schedule 80, type 304										
	8510	1/4" pipe size	Q-15	110	.145	L.F.	5.75	4.51	.91	11.17	14.10	
	8520	3/8" pipe size		109	.147		6.50	4.55	.92	11.97	15	
	8530	1/2" pipe size		106	.151		8.10	4.68	.95	13.73	17.05	
	8540	3/4" pipe size		96	.167		10.35	5.15	1.05	16.55	20.50	
	8550	1" pipe size		87	.184		13.75	5.70	1.16	20.61	25	
	8560	1-1/4" pipe size		81	.198		17.70	6.10	1.24	25.04	30	
	8570	1-1/2" pipe size		74	.216		20	6.70	1.36	28.06	33.50	
	8580	2" pipe size		58	.276		25	8.55	1.74	35.29	42.50	
	8590	2-1/2" pipe size		46	.348		51.50	10.80	2.19	64.49	75	
	8600	3" pipe size		41	.390		69	12.10	2.45	83.55	97	
	8610	4" pipe size		33	.485		108	15.05	3.05	126.10	145	
	8630	6" pipe size	Q-16	30	.800		237	25.50	3.35	265.85	305	
	8640	Schedule 80, type 316										
	8650	1/4" pipe size	Q-15	110	.145	L.F.	7.15	4.51	.91	12.57	15.65	
	8660	3/8" pipe size		109	.147		8.15	4.55	.92	13.62	16.80	
	8670	1/2" pipe size		106	.151		11.60	4.68	.95	17.23	21	
	8680	3/4" pipe size		96	.167		12.80	5.15	1.05	19	23	
	8690	1" pipe size		87	.184		17.10	5.70	1.16	23.96	28.50	
	8700	1-1/4" pipe size		81	.198		23	6.10	1.24	30.34	36	
	8710	1-1/2" pipe size		74	.216		26	6.70	1.36	34.06	40.50	
	8720	2" pipe size		58	.276		32.50	8.55	1.74	42.79	51	

15107 | Metal Pipe & Fittings

920		CREW	DAILY OUTPUT	LABOR-HOURS	UNIT	2001 BARE COSTS MAT.	LABOR	EQUIP.	TOTAL	TOTAL INCL O&P
8730	2-1/2" pipe size (R15100-070)	Q-15	46	.348	L.F.	85	10.80	2.19	97.99	112
8740	3" pipe size		41	.390		99	12.10	2.45	113.55	130
8760	4" pipe size		33	.485		137	15.05	3.05	155.10	177
8770	6" pipe size	Q-16	30	.800		370	25.50	3.35	398.85	450
8790	Schedule 160, type 304									
8800	1/2" pipe size	Q-15	96	.167	L.F.	11.40	5.15	1.05	17.60	21.50
8810	3/4" pipe size		88	.182		13.35	5.65	1.14	20.14	24.50
8820	1" pipe size		80	.200		22.50	6.20	1.26	29.96	35.50
8840	1-1/2" pipe size		67	.239		31	7.40	1.50	39.90	47
8850	2" pipe size		53	.302		44.50	9.35	1.90	55.75	65
8870	3" pipe size		37	.432		72.50	13.40	2.72	88.62	103
8900	Schedule 160, type 316									
8910	1/2" pipe size	Q-15	96	.167	L.F.	16.85	5.15	1.05	23.05	27.50
8920	3/4" pipe size		88	.182		17.70	5.65	1.14	24.49	29.50
8930	1" pipe size		80	.200		25	6.20	1.26	32.46	38
8940	1-1/4" pipe size		73	.219		35.50	6.80	1.38	43.68	51
8950	1-1/2" pipe size		67	.239		41	7.40	1.50	49.90	58
8960	2" pipe size		53	.302		56.50	9.35	1.90	67.75	78.50
8990	4" pipe size		30	.533		143	16.55	3.35	162.90	186
9100	Threading pipe labor, sst, one end, schedules 40 & 80									
9110	1/4" through 3/4" pipe size	1 Plum	61.50	.130	Ea.		4.48		4.48	6.75
9120	1" through 2" pipe size		55.90	.143			4.93		4.93	7.45
9130	2-1/2" pipe size		41.50	.193			6.65		6.65	10
9140	3" pipe size		38.50	.208			7.15		7.15	10.80
9150	3-1/2" pipe size	Q-1	68.40	.234			7.25		7.25	10.95
9160	4" pipe size		73	.219			6.80		6.80	10.25
9170	5" pipe size		40.70	.393			12.20		12.20	18.40
9180	6" pipe size		35.40	.452			14		14	21
9190	8" pipe size		22.30	.717			22		22	33.50
9200	10" pipe size		16.10	.994			31		31	46.50
9210	12" pipe size		12.30	1.301			40.50		40.50	61
9250	Welding labor per joint for stainless steel									
9260	Schedule 5 and 10									
9270	1/4" pipe size	Q-15	36	.444	Ea.		13.80	2.80	16.60	24
9280	3/8" pipe size		35	.457			14.15	2.88	17.03	24.50
9290	1/2" pipe size		35	.457			14.15	2.88	17.03	24.50
9300	3/4" pipe size		28	.571			17.70	3.59	21.29	30.50
9310	1" pipe size		25	.640			19.85	4.03	23.88	34.50
9320	1-1/4" pipe size		22	.727			22.50	4.57	27.07	39
9330	1-1/2" pipe size		21	.762			23.50	4.79	28.29	41
9340	2" pipe size		18	.889			27.50	5.60	33.10	47.50
9350	2-1/2" pipe size		14	1.143			35.50	7.20	42.70	61.50
9360	3" pipe size		13	1.231			38	7.75	45.75	66
9370	4" pipe size		12	1.333			41.50	8.40	49.90	72
9380	5" pipe size		10	1.600			49.50	10.05	59.55	86
9390	6" pipe size		9	1.778			55	11.20	66.20	95.50
9400	8" pipe size		6	2.667			82.50	16.75	99.25	143
9410	10" pipe size		4	4			124	25	149	215
9420	12" pipe size		3	5.333			165	33.50	198.50	287
9500	Schedule 40									
9510	1/4" pipe size	Q-15	35	.457	Ea.		14.15	2.88	17.03	24.50
9520	3/8" pipe size		35	.457			14.15	2.88	17.03	24.50
9530	1/2" pipe size		34	.471			14.60	2.96	17.56	25.50
9540	3/4" pipe size		28	.571			17.70	3.59	21.29	30.50
9550	1" pipe size		24	.667			20.50	4.19	24.69	35.50
9560	1-1/4" pipe size		21	.762			23.50	4.79	28.29	41

	15107 | Metal Pipe & Fittings		CREW	DAILY OUTPUT	LABOR-HOURS	UNIT	2001 BARE COSTS MAT.	LABOR	EQUIP.	TOTAL	TOTAL INCL O&P	
920	9570	1-1/2" pipe size R15100-070	Q-15	20	.800	Ea.		25	5.05	30.05	43	920
	9580	2" pipe size		17	.941			29	5.90	34.90	50.50	
	9590	2-1/2" pipe size		14	1.143			35.50	7.20	42.70	61.50	
	9600	3" pipe size		13	1.231			38	7.75	45.75	66	
	9610	4" pipe size		11	1.455			45	9.15	54.15	78	
	9620	5" pipe size		9	1.778			55	11.20	66.20	95.50	
	9630	6" pipe size		8	2			62	12.60	74.60	107	
	9640	8" pipe size		5	3.200			99	20	119	172	
	9650	10" pipe size		4	4			124	25	149	215	
	9660	12" pipe size		3	5.333			165	33.50	198.50	287	
	9750	Schedule 80										
	9760	1/4" pipe size	Q-15	28	.571	Ea.		17.70	3.59	21.29	30.50	
	9770	3/8" pipe size		28	.571			17.70	3.59	21.29	30.50	
	9780	1/2" pipe size		28	.571			17.70	3.59	21.29	30.50	
	9790	3/4" pipe size		24	.667			20.50	4.19	24.69	35.50	
	9800	1" pipe size		21	.762			23.50	4.79	28.29	41	
	9810	1-1/4" pipe size		20	.800			25	5.05	30.05	43	
	9820	1-1/2" pipe size		19	.842			26	5.30	31.30	45.50	
	9830	2" pipe size		16	1			31	6.30	37.30	54	
	9840	2-1/2" pipe size		13	1.231			38	7.75	45.75	66	
	9850	3" pipe size		12	1.333			41.50	8.40	49.90	72	
	9860	4" pipe size		8	2			62	12.60	74.60	107	
	9870	5" pipe size		6	2.667			82.50	16.75	99.25	143	
	9880	6" pipe size		5	3.200			99	20	119	172	
	9890	8" pipe size		3	5.333			165	33.50	198.50	287	
	9900	10" pipe size		2	8			248	50.50	298.50	430	
	9910	12" pipe size		1	16			495	101	596	860	
	9920	Schedule 160, 1/2" pipe size		25	.640			19.85	4.03	23.88	34.50	
	9930	3/4" pipe size		22	.727			22.50	4.57	27.07	39	
	9940	1" pipe size		19	.842			26	5.30	31.30	45.50	
	9950	1-1/4" pipe size		18	.889			27.50	5.60	33.10	47.50	
	9960	1-1/2" pipe size		17	.941			29	5.90	34.90	50.50	
	9970	2" pipe size		14	1.143			35.50	7.20	42.70	61.50	
	9980	3" pipe size		11	1.455			45	9.15	54.15	78	
	9990	4" pipe size		7	2.286			71	14.40	85.40	123	
960	0010	**PIPE, STAINLESS STEEL, FITTINGS**										960
	0100	Butt weld joint, schedule 5, type 304										
	0120	90° Elbow, long										
	0140	1/2"	Q-15	16	1	Ea.	9.55	31	6.30	46.85	64.50	
	0150	3/4"		16	1		15	31	6.30	52.30	70.50	
	0160	1"		15	1.067		10.95	33	6.70	50.65	69.50	
	0170	1-1/4"		14	1.143		17.50	35.50	7.20	60.20	80.50	
	0180	1-1/2"		13	1.231		13.75	38	7.75	59.50	81	
	0190	2"		11	1.455		18.15	45	9.15	72.30	98	
	0200	2-1/2"		8	2		26.50	62	12.60	101.10	136	
	0210	3"		6	2.667		33	82.50	16.75	132.25	180	
	0220	3-1/2"		5	3.200		175	99	20	294	365	
	0230	4"		4	4		61.50	124	25	210.50	282	
	0240	5"		3	5.333		440	165	33.50	638.50	765	
	0250	6"	Q-16	4	6		175	193	25	393	510	
	0260	8"		3	8		298	257	33.50	588.50	750	
	0270	10"		2	12		615	385	50.50	1,050.50	1,300	
	0280	12"		1.57	15.287		970	490	64	1,524	1,875	
	0320	For schedule 5, type 316, add					30%					
	0600	45° Elbow, long										

15 MECHANICAL

Important: See the Reference Section for critical supporting data - Reference Nos., Crews, & City Cost Indexes

	15107 | Metal Pipe & Fittings	CREW	DAILY OUTPUT	LABOR-HOURS	UNIT	2001 BARE COSTS MAT.	LABOR	EQUIP.	TOTAL	TOTAL INCL O&P	
960	0620 1/2"	Q-15	16	1	Ea.	13.75	31	6.30	51.05	69	960
	0630 3/4"		16	1		13.75	31	6.30	51.05	69	
	0640 1"		15	1.067		15.50	33	6.70	55.20	74.50	
	0650 1-1/4"		14	1.143		26.50	35.50	7.20	69.20	90.50	
	0660 1-1/2"		13	1.231		14.10	38	7.75	59.85	81.50	
	0670 2"		11	1.455		16.20	45	9.15	70.35	96	
	0680 2-1/2"		8	2		27	62	12.60	101.60	137	
	0690 3"		6	2.667		27.50	82.50	16.75	126.75	174	
	0700 3-1/2"		5	3.200		175	99	20	294	365	
	0710 4"		4	4		49	124	25	198	269	
	0720 5"		3	5.333		305	165	33.50	503.50	620	
	0730 6"	Q-16	4	6		105	193	25	323	435	
	0740 8"		3	8		183	257	33.50	473.50	630	
	0750 10"		2	12		410	385	50.50	845.50	1,075	
	0760 12"		1.57	15.287		655	490	64	1,209	1,525	
	0800 For schedule 5, type 316, add					25%					
	1100 Tee, straight										
	1130 1/2"	Q-15	10	1.600	Ea.	38.50	49.50	10.05	98.05	129	
	1140 3/4"		10	1.600		38.50	49.50	10.05	98.05	129	
	1150 1"		10	1.600		37	49.50	10.05	96.55	127	
	1160 1-1/4"		9	1.778		63	55	11.20	129.20	165	
	1170 1-1/2"		8	2		37	62	12.60	111.60	148	
	1180 2"		7	2.286		38.50	71	14.40	123.90	165	
	1190 2-1/2"		5	3.200		110	99	20	229	293	
	1200 3"		4	4		72	124	25	221	294	
	1210 3-1/2"		3	5.333		192	165	33.50	390.50	500	
	1220 4"		2	8		102	248	50.50	400.50	545	
	1230 5"		2	8		410	248	50.50	708.50	880	
	1240 6"	Q-16	2.60	9.231		305	297	38.50	640.50	830	
	1250 8"		2	12		700	385	50.50	1,135.50	1,400	
	1260 10"		1.80	13.333		1,100	430	56	1,586	1,900	
	1270 12"		1.40	17.143		1,400	550	72	2,022	2,425	
	1320 For schedule 5, type 316, add					25%					
	2000 Butt weld joint, schedule 10, type 304										
	2020 90° elbow, long										
	2040 1/2"	Q-15	16	1	Ea.	9	31	6.30	46.30	64	
	2050 3/4"		16	1		9	31	6.30	46.30	64	
	2060 1"		15	1.067		9	33	6.70	48.70	67.50	
	2070 1-1/4"		14	1.143		16.80	35.50	7.20	59.50	80	
	2080 1-1/2"		13	1.231		13.10	38	7.75	58.85	80.50	
	2090 2"		11	1.455		17.50	45	9.15	71.65	97.50	
	2100 2-1/2"		8	2		40	62	12.60	114.60	151	
	2110 3"		6	2.667		47.50	82.50	16.75	146.75	195	
	2120 3-1/2"		5	3.200		202	99	20	321	395	
	2130 4"		4	4		77.50	124	25	226.50	300	
	2140 5"		3	5.333		415	165	33.50	613.50	740	
	2150 6"	Q-16	4	6		168	193	25	386	505	
	2160 8"		3	8		350	257	33.50	640.50	810	
	2170 10"		2	12		755	385	50.50	1,190.50	1,475	
	2180 12"		1.57	15.287		1,175	490	64	1,729	2,075	
	2500 45° elbow, long										
	2520 1/2"	Q-15	16	1	Ea.	16.50	31	6.30	53.80	72	
	2530 3/4"		16	1		16.50	31	6.30	53.80	72	
	2540 1"		15	1.067		14.70	33	6.70	54.40	73.50	
	2550 1-1/4"		14	1.143		31.50	35.50	7.20	74.20	96	
	2560 1-1/2"		13	1.231		13.10	38	7.75	58.85	80.50	

15107 | Metal Pipe & Fittings

		CREW	DAILY OUTPUT	LABOR-HOURS	UNIT	2001 BARE COSTS MAT.	LABOR	EQUIP.	TOTAL	TOTAL INCL O&P
960 2570	2″	Q-15	11	1.455	Ea.	15.05	45	9.15	69.20	94.50
2580	2-1/2″		8	2		36	62	12.60	110.60	147
2590	3″		6	2.667		40.50	82.50	16.75	139.75	188
2600	3-1/2″		5	3.200		202	99	20	321	395
2610	4″		4	4		45.50	124	25	194.50	265
2620	5″		3	5.333		315	165	33.50	513.50	630
2630	6″	Q-16	4	6		98	193	25	316	425
2640	8″		3	8		233	257	33.50	523.50	685
2650	10″		2	12		540	385	50.50	975.50	1,225
2660	12″		1.57	15.287		825	490	64	1,379	1,725
3000	Tee, straight									
3030	1/2″	Q-15	10	1.600	Ea.	41.50	49.50	10.05	101.05	132
3040	3/4″		10	1.600		41.50	49.50	10.05	101.05	132
3050	1″		10	1.600		31.50	49.50	10.05	91.05	121
3060	1-1/4″		9	1.778		63	55	11.20	129.20	165
3070	1-1/2″		8	2		34.50	62	12.60	109.10	145
3080	2″		7	2.286		35	71	14.40	120.40	161
3090	2-1/2″		5	3.200		116	99	20	235	299
3100	3″		4	4		81	124	25	230	305
3110	3-1/2″		3	5.333		122	165	33.50	320.50	420
3120	4″		2	8		96	248	50.50	394.50	535
3130	5″		2	8		390	248	50.50	688.50	855
3140	6″	Q-16	2.60	9.231		290	297	38.50	625.50	815
3150	8″		2	12		705	385	50.50	1,140.50	1,400
3151	10″		1.80	13.333		1,175	430	56	1,661	1,975
3152	12″		1.40	17.143		1,625	550	72	2,247	2,675
3154	For schedule 10, type 316, add					25%				
3281	Butt weld joint, schedule 40, type 304									
3284	90° Elbow, long, 1/2″	Q-15	14	1.143	Ea.	15.75	35.50	7.20	58.45	79
3288	3/4″		14	1.143		15.75	35.50	7.20	58.45	79
3289	1″		12	1.333		17.25	41.50	8.40	67.15	91
3290	1-1/4″		11	1.455		25.50	45	9.15	79.65	106
3300	1-1/2″		10	1.600		20.50	49.50	10.05	80.05	109
3310	2″		9	1.778		28.50	55	11.20	94.70	127
3320	2-1/2″		7	2.286		59.50	71	14.40	144.90	188
3330	3″		6	2.667		76	82.50	16.75	175.25	227
3340	3-1/2″		5.50	2.909		315	90	18.30	423.30	505
3350	4″		5	3.200		140	99	20	259	325
3360	5″		4	4		495	124	25	644	760
3370	6″	Q-16	6	4		440	129	16.75	585.75	695
3380	8″		4	6		740	193	25	958	1,125
3390	10″		3	8		1,550	257	33.50	1,840.50	2,125
3400	12″		2	12		2,175	385	50.50	2,610.50	3,025
3410	For schedule 40, type 316, add					25%				
3460	45° Elbow, long, 1/2″	Q-15	14	1.143	Ea.	21	35.50	7.20	63.70	84.50
3470	3/4″		14	1.143		21	35.50	7.20	63.70	84.50
3480	1″		12	1.333		21	41.50	8.40	70.90	95
3490	1-1/4″		11	1.455		31	45	9.15	85.15	112
3500	1-1/2″		10	1.600		24	49.50	10.05	83.55	113
3510	2″		9	1.778		29.50	55	11.20	95.70	127
3520	2-1/2″		7	2.286		59.50	71	14.40	144.90	188
3530	3″		6	2.667		55	82.50	16.75	154.25	203
3540	3-1/2″		5.50	2.909		320	90	18.30	428.30	510
3550	4″		5	3.200		99	99	20	218	281
3560	5″		4	4		345	124	25	494	595
3570	6″	Q-16	6	4		258	129	16.75	403.75	495

15107 | Metal Pipe & Fittings		CREW	DAILY OUTPUT	LABOR-HOURS	UNIT	2001 BARE COSTS MAT.	LABOR	EQUIP.	TOTAL	TOTAL INCL O&P
960 3580	8"	Q-16	4	6	Ea.	490	193	25	708	860
3590	10"		3	8		920	257	33.50	1,210.50	1,425
3600	12"		2	12		1,275	385	50.50	1,710.50	2,025
3610	For schedule 40, type 316, add					25%				
3660	Tee, straight 1/2"	Q-15	11	1.455	Ea.	44.50	45	9.15	98.65	127
3670	3/4"		9	1.778		44.50	55	11.20	110.70	144
3680	1"		8	2		44.50	62	12.60	119.10	156
3690	1-1/4"		7	2.286		73.50	71	14.40	158.90	204
3700	1-1/2"		6.50	2.462		44.50	76.50	15.50	136.50	181
3710	2"		6	2.667		48	82.50	16.75	147.25	196
3720	2-1/2"		5	3.200		120	99	20	239	305
3730	3"		4	4		97	124	25	246	320
3740	3-1/2"		3.70	4.324		139	134	27	300	385
3750	4"		3.50	4.571		139	142	29	310	400
3760	5"		3	5.333		515	165	33.50	713.50	855
3770	6"	Q-16	4	6		445	193	25	663	805
3780	8"		3	8		1,025	257	33.50	1,315.50	1,550
3790	10"		2	12		2,175	385	50.50	2,610.50	3,000
3800	12"		1.50	16		2,200	515	67	2,782	3,250
3810	For schedule 40, type 316, add					25%				
3820	Tee, reducing on outlet, 3/4" x 1/2"	Q-15	11	1.455	Ea.	48	45	9.15	102.15	131
3822	1" x 1/2"		9	1.778		66.50	55	11.20	132.70	169
3824	1" x 3/4"		9	1.778		55.50	55	11.20	121.70	156
3826	1-1/4" x 1"		8	2		176	62	12.60	250.60	300
3828	1-1/2" x 1/2"		7	2.286		92.50	71	14.40	177.90	225
3830	1-1/2" x 3/4"		7	2.286		92.50	71	14.40	177.90	225
3832	1-1/2" x 1"		7	2.286		66.50	71	14.40	151.90	196
3834	2" x 1"		6.50	2.462		65	76.50	15.50	157	203
3836	2" x 1-1/2"		6.50	2.462		61	76.50	15.50	153	199
3838	2-1/2" x 2"		6	2.667		157	82.50	16.75	256.25	315
3840	3" x 1-1/2"		5	3.200		134	99	20	253	320
3842	3" x 2"		5	3.200		113	99	20	232	296
3844	4" x 2"		4	4		204	124	25	353	440
3846	4" x 3"		4	4		167	124	25	316	400
3848	5" x 4"		3.50	4.571		525	142	29	696	825
3850	6" x 3"	Q-16	4.60	5.217		420	168	22	610	735
3852	6" x 4"		4.60	5.217		380	168	22	570	695
3854	8" x 4"		4	6		700	193	25	918	1,100
3856	10" x 8"		3	8		1,950	257	33.50	2,240.50	2,550
3858	12" x 10"		2	12		2,400	385	50.50	2,835.50	3,275
3950	Reducer, concentric, 3/4" x 1/2"	Q-15	14	1.143		22	35.50	7.20	64.70	86
3952	1" x 3/4"		13	1.231		26	38	7.75	71.75	94.50
3954	1-1/4" x 3/4"		12	1.333		74	41.50	8.40	123.90	153
3956	1-1/4" x 1"		12	1.333		40	41.50	8.40	89.90	116
3958	1-1/2" x 3/4"		11	1.455		46	45	9.15	100.15	129
3960	1-1/2" x 1"		11	1.455		15.45	45	9.15	69.60	95
3962	2" x 1"		10	1.600		18.50	49.50	10.05	78.05	107
3964	2" x 1-1/2"		10	1.600		15.45	49.50	10.05	75	103
3966	2-1/2" x 1"		8	2		111	62	12.60	185.60	229
3968	2-1/2" x 2"		8	2		55.50	62	12.60	130.10	168
3970	3" x 1"		7	2.286		74	71	14.40	159.40	204
3972	3" x 1-1/2"		7	2.286		34.50	71	14.40	119.90	160
3974	3" x 2"		7	2.286		28	71	14.40	113.40	153
3976	4" x 2"		6	2.667		38	82.50	16.75	137.25	185
3978	4" x 3"		6	2.667		33.50	82.50	16.75	132.75	180
3980	5" x 3"		4	4		278	124	25	427	520

MECHANICAL 15

15107	Metal Pipe & Fittings	CREW	DAILY OUTPUT	LABOR-HOURS	UNIT	2001 BARE COSTS MAT.	LABOR	EQUIP.	TOTAL	TOTAL INCL O&P
960 3982	5" x 4"	Q-15	4	4	Ea.	222	124	25	371	460 960
3984	6" x 3"	Q-16	7	3.429		106	110	14.35	230.35	298
3986	6" x 4"		7	3.429		88	110	14.35	212.35	279
3988	8" x 4"		5	4.800		335	154	20	509	620
3990	8" x 6"		5	4.800		213	154	20	387	490
3992	10" x 6"		4	6		445	193	25	663	810
3994	10" x 8"		4	6		263	193	25	481	610
3995	12" x 6"		3	8		665	257	33.50	955.50	1,150
3996	12" x 8"		3	8		610	257	33.50	900.50	1,100
3997	12" x 10"		3	8		365	257	33.50	655.50	830
6000	Threaded companion flange									
6010	Stainless steel, 150 lb., type 304									
6020	1/2" diam.	1 Plum	30	.267	Ea.	45.50	9.20		54.70	64
6030	3/4" diam.		28	.286		50.50	9.85		60.35	71
6040	1" diam.		27	.296		56.50	10.20		66.70	78
6050	1-1/4" diam.	Q-1	44	.364		77	11.25		88.25	102
6060	1-1/2" diam.		40	.400		79	12.40		91.40	106
6070	2" diam.		36	.444		88	13.80		101.80	118
6080	2-1/2" diam.		28	.571		135	17.70		152.70	175
6090	3" diam.		20	.800		147	25		172	199
6110	4" diam.		12	1.333		231	41.50		272.50	315
6130	6" diam.	Q-2	14	1.714		440	55		495	570
6140	8" diam.	"	12	2		825	64.50		889.50	1,000
6150	For type 316 add					40%				
6260	Weld flanges, stainless steel, type 304									
6270	Slip on, 150 lb. (welded, front and back)									
6280	1/2" diam.	Q-15	16	1	Ea.	45.50	31	6.30	82.80	104
6290	3/4" diam.		16	1		50.50	31	6.30	87.80	110
6300	1" diam.		13	1.231		34.50	38	7.75	80.25	104
6310	1-1/4" diam.		12	1.333		77	41.50	8.40	126.90	157
6320	1-1/2" diam.		11	1.455		51	45	9.15	105.15	134
6330	2" diam.		10	1.600		66.50	49.50	10.05	126.05	159
6340	2-1/2" diam.		9	1.778		135	55	11.20	201.20	243
6350	3" diam.		8	2		147	62	12.60	221.60	268
6370	4" diam.		6	2.667		140	82.50	16.75	239.25	297
6390	6" diam.	Q-16	6	4		220	129	16.75	365.75	455
6400	8" diam.	"	4	6		825	193	25	1,043	1,225
6410	For type 316, add					40%				
6530	Weld neck 150 lb.									
6540	1/2" diam.	Q-15	35	.457	Ea.	32.50	14.15	2.88	49.53	60.50
6550	3/4" diam.		32	.500		37	15.50	3.15	55.65	67.50
6560	1" diam.		24	.667		44	20.50	4.19	68.69	83.50
6570	1-1/4" diam.		23	.696		52.50	21.50	4.38	78.38	95.50
6580	1-1/2" diam.		20	.800		61.50	25	5.05	91.55	111
6590	2" diam.		18	.889		79	27.50	5.60	112.10	134
6600	2-1/2" diam.		14	1.143		131	35.50	7.20	173.70	205
6610	3" diam.		12	1.333		149	41.50	8.40	198.90	236
6630	4" diam.		9	1.778		193	55	11.20	259.20	305
6640	5" diam.		9	1.778		256	55	11.20	322.20	375
6650	6" diam.	Q-16	9	2.667		298	85.50	11.15	394.65	470
6652	8" diam.		8	3		420	96.50	12.55	529.05	620
6654	10" diam.		7	3.429		575	110	14.35	699.35	810
6656	12" diam.		6	4		910	129	16.75	1,055.75	1,200
6670	For type 316 add					23%				
7000	Threaded joint, 150 lb., type 304									
7030	90° elbow									

15107 | Metal Pipe & Fittings

			CREW	DAILY OUTPUT	LABOR-HOURS	UNIT	2001 BARE COSTS MAT.	LABOR	EQUIP.	TOTAL	TOTAL INCL O&P	
960	7040	1/8"	1 Plum	13	.615	Ea.	12.80	21		33.80	46	960
	7050	1/4"		13	.615		12.80	21		33.80	46	
	7070	3/8"		13	.615		15.45	21		36.45	49	
	7080	1/2"		12	.667		7.90	23		30.90	43	
	7090	3/4"		11	.727		8.95	25		33.95	48	
	7100	1"		10	.800		11.90	27.50		39.40	54.50	
	7110	1-1/4"	Q-1	17	.941		33	29		62	80	
	7120	1-1/2"		16	1		20	31		51	69	
	7130	2"		14	1.143		29	35.50		64.50	85.50	
	7140	2-1/2"		11	1.455		144	45		189	227	
	7150	3"		8	2		182	62		244	295	
	7160	4"	Q-2	11	2.182		201	70		271	325	
	7180	45° elbow										
	7190	1/8"	1 Plum	13	.615	Ea.	17.60	21		38.60	51.50	
	7200	1/4"		13	.615		17.60	21		38.60	51.50	
	7210	3/8"		13	.615		19.60	21		40.60	53.50	
	7220	1/2"		12	.667		9.50	23		32.50	45	
	7230	3/4"		11	.727		11.65	25		36.65	51	
	7240	1"		10	.800		12.85	27.50		40.35	55.50	
	7250	1-1/4"	Q-1	17	.941		31	29		60	78	
	7260	1-1/2"		16	1		20.50	31		51.50	69.50	
	7270	2"		14	1.143		30.50	35.50		66	87	
	7280	2-1/2"		11	1.455		204	45		249	292	
	7290	3"		8	2		298	62		360	425	
	7300	4"	Q-2	11	2.182		254	70		324	385	
	7320	Tee, straight										
	7330	1/8"	1 Plum	9	.889	Ea.	19.70	30.50		50.20	67.50	
	7340	1/4"		9	.889		19.70	30.50		50.20	67.50	
	7350	3/8"		9	.889		22.50	30.50		53	71	
	7360	1/2"		8	1		12.15	34.50		46.65	65.50	
	7370	3/4"		7	1.143		12.75	39.50		52.25	73.50	
	7380	1"		6.50	1.231		15.25	42.50		57.75	81	
	7390	1-1/4"	Q-1	11	1.455		45	45		90	117	
	7400	1-1/2"		10	1.600		31	49.50		80.50	109	
	7410	2"		9	1.778		37	55		92	124	
	7420	2-1/2"		7	2.286		208	71		279	335	
	7430	3"		5	3.200		315	99		414	500	
	7440	4"	Q-2	7	3.429		298	110		408	495	
	7460	Coupling, straight										
	7470	1/8"	1 Plum	19	.421	Ea.	4.86	14.50		19.36	27.50	
	7480	1/4"		19	.421		5.85	14.50		20.35	28.50	
	7490	3/8"		19	.421		6.90	14.50		21.40	29.50	
	7500	1/2"		19	.421		4.81	14.50		19.31	27.50	
	7510	3/4"		18	.444		12.45	15.30		27.75	36.50	
	7520	1"		15	.533		16.50	18.35		34.85	45.50	
	7530	1-1/4"	Q-1	26	.615		14.15	19.10		33.25	44.50	
	7540	1-1/2"		24	.667		17.05	20.50		37.55	50	
	7550	2"		21	.762		27	23.50		50.50	65.50	
	7560	2-1/2"		18	.889		61.50	27.50		89	109	
	7570	3"		14	1.143		82.50	35.50		118	144	
	7580	4"	Q-2	16	1.500		123	48		171	208	
	7600	Reducer, concentric, 1/2"	1 Plum	12	.667		4.23	23		27.23	39	
	7610	3/4"		11	.727		5.50	25		30.50	44	
	7612	1"		10	.800		9.60	27.50		37.10	52	
	7614	1-1/4"	Q-1	17	.941		19.60	29		48.60	65.50	
	7616	1-1/2"		16	1		21.50	31		52.50	71	

15107 | Metal Pipe & Fittings

			CREW	DAILY OUTPUT	LABOR-HOURS	UNIT	2001 BARE COSTS MAT.	LABOR	EQUIP.	TOTAL	TOTAL INCL O&P	
960	7618	2″	Q-1	14	1.143	Ea.	34	35.50		69.50	90.50	960
	7620	2-1/2″		11	1.455		71	45		116	146	
	7622	3″		8	2		103	62		165	208	
	7624	4″	Q-2	11	2.182		148	70		218	268	
	7710	Union										
	7720	1/8″	1 Plum	12	.667	Ea.	23.50	23		46.50	60.50	
	7730	1/4″		12	.667		23.50	23		46.50	60.50	
	7740	3/8″		12	.667		27.50	23		50.50	64.50	
	7750	1/2″		11	.727		16.35	25		41.35	56	
	7760	3/4″		10	.800		23	27.50		50.50	66.50	
	7770	1″		9	.889		33	30.50		63.50	82.50	
	7780	1-1/4″	Q-1	16	1		148	31		179	209	
	7790	1-1/2″		15	1.067		160	33		193	226	
	7800	2″		13	1.231		202	38		240	280	
	7810	2-1/2″		10	1.600		435	49.50		484.50	550	
	7820	3″		7	2.286		575	71		646	740	
	7830	4″	Q-2	10	2.400		795	77		872	990	
	7838	Caps										
	7840	1/2″	1 Plum	24	.333	Ea.	4.61	11.50		16.11	22.50	
	7841	3/4″		22	.364		6.70	12.55		19.25	26.50	
	7842	1″		20	.400		10.70	13.80		24.50	33	
	7843	1-1/2″	Q-1	32	.500		28	15.50		43.50	54.50	
	7844	2″	″	28	.571		36	17.70		53.70	66	
	7845	4″	Q-2	22	1.091		141	35		176	208	
	7850	For 150 lb., type 316, add					25%					

15108 | Plastic Pipe & Fittings

			CREW	DAILY OUTPUT	LABOR-HOURS	UNIT	MAT.	LABOR	EQUIP.	TOTAL	TOTAL INCL O&P	
520	0010	**PIPE, PLASTIC** R15100-070										520
	0020	Fiberglass reinforced, couplings 10′ O.C., hangers 3 per 10′										
	0080	General service										
	0120	2″ diameter	Q-1	59	.271	L.F.	10.15	8.40		18.55	24	
	0140	3″ diameter		52	.308		12.80	9.55		22.35	28.50	
	0150	4″ diameter		48	.333		14.55	10.35		24.90	31.50	
	0160	6″ diameter		39	.410		22	12.70		34.70	43	
	0170	8″ diameter	Q-2	49	.490		33.50	15.75		49.25	61	
	0180	10″ diameter		41	.585		51	18.80		69.80	84.50	
	0190	12″ diameter		36	.667		65	21.50		86.50	104	
	0200	High strength										
	0240	2″ diameter	Q-1	58	.276	L.F.	10.45	8.55		19	24.50	
	0260	3″ diameter		51	.314		14.70	9.75		24.45	31	
	0280	4″ diameter		47	.340		17.40	10.55		27.95	35	
	0300	6″ diameter		38	.421		26.50	13.05		39.55	48.50	
	0320	8″ diameter	Q-2	48	.500		41	16.10		57.10	69.50	
	0340	10″ diameter		40	.600		60	19.30		79.30	95	
	0360	12″ diameter		36	.667		74	21.50		95.50	114	
	0550	To delete coupling & hangers, subtract										
	0560	2″ diam. to 6″ diam.					33%	56%				
	0570	8″ diam. to 12″ diam.					31%	52%				
	0600	PVC, high impact/pressure, cplgs. 10′ O.C., hangers 3 per 10′										
	1020	Schedule 80										
	1040	1/4″ diameter	1 Plum	58	.138	L.F.	2.65	4.75		7.40	10.05	
	1060	3/8″ diameter		55	.145		2.84	5		7.84	10.65	
	1070	1/2″ diameter		50	.160		2.53	5.50		8.03	11.10	
	1080	3/4″ diameter		47	.170		2.68	5.85		8.53	11.80	
	1090	1″ diameter		43	.186		3.07	6.40		9.47	13	

15 MECHANICAL

15108 | Plastic Pipe & Fittings

520

R15100-070

		CREW	DAILY OUTPUT	LABOR-HOURS	UNIT	2001 BARE COSTS MAT.	LABOR	EQUIP.	TOTAL	TOTAL INCL O&P
1100	1-1/4" diameter	1 Plum	39	.205	L.F.	3.55	7.05		10.60	14.55
1110	1-1/2" diameter		34	.235		3.84	8.10		11.94	16.45
1120	2" diameter	Q-1	55	.291		4.44	9		13.44	18.50
1130	2-1/2" diameter		52	.308		5.70	9.55		15.25	20.50
1140	3" diameter		50	.320		7.80	9.90		17.70	23.50
1150	4" diameter		46	.348		10.25	10.80		21.05	27.50
1160	5" diameter		42	.381		13.80	11.80		25.60	33
1170	6" diameter		38	.421		18	13.05		31.05	39.50
1180	8" diameter	Q-2	47	.511		25.50	16.40		41.90	53
1190	10" diameter		42	.571		38.50	18.35		56.85	70
1200	12" diameter		38	.632		50	20.50		70.50	85.50
1730	To delete coupling & hangers, subtract									
1740	1/4" diam. to 1/2" diam.					62%	80%			
1750	3/4" diam. to 1-1/4" diam.					58%	73%			
1760	1-1/2" diam. to 6" diam.					40%	57%			
1770	8" diam. to 12" diam.					34%	50%			
1800	PVC, couplings 10' O.C., hangers 3 per 10'									
1820	Schedule 40									
1860	1/2" diameter	1 Plum	54	.148	L.F.	2.23	5.10		7.33	10.15
1870	3/4" diameter		51	.157		2.28	5.40		7.68	10.65
1880	1" diameter		46	.174		2.48	6		8.48	11.75
1890	1-1/4" diameter		42	.190		2.72	6.55		9.27	12.90
1900	1-1/2" diameter		36	.222		2.85	7.65		10.50	14.70
1910	2" diameter	Q-1	59	.271		3.17	8.40		11.57	16.20
1920	2-1/2" diameter		56	.286		3.97	8.85		12.82	17.70
1930	3" diameter		53	.302		4.96	9.35		14.31	19.60
1940	4" diameter		48	.333		6.05	10.35		16.40	22.50
1950	5" diameter		43	.372		7.75	11.55		19.30	26
1960	6" diameter		39	.410		10.10	12.70		22.80	30.50
1970	8" diameter	Q-2	48	.500		13.10	16.10		29.20	39
1980	10" diameter		43	.558		25.50	17.95		43.45	55.50
1990	12" diameter		42	.571		33.50	18.35		51.85	64
2000	14" diameter		31	.774		52.50	25		77.50	95.50
2010	16" diameter		23	1.043		83	33.50		116.50	142
2340	To delete coupling & hangers, subtract									
2360	1/2" diam. to 1-1/4" diam.					65%	74%			
2370	1-1/2" diam. to 6" diam.					44%	57%			
2380	8" diam. to 12" diam.					41%	53%			
2390	14" diam. to 16" diam.					48%	45%			
2420	Schedule 80									
2440	1/4" diameter	1 Plum	58	.138	L.F.	2.10	4.75		6.85	9.45
2450	3/8" diameter		55	.145		2.10	5		7.10	9.85
2460	1/2" diameter		50	.160		2.27	5.50		7.77	10.80
2470	3/4" diameter		47	.170		2.34	5.85		8.19	11.45
2480	1" diameter		43	.186		2.57	6.40		8.97	12.45
2490	1-1/4" diameter		39	.205		2.85	7.05		9.90	13.80
2500	1-1/2" diameter		34	.235		3.01	8.10		11.11	15.55
2510	2" diameter	Q-1	55	.291		3.30	9		12.30	17.25
2520	2-1/2" diameter		52	.308		3.86	9.55		13.41	18.65
2530	3" diameter		50	.320		5.45	9.90		15.35	21
2540	4" diameter		46	.348		6.80	10.80		17.60	24
2550	5" diameter		42	.381		10	11.80		21.80	29
2560	6" diameter		38	.421		11.75	13.05		24.80	32.50
2570	8" diameter	Q-2	47	.511		15.65	16.40		32.05	42
2580	10" diameter		42	.571		31.50	18.35		49.85	62
2590	12" diameter		38	.632		40	20.50		60.50	74.50

520

15108 | Plastic Pipe & Fittings

			CREW	DAILY OUTPUT	LABOR-HOURS	UNIT	2001 BARE COSTS MAT.	LABOR	EQUIP.	TOTAL	TOTAL INCL O&P
520	2830	To delete coupling & hangers, subtract R15100-070									
	2840	1/4" diam. to 1/2" diam.					66%	80%			
	2850	3/4" diam. to 1-1/4" diam.					61%	73%			
	2860	1-1/2" diam. to 6" diam.					41%	57%			
	2870	8" diam. to 12" diam.					31%	50%			
	2900	Schedule 120									
	2910	1/2" diameter	1 Plum	50	.160	L.F.	2.66	5.50		8.16	11.20
	2950	3/4" diameter		47	.170		2.84	5.85		8.69	12
	2960	1" diameter		43	.186		3.31	6.40		9.71	13.30
	2970	1-1/4" diameter		39	.205		3.89	7.05		10.94	14.95
	2980	1-1/2" diameter		33	.242		4.26	8.35		12.61	17.30
	2990	2" diameter	Q-1	54	.296		5.05	9.20		14.25	19.45
	3000	2-1/2" diameter		52	.308		7.35	9.55		16.90	22.50
	3010	3" diameter		49	.327		9.05	10.10		19.15	25.50
	3020	4" diameter		45	.356		12.85	11		23.85	31
	3030	6" diameter		37	.432		23	13.40		36.40	45.50
	3240	To delete coupling & hangers, subtract									
	3250	1/2" diam. to 1-1/4" diam.					52%	74%			
	3260	1-1/2" diam. to 4" diam.					30%	57%			
	3270	6" diam.					17%	50%			
	3300	PVC, pressure, couplings 10' O.C., hangers 3 per 10'									
	3310	SDR 26, 160 psi									
	3350	1-1/4" diameter	1 Plum	42	.190	L.F.	2.55	6.55		9.10	12.70
	3360	1-1/2" diameter	"	36	.222		2.66	7.65		10.31	14.50
	3370	2" diameter	Q-1	59	.271		2.83	8.40		11.23	15.80
	3380	2-1/2" diameter		56	.286		3.60	8.85		12.45	17.30
	3390	3" diameter		53	.302		4.56	9.35		13.91	19.15
	3400	4" diameter		48	.333		5.65	10.35		16	22
	3420	6" diameter		39	.410		9.85	12.70		22.55	30
	3430	8" diameter	Q-2	48	.500		13.10	16.10		29.20	39
	3660	To delete coupling & hangers, subtract									
	3670	1-1/4" diam.					63%	68%			
	3680	1-1/2" diam. to 4" diam.					48%	57%			
	3690	6" diam. to 8" diam.					60%	54%			
	3720	SDR 21, 200 psi, 1/2" diameter	1 Plum	54	.148	L.F.	2.19	5.10		7.29	10.10
	3740	3/4" diameter		51	.157		2.20	5.40		7.60	10.60
	3750	1" diameter		46	.174		2.36	6		8.36	11.65
	3760	1-1/4" diameter		42	.190		2.58	6.55		9.13	12.75
	3770	1-1/2" diameter		36	.222		2.70	7.65		10.35	14.50
	3780	2" diameter	Q-1	59	.271		2.91	8.40		11.31	15.90
	3790	2-1/2" diameter		56	.286		4.29	8.85		13.14	18.05
	3800	3" diameter		53	.302		4.72	9.35		14.07	19.35
	3810	4" diameter		48	.333		5.90	10.35		16.25	22
	3830	6" diameter		39	.410		10.40	12.70		23.10	30.50
	3840	8" diameter	Q-2	48	.500		14.25	16.10		30.35	40
	4000	To delete coupling & hangers, subtract									
	4010	1/2" diam. to 3/4" diam.					71%	77%			
	4020	1" diam. to 1-1/4" diam.					63%	70%			
	4030	1-1/2" diam. to 6" diam.					44%	57%			
	4040	8" diam.					46%	54%			
	4100	DWV type, schedule 40, couplings 10' O.C., hangers 3 per 10'									
	4120	ABS									
	4140	1-1/4" diameter	1 Plum	42	.190	L.F.	2.33	6.55		8.88	12.45
	4150	1-1/2" diameter	"	36	.222		2.36	7.65		10.01	14.15
	4160	2" diameter	Q-1	59	.271		2.49	8.40		10.89	15.45
	4170	3" diameter		53	.302		3.62	9.35		12.97	18.15

	15108 | Plastic Pipe & Fittings	CREW	DAILY OUTPUT	LABOR-HOURS	UNIT	2001 BARE COSTS MAT.	LABOR	EQUIP.	TOTAL	TOTAL INCL O&P
520 4180	4″ diameter R15100-070	Q-1	48	.333	L.F.	4.07	10.35		14.42	20
4190	6″ diameter	↓	39	.410	↓	12.90	12.70		25.60	33.50
4360	To delete coupling & hangers, subtract									
4370	1-1/4″ diam.					64%	68%			
4380	1-1/2″ diam. to 6″ diam.					54%	57%			
4400	PVC									
4410	1-1/4″ diameter	1 Plum	42	.190	L.F.	2.33	6.55		8.88	12.45
4420	1-1/2″ diameter	″	36	.222		2.44	7.65		10.09	14.25
4460	2″ diameter	Q-1	59	.271		2.51	8.40		10.91	15.45
4470	3″ diameter		53	.302		4.34	9.35		13.69	18.95
4480	4″ diameter		48	.333		4.74	10.35		15.09	21
4490	6″ diameter	↓	39	.410		8.15	12.70		20.85	28
4500	8″ diameter	Q-2	48	.500	↓	17.35	16.10		33.45	43.50
4750	To delete coupling & hangers, subtract									
4760	1-1/4″ diam. to 1-1/2″ diam.					71%	64%			
4770	2″ diam. to 8″ diam.					60%	57%			
4800	PVC, clear pipe, cplgs. 10′ O.C., hangers 3 per 10′, Sched. 40									
4840	1/4″ diameter	1 Plum	59	.136	L.F.	2.49	4.67		7.16	9.80
4850	3/8″ diameter		56	.143		2.72	4.92		7.64	10.45
4860	1/2″ diameter		54	.148		3.10	5.10		8.20	11.10
4870	3/4″ diameter		51	.157		3.44	5.40		8.84	11.95
4880	1″ diameter		46	.174		4.31	6		10.31	13.80
4890	1-1/4″ diameter		42	.190		5.20	6.55		11.75	15.60
4900	1-1/2″ diameter	↓	36	.222		5.85	7.65		13.50	17.95
4910	2″ diameter	Q-1	59	.271		7.20	8.40		15.60	20.50
4920	2-1/2″ diameter		56	.286		10.60	8.85		19.45	25
4930	3″ diameter		53	.302		13.30	9.35		22.65	29
4940	3-1/2″ diameter		50	.320		20	9.90		29.90	37
4950	4″ diameter	↓	48	.333	↓	19.05	10.35		29.40	36.50
5250	To delete coupling & hangers, subtract									
5260	1/4″ diam. to 3/8″ diam.					60%	81%			
5270	1/2″ diam. to 3/4″ diam.					41%	77%			
5280	1″ diam. to 1-1/2″ diam.					26%	67%			
5290	2″ diam. to 4″ diam.					16%	58%			
5360	CPVC, couplings 10′ O.C., hangers 3 per 10′									
5380	Schedule 40									
5460	1/2″ diameter	1 Plum	54	.148	L.F.	2.88	5.10		7.98	10.85
5470	3/4″ diameter		51	.157		3.18	5.40		8.58	11.65
5480	1″ diameter		46	.174		3.75	6		9.75	13.15
5490	1-1/4″ diameter		42	.190		4.35	6.55		10.90	14.70
5500	1-1/2″ diameter	↓	36	.222		5.35	7.65		13	17.45
5510	2″ diameter	Q-1	59	.271		6.20	8.40		14.60	19.55
5520	2-1/2″ diameter		56	.286		9.95	8.85		18.80	24.50
5530	3″ diameter		53	.302		12	9.35		21.35	27.50
5540	4″ diameter		48	.333		17.35	10.35		27.70	34.50
5550	6″ diameter	↓	43	.372	↓	24.50	11.55		36.05	44.50
5730	To delete coupling & hangers, subtract									
5740	1/2″ diam. to 3/4″ diam.					37%	77%			
5750	1″ diam. to 1-1/4″ diam.					27%	70%			
5760	1-1/2″ diam. to 3″ diam.					21%	57%			
5770	4″ diam. to 6″ diam.					16%	57%			
5800	Schedule 80									
5860	1/2″ diameter	1 Plum	50	.160	L.F.	3.27	5.50		8.77	11.90
5870	3/4″ diameter		47	.170		3.36	5.85		9.21	12.55
5880	1″ diameter		43	.186		4.02	6.40		10.42	14.05
5890	1-1/4″ diameter	↓	39	.205	↓	4.76	7.05		11.81	15.90

15108 | Plastic Pipe & Fittings

	15108 Plastic Pipe & Fittings	CREW	DAILY OUTPUT	LABOR-HOURS	UNIT	2001 BARE COSTS MAT.	LABOR	EQUIP.	TOTAL	TOTAL INCL O&P
520 5900	1-1/2" diameter R15100 -070	1 Plum	34	.235	L.F.	5.90	8.10		14	18.75
5910	2" diameter	Q-1	55	.291		7.05	9		16.05	21.50
5920	2-1/2" diameter		52	.308		11.05	9.55		20.60	26.50
5930	3" diameter		50	.320		13.65	9.90		23.55	30
5940	4" diameter		46	.348		19.95	10.80		30.75	38.50
5950	6" diameter		38	.421		30.50	13.05		43.55	53
5960	8" diameter	Q-2	47	.511		64.50	16.40		80.90	96
6060	To delete couplings & hangers, subtract									
6070	1/2" diam. to 3/4" diam.					44%	77%			
6080	1" diam. to 1-1/4" diam.					32%	71%			
6090	1-1/2" diam. to 4" diam.					25%	58%			
6100	6" diam. to 8" diam.					20%	53%			
6240	CTS, 1/2" diameter	1 Plum	54	.148	L.F.	2.12	5.10		7.22	10.05
6250	3/4" diameter		51	.157		2.60	5.40		8	11
6260	1" diameter		46	.174		2.90	6		8.90	12.25
6270	1 1/4"		42	.190		3.40	6.55		9.95	13.65
6280	1 1/2" diameter		36	.222		3.92	7.65		11.57	15.85
6290	2" diameter	Q-1	59	.271		5.30	8.40		13.70	18.55
6370	To delete coupling & hangers, subtract									
6380	1/2" diam.					51%	79%			
6390	3/4" diam.					40%	76%			
6392	1" thru 2" diam.					72%	68%			
6500	Residential installation, plastic pipe									
6510	Couplings 10' O.C., strap hangers 3 per 10'									
6520	PVC, Schedule 40									
6530	1/2" diameter	1 Plum	138	.058	L.F.	.82	2		2.82	3.91
6540	3/4" diameter		128	.063		.88	2.15		3.03	4.21
6550	1" diameter		119	.067		.99	2.32		3.31	4.59
6560	1-1/4" diameter		111	.072		1.17	2.48		3.65	5.05
6570	1-1/2" diameter		104	.077		1.44	2.65		4.09	5.60
6580	2" diameter	Q-1	197	.081		1.64	2.52		4.16	5.60
6590	2-1/2" diameter		162	.099		2.81	3.06		5.87	7.70
6600	4" diameter		123	.130		5.70	4.03		9.73	12.35
6700	PVC, DWV, Schedule 40									
6720	1-1/4" diameter	1 Plum	100	.080	L.F.	1.26	2.76		4.02	5.55
6730	1-1/2" diameter	"	94	.085		1.43	2.93		4.36	6
6740	2" diameter	Q-1	178	.090		1.50	2.79		4.29	5.85
6760	4" diameter	"	110	.145		5.45	4.51		9.96	12.80
7280	Polyethylene, flexible, no couplings or hangers									
7282	Note: For labor costs add 25% to the couplings and fittings labor total.									
7300	SDR 15, 100 psi									
7310	3/4" diameter				L.F.	.17			.17	.19
7350	1" diameter					.22			.22	.24
7360	1-1/4" diameter					.41			.41	.45
7370	1-1/2" diameter					.49			.49	.54
7380	2" diameter					.80			.80	.88
7700	SIDR 9, 160 psi									
7710	1/2" diameter				L.F.	.22			.22	.24
7750	3/4" diameter					.22			.22	.24
7760	1" diameter					.34			.34	.37
7770	1-1/4" diameter					.60			.60	.66
7780	1-1/2" diameter					.84			.84	.92
7790	2" diameter					1.38			1.38	1.52
8120	SDR 9, 200 psi									
8150	3/4" diameter				L.F.	.20			.20	.22
8160	1" diameter					.32			.32	.35

15 MECHANICAL

15108 | Plastic Pipe & Fittings

		15108 Plastic Pipe & Fittings	Crew	Daily Output	Labor-Hours	Unit	2001 Bare Costs Mat.	Labor	Equip.	Total	Total Incl O&P	
520	8170	1-1/4" diameter (R15100-070)				L.F.	.49			.49	.54	520
	8420	SDR 7, 250 psi										
	8440	3/4" diameter				L.F.	.28			.28	.31	
	8450	1" diameter					.46			.46	.51	
	8460	1-1/4" diameter					.80			.80	.88	
	8800	PVC, type PSP, drain & sewer, belled end gasket jnt., no hngr.										
	8840	3" diameter				L.F.	.32			.32	.35	
	8850	4" diameter					.43			.43	.47	
	8860	6" diameter					.89			.89	.98	
	9000	Perforated										
	9040	4" diameter				L.F.	.57			.57	.63	
560	0010	**PIPE, PLASTIC, FITTINGS**										560
	0030	Epoxy resin, fiberglass reinforced, general service										
	0090	Elbow, 90°, 2"	Q-1	33.10	.483	Ea.	45.50	15		60.50	72.50	
	0100	3"		20.80	.769		64.50	24		88.50	107	
	0110	4"		16.50	.970		88	30		118	143	
	0120	6"		10.10	1.584		134	49		183	222	
	0130	8"	Q-2	9.30	2.581		248	83		331	395	
	0140	10"		8.50	2.824		310	91		401	475	
	0150	12"		7.60	3.158		445	102		547	645	
	0160	45° Elbow, same as 90°										
	0170	Elbow, 90°, flanged										
	0172	2"	Q-1	23	.696	Ea.	91	21.50		112.50	133	
	0173	3"		16	1		105	31		136	163	
	0174	4"		13	1.231		138	38		176	209	
	0176	6"		8	2		249	62		311	370	
	0177	8"	Q-2	9	2.667		450	85.50		535.50	620	
	0178	10"		7	3.429		615	110		725	840	
	0179	12"		5	4.800		835	154		989	1,150	
	0186	Elbow, 45°, flanged										
	0188	2"	Q-1	23	.696	Ea.	91.50	21.50		113	134	
	0189	3"		16	1		105	31		136	163	
	0190	4"		13	1.231		137	38		175	209	
	0192	6"		8	2		251	62		313	370	
	0193	8"	Q-2	9	2.667		450	85.50		535.50	625	
	0194	10"		7	3.429		615	110		725	840	
	0195	12"		5	4.800		835	154		989	1,150	
	0290	Tee, 2"	Q-1	20	.800		55	25		80	98	
	0300	3"		13.90	1.151		64.50	35.50		100	125	
	0310	4"		11	1.455		76.50	45		121.50	152	
	0320	6"		6.70	2.388		205	74		279	340	
	0330	8"	Q-2	6.20	3.871		237	124		361	450	
	0340	10"		5.70	4.211		380	135		515	620	
	0350	12"		5.10	4.706		460	151		611	735	
	0352	Tee, flanged										
	0354	2"	Q-1	17	.941	Ea.	124	29		153	180	
	0355	3"		10	1.600		166	49.50		215.50	258	
	0356	4"		8	2		185	62		247	297	
	0358	6"		5	3.200		365	99		464	550	
	0359	8"	Q-2	6	4		625	129		754	880	
	0360	10"		5	4.800		905	154		1,059	1,225	
	0361	12"		4	6		1,150	193		1,343	1,575	
	0365	Wye, flanged										
	0367	2"	Q-1	17	.941	Ea.	320	29		349	395	
	0368	3"		10	1.600		355	49.50		404.50	465	

15108 | Plastic Pipe & Fittings		CREW	DAILY OUTPUT	LABOR-HOURS	UNIT	2001 BARE COSTS MAT.	LABOR	EQUIP.	TOTAL	TOTAL INCL O&P
560 0369	4"	Q-1	8	2	Ea.	440	62		502	580
0371	6"		5	3.200		530	99		629	730
0372	8"	Q-2	6	4		690	129		819	950
0373	10"		5	4.800		1,175	154		1,329	1,500
0374	12"		4	6		1,425	193		1,618	1,850
0380	Couplings									
0410	2"	Q-1	33.10	.483	Ea.	9.85	15		24.85	33.50
0420	3"		20.80	.769		16.95	24		40.95	54.50
0430	4"		16.50	.970		20	30		50	67.50
0440	6"		10.10	1.584		38	49		87	116
0450	8"	Q-2	9.30	2.581		58	83		141	189
0460	10"		8.50	2.824		87.50	91		178.50	233
0470	12"		7.60	3.158		117	102		219	282
0473	High corrosion resistant couplings, add					30%				
0474	Reducer, concentric, flanged									
0475	2" x 1-1/2"	Q-1	30	.533	Ea.	132	16.55		148.55	170
0476	3" x 2"		24	.667		150	20.50		170.50	196
0477	4" x 3"		19	.842		158	26		184	214
0479	6" x 4"		15	1.067		206	33		239	277
0480	8" x 6"	Q-2	16	1.500		289	48		337	395
0481	10" x 8"		13	1.846		385	59.50		444.50	515
0482	12" x 10"		11	2.182		555	70		625	715
0486	Adapter, bell x male or female									
0488	2"	Q-1	28	.571	Ea.	17.70	17.70		35.40	46
0489	3"		20	.800		22.50	25		47.50	62.50
0491	4"		17	.941		31	29		60	78
0492	6"		12	1.333		61.50	41.50		103	130
0493	8"	Q-2	15	1.600		86	51.50		137.50	172
0494	10"	"	11	2.182		151	70		221	272
0528	Flange									
0532	2"	Q-1	46	.348	Ea.	17.40	10.80		28.20	35.50
0533	3"		32	.500		21	15.50		36.50	47
0534	4"		26	.615		29.50	19.10		48.60	61.50
0536	6"		16	1		50	31		81	102
0537	8"	Q-2	18	1.333		82	43		125	155
0538	10"		14	1.714		114	55		169	209
0539	12"		10	2.400		161	77		238	293
2100	PVC schedule 80									
2110	90° elbow, 1/2"	1 Plum	30.30	.264	Ea.	.98	9.10		10.08	14.85
2130	3/4"		26	.308		1.27	10.60		11.87	17.40
2140	1"		22.70	.352		2.03	12.15		14.18	20.50
2150	1-1/4"		20.20	.396		2.71	13.65		16.36	23.50
2160	1-1/2"		18.20	.440		2.91	15.15		18.06	26
2170	2"	Q-1	33.10	.483		3.51	15		18.51	26.50
2180	3"		20.80	.769		9.25	24		33.25	46
2190	4"		16.50	.970		19.15	30		49.15	66.50
2200	6"		10.10	1.584		54.50	49		103.50	134
2210	8"	Q-2	9.30	2.581		150	83		233	290
2250	45° elbow, 1/2"	1 Plum	30.30	.264		1.86	9.10		10.96	15.80
2270	3/4"		26	.308		2.83	10.60		13.43	19.10
2280	1"		22.70	.352		4.24	12.15		16.39	23
2290	1-1/4"		20.20	.396		5.40	13.65		19.05	26.50
2300	1-1/2"		18.20	.440		6.40	15.15		21.55	30
2310	2"	Q-1	33.10	.483		8.30	15		23.30	31.50
2320	3"		20.80	.769		21	24		45	59.50
2330	4"		16.50	.970		38	30		68	87.50 560

Important: See the Reference Section for critical supporting data - Reference Nos., Crews, & City Cost Indexes

15108 | Plastic Pipe & Fittings		CREW	DAILY OUTPUT	LABOR-HOURS	UNIT	2001 BARE COSTS MAT.	LABOR	EQUIP.	TOTAL	TOTAL INCL O&P
560 2340	6"	Q-1	10.10	1.584	Ea.	48	49		97	127
2350	8"	Q-2	9.30	2.581		92.50	83		175.50	227
2400	Tee, 1/2"	1 Plum	20.20	.396		2.79	13.65		16.44	23.50
2420	3/4"		17.30	.462		2.91	15.95		18.86	27
2430	1"		15.20	.526		3.63	18.15		21.78	31.50
2440	1-1/4"		13.50	.593		10	20.50		30.50	42
2450	1-1/2"		12.10	.661		10	23		33	45.50
2460	2"	Q-1	20	.800		12.50	25		37.50	51.50
2470	3"		13.90	1.151		17	35.50		52.50	72.50
2480	4"		11	1.455		19.65	45		64.65	89.50
2490	6"		6.70	2.388		61	74		135	180
2500	8"	Q-2	6.20	3.871		198	124		322	405
2510	Flange, socket, 150 lb., 1/2"	1 Plum	55.60	.144		5.40	4.96		10.36	13.40
2514	3/4"		47.60	.168		5.75	5.80		11.55	15.10
2518	1"		41.70	.192		6.40	6.60		13	17.05
2522	1-1/2"		33.30	.240		6.75	8.30		15.05	19.95
2526	2"	Q-1	60.60	.264		9	8.20		17.20	22.50
2530	4"		30.30	.528		19.40	16.35		35.75	46
2534	6"		18.50	.865		30.50	27		57.50	74
2538	8"	Q-2	17.10	1.404		50	45		95	123
2550	Coupling, 1/2"	1 Plum	30.30	.264		1.78	9.10		10.88	15.70
2570	3/4"		26	.308		2.44	10.60		13.04	18.70
2580	1"		22.70	.352		2.47	12.15		14.62	21
2590	1-1/4"		20.20	.396		3.77	13.65		17.42	24.50
2600	1-1/2"		18.20	.440		4.07	15.15		19.22	27.50
2610	2"	Q-1	33.10	.483		4.36	15		19.36	27.50
2620	3"		20.80	.769		12.35	24		36.35	49.50
2630	4"		16.50	.970		15.45	30		45.45	62.50
2640	6"		10.10	1.584		33.50	49		82.50	111
2650	8"	Q-2	9.30	2.581		35	83		118	164
2660	10"		8.50	2.824		41	91		132	182
2670	12"		7.60	3.158		44	102		146	202
2700	PVC (white), schedule 40, socket joints									
2760	90° elbow, 1/2"	1 Plum	33.30	.240	Ea.	.26	8.30		8.56	12.80
2770	3/4"		28.60	.280		.29	9.65		9.94	14.85
2780	1"		25	.320		.53	11		11.53	17.25
2790	1-1/4"		22.20	.360		.94	12.40		13.34	19.80
2800	1-1/2"		20	.400		1	13.80		14.80	22
2810	2"	Q-1	36.40	.440		1.56	13.65		15.21	22
2820	2-1/2"		26.70	.599		4.75	18.60		23.35	33.50
2830	3"		22.90	.699		5.70	21.50		27.20	39
2840	4"		18.20	.879		10.20	27.50		37.70	52
2850	5"		12.10	1.322		26.50	41		67.50	91
2860	6"		11.10	1.441		32.50	44.50		77	103
2870	8"	Q-2	10.30	2.330		84.50	75		159.50	206
2980	45° elbow, 1/2"	1 Plum	33.30	.240		.43	8.30		8.73	12.95
2990	3/4"		28.60	.280		.66	9.65		10.31	15.30
3000	1"		25	.320		.80	11		11.80	17.55
3010	1-1/4"		22.20	.360		1.12	12.40		13.52	20
3020	1-1/2"		20	.400		.90	13.80		14.70	22
3030	2"	Q-1	36.40	.440		1.83	13.65		15.48	22.50
3040	2-1/2"		26.70	.599		4.76	18.60		23.36	33.50
3050	3"		22.90	.699		7.40	21.50		28.90	40.50
3060	4"		18.20	.879		13.25	27.50		40.75	55.50
3070	5"		12.10	1.322		26.50	41		67.50	91
3080	6"		11.10	1.441		32.50	44.50		77	104

560

	15108 | Plastic Pipe & Fittings		CREW	DAILY OUTPUT	LABOR-HOURS	UNIT	2001 BARE COSTS MAT.	LABOR	EQUIP.	TOTAL	TOTAL INCL O&P	
560	3090	8"	Q-2	10.30	2.330	Ea.	93	75		168	215	560
	3180	Tee, 1/2"	1 Plum	22.20	.360		.32	12.40		12.72	19.10	
	3190	3/4"		19	.421		.36	14.50		14.86	22.50	
	3200	1"		16.70	.479		.70	16.50		17.20	26	
	3210	1-1/4"		14.80	.541		1.10	18.60		19.70	29	
	3220	1-1/2"		13.30	.601		1.33	20.50		21.83	33	
	3230	2"	Q-1	24.20	.661		1.93	20.50		22.43	33	
	3240	2-1/2"		17.80	.899		6.35	28		34.35	49	
	3250	3"		15.20	1.053		8.35	32.50		40.85	58.50	
	3260	4"		12.10	1.322		15.10	41		56.10	78.50	
	3270	5"		8.10	1.975		36.50	61		97.50	133	
	3280	6"		7.40	2.162		51	67		118	157	
	3290	8"	Q-2	6.80	3.529		60.50	113		173.50	238	
	3380	Coupling, 1/2"	1 Plum	33.30	.240		.17	8.30		8.47	12.70	
	3390	3/4"		28.60	.280		.24	9.65		9.89	14.80	
	3400	1"		25	.320		.41	11		11.41	17.10	
	3410	1-1/4"		22.20	.360		.56	12.40		12.96	19.35	
	3420	1-1/2"		20	.400		.60	13.80		14.40	21.50	
	3430	2"	Q-1	36.40	.440		.94	13.65		14.59	21.50	
	3440	2-1/2"		26.70	.599		2.06	18.60		20.66	30.50	
	3450	3"		22.90	.699		3.24	21.50		24.74	36	
	3460	4"		18.20	.879		5.05	27.50		32.55	46.50	
	3470	5"		12.10	1.322		11.40	41		52.40	74.50	
	3480	6"		11.10	1.441		14.75	44.50		59.25	84	
	3490	8"	Q-2	10.30	2.330		28	75		103	144	
	3600	Cap, schedule 40, PVC socket 1/2"	1 Plum	60.60	.132		.32	4.55		4.87	7.20	
	3610	3/4"		51.90	.154		.28	5.30		5.58	8.30	
	3620	1"		45.50	.176		.44	6.05		6.49	9.65	
	3630	1-1/4"		40.40	.198		.60	6.80		7.40	10.95	
	3640	1-1/2"		36.40	.220		.66	7.55		8.21	12.20	
	3650	2"	Q-1	66.10	.242		.80	7.50		8.30	12.25	
	3660	2-1/2"		48.50	.330		2.56	10.25		12.81	18.25	
	3670	3"		41.60	.385		2.80	11.90		14.70	21	
	3680	4"		33.10	.483		6.35	15		21.35	29.50	
	3690	6"		20.20	.792		15.25	24.50		39.75	54	
	3700	8"	Q-2	18.60	1.290		38.50	41.50		80	105	
	3710	Reducing insert, schedule 40, socket weld										
	3712	3/4"	1 Plum	31.50	.254	Ea.	.28	8.75		9.03	13.50	
	3713	1"		27.50	.291		.51	10		10.51	15.70	
	3715	1-1/2"		22	.364		.71	12.55		13.26	19.70	
	3716	2"	Q-1	40	.400		1.17	12.40		13.57	20	
	3717	4"		20	.800		6.25	25		31.25	44.50	
	3718	6"		12.20	1.311		15.50	40.50		56	78.50	
	3719	8"	Q-2	11.30	2.124		54	68.50		122.50	163	
	3730	Reducing insert, socket weld x female/male thread										
	3732	1/2"	1 Plum	38.30	.209	Ea.	1.25	7.20		8.45	12.25	
	3733	3/4"		32.90	.243		.77	8.40		9.17	13.50	
	3734	1"		28.80	.278		1.09	9.55		10.64	15.65	
	3736	1-1/2"		23	.348		1.95	12		13.95	20.50	
	3737	2"	Q-1	41.90	.382		1.04	11.85		12.89	19	
	3738	4"	"	20.90	.766		7.80	23.50		31.30	44.50	
	3742	Male adapter, socket weld x male thread										
	3744	1/2"	1 Plum	38.30	.209	Ea.	.24	7.20		7.44	11.10	
	3745	3/4"		32.90	.243		.36	8.40		8.76	13.05	
	3746	1"		28.80	.278		.47	9.55		10.02	14.95	
	3748	1-1/2"		23	.348		.77	12		12.77	18.95	

15108 | Plastic Pipe & Fittings		CREW	DAILY OUTPUT	LABOR-HOURS	UNIT	2001 BARE COSTS MAT.	LABOR	EQUIP.	TOTAL	TOTAL INCL O&P
560 3749	2"	Q-1	41.90	.382	Ea.	1.01	11.85		12.86	18.95
3750	4"	"	20.90	.766		5.60	23.50		29.10	42
3754	Female adapter, socket weld x female thread									
3756	1/2"	1 Plum	38.30	.209	Ea.	.39	7.20		7.59	11.30
3757	3/4"		32.90	.243		.50	8.40		8.90	13.20
3758	1"		28.80	.278		.58	9.55		10.13	15.10
3760	1-1/2"		23	.348		.77	12		12.77	18.95
3761	2"	Q-1	41.90	.382		1.04	11.85		12.89	19
3762	4"	"	20.90	.766		5.85	23.50		29.35	42.50
3800	PVC, schedule 80, socket joints									
3810	Reducing insert									
3812	3/4"	1 Plum	28.60	.280	Ea.	.78	9.65		10.43	15.40
3813	1"		25	.320		3.03	11		14.03	20
3815	1-1/2"		20	.400		4.79	13.80		18.59	26.50
3816	2"	Q-1	36.40	.440		6.80	13.65		20.45	28
3817	4"		18.20	.879		26	27.50		53.50	69.50
3818	6"		11.10	1.441		36	44.50		80.50	108
3819	8"	Q-2	10.20	2.353		56	75.50		131.50	176
3830	Reducing insert, socket weld x female/male thread									
3832	1/2"	1 Plum	34.80	.230	Ea.	2.28	7.90		10.18	14.45
3833	3/4"		29.90	.268		1.98	9.20		11.18	16.10
3834	1"		26.10	.307		3.43	10.55		13.98	19.70
3836	1-1/2"		20.90	.383		5.75	13.20		18.95	26.50
3837	2"	Q-1	38	.421		8.35	13.05		21.40	29
3838	4"	"	19	.842		23.50	26		49.50	65.50
3844	Adapter, male socket x male thread									
3846	1/2"	1 Plum	34.80	.230	Ea.	1.80	7.90		9.70	13.95
3847	3/4"		29.90	.268		1.98	9.20		11.18	16.10
3848	1"		26.10	.307		3.43	10.55		13.98	19.70
3850	1-1/2"		20.90	.383		5.75	13.20		18.95	26.50
3851	2"	Q-1	38	.421		8.35	13.05		21.40	29
3852	4"	"	19	.842		11.60	26		37.60	52.50
3860	Adapter, female socket x female thread									
3862	1/2"	1 Plum	34.80	.230	Ea.	1.44	7.90		9.34	13.55
3863	3/4"		29.90	.268		3.37	9.20		12.57	17.60
3864	1"		26.10	.307		4.97	10.55		15.52	21.50
3866	1-1/2"		20.90	.383		9.90	13.20		23.10	31
3867	2"	Q-1	38	.421		17.25	13.05		30.30	38.50
3868	4"	"	19	.842		52.50	26		78.50	97.50
3872	Union, socket joints									
3874	1/2"	1 Plum	25.80	.310	Ea.	4.97	10.70		15.67	21.50
3875	3/4"		22.10	.362		7.20	12.45		19.65	27
3876	1"		19.30	.415		6.15	14.30		20.45	28.50
3878	1-1/2"		15.50	.516		16.20	17.80		34	45
3879	2"	Q-1	28.10	.569		22	17.65		39.65	50.50
3888	Cap									
3890	1/2"	1 Plum	54.50	.147	Ea.	2.38	5.05		7.43	10.25
3891	3/4"		46.70	.171		3.33	5.90		9.23	12.55
3892	1"		41	.195		4.44	6.70		11.14	15.05
3894	1-1/2"		32.80	.244		5.35	8.40		13.75	18.60
3895	2"	Q-1	59.50	.269		14.10	8.35		22.45	28
3896	4"		30	.533		42.50	16.55		59.05	72
3897	6"		18.20	.879		108	27.50		135.50	160
3898	8"	Q-2	16.70	1.437		136	46		182	220
4500	DWV, ABS, non pressure, socket joints									
4540	1/4 Bend, 1-1/4"	1 Plum	20.20	.396	Ea.	1.90	13.65		15.55	22.50

560

MECHANICAL 15

15108 | Plastic Pipe & Fittings

			CREW	DAILY OUTPUT	LABOR-HOURS	UNIT	2001 BARE COSTS MAT.	LABOR	EQUIP.	TOTAL	TOTAL INCL O&P	
560	4560	1-1/2"	1 Plum	18.20	.440	Ea.	.85	15.15		16	24	560
	4570	2"	Q-1	33.10	.483		1.39	15		16.39	24	
	4580	3"		20.80	.769		3.31	24		27.31	39.50	
	4590	4"		16.50	.970		5.75	30		35.75	52	
	4600	6"	↓	10.10	1.584	↓	36.50	49		85.50	114	
	4650	1/8 Bend, same as 1/4 Bend										
	4800	Tee, sanitary										
	4820	1-1/4"	1 Plum	13.50	.593	Ea.	2.40	20.50		22.90	33.50	
	4830	1-1/2"	"	12.10	.661		1.25	23		24.25	36	
	4840	2"	Q-1	20	.800		1.77	25		26.77	39.50	
	4850	3"		13.90	1.151		4.50	35.50		40	59	
	4860	4"		11	1.455		11.15	45		56.15	80.50	
	4862	Tee, sanitary, reducing, 2" x 1-1/2"		22	.727		2.63	22.50		25.13	37	
	4864	3" x 2"		15.30	1.046		3.80	32.50		36.30	53	
	4868	4" x 3"	↓	12.10	1.322	↓	11.65	41		52.65	75	
	4870	Combination Y and 1/8 bend										
	4872	1-1/2"	1 Plum	12.10	.661	Ea.	4.51	23		27.51	39.50	
	4874	2"	Q-1	20	.800		5.45	25		30.45	43.50	
	4876	3"		13.90	1.151		9.65	35.50		45.15	64.50	
	4878	4"		11	1.455		18.95	45		63.95	89	
	4880	3" x 1-1/2"		15.50	1.032		11.25	32		43.25	61	
	4882	4" x 3"	↓	12.10	1.322		16.95	41		57.95	80.50	
	4900	Wye, 1-1/4"	1 Plum	13.50	.593		2.77	20.50		23.27	34	
	4902	1-1/2"	"	12.10	.661		2.83	23		25.83	37.50	
	4904	2"	Q-1	20	.800		2.90	25		27.90	40.50	
	4906	3"		13.90	1.151		6.70	35.50		42.20	61.50	
	4908	4"		11	1.455		15.80	45		60.80	85.50	
	4910	6"		6.70	2.388		60	74		134	178	
	4918	3" x 1-1/2"		15.50	1.032		6.60	32		38.60	56	
	4920	4" x 3"		12.10	1.322		10.95	41		51.95	74	
	4922	6" x 4"	↓	6.90	2.319		35.50	72		107.50	148	
	4930	Double Wye, 1-1/2"	1 Plum	9.10	.879		4.51	30.50		35.01	50.50	
	4932	2"	Q-1	16.60	.964		5.80	30		35.80	51.50	
	4934	3"		10.40	1.538		15	47.50		62.50	88.50	
	4936	4"		8.25	1.939		30.50	60		90.50	125	
	4940	2" x 1-1/2"		16.80	.952		5.80	29.50		35.30	51	
	4942	3" x 2"		10.60	1.509		11.10	47		58.10	83	
	4944	4" x 3"		8.45	1.893		32	58.50		90.50	124	
	4946	6" x 4"		7.25	2.207		67	68.50		135.50	177	
	4950	Reducer bushing, 2" x 1-1/2"		36.40	.440		.99	13.65		14.64	21.50	
	4952	3" x 1-1/2"		27.30	.586		3.57	18.15		21.72	31.50	
	4954	4" x 2"		18.20	.879		8.20	27.50		35.70	50	
	4956	6" x 4"	↓	11.10	1.441		22	44.50		66.50	91.50	
	4960	Couplings, 1-1/2"	1 Plum	18.20	.440		2.07	15.15		17.22	25.50	
	4962	2"	Q-1	33.10	.483		.66	15		15.66	23	
	4963	3"		20.80	.769		2.22	24		26.22	38.50	
	4964	4"		16.50	.970		3.40	30		33.40	49	
	4966	6"		10.10	1.584		18	49		67	94	
	4970	2" x 1-1/2"		33.30	.480		1.76	14.90		16.66	24.50	
	4972	3" x 1-1/2"		21	.762		5.05	23.50		28.55	41	
	4974	4" x 3"	↓	16.70	.958		8.70	29.50		38.20	54.50	
	4978	Closet flange, 4"	1 Plum	32	.250		3.23	8.60		11.83	16.55	
	4980	4" x 3"	"	34	.235	↓	3.03	8.10		11.13	15.60	
	5000	DWV, PVC, schedule 40, socket joints										
	5040	1/4 bend, 1-1/4"	1 Plum	20.20	.396	Ea.	1.64	13.65		15.29	22.50	
	5060	1-1/2"	"	18.20	.440	↓	.82	15.15		15.97	24	

 Important: See the Reference Section for critical supporting data - Reference Nos., Crews, & City Cost Indexes

15108 | Plastic Pipe & Fittings

			CREW	DAILY OUTPUT	LABOR-HOURS	UNIT	2001 BARE COSTS MAT.	LABOR	EQUIP.	TOTAL	TOTAL INCL O&P
560	5070	2″	Q-1	33.10	.483	Ea.	1.28	15		16.28	24
	5080	3″		20.80	.769		3.19	24		27.19	39.50
	5090	4″		16.50	.970		5.70	30		35.70	52
	5100	6″		10.10	1.584		35	49		84	113
	5105	8″	Q-2	9.30	2.581		70.50	83		153.50	203
	5106	10″	″	8.50	2.824		70.50	91		161.50	215
	5110	1/4 bend, long sweep, 1-1/2″	1 Plum	18.20	.440		2.61	15.15		17.76	26
	5112	2″	Q-1	33.10	.483		2.11	15		17.11	25
	5114	3″		20.80	.769		5.55	24		29.55	42
	5116	4″		16.50	.970		10.95	30		40.95	57.50
	5150	1/8 bend, 1-1/4″	1 Plum	20.20	.396		1.67	13.65		15.32	22.50
	5170	1-1/2″	″	18.20	.440		.92	15.15		16.07	24
	5180	2″	Q-1	33.10	.483		1.35	15		16.35	24
	5190	3″		20.80	.769		2.75	24		26.75	39
	5200	4″		16.50	.970		4.56	30		34.56	50.50
	5210	6″		10.10	1.584		32	49		81	109
	5215	8″	Q-2	9.30	2.581		47.50	83		130.50	177
	5216	10″		8.50	2.824		71.50	91		162.50	216
	5217	12″		7.60	3.158		114	102		216	279
	5250	Tee, sanitary 1-1/4″	1 Plum	13.50	.593		2.94	20.50		23.44	34
	5254	1-1/2″	″	12.10	.661		1.24	23		24.24	36
	5255	2″	Q-1	20	.800		1.69	25		26.69	39.50
	5256	3″		13.90	1.151		4.65	35.50		40.15	59
	5257	4″		11	1.455		8.85	45		53.85	78
	5259	6″		6.70	2.388		51.50	74		125.50	169
	5261	8″	Q-2	6.20	3.871		154	124		278	360
	5264	2″ x 1-1/2″	Q-1	22	.727		4.46	22.50		26.96	39
	5266	3″ x 1-1/2″		15.50	1.032		3.56	32		35.56	52.50
	5268	4″ x 3″		12.10	1.322		16.35	41		57.35	80
	5271	6″ x 4″		6.90	2.319		55.50	72		127.50	170
	5276	Tee, sanitary, reducing									
	5281	2″ x 1-1/2″ x 1-1/2″	Q-1	23	.696	Ea.	2.46	21.50		23.96	35
	5282	2″ x 1-1/2″ x 2″		22	.727		3.38	22.50		25.88	37.50
	5283	2″ x 2″ x 1-1/2″		22	.727		1.95	22.50		24.45	36
	5284	3″ x 3″ x 1-1/2″		15.50	1.032		3.96	32		35.96	53
	5285	3″ x 3″ x 2″		15.30	1.046		4.22	32.50		36.72	53.50
	5286	4″ x 4″ x 1-1/2″		12.30	1.301		13.60	40.50		54.10	76
	5287	4″ x 4″ x 2″		12.20	1.311		13.35	40.50		53.85	76
	5288	4″ x 4″ x 3″		12.10	1.322		16.35	41		57.35	80
	5291	6″ x 6″ x 4″		6.90	2.319		55.50	72		127.50	170
	5294	Tee, double sanitary									
	5295	1-1/2″	1 Plum	9.10	.879	Ea.	4.10	30.50		34.60	50
	5296	2″	Q-1	16.60	.964		4.52	30		34.52	50
	5297	3″		10.40	1.538		12.65	47.50		60.15	86
	5298	4″		8.25	1.939		20.50	60		80.50	114
	5303	Wye, reducing									
	5304	2″ x 1-1/2″ x 1-1/2″	Q-1	23	.696	Ea.	3.81	21.50		25.31	36.50
	5305	2″ x 2″ x 1-1/2″		22	.727		4.80	22.50		27.30	39.50
	5306	3″ x 3″ x 2″		15.30	1.046		9	32.50		41.50	59
	5307	4″ x 4″ x 2″		12.20	1.311		9.25	40.50		49.75	71.50
	5309	4″ x 4″ x 3″		12.10	1.322		10.80	41		51.80	74
	5314	Combination Y & 1/8 bend, 1-1/2″	1 Plum	12.10	.661		3.92	23		26.92	39
	5315	2″	Q-1	20	.800		5.05	25		30.05	43
	5317	3″		13.90	1.151		8.35	35.50		43.85	63
	5318	4″		11	1.455		16.40	45		61.40	86
	5324	Combination Y & 1/8 bend, reducing									

MECHANICAL 15

15108 | Plastic Pipe & Fittings

		CREW	DAILY OUTPUT	LABOR-HOURS	UNIT	2001 BARE COSTS MAT.	LABOR	EQUIP.	TOTAL	TOTAL INCL O&P
560 5325	2" x 2" x 1-1/2"	Q-1	22	.727	Ea.	5.60	22.50		28.10	40
5327	3" x 3" x 1-1/2"		15.50	1.032		9.75	32		41.75	59
5328	3" x 3" x 2"		15.30	1.046		6.10	32.50		38.60	55.50
5329	4" x 4" x 2"		12.20	1.311		13.55	40.50		54.05	76.50
5331	Wye, 1-1/4"	1 Plum	13.50	.593		3.13	20.50		23.63	34.50
5332	1-1/2"	"	12.10	.661		2.44	23		25.44	37
5333	2"	Q-1	20	.800		2.45	25		27.45	40
5334	3"		13.90	1.151		5.55	35.50		41.05	60
5335	4"		11	1.455		10.35	45		55.35	79.50
5336	6"		6.70	2.388		38	74		112	154
5337	8"	Q-2	6.20	3.871		39	124		163	231
5338	10"		5.70	4.211		61	135		196	271
5339	12"		5.10	4.706		90	151		241	325
5341	2" x 1-1/2"	Q-1	22	.727		4.31	22.50		26.81	38.50
5342	3" x 1-1/2"		15.50	1.032		5.70	32		37.70	55
5343	4" x 3"		12.10	1.322		6.70	41		47.70	69.50
5344	6" x 4"		6.90	2.319		38	72		110	151
5345	8" x 6"	Q-2	6.40	3.750		52	121		173	239
5347	Double wye, 1-1/2"	1 Plum	9.10	.879		5.20	30.50		35.70	51
5348	2"	Q-1	16.60	.964		6.70	30		36.70	52.50
5349	3"		10.40	1.538		17.25	47.50		64.75	91
5350	4"		8.25	1.939		35	60		95	130
5354	2" x 1-1/2"		16.80	.952		6.10	29.50		35.60	51.50
5355	3" x 2"		10.60	1.509		12.85	47		59.85	84.50
5356	4" x 3"		8.45	1.893		27.50	58.50		86	119
5357	6" x 4"		7.25	2.207		58	68.50		126.50	167
5374	Coupling, 1-1/4"	1 Plum	20.20	.396		1.58	13.65		15.23	22
5376	1-1/2"	"	18.20	.440		.53	15.15		15.68	23.50
5378	2"	Q-1	33.10	.483		.55	15		15.55	23
5380	3"		20.80	.769		1.58	24		25.58	37.50
5390	4"		16.50	.970		2.61	30		32.61	48.50
5400	6"		10.10	1.584		13.95	49		62.95	89.50
5402	8"	Q-2	9.30	2.581		31.50	83		114.50	160
5404	2" x 1-1/2"	Q-1	33.30	.480		1.37	14.90		16.27	24
5406	3" x 1-1/2"		21	.762		4.37	23.50		27.87	40.50
5408	4" x 3"		16.70	.958		7.55	29.50		37.05	53.50
5410	Reducer bushing, 2" x 1-1/4"		36.50	.438		.92	13.60		14.52	21.50
5411	2" x 1-1/2"		36.40	.440		1.37	13.65		15.02	22
5412	3" x 1-1/2"		27.30	.586		4.37	18.15		22.52	32.50
5413	3" x 2"		27.10	.590		3.65	18.30		21.95	31.50
5414	4" x 2"		18.20	.879		8.05	27.50		35.55	50
5415	4" x 3"		16.70	.958		8.40	29.50		37.90	54.50
5416	6" x 4"		11.10	1.441		22.50	44.50		67	92.50
5418	8" x 6"	Q-2	10.20	2.353		45.50	75.50		121	164
5425	Closet flange 4"	Q-1	32	.500		10	15.50		25.50	34.50
5426	4" x 3"	"	34	.471		7.20	14.60		21.80	30
5450	Solvent cement for PVC, industrial grade, per quart				Qt.	10.80			10.80	11.90
5500	CPVC, Schedule 80, threaded joints									
5540	90° Elbow, 1/4"	1 Plum	32	.250	Ea.	6.25	8.60		14.85	19.90
5560	1/2"		30.30	.264		2.45	9.10		11.55	16.45
5570	3/4"		26	.308		3.13	10.60		13.73	19.45
5580	1"		22.70	.352		4.96	12.15		17.11	24
5590	1-1/4"		20.20	.396		10.75	13.65		24.40	32.50
5600	1-1/2"		18.20	.440		12	15.15		27.15	36
5610	2"	Q-1	33.10	.483		14.45	15		29.45	38.50
5620	2-1/2"		24.20	.661		33.50	20.50		54	67.50

15108 | Plastic Pipe & Fittings

		15108 | Plastic Pipe & Fittings	CREW	DAILY OUTPUT	LABOR-HOURS	UNIT	2001 BARE COSTS MAT.	LABOR	EQUIP.	TOTAL	TOTAL INCL O&P	
560	5630	3"	Q-1	20.80	.769	Ea.	37.50	24		61.50	77.50	560
	5640	4"		16.50	.970		68	30		98	120	
	5650	6"		10.10	1.584		136	49		185	224	
	5700	45° Elbow same as 90° Elbow										
	5850	Tee, 1/4"	1 Plum	22	.364	Ea.	5.75	12.55		18.30	25	
	5870	1/2"		20.20	.396		5.75	13.65		19.40	27	
	5880	3/4"		17.30	.462		5.85	15.95		21.80	30.50	
	5890	1"		15.20	.526		7.15	18.15		25.30	35.50	
	5900	1-1/4"		13.50	.593		15.10	20.50		35.60	47.50	
	5910	1-1/2"		12.10	.661		17.30	23		40.30	53.50	
	5920	2"	Q-1	20	.800		19.25	25		44.25	58.50	
	5930	2-1/2"		16.20	.988		49	30.50		79.50	100	
	5940	3"		13.90	1.151		49	35.50		84.50	108	
	5950	4"		11	1.455		65	45		110	140	
	5960	6"		6.70	2.388		170	74		244	299	
	6000	Coupling, 1/4"	1 Plum	32	.250		3.88	8.60		12.48	17.25	
	6020	1/2"		30.30	.264		2.72	9.10		11.82	16.75	
	6030	3/4"		26	.308		3.87	10.60		14.47	20.50	
	6040	1"		22.70	.352		4.16	12.15		16.31	23	
	6050	1-1/4"		20.20	.396		5.15	13.65		18.80	26	
	6060	1-1/2"		18.20	.440		11.45	15.15		26.60	35.50	
	6070	2"	Q-1	33.10	.483		12	15		27	35.50	
	6080	2-1/2"		24.20	.661		28.50	20.50		49	62.50	
	6090	3"		20.80	.769		33.50	24		57.50	73	
	6100	4"		16.50	.970		58	30		88	110	
	6110	6"		10.10	1.584		45.50	49		94.50	124	
	6120	8"	Q-2	9.30	2.581		215	83		298	360	
	6200	CTS, 100 psi at 180°F, hot and cold water										
	6230	90° Elbow, 1/2"	1 Plum	20	.400	Ea.	.07	13.80		13.87	21	
	6250	3/4"		19	.421		.16	14.50		14.66	22	
	6251	1"		16	.500		.52	17.25		17.77	26.50	
	6252	1-1/4"		15	.533		.89	18.35		19.24	28.50	
	6253	1-1/2"		14	.571		1.43	19.70		21.13	31	
	6254	2"	Q-1	23	.696		3.08	21.50		24.58	36	
	6260	45° Elbow, 1/2"	1 Plum	20	.400		.12	13.80		13.92	21	
	6280	3/4"		19	.421		.18	14.50		14.68	22	
	6281	1"		16	.500		.48	17.25		17.73	26.50	
	6282	1-1/4"		15	.533		.94	18.35		19.29	28.50	
	6283	1-1/2"		14	.571		1.49	19.70		21.19	31	
	6284	2"	Q-1	23	.696		3.14	21.50		24.64	36	
	6290	Tee, 1/2"	1 Plum	13	.615		.10	21		21.10	32	
	6310	3/4"		12	.667		.23	23		23.23	35	
	6311	1"		11	.727		1.18	25		26.18	39.50	
	6312	1-1/4"		10	.800		1.80	27.50		29.30	43.50	
	6313	1-1/2"		10	.800		2.34	27.50		29.84	44	
	6314	2"	Q-1	17	.941		3.80	29		32.80	48	
	6320	Coupling, 1/2"	1 Plum	22	.364		.09	12.55		12.64	19	
	6340	3/4"		21	.381		.10	13.10		13.20	19.90	
	6341	1"		18	.444		.45	15.30		15.75	23.50	
	6342	1-1/4"		17	.471		.56	16.20		16.76	25	
	6343	1-1/2"		16	.500		.79	17.25		18.04	27	
	6344	2"	Q-1	28	.571		1.58	17.70		19.28	28	
	6360	Solvent cement for CPVC, commercial grade, per quart				Qt.	10.45			10.45	11.50	
	7340	PVC flange, slip-on, Sch 80 std., 1/2"	1 Plum	22	.364	Ea.	6.85	12.55		19.40	26.50	
	7350	3/4"		21	.381		7.30	13.10		20.40	28	
	7360	1"		18	.444		8.15	15.30		23.45	32	

15108 | Plastic Pipe & Fittings

		CREW	DAILY OUTPUT	LABOR-HOURS	UNIT	2001 BARE COSTS MAT.	LABOR	EQUIP.	TOTAL	TOTAL INCL O&P	
560	7370 1-1/4"	1 Plum	17	.471	Ea.	8.40	16.20		24.60	34	560
	7380 1-1/2"		16	.500		8.60	17.25		25.85	35.50	
	7390 2"	Q-1	26	.615		11.40	19.10		30.50	41.50	
	7400 2-1/2"		24	.667		14.30	20.50		34.80	46.50	
	7410 3"		18	.889		19.50	27.50		47	63	
	7420 4"		15	1.067		24.50	33		57.50	77	
	7430 6"		10	1.600		40	49.50		89.50	119	
	7440 8"	Q-2	11	2.182		74.50	70		144.50	188	
	7550 Union, schedule 40, socket joints, 1/2"	1 Plum	19	.421		4.67	14.50		19.17	27	
	7560 3/4"		18	.444		5.90	15.30		21.20	29.50	
	7570 1"		15	.533		6.75	18.35		25.10	35	
	7580 1-1/4"		14	.571		13.35	19.70		33.05	44	
	7590 1-1/2"		13	.615		15.15	21		36.15	48.50	
	7600 2"	Q-1	20	.800		20.50	25		45.50	60	
	7990 Polybutyl/polyethyl pipe, for copper fittings see 15107-460-7000										
	8000 Compression type, PVC, 160 psi cold water										
	8010 Coupling, 3/4" CTS	1 Plum	21	.381	Ea.	2.11	13.10		15.21	22	
	8020 1" CTS		18	.444		2.61	15.30		17.91	26	
	8030 1-1/4" CTS		17	.471		3.63	16.20		19.83	28.50	
	8040 1-1/2" CTS		16	.500		5	17.25		22.25	31.50	
	8050 2" CTS		15	.533		6.95	18.35		25.30	35	
	8060 Female adapter, 3/4" FPT x 3/4" CTS		23	.348		3.35	12		15.35	22	
	8070 3/4" FPT x 1" CTS		21	.381		5	13.10		18.10	25.50	
	8080 1" FPT x 1" CTS		20	.400		5.05	13.80		18.85	26.50	
	8090 1-1/4" FPT x 1-1/4" CTS		18	.444		6.60	15.30		21.90	30.50	
	8100 1-1/2" FPT x 1-1/2" CTS		16	.500		7.70	17.25		24.95	34.50	
	8110 2" FPT x 2" CTS		13	.615		11.45	21		32.45	44.50	
	8130 Male adapter, 3/4" MPT x 3/4" CTS		23	.348		3.02	12		15.02	21.50	
	8140 3/4" MPT x 1" CTS		21	.381		3.71	13.10		16.81	24	
	8150 1" MPT x 1" CTS		20	.400		3.79	13.80		17.59	25	
	8160 1-1/4" MPT x 1-1/4" CTS		18	.444		5.80	15.30		21.10	29.50	
	8170 1-1/2" MPT x 1-1/2" CTS		16	.500		6.75	17.25		24	33.50	
	8180 2" MPT x 2" CTS		13	.615		10.75	21		31.75	44	
	8200 Spigot adapter, 3/4" IPS x 3/4" CTS		23	.348		3.02	12		15.02	21.50	
	8210 3/4" IPS x 1" CTS		21	.381		3.02	13.10		16.12	23	
	8220 1" IPS x 1" CTS		20	.400		3.26	13.80		17.06	24.50	
	8230 1-1/4" IPS x 1-1/4" CTS		18	.444		5.80	15.30		21.10	29.50	
	8240 1-1/2" IPS x 1-1/2" CTS		16	.500		6.75	17.25		24	33.50	
	8250 2" IPS x 2" CTS		13	.615		10.75	21		31.75	44	
	8270 Price includes insert stiffeners										
	8280 250 psi is same price as 160 psi										
	8300 Insert type, nylon, 160 & 250 psi, cold water										
	8310 Clamp ring stainless steel, 3/4" IPS	1 Plum	115	.070	Ea.	.62	2.40		3.02	4.30	
	8320 1" IPS		107	.075		.63	2.58		3.21	4.58	
	8330 1-1/4" IPS		101	.079		.63	2.73		3.36	4.81	
	8340 1-1/2" IPS		95	.084		.64	2.90		3.54	5.10	
	8350 2" IPS		85	.094		.64	3.24		3.88	5.60	
	8370 Coupling, 3/4" IPS		22	.364		.62	12.55		13.17	19.60	
	8380 1" IPS		19	.421		.83	14.50		15.33	23	
	8390 1-1/4" IPS		18	.444		1.49	15.30		16.79	24.50	
	8400 1-1/2" IPS		17	.471		1.94	16.20		18.14	26.50	
	8410 2" IPS		16	.500		2.33	17.25		19.58	28.50	
	8430 Elbow, 90°, 3/4" IPS		22	.364		.91	12.55		13.46	19.90	
	8440 1" IPS		19	.421		1	14.50		15.50	23	
	8450 1-1/4" IPS		18	.444		1.12	15.30		16.42	24	
	8460 1-1/2" IPS		17	.471		1.33	16.20		17.53	26	

15108 | Plastic Pipe & Fittings

		15108 Plastic Pipe & Fittings	CREW	DAILY OUTPUT	LABOR-HOURS	UNIT	2001 BARE COSTS MAT.	LABOR	EQUIP.	TOTAL	TOTAL INCL O&P	
560	8470	2" IPS	1 Plum	16	.500	Ea.	1.85	17.25		19.10	28	560
	8490	Male adapter, 3/4" IPS x 3/4" MPT		25	.320		.63	11		11.63	17.35	
	8500	1" IPS x 1" MPT		21	.381		.83	13.10		13.93	20.50	
	8510	1-1/4" IPS x 1-1/4" MPT		20	.400		1.37	13.80		15.17	22.50	
	8520	1-1/2" IPS x 1-1/2" MPT		18	.444		1.81	15.30		17.11	25	
	8530	2" IPS x 2" MPT		15	.533		2.33	18.35		20.68	30	
	8550	Tee, 3/4" IPS		14	.571		1.19	19.70		20.89	31	
	8560	1" IPS		13	.615		1.94	21		22.94	34	
	8570	1-1/4" IPS		12	.667		3.50	23		26.50	38.50	
	8580	1-1/2" IPS		11	.727		4.40	25		29.40	43	
	8590	2" IPS		10	.800		5.95	27.50		33.45	48	
	8610	Insert type, PVC, 100 psi @ 180°F, hot & cold water										
	8620	Coupler, male, 3/8" CTS x 3/8" MPT	1 Plum	29	.276	Ea.	.38	9.50		9.88	14.75	
	8630	3/8" CTS x 1/2" MPT		28	.286		.38	9.85		10.23	15.25	
	8640	1/2" CTS x 1/2" MPT		27	.296		.38	10.20		10.58	15.80	
	8650	1/2" CTS x 3/4" MPT		26	.308		1.07	10.60		11.67	17.20	
	8660	3/4" CTS x 1/2" MPT		25	.320		1.07	11		12.07	17.85	
	8670	3/4" CTS x 3/4" MPT		25	.320		.45	11		11.45	17.15	
	8700	Coupling, 3/8" CTS x 1/2" CTS		25	.320		2.15	11		13.15	19	
	8710	1/2" CTS		23	.348		2.50	12		14.50	21	
	8720	1/2" CTS x stub		23	.348		2.52	12		14.52	21	
	8730	3/4" CTS		22	.364		3.96	12.55		16.51	23.50	
	8750	Elbow 90°, 3/8" CTS		25	.320		2.67	11		13.67	19.60	
	8760	1/2" CTS		23	.348		2.83	12		14.83	21	
	8770	3/4" CTS		22	.364		4.61	12.55		17.16	24	
	8800	Rings, crimp, copper, 3/8" CTS		120	.067		.11	2.30		2.41	3.59	
	8810	1/2" CTS		117	.068		.12	2.36		2.48	3.69	
	8820	3/4" CTS		115	.070		.15	2.40		2.55	3.79	
	8850	Reducer tee, 3/8" x 3/8" x 1/2" CTS		17	.471		1.46	16.20		17.66	26	
	8860	1/2" x 3/8" x 1/2" CTS		15	.533		1.25	18.35		19.60	29	
	8870	3/4" x 1/2" x 1/2" CTS		14	.571		1.31	19.70		21.01	31	
	8890	3/4" x 3/4" x 1/2" CTS		14	.571		1.24	19.70		20.94	31	
	8900	3/4" x 1/2" x 3/8" CTS		14	.571		1.09	19.70		20.79	30.50	
	8930	Tee, 3/8" CTS		17	.471		1.05	16.20		17.25	25.50	
	8940	1/2" CTS		15	.533		1.19	18.35		19.54	29	
	8950	3/4" CTS		14	.571		1.29	19.70		20.99	31	
	8960	Copper rings included in fitting price										
	9000	Flare type, assembled, acetal, hot & cold water										
	9010	Coupling, 1/4" & 3/8" CTS	1 Plum	24	.333	Ea.	2.14	11.50		13.64	19.70	
	9020	1/2" CTS		22	.364		2.45	12.55		15	21.50	
	9030	3/4" CTS		21	.381		3.61	13.10		16.71	24	
	9040	1" CTS		18	.444		4.61	15.30		19.91	28	
	9050	Elbow 90°, 1/4" CTS		26	.308		2.38	10.60		12.98	18.60	
	9060	3/8" CTS		24	.333		2.55	11.50		14.05	20	
	9070	1/2" CTS		22	.364		3.01	12.55		15.56	22	
	9080	3/4" CTS		21	.381		4.61	13.10		17.71	25	
	9090	1" CTS		18	.444		5.80	15.30		21.10	29.50	
	9110	Tee ,1/4" & 3/8" CTS		15	.533		2.62	18.35		20.97	30.50	
	9120	1/2" CTS		14	.571		3.44	19.70		23.14	33.50	
	9130	3/4" CTS		13	.615		5.25	21		26.25	38	
	9140	1" CTS		12	.667		7	23		30	42	
	9550	For plastic hangers see 15060-300-8000										
	9560	For copper/brass fittings see 15107-460-7000										
590	0010	**PIPE, HIGH DENSITY POLYETHYLENE PLASTIC (HDPE)**										590
	0020	Not incl. hangers, trenching, backfill, hoisting or digging equipment.										

MECHANICAL 15

15108 | Plastic Pipe & Fittings

		15108 Plastic Pipe & Fittings	CREW	DAILY OUTPUT	LABOR-HOURS	UNIT	2001 BARE COSTS MAT.	LABOR	EQUIP.	TOTAL	TOTAL INCL O&P	
590	0030	Standard length is 40′, add a weld for each joint.										590
	0040	Single wall										
	0050	Straight										
	0054	1″ diameter DR 11				L.F.	.28			.28	.31	
	0058	1-1/2″ diameter DR 11					.62			.62	.68	
	0062	2″ diameter DR 11					.79			.79	.87	
	0066	3″ diameter DR 11					1.72			1.72	1.89	
	0070	3″ diameter DR 17					1.15			1.15	1.27	
	0074	4″ diameter DR 11					2.84			2.84	3.12	
	0078	4″ diameter DR 17					1.91			1.91	2.10	
	0082	6″ diameter DR 11					6.15			6.15	6.75	
	0086	6″ diameter DR 17					4.13			4.13	4.54	
	0090	8″ diameter DR 11					10.40			10.40	11.45	
	0094	8″ diameter DR 26					4.69			4.69	5.15	
	0098	10″ diameter DR 11					16.20			16.20	17.85	
	0102	10″ diameter DR 26					7.25			7.25	8	
	0106	12″ diameter DR 11					23			23	25	
	0110	12″ diameter DR 26					10.25			10.25	11.25	
	0114	16″ diameter DR 11					34			34	37	
	0118	16″ diameter DR 26					15.20			15.20	16.70	
	0122	18″ diameter DR 11					43			43	47	
	0126	18″ diameter DR 26					19.20			19.20	21	
	0130	20″ diameter DR 11					53			53	58	
	0134	20″ diameter DR 26					23.50			23.50	26	
	0138	22″ diameter DR 11					64			64	70.50	
	0142	22″ diameter DR 26					28.50			28.50	31.50	
	0146	24″ diameter DR 11					76			76	84	
	0150	24″ diameter DR 26					34			34	37.50	
	0154	28″ diameter DR 17					69.50			69.50	76.50	
	0158	28″ diameter DR 26					46.50			46.50	51	
	0162	30″ diameter DR 21					65.50			65.50	72	
	0166	30″ diameter DR 26					43			43	47.50	
	0170	36″ diameter DR 26					43			43	47.50	
	0174	42″ diameter DR 26					105			105	115	
	0178	48″ diameter DR 26					132			132	146	
	0182	54″ diameter DR 26					173			173	190	
	0300	90° Elbow										
	0304	1″ diameter DR 11				Ea.	6.45			6.45	7.10	
	0308	1-1/2″ diameter DR 11					10.75			10.75	11.85	
	0312	2″ diameter DR 11					12			12	13.20	
	0316	3″ diameter DR 11					23			23	25.50	
	0320	3″ diameter DR 17					23			23	25.50	
	0324	4″ diameter DR 11					27			27	29.50	
	0328	4″ diameter DR 17					27			27	29.50	
	0332	6″ diameter DR 11					79			79	87	
	0336	6″ diameter DR 17					79			79	87	
	0340	8″ diameter DR 11					260			260	286	
	0344	8″ diameter DR 26					177			177	195	
	0348	10″ diameter DR 11					645			645	710	
	0352	10″ diameter DR 26					325			325	360	
	0356	12″ diameter DR 11					645			645	710	
	0360	12″ diameter DR 26					355			355	390	
	0364	16″ diameter DR 11					720			720	795	
	0368	16″ diameter DR 26					580			580	640	
	0372	18″ diameter DR 11					755			755	830	
	0376	18″ diameter DR 26					755			755	830	

	15108 | Plastic Pipe & Fittings	CREW	DAILY OUTPUT	LABOR-HOURS	UNIT	2001 BARE COSTS MAT.	LABOR	EQUIP.	TOTAL	TOTAL INCL O&P	
590	0380 20″ diameter DR 11				Ea.	1,125			1,125	1,225	590
	0384 20″ diameter DR 26					1,125			1,125	1,225	
	0388 22″ diameter DR 11					1,400			1,400	1,550	
	0392 22″ diameter DR 26					1,400			1,400	1,550	
	0396 24″ diameter DR 11					1,625			1,625	1,775	
	0400 24″ diameter DR 26					1,625			1,625	1,775	
	0404 28″ diameter DR 17					2,450			2,450	2,700	
	0408 28″ diameter DR 26					2,450			2,450	2,700	
	0412 30″ diameter DR 17					2,250			2,250	2,475	
	0416 30″ diameter DR 26					2,250			2,250	2,475	
	0420 36″ diameter DR 26					3,450			3,450	3,775	
	0424 42″ diameter DR 26					3,875			3,875	4,250	
	0428 48″ diameter DR 26					4,075			4,075	4,500	
	0432 54″ diameter DR 26				↓	9,475			9,475	10,400	
	0500 45° Elbow										
	0512 2″ diameter DR 11				Ea.	24			24	26.50	
	0516 3″ diameter DR 11					21			21	23	
	0520 3″ diameter DR 17					34			34	37.50	
	0524 4″ diameter DR 11					35			35	38.50	
	0528 4″ diameter DR 17					42			42	46	
	0532 6″ diameter DR 11					74			74	81.50	
	0536 6″ diameter DR 17					70			70	77	
	0540 8″ diameter DR 11					175			175	193	
	0544 8″ diameter DR 26					105			105	116	
	0548 10″ diameter DR 11					490			490	540	
	0552 10″ diameter DR 26					142			142	156	
	0556 12″ diameter DR 11					590			590	650	
	0560 12″ diameter DR 26					224			224	246	
	0564 16″ diameter DR 11					282			282	310	
	0568 16″ diameter DR 26					282			282	310	
	0572 18″ diameter DR 11					415			415	455	
	0576 18″ diameter DR 26					415			415	455	
	0580 20″ diameter DR 11					645			645	710	
	0584 20″ diameter DR 26					645			645	710	
	0588 22″ diameter DR 11					755			755	830	
	0592 22″ diameter DR 26					755			755	830	
	0596 24″ diameter DR 11					925			925	1,025	
	0600 24″ diameter DR 26					925			925	1,025	
	0604 28″ diameter DR 17					1,350			1,350	1,500	
	0608 28″ diameter DR 26					1,350			1,350	1,500	
	0612 30″ diameter DR 17					1,300			1,300	1,425	
	0616 30″ diameter DR 26					1,300			1,300	1,425	
	0620 36″ diameter DR 26					2,125			2,125	2,350	
	0624 42″ diameter DR 26					2,600			2,600	2,850	
	0628 48″ diameter DR 26					2,675			2,675	2,925	
	0632 54″ diameter DR 26				↓	4,350			4,350	4,775	
	0700 Tee										
	0704 1″ diameter DR 11				Ea.	6.45			6.45	7.10	
	0708 1-1/2″ diameter DR 11					10.75			10.75	11.85	
	0712 2″ diameter DR 11					15			15	16.50	
	0716 3″ diameter DR 11					24			24	26.50	
	0720 3″ diameter DR 17					24			24	26.50	
	0724 4″ diameter DR 11					38			38	42	
	0728 4″ diameter DR 17					38			38	42	
	0732 6″ diameter DR 11					94			94	103	
	0736 6″ diameter DR 17				↓	94			94	103	

15108 | Plastic Pipe & Fittings		CREW	DAILY OUTPUT	LABOR-HOURS	UNIT	2001 BARE COSTS MAT.	LABOR	EQUIP.	TOTAL	TOTAL INCL O&P	
590	0740 8″ diameter DR 11				Ea.	217			217	239	590
	0744 8″ diameter DR 17					197			197	217	
	0748 10″ diameter DR 11					675			675	745	
	0752 10″ diameter DR 17					242			242	266	
	0756 12″ diameter DR 11					975			975	1,075	
	0760 12″ diameter DR 17					276			276	305	
	0764 16″ diameter DR 11					645			645	710	
	0768 16″ diameter DR 17					645			645	710	
	0772 18″ diameter DR 11					685			685	755	
	0776 18″ diameter DR 17					685			685	755	
	0780 20″ diameter DR 11					850			850	935	
	0784 20″ diameter DR 17					850			850	935	
	0788 22″ diameter DR 11					1,175			1,175	1,300	
	0792 22″ diameter DR 17					1,175			1,175	1,300	
	0796 24″ diameter DR 11					1,175			1,175	1,300	
	0800 24″ diameter DR 17					1,175			1,175	1,300	
	0804 28″ diameter DR 17					2,100			2,100	2,300	
	0812 30″ diameter DR 17					2,350			2,350	2,575	
	0820 36″ diameter DR 17					4,675			4,675	5,125	
	0824 42″ diameter DR 26					5,125			5,125	5,650	
	0828 48″ diameter DR 26				▼	6,000			6,000	6,600	
	1000 Flange adptr, w/back-up ring and 1/2 cost of plated bolt set										
	1004 1″ diameter DR 11				Ea.	59			59	65	
	1008 1-1/2″ diameter DR 11					60.50			60.50	66.50	
	1012 2″ diameter DR 11					33.50			33.50	36.50	
	1016 3″ diameter DR 11					41			41	45	
	1020 3″ diameter DR 17					41			41	45	
	1024 4″ diameter DR 11					54			54	59	
	1028 4″ diameter DR 17					54			54	59	
	1032 6″ diameter DR 11					78.50			78.50	86	
	1036 6″ diameter DR 17					78.50			78.50	86	
	1040 8″ diameter DR 11					112			112	123	
	1044 8″ diameter DR 26					112			112	123	
	1048 10″ diameter DR 11					168			168	185	
	1052 10″ diameter DR 26					168			168	185	
	1056 12″ diameter DR 11					243			243	267	
	1060 12″ diameter DR 26					243			243	267	
	1064 16″ diameter DR 11					520			520	570	
	1068 16″ diameter DR 26					520			520	570	
	1072 18″ diameter DR 11					665			665	730	
	1076 18″ diameter DR 26					665			665	730	
	1080 20″ diameter DR 11					950			950	1,050	
	1084 20″ diameter DR 26					950			950	1,050	
	1088 22″ diameter DR 11					1,125			1,125	1,225	
	1092 22″ diameter DR 26					1,125			1,125	1,225	
	1096 24″ diameter DR 17					1,175			1,175	1,300	
	1100 24″ diameter DR 32.5					1,175			1,175	1,300	
	1104 28″ diameter DR 15.5					2,025			2,025	2,225	
	1108 28″ diameter DR 32.5					1,975			1,975	2,175	
	1112 30″ diameter DR 11					2,175			2,175	2,375	
	1116 30″ diameter DR 21					1,875			1,875	2,075	
	1120 36″ diameter DR 26					2,800			2,800	3,075	
	1124 42″ diameter DR 26					3,150			3,150	3,475	
	1128 48″ diameter DR 26					4,450			4,450	4,900	
	1132 54″ diameter DR 26				▼	4,975			4,975	5,475	
	1200 Reducer										

15108 | Plastic Pipe & Fittings

	15108 | Plastic Pipe & Fittings	CREW	DAILY OUTPUT	LABOR-HOURS	UNIT	2001 BARE COSTS MAT.	LABOR	EQUIP.	TOTAL	TOTAL INCL O&P
590 1204	1-1/2" x 1" diameter DR 11				Ea.	6.90			6.90	7.55
1208	2" x 1-1/2" diameter DR 11					10.75			10.75	11.85
1212	3" x 2" diameter DR 11					14			14	15.40
1216	4" x 2" diameter DR 11					21			21	23
1220	4" x 3" diameter DR 11					21			21	23
1224	6" x 4" diameter DR 11					49			49	54
1228	8" x 6" diameter DR 11					86			86	94.50
1232	10" x 8" diameter DR 11					115			115	127
1236	12" x 8" diameter DR 11					215			215	237
1240	12" x 10" diameter DR 11					154			154	169
1244	14" x 12" diameter DR 11					173			173	190
1248	16" x 14" diameter DR 11					208			208	229
1252	18" x 16" diameter DR 11					415			415	460
1256	20" x 18" diameter DR 11					565			565	620
1260	22" x 20" diameter DR 11					1,150			1,150	1,275
1264	24" x 22" diameter DR 11					740			740	810
1268	26" x 24" diameter DR 11					865			865	950
1272	28" x 24" diameter DR 11					995			995	1,100
1276	32" x 28" diameter DR 17					1,200			1,200	1,325
1280	36" x 32" diameter DR 17					1,425			1,425	1,550
4000	Welding labor per joint, not including welding machine									
4010	Pipe joint size (cost based on thickest wall for each dia.)									
4030	1" pipe size	4 Skwk	273	.117	Ea.		3.50		3.50	5.50
4040	1-1/2" pipe size		175	.183			5.45		5.45	8.55
4050	2" pipe size		128	.250			7.45		7.45	11.70
4060	3" pipe size		100	.320			9.55		9.55	15
4070	4" pipe size		77	.416			12.40		12.40	19.45
4080	6" pipe size	5 Skwk	63	.635			18.95		18.95	30
4090	8" pipe size		48	.833			25		25	39
4100	10" pipe size		40	1			30		30	47
4110	12" pipe size	6 Skwk	41	1.171			35		35	55
4120	16" pipe size		34	1.412			42		42	66
4130	18" pipe size		32	1.500			45		45	70.50
4140	20" pipe size	8 Skwk	37	1.730			51.50		51.50	81
4150	22" pipe size		35	1.829			54.50		54.50	85.50
4160	24" pipe size		34	1.882			56		56	88
4170	28" pipe size		33	1.939			58		58	91
4180	30" pipe size		32	2			59.50		59.50	93.50
4190	36" pipe size		31	2.065			61.50		61.50	96.50
4200	42" pipe size		30	2.133			63.50		63.50	100
4210	48" pipe size	9 Skwk	33	2.182			65		65	102
4220	54" pipe size	"	31	2.323			69.50		69.50	109
4300	Note: Cost for set up each time welder is moved.									
4301	Add 50% of a weld cost									
4310	Welder usually remains stationary with pipe moved through it.									
4340	Weld machine, rental per day based on dia. capacity									
4350	1" thru 2" diameter				Ea.			35		38.50
4360	3" thru 4" diameter							40		44
4370	6" thru 8" diameter							90		99
4380	10" thru 12" diameter							155		170.50
4390	16" thru 18" diameter							225		247.50
4400	20" thru 24" diameter							435		478.50
4410	28" thru 32" diameter							475		525
4420	36" diameter							495		545
4430	42" thru 54" diameter							775		855
5000	Dual wall contained pipe									

	15108 | Plastic Pipe & Fittings	CREW	DAILY OUTPUT	LABOR-HOURS	UNIT	2001 BARE COSTS MAT.	LABOR	EQUIP.	TOTAL	TOTAL INCL O&P
590 5040	Straight									
5054	1" DR 11 x 3" DR 11				L.F.	5.05			5.05	5.55
5058	1" DR 11 x 4" DR 11					5.55			5.55	6.10
5062	1-1/2" DR 11 x 4" DR 17					5.85			5.85	6.45
5066	2" DR 11 x 4" DR 17					5.90			5.90	6.50
5070	2" DR 11 x 6" DR 17					9.25			9.25	10.15
5074	3" DR 11 x 6" DR 17					10.15			10.15	11.20
5078	3" DR 11 x 6" DR 26					8.15			8.15	8.95
5082	3" DR 17 x 8" DR 11					16.80			16.80	18.45
5086	3" DR 17 x 8" DR 17					13.25			13.25	14.60
5090	4" DR 11 x 8" DR 17					15			15	16.50
5094	4" DR 17 x 8" DR 26					11.85			11.85	13.05
5098	6" DR 11 x 10" DR 17					24			24	26.50
5102	6" DR 17 x 10" DR 26					18.25			18.25	20
5106	6" DR 26 x 10" DR 26					16.85			16.85	18.50
5110	8" DR 17 x 12" DR 26					26.50			26.50	29
5114	8" DR 26 x 12" DR 32.5					22			22	24
5118	10" DR 17 x 14" DR 26					34.50			34.50	38
5122	10" DR 17 x 16" DR 26					39			39	43
5126	10" DR 26 x 16" DR 26					35.50			35.50	39
5130	12" DR 26 x 16" DR 26					40			40	44
5134	12" DR 17 x 18" DR 26					51.50			51.50	56.50
5138	12" DR 26 x 18" DR 26					46			46	50.50
5142	14" DR 26 x 20" DR 32.5					50.50			50.50	55.50
5146	16" DR 26 x 22" DR 32.5					76			76	83.50
5150	18" DR 26 x 24" DR 32.5					70			70	77
5154	20" DR 32.5 x 28" DR 32.5					90.50			90.50	99.50
5158	22" DR 32.5 x 30" DR 32.5					90.50			90.50	99.50
5162	24" DR 32.5 x 32" DR 32.5					115			115	127
5166	36" DR 32.5 x 42" DR 32.5				↓	195			195	215
5300	Force transfer coupling									
5354	1" DR 11 x 3" DR 11				Ea.	158			158	174
5358	1" DR 11 x 4" DR 17					166			166	183
5362	1-1/2" DR 11 x 4" DR 17					174			174	192
5366	2" DR 11 x 4" DR 17					182			182	200
5370	2" DR 11 x 6" DR 17					264			264	291
5374	3" DR 11 x 6" DR 17					264			264	291
5378	3" DR 11 x 6" DR 26					264			264	291
5382	3" DR 11 x 8" DR 11					279			279	305
5386	3" DR 11 x 8" DR 17					279			279	305
5390	4" DR 11 x 8" DR 17					279			279	305
5394	4" DR 17 x 8" DR 26					245			245	270
5398	6" DR 11 x 10" DR 17					355			355	390
5402	6" DR 17 x 10" DR 26					292			292	320
5406	6" DR 26 x 10" DR 26					292			292	320
5410	8" DR 17 x 12" DR 26					390			390	425
5414	8" DR 26 x 12" DR 32.5					315			315	345
5418	10" DR 17 x 14" DR 26					510			510	560
5422	10" DR 17 x 16" DR 26					510			510	560
5426	10" DR 26 x 16" DR 26					510			510	560
5430	12" DR 26 x 16" DR 26					635			635	700
5434	12" DR 17 x 18" DR 26					655			655	720
5438	12" DR 26 x 18" DR 26					655			655	720
5442	14" DR 26 x 20" DR 32.5					690			690	760
5446	16" DR 26 x 22" DR 32.5					710			710	780
5450	18" DR 26 x 24" DR 32.5				↓	770			770	845

590

15108 | Plastic Pipe & Fittings

			CREW	DAILY OUTPUT	LABOR-HOURS	UNIT	2001 BARE COSTS MAT.	LABOR	EQUIP.	TOTAL	TOTAL INCL O&P	
590	5454	20″ DR 32.5 x 28″ DR 32.5				Ea.	940			940	1,025	590
	5458	22″ DR 32.5 x 30″ DR 32.5					1,025			1,025	1,150	
	5462	24″ DR 32.5 x 32″ DR 32.5					1,125			1,125	1,225	
	5466	36″ DR 32.5 x 42″ DR 32.5				↓	1,850			1,850	2,050	
	5600	90° Elbow										
	5654	1″ DR 11 x 3″ DR 11				Ea.	131			131	144	
	5658	1″ DR 11 x 4″ DR 17					125			125	138	
	5662	1-1/2″ DR 11 x 4″ DR 17					137			137	151	
	5666	2″ DR 11 x 4″ DR 17					145			145	160	
	5670	2″ DR 11 x 6″ DR 17					184			184	203	
	5674	3″ DR 11 x 6″ DR 17					211			211	232	
	5678	3″ DR 17 x 6″ DR 26					168			168	185	
	5682	3″ DR 17 x 8″ DR 11					340			340	375	
	5686	3″ DR 17 x 8″ DR 17					265			265	292	
	5690	4″ DR 11 x 8″ DR 17					310			310	345	
	5694	4″ DR 17 x 8″ DR 26					248			248	273	
	5698	6″ DR 11 x 10″ DR 17					415			415	455	
	5702	6″ DR 17 x 10″ DR 26					320			320	350	
	5706	6″ DR 26 x 10″ DR 26					298			298	330	
	5710	8″ DR 17 x 12″ DR 26					475			475	520	
	5714	8″ DR 26 x 12″ DR 32.5					435			435	480	
	5718	10″ DR 17 x 14″ DR 26					745			745	820	
	5722	10″ DR 17 x 16″ DR 26					710			710	785	
	5726	10″ DR 26 x 16″ DR 26					660			660	725	
	5730	12″ DR 26 x 16″ DR 26					730			730	805	
	5734	12″ DR 17 x 18″ DR 26					870			870	960	
	5738	12″ DR 26 x 18″ DR 26					800			800	880	
	5742	14″ DR 26 x 20″ DR 32.5					1,050			1,050	1,150	
	5746	16″ DR 26 x 22″ DR 32.5					1,075			1,075	1,175	
	5750	18″ DR 26 x 24″ DR 32.5					1,300			1,300	1,425	
	5754	20″ DR 32.5 x 28″ DR 32.5					1,500			1,500	1,650	
	5758	22″ DR 32.5 x 30″ DR 32.5					2,050			2,050	2,250	
	5762	24″ DR 32.5 x 32″ DR 32.5					2,000			2,000	2,200	
	5766	36″ DR 32.5 x 42″ DR 32.5				↓	3,250			3,250	3,550	
	5800	45° Elbow										
	5804	1″ DR 11 x 3″ DR 11				Ea.	78.50			78.50	86	
	5808	1″ DR 11 x 4″ DR 17					78			78	85.50	
	5812	1-1/2″ DR 11 x 4″ DR 17					83.50			83.50	92	
	5816	2″ DR 11 x 4″ DR 17					91			91	100	
	5820	2″ DR 11 x 6″ DR 17					114			114	126	
	5824	3″ DR 11 x 6″ DR 17					127			127	140	
	5828	3″ DR 17 x 6″ DR 26					105			105	116	
	5832	3″ DR 17 x 8″ DR 11					194			194	214	
	5836	3″ DR 17 x 8″ DR 17					155			155	171	
	5840	4″ DR 11 x 8″ DR 17					182			182	200	
	5844	4″ DR 17 x 8″ DR 26					148			148	163	
	5848	6″ DR 11 x 10″ DR 17					240			240	264	
	5852	6″ DR 17 x 10″ DR 26					188			188	207	
	5856	6″ DR 26 x 10″ DR 26					177			177	195	
	5860	8″ DR 17 x 12″ DR 26					286			286	315	
	5864	8″ DR 26 x 12″ DR 32.5					267			267	294	
	5868	10″ DR 17 x 14″ DR 26					450			450	495	
	5872	10″ DR 17 x 16″ DR 26					440			440	480	
	5876	10″ DR 26 x 16″ DR 26					410			410	450	
	5880	12″ DR 26 x 16″ DR 26					465			465	510	
	5884	12″ DR 17 x 18″ DR 26				↓	545			545	600	

	15108 | Plastic Pipe & Fittings	CREW	DAILY OUTPUT	LABOR-HOURS	UNIT	2001 BARE COSTS MAT.	LABOR	EQUIP.	TOTAL	TOTAL INCL O&P	
590	5888 12″ DR 26 x 18″ DR 26				Ea.	510			510	560	590
	5892 14″ DR 26 x 20″ DR 32.5					655			655	720	
	5896 16″ DR 26 x 22″ DR 32.5					685			685	755	
	5900 18″ DR 26 x 24″ DR 32.5					810			810	895	
	5904 20″ DR 32.5 x 28″ DR 32.5					995			995	1,100	
	5908 22″ DR 32.5 x 30″ DR 32.5					1,275			1,275	1,400	
	5912 24″ DR 32.5 x 32″ DR 32.5					1,250			1,250	1,375	
	5916 36″ DR 32.5 x 42″ DR 32.5					1,950			1,950	2,150	
	6000 Access port with 4″ riser										
	6050 1″ DR 11 x 4″ DR 17				Ea.	171			171	188	
	6054 1-1/2″ DR 11 x 4″ DR 17					175			175	192	
	6058 2″ DR 11 x 6″ DR 17					213			213	234	
	6062 3″ DR 11 x 6″ DR 17					214			214	235	
	6066 3″ DR 17 x 6″ DR 26					210			210	231	
	6070 3″ DR 17 x 8″ DR 11					240			240	264	
	6074 3″ DR 17 x 8″ DR 17					229			229	252	
	6078 4″ DR 11 x 8″ DR 17					237			237	261	
	6082 4″ DR 17 x 8″ DR 26					227			227	250	
	6086 6″ DR 11 x 10″ DR 17					297			297	325	
	6090 6″ DR 17 x 10″ DR 26					273			273	300	
	6094 6″ DR 26 x 10″ DR 26					268			268	295	
	6098 8″ DR 17 x 12″ DR 26					287			287	315	
	6102 8″ DR 26 x 12″ DR 32.5					274			274	300	
	6200 End termination with vent plug										
	6204 1″ DR 11 x 3″ DR 11				Ea.	142			142	156	
	6208 1″ DR 11 x 4″ DR 17					151			151	167	
	6212 1-1/2″ DR 11 x 4″ DR 17					151			151	167	
	6216 2″ DR 11 x 4″ DR 17					165			165	181	
	6220 2″ DR 11 x 6″ DR 17					188			188	206	
	6224 3″ DR 11 x 6″ DR 17					188			188	206	
	6228 3″ DR 17 x 6″ DR 26					188			188	206	
	6232 3″ DR 17 x 8″ DR 11					260			260	285	
	6236 3″ DR 17 x 8″ DR 17					260			260	285	
	6240 4″ DR 11 x 8″ DR 17					260			260	285	
	6244 4″ DR 17 x 8″ DR 26					225			225	248	
	6248 6″ DR 11 x 10″ DR 17					310			310	340	
	6252 6″ DR 17 x 10″ DR 26					253			253	278	
	6256 6″ DR 26 x 10″ DR 26					253			253	278	
	6260 8″ DR 17 x 12″ DR 26					375			375	410	
	6264 8″ DR 26 x 12″ DR 32.5					300			300	330	
	6268 10″ DR 17 x 14″ DR 26					390			390	430	
	6272 10″ DR 17 x 16″ DR 26					390			390	430	
	6276 10″ DR 26 x 16″ DR 26					390			390	430	
	6280 12″ DR 26 x 16″ DR 26					390			390	430	
	6284 12″ DR 17 x 18″ DR 26					515			515	565	
	6288 12″ DR 26 x 18″ DR 26					515			515	565	
	6292 14″ DR 26 x 20″ DR 32.5					525			525	575	
	6296 16″ DR 26 x 22″ DR 32.5					605			605	665	
	6300 18″ DR 26 x 24″ DR 32.5					625			625	690	
	6304 20″ DR 32.5 x 28″ DR 32.5					800			800	880	
	6308 22″ DR 32.5 x 30″ DR 32.5					880			880	965	
	6312 24″ DR 32.5 x 32″ DR 32.5					970			970	1,075	
	6316 36″ DR 32.5 x 42″ DR 32.5					1,425			1,425	1,550	
	6600 Tee										
	6604 1″ DR 11 x 3″ DR 11				Ea.	144			144	159	
	6608 1″ DR 11 x 4″ DR 17					179			179	197	

15108 | Plastic Pipe & Fittings		CREW	DAILY OUTPUT	LABOR-HOURS	UNIT	2001 BARE COSTS MAT.	LABOR	EQUIP.	TOTAL	TOTAL INCL O&P
590 6612	1-1/2" DR 11 x 4" DR 17				Ea.	199			199	219
6616	2" DR 11 x 4" DR 17					217			217	239
6620	2" DR 11 x 6" DR 17					264			264	290
6624	3" DR 11 x 6" DR 17					305			305	335
6628	3" DR 17 x 6" DR 26					296			296	325
6632	3" DR 17 x 8" DR 11					360			360	395
6636	3" DR 17 x 8" DR 17					340			340	375
6640	4" DR 11 x 8" DR 17					370			370	410
6644	4" DR 17 x 8" DR 26					355			355	390
6648	6" DR 11 x 10" DR 17					465			465	510
6652	6" DR 17 x 10" DR 26					440			440	485
6656	6" DR 26 x 10" DR 26					435			435	480
6660	8" DR 17 x 12" DR 26					435			435	480
6664	8" DR 26 x 12" DR 32.5					570			570	625
6668	10" DR 17 x 14" DR 26					655			655	725
6672	10" DR 17 x 16" DR 26					690			690	760
6676	10" DR 26 x 16" DR 26					690			690	760
6680	12" DR 26 x 16" DR 26					745			745	815
6684	12" DR 17 x 18" DR 26					745			745	815
6688	12" DR 26 x 18" DR 26					825			825	905
6692	14" DR 26 x 20" DR 32.5					1,100			1,100	1,200
6696	16" DR 26 x 22" DR 32.5					1,325			1,325	1,475
6700	18" DR 26 x 24" DR 32.5					1,450			1,450	1,600
6704	20" DR 32.5 x 28" DR 32.5					1,825			1,825	2,025
6708	22" DR 32.5 x 30" DR 32.5					2,175			2,175	2,400
6712	24" DR 32.5 x 32" DR 32.5					2,600			2,600	2,875
6716	36" DR 32.5 x 42" DR 32.5				▼	5,300			5,300	5,850
6800	Wye									
6816	2" DR 11 x 4" DR 17				Ea.	274			274	300
6820	2" DR 11 x 6" DR 17					310			310	340
6824	3" DR 11 x 6" DR 17					315			315	345
6828	3" DR 17 x 6" DR 26					305			305	335
6832	3" DR 17 x 8" DR 11					380			380	420
6836	3" DR 17 x 8" DR 17					355			355	395
6840	4" DR 11 x 8" DR 17					390			390	430
6844	4" DR 17 x 8" DR 26					370			370	405
6848	6" DR 11 x 10" DR 17					480			480	530
6852	6" DR 17 x 10" DR 26					450			450	495
6856	6" DR 26 x 10" DR 26					445			445	490
6860	8" DR 17 x 12" DR 26					640			640	700
6864	8" DR 26 x 12" DR 32.5					615			615	680
6868	10" DR 17 x 14" DR 26					770			770	845
6872	10" DR 17 x 16" DR 26					815			815	900
6876	10" DR 26 x 16" DR 26					805			805	885
6880	12" DR 26 x 16" DR 26					905			905	995
6884	12" DR 17 x 18" DR 26					1,025			1,025	1,125
6888	12" DR 26 x 18" DR 26					1,025			1,025	1,125
6892	14" DR 26 x 20" DR 32.5					1,250			1,250	1,375
6896	16" DR 26 x 22" DR 32.5					1,800			1,800	2,000
6900	18" DR 26 x 24" DR 32.5					2,125			2,125	2,350
6904	20" DR 32.5 x 28" DR 32.5					2,675			2,675	2,925
6908	22" DR 32.5 x 30" DR 32.5					3,025			3,025	3,350
6912	24" DR 32.5 x 32" DR 32.5				▼	3,450			3,450	3,800
9000	Welding labor per joint, not including welding machine									
9010	Pipe joint size, outer pipe (cost based on the thickest walls)									
9020	Straight pipe									590

MECHANICAL 15

15108 | Plastic Pipe & Fittings

			CREW	DAILY OUTPUT	LABOR-HOURS	UNIT	2001 BARE COSTS MAT.	LABOR	EQUIP.	TOTAL	TOTAL INCL O&P	
590	9050	3″ pipe size	4 Skwk	96	.333	Ea.		9.95		9.95	15.60	590
	9060	4″ pipe size	″	77	.416			12.40		12.40	19.45	
	9070	6″ pipe size	5 Skwk	60	.667			19.90		19.90	31	
	9080	8″ pipe size	″	40	1			30		30	47	
	9090	10″ pipe size	6 Skwk	41	1.171			35		35	55	
	9100	12″ pipe size		39	1.231			36.50		36.50	57.50	
	9110	14″ pipe size		38	1.263			37.50		37.50	59	
	9120	16″ pipe size	↓	35	1.371			41		41	64.50	
	9130	18″ pipe size	8 Skwk	45	1.422			42.50		42.50	66.50	
	9140	20″ pipe size		42	1.524			45.50		45.50	71.50	
	9150	22″ pipe size		40	1.600			48		48	75	
	9160	24″ pipe size		38	1.684			50.50		50.50	79	
	9170	28″ pipe size		37	1.730			51.50		51.50	81	
	9180	30″ pipe size		36	1.778			53		53	83.50	
	9190	32″ pipe size		35	1.829			54.50		54.50	85.50	
	9200	42″ pipe size	↓	32	2	↓		59.50		59.50	93.50	
	9300	Note: Cost for set up each time welder is moved.										
	9301	Add 100% of weld labor cost										
	9310	For handling between fitting welds add 50% of weld labor cost										
	9320	Welder usually remains stationary with pipe moved through it										
	9360	Weld machine, rental per day based on dia. capacity										
	9380	3″ thru 4″ diameter				Ea.			55		60.50	
	9390	6″ thru 8″ diameter							180		198	
	9400	10″ thru 12″ diameter							270		297	
	9410	14″ thru 18″ diameter							365		402	
	9420	20″ thru 24″ diameter							650		715	
	9430	28″ thru 32″ diameter							765		840	
	9440	42″ thru 58″ diameter				↓			790		870	

15110 | Valves

			CREW	DAILY OUTPUT	LABOR-HOURS	UNIT	MAT.	LABOR	EQUIP.	TOTAL	TOTAL INCL O&P	
100	0010	**VALVES, BRASS**										100
	0030	For motorized valves, see Division 13838-200										
	0500	Gas cocks, threaded										
	0510	1/4″ size	1 Plum	26	.308	Ea.	7.80	10.60		18.40	24.50	
	0520	3/8″ size		24	.333		8.25	11.50		19.75	26.50	
	0530	1/2″ size		24	.333		7.65	11.50		19.15	26	
	0540	3/4″ size		22	.364		10.30	12.55		22.85	30.50	
	0550	1″ size		19	.421		12.60	14.50		27.10	36	
	0560	1-1/4″ size		15	.533		36	18.35		54.35	67	
	0570	1-1/2″ size		13	.615		45	21		66	81.50	
	0580	2″ size	↓	11	.727	↓	65.50	25		90.50	110	
	0670	For larger sizes use lubricated plug valve, Section 15110-600										
160	0010	**VALVES, BRONZE** R15100-090										160
	1020	Angle, 150 lb., rising stem, threaded										
	1030	1/8″ size	1 Plum	24	.333	Ea.	42	11.50		53.50	63.50	
	1040	1/4″ size		24	.333		42	11.50		53.50	63.50	
	1050	3/8″ size		24	.333		42	11.50		53.50	63.50	
	1060	1/2″ size		22	.364		42	12.55		54.55	65	
	1070	3/4″ size		20	.400		56.50	13.80		70.30	83	
	1080	1″ size		19	.421		81	14.50		95.50	111	
	1090	1-1/4″ size		15	.533		104	18.35		122.35	143	
	1100	1-1/2″ size		13	.615		135	21		156	181	
	1110	2″ size	↓	11	.727	↓	219	25		244	279	
	1380	Ball, 150 psi, threaded										
	1400	1/4″ size	1 Plum	24	.333	Ea.	6.40	11.50		17.90	24.50	
	1430	3/8″ size	↓	24	.333	↓	6.40	11.50		17.90	24.50	

15 MECHANICAL

15110 | Valves

		Description	CREW	DAILY OUTPUT	LABOR-HOURS	UNIT	2001 BARE COSTS MAT.	LABOR	EQUIP.	TOTAL	TOTAL INCL O&P
160	1450	1/2" size R15100-090	1 Plum	22	.364	Ea.	6.40	12.55		18.95	26
	1460	3/4" size		20	.400		10.60	13.80		24.40	32.50
	1470	1" size		19	.421		13.30	14.50		27.80	36.50
	1480	1-1/4" size		15	.533		22.50	18.35		40.85	52.50
	1490	1-1/2" size		13	.615		28.50	21		49.50	63.50
	1500	2" size		11	.727		35.50	25		60.50	77.50
	1510	2-1/2" diameter		9	.889		169	30.50		199.50	232
	1520	3" diameter		8	1		196	34.50		230.50	267
	1600	Butterfly, 175 psi, full port, solder or threaded ends									
	1610	Stainless steel disc and stem									
	1620	1/4" size	1 Plum	24	.333	Ea.	5.75	11.50		17.25	23.50
	1630	3/8" size		24	.333		5.75	11.50		17.25	23.50
	1640	1/2" size		22	.364		6.10	12.55		18.65	25.50
	1650	3/4" size		20	.400		9.85	13.80		23.65	32
	1660	1" size		19	.421		12.25	14.50		26.75	35.50
	1670	1-1/4" size		15	.533		19.45	18.35		37.80	49
	1680	1-1/2" size		13	.615		25	21		46	59.50
	1690	2" size		11	.727		31.50	25		56.50	72.50
	1750	Check, swing, class 150, regrinding disc, threaded									
	1800	1/8" size	1 Plum	24	.333	Ea.	22.50	11.50		34	42.50
	1830	1/4" size		24	.333		22.50	11.50		34	42.50
	1840	3/8" size		24	.333		22.50	11.50		34	42.50
	1850	1/2" size		24	.333		22.50	11.50		34	42.50
	1860	3/4" size		20	.400		30	13.80		43.80	54
	1870	1" size		19	.421		45	14.50		59.50	71.50
	1880	1-1/4" size		15	.533		62.50	18.35		80.85	96
	1890	1-1/2" size		13	.615		73.50	21		94.50	113
	1900	2" size		11	.727		108	25		133	157
	1910	2-1/2" size	Q-1	15	1.067		179	33		212	247
	1920	3" size	"	13	1.231		265	38		303	350
	2000	For 200 lb, add					5%	10%			
	2040	For 300 lb, add					15%	15%			
	2060	Check swing, 300#, sweat, 3/8" size	1 Plum	24	.333	Ea.	12.90	11.50		24.40	31.50
	2070	1/2" size		24	.333		12.90	11.50		24.40	31.50
	2080	3/4" size		20	.400		16.45	13.80		30.25	39
	2090	1" size		19	.421		23	14.50		37.50	47
	2100	1-1/4" size		15	.533		29.50	18.35		47.85	60
	2110	1-1/2 size		13	.615		39.50	21		60.50	75.50
	2120	2" size		11	.727		59	25		84	103
	2130	2-1/2" size	Q-1	15	1.067		141	33		174	205
	2140	3" size	"	13	1.231		188	38		226	264
	2850	Gate, N.R.S., soldered, 125 psi									
	2900	3/8" size	1 Plum	24	.333	Ea.	16.85	11.50		28.35	36
	2920	1/2" size		24	.333		14.75	11.50		26.25	33.50
	2940	3/4" size		20	.400		16.85	13.80		30.65	39.50
	2950	1" size		19	.421		24	14.50		38.50	48.50
	2960	1-1/4" size		15	.533		40	18.35		58.35	71
	2970	1-1/2" size		13	.615		45	21		66	81.50
	2980	2" size		11	.727		57.50	25		82.50	102
	2990	2-1/2" size	Q-1	15	1.067		145	33		178	209
	3000	3" size	"	13	1.231		189	38		227	266
	3350	Threaded, class 150									
	3410	1/4" size	1 Plum	24	.333	Ea.	25	11.50		36.50	45
	3420	3/8" size		24	.333		25	11.50		36.50	45
	3430	1/2" size		24	.333		23	11.50		34.50	42.50
	3440	3/4" size		20	.400		25.50	13.80		39.30	49

15110 | Valves

		Description	Crew	Daily Output	Labor-Hours	Unit	2001 Bare Costs Mat.	2001 Bare Costs Labor	2001 Bare Costs Equip.	2001 Bare Costs Total	Total Incl O&P	
160	3450	1" size (R15100-090)	1 Plum	19	.421	Ea.	33	14.50		47.50	58.50	160
	3460	1-1/4" size		15	.533		44.50	18.35		62.85	76.50	
	3470	1-1/2" size		13	.615		56	21		77	94	
	3480	2" size		11	.727		77	25		102	123	
	3490	2-1/2" size	Q-1	15	1.067		178	33		211	246	
	3500	3" size	"	13	1.231		251	38		289	335	
	3600	Gate, flanged, 150 lb.										
	3610	1" size	1 Plum	7	1.143	Ea.	415	39.50		454.50	520	
	3620	1-1/2" size		6	1.333		515	46		561	640	
	3630	2" size		5	1.600		765	55		820	925	
	3640	3" size	Q-1	4.50	3.556		1,375	110		1,485	1,675	
	3850	Rising stem, soldered, 300 psi										
	3900	3/8" size	1 Plum	24	.333	Ea.	37.50	11.50		49	59	
	3920	1/2" size		24	.333		35	11.50		46.50	56	
	3940	3/4" size		20	.400		44.50	13.80		58.30	70	
	3950	1" size		19	.421		55	14.50		69.50	82.50	
	3960	1-1/4" size		15	.533		76	18.35		94.35	111	
	3970	1-1/2" size		13	.615		93	21		114	134	
	3980	2" size		11	.727		137	25		162	189	
	3990	2-1/2" size	Q-1	15	1.067		330	33		363	415	
	4000	3" size	"	13	1.231		385	38		423	485	
	4250	Threaded, class 150										
	4310	1/4" size	1 Plum	24	.333	Ea.	21.50	11.50		33	41.50	
	4320	3/8" size		24	.333		21.50	11.50		33	41.50	
	4330	1/2" size		24	.333		20.50	11.50		32	40	
	4340	3/4" size		20	.400		24	13.80		37.80	47.50	
	4350	1" size		19	.421		32	14.50		46.50	57	
	4360	1-1/4" size		15	.533		43	18.35		61.35	75	
	4370	1-1/2" size		13	.615		54	21		75	91.50	
	4380	2" size		11	.727		73.50	25		98.50	119	
	4390	2-1/2" size	Q-1	15	1.067		171	33		204	238	
	4400	3" size	"	13	1.231		240	38		278	320	
	4500	For 300 psi, threaded, add					100%	15%				
	4540	For chain operated type, add					15%					
	4850	Globe, class 150, rising stem, threaded										
	4920	1/4" size	1 Plum	24	.333	Ea.	31.50	11.50		43	52	
	4940	3/8" size		24	.333		31.50	11.50		43	52	
	4950	1/2" size		24	.333		31.50	11.50		43	52	
	4960	3/4" size		20	.400		43	13.80		56.80	68	
	4970	1" size		19	.421		67	14.50		81.50	95.50	
	4980	1-1/4" size		15	.533		105	18.35		123.35	144	
	4990	1-1/2" size		13	.615		128	21		149	172	
	5000	2" size		11	.727		191	25		216	248	
	5010	2-1/2" size	Q-1	15	1.067		385	33		418	475	
	5020	3" size	"	13	1.231		550	38		588	660	
	5120	For 300 lb threaded, add					50%	15%				
	5600	Relief, pressure & temperature, self-closing, ASME, threaded										
	5640	3/4" size	1 Plum	28	.286	Ea.	65	9.85		74.85	86.50	
	5650	1" size		24	.333		94.50	11.50		106	121	
	5660	1-1/4" size		20	.400		189	13.80		202.80	229	
	5670	1-1/2" size		18	.444		360	15.30		375.30	425	
	5680	2" size		16	.500		395	17.25		412.25	460	
	5950	Pressure, poppet type, threaded										
	6000	1/2" size	1 Plum	30	.267	Ea.	13.15	9.20		22.35	28.50	
	6040	3/4" size	"	28	.286	"	44	9.85		53.85	63.50	
	6400	Pressure, water, ASME, threaded										

15110 | Valves

R15100-090

160		Crew	Daily Output	Labor-Hours	Unit	Mat. (2001 Bare Costs)	Labor (2001 Bare Costs)	Equip. (2001 Bare Costs)	Total (2001 Bare Costs)	Total Incl O&P
6440	3/4" size	1 Plum	28	.286	Ea.	43	9.85		52.85	62.50
6450	1" size		24	.333		91	11.50		102.50	117
6460	1-1/4"size		20	.400		149	13.80		162.80	185
6470	1-1/2" size		18	.444		206	15.30		221.30	249
6480	2" size		16	.500		298	17.25		315.25	350
6490	2-1/2" size		15	.533		875	18.35		893.35	990
6900	Reducing, water pressure									
6920	300 psi to 25-75 psi, threaded or sweat									
6940	1/2" size	1 Plum	24	.333	Ea.	108	11.50		119.50	136
6950	3/4" size		20	.400		108	13.80		121.80	140
6960	1" size		19	.421		167	14.50		181.50	206
6970	1-1/4" size		15	.533		300	18.35		318.35	360
6980	1-1/2" size		13	.615		450	21		471	530
6990	2" size		11	.727		655	25		680	760
7100	For built-in by-pass or 10-35 psi, add					7			7	7.70
7700	High capacity, 250 psi to 25-75 psi, threaded									
7740	1/2" size	1 Plum	24	.333	Ea.	108	11.50		119.50	136
7780	3/4" size		20	.400		108	13.80		121.80	140
7790	1" size		19	.421		167	14.50		181.50	206
7800	1-1/4" size		15	.533		300	18.35		318.35	360
7810	1-1/2" size		13	.615		450	21		471	530
7820	2" size		11	.727		655	25		680	760
7830	2-1/2" size		9	.889		870	30.50		900.50	1,000
7840	3" size		8	1		1,025	34.50		1,059.50	1,175
7850	3" flanged (iron body)	Q-1	10	1.600		1,050	49.50		1,099.50	1,225
7860	4" flanged (iron body)	"	8	2		1,750	62		1,812	2,000
7920	For higher pressure, add					25%				
8000	Silent check, bronze trim									
8010	Compact wafer type, for 125 or 150 lb. flanges									
8020	1-1/2" size	1 Plum	11	.727	Ea.	255	25		280	320
8021	2" size	"	9	.889		274	30.50		304.50	345
8022	2-1/2" size	Q-1	9	1.778		288	55		343	400
8023	3" size		8	2		320	62		382	450
8024	4" size		5	3.200		550	99		649	755
8025	5" size	Q-2	6	4		760	129		889	1,025
8026	6" size	"	5	4.800		865	154		1,019	1,175
8050	For 250 or 300 lb flanges, thru 6" no change									
8060	Full flange wafer type, 150 lb.									
8063	1-1/2" size	1 Plum	11	.727	Ea.	340	25		365	415
8064	2" size	"	9	.889		460	30.50		490.50	550
8065	2-1/2" size	Q-1	9	1.778		520	55		575	655
8066	3" size		8	2		575	62		637	725
8067	4" size		5	3.200		850	99		949	1,075
8068	5" size	Q-2	6	4		1,100	129		1,229	1,400
8069	6" size	"	5	4.800		1,450	154		1,604	1,800
8080	For 300 lb., add					40%	10%			
8100	Globe type, 150 lb.									
8110	2" size	1 Plum	9	.889	Ea.	535	30.50		565.50	635
8111	2-1/2" size	Q-1	9	1.778		660	55		715	815
8112	3" size		8	2		790	62		852	965
8113	4" size		5	3.200		1,100	99		1,199	1,350
8114	5" size	Q-2	6	4		1,350	129		1,479	1,675
8115	6" size	"	5	4.800		1,875	154		2,029	2,275
8130	For 300 lb., add					20%	10%			
8140	Screwed end type, 250 lb.									
8141	1/2" size	1 Plum	24	.333	Ea.	48	11.50		59.50	70.50

15110 | Valves

			CREW	DAILY OUTPUT	LABOR-HOURS	UNIT	2001 BARE COSTS MAT.	LABOR	EQUIP.	TOTAL	TOTAL INCL O&P	
160	8142	3/4" size (R15100-090)	1 Plum	20	.400	Ea.	48	13.80		61.80	74	160
	8143	1" size		19	.421		55	14.50		69.50	82.50	
	8144	1-1/4" size		15	.533		76	18.35		94.35	111	
	8145	1-1/2" size		13	.615		85	21		106	126	
	8146	2" size		11	.727		114	25		139	164	
	8350	Tempering, water, sweat connections										
	8400	1/2" size	1 Plum	24	.333	Ea.	40	11.50		51.50	61.50	
	8440	3/4" size	"	20	.400	"	49	13.80		62.80	75	
	8650	Threaded connections										
	8700	1/2" size	1 Plum	24	.333	Ea.	49	11.50		60.50	71.50	
	8740	3/4" size		20	.400		187	13.80		200.80	227	
	8750	1" size		19	.421		206	14.50		220.50	249	
	8760	1-1/4" size		15	.533		330	18.35		348.35	390	
	8770	1-1/2" size		13	.615		355	21		376	425	
	8780	2" size		11	.727		540	25		565	630	
200	0010	**VALVES, IRON BODY** (R15100-090)										200
	0020	For grooved joint, see Division 15107-690										
	0100	Angle, 125 lb.										
	0110	Flanged										
	0116	2" size	1 Plum	5	1.600	Ea.	630	55		685	780	
	0118	4" size	Q-1	3	5.333		1,050	165		1,215	1,400	
	0120	6" size	Q-2	3	8		2,050	257		2,307	2,650	
	0122	8" size	"	2.50	9.600		3,650	310		3,960	4,500	
	1020	Butterfly, wafer type, gear actuator, 200 lb.										
	1030	2" size	1 Plum	14	.571	Ea.	109	19.70		128.70	149	
	1040	2-1/2" size	Q-1	9	1.778		112	55		167	206	
	1050	3" size		8	2		116	62		178	222	
	1060	4" size		5	3.200		145	99		244	310	
	1070	5" size	Q-2	5	4.800		175	154		329	425	
	1080	6" size		5	4.800		198	154		352	450	
	1090	8" size		4.50	5.333		264	171		435	550	
	1100	10" size		4	6		330	193		523	655	
	1110	12" size		3	8		460	257		717	895	
	1200	Wafer type, lever actuator, 200 lb.										
	1220	2" size	1 Plum	14	.571	Ea.	76	19.70		95.70	113	
	1230	2-1/2" size	Q-1	9	1.778		79	55		134	170	
	1240	3" size		8	2		83.50	62		145.50	185	
	1250	4" size		5	3.200		102	99		201	262	
	1260	5" size	Q-2	5	4.800		141	154		295	390	
	1270	6" size		5	4.800		170	154		324	420	
	1280	8" size		4.50	5.333		251	171		422	535	
	1290	10" size		4	6		325	193		518	650	
	1300	12" size		3	8		505	257		762	945	
	1600	Gate, threaded, 125 lb.										
	1604	1-1/2"	1 Plum	13	.615	Ea.	150	21		171	196	
	1605	2"		11	.727		530	25		555	625	
	1606	3"		8	1		420	34.50		454.50	515	
	1607	4"		5	1.600		955	55		1,010	1,125	
	1650	Gate, 125 lb., N.R.S.										
	2150	Flanged										
	2200	2" size	1 Plum	5	1.600	Ea.	276	55		331	390	
	2240	2-1/2" size	Q-1	5	3.200		284	99		383	460	
	2260	3" size		4.50	3.556		320	110		430	515	
	2280	4" size		3	5.333		455	165		620	750	
	2290	5" size	Q-2	3.40	7.059		775	227		1,002	1,200	

15110 | Valves

200		CREW	DAILY OUTPUT	LABOR-HOURS	UNIT	2001 BARE COSTS MAT.	LABOR	EQUIP.	TOTAL	TOTAL INCL O&P
2300	6" size R15100-090	Q-2	3	8	Ea.	775	257		1,032	1,250
2320	8" size		2.50	9.600		1,325	310		1,635	1,950
2340	10" size		2.20	10.909		2,350	350		2,700	3,100
2360	12" size		1.70	14.118		3,225	455		3,680	4,225
2420	For 250 lb flanged, add					200%	10%			
3500	OS&Y, 125 lb., threaded									
3504	3/4" size	1 Plum	18	.444	Ea.	385	15.30		400.30	450
3505	1" size		16	.500		440	17.25		457.25	505
3506	1-1/2" size		13	.615		465	21		486	540
3507	2" size		11	.727		475	25		500	565
3508	2-1/2" size		9	.889		490	30.50		520.50	585
3510	4" size		7	1.143		790	39.50		829.50	925
3550	OS&Y, 125 lb., flanged									
3600	2" size	1 Plum	5	1.600	Ea.	209	55		264	310
3640	2-1/2" size	Q-1	5	3.200		217	99		316	390
3660	3" size		4.50	3.556		243	110		353	435
3670	3-1/2" size		3	5.333		271	165		436	550
3680	4" size		3	5.333		345	165		510	630
3690	5" size	Q-2	3.40	7.059		570	227		797	975
3700	6" size		3	8		570	257		827	1,025
3720	8" size		2.50	9.600		1,025	310		1,335	1,600
3740	10" size		2.20	10.909		1,850	350		2,200	2,575
3760	12" size		1.70	14.118		2,475	455		2,930	3,400
3900	For 250 lb, flanged, add					200%	10%			
4350	Globe, OS&Y									
4352	Class 125, threaded									
4358	4" size	Q-1	12	1.333	Ea.	1,025	41.50		1,066.50	1,200
4540	Class 125, flanged									
4550	2" size	1 Plum	5	1.600	Ea.	430	55		485	555
4560	2-1/2" size	Q-1	5	3.200		605	99		704	815
4570	3" size		4.50	3.556		525	110		635	745
4580	4" size		3	5.333		750	165		915	1,075
4590	5" size	Q-2	3.40	7.059		1,375	227		1,602	1,850
4600	6" size		3	8		1,375	257		1,632	1,900
4610	8" size		2.50	9.600		2,675	310		2,985	3,425
4612	10" size		2.20	10.909		4,200	350		4,550	5,125
4614	12" size		1.70	14.118		4,850	455		5,305	6,000
5040	Class 250, flanged									
5050	2" size	1 Plum	4.50	1.778	Ea.	645	61		706	805
5060	2-1/2" size	Q-1	4.50	3.556		840	110		950	1,100
5070	3" size		4	4		870	124		994	1,150
5080	4" size		2.70	5.926		1,275	184		1,459	1,675
5090	5" size	Q-2	3	8		2,300	257		2,557	2,925
5100	6" size		2.70	8.889		2,300	286		2,586	2,950
5110	8" size		2.20	10.909		3,875	350		4,225	4,800
5120	10" size		2	12		5,250	385		5,635	6,350
5130	12" size		1.60	15		8,175	480		8,655	9,725
5240	Valve sprocket rim w/chain, for 2" valve	1 Stpi	30	.267		71.50	9.30		80.80	92.50
5250	2-1/2" valve		27	.296		71.50	10.35		81.85	94
5260	3-1/2" valve		25	.320		71.50	11.15		82.65	95.50
5270	6" valve		20	.400		71.50	13.95		85.45	99.50
5280	8" valve		18	.444		108	15.50		123.50	143
5290	12" valve		16	.500		108	17.45		125.45	146
5300	16" valve		12	.667		147	23		170	196
5310	20" valve		10	.800		147	28		175	203
5320	36" valve		8	1		272	35		307	350

15110 | Valves

		Description	Crew	Daily Output	Labor-Hours	Unit	2001 Bare Costs Mat.	Labor	Equip.	Total	Total Incl O&P	
200	5450	Swing check, 125 lb., threaded R15100-090										200
	5500	2" size	1 Plum	11	.727	Ea.	340	25		365	415	
	5540	2-1/2" size	Q-1	15	1.067		400	33		433	490	
	5550	3" size		13	1.231		445	38		483	550	
	5560	4" size		10	1.600		710	49.50		759.50	860	
	5950	Flanged										
	5994	1" size	1 Plum	7	1.143	Ea.	78	39.50		117.50	146	
	5998	1-1/2" size		6	1.333		114	46		160	195	
	6000	2" size		5	1.600		144	55		199	241	
	6040	2-1/2" size	Q-1	5	3.200		174	99		273	340	
	6050	3" size		4.50	3.556		187	110		297	370	
	6060	4" size		3	5.333		294	165		459	575	
	6070	6" size	Q-2	3	8		505	257		762	945	
	6080	8" size		2.50	9.600		945	310		1,255	1,525	
	6090	10" size		2.20	10.909		1,625	350		1,975	2,300	
	6100	12" size		1.70	14.118		2,525	455		2,980	3,450	
	6110	18" size		1.30	18.462		16,600	595		17,195	19,100	
	6114	24" size		.75	32		30,800	1,025		31,825	35,400	
	6160	For 250 lb flanged, add					200%	20%				
	6600	Silent check, bronze trim										
	6610	Compact wafer type, for 125 or 150 lb. flanges										
	6630	1-1/2" size	1 Plum	11	.727	Ea.	118	25		143	167	
	6640	2" size	"	9	.889		140	30.50		170.50	200	
	6650	2-1/2" size	Q-1	9	1.778		154	55		209	252	
	6660	3" size		8	2		165	62		227	275	
	6670	4" size		5	3.200		210	99		309	380	
	6680	5" size	Q-2	6	4		275	129		404	500	
	6690	6" size		6	4		370	129		499	600	
	6700	8" size		4.50	5.333		660	171		831	985	
	6710	10" size		4	6		1,150	193		1,343	1,550	
	6720	12" size		3	8		2,175	257		2,432	2,800	
	6740	For 250 or 300 lb. flanges, thru 6" no change										
	6741	For 8" and 10", add				Ea.	11%	10%				
	6750	Twin disc										
	6752	2" size	1 Plum	9	.889	Ea.	168	30.50		198.50	231	
	6754	4" size	Q-1	5	3.200		282	99		381	460	
	6756	6" size	Q-2	5	4.800		415	154		569	690	
	6758	8" size		4.50	5.333		645	171		816	965	
	6760	10" size		4	6		1,025	193		1,218	1,425	
	6762	12" size		3	8		1,250	257		1,507	1,750	
	6764	18" size		1.50	16		5,050	515		5,565	6,325	
	6766	24" size		.75	32		7,475	1,025		8,500	9,775	
	6800	Full flange type, 150 lb.										
	6900	Globe type, 125 lb.										
	6911	2-1/2" size	Q-1	9	1.778	Ea.	294	55		349	410	
	6912	3" size		8	2		315	62		377	445	
	6913	4" size		5	3.200		425	99		524	620	
	6914	5" size	Q-2	6	4		550	129		679	800	
	6915	6" size		5	4.800		685	154		839	990	
	6916	8" size		4.50	5.333		1,250	171		1,421	1,625	
	6917	10" size		4	6		1,550	193		1,743	2,000	
	6918	12" size		3	8		2,625	257		2,882	3,300	
	6940	For 250 lb., add					40%	10%				
	6980	Screwed end type, 125 lb.										
	6981	1" size	1 Plum	19	.421	Ea.	50.50	14.50		65	77.50	
	6982	1-1/4" size		15	.533		65.50	18.35		83.85	99.50	

15110 | Valves

		15110 Valves	CREW	DAILY OUTPUT	LABOR-HOURS	UNIT	MAT.	LABOR	EQUIP.	TOTAL	TOTAL INCL O&P	
							2001 BARE COSTS					
200	6983	1-1/2" size (R15100-090)	1 Plum	13	.615	Ea.	81	21		102	121	200
	6984	2" size		11	.727		110	25		135	159	
300	0010	**VALVES, LINED, CORROSION RESISTANT/HIGH PURITY**										300
	2000	Butterfly 150 lb. ductile iron,										
	2010	Wafer type										
	2030	TFE lined 3" lever handle	Q-1	8	2	Ea.	950	62		1,012	1,150	
	2050	4" lever handle	"	5	3.200		1,325	99		1,424	1,600	
	2070	6" gear operated	Q-2	4.50	5.333		1,950	171		2,121	2,400	
	2080	8" gear operated		4	6		2,550	193		2,743	3,100	
	2090	10" gear operated		3.50	6.857		3,225	220		3,445	3,875	
	2100	12" gear operated		2.50	9.600		4,525	310		4,835	5,450	
	3500	Check lift, 125 lb., cast iron flanged										
	3510	Horizontal PPL or SL lined										
	3530	1" size	1 Plum	14	.571	Ea.	375	19.70		394.70	445	
	3540	1-1/2" size		11	.727		455	25		480	545	
	3550	2" size		8	1		530	34.50		564.50	635	
	3560	2-1/2" size	Q-1	5	3.200		695	99		794	915	
	3570	3" size		4.50	3.556		870	110		980	1,125	
	3590	4" size		3	5.333		1,150	165		1,315	1,500	
	3610	6" size	Q-2	3	8		1,950	257		2,207	2,550	
	3620	8" size	"	2.50	9.600		4,300	310		4,610	5,200	
	4250	Vertical PPL or SL lined										
	4270	1" size	1 Plum	14	.571	Ea.	340	19.70		359.70	400	
	4290	1-1/2" size		11	.727		420	25		445	500	
	4300	2" size		8	1		490	34.50		524.50	585	
	4310	2-1/2" size	Q-1	5	3.200		660	99		759	875	
	4320	3" size		4.50	3.556		735	110		845	975	
	4340	4" size		3	5.333		970	165		1,135	1,325	
	4360	6" size	Q-2	3	8		1,625	257		1,882	2,200	
	4370	8" size	"	2.50	9.600		2,950	310		3,260	3,725	
	5000	Clamp type, ductile iron, 150 lb. flanged										
	5010	TFE lined										
	5030	1" size, lever handle	1 Plum	9	.889	Ea.	700	30.50		730.50	815	
	5050	1-1/2" size, lever handle		6	1.333		925	46		971	1,100	
	5060	2" size, lever handle		5	1.600		1,125	55		1,180	1,300	
	5080	3" size, lever handle	Q-1	4.50	3.556		1,400	110		1,510	1,725	
	5100	4" size, gear operated	"	3	5.333		2,025	165		2,190	2,475	
	5120	6" size, gear operated	Q-2	3	8		5,725	257		5,982	6,700	
	5130	8" size, gear operated	"	2.50	9.600		6,975	310		7,285	8,125	
	6000	Diaphragm type, cast iron, 125 lb. flanged										
	6010	PTFE or VITON, lined										
	6030	1" size, handwheel operated	1 Plum	9	.889	Ea.	143	30.50		173.50	203	
	6050	1-1/2" size, handwheel operated		6	1.333		183	46		229	271	
	6060	2" size, handwheel operated		5	1.600		230	55		285	335	
	6070	2-1/2" size, handwheel operated	Q-1	5	3.200		365	99		464	550	
	6080	3" size, handwheel operated		4.50	3.556		395	110		505	600	
	6100	4" size, handwheel operated		3	5.333		645	165		810	960	
	6120	6" size, handwheel operated	Q-2	3	8		1,225	257		1,482	1,750	
	6130	8" size, handwheel operated	"	2.50	9.600		2,600	310		2,910	3,325	
500	0010	**VALVES, PLASTIC** (R15100-090)										500
	1100	Angle, PVC, threaded										
	1110	1/4" size	1 Plum	26	.308	Ea.	47.50	10.60		58.10	68	
	1120	1/2" size		26	.308		47.50	10.60		58.10	68	
	1130	3/4" size		25	.320		55.50	11		66.50	77.50	
	1140	1" size		23	.348		66.50	12		78.50	91	

15110 | Valves

		Description	Crew	Daily Output	Labor-Hours	Unit	2001 Bare Costs Mat.	2001 Bare Costs Labor	2001 Bare Costs Equip.	2001 Bare Costs Total	Total Incl O&P	
500	1150	Ball, PVC, socket or threaded, single union R15100-090										500
	1200	1/4" size	1 Plum	26	.308	Ea.	24	10.60		34.60	42.50	
	1220	3/8" size		26	.308		24	10.60		34.60	42.50	
	1230	1/2" size		26	.308		24	10.60		34.60	42.50	
	1240	3/4" size		25	.320		28.50	11		39.50	48	
	1250	1" size		23	.348		34	12		46	55.50	
	1260	1-1/4" size		21	.381		45	13.10		58.10	69.50	
	1270	1-1/2" size		20	.400		57	13.80		70.80	83.50	
	1280	2" size		17	.471		81.50	16.20		97.70	114	
	1290	2-1/2" size	Q-1	26	.615		204	19.10		223.10	253	
	1300	3" size		24	.667		204	20.50		224.50	255	
	1310	4" size		20	.800		262	25		287	325	
	1360	For PVC, flanged, add					100%	15%				
	1450	Double union 1/2" size	1 Plum	26	.308		13.80	10.60		24.40	31	
	1460	3/4" size		25	.320		16.20	11		27.20	34.50	
	1470	1" size		23	.348		17.75	12		29.75	37.50	
	1480	1-1/4" size		21	.381		27	13.10		40.10	49.50	
	1490	1-1/2" size		20	.400		30	13.80		43.80	54	
	1500	2" size		17	.471		42.50	16.20		58.70	71	
	1650	CPVC, socket or threaded, single union										
	1700	1/2" size	1 Plum	26	.308	Ea.	33.50	10.60		44.10	52.50	
	1720	3/4" size		25	.320		42	11		53	62.50	
	1730	1" size		23	.348		50	12		62	73	
	1750	1-1/4" size		21	.381		83.50	13.10		96.60	112	
	1760	1-1/2" size		20	.400		83.50	13.80		97.30	113	
	1770	2" size		17	.471		116	16.20		132.20	153	
	1780	3" size	Q-1	24	.667		345	20.50		365.50	410	
	1840	For CPVC, flanged, add					65%	15%				
	1880	For true union, socket or threaded, add					50%	5%				
	2050	Polypropylene, threaded										
	2100	1/4" size	1 Plum	26	.308	Ea.	30	10.60		40.60	49	
	2120	3/8" size		26	.308		30	10.60		40.60	49	
	2130	1/2" size		26	.308		30	10.60		40.60	49	
	2140	3/4" size		25	.320		37.50	11		48.50	57.50	
	2150	1" size		23	.348		44.50	12		56.50	67	
	2160	1-1/4" size		21	.381		64.50	13.10		77.60	91	
	2170	1-1/2" size		20	.400		74	13.80		87.80	103	
	2180	2" size		17	.471		101	16.20		117.20	136	
	2190	3" size	Q-1	24	.667		266	20.50		286.50	325	
	2200	4" size	"	20	.800		445	25		470	530	
	2550	PVC, three way, socket or threaded										
	2600	1/2" size	1 Plum	26	.308	Ea.	56.50	10.60		67.10	78	
	2640	3/4" size		25	.320		64	11		75	87	
	2650	1" size		23	.348		69.50	12		81.50	94.50	
	2660	1-1/2" size		20	.400		140	13.80		153.80	175	
	2670	2" size		17	.471		188	16.20		204.20	232	
	2680	3" size	Q-1	24	.667		410	20.50		430.50	480	
	2740	For flanged, add					60%	15%				
	3150	Ball check, PVC, socket or threaded										
	3200	1/4" size	1 Plum	26	.308	Ea.	29	10.60		39.60	48	
	3220	3/8" size		26	.308		29	10.60		39.60	48	
	3240	1/2" size		26	.308		29	10.60		39.60	48	
	3250	3/4" size		25	.320		32.50	11		43.50	52	
	3260	1" size		23	.348		40.50	12		52.50	63	
	3270	1-1/4" size		21	.381		68	13.10		81.10	95	
	3280	1-1/2" size		20	.400		68	13.80		81.80	96	

15110 | Valves

500

15110 Valves		Crew	Daily Output	Labor-Hours	Unit	2001 Bare Costs Mat.	Labor	Equip.	Total	Total Incl O&P
3290	2" size (R15100-090)	1 Plum	17	.471	Ea.	93	16.20		109.20	127
3310	3" size	Q-1	24	.667		257	20.50		277.50	315
3320	4" size	"	20	.800		365	25		390	440
3360	For PVC, flanged, add					50%	15%			
3750	CPVC, socket or threaded									
3800	1/2" size	1 Plum	26	.308	Ea.	42.50	10.60		53.10	62.50
3840	3/4" size		25	.320		50.50	11		61.50	72
3850	1" size		23	.348		60	12		72	83.50
3860	1-1/2" size		20	.400		103	13.80		116.80	134
3870	2" size		17	.471		138	16.20		154.20	177
3880	3" size	Q-1	24	.667		375	20.50		395.50	440
3920	4" size	"	20	.800		505	25		530	595
3930	For CPVC, flanged, add					40%	15%			
4340	Polypropylene, threaded									
4360	1/2" size	1 Plum	26	.308	Ea.	25	10.60		35.60	43.50
4400	3/4" size		25	.320		29	11		40	48.50
4440	1" size		23	.348		37.50	12		49.50	59
4450	1-1/2" size		20	.400		72.50	13.80		86.30	101
4460	2" size		17	.471		91	16.20		107.20	125
4500	For polypropylene flanged, add					200%	15%			
4850	Foot valve, PVC, socket or threaded									
4900	1/2" size	1 Plum	34	.235	Ea.	40.50	8.10		48.60	57
4930	3/4" size		32	.250		46	8.60		54.60	64
4940	1" size		28	.286		60	9.85		69.85	81
4950	1-1/4" size		27	.296		115	10.20		125.20	142
4960	1-1/2" size		26	.308		115	10.60		125.60	143
4970	2" size		24	.333		134	11.50		145.50	164
4980	3" size		20	.400		320	13.80		333.80	370
4990	4" size		18	.444		560	15.30		575.30	640
5000	For flanged, add					25%	10%			
5050	CPVC, socket or threaded									
5060	1/2" size	1 Plum	34	.235	Ea.	89.50	8.10		97.60	111
5070	3/4" size		32	.250		108	8.60		116.60	132
5080	1" size		28	.286		127	9.85		136.85	155
5090	1-1/4" size		27	.296		200	10.20		210.20	235
5100	1-1/2" size		26	.308		208	10.60		218.60	245
5110	2" size		24	.333		267	11.50		278.50	310
5120	3" size		20	.400		505	13.80		518.80	575
5130	4" size		18	.444		1,075	15.30		1,090.30	1,225
5140	For flanged, add					25%	10%			
5280	Needle valve, PVC, threaded									
5300	1/4" size	1 Plum	26	.308	Ea.	32.50	10.60		43.10	52
5340	3/8" size		26	.308		38	10.60		48.60	58
5360	1/2" size		26	.308		38	10.60		48.60	58
5380	For polypropylene, add					10%				
5800	Y check, PVC, socket or threaded									
5820	1/2" size	1 Plum	26	.308	Ea.	52.50	10.60		63.10	73.50
5840	3/4" size		25	.320		56.50	11		67.50	78.50
5850	1" size		23	.348		62	12		74	86
5860	1-1/4" size		21	.381		97	13.10		110.10	127
5870	1-1/2" size		20	.400		105	13.80		118.80	137
5880	2" size		17	.471		131	16.20		147.20	170
5890	2-1/2" size		15	.533		277	18.35		295.35	335
5900	3" size	Q-1	24	.667		260	20.50		280.50	315
5910	4" size	"	20	.800		455	25		480	540
5960	For PVC flanged, add					45%	15%			

500

		15110 | Valves	CREW	DAILY OUTPUT	LABOR-HOURS	UNIT	2001 BARE COSTS MAT.	LABOR	EQUIP.	TOTAL	TOTAL INCL O&P	
500	6350	Y sediment strainer, PVC, socket or threaded R15100-090										500
	6400	1/2" size	1 Plum	26	.308	Ea.	29.50	10.60		40.10	48.50	
	6440	3/4" size		24	.333		32	11.50		43.50	53	
	6450	1" size		23	.348		39	12		51	61	
	6460	1-1/4" size		21	.381		64.50	13.10		77.60	91	
	6470	1-1/2" size		20	.400		67.50	13.80		81.30	95.50	
	6480	2" size		17	.471		78.50	16.20		94.70	111	
	6490	2-1/2" size		15	.533		192	18.35		210.35	239	
	6500	3" size	Q-1	24	.667		192	20.50		212.50	242	
	6510	4" size	"	20	.800		320	25		345	390	
	6560	For PVC, flanged, add					55%	15%				
600	0010	**VALVES, SEMI-STEEL** R15100-090										600
	1020	Lubricated plug valve, threaded, 200 psi										
	1030	1/2" pipe size	1 Plum	18	.444	Ea.	61.50	15.30		76.80	90.50	
	1040	3/4" pipe size		16	.500		61.50	17.25		78.75	93.50	
	1050	1" pipe size		14	.571		79	19.70		98.70	116	
	1060	1-1/4" pipe size		12	.667		94.50	23		117.50	139	
	1070	1-1/2" pipe size		11	.727		102	25		127	150	
	1080	2" pipe size		8	1		120	34.50		154.50	184	
	1090	2-1/2" pipe size	Q-1	5	3.200		186	99		285	355	
	1100	3" pipe size	"	4.50	3.556		228	110		338	415	
	6990	Flanged, 200 psi										
	7000	2" pipe size	1 Plum	8	1	Ea.	145	34.50		179.50	211	
	7010	2-1/2" pipe size	Q-1	5	3.200		218	99		317	390	
	7020	3" pipe size		4.50	3.556		263	110		373	455	
	7030	4" pipe size		3	5.333		335	165		500	615	
	7036	5" pipe size		2.50	6.400		500	198		698	850	
	7040	6" pipe size	Q-2	3	8		655	257		912	1,100	
	7050	8" pipe size		2.50	9.600		1,175	310		1,485	1,775	
	7060	10" pipe size		2.20	10.909		1,725	350		2,075	2,425	
	7070	12" pipe size		1.70	14.118		2,975	455		3,430	3,925	
800	0010	**VALVES, STAINLESS STEEL** R15100-090										800
	1700	Check, 200 lb., threaded										
	1710	1/4" size	1 Plum	24	.333	Ea.	72	11.50		83.50	96.50	
	1720	1/2" size		22	.364		72	12.55		84.55	98	
	1730	3/4" size		20	.400		79	13.80		92.80	108	
	1750	1" size		19	.421		101	14.50		115.50	133	
	1760	1-1/2" size		13	.615		191	21		212	242	
	1770	2" size		11	.727		320	25		345	390	
	1800	150 lb., flanged										
	1810	2-1/2" size	Q-1	5	3.200	Ea.	765	99		864	990	
	1820	3" size		4.50	3.556		765	110		875	1,000	
	1830	4" size		3	5.333		1,150	165		1,315	1,525	
	1840	6" size	Q-2	3	8		2,050	257		2,307	2,650	
	1850	8" size	"	2.50	9.600		3,425	310		3,735	4,225	
	2100	Gate, OS&Y, 150 lb., flanged										
	2120	1/2" size	1 Plum	18	.444	Ea.	310	15.30		325.30	365	
	2140	3/4" size		16	.500		335	17.25		352.25	395	
	2150	1" size		14	.571		375	19.70		394.70	440	
	2155	1-1/4" size		12	.667		470	23		493	550	
	2160	1-1/2" size		11	.727		470	25		495	555	
	2170	2" size		8	1		520	34.50		554.50	625	
	2180	2-1/2" size	Q-1	5	3.200		810	99		909	1,050	
	2190	3" size		4.50	3.556		885	110		995	1,150	
	2200	4" size		3	5.333		1,225	165		1,390	1,600	

		15110 | Valves	CREW	DAILY OUTPUT	LABOR-HOURS	UNIT	2001 BARE COSTS MAT.	LABOR	EQUIP.	TOTAL	TOTAL INCL O&P	
800	2205	5" size R15100-090	Q-1	2.80	5.714	Ea.	2,200	177		2,377	2,700	800
	2210	6" size	Q-2	3	8		2,200	257		2,457	2,825	
	2220	8" size		2.50	9.600		3,400	310		3,710	4,225	
	2230	10" size		2.30	10.435		4,950	335		5,285	5,950	
	2240	12" size	↓	1.90	12.632		6,900	405		7,305	8,225	
	2260	For 300 lb., flanged, add				↓	120%	15%				
	2600	600 lb., flanged										
	2620	1/2" size	1 Plum	16	.500	Ea.	720	17.25		737.25	815	
	2640	3/4" size		14	.571		880	19.70		899.70	995	
	2650	1" size		12	.667		1,100	23		1,123	1,250	
	2660	1-1/2" size		10	.800		1,550	27.50		1,577.50	1,750	
	2670	2" size	↓	7	1.143		2,775	39.50		2,814.50	3,100	
	2680	2-1/2" size	Q-1	4	4		7,975	124		8,099	8,950	
	2690	3" size	"	3.60	4.444	↓	8,050	138		8,188	9,050	
	3100	Globe, OS&Y, 150 lb., flanged										
	3120	1/2" size	1 Plum	18	.444	Ea.	345	15.30		360.30	405	
	3140	3/4" size		16	.500		380	17.25		397.25	445	
	3150	1" size		14	.571		470	19.70		489.70	550	
	3160	1-1/2" size		11	.727		700	25		725	810	
	3170	2" size	↓	8	1		855	34.50		889.50	990	
	3180	2-1/2" size	Q-1	5	3.200		1,800	99		1,899	2,125	
	3190	3" size		4.50	3.556		1,850	110		1,960	2,200	
	3200	4" size	↓	3	5.333		2,150	165		2,315	2,625	
	3210	6" size	Q-2	3	8	↓	4,075	257		4,332	4,875	
		15120 | Piping Specialties										
120	0010	**AIR CONTROL**										120
	0030	Air separator, with strainer										
	0040	2" diameter	Q-5	6	2.667	Ea.	515	83.50		598.50	695	
	0080	2-1/2" diameter		5	3.200		585	100		685	795	
	0100	3" diameter	↓	4	4	↓	890	126		1,016	1,175	
	1000	Micro-bubble separator for total air removal, closed loop system										
	1010	Requires bladder type tank in system.										
	1020	Water (hot or chilled) or glycol system										
	1030	Threaded										
	1040	3/4" diameter	1 Stpi	20	.400	Ea.	67	13.95		80.95	94.50	
	1050	1" diameter		19	.421		75	14.65		89.65	105	
	1060	1-1/4" diameter		16	.500		103	17.45		120.45	141	
	1070	1-1/2" diameter		13	.615		135	21.50		156.50	181	
	1080	2" diameter	↓	11	.727		655	25.50		680.50	760	
	1090	2-1/2" diameter	Q-5	15	1.067		725	33.50		758.50	850	
	1100	3" diameter		13	1.231		1,050	38.50		1,088.50	1,200	
	1110	4" diameter	↓	10	1.600	↓	1,125	50		1,175	1,300	
140	0010	**AIR PURGING SCOOP** with tappings										140
	0020	for air vent and expansion tank connection										
	0100	1" pipe size, threaded	1 Stpi	19	.421	Ea.	13.25	14.65		27.90	36.50	
	0110	1-1/4" pipe size, threaded		15	.533		13.30	18.60		31.90	42.50	
	0120	1-1/2" pipe size, threaded		13	.615		29	21.50		50.50	64.50	
	0130	2" pipe size, threaded	↓	11	.727	↓	33	25.50		58.50	75	
160	0010	**AUTOMATIC AIR VENT**										160
	0020	Cast iron body, stainless steel internals, float type										
	0180	1/2" NPT inlet, 250 psi	1 Stpi	10	.800	Ea.	251	28		279	320	
	0220	3/4" NPT inlet, 250 psi	"	10	.800	"	251	28		279	320	
	0600	Forged steel body, stainless steel internals, float type										
	0640	1/2" NPT inlet, 750 psi	1 Stpi	12	.667	Ea.	770	23		793	880	

15120 | Piping Specialties

		15120 Piping Specialties	CREW	DAILY OUTPUT	LABOR-HOURS	UNIT	2001 BARE COSTS MAT.	LABOR	EQUIP.	TOTAL	TOTAL INCL O&P	
160	0680	3/4″ NPT inlet, 750 psi	1 Stpi	12	.667	Ea.	770	23		793	880	160
	1100	Formed steel body, non corrosive										
	1110	1/8″ NPT inlet 150 psi	1 Stpi	32	.250	Ea.	6.70	8.70		15.40	20.50	
	1120	1/4″ NPT inlet 150 psi	↓	32	.250	↓	22.50	8.70		31.20	38	
	1130	3/4″ NPT inlet 150 psi	↓	32	.250	↓	22.50	8.70		31.20	38	
	1300	Chrome plated brass, automatic/manual, for radiators										
	1310	1/8″ NPT inlet, nickel plated brass	1 Stpi	32	.250	Ea.	3.92	8.70		12.62	17.45	
200	0010	**CIRCUIT SETTER** Balance valve										200
	0018	Threaded										
	0019	1/2″ pipe size	1 Stpi	22	.364	Ea.	43	12.65		55.65	66.50	
	0020	3/4″ pipe size	″	20	.400	″	39	13.95		52.95	64	
220	0010	**COCKS, DRAINS & SPECIALTIES**										220
	1000	Boiler drain										
	1010	Pipe thread to hose										
	1020	Bronze										
	1030	1/2″ size	1 Stpi	36	.222	Ea.	4.10	7.75		11.85	16.20	
	1040	3/4″ size	″	34	.235	″	4.44	8.20		12.64	17.30	
	1100	Solder to hose										
	1110	Bronze										
	1120	1/2″ size	1 Stpi	46	.174	Ea.	4.10	6.05		10.15	13.65	
	1130	3/4″ size	″	44	.182	″	4.44	6.35		10.79	14.45	
	1600	With built-in vacuum breaker										
	1610	1/2″ IP or solder	1 Stpi	36	.222	Ea.	12.60	7.75		20.35	25.50	
	1630	With tamper proof vacuum breaker										
	1640	1/2″ IP or solder	1 Stpi	36	.222	Ea.	15.65	7.75		23.40	29	
	1650	3/4″ IP or solder	″	34	.235	″	15.95	8.20		24.15	30	
	3000	Cocks										
	3010	Air, lever or tee handle										
	3020	Bronze, single thread										
	3030	1/8″ size	1 Stpi	52	.154	Ea.	5.45	5.35		10.80	14.10	
	3040	1/4″ size	↓	46	.174	↓	5.90	6.05		11.95	15.65	
	3050	3/8″ size	↓	40	.200	↓	6.65	6.95		13.60	17.80	
	3060	1/2″ size	↓	36	.222	↓	6.65	7.75		14.40	19.05	
	3100	Bronze, double thread										
	3110	1/8″ size	1 Stpi	26	.308	Ea.	7.30	10.70		18	24.50	
	3120	1/4″ size	↓	22	.364	↓	7.05	12.65		19.70	27	
	3130	3/8″ size	↓	18	.444	↓	7.80	15.50		23.30	32	
	3140	1/2″ size	↓	15	.533	↓	9.25	18.60		27.85	38	
	4500	Gauge cock, brass										
	4510	1/4″ FPT	1 Stpi	24	.333	Ea.	6.15	11.60		17.75	24.50	
	4512	1/4″ MPT	″	24	.333	″	7.15	11.60		18.75	25.50	
	4600	Pigtail, steam syphon										
	4604	1/4″	1 Stpi	24	.333	Ea.	9.30	11.60		20.90	28	
	4650	Snubber valve										
	4654	1/4″	1 Stpi	22	.364	Ea.	9.40	12.65		22.05	29.50	
	4660	Nipple, black steel										
	4664	1/4″ x 3″	1 Stpi	37	.216	Ea.	.44	7.55		7.99	11.85	
250	0010	**DIELECTRIC UNIONS** Standard gaskets for water and air										250
	0020	250 psi maximum pressure										
	0280	Female IPT to sweat, straight										
	0300	1/2″ pipe size	1 Plum	24	.333	Ea.	3.15	11.50		14.65	21	
	0340	3/4″ pipe size	↓	20	.400	↓	3.15	13.80		16.95	24.50	
	0360	1″ pipe size	↓	19	.421	↓	6.10	14.50		20.60	28.50	
	0380	1-1/4″ pipe size	↓	15	.533	↓	10	18.35		28.35	38.50	
	0400	1-1/2″ pipe size	↓	13	.615	↓	15	21		36	48.50	

	15120 | Piping Specialties		CREW	DAILY OUTPUT	LABOR-HOURS	UNIT	2001 BARE COSTS MAT.	LABOR	EQUIP.	TOTAL	TOTAL INCL O&P
250	0420	2″ pipe size	1 Plum	11	.727	Ea.	21	25		46	61
	0580	Female IPT to brass pipe thread, straight									
	0600	1/2″ pipe size	1 Plum	24	.333	Ea.	5.95	11.50		17.45	24
	0640	3/4″ pipe size		20	.400		7.60	13.80		21.40	29.50
	0660	1″ pipe size		19	.421		11.40	14.50		25.90	34.50
	0680	1-1/4″ pipe size		15	.533		17.05	18.35		35.40	46.50
	0700	1-1/2″ pipe size		13	.615		23.50	21		44.50	58
	0720	2″ pipe size		11	.727		30.50	25		55.50	72
	0780	Female IPT to female IPT, straight									
	0800	1/2″ pipe size	1 Plum	24	.333	Ea.	4.75	11.50		16.25	22.50
	0840	3/4″ pipe size		20	.400		5.30	13.80		19.10	27
	0860	1″ pipe size		19	.421		7	14.50		21.50	29.50
	0880	1-1/4″ pipe size		15	.533		9.75	18.35		28.10	38.50
	0900	1-1/2″ pipe size		13	.615		14.95	21		35.95	48.50
	0920	2″ pipe size		11	.727		21.50	25		46.50	61.50
	2000	175 psi maximum pressure									
	2180	Female IPT to sweat									
	2200	1-1/2″ pipe size	1 Plum	11	.727	Ea.	38.50	25		63.50	80
	2240	2″ pipe size	″	9	.889		44	30.50		74.50	94.50
	2260	2-1/2″ pipe size	Q-1	15	1.067		48	33		81	103
	2280	3″ pipe size		14	1.143		65.50	35.50		101	126
	2300	4″ pipe size		11	1.455		174	45		219	260
	2320	5″ pipe size	Q-2	14	1.714		284	55		339	395
	2340	6″ pipe size		12	2		330	64.50		394.50	455
	2360	8″ pipe size		10	2.400		910	77		987	1,125
	2480	Female IPT to brass pipe									
	2500	1-1/2″ pipe size	1 Plum	11	.727	Ea.	58	25		83	102
	2540	2″ pipe size	″	9	.889		69	30.50		99.50	122
	2560	2-1/2″ pipe size	Q-1	15	1.067		100	33		133	160
	2580	3″ pipe size		14	1.143		116	35.50		151.50	182
	2600	4″ pipe size		11	1.455		199	45		244	287
	2620	5″ pipe size	Q-2	14	1.714		224	55		279	330
	2640	6″ pipe size		12	2		267	64.50		331.50	390
	2680	8″ pipe size		10	2.400		435	77		512	595
	2880	Female IPT to female IPT									
	2900	1-1/2″ pipe size	1 Plum	11	.727	Ea.	36	25		61	77.50
	2940	2″ pipe size	″	9	.889		38.50	30.50		69	88.50
	2960	2-1/2″ pipe size	Q-1	15	1.067		48	33		81	103
	2980	3″ pipe size		14	1.143		60	35.50		95.50	120
	3000	4″ pipe size		11	1.455		91	45		136	168
	3020	5″ pipe size	Q-2	14	1.714		105	55		160	199
	3040	6″ pipe size		12	2		141	64.50		205.50	252
	3060	8″ pipe size		10	2.400		192	77		269	325
	3380	Copper to copper									
	3400	1-1/2″ pipe size	1 Plum	11	.727	Ea.	47.50	25		72.50	90.50
	3440	2″ pipe size	″	9	.889		54	30.50		84.50	106
	3460	2-1/2″ pipe size	Q-1	15	1.067		58	33		91	114
	3480	3″ pipe size		14	1.143		72.50	35.50		108	133
	3500	4″ pipe size		11	1.455		230	45		275	320
	3520	5″ pipe size	Q-2	14	1.714		370	55		425	490
	3540	6″ pipe size		12	2		450	64.50		514.50	590
	3560	8″ pipe size		10	2.400		1,200	77		1,277	1,450
300	0010	**EXPANSION JOINTS**									
	0100	Bellows type, neoprene cover, flanged spool									
	0120	6″ face to face, 1″ diameter	1 Stpi	13	.615	Ea.	158	21.50		179.50	207
	0140	1-1/4″ diameter		11	.727		161	25.50		186.50	216

15120 | Piping Specialties

		15120 Piping Specialties	CREW	DAILY OUTPUT	LABOR-HOURS	UNIT	2001 BARE COSTS MAT.	LABOR	EQUIP.	TOTAL	TOTAL INCL O&P	
300	0160	1-1/2″ diameter	1 Stpi	10.60	.755	Ea.	161	26.50		187.50	217	300
	0180	2″ diameter	Q-5	13.30	1.203		163	38		201	236	
	0190	2-1/2″ diameter		12.40	1.290		170	40.50		210.50	248	
	0200	3″ diameter		11.40	1.404		193	44		237	279	
	0210	4″ diameter		8.40	1.905		206	60		266	320	
	0220	5″ diameter		7.60	2.105		254	66		320	380	
	0230	6″ diameter		6.80	2.353		261	74		335	400	
	0240	8″ diameter		5.40	2.963		299	93		392	470	
	0250	10″ diameter		5	3.200		400	100		500	590	
	0260	12″ diameter		4.60	3.478		455	109		564	665	
	0480	10″ face to face, 2″ diameter		13	1.231		240	38.50		278.50	325	
	0500	2-1/2″ diameter		12	1.333		253	42		295	340	
	0520	3″ diameter		11	1.455		257	45.50		302.50	350	
	0540	4″ diameter		8	2		287	63		350	410	
	0560	5″ diameter		7	2.286		390	71.50		461.50	540	
	0580	6″ diameter		6	2.667		355	83.50		438.50	520	
	0600	8″ diameter		5	3.200		425	100		525	615	
	0620	10″ diameter		4.60	3.478		455	109		564	665	
	0640	12″ diameter		4	4		520	126		646	760	
	0660	14″ diameter		3.80	4.211		650	132		782	915	
	0680	16″ diameter		2.90	5.517		745	173		918	1,075	
	0700	18″ diameter		2.50	6.400		840	201		1,041	1,225	
	0720	20″ diameter		2.10	7.619		880	239		1,119	1,325	
	0740	24″ diameter		1.80	8.889		1,000	279		1,279	1,550	
	0760	26″ diameter		1.40	11.429		1,125	360		1,485	1,775	
	0780	30″ diameter		1.20	13.333		1,225	420		1,645	1,975	
	0800	36″ diameter		1	16		1,525	500		2,025	2,425	
320	0010	**EXPANSION TANKS**										320
	1500	Plastic, corrosion resistant, see Div. 152-132										
	1505	Fiberglass and steel single / double wall storage, see Div 13201										
	1510	Tank leak detection systems, see Div 13851-350										
	2000	Steel, liquid expansion, ASME, painted, 15 gallon capacity	Q-5	17	.941	Ea.	340	29.50		369.50	415	
	2020	24 gallon capacity		14	1.143		340	36		376	430	
	2040	30 gallon capacity		12	1.333		375	42		417	480	
	2060	40 gallon capacity		10	1.600		430	50		480	550	
	2360	Galvanized										
	2370	15 gallon capacity	Q-5	17	.941	Ea.	495	29.50		524.50	590	
	2380	24 gallon capacity		14	1.143		625	36		661	740	
	2390	30 gallon capacity		12	1.333		600	42		642	725	
	3000	Steel ASME expansion, rubber diaphragm, 19 gal. cap. accept.		12	1.333		1,275	42		1,317	1,475	
	3020	31 gallon capacity		8	2		1,425	63		1,488	1,650	
	3040	61 gallon capacity		6	2.667		2,000	83.50		2,083.50	2,300	
350	0010	**FLEXIBLE CONNECTORS**, Corrugated, 7/8″ O.D., 1/2″ I.D.										350
	0050	Gas, seamless brass, steel fittings										
	0200	12″ long	1 Plum	36	.222	Ea.	9.95	7.65		17.60	22.50	
	0220	18″ long		36	.222		12.35	7.65		20	25	
	0240	24″ long		34	.235		14.60	8.10		22.70	28.50	
	0260	30″ long		34	.235		15.75	8.10		23.85	29.50	
	0280	36″ long		32	.250		17.40	8.60		26	32	
	0320	48″ long		30	.267		22	9.20		31.20	38.50	
	0340	60″ long		30	.267		26.50	9.20		35.70	43	
	0360	72″ long		30	.267		30.50	9.20		39.70	47.50	
	2000	Water, copper tubing, dielectric separators										
	2100	12″ long	1 Plum	36	.222	Ea.	8.50	7.65		16.15	21	

	15120 | Piping Specialties		CREW	DAILY OUTPUT	LABOR-HOURS	UNIT	2001 BARE COSTS MAT.	LABOR	EQUIP.	TOTAL	TOTAL INCL O&P	
350	2220	15″ long	1 Plum	36	.222	Ea.	9.45	7.65		17.10	22	350
	2240	18″ long		36	.222		10.30	7.65		17.95	23	
	2260	24″ long		34	.235		12.65	8.10		20.75	26	
370	0010	**FLEXIBLE METAL HOSE** Connectors, standard lengths										370
	0100	Bronze braided, bronze ends										
	0120	3/8″ diameter x 12″	1 Stpi	26	.308	Ea.	14.40	10.70		25.10	32	
	0140	1/2″ diameter x 12″		24	.333		13.70	11.60		25.30	32.50	
	0160	3/4″ diameter x 12″		20	.400		20	13.95		33.95	43	
	0180	1″ diameter x 18″		19	.421		26	14.65		40.65	50.50	
	0200	1-1/2″ diameter x 18″		13	.615		41	21.50		62.50	77.50	
	0220	2″ diameter x 18″		11	.727		49.50	25.50		75	93	
	1000	Carbon steel ends										
	1020	1/4″ diameter x 12″	1 Stpi	28	.286	Ea.	11	9.95		20.95	27	
	1040	3/8″ diameter x 12″		26	.308		11.30	10.70		22	28.50	
	1060	1/2″ diameter x 12″		24	.333		11	11.60		22.60	29.50	
	1080	1/2″ diameter x 24″		24	.333		14.10	11.60		25.70	33	
	1100	1/2″ diameter x 36″		24	.333		17.20	11.60		28.80	36.50	
	1120	3/4″ diameter x 12″		20	.400		14.60	13.95		28.55	37	
	1140	3/4″ diameter x 24″		20	.400		17.90	13.95		31.85	40.50	
	1160	3/4″ diameter x 36″		20	.400		21	13.95		34.95	44.50	
	1180	1″ diameter x 18″		19	.421		19.20	14.65		33.85	43	
	1200	1″ diameter x 30″		19	.421		22.50	14.65		37.15	47	
	1220	1″ diameter x 36″		19	.421		24.50	14.65		39.15	49	
	1240	1-1/4″ diameter x 18″		15	.533		19.60	18.60		38.20	49.50	
	1260	1-1/4″ diameter x 36″		15	.533		26	18.60		44.60	56.50	
	1280	1-1/2″ diameter x 18″		13	.615		21	21.50		42.50	55.50	
	1300	1-1/2″ diameter x 36″		13	.615		28	21.50		49.50	63.50	
	1320	2″ diameter x 24″		11	.727		27.50	25.50		53	69	
	1340	2″ diameter x 36″		11	.727		34	25.50		59.50	76	
	1360	2-1/2″ diameter x 24″		9	.889		39	31		70	90	
	1380	2-1/2″ diameter x 36″		9	.889		46	31		77	97.50	
	1400	3″ diameter x 24″		7	1.143		46.50	40		86.50	111	
	1420	3″ diameter x 36″		7	1.143		54	40		94	120	
	2000	Carbon steel braid, carbon steel solid ends										
	2100	1/2″ diameter x 12″	1 Stpi	24	.333	Ea.	11	11.60		22.60	29.50	
	2120	3/4″ diameter x 12″		20	.400		14.60	13.95		28.55	37	
	2140	1″ diameter x 12″		19	.421		17.40	14.65		32.05	41	
	2160	1-1/4″ diameter x 12″		15	.533		17.40	18.60		36	47	
	2180	1-1/2″ diameter x 12″		13	.615		18.80	21.50		40.30	53	
	3000	Stainless steel braid, welded on carbon steel ends										
	3100	1/2″ diameter x 12″	1 Stpi	24	.333	Ea.	16.10	11.60		27.70	35.50	
	3120	3/4″ diameter x 12″		20	.400		23	13.95		36.95	46	
	3140	3/4″ diameter x 24″		20	.400		27.50	13.95		41.45	51.50	
	3160	3/4″ diameter x 36″		20	.400		32.50	13.95		46.45	57	
	3180	1″ diameter x 12″		19	.421		26	14.65		40.65	50.50	
	3200	1″ diameter x 24″		19	.421		31.50	14.65		46.15	56.50	
	3220	1″ diameter x 36″		19	.421		37	14.65		51.65	63	
	3240	1-1/4″ diameter x 12″		15	.533		28.50	18.60		47.10	59	
	3260	1-1/4″ diameter x 24″		15	.533		36	18.60		54.60	67.50	
	3280	1-1/4″ diameter x 36″		15	.533		43.50	18.60		62.10	76	
	3300	1-1/2″ diameter x 12″		13	.615		31	21.50		52.50	67	
	3320	1-1/2″ diameter x 24″		13	.615		40	21.50		61.50	76.50	
	3340	1-1/2″ diameter x 36″		13	.615		48.50	21.50		70	86	
	3400	Metal sst braid, over corrugated stainless steel, flanged ends										
	3410	150 PSI										

	15120 | Piping Specialties		CREW	DAILY OUTPUT	LABOR-HOURS	UNIT	2001 BARE COSTS MAT.	LABOR	EQUIP.	TOTAL	TOTAL INCL O&P	
370	3420	1/2″ diameter x 12″	1 Stpi	24	.333	Ea.	20	11.60		31.60	39.50	370
	3430	1″ diameter x 12″		20	.400		32	13.95		45.95	56	
	3440	1-1/2″ diameter x 12″		15	.533		37	18.60		55.60	68.50	
	3450	2-1/2″ diameter x 9″		12	.667		54	23		77	94.50	
	3460	3″ diameter x 9″		9	.889		65	31		96	119	
	3470	4″ diameter x 9″		7	1.143		80	40		120	148	
	3480	4″ diameter x 30″		5	1.600		170	56		226	271	
	3490	4″ diameter x 36″		4.80	1.667		190	58		248	297	
	3500	6″ diameter x 11″		5	1.600		132	56		188	229	
	3510	6″ diameter x 36″		3.80	2.105		251	73.50		324.50	390	
	3520	8″ diameter x 12″		4	2		256	69.50		325.50	385	
	3530	10″ diameter x 13″		3	2.667		325	93		418	500	
	3540	12″ diameter x 14″	Q-5	4	4		655	126		781	910	
	6000	Molded rubber with helical wire reinforcement										
	6010	150 PSI										
	6020	1-1/2″ diameter x 12″	1 Stpi	15	.533	Ea.	39	18.60		57.60	70.50	
	6030	2″ diameter x 12″		12	.667		145	23		168	195	
	6040	3″ diameter x 12″		8	1		187	35		222	259	
	6050	4″ diameter x 12″		6	1.333		218	46.50		264.50	310	
	6060	6″ diameter x 18″		4	2		275	69.50		344.50	410	
	6070	8″ diameter x 24″		3	2.667		355	93		448	530	
	6080	10″ diameter x 24″		2	4		400	139		539	650	
	6090	12″ diameter x 24″	Q-5	3	5.333		455	167		622	755	
	7000	Molded teflon with stainless steel flanges										
	7010	150 PSI										
	7020	2-1/2″ diameter x 3-3/16″	Q-1	7.80	2.051	Ea.	325	63.50		388.50	455	
	7030	3″ diameter x 3-5/8″		6.50	2.462		415	76.50		491.50	575	
	7040	4″ diameter x 3-5/8″		5	3.200		450	99		549	645	
	7050	6″ diameter x 4″		4.30	3.721		645	115		760	885	
	7060	8″ diameter x 6″		3.80	4.211		860	131		991	1,150	
400	0010	**FLOAT VALVES**										400
	0020	With ball and bracket										
	0030	Single seat, threaded										
	0040	Brass body										
	0050	1/2″	1 Stpi	11	.727	Ea.	37	25.50		62.50	79	
	0060	3/4″		9	.889		44.50	31		75.50	96	
	0070	1″		7	1.143		56.50	40		96.50	123	
	0080	1-1/2″		4.50	1.778		83	62		145	185	
	0090	2″		3.60	2.222		84	77.50		161.50	210	
	0300	For condensate receivers, CI, in-line mount										
	0320	1″ inlet	1 Stpi	7	1.143	Ea.	116	40		156	187	
	0360	For condensate receiver, CI, external float, flanged tank mount										
	0370	3/4″ inlet	1 Stpi	5	1.600	Ea.	98.50	56		154.50	192	
420	0010	**FLOW CHECK CONTROL**										420
	0100	Bronze body, soldered										
	0110	3/4″ size	1 Stpi	20	.400	Ea.	34.50	13.95		48.45	59	
	0120	1″ size	″	19	.421	″	41.50	14.65		56.15	67.50	
	0200	Cast iron body, threaded										
	0210	3/4″ size	1 Stpi	20	.400	Ea.	27.50	13.95		41.45	51.50	
	0220	1″ size		19	.421		31	14.65		45.65	56	
	0230	1-1/4″ size		15	.533		37.50	18.60		56.10	69.50	
	0240	1-1/2″ size		13	.615		57.50	21.50		79	96	
	0250	2″ size		11	.727		82.50	25.50		108	130	
520	0010	**HYDRONIC HEATING CONTROL VALVES**										520
	0050	Hot water, nonelectric, thermostatic										

15120 | Piping Specialties

			CREW	DAILY OUTPUT	LABOR-HOURS	UNIT	2001 BARE COSTS MAT.	LABOR	EQUIP.	TOTAL	TOTAL INCL O&P	
520	0100	Radiator supply, 1/2″ diameter	1 Stpi	24	.333	Ea.	36	11.60		47.60	57	520
	0120	3/4″ diameter	″	20	.400	″	36.50	13.95		50.45	61.50	
	1000	Manual, radiator supply										
	1010	1/2″ pipe size, angle union	1 Stpi	24	.333	Ea.	28.50	11.60		40.10	48.50	
	1020	3/4″ pipe size, angle union	″	20	.400	″	26.50	13.95		40.45	50.50	
	1100	Radiator, balancing, straight, sweat connections										
	1110	1/2″ pipe size	1 Stpi	24	.333	Ea.	7.70	11.60		19.30	26	
	1120	3/4″ pipe size	″	20	.400	″	11.30	13.95		25.25	33.50	
	1200	Steam, radiator, supply										
	1210	1/2″ pipe size, angle union	1 Stpi	24	.333	Ea.	25.50	11.60		37.10	45.50	
	1220	3/4″ pipe size, angle union	″	20	.400	″	28	13.95		41.95	52	
	8000	System balancing and shut-off										
	8020	Butterfly, quarter turn, calibrated, threaded or solder										
	8040	Bronze, -30° F to +350° F, pressure to 175 psi										
	8060	1/2″ size	1 Stpi	22	.364	Ea.	9.80	12.65		22.45	30	
	8070	3/4″ size	″	20	.400	″	10.55	13.95		24.50	32.50	
580	0010	**MIXING VALVE** Automatic water tempering										580
	0050	3/4″ size	1 Stpi	18	.444	Ea.	335	15.50		350.50	390	
	0100	1″ size		16	.500		475	17.45		492.45	550	
	0120	1-1/4″ size		13	.615		690	21.50		711.50	790	
	0140	1-1/2″ size		10	.800		820	28		848	945	
	0160	2″ size		8	1		1,025	35		1,060	1,200	
	0180	3″ size	↓	4	2	↓	2,375	69.50		2,444.50	2,725	
610	0010	**MONOFLOW TEE FITTING**										610
	1100	For one pipe hydronic, supply and return										
	1110	Copper, soldered										
	1120	3/4″ x 1/2″ size	1 Stpi	13	.615	Ea.	12	21.50		33.50	45.50	
670	0010	**PRESSURE REGULATOR**										670
	0100	Gas appliance regulators										
	0106	Main burner and pilot applications										
	0108	Rubber seat poppet type										
	0109	1/8″ pipe size	1 Stpi	24	.333	Ea.	10.60	11.60		22.20	29.50	
	0110	1/4″ pipe size		24	.333		12.15	11.60		23.75	31	
	0112	3/8″ pipe size		24	.333		13.10	11.60		24.70	32	
	0113	1/2″ pipe size		24	.333		14.55	11.60		26.15	33.50	
	0114	3/4″ pipe size	↓	20	.400	↓	17.45	13.95		31.40	40	
	0122	Lever action type										
	0123	3/8″ pipe size	1 Stpi	24	.333	Ea.	21.50	11.60		33.10	41	
	0124	1/2″ pipe size		24	.333		21.50	11.60		33.10	41	
	0125	3/4″ pipe size		20	.400		43	13.95		56.95	68.50	
	0126	1″ pipe size	↓	19	.421	↓	43	14.65		57.65	69.50	
	0132	Double diaphragm type										
	0133	3/8″ pipe size	1 Stpi	24	.333	Ea.	24.50	11.60		36.10	44.50	
	0134	1/2″ pipe size		24	.333		37.50	11.60		49.10	58.50	
	0135	3/4″ pipe size		20	.400		64	13.95		77.95	91.50	
	0136	1″ pipe size		19	.421		78.50	14.65		93.15	108	
	0137	1-1/4″ pipe size		15	.533		255	18.60		273.60	310	
	0138	1-1/2″ pipe size		13	.615		475	21.50		496.50	555	
	0139	2″ pipe size	↓	11	.727		475	25.50		500.50	560	
	0140	2-1/2″ pipe size	Q-5	15	1.067		925	33.50		958.50	1,075	
	0141	3″ pipe size		13	1.231		925	38.50		963.50	1,075	
	0142	4″ pipe size (flanged)	↓	8	2	↓	1,700	63		1,763	1,975	
	0160	Main burner only										
	0162	Straight-thru-flow design										
	0163	1/2″ pipe size	1 Stpi	24	.333	Ea.	27.50	11.60		39.10	48	

15120 | Piping Specialties

			CREW	DAILY OUTPUT	LABOR-HOURS	UNIT	2001 BARE COSTS MAT.	LABOR	EQUIP.	TOTAL	TOTAL INCL O&P	
670	0164	3/4" pipe size	1 Stpi	20	.400	Ea.	37.50	13.95		51.45	62.50	670
	0165	1" pipe size		19	.421		56	14.65		70.65	83.50	
	0166	1-1/4" pipe size	↓	15	.533	↓	56	18.60		74.60	89.50	
	0200	Oil, light, hot water, ordinary steam, threaded										
	0220	Bronze body, 1/4" size	1 Stpi	24	.333	Ea.	122	11.60		133.60	152	
	0230	3/8" size		24	.333		132	11.60		143.60	163	
	0240	1/2" size		24	.333		163	11.60		174.60	198	
	0250	3/4" size		20	.400		195	13.95		208.95	235	
	0260	1" size		19	.421		297	14.65		311.65	345	
	0270	1-1/4" size		15	.533		400	18.60		418.60	470	
	0320	Iron body, 1/4" size		24	.333		95	11.60		106.60	123	
	0330	3/8" size		24	.333		112	11.60		123.60	142	
	0340	1/2" size		24	.333		119	11.60		130.60	149	
	0350	3/4" size		20	.400		144	13.95		157.95	179	
	0360	1" size		19	.421		188	14.65		202.65	229	
	0370	1-1/4" size	↓	15	.533	↓	261	18.60		279.60	315	
	9000	For water pressure regulators, see Div. 15110-160										
730	0010	**SLEEVES AND ESCUTCHEONS**										730
	0100	Pipe sleeve										
	0110	Steel, w/water stop, 12" long, with link seal										
	0120	2" diam. for 1/2" carrier pipe	1 Plum	8.40	.952	Ea.	24	33		57	75.50	
	0130	2-1/2" diam. for 3/4" carrier pipe		8	1		28.50	34.50		63	83.50	
	0140	2-1/2" diam. for 1" carrier pipe		8	1		26.50	34.50		61	81	
	0150	3" diam. for 1-1/4" carrier pipe		7.20	1.111		37.50	38.50		76	99	
	0160	3-1/2" diam. for 1-1/2" carrier pipe		6.80	1.176		38.50	40.50		79	103	
	0170	4" diam. for 2" carrier pipe		6	1.333		42.50	46		88.50	117	
	0180	4" diam. for 2-1/2" carrier pipe		6	1.333		42	46		88	116	
	0190	5" diam. for 3" carrier pipe		5.40	1.481		53	51		104	136	
	0200	6" diam. for 4" carrier pipe	↓	4.80	1.667		61.50	57.50		119	154	
	0210	10" diam. for 6" carrier pipe	Q-1	8	2		115	62		177	220	
	0220	12" diam. for 8" carrier pipe		7.20	2.222		165	69		234	286	
	0230	14" diam. for 10" carrier pipe		6.40	2.500		174	77.50		251.50	310	
	0240	16" diam. for 12" carrier pipe		5.80	2.759		203	85.50		288.50	355	
	0250	18" diam. for 14" carrier pipe		5.20	3.077		229	95.50		324.50	395	
	0260	24" diam. for 18" carrier pipe		4	4		380	124		504	600	
	0270	24" diam. for 20" carrier pipe		4	4		310	124		434	525	
	0280	30" diam. for 24" carrier pipe	↓	3.20	5	↓	700	155		855	1,000	
	0500	Wall sleeve										
	0510	Ductile iron with rubber gasket seal										
	0520	3"	1 Plum	8.40	.952	Ea.	96	33		129	156	
	0530	4"		7.20	1.111		120	38.50		158.50	190	
	0540	6"		6	1.333		161	46		207	247	
	0550	8"		4	2		206	69		275	330	
	0560	10"		3	2.667		231	92		323	395	
	0570	12"	↓	2.40	3.333	↓	274	115		389	475	
	5000	Escutcheon										
	5100	Split ring, pipe										
	5110	Chrome plated										
	5120	1/2"	1 Plum	160	.050	Ea.	.88	1.72		2.60	3.57	
	5130	3/4"		160	.050		.92	1.72		2.64	3.61	
	5140	1"		135	.059		.98	2.04		3.02	4.16	
	5150	1-1/2"		115	.070		1.48	2.40		3.88	5.25	
	5160	2"		100	.080		1.52	2.76		4.28	5.85	
	5170	4"		80	.100		3.35	3.45		6.80	8.90	
	5180	6"	↓	68	.118	↓	5.60	4.05		9.65	12.25	

Where can you get cost data
with just one click?

Get your Means CostWorks 2001 CD today!

FAX back the form below Toll Free: 1-800-632-6732 or Order by Phone: 1-800-334-3509

Select the individual digital cost books or packages you need.

Catalog No.	Title	Price	Cost
65061	Assemblies Cost Data	$159.95	
65011	Building Construction Cost Data	$110.95	
65111	Concrete & Masonry Cost Data	$110.95	
65031	Electrical Cost Data	$110.95	
65161	Heavy Construction Cost Data	$110.95	
65091	Interior Cost Data	$110.95	
65181	Light Commercial Cost Data	$139.95	
65021	Mechanical Cost Data	$110.95	
65151	Open Shop BCCD	$110.95	
65211	Plumbing Cost Data	$110.95	
65281	Site Work & Landscape Cost Data	$110.95	
65051	Square Foot Model Costs	$159.95	
65901	Wage Rate Utility (Available for above titles only)	$110.95	
65201	Facilities Construction Cost Data	$259.95	
65041	Repair & Remodeling Cost Data	$110.95	
65311	Building Construction Cost Data/Metric	$110.95	
65361	Heavy Construction Cost Data/Metric	$110.95	
65171	Residential Cost Data	$129.95	
65301	Facilities Maintenance & Repair Cost Data	$259.95	
PACKAGES			
Catalog No.	**Title**	**Price**	**Cost**
65511	Builder's Package	$299.99	
65521	Building Professional's Package	$349.99	
65531	Facility Manager's Package	$449.99	
65541	Design Professional's Package	$349.99	
		Subtotal	
		MA residents please add 5% sales tax	
		Shipping and Handling	$5.00
		Grand total	

Name: ______________________

Company: ______________________

Street: ______________________

City ______________ State: ______________ Zip: ______________

Phone (   ) ______________ FAX: (   ) ______________

E-Mail: ______________________

All items shipped on a single CD with software code to unlock specific selections purchased.

Buy additional keys any time you need to unlock additional cost data books!

☐ Bill me ☐ Please charge my credit card

☐ Visa ☐ MasterCard ☐ American Express ☐ Discover

Card # ______________________ Expiration Date: ______

Cardholder Signature: ______________________

RSMeans
CMDGROUP

CWCW 2001

15120 | Piping Specialties

		Description	Crew	Daily Output	Labor-Hours	Unit	2001 Bare Costs Mat.	Labor	Equip.	Total	Total Incl O&P	
790	0010	**STRAINERS, BASKET TYPE** Perforated stainless steel basket										790
	0100	Brass or monel available										
	2000	Simplex style										
	2300	Bronze body										
	2320	Screwed, 3/8″ pipe size	1 Stpi	22	.364	Ea.	98	12.65		110.65	127	
	2340	1/2″ pipe size		20	.400		100	13.95		113.95	131	
	2360	3/4″ pipe size		17	.471		103	16.40		119.40	138	
	2380	1″ pipe size		15	.533		171	18.60		189.60	216	
	2400	1-1/4″ pipe size		13	.615		201	21.50		222.50	254	
	2420	1-1/2″ pipe size		12	.667		260	23		283	320	
	2440	2″ pipe size		10	.800		395	28		423	475	
	2460	2-1/2″ pipe size	Q-5	15	1.067		565	33.50		598.50	670	
	2480	3″ pipe size	″	14	1.143		815	36		851	955	
	2600	Flanged, 2″ pipe size	1 Stpi	6	1.333		600	46.50		646.50	730	
	2620	2-1/2″ pipe size	Q-5	4.50	3.556		730	112		842	975	
	2640	3″ pipe size		3.50	4.571		1,075	143		1,218	1,400	
	2660	4″ pipe size		3	5.333		1,800	167		1,967	2,225	
	2680	5″ pipe size	Q-6	3.40	7.059		2,150	230		2,380	2,700	
	2700	6″ pipe size		3	8		3,525	260		3,785	4,275	
	2710	8″ pipe size		2.50	9.600		6,300	310		6,610	7,400	
	3600	Iron body										
	3700	Screwed, 3/8″ pipe size	1 Stpi	22	.364	Ea.	112	12.65		124.65	142	
	3720	1/2″ pipe size		20	.400		112	13.95		125.95	144	
	3740	3/4″ pipe size		17	.471		147	16.40		163.40	187	
	3760	1″ pipe size		15	.533		147	18.60		165.60	190	
	3780	1-1/4″ pipe size		13	.615		197	21.50		218.50	249	
	3800	1-1/2″ pipe size		12	.667		216	23		239	272	
	3820	2″ pipe size		10	.800		258	28		286	325	
	3840	2-1/2″ pipe size	Q-5	15	1.067		370	33.50		403.50	455	
	3860	3″ pipe size	″	14	1.143		420	36		456	515	
	4000	Flanged, 2″ pipe size	1 Stpi	6	1.333		350	46.50		396.50	455	
	4020	2-1/2″ pipe size	Q-5	4.50	3.556		475	112		587	695	
	4040	3″ pipe size		3.50	4.571		505	143		648	770	
	4060	4″ pipe size		3	5.333		760	167		927	1,100	
	4080	5″ pipe size	Q-6	3.40	7.059		1,200	230		1,430	1,675	
	4100	6″ pipe size		3	8		1,475	260		1,735	2,025	
	4120	8″ pipe size		2.50	9.600		2,725	310		3,035	3,475	
	4140	10″ pipe size		2.20	10.909		4,100	355		4,455	5,025	
	4160	12″ pipe size		1.70	14.118		4,875	460		5,335	6,075	
	4180	14″ pipe size		1.40	17.143		6,150	560		6,710	7,600	
	4200	16″ pipe size		1	24		6,425	780		7,205	8,250	
	7000	Stainless steel body										
	7200	Screwed, 1″ pipe size	1 Stpi	15	.533	Ea.	395	18.60		413.60	465	
	7210	1-1/4″ pipe size		13	.615		575	21.50		596.50	670	
	7220	1-1/2″ pipe size		12	.667		575	23		598	670	
	7240	2″ pipe size		10	.800		845	28		873	970	
	7260	2-1/2″ pipe size	Q-5	15	1.067		1,400	33.50		1,433.50	1,600	
	7280	3″ pipe size	″	14	1.143		1,850	36		1,886	2,100	
	7400	Flanged, 2″ pipe size	1 Stpi	6	1.333		1,150	46.50		1,196.50	1,350	
	7420	2-1/2″ pipe size	Q-5	4.50	3.556		2,100	112		2,212	2,500	
	7440	3″ pipe size		3.50	4.571		2,150	143		2,293	2,600	
	7460	4″ pipe size		3	5.333		3,375	167		3,542	3,975	
	7480	6″ pipe size	Q-6	3	8		5,850	260		6,110	6,850	
	7500	8″ pipe size	″	2.50	9.600		9,925	310		10,235	11,400	
	8100	Duplex style										
	8200	Bronze body										

15120 | Piping Specialties

			CREW	DAILY OUTPUT	LABOR-HOURS	UNIT	2001 BARE COSTS MAT.	LABOR	EQUIP.	TOTAL	TOTAL INCL O&P	
790	8240	Screwed, 3/4″ pipe size	1 Stpi	16	.500	Ea.	1,050	17.45		1,067.45	1,175	790
	8260	1″ pipe size		14	.571		1,050	19.90		1,069.90	1,175	
	8280	1-1/4″ pipe size		12	.667		2,050	23		2,073	2,275	
	8300	1-1/2″ pipe size		11	.727		2,050	25.50		2,075.50	2,300	
	8320	2″ pipe size		9	.889		3,250	31		3,281	3,625	
	8340	2-1/2″ pipe size	Q-5	14	1.143		3,250	36		3,286	3,625	
	8420	Flanged, 2″ pipe size	1 Stpi	6	1.333		3,675	46.50		3,721.50	4,125	
	8440	2-1/2″ pipe size	Q-5	4.50	3.556		3,775	112		3,887	4,325	
	8460	3″ pipe size		3.50	4.571		5,200	143		5,343	5,925	
	8480	4″ pipe size		3	5.333		8,200	167		8,367	9,275	
	8500	5″ pipe size	Q-6	3.40	7.059		14,300	230		14,530	16,000	
	8520	6″ pipe size	″	3	8		18,000	260		18,260	20,200	
	8700	Iron body										
	8740	Screwed, 3/4″ pipe size	1 Stpi	16	.500	Ea.	545	17.45		562.45	625	
	8760	1″ pipe size		14	.571		545	19.90		564.90	630	
	8780	1-1/4″ pipe size		12	.667		950	23		973	1,075	
	8800	1-1/2″ pipe size		11	.727		950	25.50		975.50	1,100	
	8820	2″ pipe size		9	.889		1,625	31		1,656	1,825	
	8840	2-1/2″ pipe size	Q-5	14	1.143		1,625	36		1,661	1,825	
	9000	Flanged, 2″ pipe size	1 Stpi	6	1.333		1,700	46.50		1,746.50	1,950	
	9020	2-1/2″ pipe size	Q-5	4.50	3.556		1,775	112		1,887	2,125	
	9040	3″ pipe size		3.50	4.571		2,250	143		2,393	2,700	
	9060	4″ pipe size		3	5.333		3,550	167		3,717	4,175	
	9080	5″ pipe size	Q-6	3.40	7.059		7,150	230		7,380	8,225	
	9100	6″ pipe size		3	8		7,150	260		7,410	8,275	
	9120	8″ pipe size		2.50	9.600		13,700	310		14,010	15,600	
	9140	10″ pipe size		2.20	10.909		19,100	355		19,455	21,500	
	9160	12″ pipe size		1.70	14.118		21,200	460		21,660	24,000	
	9170	14″ pipe size		1.40	17.143		25,700	560		26,260	29,100	
	9180	16″ pipe size		1	24		32,200	780		32,980	36,700	
	9700	Stainless steel body										
	9740	Screwed, 1″ pipe size	1 Stpi	14	.571	Ea.	2,775	19.90		2,794.90	3,075	
	9760	1-1/2″ pipe size		11	.727		4,200	25.50		4,225.50	4,675	
	9780	2″ pipe size		9	.889		6,775	31		6,806	7,500	
	9860	Flanged, 2″ pipe size		6	1.333		7,050	46.50		7,096.50	7,825	
	9880	2-1/2″ pipe size	Q-5	4.50	3.556		7,400	112		7,512	8,325	
	9900	3″ pipe size		3.50	4.571		11,600	143		11,743	13,000	
	9920	4″ pipe size		3	5.333		14,200	167		14,367	15,900	
	9940	6″ pipe size	Q-6	3	8		25,000	260		25,260	27,900	
	9960	8″ pipe size	″	2.50	9.600		46,300	310		46,610	51,500	
820	0010	**STRAINERS, Y TYPE** Bronze body										820
	0050	Screwed, 150 lb., 1/4″ pipe size	1 Stpi	24	.333	Ea.	11.40	11.60		23	30	
	0070	3/8″ pipe size		24	.333		15.90	11.60		27.50	35	
	0100	1/2″ pipe size		20	.400		15.90	13.95		29.85	38.50	
	0120	3/4″ pipe size		19	.421		17.40	14.65		32.05	41	
	0140	1″ pipe size		17	.471		18.70	16.40		35.10	45.50	
	0150	1-1/4″ pipe size		15	.533		42	18.60		60.60	74	
	0160	1-1/2″ pipe size		14	.571		40.50	19.90		60.40	74.50	
	0180	2″ pipe size		13	.615		53.50	21.50		75	91.50	
	0182	3″ pipe size		12	.667		286	23		309	350	
	0200	300 lb., 2-1/2″ pipe size	Q-5	17	.941		286	29.50		315.50	360	
	0220	3″ pipe size		16	1		565	31.50		596.50	675	
	0240	4″ pipe size		15	1.067		1,300	33.50		1,333.50	1,475	
	0500	For 300 lb rating 1/4″ thru 2″, add					15%					
	1000	Flanged, 150 lb., 1-1/2″ pipe size	1 Stpi	11	.727	Ea.	305	25.50		330.50	375	
	1020	2″ pipe size	″	8	1		390	35		425	485	

Important: See the Reference Section for critical supporting data - Reference Nos., Crews, & City Cost Indexes

15120 | Piping Specialties

			CREW	DAILY OUTPUT	LABOR-HOURS	UNIT	2001 BARE COSTS MAT.	LABOR	EQUIP.	TOTAL	TOTAL INCL O&P	
820	1030	2-1/2" pipe size	Q-5	5	3.200	Ea.	505	100		605	705	820
	1040	3" pipe size		4.50	3.556		620	112		732	855	
	1060	4" pipe size		3	5.333		945	167		1,112	1,275	
	1080	5" pipe size	Q-6	3.40	7.059		1,875	230		2,105	2,425	
	1100	6" pipe size		3	8		2,500	260		2,760	3,150	
	1106	8" pipe size		2.60	9.231		3,900	300		4,200	4,750	
	1500	For 300 lb rating, add					40%					
840	0010	**STRAINERS, Y TYPE** Iron body										840
	0050	Screwed, 250 lb., 1/4" pipe size	1 Stpi	20	.400	Ea.	6.65	13.95		20.60	28.50	
	0070	3/8" pipe size		20	.400		6.30	13.95		20.25	28	
	0100	1/2" pipe size		20	.400		6.65	13.95		20.60	28.50	
	0120	3/4" pipe size		18	.444		7.95	15.50		23.45	32.50	
	0140	1" pipe size		16	.500		10.75	17.45		28.20	38.50	
	0150	1-1/4" pipe size		15	.533		14.75	18.60		33.35	44	
	0160	1-1/2" pipe size		12	.667		17.75	23		40.75	54.50	
	0180	2" pipe size		8	1		27.50	35		62.50	82.50	
	0200	2-1/2" pipe size	Q-5	12	1.333		140	42		182	217	
	0220	3" pipe size		11	1.455		151	45.50		196.50	235	
	0240	4" pipe size		5	3.200		255	100		355	430	
	0500	For galvanized body, add					50%					
	1000	Flanged, 125 lb., 1-1/2" pipe size	1 Stpi	11	.727	Ea.	86	25.50		111.50	133	
	1020	2" pipe size	"	8	1		64	35		99	123	
	1030	2-1/2" pipe size	Q-5	5	3.200		72.50	100		172.50	232	
	1040	3" pipe size		4.50	3.556		84.50	112		196.50	261	
	1060	4" pipe size		3	5.333		155	167		322	425	
	1080	5" pipe size	Q-6	3.40	7.059		242	230		472	610	
	1100	6" pipe size		3	8		295	260		555	720	
	1120	8" pipe size		2.50	9.600		495	310		805	1,025	
	1140	10" pipe size		2	12		950	390		1,340	1,650	
	1160	12" pipe size		1.70	14.118		1,425	460		1,885	2,275	
	1170	14" pipe size		1.30	18.462		2,625	600		3,225	3,800	
	1180	16" pipe size		1	24		3,725	780		4,505	5,275	
	1500	For 250 lb rating, add					20%					
	2000	For galvanized body, add					50%					
	2500	For steel body, add					40%					
870	0010	**SUCTION DIFFUSERS**										870
	0100	Cast iron body with integral straightening vanes, strainer										
	1000	Flanged										
	1010	2" inlet, 1-1/2" pump side	1 Stpi	6	1.333	Ea.	213	46.50		259.50	305	
	1020	2" pump side	"	5	1.600		221	56		277	325	
	1030	3" inlet, 2" pump side	Q-5	6.50	2.462		270	77		347	415	
900	0010	**THERMOFLO INDICATOR** For balancing										900
	1000	Sweat connections, 1-1/4" pipe size	1 Stpi	12	.667	Ea.	345	23		368	415	
	1020	1-1/2" pipe size	"	10	.800	"	355	28		383	430	
920	0010	**VENTURI FLOW** Measuring device										920
	0050	1/2" diameter	1 Stpi	24	.333	Ea.	105	11.60		116.60	134	
	0100	3/4" diameter		20	.400		93	13.95		106.95	123	
	0120	1" diameter		19	.421		94	14.65		108.65	125	
	0140	1-1/4" diameter		15	.533		134	18.60		152.60	175	
	0160	1-1/2" diameter		13	.615		140	21.50		161.50	187	
	0180	2" diameter		11	.727		166	25.50		191.50	222	
	0200	2-1/2" diameter	Q-5	16	1		259	31.50		290.50	335	
	0220	3" diameter		14	1.143		335	36		371	425	
	0240	4" diameter		11	1.455		375	45.50		420.50	485	

15120 | Piping Specialties

			CREW	DAILY OUTPUT	LABOR-HOURS	UNIT	2001 BARE COSTS MAT.	LABOR	EQUIP.	TOTAL	TOTAL INCL O&P	
920	0260	5" diameter	Q-6	4	6	Ea.	585	195		780	940	920
	0280	6" diameter		3.50	6.857		660	223		883	1,050	
	0300	8" diameter		3	8		1,050	260		1,310	1,550	
	0320	10" diameter		2	12		1,925	390		2,315	2,700	
	0330	12" diameter		1.80	13.333		3,075	435		3,510	4,025	
	0340	14" diameter		1.60	15		3,250	490		3,740	4,300	
	0350	16" diameter		1.40	17.143		6,500	560		7,060	8,000	
	0500	For meter, add					1,175			1,175	1,300	
940	0010	**WATER SUPPLY METERS**										940
	1000	Detector, serves dual systems such as fire and domestic or										
	1020	process water, wide range cap., UL and FM approved										
	1100	3" mainline x 2" by-pass, 400 GPM	Q-1	3.60	4.444	Ea.	5,100	138		5,238	5,825	
	1140	4" mainline x 2" by-pass, 700 GPM	"	2.50	6.400		5,500	198		5,698	6,350	
	1180	6" mainline x 3" by-pass, 1600 GPM	Q-2	2.60	9.231		7,475	297		7,772	8,650	
	1220	8" mainline x 4" by-pass, 2800 GPM		2.10	11.429		11,300	365		11,665	13,100	
	1260	10" mainline x 6" by-pass, 4400 GPM		2	12		18,400	385		18,785	20,800	
	1300	10"x12" mainlines x 6" by-pass,5400 GPM		1.70	14.118		19,100	455		19,555	21,700	
	2000	Domestic/commercial, bronze										
	2020	Threaded										
	2060	5/8" diameter, to 20 GPM	1 Plum	16	.500	Ea.	55.50	17.25		72.75	87	
	2080	3/4" diameter, to 30 GPM		14	.571		96	19.70		115.70	136	
	2100	1" diameter, to 50 GPM		12	.667		131	23		154	179	
	2300	Threaded/flanged										
	2340	1-1/2" diameter, to 100 GPM	1 Plum	8	1	Ea.	405	34.50		439.50	495	
	2360	2" diameter, to 160 GPM	"	6	1.333	"	535	46		581	660	
	2600	Flanged, compound										
	2640	3" diameter, 320 GPM	Q-1	3	5.333	Ea.	2,425	165		2,590	2,900	
	2660	4" diameter, to 500 GPM		1.50	10.667		3,775	330		4,105	4,650	
	2680	6" diameter, to 1,000 GPM		1	16		5,425	495		5,920	6,725	
	2700	8" diameter, to 1,800 GPM		.80	20		10,700	620		11,320	12,700	
	7000	Turbine										
	7260	Flanged										
	7300	2" diameter, to 160 GPM	1 Plum	7	1.143	Ea.	690	39.50		729.50	820	
	7320	3" diameter, to 450 GPM	Q-1	3.60	4.444		995	138		1,133	1,300	
	7340	4" diameter, to 650 GPM	"	2.50	6.400		1,700	198		1,898	2,175	
	7360	6" diameter, to 1800 GPM	Q-2	2.60	9.231		2,850	297		3,147	3,575	
	7380	8" diameter, to 2500 GPM		2.10	11.429		4,850	365		5,215	5,875	
	7400	10" diameter, to 5500 GPM		1.70	14.118		6,575	455		7,030	7,925	

15140 | Domestic Water Piping

			CREW	DAILY OUTPUT	LABOR-HOURS	UNIT	MAT.	LABOR	EQUIP.	TOTAL	TOTAL INCL O&P	
100	0010	**BACKFLOW PREVENTER** Includes valves										100
	0020	and four test cocks, corrosion resistant, automatic operation										
	1000	Double check principle										
	1080	Threaded, with gate valves										
	1100	3/4" pipe size	1 Plum	16	.500	Ea.	595	17.25		612.25	680	
	1120	1" pipe size		14	.571		620	19.70		639.70	710	
	1140	1-1/2" pipe size		10	.800		705	27.50		732.50	815	
	1160	2" pipe size		7	1.143		870	39.50		909.50	1,025	
	1300	Flanged, valves are OS&Y										
	1370	1" pipe size	1 Plum	5	1.600	Ea.	650	55		705	800	
	1374	1-1/2" pipe size		5	1.600		740	55		795	900	
	1378	2" pipe size		4.80	1.667		905	57.50		962.50	1,075	
	1380	3" pipe size	Q-1	4.50	3.556		1,450	110		1,560	1,775	
	1400	4" pipe size	"	3	5.333		2,100	165		2,265	2,550	
	1420	6" pipe size	Q-2	3	8		3,300	257		3,557	4,025	
	1430	8" pipe size	"	2	12		5,725	385		6,110	6,875	

Important: See the Reference Section for critical supporting data - Reference Nos., Crews, & City Cost Indexes

	15140 | Domestic Water Piping		CREW	DAILY OUTPUT	LABOR-HOURS	UNIT	2001 BARE COSTS MAT.	LABOR	EQUIP.	TOTAL	TOTAL INCL O&P	
100	4000	Reduced pressure principle										100
	4100	Threaded, valves are ball										
	4120	3/4" pipe size	1 Plum	16	.500	Ea.	685	17.25		702.25	780	
	4140	1" pipe size		14	.571		730	19.70		749.70	830	
	4150	1-1/4" pipe size		12	.667		1,025	23		1,048	1,150	
	4160	1-1/2" pipe size		10	.800		1,025	27.50		1,052.50	1,175	
	4180	2" pipe size	↓	7	1.143	↓	1,050	39.50		1,089.50	1,200	
	5000	Flanged, bronze, valves are OS&Y										
	5060	2-1/2" pipe size	Q-1	5	3.200	Ea.	2,725	99		2,824	3,150	
	5080	3" pipe size		4.50	3.556		3,375	110		3,485	3,900	
	5100	4" pipe size	↓	3	5.333		4,800	165		4,965	5,525	
	5120	6" pipe size	Q-2	3	8	↓	10,200	257		10,457	11,600	
	5600	Flanged, iron, valves are OS&Y										
	5660	2-1/2" pipe size	Q-1	5	3.200	Ea.	1,825	99		1,924	2,150	
	5680	3" pipe size		4.50	3.556		1,950	110		2,060	2,300	
	5700	4" pipe size	↓	3	5.333		2,575	165		2,740	3,075	
	5720	6" pipe size	Q-2	3	8		4,050	257		4,307	4,850	
	5740	8" pipe size		2	12		8,175	385		8,560	9,575	
	5760	10" pipe size	↓	1	24	↓	11,200	770		11,970	13,500	
600	0010	**VACUUM BREAKERS** Hot or cold water										600
	1030	Anti-siphon, brass										
	1040	1/4" size	1 Plum	24	.333	Ea.	18.65	11.50		30.15	38	
	1050	3/8" size		24	.333		18.65	11.50		30.15	38	
	1060	1/2" size		24	.333		21	11.50		32.50	41	
	1080	3/4" size		20	.400		25	13.80		38.80	48.50	
	1100	1" size		19	.421		39	14.50		53.50	65	
	1120	1-1/4" size		15	.533		68.50	18.35		86.85	103	
	1140	1-1/2" size		13	.615		80.50	21		101.50	121	
	1160	2" size		11	.727		125	25		150	176	
	1180	2-1/2" size		9	.889		360	30.50		390.50	440	
	1200	3" size	↓	7	1.143	↓	480	39.50		519.50	585	
	1300	For polished chrome, (1/4" thru 1"), add					50%					
	1900	Vacuum relief, water service, bronze										
	2000	1/2" size	1 Plum	30	.267	Ea.	17.70	9.20		26.90	33.50	
	2040	3/4" size	"	28	.286	"	22	9.85		31.85	39.50	
700	0010	**VACUUM BREAKERS**										700
	0011	See also backflow preventers 15140-100										
	1000	Anti-siphon continuous pressure type										
	1010	Max. 150 PSI - 210°F										
	1020	Bronze body										
	1030	1/2" size	1 Stpi	24	.333	Ea.	80.50	11.60		92.10	106	
	1040	3/4" size		20	.400		80.50	13.95		94.45	110	
	1050	1" size		19	.421		83.50	14.65		98.15	114	
	1060	1-1/4" size		15	.533		164	18.60		182.60	209	
	1070	1-1/2" size		13	.615		202	21.50		223.50	256	
	1080	2" size	↓	11	.727	↓	209	25.50		234.50	269	
	1200	Max. 125 PSI with atmospheric vent										
	1210	Brass, in-line construction										
	1220	1/4" size	1 Stpi	24	.333	Ea.	30.50	11.60		42.10	51.50	
	1230	3/8" size	"	24	.333		30.50	11.60		42.10	51.50	
	1260	For polished chrome finish, add				↓	13%					
	2000	Anti-siphon, non-continuous pressure type										
	2010	Hot or cold water 125 PSI - 210° F										
	2020	Bronze body										
	2030	1/4" size	1 Stpi	24	.333	Ea.	18.65	11.60		30.25	38	

	15140 | Domestic Water Piping		Crew	Daily Output	Labor-Hours	Unit	2001 Bare Costs Mat.	Labor	Equip.	Total	Total Incl O&P	
700	2040	3/8" size	1 Stpi	24	.333	Ea.	18.65	11.60		30.25	38	700
	2050	1/2" size		24	.333		21	11.60		32.60	41	
	2060	3/4" size		20	.400		25	13.95		38.95	48.50	
	2070	1" size		19	.421		39	14.65		53.65	65	
	2080	1-1/4" size		15	.533		68.50	18.60		87.10	104	
	2090	1-1/2" size		13	.615		80.50	21.50		102	121	
	2100	2" size		11	.727		125	25.50		150.50	177	
	2110	2-1/2" size		8	1		360	35		395	450	
	2120	3" size		6	1.333		480	46.50		526.50	595	
	2150	For polished chrome finish, add					50%					
800	0010	**WATER HAMMER ARRESTORS / SHOCK ABSORBERS**										800
	0490	Copper										
	0500	3/4" male I.P.S. For 1 to 11 fixtures	1 Plum	12	.667	Ea.	16.05	23		39.05	52	
	0600	1" male I.P.S., For 12 to 32 fixtures		8	1		41	34.50		75.50	97	
	0700	1-1/4" male I.P.S. For 33 to 60 fixtures		8	1		47.50	34.50		82	105	
	0800	1-1/2" male I.P.S. For 61 to 113 fixtures		8	1		65	34.50		99.50	124	
	0900	2" male I.P.S.For 114 to 154 fixtures		8	1		104	34.50		138.50	166	
	1000	2-1/2" male I.P.S. For 155 to 330 fixtures		4	2		231	69		300	360	
	4000	Bellows type										
	4010	3/4" FNPT, to 11 fixture units	1 Plum	10	.800	Ea.	68.50	27.50		96	117	
	4020	1" FNPT, to 32 fixture units		8.80	.909		138	31.50		169.50	200	
	4030	1" FNPT, to 60 fixture units		8.80	.909		206	31.50		237.50	275	
	4040	1" FNPT, to 113 fixture units		8.80	.909		515	31.50		546.50	620	
	4050	1" FNPT, to 154 fixture units		6.60	1.212		585	42		627	705	
	4060	1-1/2" FNPT, to 300 fixture units		5	1.600		720	55		775	875	

	15155 | Drainage Specialties		Crew	Daily Output	Labor-Hours	Unit	Mat.	Labor	Equip.	Total	Total Incl O&P	
160	0010	**CLEANOUTS**										160
	0060	Floor type										
	0080	Round or square, scoriated nickel bronze top										
	0100	2" pipe size	1 Plum	10	.800	Ea.	73.50	27.50		101	123	
	0120	3" pipe size		8	1		110	34.50		144.50	173	
	0140	4" pipe size		6	1.333		110	46		156	191	
	0160	5" pipe size		4	2		140	69		209	258	
	0180	6" pipe size	Q-1	6	2.667		140	82.50		222.50	279	
	0200	8" pipe size	"	4	4		247	124		371	460	
	0340	Recessed for tile, same price										
	0980	Round top, recessed for terrazzo										
	1000	2" pipe size	1 Plum	9	.889	Ea.	73.50	30.50		104	127	
	1080	3" pipe size		6	1.333		110	46		156	191	
	1100	4" pipe size		4	2		110	69		179	225	
	1120	5" pipe size	Q-1	6	2.667		140	82.50		222.50	279	
	1140	6" pipe size		5	3.200		140	99		239	305	
	1160	8" pipe size		4	4		247	124		371	460	
	2000	Round scoriated nickel bronze top, extra heavy duty										
	2060	2" pipe size	1 Plum	9	.889	Ea.	100	30.50		130.50	156	
	2080	3" pipe size		6	1.333		136	46		182	220	
	2100	4" pipe size		4	2		136	69		205	254	
	2120	5" pipe size	Q-1	6	2.667		166	82.50		248.50	310	
	2140	6" pipe size		5	3.200		166	99		265	335	
	2160	8" pipe size		4	4		274	124		398	485	
	4000	Wall type, square smooth cover, over wall frame										
	4060	2" pipe size	1 Plum	14	.571	Ea.	120	19.70		139.70	162	
	4080	3" pipe size		12	.667		130	23		153	177	
	4100	4" pipe size		10	.800		139	27.50		166.50	195	

15155 | Drainage Specialties

			CREW	DAILY OUTPUT	LABOR-HOURS	UNIT	2001 BARE COSTS MAT.	LABOR	EQUIP.	TOTAL	TOTAL INCL O&P	
160	4120	5″ pipe size	1 Plum	9	.889	Ea.	202	30.50		232.50	268	160
	4140	6″ pipe size		8	1		224	34.50		258.50	298	
	4160	8″ pipe size	Q-1	11	1.455		298	45		343	400	
	5000	Extension, C.I.;bronze countersunk plug, 8″ long										
	5040	2″ pipe size	1 Plum	16	.500	Ea.	62.50	17.25		79.75	94.50	
	5060	3″ pipe size		14	.571		72.50	19.70		92.20	109	
	5080	4″ pipe size		13	.615		74.50	21		95.50	114	
	5100	5″ pipe size		12	.667		117	23		140	163	
	5120	6″ pipe size		11	.727		214	25		239	273	
170	0010	**CLEANOUT TEE**										170
	0100	Cast iron, B&S, with countersunk plug										
	0200	2″ pipe size	1 Plum	4	2	Ea.	100	69		169	214	
	0220	3″ pipe size		3.60	2.222		110	76.50		186.50	237	
	0240	4″ pipe size		3.30	2.424		136	83.50		219.50	276	
	0260	5″ pipe size	Q-1	5.50	2.909		295	90		385	460	
	0280	6″ pipe size	″	5	3.200		365	99		464	555	
	0300	8″ pipe size	Q-3	5	6.400		420	210		630	780	
	0500	For round smooth access cover, same price										
	0600	For round scoriated access cover, same price										
	0700	For square smooth access cover, add					60%					
	2000	Cast iron, no hub										
	2010	Cleanout tee, with 2 couplings										
	2012	2″	Q-1	22	.727	Ea.	14.20	22.50		36.70	49.50	
	2014	3″		19	.842		19.35	26		45.35	61	
	2016	4″		16.50	.970		28	30		58	76.50	
	2018	6″	Q-2	20	1.200		69	38.50		107.50	134	
	2040	Cleanout plug, no hub										
	2042	2″	Q-1	44	.364	Ea.	5.40	11.25		16.65	23	
	2046	3″		38	.421		8.40	13.05		21.45	29	
	2048	4″		33	.485		13.75	15.05		28.80	37.50	
	2050	6″	Q-2	40	.600		49.50	19.30		68.80	83.50	
	2052	8″	″	33	.727		69	23.50		92.50	112	
	4000	Plastic, tees and adapters. Add plugs										
	4010	ABS, DWV										
	4020	Cleanout tee, 1-1/2″ pipe size	1 Plum	15	.533	Ea.	3.82	18.35		22.17	31.50	
	4030	2″ pipe size	Q-1	27	.593		5.05	18.35		23.40	33	
	4040	3″ pipe size		21	.762		10.10	23.50		33.60	46.50	
	4050	4″ pipe size		16	1		18.85	31		49.85	68	
	4100	Cleanout plug, 1-1/2″ pipe size	1 Plum	32	.250		.79	8.60		9.39	13.85	
	4110	2″ pipe size	Q-1	56	.286		.92	8.85		9.77	14.35	
	4120	3″ pipe size		36	.444		1.54	13.80		15.34	22.50	
	4130	4″ pipe size		30	.533		3.23	16.55		19.78	28.50	
	4180	Cleanout adapter fitting, 1-1/2″ pipe size	1 Plum	32	.250		1.24	8.60		9.84	14.35	
	4190	2″ pipe size	Q-1	56	.286		1.73	8.85		10.58	15.25	
	4200	3″ pipe size		36	.444		4.91	13.80		18.71	26.50	
	4210	4″ pipe size		30	.533		8.05	16.55		24.60	34	
	5000	PVC, DWV										
	5010	Cleanout tee, 1-1/2″ pipe size	1 Plum	15	.533	Ea.	1.24	18.35		19.59	29	
	5020	2″ pipe size	Q-1	27	.593		2.37	18.35		20.72	30	
	5030	3″ pipe size		21	.762		4.65	23.50		28.15	40.50	
	5040	4″ pipe size		16	1		8.90	31		39.90	57	
	5090	Cleanout plug, 1-1/2″ pipe size	1 Plum	32	.250		.72	8.60		9.32	13.80	
	5100	2″ pipe size	Q-1	56	.286		.89	8.85		9.74	14.35	
	5110	3″ pipe size		36	.444		1.89	13.80		15.69	23	
	5120	4″ pipe size		30	.533		2.79	16.55		19.34	28	

15155 | Drainage Specialties

		15155 Drainage Specialties	CREW	DAILY OUTPUT	LABOR-HOURS	UNIT	2001 BARE COSTS MAT.	LABOR	EQUIP.	TOTAL	TOTAL INCL O&P	
170	5130	6" pipe size	Q-1	24	.667	Ea.	8.55	20.50		29.05	40.50	170
	5170	Cleanout adapter fitting, 1-1/2" pipe size	1 Plum	32	.250		1.07	8.60		9.67	14.20	
	5180	2" pipe size	Q-1	56	.286		1.50	8.85		10.35	15	
	5190	3" pipe size		36	.444		4.25	13.80		18.05	25.50	
	5200	4" pipe size		30	.533		6.95	16.55		23.50	32.50	
	5210	6" pipe size		24	.667		15.60	20.50		36.10	48	
	5300	Cleanout tee, with plug										
	5310	4"	Q-1	14	1.143	Ea.	13.75	35.50		49.25	68.50	
	5320	6"	"	8	2		33	62		95	130	
	5330	8"	Q-2	11	2.182		44.50	70		114.50	155	
	5340	12"	"	10	2.400		121	77		198	249	
	5400	Polypropylene, Schedule 40										
	5410	Cleanout tee with plug										
	5420	1-1/2"	1 Plum	10	.800	Ea.	13.20	27.50		40.70	56	
	5430	2"	Q-1	17	.941		18.60	29		47.60	64.50	
	5440	3"		11	1.455		39	45		84	111	
	5450	4"		9	1.778		53	55		108	142	
300	0010	DRAINS (A8.1-310)										300
	0140	Cornice, C.I., 45° or 90° outlet										
	0180	1-1/2" & 2" pipe size	Q-1	14	1.143	Ea.	91	35.50		126.50	154	
	0200	3" and 4" pipe size	"	12	1.333		124	41.50		165.50	199	
	0260	For galvanized body, add					24.50			24.50	27	
	0280	For polished bronze dome, add					21			21	23	
	0400	Deck, auto park, C.I., 13" top										
	0440	3", 4", 5", and 6" pipe size	Q-1	8	2	Ea.	540	62		602	685	
	0480	For galvanized body, add				"	255			255	281	
	0800	Promenade, heelproof grate, C.I., 14" top										
	0840	2", 3", and 4" pipe size	Q-1	10	1.600	Ea.	207	49.50		256.50	305	
	0860	5" and 6" pipe size		9	1.778		255	55		310	365	
	0880	8" pipe size		8	2		305	62		367	430	
	0940	For galvanized body, add					134			134	147	
	0960	For polished bronze top, add					139			139	153	
	1200	Promenade, heelproof grate, C.I., lateral, 14" top										
	1240	2", 3" and 4" pipe size	Q-1	10	1.600	Ea.	270	49.50		319.50	370	
	1260	5" and 6" pipe size		9	1.778		320	55		375	435	
	1280	8" pipe size		8	2		365	62		427	500	
	1340	For galvanized body, add					134			134	147	
	1360	For polished bronze top, add					140			140	154	
	1500	Promenade, slotted grate, C.I., 11" top										
	1540	2", 3", 4", 5", and 6" pipe size	Q-1	12	1.333	Ea.	185	41.50		226.50	267	
	1600	For galvanized body, add					75.50			75.50	83	
	1640	For polished bronze top, add					110			110	121	
	2000	Floor, medium duty, C.I., deep flange, 7" dia top										
	2040	2" and 3" pipe size	Q-1	12	1.333	Ea.	75.50	41.50		117	146	
	2080	For galvanized body, add					31.50			31.50	35	
	2120	For polished bronze top, add					38			38	42	
	2160	Heavy duty, C.I., 12" dia anti-tilt grate										
	2180	2", 3", 4", 5" and 6" pipe size	Q-1	10	1.600	Ea.	218	49.50		267.50	315	
	2220	For galvanized body, add					97.50			97.50	107	
	2240	For polished bronze top, add					111			111	122	
	2300	X-Heavy duty, C.I., 15" antitilt grate										
	2320	4", 5", 6", and 8" pipe size	Q-1	8	2	Ea.	465	62		527	610	
	2360	For galvanized body, add					159			159	174	
	2380	For polished bronze top, add					186			186	204	
	2400	Heavy duty, with sediment bucket, C.I., 12" dia. loose grate										

15 MECHANICAL

15155 | Drainage Specialties

300			CREW	DAILY OUTPUT	LABOR-HOURS	UNIT	2001 BARE COSTS MAT.	LABOR	EQUIP.	TOTAL	TOTAL INCL O&P	300
	2420	2″, 3″, 4″, 5″, and 6″ pipe size	A8.1 -310	Q-1	9	1.778	Ea.	259	55		314	370
	2440	For galvanized body, add						144			144	159
	2460	For polished bronze top, add						107			107	118
	2500	Heavy duty, cleanout & trap w/bucket, C.I., 15″ top										
	2540	2″, 3″, and 4″ pipe size		Q-1	6	2.667	Ea.	1,975	82.50		2,057.50	2,300
	2560	For galvanized body, add						555			555	610
	2580	For polished bronze top, add						615			615	675
	2600	Medium duty, with perforated SS basket, C.I., body,										
	2610	18″ top for refuse container washing area										
	2620	2″ thru 6″ pipe size		Q-1	4	4	Ea.	1,450	124		1,574	1,775
	2630	Acid resistant										
	2638	PVC										
	2640	2″, 3″ and 4″ pipe size		Q-1	16	1	Ea.	289	31		320	365
	2644	Cast iron, epoxy coated										
	2646	2″, 3″ and 4″ pipe size		Q-1	14	1.143	Ea.	123	35.50		158.50	190
	2650	PVC or ABS thermoplastic										
	2660	3″ and 4″ pipe size		Q-1	16	1	Ea.	53	31		84	106
	2680	Extra heavy duty, oil intercepting, gas seal cone,										
	2690	with cleanout, loose grate, C.I., body 16″ top										
	2700	3″ and 4″ diameter outlet, 4″ slab depth		Q-1	4	4	Ea.	2,250	124		2,374	2,650
	2720	3″ and 4″ diameter outlet, 8″ slab depth			3	5.333		2,450	165		2,615	2,925
	2740	3″ & 4″ dia. outlet, 10″-12″ slab depth, 16″top			2	8		2,525	248		2,773	3,150
	2780	Shower, with strainer, uniform diam. trap, bronze top										
	2800	1-1/2″, 2″ and 3″ pipe size		Q-1	8	2	Ea.	159	62		221	269
	2820	4″ pipe size		″	7	2.286		175	71		246	300
	2840	For galvanized body, add						62			62	68.50
	2860	With strainer, backwater valve, drum trap										
	2880	1-1/2″, 2″,& 3″ pipe size		Q-1	8	2	Ea.	143	62		205	251
	2890	4″ pipe size		″	7	2.286		199	71		270	325
	2900	For galvanized body, add						62			62	68
	2910	Prison cell, vandal-proof, 1-1/2″, and 2″ diam. pipe		Q-1	12	1.333		152	41.50		193.50	230
	2920	3″ pipe size		″	10	1.600		170	49.50		219.50	262
	2930	Trap drain, light duty, backwater valve C.I. top										
	2950	8″ diameter top, 2″ pipe size		Q-1	12	1.333	Ea.	144	41.50		185.50	222
	2960	10″ diameter top, 3″ pipe size			10	1.600		199	49.50		248.50	294
	2970	12″ diameter top, 4″ pipe size			8	2		280	62		342	405
	2990	Pool main discharge, polished bronze top, C.I. body										
	3000	8″ diam. grate, 2″, 3″ & 4″ pipe size		Q-1	8	2	Ea.	290	62		352	415
	3100	12″ diam. grate, 4″, 5″ & 6″ pipe size		″	8	2		485	62		547	630
	3160	For galvanized body, add						116			116	127
	3200	Gutter, bronze body, polished bronze top										
	3220	9″ grate, 1-1/2″ & 2″ pipe size		Q-1	8	2	Ea.	68	62		130	168
	3240	10″ grate, 2″, 2-1/2″ & 3″ pipe size		″	7	2.286	″	84.50	71		155.50	200
	3380	Overflow, 5″ dome & 6″ high pipe polished bronze,										
	3400	cast iron body, 2″, 3″, & 4″ pipe size		Q-1	10	1.600	Ea.	365	49.50		414.50	475
	3580	Recirculating inlet, 6″ sq face polished bronze,										
	3600	C.I. body, 2″, 2-1/2″ & 3″ pipe size		Q-1	10	1.600	Ea.	51	49.50		100.50	131
	3620	For galvanized body, add					″	22.50			22.50	24.50
	3860	Roof, flat metal deck, C.I. body, 12″ C.I. dome										
	3880	2″ pipe size		Q-1	15	1.067	Ea.	108	33		141	168
	3890	3″ pipe size			14	1.143		147	35.50		182.50	216
	3900	4″ pipe size			13	1.231		147	38		185	220
	3910	5″ pipe size			12	1.333		108	41.50		149.50	181
	3920	6″ pipe size			10	1.600		255	49.50		304.50	355
	4280	Integral expansion joint, C.I. body, 12″ C.I. dome										
	4300	2″ pipe size		Q-1	8	2	Ea.	221	62		283	335

MECHANICAL 15

15155 | Drainage Specialties

		15155 Drainage Specialties	CREW	DAILY OUTPUT	LABOR-HOURS	UNIT	2001 BARE COSTS MAT.	LABOR	EQUIP.	TOTAL	TOTAL INCL O&P	
300	4320	3″ pipe size (A8.1 -310)	Q-1	7	2.286	Ea.	269	71		340	400	300
	4340	4″ pipe size		6	2.667		285	82.50		367.50	440	
	4360	5″ pipe size		4	4		245	124		369	455	
	4380	6″ pipe size		3	5.333		485	165		650	780	
	4400	8″ pipe size		3	5.333		640	165		805	955	
	4440	For galvanized body, add					86.50			86.50	95	
	4480	Flexible neoprene bellows, no-hub connection										
	4500	3″ and 4″ pipe size	Q-1	5	3.200	Ea.	64	99		163	221	
	4520	6″ and 8″ pipe size	″	4	4	″	142	124		266	345	
	4620	Main, all aluminum, 12″ low profile dome										
	4640	2″, 3″ and 4″ pipe size	Q-1	14	1.143	Ea.	164	35.50		199.50	234	
	4660	5″ and 6″ pipe size		13	1.231		215	38		253	295	
	4680	8″ pipe size		10	1.600		268	49.50		317.50	370	
	4690	Main, CI body, 12″ poly. dome, 2″, 3″, & 4″ pipe		8	2		121	62		183	228	
	4710	5″ and 6″ pipe size		6	2.667		173	82.50		255.50	315	
	4720	8″ pipe size		4	4		225	124		349	435	
	4730	For underdeck clamp, add		22	.727		84	22.50		106.50	127	
	4740	For vandalproof dome, add					21			21	23.50	
	4750	For galvanized body, add					144			144	159	
	4760	Main, PVC body and dome, 2″ pipe size	Q-1	14	1.143		29	35.50		64.50	85.50	
	4780	3″ pipe size		14	1.143		29	35.50		64.50	85.50	
	4800	4″ pipe size		14	1.143		29	35.50		64.50	85.50	
	4820	For underdeck clamp, add		24	.667		16	20.50		36.50	48.50	
	4900	Terrace planting area, with perforated overflow, C.I.										
	4920	2″, 3″ and 4″ pipe size	Q-1	8	2	Ea.	229	62		291	345	
	4980	Scupper floor, oblique strainer, C.I.										
	5000	6″ x 7″ top, 2″, 3″ and 4″ pipe size	Q-1	16	1	Ea.	106	31		137	164	
	5100	8″ x 12″ top, 5″ and 6″ pipe size	″	14	1.143		206	35.50		241.50	281	
	5160	For galvanized body, add					40%					
	5200	For polished bronze strainer, add					85%					
	5980	Trench, floor, hvy duty, modular, C.I., 12″ x 12″ top										
	6000	2″, 3″, 4″, 5″, & 6″ pipe size	Q-1	8	2	Ea.	335	62		397	465	
	6100	For polished bronze top, add					164			164	181	
	6200	For 12″ extension section, C.I. top, add					335			335	370	
	6240	For 12″ extension section, polished bronze top, add					157			157	172	
	6960	Backwater valve, soil pipe, C.I. body										
	6980	Bronze gate and automatic flapper valves										
	7000	3″ and 4″ pipe size	Q-1	13	1.231	Ea.	665	38		703	790	
	7100	5″ and 6″ pipe size	″	13	1.231	″	1,075	38		1,113	1,225	
	7240	Bronze flapper valve, bolted cover										
	7260	2″ pipe size	Q-1	16	1	Ea.	224	31		255	293	
	7280	3″ pipe size		14.50	1.103		350	34		384	435	
	7300	4″ pipe size		13	1.231		430	38		468	535	
	7320	5″ pipe size	Q-2	18	1.333		515	43		558	635	
	7340	6″ pipe size	″	17	1.412		620	45.50		665.50	750	
	7360	8″ pipe size	Q-3	10	3.200		870	105		975	1,125	
	7380	10″ pipe size	″	9	3.556		1,425	117		1,542	1,725	
	7500	For threaded cover, same cost										
	7540	Revolving disk type, same cost as flapper type										
340	0010	**FLOOR RECEPTORS** For connection to 2″, 3″ & 4″ diameter pipe										340
	0200	12-1/2″ square top, 25 sq in open area	Q-1	10	1.600	Ea.	365	49.50		414.50	475	
	0300	For grate with 4″ diam. x 3-3/4″ high funnel, add					51.50			51.50	56.50	
	0400	For grate with 6″ diameter x 6″ high funnel, add					66.50			66.50	73	
	0500	For full hinged grate with open center, add					48			48	53	
	0600	For aluminum bucket, add					42.50			42.50	47	

15 MECHANICAL

15155 | Drainage Specialties

		15155 Drainage Specialties	CREW	DAILY OUTPUT	LABOR-HOURS	UNIT	2001 BARE COSTS MAT.	LABOR	EQUIP.	TOTAL	TOTAL INCL O&P	
340	0700	For acid-resisting bucket, add				Ea.	81			81	89.50	340
	0800	For galvanized bucket, add					53			53	58.50	
	0900	For stainless steel mesh bucket liner, add					64			64	70.50	
	1000	For bronze antisplash dome strainer, add					31			31	34	
	1100	For partial solid cover, add					15.60			15.60	17.15	
	1200	For trap primer connection, add					25.50			25.50	28.50	
	2000	12-5/8″ diameter top, 40 sq. in. open area	Q-1	10	1.600		228	49.50		277.50	325	
	2100	For options, add same prices as square top										
	3000	8″ x 4″ rectangular top, 7.5 sq. in. open area	Q-1	14	1.143	Ea.	240	35.50		275.50	320	
	3100	For trap primer connections, add					25.50			25.50	28.50	
	4000	24″ x 16″ rectangular top, 70 sq. in. open area	Q-1	4	4		1,225	124		1,349	1,525	
	4100	For trap primer connection, add					51			51	56	
400	0010	**INTERCEPTORS**										400
	0150	Grease, cast iron, 4 GPM, 8 lb. fat capacity	1 Plum	4	2	Ea.	445	69		514	595	
	0200	7 GPM, 14 lb. fat capacity		4	2		615	69		684	785	
	1000	10 GPM, 20 lb. fat capacity		4	2		725	69		794	905	
	1040	15 GPM, 30 lb. fat capacity		4	2		1,075	69		1,144	1,275	
	1060	20 GPM, 40 lb. fat capacity		3	2.667		1,325	92		1,417	1,600	
	1080	25 GPM, 50 lb. fat capacity	Q-1	3.50	4.571		1,475	142		1,617	1,850	
	1100	35 GPM, 70 lb. fat capacity		3	5.333		1,825	165		1,990	2,250	
	1110	50 GPM, 100 lb. fat capacity		2	8		1,825	248		2,073	2,375	
	1120	Fabricated steel, 50 GPM, 100 lb. fat capacitv		2	8		2,425	248		2,673	3,025	
	1140	75 GPM, 150 lb. fat capacity		2	8		4,875	248		5,123	5,725	
	1160	100 GPM, 200 lb. fat capacity		2	8		5,475	248		5,723	6,400	
	1180	150 GPM, 300 lb. fat capacity		2	8		6,100	248		6,348	7,075	
	1200	200 GPM, 400 lb. fat capacity		1.50	10.667		8,825	330		9,155	10,200	
	1220	250 GPM, 500 lb. fat capacity		1.30	12.308		10,300	380		10,680	11,900	
	1240	300 GPM, 600 lb. fat capacity		1	16		12,100	495		12,595	14,100	
	1260	400 GPM, 800 lb. fat capacity	Q-2	1.20	20		14,800	645		15,445	17,300	
	1280	500 GPM, 1000 lb. fat capacity	″	1	24		17,600	770		18,370	20,500	
	1560	For chemical add-port, add					97			97	106	
	1580	For seepage pan, add					7%					
	3000	Hair, cast iron, 1-1/4″ and 1-1/2″ pipe connection	1 Plum	8	1	Ea.	165	34.50		199.50	234	
	3100	For chrome-plated cast iron, add					103			103	113	
	3200	For polished bronze, add					305			305	335	
	4000	Oil, fabricated steel, 10 GPM, 2″ pipe size	1 Plum	4	2		990	69		1,059	1,200	
	4100	15 GPM, 2″ or 3″ pipe size		4	2		1,350	69		1,419	1,600	
	4120	20 GPM, 2″ or 3″ pipe size		3	2.667		1,650	92		1,742	1,950	
	4140	25 GPM, 2″ or 3″ pipe size	Q-1	3.50	4.571		1,775	142		1,917	2,200	
	4160	35 GPM, 2″, 3″, or 4″ pipe size		3	5.333		2,175	165		2,340	2,650	
	4180	50 GPM, 2″, 3″, or 4″ pipe size		2	8		2,925	248		3,173	3,575	
	4200	75 GPM, 3″ pipe size		2	8		5,050	248		5,298	5,950	
	4220	100 GPM, 3″ pipe size		2	8		5,475	248		5,723	6,400	
	4240	150 GPM, 4″ pipe size		2	8		6,800	248		7,048	7,850	
	4260	200 GPM, 4″ pipe size		1.50	10.667		9,600	330		9,930	11,100	
	4280	250 GPM, 5″ pipe size		1.30	12.308		11,100	380		11,480	12,800	
	4300	300 GPM, 5″ pipe size		1	16		12,600	495		13,095	14,600	
	4320	400 GPM, 6″ pipe size	Q-2	1.20	20		16,200	645		16,845	18,800	
	4340	500 GPM, 6″ pipe size	″	1	24		20,400	770		21,170	23,600	
	6000	Solids, precious metals recovery, C.I., 1-1/4″ to 2″ pipe	1 Plum	4	2		251	69		320	380	
	6100	Dental Lab., large, C.I., 1-1/2″ to 2″ pipe	″	3	2.667		880	92		972	1,100	
680	0010	**SEPARATORS** Entrainment eliminator, steel body, 150 PSIG										680
	0100	1/4″ size	1 Stpi	24	.333	Ea.	134	11.60		145.60	165	
	0120	1/2″ size		24	.333		144	11.60		155.60	176	
	0140	3/4″ size		20	.400		155	13.95		168.95	192	

15155 | Drainage Specialties

		15155 Drainage Specialties	CREW	DAILY OUTPUT	LABOR-HOURS	UNIT	2001 BARE COSTS MAT.	LABOR	EQUIP.	TOTAL	TOTAL INCL O&P	
680	0160	1" size	1 Stpi	19	.421	Ea.	206	14.65		220.65	249	680
	0180	1-1/4" size		15	.533		237	18.60		255.60	289	
	0200	1-1/2" size		13	.615		268	21.50		289.50	330	
	0220	2" size		11	.727		299	25.50		324.50	370	
740	0010	**SINK WASTE TREATMENT** System for commercial kitchens										740
	0100	includes clock timer, & fittings										
	0200	System less chemical, wall mounted cabinet	1 Plum	16	.500	Ea.	216	17.25		233.25	264	
	2000	Chemical, 1 gallon, add					38			38	42	
	2100	6 gallons, add					126			126	139	
	2200	15 gallons, add					345			345	380	
	2300	30 gallons, add					645			645	710	
	2400	55 gallons, add					1,100			1,100	1,200	
780	0010	**TRAPS**										780
	0030	Cast iron, service weight										
	0050	Running P trap, without vent										
	1100	2"	Q-1	16	1	Ea.	17.85	31		48.85	66.50	
	1140	3"		14	1.143		29	35.50		64.50	85.50	
	1150	4"		13	1.231		52.50	38		90.50	116	
	1160	6"	Q-2	17	1.412		243	45.50		288.50	335	
	1180	Running trap, single hub, with vent										
	2080	3" pipe size, 3" vent	Q-1	14	1.143	Ea.	44	35.50		79.50	102	
	2120	4" pipe size, 4" vent	"	13	1.231		60	38		98	124	
	2140	5" pipe size, 4" vent	Q-2	11	2.182		96	70		166	211	
	2160	6" pipe size, 4" vent		10	2.400		258	77		335	400	
	2180	6" pipe size, 6" vent		8	3		281	96.50		377.50	455	
	2200	8" pipe size, 4" vent	Q-3	10	3.200		850	105		955	1,100	
	2220	8" pipe size, 6" vent	"	8	4		920	131		1,051	1,225	
	2300	For double hub, vent, add					10%	20%				
	2800	S trap,										
	2850	4" pipe size	Q-1	13	1.231	Ea.	96	38		134	164	
	3000	P trap, B&S, 2" pipe size		16	1		13.20	31		44.20	61.50	
	3040	3" pipe size		14	1.143		19.50	35.50		55	75	
	3060	4" pipe size		13	1.231		28.50	38		66.50	89	
	3080	5" pipe size	Q-2	18	1.333		60	43		103	131	
	3100	6" pipe size	"	17	1.412		88	45.50		133.50	165	
	3120	8" pipe size	Q-3	11	2.909		250	95.50		345.50	420	
	3130	10" pipe size	"	10	3.200		495	105		600	700	
	3150	P trap, no hub, 1-1/2" pipe size	Q-1	17	.941		7.75	29		36.75	52.50	
	3160	2" pipe size		16	1		7.45	31		38.45	55	
	3170	3" pipe size		14	1.143		16.45	35.50		51.95	71.50	
	3180	4" pipe size		13	1.231		28.50	38		66.50	89	
	3190	6" pipe size	Q-2	17	1.412		69	45.50		114.50	145	
	3350	Deep seal trap, B&S										
	3400	1-1/4" pipe size	Q-1	14	1.143	Ea.	17.10	35.50		52.60	72.50	
	3410	1-1/2" pipe size		14	1.143		17.10	35.50		52.60	72.50	
	3420	2" pipe size		14	1.143		17.05	35.50		52.55	72.50	
	3440	3" pipe size		12	1.333		27.50	41.50		69	93	
	3460	4" pipe size		11	1.455		44	45		89	116	
	3500	For trap primer connection, add					11.65			11.65	12.85	
	3540	For trap with floor cleanout, add					70%	5%				
	3580	For trap with adjustable cleanout, add	Q-1	10	1.600	Ea.	63	49.50		112.50	145	
	3800	Drum trap, 4" x 5", 1-1/2" tapping	Q-2	17	1.412		13.50	45.50		59	83.50	
	3820	2" tapping	"	17	1.412		13.50	45.50		59	83.50	
	3840	For galvanized, add					60%					
	4700	Copper, drainage, drum trap										
	4800	3" x 5" solid, 1-1/2" pipe size	1 Plum	16	.500	Ea.	16.70	17.25		33.95	44.50	

15155 | Drainage Specialties

780

	15155 Drainage Specialties	CREW	DAILY OUTPUT	LABOR-HOURS	UNIT	2001 BARE COSTS MAT.	LABOR	EQUIP.	TOTAL	TOTAL INCL O&P
4840	3″ x 6″ swivel, 1-1/2″ pipe size	1 Plum	16	.500	Ea.	24.50	17.25		41.75	53
5100	P trap, standard pattern									
5200	1-1/4″ pipe size	1 Plum	18	.444	Ea.	11.90	15.30		27.20	36
5240	1-1/2″ pipe size		17	.471		11.90	16.20		28.10	37.50
5260	2″ pipe size		15	.533		18.35	18.35		36.70	47.50
5280	3″ pipe size		11	.727		44.50	25		69.50	87
5340	With cleanout and slip joint									
5360	1-1/4″ pipe size	1 Plum	18	.444	Ea.	16.85	15.30		32.15	41.50
5400	1-1/2″ pipe size		17	.471		17.20	16.20		33.40	43.50
5420	2″ pipe size		15	.533		28	18.35		46.35	58
5460	For swivel, add					6.75			6.75	7.45
5750	Chromed brass, tubular, P trap, without cleanout, 20 Ga.									
5800	1-1/4″ pipe size	1 Plum	18	.444	Ea.	11.25	15.30		26.55	35.50
5840	1-1/2″ pipe size	″	17	.471	″	13.40	16.20		29.60	39.50
5900	With cleanout, 20 Ga.									
5940	1-1/4″ pipe size	1 Plum	18	.444	Ea.	22	15.30		37.30	47
6000	1-1/2″ pipe size	″	17	.471	″	27	16.20		43.20	54
6350	S trap, without cleanout, 20 Ga.									
6400	1-1/4″ pipe size	1 Plum	18	.444	Ea.	24.50	15.30		39.80	49.50
6440	1-1/2″ pipe size	″	17	.471	″	28	16.20		44.20	55
6550	With cleanout, 20 Ga.									
6600	1-1/4″ pipe size	1 Plum	18	.444	Ea.	41.50	15.30		56.80	69
6640	1-1/2″ pipe size	″	17	.471		45.50	16.20		61.70	74.50
6660	Corrosion resistant, glass, P trap, 1-1/2″ pipe size	Q-1	17	.941		38	29		67	86
6670	2″ pipe size		16	1		50	31		81	102
6680	3″ pipe size		14	1.143		103	35.50		138.50	167
6690	4″ pipe size		13	1.231		153	38		191	227
6700	6″ pipe size	Q-2	17	1.412		590	45.50		635.50	715
6710	ABS DWV P trap, solvent weld joint									
6720	1-1/2″ pipe size	1 Plum	18	.444	Ea.	31	15.30		46.30	57
6722	2″ pipe size		17	.471		50	16.20		66.20	79.50
6724	3″ pipe size		15	.533		228	18.35		246.35	279
6726	4″ pipe size		14	.571		470	19.70		489.70	545
6732	PVC DWV P trap, solvent weld joint									
6733	1-1/2″ pipe size	1 Plum	18	.444	Ea.	2.39	15.30		17.69	25.50
6734	2″ pipe size		17	.471		3.26	16.20		19.46	28
6735	3″ pipe size		15	.533		16.20	18.35		34.55	45.50
6736	4″ pipe size		14	.571		39	19.70		58.70	72.50
6760	PP DWV, dilution trap, 1-1/2″ pipe size		16	.500		101	17.25		118.25	137
6770	P trap, 1-1/2″ pipe size		17	.471		27	16.20		43.20	54
6780	2″ pipe size		16	.500		37	17.25		54.25	66.50
6790	3″ pipe size		14	.571		85	19.70		104.70	123
6800	4″ pipe size		13	.615		109	21		130	151
6810	Running trap, 1-1/2″ pipe size		16	.500		28.50	17.25		45.75	57.50
6820	2″ pipe size		15	.533		40	18.35		58.35	71.50
6830	S trap, 1-1/2″ pipe size		16	.500		29	17.25		46.25	58
6840	2″ pipe size		15	.533		36.50	18.35		54.85	67.50
6850	Universal trap, 1-1/2″ pipe size		14	.571		82.50	19.70		102.20	121
6860	PVC DWV hub x hub, basin trap, 1-1/4″ pipe size		18	.444		4.22	15.30		19.52	27.50
6870	Sink P trap, 1-1/2″ pipe size		18	.444		4.22	15.30		19.52	27.50
6880	Tubular S trap, 1-1/2″ pipe size		17	.471		10	16.20		26.20	35.50
6890	PVC sch. 40 DWV, drum trap									
6900	1-1/2″ pipe size	1 Plum	16	.500	Ea.	12.40	17.25		29.65	39.50
6910	P trap, 1-1/2″ pipe size		18	.444		2.39	15.30		17.69	25.50
6920	2″ pipe size		17	.471		3.26	16.20		19.46	28
6930	3″ pipe size		15	.533		16.20	18.35		34.55	45.50

780

15100 | Building Services Piping

15 MECHANICAL

		15155 | Drainage Specialties	Crew	Daily Output	Labor-Hours	Unit	2001 Bare Costs Mat.	Labor	Equip.	Total	Total Incl O&P	
780	6940	4" pipe size	1 Plum	14	.571	Ea.	39	19.70		58.70	72.50	780
	6950	P trap w/clean out, 1-1/2" pipe size		18	.444		5.80	15.30		21.10	29.50	
	6960	2" pipe size		17	.471		9.85	16.20		26.05	35.50	
	6970	P trap adjustable, 1-1/2" pipe size		17	.471		3.12	16.20		19.32	28	
	6980	P trap adj. w/union & cleanout, 1-1/2" pipe size		16	.500		10.30	17.25		27.55	37.50	
	7000	Trap primer, flow through type, 1/2" diameter		24	.333		41	11.50		52.50	62.50	
	7100	With sediment strainer		22	.364		41	12.55		53.55	64	
	7450	Trap primer distribution unit										
	7500	2 openings	1 Plum	18	.444	Ea.	23.50	15.30		38.80	49	
	7540	3 opening		17	.471		25.50	16.20		41.70	52.50	
	7560	4 opening		16	.500		27	17.25		44.25	55.50	
	7850	Trap primer manifold										
	7900	2 outlet	1 Plum	18	.444	Ea.	51	15.30		66.30	79.50	
	7940	4 outlet		16	.500		79.50	17.25		96.75	114	
	7960	6 outlet		15	.533		100	18.35		118.35	138	
	7980	8 outlet		13	.615		125	21		146	170	
940	0010	**VENT FLASHING**										940
	1000	Aluminum with lead ring										
	1020	1-1/4" pipe	1 Plum	20	.400	Ea.	6.40	13.80		20.20	28	
	1030	1-1/2" pipe		20	.400		6.40	13.80		20.20	28	
	1040	2" pipe		18	.444		6.85	15.30		22.15	30.50	
	1050	3" pipe		17	.471		7.60	16.20		23.80	33	
	1060	4" pipe		16	.500		9.15	17.25		26.40	36	
	1350	Copper with neoprene ring										
	1400	1-1/4" pipe	1 Plum	20	.400	Ea.	14.05	13.80		27.85	36.50	
	1430	1-1/2" pipe		20	.400		14.05	13.80		27.85	36.50	
	1440	2" pipe		18	.444		14.85	15.30		30.15	39.50	
	1450	3" pipe		17	.471		17.40	16.20		33.60	43.50	
	1460	4" pipe		16	.500		19.20	17.25		36.45	47	
	2000	Galvanized with neoprene ring										
	2020	1-1/4" pipe	1 Plum	20	.400	Ea.	5.10	13.80		18.90	26.50	
	2030	1-1/2" pipe		20	.400		5.10	13.80		18.90	26.50	
	2040	2" pipe		18	.444		5.40	15.30		20.70	29	
	2050	3" pipe		17	.471		5.85	16.20		22.05	31	
	2060	4" pipe		16	.500		7.60	17.25		24.85	34.50	
	2980	Neoprene, one piece										
	3000	1-1/4" pipe	1 Plum	24	.333	Ea.	4.92	11.50		16.42	23	
	3030	1-1/2" pipe		24	.333		4.92	11.50		16.42	23	
	3040	2" pipe		23	.348		5.15	12		17.15	24	
	3050	3" pipe		21	.381		6.25	13.10		19.35	26.50	
	3060	4" pipe		20	.400		7.90	13.80		21.70	29.50	

		15180 | Heating and Cooling Piping	Crew	Daily Output	Labor-Hours	Unit	Mat.	Labor	Equip.	Total	Total Incl O&P	
100	0010	**ANTI-FREEZE** Inhibited										100
	0900	Ethylene glycol concentrated										
	1000	55 gallon drums, small quantities				Gal.	8.75			8.75	9.60	
	1200	Large quantities					8.20			8.20	9	
	2000	Propylene glycol, for solar heat, small quantities					9.95			9.95	10.90	
	2100	Large quantities					8.80			8.80	9.70	
200	0010	**PUMPS, CIRCULATING** Heated or chilled water application										200
	0600	Bronze, sweat connections, 1/40 HP, in line										
	0640	3/4" size	Q-1	16	1	Ea.	114	31		145	172	
	1000	Flange connection, 3/4" to 1-1/2" size										
	1040	1/12 HP	Q-1	6	2.667	Ea.	310	82.50		392.50	465	
	1060	1/8 HP		6	2.667		535	82.50		617.50	710	

	15180 | Heating and Cooling Piping		CREW	DAILY OUTPUT	LABOR-HOURS	UNIT	2001 BARE COSTS MAT.	LABOR	EQUIP.	TOTAL	TOTAL INCL O&P	
200	1100	1/3 HP	Q-1	6	2.667	Ea.	595	82.50		677.50	780	200
	1140	2" size, 1/6 HP		5	3.200		765	99		864	990	
	1180	2-1/2" size, 1/4 HP		5	3.200		995	99		1,094	1,250	
	1220	3" size, 1/4 HP		4	4		1,050	124		1,174	1,325	
	1260	1/3 HP		4	4		1,275	124		1,399	1,575	
	1300	1/2 HP		4	4		1,300	124		1,424	1,600	
	1340	3/4 HP		4	4		1,425	124		1,549	1,725	
	1380	1 HP		4	4		2,300	124		2,424	2,700	
	2000	Cast iron, flange connection										
	2040	3/4" to 1-1/2" size, in line, 1/12 HP	Q-1	6	2.667	Ea.	202	82.50		284.50	345	
	2060	1/8 HP		6	2.667		340	82.50		422.50	495	
	2100	1/3 HP		6	2.667		380	82.50		462.50	540	
	2140	2" size, 1/6 HP		5	3.200		415	99		514	605	
	2180	2-1/2" size, 1/4 HP		5	3.200		540	99		639	745	
	2220	3" size, 1/4 HP		4	4		550	124		674	790	
	2260	1/3 HP		4	4		745	124		869	1,000	
	2300	1/2 HP		4	4		770	124		894	1,025	
	2340	3/4 HP		4	4		890	124		1,014	1,175	
	2380	1 HP		4	4		1,275	124		1,399	1,600	
	2600	For non-ferrous impeller, add					3%					
	3000	High head, bronze impeller										
	3030	1-1/2" size 1/2 HP	Q-1	5	3.200	Ea.	660	99		759	875	
	3040	1-1/2" size 3/4 HP		5	3.200		705	99		804	925	
	3050	2" size 1 HP		4	4		845	124		969	1,125	
	3090	2" size 1-1/2 HP		4	4		1,050	124		1,174	1,325	
	4000	Close coupled, end suction, bronze impeller										
	4040	1-1/2" size, 1-1/2 HP, to 40 GPM	Q-1	3	5.333	Ea.	1,050	165		1,215	1,425	
	4090	2" size, 2 HP, to 50 GPM		3	5.333		1,075	165		1,240	1,450	
	4100	2" size, 3 HP, to 90 GPM		2.30	6.957		1,200	216		1,416	1,625	
	4190	2-1/2" size, 3 HP, to 150 GPM		2	8		1,275	248		1,523	1,800	
	4300	3" size, 5 HP, to 225 GPM		1.80	8.889		1,400	276		1,676	1,950	
	4410	3" size, 10 HP, to 350 GPM		1.60	10		2,000	310		2,310	2,675	
	4420	4" size, 7-1/2 HP, to 350 GPM		1.60	10		1,950	310		2,260	2,625	
	4520	4" size, 10 HP, to 600 GPM	Q-2	1.70	14.118		2,400	455		2,855	3,300	
	4530	5" size, 15 HP, to 1000 GPM		1.70	14.118		2,875	455		3,330	3,850	
	4610	5" size, 20 HP, to 1350 GPM		1.50	16		3,175	515		3,690	4,275	
	4620	5" size, 25 HP, to 1550 GPM		1.50	16		3,775	515		4,290	4,950	
	5000	Base mounted, bronze impeller, coupling guard										
	5040	1-1/2" size, 1-1/2 HP, to 40 GPM	Q-1	2.30	6.957	Ea.	1,500	216		1,716	1,975	
	5090	2" size, 2 HP, to 50 GPM		2.30	6.957		1,700	216		1,916	2,200	
	5100	2" size, 3 HP, to 90 GPM		2	8		1,750	248		1,998	2,300	
	5190	2-1/2" size, 3 HP, to 150 GPM		1.80	8.889		1,800	276		2,076	2,400	
	5300	3" size, 5 HP, to 225 GPM		1.60	10		1,900	310		2,210	2,575	
	5410	4" size, 5 HP, to 350 GPM		1.50	10.667		2,075	330		2,405	2,775	
	5420	4" size, 7-1/2 HP, to 350 GPM		1.50	10.667		2,325	330		2,655	3,050	
	5520	5" size, 10 HP, to 600 GPM	Q-2	1.60	15		2,375	480		2,855	3,325	
	5530	5" size, 15 HP, to 1000 GPM		1.60	15		3,275	480		3,755	4,325	
	5610	6" size, 20 HP, to 1350 GPM		1.40	17.143		3,600	550		4,150	4,800	
	5620	6" size, 25 HP, to 1550 GPM		1.40	17.143		4,600	550		5,150	5,900	
	5800	The above pump capacities are based on 1800 RPM,										
	5810	at a 60 foot head. Increasing the RPM										
	5820	or decreasing the head will increase the GPM.										

	15195 | Fuel Systems											
490	0010	FUEL OIL SPECIALTIES										490
	0020	Foot valve, single poppet, metal to metal construction										

15100 | Building Services Piping

	15195 | Fuel Systems		CREW	DAILY OUTPUT	LABOR-HOURS	UNIT	2001 BARE COSTS MAT.	LABOR	EQUIP.	TOTAL	TOTAL INCL O&P	
490	0040	Bevel seat, 1/2" diameter	1 Stpi	20	.400	Ea.	36.50	13.95		50.45	61.50	490
	0060	3/4" diameter		18	.444		38.50	15.50		54	65.50	
	1000	Oil filters, 3/8" IPT., 20 gal. per hour		20	.400		17.25	13.95		31.20	40	
	1800	Pump and motor sets										
	1810	Light fuel and diesel oils, 100 PSI										
	1820	25 GPH 1/4 HP	Q-5	6	2.667	Ea.	765	83.50		848.50	965	
	1850	160 GPH, 1/3 HP	"	4	4	"	850	126		976	1,125	
	2000	Remote tank gauging system, self contained										
	2020	30' transmission line										
	2100	30" pointer travel	1 Stpi	2	4	Ea.	1,750	139		1,889	2,125	
	2120	5" pointer travel	"	2.50	3.200		1,200	112		1,312	1,500	
	2500	For each additional 10' transmission line, add					43			43	47	
	3000	Valve, ball check, globe type, 3/8" diameter	1 Stpi	24	.333		6.30	11.60		17.90	24.50	
	3500	Fusible, 3/8" diameter		24	.333		5.15	11.60		16.75	23	
	3600	1/2" diameter		24	.333		13.40	11.60		25	32.50	
	3610	3/4" diameter		20	.400		26.50	13.95		40.45	50.50	
	3620	1" diameter		19	.421		81	14.65		95.65	111	
	4000	Nonfusible, 3/8" diameter		24	.333		9.55	11.60		21.15	28	
	4500	Shutoff, gate type, lever handle, spring-fusible kit										
	4520	1/4" diameter	1 Stpi	14	.571	Ea.	15.15	19.90		35.05	46.50	
	4540	3/8" diameter		12	.667		14.05	23		37.05	50.50	
	4560	1/2" diameter		10	.800		19.95	28		47.95	64	
	4570	3/4" diameter		8	1		31.50	35		66.50	87.50	
	5000	Vent alarm, whistling signal					26			26	28.50	
	5500	Vent protector/breather, 1-1/4" diameter	1 Stpi	32	.250		7.35	8.70		16.05	21	
605	0010	**METERS**										605
	4000	Residential										
	4010	Gas meter, residential, 3/4" pipe size	1 Plum	14	.571	Ea.	115	19.70		134.70	157	
	4020	Gas meter, residential, 1" pipe size		12	.667		138	23		161	187	
	4030	Gas meter, residential, 1-1/4" pipe size		10	.800		138	27.50		165.50	194	

15 MECHANICAL

15200 | Process Piping

	15210 | Process Air/Gas Piping		CREW	DAILY OUTPUT	LABOR-HOURS	UNIT	2001 BARE COSTS MAT.	LABOR	EQUIP.	TOTAL	TOTAL INCL O&P	
100	0010	**COMPRESSORS**										100
	5250	Air, reciprocating air cooled, splash lubricated, tank mounted										
	5300	Single stage, 1 phase, 140 psi										
	5303	1/2 HP, 30 gal tank	1 Stpi	3	2.667	Ea.	1,050	93		1,143	1,325	
	5305	3/4 HP, 30 gal tank		2.60	3.077		1,075	107		1,182	1,350	
	5307	1 HP, 30 gal tank		2.20	3.636		1,375	127		1,502	1,700	
	5309	2 HP, 30 gal tank	Q-5	4	4		1,750	126		1,876	2,125	
	5310	3 HP, 30 gal vertical tank		3.60	4.444		2,125	139		2,264	2,550	
	5314	3 HP, 60 gal tank		3.50	4.571		2,825	143		2,968	3,325	
	5320	5 HP, 60 gal vertical tank		3.20	5		2,900	157		3,057	3,400	
	5330	5 HP, 80 gal vertical tank		3	5.333		3,050	167		3,217	3,600	
	5340	7.5 HP, 80 gal tank		2.60	6.154		5,225	193		5,418	6,050	
	5600	2 stage pkg., 3 phase										
	5650	6 CFM at 125 psi 1-1/2 HP, 60 gal tank	Q-5	3	5.333	Ea.	2,275	167		2,442	2,750	

15210 | Process Air/Gas Piping

			CREW	DAILY OUTPUT	LABOR-HOURS	UNIT	2001 BARE COSTS MAT.	LABOR	EQUIP.	TOTAL	TOTAL INCL O&P	
100	5670	10.9 CFM at 125 psi, 3 HP, 80 gal tank	Q-5	1.50	10.667	Ea.	2,725	335		3,060	3,500	100
	5680	38.7 CFM at 125 psi, 10 HP, 120 gal tank		.60	26.667		4,900	835		5,735	6,650	
	5690	105 CFM at 125 psi, 25 HP, 250 gal tank	Q-6	.60	40		9,425	1,300		10,725	12,400	
	5800	With single stage pump										
	5850	8.3 CFM at 125 psi, 2 HP, 80 gal tank	Q-6	3.50	6.857	Ea.	2,575	223		2,798	3,150	
	5860	38.7 CFM at 125 psi, 10 HP, 120 gal tank	"	.90	26.667	"	5,425	865		6,290	7,275	
	6000	Reciprocating, 2 stage, tank mtd, 3 Ph., Cap. rated @175 PSIG										
	6050	Pressure lubcatd, hvy duty, 9.7 CFM, 3 HP, 120 Gal. tank	Q-5	1.30	12.308	Ea.	4,150	385		4,535	5,125	
	6054	5 CFM, 1-1/2 HP, 80 gal tank		2.80	5.714		2,425	179		2,604	2,950	
	6056	6.4 CFM, 2 HP, 80 gal tank		2	8		2,575	251		2,826	3,200	
	6058	8.1 CFM, 3 HP, 80 gal tank		1.70	9.412		2,675	295		2,970	3,400	
	6059	14.8 CFM, 5 HP, 80 gal tank		1	16		2,750	500		3,250	3,800	
	6060	16.5 CFM, 5 HP, 120 gal. tank		1	16		4,200	500		4,700	5,375	
	6063	13 CFM, 6 HP, 80 gal tank		.90	17.778		4,425	560		4,985	5,725	
	6066	19.8 CFM, 7.5 HP, 80 gal tank		.80	20		4,675	630		5,305	6,075	
	6070	25.8 CFM, 7-1/2 HP, 120 gal. tank		.80	20		6,250	630		6,880	7,825	
	6078	34.8 CFM, 10 HP, 80 gal. tank		.70	22.857		5,925	715		6,640	7,575	
	6080	34.8 CFM, 10 HP, 120 gal. tank		.60	26.667		6,275	835		7,110	8,175	
	6090	53.7 CFM, 15 HP, 120 gal. tank	Q-6	.80	30		7,800	975		8,775	10,100	
	6100	76.7 CFM, 20 HP, 120 gal. tank		.70	34.286		9,650	1,125		10,775	12,300	
	6104	76.7 CFM, 20 HP, 240 gal tank		.68	35.294		9,450	1,150		10,600	12,100	
	6110	90.1 CFM, 25 HP, 120 gal. tank		.63	38.095		10,100	1,250		11,350	13,000	
	6120	101 CFM, 30 HP, 120 gal. tank		.57	42.105		10,900	1,375		12,275	14,100	
	6130	101 CFM, 30 HP, 250 gal. tank		.52	46.154		11,800	1,500		13,300	15,200	
	6200	Oil-less, 13.6 CFM, 5 HP, 120 gal. tank	Q-5	.88	18.182		12,400	570		12,970	14,500	
	6210	13.6 CFM, 5 HP, 250 gal. tank		.80	20		13,300	630		13,930	15,600	
	6220	18.2 CFM, 7.5 HP, 120 gal. tank		.73	21.918		12,600	690		13,290	14,900	
	6230	18.2 CFM, 7.5 HP, 250 gal. tank		.67	23.881		13,400	750		14,150	15,900	
	6250	30.5 CFM, 10 HP, 120 gal. tank		.57	28.070		14,700	880		15,580	17,400	
	6260	30.5 CFM, 10 HP, 250 gal. tank		.53	30.189		15,500	945		16,445	18,500	
	6270	41.3 CFM, 15 HP, 120 gal. tank	Q-6	.70	34.286		15,900	1,125		17,025	19,200	
	6280	41.3 CFM, 15 HP, 250 gal. tank	"	.67	35.821		16,700	1,175		17,875	20,200	
200	0010	**COMPRESSOR ACCESSORIES**										200
	0100	Aftercooler										
	0130	Air cooled, based on 100 psig and 250°F										
	0140	Tank mounted single stage, 20 SCFM	Q-5	7.50	2.133	Ea.	625	67		692	790	
	0150	46 SCFM		7	2.286		730	71.50		801.50	910	
	0160	77 SCFM		6.50	2.462		1,225	77		1,302	1,475	
	0170	Floor mounted horizontal draft, 36 SCFM		6.50	2.462		645	77		722	825	
	0180	78 SCFM		6	2.667		800	83.50		883.50	1,000	
	0190	190 SCFM		5.70	2.807		1,225	88		1,313	1,475	
	0200	260 SCFM		5.20	3.077		1,375	96.50		1,471.50	1,675	
	0210	440 SCFM		4	4		1,850	126		1,976	2,225	
	0220	650 SCFM		3.50	4.571		2,000	143		2,143	2,425	
	0230	Floor mounted vertical draft, 850 SCFM		2.50	6.400		3,450	201		3,651	4,100	
	0260	2060 SCFM		2	8		5,675	251		5,926	6,625	
	0280	3640 SCFM		1.70	9.412		13,900	295		14,195	15,700	
	0300	Separator with automatic float trap										
	0330	600 lb/hr drain cap. 125 psi 3/4" conn.	Q-5	7.50	2.133	Ea.	315	67		382	445	
	0340	1200 lb/hr drain cap. 125 psi 1" conn.		7	2.286		330	71.50		401.50	475	
	0350	1200 lb/hr drain cap. 125 psi 1-1/2" conn.		6	2.667		375	83.50		458.50	540	
	0360	1200 lb/hr drain cap. 125 psi 2" conn.		5	3.200		405	100		505	595	
	0370	1500 CFM at 200 psi 2-1/2" conn.		4	4		960	126		1,086	1,250	
	0380	4000 CFM at 200 psi 4" conn.		3	5.333		1,050	167		1,217	1,400	
	0400	Water cooled, fixed tube at 250°F										
	0430	52 CFM	Q-5	5	3.200	Ea.	405	100		505	595	

15210 | Process Air/Gas Piping

		Description	CREW	DAILY OUTPUT	LABOR-HOURS	UNIT	2001 BARE COSTS MAT.	LABOR	EQUIP.	TOTAL	TOTAL INCL O&P	
200	0440	104 CFM	Q-5	4	4	Ea.	470	126		596	710	200
	0450	182 CFM		3.50	4.571		730	143		873	1,025	
	0460	312 CFM		3	5.333		815	167		982	1,150	
	0480	494 CFM		2.80	5.714		1,700	179		1,879	2,150	
	0490	702 CFM		2.40	6.667		2,275	209		2,484	2,825	
	0500	806 CFM		2.20	7.273		2,675	228		2,903	3,300	
	0510	1092 CFM		2	8		3,325	251		3,576	4,050	
	0520	1430 CFM		1.50	10.667		3,925	335		4,260	4,800	
	0600	Separator, horizontal										
	0610	100 CFM	Q-5	15	1.067	Ea.	221	33.50		254.50	294	
	0620	350 CFM		10	1.600		320	50		370	425	
	0630	900 CFM		7	2.286		775	71.50		846.50	965	
	0640	1500 CFM		5	3.200		860	100		960	1,100	
	0650	2800 CFM		3.50	4.571		1,275	143		1,418	1,625	
	0700	Dryers, air, deliquescent at 90 psi										
	0730	21 CFM	Q-5	14	1.143	Ea.	655	36		691	780	
	0740	63 CFM		12	1.333		1,200	42		1,242	1,400	
	0750	140 CFM		8	2		2,150	63		2,213	2,450	
	0760	335 CFM		5.50	2.909		2,900	91.50		2,991.50	3,325	
	0770	525 CFM		5	3.200		3,800	100		3,900	4,325	
	0780	760 CFM		3	5.333		5,300	167		5,467	6,075	
	0790	1050 CFM		2.50	6.400		6,575	201		6,776	7,550	
	0800	1340 CFM		2	8		7,750	251		8,001	8,900	
	0810	1700 CFM		1.50	10.667		9,850	335		10,185	11,300	
	0820	2550 CFM	Q-6	2	12		12,800	390		13,190	14,700	
	0830	3550 CFM		1.50	16		16,100	520		16,620	18,500	
	0840	4700 CFM		1	24		20,000	780		20,780	23,200	
	0850	6100 CFM	Q-7	1	32		23,400	1,075		24,475	27,300	
	0860	8400 CFM		.80	40		29,000	1,325		30,325	33,900	
	0870	10200 CFM		.60	53.333		31,500	1,775		33,275	37,400	
	0880	12100 CFM		.50	64		33,800	2,125		35,925	40,300	
	0890	14200 CFM		.40	80		35,600	2,650		38,250	43,200	
	0950	Desiccant 33 lb. bag					30.50			30.50	33.50	
	0960	55 lb. bag					50			50	55	
	0980	2200 lb. bulk bag 1 or 2					2,000			2,000	2,200	
	0990	2200 lb. bulk bag 3 or 4					1,950			1,950	2,150	
	1250	Refrigeration, 50°F pressure dewpoint										
	1280	20 CFM	Q-5	7	2.286	Ea.	1,250	71.50		1,321.50	1,475	
	1290	60 CFM		6	2.667		1,775	83.50		1,858.50	2,075	
	1300	130 CFM		5.50	2.909		2,300	91.50		2,391.50	2,675	
	1310	150 CFM		5	3.200		4,825	100		4,925	5,450	
	1320	200 CFM		4.50	3.556		4,950	112		5,062	5,625	
	1330	280 CFM		4	4		5,425	126		5,551	6,150	
	1340	350 CFM		3.50	4.571		5,900	143		6,043	6,700	
	1350	450 CFM		3	5.333		6,775	167		6,942	7,700	
	1360	520 CFM		2.80	5.714		7,475	179		7,654	8,500	
	1370	650 CFM		2.40	6.667		8,225	209		8,434	9,375	
	1380	800 CFM		2.20	7.273		8,775	228		9,003	10,000	
	1390	1000 CFM		2	8		11,300	251		11,551	12,800	
	1400	1400 CFM		1.50	10.667		13,300	335		13,635	15,100	
	1410	2000 CFM		1.20	13.333		16,200	420		16,620	18,400	
	1420	2400 CFM		1	16		23,400	500		23,900	26,500	
	1430	3000 CFM	Q-6	1.30	18.462		31,700	600		32,300	35,800	
	1440	3600 CFM		1.20	20		35,700	650		36,350	40,300	
	1450	4800 CFM		1	24		41,800	780		42,580	47,100	
	1460	6000 CFM		.80	30		51,500	975		52,475	58,000	

15210	Process Air/Gas Piping	CREW	DAILY OUTPUT	LABOR-HOURS	UNIT	2001 BARE COSTS MAT.	LABOR	EQUIP.	TOTAL	TOTAL INCL O&P
200 1470	7200 CFM	Q-6	.50	48	Ea.	60,000	1,550		61,550	68,500
1500	Package includes dryer & filtration systems									
1530	150 CFM	Q-5	4.20	3.810	Ea.	7,050	120		7,170	7,925
1540	200 CFM		3.90	4.103		7,200	129		7,329	8,125
1550	280 CFM		3.50	4.571		8,250	143		8,393	9,300
1560	350 CFM		3.20	5		9,275	157		9,432	10,400
1570	450 CFM		3	5.333		10,300	167		10,467	11,600
1580	520 CFM		2.50	6.400		10,900	201		11,101	12,300
1590	650 CFM		2.20	7.273		11,300	228		11,528	12,800
1600	800 CFM		2	8		12,000	251		12,251	13,600
1610	1000 CFM		1.80	8.889		15,400	279		15,679	17,300
1620	1200 CFM		1.40	11.429		17,300	360		17,660	19,500
1630	1400 CFM		1.20	13.333		18,900	420		19,320	21,300
1640	2000 CFM		1	16		21,700	500		22,200	24,700
1650	2400 CFM		.70	22.857		28,000	715		28,715	31,900
2000	Regenerative adsorption with desiccant									
2030	50 CFM	Q-5	5	3.200	Ea.	4,250	100		4,350	4,825
2040	75 CFM		4.50	3.556		4,525	112		4,637	5,175
2050	100 CFM		4.20	3.810		4,975	120		5,095	5,650
2060	125 CFM		4	4		5,400	126		5,526	6,150
2070	170 CFM		3.70	4.324		5,850	136		5,986	6,625
2080	240 CFM		3.40	4.706		7,100	148		7,248	8,025
2090	350 CFM		3	5.333		7,600	167		7,767	8,625
2100	520 CFM		2.50	6.400		9,100	201		9,301	10,300
2110	650 CFM		2	8		10,400	251		10,651	11,900
2120	800 CFM		1.50	10.667		13,100	335		13,435	14,900
2130	1000 CFM		1.20	13.333		15,200	420		15,620	17,400
2140	1200 CFM		1	16		16,000	500		16,500	18,400
2150	1600 CFM		.80	20		20,000	630		20,630	23,000
2160	2000 CFM		.70	22.857		22,800	715		23,515	26,200
2170	2500 CFM		.60	26.667		28,200	835		29,035	32,300
2180	3000 CFM	Q-6	.90	26.667		29,500	865		30,365	33,700
2190	3500 CFM		.70	34.286		42,700	1,125		43,825	48,600
2200	4000 CFM		.50	48		43,900	1,550		45,450	50,500
2300	Desiccant, 50 lb. bag					122			122	134
2330	350 lb. drum					845			845	930
3000	Filters Air									
3100	Adsorptive, odors by glass fiber & carbon, 90 psi									
3140	110 CFM	Q-5	8	2	Ea.	835	63		898	1,025
3150	190 CFM		7.50	2.133		950	67		1,017	1,150
3160	290 CFM		7	2.286		1,175	71.50		1,246.50	1,400
3170	380 CFM		6	2.667		1,600	83.50		1,683.50	1,875
3180	570 CFM		5	3.200		2,250	100		2,350	2,625
3190	870 CFM		4.50	3.556		2,550	112		2,662	3,000
3200	1160 CFM		4	4		3,275	126		3,401	3,800
3210	1450 CFM		3.50	4.571		3,850	143		3,993	4,450
3220	1740 CFM		3	5.333		4,300	167		4,467	4,975
3260	Coalescent system, liquids & solids 90 psi high purity									
3300	110 CFM	Q-5	7	2.286	Ea.	3,125	71.50		3,196.50	3,525
3310	190 CFM		6.70	2.388		3,400	75		3,475	3,850
3320	290 CFM		6.50	2.462		3,575	77		3,652	4,075
3330	380 CFM		6	2.667		4,200	83.50		4,283.50	4,750
3340	570 CFM		4.50	3.556		5,275	112		5,387	5,975
3350	870 CFM		4	4		5,575	126		5,701	6,325
3360	1160 CFM		3.50	4.571		6,350	143		6,493	7,200
3370	1450 CFM		3	5.333		7,150	167		7,317	8,100 200

15210 | Process Air/Gas Piping

			CREW	DAILY OUTPUT	LABOR-HOURS	UNIT	2001 BARE COSTS MAT.	LABOR	EQUIP.	TOTAL	TOTAL INCL O&P	
200	3380	1740 CFM	Q-5	2.50	6.400	Ea.	7,550	201		7,751	8,625	200
	3390	2030 CFM	↓	2	8	↓	7,825	251		8,076	8,975	
	3500	Combination air line filter, regulator and lubricator										
	3510	1/4" NPT	1 Stpi	16	.500	Ea.	105	17.45		122.45	142	
	3520	3/8" NPT		16	.500		117	17.45		134.45	155	
	3530	1/2" NPT		15	.533		117	18.60		135.60	156	
	3540	3/4" NPT		14	.571		235	19.90		254.90	289	
	3550	1" NPT	↓	12	.667	↓	235	23		258	294	
	4000	Couplers, air line, sleeve type										
	4010	Female, connection size NPT										
	4020	1/4"	1 Stpi	38	.211	Ea.	3.92	7.35		11.27	15.35	
	4030	3/8"		36	.222		4.89	7.75		12.64	17.10	
	4040	1/2"		35	.229		10.25	7.95		18.20	23.50	
	4050	3/4"	↓	34	.235	↓	11.75	8.20		19.95	25.50	
	4100	Male										
	4110	1/4"	1 Stpi	38	.211	Ea.	3.88	7.35		11.23	15.30	
	4120	3/8"		36	.222		4.89	7.75		12.64	17.10	
	4130	1/2"		35	.229		9.70	7.95		17.65	22.50	
	4140	3/4"	↓	34	.235	↓	11.45	8.20		19.65	25	
	4150	Coupler, combined male and female halves										
	4160	1/2"	1 Stpi	17	.471	Ea.	19.95	16.40		36.35	47	
	4170	3/4"	"	15	.533	"	23	18.60		41.60	53.50	
900	0010	**VACUUM PUMPS**										900
	0100	Medical, with receiver										
	0110	Duplex										
	0120	20 SCFM	Q-1	1.14	14.035	Ea.	14,100	435		14,535	16,200	
	0130	60 SCFM, 10 HP	Q-2	1.20	20	"	24,000	645		24,645	27,300	
	0200	Triplex										
	0220	180 SCFM	Q-2	.86	27.907	Ea.	46,400	895		47,295	52,500	
	0300	Dental oral										
	0310	Duplex										
	0330	165 SCFM with 77 Gal separator	Q-2	1.30	18.462	Ea.	42,000	595		42,595	47,100	

15220 | Water Process/Waste Piping

			CREW	DAILY OUTPUT	LABOR-HOURS	UNIT	MAT.	LABOR	EQUIP.	TOTAL	TOTAL INCL O&P	
100	0010	**LABORATORY EQUIPMENT** Corrosion resistant										100
	7000	Tanks, covers included										
	7400	Kaynar (PVDF), FRP casing, high chemical resistance										
	7410	Temperature range to 230°F										
	7420	7 gallon, 12" x 12" x 12"	Q-1	20	.800	Ea.	550	25		575	645	
	7430	15 gallon, 18" x 12" x 18"		17	.941		885	29		914	1,025	
	7440	30 gallon, 24" x 18" x 18"	↓	12	1.333	↓	1,275	41.50		1,316.50	1,475	
	7500	Polyethylene, neutralization & dilution										
	7510	Continuous use to 180°F, includes 4" inlet & outlet										
	7530	30 gallon, upright cylinder	Q-1	10	1.600	Ea.	260	49.50		309.50	360	
	7550	55 gallon, upright cylinder		5	3.200		325	99		424	510	
	7570	100 gallon, upright cylinder		4	4		565	124		689	810	
	7590	250 gallon, upright cylinder	↓	3	5.333	↓	1,025	165		1,190	1,375	
	7650	Polyethylene with ultra violet light inhibitor										
	7660	Continuous use to 180°F, incl. molded leg supports & ftngs										
	7680	55 gallon 24" O.D. x 36" long	Q-1	10	1.600	Ea.	244	49.50		293.50	345	
	7700	200 gallon 33" O.D. x 65" long		6	2.667		435	82.50		517.50	605	
	7720	300 gallon 38" O.D. x 72" long		5	3.200		490	99		589	690	
	7740	500 gallon 48" O.D. x 75" long	↓	3	5.333	↓	595	165		760	905	
	7800	Polyethylene liner, fiberglass casing										
	7810	Continuous service to 220°F										
	7830	2 gallon, 8" x 8" x 8"	Q-1	20	.800	Ea.	171	25		196	226	

15220 | Water Process/Waste Piping

		15220 Water Process/Waste Piping	CREW	DAILY OUTPUT	LABOR-HOURS	UNIT	2001 BARE COSTS MAT.	LABOR	EQUIP.	TOTAL	TOTAL INCL O&P	
100	7850	7 gallon, 12″ x 12″ x 12″	Q-1	20	.800	Ea.	207	25		232	265	100
	7870	15 gallon, 24″ x 12″ x 12″		17	.941		340	29		369	420	
	8010	30 gallon, 24″ x 18″ x 18″		12	1.333		495	41.50		536.50	605	
	8070	45 gallon, 24″ x 18″ x 24″		10	1.600		580	49.50		629.50	710	
	8080	90 gallon, 36″ x 24″ x 24″		8	2		1,125	62		1,187	1,325	
	8150	Polyethylene, heavy duty walls										
	8160	Continuous service to 180°F										
	8180	5 gallon, 11″ I.D. x 15″ deep	Q-1	20	.800	Ea.	65.50	25		90.50	110	
	8210	15 gallon, 14″ I.D. x 27″ deep		17	.941		96.50	29		125.50	150	
	8230	55 gallon, 22″ I.D. x 36″ deep		10	1.600		204	49.50		253.50	300	
	8250	100 gallon 28″ I.D. x 42″ deep		8	2		440	62		502	580	
	8270	200 gallon 36″ I.D. x 48″ deep		6	2.667		555	82.50		637.50	735	
	8290	360 gallon 48″ I.D. x 48″ deep		5	3.200		905	99		1,004	1,150	
	8350	With fiberglass casing										
	8420	55 gallon tank size	Q-1	20	.800	Ea.	495	25		520	585	
	8440	100 gallon tank size		16	1		765	31		796	885	
	8460	200 gallon tank size		12	1.333		1,100	41.50		1,141.50	1,275	
	8480	360 gallon tank size		10	1.600		1,650	49.50		1,699.50	1,875	
		15230 Industrial Process Piping										
500	0010	**PUMPS, GENERAL UTILITY** With motor										500
	0200	Multi-stage, horizontal split, for boiler feed applications										
	0300	Two stage, 3″ discharge x 4″ suction, 75 HP	Q-7	.30	106	Ea.	14,900	3,550		18,450	21,800	
	0340	Four stage, 3″ discharge x 4″ suction, 150 HP	"	.18	177	"	24,500	5,900		30,400	35,900	
	2000	Single stage										
	2060	End suction, 1″D. x 2″S., 3 HP	Q-1	.50	32	Ea.	3,650	990		4,640	5,500	
	2100	1-1/2″D. x 3″S., 10 HP	"	.40	40		3,925	1,250		5,175	6,200	
	2140	2″D. x 3″S., 15 HP	Q-2	.60	40		4,275	1,275		5,550	6,650	
	2180	3″D. x 4″S., 20 HP		.50	48		4,550	1,550		6,100	7,325	
	2220	4″D. x 6″S., 30 HP		.40	60		4,925	1,925		6,850	8,325	
	3000	Double suction, 2″D. x 2-1/2″S., 10 HP	Q-1	.30	53.333		5,050	1,650		6,700	8,050	
	3060	3″D. x 4″S., 15 HP	Q-2	.46	52.174		5,200	1,675		6,875	8,250	
	3100	4″D. x 5″S., 30 HP		.40	60		7,050	1,925		8,975	10,700	
	3140	5″D. x 6″S., 50 HP		.33	72.727		7,575	2,350		9,925	11,900	
	3180	6″D. x 8″S., 60 HP	Q-3	.30	106		8,475	3,500		11,975	14,600	
	3190	75 HP, to 2500 GPM		.28	114		10,500	3,750		14,250	17,200	
	3220	100 HP, to 3000 GPM		.26	123		13,300	4,050		17,350	20,800	
	3240	150 HP, to 4000 GPM		.24	133		17,900	4,375		22,275	26,300	
	4000	Centrifugal, end suction, mounted on base										
	4010	Horizontal mounted, with drip proof motor, rated @ 100′ head										
	4020	Vertical split case, single stage										
	4040	100 GPM, 5 HP, 1-1/2″ discharge	Q-1	1.70	9.412	Ea.	1,950	292		2,242	2,600	
	4050	200 GPM, 10 HP, 2″ discharge		1.30	12.308		2,325	380		2,705	3,125	
	4060	250 GPM, 10 HP, 3″ discharge		1.28	12.500		2,425	390		2,815	3,250	
	4070	300 GPM, 15 HP, 2″ discharge	Q-2	1.56	15.385		2,625	495		3,120	3,650	
	4080	500 GPM, 20 HP, 4″ discharge		1.44	16.667		3,100	535		3,635	4,200	
	4090	750 GPM, 30 HP, 4″ discharge		1.20	20		4,150	645		4,795	5,550	
	4100	1050 GPM, 40 HP, 5″ discharge		1	24		4,475	770		5,245	6,100	
	4110	1500 GPM, 60 HP, 6″ discharge		.60	40		6,525	1,275		7,800	9,125	
	4120	2000 GPM, 75 HP, 6″ discharge		.50	48		7,500	1,550		9,050	10,600	
	4130	3000 GPM, 100 HP, 8″ discharge		.40	60		11,300	1,925		13,225	15,300	
	4200	Horizontal split case, single stage										
	4210	100 GPM, 7.5 HP, 1-1/2″ discharge	Q-1	1.70	9.412	Ea.	3,200	292		3,492	3,950	
	4220	250 GPM, 15 HP, 2-1/2″ discharge	"	1.30	12.308		4,175	380		4,555	5,175	
	4230	500 GPM, 20 HP, 4″ discharge	Q-2	1.60	15		5,050	480		5,530	6,275	
	4240	750 GPM, 25 HP, 5″ discharge		1.54	15.584		5,225	500		5,725	6,500	

15230 | Industrial Process Piping

		15230 Industrial Process Piping	CREW	DAILY OUTPUT	LABOR-HOURS	UNIT	2001 BARE COSTS MAT.	LABOR	EQUIP.	TOTAL	TOTAL INCL O&P	
500	4250	1000 GPM, 40 HP, 5″ discharge	Q-2	1.20	20	Ea.	6,825	645		7,470	8,500	500
	4260	1500 GPM, 50 HP, 6″ discharge	Q-3	1.42	22.535		8,600	740		9,340	10,600	
	4270	2000 GPM, 75 HP, 8″ discharge		1.14	28.070		11,700	920		12,620	14,200	
	4280	3000 GPM, 100 HP, 10″ discharge		.96	33.333		13,000	1,100		14,100	16,000	
	4290	3500 GPM, 150 HP, 10″ discharge		.86	37.209		16,300	1,225		17,525	19,800	
	4300	4000 GPM, 200 HP, 10″ discharge		.66	48.485		17,900	1,600		19,500	22,100	
	4330	Horizontal split case, two stage, 500′ head										
	4340	100 GPM, 40 HP, 1-1/2″ discharge	Q-2	1.70	14.118	Ea.	4,150	455		4,605	5,225	
	4350	200 GPM, 50 HP, 1-1/2″ discharge	″	1.44	16.667		4,600	535		5,135	5,875	
	4360	300 GPM, 75 HP, 2″ discharge	Q-3	1.57	20.382		4,900	670		5,570	6,400	
	4370	400 GPM, 100 HP, 3″ discharge		1.14	28.070		6,375	920		7,295	8,425	
	4380	800 GPM, 200 HP, 4″ discharge		.86	37.209		15,000	1,225		16,225	18,400	
	5000	Centrifugal, in-line										
	5006	Vertical mount, iron body, 125 lb. flgd, 3550 RPM TEFC mtr										
	5010	Single stage										
	5020	50 GPM, 3 HP, 1-1/2″ discharge	Q-1	2.30	6.957	Ea.	510	216		726	885	
	5030	75 GPM, 5 HP, 1-1/2″ discharge		1.60	10		865	310		1,175	1,425	
	5040	100 GPM, 7.5 HP, 1-1/2″ discharge		1.30	12.308		930	380		1,310	1,600	
	5050	125 GPM, 10 HP, 1-1/2″ discharge	Q-2	1.70	14.118		1,150	455		1,605	1,950	
	5060	150 GPM, 15 HP, 1-1/2″ discharge		1.60	15		1,475	480		1,955	2,350	
	5070	200 GPM, 30 HP, 1-1/2″ discharge		1.50	16		2,425	515		2,940	3,425	
	5080	250 GPM, 40 HP, 2″ discharge		1.40	17.143		3,200	550		3,750	4,350	
	5090	300 GPM, 50 HP, 3″ discharge		1.30	18.462		3,875	595		4,470	5,175	
	5100	400 GPM, 75 HP, 3″ discharge	Q-3	.60	53.333		6,325	1,750		8,075	9,600	
	5110	600 GPM, 100 HP, 3″ discharge	″	.50	64		8,475	2,100		10,575	12,500	
	8000	Vertical submerged, with non-submerged motor										
	8060	1″D., 3 HP	Q-1	.50	32	Ea.	5,275	990		6,265	7,300	
	8100	1-1/2″D., 10 HP	″	.30	53.333		5,525	1,650		7,175	8,575	
	8140	2″D., 15 HP	Q-2	.40	60		5,675	1,925		7,600	9,150	
	8180	3″D., 25 HP		.30	80		5,800	2,575		8,375	10,300	
	8220	4″D., 30 HP		.20	120		6,500	3,850		10,350	13,000	
600	0010	**PUMPS MISCELLANEOUS**										600
	0020	Water pump, portable, gasoline powered										
	0100	6000 GPH, 2″ discharge	Q-1	11	1.455	Ea.	785	45		830	935	
	0110	8000 GPH, 2″ discharge		10.50	1.524		785	47		832	935	
	0120	10,000 GPH, 2″ discharge		10	1.600		1,275	49.50		1,324.50	1,475	
	0500	Pump, propylene body, housing and impeller										
	0510	22 GPM, 1/3 HP, 40′ HD	1 Plum	5	1.600	Ea.	485	55		540	620	
	0520	33 GPM, 1/2 HP, 40′ HD	Q-1	5	3.200		495	99		594	695	
	0530	53 GPM, 3/4 HP, 40′ HD	″	4	4		530	124		654	765	
	0600	Rotary pump, CI										
	0610	10 GPM, 1/2 HP, 1″ discharge	Q-1	5	3.200	Ea.	910	99		1,009	1,150	
	0614	10 GPM, 3/4 HP, 1″ discharge		4.50	3.556		910	110		1,020	1,175	
	0618	10 GPM 1 HP, 1″ discharge		4	4		945	124		1,069	1,225	
	0620	25 GPM, 1 HP, 1-1/4″ discharge		3.80	4.211		945	131		1,076	1,250	
	0624	25 GPM, 1.5 HP, 1-1/4″ discharge		3.60	4.444		1,025	138		1,163	1,325	
	0628	25 GPM, 2 HP, 1-1/4″ discharge		3.20	5		1,025	155		1,180	1,350	
	0700	Fuel oil pump, 2 stage, 3450 RPM										
	0710	1/4 HP, 100 PSI	Q-1	8	2	Ea.	1,550	62		1,612	1,800	
	0800	Hydraulic oil pump										
	0810	75 HP, 3 PH, 480 V	Q-3	1.20	26.667	Ea.	5,075	875		5,950	6,900	
	1000	Turbine pump, CI										
	1010	50 GPM, 2 HP, 3″ discharge	Q-1	.80	20	Ea.	1,050	620		1,670	2,075	
	1020	100 GPM, 3 HP, 4″ discharge	Q-2	.96	25		1,575	805		2,380	2,950	
	1030	250 GPM, 15 HP, 6″ discharge	″	.94	25.532		3,150	820		3,970	4,725	

15200 | Process Piping

	15230 | Industrial Process Piping		CREW	DAILY OUTPUT	LABOR-HOURS	UNIT	2001 BARE COSTS MAT.	LABOR	EQUIP.	TOTAL	TOTAL INCL O&P	
600	1040	500 GPM, 25 HP, 6″ discharge	Q-3	1.30	24.615	Ea.	5,250	810		6,060	7,000	600
	1050	1000 GPM, 50 HP, 8″ discharge		1.14	28.070		8,400	920		9,320	10,700	
	1060	2000 GPM, 100 HP, 10″ discharge		1	32		9,450	1,050		10,500	12,000	
	1070	3000 GPM, 150 HP, 10″ discharge		.80	40		10,500	1,325		11,825	13,600	
	1080	4000 GPM, 200 HP, 12″ discharge		.70	45.714		15,800	1,500		17,300	19,600	
	1090	6000 GPM, 300 HP, 14″ discharge		.60	53.333		17,900	1,750		19,650	22,400	
	1100	10,000 GPM, 300 HP, 18″ discharge		.58	55.172		19,500	1,800		21,300	24,100	

15400 | Plumbing Fixtures & Equipment

	15410 | Plumbing Fixtures		CREW	DAILY OUTPUT	LABOR-HOURS	UNIT	2001 BARE COSTS MAT.	LABOR	EQUIP.	TOTAL	TOTAL INCL O&P	
040	0010	**FIXTURES** Includes trim fittings unless otherwise noted (R15100-410)										040
	0080	For rough-in, supply, waste, and vent, see add for each type										
	0120	For electric water coolers, see division 15413										
	0160	For color, unless otherwise noted, add				Ea.	20%					
200	0010	**CARRIERS/SUPPORTS** For plumbing fixtures										200
	0500	Drinking fountain, wall mounted										
	0600	Plate type with studs, top back plate	1 Plum	7	1.143	Ea.	26.50	39.50		66	88.50	
	0700	Top front and back plate		7	1.143		33	39.50		72.50	95.50	
	0800	Top & bottom, front & back plates, w/bearing jacks		7	1.143		47.50	39.50		87	112	
	3000	Lavatory, concealed arm										
	3050	Floor mounted, single										
	3100	High back fixture	1 Plum	6	1.333	Ea.	161	46		207	247	
	3200	Flat slab fixture		6	1.333		188	46		234	276	
	3220	Paraplegic		6	1.333		136	46		182	220	
	3250	Floor mounted, back to back										
	3300	High back fixtures	1 Plum	5	1.600	Ea.	230	55		285	335	
	3400	Flat slab fixtures		5	1.600		282	55		337	395	
	3430	Paraplegic		5	1.600		191	55		246	293	
	3500	Wall mounted, in stud or masonry										
	3600	High back fixture	1 Plum	6	1.333	Ea.	75.50	46		121.50	153	
	3700	Flat slab fixture	″	6	1.333	″	122	46		168	204	
	4000	Exposed arm type, floor mounted										
	4100	Single high back or flat slab fixture	1 Plum	6	1.333	Ea.	182	46		228	270	
	4200	Back to back, high back or flat slab fixtures		5	1.600		360	55		415	480	
	4300	Wall mounted, High back or flat slab lavatory		6	1.333		159	46		205	245	
	4600	Sink, floor mounted										
	4650	Exposed arm system										
	4700	Single heavy fixture	1 Plum	5	1.600	Ea.	298	55		353	410	
	4750	Single heavy sink with slab		5	1.600		201	55		256	305	
	4800	Back to back, standard fixtures		5	1.600		230	55		285	335	
	4850	Back to back, heavy fixtures		5	1.600		268	55		323	380	
	4900	Back to back, heavy sink with slab		5	1.600		268	55		323	380	
	4950	Exposed offset arm system										
	5000	Single heavy deep fixture	1 Plum	5	1.600	Ea.	256	55		311	365	
	5100	Plate type system										
	5200	With bearing jacks, single fixture	1 Plum	5	1.600	Ea.	271	55		326	380	
	5300	With exposed arms, single heavy fixture		5	1.600		350	55		405	465	
	5400	Wall mounted, exposed arms, single heavy fixture		5	1.600		126	55		181	222	

15410 | Plumbing Fixtures

	15410 Plumbing Fixtures		CREW	DAILY OUTPUT	LABOR-HOURS	UNIT	2001 BARE COSTS MAT.	LABOR	EQUIP.	TOTAL	TOTAL INCL O&P	
200	6000	Urinal, floor mounted, 2″ or 3″ coupling, blowout type	1 Plum	6	1.333	Ea.	171	46		217	258	200
	6100	With fixture or hanger bolts, blowout or washout		6	1.333		122	46		168	204	
	6200	With bearing plate		6	1.333		136	46		182	220	
	6300	Wall mounted, plate type system	↓	6	1.333	↓	97	46		143	176	
	6980	Water closet, siphon jet										
	7000	Horizontal, adjustable, caulk										
	7040	Single, 4″ pipe size	1 Plum	6	1.333	Ea.	205	46		251	296	
	7050	4″ pipe size, paraplegic		6	1.333		205	46		251	296	
	7060	5″ pipe size		6	1.333		271	46		317	370	
	7100	Double, 4″ pipe size		5	1.600		385	55		440	505	
	7110	4″ pipe size, paraplegic		5	1.600		205	55		260	310	
	7120	5″ pipe size	↓	5	1.600	↓	271	55		326	380	
	7160	Horizontal, adjustable, extended, caulk										
	7180	Single, 4″ pipe size	1 Plum	6	1.333	Ea.	205	46		251	296	
	7200	5″ pipe size		6	1.333		365	46		411	475	
	7240	Double, 4″ pipe size		5	1.600		277	55		332	390	
	7260	5″ pipe size	↓	5	1.600	↓	365	55		420	490	
	7400	Vertical, adjustable, caulk or thread										
	7440	Single, 4″ pipe size	1 Plum	6	1.333	Ea.	242	46		288	335	
	7460	5″ pipe size		6	1.333		305	46		351	405	
	7480	6″ pipe size		5	1.600		360	55		415	480	
	7520	Double, 4″ pipe size		5	1.600		420	55		475	550	
	7540	5″ pipe size		5	1.600		485	55		540	620	
	7560	6″ pipe size	↓	4	2	↓	535	69		604	695	
	7600	Vertical, adjustable, extended, caulk										
	7620	Single, 4″ pipe size	1 Plum	6	1.333	Ea.	242	46		288	335	
	7640	5″ pipe size		6	1.333		305	46		351	405	
	7680	6″ pipe size		5	1.600		360	55		415	480	
	7720	Double, 4″ pipe size		5	1.600		420	55		475	550	
	7740	5″ pipe size		5	1.600		485	55		540	620	
	7760	6″ pipe size	↓	4	2	↓	535	69		604	695	
	7780	Water closet, blow out										
	7800	Vertical offset, caulk or thread										
	7820	Single, 4″ pipe size	1 Plum	6	1.333	Ea.	232	46		278	325	
	7840	Double, 4″ pipe size	″	5	1.600	″	395	55		450	520	
	7880	Vertical offset, extended, caulk										
	7900	Single, 4″ pipe size	1 Plum	6	1.333	Ea.	291	46		337	390	
	7920	Double, 4″ pipe size	″	5	1.600	″	455	55		510	585	
	7960	Vertical, for floor mounted back-outlet										
	7980	Single, 4″ thread, 2″ vent	1 Plum	6	1.333	Ea.	187	46		233	276	
	8000	Double, 4″ thread, 2″ vent	″	6	1.333	″	335	46		381	440	
	8040	Vertical, for floor mounted back-outlet, extended										
	8060	Single, 4″ caulk, 2″ vent	1 Plum	6	1.333	Ea.	179	46		225	267	
	8080	Double, 4″ caulk, 2″ vent	″	6	1.333	″	278	46		324	375	
	8200	Water closet, residential										
	8220	Vertical centerline, floor mount										
	8240	Single, 3″ caulk, 2″ or 3″ vent	1 Plum	6	1.333	Ea.	185	46		231	274	
	8260	4″ caulk, 2″ or 4″ vent		6	1.333		233	46		279	325	
	8280	3″ copper sweat, 3″ vent		6	1.333		185	46		231	274	
	8300	4″ copper sweat, 4″ vent	↓	6	1.333	↓	233	46		279	325	
	8400	Vertical offset, floor mount										
	8420	Single, 3″ or 4″ caulk, vent	1 Plum	4	2	Ea.	232	69		301	360	
	8440	3″ or 4″ copper sweat, vent		5	1.600		232	55		287	340	
	8460	Double, 3″ or 4″ caulk, vent		4	2		395	69		464	540	
	8480	3″ or 4″ copper sweat, vent	↓	5	1.600	↓	395	55		450	520	
	9000	Water cooler (electric), floor mounted										

	15410 | Plumbing Fixtures		Crew	Daily Output	Labor-Hours	Unit	2001 Bare Costs Mat.	Labor	Equip.	Total	Total Incl O&P	
200	9100	Plate type with bearing plate, single	1 Plum	6	1.333	Ea.	108	46		154	189	200
	9120	Plate type with bearing plate, double		4	2		159	69		228	278	
	9140	Plate type with bearing plate, back to back		4	2		151	69		220	270	
300	0010	**FAUCETS/FITTINGS**										300
	0150	Bath, faucets, diverter spout combination, sweat	1 Plum	8	1	Ea.	102	34.50		136.50	164	
	0200	For integral stops, IPS unions, add					68.50			68.50	75.50	
	0300	Three valve combinations, spout, head, arm, flange, sweat	1 Plum	6	1.333		65.50	46		111.50	142	
	0400	For integral stops, IPS unions, add					91			91	100	
	0420	Bath, press-bal mix valve w/diverter, spout, shower hd, arm/flange	1 Plum	8	1		102	34.50		136.50	164	
	0500	Drain, central lift, 1-1/2" IPS male		20	.400		35.50	13.80		49.30	60.50	
	0600	Trip lever, 1-1/2" IPS male		20	.400		36.50	13.80		50.30	61.50	
	0700	Pop up, 1-1/2" IPS male		18	.444		43	15.30		58.30	70.50	
	0800	Chain and stopper, 1-1/2" IPS male		24	.333		29	11.50		40.50	49	
	0810	Bidet										
	0812	Fitting, over the rim, swivel spray/pop-up drain	1 Plum	8	1	Ea.	126	34.50		160.50	190	
	0840	Flush valves, with vacuum breaker										
	0850	Water closet										
	0860	Exposed, rear spud	1 Plum	8	1	Ea.	106	34.50		140.50	169	
	0870	Top spud		8	1		97.50	34.50		132	159	
	0880	Concealed, rear spud		8	1		139	34.50		173.50	205	
	0890	Top spud		8	1		138	34.50		172.50	204	
	0900	Wall hung		8	1		123	34.50		157.50	187	
	0920	Urinal										
	0930	Exposed, stall	1 Plum	8	1	Ea.	97.50	34.50		132	159	
	0940	Wall, (washout)		8	1		97.50	34.50		132	159	
	0950	Pedestal, top spud		8	1		99	34.50		133.50	161	
	0960	Concealed, stall		8	1		112	34.50		146.50	175	
	0970	Wall (washout)		8	1		115	34.50		149.50	179	
	0971	Automatic flush sensor and operator for										
	0972	urinals or water closets	1 Plum	16	.500	Ea.	320	17.25		337.25	375	
	1000	Kitchen sink faucets, top mount, cast spout		10	.800		48	27.50		75.50	94.50	
	1100	For spray, add		24	.333		12	11.50		23.50	30.50	
	1110	For basket strainer w/tail piece, add					11.75			11.75	12.90	
	1200	Wall type, swing tube spout	1 Plum	10	.800		59.50	27.50		87	107	
	1240	For soap dish, add					4.35			4.35	4.79	
	1250	For basket strainer w/tail piece, add					11.75			11.75	12.90	
	2000	Laundry faucets, shelf type, IPS or copper unions	1 Plum	12	.667		39.50	23		62.50	78	
	2100	Lavatory faucet, centerset, without drain		10	.800		35	27.50		62.50	79.50	
	2200	For pop-up drain, add		16	.500		14.20	17.25		31.45	41.50	
	2250	For acrylic handles, add					7.10			7.10	7.80	
	2400	Concealed, 12" centers	1 Plum	10	.800		76.50	27.50		104	126	
	2450	For pop-up drain, add		16	.500		14.20	17.25		31.45	41.50	
	2600	Shelfback, 4" to 6" centers, 17 Ga. tailpiece		10	.800		60.50	27.50		88	108	
	2650	For pop-up drain, add		16	.500		14.20	17.25		31.45	41.50	
	2700	For shampoo faucet with supply tube, add		24	.333		38	11.50		49.50	59.50	
	2800	Self-closing, center set		10	.800		102	27.50		129.50	155	
	2810	Automatic sensor and operator, with faucet head		6.15	1.301		284	45		329	385	
	2850	Medical, bedpan cleanser, with pedal valve,		12	.667		450	23		473	530	
	2860	With screwdriver stop valve		12	.667		234	23		257	293	
	2870	With self-closing spray valve		12	.667		176	23		199	228	
	2900	Faucet, gooseneck spout, wrist handles, grid drain		10	.800		138	27.50		165.50	194	
	2940	Mixing valve, knee action, screwdriver stops		4	2		520	69		589	680	
	3000	Service sink faucet, cast spout, pail hook, hose end		14	.571		67.50	19.70		87.20	104	
	4000	Shower by-pass valve with union		18	.444		47.50	15.30		62.80	75	
	4100	Shower arm with flange and head		22	.364		13.50	12.55		26.05	34	

	15410 | Plumbing Fixtures		CREW	DAILY OUTPUT	LABOR-HOURS	UNIT	2001 BARE COSTS MAT.	LABOR	EQUIP.	TOTAL	TOTAL INCL O&P	
300	4140	Shower, hand held, pin mount, massage action, chrome	1 Plum	22	.364	Ea.	63.50	12.55		76.05	89	300
	4142	Polished brass		22	.364		121	12.55		133.55	152	
	4144	Shower, hand held, wall mtd, adj. spray, 2 wall mounts, chrome		20	.400		128	13.80		141.80	162	
	4146	Polished brass		20	.400		235	13.80		248.80	280	
	4148	Shower, hand held head, bar mounted 24", adj. spray, chrome		20	.400		159	13.80		172.80	196	
	4150	Polished brass		20	.400		350	13.80		363.80	405	
	4200	Shower thermostatic mixing valve, concealed		8	1		226	34.50		260.50	300	
	4300	For inlet strainer, check, and stops, add					54			54	59.50	
	5000	Sillcock, compact, brass, IPS or copper to hose	1 Plum	24	.333		4.94	11.50		16.44	23	
	6000	Stop and waste valves, bronze										
	6100	Angle, solder end 1/2"	1 Plum	24	.333	Ea.	4.18	11.50		15.68	22	
	6110	3/4"		20	.400		4.59	13.80		18.39	26	
	6300	Straightway, solder end 3/8"		24	.333		3.59	11.50		15.09	21.50	
	6310	1/2"		24	.333		2.32	11.50		13.82	19.90	
	6320	3/4"		20	.400		2.68	13.80		16.48	24	
	6330	1"		19	.421		4	14.50		18.50	26.50	
	6400	Straightway, threaded 3/8"		24	.333		4.07	11.50		15.57	22	
	6410	1/2"		24	.333		3.13	11.50		14.63	21	
	6420	3/4"		20	.400		3.73	13.80		17.53	25	
	6430	1"		19	.421		10.50	14.50		25	33.50	
	7800	Water closet, wax gasket		96	.083		1.24	2.87		4.11	5.70	
	7820	Gasket toilet tank to bowl		32	.250		1.19	8.60		9.79	14.30	
	8000	Water supply stops, polished chrome plate										
	8200	Angle, 3/8"	1 Plum	24	.333	Ea.	4.73	11.50		16.23	22.50	
	8300	1/2"		22	.364		4.73	12.55		17.28	24	
	8400	Straight, 3/8"		26	.308		7.05	10.60		17.65	24	
	8500	1/2"		24	.333		7.05	11.50		18.55	25	
	8600	Water closet, angle, w/flex riser, 3/8"		24	.333		11.05	11.50		22.55	29.50	
	9100	Miscellaneous										
	9720	Teflon tape, 1/2" x 520" roll				Ea.	1.09			1.09	1.20	
800	0010	**TOILET SEATS**										800
	0100	Molded composition, white										
	0150	Industrial, w/o cover, open front, regular bowl	1 Plum	24	.333	Ea.	14.40	11.50		25.90	33	
	0200	With self-sustaining hinge		24	.333		15.85	11.50		27.35	35	
	0220	With self-sustaining check hinge		24	.333		15.85	11.50		27.35	35	
	0240	Extra heavy, with check hinge		24	.333		28.50	11.50		40	49	
	0260	Elongated bowl, same price										
	0300	Junior size, w/o cover, open front	1 Plum	24	.333	Ea.	29.50	11.50		41	50	
	0320	Regular primary bowl, open front		24	.333		29.50	11.50		41	50	
	0340	Regular baby bowl, open front, check hinge		24	.333		29.50	11.50		41	50	
	0380	Open back & front, w/o cover, reg. or elongated bowl		24	.333		28	11.50		39.50	48.50	
	0400	Residential										
	0420	Regular bowl, w/cover, closed front	1 Plum	24	.333	Ea.	27.50	11.50		39	47.50	
	0440	Open front	"	24	.333	"	27	11.50		38.50	47.50	
	0460	Elongated bowl, add					25%					
	0500	Self-raising hinge, w/o cover, open front										
	0520	Regular bowl	1 Plum	24	.333	Ea.	89	11.50		100.50	115	
	0540	Elongated bowl	"	24	.333	"	92	11.50		103.50	118	
	0700	Molded wood, white, with cover										
	0720	Closed front, regular bowl, square back	1 Plum	24	.333	Ea.	9.45	11.50		20.95	28	
	0740	Extended back		24	.333		13.25	11.50		24.75	32	
	0780	Elongated bowl, square back		24	.333		12.35	11.50		23.85	31	
	0800	Open front		24	.333		13.25	11.50		24.75	32	
	0850	Decorator styles										
	0870	Cane top	1 Plum	24	.333	Ea.	22	11.50		33.50	41.50	
	0890	Vinyl top, patterned		24	.333		13.10	11.50		24.60	32	

	15410 | Plumbing Fixtures	CREW	DAILY OUTPUT	LABOR-HOURS	UNIT	2001 BARE COSTS MAT.	LABOR	EQUIP.	TOTAL	TOTAL INCL O&P
800	0900 Vinyl padded, plain colors, regular bowl	1 Plum	24	.333	Ea.	13.10	11.50		24.60	32
	0930 Elongated bowl	↓	24	.333	↓	18.85	11.50		30.35	38
	1000 Solid plastic, white									
	1030 Industrial, w/o cover, open front, regular bowl	1 Plum	24	.333	Ea.	17.90	11.50		29.40	37
	1080 Extra heavy, concealed check hinge		24	.333		14.40	11.50		25.90	33
	1100 Self-sustaining hinge		24	.333		16.05	11.50		27.55	35
	1150 Elongated bowl		24	.333		18.55	11.50		30.05	38
	1170 Concealed check		24	.333		14.45	11.50		25.95	33.50
	1190 Self-sustaining hinge, concealed check		24	.333		37.50	11.50		49	59
	1220 Residential, with cover, closed front, regular bowl		24	.333		24	11.50		35.50	44
	1240 Elongated bowl		24	.333		32	11.50		43.50	53
	1260 Open front, regular bowl		24	.333		28.50	11.50		40	49
	1280 Elongated bowl	↓	24	.333	↓	35.50	11.50		47	56.50
900	0010 **WASHER DRYER ACCESSORIES**									
	1020 Valves ball type single lever									
	1030 1/2″ diam., IPS	1 Plum	21	.381	Ea.	25	13.10		38.10	47.50
	1040 1/2″ diam., solder	″	21	.381	″	25	13.10		38.10	47.50
	1050 Recessed box, 16 ga., two hose valves and drain									
	1060 1/2″ size, 1-1/2″ drain	1 Plum	18	.444	Ea.	39	15.30		54.30	66
	1070 1/2″ size, 2″ drain	″	17	.471	″	42	16.20		58.20	70.50
	1080 With grounding electric receptacle									
	1090 1/2″ size, 1-1/2″ drain	1 Plum	18	.444	Ea.	45.50	15.30		60.80	73.50
	1100 1/2″ size, 2″ drain	″	17	.471	″	42	16.20		58.20	70.50
	1110 With grounding and dryer receptacle									
	1120 1/2″ size, 1-1/2″ drain	1 Plum	18	.444	Ea.	52.50	15.30		67.80	81
	1130 1/2″ size, 2″ drain	″	17	.471	″	56.50	16.20		72.70	87
	1140 Recessed box 16 ga., ball valves with single lever and drain									
	1150 1/2″ size, 1-1/2″ drain	1 Plum	19	.421	Ea.	68.50	14.50		83	97
	1160 1/2″ size, 2″ drain	″	18	.444	″	68.50	15.30		83.80	98.50
	1170 With grounding electric receptacle									
	1180 1/2″ size, 1-1/2″ drain	1 Plum	19	.421	Ea.	70	14.50		84.50	99
	1190 1/2″ size, 2″ drain	″	18	.444	″	75	15.30		90.30	106
	1200 With grounding and dryer receptacles									
	1210 1/2″ size, 1-1/2″ drain	1 Plum	19	.421	Ea.	76.50	14.50		91	106
	1220 1/2″ size, 2″ drain	″	18	.444		82	15.30		97.30	114
	1230 For duplex grounding receptacle add				↓	19.50			19.50	21.50
	1300 Recessed box, 20 ga., two hose valves and drain (economy type)									
	1310 1/2″ size, 1-1/2″ drain	1 Plum	19	.421	Ea.	32	14.50		46.50	57.50
	1320 1/2″ size, 2″ drain		18	.444		32.50	15.30		47.80	59
	1330 Box with drain only		24	.333		17.55	11.50		29.05	36.50
	1340 1/2″ size, 1-1/2″ ABS/PVC drain		19	.421		20	14.50		34.50	44
	1350 1/2″ size, 2″ ABS/PVC drain		18	.444		15.40	15.30		30.70	40
	1352 Box with drain only		24	.333		17.85	11.50		29.35	37
	1360 1/2″ size, 2″ drain ABS/PVC, 15 A receptacle	↓	24	.333	↓	42	11.50		53.50	63.50
	1400 Wall mounted									
	1410 1/2″ size, 1-1/2″ plastic drain	1 Plum	19	.421	Ea.	17.85	14.50		32.35	41.50
	1420 1/2″ size, 2″ plastic drain	″	18	.444	″	14.90	15.30		30.20	39.50
	1500 Dryer vent kit, recessed, 16 ga.									
	1510 With door, electrical outlets, flex duct	1 Plum	20	.400	Ea.	120	13.80		133.80	153
	1980 Rough-in, supply, waste, and vent for washer boxes	″	3.46	2.310	″	45.50	79.50		125	170

	15411 | Commercial/Indust Fixtures	CREW	DAILY OUTPUT	LABOR-HOURS	UNIT	MAT.	LABOR	EQUIP.	TOTAL	TOTAL INCL O&P
400	0010 **HOT WATER DISPENSERS**									
	0160 Commercial, 100 cup, 11.3 amp	1 Plum	14	.571	Ea.	305	19.70		324.70	365

	15411 | Commercial/Indust Fixtures		CREW	DAILY OUTPUT	LABOR-HOURS	UNIT	2001 BARE COSTS MAT.	LABOR	EQUIP.	TOTAL	TOTAL INCL O&P	
400	3180	Household, 60 cup	1 Plum	14	.571	Ea.	154	19.70		173.70	199	400
500	0010	**HYDRANTS**										500
	0050	Wall type, moderate climate, bronze, encased										
	0200	3/4" IPS connection	1 Plum	16	.500	Ea.	226	17.25		243.25	275	
	0300	1" IPS connection	"	14	.571		226	19.70		245.70	279	
	0500	Anti-siphon type					30.50			30.50	33.50	
	1000	Non-freeze, bronze, exposed										
	1100	3/4" IPS connection, 4" to 9" thick wall	1 Plum	14	.571	Ea.	153	19.70		172.70	199	
	1120	10" to 14" thick wall		12	.667		166	23		189	218	
	1140	15" to 19" thick wall		12	.667		185	23		208	239	
	1160	20" to 24" thick wall		10	.800		201	27.50		228.50	263	
	1200	For 1" IPS connection, add					15%	10%				
	1240	For 3/4" adapter type vacuum breaker, add				Ea.	19.25			19.25	21	
	1280	For anti-siphon type				"	40.50			40.50	44.50	
	2000	Non-freeze bronze, encased										
	2100	3/4" IPS connection, 5" to 9" thick wall	1 Plum	14	.571	Ea.	264	19.70		283.70	320	
	2120	10" to 14" thick wall		12	.667		277	23		300	340	
	2140	15" to 19" thick wall		12	.667		297	23		320	360	
	2160	20" to 24" thick wall		10	.800		310	27.50		337.50	380	
	2200	For 1" IPS connection, add					10%	10%				
	2280	Anti-siphon type				Ea.	124			124	136	
	3000	Ground box type, bronze frame, 3/4" IPS connection										
	3080	Non-freeze, all bronze, polished face, set flush										
	3100	2 feet depth of bury	1 Plum	8	1	Ea.	282	34.50		316.50	360	
	3120	3 feet depth of bury		8	1		305	34.50		339.50	385	
	3140	4 feet depth of bury		8	1		325	34.50		359.50	405	
	3160	5 feet depth of bury		7	1.143		350	39.50		389.50	440	
	3180	6 feet depth of bury		7	1.143		370	39.50		409.50	465	
	3200	7 feet depth of bury		6	1.333		390	46		436	500	
	3220	8 feet depth of bury		5	1.600		415	55		470	540	
	3240	9 feet depth of bury		4	2		435	69		504	585	
	3260	10 feet depth of bury		4	2		455	69		524	605	
	3400	For 1" IPS connection, add					15%	10%				
	3450	For 1-1/4" IPS connection, add					325%	14%				
	3500	For 1-1/2" connection, add					370%	18%				
	3550	For 2" connection, add					445%	24%				
	3600	For tapped drain port in box, add					25.50			25.50	28.50	
	4000	Non-freeze, CI body, bronze frame & scoriated cover										
	4010	with hose storage										
	4100	2 feet depth of bury	1 Plum	7	1.143	Ea.	535	39.50		574.50	645	
	4120	3 feet depth of bury		7	1.143		555	39.50		594.50	670	
	4140	4 feet depth of bury		7	1.143		575	39.50		614.50	690	
	4160	5 feet depth of bury		6.50	1.231		585	42.50		627.50	710	
	4180	6 feet depth of bury		6	1.333		595	46		641	725	
	4200	7 feet depth of bury		5.50	1.455		615	50		665	755	
	4220	8 feet depth of bury		5	1.600		640	55		695	785	
	4240	9 feet depth of bury		4.50	1.778		660	61		721	820	
	4260	10 feet depth of bury		4	2		680	69		749	855	
	4280	For 1" IPS connection, add					117			117	129	
	4300	For tapped drain port in box, add					25.50			25.50	28.50	
	5000	Moderate climate, all bronze, polished face										
	5020	and scoriated cover, set flush										
	5100	3/4" IPS connection	1 Plum	16	.500	Ea.	201	17.25		218.25	247	
	5120	1" IPS connection	"	14	.571		200	19.70		219.70	250	
	5200	For tapped drain port in box, add					25.50			25.50	28.50	

15411 | Commercial/Indust Fixtures

			CREW	DAILY OUTPUT	LABOR-HOURS	UNIT	2001 BARE COSTS MAT.	LABOR	EQUIP.	TOTAL	TOTAL INCL O&P	
500	6000	Ground post type, all non-freeze, all bronze, aluminum casing										500
	6010	guard, exposed head, 3/4″ IPS connection										
	6100	2 feet depth of bury	1 Plum	8	1	Ea.	283	34.50		317.50	360	
	6120	3 feet depth of bury		8	1		305	34.50		339.50	385	
	6140	4 feet depth of bury		8	1		330	34.50		364.50	415	
	6160	5 feet depth of bury		7	1.143		355	39.50		394.50	450	
	6180	6 feet depth of bury		7	1.143		375	39.50		414.50	475	
	6200	7 feet depth of bury		6	1.333		405	46		451	515	
	6220	8 feet depth of bury		5	1.600		430	55		485	555	
	6240	9 feet depth of bury		4	2		455	69		524	605	
	6260	10 feet depth of bury		4	2		480	69		549	635	
	6300	For 1″ IPS connection, add					40%	10%				
	6350	For 1-1/4″ IPS connection, add					140%	14%				
	6400	For 1-1/2″ IPS connection, add					225%	18%				
	6450	For 2″ IPS connection, add					315%	24%				
700	0010	**URINALS** A8.1 -450										700
	0100	For automatic flush see 15410-300-0972										
	3000	Wall hung, vitreous china, with hanger & self-closing valve										
	3100	Siphon jet type	Q-1	3	5.333	Ea.	279	165		444	555	
	3120	Blowout type		3	5.333		320	165		485	600	
	3300	Rough-in, supply, waste & vent		2.83	5.654		80.50	175		255.50	355	
	5000	Stall type, vitreous china, includes valve		2.50	6.400		450	198		648	795	
	5100	3″ seam cover, add		12	1.333		118	41.50		159.50	193	
	5200	6″ seam cover, add		12	1.333		164	41.50		205.50	243	
	6980	Rough-in, supply, waste and vent R15100 -410		1.99	8.040		99	249		348	485	
	8000	Waterless (no flush) urinal										
	8010	Wall hung										
	8020	Standard unit	Q-1	21.30	.751	Ea.	480	23.50		503.50	560	
	8030	ADA compliant unit	″	21.30	.751		500	23.50		523.50	585	
	8070	For solid color, add					60			60	66	
	8080	For 2″ brass flange, (new const.), add	Q-1	96	.167		24	5.15		29.15	34.50	
	8090	Rough-in, supply, waste & vent	″	3.86	4.145		70	129		199	271	
	8100	Trap liquid										
	8110	1 quart				Ea.	16			16	17.60	
	8120	1 gallon				″	64			64	70.50	
800	0010	**WASH CENTER** Prefabricated,stainless steel, semirecessed										800
	0050	Lavatory, storage cabinet, mirror, light & switch, electric										
	0060	outlet, towel dispenser, waste receptacle & trim										
	0100	Foot water valve, cup & soap dispenser,16″ W x 54-3/4″ H	Q-1	8	2	Ea.	1,925	62		1,987	2,200	
	0200	Handicap, wrist blade handles, 17″ W x 66-1/2″ H		8	2		900	62		962	1,075	
	0220	20″ W x 67-3/8″ H		8	2		2,150	62		2,212	2,475	
	0300	Push button metering & thermostatic mixing valves										
	0320	Handicap 17″ W x 27-1/2″ H	Q-1	8	2	Ea.	1,075	62		1,137	1,300	
	0400	Rough-in, supply, waste and vent	″	2.10	7.619	″	50	236		286	410	
840	0010	**WASH FOUNTAINS** Rigging not included A8.1 -560										840
	1900	Group, foot control										
	2000	Precast terrazzo, circular, 36″ diam., 5 or 6 persons	Q-2	3	8	Ea.	2,275	257		2,532	2,900	
	2100	54″ diameter for 8 or 10 persons		2.50	9.600		2,750	310		3,060	3,500	
	2400	Semi-circular, 36″ diam. for 3 persons		3	8		2,050	257		2,307	2,650	
	2500	54″ diam. for 4 or 5 persons		2.50	9.600		2,525	310		2,835	3,250	
	2700	Quarter circle (corner), 54″ for 3 persons		3.50	6.857		2,525	220		2,745	3,100	
	3000	Stainless steel, circular, 36″ diameter		3.50	6.857		2,650	220		2,870	3,225	
	3100	54″ diameter		2.80	8.571		3,475	276		3,751	4,250	
	3400	Semi-circular, 36″ diameter		3.50	6.857		2,275	220		2,495	2,825	

15411 | Commercial/Indust Fixtures

	Line	Description	Crew	Daily Output	Labor-Hours	Unit	2001 Bare Costs Mat.	Labor	Equip.	Total	Total Incl O&P	
840	3500	54″ diameter (A8.1-560)	Q-2	2.80	8.571	Ea.	3,000	276		3,276	3,725	840
	5000	Thermoplastic, pre-assembled, circular, 36″ diameter		6	4		2,050	129		2,179	2,450	
	5100	54″ diameter		4	6		2,400	193		2,593	2,950	
	5400	Semi-circular, 36″ diameter		6	4		1,925	129		2,054	2,300	
	5600	54″ diameter		4	6		2,275	193		2,468	2,800	
	5610	Group, infrared control, barrier free										
	5614	Precast terrazzo										
	5620	Semi-circular 36″ diam. for 3 persons	Q-2	3	8	Ea.	3,000	257		3,257	3,700	
	5630	46″ diam. for 4 persons		2.80	8.571		3,350	276		3,626	4,125	
	5640	Circular, 54″ diam. for 8 persons, button control		2.50	9.600		5,150	310		5,460	6,150	
	5700	Rough-in, supply, waste and vent for above wash fountains	Q-1	1.82	8.791		82.50	273		355.50	500	
	6200	Duo for small washrooms, stainless steel		2	8		1,550	248		1,798	2,100	
	6400	Bowl with backsplash		2	8		1,325	248		1,573	1,825	
	6500	Rough-in, supply, waste & vent for duo fountains		2.02	7.921		46	246		292	420	
900	0010	**WATER CHILLERS REMOTE** 80° F inlet										900
	0100	Air cooled, 50° F outlet, 115V, 4.1 GPH	1 Plum	6	1.333	Ea.	520	46		566	640	
	0200	5.7 GPH		5.50	1.455		730	50		780	880	
	0300	8.0 GPH		5	1.600		560	55		615	700	
	0400	10.0 GPH		4.50	1.778		680	61		741	840	
	0500	13.4 GPH		4	2		1,025	69		1,094	1,225	
	0700	29 GPH	Q-1	5	3.200		1,250	99		1,349	1,525	
	1000	230V, 32 GPH	″	5	3.200		1,425	99		1,524	1,700	
	1200	For remote grill, add	1 Plum	16	.500		61.50	17.25		78.75	93.50	

15412 | Drinking Fountains

	Line	Description	Crew	Daily Output	Labor-Hours	Unit	2001 Bare Costs Mat.	Labor	Equip.	Total	Total Incl O&P	
200	0010	**DRINKING FOUNTAIN** For connection to cold water supply										200
	0800	For remote water chiller, see division 15411										
	1000	Wall mounted, non-recessed (R15100-410)										
	1200	Aluminum,										
	1280	Dual bubbler type (R15100-430)	1 Plum	3.20	2.500	Ea.	935	86		1,021	1,150	
	1400	Bronze, with no back		4	2		990	69		1,059	1,200	
	1600	Cast iron, enameled, low back, single bubbler (A8.1-420)		4	2		425	69		494	575	
	1640	Dual bubbler type		3.20	2.500		790	86		876	1,000	
	1680	Triple bubbler type		3.20	2.500		1,050	86		1,136	1,275	
	1800	Cast aluminum, enameled, for correctional institutions		4	2		595	69		664	760	
	2000	Fiberglass, 12″ back, single bubbler unit		4	2		530	69		599	690	
	2040	Dual bubbler		3.20	2.500		770	86		856	975	
	2080	Triple bubbler		3.20	2.500		995	86		1,081	1,225	
	2200	Polymarble, no back, single bubbler		4	2		475	69		544	630	
	2240	Dual bubbler		3.20	2.500		815	86		901	1,025	
	2280	Triple bubbler		3.20	2.500		1,075	86		1,161	1,325	
	2400	Precast stone, no back		4	2		480	69		549	635	
	2700	Stainless steel, single bubbler, no back		4	2		795	69		864	980	
	2740	With back		4	2		945	69		1,014	1,150	
	2780	Dual handle & wheelchair projection type		4	2		495	69		564	650	
	2820	Dual level for handicapped type		3.20	2.500		1,000	86		1,086	1,225	
	2840	Vandal resistant type		4	2		420	69		489	570	
	3300	Vitreous china										
	3340	7″ back	1 Plum	4	2	Ea.	405	69		474	550	
	3940	For vandal-resistant bottom plate, add					50.50			50.50	56	
	3960	For freeze-proof valve system, add	1 Plum	2	4		400	138		538	650	
	3980	For rough-in, supply and waste, add	″	2.21	3.620		35	125		160	227	
	4000	Wall mounted, semi-recessed										
	4200	Poly-marble, single bubbler	1 Plum	4	2	Ea.	645	69		714	815	
	4600	Stainless steel, satin finish, single bubbler		4	2		755	69		824	935	

	15412 | Drinking Fountains			Crew	Daily Output	Labor-Hours	Unit	2001 Bare Costs Mat.	Labor	Equip.	Total	Total Incl O&P	
200	4900	Vitreous china, single bubbler	R15100-410	1 Plum	4	2	Ea.	435	69		504	580	200
	5980	For rough-in, supply and waste, add			1.83	4.372		35	151		186	266	
	6000	Wall mounted, fully recessed	R15100-430										
	6400	Poly-marble, single bubbler		1 Plum	4	2	Ea.	725	69		794	905	
	6440	For water glass filler, add	A8.1-420					123			123	135	
	6800	Stainless steel, single bubbler		1 Plum	4	2		635	69		704	800	
	6900	Fountain and cuspidor combination			2	4		1,300	138		1,438	1,625	
	7560	For freeze-proof valve system, add			2	4		495	138		633	755	
	7580	For rough-in, supply and waste, add			1.83	4.372		35	151		186	266	
	7600	Floor mounted, pedestal type											
	7700	Aluminum, architectural style, C.I. base		1 Plum	2	4	Ea.	440	138		578	690	
	7780	Wheelchair handicap unit			2	4		1,025	138		1,163	1,350	
	8000	Bronze, architectural style			2	4		1,125	138		1,263	1,450	
	8040	Enameled steel cylindrical column style			2	4		955	138		1,093	1,250	
	8200	Precast stone/concrete, cylindrical column			1	8		735	276		1,011	1,225	
	8240	Wheelchair handicap unit			1	8		1,075	276		1,351	1,600	
	8400	Stainless steel, architectural style			2	4		1,125	138		1,263	1,450	
	8600	Enameled iron, heavy duty service, 2 bubblers			2	4		1,050	138		1,188	1,350	
	8660	4 bubblers			2	4		1,125	138		1,263	1,425	
	8880	For freeze-proof valve system, add			2	4		320	138		458	560	
	8900	For rough-in, supply and waste, add			1.83	4.372		35	151		186	266	
	9100	Deck mounted											
	9500	Stainless steel, circular receptor		1 Plum	4	2	Ea.	281	69		350	415	
	9540	14" x 9" receptor			4	2		240	69		309	370	
	9580	25" x 17" deep receptor, with water glass filler			3	2.667		266	92		358	430	
	9760	White enameled steel, 14" x 9" receptor			4	2		240	69		309	370	
	9860	White enameled cast iron, 24" x 16" receptor			3	2.667		230	92		322	390	
	9980	For rough-in, supply and waste, add			1.83	4.372		35	151		186	266	

	15413 | Electric Water Coolers			Crew	Daily Output	Labor-Hours	Unit	Mat.	Labor	Equip.	Total	Total Incl O&P	
900	0010	**WATER COOLER**	A8.1-460										900
	0030	See line 15413-900-9800 for rough-in, waste & vent											
	0040	for all water coolers	R15100-430										
	0100	Wall mounted, non-recessed											
	0140	4 GPH		Q-1	4	4	Ea.	475	124		599	705	
	0160	8 GPH, barrier free, sensor operated			4	4		620	124		744	870	
	0180	8.2 GPH			4	4		540	124		664	775	
	0220	14.3 GPH			4	4		575	124		699	815	
	0600	For hot and cold water, add						138			138	151	
	0640	For stainless steel cabinet, add						67			67	74	
	1000	Dual height, 8.2 GPH		Q-1	3.80	4.211		780	131		911	1,050	
	1040	14.3 GPH		"	3.80	4.211		815	131		946	1,100	
	1240	For stainless steel cabinet, add						112			112	123	
	2600	Wheelchair type, 8 GPH		Q-1	4	4		1,375	124		1,499	1,700	
	3000	Simulated recessed, 8 GPH			4	4		690	124		814	945	
	3040	11.5 GPH			4	4		735	124		859	990	
	3200	For glass filler, add						73.50			73.50	81	
	3240	For stainless steel cabinet, add						48			48	53	
	3300	Semi-recessed, 8.1 GPH		Q-1	4	4		785	124		909	1,050	
	3320	12 GPH		"	4	4		830	124		954	1,100	
	3340	For glass filler, add						73.50			73.50	81	
	3360	For stainless steel cabinet, add						48			48	53	
	3400	Full recessed, stainless steel, 8 GPH		Q-1	3.50	4.571		1,200	142		1,342	1,550	
	3420	11.5 GPH		"	3.50	4.571		1,250	142		1,392	1,600	
	3460	For glass filler, add						73.50			73.50	81	
	3600	For mounting can only						147			147	162	

		15413 | Electric Water Coolers	CREW	DAILY OUTPUT	LABOR-HOURS	UNIT	2001 BARE COSTS MAT.	LABOR	EQUIP.	TOTAL	TOTAL INCL O&P
900	4600	Floor mounted, flush-to-wall (A8.1-460)									
	4640	4 GPH	1 Plum	3	2.667	Ea.	480	92		572	670
	4680	8.2 GPH (R15100-430)		3	2.667		540	92		632	730
	4720	14.3 GPH		3	2.667		565	92		657	765
	4960	For hot and cold water, add					250			250	275
	4980	For stainless steel cabinet, add					96			96	106
	5000	Dual height, 8.2 GPH	1 Plum	2	4		845	138		983	1,150
	5040	14.3 GPH	"	2	4		860	138		998	1,150
	5120	For stainless steel cabinet, add					144			144	158
	5600	Explosion Proof, 16 GPH	1 Plum	3	2.667		1,300	92		1,392	1,575
	5640	For stainless steel cabinet, add					96			96	106
	6000	Refrigerator Compartment Type, 4.5 GPH	1 Plum	3	2.667		1,050	92		1,142	1,325
	6600	Bottle Supply Type, 1.0 GPH		4	2		281	69		350	415
	6640	Hot and cold, 1.0 GPH		4	2		380	69		449	525
	9800	For supply, waste & vent, all coolers		2.21	3.620		35	125		160	227

		15414 | Emergency Fixtures	CREW	DAILY OUTPUT	LABOR-HOURS	UNIT	MAT.	LABOR	EQUIP.	TOTAL	TOTAL INCL O&P
200	0010	**INDUSTRIAL SAFETY FIXTURES** Rough-in not included									
	1000	Eye wash fountain									
	1400	Plastic bowl, pedestal mounted	Q-1	4	4	Ea.	182	124		306	390
	1600	Unmounted		4	4		131	124		255	330
	1800	Wall mounted		4	4		134	124		258	335
	2000	Stainless steel, pedestal mounted		4	4		273	124		397	485
	2200	Unmounted		4	4		182	124		306	390
	2400	Wall mounted		4	4		174	124		298	380
	3000	Eye wash, portable, self-contained					375			375	410
	4000	Eye and face wash, combination fountain									
	4200	Stainless steel, pedestal mounted	Q-1	4	4	Ea.	390	124		514	615
	4400	Unmounted		4	4		294	124		418	510
	4600	Wall mounted		4	4		300	124		424	515
	5000	Shower, single head, drench, ball valve, pull, freestanding		4	4		255	124		379	465
	5200	Horizontal or vertical supply		4	4		232	124		356	445
	6000	Multi-nozzle, eye/face wash combination		4	4		480	124		604	715
	6400	Multi-nozzle, 12 spray, shower only		4	4		1,100	124		1,224	1,375
	6600	For freeze-proof, add		6	2.667		268	82.50		350.50	420
	8000	Walk-thru decontamination with eye-face wash		2	8		1,725	248		1,973	2,275
	8200	For freeze proof, add		4	4		440	124		564	670

		15417 | Institutional Fixtures	CREW	DAILY OUTPUT	LABOR-HOURS	UNIT	MAT.	LABOR	EQUIP.	TOTAL	TOTAL INCL O&P
700	0010	**PRISON/INSTITUTION FIXTURES,** Stainless steel									
	1000	Lavatory, wall hung, push button filler valve									
	1100	Rectangular bowl	Q-1	8	2	Ea.	715	62		777	880
	1200	Oval bowl		8	2		710	62		772	875
	1240	Oval bowl, corner mount		8	2		820	62		882	995
	1300	For lavatory rough-in, supply, waste and vent		1.50	10.667		64.50	330		394.50	570
	1700	Service sink, with soap dish									
	1740	24" x 19" size	Q-1	3	5.333	Ea.	1,300	165		1,465	1,675
	1790	For sink rough-in, supply, waste and vent	"	.89	17.978	"	130	555		685	985
	1800	Shower cabinet, unitized									
	1840	36" x 36" x 88"	Q-1	2.20	7.273	Ea.	3,200	225		3,425	3,850
	1900	Shower package for built-in									
	1940	Hot & cold valves, recessed soap dish	Q-1	6	2.667	Ea.	470	82.50		552.50	640
	2000	Urinal, back supply and flush									
	2200	Wall hung	Q-1	4	4	Ea.	1,625	124		1,749	1,950
	2240	Stall		2.50	6.400		1,175	198		1,373	1,600

	15417 | Institutional Fixtures	CREW	DAILY OUTPUT	LABOR-HOURS	UNIT	2001 BARE COSTS MAT.	LABOR	EQUIP.	TOTAL	TOTAL INCL O&P	
700	2300 For urinal rough-in, supply, waste and vent	Q-1	1.49	10.738	Ea.	96	335		431	610	700
	3000 Water closet, integral seat, back supply and flush										
	3300 Wall hung, wall outlet	Q-1	5.80	2.759	Ea.	730	85.50		815.50	935	
	3400 Floor mount, wall outlet		5.80	2.759		835	85.50		920.50	1,050	
	3440 Floor mount, floor outlet		5.80	2.759		915	85.50		1,000.50	1,125	
	3480 For recessed tissue holder, add					64			64	70.50	
	3500 For water closet rough-in, supply, waste and vent	Q-1	1.19	13.445		119	415		534	760	
	5000 Water closet and lavatory units, push button filler valves,										
	5010 soap & paper holders, seat										
	5300 Wall hung	Q-1	5	3.200	Ea.	1,650	99		1,749	1,975	
	5400 Floor mount		5	3.200		1,650	99		1,749	1,975	
	6300 For unit rough-in, supply, waste and vent		1	16		131	495		626	895	

	15418 | Resi/Comm/Industrial Fixtures	CREW	DAILY OUTPUT	LABOR-HOURS	UNIT	MAT.	LABOR	EQUIP.	TOTAL	TOTAL INCL O&P	
100	0010 BATHS (A8.1 -410)										100
	0100 Tubs, recessed porcelain enamel on cast iron, with trim										
	0140 42" x 37"	Q-1	5	3.200	Ea.	710	99		809	930	
	0180 48" x 42"		4	4		1,050	124		1,174	1,325	
	0220 72" x 36"		3	5.333		1,175	165		1,340	1,550	
	0300 Mat bottom, 4' long		5.50	2.909		910	90		1,000	1,125	
	0340 4'-6" long		5	3.200		615	99		714	825	
	0380 5' long		4.40	3.636		360	113		473	565	
	0420 5'-6" long		4	4		870	124		994	1,150	
	0480 Above floor drain, 5' long		4	4		585	124		709	825	
	0560 Corner 48" x 44"		4.40	3.636		1,250	113		1,363	1,575	
	0750 For color, add					25%					
	0760 For designer colors & trim, add					60%					
	2000 Enameled formed steel, 4'-6" long	Q-1	5.80	2.759	Ea.	277	85.50		362.50	435	
	2200 5' long		5.50	2.909		261	90		351	425	
	2300 Above floor drain, 5' long		5.50	2.909		242	90		332	400	
	2350 For color, add					10%					
	4000 Soaking, acrylic with pop-up drain, 60" x 32" x 21" deep	Q-1	5.50	2.909	Ea.	730	90		820	940	
	4100 60" x 48" x 18-1/2" deep		5	3.200		590	99		689	800	
	4200 72" x 45" x 21-1/2" deep		4.80	3.333		775	103		878	1,000	
	4310 For color, add					5%					
	4311 For designer colors & trim, add					20%					
	4600 Module tub & showerwall surround, molded fiberglass										
	4610 5' long x 34" wide x 76" high	Q-1	4	4	Ea.	610	124		734	860	
	4620 For color add					10%					
	4621 For designer colors and trim add					25%					
	4750 Handicap with 1-1/2" OD grab bar, antiskid bottom										
	4760 60" x 32-3/4" x 72" high	Q-1	4	4	Ea.	655	124		779	905	
	4770 60" x 30" x 71" high with molded seat	"	3.50	4.571	"	815	142		957	1,125	
	5000 Hospital type, with trim (see 15410-300)										
	5050 Bathing pool, porcelain enamel on cast iron, grab bars										
	5060 pop-up drain, 72" x 36"	Q-1	3	5.333	Ea.	2,800	165		2,965	3,325	
	5100 Perineal (sitz), vitreous china		3	5.333		815	165		980	1,150	
	5120 Pedestal, vitreous china		8	2		127	62		189	234	
	5180 Pier tub, porcelain enamel on cast iron, 66-3/4" x 30"		3	5.333		2,025	165		2,190	2,475	
	5200 Base, porcelain enamel on cast iron		8	2		955	62		1,017	1,150	
	5300 Whirlpool, porcelain enamel on cast iron, 72" x 36"		1	16		2,700	495		3,195	3,725	
	5310 For color add					5%					
	5311 For designer colors and trim add					15%					
	6000 Whirlpool, bath with vented overflow, molded fiberglass										
	6100 66" x 48" x 24"	Q-1	1	16	Ea.	2,050	495		2,545	3,000	
	6400 72" x 36" x 24"		1	16		2,000	495		2,495	2,950	

	15418 | Resi/Comm/Industrial Fixtures		Ref.	Crew	Daily Output	Labor-Hours	Unit	2001 Bare Costs Mat.	Labor	Equip.	Total	Total Incl O&P	
100	6500	60″ x 30″ x 21″	A8.1 -410	Q-1	1	16	Ea.	1,750	495		2,245	2,675	100
	6600	72″ x 42″ x 22″			1	16		2,825	495		3,320	3,850	
	6700	83″ x 65″			.30	53.333		3,850	1,650		5,500	6,750	
	6710	For color add						10%					
	6711	For designer colors and trim add						25%					
	7000	Redwood hot tub system											
	7050	4′ diameter x 4′ deep		Q-1	1	16	Ea.	1,050	495		1,545	1,900	
	7100	5′ diameter x 4′ deep			1	16		1,425	495		1,920	2,300	
	7150	6′ diameter x 4′ deep			.80	20		1,725	620		2,345	2,825	
	7200	8′ diameter x 4′ deep			.80	20		2,425	620		3,045	3,600	
	9600	Rough-in, supply, waste and vent, for all above tubs, add			2.07	7.729		101	240		341	470	
200	0010	**BIDET**											200
	0180	Vitreous china, with trim on fixture		Q-1	5	3.200	Ea.	505	99		604	705	
	0200	With trim for wall mounting		"	5	3.200		385	99		484	575	
	9590	For color add						40%					
	9591	For designer colors and trim add						50%					
	9600	For rough-in, supply, waste and vent, add		Q-1	1.78	8.989		89	279		368	520	
400	0010	**LAUNDRY SINKS** With trim	A8.1 -432										400
	0020	Porcelain enamel on cast iron, black iron frame											
	0050	24″ x 20″, single compartment		Q-1	6	2.667	Ea.	340	82.50		422.50	495	
	0100	24″ x 23″, single compartment			6	2.667		365	82.50		447.50	530	
	0200	48″ x 20″, double compartment			5	3.200		640	99		739	850	
	2000	Molded stone, on wall hanger or legs											
	2020	22″ x 23″, single compartment		Q-1	6	2.667	Ea.	117	82.50		199.50	254	
	2100	45″ x 21″, double compartment		"	5	3.200	"	243	99		342	415	
	3000	Plastic, on wall hanger or legs											
	3020	18″ x 23″, single compartment		Q-1	6.50	2.462	Ea.	82	76.50		158.50	206	
	3100	20″ x 24″, single compartment			6.50	2.462		108	76.50		184.50	234	
	3200	36″ x 23″, double compartment			5.50	2.909		127	90		217	276	
	3300	40″ x 24″, double compartment			5.50	2.909		187	90		277	340	
	5000	Stainless steel, counter top, 22″ x 17″ single compartment			6	2.667		270	82.50		352.50	420	
	5100	22″ x 22″, single compartment			6	2.667		340	82.50		422.50	500	
	5200	33″ x 22″, double compartment			5	3.200		315	99		414	495	
	9600	Rough-in, supply, waste and vent, for all laundry sinks			2.14	7.477		63	232		295	420	
450	0010	**LAVATORIES** With trim, white unless noted otherwise	A8.1 -433										450
	0500	Vanity top, porcelain enamel on cast iron											
	0600	20″ x 18″	R15100 -410	Q-1	6.40	2.500	Ea.	156	77.50		233.50	289	
	0640	33″ x 19″ oval			6.40	2.500		335	77.50		412.50	485	
	0680	20″ x 17″ oval			6.40	2.500		156	77.50		233.50	289	
	0720	19″ round			6.40	2.500		144	77.50		221.50	276	
	0760	20″ x 12″ triangular bowl			6.40	2.500		162	77.50		239.50	296	
	0860	For color, add						25%					
	0861	For designer colors and trim, add						70%					
	1000	Cultured marble, 19″ x 17″, single bowl		Q-1	6.40	2.500	Ea.	101	77.50		178.50	228	
	1040	25″ x 19″, single bowl			6.40	2.500		128	77.50		205.50	258	
	1080	31″ x 19″, single bowl			6.40	2.500		142	77.50		219.50	273	
	1120	25″ x 22″, single bowl			6.40	2.500		137	77.50		214.50	268	
	1160	37″ x 22″, single bowl			6.40	2.500		168	77.50		245.50	300	
	1200	49″ x 22″, single bowl			6.40	2.500		208	77.50		285.50	345	
	1580	For color, same price											
	1900	Stainless steel, self-rimming, 25″ x 22″, single bowl, ledge		Q-1	6.40	2.500	Ea.	162	77.50		239.50	295	
	1960	17″ x 22″, single bowl			6.40	2.500		157	77.50		234.50	290	
	2040	18-3/4″ round			6.40	2.500		325	77.50		402.50	470	
	2600	Steel, enameled, 20″ x 17″, single bowl			5.80	2.759		84.50	85.50		170	222	

	15418 | Resi/Comm/Industrial Fixtures		CREW	DAILY OUTPUT	LABOR-HOURS	UNIT	2001 BARE COSTS MAT.	LABOR	EQUIP.	TOTAL	TOTAL INCL O&P	
450	2660	19″ round (A8.1 -433)	Q-1	5.80	2.759	Ea.	82.50	85.50		168	220	450
	2720	18″ round		5.80	2.759		71.50	85.50		157	208	
	2860	For color, add (R15100 -410)					10%					
	2861	For designer colors and trim, add					20%					
	2900	Vitreous china, 20″ x 16″, single bowl	Q-1	5.40	2.963	Ea.	168	92		260	325	
	2960	20″ x 17″, single bowl		5.40	2.963		102	92		194	252	
	3020	19″ round, single bowl		5.40	2.963		101	92		193	250	
	3080	19″ x 16″, single bowl		5.40	2.963		168	92		260	325	
	3140	17″ x 14″, single bowl		5.40	2.963		132	92		224	284	
	3200	22″ x 13″, single bowl		5.40	2.963		176	92		268	335	
	3560	For color, add					50%					
	3561	For designer colors and trim, add					100%					
	3580	Rough-in, supply, waste and vent for all above lavatories	Q-1	2.30	6.957	Ea.	55.50	216		271.50	385	
	4000	Wall hung										
	4040	Porcelain enamel on cast iron, 16″ x 14″, single bowl	Q-1	8	2	Ea.	253	62		315	375	
	4060	18″ x 15″ single bowl		8	2		299	62		361	425	
	4120	19″ x 17″, single bowl		8	2		208	62		270	325	
	4180	20″ x 18″, single bowl		8	2		184	62		246	296	
	4240	22″ x 19″, single bowl		8	2		335	62		397	465	
	4580	For color, add					30%					
	4581	For designer colors and trim, add					75%					
	6000	Vitreous china, 18″ x 15″, single bowl with backsplash	Q-1	7	2.286	Ea.	136	71		207	256	
	6060	19″ x 17″, single bowl		7	2.286		120	71		191	238	
	6120	24″ x 20″, single bowl		7	2.286		236	71		307	365	
	6180	19″ x 19″, corner style		8	2		360	62		422	495	
	6210	27″ x 20″, wheelchair type		7	2.286		206	71		277	335	
	6500	For color, add					30%					
	6501	For designer colors and trim, add					50%					
	6700	Hospital type, without trim (see 15410-300)										
	6710	20″ x 18″, contoured splash shield	Q-1	8	2	Ea.	101	62		163	205	
	6720	26″ x 20″, patient,corner side deck & back		7	2.286		335	71		406	475	
	6730	28″ x 20″, surgeon, side decks		8	2		405	62		467	540	
	6740	28″ x 22″, surgeon scrub-up, deep bowl		8	2		555	62		617	705	
	6750	20″ x 27″, patient, wheelchair		7	2.286		267	71		338	400	
	6760	30″ x 22″, all purpose		7	2.286		525	71		596	685	
	6770	30″ x 22″, plaster work		7	2.286		525	71		596	685	
	6820	20″ x 24″ clinic service, liquid/solid waste		6	2.667		660	82.50		742.50	850	
	6960	Rough-in, supply, waste and vent for above lavatories		1.66	9.639		187	299		486	655	
500	0010	**SHOWERS** (A8.1 -440)										500
	1500	Stall, with drain only. Add for valve and door/curtain										
	1510	Baked enamel, molded stone receptor, 30″ square	Q-1	2	8	Ea.	340	248		588	750	
	1520	32″ square		2	8		315	248		563	720	
	1530	36″ square		2	8		410	248		658	825	
	1540	Terrazzo receptor, 32″ square		2	8		990	248		1,238	1,475	
	1560	36″ square		1.80	8.889		820	276		1,096	1,325	
	1580	36″ corner angle		1.80	8.889		750	276		1,026	1,250	
	1600	For color, add					10%					
	1601	For designer colors and trim, add					15%					
	1604	For thermostatic valve add				Ea.	226			226	249	
	3000	Fiberglass, one piece, with 3 walls, 32″ x 32″ square	Q-1	2.40	6.667		291	207		498	630	
	3100	36″ x 36″ square	″	2.40	6.667		335	207		542	675	
	3200	Handicap, 1-1/2″ O.D. grab bars, nonskid floor										
	3210	48″ x 34-1/2″ x 72″ corner seat	Q-1	2.40	6.667	Ea.	340	207		547	685	
	3220	60″ x 34-1/2″ x 72″ corner seat		2	8		565	248		813	995	
	3230	48″ x 34-1/2″ x 72″ fold up seat		2.20	7.273		535	225		760	925	
	3250	64″ x 65-3/4″ x 81-1/2″ fold. seat, whlchr.		1.80	8.889		2,675	276		2,951	3,375	

	15418 | Resi/Comm/Industrial Fixtures	CREW	DAILY OUTPUT	LABOR-HOURS	UNIT	2001 BARE COSTS MAT.	LABOR	EQUIP.	TOTAL	TOTAL INCL O&P	
500	3260 For thermostatic valve add (A8.1 -440)				Ea.	226			226	249	500
	4000 Polypropylene, with molded-stone floor, 30″ x 30″	Q-1	2	8		370	248		618	785	
	4100 32″ x 32″	″	2	8		360	248		608	770	
	4110 For thermostatic valve add					226			226	249	
	4200 Rough-in, supply, waste and vent for above showers	Q-1	2.05	7.805		58	242		300	430	
	4520 Showers, fiberglass receptor only	1 Plum	8	1		155	34.50		189.50	223	
	5000 Built-in, head, arm, 4 GPM valve		4	2		73.50	69		142.50	185	
	5200 Head, arm, by-pass, integral stops, handles		3.60	2.222		189	76.50		265.50	325	
	5500 Head, water economizer, 3.0 GPM		24	.333		52	11.50		63.50	74.50	
	5800 Mixing valve, built-in		6	1.333		95.50	46		141.50	175	
	5900 Exposed		6	1.333		415	46		461	530	
	5950 Module, handicap, SS panel, fixed & hand held head, control										
	5960 valves, grab bar, curtain & rod, folding seat	1 Plum	4	2	Ea.	1,325	69		1,394	1,550	
	6000 Group, w/pressure balancing valve, rough-in and rigging not included										
	6800 Column, 6 heads, no receptors, less partitions	Q-1	3	5.333	Ea.	1,450	165		1,615	1,825	
	6900 With stainless steel partitions		1	16		3,350	495		3,845	4,450	
	7600 5 heads, no receptors, less partitions		3	5.333		1,200	165		1,365	1,575	
	7620 4 heads (1 handicap) no receptors, less partitions		3	5.333		1,675	165		1,840	2,100	
	7700 With stainless steel partitions		1	16		1,200	495		1,695	2,075	
	8000 Wall, 2 heads, no receptors, less partitions		4	4		720	124		844	975	
	8100 With stainless steel partitions		2	8		1,675	248		1,923	2,225	
600	0010 **SINKS** With faucets and drain (A8.1 -431)										600
	0050 Laboratory sinks, corrosion resistant										
	1000 Polyethylene, single sink, bench mounted, with (A8.1 -434)										
	1020 plug & waste fitting with 1-1/2″ straight threads										
	1030 2 drainboards, backnut & strainer										
	1050 18-1/2″ x 15-1/2″ x 12-1/2″ sink, 54″ x 24″ O.D.	Q-1	3	5.333	Ea.	780	165		945	1,100	
	1100 Single drainboard, backnut & strainer										
	1130 18-1/2″ x 15-1/2″ x 12-1/2″ sink, 47″ x 24″ O.D.	Q-1	3	5.333	Ea.	815	165		980	1,150	
	1150 18-1/2″ x 15-1/2″ x 12-1/2″ sink, 70″ x 24″ O.D.	″	3	5.333	″	830	165		995	1,175	
	1290 Flanged 1-1/4″ wide, rectangular with strainer										
	1300 plug & waste fitting, 1-1/2″ straight threads										
	1320 12″ x 12″ x 8″ sink, 14-1/2″ x 14-1/2″ O.D.	Q-1	4	4	Ea.	260	124		384	475	
	1340 16″ x 16″ x 8″ sink, 18-1/2″ x 18-1/2″ O.D.		4	4		222	124		346	430	
	1360 21″ x 18″ x 10″ sink, 23-1/2″ x 20-1/2″ O.D.		4	4		270	124		394	485	
	1490 For rough-in, supply, waste & vent, add		2.02	7.921		68.50	246		314.50	445	
	1600 Polypropylene										
	1620 Cup sink, oval, integral strainers										
	1640 6″ x 3″ I.D., 7″ x 4″ O.D.	Q-1	6	2.667	Ea.	82.50	82.50		165	216	
	1660 9″ x 3″ I.D., 10″ x 4-1/2″ O.D.	″	6	2.667	″	93	82.50		175.50	227	
	1720 Overflow standpipe										
	1740 1-1/2″ diam. x 11″ long				Ea.	24.50			24.50	27	
	1980 For rough-in, supply, waste & vent, add	Q-1	1.70	9.412		70	292		362	515	
	2000 Kitchen, counter top style, P.E. on C.I., 24″ x 21″ single bowl		5.60	2.857		186	88.50		274.50	340	
	2100 31″ x 22″ single bowl		5.60	2.857		325	88.50		413.50	495	
	2200 32″ x 21″ double bowl		4.80	3.333		267	103		370	450	
	2300 42″ x 21″ double bowl		4.80	3.333		440	103		543	640	
	2310 For color, add					20%					
	2311 For designer colors and trim, add					50%					
	3000 Stainless steel, self rimming, 19″ x 18″ single bowl	Q-1	5.60	2.857	Ea.	280	88.50		368.50	445	
	3100 25″ x 22″ single bowl		5.60	2.857		310	88.50		398.50	475	
	3200 33″ x 22″ double bowl		4.80	3.333		445	103		548	645	
	3300 43″ x 22″ double bowl		4.80	3.333		510	103		613	715	
	3400 22″ x 43″ triple bowl		4.40	3.636		760	113		873	1,000	
	3500 Corner double bowl each 14″ x 16″		4.80	3.333		490	103		593	695	

Important: See the Reference Section for critical supporting data - Reference Nos., Crews, & City Cost Indexes

15418	Resi/Comm/Industrial Fixtures		CREW	DAILY OUTPUT	LABOR-HOURS	UNIT	2001 BARE COSTS MAT.	LABOR	EQUIP.	TOTAL	TOTAL INCL O&P	
600	4000	Steel, enameled, with ledge, 24" x 21" single bowl (A8.1-431)	Q-1	5.60	2.857	Ea.	99.50	88.50		188	243	600
	4100	32" x 21" double bowl		4.80	3.333		126	103		229	294	
	4960	For color sinks except stainless steel, add (A8.1-434)					10%					
	4961	For designer colors and trim add					20%					
	4980	For rough-in, supply, waste and vent, counter top sinks	Q-1	2.14	7.477		63	232		295	420	
	5000	Kitchen, raised deck, P.E. on C.I.										
	5100	32" x 21", dual level, double bowl	Q-1	2.60	6.154	Ea.	360	191		551	685	
	5200	42" x 21", double bowl & disposer well	"	2.20	7.273		535	225		760	930	
	5700	For color, add					20%					
	5701	For designer colors and trim add					50%					
	5790	For rough-in, supply, waste & vent, sinks	Q-1	1.85	8.649		63	268		331	475	
	6650	Service, floor, corner, P.E. on C.I., 28" x 28"	"	4.40	3.636		475	113		588	690	
	6750	Vinyl coated rim guard, add					61.50			61.50	68	
	6760	Mop sink, molded stone, 24" x 36"	1 Plum	3.33	2.402		187	83		270	330	
	6770	Mop sink, molded stone, 24" x 36", w/rim 3 sides	"	3.33	2.402		214	83		297	360	
	6790	For rough-in, supply, waste & vent, floor service sinks	Q-1	1.64	9.756		127	300		427	595	
	7000	Service, wall, P.E. on C.I., roll rim, 22" x 18"		4	4		405	124		529	630	
	7100	24" x 20"		4	4		445	124		569	675	
	7600	For stainless steel rim guard, add					79.50			79.50	87.50	
	7800	For stainless steel rim guard, front only, add					31			31	34	
	8600	Vitreous china, 22" x 20"	Q-1	4	4		380	124		504	605	
	8960	For stainless steel rim guard, front or side, add					31			31	34	
	8980	For rough-in, supply, waste & vent, wall service sinks	Q-1	1.30	12.308		245	380		625	845	
900	0010	**WATER CLOSETS** (A8.1-470)										900
	0020	For seats, see 15410-800										
	0030	For automatic flush, see 15410-300-0972 (A8.1-510)										
	0150	Tank type, vitreous china, incl. seat, supply pipe w/stop										
	0200	Wall hung, one piece (R15100-410)	Q-1	5.30	3.019	Ea.	425	93.50		518.50	610	
	0400	Two piece, close coupled		5.30	3.019		480	93.50		573.50	665	
	0960	For rough-in, supply, waste, vent and carrier		2.73	5.861		251	182		433	550	
	1000	Floor mounted, one piece		5.30	3.019		480	93.50		573.50	670	
	1020	One piece, low profile		5.30	3.019		535	93.50		628.50	730	
	1050	One piece combination		5.30	3.019		610	93.50		703.50	810	
	1100	Two piece, close coupled, water saver		5.30	3.019		133	93.50		226.50	288	
	1150	With wall outlet		5.30	3.019		440	93.50		533.50	625	
	1200	With 18" high bowl		5.30	3.019		355	93.50		448.50	530	
	1960	For color, add					30%					
	1961	For designer colors and trim, add					55%					
	1980	For rough-in, supply, waste and vent	Q-1	3.05	5.246	Ea.	114	163		277	370	
	3000	Bowl only, with flush valve, seat										
	3100	Wall hung	Q-1	5.80	2.759	Ea.	335	85.50		420.50	500	
	3150	Hospital type, slotted rim for bed pan										
	3160	Elongated bowl	Q-1	5.80	2.759	Ea.	335	85.50		420.50	495	
	3200	For rough-in, supply, waste and vent, single WC		2.56	6.250		258	194		452	575	
	3300	Floor mounted		5.80	2.759		300	85.50		385.50	460	
	3350	With wall outlet		5.80	2.759		450	85.50		535.50	625	
	3360	Hospital type, slotted rim for bed pan										
	3370	Elongated bowl	Q-1	5	3.200	Ea.	310	99		409	490	
	3380	Elongated bowl, 18" high		5	3.200		310	99		409	495	
	3390	Water closet bowl w/auto self flush, self clean, self sanitizing		4.70	3.404		1,500	106		1,606	1,800	
	3400	For rough-in, supply, waste and vent, single WC		2.84	5.634		120	175		295	395	
	3500	Gang side by side carrier system, rough-in, supply, waste & vent										
	3510	For single hook-up	Q-1	1.97	8.122	Ea.	305	252		557	715	
	3520	For each additional hook-up, add	"	2.14	7.477	"	280	232		512	660	
	3550	Gang back to back carrier system, rough-in, supply, waste & vent										

15418 | Resi/Comm/Industrial Fixtures

			Crew	Daily Output	Labor-Hours	Unit	2001 Bare Costs Mat.	Labor	Equip.	Total	Total Incl O&P	
900	3560	For pair hook-up (A8.1 -470)	Q-1	1.76	9.091	Pr.	490	282		772	965	900
	3570	For each additional pair hook-up, add	"	1.81	8.840	"	465	274		739	925	
	4000	Water conserving systems (A8.1 -510)										
	4900	2 quart flush, residential	Q-1	5.40	2.963	Ea.	450	92		542	635	
	4980	For rough-in, supply, waste and vent (R15100 -410)		1.94	8.247		130	256		386	530	
	5100	2 quart flush		4.60	3.478		470	108		578	680	
	5200	For remote valve, add		24	.667		150	20.50		170.50	197	
	5300	For residential air compressor		6	2.667		685	82.50		767.50	875	
	5400	For light industrial air compressor		4	4		1,125	124		1,249	1,425	
	5500	For heavy duty industrial air compressor		1	16		1,700	495		2,195	2,625	
	5600	For dual compressor alternator		20	.800		565	25		590	660	

15440 | Plumbing Pumps

			Crew	Daily Output	Labor-Hours	Unit	2001 Bare Costs Mat.	Labor	Equip.	Total	Total Incl O&P	
240	0010	**PUMPS, PRESSURE BOOSTER SYSTEM**										240
	0200	Pump system, with diaphragm tank, control, press. switch										
	0300	1 HP pump	Q-1	1.30	12.308	Ea.	3,050	380		3,430	3,925	
	0400	1-1/2 HP pump		1.25	12.800		3,075	395		3,470	3,975	
	0420	2 HP pump		1.20	13.333		3,150	415		3,565	4,075	
	0440	3 HP pump		1.10	14.545		3,200	450		3,650	4,175	
	0460	5 HP pump	Q-2	1.50	16		3,525	515		4,040	4,650	
	0480	7-1/2 HP pump		1.42	16.901		3,925	545		4,470	5,150	
	0500	10 HP pump		1.34	17.910		4,125	575		4,700	5,400	
	1000	Pump/ energy storage system, diaphragm tank, 3 HP pump										
	1100	motor, PRV, switch, gauge, control center, flow switch										
	1200	125 lb. working pressure	Q-2	.70	34.286	Ea.	8,850	1,100		9,950	11,400	
	1300	250 lb. working pressure	"	.64	37.500	"	9,675	1,200		10,875	12,400	
400	0010	**PUMPS, GRINDER SYSTEM** Complete, incl. check valve, tank, std.										400
	0020	controls incl. alarm/disconnect panel w/wire. Excavation not included										
	0260	Simplex, 9 GPM at 60 PSIG, 70 gal. tank				Ea.	2,150			2,150	2,375	
	0300	For manway, 26″ I.D., 18″ high, add					405			405	445	
	0340	26″ I.D., 36″ high, add					475			475	525	
	0380	43″ I.D., 4′ high, add					525			525	575	
	0600	Simplex, 9 GPM at 60 PSIG, 150 gal. tank					2,450			2,450	2,700	
	0660	For manway, add										
	0700	26″ I.D., 36″ high, add				Ea.	490			490	540	
	0740	26″ I.D., 4′ high, add					570			570	625	
	2000	Duplex, 18 GPM at 60 PSIG, 150 gal. tank					4,050			4,050	4,450	
	2060	For manway 43″ I.D., 4′ high, add					1,300			1,300	1,425	
	2400	For core only					1,625			1,625	1,775	
	2410	For salt water core, add					156			156	172	
	2600	For addit. manway 5′ to 10′ high, 60 gal tank, per ft., add				V.L.F.	145			145	159	
	2620	For manways 5′ to 10′, 120 gal tank, add				Ea.	158			158	173	
800	0010	**PUMPS, SEWAGE EJECTOR** With operating and level controls										800
	0100	Simplex system incl. tank, cover, pump 15′ head										
	0500	37 gal PE tank, 12 GPM, 1/2 HP, 2″ discharge	Q-1	3.20	5	Ea.	365	155		520	635	
	0510	3″ discharge		3.10	5.161		390	160		550	670	
	0530	87 GPM, .7 HP, 2″ discharge		3.20	5		560	155		715	850	
	0540	3″ discharge		3.10	5.161		610	160		770	910	
	0600	45 gal. coated stl tank, 12 GPM, 1/2 HP, 2″ discharge		3	5.333		645	165		810	960	
	0610	3″ discharge		2.90	5.517		675	171		846	1,000	
	0630	87 GPM, .7 HP, 2″ discharge		3	5.333		830	165		995	1,150	
	0640	3″ discharge		2.90	5.517		875	171		1,046	1,225	
	0660	134 GPM, 1 HP, 2″ discharge		2.80	5.714		895	177		1,072	1,250	
	0680	3″ discharge		2.70	5.926		945	184		1,129	1,325	

Important: See the Reference Section for critical supporting data - Reference Nos., Crews, & City Cost Indexes

	15440 | Plumbing Pumps		CREW	DAILY OUTPUT	LABOR-HOURS	UNIT	2001 BARE COSTS MAT.	LABOR	EQUIP.	TOTAL	TOTAL INCL O&P	
800	0700	70 gal. PE tank, 12 GPM, 1/2 HP, 2" discharge	Q-1	2.60	6.154	Ea.	730	191		921	1,100	800
	0710	3" discharge		2.40	6.667		780	207		987	1,175	
	0730	87 GPM, 0.7 HP, 2" discharge		2.50	6.400		980	198		1,178	1,375	
	0740	3" discharge		2.30	6.957		1,000	216		1,216	1,425	
	0760	134 GPM, 1 HP, 2" discharge		2.20	7.273		1,025	225		1,250	1,475	
	0770	3" discharge		2	8		1,100	248		1,348	1,575	
	0800	75 gal. coated stl. tank, 12 GPM, 1/2 HP, 2 " discharge		2.40	6.667		735	207		942	1,125	
	0810	3" discharge		2.20	7.273		765	225		990	1,175	
	0830	87 GPM, .7 HP, 2" discharge		2.30	6.957		925	216		1,141	1,350	
	0840	3" discharge		2.10	7.619		970	236		1,206	1,425	
	0860	134 GPM, 1 HP, 2" discharge		2	8		990	248		1,238	1,475	
	0880	3" discharge	↓	1.80	8.889	↓	1,050	276		1,326	1,575	
	1040	Duplex system incl. tank, covers, pumps										
	1060	110 gal. fiberglass tank, 24 GPM, 1/2 HP, 2" discharge	Q-1	1.60	10	Ea.	1,425	310		1,735	2,050	
	1080	3" discharge		1.40	11.429		1,475	355		1,830	2,150	
	1100	174 GPM, .7 HP, 2" discharge		1.50	10.667		1,875	330		2,205	2,550	
	1120	3" discharge		1.30	12.308		1,950	380		2,330	2,700	
	1140	268 GPM, 1 HP, 2" discharge		1.20	13.333		2,025	415		2,440	2,850	
	1160	3" discharge	↓	1	16		2,100	495		2,595	3,050	
	1260	135 gal. coated stl. tank, 24 GPM, 1/2 HP, 2" discharge	Q-2	1.70	14.118		1,450	455		1,905	2,275	
	2000	3" discharge		1.60	15		1,525	480		2,005	2,400	
	2640	174 GPM, .7 HP, 2" discharge		1.60	15		1,900	480		2,380	2,800	
	2660	3" discharge		1.50	16		2,000	515		2,515	2,975	
	2700	268 GPM, 1 HP, 2" discharge		1.30	18.462		2,050	595		2,645	3,150	
	3040	3" discharge		1.10	21.818		2,150	700		2,850	3,425	
	3060	275 gal. coated stl. tank, 24 GPM, 1/2 HP, 2" discharge		1.50	16		1,800	515		2,315	2,750	
	3080	3" discharge		1.40	17.143		1,825	550		2,375	2,825	
	3100	174 GPM, .7 HP, 2" discharge		1.40	17.143		2,325	550		2,875	3,375	
	3120	3" discharge		1.30	18.462		2,350	595		2,945	3,475	
	3140	268 GPM, 1 HP, 2" discharge		1.10	21.818		2,550	700		3,250	3,850	
	3160	3" discharge	↓	.90	26.667	↓	2,650	855		3,505	4,200	
	3260	Pump system accessories, add										
	3300	Alarm horn and lights, 115V mercury switch	Q-1	8	2	Ea.	82	62		144	184	
	3340	Switch, mag. contactor, alarm bell, light, 3 level control		5	3.200		390	99		489	580	
	3380	Alternator, mercury switch activated	↓	4	4	↓	785	124		909	1,050	
900	0010	**PUMPS, PEDESTAL SUMP** With float control										900
	0400	Molded PVC base, 21 GPM at 15' head, 1/3 HP	1 Plum	5	1.600	Ea.	86.50	55		141.50	178	
	0800	Iron base, 21 GPM at 15' head, 1/3 HP		5	1.600		110	55		165	204	
	1200	Solid brass, 21 GPM at 15' head, 1/3 HP	↓	5	1.600	↓	179	55		234	280	
940	0010	**PUMPS, SUBMERSIBLE** Sump										940
	7000	Sump pump, automatic										
	7100	Plastic, 1-1/4" discharge, 1/4 HP	1 Plum	6	1.333	Ea.	105	46		151	186	
	7140	1/3 HP		5	1.600		127	55		182	223	
	7160	1/2 HP		5	1.600		153	55		208	251	
	7180	1-1/2" discharge, 1/2 HP		4	2		171	69		240	292	
	7500	Cast iron, 1-1/4" discharge, 1/4 HP		6	1.333		125	46		171	207	
	7540	1/3 HP		6	1.333		140	46		186	224	
	7560	1/2 HP	↓	5	1.600	↓	176	55		231	276	

	15450 | Potable Water Tanks		CREW	DAILY OUTPUT	LABOR-HOURS	UNIT	MAT.	LABOR	EQUIP.	TOTAL	TOTAL INCL O&P	
900	0010	**WATER HEATER STORAGE TANKS** 125 psi ASME										900
	2000	Galvanized steel, 15 gal., 14" diam., 26" LOA	1 Plum	12	.667	Ea.	490	23		513	570	
	2060	30 gal., 14" diam. x 49" LOA		11	.727		580	25		605	680	
	2080	80 gal., 20" diam. x 64" LOA	↓	9	.889	↓	900	30.50		930.50	1,025	

15450 | Potable Water Tanks

		15450 Potable Water Tanks	CREW	DAILY OUTPUT	LABOR-HOURS	UNIT	2001 BARE COSTS MAT.	LABOR	EQUIP.	TOTAL	TOTAL INCL O&P	
900	2100	135 gal., 24″ diam. x 75″ LOA	1 Plum	6	1.333	Ea.	1,325	46		1,371	1,550	900
	2120	240 gal., 30″ diam. x 86″ LOA		4	2		2,275	69		2,344	2,600	
	2140	300 gal., 36″ diam. x 76″ LOA		3	2.667		3,700	92		3,792	4,225	
	2160	400 gal., 36″ diam. x 100″ LOA	Q-1	4	4		4,550	124		4,674	5,200	
	2180	500 gal., 36″ diam., x 126″ LOA	″	3	5.333		5,625	165		5,790	6,450	
	3000	Glass lined, P.E., 80 gal., 20″ diam. x 60″ LOA	1 Plum	9	.889		1,600	30.50		1,630.50	1,800	
	3060	140 gal., 24″ diam. x 80″ LOA		6	1.333		2,275	46		2,321	2,575	
	3080	225 gal., 30″ diam. x 78″ LOA		4	2		2,650	69		2,719	3,025	
	3100	325 gal., 36″ diam. x 81″ LOA		3	2.667		3,400	92		3,492	3,900	
	3120	460 gal., 42″ diam. x 84″ LOA	Q-1	4	4		3,575	124		3,699	4,125	
	3140	605 gal., 48″ diam. x 87″ LOA		3	5.333		6,850	165		7,015	7,775	
	3160	740 gal., 54″ diam. x 91″ LOA		3	5.333		7,700	165		7,865	8,725	
	3180	940 gal., 60″ diam. x 93″ LOA		2.50	6.400		8,900	198		9,098	10,100	
	3200	1330 gal., 66″ diam. x 107″ LOA		2	8		12,200	248		12,448	13,800	
	3220	1615 gal., 72″ diam. x 110″ LOA		1.50	10.667		13,900	330		14,230	15,700	
	3240	2285 gal., 84″ diam. x 128″ LOA		1	16		17,100	495		17,595	19,600	
	3260	3440 gal., 96″ diam. x 157″ LOA	Q-2	1.50	16		25,400	515		25,915	28,700	

15460 | Domestic Water Cond Equipment

		15460 Domestic Water Cond Equipment	CREW	DAILY OUTPUT	LABOR-HOURS	UNIT	MAT.	LABOR	EQUIP.	TOTAL	TOTAL INCL O&P	
900	0010	**WATER SOFTENER**										900
	5800	Softener systems, automatic, intermediate sizes										
	5820	available, may be used in multiples.										
	6000	Hardness capacity between regenerations and flow										
	6100	150,000 grains, 37 GPM cont., 51 GPM peak	Q-1	1.20	13.333	Ea.	3,575	415		3,990	4,550	
	6200	300,000 grains, 81 GPM cont., 113 GPM peak		1	16		5,300	495		5,795	6,575	
	6300	750,000 grains, 160 GPM cont., 230 GPM peak		.80	20		6,900	620		7,520	8,500	
	6400	900,000 grains, 185 GPM cont., 270 GPM peak		.70	22.857		11,100	710		11,810	13,300	
	8000	Water treatment, salts, 50 lb. bag										
	8020	Salt, water softener, bag, pelletized				Lb.	.12			.12	.13	
	8030	Salt, water softener, bag, crystal rock salt				″	.10			.10	.11	

15470 | Domstc Water Filt Equip

		15470 Domstc Water Filt Equip	CREW	DAILY OUTPUT	LABOR-HOURS	UNIT	MAT.	LABOR	EQUIP.	TOTAL	TOTAL INCL O&P	
400	0010	**WATER FILTERS** Purification and treatment										400
	1000	Cartridge style, dirt and rust type	1 Plum	12	.667	Ea.	27	23		50	64	
	1200	Replacement cartridge		32	.250		.83	8.60		9.43	13.90	
	1600	Taste and odor type		12	.667		26.50	23		49.50	64	
	1700	Replacement cartridge		32	.250		1.24	8.60		9.84	14.35	
	3000	Central unit, dirt/rust/odor/taste/scale		4	2		610	69		679	775	
	3100	Replacement cartridge, standard		20	.400		120	13.80		133.80	153	
	3600	Replacement cartridge, heavy duty		20	.400		131	13.80		144.80	165	
	8000	Commercial, fully automatic or push button automatic										
	8200	Iron removal, 660 GPH, 1″ pipe size	Q-1	1.50	10.667	Ea.	1,575	330		1,905	2,250	
	8240	1500 GPH, 1-1/4″ pipe size		1	16		2,675	495		3,170	3,675	
	8280	2340 GPH, 1-1/2″ pipe size		.80	20		2,925	620		3,545	4,150	
	8320	3420 GPH, 2″ pipe size		.60	26.667		5,400	825		6,225	7,200	
	8360	4620 GPH, 2-1/2″ pipe size		.50	32		8,575	990		9,565	11,000	
	8500	Neutralizer for acid water, 780 GPH, 1″ pipe size		1.50	10.667		1,525	330		1,855	2,175	
	8540	1140 GPH, 1-1/4″ pipe size		1	16		1,725	495		2,220	2,650	
	8580	1740 GPH, 1-1/2″ pipe size		.80	20		2,500	620		3,120	3,675	
	8620	2520 GPH, 2″ pipe size		.60	26.667		3,250	825		4,075	4,825	
	8660	3480 GPH, 2-1/2″ pipe size		.50	32		5,400	990		6,390	7,450	
	8800	Sediment removal, 780 GPH, 1″ pipe size		1.50	10.667		1,450	330		1,780	2,100	
	8840	1140 GPH, 1-1/4″ pipe size		1	16		1,750	495		2,245	2,675	
	8880	1740 GPH, 1-1/2″ pipe size		.80	20		2,325	620		2,945	3,475	

Important: See the Reference Section for critical supporting data - Reference Nos., Crews, & City Cost Indexes

15470 | Domstc Water Filt Equip

			CREW	DAILY OUTPUT	LABOR-HOURS	UNIT	2001 BARE COSTS MAT.	LABOR	EQUIP.	TOTAL	TOTAL INCL O&P	
400	8920	2520 GPH, 2" pipe size	Q-1	.60	26.667	Ea.	3,350	825		4,175	4,925	400
	8960	3480 GPH, 2-1/2" pipe size		.50	32		5,250	990		6,240	7,275	
	9200	Taste and odor removal, 660 GPH, 1" pipe size		1.50	10.667		2,050	330		2,380	2,750	
	9240	1500 GPH, 1-1/4" pipe size		1	16		3,550	495		4,045	4,650	
	9280	2340 GPH, 1-1/2" pipe size		.80	20		3,400	620		4,020	4,675	
	9320	3420 GPH, 2" pipe size		.60	26.667		6,100	825		6,925	7,975	
	9360	4620 GPH, 2-1/2" pipe size		.50	32		9,600	990		10,590	12,100	

15480 | Domestic Water Heaters

			CREW	DAILY OUTPUT	LABOR-HOURS	UNIT	MAT.	LABOR	EQUIP.	TOTAL	TOTAL INCL O&P	
200	0010	WATER HEATERS A8.1-110										200
	0020	For solar, see division 13600										
	1000	Residential, elec., glass lined tank, 5 yr, 10 gal., sgl. element A8.1-120	1 Plum	2.30	3.478	Ea.	195	120		315	395	
	1040	20 gallon, single element		2.20	3.636		228	125		353	440	
	1060	30 gallon, double element A8.1-130		2.20	3.636		246	125		371	460	
	1080	40 gallon, double element		2	4		265	138		403	500	
	1100	52 gallon, double element A8.1-160		2	4		300	138		438	545	
	1120	66 gallon, double element		1.80	4.444		410	153		563	680	
	1140	80 gallon, double element A8.1-170		1.60	5		465	172		637	770	
	1180	120 gallon, double element		1.40	5.714		680	197		877	1,050	
	2000	Gas fired, glass lined tank, 5 yr, vent not incl., 20 gallon		2.10	3.810		188	131		319	405	
	2040	30 gallon		2	4		330	138		468	575	
	2060	40 gallon		1.90	4.211		350	145		495	605	
	2080	50 gallon		1.80	4.444		480	153		633	755	
	2100	75 gallon		1.50	5.333		525	184		709	855	
	2120	100 gallon		1.30	6.154		790	212		1,002	1,175	
	3000	Oil fired, glass lined tank, 5 yr, vent not included, 30 gal. A8.1-180		2	4		760	138		898	1,050	
	3040	50 gallon		1.80	4.444		1,025	153		1,178	1,350	
	3060	70 gallon		1.50	5.333		1,350	184		1,534	1,775	
	3080	85 gallon		1.40	5.714		3,425	197		3,622	4,075	
	4000	Commercial, 100° rise. NOTE: for each size tank, a range of										
	4010	heaters between the ones shown are available										
	4020	Electric										
	4100	5 gal., 3 kW, 12 GPH, 208V	1 Plum	2	4	Ea.	1,300	138		1,438	1,625	
	4120	10 gal., 6 kW, 25 GPH, 208V		2	4		1,425	138		1,563	1,775	
	4140	50 gal., 9 kW, 37 GPH, 208V		1.80	4.444		1,975	153		2,128	2,400	
	4160	50 gal., 36 kW, 148 GPH, 208V		1.80	4.444		3,000	153		3,153	3,550	
	4180	80 gal., 12 kW, 49 GPH, 208V		1.50	5.333		2,425	184		2,609	2,950	
	4200	80 gal., 36 kW, 148 GPH, 208V		1.50	5.333		3,350	184		3,534	3,975	
	4220	100 gal., 36 kW, 148 GPH, 208V		1.20	6.667		3,500	230		3,730	4,200	
	4240	120 gal., 36 kW, 148 GPH, 208V		1.20	6.667		3,650	230		3,880	4,350	
	4260	150 gal., 15 kW , 61 GPH, 480V		1	8		8,575	276		8,851	9,850	
	4280	150 gal., 120 kW, 490 GPH, 480V		1	8		12,000	276		12,276	13,600	
	4300	200 gal., 15 kW, 61 GPH, 480V	Q-1	1.70	9.412		9,275	292		9,567	10,600	
	4320	200 gal., 120 kW , 490 GPH, 480V		1.70	9.412		12,700	292		12,992	14,400	
	4340	250 gal., 15 kW, 61 GPH, 480V		1.50	10.667		9,575	330		9,905	11,000	
	4360	250 gal., 150 kW, 615 GPH, 480V		1.50	10.667		13,900	330		14,230	15,800	
	4380	300 gal., 30 kW, 123 GPH, 480V		1.30	12.308		10,500	380		10,880	12,200	
	4400	300 gal., 180 kW, 738 GPH, 480V		1.30	12.308		15,200	380		15,580	17,400	
	4420	350 gal., 30 kW, 123 GPH, 480V		1.10	14.545		11,100	450		11,550	13,000	
	4440	350 gal, 180 kW, 738 GPH, 480V		1.10	14.545		15,700	450		16,150	18,000	
	4460	400 gal., 30 kW, 123 GPH, 480V		1	16		12,600	495		13,095	14,700	
	4480	400 gal., 210 kW, 860 GPH, 480V		1	16		18,100	495		18,595	20,700	
	4500	500 gal., 30 kW, 123 GPH, 480V		.80	20		14,700	620		15,320	17,100	
	4520	500 gal., 240 kW, 984 GPH, 480V		.80	20		21,700	620		22,320	24,700	
	4540	600 gal., 30 kW, 123 GPH, 480V	Q-2	1.20	20		16,900	645		17,545	19,500	

15480 | Domestic Water Heaters			CREW	DAILY OUTPUT	LABOR-HOURS	UNIT	2001 BARE COSTS MAT.	LABOR	EQUIP.	TOTAL	TOTAL INCL O&P	
200 4560	600 gal., 300 kW, 1230 GPH, 480V	A8.1 -110	Q-2	1.20	20	Ea.	25,600	645		26,245	29,100	200
4580	700 gal., 30 kW, 123 GPH, 480V			1	24		17,700	770		18,470	20,700	
4600	700 gal., 300 kW, 1230 GPH, 480V	A8.1 -120		1	24		26,600	770		27,370	30,500	
4620	800 gal., 60 kW, 245 GPH, 480V			.90	26.667		19,200	855		20,055	22,400	
4640	800 gal., 300 kW, 1230 GPH, 480V	A8.1 -130		.90	26.667		27,200	855		28,055	31,300	
4660	1000 gal., 60 kW, 245 GPH, 480V			.70	34.286		21,000	1,100		22,100	24,800	
4680	1000 gal., 480 kW, 1970 GPH, 480V	A8.1 -160		.70	34.286		21,500	1,100		22,600	25,400	
4700	1250 gal., 60 kW, 245 GPH, 480V			.60	40		23,700	1,275		24,975	28,100	
4720	1250 gal., 480 kW, 1970 GPH, 480V	A8.1 -170		.60	40		36,000	1,275		37,275	41,600	
4740	1500 gal., 60 kW, 245 GPH, 480V			.50	48		31,300	1,550		32,850	36,700	
4760	1500 gal., 480 kW, 1970 GPH, 480V	A8.1 -180		.50	48		43,400	1,550		44,950	50,000	
4770	2000 gal., 480 kW, 1970 GPH, 480V			.48	50		24,500	1,600		26,100	29,400	
5400	Modulating step control, 2-5 steps		1 Elec	5.30	1.509		1,125	52		1,177	1,325	
5440	6-10 steps			3.20	2.500		1,450	86		1,536	1,725	
5460	11-15 steps			2.70	2.963		1,775	102		1,877	2,100	
5480	16-20 steps			1.60	5		2,075	172		2,247	2,550	
6000	Gas fired, flush jacket, std. controls, vent not incl.											
6040	75 MBH input, 73 GPH		1 Plum	1.40	5.714	Ea.	1,250	197		1,447	1,675	
6060	98 MBH input, 95 GPH			1.40	5.714		2,175	197		2,372	2,700	
6080	120 MBH input, 110 GPH			1.20	6.667		2,375	230		2,605	2,950	
6100	120 MBH input, 115 GPH			1.10	7.273		1,975	251		2,226	2,550	
6120	140 MBH input, 130 GPH			1	8		2,500	276		2,776	3,175	
6140	155 MBH input, 150 GPH			.80	10		2,925	345		3,270	3,725	
6160	180 MBH input, 170 GPH			.70	11.429		2,575	395		2,970	3,425	
6180	200 MBH input, 192 GPH			.60	13.333		2,600	460		3,060	3,550	
6200	250 MBH input, 245 GPH			.50	16		2,750	550		3,300	3,850	
6220	260 MBH input, 250 GPH		Q-1	.80	20		2,925	620		3,545	4,150	
6240	360 MBH input, 360 GPH			.80	20		3,450	620		4,070	4,700	
6260	500 MBH input, 480 GPH			.70	22.857		4,775	710		5,485	6,325	
6280	725 MBH input, 690 GPH			.60	26.667		8,150	825		8,975	10,200	
6900	For low water cutoff, add		1 Plum	8	1		198	34.50		232.50	270	
6960	For bronze body hot water circulator, add		"	4	2		1,100	69		1,169	1,300	
8000	Oil fired, flush jacket, std. controls, vent not incl.											
8060	103 MBH gross output, 116 GPH		1 Plum	1.10	7.273	Ea.	895	251		1,146	1,375	
8080	122 MBH gross output, 141 GPH			1	8		1,400	276		1,676	1,975	
8100	137 MBH gross output, 161 GPH			.80	10		1,500	345		1,845	2,175	
8120	168 MBH gross output, 192 GPH			.60	13.333		1,675	460		2,135	2,550	
8140	195 MBH gross output, 224 GPH			.50	16		1,825	550		2,375	2,850	
8160	225 MBH gross output, 256 GPH		Q-1	.80	20		2,125	620		2,745	3,275	
8180	262 MBH gross output, 315 GPH			.70	22.857		2,225	710		2,935	3,525	
8200	315 MBH gross output, 409 GPH			.70	22.857		2,725	710		3,435	4,075	
8220	420 MBH gross output, 504 GPH			.60	26.667		3,100	825		3,925	4,675	
8240	525 MBH gross output, 630 GPH			.50	32		4,300	990		5,290	6,225	
8260	630 MBH gross output, 756 GPH			.50	32		4,475	990		5,465	6,425	
8280	735 MBH gross output, 880 GPH			.40	40		4,975	1,250		6,225	7,350	
8300	840 MBH gross output, 1000 GPH			.40	40		5,725	1,250		6,975	8,175	
8320	1050 MBH gross output, 1260 GPH		Q-2	.60	40		6,500	1,275		7,775	9,075	
8340	1365 MBH gross output, 1640 GPH			.50	48		7,975	1,550		9,525	11,100	
8360	1680 MBH gross output, 2000 GPH			.40	60		8,750	1,925		10,675	12,500	
8380	2310 MBH gross output, 2780 GPH			.30	80		11,700	2,575		14,275	16,700	
8400	2835 MBH gross output, 3400 GPH			.30	80		13,800	2,575		16,375	19,100	
8420	3150 MBH gross output, 3780 GPH			.20	120		15,200	3,850		19,050	22,500	
8500	Oil fired, polymer lined											
8510	400 MBH, 125 gallon		Q-2	1	24	Ea.	13,000	770		13,770	15,500	
8520	400 MBH, 600 gallon			.80	30		23,000	965		23,965	26,800	
8530	800 MBH, 400 gallon			.67	35.821		21,700	1,150		22,850	25,600	200

Important: See the Reference Section for critical supporting data - Reference Nos., Crews, & City Cost Indexes

15480 | Domestic Water Heaters

	Line	Description	Ref.	Crew	Daily Output	Labor-Hours	Unit	2001 Bare Costs Mat.	Labor	Equip.	Total	Total Incl O&P	
200	8540	800 MBH, 600 gallon	A8.1-110	Q-2	.60	40	Ea.	27,800	1,275		29,075	32,600	200
	8550	1000 MBH, 600 gallon			.50	48		28,700	1,550		30,250	33,900	
	8560	1200 MBH, 900 gallon	A8.1-120		.40	60		32,700	1,925		34,625	38,900	
	8900	For low water cutoff, add		1 Plum	8	1		198	34.50		232.50	270	
	8960	For bronze body hot water circulator, add	A8.1-130	"	4	2		745	69		814	920	
	8965	Point of use, electric, glass lined											
	8969	Energy saver	A8.1-160										
	8970	2.5 gal. single element		1 Plum	2.80	2.857	Ea.	159	98.50		257.50	325	
	8971	4 gal. single element	A8.1-170		2.80	2.857		174	98.50		272.50	340	
	8974	6 gal. single element			2.50	3.200		188	110		298	375	
	8975	10 gal. single element	A8.1-180		2.50	3.200		195	110		305	380	
	8976	15 gal. single element			2.40	3.333		210	115		325	405	
	8977	20 gal. single element			2.40	3.333		228	115		343	425	
	8978	30 gal. single element			2.30	3.478		253	120		373	460	
	8979	40 gal. single element			2.20	3.636		275	125		400	495	
	8988	Commercial (ASHRAE energy std. 90)											
	8989	6 gallon		1 Plum	2.50	3.200	Ea.	320	110		430	520	
	8990	10 gallon			2.50	3.200		345	110		455	540	
	8991	15 gallon			2.40	3.333		355	115		470	565	
	8992	20 gallon			2.40	3.333		385	115		500	595	
	8993	30 gallon			2.30	3.478		810	120		930	1,075	
	8995	Under the sink, copper, w/bracket											
	8996	2.5 gallon		1 Plum	4	2	Ea.	125	69		194	242	
700	0010	**TEMPERATURE MAINTENANCE CABLE**											700
	0040	Components											
	0080	Heating cable											
	0100	208 V											
	0120	105 °F		Q-1	1,060.80	.015	L.F.	8	.47		8.47	9.50	
	0130	115 °F			1,060.80	.015		8	.47		8.47	9.50	
	0140	125 °F			1,060.80	.015		8	.47		8.47	9.50	
	0150	140 °F			1,060.80	.015		8	.47		8.47	9.50	
	0200	120 V											
	0220	125 °F		Q-1	1,060.80	.015	L.F.	5.15	.47		5.62	6.35	
	0300	Power kit w/1 end seal		1 Elec	48.80	.164	Ea.	44	5.60		49.60	56.50	
	0310	Splice kit			35.50	.225		52	7.75		59.75	68.50	
	0320	End seal			160	.050		6.95	1.72		8.67	10.20	
	0330	Tee kit w/1 end seal			26.80	.299		55	10.25		65.25	76	
	0340	Powered splice w/2 end seals			20	.400		52	13.70		65.70	78	
	0350	Powered tee kit w/3 end seals			18	.444		67	15.25		82.25	96.50	
	0360	Cross kit w/2 end seals			18.60	.430		69	14.75		83.75	98	
	0500	Recommended thickness of fiberglass insulation											
	0510	Pipe size											
	0520	1/2" - 1" use 1" insulation											
	0530	1-1/4" - 2" use 1-1/2" insulation											
	0540	2-1/2" - 6" use 2" insulation											
	0560	NOTE: For pipe sizes 1-1/4" and smaller use 1/4" larger diameter											
	0570	insulation to allow room for installation over cable.											
940	0010	**WATER HEATER PACKAGED SYSTEMS**											940
	1000	Car wash package, continuous duty, high recovery, gas fired											
	1040	100° rise, 180 MBH input, 174 GPH		1 Plum	3	2.667	Ea.	2,875	92		2,967	3,325	
	1060	280 MBH input, 270 GPH			2.50	3.200		3,075	110		3,185	3,550	
	1080	400 MBH input, 386 GPH			2.50	3.200		3,525	110		3,635	4,075	
	1100	480 MBH input, 464 GPH			2	4		3,725	138		3,863	4,300	
	1120	605 MBH input, 584 GPH			1.50	5.333		4,250	184		4,434	4,950	
	1140	700 MBH input, 676 GPH			1	8		4,425	276		4,701	5,300	

15400 | Plumbing Fixtures & Equipment

	15480 | Domestic Water Heaters		CREW	DAILY OUTPUT	LABOR-HOURS	UNIT	2001 BARE COSTS MAT.	LABOR	EQUIP.	TOTAL	TOTAL INCL O&P	
940	1160	1000 MBH input, 966 GPH	Q-1	1.60	10	Ea.	4,925	310		5,235	5,900	940
	1180	1200 MBH input, 1159 GPH		1.40	11.429		6,700	355		7,055	7,900	
	1200	1400 MBH input, 1353 GPH		1.20	13.333		7,300	415		7,715	8,650	
	3000	Combination dishwasher & general purpose, 2 temp. gas fired										
	3040	124 GPH @ 140° rise; 434 GPH @ 40° rise	1 Plum	2	4	Ea.	2,925	138		3,063	3,425	
	3060	193 GPH @ 140° rise; 677 GPH @ 40° rise		1.60	5		3,350	172		3,522	3,950	
	3080	276 GPH @ 140° rise; 969 GPH @ 40° rise		1.40	5.714		3,825	197		4,022	4,500	
	3100	330 GPH @ 140° rise; 1160 GPH @ 40°rise		1.20	6.667		4,050	230		4,280	4,800	
	3120	417 GPH @ 140° rise; 1460 GPH @ 40°rise		1	8		4,575	276		4,851	5,450	
	3140	481 GPH @ 140° rise; 1685 GPH @ 40°rise	Q-1	1.40	11.429		5,275	355		5,630	6,325	
	3160	688 GPH @ 140° rise; 2410 GPH @ 40°rise		1.20	13.333		5,775	415		6,190	6,975	
	3180	828 GPH @ 140° rise; 2790 GPH @ 40°rise		1	16		7,550	495		8,045	9,075	
	3200	965 GPH @ 140° rise; 3370 GPH @ 40°rise		.80	20		8,175	620		8,795	9,900	
	3960	For unit base, add					226			226	249	
	5000	Coin laundry units, gas fired, 100° rise										
	5020	Single heater,										
	5040	280 MBH input, 270 GPH	1 Plum	1.60	5	Ea.	3,350	172		3,522	3,950	
	5060	400 MBH input, 386 GPH		1.30	6.154		4,425	212		4,637	5,175	
	5080	480 MBH input, 464 GPH		1	8		4,650	276		4,926	5,550	
	5100	605 MBH input, 584 GPH	Q-1	1.40	11.429		5,475	355		5,830	6,550	
	5120	700 MBH input, 676 GPH		1.20	13.333		5,775	415		6,190	6,975	
	5140	1000 MBH input, 966 GPH		1	16		6,275	495		6,770	7,650	
	5160	1200 MBH input, 1159 GPH		.90	17.778		8,150	550		8,700	9,800	
	5180	1400 MBH input, 1353 GPH		.70	22.857		9,050	710		9,760	11,100	
	6000	Multiple heater										
	6040	560 MBH input, 540 GPH	1 Plum	1	8	Ea.	5,300	276		5,576	6,275	
	6060	800 MBH input, 772 GPH	Q-1	1.30	12.308		6,875	380		7,255	8,150	
	6080	960 MBH input, 928 GPH		1.20	13.333		7,775	415		8,190	9,175	
	6100	1210 MBH input, 1168 GPH		1	16		8,725	495		9,220	10,400	
	6120	1400 MBH input, 1352 GPH		1	16		9,125	495		9,620	10,800	
	6140	1700 MBH input, 1642 GPH		.80	20		10,400	620		11,020	12,400	
	6160	2000 MBH input, 1932 GPH		.70	22.857		12,600	710		13,310	15,000	
	6180	2400 MBH input, 2318 GPH	Q-2	.90	26.667		14,200	855		15,055	16,900	
	6200	2600 MBH input, 2512 GPH		.80	30		15,800	965		16,765	18,900	
	6220	2800 MBH input, 2706 GPH		.70	34.286		17,800	1,100		18,900	21,300	
	6800	Standard system sizing is based on										
	6820	80% of the washers operating at one time.										

15500 | Heat Generation Equipment

	15510 | Heating Boilers and Accessories		CREW	DAILY OUTPUT	LABOR-HOURS	UNIT	2001 BARE COSTS MAT.	LABOR	EQUIP.	TOTAL	TOTAL INCL O&P	
120	0010	**BURNERS**										120
	0990	Residential, conversion, gas fired, LP or natural										
	1000	Gun type, atmospheric input 72 to 200 MBH	Q-1	2.50	6.400	Ea.	655	198		853	1,025	
	1020	120 to 360 MBH	"	2	8	"	725	248		973	1,175	
	3000	Flame retention oil fired assembly, input										
	3020	.50 to 2.25 GPH	Q-1	2.40	6.667	Ea.	291	207		498	630	
	3040	2.0 to 5.0 GPH	"	2	8		525	248		773	955	
	4600	Gas safety, shut off valve, 3/4" threaded	1 Stpi	20	.400		272	13.95		285.95	320	

Important: See the Reference Section for critical supporting data - Reference Nos., Crews, & City Cost Indexes

15510 | Heating Boilers and Accessories

		15510 | Heating Boilers and Accessories	CREW	DAILY OUTPUT	LABOR-HOURS	UNIT	2001 BARE COSTS MAT.	LABOR	EQUIP.	TOTAL	TOTAL INCL O&P	
120	4610	1" threaded	1 Stpi	19	.421	Ea.	274	14.65		288.65	320	120
	4620	1-1/4" threaded		15	.533		298	18.60		316.60	360	
	4630	1-1/2" threaded		13	.615		320	21.50		341.50	390	
	4640	2" threaded	↓	11	.727		360	25.50		385.50	435	
	4650	2-1/2" threaded	Q-1	15	1.067		415	33		448	505	
	4660	3" threaded		13	1.231		570	38		608	685	
	4670	4" flanged	↓	3	5.333		2,150	165		2,315	2,625	
	4680	6" flanged	Q-2	3	8	↓	5,000	257		5,257	5,900	
300	0010	**BOILERS, ELECTRIC, ASME** Standard controls and trim (A8.3-110)										300
	1000	Steam, 6 KW, 20.5 MBH	Q-19	1.20	20	Ea.	7,900	645		8,545	9,675	
	1040	9 KW, 30.7 MBH		1.20	20		8,250	645		8,895	10,100	
	1060	18 KW, 61.4 MBH		1.20	20		8,325	645		8,970	10,100	
	1080	24 KW, 81.8 MBH		1.10	21.818		8,400	705		9,105	10,300	
	1120	36 KW, 123 MBH		1.10	21.818		8,575	705		9,280	10,500	
	2000	Hot water, 6 KW, 20.5 MBH		1.30	18.462		3,400	595		3,995	4,625	
	2020	9 KW, 30.7 MBH		1.30	18.462		3,425	595		4,020	4,650	
	2040	24 KW, 82 MBH		1.20	20		3,625	645		4,270	4,975	
	2060	30 KW, 103 MBH		1.20	20		3,725	645		4,370	5,075	
	2070	36 KW, 123 MBH		1.20	20		3,950	645		4,595	5,325	
	2080	48 KW, 164 MBH	↓	1.10	21.818	↓	3,975	705		4,680	5,425	
400	0010	**BOILERS, GAS FIRED** Natural or propane, standard controls										400
	1000	Cast iron, with insulated jacket										
	2000	Steam, gross output, 81 MBH	Q-7	1.40	22.857	Ea.	1,125	760		1,885	2,400	
	2020	102 MBH		1.30	24.615		1,300	820		2,120	2,650	
	2040	122 MBH		1	32		1,400	1,075		2,475	3,150	
	2060	163 MBH		.90	35.556		1,725	1,175		2,900	3,675	
	2080	203 MBH		.90	35.556		1,900	1,175		3,075	3,875	
	2100	240 MBH		.85	37.647		1,900	1,250		3,150	4,000	
	2120	280 MBH		.80	40		2,425	1,325		3,750	4,675	
	2140	320 MBH		.70	45.714		2,650	1,525		4,175	5,200	
	3000	Hot water, gross output, 80 MBH		1.46	21.918		1,000	730		1,730	2,225	
	3020	100 MBH		1.35	23.704		1,175	790		1,965	2,500	
	3040	122 MBH		1.10	29.091		1,275	965		2,240	2,850	
	3060	163 MBH		1	32		1,600	1,075		2,675	3,350	
	3080	203 MBH		1	32		1,775	1,075		2,850	3,575	
	3100	240 MBH		.95	33.684		1,800	1,125		2,925	3,675	
	3120	280 MBH		.90	35.556		2,250	1,175		3,425	4,250	
	3140	320 MBH	↓	.80	40	↓	2,475	1,325		3,800	4,700	
	6000	Hot water, including burner & one zone valve, gross output										
	6010	51.2 MBH	Q-6	2	12	Ea.	1,575	390		1,965	2,325	
	6020	72 MBH		2	12		1,750	390		2,140	2,525	
	6040	89 MBH		1.90	12.632		1,775	410		2,185	2,600	
	6060	105 MBH		1.80	13.333		2,000	435		2,435	2,850	
	6080	132 MBH		1.70	14.118		2,275	460		2,735	3,200	
	6100	155 MBH		1.50	16		2,650	520		3,170	3,675	
	6110	186 MBH	↓	1.40	17.143	↓	3,200	560		3,760	4,350	
	7000	For tankless water heater on smaller gas units, add					10%					
	7050	For additional zone valves up to 312 MBH add				Ea.	101			101	111	
	7990	Special feature gas fired boilers										
	8000	Pulse combustion, standard controls / trim, 44,000 BTU	Q-5	1.50	10.667	Ea.	2,275	335		2,610	3,000	
	8050	88,000 BTU		1.40	11.429		2,700	360		3,060	3,500	
	8080	134,000 BTU	↓	1.20	13.333	↓	3,500	420		3,920	4,450	
	9000	Wall hung, C.I., sealed combustion, direct vent										
	9010	Packaged, net output										
	9020	45,000 BTUH	Q-6	1.33	18.045	Ea.	1,375	585		1,960	2,375	
	9030	60,800 BTUH	↓	1.33	18.045	↓	1,525	585		2,110	2,550	

15510 | Heating Boilers and Accessories

		15510 Heating Boilers and Accessories	CREW	DAILY OUTPUT	LABOR-HOURS	UNIT	2001 BARE COSTS MAT.	LABOR	EQUIP.	TOTAL	TOTAL INCL O&P	
400	9040	77,000 BTUH	Q-6	1.33	18.045	Ea.	1,650	585		2,235	2,700	400
500	0010	**BOILERS, OIL FIRED** Standard controls, flame retention burner										500
	1000	Cast iron, with insulated flush jacket										
	2000	Steam, gross output, 109 MBH	Q-7	1.20	26.667	Ea.	1,375	885		2,260	2,875	
	2020	144 MBH		1.10	29.091		1,600	965		2,565	3,225	
	2040	173 MBH		1	32		1,750	1,075		2,825	3,500	
	2060	207 MBH	↓	.90	35.556	↓	1,875	1,175		3,050	3,850	
	3000	Hot water, same price as steam										
	4000	For tankless coil in smaller sizes, add					15%					
	5000	Steel, insulated jacket, burner										
	6000	Steam, full water leg construction, gross output										
	6020	144 MBH	Q-6	1.33	18.045	Ea.	3,275	585		3,860	4,475	
	6040	198 MBH	"	1.20	20	"	3,550	650		4,200	4,875	
	6400	Larger sizes are same as steel, gas fired										
	7000	Hot water, gross output, 103 MBH	Q-6	1.60	15	Ea.	965	490		1,455	1,800	
	7020	122 MBH		1.45	16.506		2,100	535		2,635	3,100	
	7040	137 MBH		1.36	17.595		2,450	570		3,020	3,575	
	7060	168 MBH		1.30	18.405		2,625	600		3,225	3,800	
	7080	225 MBH	↓	1.22	19.704	↓	2,725	640		3,365	3,975	
	7340	For tankless coil in steam or hot water, add					7%					
880	0010	**SWIMMING POOL HEATERS** Not including wiring, external										880
	0020	piping, base or pad,										
	0060	Gas fired, input, 115 MBH	Q-6	3	8	Ea.	1,525	260		1,785	2,075	
	0100	135 MBH		2	12		1,725	390		2,115	2,475	
	0160	155 MBH		1.50	16		1,825	520		2,345	2,775	
	0180	175 MBH		1.30	18.462		2,025	600		2,625	3,125	
	0200	190 MBH		1	24		2,350	780		3,130	3,775	
	0220	235 MBH		.70	34.286		2,875	1,125		4,000	4,850	
	0240	285 MBH		.60	40		3,700	1,300		5,000	6,050	
	0260	365 MBH		.50	48		4,175	1,550		5,725	6,925	
	0280	500 MBH		.40	60		5,700	1,950		7,650	9,225	
	0300	600 MBH		.35	68.571		6,550	2,225		8,775	10,600	
	0320	800 MBH		.33	72.727		7,825	2,375		10,200	12,200	
	0360	1000 MBH		.22	109		9,125	3,550		12,675	15,500	
	0370	1,200 MBH		.21	114		11,400	3,725		15,125	18,100	
	0380	1,500 MBH		.19	126		14,100	4,100		18,200	21,700	
	0400	1,800 MBH		.14	171		15,900	5,575		21,475	25,900	
	0410	2,450 MBH		.13	184		21,200	6,000		27,200	32,400	
	0420	3,000 MBH		.11	218		25,800	7,100		32,900	39,100	
	0440	3,750 MBH		.09	266		32,100	8,675		40,775	48,400	
	1000	Oil fired, gross output, 97 MBH		1.70	14.118		1,750	460		2,210	2,625	
	1020	118 MBH		1.70	14.118		1,775	460		2,235	2,650	
	1040	134 MBH		1.60	15		1,900	490		2,390	2,800	
	1060	161 MBH		1.30	18.462		2,300	600		2,900	3,425	
	1080	187 MBH		1.20	20		2,425	650		3,075	3,650	
	1100	214 MBH		1.10	21.818		2,775	710		3,485	4,125	
	1120	262 MBH		.89	26.966		3,125	875		4,000	4,775	
	1140	340 MBH		.68	35.294		4,150	1,150		5,300	6,275	
	1160	420 MBH		.58	41.379		4,750	1,350		6,100	7,250	
	1180	525 MBH		.46	52.174		6,075	1,700		7,775	9,275	
	1200	630 MBH		.41	58.537		6,775	1,900		8,675	10,400	
	1220	735 MBH		.35	68.571		8,350	2,225		10,575	12,600	
	1240	840 MBH		.33	72.727		9,550	2,375		11,925	14,100	
	1260	1,050 MBH	↓	.27	88.889	↓	11,600	2,900		14,500	17,200	

15 MECHANICAL

15500 | Heat Generation Equipment

	15510 | Heating Boilers and Accessories		CREW	DAILY OUTPUT	LABOR-HOURS	UNIT	2001 BARE COSTS MAT.	LABOR	EQUIP.	TOTAL	TOTAL INCL O&P	
880	1280	1,365 MBH	Q-6	.21	114	Ea.	14,900	3,725		18,625	22,000	880
	1300	1,680 MBH		.18	133		17,600	4,325		21,925	25,900	
	1320	2,310 MBH		.13	184		22,700	6,000		28,700	34,000	
	1340	2,835 MBH		.11	218		28,200	7,100		35,300	41,700	
	1360	3,150 MBH		.09	266		31,700	8,675		40,375	48,000	
	2000	Electric, 12 KW, 4,800 gallon pool	Q-19	3	8		1,525	259		1,784	2,075	
	2020	15 KW, 7,200 gallon pool		2.80	8.571		1,525	277		1,802	2,100	
	2040	24 KW, 9,600 gallon pool		2.40	10		2,050	325		2,375	2,725	
	2060	30 KW, 12,000 gallon pool		2	12		2,125	390		2,515	2,900	
	2080	35 KW, 14,400 gallon pool		1.60	15		2,125	485		2,610	3,050	
	2100	55 KW, 24,000 gallon pool		1.20	20		2,900	645		3,545	4,175	
	9000	To select pool heater: 12 BTUH x S.F. pool area										
	9010	X temperature differential =required output										
	9050	For electric, KW = gallons x 2.5 divided by 1000										
	9100	For family home type pool, double the										
	9110	Rated gallon capacity = 1/2°F rise per hour										

	15550 | Breechings, Chimneys & Stacks		CREW	DAILY OUTPUT	LABOR-HOURS	UNIT	MAT.	LABOR	EQUIP.	TOTAL	TOTAL INCL O&P	
200	0010	**DRAFT CONTROLS**										200
	5000	Vent damper, bi-metal, gas, 3" diameter	Q-9	24	.667	Ea.	31.50	20		51.50	65.50	
	5010	4" diameter		24	.667		35.50	20		55.50	70	
	5101	Electric, automatic, gas, 4" diameter		24	.667		137	20		157	181	
	5250	Automatic, oil, 4" diameter		24	.667		153	20		173	199	
440	0010	**VENT CHIMNEY** Prefab metal, U.L. listed										440
	0020	Gas, double wall, galvanized steel										
	0080	3" diameter	Q-9	72	.222	V.L.F.	3.48	6.70		10.18	14.20	
	0100	4" diameter		68	.235		4.35	7.10		11.45	15.75	
	0120	5" diameter		64	.250		5.05	7.55		12.60	17.25	
	0140	6" diameter		60	.267		6.25	8.05		14.30	19.30	
	0160	7" diameter		56	.286		8.65	8.65		17.30	23	
	0180	8" diameter		52	.308		9.65	9.30		18.95	25	
	0200	10" diameter		48	.333		20.50	10.05		30.55	38	
	0220	12" diameter		44	.364		27	11		38	47	
	0240	14" diameter		42	.381		46	11.50		57.50	68.50	
	0260	16" diameter		40	.400		62	12.10		74.10	86.50	
	0280	18" diameter		38	.421		80	12.70		92.70	108	
	0300	20" diameter	Q-10	36	.667		94.50	21		115.50	136	
	0320	22" diameter		34	.706		120	22		142	166	
	0340	24" diameter		32	.750		147	23.50		170.50	199	
	0600	For 4", 5" and 6" oval, add					50%					
	0650	Gas, double wall, galvanized steel, fittings										
	0660	Elbow 45°, 3" diameter	Q-9	36	.444	Ea.	8.40	13.40		21.80	30	
	0670	4" diameter		34	.471		10.20	14.20		24.40	33.50	
	0680	5" diameter		32	.500		12	15.10		27.10	36.50	
	0690	6" diameter		30	.533		14.35	16.10		30.45	41	
	0700	7" diameter		28	.571		20.50	17.25		37.75	49.50	
	0710	8" diameter		26	.615		24	18.60		42.60	54.50	
	0720	10" diameter		24	.667		59	20		79	96	
	0730	12" diameter		22	.727		75	22		97	117	
	0740	14" diameter		21	.762		105	23		128	151	
	0750	16" diameter		20	.800		131	24		155	183	
	0760	18" diameter		19	.842		177	25.50		202.50	234	
	0770	20" diameter	Q-10	18	1.333		194	42		236	278	
	0780	22" diameter		17	1.412		292	44		336	390	
	0790	24" diameter		16	1.500		380	47		427	495	

440	15550 | Breechings, Chimneys & Stacks	CREW	DAILY OUTPUT	LABOR-HOURS	UNIT	2001 BARE COSTS MAT.	LABOR	EQUIP.	TOTAL	TOTAL INCL O&P
0916	Adjustable length									
0918	3″ diameter, to 12″	Q-9	36	.444	Ea.	7.05	13.40		20.45	28.50
0920	4″ diameter, to 12″		34	.471		8.35	14.20		22.55	31
0924	6″ diameter, to 12″		30	.533		10.80	16.10		26.90	37
0928	8″ diameter, to 12″		26	.615		15.85	18.60		34.45	46
0930	10″ diameter, to 18″		24	.667		59.50	20		79.50	96.50
0932	12″ diameter, to 18″		22	.727		72.50	22		94.50	114
0936	16″ diameter, to 18″		20	.800		140	24		164	192
0938	18″ diameter, to 18″		19	.842		167	25.50		192.50	224
0944	24″ diameter, to 18″	Q-10	16	1.500		299	47		346	405
0950	Elbow 90°, adjustable, 3″ diameter	Q-9	36	.444		14.40	13.40		27.80	36.50
0960	4″ diameter		34	.471		17.50	14.20		31.70	41.50
0970	5″ diameter		32	.500		20.50	15.10		35.60	46
0980	6″ diameter		30	.533		24.50	16.10		40.60	52
0990	7″ diameter		28	.571		38.50	17.25		55.75	69
1010	8″ diameter		26	.615		41	18.60		59.60	73.50
1020	Wall thimble, 4 to 7″ adjustable, 3″ diameter		36	.444		9.10	13.40		22.50	30.50
1022	4″ diameter		34	.471		10.15	14.20		24.35	33
1024	5″ diameter		32	.500		11.40	15.10		26.50	36
1026	6″ diameter		30	.533		14	16.10		30.10	40.50
1028	7″ diameter		28	.571		16.25	17.25		33.50	44.50
1030	8″ diameter		26	.615		16.85	18.60		35.45	47
1040	Roof flashing, 3″ diameter		36	.444		5.20	13.40		18.60	26
1050	4″ diameter		34	.471		5.85	14.20		20.05	28.50
1060	5″ diameter		32	.500		9.10	15.10		24.20	33.50
1070	6″ diameter		30	.533		12.75	16.10		28.85	39
1080	7″ diameter		28	.571		16.10	17.25		33.35	44
1090	8″ diameter		26	.615		17.50	18.60		36.10	48
1100	10″ diameter		24	.667		47.50	20		67.50	83.50
1110	12″ diameter		22	.727		61	22		83	101
1120	14″ diameter		20	.800		75	24		99	120
1130	16″ diameter		18	.889		92.50	27		119.50	144
1140	18″ diameter		16	1		125	30		155	185
1150	20″ diameter	Q-10	18	1.333		163	42		205	244
1160	22″ diameter		14	1.714		206	53.50		259.50	310
1170	24″ diameter		12	2		247	62.50		309.50	370
1200	Tee, 3″ diameter	Q-9	27	.593		21.50	17.90		39.40	51
1210	4″ diameter		26	.615		22	18.60		40.60	53
1220	5″ diameter		25	.640		24	19.35		43.35	56.50
1230	6″ diameter		24	.667		26	20		46	59.50
1240	7″ diameter		23	.696		28.50	21		49.50	64
1250	8″ diameter		22	.727		39	22		61	77
1260	10″ diameter		21	.762		106	23		129	153
1270	12″ diameter		20	.800		128	24		152	178
1280	14″ diameter		18	.889		208	27		235	271
1290	16″ diameter		16	1		295	30		325	370
1300	18″ diameter		14	1.143		370	34.50		404.50	460
1310	20″ diameter	Q-10	17	1.412		500	44		544	620
1320	22″ diameter		13	1.846		620	58		678	775
1330	24″ diameter		12	2		705	62.50		767.50	875
1460	Tee cap, 3″ diameter	Q-9	45	.356		1.53	10.75		12.28	18.30
1470	4″ diameter		42	.381		1.69	11.50		13.19	19.60
1480	5″ diameter		40	.400		2.28	12.10		14.38	21
1490	6″ diameter		37	.432		2.63	13.05		15.68	23
1500	7″ diameter		35	.457		4.29	13.80		18.09	26
1510	8″ diameter		34	.471		5	14.20		19.20	27.50

15550 | Breechings, Chimneys & Stacks

			CREW	DAILY OUTPUT	LABOR-HOURS	UNIT	2001 BARE COSTS MAT.	LABOR	EQUIP.	TOTAL	TOTAL INCL O&P	
440	1520	10″ diameter	Q-9	32	.500	Ea.	9.60	15.10		24.70	34	440
	1530	12″ diameter		30	.533		12.80	16.10		28.90	39	
	1540	14″ diameter		28	.571		26.50	17.25		43.75	55.50	
	1550	16″ diameter		25	.640		30.50	19.35		49.85	63.50	
	1560	18″ diameter		24	.667		34.50	20		54.50	69	
	1570	20″ diameter	Q-10	27	.889		37	28		65	83.50	
	1580	22″ diameter		22	1.091		42.50	34		76.50	99.50	
	1590	24″ diameter		21	1.143		48	36		84	108	
	1750	Top, 3″ diameter	Q-9	46	.348		8	10.50		18.50	25	
	1760	4″ diameter		44	.364		8.40	11		19.40	26	
	1770	5″ diameter		42	.381		11.55	11.50		23.05	30.50	
	1780	6″ diameter		40	.400		14.65	12.10		26.75	35	
	1790	7″ diameter		38	.421		22	12.70		34.70	43.50	
	1800	8″ diameter		36	.444		27.50	13.40		40.90	50.50	
	1810	10″ diameter		34	.471		66	14.20		80.20	94.50	
	1820	12″ diameter		32	.500		99.50	15.10		114.60	133	
	1830	14″ diameter		30	.533		113	16.10		129.10	149	
	1840	16″ diameter		28	.571		204	17.25		221.25	251	
	1850	18″ diameter		26	.615		370	18.60		388.60	435	
	1860	20″ diameter	Q-10	28	.857		400	27		427	480	
	1870	22″ diameter		22	1.091		545	34		579	655	
	1880	24″ diameter		20	1.200		680	37.50		717.50	805	
	1882	26″ diameter		19	1.263		1,025	39.50		1,064.50	1,175	
	1884	28″ diameter		18	1.333		1,175	42		1,217	1,350	
	1886	30″ diameter		17	1.412		1,300	44		1,344	1,525	
	1888	32″ diameter		16	1.500		1,375	47		1,422	1,600	
	1890	34″ diameter		15	1.600		1,475	50		1,525	1,700	
	1892	36″ diameter		14	1.714		1,575	53.50		1,628.50	1,825	
	1900	Gas, double wall, galvanized steel, oval										
	1904	4″ x 1′	Q-9	68	.235	V.L.F.	11.20	7.10		18.30	23.50	
	1906	5″		64	.250		23	7.55		30.55	36.50	
	1908	5″/6″ x 1′		60	.267		23	8.05		31.05	38	
	1910	Oval fittings										
	1912	Adjustable length										
	1914	4″ diameter to 12″ lg	Q-9	34	.471	Ea.	11.85	14.20		26.05	35	
	1916	5″ diameter to 12″ lg		32	.500		33	15.10		48.10	59.50	
	1918	5″/6″ diameter to 12″		30	.533		30	16.10		46.10	58	
	1920	Elbow 45°										
	1922	4″	Q-9	34	.471	Ea.	21.50	14.20		35.70	46	
	1924	5″		32	.500		41	15.10		56.10	68.50	
	1926	5″/6″		30	.533		39	16.10		55.10	67.50	
	1930	Elbow 45°, flat										
	1932	4″	Q-9	34	.471	Ea.	21.50	14.20		35.70	46	
	1934	5″		32	.500		41	15.10		56.10	68.50	
	1936	5″/6″		30	.533		41	16.10		57.10	70	
	1940	Top										
	1942	4″	Q-9	44	.364	Ea.	16.10	11		27.10	34.50	
	1944	5″		42	.381		24.50	11.50		36	45	
	1946	5″/6″		40	.400		24.50	12.10		36.60	45.50	
	1950	Adjustable flashing										
	1952	4″	Q-9	34	.471	Ea.	9.70	14.20		23.90	32.50	
	1954	5″		32	.500		28.50	15.10		43.60	55	
	1956	5″/6″		30	.533		19.20	16.10		35.30	46	
	1960	Tee										
	1962	4″	Q-9	26	.615	Ea.	32	18.60		50.60	64	
	1964	5″		25	.640		61.50	19.35		80.85	98	

MECHANICAL 15

15550	Breechings, Chimneys & Stacks	CREW	DAILY OUTPUT	LABOR-HOURS	UNIT	2001 BARE COSTS MAT.	LABOR	EQUIP.	TOTAL	TOTAL INCL O&P
440 1966	5"/6"	Q-9	24	.667	Ea.	62	20		82	99.50 440
1970	Tee with short snout									
1972	4"	Q-9	26	.615	Ea.	32	18.60		50.60	64
3200	All fuel, pressure tight, double wall, U.L. listed, 1400°F.									
3210	304 stainless steel liner, aluminized steel outer jacket									
3220	6" diameter	Q-9	60	.267	L.F.	35.50	8.05		43.55	51.50
3221	8" diameter		52	.308		43.50	9.30		52.80	62.50
3222	10" diameter		48	.333		48.50	10.05		58.55	69
3223	12" diameter		44	.364		55.50	11		66.50	78
3224	14" diameter		42	.381		63.50	11.50		75	88
3225	16" diameter		40	.400		67.50	12.10		79.60	92.50
3226	18" diameter		38	.421		76	12.70		88.70	103
3227	20" diameter	Q-10	36	.667		87	21		108	128
3228	24" diameter	"	32	.750		112	23.50		135.50	160
3260	For 316 stainless steel liner add					6%				
3280	All fuel, pressure tight, double wall fittings									
3284	304 stainless steel inner, aluminized steel jacket									
3288	Adjustable 18"/30" section									
3291	5" diameter	Q-9	32	.500	Ea.	91	15.10		106.10	124
3292	6" diameter		30	.533		115	16.10		131.10	152
3293	8" diameter		26	.615		133	18.60		151.60	175
3294	10" diameter		24	.667		149	20		169	195
3295	12" diameter		22	.727		170	22		192	221
3296	14" diameter		21	.762		196	23		219	252
3297	16" diameter		20	.800		204	24		228	262
3298	18" diameter		19	.842		251	25.50		276.50	315
3299	20" diameter	Q-10	18	1.333		290	42		332	385
3300	24" diameter	"	16	1.500		375	47		422	485
3326	For 316 stainless steel liner, add					15%				
3350	Elbow 90° fixed									
3353	5" diameter	Q-9	32	.500	Ea.	170	15.10		185.10	211
3354	6" diameter		30	.533		200	16.10		216.10	245
3355	8" diameter		26	.615		221	18.60		239.60	273
3356	10" diameter		24	.667		254	20		274	310
3357	12" diameter		22	.727		287	22		309	350
3358	14" diameter		21	.762		335	23		358	405
3359	16" diameter		20	.800		375	24		399	455
3360	18" diameter		19	.842		440	25.50		465.50	520
3361	20" diameter	Q-10	18	1.333		490	42		532	600
3362	24" diameter	"	16	1.500		605	47		652	740
3380	For 316 stainless steel liner, add					20%				
3400	Elbow 45°									
3404	6" diameter	Q-9	30	.533	Ea.	127	16.10		143.10	164
3405	8" diameter		26	.615		143	18.60		161.60	187
3406	10" diameter		24	.667		161	20		181	208
3407	12" diameter		22	.727		184	22		206	236
3408	14" diameter		21	.762		211	23		234	268
3409	16" diameter		20	.800		238	24		262	299
3410	18" diameter		19	.842		269	25.50		294.50	335
3411	20" diameter	Q-10	18	1.333		310	42		352	405
3412	24" diameter	"	16	1.500		395	47		442	510
3430	For 316 stainless steel liner, add					20%				
3450	Tee 90°									
3454	6" diameter	Q-9	24	.667	Ea.	132	20		152	176
3455	8" diameter		22	.727		152	22		174	201
3456	10" diameter		21	.762		178	23		201	231

15 MECHANICAL

Important: See the Reference Section for critical supporting data - Reference Nos., Crews, & City Cost Indexes

15550 | Breechings, Chimneys & Stacks		CREW	DAILY OUTPUT	LABOR-HOURS	UNIT	2001 BARE COSTS MAT.	LABOR	EQUIP.	TOTAL	TOTAL INCL O&P
440	3457 12″ diameter	Q-9	20	.800	Ea.	208	24		232	266
	3458 14″ diameter		18	.889		245	27		272	310
	3459 16″ diameter		16	1		275	30		305	350
	3460 18″ diameter	↓	14	1.143		325	34.50		359.50	410
	3461 20″ diameter	Q-10	17	1.412		365	44		409	470
	3462 24″ diameter	″	12	2		450	62.50		512.50	590
	3480 For Tee Cap, add					35%	20%			
	3500 For 316 stainless steel liner, add				↓	22%				
	3520 Plate support, galvanized									
	3524 6″ diameter	Q-9	26	.615	Ea.	83	18.60		101.60	120
	3525 8″ diameter		22	.727		93.50	22		115.50	137
	3526 10″ diameter		20	.800		102	24		126	150
	3527 12″ diameter		18	.889		106	27		133	159
	3528 14″ diameter		17	.941		128	28.50		156.50	185
	3529 16″ diameter		16	1		136	30		166	196
	3530 18″ diameter	↓	15	1.067		143	32		175	207
	3531 20″ diameter	Q-10	16	1.500		151	47		198	239
	3532 24″ diameter	″	14	1.714	↓	159	53.50		212.50	258
	3570 Bellows, lined, 316 stainless steel only									
	3574 6″ diameter	Q-9	30	.533	Ea.	284	16.10		300.10	340
	3575 8″ diameter		26	.615		310	18.60		328.60	370
	3576 10″ diameter		24	.667		340	20		360	405
	3577 12″ diameter		22	.727		375	22		397	450
	3578 14″ diameter		21	.762		425	23		448	500
	3579 16″ diameter		20	.800		470	24		494	555
	3580 18″ diameter	↓	19	.842		525	25.50		550.50	615
	3581 20″ diameter	Q-10	18	1.333	↓	590	42		632	715
	3600 Ventilated roof thimble, 304 stainless steel									
	3620 6″ diameter	Q-9	26	.615	Ea.	191	18.60		209.60	239
	3624 8″ diameter		22	.727		205	22		227	259
	3625 10″ diameter		20	.800		206	24		230	265
	3626 12″ diameter		18	.889		222	27		249	286
	3627 14″ diameter		17	.941		239	28.50		267.50	305
	3628 16″ diameter		16	1		254	30		284	325
	3629 18″ diameter	↓	15	1.067		274	32		306	350
	3630 20″ diameter	Q-10	16	1.500		293	47		340	400
	3631 24″ diameter	″	14	1.714	↓	330	53.50		383.50	450
	3650 For 316 stainless steel, add					10%				
	3670 Exit cone, 316 stainless steel only									
	3674 6″ diameter	Q-9	46	.348	Ea.	125	10.50		135.50	154
	3675 8″ diameter		42	.381		131	11.50		142.50	163
	3676 10″ diameter		40	.400		142	12.10		154.10	175
	3677 12″ diameter		38	.421		149	12.70		161.70	184
	3678 14″ diameter		37	.432		166	13.05		179.05	202
	3679 16″ diameter		36	.444		182	13.40		195.40	222
	3680 18″ diameter	↓	35	.457		224	13.80		237.80	269
	3681 20″ diameter	Q-10	28	.857		267	27		294	335
	3682 24″ diameter	″	26	.923	↓	345	29		374	425
	3720 Roof support assy., incl. 30″ pipe sect, 304 st. st.									
	3724 6″ diameter	Q-9	25	.640	Ea.	218	19.35		237.35	270
	3725 8″ diameter		21	.762		238	23		261	298
	3726 10″ diameter		19	.842		261	25.50		286.50	325
	3727 12″ diameter		17	.941		278	28.50		306.50	350
	3728 14″ diameter		16	1		310	30		340	385
	3729 16″ diameter		15	1.067		330	32		362	415
	3730 18″ diameter	↓	14	1.143	↓	355	34.50		389.50	445

	15550 | Breechings, Chimneys & Stacks		CREW	DAILY OUTPUT	LABOR-HOURS	UNIT	2001 BARE COSTS MAT.	LABOR	EQUIP.	TOTAL	TOTAL INCL O&P	
440	3731	20″ diameter	Q-10	15	1.600	Ea.	380	50		430	500	440
	3732	24″ diameter	″	13	1.846		425	58		483	560	
	3750	For 316 stainless steel, add					20%					
	3770	Stack cap, 316 stainless steel only										
	3773	5″ diameter	Q-9	48	.333	Ea.	119	10.05		129.05	147	
	3774	6″ diameter		46	.348		140	10.50		150.50	170	
	3775	8″ diameter		42	.381		162	11.50		173.50	196	
	3776	10″ diameter		40	.400		190	12.10		202.10	228	
	3777	12″ diameter		38	.421		221	12.70		233.70	263	
	3778	14″ diameter		37	.432		260	13.05		273.05	305	
	3779	16″ diameter		36	.444		297	13.40		310.40	345	
	3780	18″ diameter		35	.457		335	13.80		348.80	390	
	3781	20″ diameter	Q-10	28	.857		385	27		412	460	
	3782	24″ diameter	″	26	.923		460	29		489	555	
	3800	Round storm collar										
	3806	5″ diameter	Q-9	32	.500	Ea.	25	15.10		40.10	51	
	7800	All fuel, double wall, stainless steel, 6″ diameter		60	.267	V.L.F.	28.50	8.05		36.55	44	
	7802	7″ diameter		56	.286		37	8.65		45.65	54	
	7804	8″ diameter		52	.308		43	9.30		52.30	62	
	7806	10″ diameter		48	.333		63	10.05		73.05	85	
	7808	12″ diameter		44	.364		84.50	11		95.50	110	
	7810	14″ diameter		42	.381		111	11.50		122.50	140	
	8000	All fuel, double wall, stainless steel fittings										
	8010	Roof support 6″ diameter	Q-9	30	.533	Ea.	73.50	16.10		89.60	106	
	8020	7″ diameter		28	.571		83.50	17.25		100.75	118	
	8030	8″ diameter		26	.615		90	18.60		108.60	128	
	8040	10″ diameter		24	.667		119	20		139	162	
	8050	12″ diameter		22	.727		143	22		165	191	
	8060	14″ diameter		21	.762		181	23		204	235	
	8100	Elbow 15°, 6″ diameter		30	.533		64	16.10		80.10	95.50	
	8120	7″ diameter		28	.571		83.50	17.25		100.75	118	
	8140	8″ diameter		26	.615		82.50	18.60		101.10	119	
	8160	10″ diameter		24	.667		108	20		128	149	
	8180	12″ diameter		22	.727		131	22		153	178	
	8200	14″ diameter		21	.762		156	23		179	208	
	8300	Insulated tee with insulated tee cap, 6″ diameter		30	.533		121	16.10		137.10	158	
	8340	7″ diameter		28	.571		159	17.25		176.25	202	
	8360	8″ diameter		26	.615		179	18.60		197.60	226	
	8380	10″ diameter		24	.667		252	20		272	310	
	8400	12″ diameter		22	.727		355	22		377	425	
	8420	14″ diameter		21	.762		465	23		488	545	
	8500	Joist shield, 6″ diameter		30	.533		37	16.10		53.10	65.50	
	8510	7″ diameter		28	.571		40	17.25		57.25	70.50	
	8520	8″ diameter		26	.615		49.50	18.60		68.10	83	
	8530	10″ diameter		24	.667		66.50	20		86.50	105	
	8540	12″ diameter		22	.727		83	22		105	126	
	8550	14″ diameter		21	.762		103	23		126	149	
	8600	Round top, 6″ diameter		30	.533		41	16.10		57.10	70	
	8620	7″ diameter		28	.571		56	17.25		73.25	88	
	8640	8″ diameter		26	.615		75	18.60		93.60	111	
	8660	10″ diameter		24	.667		135	20		155	180	
	8680	12″ diameter		22	.727		193	22		215	246	
	8700	14″ diameter		21	.762		258	23		281	320	
	8800	Adjustable roof flashing, 6″ diameter		30	.533		49	16.10		65.10	78.50	
	8820	7″ diameter		28	.571		56	17.25		73.25	88	
	8840	8″ diameter		26	.615		60.50	18.60		79.10	95	

	15550 | Breechings, Chimneys & Stacks		CREW	DAILY OUTPUT	LABOR-HOURS	UNIT	2001 BARE COSTS MAT.	LABOR	EQUIP.	TOTAL	TOTAL INCL O&P	
440	8860	10″ diameter	Q-9	24	.667	Ea.	78	20		98	117	440
	8880	12″ diameter		22	.727		100	22		122	145	
	8900	14″ diameter		21	.762		125	23		148	174	
	9000	High temp. (2000° F), steel jacket, acid resist. refractory lining										
	9010	11 ga. galvanized jacket, U.L. listed										
	9020	Straight section, 48″ long, 10″ diameter	Q-10	13.30	1.805	Ea.	335	56.50		391.50	455	
	9030	12″ diameter		11.20	2.143		360	67		427	505	
	9040	18″ diameter		7.40	3.243		510	102		612	715	
	9050	24″ diameter		4.60	5.217		695	163		858	1,025	
	9120	Tee section, 10″ diameter		4.40	5.455		705	171		876	1,050	
	9130	12″ diameter		3.70	6.486		745	203		948	1,125	
	9140	18″ diameter		2.40	10		1,000	315		1,315	1,575	
	9150	24″ diameter		1.50	16		1,400	500		1,900	2,300	
	9220	Cleanout pier section, 10″ diameter		3.50	6.857		545	215		760	930	
	9230	12″ diameter		2.50	9.600		570	300		870	1,100	
	9240	18″ diameter		1.90	12.632		730	395		1,125	1,425	
	9250	24″ diameter		1.30	18.462		925	580		1,505	1,925	
	9320	For drain, add					53%					
	9330	Elbow, 30° and 45°, 10″ diameter	Q-10	6.60	3.636		460	114		574	680	
	9340	12″ diameter		5.60	4.286		495	134		629	750	
	9350	18″ diameter		3.70	6.486		745	203		948	1,125	
	9360	24″ diameter		2.30	10.435		1,100	325		1,425	1,725	
	9430	For 60° and 90° elbow, add					112%					
	9440	End cap, 10″ diameter	Q-10	26	.923		790	29		819	910	
	9450	12″ diameter		22	1.091		900	34		934	1,050	
	9460	18″ diameter		15	1.600		1,125	50		1,175	1,300	
	9470	24″ diameter		9	2.667		1,375	83.50		1,458.50	1,625	
	9540	Increaser (1 diameter), 10″ diameter		6.60	3.636		390	114		504	605	
	9550	12″ diameter		5.60	4.286		420	134		554	665	
	9560	18″ diameter		3.70	6.486		610	203		813	985	
	9570	24″ diameter		2.30	10.435		880	325		1,205	1,475	
	9600	For expansion joints, add to straight section					7%					
	9610	For 1/4″ hot rolled steel jacket, add					157%					
	9620	For 2950°F very high temperature, add					89%					
	9630	26 ga. aluminized jacket, straight section, 48″ long										
	9640	10″ diameter	Q-10	15.30	1.569	V.L.F.	197	49		246	292	
	9650	12″ diameter		12.90	1.860		208	58.50		266.50	320	
	9660	18″ diameter		8.50	2.824		300	88.50		388.50	465	
	9670	24″ diameter		5.30	4.528		430	142		572	695	
	9700	Accessories (all models)										
	9710	Guy band, 10″ diameter	Q-10	32	.750	Ea.	61	23.50		84.50	104	
	9720	12″ diameter		30	.800		64.50	25		89.50	110	
	9730	18″ diameter		26	.923		83.50	29		112.50	137	
	9740	24″ diameter		24	1		124	31.50		155.50	185	
	9810	Draw band, galv. stl., 11 gauge, 10″ diameter		32	.750		53	23.50		76.50	95	
	9820	12″ diameter		30	.800		57	25		82	101	
	9830	18″ diameter		26	.923		71.50	29		100.50	123	
	9840	24″ diameter		24	1		90.50	31.50		122	148	
	9910	Draw band, aluminized stl., 26 gauge, 10″ diameter		32	.750		15.20	23.50		38.70	53	
	9920	12″ diameter		30	.800		15.20	25		40.20	55	
	9930	18″ diameter		26	.923		19	29		48	65.50	
	9940	24″ diameter		24	1		23	31.50		54.50	73.50	

15700 | Heating/Ventilating/Air Conditioning Equipment

	15705 | Curbs/Pads/Stands Prefab		CREW	DAILY OUTPUT	LABOR-HOURS	UNIT	2001 BARE COSTS MAT.	LABOR	EQUIP.	TOTAL	TOTAL INCL O&P	
600	0010	CURBS/PADS PREFABRICATED										600
	6000	Pad, fiberglass reinforced concrete with polystyrene foam core										
	6050	Condenser, 2″ thick, 20″ x 38″	1 Shee	8	1	Ea.	17.75	33.50		51.25	71.50	
	6220	30″ x 36″	″	8	1		29.50	33.50		63	84	
	6340	36″ x 54″	Q-9	6	2.667		53	80.50		133.50	183	

	15710 | Heat Exchangers		CREW	DAILY OUTPUT	LABOR-HOURS	UNIT	MAT.	LABOR	EQUIP.	TOTAL	TOTAL INCL O&P	
900	0010	HEAT EXCHANGERS										900
	0016	Shell & tube type, 4 pass, 3/4″ O.D. copper tubes,										
	0020	C.I. heads, C.I. tube sheet, steel shell										
	0100	Hot water 40°F to 180°F, by steam at 10 PSI										
	0120	8 GPM	Q-5	6	2.667	Ea.	895	83.50		978.50	1,100	
	0140	10 GPM		5	3.200		1,350	100		1,450	1,625	
	0160	40 GPM		4	4		2,075	126		2,201	2,500	
	0500	For bronze head and tube sheet, add					50%					
	1000	Hot water 40°F to 140°F, by water at 200°F										
	1020	7 GPM	Q-5	6	2.667	Ea.	1,100	83.50		1,183.50	1,325	
	1040	16 GPM		5	3.200		1,550	100		1,650	1,850	
	1060	34 GPM		4	4		2,350	126		2,476	2,800	
	1080	55 GPM		3	5.333		3,400	167		3,567	4,000	
	1100	74 GPM		1.50	10.667		4,225	335		4,560	5,150	
	1120	86 GPM		1.40	11.429		5,650	360		6,010	6,775	
	1140	112 GPM	Q-6	2	12		7,025	390		7,415	8,350	
	1160	126 GPM		1.80	13.333		8,775	435		9,210	10,300	
	1180	152 GPM		1	24		11,200	780		11,980	13,500	
	3000	Plate type,										
	3100	400 GPM	Q-6	.80	30	Ea.	24,500	975		25,475	28,500	
	3120	800 GPM	″	.50	48		43,100	1,550		44,650	49,800	
	3140	1200 GPM	Q-7	.34	94.118		88,500	3,125		91,625	102,000	
	3160	1800 GPM	″	.24	133		133,000	4,425		137,425	152,500	
	8000	Heat pipe type, glycol, 50% efficient										
	8010	100 MBH, 1700 CFM	1 Stpi	.80	10	Ea.	15,700	350		16,050	17,800	
	8020	160 MBH, 2700 CFM		.60	13.333		19,800	465		20,265	22,500	
	8030	620 MBH, 4000 CFM		.40	20		26,900	695		27,595	30,700	

	15730 | Unitary Air Conditioning Equip		CREW	DAILY OUTPUT	LABOR-HOURS	UNIT	MAT.	LABOR	EQUIP.	TOTAL	TOTAL INCL O&P	
500	0010	**PACKAGED TERMINAL AIR CONDITIONER** Cabinet, wall sleeve,										500
	0100	louver, electric heat, thermostat, manual changeover, 208 V										
	0200	6,000 BTUH cooling, 8800 BTU heat	Q-5	6	2.667	Ea.	915	83.50		998.50	1,125	
	0220	9,000 BTUH cooling, 13,900 BTU heat		5	3.200		935	100		1,035	1,175	
	0240	12,000 BTUH cooling, 13,900 BTU heat		4	4		970	126		1,096	1,275	
	0260	15,000 BTUH cooling, 13,900 BTU heat		3	5.333		1,050	167		1,217	1,400	
800	0010	WINDOW UNIT AIR CONDITIONERS										800
	4000	Portable/window, 15 amp 125V grounded receptacle required										
	4060	5000 BTUH	1 Carp	8	1	Ea.	296	29		325	370	
	4340	6000 BTUH		8	1		335	29		364	415	
	4480	8000 BTUH		6	1.333		385	39		424	485	
	4500	10,000 BTUH		6	1.333		455	39		494	560	
	4520	12,000 BTUH	L-2	8	2		495	51.50		546.50	625	
	4600	Window/thru-the-wall, 15 amp 230V grounded receptacle required										
	4780	17,000 BTUH	L-2	6	2.667	Ea.	715	68.50		783.50	890	
	4940	25,000 BTUH		4	4		1,025	103		1,128	1,275	
	4960	29,000 BTUH		4	4		1,200	103		1,303	1,475	

Important: See the Reference Section for critical supporting data - Reference Nos., Crews, & City Cost Indexes

15740 | Heat Pumps

		15740 Heat Pumps	Crew	Daily Output	Labor-Hours	Unit	2001 Bare Costs Mat.	Labor	Equip.	Total	Total Incl O&P	
100	0010	**HEAT PUMPS** (Not including interconnecting tubing)										100
	1000	Air to air, split system, not including curbs, pads, or ductwork										
	1010	For curbs/pads see division 15705										
	1015	1.5 ton cooling, 7 MBH heat @ 0°F	Q-5	1.22	13.115	Ea.	1,875	410		2,285	2,675	
	1020	2 ton cooling, 8.5 MBH heat @ 0°F		1.20	13.333		2,100	420		2,520	2,925	
	1030	2.5 ton cooling, 10 MBH heat @ 0°F		1	16		2,375	500		2,875	3,375	
	1040	3 ton cooling, 13 MBH heat @ 0°F		.80	20		2,700	630		3,330	3,900	
	1050	3.5 ton cooling, 18 MBH heat @ 0°F		.75	21.333		3,275	670		3,945	4,600	
	1054	4 ton cooling, 24 MBH heat @ 0°F		.60	26.667		3,650	835		4,485	5,300	
	1060	5 ton cooling, 27 MBH heat @ 0°F		.50	32		4,275	1,000		5,275	6,225	
	1300	Supplementary electric heat coil, included										
	1500	Single package, not including curbs, pads, or plenums										
	1502	1/2 ton cooling, supplementary heat not incl.	Q-5	8	2	Ea.	1,200	63		1,263	1,425	
	1504	3/4 ton cooling, supplementary heat not incl.		6	2.667		1,225	83.50		1,308.50	1,475	
	1506	1 ton cooling, supplementary heat not incl.		4	4		1,325	126		1,451	1,650	
	1510	1.5 ton cooling, 5 MBH heat @ 0°F		1.55	10.323		2,300	325		2,625	3,025	
	1520	2 ton cooling, 6.5 MBH heat @ 0°F		1.50	10.667		2,525	335		2,860	3,275	
	1540	2.5 ton cooling, 8 MBH heat @ 0°F		1.40	11.429		2,875	360		3,235	3,725	
	1560	3 ton cooling, 10 MBH heat @ 0°F		1.20	13.333		3,025	420		3,445	3,950	
	1570	3.5 ton cooling, 11 MBH heat @ 0°F		1	16		3,425	500		3,925	4,525	
	1580	4 ton cooling, 13 MBH heat @ 0°F		.96	16.667		3,800	525		4,325	4,975	
	1620	5 ton cooling, 27 MBH heat @ 0°F		.65	24.615		4,375	770		5,145	6,000	
	1640	7.5 ton cooling, 35 MBH heat @ 0°F		.40	40		8,225	1,250		9,475	11,000	
	2000	Water source to air, single package										
	2100	1 ton cooling, 13 MBH heat @ 75°F	Q-5	2	8	Ea.	965	251		1,216	1,450	
	2120	1.5 ton cooling, 17 MBH heat @ 75°F		1.80	8.889		1,150	279		1,429	1,700	
	2140	2 ton cooling, 19 MBH heat @ 75°F		1.70	9.412		1,225	295		1,520	1,775	
	2160	2.5 ton cooling, 25 MBH heat @ 75°F		1.60	10		1,300	315		1,615	1,900	
	2180	3 ton cooling, 27 MBH heat @ 75°F		1.40	11.429		1,350	360		1,710	2,025	
	2190	3.5 ton cooling, 29 MBH heat @ 75°F		1.30	12.308		1,450	385		1,835	2,150	
	2200	4 ton cooling, 31 MBH heat @ 75°F		1.20	13.333		1,750	420		2,170	2,550	
	2220	5 ton cooling, 29 MBH heat @ 75°F		.90	17.778		2,125	560		2,685	3,200	
	3960	For supplementary heat coil, add					10%					

15766 | Fin Tube Radiation

		15766 Fin Tube Radiation	Crew	Daily Output	Labor-Hours	Unit	2001 Bare Costs Mat.	Labor	Equip.	Total	Total Incl O&P	
190	0010	**HYDRONIC HEATING** Terminal units, not incl. main supply pipe										190
	1000	Radiation										
	1100	Panel, baseboard, C.I., including supports, no covers	Q-5	46	.348	L.F.	21	10.90		31.90	39.50	
	1310	Baseboard, pkgd, 1/2″ copper tube, alum. fin, 7″ high		60	.267		5.05	8.35		13.40	18.25	
	1320	3/4″ copper tube, alum. fin, 7″ high		58	.276		5.35	8.65		14	18.95	
	1381	Rough in baseboard panel & fin tube, supply & balance valves	1 Stpi	1.06	7.547	Ea.	85.50	263		348.50	490	
	1500	Note: fin tube may also require corners, caps, etc.										
	1990	Convector unit, floor recessed, flush, with trim										
	2000	for under large glass wall areas, no damper	Q-5	20	.800	L.F.	24.50	25		49.50	65	
	2100	For unit with damper, add				″	9.95			9.95	10.95	
	3000	Radiators, cast iron										
	3100	Free standing or wall hung, 6 tube, 25″ high	Q-5	96	.167	Section	21.50	5.25		26.75	31.50	
	3150	4 tube 25″ high		96	.167		15.90	5.25		21.15	25.50	
	3200	4 tube, 19″ high		96	.167		14.80	5.25		20.05	24	
	3250	Adj. brackets, 2 per wall radiator up to 30 sections	1 Stpi	32	.250	Ea.	32	8.70		40.70	48	
	3500	Recessed, 20″ high x 5″ deep, without grille	Q-5	60	.267	Section	17.60	8.35		25.95	32	
	3600	For inlet grille, add				″	1.25			1.25	1.38	
	9500	To convert SFR to BTU rating: Hot water, 150 x SFR										
	9510	Forced hot water, 180 x SFR; steam, 240 x SFR										

15768 | Infrared Heaters

		15768 Infrared Heaters	CREW	DAILY OUTPUT	LABOR-HOURS	UNIT	2001 BARE COSTS MAT.	LABOR	EQUIP.	TOTAL	TOTAL INCL O&P	
600	0010	**INFRA-RED UNIT**										600
	0020	Gas fired, unvented, electric ignition, 100% shutoff. Piping and										
	0030	wiring not included										
	0060	Input, 15 MBH	Q-5	7	2.286	Ea.	310	71.50		381.50	450	
	0100	30 MBH		6	2.667		355	83.50		438.50	515	
	0120	45 MBH		5	3.200		390	100		490	575	
	0140	50 MBH		4.50	3.556		420	112		532	635	
	0160	60 MBH		4	4		455	126		581	690	
	0180	75 MBH		3	5.333		485	167		652	790	
	0200	90 MBH		2.50	6.400		545	201		746	905	
	0220	105 MBH		2	8		680	251		931	1,125	
	0240	120 MBH	↓	2	8	↓	805	251		1,056	1,275	
	2000	Electric, single or three phase										
	2050	6 kW, 20,478 BTU	1 Elec	2.30	3.478	Ea.	345	119		464	560	
	2100	13.5 KW, 40,956 BTU		2.20	3.636		535	125		660	775	
	2150	24 KW, 81,912 BTU	↓	2	4	↓	1,175	137		1,312	1,475	
	3000	Oil fired, pump, controls, fusible valve, oil supply tank										
	3050	91,000 BTU	Q-5	2.50	6.400	Ea.	5,100	201		5,301	5,925	
	3080	105.000 BTU		2.25	7.111		5,350	223		5,573	6,200	
	3110	119,000 BTU	↓	2	8	↓	5,625	251		5,876	6,575	

15770 | Floor-Heating & Snow-Melting Eq.

		15770 Floor-Heating & Snow-Melting Eq.	CREW	DAILY OUTPUT	LABOR-HOURS	UNIT	MAT.	LABOR	EQUIP.	TOTAL	TOTAL INCL O&P	
200	0010	**ELECTRIC HEATING**, equipments are not including wires & conduits										200
	0200	Snow melting for paved surface embedded mat heaters & controls	1 Elec	130	.062	S.F.	6.10	2.11		8.21	9.85	
	0400	Cable heating, radiant heat plaster, no controls, in South		130	.062		7.25	2.11		9.36	11.15	
	0600	In North		90	.089		7.25	3.05		10.30	12.55	
	0800	Cable on 1/2" board not incl. controls, tract housing		90	.089		6.10	3.05		9.15	11.25	
	1000	Custom housing		80	.100	↓	7.10	3.43		10.53	12.90	
	1100	Rule of thumb: Baseboard units, including control		4.40	1.818	kW	75	62.50		137.50	176	
	1200	Duct heaters, including controls		5.30	1.509	"	61.50	52		113.50	146	
	1300	Baseboard heaters, 2' long, 375 watt		8	1	Ea.	34	34.50		68.50	88.50	
	1400	3' long, 500 watt		8	1		41	34.50		75.50	96	
	1600	4' long, 750 watt		6.70	1.194		48	41		89	114	
	1800	5' long, 935 watt		5.70	1.404		57	48		105	135	
	2000	6' long, 1125 watt		5	1.600		65	55		120	154	
	2200	7' long, 1310 watt		4.40	1.818		76	62.50		138.50	177	
	2400	8' long, 1500 watt		4	2		81	68.50		149.50	191	
	2600	9' long, 1680 watt		3.60	2.222		134	76		210	261	
	2800	10' long, 1875 watt	↓	3.30	2.424	↓	134	83		217	271	
	2950	Wall heaters with fan, 120 to 277 volt										
	2970	surface mounted, residential, 750 watt	1 Elec	7	1.143	Ea.	105	39		144	175	
	2980	1000 watt		7	1.143		107	39		146	177	
	2990	1250 watt		6	1.333		120	45.50		165.50	201	
	3000	1500 watt		5	1.600		120	55		175	214	
	3160	Recessed, residential, 750 watt		6	1.333		84	45.50		129.50	161	
	3170	1000 watt		6	1.333		85	45.50		130.50	162	
	3180	1250 watt		5	1.600		94	55		149	185	
	3190	1500 watt		4	2		94	68.50		162.50	205	
	3600	Thermostats, integral		16	.500		25	17.15		42.15	53	
	3800	Line voltage, 1 pole		8	1		25	34.50		59.50	78.50	
	3810	2 pole		8	1		25	34.50		59.50	78.50	
	3820	Low voltage, 1 pole	↓	8	1	↓	25	34.50		59.50	78.50	
	4000	Heat trace system, 400 degree										
	4020	115V, 2.5 watts per L.F.	1 Elec	530	.015	L.F.	5.80	.52		6.32	7.15	
	4030	5 watts per L.F.		530	.015		5.80	.52		6.32	7.15	
	4050	10 watts per L.F.	↓	530	.015	↓	7.35	.52		7.87	8.85	

15 MECHANICAL

Important: See the Reference Section for critical supporting data - Reference Nos., Crews, & City Cost Indexes

	15770 | Floor-Heating & Snow-Melting Eq.		CREW	DAILY OUTPUT	LABOR-HOURS	UNIT	2001 BARE COSTS MAT.	LABOR	EQUIP.	TOTAL	TOTAL INCL O&P	
200	4060	220V, 4 watts per L.F.	1 Elec	530	.015	L.F.	5.80	.52		6.32	7.15	200
	4080	480V, 8 watts per L.F.		530	.015		5.80	.52		6.32	7.15	
	4200	Heater raceway										
	4220	5/8" W x 3/8" H	1 Elec	200	.040	L.F.	3.95	1.37		5.32	6.40	
	4240	5/8" W x 1/2" H	"	190	.042	"	5.50	1.44		6.94	8.20	
	4260	Heat transfer cement										
	4280	1 gallon				Ea.	50			50	55	
	4300	5 gallon				"	205			205	226	
	4320	Snap band, clamp										
	4340	3/4" pipe size	1 Elec	470	.017	Ea.		.58		.58	.87	
	4360	1" pipe size		444	.018			.62		.62	.92	
	4380	1-1/4" pipe size		400	.020			.69		.69	1.02	
	4400	1-1/2" pipe size		355	.023			.77		.77	1.15	
	4420	2" pipe size		320	.025			.86		.86	1.28	
	4440	3" pipe size		160	.050			1.72		1.72	2.56	
	4460	4" pipe size		100	.080			2.74		2.74	4.10	
	4480	Single pole thermostat NEMA 4, 30 amp		8	1		500	34.50		534.50	600	
	4500	NEMA 7, 30 amp		7	1.143		590	39		629	705	
	4520	Double pole, NEMA 4, 30 amp		7	1.143		500	39		539	610	
	4540	NEMA 7, 30 amp		6	1.333		590	45.50		635.50	715	
	4560	Thermostat/contactor combination, NEMA 4										
	4580	30 amp 4 pole	1 Elec	3.60	2.222	Ea.	525	76		601	695	
	4600	50 amp 4 pole		3	2.667		530	91.50		621.50	720	
	4620	75 amp 3 pole		2.50	3.200		800	110		910	1,050	
	4640	75 amp 4 pole		2.30	3.478		1,150	119		1,269	1,425	
	4680	Control transformer, 50 VA		4	2		81	68.50		149.50	191	
	4700	75 VA		3.10	2.581		88	88.50		176.50	229	
	4720	Expediter fitting		11	.727		20	25		45	59	
	5000	Radiant heating ceiling panels, 2' x 4', 500 watt		16	.500		183	17.15		200.15	227	
	5050	750 watt		16	.500		202	17.15		219.15	248	
	5200	For recessed plaster frame, add		32	.250		36	8.60		44.60	52.50	
	5300	Infra-red quartz heaters, 120 volts, 1000 watts		6.70	1.194		125	41		166	199	
	5350	1500 watt		5	1.600		125	55		180	220	
	5400	240 volts, 1500 watt		5	1.600		125	55		180	220	
	5450	2000 watt		4	2		125	68.50		193.50	240	
	5500	3000 watt		3	2.667		142	91.50		233.50	293	
	5550	4000 watt		2.60	3.077		252	106		358	435	
	5570	Modulating control		.80	10		66	345		411	585	
	5600	Unit heaters, heavy duty, with fan & mounting bracket										
	5650	Single phase, 208-240-277 volt, 3 kW	1 Elec	3.20	2.500	Ea.	350	86		436	515	
	5700	4 kW		2.80	2.857		360	98		458	540	
	5750	5 kW		2.40	3.333		360	114		474	565	
	5800	7 kW		1.90	4.211		560	144		704	830	
	5850	10 kW		1.30	6.154		610	211		821	985	
	5900	13 kW		1	8		980	274		1,254	1,475	
	5950	15 kW		.90	8.889		1,050	305		1,355	1,600	
	6000	480 volt, 3 kW		3.30	2.424		460	83		543	630	
	6020	4 kW		3	2.667		525	91.50		616.50	715	
	6040	5 kW		2.60	3.077		525	106		631	740	
	6060	7 kW		2	4		650	137		787	920	
	6080	10 kW		1.40	5.714		650	196		846	1,000	
	6100	13 kW		1.10	7.273		1,075	249		1,324	1,550	
	6120	15 kW		1	8		1,075	274		1,349	1,575	
	6140	20 kW		.90	8.889		1,375	305		1,680	1,975	
	6300	3 phase, 208-240 volt, 5 kW		2.40	3.333		355	114		469	560	
	6320	7 kW		1.90	4.211		545	144		689	815	

	15770 | Floor-Heating & Snow-Melting Eq.		CREW	DAILY OUTPUT	LABOR-HOURS	UNIT	2001 BARE COSTS MAT.	LABOR	EQUIP.	TOTAL	TOTAL INCL O&P	
200	6340	10 kW	1 Elec	1.30	6.154	Ea.	590	211		801	965	200
	6360	15 kW		.90	8.889		980	305		1,285	1,525	
	6380	20 kW		.70	11.429		1,400	390		1,790	2,125	
	6400	25 kW		.50	16		1,675	550		2,225	2,675	
	6500	480 volt, 5 kW		2.60	3.077		500	106		606	710	
	6520	7 kW		2	4		625	137		762	895	
	6540	10 kW		1.40	5.714		660	196		856	1,025	
	6560	13 kW		1.10	7.273		1,075	249		1,324	1,550	
	6580	15 kW		1	8		1,075	274		1,349	1,575	
	6600	20 kW		.90	8.889		1,400	305		1,705	1,975	
	6800	Vertical discharge heaters, with fan										
	6820	Single phase, 208-240-277 volt, 10 kW	1 Elec	1.30	6.154	Ea.	1,150	211		1,361	1,575	
	6840	15 kW		.90	8.889		1,400	305		1,705	1,975	
	6900	3 phase, 208-240 volt, 10 kW		1.30	6.154		1,075	211		1,286	1,525	
	6920	15 kW		.90	8.889		1,225	305		1,530	1,800	
	6940	20 kW		.70	11.429		1,450	390		1,840	2,175	
	7100	480 volt, 10 kW		1.40	5.714		1,275	196		1,471	1,700	
	7120	15 kW		1	8		1,400	274		1,674	1,950	
	7140	20 kW		.90	8.889		1,700	305		2,005	2,325	
	7160	25 kW		.60	13.333		1,950	455		2,405	2,825	
	7900	Cabinet convector heaters, 240 volt										
	7920	2' long, 1000 watt	1 Elec	5.30	1.509	Ea.	445	52		497	570	
	7940	1500 watt		5.30	1.509		470	52		522	600	
	7960	2000 watt		5.30	1.509		500	52		552	630	
	7980	3' long, 1500 watt		4.60	1.739		515	59.50		574.50	655	
	8000	2250 watt		4.60	1.739		520	59.50		579.50	660	
	8020	3000 watt		4.60	1.739		575	59.50		634.50	725	
	8040	4' long, 2000 watt		4	2		585	68.50		653.50	745	
	8060	3000 watt		4	2		615	68.50		683.50	775	
	8080	4000 watt		4	2		645	68.50		713.50	810	
	8100	Available also in 208 or 277 volt										
	8200	Cabinet unit heaters, 120 to 277 volt, 1 pole,										
	8220	wall mounted, 2 kW	1 Elec	4.60	1.739	Ea.	1,600	59.50		1,659.50	1,850	
	8230	3 kW		4.60	1.739		1,650	59.50		1,709.50	1,925	
	8240	4 kW		4.40	1.818		1,725	62.50		1,787.50	2,000	
	8250	5 kW		4.40	1.818		1,800	62.50		1,862.50	2,075	
	8260	6 kW		4.20	1.905		1,825	65.50		1,890.50	2,100	
	8270	8 kW		4	2		1,925	68.50		1,993.50	2,225	
	8280	10 kW		3.80	2.105		1,950	72		2,022	2,250	
	8290	12 kW		3.50	2.286		2,100	78.50		2,178.50	2,425	
	8300	13.5 kW		2.90	2.759		2,400	94.50		2,494.50	2,800	
	8310	16 kW		2.70	2.963		2,400	102		2,502	2,800	
	8320	20 kW		2.30	3.478		2,650	119		2,769	3,100	
	8330	24 kW		1.90	4.211		2,950	144		3,094	3,450	
	8350	Recessed, 2 kW		4.40	1.818		1,800	62.50		1,862.50	2,075	
	8370	3 kW		4.40	1.818		1,800	62.50		1,862.50	2,075	
	8380	4 kW		4.20	1.905		1,850	65.50		1,915.50	2,125	
	8390	5 kW		4.20	1.905		1,925	65.50		1,990.50	2,225	
	8400	6 kW		4	2		1,950	68.50		2,018.50	2,225	
	8410	8 kW		3.80	2.105		2,075	72		2,147	2,375	
	8420	10 kW		3.50	2.286		2,125	78.50		2,203.50	2,450	
	8430	12 kW		2.90	2.759		2,250	94.50		2,344.50	2,625	
	8440	13.5 kW		2.70	2.963		2,550	102		2,652	2,950	
	8450	16 kW		2.30	3.478		2,575	119		2,694	3,025	
	8460	20 kW		1.90	4.211		2,800	144		2,944	3,300	
	8470	24 kW		1.60	5		3,075	172		3,247	3,625	

Important: See the Reference Section for critical supporting data - Reference Nos., Crews, & City Cost Indexes

	15770 | Floor-Heating & Snow-Melting Eq.		CREW	DAILY OUTPUT	LABOR-HOURS	UNIT	2001 BARE COSTS MAT.	LABOR	EQUIP.	TOTAL	TOTAL INCL O&P	
200	8490	Ceiling mounted, 2 kW	1 Elec	3.20	2.500	Ea.	1,575	86		1,661	1,875	200
	8510	3 kW		3.20	2.500		1,650	86		1,736	1,925	
	8520	4 kW		3	2.667		1,700	91.50		1,791.50	2,000	
	8530	5 kW		3	2.667		1,775	91.50		1,866.50	2,075	
	8540	6 kW		2.80	2.857		1,800	98		1,898	2,150	
	8550	8 kW		2.40	3.333		1,925	114		2,039	2,275	
	8560	10 kW		2.20	3.636		1,950	125		2,075	2,325	
	8570	12 kW		2	4		2,100	137		2,237	2,500	
	8580	13.5 kW		1.50	5.333		2,225	183		2,408	2,725	
	8590	16 kW		1.30	6.154		2,400	211		2,611	2,975	
	8600	20 kW		.90	8.889		2,650	305		2,955	3,375	
	8610	24 kW	↓	.60	13.333	↓	2,925	455		3,380	3,900	
	8630	208 to 480 V, 3 pole										
	8650	Wall mounted, 2 kW	1 Elec	4.60	1.739	Ea.	1,600	59.50		1,659.50	1,850	
	8670	3 kW		4.60	1.739		1,650	59.50		1,709.50	1,925	
	8680	4 kW		4.40	1.818		1,725	62.50		1,787.50	2,000	
	8690	5 kW		4.40	1.818		1,800	62.50		1,862.50	2,075	
	8700	6 kW		4.20	1.905		1,800	65.50		1,865.50	2,100	
	8710	8 kW		4	2		1,925	68.50		1,993.50	2,225	
	8720	10 kW		3.80	2.105		1,975	72		2,047	2,275	
	8730	12 kW		3.50	2.286		2,100	78.50		2,178.50	2,425	
	8740	13.5 kW		2.90	2.759		2,400	94.50		2,494.50	2,775	
	8750	16 kW		2.70	2.963		2,400	102		2,502	2,800	
	8760	20 kW		2.30	3.478		2,650	119		2,769	3,100	
	8770	24 kW		1.90	4.211		2,950	144		3,094	3,450	
	8790	Recessed, 2 kW		4.40	1.818		2,100	62.50		2,162.50	2,400	
	8810	3 kW		4.40	1.818		1,775	62.50		1,837.50	2,050	
	8820	4 kW		4.20	1.905		1,850	65.50		1,915.50	2,125	
	8830	5 kW		4.20	1.905		1,925	65.50		1,990.50	2,200	
	8840	6 kW		4	2		1,950	68.50		2,018.50	2,225	
	8850	8 kW		3.80	2.105		2,075	72		2,147	2,375	
	8860	10 kW		3.50	2.286		2,125	78.50		2,203.50	2,450	
	8870	12 kW		2.90	2.759		2,225	94.50		2,319.50	2,600	
	8880	13.5 kW		2.70	2.963		2,550	102		2,652	2,950	
	8890	16 kW		2.30	3.478		2,575	119		2,694	3,025	
	8900	20 kW		1.90	4.211		2,800	144		2,944	3,300	
	8920	24 kW		1.60	5		3,075	172		3,247	3,625	
	8940	Ceiling mount, 2 kW		3.20	2.500		1,575	86		1,661	1,875	
	8950	3 kW		3.20	2.500		1,650	86		1,736	1,925	
	8960	4 kW		3	2.667		1,700	91.50		1,791.50	2,000	
	8970	5 kW		3	2.667		1,775	91.50		1,866.50	2,075	
	8980	6 kW		2.80	2.857		1,800	98		1,898	2,150	
	8990	8 kW		2.40	3.333		1,925	114		2,039	2,275	
	9000	10 kW		2.20	3.636		1,950	125		2,075	2,325	
	9020	13.5 kW		1.50	5.333		2,100	183		2,283	2,575	
	9030	16 kW		1.30	6.154		2,400	211		2,611	2,975	
	9040	20 kW		.90	8.889		2,650	305		2,955	3,375	
	9060	24 kW	↓	.60	13.333	↓	2,950	455		3,405	3,900	
500	0010	**RADIANT FLOOR HEATING**										500
	0100	Tubing, PEX (cross-linked polyethylene)										
	0110	Oxygen barrier type for systems with ferrous materials										
	0120	1/2"	Q-5	800	.020	L.F.	.49	.63		1.12	1.49	
	0130	3/4"		535	.030		.62	.94		1.56	2.10	
	0140	1"	↓	400	.040	↓	1.25	1.26		2.51	3.28	
	0200	Non barrier type for ferrous free systems										
	0210	1/2"	Q-5	800	.020	L.F.	.27	.63		.90	1.25	

MECHANICAL 15

15770 | Floor-Heating & Snow-Melting Eq.		CREW	DAILY OUTPUT	LABOR-HOURS	UNIT	2001 BARE COSTS MAT.	LABOR	EQUIP.	TOTAL	TOTAL INCL O&P
500	0220 3/4"	Q-5	535	.030	L.F.	.44	.94		1.38	1.90 500
	0230 1"	↓	400	.040	↓	.77	1.26		2.03	2.75
	1000 Manifolds									
	1110 Brass									
	1120 Valved									
	1130 1", 2 circuit	Q-5	13	1.231	Ea.	36	38.50		74.50	98
	1140 1", 3 circuit		11	1.455		49	45.50		94.50	123
	1150 1", 4 circuit		9	1.778		66	56		122	157
	1160 1-1/4", 3 circuit		10	1.600		55.50	50		105.50	137
	1170 1-1/4", 4 circuit	↓	8	2	↓	67.50	63		130.50	170
	1300 Valveless									
	1310 1", 2 circuit	Q-5	14	1.143	Ea.	18	36		54	74
	1320 1", 3 circuit		12	1.333		24	42		66	89.50
	1330 1", 4 circuit		10	1.600		31	50		81	110
	1340 1-1/4", 3 circuit		11	1.455		30	45.50		75.50	102
	1350 1-1/4", 4 circuit	↓	9	1.778	↓	30	56		86	117
	1400 Manifold connector kit									
	1410 Mounting brackets, couplings, end cap, air vent, valve, plug									
	1420 1"	Q-5	8	2	Ea.	61.50	63		124.50	163
	1430 1-1/4"	"	7	2.286	"	75	71.50		146.50	191
	1610 Copper, (cut to size)									
	1620 1" x 1/2" x 72" Lg, 2" sweat drops, 24 circuit	Q-5	3.33	4.805	Ea.	62.50	151		213.50	297
	1630 1-1/4" x 1/2" x 72" Lg, 2" sweat drops, 24 circuit		3.20	5		73	157		230	315
	1640 1-1/2" x 1/2" x 72" Lg, 2" sweat drops, 24 circuit		3	5.333		88.50	167		255.50	350
	1650 1-1/4" x 3/4" x 72" Lg, 2" sweat drops, 24 circuit		3.10	5.161		78	162		240	330
	1660 1-1/2" x 3/4" x 72" Lg, 2" sweat drops, 24 circuit	↓	2.90	5.517	↓	78	173		251	345
	1710 Copper, (modular)									
	1720 1" x 1/2", 2" sweat drops, 2 circuit	Q-5	17.50	.914	Ea.	5.85	28.50		34.35	50
	1730 1" x 1/2", 2" sweat drops, 3 circuit		14.50	1.103		8.80	34.50		43.30	62
	1740 1" x 1/2", 2" sweat drops, 4 circuit		12.30	1.301		10.15	41		51.15	72.50
	1750 1-1/4" x 1/2", 2" sweat drops, 2 circuit		17	.941		6.75	29.50		36.25	52
	1760 1-1/4" x 1/2", 2" sweat drops, 3 circuit		14	1.143		10.15	36		46.15	65
	1770 1-1/4" x 1/2", 2" sweat drops, 4 circuit	↓	11.80	1.356	↓	13.50	42.50		56	79
	3000 Valves									
	3110 Motorized zone valve operator, 24V	Q-5	30	.533	Ea.	37	16.75		53.75	66
	3120 Motorized zone valve with operator complete, 24V									
	3130 3/4"	Q-5	35	.457	Ea.	64.50	14.35		78.85	92.50
	3140 1"		32	.500		73.50	15.70		89.20	105
	3150 1-1/4"		29.60	.541		90	16.95		106.95	125
	3200 Flow control/isolation valve, 1/2"	↓	32	.500	↓	20.50	15.70		36.20	46
	3300 Thermostatic mixing valves									
	3310 1/2"	Q-5	26.60	.601	Ea.	67.50	18.90		86.40	103
	3320 3/4"		25	.640		82.50	20		102.50	122
	3330 1"	↓	21.40	.748	↓	88	23.50		111.50	132
	3400 3 Way mixing / diverting valve, manual, brass									
	3410 1/2"	Q-5	21.30	.751	Ea.	86.50	23.50		110	131
	3420 3/4"		20	.800		88.50	25		113.50	136
	3430 1"	↓	17.80	.899	↓	104	28		132	157
	3500 4 Way mixing / diverting valve, manual, brass									
	3510 1/2"	Q-5	16	1	Ea.	92.50	31.50		124	149
	3520 3/4"		15	1.067		104	33.50		137.50	165
	3530 1"		13.30	1.203		116	38		154	184
	3540 1-1/4"		11.40	1.404		127	44		171	206
	3550 1-1/2"		10.60	1.509		170	47.50		217.50	259
	3560 2"		10	1.600		184	50		234	278
	3570 2-1/2"	↓	8	2	↓	505	63		568	650

15700 | Heating/Ventilating/Air Conditioning Equipment

15770 | Floor-Heating & Snow-Melting Eq.

			CREW	DAILY OUTPUT	LABOR-HOURS	UNIT	2001 BARE COSTS MAT.	LABOR	EQUIP.	TOTAL	TOTAL INCL O&P	
500	3800	Motor control, 3 or 4 way, for valves up to 1-1/2"	Q-5	30	.533	Ea.	124	16.75		140.75	162	500
	5000	Radiant floor heating, zone control panel										
	5110	6 Zone control										
	5120	For zone valves	Q-5	16	1	Ea.	146	31.50		177.50	209	
	5130	For circulators	"	16	1	"	156	31.50		187.50	220	
	7000	PEX tubing fittings, compression type										
	7110	PEX tubing to brass valve conn., 1/2"	Q-5	54	.296	Ea.	2.12	9.30		11.42	16.40	
	7200	PEX x male NPT										
	7210	1/2" x 1/2"	Q-5	54	.296	Ea.	4	9.30		13.30	18.45	
	7220	3/4" x 3/4"	↓	46	.348	↓	6.75	10.90		17.65	24	
	7230	1" x 1"	↓	40	.400	↓	15	12.55		27.55	35.50	
	7300	PEX coupling										
	7310	1/2" x 1/2"	Q-5	50	.320	Ea.	4.88	10.05		14.93	20.50	
	7320	3/4" x 3/4"	↓	42	.381	↓	14.05	11.95		26	33.50	
	7330	1" x 1"	↓	36	.444	↓	18	13.95		31.95	41	
	7400	PEX x copper sweat										
	7410	1/2" x 1/2"	Q-5	38	.421	Ea.	3.50	13.20		16.70	24	
	7420	3/4" x 3/4"	↓	34	.471	↓	7.50	14.75		22.25	31	
	7430	1" x 1"	↓	29	.552	↓	16.50	17.30		33.80	44	

15800 | Air Distribution

15830 | Fans

			CREW	DAILY OUTPUT	LABOR-HOURS	UNIT	2001 BARE COSTS MAT.	LABOR	EQUIP.	TOTAL	TOTAL INCL O&P	
100	0010	**FANS**										100
	2500	Ceiling fan, right angle, extra quiet, 0.10" S.P.										
	2520	95 CFM	Q-20	20	1	Ea.	110	31		141	169	
	2540	210 CFM		19	1.053		130	32.50		162.50	193	
	2560	385 CFM	↓	18	1.111		165	34.50		199.50	235	
	2640	For wall or roof cap, add	1 Shee	16	.500		110	16.80		126.80	147	
	2660	For straight thru fan, add					10%					
	2680	For speed control switch, add	1 Elec	16	.500	↓	60	17.15		77.15	91.50	
	6650	Residential, bath exhaust, grille, back draft damper										
	6660	50 CFM	Q-20	24	.833	Ea.	28.50	26		54.50	71	
	6670	110 CFM		22	.909		47	28		75	95	
	6680	Light combination, squirrel cage, 100 watt, 70 CFM	↓	24	.833	↓	58	26		84	104	
	6700	Light/heater combination, ceiling mounted										
	6710	70 CFM, 1450 watt	Q-20	24	.833	Ea.	70.50	26		96.50	117	
	6800	Heater combination, recessed, 70 CFM		24	.833		33.50	26		59.50	76	
	6820	With 2 infrared bulbs	↓	23	.870	↓	50.50	27		77.50	97	
	6940	Residential roof jacks and wall caps										
	6944	Wall cap with back draft damper										
	6946	3" & 4" dia. round duct	1 Shee	11	.727	Ea.	12.05	24.50		36.55	51	
	6948	6" dia. round duct	"	11	.727	"	29	24.50		53.50	69.50	
	6958	Roof jack with bird screen and back draft damper										
	6960	3" & 4" dia. round duct	1 Shee	11	.727	Ea.	11.60	24.50		36.10	50.50	
	6962	3-1/4" x 10" rectangular duct	"	10	.800	"	21.50	27		48.50	65	
	6980	Transition										
	6982	3-1/4" x 10" to 6" dia. round	1 Shee	20	.400	Ea.	13.05	13.40		26.45	35	

15860 | Air Cleaning Devices

			CREW	DAILY OUTPUT	LABOR-HOURS	UNIT	2001 BARE COSTS MAT.	LABOR	EQUIP.	TOTAL	TOTAL INCL O&P	
500	0010	EXHAUST SYSTEMS A8.5-110										500
	0500	Engine exhaust, garage, in-floor system										
	0510	Single tube outlet assemblies										
	0520	For transite pipe ducting, self-storing tube										
	0530	3" tubing adapter plate	1 Shee	16	.500	Ea.	162	16.80		178.80	204	
	0540	4" tubing adapter plate		16	.500		162	16.80		178.80	204	
	0550	5" tubing adapter plate		16	.500		162	16.80		178.80	204	
	0600	For vitrified tile ducting										
	0610	3" tubing adapter plate, self-storing tube	1 Shee	16	.500	Ea.	162	16.80		178.80	204	
	0620	4" tubing adapter plate, self-storing tube		16	.500		162	16.80		178.80	204	
	0660	5" tubing adapter plate, self-storing tube		16	.500		162	16.80		178.80	204	
	0800	Two tube outlet assemblies										
	0810	For transite pipe ducting, self-storing tube										
	0820	3" tubing, dual exhaust adapter plate	1 Shee	16	.500	Ea.	162	16.80		178.80	204	
	0850	For vitrified tile ducting										
	0860	3" tubing, dual exhaust, self-storing tube	1 Shee	16	.500	Ea.	162	16.80		178.80	204	
	0870	3" tubing, double outlet, non-storing tubes	"	16	.500	"	162	16.80		178.80	204	
	0900	Accessories for metal tubing, (overhead systems also)										
	0910	Adapters, for metal tubing end										
	0920	3" tail pipe type				Ea.	27			27	29.50	
	0930	4" tail pipe type					27			27	29.50	
	0940	5" tail pipe type					27			27	29.50	
	0990	5" diesel stack type					67.50			67.50	74.50	
	1000	6" diesel stack type					67.50			67.50	74.50	
	1100	Bullnose (guide) required for in-floor assemblies										
	1110	3" tubing size				Ea.	13.50			13.50	14.85	
	1120	4" tubing size					13.50			13.50	14.85	
	1130	5" tubing size					13.50			13.50	14.85	
	1150	Plain rings, for tubing end										
	1160	3" tubing size				Ea.	9			9	9.90	
	1170	4" tubing size				"	9			9	9.90	
	1200	Tubing, galvanized, flexible, (for overhead systems also)										
	1210	3" ID				L.F.	4.82			4.82	5.30	
	1220	4" ID					5.75			5.75	6.35	
	1230	5" ID					6.75			6.75	7.45	
	1240	6" ID					7.65			7.65	8.40	
	1250	Stainless steel, flexible, (for overhead sys, also)										
	1260	3" ID				L.F.	10			10	11	
	1270	4" ID					13			13	14.30	
	1280	5" ID					15.70			15.70	17.30	
	1290	6" ID					18.70			18.70	20.50	
	1500	Engine exhaust, garage, ovrhd. compon., for neoprene tubing										
	1510	Alternate metal tubing & accessories see above										
	1550	Adapters, for neoprene tubing end										
	1560	3" tail pipe, adjustable, neoprene				Ea.	27			27	29.50	
	1570	3" tail pipe, heavy wall neoprene					38			38	41.50	
	1580	4" tail pipe, heavy wall neoprene					39.50			39.50	43.50	
	1590	5" tail pipe, heavy wall neoprene					58.50			58.50	64.50	
	1650	Connectors, tubing										
	1660	3" interior, aluminum				Ea.	16.20			16.20	17.80	
	1670	4" interior, aluminum					27			27	29.50	
	1710	5" interior, neoprene					27			27	29.50	
	1750	3" spiralock, neoprene					9			9	9.90	
	1760	4" spiralock, neoprene					11.70			11.70	12.85	
	1780	Y for 3" ID tubing, neoprene, dual exhaust					19.80			19.80	22	
	1790	Y for 4" ID tubing, aluminum, dual exhaust					40.50			40.50	44.50	

Important: See the Reference Section for critical supporting data - Reference Nos., Crews, & City Cost Indexes

15860 | Air Cleaning Devices

			CREW	DAILY OUTPUT	LABOR-HOURS	UNIT	2001 BARE COSTS MAT.	LABOR	EQUIP.	TOTAL	TOTAL INCL O&P	
500	1850	Elbows, aluminum, splice into tubing for strap A8.5-110										500
	1860	3″ neoprene tubing size				Ea.	23.50			23.50	25.50	
	1870	4″ neoprene tubing size					31.50			31.50	34.50	
	1900	Flange assemblies, connect tubing to overhead duct					29			29	31.50	
	2000	Hardware and accessories										
	2020	Cable, galvanized, 1/8″ diameter				L.F.	.27			.27	.30	
	2040	Cleat, tie down cable or rope				Ea.	3.60			3.60	3.96	
	2060	Pulley					4.50			4.50	4.95	
	2080	Pulley hook, universal					3.60			3.60	3.96	
	2100	Rope, nylon, 1/4″ diameter				L.F.	.26			.26	.29	
	2120	Winch, 1″ diameter				Ea.	72			72	79	
	2150	Lifting strap, mounts on neoprene										
	2160	3″ tubing size				Ea.	10.80			10.80	11.90	
	2170	4″ tubing size					10.80			10.80	11.90	
	2180	5″ tubing size					10.80			10.80	11.90	
	2190	6″ tubing size					10.80			10.80	11.90	
	2200	Tubing, neoprene, 11′ lengths										
	2210	3″ ID				L.F.	6.10			6.10	6.70	
	2220	4″ ID					8.90			8.90	9.80	
	2230	5″ ID					12.95			12.95	14.20	
	2500	Engine exhaust, thru-door outlet										
	2510	3″ tube size	1 Carp	16	.500	Ea.	31.50	14.60		46.10	57.50	
	2530	4″ tube size	"	16	.500	"	31.50	14.60		46.10	57.50	
	3000	Tubing, exhaust, flex hose, with										
	3010	coupler, damper and tail pipe adapter										
	3020	Neoprene										
	3040	3″ x 20′	1 Shee	6	1.333	Ea.	185	44.50		229.50	272	
	3050	4″ x 15′		5.40	1.481		203	49.50		252.50	300	
	3060	4″ x 20′		5	1.600		248	53.50		301.50	355	
	3070	5″ x 15′		4.40	1.818		300	61		361	425	
	3100	Galvanized										
	3110	3″ x 20′	1 Shee	6	1.333	Ea.	149	44.50		193.50	232	
	3120	4″ x 17′		5.60	1.429		155	48		203	244	
	3130	4″ x 20′		5	1.600		172	53.50		225.50	272	
	3140	5″ x 17′		4.60	1.739		190	58.50		248.50	299	
	7500	Welding fume elimination accessories for garage exhaust systems										
	7600	Cut off (blast gate)										
	7610	3″ tubing size, 3″ x 6″ opening	1 Shee	24	.333	Ea.	16.20	11.20		27.40	35	
	7620	4″ tubing size, 4″ x 8″ opening		24	.333		18	11.20		29.20	37	
	7630	5″ tubing size, 5″ x 10″ opening		24	.333		21.50	11.20		32.70	41.50	
	7640	6″ tubing size		24	.333		25	11.20		36.20	45	
	7650	8″ tubing size		24	.333		36	11.20		47.20	57	
	7700	Hoods, magnetic, with handle & screen										
	7710	3″ tubing size, 3″ x 6″ opening	1 Shee	24	.333	Ea.	67.50	11.20		78.70	92	
	7720	4″ tubing size, 4″ x 8″ opening		24	.333		67.50	11.20		78.70	92	
	7730	5″ tubing size, 5″ x 10″ opening		24	.333		67.50	11.20		78.70	92	

15900 | HVAC Instrumentation & Controls

	15905 | HVAC Instrumentation		CREW	DAILY OUTPUT	LABOR-HOURS	UNIT	2001 BARE COSTS MAT.	LABOR	EQUIP.	TOTAL	TOTAL INCL O&P	
960	0010	WATER LEVEL CONTROLS										960
	3000	Low water cut-off for hot water boiler, 50 psi maximum										
	3100	1″ top & bottom equalizing pipes, manual reset	1 Stpi	14	.571	Ea.	198	19.90		217.90	248	
	3200	1″ top & bottom equalizing pipes		14	.571		198	19.90		217.90	248	
	3300	2-1/2″ side connection for nipple-to-boiler		14	.571		175	19.90		194.90	222	

15950 | Testing/Adjusting/Balancing

	15955 | HVAC Test/Adjust/Balance		CREW	DAILY OUTPUT	LABOR-HOURS	UNIT	2001 BARE COSTS MAT.	LABOR	EQUIP.	TOTAL	TOTAL INCL O&P	
700	0010	PIPING, TESTING										700
	0100	Nondestructive testing										
	0110	Nondestructive hydraulic pressure test										
	0120	1″ - 4″ pipe										
	0140	0 - 250 L.F.	1 Stpi	.75	10.667	Ea.		370		370	560	
	0160	250 - 500 L.F.	″	.64	12.500			435		435	660	
	0180	500 - 1000 L.F.	Q-5	1	16			500		500	760	
	0200	1000 - 2000 L.F.	″	.53	30.189			945		945	1,425	
	0300	6″ - 10″ pipe										
	0320	0 - 250 L.F.	Q-5	1	16	Ea.		500		500	760	
	0340	250 - 500 L.F.		.33	48.485			1,525		1,525	2,300	
	0360	500 - 1000 L.F.		.21	76.190			2,400		2,400	3,600	
	0380	1000 - 2000 L.F.		.11	145			4,575		4,575	6,900	
	1000	Pneumatic pressure test, includes soaping joints										
	1120	1″ - 4″ pipe										
	1140	0 - 250 L.F.	Q-5	.90	17.778	Ea.	6.30	560		566.30	845	
	1160	250 - 500 L.F.		.55	29.091		12.60	915		927.60	1,400	
	1180	500 - 1000 L.F.		.45	35.556		18.90	1,125		1,143.90	1,700	
	1200	1000 - 2000 L.F.		.23	69.565		25	2,175		2,200	3,325	
	1300	6″ - 10″ pipe										
	1320	0 - 250 L.F.	Q-5	.45	35.556	Ea.	6.30	1,125		1,131.30	1,675	
	1340	250 - 500 L.F.		.29	55.172		12.60	1,725		1,737.60	2,650	
	1360	500 - 1000 L.F.		.18	88.889		25	2,800		2,825	4,225	
	1380	1000 - 2000 L.F.		.09	177		31.50	5,575		5,606.50	8,450	
	2000	X-Ray of welds										
	2110	2″ diam.	1 Stpi	8	1	Ea.	3	35		38	56	
	2120	3″ diam.		8	1		3	35		38	56	
	2130	4″ diam.		8	1		4.50	35		39.50	57.50	
	2140	6″ diam.		8	1		4.50	35		39.50	57.50	
	2150	8″ diam.		6.60	1.212		4.50	42		46.50	69	
	2160	10″ diam.		6	1.333		6	46.50		52.50	76.50	
	3000	Liquid penetration of welds										
	3110	2″ diam.	1 Stpi	14	.571	Ea.	1	19.90		20.90	31	
	3120	3″ diam.		13.60	.588		1	20.50		21.50	32	
	3130	4″ diam.		13.40	.597		1	21		22	32.50	
	3140	6″ diam.		13.20	.606		1	21		22	33	
	3150	8″ diam.		13	.615		1.50	21.50		23	34	
	3160	10″ diam.		12.80	.625		1.50	22		23.50	34.50	

For information about Means Estimating Seminars, see yellow pages 11 and 12 in back of book

Division 16 Electrical

Estimating Tips

16060 Grounding & Bonding

- When taking off grounding system, identify separately the type and size of wire and list each unique type of ground connection.

16100 Wiring Methods

- Conduit should be taken off in three main categories: power distribution, branch power, and branch lighting, so the estimator can concentrate on systems and components, therefore making it easier to ensure all items have been accounted for.
- For cost modifications for elevated conduit installation, add the percentages to labor according to the height of installation and only the quantities exceeding the different height levels, not to the total conduit quantities.
- Remember that aluminum wiring of equal ampacity is larger in diameter than copper and may require larger conduit.
- If more than three wires at a time are being pulled, deduct percentages from the labor hours of that grouping of wires.
- The estimator should take the weights of materials into consideration when completing a takeoff. Topics to consider include: How will the materials be supported? What methods of support are available? How high will the support structure have to reach? Will the final support structure be able to withstand the total burden? Is the support material included or separate from the fixture, equipment and material specified?

16200 Electrical Power

- Do not overlook the costs for equipment used in the installation. If scaffolding or highlifts are available in the field, contractors may use them in lieu of the proposed ladders and rolling staging.

16400 Low-Voltage Distribution

- Supports and concrete pads may be shown on drawings for the larger equipment, or the support system may be just a piece of plywood for the back of a panelboard. In either case, it must be included in the costs.

16500 Lighting

- Fixtures should be taken off room by room, using the fixture schedule, specifications, and the ceiling plan. For large concentrations of lighting fixtures in the same area deduct the percentages from labor hours.

16700 Communications
16800 Sound & Video

- When estimating material costs for special systems, it is always prudent to obtain manufacturers' quotations for equipment prices and special installation requirements which will affect the total costs.

Reference Numbers

Reference numbers are shown in bold squares at the beginning of some major classifications. These numbers refer to related items in the Reference Section. The reference information may be an estimating procedure, an alternate pricing method or technical information.

Note: Not all subdivisions listed here necessarily appear in this publication.

16132 Conduit & Tubing			CREW	DAILY OUTPUT	LABOR-HOURS	UNIT	2001 BARE COSTS MAT.	LABOR	EQUIP.	TOTAL	TOTAL INCL O&P	
260	0010	CUTTING AND DRILLING										260
	0100	Hole drilling to 10′ high, concrete wall										
	0110	8″ thick, 1/2″ pipe size	1 Elec	12	.667	Ea.		23		23	34	
	0120	3/4″ pipe size		12	.667			23		23	34	
	0130	1″ pipe size		9.50	.842			29		29	43	
	0140	1-1/4″ pipe size		9.50	.842			29		29	43	
	0150	1-1/2″ pipe size		9.50	.842			29		29	43	
	0160	2″ pipe size		4.40	1.818			62.50		62.50	93	
	0170	2-1/2″ pipe size		4.40	1.818			62.50		62.50	93	
	0180	3″ pipe size		4.40	1.818			62.50		62.50	93	
	0190	3-1/2″ pipe size		3.30	2.424			83		83	124	
	0200	4″ pipe size		3.30	2.424			83		83	124	
	0500	12″ thick, 1/2″ pipe size		9.40	.851			29		29	43.50	
	0520	3/4″ pipe size		9.40	.851			29		29	43.50	
	0540	1″ pipe size		7.30	1.096			37.50		37.50	56	
	0560	1-1/4″ pipe size		7.30	1.096			37.50		37.50	56	
	0570	1-1/2″ pipe size		7.30	1.096			37.50		37.50	56	
	0580	2″ pipe size		3.60	2.222			76		76	114	
	0590	2-1/2″ pipe size		3.60	2.222			76		76	114	
	0600	3″ pipe size		3.60	2.222			76		76	114	
	0610	3-1/2″ pipe size		2.80	2.857			98		98	146	
	0630	4″ pipe size		2.50	3.200			110		110	164	
	0650	16″ thick, 1/2″ pipe size		7.60	1.053			36		36	54	
	0670	3/4″ pipe size		7	1.143			39		39	58.50	
	0690	1″ pipe size		6	1.333			45.50		45.50	68.50	
	0710	1-1/4″ pipe size		5.50	1.455			50		50	74.50	
	0730	1-1/2″ pipe size		5.50	1.455			50		50	74.50	
	0750	2″ pipe size		3	2.667			91.50		91.50	137	
	0770	2-1/2″ pipe size		2.70	2.963			102		102	152	
	0790	3″ pipe size		2.50	3.200			110		110	164	
	0810	3-1/2″ pipe size		2.30	3.478			119		119	178	
	0830	4″ pipe size		2	4			137		137	205	
	0850	20″ thick, 1/2″ pipe size		6.40	1.250			43		43	64	
	0870	3/4″ pipe size		6	1.333			45.50		45.50	68.50	
	0890	1″ pipe size		5	1.600			55		55	82	
	0910	1-1/4″ pipe size		4.80	1.667			57		57	85.50	
	0930	1-1/2″ pipe size		4.60	1.739			59.50		59.50	89	
	0950	2″ pipe size		2.70	2.963			102		102	152	
	0970	2-1/2″ pipe size		2.40	3.333			114		114	171	
	0990	3″ pipe size		2.20	3.636			125		125	186	
	1010	3-1/2″ pipe size		2	4			137		137	205	
	1030	4″ pipe size		1.70	4.706			161		161	241	
	1050	24″ thick, 1/2″ pipe size		5.50	1.455			50		50	74.50	
	1070	3/4″ pipe size		5.10	1.569			54		54	80.50	
	1090	1″ pipe size		4.30	1.860			64		64	95.50	
	1110	1-1/4″ pipe size		4	2			68.50		68.50	102	
	1130	1-1/2″ pipe size		4	2			68.50		68.50	102	
	1150	2″ pipe size		2.40	3.333			114		114	171	
	1170	2-1/2″ pipe size		2.20	3.636			125		125	186	
	1190	3″ pipe size		2	4			137		137	205	
	1210	3-1/2″ pipe size		1.80	4.444			152		152	228	
	1230	4″ pipe size		1.50	5.333			183		183	273	
	1500	Brick wall, 8″ thick, 1/2″ pipe size		18	.444			15.25		15.25	23	
	1520	3/4″ pipe size		18	.444			15.25		15.25	23	
	1540	1″ pipe size		13.30	.601			20.50		20.50	31	
	1560	1-1/4″ pipe size	↓	13.30	.601	↓		20.50		20.50	31	

16132 | Conduit & Tubing

			CREW	DAILY OUTPUT	LABOR-HOURS	UNIT	2001 BARE COSTS MAT.	LABOR	EQUIP.	TOTAL	TOTAL INCL O&P	
260	1580	1-1/2" pipe size	1 Elec	13.30	.601	Ea.		20.50		20.50	31	260
	1600	2" pipe size		5.70	1.404			48		48	72	
	1620	2-1/2" pipe size		5.70	1.404			48		48	72	
	1640	3" pipe size		5.70	1.404			48		48	72	
	1660	3-1/2" pipe size		4.40	1.818			62.50		62.50	93	
	1680	4" pipe size		4	2			68.50		68.50	102	
	1700	12" thick, 1/2" pipe size		14.50	.552			18.90		18.90	28.50	
	1720	3/4" pipe size		14.50	.552			18.90		18.90	28.50	
	1740	1" pipe size		11	.727			25		25	37	
	1760	1-1/4" pipe size		11	.727			25		25	37	
	1780	1-1/2" pipe size		11	.727			25		25	37	
	1800	2" pipe size		5	1.600			55		55	82	
	1820	2-1/2" pipe size		5	1.600			55		55	82	
	1840	3" pipe size		5	1.600			55		55	82	
	1860	3-1/2" pipe size		3.80	2.105			72		72	108	
	1880	4" pipe size		3.30	2.424			83		83	124	
	1900	16" thick, 1/2" pipe size		12.30	.650			22.50		22.50	33.50	
	1920	3/4" pipe size		12.30	.650			22.50		22.50	33.50	
	1940	1" pipe size		9.30	.860			29.50		29.50	44	
	1960	1-1/4" pipe size		9.30	.860			29.50		29.50	44	
	1980	1-1/2" pipe size		9.30	.860			29.50		29.50	44	
	2000	2" pipe size		4.40	1.818			62.50		62.50	93	
	2010	2-1/2" pipe size		4.40	1.818			62.50		62.50	93	
	2030	3" pipe size		4.40	1.818			62.50		62.50	93	
	2050	3-1/2" pipe size		3.30	2.424			83		83	124	
	2070	4" pipe size		3	2.667			91.50		91.50	137	
	2090	20" thick, 1/2" pipe size		10.70	.748			25.50		25.50	38.50	
	2110	3/4" pipe size		10.70	.748			25.50		25.50	38.50	
	2130	1" pipe size		8	1			34.50		34.50	51	
	2150	1-1/4" pipe size		8	1			34.50		34.50	51	
	2170	1-1/2" pipe size		8	1			34.50		34.50	51	
	2190	2" pipe size		4	2			68.50		68.50	102	
	2210	2-1/2" pipe size		4	2			68.50		68.50	102	
	2230	3" pipe size		4	2			68.50		68.50	102	
	2250	3-1/2" pipe size		3	2.667			91.50		91.50	137	
	2270	4" pipe size		2.70	2.963			102		102	152	
	2290	24" thick, 1/2" pipe size		9.40	.851			29		29	43.50	
	2310	3/4" pipe size		9.40	.851			29		29	43.50	
	2330	1" pipe size		7.10	1.127			38.50		38.50	57.50	
	2350	1-1/4" pipe size		7.10	1.127			38.50		38.50	57.50	
	2370	1-1/2" pipe size		7.10	1.127			38.50		38.50	57.50	
	2390	2" pipe size		3.60	2.222			76		76	114	
	2410	2-1/2" pipe size		3.60	2.222			76		76	114	
	2430	3" pipe size		3.60	2.222			76		76	114	
	2450	3-1/2" pipe size		2.80	2.857			98		98	146	
	2470	4" pipe size		2.50	3.200			110		110	164	
	3000	Knockouts to 8' high, metal boxes & enclosures										
	3020	With hole saw, 1/2" pipe size	1 Elec	53	.151	Ea.		5.20		5.20	7.75	
	3040	3/4" pipe size		47	.170			5.85		5.85	8.70	
	3050	1" pipe size		40	.200			6.85		6.85	10.25	
	3060	1-1/4" pipe size		36	.222			7.60		7.60	11.40	
	3070	1-1/2" pipe size		32	.250			8.60		8.60	12.80	
	3080	2" pipe size		27	.296			10.15		10.15	15.15	
	3090	2-1/2" pipe size		20	.400			13.70		13.70	20.50	
	4010	3" pipe size		16	.500			17.15		17.15	25.50	
	4030	3-1/2" pipe size		13	.615			21		21	31.50	

16100 | Wiring Methods

16132 | Conduit & Tubing

			CREW	DAILY OUTPUT	LABOR-HOURS	UNIT	2001 BARE COSTS MAT.	LABOR	EQUIP.	TOTAL	TOTAL INCL O&P	
260	4050	4″ pipe size	1 Elec	11	.727	Ea.		25		25	37	260
	4070	With hand punch set, 1/2″ pipe size		40	.200			6.85		6.85	10.25	
	4090	3/4″ pipe size		32	.250			8.60		8.60	12.80	
	4110	1″ pipe size		30	.267			9.15		9.15	13.65	
	4130	1-1/4″ pipe size		28	.286			9.80		9.80	14.65	
	4150	1-1/2″ pipe size		26	.308			10.55		10.55	15.75	
	4170	2″ pipe size		20	.400			13.70		13.70	20.50	
	4190	2-1/2″ pipe size		17	.471			16.15		16.15	24	
	4200	3″ pipe size		15	.533			18.30		18.30	27.50	
	4220	3-1/2″ pipe size		12	.667			23		23	34	
	4240	4″ pipe size		10	.800			27.50		27.50	41	
	4260	With hydraulic punch, 1/2″ pipe size		44	.182			6.25		6.25	9.30	
	4280	3/4″ pipe size		38	.211			7.20		7.20	10.80	
	4300	1″ pipe size		38	.211			7.20		7.20	10.80	
	4320	1-1/4″ pipe size		38	.211			7.20		7.20	10.80	
	4340	1-1/2″ pipe size		38	.211			7.20		7.20	10.80	
	4360	2″ pipe size		32	.250			8.60		8.60	12.80	
	4380	2-1/2″ pipe size		27	.296			10.15		10.15	15.15	
	4400	3″ pipe size		23	.348			11.95		11.95	17.80	
	4420	3-1/2″ pipe size		20	.400			13.70		13.70	20.50	
	4440	4″ pipe size		18	.444			15.25		15.25	23	

16139 | Residential Wiring

			CREW	DAILY OUTPUT	LABOR-HOURS	UNIT	MAT.	LABOR	EQUIP.	TOTAL	TOTAL INCL O&P	
700	0010	**RESIDENTIAL WIRING**										700
	0020	20′ avg. runs and #14/2 wiring incl. unless otherwise noted										
	1000	Service & panel, includes 24′ SE-AL cable, service eye, meter,										
	1010	Socket, panel board, main bkr., ground rod, 15 or 20 amp										
	1020	1-pole circuit breakers, and misc. hardware										
	1100	100 amp, with 10 branch breakers	1 Elec	1.19	6.723	Ea.	395	231		626	780	
	1110	With PVC conduit and wire		.92	8.696		430	298		728	915	
	1120	With RGS conduit and wire		.73	10.959		550	375		925	1,175	
	1150	150 amp, with 14 branch breakers		1.03	7.767		615	266		881	1,075	
	1170	With PVC conduit and wire		.82	9.756		680	335		1,015	1,250	
	1180	With RGS conduit and wire		.67	11.940		895	410		1,305	1,600	
	1200	200 amp, with 18 branch breakers	2 Elec	1.80	8.889		805	305		1,110	1,350	
	1220	With PVC conduit and wire		1.46	10.959		870	375		1,245	1,525	
	1230	With RGS conduit and wire		1.24	12.903		1,150	445		1,595	1,925	
	1800	Lightning surge suppressor for above services, add	1 Elec	32	.250		39.50	8.60		48.10	56.50	
	2000	Switch devices										
	2100	Single pole, 15 amp, Ivory, with a 1-gang box, cover plate,										
	2110	Type NM (Romex) cable	1 Elec	17.10	.468	Ea.	6.85	16.05		22.90	31.50	
	2120	Type MC (BX) cable		14.30	.559		16.95	19.20		36.15	47	
	2130	EMT & wire		5.71	1.401		17.65	48		65.65	91	
	2150	3-way, #14/3, type NM cable		14.55	.550		10.80	18.85		29.65	40	
	2170	Type MC cable		12.31	.650		21.50	22.50		44	57	
	2180	EMT & wire		5	1.600		20	55		75	104	
	2200	4-way, #14/3, type NM cable		14.55	.550		23.50	18.85		42.35	54	
	2220	Type MC cable		12.31	.650		34	22.50		56.50	71	
	2230	EMT & wire		5	1.600		32.50	55		87.50	118	
	2250	S.P., 20 amp, #12/2, type NM cable		13.33	.600		12.05	20.50		32.55	44	
	2270	Type MC cable		11.43	.700		21	24		45	59	
	2280	EMT & wire		4.85	1.649		23.50	56.50		80	111	
	2290	S.P. rotary dimmer, 600W, no wiring		17	.471		15.70	16.15		31.85	41.50	
	2300	S.P. rotary dimmer, 600W, type NM cable		14.55	.550		18.60	18.85		37.45	48.50	
	2320	Type MC cable		12.31	.650		28.50	22.50		51	65	

700	16139 | Residential Wiring	CREW	DAILY OUTPUT	LABOR-HOURS	UNIT	2001 BARE COSTS MAT.	LABOR	EQUIP.	TOTAL	TOTAL INCL O&P
2330	EMT & wire	1 Elec	5	1.600	Ea.	30	55		85	116
2350	3-way rotary dimmer, type NM cable		13.33	.600		16	20.50		36.50	48
2370	Type MC cable		11.43	.700		26	24		50	64.50
2380	EMT & wire		4.85	1.649		27.50	56.50		84	115
2400	Interval timer wall switch, 20 amp, 1-30 min., #12/2									
2410	Type NM cable	1 Elec	14.55	.550	Ea.	29.50	18.85		48.35	60.50
2420	Type MC cable		12.31	.650		36	22.50		58.50	73.50
2430	EMT & wire		5	1.600		41	55		96	127
2500	Decorator style									
2510	S.P., 15 amp, type NM cable	1 Elec	17.10	.468	Ea.	10.40	16.05		26.45	35.50
2520	Type MC cable		14.30	.559		20.50	19.20		39.70	51
2530	EMT & wire		5.71	1.401		21	48		69	95
2550	3-way, #14/3, type NM cable		14.55	.550		14.35	18.85		33.20	44
2570	Type MC cable		12.31	.650		25	22.50		47.50	61
2580	EMT & wire		5	1.600		23.50	55		78.50	108
2600	4-way, #14/3, type NM cable		14.55	.550		27	18.85		45.85	57.50
2620	Type MC cable		12.31	.650		37.50	22.50		60	75
2630	EMT & wire		5	1.600		36	55		91	122
2650	S.P., 20 amp, #12/2, type NM cable		13.33	.600		15.60	20.50		36.10	47.50
2670	Type MC cable		11.43	.700		24.50	24		48.50	63
2680	EMT & wire		4.85	1.649		27	56.50		83.50	114
2700	S.P., slide dimmer, type NM cable		17.10	.468		25.50	16.05		41.55	52.50
2720	Type MC cable		14.30	.559		36	19.20		55.20	68
2730	EMT & wire		5.71	1.401		37.50	48		85.50	113
2750	S.P., touch dimmer, type NM cable		17.10	.468		21.50	16.05		37.55	47.50
2770	Type MC cable		14.30	.559		31.50	19.20		50.70	63.50
2780	EMT & wire		5.71	1.401		33	48		81	108
2800	3-way touch dimmer, type NM cable		13.33	.600		39	20.50		59.50	73.50
2820	Type MC cable		11.43	.700		49	24		73	90
2830	EMT & wire		4.85	1.649		50.50	56.50		107	141
3000	Combination devices									
3100	S.P. switch/15 amp recpt., Ivory, 1-gang box, plate									
3110	Type NM cable	1 Elec	11.43	.700	Ea.	14.65	24		38.65	52
3120	Type MC cable		10	.800		25	27.50		52.50	68
3130	EMT & wire		4.40	1.818		26.50	62.50		89	122
3150	S.P. switch/pilot light, type NM cable		11.43	.700		15.30	24		39.30	53
3170	Type MC cable		10	.800		25.50	27.50		53	69
3180	EMT & wire		4.43	1.806		27	62		89	122
3190	2-S.P. switches, 2-#14/2, no wiring		14	.571		5.45	19.60		25.05	35.50
3200	2-S.P. switches, 2-#14/2, type NM cables		10	.800		16.85	27.50		44.35	59.50
3220	Type MC cable		8.89	.900		33.50	31		64.50	83
3230	EMT & wire		4.10	1.951		28	67		95	131
3250	3-way switch/15 amp recpt., #14/3, type NM cable		10	.800		21.50	27.50		49	64.50
3270	Type MC cable		8.89	.900		32	31		63	81
3280	EMT & wire		4.10	1.951		30.50	67		97.50	134
3300	2-3 way switches, 2-#14/3, type NM cables		8.89	.900		28.50	31		59.50	77.50
3320	Type MC cable		8	1		46.50	34.50		81	102
3330	EMT & wire		4	2		35	68.50		103.50	141
3350	S.P. switch/20 amp recpt., #12/2, type NM cable		10	.800		25.50	27.50		53	69
3370	Type MC cable		8.89	.900		32	31		63	81.50
3380	EMT & wire		4.10	1.951		37	67		104	141
3400	Decorator style									
3410	S.P. switch/15 amp recpt., type NM cable	1 Elec	11.43	.700	Ea.	18.20	24		42.20	56
3420	Type MC cable		10	.800		28.50	27.50		56	72
3430	EMT & wire		4.40	1.818		30	62.50		92.50	126
3450	S.P. switch/pilot light, type NM cable		11.43	.700		18.85	24		42.85	56.50

	16139 | Residential Wiring		CREW	DAILY OUTPUT	LABOR-HOURS	UNIT	2001 BARE COSTS MAT.	LABOR	EQUIP.	TOTAL	TOTAL INCL O&P	
700	3470	Type MC cable	1 Elec	10	.800	Ea.	29	27.50		56.50	73	700
	3480	EMT & wire		4.40	1.818		30.50	62.50		93	127	
	3500	2-S.P. switches, 2-#14/2, type NM cables		10	.800		20.50	27.50		48	63.50	
	3520	Type MC cable		8.89	.900		37	31		68	87	
	3530	EMT & wire		4.10	1.951		31.50	67		98.50	135	
	3550	3-way/15 amp recpt., #14/3, type NM cable		10	.800		25	27.50		52.50	68.50	
	3570	Type MC cable		8.89	.900		35.50	31		66.50	85	
	3580	EMT & wire		4.10	1.951		34	67		101	138	
	3650	2-3 way switches, 2-#14/3, type NM cables		8.89	.900		32	31		63	81.50	
	3670	Type MC cable		8	1		50	34.50		84.50	106	
	3680	EMT & wire		4	2		38.50	68.50		107	145	
	3700	S.P. switch/20 amp recpt., #12/2, type NM cable		10	.800		29	27.50		56.50	73	
	3720	Type MC cable		8.89	.900		36	31		67	85.50	
	3730	EMT & wire		4.10	1.951		40.50	67		107.50	145	
	4000	Receptacle devices										
	4010	Duplex outlet, 15 amp recpt., Ivory, 1-gang box, plate										
	4015	Type NM cable	1 Elec	14.55	.550	Ea.	5.40	18.85		24.25	34	
	4020	Type MC cable		12.31	.650		15.50	22.50		38	50.50	
	4030	EMT & wire		5.33	1.501		16.20	51.50		67.70	95	
	4050	With #12/2, type NM cable		12.31	.650		6.70	22.50		29.20	41	
	4070	Type MC cable		10.67	.750		15.75	25.50		41.25	56	
	4080	EMT & wire		4.71	1.699		18.05	58.50		76.55	107	
	4100	20 amp recpt., #12/2, type NM cable		12.31	.650		9.95	22.50		32.45	44.50	
	4120	Type MC cable		10.67	.750		19	25.50		44.50	59.50	
	4130	EMT & wire		4.71	1.699		21.50	58.50		80	111	
	4140	For GFI see line 4300 below										
	4150	Decorator style, 15 amp recpt., type NM cable	1 Elec	14.55	.550	Ea.	8.95	18.85		27.80	38	
	4170	Type MC cable		12.31	.650		19.05	22.50		41.55	54.50	
	4180	EMT & wire		5.33	1.501		19.75	51.50		71.25	98.50	
	4200	With #12/2, type NM cable		12.31	.650		10.25	22.50		32.75	45	
	4220	Type MC cable		10.67	.750		19.30	25.50		44.80	60	
	4230	EMT & wire		4.71	1.699		21.50	58.50		80	111	
	4250	20 amp recpt. #12/2, type NM cable		12.31	.650		13.50	22.50		36	48.50	
	4270	Type MC cable		10.67	.750		22.50	25.50		48	63.50	
	4280	EMT & wire		4.71	1.699		25	58.50		83.50	115	
	4300	GFI, 15 amp recpt., type NM cable		12.31	.650		33	22.50		55.50	70	
	4320	Type MC cable		10.67	.750		43	25.50		68.50	86	
	4330	EMT & wire		4.71	1.699		44	58.50		102.50	136	
	4350	GFI with #12/2, type NM cable		10.67	.750		34.50	25.50		60	76.50	
	4370	Type MC cable		9.20	.870		43.50	30		73.50	92.50	
	4380	EMT & wire		4.21	1.900		45.50	65		110.50	148	
	4400	20 amp recpt., #12/2 type NM cable		10.67	.750		36	25.50		61.50	78	
	4420	Type MC cable		9.20	.870		45	30		75	94	
	4430	EMT & wire		4.21	1.900		47.50	65		112.50	150	
	4500	Weather-proof cover for above receptacles, add		32	.250		4.40	8.60		13	17.65	
	4550	Air conditioner outlet, 20 amp-240 volt recpt.										
	4560	30' of #12/2, 2 pole circuit breaker										
	4570	Type NM cable	1 Elec	10	.800	Ea.	40.50	27.50		68	85.50	
	4580	Type MC cable		9	.889		52	30.50		82.50	103	
	4590	EMT & wire		4	2		51.50	68.50		120	159	
	4600	Decorator style, type NM cable		10	.800		44	27.50		71.50	89.50	
	4620	Type MC cable		9	.889		56	30.50		86.50	107	
	4630	EMT & wire		4	2		55	68.50		123.50	163	
	4650	Dryer outlet, 30 amp-240 volt recpt., 20' of #10/3										
	4660	2 pole circuit breaker										
	4670	Type NM cable	1 Elec	6.41	1.248	Ea.	47.50	43		90.50	117	

Important: See the Reference Section for critical supporting data - Reference Nos., Crews, & City Cost Indexes

	16139 | Residential Wiring	CREW	DAILY OUTPUT	LABOR-HOURS	UNIT	2001 BARE COSTS MAT.	LABOR	EQUIP.	TOTAL	TOTAL INCL O&P
700	4680 Type MC cable	1 Elec	5.71	1.401	Ea.	56	48		104	134
	4690 EMT & wire	↓	3.48	2.299	↓	53	79		132	176
	4700 Range outlet, 50 amp-240 volt recpt., 30′ of #8/3									
	4710 Type NM cable	1 Elec	4.21	1.900	Ea.	74	65		139	179
	4720 Type MC cable		4	2		101	68.50		169.50	213
	4730 EMT & wire		2.96	2.703		69	92.50		161.50	214
	4750 Central vacuum outlet, Type NM cable		6.40	1.250		43.50	43		86.50	112
	4770 Type MC cable		5.71	1.401		59	48		107	137
	4780 EMT & wire	↓	3.48	2.299	↓	53.50	79		132.50	177
	4800 30 amp-110 volt locking recpt., #10/2 circ. bkr.									
	4810 Type NM cable	1 Elec	6.20	1.290	Ea.	51.50	44.50		96	123
	4830 EMT & wire	″	3.20	2.500	″	62.50	86		148.50	197
	4900 Low voltage outlets									
	4910 Telephone recpt., 20′ of 4/C phone wire	1 Elec	26	.308	Ea.	6.70	10.55		17.25	23
	4920 TV recpt., 20′ of RG59U coax wire, F type connector	″	16	.500	″	11.15	17.15		28.30	38
	4950 Door bell chime, transformer, 2 buttons, 60′ of bellwire									
	4970 Economy model	1 Elec	11.50	.696	Ea.	48.50	24		72.50	89
	4980 Custom model		11.50	.696		83.50	24		107.50	128
	4990 Luxury model, 3 buttons	↓	9.50	.842	↓	228	29		257	294
	6000 Lighting outlets									
	6050 Wire only (for fixture), type NM cable	1 Elec	32	.250	Ea.	3.77	8.60		12.37	16.95
	6070 Type MC cable		24	.333		11.60	11.45		23.05	30
	6080 EMT & wire		10	.800		11.40	27.50		38.90	53.50
	6100 Box (4″), and wire (for fixture), type NM cable		25	.320		7.65	11		18.65	25
	6120 Type MC cable		20	.400		15.50	13.70		29.20	37.50
	6130 EMT & wire	↓	11	.727	↓	15.30	25		40.30	54
	6200 Fixtures (use with lines 6050 or 6100 above)									
	6210 Canopy style, economy grade	1 Elec	40	.200	Ea.	25.50	6.85		32.35	38.50
	6220 Custom grade		40	.200		46	6.85		52.85	61
	6250 Dining room chandelier, economy grade		19	.421		76	14.45		90.45	105
	6260 Custom grade		19	.421		225	14.45		239.45	270
	6270 Luxury grade		15	.533		495	18.30		513.30	575
	6310 Kitchen fixture (fluorescent), economy grade		30	.267		51.50	9.15		60.65	70
	6320 Custom grade		25	.320		161	11		172	193
	6350 Outdoor, wall mounted, economy grade		30	.267		27	9.15		36.15	43
	6360 Custom grade		30	.267		101	9.15		110.15	125
	6370 Luxury grade		25	.320		227	11		238	266
	6410 Outdoor PAR floodlights, 1 lamp, 150 watt		20	.400		20	13.70		33.70	42.50
	6420 2 lamp, 150 watt each		20	.400		33.50	13.70		47.20	57.50
	6430 For infrared security sensor, add		32	.250		87	8.60		95.60	109
	6450 Outdoor, quartz-halogen, 300 watt flood		20	.400		38	13.70		51.70	62.50
	6600 Recessed downlight, round, pre-wired, 50 or 75 watt trim		30	.267		35	9.15		44.15	52
	6610 With shower light trim		30	.267		42	9.15		51.15	59.50
	6620 With wall washer trim		28	.286		52.50	9.80		62.30	72.50
	6630 With eye-ball trim	↓	28	.286		49	9.80		58.80	68.50
	6640 For direct contact with insulation, add					1.60			1.60	1.76
	6700 Porcelain lamp holder	1 Elec	40	.200		3.50	6.85		10.35	14.10
	6710 With pull switch		40	.200		3.75	6.85		10.60	14.40
	6750 Fluorescent strip, 1-20 watt tube, wrap around diffuser, 24″		24	.333		51.50	11.45		62.95	73.50
	6770 2-40 watt tubes, 48″		20	.400		78	13.70		91.70	107
	6780 With residential ballast		20	.400		89.50	13.70		103.20	119
	6800 Bathroom heat lamp, 1-250 watt		28	.286		30.50	9.80		40.30	48
	6810 2-250 watt lamps	↓	28	.286	↓	51	9.80		60.80	70.50
	6820 For timer switch, see line 2400									
	6900 Outdoor post lamp, incl. post, fixture, 35′ of #14/2									
	6910 Type NMC cable	1 Elec	3.50	2.286	Ea.	181	78.50		259.50	315

16139 | Residential Wiring

		16139 Residential Wiring	CREW	DAILY OUTPUT	LABOR-HOURS	UNIT	2001 BARE COSTS MAT.	LABOR	EQUIP.	TOTAL	TOTAL INCL O&P	
700	6920	Photo-eye, add	1 Elec	27	.296	Ea.	29	10.15		39.15	47	700
	6950	Clock dial time switch, 24 hr., w/enclosure, type NM cable		11.43	.700		51	24		75	92	
	6970	Type MC cable		11	.727		61	25		86	104	
	6980	EMT & wire	↓	4.85	1.649	↓	61.50	56.50		118	152	
	7000	Alarm systems										
	7050	Smoke detectors, box, #14/3, type NM cable	1 Elec	14.55	.550	Ea.	28.50	18.85		47.35	59.50	
	7070	Type MC cable		12.31	.650		37	22.50		59.50	74	
	7080	EMT & wire	↓	5	1.600		35.50	55		90.50	121	
	7090	For relay output to security system, add				↓	11.75			11.75	12.95	
	8000	Residential equipment										
	8050	Disposal hook-up, incl. switch, outlet box, 3′ of flex										
	8060	20 amp-1 pole circ. bkr., and 25′ of #12/2										
	8070	Type NM cable	1 Elec	10	.800	Ea.	19.70	27.50		47.20	62.50	
	8080	Type MC cable		8	1		30	34.50		64.50	84	
	8090	EMT & wire	↓	5	1.600	↓	32.50	55		87.50	118	
	8100	Trash compactor or dishwasher hook-up, incl. outlet box,										
	8110	3′ of flex, 15 amp-1 pole circ. bkr., and 25′ of #14/2										
	8120	Type NM cable	1 Elec	10	.800	Ea.	13.70	27.50		41.20	56	
	8130	Type MC cable		8	1		26	34.50		60.50	79.50	
	8140	EMT & wire	↓	5	1.600	↓	26.50	55		81.50	112	
	8150	Hot water sink dispensor hook-up, use line 8100										
	8200	Vent/exhaust fan hook-up, type NM cable	1 Elec	32	.250	Ea.	3.77	8.60		12.37	16.95	
	8220	Type MC cable		24	.333		11.60	11.45		23.05	30	
	8230	EMT & wire	↓	10	.800	↓	11.40	27.50		38.90	53.50	
	8250	Bathroom vent fan, 50 CFM (use with above hook-up)										
	8260	Economy model	1 Elec	15	.533	Ea.	21.50	18.30		39.80	51	
	8270	Low noise model		15	.533		30	18.30		48.30	60.50	
	8280	Custom model	↓	12	.667	↓	111	23		134	156	
	8300	Bathroom or kitchen vent fan, 110 CFM										
	8310	Economy model	1 Elec	15	.533	Ea.	56	18.30		74.30	89	
	8320	Low noise model	"	15	.533	"	75	18.30		93.30	110	
	8350	Paddle fan, variable speed (w/o lights)										
	8360	Economy model (AC motor)	1 Elec	10	.800	Ea.	100	27.50		127.50	151	
	8370	Custom model (AC motor)		10	.800		173	27.50		200.50	231	
	8380	Luxury model (DC motor)		8	1		340	34.50		374.50	425	
	8390	Remote speed switch for above, add	↓	12	.667	↓	23	23		46	59.50	
	8500	Whole house exhaust fan, ceiling mount, 36″, variable speed										
	8510	Remote switch, incl. shutters, 20 amp-1 pole circ. bkr.										
	8520	30′ of #12/2, type NM cable	1 Elec	4	2	Ea.	550	68.50		618.50	705	
	8530	Type MC cable		3.50	2.286		560	78.50		638.50	735	
	8540	EMT & wire	↓	3	2.667	↓	565	91.50		656.50	755	
	8600	Whirlpool tub hook-up, incl. timer switch, outlet box										
	8610	3′ of flex, 20 amp-1 pole GFI circ. bkr.										
	8620	30′ of #12/2, type NM cable	1 Elec	5	1.600	Ea.	75.50	55		130.50	165	
	8630	Type MC cable		4.20	1.905		82.50	65.50		148	188	
	8640	EMT & wire	↓	3.40	2.353	↓	84.50	80.50		165	213	
	8650	Hot water heater hook-up, incl. 1-2 pole circ. bkr., box;										
	8660	3′ of flex, 20′ of #10/2, type NM cable	1 Elec	5	1.600	Ea.	21	55		76	105	
	8670	Type MC cable		4.20	1.905		34	65.50		99.50	135	
	8680	EMT & wire	↓	3.40	2.353	↓	28.50	80.50		109	152	
	9000	Heating/air conditioning										
	9050	Furnace/boiler hook-up, incl. firestat, local on-off switch										
	9060	Emergency switch, and 40′ of type NM cable	1 Elec	4	2	Ea.	43	68.50		111.50	150	
	9070	Type MC cable		3.50	2.286		59.50	78.50		138	183	
	9080	EMT & wire	↓	1.50	5.333	↓	61	183		244	340	
	9100	Air conditioner hook-up, incl. local 60 amp disc. switch										

16100 | Wiring Methods

	16139 | Residential Wiring		Crew	Daily Output	Labor-Hours	Unit	2001 Bare Costs Mat.	Labor	Equip.	Total	Total Incl O&P
700	9110	3′ sealtite, 40 amp, 2 pole circuit breaker									
	9130	40′ of #8/2, type NM cable	1 Elec	3.50	2.286	Ea.	140	78.50		218.50	271
	9140	Type MC cable		3	2.667		185	91.50		276.50	340
	9150	EMT & wire		1.30	6.154		148	211		359	480
	9200	Heat pump hook-up, 1-40 & 1-100 amp 2 pole circ. bkr.									
	9210	Local disconnect switch, 3′ sealtite									
	9220	40′ of #8/2 & 30′ of #3/2									
	9230	Type NM cable	1 Elec	1.30	6.154	Ea.	345	211		556	695
	9240	Type MC cable		1.08	7.407		455	254		709	880
	9250	EMT & wire		.94	8.511		345	292		637	815
	9500	Thermostat hook-up, using low voltage wire									
	9520	Heating only	1 Elec	24	.333	Ea.	5.70	11.45		17.15	23.50
	9530	Heating/cooling	″	20	.400	″	7.05	13.70		20.75	28.50

	16150 | Wiring Connections		Crew	Daily Output	Labor-Hours	Unit	Mat.	Labor	Equip.	Total	Total Incl O&P
275	0010	**MOTOR CONNECTIONS**									
	0020	Flexible conduit and fittings, 115 volt, 1 phase, up to 1 HP motor	1 Elec	8	1	Ea.	4.39	34.50		38.89	56
	0050	2 HP motor		6.50	1.231		4.49	42		46.49	68
	0100	3 HP motor		5.50	1.455		7.50	50		57.50	82.50
	0120	230 volt, 10 HP motor, 3 phase		4.20	1.905		8.10	65.50		73.60	106
	0150	15 HP motor		3.30	2.424		15.55	83		98.55	141
	0200	25 HP motor		2.70	2.963		17.75	102		119.75	172
	0400	50 HP motor		2.20	3.636		46.50	125		171.50	238
	0600	100 HP motor		1.50	5.333		117	183		300	400

16400 | Low-Voltage Distribution

	16420 | Enclosed Controllers		Crew	Daily Output	Labor-Hours	Unit	2001 Bare Costs Mat.	Labor	Equip.	Total	Total Incl O&P
220	0010	**CONTROL STATIONS**									
	0050	NEMA 1, heavy duty, stop/start	1 Elec	8	1	Ea.	113	34.50		147.50	175
	0100	Stop/start, pilot light		6.20	1.290		153	44.50		197.50	234
	0200	Hand/off/automatic		6.20	1.290		83.50	44.50		128	158
	0400	Stop/start/reverse		5.30	1.509		153	52		205	246
	0500	NEMA 7, heavy duty, stop/start		6	1.333		188	45.50		233.50	276
	0600	Stop/start, pilot light		4	2		230	68.50		298.50	355

	16440 | Swbds, Panels & Cont Centers		Crew	Daily Output	Labor-Hours	Unit	Mat.	Labor	Equip.	Total	Total Incl O&P
660	0010	**MOTOR STARTERS & CONTROLS**									
	0050	Magnetic, FVNR, with enclosure and heaters, 480 volt									
	0080	2 HP, size 00	1 Elec	3.50	2.286	Ea.	167	78.50		245.50	300
	0100	5 HP, size 0		2.30	3.478		202	119		321	400
	0200	10 HP, size 1		1.60	5		227	172		399	505
	0300	25 HP, size 2	2 Elec	2.20	7.273		425	249		674	840
	0400	50 HP, size 3		1.80	8.889		695	305		1,000	1,225
	0500	100 HP, size 4		1.20	13.333		1,550	455		2,005	2,375
	0600	200 HP, size 5		.90	17.778		3,600	610		4,210	4,875
	0610	400 HP, size 6		.80	20		10,000	685		10,685	12,000

16440 | Swbds, Panels & Cont Centers

			CREW	DAILY OUTPUT	LABOR-HOURS	UNIT	2001 BARE COSTS MAT.	LABOR	EQUIP.	TOTAL	TOTAL INCL O&P	
660	0620	NEMA 7, 5 HP, size 0	1 Elec	1.60	5	Ea.	865	172		1,037	1,200	660
	0630	10 HP, size 1	"	1.10	7.273		905	249		1,154	1,375	
	0640	25 HP, size 2	2 Elec	1.80	8.889		1,450	305		1,755	2,050	
	0650	50 HP, size 3		1.20	13.333		2,150	455		2,605	3,050	
	0660	100 HP, size 4		.90	17.778		3,525	610		4,135	4,775	
	0670	200 HP, size 5		.50	32		8,250	1,100		9,350	10,700	
	0700	Combination, with motor circuit protectors, 5 HP, size 0	1 Elec	1.80	4.444		655	152		807	950	
	0800	10 HP, size 1	"	1.30	6.154		680	211		891	1,075	
	0900	25 HP, size 2	2 Elec	2	8		955	274		1,229	1,450	
	1000	50 HP, size 3		1.32	12.121		1,375	415		1,790	2,150	
	1200	100 HP, size 4		.80	20		3,000	685		3,685	4,325	
	1220	NEMA 7, 5 HP, size 0	1 Elec	1.30	6.154		1,525	211		1,736	2,000	
	1230	10 HP, size 1	"	1	8		1,575	274		1,849	2,125	
	1240	25 HP, size 2	2 Elec	1.32	12.121		2,075	415		2,490	2,925	
	1250	50 HP, size 3		.80	20		3,425	685		4,110	4,800	
	1260	100 HP, size 4		.60	26.667		5,325	915		6,240	7,225	
	1270	200 HP, size 5		.40	40		11,500	1,375		12,875	14,800	
	1400	Combination, with fused switch, 5 HP, size 0	1 Elec	1.80	4.444		500	152		652	780	
	1600	10 HP, size 1	"	1.30	6.154		535	211		746	905	
	1800	25 HP, size 2	2 Elec	2	8		870	274		1,144	1,375	
	2000	50 HP, size 3		1.32	12.121		1,475	415		1,890	2,250	
	2200	100 HP, size 4		.80	20		2,575	685		3,260	3,850	
	3500	Magnetic FVNR with NEMA 12, enclosure & heaters, 480 volt										
	3600	5 HP, size 0	1 Elec	2.20	3.636	Ea.	190	125		315	395	
	3700	10 HP, size 1	"	1.50	5.333		286	183		469	590	
	3800	25 HP, size 2	2 Elec	2	8		535	274		809	1,000	
	3900	50 HP, size 3		1.60	10		825	345		1,170	1,425	
	4000	100 HP, size 4		1	16		1,975	550		2,525	3,000	
	4100	200 HP, size 5		.80	20		4,725	685		5,410	6,225	
	4200	Combination, with motor circuit protectors, 5 HP, size 0	1 Elec	1.70	4.706		630	161		791	935	
	4300	10 HP, size 1	"	1.20	6.667		655	229		884	1,050	
	4400	25 HP, size 2	2 Elec	1.80	8.889		980	305		1,285	1,525	
	4500	50 HP, size 3		1.20	13.333		1,600	455		2,055	2,425	
	4600	100 HP, size 4		.74	21.622		3,600	740		4,340	5,075	
	4700	Combination, with fused switch, 5 HP, size 0	1 Elec	1.70	4.706		610	161		771	910	
	4800	10 HP, size 1	"	1.20	6.667		635	229		864	1,050	
	4900	25 HP, size 2	2 Elec	1.80	8.889		970	305		1,275	1,525	
	5000	50 HP, size 3		1.20	13.333		1,550	455		2,005	2,375	
	5100	100 HP, size 4		.74	21.622		3,150	740		3,890	4,550	
	5200	Factory installed controls, adders to size 0 thru 5										
	5300	Start-stop push button	1 Elec	32	.250	Ea.	40	8.60		48.60	57	
	5400	Hand-off-auto-selector switch		32	.250		40	8.60		48.60	57	
	5500	Pilot light		32	.250		74.50	8.60		83.10	95	
	5600	Start-stop-pilot		32	.250		115	8.60		123.60	139	
	5700	Auxiliary contact, NO or NC		32	.250		55	8.60		63.60	73.50	
	5800	NO-NC		32	.250		110	8.60		118.60	134	

For information about Means Estimating Seminars, see yellow pages 11 and 12 in back of book

16 ELECTRICAL

Division 17
Square Foot & Cubic Foot Costs

Estimating Tips

- The cost figures in Division 17 were derived from approximately 10,200 projects contained in the Means database of completed construction projects, and include the contractor's overhead and profit, but do not generally include architectural fees or land costs. The figures have been adjusted to January of the current year. New projects are added to our files each year, and outdated projects are discarded. For this reason, certain costs may not show a uniform annual progression. In no case are all subdivisions of a project listed.
- These projects were located throughout the U.S. and reflect a tremendous variation in square foot (S.F.) and cubic foot (C.F.) costs. This is due to differences, not only in labor and material costs, but also in individual owners' requirements. For instance, a bank in a large city would have different features than one in a rural area. This is true of all the different types of buildings analyzed. Therefore, caution should be exercised when using Division 17 costs. For example, for court houses, costs in the database are local court house costs and will not apply to the larger, more elaborate federal court houses. As a general rule, the projects in the 1/4 column do not include any site work or equipment, while the projects in the 3/4 column may include both equipment and site work. The median figures do not generally include site work.
- None of the figures "go with" any others. All individual cost items were computed and tabulated separately. Thus the sum of the median figures for Plumbing, HVAC and Electrical will not normally total up to the total Mechanical and Electrical costs arrived at by separate analysis and tabulation of the projects.
- Each building was analyzed as to total and component costs and percentages. The figures were arranged in ascending order with the results tabulated as shown. The 1/4 column shows that 25% of the projects had lower costs, 75% higher. The 3/4 column shows that 75% of the projects had lower costs, 25% had higher. The median column shows that 50% of the projects had lower costs, 50% had higher.
- There are two times when square foot costs are useful. The first is in the conceptual stage when no details are available. Then square foot costs make a useful starting point. The second is after the bids are in and the costs can be worked back into their appropriate units for information purposes. As soon as details become available in the project design, the square foot approach should be discontinued and the project priced as to its particular components. When more precision is required or for estimating the replacement cost of specific buildings, the current edition of *Means Square Foot Costs* should be used.
- In using the figures in Division 17, it is recommended that the median column be used for preliminary figures if no additional information is available. The median figures, when multiplied by the total city construction cost index figures (see City Cost Indexes) and then multiplied by the project size modifier in Reference Number R17100-100, should present a fairly accurate base figure, which would then have to be adjusted in view of the estimator's experience, local economic conditions, code requirements and the owner's particular requirements. There is no need to factor the percentage figures, as these should remain constant from city to city. All tabulations mentioning air conditioning had at least partial air conditioning.
- The editors of this book would greatly appreciate receiving cost figures on one or more of your recent projects which would then be included in the averages for next year. All cost figures received will be kept confidential except that they will be averaged with other similar projects to arrive at S.F. and C.F. cost figures for next year's book. See the last page of the book for details and the discount available for submitting one or more of your projects.

17100 | S.F. & C.F. Costs

				UNIT COSTS			% OF TOTAL			
			UNIT	1/4	MEDIAN	3/4	1/4	MEDIAN	3/4	
010	0010	**APARTMENTS** Low Rise (1 to 3 story) R17100-100	S.F.	43.95	55.05	73.60				010
	0020	Total project cost	C.F.	3.94	5.25	6.45				
	2720	Plumbing	S.F.	3.39	4.40	5.55	6.70%	9%	10.10%	
	2770	Heating, ventilating, air conditioning		2.18	2.69	3.95	4.20%	5.80%	7.70%	
	2900	Electrical		2.53	3.30	4.52	5.20%	6.70%	8.40%	
	3100	Total: Mechanical & Electrical		8.75	11.15	14	16%	18.20%	23%	
	9000	Per apartment unit, total cost	Apt.	40,500	61,700	92,200				
	9500	Total: Mechanical & Electrical	"	7,700	12,200	15,900				
020	0010	**APARTMENTS** Mid Rise (4 to 7 story)	S.F.	57.15	69.50	85				020
	0020	Total project costs	C.F.	4.55	6.30	8.60				
	2720	Plumbing	S.F.	3.42	5.45	6.05	6.30%	7.40%	10%	
	2900	Electrical		4.02	5.50	6.40	6.70%	7.50%	8.90%	
	3100	Total: Mechanical & Electrical		11.60	15.70	19.25	18.80%	21.60%	25.10%	
	9000	Per apartment unit, total cost	Apt.	65,900	77,800	128,700				
	9500	Total: Mechanical & Electrical	"	12,400	14,400	18,200				
030	0010	**APARTMENTS** High Rise (8 to 24 story)	S.F.	66.05	79.70	97.40				030
	0020	Total project costs	C.F.	5.62	7.85	9.55				
	2720	Plumbing	S.F.	4.87	5.75	7.05	7%	9.10%	10.60%	
	2900	Electrical		4.54	5.75	7.75	6.40%	7.70%	8.90%	
	3100	Total: Mechanical & Electrical		13.60	17.35	20.90	18%	22.30%	24.40%	
	9000	Per apartment unit, total cost	Apt.	68,400	74,200	104,900				
	9500	Total: Mechanical & Electrical	"	14,800	16,900	18,000				
040	0010	**AUDITORIUMS**	S.F.	67.95	92.05	124				040
	0020	Total project costs	C.F.	4.30	6	9				
	2720	Plumbing	S.F.	4.35	6	7.60	6.30%	7.20%	9.05%	
	2900	Electrical		5.55	7.65	9.55	6.80%	9%	10.60%	
	3100	Total: Mechanical & Electrical		10.60	14.25	24.80	14.40%	18.50%	23.60%	
050	0010	**AUTOMOTIVE SALES**	S.F.	49.50	56.95	83.85				050
	0020	Total project costs	C.F.	3.30	4.02	5.20				
	2720	Plumbing	S.F.	2.48	4.02	4.39	4.70%	6.50%	7%	
	2770	Heating, ventilating, air conditioning		3.58	5.55	5.99	6.40%	10.30%	10.40%	
	2900	Electrical		3.94	6.13	6.95	7.30%	10%	12.40%	
	3100	Total: Mechanical & Electrical		8.90	12.85	16.05	17%	20.40%	26%	
060	0010	**BANKS**	S.F.	98.60	123	156				060
	0020	Total project costs	C.F.	7.10	9.65	12.65				
	2720	Plumbing	S.F.	3.14	4.49	6.50	2.80%	3.90%	4.90%	
	2770	Heating, ventilating, air conditioning		6.10	7.95	10.55	5%	7.20%	8.50%	
	2900	Electrical		9.55	12.60	16.40	8.30%	10.30%	12.20%	
	3100	Total: Mechanical & Electrical		22.65	30.50	37	16.60%	20.40%	24.40%	
130	0010	**CHURCHES**	S.F.	66.20	83.15	108				130
	0020	Total project costs	C.F.	4.16	5.25	6.90				
	2720	Plumbing	S.F.	2.61	3.65	5.40	3.60%	5%	6.30%	
	2770	Heating, ventilating, air conditioning		6.10	7.95	11.70	7.60%	10%	12.20%	
	2900	Electrical		5.50	7.60	9.85	7.30%	8.80%	10.90%	
	3100	Total: Mechanical & Electrical		14.30	21.45	31.15	19.60%	22.90%	25.40%	
150	0010	**CLUBS, COUNTRY**	S.F.	69.85	82.85	109				150
	0020	Total project costs	C.F.	5.80	7.05	9.75				
	2720	Plumbing	S.F.	4.62	6.45	14.70	6.40%	9.60%	10%	
	2900	Electrical		5.65	8.10	10.65	7%	9.30%	11.40%	
	3100	Total: Mechanical & Electrical		12.20	15.70	29.20	16.10%	22.30%	30.90%	
170	0010	**CLUBS, SOCIAL** Fraternal	S.F.	56.45	80.65	106				170
	0020	Total project costs	C.F.	3.39	5.45	6.30				
	2720	Plumbing	S.F.	3.25	4.50	5.45	5.20%	6.80%	8.30%	
	2770	Heating, ventilating, air conditioning		5.20	6.30	8.10	8.70%	11%	14.40%	
	2900	Electrical		4.55	7.10	8.60	7.30%	9.50%	11.40%	
	3100	Total: Mechanical & Electrical		12.15	16.50	23.60	19.30%	23%	33.10%	

See R17100-100 and City Cost Indexes in the Reference Section

17100 | S.F. & C.F. Costs

			UNIT	UNIT COSTS 1/4	UNIT COSTS MEDIAN	UNIT COSTS 3/4	% OF TOTAL 1/4	% OF TOTAL MEDIAN	% OF TOTAL 3/4	
180	0010	**CLUBS, Y.M.C.A.** R17100-100	S.F.	65.80	83.10	111				180
	0020	Total project costs	C.F.	3.32	5.55	8.30				
	2720	Plumbing	S.F.	4.79	9.05	10.15	6%	9.70%	16%	
	2900	Electrical		5.45	7.20	10.60	6.40%	9.20%	10.80%	
	3100	Total: Mechanical & Electrical		11.95	17.10	25.50	16.20%	21.90%	28.50%	
190	0010	**COLLEGES** Classrooms & Administration	S.F.	82.30	109	146				190
	0020	Total project costs	C.F.	6	8.35	13.55				
	2720	Plumbing	S.F.	4.03	7.90	14.65	4.80%	8.90%	11.80%	
	2900	Electrical		6.75	10.40	12.80	8.40%	10%	12.10%	
	3100	Total: Mechanical & Electrical		14.50	26.40	40.85	14.20%	24.70%	32.50%	
210	0010	**COLLEGES** Science, Engineering, Laboratories	S.F.	131	159	193				210
	0020	Total project costs	C.F.	7.75	11.35	12.85				
	2900	Electrical	S.F.	11.20	16.15	24.35	7.10%	9.60%	15.40%	
	3100	Total: Mechanical & Electrical	"	41.30	49.20	76.15	28%	31.60%	39.80%	
230	0010	**COLLEGES** Student Unions	S.F.	86.30	121	142				230
	0020	Total project costs	C.F.	4.82	6.30	8.15				
	3100	Total: Mechanical & Electrical	S.F.	22.70	32.45	35.15	17.40%	23.90%	28.90%	
250	0010	**COMMUNITY CENTERS**	"	64.50	86.60	115				250
	0020	Total project costs	C.F.	4.65	6.65	8.60				
	2720	Plumbing	S.F.	3.46	5.90	8.60	4.90%	7.10%	9.50%	
	2770	Heating, ventilating, air conditioning		5.70	8.20	11.30	7.70%	10.80%	13%	
	2900	Electrical		5.70	7.70	11.70	7.40%	9.40%	11.50%	
	3100	Total: Mechanical & Electrical		21.15	25.25	36.40	25.10%	27.90%	34.40%	
280	0010	**COURT HOUSES**	S.F.	102	118	136				280
	0020	Total project costs	C.F.	7.80	9.45	13.10				
	2720	Plumbing	S.F.	4.89	6.82	9.80	5.90%	7.40%	9.60%	
	2900	Electrical		9	10.95	14.35	8.40%	9.80%	11.60%	
	3100	Total: Mechanical & Electrical		20.20	27.70	35.55	20.10%	25.70%	28.70%	
300	0010	**DEPARTMENT STORES**	S.F.	38.05	51.45	64.95				300
	0020	Total project costs	C.F.	2	2.64	3.59				
	2720	Plumbing	S.F.	1.24	1.50	2.27	3.10%	4.20%	5.90%	
	2770	Heating, ventilating, air conditioning		3.46	5.35	8.05	8.30%	12.40%	14.80%	
	2900	Electrical		4.37	6	7.05	9.10%	12.20%	14.90%	
	3100	Total: Mechanical & Electrical		7.70	9.95	16.10	20%	19%	26.50%	
310	0010	**DORMITORIES** Low Rise (1 to 3 story)	S.F.	62.20	88.75	107				310
	0020	Total project costs	C.F.	4.31	6.60	9.90				
	2720	Plumbing	S.F.	4.34	5.80	7.35	8%	9%	9.60%	
	2770	Heating, ventilating, air conditioning		4.59	5.50	7.30	4.60%	8%	10%	
	2900	Electrical		4.59	7	8.75	6.60%	8.90%	9.60%	
	3100	Total: Mechanical & Electrical		24.10	25.25	26.30	22.50%	24.10%	29.10%	
	9000	Per bed, total cost	Bed	30,600	33,900	72,800				
320	0010	**DORMITORIES** Mid Rise (4 to 8 story)	S.F.	88.60	116	143				320
	0020	Total project costs	C.F.	9.75	10.70	12.85				
	2900	Electrical	S.F.	6.55	10.30	13.15	7.40%	8.90%	10.40%	
	3100	Total: Mechanical & Electrical	"	19.65	26.30	34.40	18.10%	19.50%	30.60%	
	9000	Per bed, total cost	Bed	12,600	28,800	59,200				
340	0010	**FACTORIES**	S.F.	33.80	49.85	77.15				340
	0020	Total project costs	C.F.	2.15	3.19	5.30				
	2720	Plumbing	S.F.	1.99	3.36	5.70	4.60%	6.20%	8.70%	
	2770	Heating, ventilating, air conditioning		3.52	5.05	6.80	5.20%	8.50%	11.40%	
	2900	Electrical		4.11	6.55	10.05	8.10%	10.50%	14.20%	
	3100	Total: Mechanical & Electrical		9.90	15.90	24.15	20.40%	28.10%	34.50%	
360	0010	**FIRE STATIONS**	S.F.	64.80	87.05	112				360
	0020	Total project costs	C.F.	3.89	5.30	7.05				

17100 | S.F. & C.F. Costs

			UNIT	UNIT COSTS			% OF TOTAL			
				1/4	MEDIAN	3/4	1/4	MEDIAN	3/4	
360	2720	Plumbing (R17100-100)	S.F.	4.21	6.20	9.15	6.20%	7.70%	9.70%	360
	2770	Heating, ventilating, air conditioning		3.63	5.95	9.10	4.90%	7.50%	9.30%	
	2900	Electrical		4.61	7.95	10.70	7%	9.20%	11%	
	3100	Total: Mechanical & Electrical	↓	20.45	25.25	29.95	18.80%	23.80%	27.30%	
370	0010	**FRATERNITY HOUSES** and Sorority Houses	S.F.	66.50	85.65	121				370
	0020	Total project costs	C.F.	6.35	6.90	8.85				
	2720	Plumbing	S.F.	4.98	5.75	10.55	5.20%	8%	10.90%	
	2900	Electrical		4.39	9.50	11.60	6.60%	9.90%	10.70%	
	3100	Total: Mechanical & Electrical	↓	12.15	17.25	20.75	14.60%	19.90%	20.70%	
380	0010	**FUNERAL HOMES**	S.F.	69.80	95.70	174				380
	0020	Total project costs	C.F.	7.15	7.95	15.30				
	2900	Electrical	S.F.	3.10	5.70	6.70	3.60%	4.40%	11%	
	3100	Total: Mechanical & Electrical	"	11.15	16.65	21.55	12.90%	17.80%	18.80%	
390	0010	**GARAGES, COMMERCIAL** (Service)	S.F.	38.30	60.80	83.60				390
	0020	Total project costs	C.F.	2.43	3.64	5.25				
	2720	Plumbing	S.F.	2.58	3.88	7.60	4.70%	7.40%	11%	
	2730	Heating & ventilating		3.61	4.92	6.80	5.30%	6.90%	9.60%	
	2900	Electrical		3.65	5.60	7.75	7.20%	9%	11.20%	
	3100	Total: Mechanical & Electrical	↓	8.55	15.95	23.55	14.10%	22.50%	26.90%	
400	0010	**GARAGES, MUNICIPAL** (Repair)	S.F.	58.40	77.75	111				400
	0020	Total project costs	C.F.	3.65	4.63	7.55				
	2720	Plumbing	S.F.	2.62	5.05	9.50	3.60%	6.70%	8%	
	2730	Heating & ventilating		4.48	6.50	12.50	6.20%	7.40%	13.50%	
	2900	Electrical		4.31	6.35	9.80	6.30%	8%	11.70%	
	3100	Total: Mechanical & Electrical	↓	11	19.75	32.55	15.20%	25.50%	32.70%	
410	0010	**GARAGES, PARKING**	S.F.	22	31.90	56.95				410
	0020	Total project costs	C.F.	2.13	2.90	4.22				
	2720	Plumbing	S.F.	.59	1	1.54	2.60%	3.40%	3.90%	
	2900	Electrical		.93	1.41	2.07	4.30%	5.20%	6.30%	
	3100	Total: Mechanical & Electrical	↓	1.24	3.64	4.67	6.50%	9.40%	12.80%	
	9000	Per car, total cost	Car	9,600	12,000	15,600				
430	0010	**GYMNASIUMS**	S.F.	63	80.45	103				430
	0020	Total project costs	C.F.	3.15	4	5.25				
	2720	Plumbing	S.F.	4	4.93	5.90	5.40%	7.30%	7.90%	
	2770	Heating, ventilating, air conditioning	"	4.30	6.55	13.15	9%	11.10%	22.60%	
	2900	Electrical	S.F.	4.79	6	8.05	6.60%	8.30%	10.70%	
	3100	Total: Mechanical & Electrical	↓	14.80	20.40	27.35	20.60%	26.20%	29.40%	
460	0010	**HOSPITALS**	↓	130	152	225				460
	0020	Total project costs	C.F.	9.20	11.35	16.35				
	2720	Plumbing	S.F.	10.85	14.60	18.80	7.80%	9.40%	11.80%	
	2770	Heating, ventilating, air conditioning		15.35	20.35	27.40	8.40%	14.60%	17%	
	2900	Electrical		13.05	16.95	26.70	9.90%	12%	14.50%	
	3100	Total: Mechanical & Electrical	↓	37.15	49.25	80.70	26.60%	33.10%	39%	
	9000	Per bed or person, total cost	Bed	46,800	99,300	156,600				
480	0010	**HOUSING** For the Elderly	S.F.	59.65	75.25	92.45				480
	0020	Total project costs	C.F.	4.23	5.90	7.55				
	2720	Plumbing	S.F.	4.43	5.80	7.75	8.10%	9.70%	10.90%	
	2730	Heating, ventilating, air conditioning		2.27	3.22	4.81	3.30%	5.60%	7.20%	
	2900	Electrical		4.44	6	7.70	7.30%	8.50%	10.50%	
	3100	Total: Mechanical & Electrical	↓	15.30	17.95	23.75	18.10%	22%	29.10%	
	9000	Per rental unit, total cost	Unit	55,500	64,500	72,100				
	9500	Total: Mechanical & Electrical	"	12,200	14,200	16,600				
500	0010	**HOUSING** Public (Low Rise)	S.F.	51.85	69.70	90.60				500
	0020	Total project costs	C.F.	3.98	5.55	6.90				
	2720	Plumbing	S.F.	3.40	4.78	6.05	6.80%	9%	11.60%	
	2730	Heating, ventilating, air conditioning		1.82	3.53	3.86	4.20%	6%	6.40%	
	2900	Electrical		3.03	4.52	6.25	5.10%	6.60%	8.30%	
	3100	Total: Mechanical & Electrical	↓	14.40	18.55	20.75	14.50%	17.60%	26.50%	

See R17100-100 and City Cost Indexes in the Reference Section

		17100 | S.F. & C.F. Costs	UNIT	UNIT COSTS 1/4	MEDIAN	3/4	% OF TOTAL 1/4	MEDIAN	3/4	
500	9000	Per apartment, total cost R17100-100	Apt.	55,000	62,600	78,600				500
	9500	Total: Mechanical & Electrical	"	10,000	13,000	15,000				
510	0010	**ICE SKATING RINKS**	S.F.	44.60	77.40	110				510
	0020	Total project costs	C.F.	3.15	3.22	3.71				
	2720	Plumbing	S.F.	1.60	3	3.07	3.10%	5.60%	6.70%	
	2900	Electrical		4.59	7.04	7.45	6.70%	15%	15.80%	
	3100	Total: Mechanical & Electrical	↓	7.70	11.10	13.85	9.90%	25.90%	29.80%	
520	0010	**JAILS**	S.F.	132	168	217				520
	0020	Total project costs	C.F.	12.30	15.10	19.25				
	2720	Plumbing	S.F.	13.30	16.85	22.25	7%	8.90%	13.80%	
	2770	Heating, ventilating, air conditioning		11.75	15.71	30.35	7.50%	9.40%	17.70%	
	2900	Electrical		14.35	18.45	22.80	8.10%	11.40%	12.40%	
	3100	Total: Mechanical & Electrical	↓	36.65	65	76.90	29.20%	31.10%	34.10%	
530	0010	**LIBRARIES**	S.F.	77.80	99.15	127				530
	0020	Total project costs	C.F.	5.45	6.80	8.85				
	2720	Plumbing	S.F.	3.15	4.45	6	3.60%	4.90%	5.70%	
	2770	Heating, ventilating, air conditioning		6.75	11.40	14.85	8%	11%	14.60%	
	2900	Electrical		8	10.25	12.85	8.30%	11%	12.10%	
	3100	Total: Mechanical & Electrical	↓	23.05	31.55	39.50	18.90%	25.30%	27.60%	
550	0010	**MEDICAL CLINICS**	S.F.	76.55	94.80	119				550
	0020	Total project costs	C.F.	5.70	7.40	9.85				
	2720	Plumbing	S.F.	5.15	7.25	9.70	6.10%	8.40%	10%	
	2770	Heating, ventilating, air conditioning		6.25	8.05	11.85	6.70%	9%	11.30%	
	2900	Electrical		6.50	9.30	12.30	8.10%	10%	12.20%	
	3100	Total: Mechanical & Electrical	↓	20.45	28.80	40.25	22%	27.60%	34.30%	
570	0010	**MEDICAL OFFICES**	S.F.	71.85	89.05	110				570
	0020	Total project costs	C.F.	5.35	7.35	10.05				
	2720	Plumbing	S.F.	4.03	6.20	8.45	5.70%	6.80%	8.60%	
	2770	Heating, ventilating, air conditioning		4.87	7.15	9.30	6.20%	8%	9.70%	
	2900	Electrical		5.70	8.30	11.60	7.60%	9.80%	11.40%	
	3100	Total: Mechanical & Electrical	↓	14.20	20.35	30.25	18.50%	22%	24.90%	
590	0010	**MOTELS**	S.F.	46.10	68.25	88				590
	0020	Total project costs	C.F.	4.02	5.65	9.25				
	2720	Plumbing	S.F.	4.67	5.95	7.10	9.40%	10.60%	12.50%	
	2770	Heating, ventilating, air conditioning		2.85	4.24	7.60	5.60%	5.60%	10%	
	2900	Electrical		4.35	5.55	7.25	7.10%	8.20%	10.40%	
	3100	Total: Mechanical & Electrical	↓	14.80	18.50	31.75	18.50%	21%	24.40%	
	9000	Per rental unit, total cost	Unit	23,500	44,600	48,200				
	9500	Total: Mechanical & Electrical	"	4,600	6,900	8,000				
600	0010	**NURSING HOMES**	S.F.	69.25	91.55	112				600
	0020	Total project costs	C.F.	5.55	7.10	9.70				
	2720	Plumbing	S.F.	6.50	8.30	11.50	9.40%	10.70%	14.20%	
	2770	Heating, ventilating, air conditioning		6.45	9.30	11.50	9.30%	11.40%	11.80%	
	2900	Electrical		7.15	8.95	12	9.70%	11%	13%	
	3100	Total: Mechanical & Electrical	↓	17.05	23.85	34.95	26%	29.90%	30.50%	
	9000	Per bed or person, total cost	Bed	30,000	36,900	49,100				
610	0010	**OFFICES** Low Rise (1 to 4 story)	S.F.	58.35	74.30	98.95				610
	0020	Total project costs	C.F.	4.22	5.90	8				
	2720	Plumbing	S.F.	2.22	3.36	4.76	3.70%	4.50%	6.10%	
	2770	Heating, ventilating, air conditioning		4.80	6.65	9.80	7.20%	10.50%	11.90%	
	2900	Electrical		4.94	6.85	9.55	7.50%	9.60%	11.10%	
	3100	Total: Mechanical & Electrical	↓	11.65	16.15	23.60	18%	21.80%	26.50%	
620	0010	**OFFICES** Mid Rise (5 to 10 story)	S.F.	64.35	78.05	106				620
	0020	Total project costs	C.F.	4.50	5.70	8.25				
	2720	Plumbing	S.F.	1.95	3.02	4.34	2.80%	3.70%	4.50%	
	2770	Heating, ventilating, air conditioning		4.90	7	11.15	7.60%	9.40%	11%	
	2900	Electrical		4.78	6.10	9.25	6.50%	8.20%	10%	
	3100	Total: Mechanical & Electrical	↓	12.20	15.60	30.95	17.90%	22.30%	29.90%	

For expanded coverage of these items see *Means Square Foot Costs 2001*

17100 | S.F. & C.F. Costs

			UNIT	UNIT COSTS			% OF TOTAL			
				1/4	MEDIAN	3/4	1/4	MEDIAN	3/4	
630	0010	**OFFICES** High Rise (11 to 20 story) R17100-100	S.F.	77.90	99.85	123				630
	0020	Total project costs	C.F.	4.98	6.90	9.90				
	2900	Electrical	S.F.	3.98	5.75	8.70	5.80%	7%	10.50%	
	3100	Total: Mechanical & Electrical	"	14.55	18.60	34.10	15.40%	17%	34.10%	
640	0010	**POLICE STATIONS**	S.F.	93.30	125	157				640
	0020	Total project costs	C.F.	7.55	9.15	12.70				
	2720	Plumbing	S.F.	5.30	8.55	11.95	5.60%	6.90%	11.30%	
	2770	Heating, ventilating, air conditioning		8.30	11	14.95	7%	10.70%	12%	
	2900	Electrical		10.35	15.50	19.65	9.70%	11.90%	14.90%	
	3100	Total: Mechanical & Electrical	↓	32.95	39.35	55.40	25.10%	31.30%	36.40%	
650	0010	**POST OFFICES**	S.F.	72.70	92	115				650
	0020	Total project costs	C.F.	4.44	5.70	6.75				
	2720	Plumbing	S.F.	3.37	4.33	5.25	4.40%	5.60%	5.60%	
	2770	Heating, ventilating, air conditioning		5.30	6.50	7.25	7%	8.40%	10.20%	
	2900	Electrical		6.15	8.55	10.10	7.20%	9.60%	11%	
	3100	Total: Mechanical & Electrical	↓	10.85	21	26.50	16.40%	20.40%	22.30%	
660	0010	**POWER PLANTS**	S.F.	542	691	1,264				660
	0020	Total project costs	C.F.	14.31	28.95	66.70				
	2900	Electrical	S.F.	36.70	77.90	116	9.50%	12.80%	18.40%	
	8100	Total: Mechanical & Electrical	"	91.47	298	667	32.50%	32.60%	52.60%	
670	0010	**RELIGIOUS EDUCATION**	S.F.	58.75	76.25	88.90				670
	0020	Total project costs	C.F.	3.35	4.80	6.20				
	2720	Plumbing	S.F.	2.51	3.57	5.03	4.30%	4.90%	8.40%	
	2770	Heating, ventilating, air conditioning		5.90	6.65	7.35	9.80%	10%	11.80%	
	2900	Electrical		4.79	6.35	8.35	7.90%	9.10%	10.30%	
	3100	Total: Mechanical & Electrical	↓	12.40	21.50	25.55	17.60%	20.50%	22%	
690	0010	**RESEARCH** Laboratories and facilities	S.F.	90.25	130	189				690
	0020	Total project costs	C.F.	6.75	11.30	15.20				
	2720	Plumbing	S.F.	9.05	11.85	18.65	7.30%	9.20%	13.30%	
	2770	Heating, ventilating, air conditioning		8.13	27.30	32.30	7.20%	16.50%	17.50%	
	2900	Electrical		10.40	17.45	29.70	9.60%	12%	15.80%	
	3100	Total: Mechanical & Electrical	↓	24	45.55	86.75	23.90%	35.60%	41.70%	
700	0010	**RESTAURANTS**	S.F.	86.90	112	145				700
	0020	Total project costs	C.F.	7.25	9.60	12.65				
	2720	Plumbing	S.F.	6.90	8.40	12	6.10%	8.30%	9.10%	
	2770	Heating, ventilating, air conditioning		9.15	12.20	15.90	9.40%	12.20%	13%	
	2900	Electrical		9.25	11.45	14.85	8.50%	10.60%	11.60%	
	3100	Total: Mechanical & Electrical	↓	28.95	30.20	38.20	21.10%	24%	29.50%	
	9000	Per seat unit, total cost	Seat	3,200	4,300	5,400				
	9500	Total: Mechanical & Electrical	"	675	960	1,263				
720	0010	**RETAIL STORES**	S.F.	40.55	54.65	71.85				720
	0020	Total project costs	C.F.	2.76	3.94	5.45				
	2720	Plumbing	S.F.	1.48	2.48	4.25	3.20%	4.60%	6.80%	
	2770	Heating, ventilating, air conditioning		3.20	4.39	6.60	6.70%	8.70%	10.10%	
	2900	Electrical		3.66	4.96	7.20	7.30%	9.90%	11.60%	
	3100	Total: Mechanical & Electrical	↓	9.80	12.55	16.30	17.10%	21.40%	23.80%	
740	0010	**SCHOOLS** Elementary	S.F.	64.55	79.60	96				740
	0020	Total project costs	C.F.	4.37	5.60	7.25				
	2720	Plumbing	S.F.	3.74	5.40	7.25	5.70%	7.10%	9.40%	
	2730	Heating, ventilating, air conditioning		5.75	9.15	12.80	8.10%	10.80%	15.20%	
	2900	Electrical		6	7.65	9.80	8.40%	10%	11.60%	
	3100	Total: Mechanical & Electrical	↓	20.80	23.05	32.55	22.20%	28%	30.40%	
	9000	Per pupil, total cost	Ea.	6,400	10,700	34,700				
	9500	Total: Mechanical & Electrical	"	2,175	2,750	11,000				
760	0010	**SCHOOLS** Junior High & Middle	S.F.	65.55	81.10	95				760
	0020	Total project costs	C.F.	4.31	5.65	6.35				
	2720	Plumbing	S.F.	4.35	4.84	6.40	5.60%	6.90%	8.10%	
	2770	Heating, ventilating, air conditioning		5.05	9.70	13.55	8.70%	12.70%	17.40%	
	2900	Electrical		6.35	7.95	9.60	7.80%	9.40%	10.60%	
	3100	Total: Mechanical & Electrical	↓	18.30	23	30.55	23.30%	25.70%	27.30%	

See R17100-100 and City Cost Indexes in the Reference Section

	17100 | S.F. & C.F. Costs		UNIT	UNIT COSTS 1/4	UNIT COSTS MEDIAN	UNIT COSTS 3/4	% OF TOTAL 1/4	% OF TOTAL MEDIAN	% OF TOTAL 3/4	
760	9000	Per pupil, total cost R17100-100	Ea.	8,300	9,500	11,800				760
780	0010	**SCHOOLS** Senior High	S.F.	70.80	81.10	114				780
	0020	Total project costs	C.F.	4.54	5.90	8.30				
	2720	Plumbing	S.F.	3.66	6.70	11	5%	6.90%	12%	
	2770	Heating, ventilating, air conditioning	↓	8.20	9.30	17.90	8.90%	11%	15%	
	2900	Electrical	↓	6.95	9.15	15.25	8.30%	10.10%	12.60%	
	3100	Total: Mechanical & Electrical	↓	19	25.05	46.85	19.80%	23.10%	27.80%	
	9000	Per pupil, total cost	Ea.	6,800	11,500	17,200				
800	0010	**SCHOOLS** Vocational	S.F.	57.80	80.75	103				800
	0020	Total project costs	C.F.	3.54	5.10	7.15				
	2720	Plumbing	S.F.	3.71	5.65	8.15	5.40%	7%	8.50%	
	2770	Heating, ventilating, air conditioning	↓	5.20	9.65	16.15	8.80%	11.90%	14.60%	
	2900	Electrical	↓	5.95	8.25	11.65	8.40%	11.20%	13.80%	
	3100	Total: Mechanical & Electrical	↓	15.65	20.95	39.80	21.70%	27.30%	33.10%	
	9000	Per pupil, total cost	Ea.	8,100	21,600	32,300				
830	0010	**SPORTS ARENAS**	S.F.	49.25	67.70	99.35				830
	0020	Total project costs	C.F.	2.75	4.88	6.35				
	2720	Plumbing	S.F.	2.62	4.36	8.20	4.50%	6.30%	9.40%	
	2770	Heating, ventilating, air conditioning	↓	5.35	7.50	9.75	5.80%	10.20%	13.50%	
	2900	Electrical	↓	4.35	6.85	8.60	7.70%	9.80%	12.30%	
	3100	Total: Mechanical & Electrical	↓	12.80	22.75	29.35	13.40%	22.50%	30.80%	
850	0010	**SUPERMARKETS**	S.F.	46.75	54.85	63.65				850
	0020	Total project costs	C.F.	2.61	3.15	4.77				
	2720	Plumbing	S.F.	2.61	3.29	3.83	5.40%	6%	7.50%	
	2770	Heating, ventilating, air conditioning	↓	3.85	4.72	5.60	8.60%	8.60%	9.60%	
	2900	Electrical	↓	5.45	6.75	8	10.40%	12.40%	13.60%	
	3100	Total: Mechanical & Electrical	↓	15.30	16.50	22.85	17.40%	20.40%	28.40%	
860	0010	**SWIMMING POOLS**	S.F.	82.65	127	177				860
	0020	Total project costs	C.F.	6.10	7.55	8.25				
	2720	Plumbing	S.F.	7	8	11.15	9.70%	12.30%	12.30%	
	2900	Electrical	↓	5.70	8.85	11.50	7.50%	8%	8%	
	3100	Total: Mechanical & Electrical	↓	12.05	25.60	47.40	17.70%	24.90%	31.60%	
870	0010	**TELEPHONE EXCHANGES**	S.F.	101	147	187				870
	0020	Total project costs	C.F.	6.22	9.55	13.55				
	2720	Plumbing	S.F.	4.25	6.55	9.60	4.50%	5.80%	6.90%	
	2770	Heating, ventilating, air conditioning	↓	8.65	19.45	24.65	11.80%	11.80%	18.40%	
	2900	Electrical	↓	10.25	15.85	28.75	10.90%	14%	17.80%	
	3100	Total: Mechanical & Electrical	↓	22	30.25	75	27.30%	33.40%	44.40%	
910	0010	**THEATERS**	S.F.	63.30	81.20	120				910
	0020	Total project costs	C.F.	2.92	4.32	6.35				
	2720	Plumbing	S.F.	1.97	2.30	6.85	2.90%	4.70%	6.80%	
	2770	Heating, ventilating, air conditioning	↓	6.15	7.45	9.20	8%	12.20%	13.40%	
	2900	Electrical	↓	5.55	7.45	15.20	8%	10%	12.40%	
	3100	Total: Mechanical & Electrical	↓	14.25	21.30	43.70	22.90%	26.60%	27.50%	
940	0010	**TOWN HALLS** City Halls & Municipal Buildings	S.F.	70.65	89.30	117				940
	0020	Total project costs	C.F.	5.30	7.40	9.90				
	2720	Plumbing	S.F.	2.51	5.20	8.60	4.20%	6.10%	7.90%	
	2770	Heating, ventilating, air conditioning	↓	5.35	10.60	15.45	7%	9%	13.50%	
	2900	Electrical	↓	6.35	8.55	12.10	8.20%	9.50%	11.70%	
	3100	Total: Mechanical & Electrical	↓	22.40	23.25	29.50	22%	25.10%	30.40%	
970	0010	**WAREHOUSES** And Storage Buildings	S.F.	26.35	36.80	56.70				970
	0020	Total project costs	C.F.	1.33	2.16	3.58				
	2720	Plumbing	S.F.	.88	1.58	2.95	2.90%	4.80%	6.60%	
	2730	Heating, ventilating, air conditioning	↓	1	2.83	3.80	2.40%	5%	8.90%	
	2900	Electrical	↓	1.56	2.94	4.86	5%	7.30%	10%	
	3100	Total: Mechanical & Electrical	↓	4.35	6.40	14.60	12.80%	18.40%	26.20%	

17100 | S.F., C.F. and % of Total Costs

		17100 | S.F. & C.F. Costs	UNIT	UNIT COSTS 1/4	UNIT COSTS MEDIAN	UNIT COSTS 3/4	% OF TOTAL 1/4	% OF TOTAL MEDIAN	% OF TOTAL 3/4	
990	0010	**WAREHOUSE & OFFICES** Combination R17100-100	S.F.	31.70	42.50	56.40				990
	0020	Total project costs	C.F.	1.64	2.41	3.56				
	2720	Plumbing	S.F.	1.23	2.11	3.33	3.60%	4.70%	6.30%	
	2770	Heating, ventilating, air conditioning		1.95	3.08	4.35	5%	5.60%	9.60%	
	2900	Electrical	S.F.	2.15	3.17	4.97	5.70%	7.70%	10%	
	3100	Total: Mechanical & Electrical	↓	5.05	8.40	10.50	13.80%	18.60%	23%	

For information about Means Estimating Seminars, see yellow pages 11 and 12 in back of book

17 SQUARE FOOT

See R17100-100 and City Cost Indexes in the Reference Section

Assemblies Section

Table of Contents

How to Use the Assemblies Cost Tables

The following is a detailed explanation of a sample Assemblies Cost Table. Most Assembly Tables are separated into three parts: 1) an illustration of the system to be estimated; 2) the components and related costs of a typical system; and 3) the costs for similar systems with dimensional and/or size variations. For costs of the components that comprise these systems or "assemblies" refer to the Unit Price Section. Next to each bold number below is the item being described with the appropriate component of the sample entry following in parenthesis. In most cases, if the work is to be subcontracted, the general contractor will need to add an additional markup (R.S. Means suggests using 10%) to the "Total" figures.

1 System/Line Numbers (A8.3-151-1760)

Each Assemblies Cost Line has been assigned a unique identification number based on the UniFormat classification system.

UniFormat Division

8.3 151 1760

Means Subdivision
Means Major Classification
Means Individual Line Number

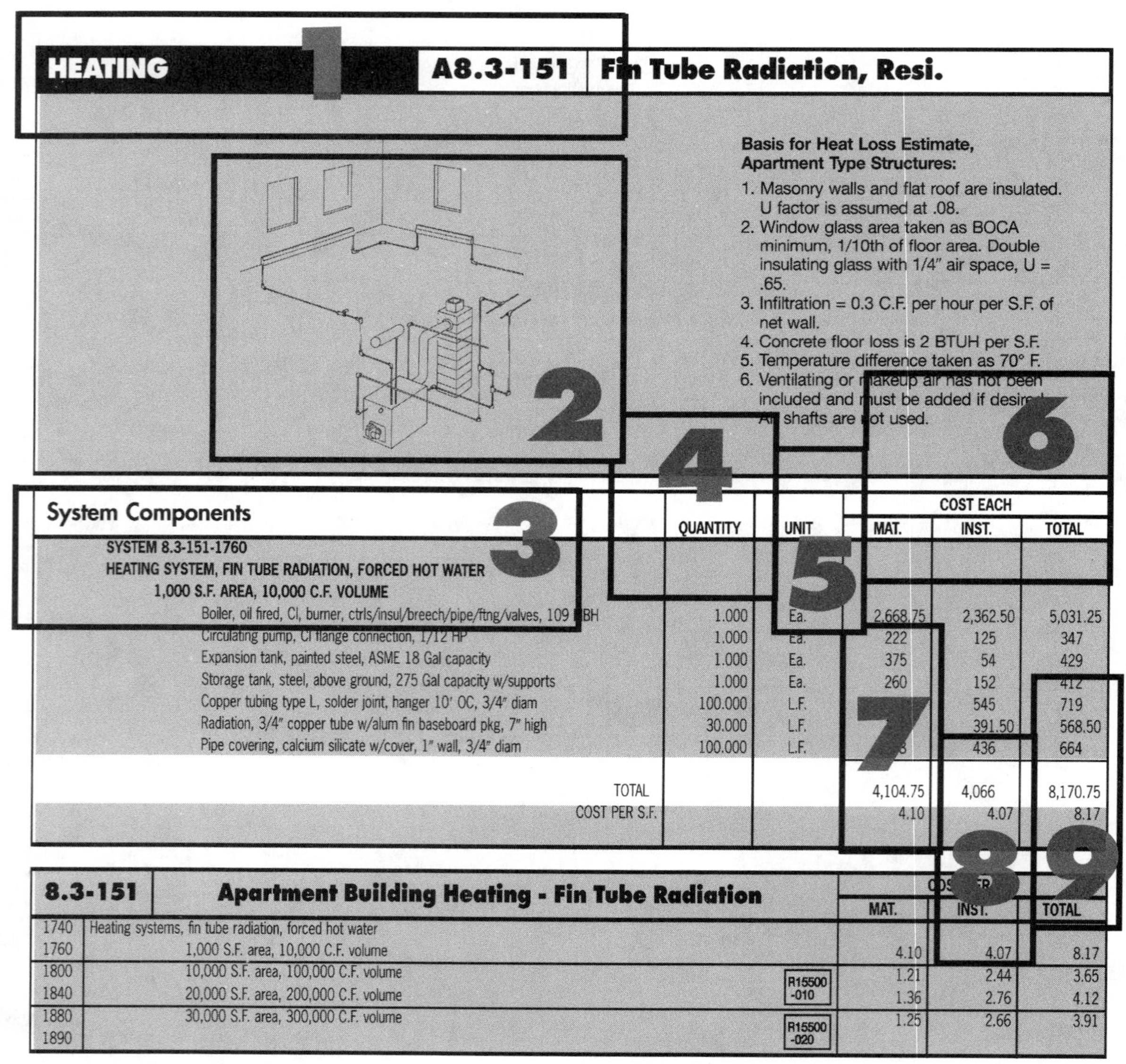

HEATING | **A8.3-151** | **Fin Tube Radiation, Resi.**

Basis for Heat Loss Estimate, Apartment Type Structures:

1. Masonry walls and flat roof are insulated. U factor is assumed at .08.
2. Window glass area taken as BOCA minimum, 1/10th of floor area. Double insulating glass with 1/4" air space, U = .65.
3. Infiltration = 0.3 C.F. per hour per S.F. of net wall.
4. Concrete floor loss is 2 BTUH per S.F.
5. Temperature difference taken as 70° F.
6. Ventilating or makeup air has not been included and must be added if desired. Air shafts are not used.

System Components	QUANTITY	UNIT	COST EACH MAT.	COST EACH INST.	COST EACH TOTAL
SYSTEM 8.3-151-1760 HEATING SYSTEM, FIN TUBE RADIATION, FORCED HOT WATER 1,000 S.F. AREA, 10,000 C.F. VOLUME					
Boiler, oil fired, CI, burner, ctrls/insul/breech/pipe/ftng/valves, 109 MBH	1.000	Ea.	2,668.75	2,362.50	5,031.25
Circulating pump, CI flange connection, 1/12 HP	1.000	Ea.	222	125	347
Expansion tank, painted steel, ASME 18 Gal capacity	1.000	Ea.	375	54	429
Storage tank, steel, above ground, 275 Gal capacity w/supports	1.000	Ea.	260	152	412
Copper tubing type L, solder joint, hanger 10' OC, 3/4" diam	100.000	L.F.	[illegible]	545	719
Radiation, 3/4" copper tube w/alum fin baseboard pkg, 7" high	30.000	L.F.	[illegible]	391.50	568.50
Pipe covering, calcium silicate w/cover, 1" wall, 3/4" diam	100.000	L.F.	[illegible]	436	664
TOTAL			4,104.75	4,066	8,170.75
COST PER S.F.			4.10	4.07	8.17

8.3-151	Apartment Building Heating - Fin Tube Radiation		CO[illegible]ER MAT.	INST.	TOTAL
1740	Heating systems, fin tube radiation, forced hot water				
1760	1,000 S.F. area, 10,000 C.F. volume		4.10	4.07	8.17
1800	10,000 S.F. area, 100,000 C.F. volume	R15500 -010	1.21	2.44	3.65
1840	20,000 S.F. area, 200,000 C.F. volume		1.36	2.76	4.12
1880	30,000 S.F. area, 300,000 C.F. volume	R15500 -020	1.25	2.66	3.91
1890					

2 *Illustration*

At the top of most assembly pages is an illustration, a brief description, and the design criteria used to develop the cost.

3 *System Components*

The components of a typical system are listed separately to show what has been included in the development of the total system price. The table below contains prices for other similar systems with dimensional and/or size variations.

4 *Quantity*

This is the number of line item units required for one system unit. For example, we assume that it will take 30 linear feet of radiation for the 1000 S.F. area shown.

5 *Unit of Measure for Each Item*

The abbreviated designation indicates the unit of measure, as defined by industry standards, upon which the price of the component is based. For example, baseboard radiation is priced by the linear foot. For a complete listing of abbreviations, see the Reference Section.

6 *Unit of Measure for Each System (Each)*

Costs shown in the three right hand columns have been adjusted by the component quantity and unit of measure for the entire system. In this example, "Cost Each" is the unit of measure for this system or "assembly."

7 *Materials (4,104.75)*

This column contains the Materials Cost of each component. These cost figures are bare costs plus 10% for profit.

8 *Installation (4,066.00)*

Installation includes labor and equipment plus the installing contractor's overhead and profit. Equipment costs are the bare rental costs plus 10% for profit. The labor overhead and profit is defined on the inside back cover of this book.

9 *Total (8,170.75)*

The figure in this column is the sum of the material and installation costs.

Material Cost	+	Installation Cost	=	Total
$4,104.75	+	$4,066.00	=	$8,170.75

Division 8
Mechanical

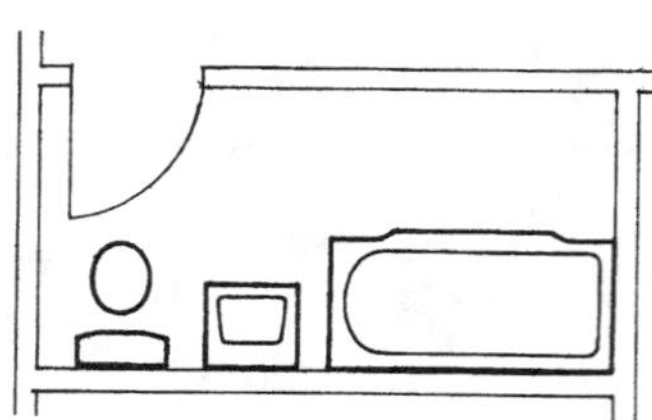

Example of Plumbing Cost Calculations: The bathroom system includes the individual fixtures such as bathtub, lavatory, shower and water closet. These fixtures are listed below as separate items merely as a checklist.

8.1-010 Plumbing Systems 20 Unit, 2 Story Apartment Building

	FIXTURE	SYSTEM	LINE	QUANTITY	UNIT	COST EACH		
						MAT.	INST.	TOTAL
0440	Bathroom	A8.1-630	3640	20	Ea.	23,900	29,700	53,600
0480	Bathtub							
0520	Booster pump[1]	not req'd.						
0560	Drinking fountain							
0600	Garbage disposal[1]	not incl.						
0660								
0680	Grease interceptor							
0720	Water heater	A8.1-170	2140	1	Ea.	3,800	2,025	5,825
0760	Kitchen sink	A8.1-431	1960	20	Ea.	11,300	10,500	21,800
0800	Laundry sink	A8.1-432	1840	4	Ea.	3,100	2,075	5,175
0840	Lavatory							
0900								
0920	Roof drain, 1 floor	A8.1-310	4200	2	Ea.	810	1,075	1,885
0960	Roof drain, add'l floor	A8.1-310	4240	20	L.F.	175	272	447
1000	Service sink	A8.1-434	4300	1	Ea.	730	725	1,455
1040	Sewage ejector[1]	not req'd.						
1080	Shower							
1100								
1160	Sump pump							
1200	Urinal							
1240	Water closet							
1320								
1360	**SUB TOTAL**					43,800	46,300	90,100
1480	Water controls	R8.1-031		10%[2]		4,375	4,625	9,000
1520	Pipe & fittings[3]	R8.1-031		30%[2]		13,100	13,900	27,000
1560	Other							
1600	Quality/complexity	R8.1-031		15%[2]		6,575	6,950	13,525
1680								
1720	**TOTAL**					68,000	72,000	140,000
1741								

[1]**Note:** Cost for items such as booster pumps, backflow preventers, sewage ejectors, water meters, etc., may be obtained from Sections 151, 152 and 154.

Water controls, pipe and fittings, and the Quality/Complexity factors come from Table R15100-040.

[2]Percentage of subtotal.

[3]Long easily discernable runs of pipe would be more accurately priced from Section 151. If this is done, reduce the miscellaneous percentage in proportion.

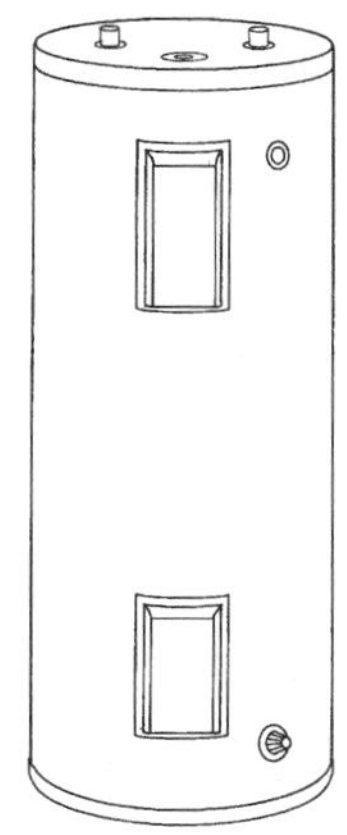

Installation includes piping and fittings within 10′ of heater. Electric water heaters do not require venting.

1 Kilowatt hour will raise:			
Gallons of Water	Degrees F	Gallons of Water	Degrees F
4.1	100°	6.8	60°
4.5	90°	8.2	50°
5.1	80°	10.0	40°
5.9	70°		

System Components	QUANTITY	UNIT	COST EACH MAT.	COST EACH INST.	COST EACH TOTAL
SYSTEM 8.1-110-1780					
ELECTRIC WATER HEATER, RESIDENTIAL, 100° F RISE					
10 GALLON TANK, 7 GPH					
Water heater, residential electric, glass lined tank, 10 Gal	1.000	Ea.	214	181	395
Copper tubing, type L, solder joint, hanger 10′ OC 1/2″ diam	30.000	L.F.	41.40	154.50	195.90
Wrought copper 90° elbow for solder joints 1/2″ diam.	4.000	Ea.	.96	84	84.96
Wrought copper Tee for solder joints, 1/2″ diam	2.000	Ea.	.80	64	64.80
Wrought copper union for soldered joints, 1/2″ diam	2.000	Ea.	4.62	44	48.62
Valve, gate, bronze, 125 lb, NRS, soldered 1/2″ diam	2.000	Ea.	32.50	34.70	67.20
Relief valve, bronze, press & temp, self-close, 3/4″ IPS	1.000	Ea.	71.50	14.85	86.35
Wrought copper adapter, CTS to MPT 3/4″ IPS	1.000	Ea.	.84	24.50	25.34
Copper tubing, type L, solder joints, 3/4″ diam	1.000	L.F.	1.74	5.45	7.19
Wrought copper 90° elbow for solder joints 3/4″ diam	1.000	Ea.	.53	22	22.53
TOTAL			368.89	629	997.89

8.1-110	Electric Water Heaters - Residential Systems	COST EACH MAT.	COST EACH INST.	COST EACH TOTAL
1760	Electric water heater, residential, 100° F rise			
1780	10 gallon tank, 7 GPH	370	630	1,000
1820	20 gallon tank, 7 GPH R15100 -120	425	670	1,095
1860	30 gallon tank, 7 GPH	485	700	1,185
1900	40 gallon tank, 8 GPH	550	770	1,320
1940	52 gallon tank, 10 GPH	595	780	1,375
1980	66 gallon tank, 13 GPH	885	885	1,770
2020	80 gallon tank, 16 GPH	950	930	1,880
2060	120 gallon tank, 23 GPH	1,450	1,100	2,550

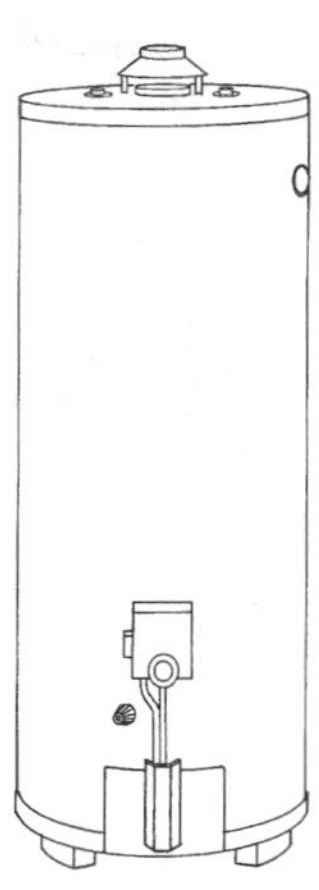

Installation includes piping and fittings within 10′ of heater. Gas heaters require vent piping (not included with these units).

System Components	QUANTITY	UNIT	COST EACH MAT.	COST EACH INST.	COST EACH TOTAL
SYSTEM 8.1-120-2220					
GAS FIRED WATER HEATER, RESIDENTIAL, 100° F RISE					
20 GALLON, 25 GPH					
Water heater, residential, gas, glass lined tank, 20 Gal	1.000	Ea.	207	198	405
Copper tubing, type L, solder joint, hanger 10′ OC, 3/4″ diam	32.000	L.F.	55.68	174.40	230.08
Wrought copper 90° elbow for solder joints 3/4″ diam	5.000	Ea.	2.65	110	112.65
Wrought copper Tee for solder joints, 3/4″ diam	2.000	Ea.	1.96	69	70.96
Wrought copper union for soldered joints, 3/4″ diam	2.000	Ea.	5.78	46	51.78
Valve, bronze, 125 lb, NRS, soldered 3/4″ diam	2.000	Ea.	37.10	42	79.10
Relief valve, press & temp, bronze, self-close, 3/4″ IPS	1.000	Ea.	71.50	14.85	86.35
Wrought copper, adapter, CTS to MPT 3/4″ IPS	1.000	Ea.	.84	24.50	25.34
Pipe steel black, schedule 40, threaded, 1/2″ diam	10.000	L.F.	13.50	66	79.50
Pipe, 90° elbow, malleable iron black, 150 lb, threaded, 1/2″ diam	2.000	Ea.	2.66	55	57.66
Pipe, union with brass seat, malleable iron black, 1/2″ diam	1.000	Ea.	5.90	29.50	35.40
Valve, gas stop w/o check, brass, 1/2″ IPS	1.000	Ea.	8.45	17.35	25.80
TOTAL			413.02	846.60	1,259.62

8.1-120	Gas Fired Water Heaters - Residential Systems	COST EACH MAT.	COST EACH INST.	COST EACH TOTAL
2200	Gas fired water heater, residential, 100° F rise			
2220	20 gallon tank, 25 GPH	415	845	1,260
2260	30 gallon tank, 32 GPH (R15100 -120)	575	860	1,435
2300	40 gallon tank, 32 GPH	635	965	1,600
2340	50 gallon tank, 63 GPH	785	965	1,750
2380	75 gallon tank, 63 GPH	905	1,075	1,980
2420	100 gallon tank, 63 GPH	1,200	1,125	2,325

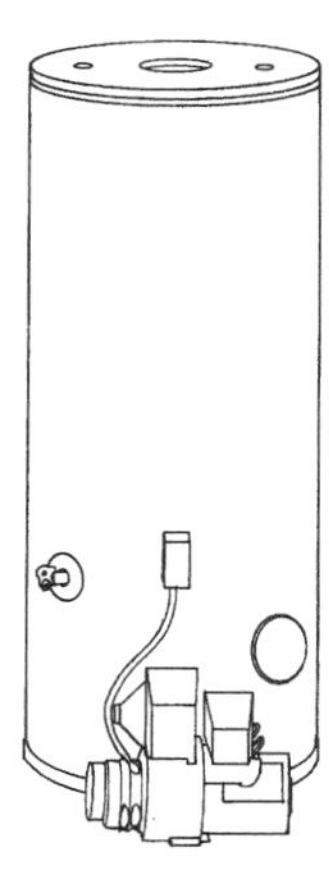

Installation includes piping and fittings within 10′ of heater. Oil fired heaters require vent piping (not included in these prices).

System Components	QUANTITY	UNIT	COST EACH MAT.	COST EACH INST.	COST EACH TOTAL
SYSTEM 8.1-130-2220					
OIL FIRED WATER HEATER, RESIDENTIAL, 100° F RISE					
30 GALLON TANK, 103 GPH					
Water heater, residential, oil glass lined tank, 30 Gal	1.000	Ea.	840	208	1,048
Copper tubing, type L, solder joint, hanger 10′ O.C. 3/4″ diam	33.000	L.F.	57.42	179.85	237.27
Wrought copper 90° elbow for solder joints 3/4″ diam	5.000	Ea.	2.65	110	112.65
Wrought copper Tee for solder joints, 3/4″ diam	2.000	Ea.	1.96	69	70.96
Wrought copper union for soldered joints, 3/4″ diam	2.000	Ea.	5.78	46	51.78
Valve, gate, bronze, 125 lb, NRS, soldered 3/4″ diam	2.000	Ea.	37.10	42	79.10
Relief valve, bronze, press & temp, self-close, 3/4″ IPS	1.000	Ea.	71.50	14.85	86.35
Wrought copper adapter, CTS to MPT, 3/4″ IPS	1.000	Ea.	.84	24.50	25.34
Copper tubing, type L, solder joint, hanger 10′ OC 3/8″ diam	10.000	L.F.	12.40	49.50	61.90
Wrought copper 90° elbow for solder joints 3/8″ diam	2.000	Ea.	1.32	37.80	39.12
Valve, globe, fusible, 3/8″ diam	1.000	Ea.	5.65	17.55	23.20
TOTAL			1,036.62	799.05	1,835.67

8.1-130	Oil Fired Water Heaters - Residential Systems	COST EACH MAT.	COST EACH INST.	COST EACH TOTAL
2200	Oil fired water heater, residential, 100° F rise			
2220	30 gallon tank, 103 GPH	1,025	800	1,825
2260	50 gallon tank, 145 GPH R15100 -120	1,375	895	2,270
2300	70 gallon tank, 164 GPH	1,800	1,000	2,800
2340	85 gallon tank, 181 GPH	4,075	1,025	5,100

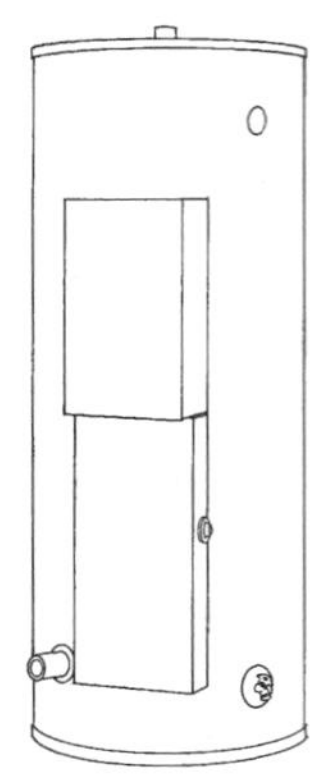

Systems below include piping and fittings within 10′ of heater. Electric water heaters do not require venting.

System Components	QUANTITY	UNIT	COST EACH MAT.	COST EACH INST.	COST EACH TOTAL
SYSTEM 8.1-160-1820					
ELECTRIC WATER HEATER, COMMERCIAL, 100° F RISE					
50 GALLON TANK, 9 KW, 37 GPH					
Water heater, commercial, electric, 50 Gal, 9 KW, 37 GPH	1.000	Ea.	2,175	231	2,406
Copper tubing, type L, solder joint, hanger 10′ OC, 3/4″ diam	34.000	L.F.	59.16	185.30	244.46
Wrought copper 90° elbow for solder joints 3/4″ diam	5.000	Ea.	2.65	110	112.65
Wrought copper Tee for solder joints, 3/4″ diam	2.000	Ea.	1.96	69	70.96
Wrought copper union for soldered joints, 3/4″ diam	2.000	Ea.	5.78	46	51.78
Valve, gate, bronze, 125 lb, NRS, soldered 3/4″ diam	2.000	Ea.	37.10	42	79.10
Relief valve, bronze, press & temp, self-close, 3/4″ IPS	1.000	Ea.	71.50	14.85	86.35
Wrought copper adapter, copper tubing to male, 3/4″ IPS	1.000	Ea.	.84	24.50	25.34
TOTAL			2,353.99	722.65	3,076.64

8.1-160	Electric Water Heaters - Commercial Systems		COST EACH MAT.	COST EACH INST.	COST EACH TOTAL
1800	Electric water heater, commercial, 100° F rise				
1820	50 gallon tank, 9 KW 37 GPH		2,350	725	3,075
1860	80 gal, 12 KW 49 GPH	R15100 -110	2,950	890	3,840
1900	36 KW 147 GPH		4,025	965	4,990
1940	120 gal, 36 KW 147 GPH	R15100 -120	4,325	1,050	5,375
1980	150 gal, 120 KW 490 GPH		13,500	1,125	14,625
2020	200 gal, 120 KW 490 GPH		14,300	1,150	15,450
2060	250 gal, 150 KW 615 GPH		15,800	1,350	17,150
2100	300 gal, 180 KW 738 GPH		17,300	1,400	18,700
2140	350 gal, 30 KW 123 GPH		12,800	1,525	14,325
2180	180 KW 738 GPH		17,800	1,525	19,325
2220	500 gal, 30 KW 123 GPH		16,700	1,800	18,500
2260	240 KW 984 GPH		24,300	1,800	26,100
2300	700 gal, 30 KW 123 GPH		20,100	2,050	22,150
2340	300 KW 1230 GPH		29,900	2,050	31,950
2380	1000 gal, 60 KW 245 GPH		24,200	2,850	27,050
2420	480 KW 1970 GPH		24,800	2,875	27,675
2460	1500 gal, 60 KW 245 GPH		35,500	3,525	39,025
2500	480 KW 1970 GPH		48,800	3,525	52,325

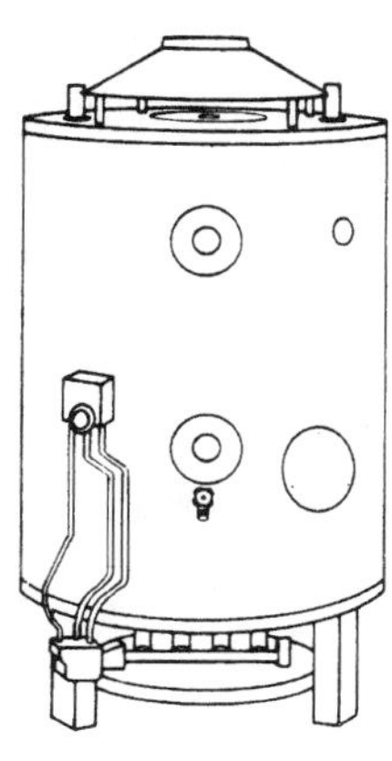

Units may be installed in multiples for increased capacity.

Included below is the heater with self-energizing gas controls, safety pilots, insulated jacket, hi-limit aquastat and pressure relief valve.

Installation includes piping and fittings within 10′ of heater. Gas heaters require vent piping (not included in these prices).

System Components	QUANTITY	UNIT	COST EACH MAT.	COST EACH INST.	COST EACH TOTAL
SYSTEM 8.1-170-1780					
GAS FIRED WATER HEATER, COMMERCIAL, 100° F RISE					
75.5 MBH INPUT, 63 GPH					
Water heater, commercial, gas, 75.5 MBH, 63 GPH	1.000	Ea.	1,375	297	1,672
Copper tubing, type L, solder joint, hanger 10′ OC, 1-1/4″ diam	30.000	L.F.	84	214.50	298.50
Wrought copper 90° elbow for solder joints 1-1/4″ diam	4.000	Ea.	7.44	110	117.44
Wrought copper Tee for solder joints, 1-1/4″ diam	2.000	Ea.	8.62	92	100.62
Wrought copper union for soldered joints, 1-1/4″ diam	2.000	Ea.	17.80	59	76.80
Valve, gate, bronze, 125 lb, NRS, soldered 1-1/4″ diam	2.000	Ea.	87	55	142
Relief valve, bronze, press & temp, self-close, 3/4″ IPS	1.000	Ea.	71.50	14.85	86.35
Copper tubing, type L, solder joints, 3/4″ diam	8.000	L.F.	13.92	43.60	57.52
Wrought copper 90° elbow for solder joints 3/4″ diam	1.000	Ea.	.53	22	22.53
Wrought copper, adapter, CTS to MPT, 3/4″ IPS	1.000	Ea.	.84	24.50	25.34
Pipe steel black, schedule 40, threaded, 3/4″ diam	10.000	L.F.	15	68	83
Pipe, 90° elbow, malleable iron black, 150 lb threaded, 3/4″ diam	2.000	Ea.	3.24	59	62.24
Pipe, union with brass seat, malleable iron black, 3/4″ diam	1.000	Ea.	6.80	32	38.80
Valve, gas stop w/o check, brass, 3/4″ IPS	1.000	Ea.	11.35	18.90	30.25
TOTAL			1,703.04	1,110.35	2,813.39

8.1-170	Gas Fired Water Heaters - Commercial Systems		COST EACH MAT.	COST EACH INST.	COST EACH TOTAL
1760	Gas fired water heater, commercial, 100° F rise				
1780	75.5 MBH input, 63 GPH		1,700	1,100	2,800
1860	100 MBH input, 91 GPH	R15100 -110	2,925	1,150	4,075
1980	155 MBH input, 150 GPH		3,525	1,350	4,875
2060	200 MBH input, 192 GPH	R15100 -120	3,275	1,625	4,900
2140	300 MBH input, 278 GPH		3,800	2,025	5,825
2180	390 MBH input, 374 GPH		4,450	2,050	6,500
2220	500 MBH input, 480 GPH		5,950	2,200	8,150
2260	600 MBH input, 576 GPH		9,650	2,375	12,025

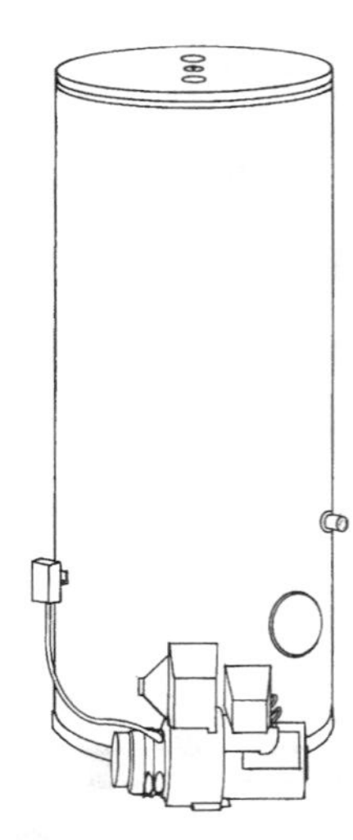

Units may be installed in multiples for increased capacity.

Included below is the heater, wired-in flame retention burners, cadmium cell primary controls, hi-limit controls, ASME pressure relief valves, draft controls, and insulated jacket.

Oil fired water heater systems include piping and fittings within 10′ of heater. Oil fired heaters require vent piping (not included in these systems).

System Components	QUANTITY	UNIT	COST EACH MAT.	COST EACH INST.	COST EACH TOTAL
SYSTEM 8.1-180-1820					
OIL FIRED WATER HEATER, COMMERCIAL, 100° F RISE					
103 MBH OUTPUT, 116 GPH					
Water heater, commercial, oil, 103 MBH	1.000	Ea.	985	380	1,365
Copper tubing, type L, solder joint, hanger 10′ OC, 3/4″ diam	34.000	L.F.	59.16	185.30	244.46
Wrought copper 90° elbow for solder joints 3/4″ diam	5.000	Ea.	2.65	110	112.65
Wrought copper Tee for solder joints, 3/4″ diam	2.000	Ea.	1.96	69	70.96
Wrought copper union for soldered joints, 3/4″ diam	2.000	Ea.	5.78	46	51.78
Valve, bronze, 125 lb, NRS, soldered 3/4″ diam	2.000	Ea.	37.10	42	79.10
Relief valve, bronze, press & temp, self-close, 3/4″ IPS	1.000	Ea.	71.50	14.85	86.35
Wrought copper adapter, copper tubing to male, 3/4″ IPS	1.000	Ea.	.84	24.50	25.34
Copper tubing, type L, solder joint, hanger 10′ OC, 3/8″ diam	10.000	L.F.	12.40	49.50	61.90
Wrought copper 90° elbow for solder joints 3/8″ diam	2.000	Ea.	1.32	37.80	39.12
Valve, globe, fusible, 3/8″ IPS	1.000	Ea.	5.65	17.55	23.20
TOTAL			1,183.36	976.50	2,159.86

8.1-180	Oil Fired Water Heaters - Commercial Systems		COST EACH MAT.	COST EACH INST.	COST EACH TOTAL
1800	Oil fired water heater, commercial, 100° F rise				
1820	103 MBH output, 116 GPH		1,175	975	2,150
1900	134 MBH output, 161 GPH	R15100 -110	1,950	1,250	3,200
1940	161 MBH output, 192 GPH		2,175	1,425	3,600
1980	187 MBH output, 224 GPH	R15100 -120	2,425	1,650	4,075
2060	262 MBH output, 315 GPH		2,850	1,900	4,750
2100	341 MBH output, 409 GPH		3,500	2,025	5,525
2140	420 MBH output, 504 GPH		3,925	2,200	6,125
2180	525 MBH output, 630 GPH		5,350	2,475	7,825
2220	630 MBH output, 756 GPH		5,550	2,475	8,025
2260	735 MBH output, 880 GPH		6,300	2,875	9,175
2300	840 MBH output, 1000 GPH		7,125	2,875	10,000
2340	1050 MBH output, 1260 GPH		7,950	2,950	10,900
2380	1365 MBH output, 1640 GPH		9,675	3,350	13,025
2420	1680 MBH output, 2000 GPH		10,900	4,200	15,100
2460	2310 MBH output, 2780 GPH		14,000	5,175	19,175
2500	2835 MBH output, 3400 GPH		16,400	5,175	21,575

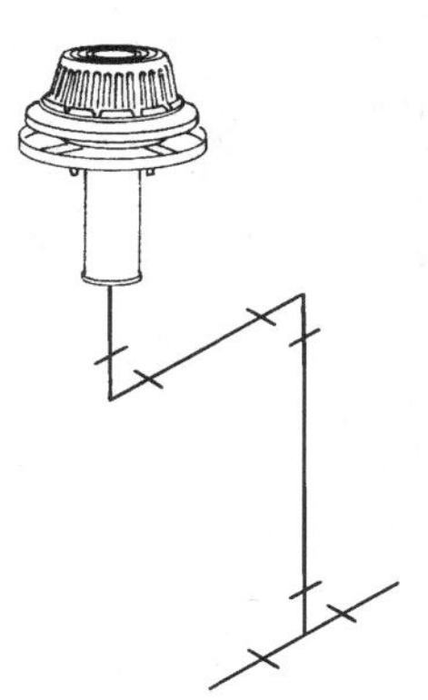

Design Assumptions: Vertical conductor size is based on a maximum rate of rainfall of 4″ per hour. To convert roof area to other rates multiply "Max. S.F. Roof Area" shown by four and divide the result by desired local rate. The answer is the local roof area that may be handled by the indicated pipe diameter.

Basic cost is for roof drain, 10′ of vertical leader and 10′ of horizontal, plus connection to the main.

Pipe Dia.	Max. S.F. Roof Area	Gallons per Min.
2″	544	23
3″	1610	67
4″	3460	144
5″	6280	261
6″	10,200	424
8″	22,000	913

System Components

System Components	QUANTITY	UNIT	COST EACH MAT.	COST EACH INST.	COST EACH TOTAL
SYSTEM 8.1-310-1880					
ROOF DRAIN, DWV PVC PIPE, 2″ DIAM., 10′ HIGH					
Drain, roof, main, PVC, dome type 2″ pipe size	1.000	Ea.	32	53.50	85.50
Clamp, roof drain, underdeck	1.000	Ea.	17.60	31	48.60
Pipe, Tee, PVC DWV, schedule 40, 2″ pipe size	1.000	Ea.	1.86	37.50	39.36
Pipe, PVC, DWV, schedule 40, 2″ diam.	20.000	L.F.	55.20	254	309.20
Pipe, elbow, PVC schedule 40, 2″ diam.	2.000	Ea.	3.44	41	44.44
TOTAL			110.10	417	527.10

8.1-310 Roof Drain Systems

8.1-310	Roof Drain Systems	COST PER SYSTEM MAT.	INST.	TOTAL
1880	Roof drain, DWV PVC, 2″ diam., piping, 10′ high	110	415	525
1920	For each additional foot add	2.76	12.70	15.46
1960	3″ diam., 10′ high	163	485	648
2000	For each additional foot add	4.78	14.15	18.93
2040	4″ diam., 10′ high	186	545	731
2080	For each additional foot add	5.20	15.60	20.80
2120	5″ diam., 10′ high	395	630	1,025
2160	For each additional foot add	8.50	17.40	25.90
2200	6″ diam., 10′ high	585	700	1,285
2240	For each additional foot add	8.95	19.20	28.15
2280	8″ diam., 10′ high	1,375	1,200	2,575
2320	For each additional foot add	14.40	24.50	38.90
3940	C.I., soil, single hub, service wt., 2″ diam. piping, 10′ high	247	455	702
3980	For each additional foot add	5.20	11.90	17.10
4120	3″ diam., 10′ high	310	495	805
4160	For each additional foot add	5.10	12.50	17.60
4200	4″ diam., 10′ high	405	540	945
4240	For each additional foot add	8.75	13.60	22.35
4280	5″ diam., 10′ high	450	595	1,045
4320	For each additional foot add	11.65	15.35	27
4360	6″ diam., 10′ high	610	635	1,245
4400	For each additional foot add	10.05	15.95	26
4440	8″ diam., 10′ high	1,500	1,300	2,800
4480	For each additional foot add	22	27	49
6040	Steel galv. sch 40 threaded, 2″ diam. piping, 10′ high	246	440	686
6080	For each additional foot add	3.34	11.70	15.04
6120	3″ diam., 10′ high	460	635	1,095
6160	For each additional foot add	6.15	17.40	23.55
6200	4″ diam., 10′ high	650	820	1,470
6240	For each additional foot add	8.85	21	29.85

8.1-310	Roof Drain Systems	COST PER SYSTEM		
		MAT.	INST.	TOTAL
6280	5″ diam., 10′ high	775	1,050	1,825
6320	For each additional foot add	16.40	29	45.40
6360	6″ diam, 10′ high	1,425	1,275	2,700
6400	For each additional foot add	22.50	37.50	60
6440	8″ diam., 10′ high	2,900	1,725	4,625
6480	For each additional foot add	36.50	43	79.50

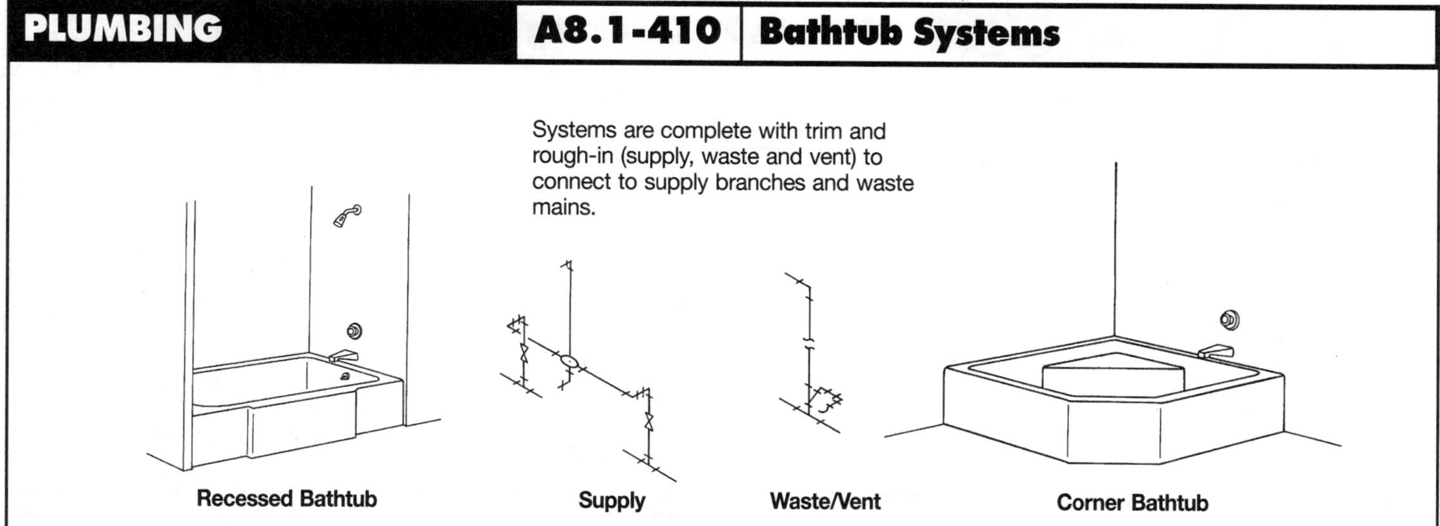
Systems are complete with trim and rough-in (supply, waste and vent) to connect to supply branches and waste mains.
Recessed Bathtub
Supply
Waste/Vent
Corner Bathtub

System Components

System Components	QUANTITY	UNIT	COST EACH MAT.	COST EACH INST.	COST EACH TOTAL
SYSTEM 8.1-410-1960					
BATHTUB RECESSED, PORCELAIN ENAMEL ON CAST IRON, 42″ X 37″					
Bath tub, porcelain enamel on cast iron, w/fittings, 42″ x 37″	1.000	Ea.	780	150	930
Pipe, steel, galvanized, schedule 40, threaded, 1-1/4″ diam	4.000	L.F.	9.48	33.60	43.08
Pipe, CI, soil, no hub, w/coupling 10′ OC, hanger 5′ OC, 4″ diam	3.000	L.F.	26.85	38.70	65.55
Combination Y and 1/8 bend for C.I. soil pipe, no hub, 4″ pipe size	1.000	Ea.	21.50		21.50
Drum trap, 3″ x 5″, copper, 1-1/2″ diam	1.000	Ea.	18.35	26	44.35
Copper tubing, type L, solder joints, hangers 10′ OC 1/2″ diam	10.000	L.F.	13.80	51.50	65.30
Wrought copper 90° elbow for solder joints, 1/2″ diam	2.000	Ea.	.48	42	42.48
Wrought copper Tee for solder joints, 1/2″ diam	2.000	Ea.	.80	64	64.80
Stop, angle supply, 1/2″ diam	2.000	Ea.	10.40	37.80	48.20
Copper tubing type DWV, solder joint, hanger 10′ OC 1-1/2″ diam	3.000	L.F.	8.37	23.10	31.47
Pipe coupling, standard, C.I. soil no hub, 4″ pipe size	2.000	Ea.	11.90	45	56.90
TOTAL			901.93	511.70	1,413.63

8.1-410	Bathtub Systems	COST EACH MAT.	COST EACH INST.	COST EACH TOTAL
1960	Bathtub, recessed, P.E. on CI., 42″ x 37″	900	510	1,410
2000	48″ x 42″	1,275	550	1,825
2040	72″ x 36″ (R15100-410)	1,425	610	2,035
2080	Mat bottom, 5′ long	515	530	1,045
2120	5′-6″ long	1,075	550	1,625
2160	Corner, 48″ x 42″	1,525	530	2,055
2200	Formed steel, enameled, 4′-6″ long	425	490	915
2240	5′ long	410	500	910

A8.1-420 Drinking Fountain System

Systems are complete with trim and rough-in (supply, waste and vent) to connect to supply branches and waste mains.

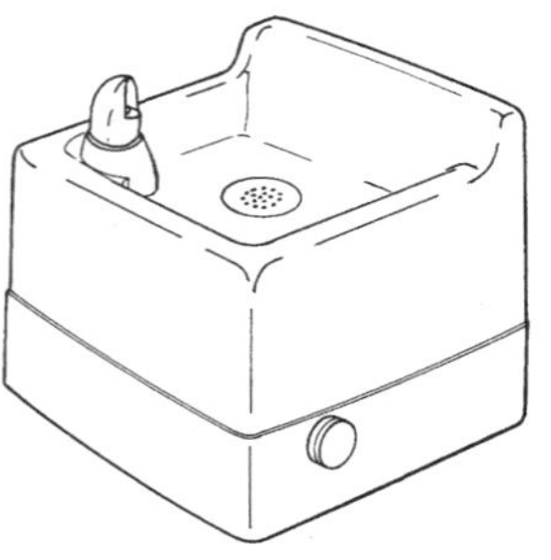
Wall Mounted, No Back

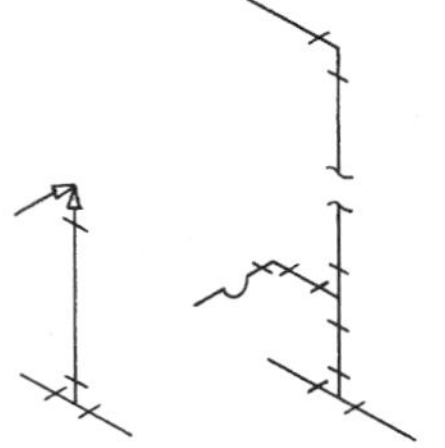
Supply Waste/Vent

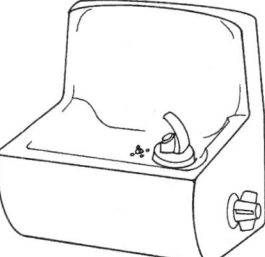

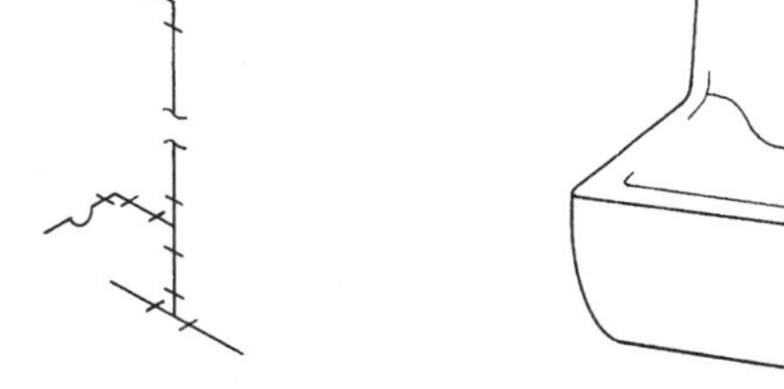
Wall Mounted, Low Back

System Components	QUANTITY	UNIT	COST EACH MAT.	COST EACH INST.	COST EACH TOTAL
SYSTEM 8.1-420-1800					
DRINKING FOUNTAIN, ONE BUBBLER, WALL MOUNTED					
NON RECESSED, BRONZE, NO BACK					
Drinking fountain, wall mount, bronze, 1 bubbler	1.000	Ea.	1,100	104	1,204
Copper tubing, type L, solder joint, hanger 10′ OC 3/8″ diam	5.000	L.F.	6.20	24.75	30.95
Stop, supply, straight, chrome, 3/8″ diam	1.000	Ea.	3.95	17.35	21.30
Wrought copper 90° elbow for solder joints 3/8″ diam	1.000	Ea.	.66	18.90	19.56
Wrought copper Tee for solder joints, 3/8″ diam	1.000	Ea.	1.13	29.50	30.63
Copper tubing, type DWV, solder joint, hanger 10′ OC 1-1/4″ diam	4.000	L.F.	9.44	27.80	37.24
P trap, standard, copper drainage, 1-1/4″ diam	1.000	Ea.	13.10	23	36.10
Wrought copper, DWV, Tee, sanitary, 1-1/4″ diam	1.000	Ea.	4.26	46	50.26
TOTAL			1,138.74	291.30	1,430.04

8.1-420	Drinking Fountain Systems	COST EACH MAT.	COST EACH INST.	COST EACH TOTAL
1740	Drinking fountain, one bubbler, wall mounted			
1760	Non recessed			
1800	Bronze, no back (R15100-410)	1,150	291	1,441
1840	Cast iron, enameled, low back	510	291	801
1880	Fiberglass, 12″ back	625	291	916
1920	Stainless steel, no back	915	291	1,206
1960	Semi-recessed, poly marble	750	291	1,041
2040	Stainless steel	870	291	1,161
2080	Vitreous china	515	291	806
2120	Full recessed, poly marble	840	291	1,131
2200	Stainless steel	735	291	1,026
2240	Floor mounted, pedestal type, aluminum	520	395	915
2320	Bronze	1,300	395	1,695
2360	Stainless steel	1,300	395	1,695

Systems are complete with trim and rough-in (supply, waste and vent) to connect to supply branches and waste mains.

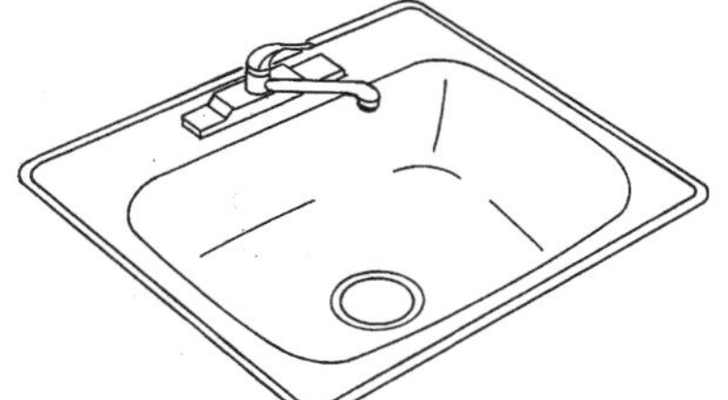

Countertop Single Bowl

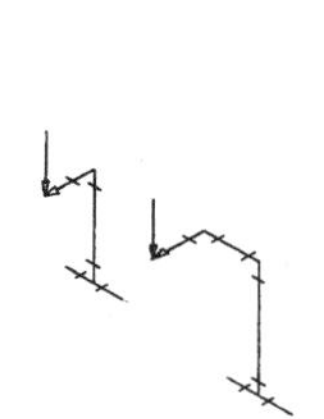

Supply

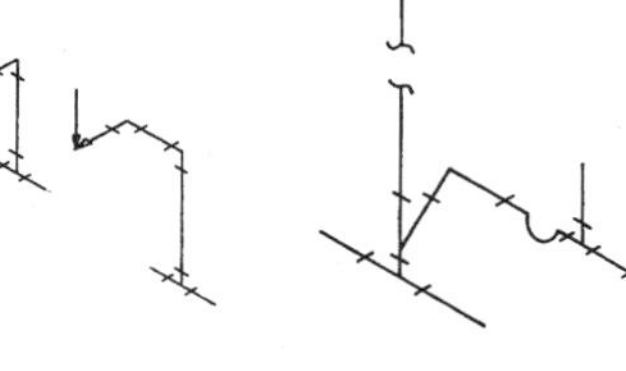

Waste/Vent

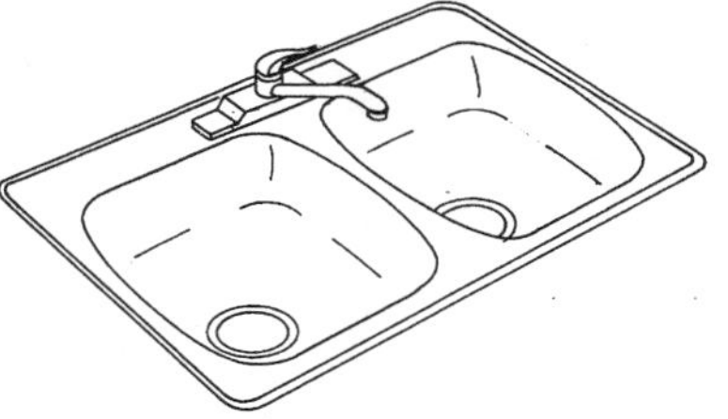

Countertop Double Bowl

System Components	QUANTITY	UNIT	COST EACH MAT.	INST.	TOTAL
SYSTEM 8.1-431-1720					
KITCHEN SINK W/TRIM, COUNTERTOP, P.E. ON C.I., 24″ X 21″, SINGLE BOWL					
Kitchen sink, counter top, PE on CI, 1 bowl, 24″ x 21″ OD	1.000	Ea.	204	134	338
Pipe, steel, galvanized, schedule 40, threaded, 1-1/4″ diam	4.000	L.F.	9.48	33.60	43.08
Copper tubing, type DWV, solder, hangers 10′ OC 1-1/2″ diam	6.000	L.F.	16.74	46.20	62.94
Wrought copper, DWV, Tee, sanitary, 1-1/2″ diam	1.000	Ea.	5.30	52	57.30
P trap, standard, copper, 1-1/2″ diam	1.000	Ea.	13.10	24.50	37.60
Copper tubing, type L, solder joints, hangers 10′ OC 1/2″ diam	10.000	L.F.	13.80	51.50	65.30
Wrought copper 90° elbow for solder joints 1/2″ diam	2.000	Ea.	.48	42	42.48
Wrought copper Tee for solder joints, 1/2″ diam	2.000	Ea.	.80	64	64.80
Stop, angle supply, chrome, 1/2″ CTS	2.000	Ea.	10.40	37.80	48.20
TOTAL			274.10	485.60	759.70

8.1-431	Kitchen Sink Systems		COST EACH MAT.	INST.	TOTAL
1720	Kitchen sink w/trim, countertop, PE on CI, 24″x21″, single bowl		274	485	759
1760	30″ x 21″ single bowl		430	485	915
1800	32″ x 21″ double bowl	R15155-210	370	525	895
1840	42″ x 21″ double bowl		565	530	1,095
1880	Stainless steel, 19″ x 18″ single bowl	R15155-220	380	485	865
1920	25″ x 22″ single bowl		410	485	895
1960	33″ x 22″ double bowl	R15100-120	565	525	1,090
2000	43″ x 22″ double bowl		640	530	1,170
2040	44″ x 22″ triple bowl		915	550	1,465
2080	44″ x 24″ corner double bowl		620	530	1,150
2120	Steel, enameled, 24″ x 21″ single bowl		179	485	664
2160	32″ x 21″ double bowl		214	525	739
2240	Raised deck, PE on CI, 32″ x 21″, dual level, double bowl		475	665	1,140
2280	42″ x 21″ dual level, triple bowl		670	720	1,390

Systems are complete with trim and rough-in (supply, waste and vent) to connect to supply branches and waste mains.

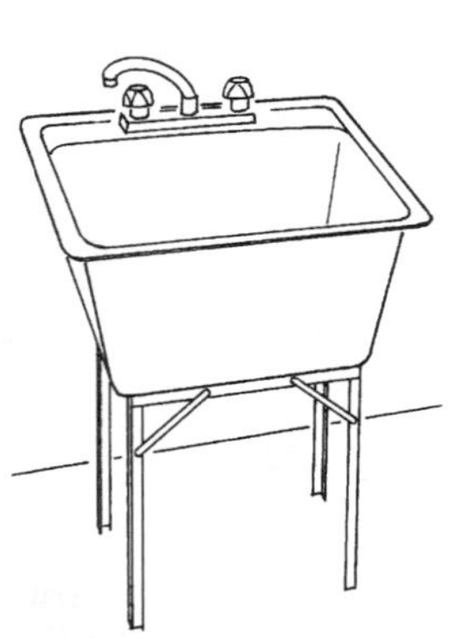
Single Compartment Sink

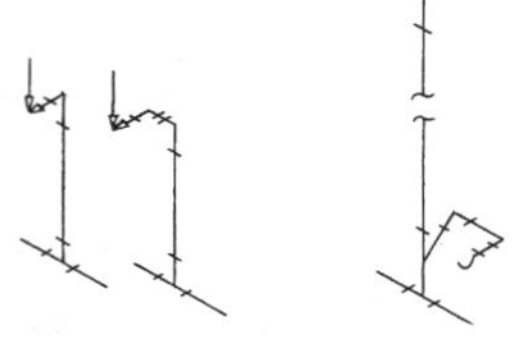
Supply Waste/Vent

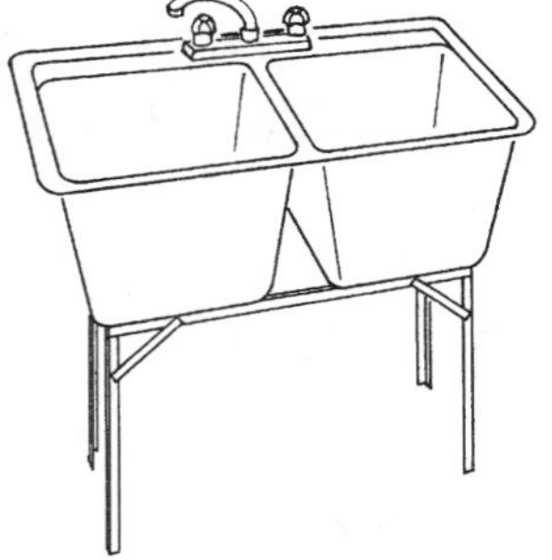
Double Compartment Sink

System Components	QUANTITY	UNIT	COST EACH MAT.	COST EACH INST.	COST EACH TOTAL
SYSTEM 8.1-432-1760					
LAUNDRY SINK W/TRIM, PE ON CI, BLACK IRON FRAME					
24" X 20" OD, SINGLE COMPARTMENT					
Laundry sink PE on CI w/trim & frame, 24" x 20" OD, 1 compartment	1.000	Ea.	370	125	495
Pipe, steel, galvanized, schedule 40, threaded, 1-1/4" diam	4.000	L.F.	9.48	33.60	43.08
Copper tubing, type DWV, solder joint, hanger 10' OC 1-1/2"diam	6.000	L.F.	16.74	46.20	62.94
Wrought copper, DWV, Tee, sanitary, 1-1/2" diam	1.000	Ea.	5.30	52	57.30
P trap, standard, copper, 1-1/2" diam	1.000	Ea.	13.10	24.50	37.60
Copper tubing type L, solder joints, hangers 10' OC, 1/2" diam	10.000	L.F.	13.80	51.50	65.30
Wrought copper 90° elbow for solder joints 1/2" diam	2.000	Ea.	.48	42	42.48
Wrought copper Tee for solder joints, 1/2" diam	2.000	Ea.	.80	64	64.80
Stop, angle supply, 1/2" diam	2.000	Ea.	10.40	37.80	48.20
TOTAL			440.10	476.60	916.70

8.1-432	Laundry Sink Systems	COST EACH MAT.	COST EACH INST.	COST EACH TOTAL
1740	Laundry sink w/trim, PE on CI, black iron frame			
1760	24" x 20", single compartment	440	475	915
1800	24" x 23" single compartment (R15100-410)	475	475	950
1840	48" x 21" double compartment	775	515	1,290
1920	Molded stone, on wall, 22" x 21" single compartment	199	475	674
1960	45"x 21" double compartment	345	515	860
2040	Plastic, on wall or legs, 18" x 23" single compartment	161	465	626
2080	20" x 24" single compartment	189	465	654
2120	36" x 23" double compartment	216	505	721
2160	40" x 24" double compartment	282	505	787
2200	Stainless steel, countertop, 22" x 17" single compartment	365	475	840
2240	19" x 22" single compartment	445	475	920
2280	33" x 22" double compartment	420	515	935

Systems are complete with trim and rough-in (supply, waste and vent) to connect to supply branches and waste mains.

Vanity Top

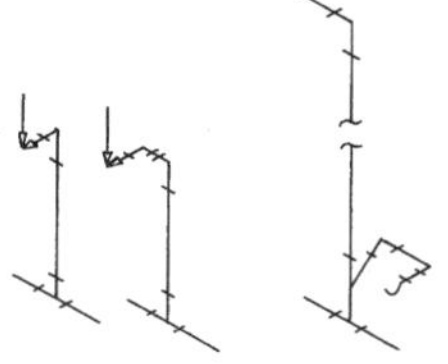
Supply Waste/Vent

Wall Hung

System Components

System Components	QUANTITY	UNIT	COST EACH MAT.	COST EACH INST.	COST EACH TOTAL
SYSTEM 8.1-433-1560					
LAVATORY W/TRIM, VANITY TOP, P.E. ON C.I., 20″ X 18″					
Lavatory w/trim, PE on CI, white, vanity top, 20″ x 18″ oval	1.000	Ea.	172	117	289
Pipe, steel, galvanized, schedule 40, threaded, 1-1/4″ diam	4.000	L.F.	9.48	33.60	43.08
Copper tubing type DWV, solder joint, hanger 10′OC 1-1/4″ diam	4.000	L.F.	9.44	27.80	37.24
Wrought copper DWV, Tee, sanitary, 1-1/4″ diam	1.000	Ea.	4.26	46	50.26
P trap w/cleanout, 20 ga, 1-1/4″ diam	1.000	Ea.	13.10	23	36.10
Copper tubing type L, solder joint, hanger 10′ OC 1/2″ diam	10.000	L.F.	13.80	51.50	65.30
Wrought copper 90° elbow for solder joints 1/2″ diam	2.000	Ea.	.48	42	42.48
Wrought copper Tee for solder joints, 1/2″ diam	2.000	Ea.	.80	64	64.80
Stop, chrome, angle supply, 1/2″ diam	2.000	Ea.	10.40	37.80	48.20
TOTAL			233.76	442.70	676.46

8.1-433	Lavatory Systems	COST EACH MAT.	COST EACH INST.	COST EACH TOTAL
1560	Lavatory w/trim, vanity top, PE on CI, 20″ x 18″, Vanity top by others.	234	445	679
1600	19″ x 16″ oval	234	445	679
1640	18″ round	221	445	666
1680	Cultured marble, 19″ x 17″	173	445	618
1720	25″ x 19″	203	445	648
1760	Stainless, self-rimming, 25″ x 22″	240	445	685
1800	17″ x 22″ R15100 -410	235	445	680
1840	Steel enameled, 20″ x 17″	155	455	610
1880	19″ round	153	455	608
1920	Vitreous china, 20″ x 16″	246	465	711
1960	19″ x 16″	246	465	711
2000	22″ x 13″	256	465	721
2040	Wall hung, PE on CI, 18″ x 15″	535	490	1,025
2080	19″ x 17″	525	490	1,015
2120	20″ x 18″	410	490	900
2160	Vitreous china, 18″ x 15″	355	500	855
2200	19″ x 17″	340	500	840
2240	24″ x 20″	465	500	965

Corrosion resistant laboratory sink systems are complete with trim and rough-in (supply, waste and vent) to connect to supply branches and waste mains.

Laboratory Sink

Supply

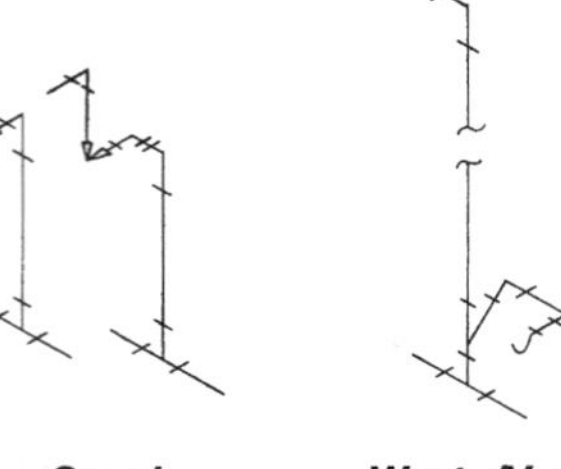

Waste/Vent

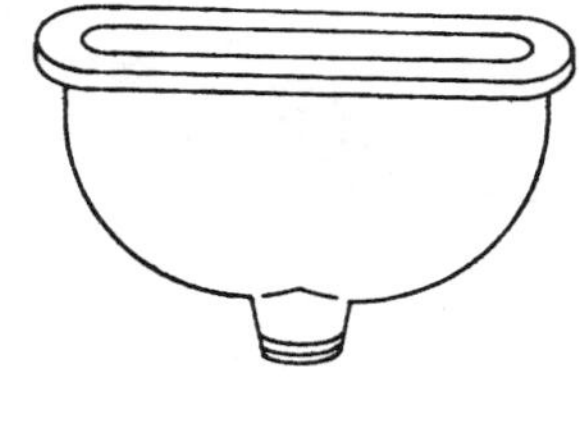

Polypropylene Cup Sink

System Components	QUANTITY	UNIT	COST EACH MAT.	INST.	TOTAL
SYSTEM 8.1-434-1600					
LABORATORY SINK W/TRIM, POLYETHYLENE, SINGLE BOWL					
DOUBLE DRAINBOARD, 54″ X 24″ OD					
Sink w/trim, polyethylene, 1 bowl, 2 drainboards 54″ x 24″ OD	1.000	Ea.	855	250	1,105
Pipe, polypropylene, schedule 40, acid resistant 1-1/2″ diam	10.000	L.F.	10.50	110	120.50
Tee, sanitary, polypropylene, acid resistant, 1-1/2″ diam	1.000	Ea.	16.90	41.50	58.40
P trap, polypropylene, acid resistant, 1-1/2″ diam	1.000	Ea.	29.50	24.50	54
Copper tubing type L, solder joint, hanger 10′ OC 1/2″ diam	10.000	L.F.	13.80	51.50	65.30
Wrought copper 90° elbow for solder joints 1/2″ diam	2.000	Ea.	.48	42	42.48
Wrought copper Tee for solder joints, 1/2″ diam	2.000	Ea.	.80	64	64.80
Stop, angle supply, chrome, 1/2″ diam	2.000	Ea.	10.40	37.80	48.20
TOTAL			937.38	621.30	1,558.68

8.1-434	Laboratory Sink Systems	COST EACH MAT.	INST.	TOTAL
1580	Laboratory sink w/trim, polyethylene, single bowl,			
1600	Double drainboard, 54″ x 24″ O.D.	935	620	1,555
1640	Single drainboard, 47″ x 24″O.D.	975	620	1,595
1680	70″ x 24″ O.D.	995	620	1,615
1760	Flanged, 14-1/2″ x 14-1/2″ O.D.	370	560	930
1800	18-1/2″ x 18-1/2″ O.D.	325	560	885
1840	23-1/2″ x 20-1/2″ O.D.	380	560	940
1920	Polypropylene, cup sink, oval, 7″ x 4″ O.D.	168	495	663
1960	10″ x 4-1/2″ O.D.	179	495	674

Important: See the Reference Section for critical supporting data - Reference Nos., Crews, & City Cost Indexes

Corrosion resistant laboratory sink systems are complete with trim and rough–in (supply, waste and vent) to connect to supply branches and waste mains.

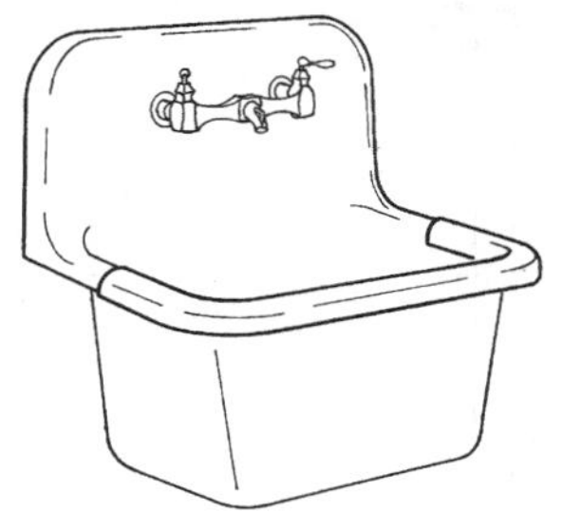

Wall Hung

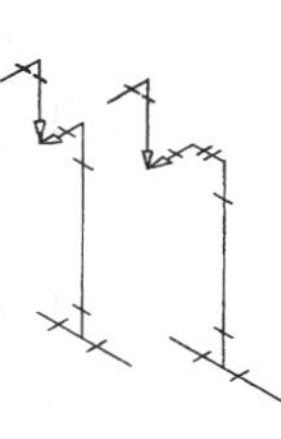

Supply

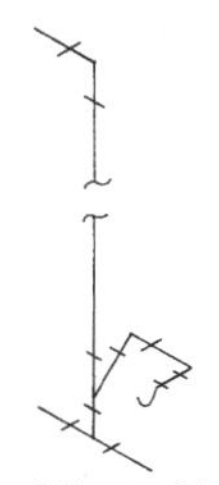

Waste/Vent

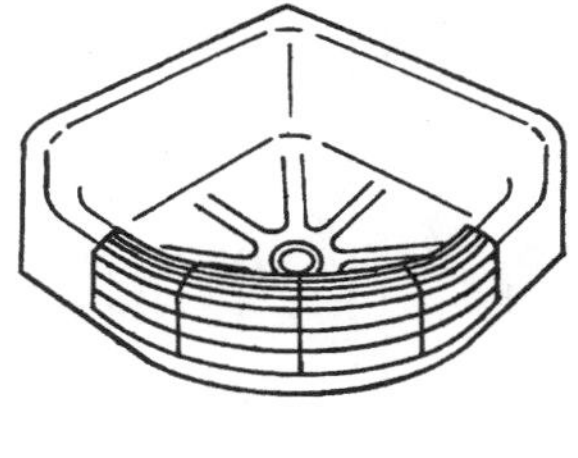

Corner, Floor

System Components	QUANTITY	UNIT	COST EACH MAT.	COST EACH INST.	COST EACH TOTAL
SYSTEM 8.1-434-4260					
SERVICE SINK, PE ON CI, CORNER FLOOR, 28″X28″, W/RIM GUARD & TRIM					
Service sink, corner floor, PE on CI, 28″ x 28″, w/rim guard & trim	1.000	Ea.	520	170	690
Copper tubing type DWV, solder joint, hanger 10′OC 3″ diam	6.000	L.F.	34.50	77.40	111.90
Copper tubing type DWV, solder joint, hanger 10′OC 2″ diam	4.000	L.F.	14.04	37.80	51.84
Wrought copper DWV, Tee, sanitary, 3″ diam	1.000	Ea.	22	107	129
P trap with cleanout & slip joint, copper 3″ diam	1.000	Ea.	49	38	87
Copper tubing, type L, solder joints, hangers 10′ OC, 1/2″ diam	10.000	L.F.	13.80	51.50	65.30
Wrought copper 90° elbow for solder joints 1/2″ diam	2.000	Ea.	.48	42	42.48
Wrought copper Tee for solder joints, 1/2″ diam	2.000	Ea.	.80	64	64.80
Stop, angle supply, chrome, 1/2″ diam	2.000	Ea.	10.40	37.80	48.20
TOTAL			665.02	625.50	1,290.52

8.1-434	Service Sink Systems		COST EACH MAT.	COST EACH INST.	COST EACH TOTAL
4260	Service sink w/trim, PE on CI, corner floor, 28″ x 28″, w/rim guard	R15100 -410	665	625	1,290
4300	Wall hung w/rim guard, 22″ x 18″		730	725	1,455
4340	24″ x 20″		775	725	1,500
4380	Vitreous china, wall hung 22″ x 20″		705	725	1,430

Systems are complete with trim and rough-in (supply, waste and vent) for connection to supply branches and waste mains.

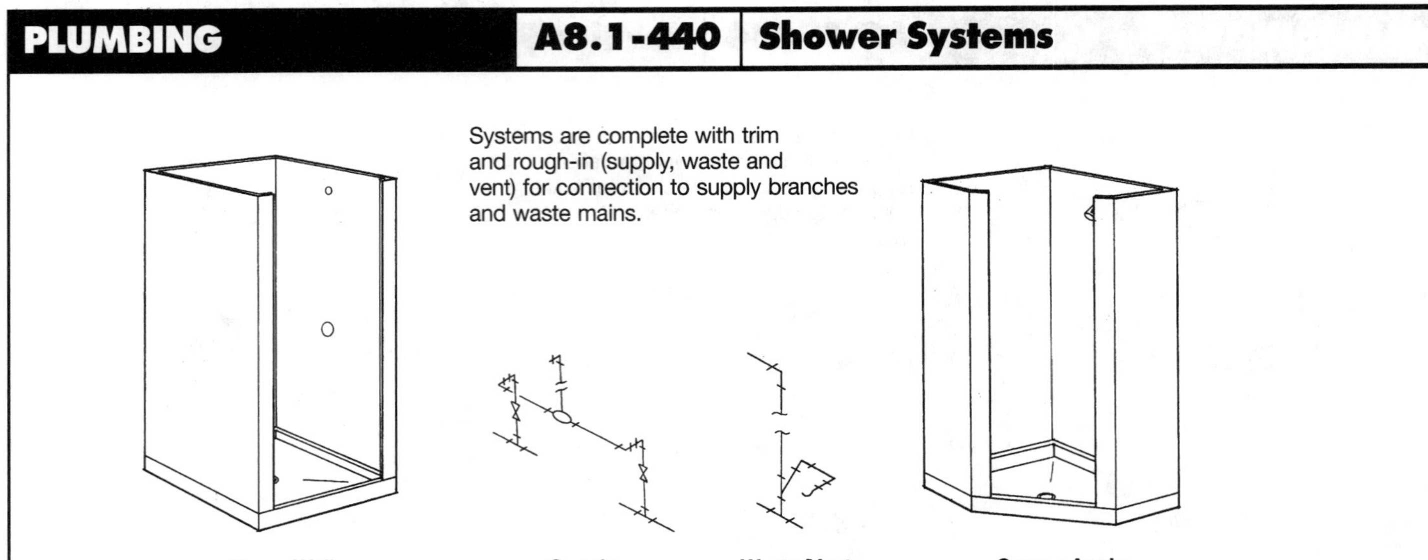

System Components	QUANTITY	UNIT	COST EACH MAT.	INST.	TOTAL
SYSTEM 8.1-440-1560					
SHOWER, STALL, BAKED ENAMEL, MOLDED STONE RECEPTOR, 30″ SQUARE					
Shower stall, enameled steel, molded stone receptor, 30″ square	1.000	Ea.	375	375	750
Copper tubing type DWV, solder joints, hangers 10′OC, 2″ diam	6.000	L.F.	16.74	46.20	62.94
Wrought copper DWV, Tee, sanitary, 2″ diam	1.000	Ea.	5.30	52	57.30
Trap, standard, copper, 2″ diam	1.000	Ea.	13.10	24.50	37.60
Copper tubing type L, solder joint, hanger 10′ OC 1/2″ diam	16.000	L.F.	22.08	82.40	104.48
Wrought copper 90° elbow for solder joints 1/2″ diam	3.000	Ea.	.72	63	63.72
Wrought copper Tee for solder joints, 1/2″ diam	2.000	Ea.	.80	64	64.80
Stop and waste, straightway, bronze, solder joint 1/2″ diam	2.000	Ea.	5.10	34.70	39.80
TOTAL			438.84	741.80	1,180.64

8.1-440	Shower Systems	COST EACH MAT.	INST.	TOTAL
1560	Shower, stall, baked enamel, molded stone receptor, 30″ square	440	740	1,180
1600	32″ square	410	740	1,150
1640	Terrazzo receptor, 32″ square (R15100 -410)	1,175	740	1,915
1680	36″ square	970	780	1,750
1720	36″ corner angle	890	780	1,670
1800	Fiberglass one piece, three walls, 32″ square	385	675	1,060
1840	36″ square	430	675	1,105
1880	Polypropylene, molded stone receptor, 30″ square	475	740	1,215
1920	32″ square	460	740	1,200
1960	Built-in head, arm, bypass, stops and handles	88	193	281

Important: See the Reference Section for critical supporting data - Reference Nos., Crews, & City Cost Indexes

Systems are complete with trim, flush valve and rough-in (supply, waste and vent) for connection to supply branches and waste mains.

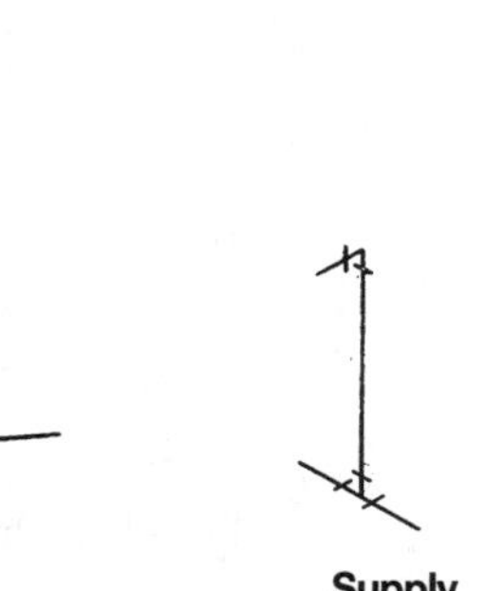

Stall Type

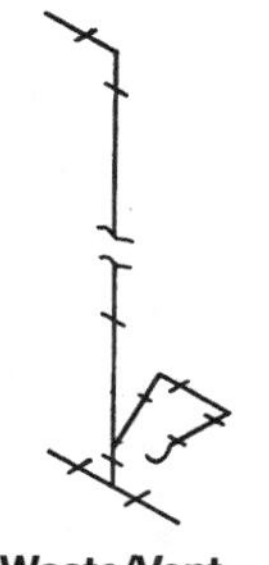

Supply

Waste/Vent

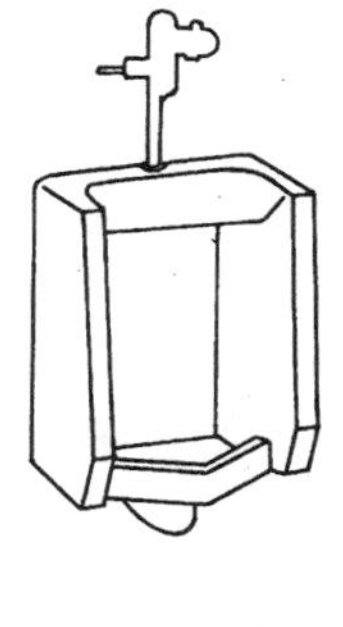

Wall Hung

System Components	QUANTITY	UNIT	COST EACH MAT.	COST EACH INST.	COST EACH TOTAL
SYSTEM 8.1-450-2000					
URINAL, VITREOUS CHINA, WALL HUNG					
Urinal, wall hung, vitreous china, incl. hanger	1.000	Ea.	305	250	555
Pipe, steel, galvanized, schedule 40, threaded, 1-1/2″ diam	5.000	L.F.	13.15	46.75	59.90
Copper tubing type DWV, solder joint, hangers 10′OC, 2″ diam	3.000	L.F.	10.53	28.35	38.88
Combination Y & 1/8 bend for CI soil pipe, no hub, 3″ diam	1.000	Ea.	9		9
Pipe, CI, no hub, cplg 10′ OC, hanger 5′ OC, 3″ diam	4.000	L.F.	29	46.80	75.80
Pipe coupling standard, CI soil, no hub, 3″ diam	3.000	Ea.	9.98	39.40	49.38
Copper tubing type L, solder joint, hanger 10′ OC 3/4″ diam	5.000	L.F.	8.70	27.25	35.95
Wrought copper 90° elbow for solder joints 3/4″ diam	1.000	Ea.	.53	22	22.53
Wrought copper Tee for solder joints, 3/4″ diam	1.000	Ea.	.98	34.50	35.48
TOTAL			391.08	536.55	927.63

8.1-450	Urinal Systems		COST EACH MAT.	COST EACH INST.	COST EACH TOTAL
2000	Urinal, vitreous china, wall hung	R15100 -410	390	535	925
2040	Stall type		605	590	1,195

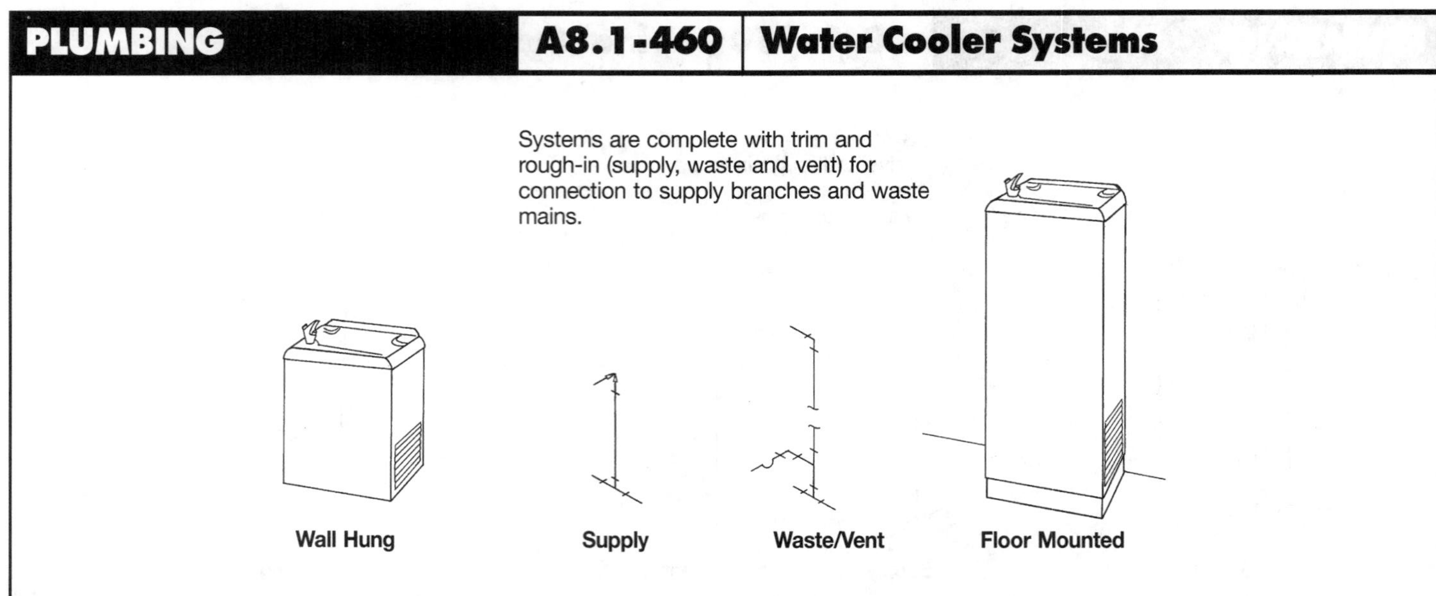
Systems are complete with trim and rough-in (supply, waste and vent) for connection to supply branches and waste mains.
Wall Hung
Supply
Waste/Vent
Floor Mounted

System Components	QUANTITY	UNIT	COST EACH MAT.	COST EACH INST.	COST EACH TOTAL
SYSTEM 8.1-460-1840					
WATER COOLER, ELECTRIC, SELF CONTAINED, WALL HUNG, 8.2 GPH					
Water cooler, wall mounted, 8.2 GPH	1.000	Ea.	590	187	777
Copper tubing type DWV, solder joint, hanger 10'OC 1-1/4" diam	4.000	L.F.	9.44	27.80	37.24
Wrought copper DWV, Tee, sanitary 1-1/4" diam	1.000	Ea.	4.26	46	50.26
P trap, copper drainage, 1-1/4" diam	1.000	Ea.	13.10	23	36.10
Copper tubing type L, solder joint, hanger 10' OC 3/8" diam	5.000	L.F.	6.20	24.75	30.95
Wrought copper 90° elbow for solder joints 3/8" diam	1.000	Ea.	.66	18.90	19.56
Wrought copper Tee for solder joints, 3/8" diam	1.000	Ea.	1.13	29.50	30.63
Stop and waste, straightway, bronze, solder, 3/8" diam	1.000	Ea.	3.95	17.35	21.30
TOTAL			628.74	374.30	1,003.04

8.1-460	Water Cooler Systems		COST EACH MAT.	COST EACH INST.	COST EACH TOTAL
1840	Water cooler, electric, wall hung, 8.2 GPH	R15100-410	630	375	1,005
1880	Dual height, 14.3 GPH		935	385	1,320
1920	Wheelchair type, 7.5 G.P.H.	♿	1,575	375	1,950
1960	Semi recessed, 8.1 G.P.H.		900	375	1,275
2000	Full recessed, 8 G.P.H.	R15100-430	1,375	400	1,775
2040	Floor mounted, 14.3 G.P.H.		665	325	990
2080	Dual height, 14.3 G.P.H.		985	395	1,380
2120	Refrigerated compartment type, 1.5 G.P.H.		1,225	325	1,550

Systems are complete with trim seat and rough-in (supply, waste and vent) for connection to supply branches and waste mains.

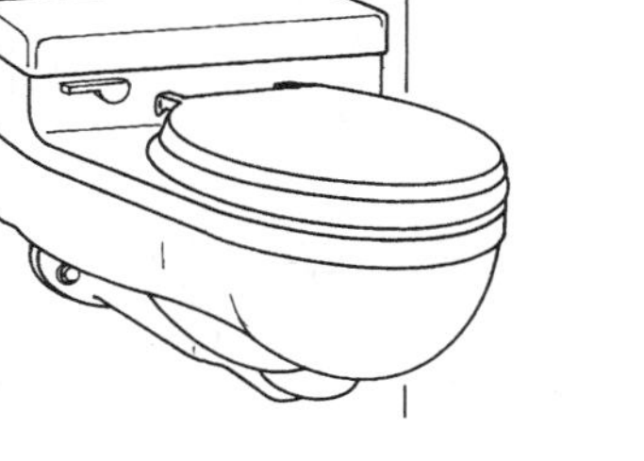

One Piece Wall Hung

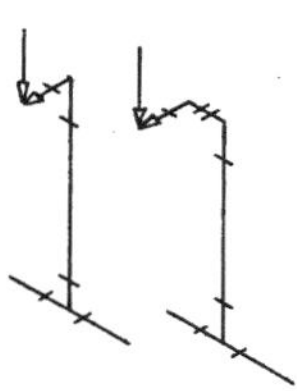

Supply

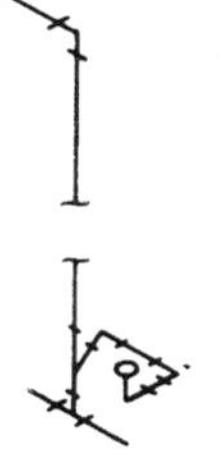

Waste/Vent

Floor Mount

System Components	QUANTITY	UNIT	COST EACH MAT.	COST EACH INST.	COST EACH TOTAL
SYSTEM 8.1-470-1840					
WATER CLOSET, VITREOUS CHINA, ELONGATED					
TANK TYPE, WALL HUNG, ONE PIECE					
Wtr closet tank type vit china wall hung 1 pc w/seat supply & stop	1.000	Ea.	470	141	611
Pipe steel galvanized, schedule 40, threaded, 2″ diam	4.000	L.F.	13.36	46.80	60.16
Pipe, CI soil, no hub, cplg 10′ OC, hanger 5′ OC, 4″ diam	2.000	L.F.	17.90	25.80	43.70
Pipe, coupling, standard coupling, CI soil, no hub, 4″ diam	2.000	Ea.	11.90	45	56.90
Copper tubing type L, solder joint, hanger 10′OC, 1/2″ diam	6.000	L.F.	8.28	30.90	39.18
Wrought copper 90° elbow for solder joints 1/2″ diam	2.000	Ea.	.48	42	42.48
Wrought copper Tee for solder joints 1/2″ diam	1.000	Ea.	.40	32	32.40
Support/carrier, for water closet, siphon jet, horiz, single, 4″ waste	1.000	Ea.	226	69.50	295.50
TOTAL			748.32	433	1,181.32

8.1-470	Water Closet Systems		COST EACH MAT.	COST EACH INST.	COST EACH TOTAL
1800	Water closet, vitreous china, elongated				
1840	Tank type, wall hung, one piece		750	435	1,185
1880	Close coupled two piece	R15100 -410	805	435	1,240
1920	Floor mount, one piece		675	470	1,145
1960	One piece low profile	R15100 -420	735	470	1,205
2000	Two piece close coupled		291	470	761
2040	Bowl only with flush valve				
2080	Wall hung		670	490	1,160
2120	Floor mount		480	475	955

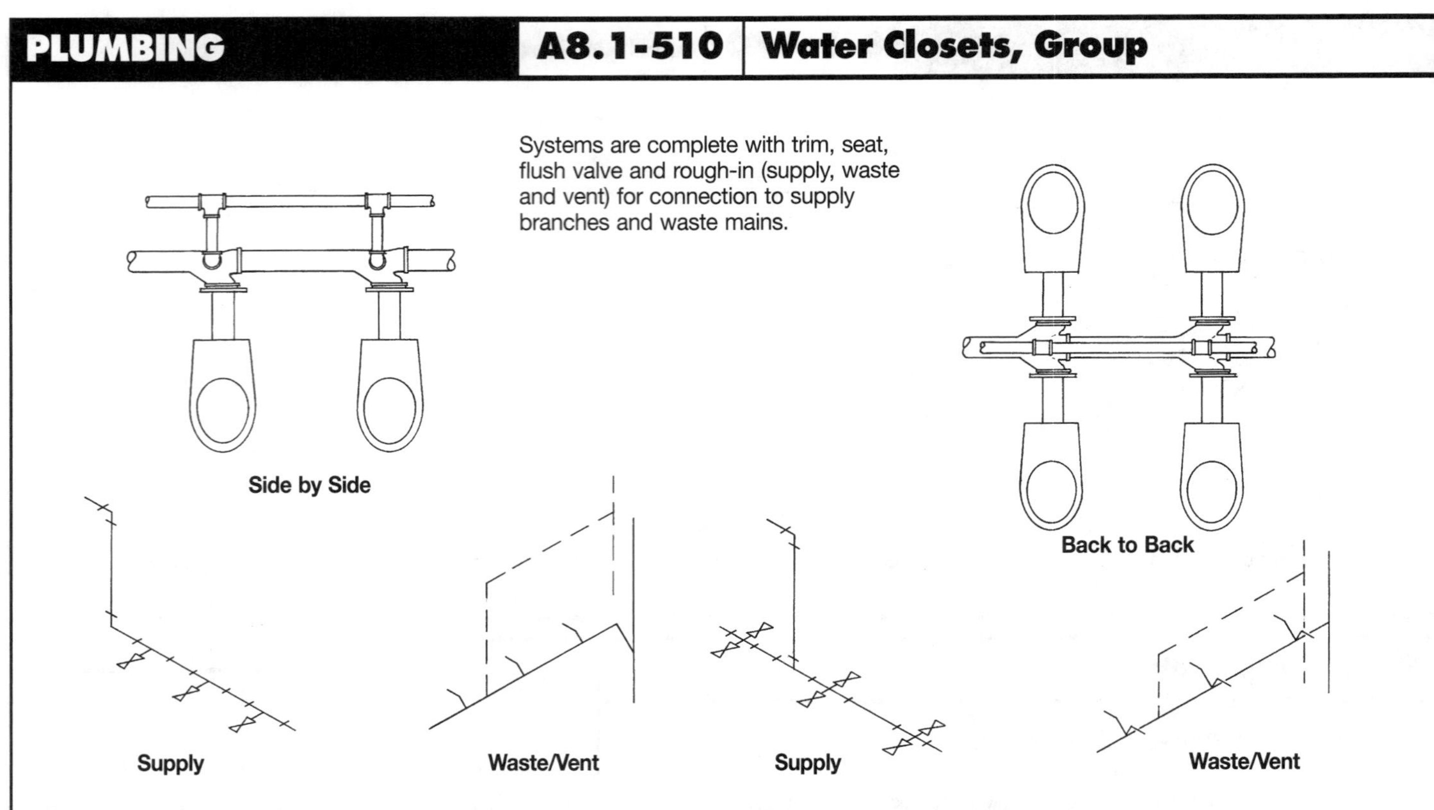
Systems are complete with trim, seat, flush valve and rough-in (supply, waste and vent) for connection to supply branches and waste mains.
Side by Side
Back to Back
Supply
Waste/Vent
Supply
Waste/Vent

System Components	QUANTITY	UNIT	COST EACH MAT.	COST EACH INST.	COST EACH TOTAL
SYSTEM 8.1-510-1760					
WATER CLOSETS, BATTERY MOUNT, WALL HUNG, SIDE BY SIDE, FIRST CLOSET					
Water closet, bowl only w/flush valve, seat, wall hung	1.000	Ea.	370	129	499
Pipe, CI soil, no hub, cplg 10′ OC, hanger 5′ OC, 4″ diam	3.000	L.F.	26.85	38.70	65.55
Coupling, standard, CI, soil, no hub, 4″ diam	2.000	Ea.	11.90	45	56.90
Copper tubing type L, solder joints, hangers 10′ OC, 1″ diam	6.000	L.F.	13.38	36.60	49.98
Copper tubing, type DWV, solder joints, hangers 10′OC, 2″ diam	6.000	L.F.	21.06	56.70	77.76
Wrought copper 90° elbow for solder joints 1″ diam	1.000	Ea.	1.23	26	27.23
Wrought copper Tee for solder joints, 1″ diam	1.000	Ea.	2.85	41.50	44.35
Support/carrier, siphon jet, horiz, adjustable single, 4″ pipe	1.000	Ea.	226	69.50	295.50
Valve, gate, bronze, 125 lb, NRS, soldered 1″ diam	1.000	Ea.	26.50	22	48.50
Wrought copper, DWV, 90° elbow, 2″ diam	1.000	Ea.	4.21	41.50	45.71
TOTAL			703.98	506.50	1,210.48

8.1-510	Water Closets, Group		COST EACH MAT.	COST EACH INST.	COST EACH TOTAL
1760	Water closets, battery mount, wall hung, side by side, first closet	R15100 -410	705	505	1,210
1800	Each additional water closet, add		670	480	1,150
3000	Back to back, first pair of closets	R15100 -420	1,275	685	1,960
3100	Each additional pair of closets, back to back		1,250	675	1,925

Systems are complete with trim, flush valve and rough-in (supply, waste and vent) for connection to supply branches and waste mains.

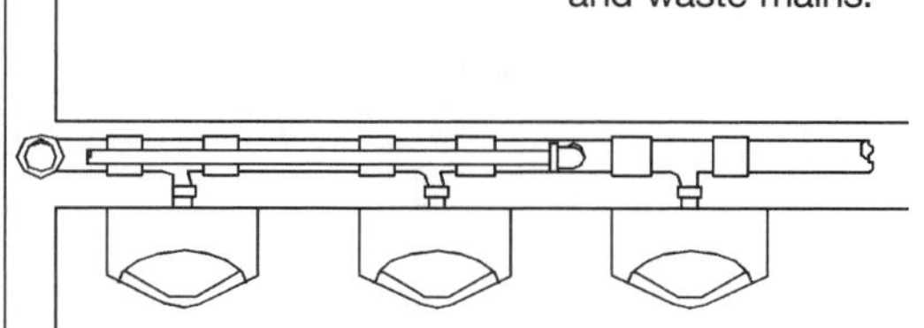

Side by Side

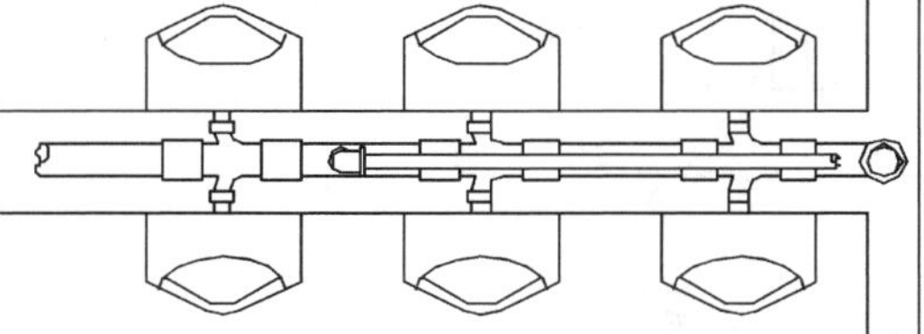

Back to Back

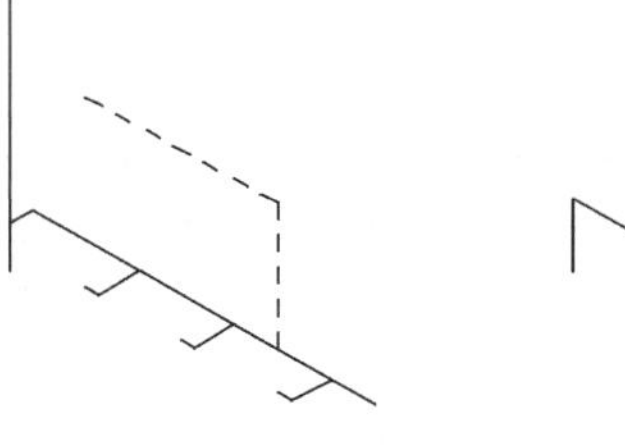

Waste/Vent

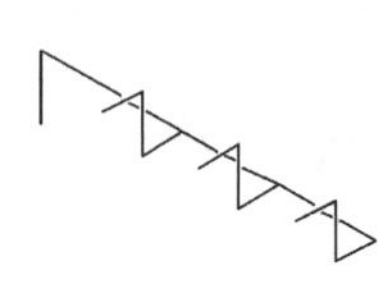

Supply

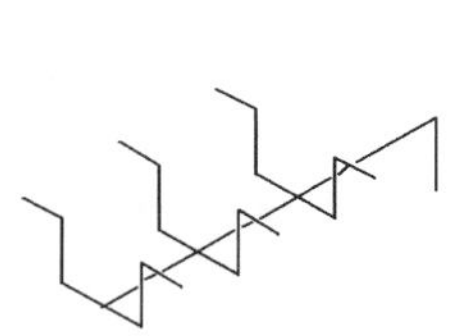

Supply

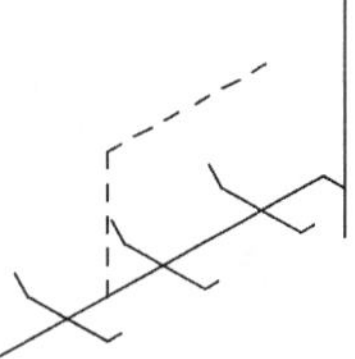

Waste/Vent

System Components	QUANTITY	UNIT	COST EACH MAT.	COST EACH INST.	COST EACH TOTAL
SYSTEM 8.1-530-1760					
URINALS, BATTERY MOUNT, WALL HUNG, SIDE BY SIDE, FIRST URINAL					
Urinal, wall hung, vitreous china, with hanger & trim	1.000	Ea.	305	250	555
No hub cast iron soil pipe, 3″ diameter	4.000	L.F.	29	46.80	75.80
No hub cast iron sanitary tee, 3″ diameter	1.000	Ea.	9		9
No hub coupling, 3″ diameter	2.000	Ea.	9.98	39.40	49.38
Copper tubing, type L, 3/4″ diameter	5.000	L.F.	8.70	27.25	35.95
Copper tubing, type DWV, 2″ diameter	2.000	L.F.	7.02	18.90	25.92
Copper 90° elbow, 3/4″ diameter	2.000	Ea.	1.06	44	45.06
Copper 90° elbow, type DWV, 2″ diameter	2.000	Ea.	4.21	41.50	45.71
Galvanized steel pipe, 1-1/2″ diameter	5.000	L.F.	13.15	46.75	59.90
Cast iron drainage elbow, 90°, 1-1/2″ diameter	1.000	Ea.	.98	34.50	35.48
TOTAL			388.10	549.10	937.20

8.1-530	Urinal Systems, Battery Mount		COST EACH MAT.	COST EACH INST.	COST EACH TOTAL
1760	Urinals, battery mount, side by side, first urinal		390	550	940
1800	Each additional urinal, add		370	535	905
2000	Back to back, first pair of urinals	R15100 -410	780	900	1,680
2100	Each additional pair of urinals, back to back		685	710	1,395

Systems are complete with trim, flush valve and rough-in (supply, waste and vent) for connection to supply branches and waste mains.

Circular Fountain

Supply

Waste/Vent

Semi-Circular Fountain

System Components

System Components	QUANTITY	UNIT	COST EACH MAT.	COST EACH INST.	COST EACH TOTAL
SYSTEM 8.1-560-1760					
GROUP WASH FOUNTAIN, PRECAST TERRAZZO					
CIRCULAR, 36″ DIAMETER					
Wash fountain, group, precast terrazzo, foot control 36″ diam	1.000	Ea.	2,500	390	2,890
Copper tubing type DWV, solder joint, hanger 10′OC, 2″ diam	10.000	L.F.	35.10	94.50	129.60
P trap, standard, copper, 2″ diam	1.000	Ea.	20	27.50	47.50
Wrought copper, Tee, sanitary, 2″ diam	1.000	Ea.	6.20	59.50	65.70
Copper tubing type L, solder joint, hanger 10′ OC 1/2″ diam	20.000	L.F.	27.60	103	130.60
Wrought copper 90° elbow for solder joints 1/2″ diam	3.000	Ea.	.72	63	63.72
Wrought copper Tee for solder joints, 1/2″ diam	2.000	Ea.	.80	64	64.80
TOTAL			2,590.42	801.50	3,391.92

8.1-560 Group Wash Fountain Systems

	8.1-560 Group Wash Fountain Systems	COST EACH MAT.	COST EACH INST.	COST EACH TOTAL
1740	Group wash fountain, precast terrazzo			
1760	Circular, 36″ diameter	2,600	800	3,400
1800	54″ diameter	3,125	875	4,000
1840	Semi-circular, 36″ diameter (R15100 -410)	2,350	800	3,150
1880	54″ diameter	2,875	875	3,750
1960	Stainless steel, circular, 36″ diameter	3,000	745	3,745
2000	54″ diameter	3,925	825	4,750
2040	Semi-circular, 36″ diameter	2,600	745	3,345
2080	54″ diameter	3,400	825	4,225
2160	Thermoplastic, circular, 36″ diameter	2,350	605	2,955
2200	54″ diameter	2,750	705	3,455
2240	Semi-circular, 36″ diameter	2,200	605	2,805
2280	54″ diameter	2,600	705	3,305

Important: See the Reference Section for critical supporting data - Reference Nos., Crews, & City Cost Indexes

Systems are complete with trim, flush valve and rough-in (supply, waste and vent) for connection to supply branches and waste mains.

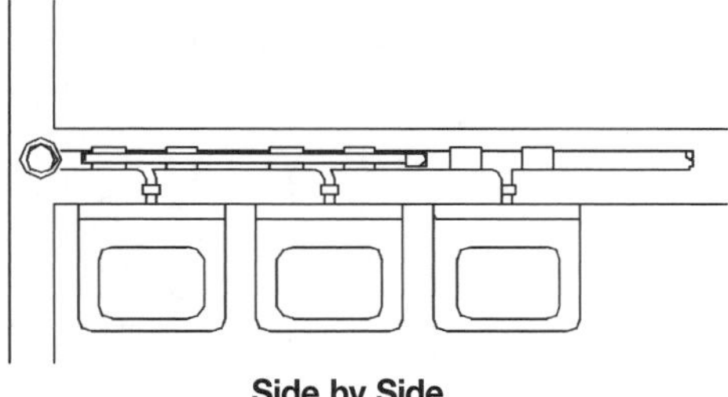

Side by Side

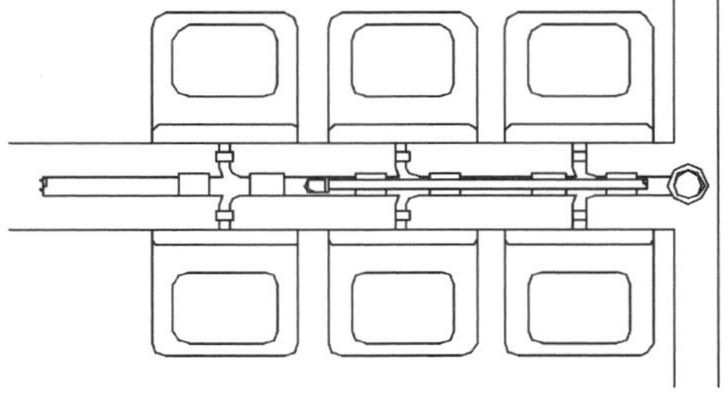

Back to Back

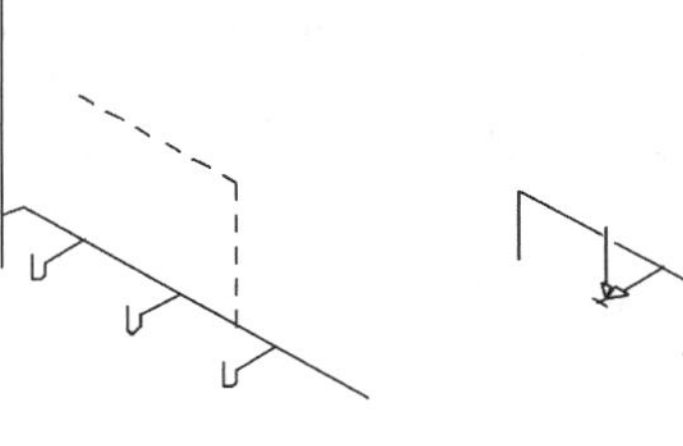

Waste/Vent

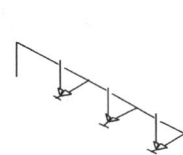

Supply
(Two supply systems required)

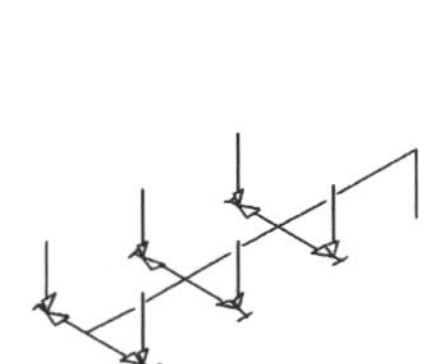

Supply
(Two supply systems required)

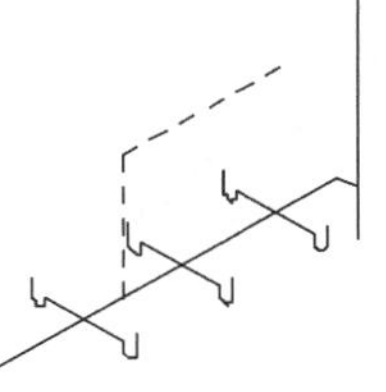

Waste/Vent

System Components	QUANTITY	UNIT	COST EACH MAT.	COST EACH INST.	COST EACH TOTAL
SYSTEM 8.1-580-1760					
LAVATORIES, BATTERY MOUNT, WALL HUNG, SIDE BY SIDE, FIRST LAVATORY					
Lavatory w/trim wall hung PE on CI 20″ x 18″	1.000	Ea.	202	93.50	295.50
Stop, chrome, angle supply, 3/8″ diameter	2.000	Ea.	10.40	34.70	45.10
Concealed arm support	1.000	Ea.	177	69.50	246.50
P trap w/cleanout, 20 ga. C.P., 1-1/4″ diameter	1.000	Ea.	24	23	47
Copper tubing, type L, 1/2″ diameter	10.000	L.F.	13.80	51.50	65.30
Copper tubing, type DWV, 1-1/4″ diameter	4.000	L.F.	9.44	27.80	37.24
Copper 90° elbow, 1/2″ diameter	2.000	Ea.	.48	42	42.48
Copper tee, 1/2″ diameter	2.000	Ea.	.80	64	64.80
DWV copper sanitary tee, 1-1/4″ diameter	2.000	Ea.	8.52	92	100.52
Galvanized steel pipe, 1-1/4″ diameter	4.000	L.F.	9.48	33.60	43.08
Black cast iron 90° elbow, 1-1/4″ diameter	1.000	Ea.	4.08	34	38.08
TOTAL			460	565.60	1,025.60

8.1-580	Lavatory Systems, Battery Mount		COST EACH MAT.	COST EACH INST.	COST EACH TOTAL
1760	Lavatories, battery mount, side by side, first lavatory		460	565	1,025
1800	Each additional lavatory, add		435	390	825
2000	Back to back, first pair of lavatories	R15100-410	785	900	1,685
2100	Each additional pair of lavatories, back to back		765	740	1,505

A8.1-620 Two Fixture Bathrooms

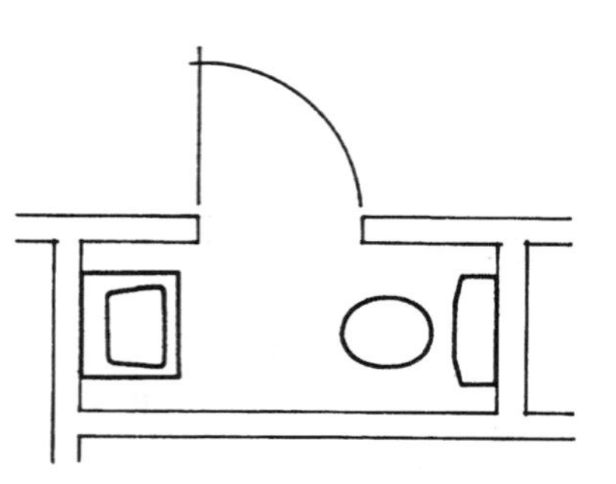

*Common wall is with an adjacent bathroom

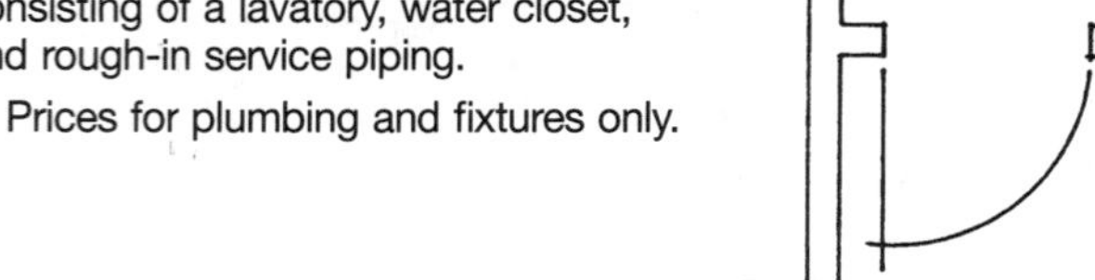

Two Fixture Bathroom Systems consisting of a lavatory, water closet, and rough-in service piping.

- Prices for plumbing and fixtures only.

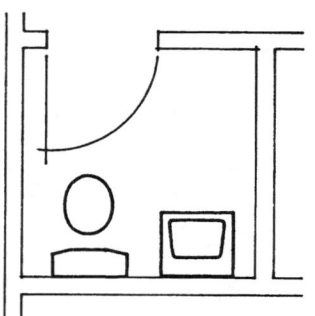

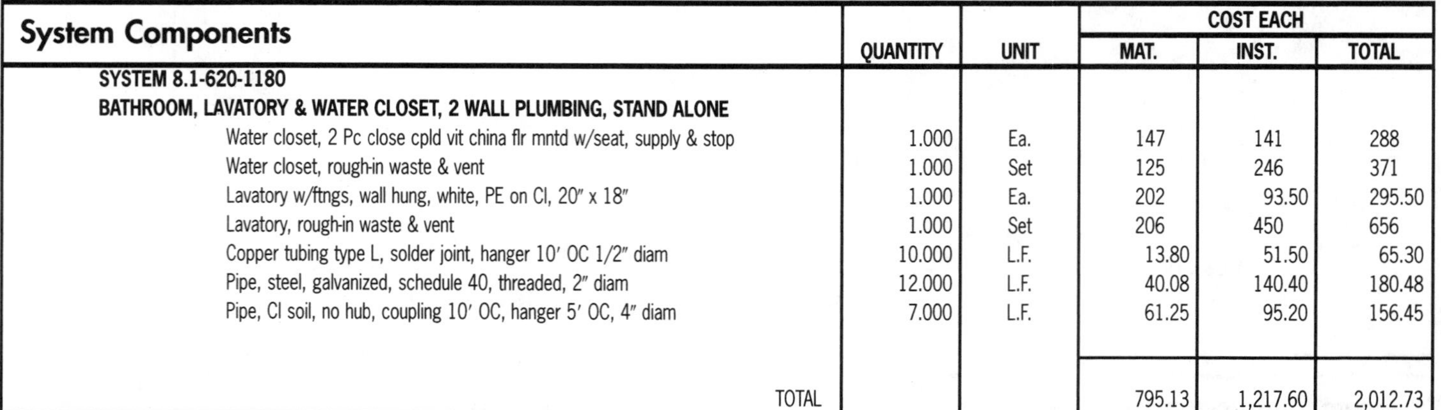

System Components	QUANTITY	UNIT	COST EACH MAT.	COST EACH INST.	COST EACH TOTAL
SYSTEM 8.1-620-1180					
BATHROOM, LAVATORY & WATER CLOSET, 2 WALL PLUMBING, STAND ALONE					
Water closet, 2 Pc close cpld vit china flr mntd w/seat, supply & stop	1.000	Ea.	147	141	288
Water closet, rough-in waste & vent	1.000	Set	125	246	371
Lavatory w/ftngs, wall hung, white, PE on CI, 20″ x 18″	1.000	Ea.	202	93.50	295.50
Lavatory, rough-in waste & vent	1.000	Set	206	450	656
Copper tubing type L, solder joint, hanger 10′ OC 1/2″ diam	10.000	L.F.	13.80	51.50	65.30
Pipe, steel, galvanized, schedule 40, threaded, 2″ diam	12.000	L.F.	40.08	140.40	180.48
Pipe, CI soil, no hub, coupling 10′ OC, hanger 5′ OC, 4″ diam	7.000	L.F.	61.25	95.20	156.45
TOTAL			795.13	1,217.60	2,012.73

8.1-620	Two Fixture Bathroom, Two Wall Plumbing	COST EACH MAT.	COST EACH INST.	COST EACH TOTAL
1180	Bathroom, lavatory & water closet, 2 wall plumbing, stand alone	795	1,225	2,020
1200	Share common plumbing wall*	750	1,050	1,800

8.1-620	Two Fixture Bathroom, One Wall Plumbing		COST EACH MAT.	COST EACH INST.	COST EACH TOTAL
2220	Bathroom, lavatory & water closet, one wall plumbing, stand alone	R15100-410	760	1,100	1,860
2240	Share common plumbing wall*		680	930	1,610
2260		R15100-420			
2280					

Important: See the Reference Section for critical supporting data - Reference Nos., Crews, & City Cost Indexes

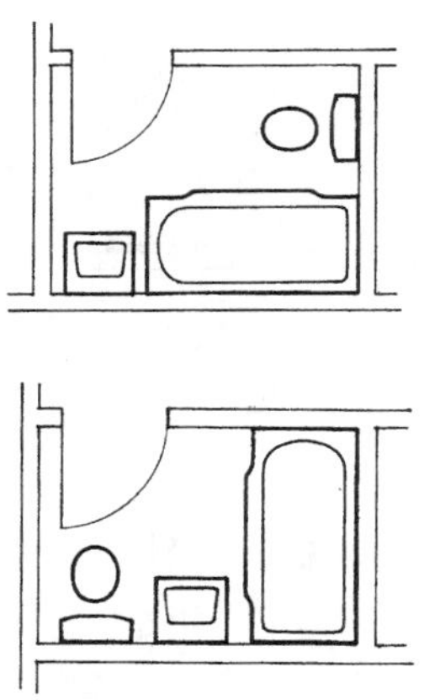

Three Fixture Bathroom Systems consisting of a lavatory, water closet, bathtub or shower and rough-in service piping.

- Prices for plumbing and fixtures only.

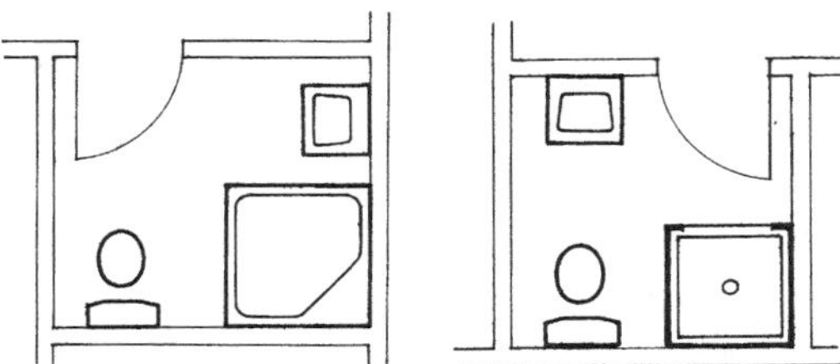

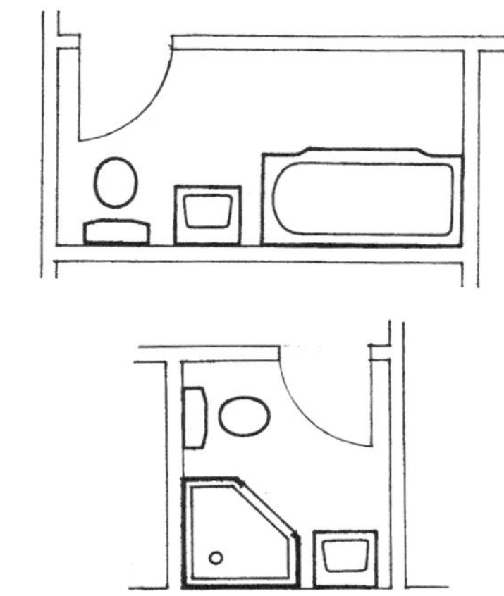

*Common wall is with an adjacent bathroom

System Components	QUANTITY	UNIT	COST EACH MAT.	COST EACH INST.	COST EACH TOTAL
SYSTEM 8.1-630-1170					
BATHROOM, LAVATORY, WATER CLOSET & BATHTUB					
ONE WALL PLUMBING, STAND ALONE					
Wtr closet, 2 pc close cpld vit china flr mntd w/seat supply & stop	1.000	Ea.	147	141	288
Water closet, rough-in waste & vent	1.000	Set	125	246	371
Lavatory w/ftngs, wall hung, white, PE on CI, 20" x 18"	1.000	Ea.	202	93.50	295.50
Lavatory, rough-in waste & vent	1.000	Set	206	450	656
Bathtub, white PE on CI, w/ftgs, mat bottom, recessed, 5' long	1.000	Ea.	395	170	565
Baths, rough-in waste and vent	1.000	Set	99.90	324	423.90
TOTAL			1,174.90	1,424.50	2,599.40

8.1-630	Three Fixture Bathroom, One Wall Plumbing		COST EACH MAT.	COST EACH INST.	COST EACH TOTAL
1150	Bathroom, three fixture, one wall plumbing	R15100 -410			
1160	Lavatory, water closet & bathtub				
1170	Stand alone	R15100 -420	1,175	1,425	2,600
1180	Share common plumbing wall *		995	1,025	2,020

8.1-630	Three Fixture Bathroom, Two Wall Plumbing	COST EACH MAT.	COST EACH INST.	COST EACH TOTAL
2130	Bathroom, three fixture, two wall plumbing			
2140	Lavatory, water closet & bathtub			
2160	Stand alone	1,175	1,450	2,625
2180	Long plumbing wall common *	1,050	1,150	2,200
3610	Lavatory, bathtub & water closet			
3620	Stand alone	1,250	1,650	2,900
3640	Long plumbing wall common *	1,200	1,475	2,675
4660	Water closet, corner bathtub & lavatory			
4680	Stand alone	2,200	1,450	3,650
4700	Long plumbing wall common *	2,075	1,100	3,175
6100	Water closet, stall shower & lavatory			
6120	Stand alone	1,175	1,875	3,050
6140	Long plumbing wall common *	1,125	1,750	2,875
7060	Lavatory, corner stall shower & water closet			
7080	Stand alone	1,575	1,700	3,275
7100	Short plumbing wall common *	1,450	1,225	2,675

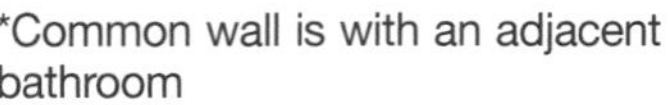

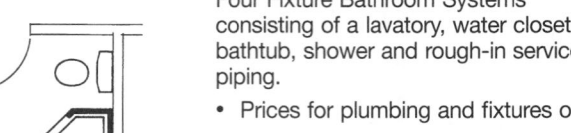

Four Fixture Bathroom Systems consisting of a lavatory, water closet, bathtub, shower and rough-in service piping.

- Prices for plumbing and fixtures only.

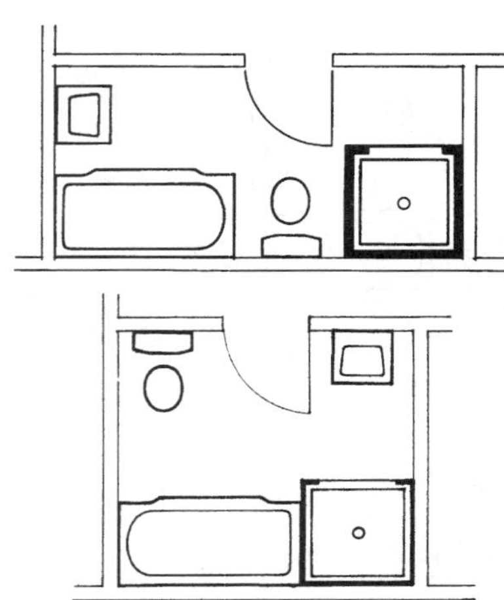

*Common wall is with an adjacent bathroom

System Components	QUANTITY	UNIT	COST EACH MAT.	INST.	TOTAL
SYSTEM 8.1-640-1160					
BATHROOM, BATHTUB, WATER CLOSET, STALL SHOWER & LAVATORY					
TWO WALL PLUMBING, STAND ALONE					
Wtr closet, 2 pc close cpld vit china flr mntd w/seat supply & stop	1.000	Ea.	147	141	288
Water closet, rough-in waste & vent	1.000	Set	125	246	371
Lavatory w/ftngs, wall hung, white PE on CI, 20″ x 18″	1.000	Ea.	202	93.50	295.50
Lavatory, rough-in waste & vent	1.000	Set	20.60	45	65.60
Bathtub, white PE on CI, w/ftgs, mat bottom, recessed, 5′ long	1.000	Ea.	395	170	565
Baths, rough-in waste and vent	1.000	Set	111	360	471
Shower stall, bkd enam, molded stone receptor, door & trim 32″ sq.	1.000	Ea.	345	375	720
Shower stall, rough-in supply, waste & vent	1.000	Set	64	365	429
TOTAL			1,409.60	1,795.50	3,205.10

8.1-640	Four Fixture Bathroom, Two Wall Plumbing		COST EACH MAT.	INST.	TOTAL
1140	Bathroom, four fixture, two wall plumbing	R15100 -410			
1150	Bathtub, water closet, stall shower & lavatory				
1160	Stand alone	R15100 -420	1,400	1,800	3,200
1180	Long plumbing wall common *		1,300	1,425	2,725
2260	Bathtub, lavatory, corner stall shower & water closet				
2280	Stand alone		1,900	1,825	3,725
2320	Long plumbing wall common *		1,775	1,475	3,250
3620	Bathtub, stall shower, lavatory & water closet				
3640	Stand alone		1,600	2,200	3,800
3660	Long plumbing wall (opp. door) common *		1,475	1,850	3,325

8.1-640	Four Fixture Bathroom, Three Wall Plumbing	COST EACH MAT.	INST.	TOTAL
4680	Bathroom, four fixture, three wall plumbing			
4700	Bathtub, stall shower, lavatory & water closet			
4720	Stand alone	2,125	2,425	4,550
4760	Long plumbing wall (opposite door) common *	2,075	2,275	4,350

A8.1-650 Five Fixture Bathrooms

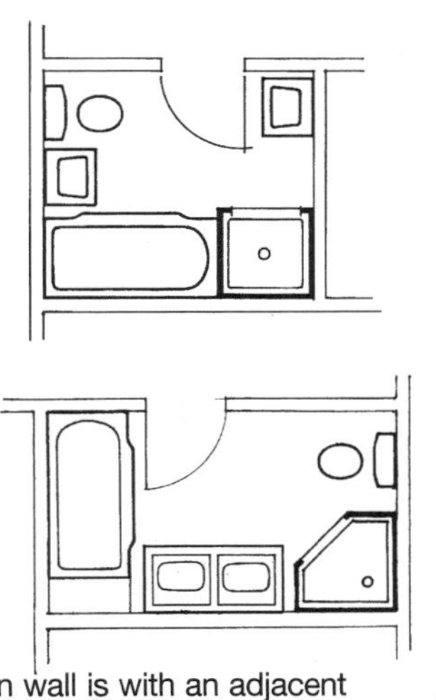

Five Fixture Bathroom Systems consisting of two lavatories, a water closet, bathtub, shower and rough-in service piping.

- Prices for plumbing and fixtures only.

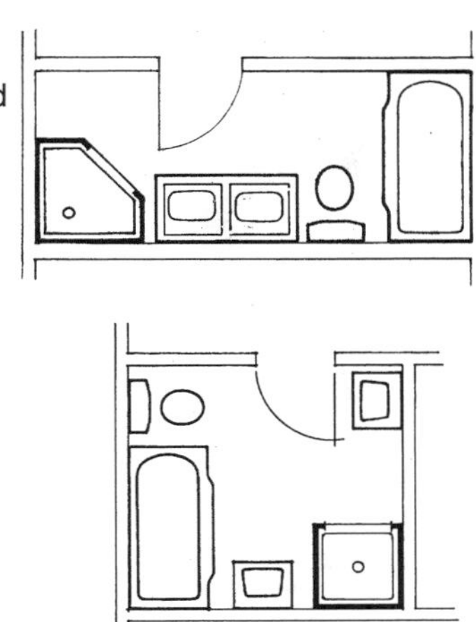

*Common wall is with an adjacent bathroom.

System Components	QUANTITY	UNIT	COST EACH MAT.	COST EACH INST.	COST EACH TOTAL
SYSTEM 8.1-650-1360					
BATHROOM, BATHTUB, WATER CLOSET, STALL SHOWER & TWO LAVATORIES					
TWO WALL PLUMBING, STAND ALONE					
Wtr closet, 2 pc close cpld vit china flr mntd incl seat,supply & stop	1.000	Ea.	147	141	288
Water closet, rough-in waste & vent	1.000	Set	125	246	371
Lavatory w/ftngs, wall hung, white PE on CI, 20″ x 18″	2.000	Ea.	404	187	591
Lavatory, rough-in waste & vent	2.000	Set	412	900	1,312
Bathtub, white PE on CI, w/ftgs, mat bottom, recessed, 5′ long	1.000	Ea.	395	170	565
Baths, rough-in waste and vent	1.000	Set	111	360	471
Shower stall, bkd enam molded stone receptor, door & ftng, 32″ sq.	1.000	Ea.	345	375	720
Shower stall, rough-in supply, waste & vent,	1.000	Set	64	365	429
TOTAL			2,003	2,744	4,747

8.1-650	Five Fixture Bathroom, Two Wall Plumbing		COST EACH MAT.	COST EACH INST.	COST EACH TOTAL
1320	Bathroom, five fixture, two wall plumbing	R15100 -410			
1340	Bathtub, water closet, stall shower & two lavatories				
1360	Stand alone	R15100 -420	2,000	2,750	4,750
1400	One short plumbing wall common,*		1,900	2,375	4,275
1500	Bathtub, two lavatories, corner stall shower & water closet				
1520	Stand alone		2,475	2,775	5,250
1540	Long plumbing wall common*		2,225	2,225	4,450

8.1-650	Five Fixture Bathroom, Three Wall Plumbing	COST EACH MAT.	COST EACH INST.	COST EACH TOTAL
2360	Bathroom, five fixture, three wall plumbing			
2380	Water closet, bathtub, two lavatories & stall shower			
2400	Stand alone	2,475	2,775	5,250
2440	One short plumbing wall common *	2,375	2,425	4,800

8.1-650	Five Fixture Bathroom, One Wall Plumbing	COST EACH MAT.	COST EACH INST.	COST EACH TOTAL
4080	Bathroom, five fixture, one wall plumbing			
4100	Bathtub, two lavatories, corner stall shower & water closet			
4120	Stand alone	2,350	2,525	4,875
4160	Share common wall *	1,975	1,725	3,700

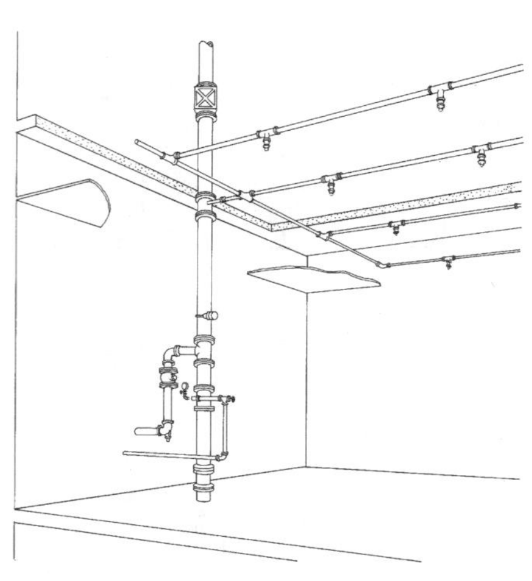

Wet Pipe System. A system employing automatic sprinklers attached to a piping system containing water and connected to a water supply so that water discharges immediately from sprinklers opened by heat from a fire.

All areas are assumed to be open.

System Components	QUANTITY	UNIT	COST EACH MAT.	COST EACH INST.	COST EACH TOTAL
SYSTEM 8.2-110-0580					
WET PIPE SPRINKLER, STEEL, BLACK, SCH. 40 PIPE					
LIGHT HAZARD, ONE FLOOR, 2000 S.F.					
Valve, gate, iron body, 125 lb, OS&Y, flanged, 4″ diam	1.000	Ea.	285	187.50	472.50
Valve, swing check, bronze, 125 lb, regrinding disc, 2-1/2″ pipe size	1.000	Ea.	147.75	37.50	185.25
Valve, angle, bronze, 150 lb, rising stem, threaded, 2″ diam	1.000	Ea.	180.75	28.50	209.25
*Alarm valve, 2-1/2″ pipe size	1.000	Ea.	615	187.50	802.50
Alarm, water motor, complete with gong	1.000	Ea.	137.25	78	215.25
Valve, swing check, w/balldrip CI with brass trim 4″ pipe size	1.000	Ea.	116.25	187.50	303.75
Pipe, steel, black, schedule 40, 4″ diam	10.000	L.F.	64.88	172.43	237.31
*Flow control valve, trim & gauges, 4″ pipe size	1.000	Set	1,237.50	427.50	1,665
Fire alarm horn, electric	1.000	Ea.	30	45.75	75.75
Pipe, steel, black, schedule 40, threaded, cplg & hngr 10′OC, 2-1/2″ diam	20.000	L.F.	75.75	225	300.75
Pipe, steel, black, schedule 40, threaded, cplg & hngr 10′OC, 2″ diam	12.500	L.F.	29.81	109.69	139.50
Pipe, steel, black, schedule 40, threaded, cplg & hngr 10′OC, 1-1/4″ diam	37.500	L.F.	63	236.25	299.25
Pipe steel, black, schedule 40, threaded cplg & hngr 10′OC, 1″ diam	112.000	L.F.	160.44	659.40	819.84
Pipe Tee, malleable iron black, 150 lb threaded, 4″ pipe size	2.000	Ea.	174	280.50	454.50
Pipe Tee, malleable iron black, 150 lb threaded, 2-1/2″ pipe size	2.000	Ea.	48.75	124.50	173.25
Pipe Tee, malleable iron black, 150 lb threaded, 2″ pipe size	1.000	Ea.	11.33	51	62.33
Pipe Tee, malleable iron black, 150 lb threaded, 1-1/4″ pipe size	5.000	Ea.	26.63	200.63	227.26
Pipe Tee, malleable iron black, 150 lb threaded, 1″ pipe size	4.000	Ea.	13.17	156	169.17
Pipe 90° elbow, malleable iron black, 150 lb threaded, 1″ pipe size	6.000	Ea.	12.69	144	156.69
Sprinkler head, standard spray, brass 135°-286°F 1/2″ NPT, 3/8″ orifice	12.000	Ea.	79.20	312	391.20
Valve, gate, bronze, NRS, class 150, threaded, 1″ pipe size	1.000	Ea.	27.38	16.50	43.88
*Standpipe connection, wall, single, flush w/plug & chain 2-1/2″x2-1/2″	1.000	Ea.	64.88	112.50	177.38
TOTAL			3,601.41	3,980.15	7,581.56
COST PER S.F.			1.80	1.99	3.79

*Not included in systems under 2000 S.F.

8.2-110	Wet Pipe Sprinkler Systems		COST PER S.F. MAT.	COST PER S.F. INST.	COST PER S.F. TOTAL
0520	Wet pipe sprinkler systems, steel, black, sch. 40 pipe				
0530	Light hazard, one floor, 500 S.F.		1.11	1.90	3.01
0560	1000 S.F.	R10520-110	1.72	1.97	3.69
0580	2000 S.F.		1.80	1.99	3.79
0600	5000 S.F.	R10520-120	.86	1.41	2.27
0620	10,000 S.F.		.53	1.16	1.69
0640	50,000 S.F.	R10520-130	.36	1.08	1.44
0660	Each additional floor, 500 S.F.		.54	1.61	2.15

8.2-110	Wet Pipe Sprinkler Systems	COST PER S.F.		
		MAT.	INST.	TOTAL
0680	1000 S.F.	.51	1.50	2.01
0700	2000 S.F.	.47	1.36	1.83
0720	5000 S.F.	.33	1.15	1.48
0740	10,000 S.F.	.30	1.06	1.36
0760	50,000 S.F.	.26	.86	1.12
1000	Ordinary hazard, one floor, 500 S.F.	1.21	2.04	3.25
1020	1000 S.F.	1.70	1.95	3.65
1040	2000 S.F.	1.85	2.10	3.95
1060	5000 S.F.	.97	1.53	2.50
1080	10,000 S.F.	.66	1.55	2.21
1100	50,000 S.F.	.60	1.54	2.14
1140	Each additional floor, 500 S.F.	.64	1.82	2.46
1160	1000 S.F.	.49	1.49	1.98
1180	2000 S.F.	.54	1.53	2.07
1200	5000 S.F.	.54	1.43	1.97
1220	10,000 S.F.	.43	1.45	1.88
1240	50,000 S.F.	.46	1.36	1.82
1500	Extra hazard, one floor, 500 S.F.	3.23	3.14	6.37
1520	1000 S.F.	2.12	2.73	4.85
1540	2000 S.F.	1.96	2.78	4.74
1560	5000 S.F.	1.21	2.45	3.66
1580	10,000 S.F.	1.12	2.33	3.45
1600	50,000 S.F.	1.05	2.26	3.31
1660	Each additional floor, 500 S.F.	.83	2.24	3.07
1680	1000 S.F.	.83	2.12	2.95
1700	2000 S.F.	.72	2.14	2.86
1720	5000 S.F.	.60	1.88	2.48
1740	10,000 S.F.	.71	1.76	2.47
1760	50,000 S.F.	.68	1.66	2.34
2020	Grooved steel, black sch. 40 pipe, light hazard, one floor, 2000 S.F.	1.83	1.66	3.49
2060	10,000 S.F.	.70	1.02	1.72
2100	Each additional floor, 2000 S.F.	.52	1.09	1.61
2150	10,000 S.F.	.34	.87	1.21
2200	Ordinary hazard, one floor, 2000 S.F.	1.87	1.79	3.66
2250	10,000 S.F.	.65	1.31	1.96
2300	Each additional floor, 2000 S.F.	.56	1.22	1.78
2350	10,000 S.F.	.42	1.21	1.63
2400	Extra hazard, one floor, 2000 S.F.	2.01	2.26	4.27
2450	10,000 S.F.	.96	1.70	2.66
2500	Each additional floor, 2000 S.F.	.79	1.74	2.53
2550	10,000 S.F.	.66	1.52	2.18
3050	Grooved steel black sch. 10 pipe, light hazard, one floor, 2000 S.F.	1.81	1.65	3.46
3100	10,000 S.F.	.56	.96	1.52
3150	Each additional floor, 2000 S.F.	.50	1.08	1.58
3200	10,000 S.F.	.33	.86	1.19
3250	Ordinary hazard, one floor, 2000 S.F.	1.85	1.78	3.63
3300	10,000 S.F.	.64	1.27	1.91
3350	Each additional floor, 2000 S.F.	.54	1.21	1.75
3400	10,000 S.F.	.41	1.17	1.58
3450	Extra hazard, one floor, 2000 S.F.	1.99	2.26	4.25
3500	10,000 S.F.	.92	1.69	2.61
3550	Each additional floor, 2000 S.F.	.77	1.74	2.51
3600	10,000 S.F.	.65	1.51	2.16
4050	Copper tubing, type M, light hazard, one floor, 2000 S.F.	1.72	1.66	3.38
4100	10,000 S.F.	.55	.99	1.54
4150	Each additional floor, 2000 S.F.	.42	1.11	1.53
4200	10,000 S.F.	.32	.89	1.21
4250	Ordinary hazard, one floor, 2000 S.F.	1.77	1.86	3.63
4300	10,000 S.F.	.64	1.20	1.84
4350	Each additional floor, 2000 S.F.	.47	1.23	1.70

8.2-110	Wet Pipe Sprinkler Systems	COST PER S.F.		
		MAT.	INST.	TOTAL
4400	10,000 S.F.	.40	1.07	1.47
4450	Extra hazard, one floor, 2000 S.F.	1.89	2.29	4.18
4500	10,000 S.F.	1.08	1.84	2.92
4550	Each additional floor, 2000 S.F.	.67	1.77	2.44
4600	10,000 S.F.	.68	1.65	2.33
5050	Copper tubing, type M, T-drill system, light hazard, one floor			
5060	2000 S.F.	1.72	1.53	3.25
5100	10,000 S.F.	.53	.83	1.36
5150	Each additional floor, 2000 S.F.	.42	.98	1.40
5200	10,000 S.F.	.30	.73	1.03
5250	Ordinary hazard, one floor, 2000 S.F.	1.73	1.56	3.29
5300	10,000 S.F.	.62	1.06	1.68
5350	Each additional floor, 2000 S.F.	.42	.99	1.41
5400	10,000 S.F.	.39	.96	1.35
5450	Extra hazard, one floor, 2000 S.F.	1.80	1.87	3.67
5500	10,000 S.F.	.91	1.33	2.24
5550	Each additional floor, 2000 S.F.	.62	1.38	2
5600	10,000 S.F.	.51	1.14	1.65

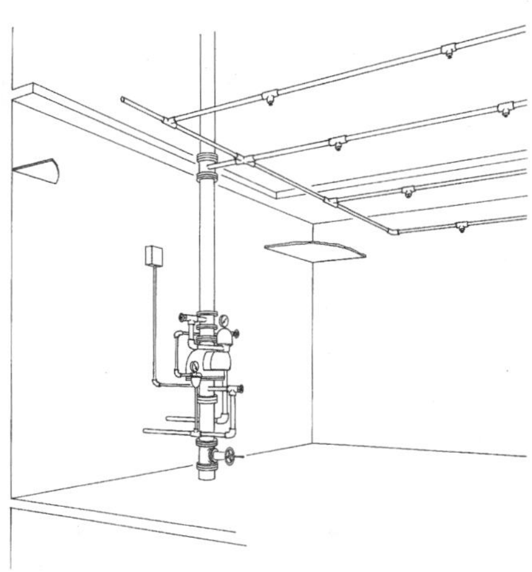

Dry Pipe System: A system employing automatic sprinklers attached to a piping system containing air under pressure, the release of which as from the opening of sprinklers permits the water pressure to open a valve known as a "dry pipe valve". The water then flows into the piping system and out the opened sprinklers.

All areas are assumed to be open.

System Components	QUANTITY	UNIT	COST EACH MAT.	COST EACH INST.	COST EACH TOTAL
SYSTEM 8.2-120-0580					
DRY PIPE SPRINKLER, STEEL BLACK, SCH. 40 PIPE					
LIGHT HAZARD, ONE FLOOR, 2000 S.F.					
Valve, gate, iron body 125 lb, OS&Y, flanged, 4″ pipe size	1.000	Ea.	285	187.50	472.50
Valve, swing check, bronze, 125 lb, regrinding disc, 2-1/2″ pipe size	1.000	Ea.	147.75	37.50	185.25
Valve, angle, bronze, 150 lb, rising stem, threaded, 2″ pipe size	1.000	Ea.	180.75	28.50	209.25
*Alarm valve, 2-1/2″ pipe size	1.000	Ea.	615	187.50	802.50
Alarm, water motor, complete with gong,	1.000	Ea.	137.25	78	215.25
Fire alarm horn, electric	1.000	Ea.	30	45.75	75.75
Valve swing check w/balldrip CI with brass trim, 4″ pipe size	1.000	Ea.	116.25	187.50	303.75
Pipe, steel, black, schedule 40, 4″ diam	10.000	L.F.	64.88	172.43	237.31
Dry pipe valve, trim & gauges, 4″ pipe size	1.000	Ea.	975	562.50	1,537.50
Pipe, steel, black, schedule 40, threaded, cplg & hngr 10′OC 2-1/2″ diam	20.000	L.F.	75.75	225	300.75
Pipe, steel, black, schedule 40, threaded, cplg & hngr 10′OC 2″ diam	12.500	L.F.	29.81	109.69	139.50
Pipe, steel, black, schedule 40, threaded, cplg & hngr 10′OC 1-1/4″ diam	37.500	L.F.	63	236.25	299.25
Pipe, steel, black, schedule 40, threaded, cplg & hngr 10′OC 1″ diam	112.000	L.F.	160.44	659.40	819.84
Pipe Tee, malleable iron black, 150 lb threaded, 4″ pipe size	2.000	Ea.	174	280.50	454.50
Pipe Tee, malleable iron black, 150 lb threaded, 2-1/2″ pipe size	2.000	Ea.	48.75	124.50	173.25
Pipe Tee, malleable iron black, 150 lb threaded, 2″ pipe size	1.000	Ea.	11.33	51	62.33
Pipe Tee, malleable iron black, 150 lb threaded, 1-1/4″ pipe size	5.000	Ea.	26.63	200.63	227.26
Pipe Tee, malleable iron black, 150 lb threaded, 1″ pipe size	4.000	Ea.	13.17	156	169.17
Pipe 90° elbow malleable iron black, 150 lb threaded, 1″ pipe size	6.000	Ea.	12.69	144	156.69
Sprinkler head dry 1/2″ orifice 1″ NPT, 3″ to 4-3/4″ length	12.000	Ea.	456	360	816
Air compressor, 200 Gal sprinkler system capacity, 1/3 HP	1.000	Ea.	577.50	240	817.50
*Standpipe connection, wall, flush, brs w/plug & chain 2-1/2″x2-1/2″	1.000	Ea.	64.88	112.50	177.38
Valve gate bronze, 300 psi, NRS, class 150, threaded, 1″ pipe size	1.000	Ea.	27.38	16.50	43.88
TOTAL			4,293.21	4,403.15	8,696.36
COST PER S.F.			2.15	2.20	4.35

*Not included in systems under 2000 S.F.

8.2-120	Dry Pipe Sprinkler Systems		COST PER S.F. MAT.	COST PER S.F. INST.	COST PER S.F. TOTAL
0520	Dry pipe sprinkler systems, steel, black, sch. 40 pipe				
0530	Light hazard, one floor, 500 S.F.		4.45	3.72	8.17
0560	1000 S.F.	R10520-110	2.41	2.18	4.59
0580	2000 S.F.		2.15	2.20	4.35
0600	5000 S.F.	R10520-120	1.07	1.50	2.57
0620	10,000 S.F.		.70	1.22	1.92
0640	50,000 S.F.	R10520-130	.49	1.10	1.59
0660	Each additional floor, 500 S.F.		.77	1.79	2.56

8.2-120	Dry Pipe Sprinkler Systems	COST PER S.F.		
		MAT.	INST.	TOTAL
0680	1000 S.F.	.64	1.48	2.12
0700	2000 S.F.	.61	1.37	1.98
0720	5000 S.F.	.48	1.17	1.65
0740	10,000 S.F.	.43	1.08	1.51
0760	50,000 S.F.	.40	.99	1.39
1000	Ordinary hazard, one floor, 500 S.F.	4.51	3.77	8.28
1020	1000 S.F.	2.45	2.21	4.66
1040	2000 S.F.	2.20	2.31	4.51
1060	5000 S.F.	1.22	1.62	2.84
1080	10,000 S.F.	.89	1.61	2.50
1100	50,000 S.F.	.81	1.61	2.42
1140	Each additional floor, 500 S.F.	.83	1.84	2.67
1160	1000 S.F.	.72	1.66	2.38
1180	2000 S.F.	.73	1.55	2.28
1200	5000 S.F.	.66	1.31	1.97
1220	10,000 S.F.	.57	1.27	1.84
1240	50,000 S.F.	.58	1.18	1.76
1500	Extra hazard, one floor, 500 S.F.	6.10	4.69	10.79
1520	1000 S.F.	3.53	3.38	6.91
1540	2000 S.F.	2.39	2.93	5.32
1560	5000 S.F.	1.39	2.23	3.62
1580	10,000 S.F.	1.37	2.12	3.49
1600	50,000 S.F.	1.40	2.06	3.46
1660	Each additional floor, 500 S.F.	1.11	2.28	3.39
1680	1000 S.F.	1.11	2.16	3.27
1700	2000 S.F.	1	2.18	3.18
1720	5000 S.F.	.84	1.89	2.73
1740	10,000 S.F.	.98	1.75	2.73
1760	50,000 S.F.	.99	1.68	2.67
2020	Grooved steel, black, sch. 40 pipe, light hazard, one floor, 2000 S.F.	2.13	1.86	3.99
2060	10,000 S.F.	.74	1.03	1.77
2100	Each additional floor, 2000 S.F.	.66	1.10	1.76
2150	10,000 S.F.	.47	.89	1.36
2200	Ordinary hazard, one floor, 2000 S.F.	2.22	2	4.22
2250	10,000 S.F.	.88	1.37	2.25
2300	Each additional floor, 2000 S.F.	.75	1.24	1.99
2350	10,000 S.F.	.61	1.23	1.84
2400	Extra hazard, one floor, 2000 S.F.	2.45	2.49	4.94
2450	10,000 S.F.	1.28	1.77	3.05
2500	Each additional floor, 2000 S.F.	1.07	1.78	2.85
2550	10,000 S.F.	.94	1.55	2.49
3050	Grooved steel black sch. 10 pipe, light hazard, one floor, 2000 S.F.	2.11	1.85	3.96
3100	10,000 S.F.	.73	1.02	1.75
3150	Each additional floor, 2000 S.F.	.64	1.09	1.73
3200	10,000 S.F.	.46	.88	1.34
3250	Ordinary hazard, one floor, 2000 S.F.	2.20	1.99	4.19
3300	10,000 S.F.	.87	1.33	2.20
3350	Each additional floor, 2000 S.F.	.73	1.23	1.96
3400	10,000 S.F.	.60	1.19	1.79
3450	Extra hazard, one floor, 2000 S.F.	2.43	2.49	4.92
3500	10,000 S.F.	1.24	1.76	3
3550	Each additional floor, 2000 S.F.	1.05	1.78	2.83
3600	10,000 S.F.	.93	1.54	2.47
4050	Copper tubing, type M, light hazard, one floor, 2000 S.F.	2.02	1.86	3.88
4100	10,000 S.F.	.72	1.05	1.77
4150	Each additional floor, 2000 S.F.	.56	1.12	1.68
4200	10,000 S.F.	.45	.91	1.36
4250	Ordinary hazard, one floor, 2000 S.F.	2.12	2.07	4.19
4300	10,000 S.F.	.87	1.26	2.13
4350	Each additional floor, 2000 S.F.	.70	1.29	1.99

8.2-120	Dry Pipe Sprinkler Systems	COST PER S.F.		
		MAT.	INST.	TOTAL
4400	10,000 S.F.	.59	1.09	1.68
4450	Extra hazard, one floor, 2000 S.F.	2.33	2.52	4.85
4500	10,000 S.F.	1.41	1.91	3.32
4550	Each additional floor, 2000 S.F.	.95	1.81	2.76
4600	10,000 S.F.	.96	1.68	2.64
5050	Copper tubing, type M, T-drill system, light hazard, one floor			
5060	2000 S.F.	2.02	1.73	3.75
5100	10,000 S.F.	.70	.89	1.59
5150	Each additional floor, 2000 S.F.	.56	.99	1.55
5200	10,000 S.F.	.43	.75	1.18
5250	Ordinary hazard, one floor, 2000 S.F.	2.08	1.77	3.85
5300	10,000 S.F.	.85	1.12	1.97
5350	Each additional floor, 2000 S.F.	.61	1.01	1.62
5400	10,000 S.F.	.55	.93	1.48
5450	Extra hazard, one floor, 2000 S.F.	2.24	2.10	4.34
5500	10,000 S.F.	1.24	1.40	2.64
5550	Each additional floor, 2000 S.F.	.86	1.39	2.25
5600	10,000 S.F.	.79	1.17	1.96

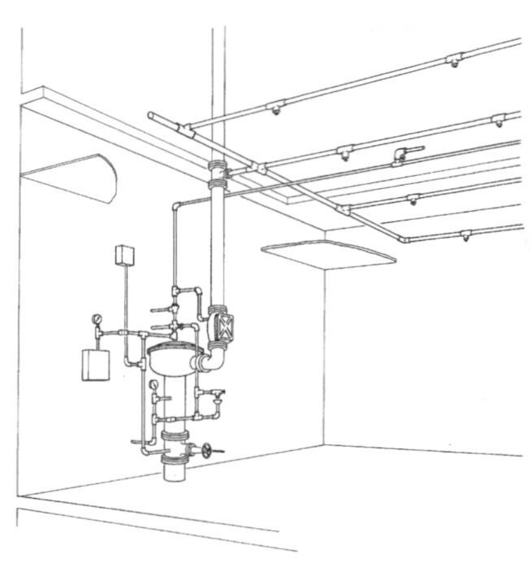

Pre-Action System: A system employing automatic sprinklers attached to a piping system containing air that may or may not be under pressure, with a supplemental heat responsive system of generally more sensitive characteristics than the automatic sprinklers themselves, installed in the same areas as the sprinklers. Actuation of the heat responsive system, as from a fire, opens a valve which permits water to flow into the sprinkler piping system and to be discharged from those sprinklers which were opened by heat from the fire.

All areas are assumed to be open.

System Components

System Components	QUANTITY	UNIT	COST EACH MAT.	COST EACH INST.	COST EACH TOTAL
SYSTEM 8.2-130-0580					
PREACTION SPRINKLER SYSTEM, STEEL BLACK SCH. 40 PIPE					
LIGHT HAZARD, 1 FLOOR, 2000 S.F.					
Valve, gate, iron body 125 lb, OS&Y, flanged, 4″ pipe size	1.000	Ea.	285	187.50	472.50
*Valve, swing check w/ball drip CI with brass trim 4″ pipe size	1.000	Ea.	116.25	187.50	303.75
Valve, swing check, bronze, 125 lb, regrinding disc, 2-1/2″ pipe size	1.000	Ea.	147.75	37.50	185.25
Valve, angle, bronze, 150 lb, rising stem, threaded, 2″ pipe size	1.000	Ea.	180.75	28.50	209.25
*Alarm valve, 2-1/2″ pipe size	1.000	Ea.	615	187.50	802.50
Alarm, water motor, complete with gong	1.000	Ea.	137.25	78	215.25
Fire alarm horn, electric	1.000	Ea.	30	45.75	75.75
Thermostatic release for release line	2.000	Ea.	438	31.50	469.50
Pipe, steel, black, schedule 40, 4″ diam	10.000	L.F.	64.88	172.43	237.31
Dry pipe valve, trim & gauges, 4″ pipe size	1.000	Ea.	975	562.50	1,537.50
Pipe, steel, black, schedule 40, threaded, cplg & hngr 10′OC 2-1/2″ diam	20.000	L.F.	75.75	225	300.75
Pipe steel black, schedule 40, threaded, cplg & hngr 10′OC 2″ diam	12.500	L.F.	29.81	109.69	139.50
Pipe, steel, black, schedule 40, threaded, cplg & hngr 10′OC 1-1/4″ diam	37.500	L.F.	63	236.25	299.25
Pipe, steel, black, schedule 40, threaded, cplg & hngr 10′OC 1″ diam	112.000	L.F.	160.44	659.40	819.84
Pipe, Tee, malleable iron, black, 150 lb threaded, 4″ diam	2.000	Ea.	174	280.50	454.50
Pipe, Tee, malleable iron, black, 150 lb threaded, 2-1/2″ pipe size	2.000	Ea.	48.75	124.50	173.25
Pipe, Tee, malleable iron, black, 150 lb threaded, 2″ pipe size	1.000	Ea.	11.33	51	62.33
Pipe, Tee, malleable iron, black, 150 lb threaded, 1-1/4″ pipe size	5.000	Ea.	26.63	200.63	227.26
Pipe, Tee, malleable iron, black, 150 lb threaded, 1″ pipe size	4.000	Ea.	13.17	156	169.17
Pipe, 90° elbow, malleable iron, blk, 150 lb threaded, 1″ pipe size	6.000	Ea.	12.69	144	156.69
Sprinkler head, std spray, brass 135°-286°F 1/2″ NPT, 3/8″ orifice	12.000	Ea.	79.20	312	391.20
Air compressor auto complete 200 Gal sprinkler sys cap, 1/3 HP	1.000	Ea.	577.50	240	817.50
*Standpipe conn.,wall,flush,brass w/plug & chain 2-1/2″ x 2-1/2″	1.000	Ea.	64.88	112.50	177.38
Valve, gate, bronze, 300 psi, NRS, class 150, threaded, 1″ pipe size	1.000	Ea.	27.38	16.50	43.88
TOTAL			4,354.41	4,386.65	8,741.06
COST PER S.F.			2.18	2.19	4.37

*Not included in systems under 2000 S.F.

8.2-130	Preaction Sprinkler Systems		COST PER S.F. MAT.	COST PER S.F. INST.	COST PER S.F. TOTAL
0520	Preaction sprinkler systems, steel, black, sch. 40 pipe				
0530	Light hazard, one floor, 500 S.F.		4.42	2.97	7.39
0560	1000 S.F.	R10520-110	2.50	2.23	4.73
0580	2000 S.F.		2.18	2.19	4.37
0600	5000 S.F.	R10520-120	1.10	1.49	2.59
0620	10,000 S.F.		.72	1.21	1.93

8.2-130	Preaction Sprinkler Systems	COST PER S.F.		
		MAT.	INST.	TOTAL
0640	50,000 S.F. R10520-130	.51	1.09	1.60
0660	Each additional floor, 500 S.F.	.97	1.60	2.57
0680	1000 S.F.	.72	1.48	2.20
0700	2000 S.F.	.69	1.38	2.07
0720	5000 S.F.	.51	1.16	1.67
0740	10,000 S.F.	.45	1.07	1.52
0760	50,000 S.F.	.45	1.01	1.46
1000	Ordinary hazard, one floor, 500 S.F.	1.66	2.18	3.84
1020	1000 S.F.	2.48	2.21	4.69
1040	2000 S.F.	2.34	2.31	4.65
1060	5000 S.F.	1.21	1.61	2.82
1080	10,000 S.F.	.85	1.60	2.45
1100	50,000 S.F.	.76	1.58	2.34
1140	Each additional floor, 500 S.F.	1.08	1.85	2.93
1160	1000 S.F.	.71	1.51	2.22
1180	2000 S.F.	.65	1.54	2.19
1200	5000 S.F.	.71	1.42	2.13
1220	10,000 S.F.	.58	1.46	2.04
1240	50,000 S.F.	.60	1.37	1.97
1500	Extra hazard, one floor, 500 S.F.	6.05	4.12	10.17
1520	1000 S.F.	3.35	3.08	6.43
1540	2000 S.F.	2.33	2.91	5.24
1560	5000 S.F.	1.39	2.39	3.78
1580	10,000 S.F.	1.30	2.35	3.65
1600	50,000 S.F.	1.30	2.28	3.58
1660	Each additional floor, 500 S.F.	1.27	2.27	3.54
1680	1000 S.F.	1.05	2.14	3.19
1700	2000 S.F.	.94	2.16	3.10
1720	5000 S.F.	.78	1.89	2.67
1740	10,000 S.F.	.86	1.77	2.63
1760	50,000 S.F.	.83	1.66	2.49
2020	Grooved steel, black, sch. 40 pipe, light hazard, one floor, 2000 S.F.	2.21	1.87	4.08
2060	10,000 S.F.	.76	1.02	1.78
2100	Each additional floor of 2000 S.F.	.74	1.11	1.85
2150	10,000 S.F.	.49	.88	1.37
2200	Ordinary hazard, one floor, 2000 S.F.	2.25	2	4.25
2250	10,000 S.F.	.84	1.36	2.20
2300	Each additional floor, 2000 S.F.	.78	1.24	2.02
2350	10,000 S.F.	.57	1.22	1.79
2400	Extra hazard, one floor, 2000 S.F.	2.39	2.47	4.86
2450	10,000 S.F.	1.16	1.75	2.91
2500	Each additional floor, 2000 S.F.	1.01	1.76	2.77
2550	10,000 S.F.	.81	1.53	2.34
3050	Grooved steel, black, sch. 10 pipe light hazard, one floor, 2000 S.F.	2.19	1.86	4.05
3100	10,000 S.F.	.75	1.01	1.76
3150	Each additional floor, 2000 S.F.	.72	1.10	1.82
3200	10,000 S.F.	.48	.87	1.35
3250	Ordinary hazard, one floor, 2000 S.F.	2.20	1.86	4.06
3300	10,000 S.F.	.72	1.31	2.03
3350	Each additional floor, 2000 S.F.	.76	1.23	1.99
3400	10,000 S.F.	.56	1.18	1.74
3450	Extra hazard, one floor, 2000 S.F.	2.37	2.47	4.84
3500	10,000 S.F.	1.11	1.74	2.85
3550	Each additional floor, 2000 S.F.	.99	1.76	2.75
3600	10,000 S.F.	.80	1.52	2.32
4050	Copper tubing, type M, light hazard, one floor, 2000 S.F.	2.10	1.87	3.97
4100	10,000 S.F.	.74	1.04	1.78
4150	Each additional floor, 2000 S.F.	.64	1.13	1.77
4200	10,000 S.F.	.36	.89	1.25
4250	Ordinary hazard, one floor, 2000 S.F.	2.15	2.07	4.22

8.2-130	Preaction Sprinkler Systems	COST PER S.F.		
		MAT.	INST.	TOTAL
4300	10,000 S.F.	.83	1.25	2.08
4350	Each additional floor, 2000 S.F.	.63	1.15	1.78
4400	10,000 S.F.	.51	.99	1.50
4450	Extra hazard, one floor, 2000 S.F.	2.27	2.50	4.77
4500	10,000 S.F.	1.27	1.89	3.16
4550	Each additional floor, 2000 S.F.	.89	1.79	2.68
4600	10,000 S.F.	.83	1.66	2.49
5050	Copper tubing, type M, T-drill system, light hazard, one floor			
5060	2000 S.F.	2.10	1.74	3.84
5100	10,000 S.F.	.72	.88	1.60
5150	Each additional floor, 2000 S.F.	.64	1	1.64
5200	10,000 S.F.	.45	.74	1.19
5250	Ordinary hazard, one floor, 2000 S.F.	2.11	1.77	3.88
5300	10,000 S.F.	.81	1.11	1.92
5350	Each additional floor, 2000 S.F.	.65	1.02	1.67
5400	10,000 S.F.	.54	.97	1.51
5450	Extra hazard, one floor, 2000 S.F.	2.18	2.08	4.26
5500	10,000 S.F.	1.10	1.38	2.48
5550	Each additional floor, 2000 S.F.	.80	1.37	2.17
5600	10,000 S.F.	.66	1.15	1.81

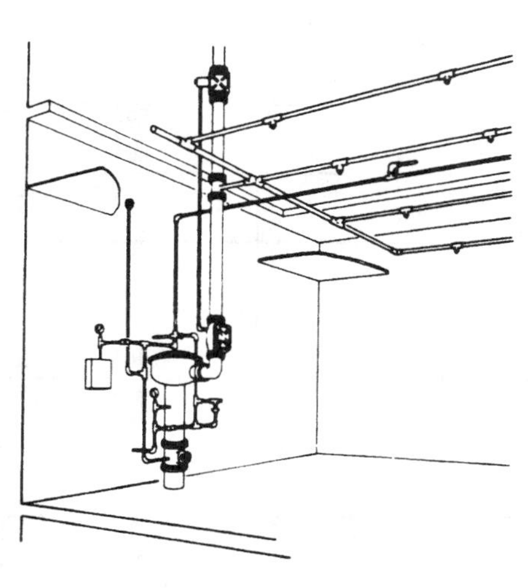

Deluge System: A system employing open sprinklers attached to a piping system connected to a water supply through a valve which is opened by the operation of a heat responsive system installed in the same areas as the sprinklers. When this valve opens, water flows into the piping system and discharges from all sprinklers attached thereto.

All areas are assumed to be open.

System Components	QUANTITY	UNIT	COST EACH MAT.	COST EACH INST.	COST EACH TOTAL
SYSTEM 8.2-140-0580					
DELUGE SPRINKLER SYSTEM, STEEL BLACK SCH. 40 PIPE					
LIGHT HAZARD, 1 FLOOR, 2000 S.F.					
Valve, gate, iron body 125 lb, OS&Y, flanged, 4″ pipe size	1.000	Ea.	285	187.50	472.50
Valve, swing check w/ball drip, CI w/brass ftngs, 4″ pipe size	1.000	Ea.	116.25	187.50	303.75
Valve, swing check, bronze, 125 lb, regrinding disc, 2-1/2″ pipe size	1.000	Ea.	147.75	37.50	185.25
Valve, angle, bronze, 150 lb, rising stem, threaded, 2″ pipe size	1.000	Ea.	180.75	28.50	209.25
*Alarm valve, 2-1/2″ pipe size	1.000	Ea.	615	187.50	802.50
Alarm, water motor, complete with gong	1.000	Ea.	137.25	78	215.25
Fire alarm horn, electric	1.000	Ea.	30	45.75	75.75
Thermostatic release for release line	2.000	Ea.	438	31.50	469.50
Pipe, steel, black, schedule 40, 4″ diam	10.000	L.F.	64.88	172.43	237.31
Deluge valve trim, pressure relief, emergency release, gauge, 4″ pipe size	1.000	Ea.	1,162.50	562.50	1,725
Deluge system pressured monitoring panel, 120V	1.000	Ea.	648.75	17.25	666
Pipe, steel, black, schedule 40, threaded, cplg & hngr 10′OC 2-1/2″ diam	20.000	L.F.	75.75	225	300.75
Pipe, steel, black, schedule 40, threaded, cplg & hngr 10′OC 2″ diam	12.500	L.F.	29.81	109.69	139.50
Pipe, steel, black, schedule 40, threaded, cplg & hngr 10′OC 1-1/4″ diam	37.500	L.F.	63	236.25	299.25
Pipe, steel, black, schedule 40, threaded, cplg & hngr 10′OC 1″ diam	112.000	L.F.	160.44	659.40	819.84
Pipe, Tee, malleable iron, black, 150 lb threaded, 4″ pipe size	2.000	Ea.	174	280.50	454.50
Pipe, Tee, malleable iron, black, 150 lb threaded, 2-1/2″ pipe size	2.000	Ea.	48.75	124.50	173.25
Pipe, Tee, malleable iron, black, 150 lb threaded, 2″ pipe size	1.000	Ea.	11.33	51	62.33
Pipe, Tee, malleable iron, black, 150 lb threaded, 1-1/4″ pipe size	5.000	Ea.	26.63	200.63	227.26
Pipe, Tee, malleable iron, black, 150 lb threaded, 1″ pipe size	4.000	Ea.	13.17	156	169.17
Pipe, 90° elbow, malleable iron, black, 150 lb threaded 1″ pipe size	6.000	Ea.	12.69	144	156.69
Sprinkler head, std spray, brass 135°-286°F 1/2″ NPT, 3/8″ orifice	9.720	Ea.	79.20	312	391.20
Air compressor, auto, complete, 200 Gal sprinkler sys cap, 1/3 HP	1.000	Ea.	577.50	240	817.50
*Standpipe connection, wall, flush w/plug & chain 2-1/2″x2-1/2″	1.000	Ea.	64.88	112.50	177.38
Valve, gate, bronze, 300 psi, NRS, class 150, threaded, 1″ pipe size	1.000	Ea.	27.38	16.50	43.88
TOTAL			5,190.66	4,403.90	9,594.56
COST PER S.F.			2.60	2.20	4.80

*Not included in systems under 2000 S.F.

8.2-140	Deluge Sprinkler Systems		COST PER S.F. MAT.	COST PER S.F. INST.	COST PER S.F. TOTAL
0520	Deluge sprinkler systems, steel, black, sch. 40 pipe				
0530	Light hazard, one floor, 500 S.F.		5.65	3	8.65
0560	1000 S.F.	R10520 -110	3.10	2.15	5.25
0580	2000 S.F.		2.60	2.20	4.80

MECHANICAL 8

8.2-140	Deluge Sprinkler Systems		COST PER S.F. MAT.	INST.	TOTAL
0600	5000 S.F.	R10520-120	1.27	1.49	2.76
0620	10,000 S.F.		.80	1.21	2.01
0640	50,000 S.F.	R10520-130	.52	1.09	1.61
0660	Each additional floor, 500 S.F.		.97	1.60	2.57
0680	1000 S.F.		.72	1.48	2.20
0700	2000 S.F.		.69	1.38	2.07
0720	5000 S.F.		.51	1.16	1.67
0740	10,000 S.F.		.45	1.07	1.52
0760	50,000 S.F.		.45	1.01	1.46
1000	Ordinary hazard, one floor, 500 S.F.		6.05	3.44	9.49
1020	1000 S.F.		3.10	2.23	5.33
1040	2000 S.F.		2.75	2.32	5.07
1060	5000 S.F.		1.38	1.61	2.99
1080	10,000 S.F.		.93	1.60	2.53
1100	50,000 S.F.		.80	1.60	2.40
1140	Each additional floor, 500 S.F.		1.08	1.85	2.93
1160	1000 S.F.		.71	1.51	2.22
1180	2000 S.F.		.65	1.54	2.19
1200	5000 S.F.		.65	1.30	1.95
1220	10,000 S.F.		.58	1.28	1.86
1240	50,000 S.F.		.57	1.26	1.83
1500	Extra hazard, one floor, 500 S.F.		7.25	4.15	11.40
1520	1000 S.F.		4.17	3.21	7.38
1540	2000 S.F.		2.74	2.92	5.66
1560	5000 S.F.		1.48	2.20	3.68
1580	10,000 S.F.		1.34	2.15	3.49
1600	50,000 S.F.		1.33	2.09	3.42
1660	Each additional floor, 500 S.F.		1.27	2.27	3.54
1680	1000 S.F.		1.05	2.14	3.19
1700	2000 S.F.		.94	2.16	3.10
1720	5000 S.F.		.78	1.89	2.67
1740	10,000 S.F.		.87	1.83	2.70
1760	50,000 S.F.		.86	1.77	2.63
2000	Grooved steel, black, sch. 40 pipe, light hazard, one floor				
2020	2000 S.F.		2.63	1.88	4.51
2060	10,000 S.F.		.85	1.03	1.88
2100	Each additional floor, 2,000 S.F.		.74	1.11	1.85
2150	10,000 S.F.		.49	.88	1.37
2200	Ordinary hazard, one floor, 2000 S.F.		2.25	2	4.25
2250	10,000 S.F.		.92	1.36	2.28
2300	Each additional floor, 2000 S.F.		.78	1.24	2.02
2350	10,000 S.F.		.57	1.22	1.79
2400	Extra hazard, one floor, 2000 S.F.		2.80	2.48	5.28
2450	10,000 S.F.		1.25	1.76	3.01
2500	Each additional floor, 2000 S.F.		1.01	1.76	2.77
2550	10,000 S.F.		.81	1.53	2.34
3000	Grooved steel, black, sch. 10 pipe, light hazard, one floor				
3050	2000 S.F.		2.32	1.79	4.11
3100	10,000 S.F.		.83	1.01	1.84
3150	Each additional floor, 2000 S.F.		.72	1.10	1.82
3200	10,000 S.F.		.48	.87	1.35
3250	Ordinary hazard, one floor, 2000 S.F.		2.64	2	4.64
3300	10,000 S.F.		.80	1.31	2.11
3350	Each additional floor, 2000 S.F.		.76	1.23	1.99
3400	10,000 S.F.		.56	1.18	1.74
3450	Extra hazard, one floor, 2000 S.F.		2.78	2.48	5.26
3500	10,000 S.F.		1.19	1.74	2.93
3550	Each additional floor, 2000 S.F.		.99	1.76	2.75
3600	10,000 S.F.		.80	1.52	2.32
4000	Copper tubing, type M, light hazard, one floor				

8.2-140	Deluge Sprinkler Systems	COST PER S.F. MAT.	INST.	TOTAL
4050	2000 S.F.	2.51	1.88	4.39
4100	10,000 S.F.	.82	1.04	1.86
4150	Each additional floor, 2000 S.F.	.64	1.12	1.76
4200	10,000 S.F.	.36	.89	1.25
4250	Ordinary hazard, one floor, 2000 S.F.	2.56	2.08	4.64
4300	10,000 S.F.	.91	1.25	2.16
4350	Each additional floor, 2000 S.F.	.63	1.15	1.78
4400	10,000 S.F.	.51	.99	1.50
4450	Extra hazard, one floor, 2000 S.F.	2.68	2.51	5.19
4500	10,000 S.F.	1.37	1.90	3.27
4550	Each additional floor, 2000 S.F.	.89	1.79	2.68
4600	10,000 S.F.	.83	1.66	2.49
5000	Copper tubing, type M, T-drill system, light hazard, one floor			
5050	2000 S.F.	2.51	1.75	4.26
5100	10,000 S.F.	.80	.88	1.68
5150	Each additional floor, 2000 S.F.	.69	1.01	1.70
5200	10,000 S.F.	.45	.74	1.19
5250	Ordinary hazard, one floor, 2000 S.F.	2.52	1.78	4.30
5300	10,000 S.F.	.89	1.11	2
5350	Each additional floor, 2000 S.F.	.64	1.01	1.65
5400	10,000 S.F.	.54	.97	1.51
5450	Extra hazard, one floor, 2000 S.F.	2.59	2.09	4.68
5500	10,000 S.F.	1.18	1.38	2.56
5550	Each additional floor, 2000 S.F.	.80	1.37	2.17
5600	10,000 S.F.	.66	1.15	1.81

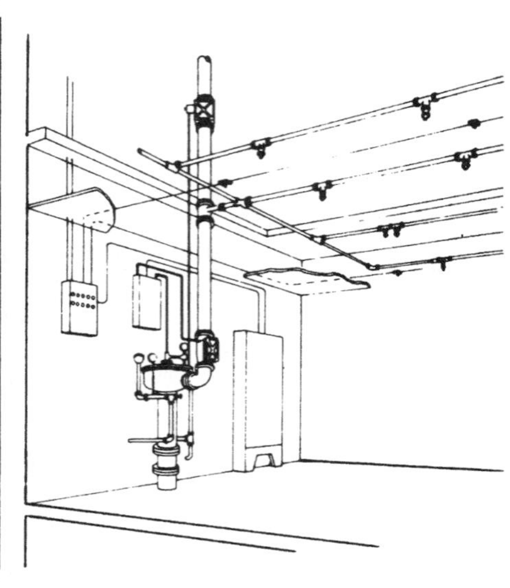

Firecycle is a fixed fire protection sprinkler system utilizing water as its extinguishing agent. It is a time delayed, recycling, preaction type which automatically shuts the water off when heat is reduced below the detector operating temperature and turns the water back on when that temperature is exceeded.

The system senses a fire condition through a closed circuit electrical detector system which controls water flow to the fire automatically. Batteries supply up to 90 hour emergency power supply for system operation. The piping system is dry (until water is required) and is monitored with pressurized air. Should any leak in the system piping occur, an alarm will sound, but water will not enter the system until heat is sensed by a Firecycle detector.

All areas are assumed to be open.

MECHANICAL 8

System Components	QUANTITY	UNIT	COST EACH		
			MAT.	INST.	TOTAL
SYSTEM 8.2-150-0580					
FIRECYCLE SPRINKLER SYSTEM, STEEL BLACK SCH. 40 PIPE					
LIGHT HAZARD, ONE FLOOR, 2000 S.F.					
Valve, gate, iron body 125 lb, OS&Y, flanged, 4″ pipe size	1.000	Ea.	285	187.50	472.50
Valve, angle, bronze, 150 lb, rising stem, threaded, 2″ pipe size	1.000	Ea.	180.75	28.50	209.25
Valve, swing check, bronze, 125 lb, regrinding disc, 2-1/2″ pipe size	1.000	Ea.	147.75	37.50	185.25
*Alarm valve, 2-1/2″ pipe size	1.000	Ea.	615	187.50	802.50
Alarm, water motor, complete with gong	1.000	Ea.	137.25	78	215.25
Pipe, steel, black, schedule 40, 4″ diam	10.000	L.F.	64.88	172.43	237.31
Fire alarm, horn, electric	1.000	Ea.	30	45.75	75.75
Pipe, steel, black, schedule 40, threaded, cplg & hngr 10′OC 2-1/2″ diam	20.000	L.F.	75.75	225	300.75
Pipe, steel, black, schedule 40, threaded, cplg & hngr 10′OC 2″ diam	12.500	L.F.	29.81	109.69	139.50
Pipe, steel, black, schedule 40, threaded, cplg & hngr 10′OC 1-1/4″ diam	37.500	L.F.	63	236.25	299.25
Pipe, steel, black, schedule 40, threaded, cplg & hngr 10′OC 1″ diam	112.000	L.F.	160.44	659.40	819.84
Pipe, Tee, malleable iron, black, 150 lb threaded, 4″ pipe size	2.000	Ea.	174	280.50	454.50
Pipe, Tee, malleable iron, black, 150 lb threaded, 2-1/2″ pipe size	2.000	Ea.	48.75	124.50	173.25
Pipe, Tee, malleable iron, black, 150 lb threaded, 2″ pipe size	1.000	Ea.	11.33	51	62.33
Pipe, Tee, malleable iron, black, 150 lb threaded, 1-1/4″ pipe size	5.000	Ea.	26.63	200.63	227.26
Pipe, Tee, malleable iron, black, 150 lb threaded, 1″ pipe size	4.000	Ea.	13.17	156	169.17
Pipe, 90° elbow, malleable iron, black, 150 lb threaded, 1″ pipe size	6.000	Ea.	12.69	144	156.69
Sprinkler head std spray, brass 135°-286°F 1/2″ NPT, 3/8″ orifice	12.000	Ea.	79.20	312	391.20
Firecycle controls, incls panel, battery, solenoid valves, press switches	1.000	Ea.	6,412.50	1,200	7,612.50
Detector, firecycle system	2.000	Ea.	472.50	39	511.50
Firecycle pkg, swing check & flow control valves w/trim 4″ pipe size	1.000	Ea.	1,743.75	562.50	2,306.25
Air compressor, auto, complete, 200 Gal sprinkler sys cap, 1/3 HP	1.000	Ea.	577.50	240	817.50
*Standpipe connection, wall, flush, brass w/plug & chain 2-1/2″x2-1/2″	1.000	Ea.	64.88	112.50	177.38
Valve, gate, bronze, 300 psi, NRS, class 150, threaded, 1″ diam	1.000	Ea.	27.38	16.50	43.88
TOTAL			11,453.91	5,406.65	16,860.56
COST PER S.F.			5.73	2.70	8.43
*Not included in systems under 2000 S.F.					

8.2-150	Firecycle Sprinkler Systems	COST PER S.F.		
		MAT.	INST.	TOTAL
0520	Firecycle sprinkler systems, steel black sch. 40 pipe			
0530	Light hazard, one floor, 500 S.F.	18.90	5.90	24.80

8.2-150	Firecycle Sprinkler Systems		COST PER S.F.	
		MAT.	INST.	TOTAL
0560	1000 S.F. R10520-110	9.70	3.68	13.38
0580	2000 S.F.	5.75	2.70	8.45
0600	5000 S.F. R10520-120	2.53	1.70	4.23
0620	10,000 S.F.	1.46	1.32	2.78
0640	50,000 S.F. R10520-130	.68	1.10	1.78
0660	Each additional floor of 500 S.F.	1	1.61	2.61
0680	1000 S.F.	.74	1.48	2.22
0700	2000 S.F.	.59	1.37	1.96
0720	5000 S.F.	.52	1.17	1.69
0740	10,000 S.F.	.49	1.08	1.57
0760	50,000 S.F.	.47	1.01	1.48
1000	Ordinary hazard, one floor, 500 S.F.	19.05	6.15	25.20
1020	1000 S.F.	9.70	3.66	13.36
1040	2000 S.F.	5.80	2.82	8.62
1060	5000 S.F.	2.64	1.82	4.46
1080	10,000 S.F.	1.59	1.71	3.30
1100	50,000 S.F.	1.05	1.79	2.84
1140	Each additional floor, 500 S.F.	1.11	1.86	2.97
1160	1000 S.F.	.73	1.51	2.24
1180	2000 S.F.	.75	1.41	2.16
1200	5000 S.F.	.66	1.31	1.97
1220	10,000 S.F.	.57	1.27	1.84
1240	50,000 S.F.	.59	1.19	1.78
1500	Extra hazard, one floor, 500 S.F.	20.50	7.05	27.55
1520	1000 S.F.	10.50	4.52	15.02
1540	2000 S.F.	5.90	3.42	9.32
1560	5000 S.F.	2.74	2.41	5.15
1580	10,000 S.F.	2.02	2.45	4.47
1600	50,000 S.F.	1.61	2.69	4.30
1660	Each additional floor, 500 S.F.	1.30	2.28	3.58
1680	1000 S.F.	1.07	2.14	3.21
1700	2000 S.F.	.96	2.16	3.12
1720	5000 S.F.	.79	1.90	2.69
1740	10,000 S.F.	.90	1.78	2.68
1760	50,000 S.F.	.88	1.71	2.59
2020	Grooved steel, black, sch. 40 pipe, light hazard, one floor			
2030	2000 S.F.	5.75	2.38	8.13
2060	10,000 S.F.	1.62	1.59	3.21
2100	Each additional floor, 2000 S.F.	.76	1.11	1.87
2150	10,000 S.F.	.53	.89	1.42
2200	Ordinary hazard, one floor, 2000 S.F.	5.80	2.51	8.31
2250	10,000 S.F.	1.71	1.55	3.26
2300	Each additional floor, 2000 S.F.	.80	1.24	2.04
2350	10,000 S.F.	.61	1.23	1.84
2400	Extra hazard, one floor, 2000 S.F.	5.95	2.98	8.93
2450	10,000 S.F.	1.88	1.85	3.73
2500	Each additional floor, 2000 S.F.	1.03	1.76	2.79
2550	10,000 S.F.	.85	1.54	2.39
3050	Grooved steel, black, sch. 10 pipe light hazard, one floor,			
3060	2000 S.F.	5.75	2.37	8.12
3100	10,000 S.F.	1.49	1.12	2.61
3150	Each additional floor, 2000 S.F.	.74	1.10	1.84
3200	10,000 S.F.	.52	.88	1.40
3250	Ordinary hazard, one floor, 2000 S.F.	5.80	2.50	8.30
3300	10,000 S.F.	1.57	1.43	3
3350	Each additional floor, 2000 S.F.	.78	1.23	2.01
3400	10,000 S.F.	.60	1.19	1.79
3450	Extra hazard, one floor, 2000 S.F.	5.90	2.98	8.88
3500	10,000 S.F.	1.84	1.84	3.68
3550	Each additional floor, 2000 S.F.	1.01	1.76	2.77

8.2-150	Firecycle Sprinkler Systems	COST PER S.F.		
		MAT.	INST.	TOTAL
3600	10,000 S.F.	.84	1.53	2.37
4060	Copper tubing, type M, light hazard, one floor, 2000 S.F.	5.65	2.38	8.03
4100	10,000 S.F.	1.48	1.15	2.63
4150	Each additional floor, 2000 S.F.	.66	1.13	1.79
4200	10,000 S.F.	.51	.91	1.42
4250	Ordinary hazard, one floor, 2000 S.F.	5.70	2.58	8.28
4300	10,000 S.F.	1.57	1.36	2.93
4350	Each additional floor, 2000 S.F.	.65	1.15	1.80
4400	10,000 S.F.	.54	.99	1.53
4450	Extra hazard, one floor, 2000 S.F.	5.80	3.01	8.81
4500	10,000 S.F.	2.07	2.01	4.08
4550	Each additional floor, 2000 S.F.	.91	1.79	2.70
4600	10,000 S.F.	.87	1.67	2.54
5060	Copper tubing, type M, T-drill system, light hazard, one floor 2000 S.F.	5.65	2.25	7.90
5100	10,000 S.F.	1.46	.99	2.45
5150	Each additional floor, 2000 S.F.	.75	1.05	1.80
5200	10,000 S.F.	.49	.75	1.24
5250	Ordinary hazard, one floor, 2000 S.F.	5.65	2.28	7.93
5300	10,000 S.F.	1.55	1.22	2.77
5350	Each additional floor, 2000 S.F.	.66	1.01	1.67
5400	10,000 S.F.	.58	.98	1.56
5450	Extra hazard, one floor, 2000 S.F.	5.75	2.59	8.34
5500	10,000 S.F.	1.83	1.48	3.31
5550	Each additional floor, 2000 S.F.	.82	1.37	2.19
5600	10,000 S.F.	.70	1.16	1.86

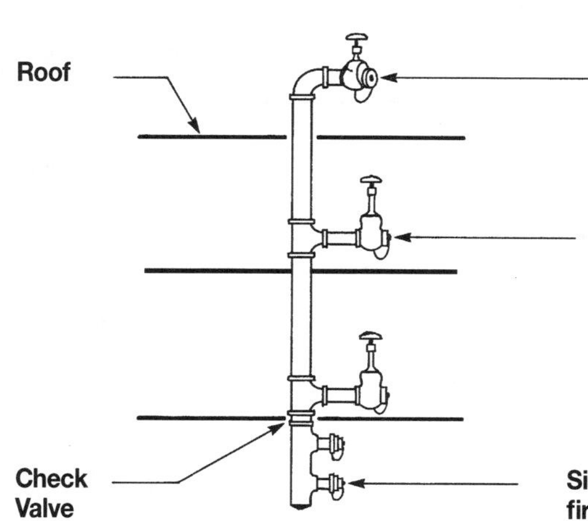

System Components	QUANTITY	UNIT	COST PER FLOOR MAT.	INST.	TOTAL
SYSTEM 8.2-310-0560					
WET STANDPIPE RISER, CLASS I, STEEL, BLACK, SCH. 40 PIPE, 10′ HEIGHT					
4″ DIAMETER PIPE, ONE FLOOR					
Pipe, steel, black, schedule 40, threaded, 4″ diam	20.000	L.F.	188	420	608
Pipe, Tee, malleable iron, black, 150 lb threaded, 4″ pipe size	2.000	Ea.	232	374	606
Pipe, 90° elbow, malleable iron, black, 150 lb threaded 4″ pipe size	1.000	Ea.	73.50	125	198.50
Pipe, nipple, steel, black, schedule 40, 2-1/2″ pipe size x 3″ long	2.000	Ea.	10.30	94	104.30
Fire valve, gate, 300 lb, brass w/handwheel, 2-1/2″ pipe size	1.000	Ea.	104	59.50	163.50
Fire valve, pressure restricting, adj, rgh brs, 2-1/2″ pipe size	1.000	Ea.	214	119	333
Valve, swing check, w/ball drip, CI w/brs ftngs, 4″ pipe size	1.000	Ea.	155	250	405
Standpipe conn wall dble flush brs w/plugs & chains 2-1/2″x2-1/2″x4″	1.000	Ea.	291	150	441
Valve, swing check, bronze, 125 lb, regrinding disc, 2-1/2″ pipe size	1.000	Ea.	197	50	247
Roof manifold, fire, w/valves & caps, horiz/vert brs 2-1/2″x2-1/2″x4″	1.000	Ea.	111	156	267
Fire, hydrolator, vent & drain, 2-1/2″ pipe size	1.000	Ea.	45	34.50	79.50
Valve, gate, iron body 125 lb, OS&Y, threaded, 4″ pipe size	1.000	Ea.	380	250	630
TOTAL			2,000.80	2,082	4,082.80

8.2-310	Wet Standpipe Risers, Class I		COST PER FLOOR MAT.	INST.	TOTAL
0550	Wet standpipe risers, Class I, steel black sch. 40, 10′ height				
0560	4″ diameter pipe, one floor		2,000	2,075	4,075
0580	Additional floors	A082 -390	475	645	1,120
0600	6″ diameter pipe, one floor		3,475	3,700	7,175
0620	Additional floors	R10520 -310	850	1,050	1,900
0640	8″ diameter pipe, one floor		5,150	4,475	9,625
0660	Additional floors	R10520 -320	1,150	1,275	2,425
0680					

8.2-310	Wet Standpipe Risers, Class II	COST PER FLOOR MAT.	INST.	TOTAL
1030	Wet standpipe risers, Class II, steel black sch. 40, 10′ height			
1040	2″ diameter pipe, one floor	845	745	1,590
1060	Additional floors	278	289	567
1080	2-1/2″ diameter pipe, one floor	1,100	1,100	2,200
1100	Additional floors	310	335	645
1120				

8.2-310	Wet Standpipe Risers, Class III	COST PER FLOOR		
		MAT.	INST.	TOTAL
1530	Wet standpipe risers, Class III, steel black sch. 40, 10′ height			
1540	4″ diameter pipe, one floor	2,050	2,075	4,125
1560	Additional floors	400	540	940
1580	6″ diameter pipe, one floor	3,525	3,700	7,225
1600	Additional floors	880	1,050	1,930
1620	8″ diameter pipe, one floor	5,225	4,475	9,700
1640	Additional floors	1,175	1,275	2,450

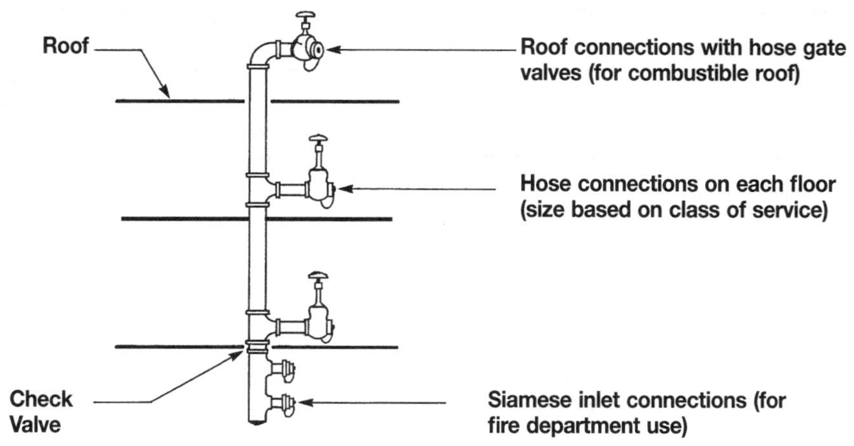

Roof

Roof connections with hose gate valves (for combustible roof)

Hose connections on each floor (size based on class of service)

Check Valve

Siamese inlet connections (for fire department use)

System Components	QUANTITY	UNIT	COST PER FLOOR MAT.	INST.	TOTAL
SYSTEM 8.2-320-0540					
DRY STANDPIPE RISER, CLASS I, PIPE, STEEL, BLACK, SCH 40, 10′ HEIGHT					
4″ DIAMETER PIPE, ONE FLOOR					
Pipe, steel, black, schedule 40, threaded, 4″ diam	20.000	L.F.	188	420	608
Pipe, Tee, malleable iron, black, 150 lb threaded, 4″ pipe size	2.000	Ea.	232	374	606
Pipe, 90° elbow, malleable iron, black, 150 lb threaded 4″ pipe size	1.000	Ea.	73.50	125	198.50
Pipe, nipple, steel, black, schedule 40, 2-1/2″ pipe size x 3″ long	2.000	Ea.	10.30	94	104.30
Fire valve gate NRS 300 lb, brass w/handwheel, 2-1/2″ pipe size	1.000	Ea.	104	59.50	163.50
Fire valve, pressure restricting, adj, rgh brs, 2-1/2″ pipe size	1.000	Ea.	107	59.50	166.50
Standpipe conn wall dble flush brs w/plugs & chains 2-1/2″x2-1/2″x4″	1.000	Ea.	291	150	441
Valve swing check w/ball drip CI w/brs ftngs, 4″pipe size	1.000	Ea.	155	250	405
Roof manifold, fire, w/valves & caps, horiz/vert brs 2-1/2″x2-1/2″x4″	1.000	Ea.	111	156	267
TOTAL			1,271.80	1,688	2,959.80

8.2-320	Dry Standpipe Risers, Class I		COST PER FLOOR MAT.	INST.	TOTAL
0530	Dry standpipe riser, Class I, steel black sch. 40, 10′ height				
0540	4″ diameter pipe, one floor		1,275	1,700	2,975
0560	Additional floors	A082 -390	430	610	1,040
0580	6″ diameter pipe, one floor		2,525	2,950	5,475
0600	Additional floors	R10520 -310	805	1,000	1,805
0620	8″ diameter pipe, one floor		3,725	3,600	7,325
0640	Additional floors	R10520 -320	1,100	1,225	2,325
0660					

8.2-320	Dry Standpipe Risers, Class II	COST PER FLOOR MAT.	INST.	TOTAL
1030	Dry standpipe risers, Class II, steel black sch. 40, 10′ height			
1040	2″ diameter pipe, one floor	690	780	1,470
1060	Additional floor	233	254	487
1080	2-1/2″ diameter pipe, one floor	825	910	1,735
1100	Additional floors	263	300	563
1120				

8.2-320	Dry Standpipe Risers, Class III	COST PER FLOOR MAT.	INST.	TOTAL
1530	Dry standpipe risers, Class III, steel black sch. 40, 10′ height			
1540	4″ diameter pipe, one floor	1,300	1,650	2,950

8.2-320	Dry Standpipe Risers, Class III	COST PER FLOOR		
		MAT.	INST.	TOTAL
1560	Additional floors	360	550	910
1580	6″ diameter pipe, one floor	2,550	2,950	5,500
1600	Additional floors	835	1,000	1,835
1620	8″ diameter pipe, one floor	3,750	3,600	7,350
1640	Additional floor	1,125	1,225	2,350

8.2-390	Standpipe Equipment	COST EACH		
		MAT.	INST.	TOTAL
0100	Adapters, reducing, 1 piece, FxM, hexagon, cast brass, 2-1/2" x 1-1/2"	30.50		30.50
0200	Pin lug, 1-1/2" x 1"	9.70		9.70
0250	3" x 2-1/2"	38		38
0300	For polished chrome, add 75% mat.			
0400	Cabinets, D.S. glass in door, recessed, steel box, not equipped			
0500	Single extinguisher, steel door & frame	62	94	156
0550	Stainless steel door & frame	109	94	203
0600	Valve, 2-1/2" angle, steel door & frame	67.50	62.50	130
0650	Aluminum door & frame	90	62.50	152.50
0700	Stainless steel door & frame	118	62.50	180.50
0750	Hose rack assy, 2-1/2" x 1-1/2" valve & 100' hose, steel door & frame	137	125	262
0800	Aluminum door & frame	203	125	328
0850	Stainless steel door & frame	264	125	389
0900	Hose rack assy,& extinguisher,2-1/2"x1-1/2" valve & hose,steel door & frame	171	150	321
0950	Aluminum	259	150	409
1000	Stainless steel	305	150	455
1550	Compressor, air, dry pipe system, automatic, 200 gal., 1/3 H.P.	770	320	1,090
1600	520 gal., 1 H.P.	805	320	1,125
1650	Alarm, electric pressure switch (circuit closer)	144	16.05	160.05
2500	Couplings, hose, rocker lug, cast brass, 1-1/2"	28.50		28.50
2550	2-1/2"	37.50		37.50
3000	Escutcheon plate, for angle valves, polished brass, 1-1/2"	11.30		11.30
3050	2-1/2"	27		27
3500	Fire pump, electric, w/controller, fittings, relief valve			
3550	4" pump, 30 H.P., 500 G.P.M.	14,900	2,350	17,250
3600	5" pump, 40 H.P., 1000 G.P.M.	21,800	2,650	24,450
3650	5" pump, 100 H.P., 1000 G.P.M.	24,400	2,950	27,350
3700	For jockey pump system, add	2,675	375	3,050
5000	Hose, per linear foot, synthetic jacket, lined,			
5100	300 lb. test, 1-1/2" diameter	1.58	.29	1.87
5150	2-1/2" diameter	2.64	.34	2.98
5200	500 lb. test, 1-1/2" diameter	1.64	.29	1.93
5250	2-1/2" diameter	2.85	.34	3.19
5500	Nozzle, plain stream, polished brass, 1-1/2" x 10"	23.50		23.50
5550	2-1/2" x 15" x 13/16" or 1-1/2"	47.50		47.50
5600	Heavy duty combination adjustable fog and straight stream w/handle 1-1/2"	273		273
5650	2-1/2" direct connection	390		390
6000	Rack, for 1-1/2" diameter hose 100 ft. long, steel	35.50	37.50	73
6050	Brass	55	37.50	92.50
6500	Reel, steel, for 50 ft. long 1-1/2" diameter hose	74	53.50	127.50
6550	For 75 ft. long 2-1/2" diameter hose	119	53.50	172.50
7050	Siamese, w/plugs & chains, polished brass, sidewalk, 4" x 2-1/2" x 2-1/2"	340	300	640
7100	6" x 2-1/2" x 2-1/2"	565	375	940
7200	Wall type, flush, 4" x 2-1/2" x 2-1/2"	291	150	441
7250	6" x 2-1/2" x 2-1/2"	395	163	558
7300	Projecting, 4" x 2-1/2" x 2-1/2"	266	150	416
7350	6" x 2-1/2" x 2-1/2"	435	163	598
7400	For chrome plate, add 15% mat.			
8000	Valves, angle, wheel handle, 300 Lb., rough brass, 1-1/2"	29	34.50	63.50
8050	2-1/2"	50.50	59.50	110
8100	Combination pressure restricting, 1-1/2"	46	34.50	80.50
8150	2-1/2"	98	59.50	157.50
8200	Pressure restricting, adjustable, satin brass, 1-1/2"	76	34.50	110.50
8250	2-1/2"	107	59.50	166.50
8300	Hydrolator, vent and drain, rough brass, 1-1/2"	45	34.50	79.50
8350	2-1/2"	45	34.50	79.50
8400	Cabinet assy, incls. adapter, rack, hose, and nozzle	435	226	661

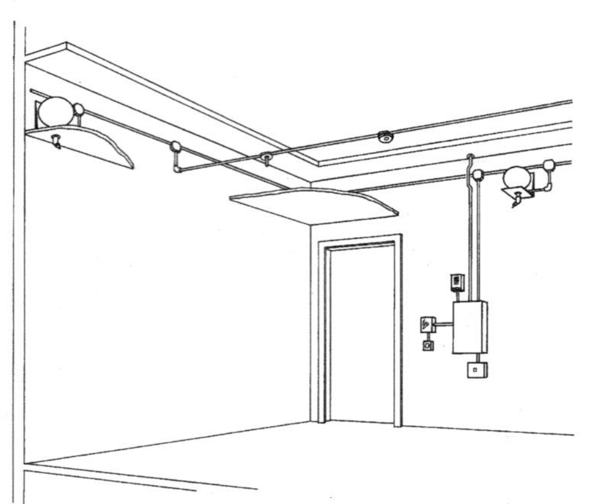

General: Automatic fire protection (suppression) systems other than water sprinklers may be desired for special environments, high risk areas, isolated locations or unusual hazards. Some typical applications would include:

Paint dip tanks
Securities vaults
Electronic data processing
Tape and data storage
Transformer rooms
Spray booths
Petroleum storage
High rack storage

Piping and wiring costs are dependent on the individual application and must be added to the component costs shown below.

All areas are assumed to be open.

8.2-810	Unit Components	COST EACH		
		MAT.	INST.	TOTAL
0020	Detectors with brackets			
0040	Fixed temperature heat detector	31	51	82
0060	Rate of temperature rise detector	37	51	88
0080	Ion detector (smoke) detector	82.50	66	148.50
0200	Extinguisher agent			
0240	200 lb FM200, container	6,375	187	6,562
0280	75 lb carbon dioxide cylinder	1,050	125	1,175
0320	Dispersion nozzle			
0340	FM200 1-1/2" dispersion nozzle	55	29.50	84.50
0380	Carbon dioxide 3" x 5" dispersion nozzle	55	23	78
0420	Control station			
0440	Single zone control station with batteries	1,400	410	1,810
0470	Multizone (4) control station with batteries	2,700	820	3,520
0490				
0500	Electric mechanical release	138	206	344
0520				
0550	Manual pull station	49.50	69.50	119
0570				
0640	Battery standby power 10" x 10" x 17"	760	102	862
0700				
0740	Bell signalling device	54.50	51	105.50

8.2-810	FM200 Systems	COST PER C.F.		
		MAT.	INST.	TOTAL
0820	Average FM200 system, minimum			1.38
0840	Maximum			2.75

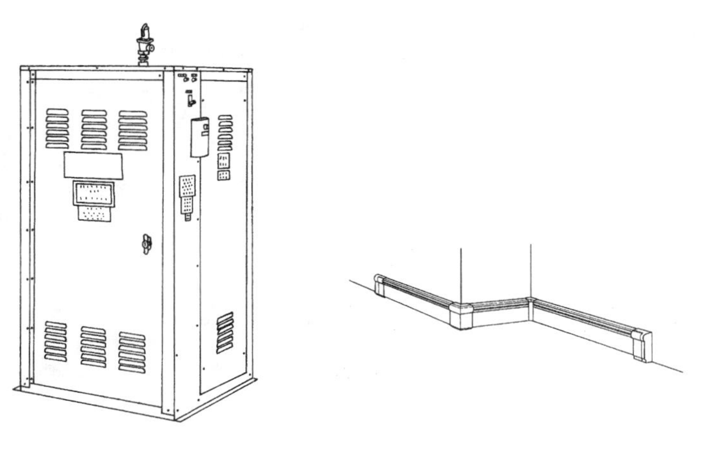

Small Electric Boiler
System Considerations:

1. Terminal units are fin tube baseboard radiation rated at 720 BTU/hr with 200° water temperature or 820 BTU/hr steam.
2. Primary use being for residential or smaller supplementary areas, the floor levels are based on 7-1/2′ ceiling heights.
3. All distribution piping is copper for boilers through 205 MBH. All piping for larger systems is steel pipe.

System Components	QUANTITY	UNIT	COST EACH		
			MAT.	INST.	TOTAL
SYSTEM 8.3-110-1120					
SMALL HEATING SYSTEM, HYDRONIC, ELECTRIC BOILER					
1,480 S.F., 61 MBH, STEAM, 1 FLOOR					
Boiler, electric steam, std cntrls, trim, ftngs and valves, 18 KW, 61.4 MBH	1.000	Ea.	10,065	1,072.50	11,137.50
Copper tubing type L, solder joint, hanger 10′OC, 1-1/4″ diam	160.000	L.F.	448	1,144	1,592
Radiation, 3/4″ copper tube w/alum fin baseboard pkg 7″ high	60.000	L.F.	354	783	1,137
Rough in baseboard panel or fin tube with valves & traps	10.000	Set	940	3,950	4,890
Pipe covering, calcium silicate w/cover 1″ wall 1-1/4″ diam	160.000	L.F.	348.80	718.40	1,067.20
Low water cut-off, quick hookup, in gage glass tappings	1.000	Ea.	161	26.50	187.50
TOTAL			12,316.80	7,694.40	20,011.20
COST PER S.F.			8.32	5.20	13.52

8.3-110	Small Heating Systems, Hydronic, Electric Boilers	COST PER S.F.		
		MAT.	INST.	TOTAL
1100	Small heating systems, hydronic, electric boilers			
1120	Steam, 1 floor, 1480 S.F., 61 M.B.H.	8.30	5.20	13.50
1160	3,000 S.F., 123 M.B.H.	4.80	4.55	9.35
1200	5,000 S.F., 205 M.B.H.	3.60	4.20	7.80
1240	2 floors, 12,400 S.F., 512 M.B.H.	2.56	4.17	6.73
1280	3 floors, 24,800 S.F., 1023 M.B.H.	2.21	4.13	6.34
1360	Hot water, 1 floor, 1,000 S.F., 41 M.B.H.	7.75	2.91	10.66
1400	2,500 S.F., 103 M.B.H.	4.60	5.20	9.80
1440	2 floors, 4,850 S.F., 205 M.B.H.	4	6.20	10.20
1480	3 floors, 9,700 S.F., 410 M.B.H.	3.71	6.45	10.16

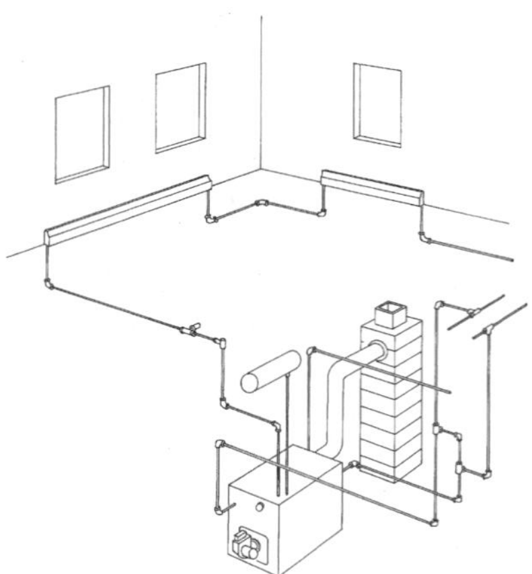

Basis for Heat Loss Estimate, Apartment Type Structures:

1. Masonry walls and flat roof are insulated. U factor is assumed at .08.
2. Window glass area taken as BOCA minimum, 1/10th of floor area. Double insulating glass with 1/4″ air space, U = .65.
3. Infiltration = 0.3 C.F. per hour per S.F. of net wall.
4. Concrete floor loss is 2 BTUH per S.F.
5. Temperature difference taken as 70° F.
6. Ventilating or makeup air has not been included and must be added if desired. Air shafts are not used.

System Components	QUANTITY	UNIT	COST EACH MAT.	COST EACH INST.	COST EACH TOTAL
SYSTEM 8.3-151-1760					
HEATING SYSTEM, FIN TUBE RADIATION, FORCED HOT WATER					
1,000 S.F. AREA, 10,000 C.F. VOLUME					
Boiler, oil fired, CI, burner, ctrls/insul/breech/pipe/ftng/valves, 109 MBH	1.000	Ea.	2,668.75	2,362.50	5,031.25
Circulating pump, CI flange connection, 1/12 HP	1.000	Ea.	222	125	347
Expansion tank, painted steel, ASME 18 Gal capacity	1.000	Ea.	375	54	429
Storage tank, steel, above ground, 275 Gal capacity w/supports	1.000	Ea.	260	152	412
Copper tubing type L, solder joint, hanger 10′ OC, 3/4″ diam	100.000	L.F.	174	545	719
Radiation, 3/4″ copper tube w/alum fin baseboard pkg, 7″ high	30.000	L.F.	177	391.50	568.50
Pipe covering, calcium silicate w/cover, 1″ wall, 3/4″ diam	100.000	L.F.	228	436	664
TOTAL			4,104.75	4,066	8,170.75
COST PER S.F.			4.10	4.07	8.17

8.3-151	Apartment Building Heating - Fin Tube Radiation	COST PER S.F. MAT.	COST PER S.F. INST.	COST PER S.F. TOTAL
1740	Heating systems, fin tube radiation, forced hot water			
1760	1,000 S.F. area, 10,000 C.F. volume	4.10	4.07	8.17
1800	10,000 S.F. area, 100,000 C.F. volume	1.21	2.44	3.65

Many styles of active solar energy systems exist. Those shown on the following page represent the majority of systems now being installed in different regions of the country.

The five active domestic hot water (DHW) systems which follow are typically specified as two or three panel systems with additional variations being the type of glazing and size of the storage tanks. Various combinations have been costed for the user's evaluation and comparison. The basic specifications from which the following systems were developed satisfy the construction detail requirements specified in the HUD Intermediate Minimum Property Standards (IMPS). If these standards are not complied with, the renewable energy system's costs could be significantly lower than shown.

To develop the system's specifications and costs it was necessary to make a number of assumptions about the systems. Certain systems are more appropriate to one climatic region than another or the systems may require modifications to be usable in particular locations. Specific instances in which the systems are not appropriate throughout the country as specified, or in which modification will be needed include the following:

- The freeze protection mechanisms provided in the DHW systems vary greatly. In harsh climates, where freeze is a major concern, a closed-loop indirect collection system may be more appropriate than a direct collection system.
- The thermosyphon water heater system described cannot be used when temperatures drop below 32° F.
- In warm climates it may be necessary to modify the systems installed to prevent overheating.

For each renewable resource (solar) system a schematic diagram and descriptive summary of the system is provided along with a list of all the components priced as part of the system. The costs were developed based on these specifications.

Considerations affecting costs which may increase or decrease beyond the estimates presented here include the following:

- Special structural qualities (allowance for earthquake, future expansion, high winds, and unusual spans or shapes);
- Isolated building site or rough terrain that would affect the transportation of personnel, material, or equipment;
- Unusual climatic conditions during the construction process;
- Substitution of other materials or system components for those used in the system specifications.

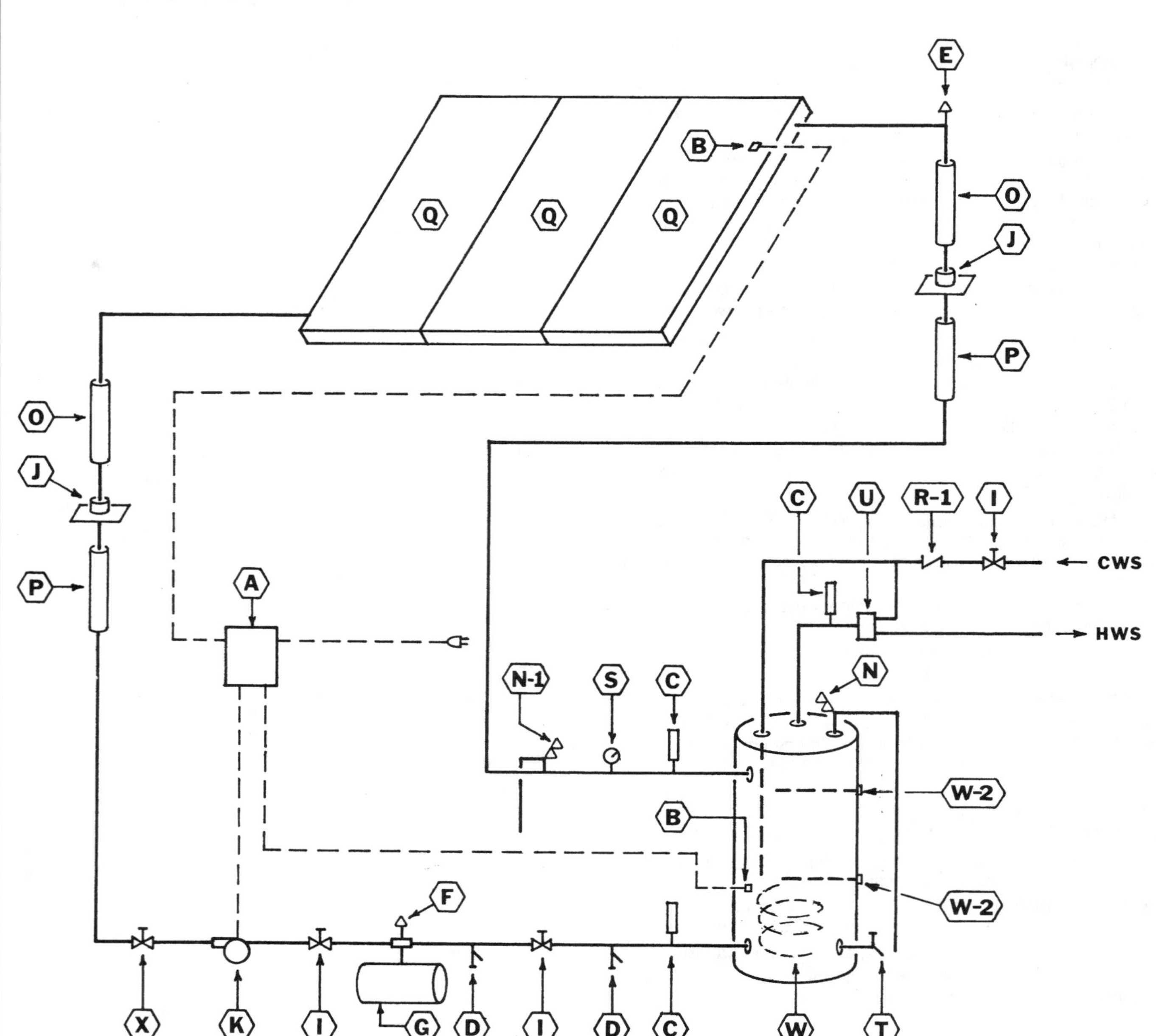

In this closed-loop indirect collection system, fluid with a low freezing temperature, such as propylene glycol, transports heat from the collectors to water storage. The transfer fluid is contained in a closed-loop consisting of collectors, supply and return piping, and a heat exchanger immersed in the storage tank. A typical two-or-three panel system contains 5 to 6 gallons of heat transfer fluid.

When the collectors become approximately 20° F warmer than the storage temperature, a controller activates the circulator. The circulator moves the fluid continuously through the collectors until the temperature difference between the collectors and storage is such that heat collection no longer occurs; at that point, the circulator shuts off. Since the heat transfer fluid has a very low freezing temperature, there is no need for it to be drained from the collectors between periods of collection.

System Components

System Components		QUANTITY	UNIT	COST EACH MAT.	COST EACH INST.	COST EACH TOTAL
	SYSTEM 8.3-610-2760					
	SOLAR, CLOSED LOOP, HOT WATER SYSTEM, IMMERSED HEAT EXCHANGER					
	3/4" TUBING, THREE 3' X 7' BLACK CHROME COLLECTORS					
A, B	Differential controller, 2 sensors, thermostat, solar energy system	1.000	Ea.	85.50	34.50	120
C	Thermometer 2" dial	3.000	Ea.	63	78	141
D, T	Fill & drain valves, brass, 3/4" connection	3.000	Ea.	16.35	52.05	68.40
E	Air vent, manual, 1/8" fitting	1.000	Ea.	2.13	13	15.13
F	Air purger	1.000	Ea.	45	34.50	79.50
G	Expansion tank	1.000	Ea.	64	13	77
I	Valve, gate, bronze, NRS, soldered 3/4" diam	3.000	Ea.	55.65	63	118.65
J	Neoprene vent flashing	2.000	Ea.	17.30	42	59.30
K	Circulator, solar heated liquid, 1/25 HP	1.000	Ea.	153	53.50	206.50
N-1, N	Relief valve, temp & press 150 psi 210°F self-closing 3/4" IPS	2.000	Ea.	17.30	27.70	45
O	Pipe covering, urethane ultraviolet cover, 1" wall, 3/4" diam	20.000	L.F.	30.80	80.80	111.60
P	Pipe covering, fiberglass, all service jacket, 1" wall, 3/4" diam	50.000	L.F.	66.50	161	227.50
	Roof clamps for solar energy collector panel	3.000	Set	5.31	32.10	37.41
Q	Collector panel solar blk chrome on copper, 1/8" temp glass, 3'x7'	3.000	Ea.	1,800	237	2,037
R-1	Valve, swing check, bronze, regrinding disc, 3/4" diam, soldered	1.000	Ea.	33	21	54
S	Pressure gauge, 60 psi, 2-1/2" dial	1.000	Ea.	23	13	36
U	Valve, water tempering, bronze, sweat connections, 3/4" diam	1.000	Ea.	54	21	75
W-2, W	Tank, water storage immersed heat exchr elec elem 2"x2# insul 120 Gal	1.000	Ea.	925	297	1,222
X	Valve, globe, bronze, rising stem, 3/4" diam, soldered	1.000	Ea.	47	21	68
	Copper tubing type L, solder joint, hanger 10' OC 3/4" diam	20.000	L.F.	34.80	109	143.80
	Copper tubing, type M, solder joint, hanger 10' OC 3/4" diam	70.000	L.F.	102.90	374.50	477.40
	Sensor wire, #22-2 conductor multistranded	.500	C.L.F.	4.13	20.50	24.63
	Solar energy heat transfer fluid, propylene glycol, anti-freeze	6.000	Gal.	58.20	89.10	147.30
	Wrought copper fittings & solder, 3/4" diam	76.000	Ea.	40.28	1,672	1,712.28
	TOTAL			3,744.15	3,560.25	7,304.40

8.3-610	Solar, Closed Loop, Hot Water Systems		COST EACH MAT.	COST EACH INST.	COST EACH TOTAL
2550	Solar, closed loop, hot water system, immersed heat exchanger				
2560	3/8" tubing, 3 ea. 4' x 4'-4" vacuum tube collectors, 80 gal. tank		4,550	3,225	7,775
2580	1/2" tubing, 4 ea. 4' x 4'-4" vacuum tube collectors, 80 gal. tank		5,450	3,500	8,950
2600	120 gal. tank		5,550	3,525	9,075
2640	2 ea. 3'x7' black chrome collectors, 80 gal. tank	R13600 -610	2,925	3,325	6,250
2660	120 gal. tank		3,050	3,325	6,375
2700	2 ea. 3'x7' flat black collectors, 120 gal. tank		2,800	3,350	6,150
2720	3 ea. 3'x7' flat black collectors, 120 gal. tank		3,275	3,425	6,700
2760	3/4" tubing, 3 ea. 3'x7' black chrome collectors, 120 gal. tank		3,750	3,550	7,300
2780	3 ea. 3'x7' flat black collectors, 120 gal. tank		3,375	3,575	6,950
2800	2 ea. 4'x9' flat black w/plastic glazing collectors 120 gal. tank		3,025	3,575	6,600
2840	1" tubing, 4 ea. 2'x9' plastic absorber & glazing collectors 120 gal. tank		4,275	4,000	8,275
2860	4 ea. 3'x7' black chrome collectors, 120 gal. tank		4,500	4,000	8,500
2880	4 ea. 3'x7' flat black absorber collectors, 120 gal. tank		4,000	4,025	8,025

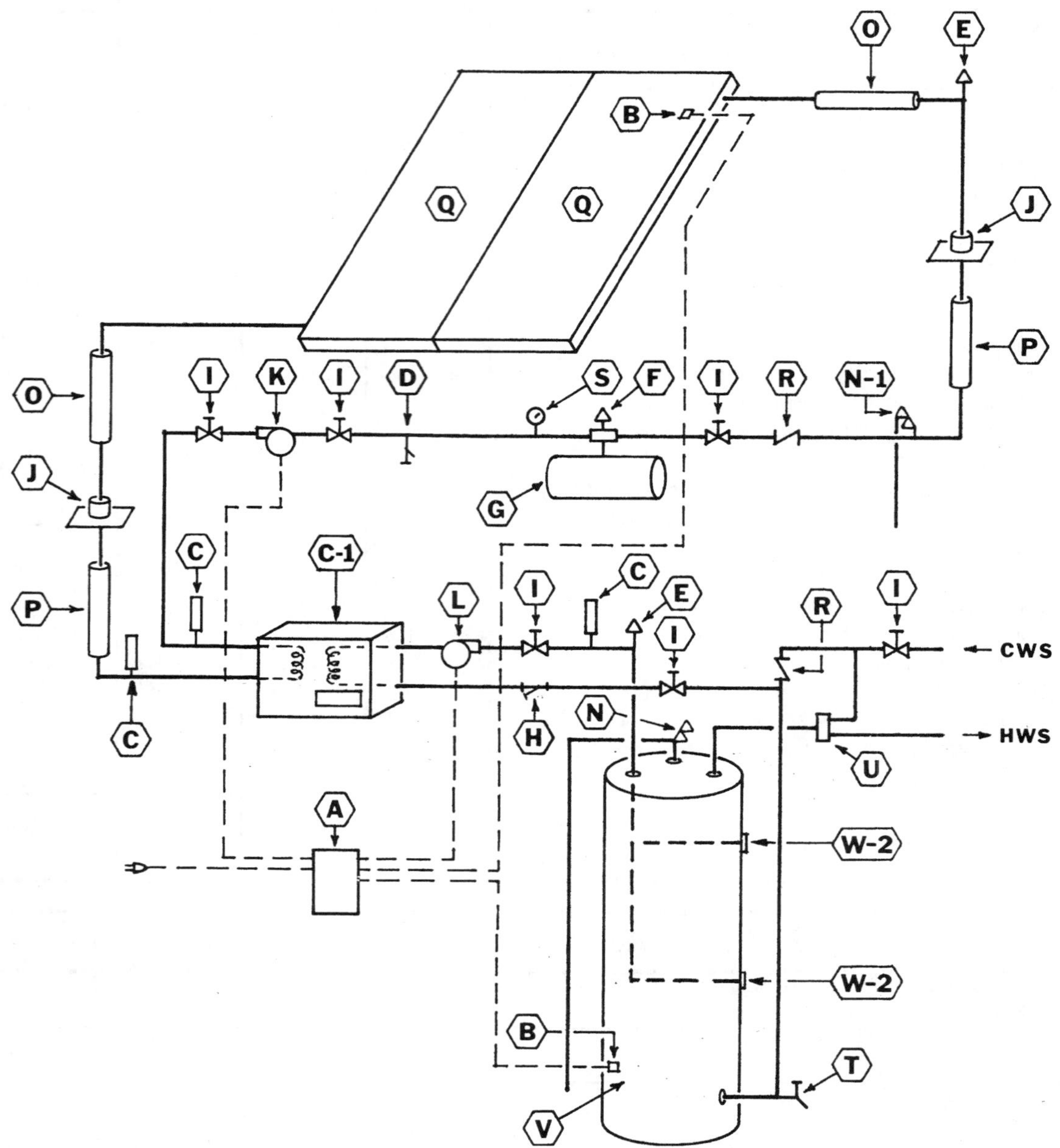

In this closed-loop indirect collection system, fluid with a low freezing temperature, such as propylene glycol, transports heat from the collectors to water storage. The transfer fluid is contained in a closed-loop consisting of collectors, supply and return piping, and a remote heat exchanger. The heat exchanger transfers heat energy from the fluid in the collector loop to potable water circulated in a storage loop. A typical two-or-three panel system contains 5 to 6 gallons of heat transfer fluid.

When the collectors become approximately 20° F warmer than the storage temperature, a controller activates the circulator on the collector and storage loops. The circulators will move the fluid and potable water through the heat exchanger until heat collection no longer occurs. At that point, the system shuts down. Since the heat transfer medium is a fluid with a very low freezing temperature, there is no need for it to be drained from the system between periods of collection.

Important: See the Reference Section for critical supporting data - Reference Nos., Crews, & City Cost Indexes

System Components		QUANTITY	UNIT	COST EACH MAT.	INST.	TOTAL
	SYSTEM 8.3-620-2760					
	SOLAR, CLOSED LOOP, ADD-ON HOT WATER SYS., EXTERNAL HEAT EXCHANGER					
	3/4" TUBING, TWO 3'X7' BLACK CHROME COLLECTORS					
A,B,G,L,K,M	Heat exchanger fluid-fluid pkg incl 2 circulators, expansion tank, Check valve, relief valve, controller, hi temp cutoff, & 2 sensors	1.000	Ea.	765	300	1,065
C	Thermometer, 2" dial	3.000	Ea.	63	78	141
D, T	Fill & drain valve, brass, 3/4" connection	1.000	Ea.	5.45	17.35	22.80
E	Air vent, manual, 1/8" fitting	2.000	Ea.	4.26	26	30.26
F	Air purger	1.000	Ea.	45	34.50	79.50
H	Strainer, Y type, bronze body, 3/4" IPS	1.000	Ea.	19.15	22	41.15
I	Valve, gate, bronze, NRS, soldered 3/4" diam	6.000	Ea.	111.30	126	237.30
J	Neoprene vent flashing	2.000	Ea.	17.30	42	59.30
N-1, N	Relief valve temp & press, 150 psi 210°F self-closing 3/4" IPS	1.000	Ea.	8.65	13.85	22.50
O	Pipe covering, urethane, ultraviolet cover, 1" wall 3/4" diam	20.000	L.F.	30.80	80.80	111.60
P	Pipe covering, fiberglass, all service jacket, 1" wall, 3/4" diam	50.000	L.F.	66.50	161	227.50
Q	Collector panel solar energy blk chrome on copper, 1/8" temp glass 3'x7'	2.000	Ea.	1,200	158	1,358
	Roof clamps for solar energy collector panels	2.000	Set	3.54	21.40	24.94
R	Valve, swing check, bronze, regrinding disc, 3/4" diam	2.000	Ea.	66	42	108
S	Pressure gauge, 60 psi, 2" dial	1.000	Ea.	23	13	36
U	Valve, water tempering, bronze, sweat connections, 3/4" diam	1.000	Ea.	54	21	75
W-2, V	Tank water storage w/heating element, drain, relief valve, existing	1.000	Ea.			
	Copper tubing type L, solder joint, hanger 10' OC 3/4" diam	20.000	L.F.	34.80	109	143.80
	Copper tubing, type M, solder joint, hanger 10' OC 3/4" diam	70.000	L.F.	102.90	374.50	477.40
	Sensor wire, #22-2 conductor multistranded	.500	C.L.F.	4.13	20.50	24.63
	Solar energy heat transfer fluid, propylene glycol anti-freeze	6.000	Gal.	58.20	89.10	147.30
	Wrought copper fittings & solder, 3/4" diam	76.000	Ea.	40.28	1,672	1,712.28
	Total			2,723.26	3,422	6,145.26

8.3-620	Solar, Closed Loop, Add-On Hot Water Systems		COST EACH MAT.	INST.	TOTAL
2550	Solar, closed loop, add-on hot water system, external heat exchanger				
2570	3/8" tubing, 3 ea. 4' x 4'-4" vacuum tube collectors		4,250	3,200	7,450
2580	1/2" tubing, 4 ea 4 x 4'-4" vacuum tube collectors	R13600 -610	5,150	3,475	8,625
2600	2 ea 3'x7' black chrome collectors		2,625	3,275	5,900
2620	3 ea 3'x7' black chrome collectors		3,225	3,350	6,575
2640	2 ea 3'x7' flat black collectors		2,375	3,275	5,650
2660	3 ea 3'x7' flat black collectors		2,850	3,375	6,225
2700	3/4" tubing, 3 ea 3'x7' black chrome collectors		3,325	3,500	6,825
2720	3 ea 3'x7' flat black absorber plate collectors		2,950	3,525	6,475
2740	2 ea 4'x9' flat black w/plastic glazing collectors		2,600	3,525	6,125
2760	2 ea 3'x7' black chrome collectors		2,725	3,425	6,150
2780	1" tubing,4 ea 2'x9' plastic absorber & glazing collectors		3,875	3,975	7,850
2800	4 ea 3'x7' black chrome absorber collectors		4,100	3,975	8,075
2820	4 ea 3'x7' flat black absorber collectors		3,600	4,000	7,600

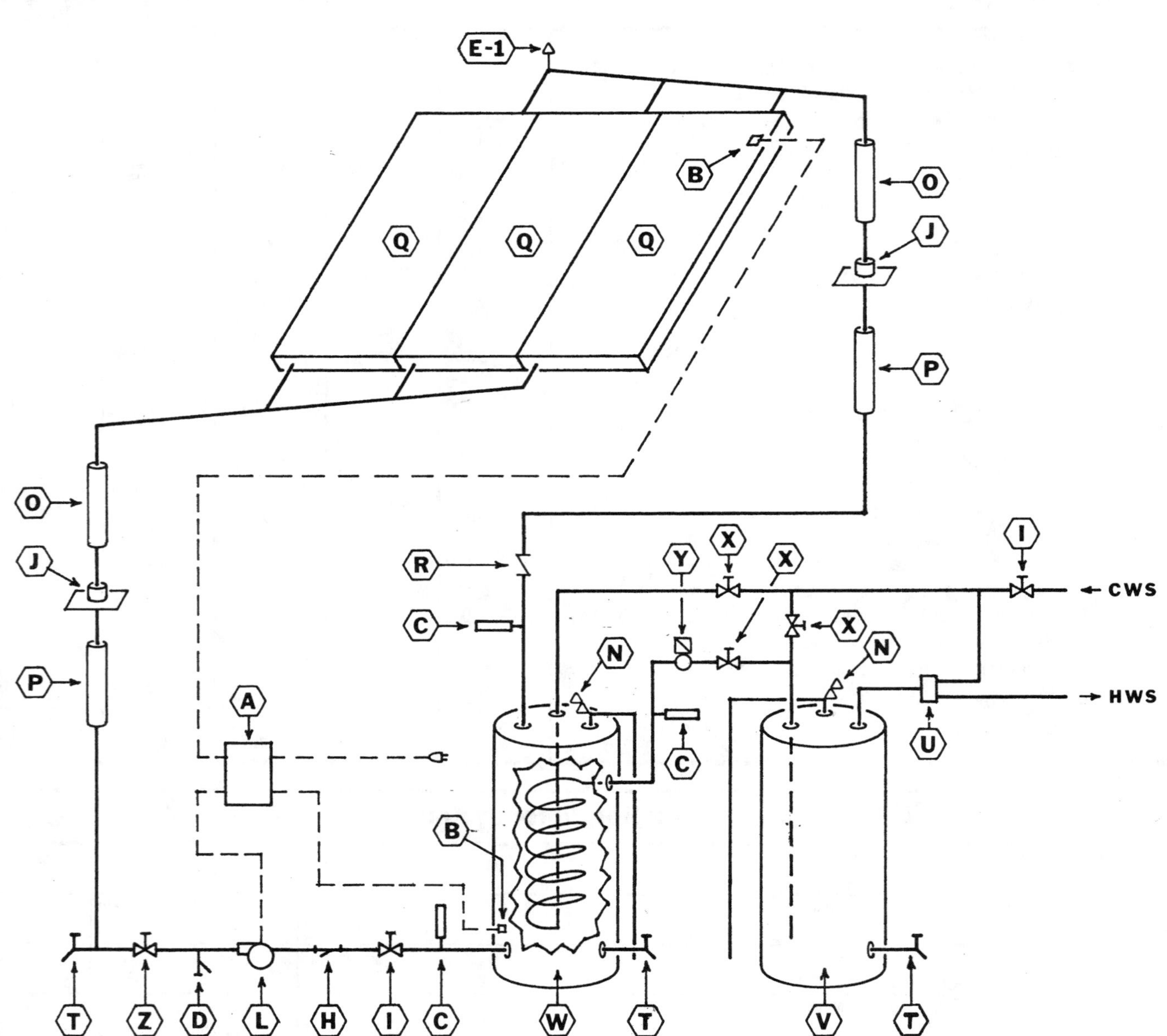

In the drainback indirect-collection system, the heat transfer fluid is distilled water contained in a loop consisting of collectors, supply and return piping, and an unpressurized holding tank. A large heat exchanger containing incoming potable water is immersed in the holding tank. When a controller activates solar collection, the distilled water is pumped through the collectors and heated and pumped back down to the holding tank. When the temperature differential between the water in the collectors and water in storage is such that collection no longer occurs, the pump turns off and gravity causes the distilled water in the collector loop to drain back to the holding tank. All the loop piping is pitched so that the water can drain out of the collectors and piping and not freeze there. As hot water is needed in the home, incoming water first flows through the holding tank with the immersed heat exchanger and is warmed and then flows through a conventional heater for any supplemental heating that is necessary.

A8.3-630 Drain Back, Immersed Exch.

System Components

System Components		QUANTITY	UNIT	COST EACH MAT.	COST EACH INST.	COST EACH TOTAL
SYSTEM 8.3-630-2760						
SOLAR, DRAINBACK, ADD ON, HOT WATER, IMMERSED HEAT EXCHANGER						
3/4″ TUBING, THREE EA 3′X7′ BLACK CHROME COLLECTOR						
A, B	Differential controller 2 sensors, thermostat, solar energy system	1.000	Ea.	85.50	34.50	120
C	Thermometer 2″ dial	3.000	Ea.	63	78	141
D, T	Fill & drain valve, brass, 3/4″ connection	1.000	Ea.	5.45	17.35	22.80
E-1	Automatic air vent 1/8″ fitting	1.000	Ea.	9.70	13	22.70
H	Strainer, Y type, bronze body, 3/4″ IPS	1.000	Ea.	19.15	22	41.15
I	Valve, gate, bronze, NRS, soldered 3/4″ diam	2.000	Ea.	37.10	42	79.10
J	Neoprene vent flashing	2.000	Ea.	17.30	42	59.30
L	Circulator, solar heated liquid, 1/20 HP	1.000	Ea.	161	62.50	223.50
N	Relief valve temp. & press. 150 psi 210°F self-closing 3/4″ IPS	1.000	Ea.	8.65	13.85	22.50
O	Pipe covering, urethane, ultraviolet cover, 1″ wall, 3/4″ diam	20.000	L.F.	30.80	80.80	111.60
P	Pipe covering, fiberglass, all service jacket, 1″ wall, 3/4″ diam	50.000	L.F.	66.50	161	227.50
Q	Collector panel solar energy blk chrome on copper, 1/8″ temp glas 3′x7′	3.000	Ea.	1,800	237	2,037
	Roof clamps for solar energy collector panels	3.000	Set	5.31	32.10	37.41
R	Valve, swing check, bronze, regrinding disc, 3/4″ diam	1.000	Ea.	33	21	54
U	Valve, water tempering, bronze sweat connections, 3/4″ diam	1.000	Ea.	54	21	75
W	Tank, water storage immersed heat exchr elec 2″x1/2# insul 120 gal	1.000	Ea.	925	297	1,222
V	Tank, water storage w/heating element, drain, relief valve, existing	1.000	Ea.			
X	Valve, globe, bronze, rising stem, 3/4″ diam, soldered	3.000	Ea.	141	63	204
Y	Flow control valve	1.000	Ea.	54.50	18.90	73.40
Z	Valve, ball, bronze, solder 3/4″ diam, solar loop flow control	1.000	Ea.	11.65	21	32.65
	Copper tubing, type L, solder joint, hanger 10′ OC 3/4″ diam	20.000	L.F.	34.80	109	143.80
	Copper tubing, type M, solder joint, hanger 10′ OC 3/4″ diam	70.000	L.F.	102.90	374.50	477.40
	Sensor wire, #22-2 conductor, multistranded	.500	C.L.F.	4.13	20.50	24.63
	Wrought copper fittings & solder, 3/4″ diam	76.000	Ea.	40.28	1,672	1,712.28
	Total			3,710.72	3,454	7,164.72

8.3-630	Solar, Drainback, Hot Water Systems	COST EACH MAT.	COST EACH INST.	COST EACH TOTAL
2550	Solar, drainback, hot water, immersed heat exchanger			
2560	3/8″ tubing, 3 ea. 4′ x 4′-4″ vacuum tube collectors	4,500	3,100	7,600
2580	1/2″ tubing, 4 ea 4′ x 4′-4″ vacuum tube collectors, 80 gal tank R13600-610	5,400	3,375	8,775
2600	120 gal tank	5,500	3,425	8,925
2640	2 ea. 3′x7′ blk chrome collectors, 80 gal tank	2,875	3,175	6,050
2660	3 ea. 3′x7′ blk chrome collectors, 120 gal tank	3,575	3,300	6,875
2700	2 ea. 3′x7′ flat blk collectors, 120 gal tank	2,725	3,225	5,950
2720	3 ea. 3′x7′ flat blk collectors, 120 gal tank	3,200	3,325	6,525
2760	3/4″ tubing, 3 ea 3′x7′ black chrome collectors, 120 gal tank	3,700	3,450	7,150
2780	3 ea. 3′x7′ flat black absorber collectors, 120 gal tank	3,325	3,475	6,800
2800	2 ea. 4′x9′ flat blk w/plastic glazing collectors 120 gal tank	3,000	3,475	6,475
2840	1″ tubing, 4 ea. 2′x9′ plastic absorber & glazing collectors, 120 gal tank	4,300	3,900	8,200
2860	4 ea. 3′x7′ black chrome absorber collectors, 120 gal tank	4,525	3,925	8,450
2880	4 ea. 3′x7′ flat black absorber collectors, 120 gal tank	4,025	3,950	7,975

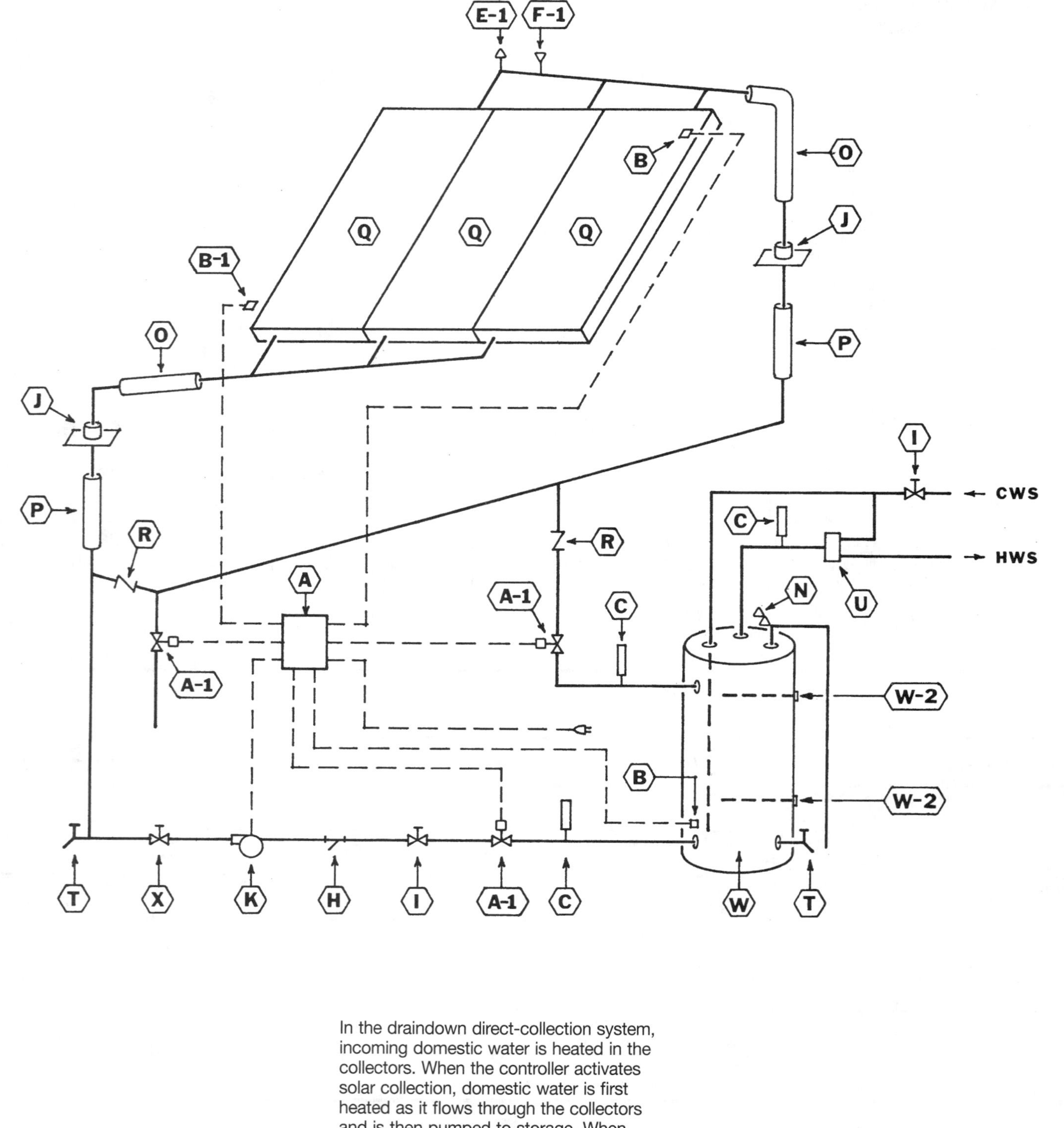

In the draindown direct-collection system, incoming domestic water is heated in the collectors. When the controller activates solar collection, domestic water is first heated as it flows through the collectors and is then pumped to storage. When conditions are no longer suitable for heat collection, the pump shuts off and the water in the loop drains down and out of the system by means of solenoid valves and properly pitched piping.

System Components

	System Components	QUANTITY	UNIT	COST EACH MAT.	COST EACH INST.	COST EACH TOTAL
	SYSTEM 8.3-640-2760					
	SOLAR, DRAINDOWN, HOT WATER, DIRECT COLLECTION					
	3/4" TUBING, THREE 3'X7' BLACK CHROME COLLECTORS					
A, B	Differential controller, 2 sensors, thermostat, solar energy system	1.000	Ea.	85.50	34.50	120
A-1	Solenoid valve, solar heating loop, brass, 3/4" diam, 24 volts	3.000	Ea.	699	138	837
B-1	Solar energy sensor, freeze prevention	1.000	Ea.	22	13	35
C	Thermometer, 2" dial	3.000	Ea.	63	78	141
E-1	Vacuum relief valve, 3/4" diam	1.000	Ea.	27.50	13	40.50
F-1	Air vent, automatic, 1/8" fitting	1.000	Ea.	9.70	13	22.70
H	Strainer, Y type, bronze body, 3/4" IPS	1.000	Ea.	19.15	22	41.15
I	Valve, gate, bronze, NRS, soldered, 3/4" diam	2.000	Ea.	37.10	42	79.10
J	Vent flashing neoprene	2.000	Ea.	17.30	42	59.30
K	Circulator, solar heated liquid, 1/25 HP	1.000	Ea.	153	53.50	206.50
N	Relief valve temp & press 150 psi 210°F self-closing 3/4" IPS	1.000	Ea.	8.65	13.85	22.50
O	Pipe covering, urethane, ultraviolet cover, 1" wall, 3/4" diam	20.000	L.F.	30.80	80.80	111.60
P	Pipe covering, fiberglass, all service jacket, 1" wall, 3/4" diam	50.000	L.F.	66.50	161	227.50
	Roof clamps for solar energy collector panels	3.000	Set	5.31	32.10	37.41
Q	Collector panel solar energy blk chrome on copper, 1/8" temp glass 3'x7'	3.000	Ea.	1,800	237	2,037
R	Valve, swing check, bronze, regrinding disc, 3/4" diam, soldered	2.000	Ea.	66	42	108
T	Drain valve, brass, 3/4" connection	2.000	Ea.	10.90	34.70	45.60
U	Valve, water tempering, bronze, sweat connections, 3/4" diam	1.000	Ea.	54	21	75
W-2, W	Tank, water storage elec elem 2"x1/2# insul 120 gal	1.000	Ea.	925	297	1,222
X	Valve, globe, bronze, rising stem, 3/4" diam, soldered	1.000	Ea.	47	21	68
	Copper tubing, type L, solder joints, hangers 10' OC 3/4" diam	20.000	L.F.	34.80	109	143.80
	Copper tubing, type M, solder joints, hangers 10' OC 3/4" diam	70.000	L.F.	102.90	374.50	477.40
	Sensor wire, #22-2 conductor, multistranded	.500	C.L.F.	4.13	20.50	24.63
	Wrought copper fittings & solder, 3/4" diam	76.000	Ea.	40.28	1,672	1,712.28
	Total			4,329.52	3,565.45	7,894.97

8.3-640	Solar, Draindown, Hot Water Systems	COST EACH MAT.	COST EACH INST.	COST EACH TOTAL
2550	Solar, draindown, hot water			
2560	3/8" tubing, 3 ea. 4' x 4'-4" vacuum tube collectors, 80 gal tank	5,100	3,400	8,500
2580	1/2" tubing, 4 ea. 4' x 4'-4" vacuum tube collectors, 80 gal tank (R13600-610)	6,025	3,500	9,525
2600	120 gal tank	6,150	3,525	9,675
2640	2 ea 3'x7' black chrome collectors, 80 gal tank	3,500	3,300	6,800
2660	3 ea 3'x7' black chrome collectors, 120 gal tank	4,225	3,425	7,650
2700	2 ea 3'x7' flat black collectors, 120 gal tank	3,375	3,350	6,725
2720	3 ea 3'x7' flat black collectors, 120 gal tank	3,825	3,425	7,250
2760	3/4" tubing, 3 ea 3'x7' black chrome collectors, 120 gal tank	4,325	3,575	7,900
2780	3 ea 3'x7' flat collectors, 120 gal tank	3,950	3,575	7,525
2800	2 ea. 4'x9' flat black & plastic glazing collectors, 120 gal tank	3,600	3,600	7,200
2840	1" tubing, 4 ea. 2'x9' plastic absorber & glazing collectors, 120 gal tank	4,875	4,025	8,900
2860	4 ea 3'x7' black chrome absorber collectors, 120 gal tank	5,100	4,025	9,125
2880	4 ea 3'x7' flat black absorber collectors, 120 gal tank	4,600	4,050	8,650

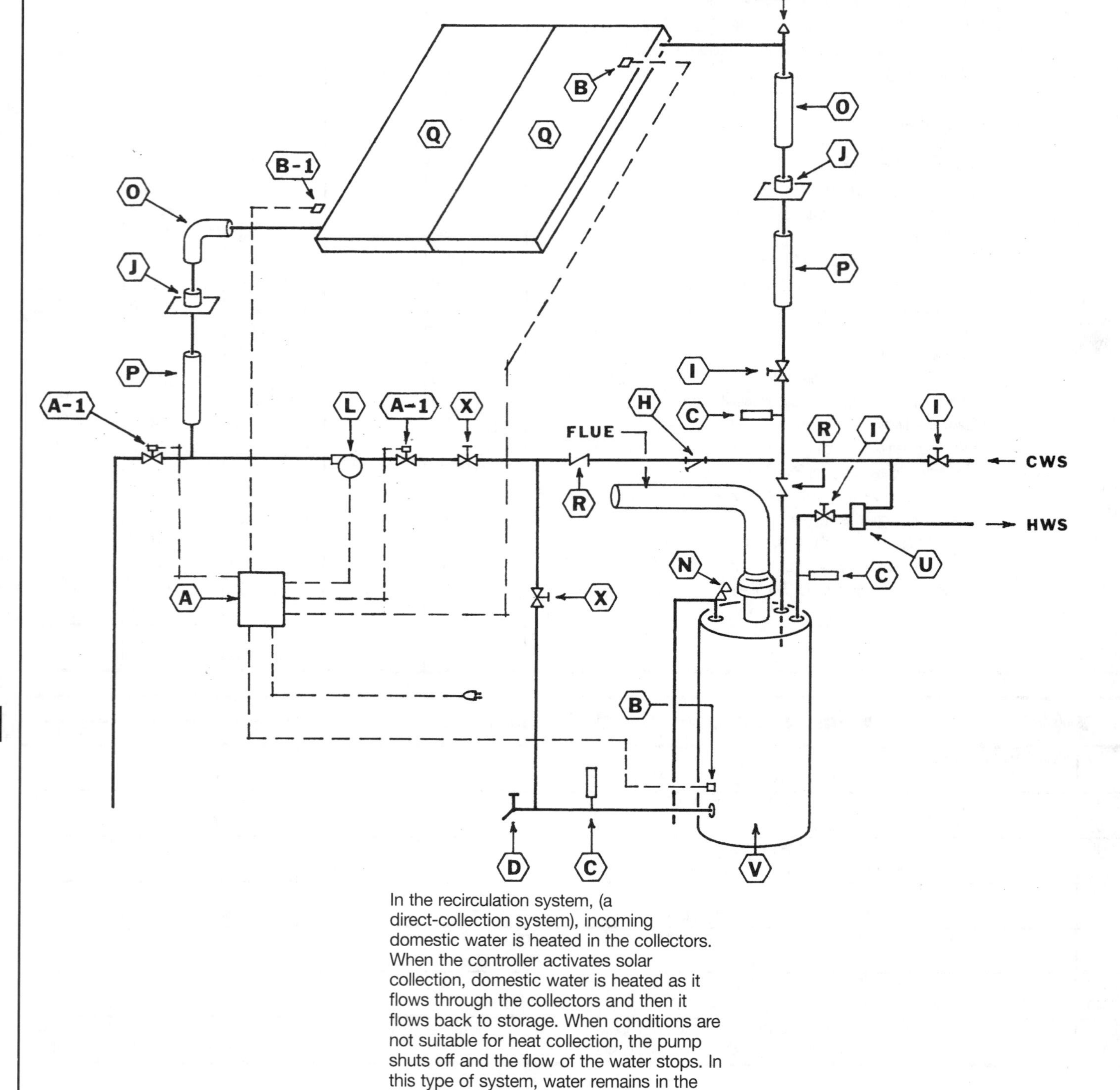

In the recirculation system, (a direct-collection system), incoming domestic water is heated in the collectors. When the controller activates solar collection, domestic water is heated as it flows through the collectors and then it flows back to storage. When conditions are not suitable for heat collection, the pump shuts off and the flow of the water stops. In this type of system, water remains in the collector loop at all times. A "frost sensor" at the collector activates the circulation of warm water from storage through the collectors when protection from freezing is required.

System Components

System Components	QUANTITY	UNIT	COST EACH MAT.	COST EACH INST.	COST EACH TOTAL
SYSTEM 8.3-650-2820					
SOLAR, RECIRCULATION, HOT WATER					
3/4″ TUBING, TWO 3′X7′ BLACK CHROME COLLECTORS					
A, B Differential controller 2 sensors, thermostat, for solar energy system	1.000	Ea.	85.50	34.50	120
A-1 Solenoid valve, solar heating loop, brass, 3/4″ IPS, 24 volts	2.000	Ea.	466	92	558
B-1 Solar energy sensor freeze prevention	1.000	Ea.	22	13	35
C Thermometer, 2″ dial	3.000	Ea.	63	78	141
D Drain valve, brass, 3/4″ connection	1.000	Ea.	5.45	17.35	22.80
F-1 Air vent, automatic, 1/8″ fitting	1.000	Ea.	9.70	13	22.70
H Strainer, Y type, bronze body, 3/4″ IPS	1.000	Ea.	19.15	22	41.15
I Valve, gate, bronze, 125 lb, soldered 3/4″ diam	3.000	Ea.	55.65	63	118.65
J Vent flashing, neoprene	2.000	Ea.	17.30	42	59.30
L Circulator, solar heated liquid, 1/20 HP	1.000	Ea.	161	62.50	223.50
N Relief valve, temp & press 150 psi 210°F self-closing 3/4″ IPS	2.000	Ea.	17.30	27.70	45
O Pipe covering, urethane, ultraviolet cover, 1″ wall, 3/4″ diam	20.000	L.F.	30.80	80.80	111.60
P Pipe covering, fiberglass, all service jacket, 1″ wall 3/4″ diam	50.000	L.F.	66.50	161	227.50
Q Collector panel solar energy blk chrome on copper, 1/8″ temp glass 3′x7′	2.000	Ea.	1,200	158	1,358
Roof clamps for solar energy collector panels	2.000	Set	3.54	21.40	24.94
R Valve, swing check, bronze, 125 lb, regrinding disc, soldered 3/4″ diam	2.000	Ea.	66	42	108
U Valve, water tempering, bronze, sweat connections, 3/4″ diam	1.000	Ea.	54	21	75
V Tank, water storage, w/heating element, drain, relief valve, existing	1.000	Ea.			
X Valve, globe, bronze, 125 lb, 3/4″ diam	2.000	Ea.	94	42	136
Copper tubing, type L, solder joints, hangers 10′ OC 3/4″ diam	20.000	L.F.	34.80	109	143.80
Copper tubing, type M, solder joints, hangers 10′ OC 3/4″ diam	70.000	L.F.	102.90	374.50	477.40
Wrought copper fittings & solder, 3/4″ diam	76.000	Ea.	40.28	1,672	1,712.28
Sensor wire, #22-2 conductor, multistranded	.500	C.L.F.	4.13	20.50	24.63
Total			2,619	3,167.25	5,786.25

8.3-650 Solar Recirculation, Domestic Hot Water Systems

8.3-650	Solar Recirculation, Domestic Hot Water Systems	COST EACH MAT.	COST EACH INST.	COST EACH TOTAL
2550	Solar recirculation, hot water			
2560	3/8″ tubing, 3 ea. 4′ x 4′-4″ vacuum tube collectors	4,125	2,950	7,075
2580	1/2″ tubing, 4 ea. 4′ x 4′-4″ vacuum tube collectors (R13600 -610)	5,025	3,225	8,250
2640	2 ea. 3′x7′ black chrome collectors	2,500	3,025	5,525
2660	3 ea. 3′x7′ black chrome collectors	3,100	3,100	6,200
2700	2 ea. 3′x7′ flat black collectors	2,250	3,025	5,275
2720	3 ea. 3′x7′ flat black collectors	2,725	3,125	5,850
2760	3/4″ tubing, 3 ea. 3′x7′ black chrome collectors	3,225	3,250	6,475
2780	3 ea. 3′x7′ flat black absorber plate collectors	2,850	3,275	6,125
2800	2 ea. 4′x9′ flat black w/plastic glazing collectors	2,500	3,275	5,775
2820	2 ea. 3′x7′ black chrome collectors	2,625	3,175	5,800
2840	1″ tubing, 4 ea. 2′x9′ black plastic absorber & glazing collectors	3,800	3,700	7,500
2860	4 ea. 3′x7′ black chrome absorber collectors	4,025	3,725	7,750
2880	4 ea. 3′x7′ flat black absorber collectors	3,525	3,750	7,275

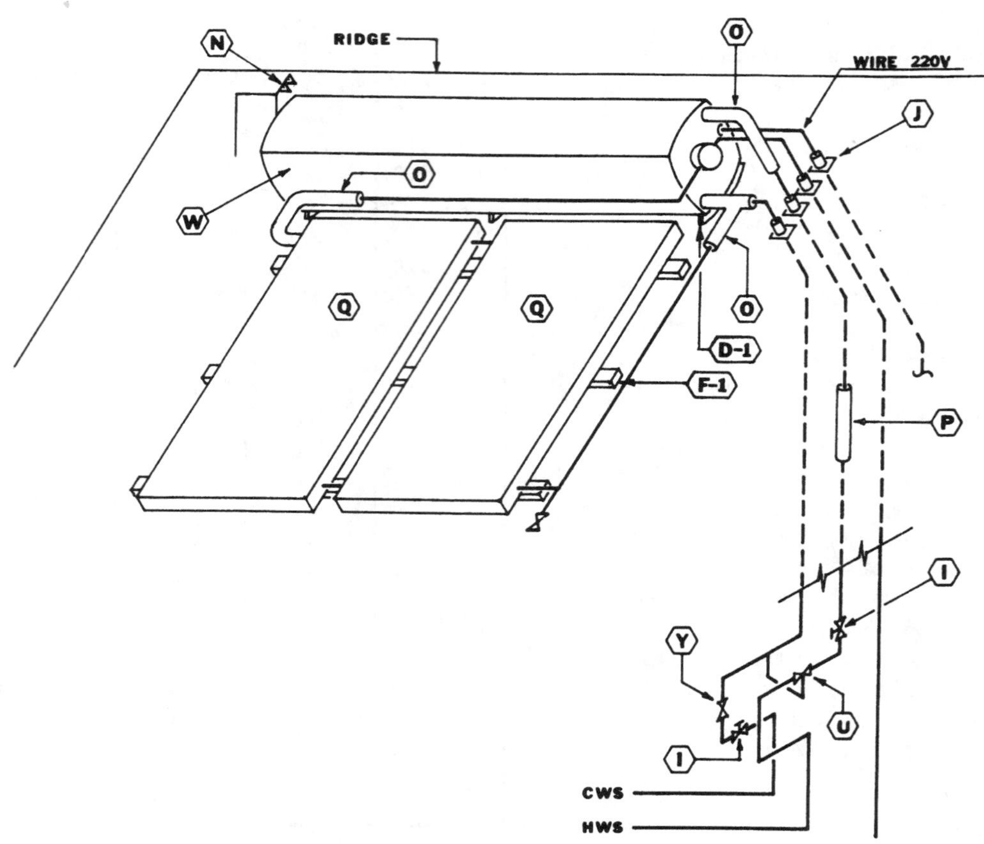

The thermosyphon domestic hot water system, a direct collection system, operates under city water pressure and does not require pumps for system operation. An insulated water storage tank is located above the collectors. As the sun heats the collectors, warm water in them rises by means of natural convection; the colder water in the storage tank flows into the collectors by means of gravity. As long as the sun is shining the water continues to flow through the collectors and to become warmer.

To prevent freezing, the system must be drained or the collectors covered with an insulated lid when the temperature drops below 32°F.

A8.3-660 Thermosyphon, Hot Water

System Components

System Components	QUANTITY	UNIT	COST EACH MAT.	COST EACH INST.	COST EACH TOTAL
SYSTEM 8.3-660-0940					
SOLAR, THERMOSYPHON, WATER HEATER					
3/4" TUBING, TWO 3'X7' BLACK CHROME COLLECTORS					
D-1 Framing lumber, fir, 2" x 6" x 8', tank cradle	.008	M.B.F.	5.20	4.68	9.88
F-1 Framing lumber, fir, 2" x 4" x 24', sleepers	.016	M.B.F.	10.56	14.08	24.64
I Valve, gate, bronze, 125 lb, soldered 1/2" diam	2.000	Ea.	32.50	34.70	67.20
J Vent flashing, neoprene	4.000	Ea.	34.60	84	118.60
O Pipe covering, urethane, ultraviolet cover, 1" wall, 1/2" diam	40.000	L.F.	47.20	158.40	205.60
P Pipe covering fiberglass all service jacket 1" wall 1/2" diam	160.000	L.F.	187.20	494.40	681.60
Q Collector panel solar, blk chrome on copper, 3/16" temp glass 3'-8" x 6'	2.000	Ea.	950	166	1,116
Y Flow control valve, globe, bronze, 125#, soldered, 1/2" diam	1.000	Ea.	34.50	17.35	51.85
U Valve, water tempering, bronze, sweat connections, 1/2" diam	1.000	Ea.	44	17.35	61.35
W Tank, water storage, solar energy system, 80 Gal, 2" x 1/2 lb insul	1.000	Ea.	820	260	1,080
Copper tubing type L, solder joints, hangers 10' OC 1/2" diam	150.000	L.F.	207	772.50	979.50
Copper tubing type M, solder joints, hangers 10' OC 1/2" diam	50.000	L.F.	58	247.50	305.50
Sensor wire, #22-2 gauge multistranded	.500	C.L.F.	4.13	20.50	24.63
Wrought copper fittings & solder, 1/2" diam	75.000	Ea.	18	1,575	1,593
Total			2,452.89	3,866.46	6,319.35

8.3-660	Thermosyphon, Hot Water		MAT.	INST.	TOTAL
0960	Solar, thermosyphon, hot water, two collector system	R13600 -610			
0970					

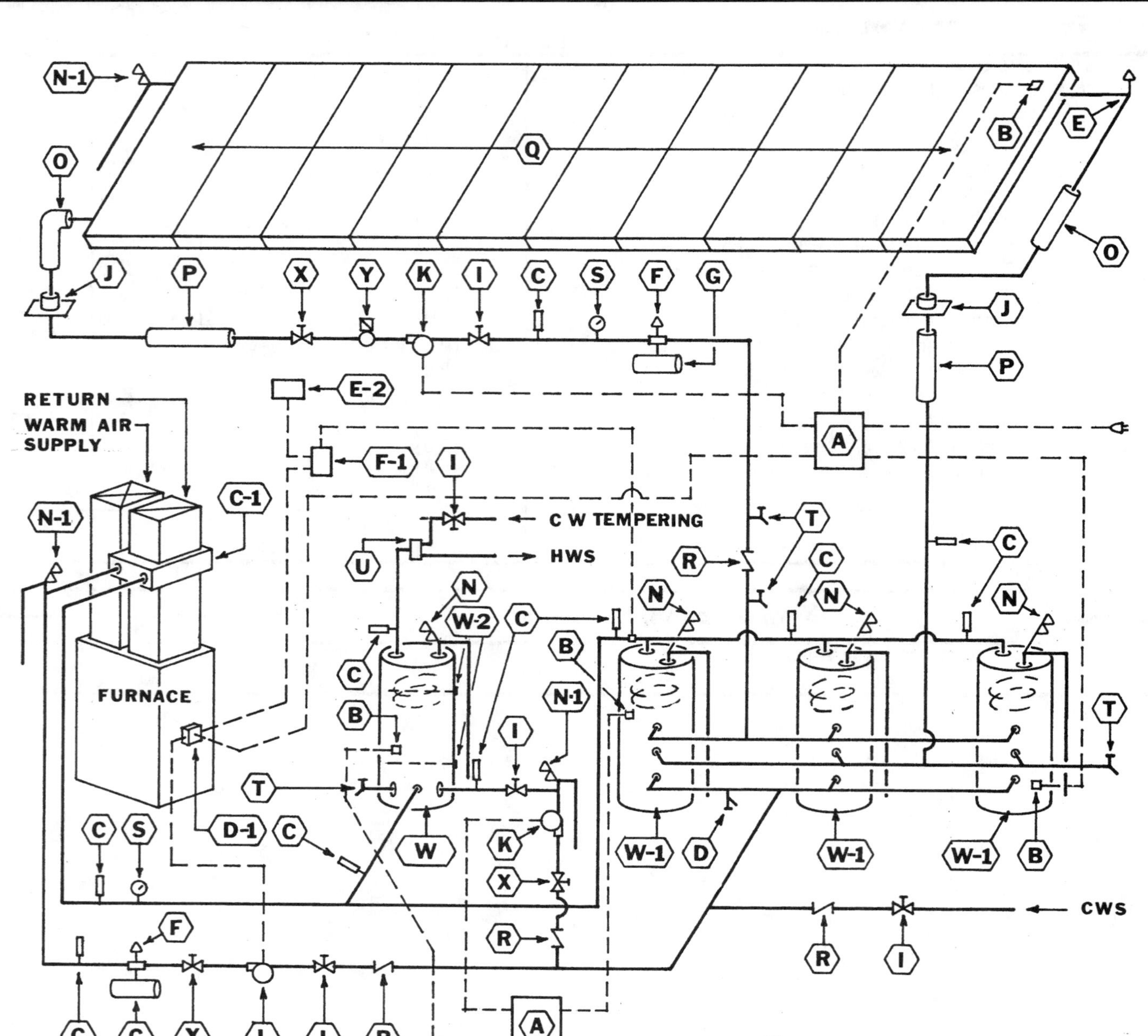

In this closed-loop indirect collection system, fluid with a low freezing temperature, propylene glycol, transports heat from the collectors to water storage. The transfer fluid is contained in a closed loop consisting of collectors, supply and return piping, and a heat exchanger immersed in the storage tank.

When the collectors become approximately 20° F warmer than the storage temperature, the controller activates the circulator. The circulator moves the fluid continuously until the temperature difference between fluid in the collectors and storage is such that the collection will no longer occur and then the circulator turns off. Since the heat transfer fluid has a very low freezing temperature, there is no need for it to be drained from the collectors between periods of collection.

System Components

		QUANTITY	UNIT	COST EACH MAT.	COST EACH INST.	COST EACH TOTAL
SYSTEM 8.3-710-2750						
SOLAR, CLOSED LOOP, SPACE/HOT WATER						
1″ TUBING, TEN 3′X7′ BLK CHROME ON COPPER ABSORBER COLLECTORS						
A, B	Differential controller 2 sensors, thermostat, solar energy system	2.000	Ea.	171	69	240
C	Thermometer, 2″ dial	10.000	Ea.	210	260	470
C-1	Heat exchanger, solar energy system, fluid to air, up flow, 80 MBH	1.000	Ea.	460	250	710
D, T	Fill & drain valves, brass, 3/4″ connection	5.000	Ea.	27.25	86.75	114
D-1	Fan center	1.000	Ea.	119	91	210
E	Air vent, manual, 1/8″ fitting	1.000	Ea.	2.13	13	15.13
E-2	Thermostat, 2 stage for sensing room temperature	1.000	Ea.	119	52	171
F	Air purger	2.000	Ea.	90	69	159
F-1	Controller, liquid temperature, solar energy system	1.000	Ea.	101	83	184
G	Expansion tank	2.000	Ea.	128	26	154
I	Valve, gate, bronze, 125 lb, soldered, 1″ diam	5.000	Ea.	132.50	110	242.50
J	Vent flashing, neoprene	2.000	Ea.	17.30	42	59.30
K	Circulator, solar heated liquid, 1/25 HP	2.000	Ea.	306	107	413
L	Circulator, solar heated liquid, 1/20 HP	1.000	Ea.	161	62.50	223.50
N	Relief valve, temp & pressure 150 psi 210°F self-closing	4.000	Ea.	34.60	55.40	90
N-1	Relief valve, pressure poppet, bronze, 30 psi, 3/4″ IPS	3.000	Ea.	145.50	44.55	190.05
O	Pipe covering, urethane, ultraviolet cover, 1″ wall, 1″ diam	50.000	L.F.	77	202	279
P	Pipe covering, fiberglass, all service jacket, 1″ wall, 1″ diam	60.000	L.F.	79.80	193.20	273
Q	Collector panel solar energy blk chrome on copper 1/8″ temp glass 3′x7′	10.000	Ea.	6,000	790	6,790
	Roof clamps for solar energy collector panel	10.000	Set	17.70	107	124.70
R	Valve, swing check, bronze, 125 lb, regrinding disc, 3/4″ & 1″ diam	4.000	Ea.	198	88	286
S	Pressure gage, 0-60 psi, for solar energy system	2.000	Ea.	46	26	72
U	Valve, water tempering, bronze, sweat connections, 3/4″ diam	1.000	Ea.	54	21	75
W-2, W-1, W	Tank, water storage immersed heat xchr elec elem 2″x1/2# ins 120 gal	4.000	Ea.	3,700	1,188	4,888
X	Valve, globe, bronze, 125 lb, rising stem, 1″ diam	3.000	Ea.	220.50	66	286.50
Y	Valve, flow control	1.000	Ea.	54.50	18.90	73.40
	Copper tubing type M, solder joint, hanger 10′ OC, 1″ diam	110.000	L.F.	205.70	654.50	860.20
	Copper tubing, type L, solder joint, hanger 10′ OC 3/4″ diam	20.000	L.F.	34.80	109	143.80
	Wrought copper fittings & solder, 3/4″ & 1″ diam	121.000	Ea.	148.83	3,146	3,294.83
	Sensor, wire, #22-2 conductor, multistranded	.700	C.L.F.	5.78	28.70	34.48
	Ductwork, galvanized steel, for heat exchanger	8.000	Lb.	11.60	39.52	51.12
	Solar energy heat transfer fluid propylene glycol, anti-freeze	25.000	Gal.	242.50	371.25	613.75
	TOTAL			13,320.99	8,470.27	21,791.26

8.3-710	Solar, Closed Loop, Space/Hot Water Systems	COST EACH MAT.	COST EACH INST.	COST EACH TOTAL
2540	Solar, closed loop, space/hot water			
2550	1/2″ tubing, 12 ea. 4′x4′4″ vacuum tube collectors	17,900	7,925	25,825
2600	3/4″ tubing, 12 ea. 4′x4′4″ vacuum tube collectors R13600-610	18,200	8,150	26,350
2650	10 ea. 3′ x 7′ black chrome absorber collectors	13,100	7,925	21,025
2700	10 ea. 3′ x 7′ flat black absorber collectors	11,800	7,950	19,750
2750	1″ tubing, 10 ea. 3′ x 7′ black chrome absorber collectors	13,300	8,475	21,775
2800	10 ea. 3′ x 7′ flat black absorber collectors	12,100	8,525	20,625
2850	6 ea. 4′ x 9′ flat black w/plastic glazing collectors	10,600	8,475	19,075
2900	12 ea. 2′ x 9′ plastic absorber and glazing collectors	13,900	8,625	22,525

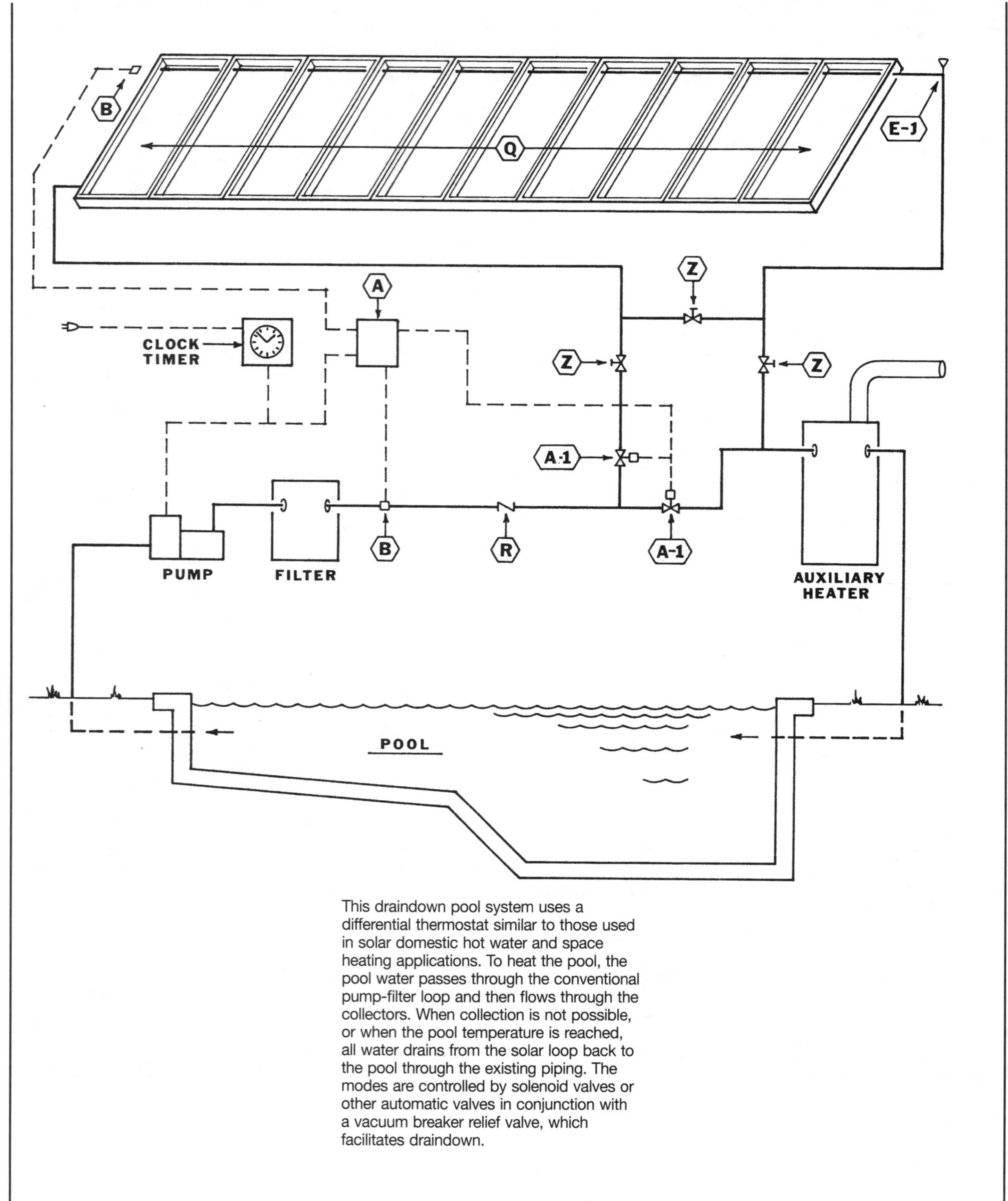

This draindown pool system uses a differential thermostat similar to those used in solar domestic hot water and space heating applications. To heat the pool, the pool water passes through the conventional pump-filter loop and then flows through the collectors. When collection is not possible, or when the pool temperature is reached, all water drains from the solar loop back to the pool through the existing piping. The modes are controlled by solenoid valves or other automatic valves in conjunction with a vacuum breaker relief valve, which facilitates draindown.

System Components

System Components	QUANTITY	UNIT	COST EACH MAT.	INST.	TOTAL
SYSTEM 8.3-720-2640					
SOLAR SWIMMING POOL HEATER, ROOF MOUNTED COLLECTORS					
TEN 4′ X 10′ FULLY WETTED UNGLAZED PLASTIC ABSORBERS					
A Differential thermostat/controller, 110V, adj pool pump system	1.000	Ea.	282	208	490
A-1 Solenoid valve, PVC, normally 1 open 1 closed (included)	2.000	Ea.			
B Sensor, thermistor type (included)	2.000	Ea.			
E-1 Valve, vacuum relief	1.000	Ea.	27.50	13	40.50
Q Collector panel, solar energy, plastic, liquid full wetted, 4′ x 10′	10.000	Ea.	2,150	1,500	3,650
R Valve, ball check, PVC, socket, 1-1/2″ diam	1.000	Ea.	75	21	96
Z Valve, ball, PVC, socket, 1-1/2″ diam	3.000	Ea.	187.50	63	250.50
Pipe, PVC, sch 40, 1-1/2″ diam	80.000	L.F.	251.20	924	1,175.20
Pipe fittings, PVC sch 40, socket joint, 1-1/2″ diam	10.000	Ea.	11	210	221
Sensor wire, #22-2 conductor, multistranded	.500	C.L.F.	4.13	20.50	24.63
Roof clamps for solar energy collector panels	10.000	Set	17.70	107	124.70
Roof strap, teflon for solar energy collector panels	26.000	L.F.	271.70	52.78	324.48
TOTAL			3,277.73	3,119.28	6,397.01

8.3-720 Solar Swimming Pool Heater Systems

8.3-720	Solar Swimming Pool Heater Systems	COST EACH MAT.	INST.	TOTAL
2530	Solar swimming pool heater systems, roof mounted collectors			
2540	10 ea. 3′x7′ black chrome absorber, 1/8″ temp. glass	7,125	2,400	9,525
2560	10 ea. 4′x8′ black chrome absorber, 3/16″ temp. glass (R13600-610)	8,225	2,875	11,100
2580	10 ea. 3′8″x6′ flat black absorber, 3/16″ temp. glass	5,875	2,450	8,325
2600	10 ea. 4′x9′ flat black absorber, plastic glazing	6,525	2,975	9,500
2620	10 ea. 2′x9′ rubber absorber, plastic glazing	12,000	3,225	15,225
2640	10 ea. 4′x10′ fully wetted unglazed plastic absorber	3,275	3,125	6,400
2660	Ground mounted collectors			
2680	10 ea. 3′x7′ black chrome absorber, 1/8″ temp. glass	7,225	2,775	10,000
2700	10 ea. 4′x8′ black chrome absorber, 3/16″ temp glass	8,325	3,250	11,575
2720	10 ea. 3′8″x6′ flat blk absorber, 3/16″ temp. glass	5,975	2,825	8,800
2740	10 ea. 4′x9′ flat blk absorber, plastic glazing	6,625	3,350	9,975
2760	10 ea. 2′x9′ rubber absorber, plastic glazing	12,100	3,500	15,600
2780	10 ea. 4′x10′ fully wetted unglazed plastic absorber	3,375	3,500	6,875

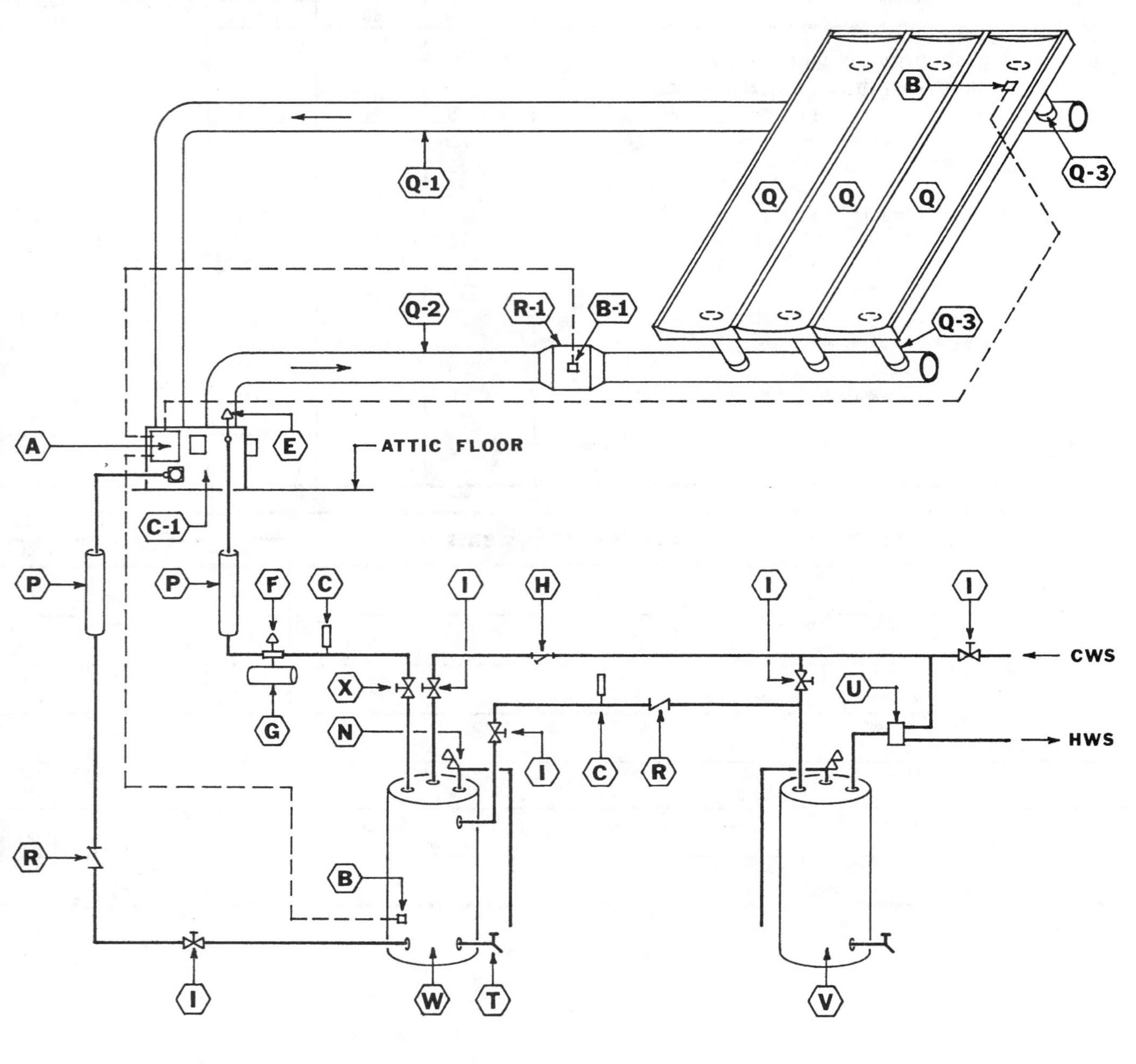

This domestic hot water pre-heat system includes heat exchanger with a circulating pump, blower, air-to-water, coil and controls, mounted in the upper collector manifold. Heat from the hot air coming out of the collectors is transferred through the heat exchanger. For each degree of DHW preheating gained, one degree less heating is needed from the fuel fired water heater. The system is simple, inexpensive to operate and can provide a substantial portion of DHW requirements for modest additional cost.

System Components

	System Components	QUANTITY	UNIT	COST EACH MAT.	COST EACH INST.	COST EACH TOTAL
	SYSTEM 8.3-730-2560					
	SOLAR, HOT WATER, AIR TO WATER HEAT EXCHANGE					
	THREE COLLECTORS, OPTICAL BLACK ON ALUMINUM, 10′X2′,80GAL TANK					
A, B	Differential controller, 2 sensors, thermostat, solar energy system	1.000	Ea.	85.50	34.50	120
C	Thermometer, 2″ dial	2.000	Ea.	42	52	94
C-1	Heat exchanger, air to fluid, up flow 70 MBH	1.000	Ea.	345	214	559
E	Air vent, manual, for solar energy system 1/8″ fitting	1.000	Ea.	2.13	13	15.13
F	Air purger	1.000	Ea.	45	34.50	79.50
G	Expansion tank	1.000	Ea.	64	13	77
H	Strainer,Y type, bronze body, 1/2″ IPS	1.000	Ea.	17.50	21	38.50
I	Valve, gate, bronze, 125 lb, NRS, soldered 1/2″ diam	5.000	Ea.	81.25	86.75	168
N	Relief valve, temp & pressure solar 150 psi 210°F self-closing	1.000	Ea.	8.65	13.85	22.50
P	Pipe covering, fiberglass, all service jacket, 1″ wall, 1/2″ diam	60.000	L.F.	70.20	185.40	255.60
Q	Collector panel solar energy, air, black on alum plate, flush mount, 10′x2′	30.000	L.F.	4,830	232.50	5,062.50
B-1, R-1	Shutter damper with shutter motor	1.000	Ea.	156.50	108.50	265
R	Backflow preventer for solar energy system, 1/2″ diam	2.000	Ea.	128	52	180
T	Drain valve, brass, 3/4″ connection	1.000	Ea.	5.45	17.35	22.80
U	Valve, water tempering, bronze, sweat connections, 1/2″ diam	1.000	Ea.	44	17.35	61.35
W	Tank, water storage, solar energy system, 80 Gal, 2″ x 2 lb insul	1.000	Ea.	820	260	1,080
V	Tank, water storage, w/heating element, drain, relief valve, existing	1.000	System			
X	Valve, globe, bronze, 125 lb, rising stem, 1/2″ diam	1.000	Ea.	34.50	17.35	51.85
	Copper tubing type M, solder joints, hangers 10′ OC 1/2″ diam	50.000	L.F.	58	247.50	305.50
	Copper tubing type L, solder joints, hangers 10′ OC 1/2″ diam	10.000	L.F.	13.80	51.50	65.30
	Wrought copper fittings & solder, 1/2″ diam	10.000	Ea.	2.40	210	212.40
	Sensor wire, #22-2 conductor multistranded	.500	C.L.F.	4.13	20.50	24.63
Q-1, Q-2	Ductwork, fiberglass, aluminized jacket, 1-1/2″ thick, 8″ diam	32.000	S.F.	143.36	132.48	275.84
Q-3	Manifold for flush mount solar energy collector panels, air	6.000	L.F.	177	27.96	204.96
	TOTAL			7,178.37	2,062.99	9,241.36

8.3-730	Solar Hot Water, Air To Water Heat Exchange	COST EACH MAT.	COST EACH INST.	COST EACH TOTAL
2550	Solar hot water, air to water heat exchange			
2560	Three collectors, optical black on aluminum, 10′ x 2′, 80 Gal tank	7,175	2,075	9,250
2580	Four collectors, optical black on aluminum, 10′ x 2′, 80 Gal tank R13600-610	8,875	2,175	11,050
2600	Four collectors, optical black on aluminum, 10′ x 2′, 120 Gal tank	9,000	2,225	11,225

A8.3-740 Air to Water Heat Exchange

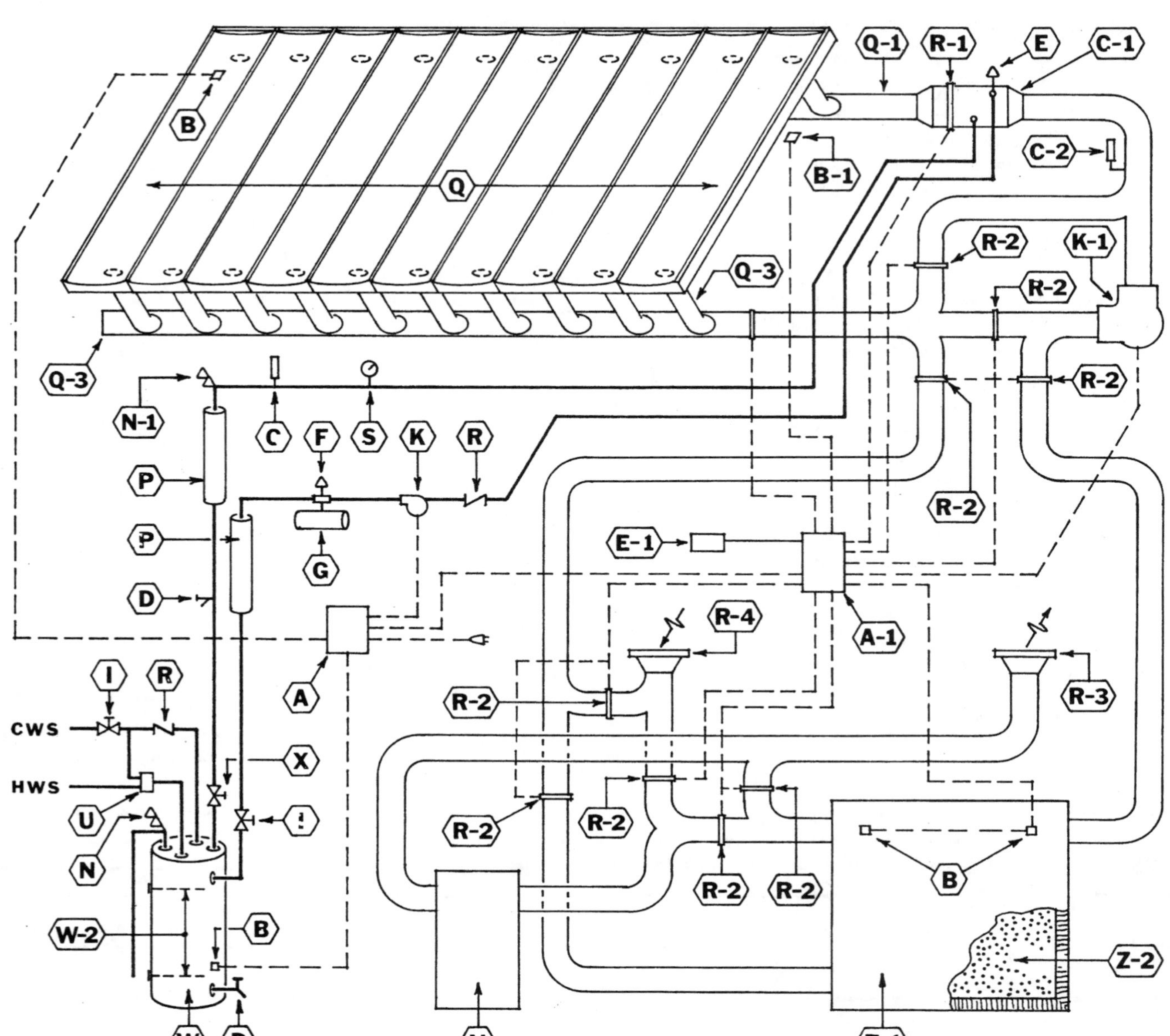

The complete Solar Air Heating System provides maximum savings of conventional fuel with both space heating and year-round domestic hot water heating. It allows for the home air conditioning to operate simultaneously and independently from the solar domestic water heating in summer. The system's modes of operation are:

Mode 1: The building is heated directly from the collectors with air circulated by the Solar Air Mover.

Mode 2: When heat is not needed in the building, the dampers change within the air mover to circulate the air from the collectors to the rock storage bin.

Mode 3: When heat is not available from the collector array and is available in rock storage, the air mover draws heated air from rock storage and directs it into the building. When heat is not available from the collectors or the rock storage bin, the auxiliary heating unit will provide heat for the building. The size of the collector array is typically 25% the size of the main floor area.

System Components

System Components		QUANTITY	UNIT	COST EACH MAT.	COST EACH INST.	COST EACH TOTAL
SYSTEM 8.3-740-0510						
SOLAR, SPACE/HOT WATER, AIR TO WATER HEAT EXCHANGE						
A, B	Differential controller 2 sensors thermos., solar energy sys liquid loop	1.000	Ea.	85.50	34.50	120
A-1, B	Differential controller 2 sensors 6 station solar energy sys air loop	1.000	Ea.	222	139	361
B-1	Solar energy sensor, freeze prevention	1.000	Ea.	22	13	35
C	Thermometer for solar energy system, 2" dial	2.000	Ea.	42	52	94
C-1	Heat exchanger, solar energy system, air to fluid, up flow, 70 MBH	1.000	Ea.	345	214	559
D	Drain valve, brass, 3/4" connection	2.000	Ea.	10.90	34.70	45.60
E	Air vent, manual, for solar energy system 1/8" fitting	1.000	Ea.	2.13	13	15.13
E-1	Thermostat, 2 stage for sensing room temperature	1.000	Ea.	119	52	171
F	Air purger	1.000	Ea.	45	34.50	79.50
G	Expansion tank, for solar energy system	1.000	Ea.	64	13	77
I	Valve, gate, bronze, 125 lb, NRS, soldered 3/4" diam	2.000	Ea.	37.10	42	79.10
K	Circulator, solar heated liquid, 1/25 HP	1.000	Ea.	153	53.50	206.50
N	Relief valve temp & press 150 psi 210°F self-closing, 3/4" IPS	1.000	Ea.	8.65	13.85	22.50
N-1	Relief valve, pressure, poppet, bronze, 30 psi, 3/4" IPS	1.000	Ea.	48.50	14.85	63.35
P	Pipe covering, fiberglass, all service jacket, 1" wall, 3/4" diam	60.000	L.F.	79.80	193.20	273
Q-3	Manifold for flush mount solar energy collector panels	20.000	L.F.	590	93.20	683.20
Q	Collector panel solar energy, air, black on alum. plate, flush mount 10'x2'	100.000	L.F.	16,100	775	16,875
R	Valve, swing check, bronze, 125 lb, regrinding disc, 3/4" diam	2.000	Ea.	66	42	108
S	Pressure gage, 2" dial, for solar energy system	1.000	Ea.	23	13	36
U	Valve, water tempering, bronze, sweat connections, 3/4" diam	1.000	Ea.	54	21	75
W-2, W	Tank, water storage, solar, elec element 2"x1/2# insul, 80 Gal	1.000	Ea.	820	260	1,080
X	Valve, globe, bronze, 125 lb, soldered, 3/4" diam	1.000	Ea.	47	21	68
	Copper tubing type L, solder joints, hangers 10' OC 3/4" diam	10.000	L.F.	17.40	54.50	71.90
	Copper tubing type M, solder joints, hangers 10' OC 3/4" diam	60.000	L.F.	88.20	321	409.20
	Wrought copper fittings & solder, 3/4" diam	26.000	Ea.	13.78	572	585.78
	Sensor wire, #22-2 conductor multistranded	1.200	C.L.F.	9.90	49.20	59.10
Q-1	Duct work, rigid fiberglass, rectangular	400.000	S.F.	364	1,328	1,692
	Ductwork, transition pieces	8.000	Ea.	118.40	248	366.40
R-2	Shutter/damper for solar heater circulator	9.000	Ea.	445.50	558	1,003.50
K-1	Shutter motor for solar heater circulator blower	9.000	Ea.	963	418.50	1,381.50
K-1	Fan, solar energy heated air circulator, space & DHW system	1.000	Ea.	1,550	1,500	3,050
Z-1	Tank, solar energy air storage, 6'-3"H 7'x7' = 306 CF/2000 Gal	1.000	Ea.	8,950	965	9,915
Z-2	Crushed stone 1-1/2"	11.000	C.Y.	225.50	54.45	279.95
C-2	Themometer, remote probe, 2" dial	1.000	Ea.	32	52	84
R-1	Solenoid valve	1.000	Ea.	233	46	279
V, R-4, R-3	Furnace, supply diffusers, return grilles, existing	1.000	Ea.	8,950	965	9,915
	Total			31,769.76	8,254.50	40,024.26

8.3-740	Air To Water Heat Exchange		COST EACH MAT.	COST EACH INST.	COST EACH TOTAL
1210	Solar, air to water heat exchange for space / hot water heating	R13600 -610			
1220					

A8.5-110 Garage Exhaust

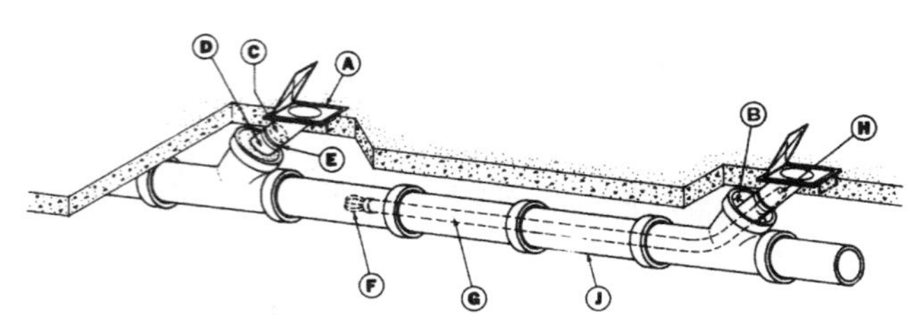

Vitrified Clay Garage Exhaust System

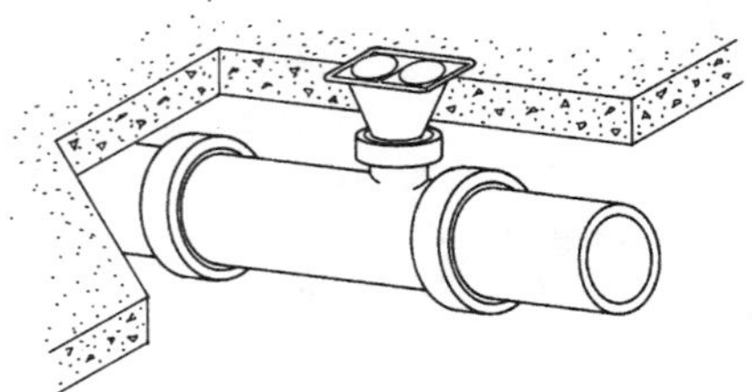
Dual Exhaust System

System Components

System Components		QUANTITY	UNIT	COST EACH MAT.	COST EACH INST.	COST EACH TOTAL
	SYSTEM 8.5-110-1040					
	GARAGE, EXHAUST, SINGLE OUTLET, 3″ EXHAUST, CARS & LIGHT TRUCKS					
A	Outlet top assy. for engine exhaust system with adapters and ftngs, 3″ diam	1.000	Ea.	178	26	204
F	Bullnose (guide) for engine exhaust system, 3″ diam	1.000	Ea.	14.85		14.85
G	Galvanized flexible tubing for engine exhaust system, 3″ diam	8.000	L.F.	42.40		42.40
H	Adapter for metal tubing end of engine exhaust system, 3″ tail pipe	1.000	Ea.	29.50		29.50
J	Pipe, sewer, vitrified clay, premium joint, 8″ diam.	18.000	L.F.	73.26	118.44	191.70
	Excavate utility trench w/chain trencher, 12 H.P., oper. walking	18.000	L.F.		23.40	23.40
	Backfill utility trench by hand, incl. compaction, 8″ wide 24″ deep	18.000	L.F.		32.22	32.22
	Stand for blower, concrete over polystyrene core, 6″ high	1.000	Ea.	2.38	12.95	15.33
	AC&V duct spiral reducer 10″x8″	1.000	Ea.	7.80	23.50	31.30
	AC&V duct spiral reducer 12″x10″	1.000	Ea.	11.10	31	42.10
	AC&V duct spiral preformed 45° elbow 8″ diam	1.000	Ea.	3.64	26.50	30.14
	AC&V utility fan, belt drive, 3 phase, 2000 CFM, 1 HP	2.000	Ea.	1,130	414	1,544
	Safety switch, heavy duty fused, 240V, 3 pole, 30 amp	1.000	Ea.	120	128	248
	Total			1,612.93	836.01	2,448.94

8.5-110 Garage Exhaust Systems

8.5-110	Garage Exhaust Systems	COST PER BAY MAT.	COST PER BAY INST.	COST PER BAY TOTAL
1040	Garage, single exhaust, 3″ outlet, cars & light trucks, one bay	1,625	835	2,460
1060	Additional bays up to seven bays	305	122	427
1500	4″ outlet, trucks, one bay	1,625	835	2,460
1520	Additional bays up to six bays	315	122	437
1600	5″ outlet, diesel trucks, one bay	1,675	835	2,510
1650	Additional single bays up to six	375	142	517
1700	Two adjoining bays	1,675	835	2,510
2000	Dual exhaust, 3″ outlets, pair of adjoining bays	1,900	930	2,830
2100	Additional pairs of adjoining bays	570	142	712

Division 12
Site Work

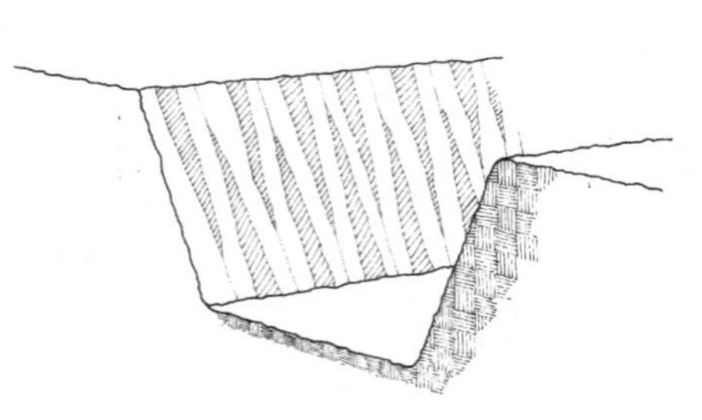

Trenching Systems are shown on a cost per linear foot basis. The systems include: excavation; backfill and removal of spoil; and compaction for various depths and trench bottom widths. The backfill has been reduced to accommodate a pipe of suitable diameter and bedding.

The slope for trench sides varies from 0:1 to 2:1.

The Expanded System Listing shows Trenching Systems that range from 2′ to 12′ in width. Depths range from 2′ to 25′.

System Components	QUANTITY	UNIT	COST PER L.F. EQUIP.	LABOR	TOTAL
SYSTEM 12.3-110-1310					
TRENCHING, BACKHOE, 0 TO 1 SLOPE, 2′ WIDE, 2′ DP, 3/8 C.Y. BUCKET					
Excavation, trench, hyd. backhoe, track mtd., 3/8 C.Y. bucket	.174	C.Y.	.23	.76	.99
Backfill and load spoil, from stockpile	.174	C.Y.	.11	.22	.33
Compaction by rammer tamper, 8″ lifts, 4 passes	.014	C.Y.	.01	.03	.04
Remove excess spoil, 6 C.Y. dump truck, 2 mile roundtrip	.160	C.Y.	.65	.47	1.12
Total			1	1.48	2.48

12.3-110	Trenching	COST PER L.F. EQUIP.	LABOR	TOTAL
1310	Trenching, backhoe, 0 to 1 slope, 2′ wide, 2′ deep, 3/8 C.Y. bucket	1	1.48	2.48
1320	3′ deep, 3/8 C.Y. bucket	1.22	2.16	3.38
1330	4′ deep, 3/8 C.Y. bucket	1.43	2.86	4.29
1340	6′ deep, 3/8 C.Y. bucket	1.84	3.64	5.48
1350	8′ deep, 1/2 C.Y. bucket	2.09	4.62	6.71
1360	10′ deep, 1 C.Y. bucket	3.13	5.55	8.68
1400	4′ wide, 2′ deep, 3/8 C.Y. bucket	2.10	2.99	5.09
1410	3′ deep, 3/8 C.Y. bucket	2.96	4.44	7.40
1420	4′ deep, 1/2 C.Y. bucket	2.84	4.45	7.29
1430	6′ deep, 1/2 C.Y. bucket	3.67	6.75	10.42
1440	8′ deep, 1/2 C.Y. bucket	5.60	8.85	14.45
1450	10′ deep, 1 C.Y. bucket	6.60	10.95	17.55
1460	12′ deep, 1 C.Y. bucket	8.25	14.05	22.30
1470	15′ deep, 1-1/2 C.Y. bucket	7.25	12	19.25
1480	18′ deep, 2-1/2 C.Y. bucket	9.85	17.15	27
1520	6′ wide, 6′ deep, 5/8 C.Y. bucket	7.40	9.50	16.90
1530	8′ deep, 3/4 C.Y. bucket	8.55	11.45	20
1540	10′ deep, 1 C.Y. bucket	9.80	13.80	23.60
1550	12′ deep, 1-1/4 C.Y. bucket	8.70	10.85	19.55
1560	16′ deep, 2 C.Y. bucket	14.20	12.35	26.55
1570	20′ deep, 3-1/2 C.Y. bucket	15.15	22	37.15
1580	24′ deep, 3-1/2 C.Y. bucket	24	37	61
1640	8′ wide, 12′ deep, 1-1/4 C.Y. bucket	13.30	14.85	28.15
1650	15′ deep, 1-1/2 C.Y. bucket	15.40	18.25	33.65
1660	18′ deep, 2-1/2 C.Y. bucket	22	18.45	40.45
1680	24′ deep, 3-1/2 C.Y. bucket	33.50	49.50	83
1730	10′ wide, 20′ deep, 3-1/2 C.Y. bucket	28	37.50	65.50
1740	24′ deep, 3-1/2 C.Y. bucket	42.50	62	104.50
1780	12′ wide, 20′ deep, 3-1/2 C.Y. bucket	34	45	79
1790	25′ deep, bucket	57	83	140
1800	1/2 to 1 slope, 2′ wide, 2′ deep, 3/8 C.Y. bucket	1.34	1.90	3.24
1810	3′ deep, 3/8 C.Y. bucket	1.84	3.14	4.98
1820	4′ deep, 3/8 C.Y. bucket	2.31	4.58	6.89
1840	6′ deep, 3/8 C.Y. bucket	3.44	7.05	10.49
1860	8′ deep, 1/2 C.Y. bucket	4.67	10.75	15.42
1880	10′ deep, 1 C.Y. bucket	8.40	15.05	23.45

12.3-110	Trenching	COST PER L.F.		
		EQUIP.	LABOR	TOTAL
2300	4' wide, 2' deep, 3/8 C.Y. bucket	2.23	3.16	5.39
2310	3' deep, 3/8 C.Y. bucket	3.67	5.25	8.92
2320	4' deep, 1/2 C.Y. bucket	3.69	5.45	9.14
2340	6' deep, 1/2 C.Y. bucket	5.20	9.45	14.65
2360	8' deep, 1/2 C.Y. bucket	8.45	13.05	21.50
2380	10' deep, 1 C.Y. bucket	11.60	19.25	30.85
2400	12' deep, 1 C.Y. bucket	12.65	20.50	33.15
2430	15' deep, 1-1/2 C.Y. bucket	15.65	25.50	41.15
2460	18' deep, 2-1/2 C.Y. bucket	27.50	30.50	58
2840	6' wide, 6' deep, 5/8 C.Y. bucket	8.25	10.50	18.75
2860	8' deep, 3/4 C.Y. bucket	11.80	15.75	27.55
2880	10' deep, 1 C.Y. bucket	14.55	20.50	35.05
2900	12' deep, 1-1/4 C.Y. bucket	13.75	17.30	31.05
2940	16' deep, 2 C.Y. bucket	21	27.50	48.50
2980	20' deep, 3-1/2 C.Y. bucket	32	47.50	79.50
3020	24' deep, 3-1/2 C.Y. bucket	57.50	88.50	146
3100	8' wide, 12' deep, 1-1/4 C.Y. bucket	19.20	21	40.20
3120	15' deep, 1-1/2 C.Y. bucket	24	28.50	52.50
3140	18' deep, 2-1/2 C.Y. bucket	37.50	31	68.50
3180	24' deep, 3-1/2 C.Y. bucket	67	99.50	166.50
3270	10' wide, 20' deep, 3-1/2 C.Y. bucket	42	59.50	101.50
3280	24' deep, 3-1/2 C.Y. bucket	77	112	189
3370	12' wide, 20' deep, 3-1/2 C.Y. bucket	48	66	114
3380	25' deep, 3-1/2 C.Y. bucket	90	130	220
3500	1 to 1 slope, 2' wide, 2' deep, 3/8 C.Y. bucket	1.56	2.22	3.78
3520	3' deep, 3/8 C.Y. bucket	2.38	3.88	6.26
3540	4' deep, 3/8 C.Y. bucket	3.01	5.85	8.86
3560	6' deep, 3/8 C.Y. bucket	4.63	9.30	13.93
3580	8' deep, 1/2 C.Y. bucket	6.45	14.65	21.10
3600	10' deep, 1 C.Y. bucket	13.75	24	37.75
3800	4' wide, 2' deep, 3/8 C.Y. bucket	2.66	3.80	6.46
3820	3' deep, 3/8 C.Y. bucket	4.69	6.65	11.34
3840	4' deep, 1/2 C.Y. bucket	4.45	6.10	10.55
3860	6' deep, 1/2 C.Y. bucket	6.45	11.35	17.80
3880	8' deep, 1/2 C.Y. bucket	11.30	17.20	28.50
3900	10' deep, 1 C.Y. bucket	15.25	24.50	39.75
3920	12' deep, 1 C.Y. bucket	21.50	36	57.50
3940	15' deep, 1-1/2 C.Y. bucket	21.50	33.50	55
3960	18' deep, 2-1/2 C.Y. bucket	39	40	79
4030	6' wide, 6' deep, 5/8 C.Y. bucket	10.05	12.80	22.85
4040	8' deep, 3/4 C.Y. bucket	14.85	20.50	35.35
4050	10' deep, 1 C.Y. bucket	19	28.50	47.50
4060	12' deep, 1-1/4 C.Y. bucket	18.55	25.50	44.05
4070	16' deep, 2 C.Y. bucket	37	36	73
4080	20' deep, 3-1/2 C.Y. bucket	46.50	75	121.50
4090	24' deep, 3-1/2 C.Y. bucket	87.50	146	233.50
4500	8' wide, 12' deep, 1-1/4 C.Y. bucket	25	29	54
4550	15' deep, 1-1/2 C.Y. bucket	32	41	73
4600	18' deep, 2-1/2 C.Y. bucket	51	47.50	98.50
4650	24' deep, 3-1/2 C.Y. bucket	97	156	253
4800	10' wide, 20' deep, 3-1/2 C.Y. bucket	60	87	147
4850	24' deep, 3-1/2 C.Y. bucket	105	165	270
4950	12' wide, 20' deep, 3-1/2 C.Y. bucket	66	93	159
4980	25' deep, 3-1/2 C.Y. bucket	128	201	329
5000	1-1/2 to 1 slope, 2' wide, 2' deep, 3/8 C.Y. bucket	1.81	2.57	4.38
5020	3' deep, 3/8 C.Y. bucket	2.90	4.60	7.50
5040	4' deep, 3/8 C.Y. bucket	3.68	6.85	10.53
5060	6' deep, 3/8 C.Y. bucket	5.65	10.95	16.60
5080	8' deep, 1/2 C.Y. bucket	7.90	17.30	25.20
5100	10' deep, 1 C.Y. bucket	15.10	25	40.10

12.3-110	Trenching	COST PER L.F.		
		EQUIP.	LABOR	TOTAL
5300	4′ wide, 2′ deep, 3/8 C.Y. bucket	2.52	3.57	6.09
5320	3′ deep, 3/8 C.Y. bucket	4.57	6.50	11.07
5340	4′ deep, 1/2 C.Y. bucket	5.30	7.15	12.45
5360	6′ deep, 1/2 C.Y. bucket	7.60	12.85	20.45
5380	8′ deep, 1/2 C.Y. bucket	13.45	19.55	33
5400	10′ deep, 1 C.Y. bucket	18.30	28	46.30
5420	12′ deep, 1 C.Y. bucket	26.50	42	68.50
5450	15′ deep, 1-1/2 C.Y. bucket	26	37	63
5480	18′ deep, 2-1/2 C.Y. bucket	48	43.50	91.50
5660	6′ wide, 6′ deep, 5/8 C.Y. bucket	11.65	14.30	25.95
5680	8′ deep, 3/4 C.Y. bucket	17.30	23	40.30
5700	10′ deep, 1 C.Y. bucket	22	31.50	53.50
5720	12′ deep, 1-1/4 C.Y. bucket	21.50	27	48.50
5760	16′ deep, 2 C.Y. bucket	44.50	38.50	83
5800	20′ deep, 3-1/2 C.Y. bucket	56.50	85.50	142
5840	24′ deep, 3-1/2 C.Y. bucket	107	166	273
6020	8′ wide, 12′ deep, 1-1/4 C.Y. bucket	29	31.50	60.50
6050	15′ deep, 1-1/2 C.Y. bucket	37.50	45	82.50
6080	18′ deep, 2-1/2 C.Y. bucket	61	51	112
6140	24′ deep, 3-1/2 C.Y. bucket	117	175	292
6300	10′ wide, 20′ deep, 3-1/2 C.Y. bucket	71	97	168
6350	24′ deep, 3-1/2 C.Y. bucket	125	184	309
6450	12′ wide, 20′ deep, 3-1/2 C.Y. bucket	78.50	103	181.50
6480	25′ deep, 3-1/2 C.Y. bucket	152	222	374
6600	2 to 1 slope, 2′ wide, 2′ deep, 3/8 C.Y. bucket	2.57	2.38	4.95
6620	3′ deep, 3/8 C.Y. bucket	3.35	5.10	8.45
6640	4′ deep, 3/8 C.Y. bucket	4.22	7.65	11.87
6660	6′ deep, 3/8 C.Y. bucket	5.85	11.60	17.45
6680	8′ deep, 1/2 C.Y. bucket	9	19.50	28.50
6700	10′ deep, 1 C.Y. bucket	17.20	28	45.20
6900	4′ wide, 2′ deep, 3/8 C.Y. bucket	2.52	3.57	6.09
6920	3′ deep, 3/8 C.Y. bucket	4.69	6.65	11.34
6940	4′ deep, 1/2 C.Y. bucket	5.90	7.65	13.55
6960	6′ deep, 1/2 C.Y. bucket	8.45	13.95	22.40
6980	8′ deep, 1/2 C.Y. bucket	14.95	21.50	36.45
7000	10′ deep, 1 C.Y. bucket	20.50	31	51.50
7050	15′ deep, 1-1/2 C.Y. bucket	29	41.50	70.50
7080	18′ deep, 2-1/2 C.Y. bucket	54.50	49.50	104
7260	6′ wide, 6′ deep, 5/8 C.Y. bucket	12.85	15.25	28.10
7280	8′ deep, 3/4 C.Y. bucket	19	24.50	43.50
7300	10′ deep, 3/4 C.Y. bucket	25	35	60
7360	16′ deep, 2 C.Y. bucket	49.50	43	92.50
7400	20′ deep, 3-1/2 C.Y. bucket	63.50	95.50	159
7440	24′ deep, 3-1/2 C.Y. bucket	120	187	307
7620	8′ wide, 12′ deep, 1-1/4 C.Y. bucket	32.50	34.50	67
7650	15′ deep, 1-1/2 C.Y. bucket	42	49.50	91.50
7680	18′ deep, 2-1/2 C.Y. bucket	68	56.50	124.50
7740	24′ deep, 3-1/2 C.Y. bucket	131	195	326
7920	10′ wide, 20′ deep, 3-1/2 C.Y. bucket	79	106	185
7940	24′ deep, 3-1/2 C.Y. bucket	139	203	342
8060	12′ wide, 20′ deep, 3-1/2 C.Y. bucket	87	112	199
8080	25′ deep, 3-1/2 C.Y. bucket	168	243	411

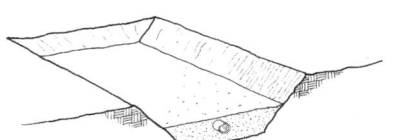

The Pipe Bedding System is shown for various pipe diameters. Compacted bank sand is used for pipe bedding and to fill 12″ over the pipe. No backfill is included. Various side slopes are shown to accommodate different soil conditions. Pipe sizes vary from 6″ to 84″ diameter.

System Components	QUANTITY	UNIT	COST PER L.F. MAT.	COST PER L.F. INST.	COST PER L.F. TOTAL
SYSTEM 12.3-310-1440					
PIPE BEDDING, SIDE SLOPE 0 TO 1, 1′ WIDE, PIPE SIZE 6″ DIAMETER					
Borrow, bank sand, 2 mile haul, machine spread	.067	C.Y.	.26	.33	.59
Compaction, vibrating plate	.067	C.Y.		.12	.12
TOTAL			.26	.45	.71

12.3-310	Pipe Bedding	COST PER L.F. MAT.	COST PER L.F. INST.	COST PER L.F. TOTAL
1440	Pipe bedding, side slope 0 to 1, 1′ wide, pipe size 6″ diameter	.26	.45	.71
1460	2′ wide, pipe size 8″ diameter	.58	.99	1.57
1480	Pipe size 10″ diameter	.59	1.02	1.61
1500	Pipe size 12″ diameter	.61	1.06	1.67
1520	3′ wide, pipe size 14″ diameter	1	1.73	2.73
1540	Pipe size 15″ diameter	1.02	1.75	2.77
1560	Pipe size 16″ diameter	1.03	1.77	2.80
1580	Pipe size 18″ diameter	1.05	1.81	2.86
1600	4′ wide, pipe size 20″ diameter	1.52	2.61	4.13
1620	Pipe size 21″ diameter	1.53	2.64	4.17
1640	Pipe size 24″ diameter	1.57	2.70	4.27
1660	Pipe size 30″ diameter	1.61	2.77	4.38
1680	6′ wide, pipe size 32″ diameter	2.81	4.85	7.66
1700	Pipe size 36″ diameter	2.89	4.98	7.87
1720	7′ wide, pipe size 48″ diameter	3.76	6.50	10.26
1740	8′ wide, pipe size 60″ diameter	4.69	8.05	12.74
1760	10′ wide, pipe size 72″ diameter	6.75	11.65	18.40
1780	12′ wide, pipe size 84″ diameter	9.20	15.85	25.05
2140	Side slope 1/2 to 1, 1′ wide, pipe size 6″ diameter	.55	.95	1.50
2160	2′ wide, pipe size 8″ diameter	.91	1.57	2.48
2180	Pipe size 10″ diameter	.99	1.70	2.69
2200	Pipe size 12″ diameter	1.06	1.84	2.90
2220	3′ wide, pipe size 14″ diameter	1.52	2.62	4.14
2240	Pipe size 15″ diameter	1.56	2.68	4.24
2260	Pipe size 16″ diameter	1.61	2.77	4.38
2280	Pipe size 18″ diameter	1.70	2.94	4.64
2300	4′ wide, pipe size 20″ diameter	2.24	3.87	6.11
2320	Pipe size 21″ diameter	2.30	3.96	6.26
2340	Pipe size 24″ diameter	2.46	4.25	6.71
2360	Pipe size 30″ diameter	2.76	4.76	7.52
2380	6′ wide, pipe size 32″ diameter	4.07	7	11.07
2400	Pipe size 36″ diameter	4.36	7.50	11.86
2420	7′ wide, pipe size 48″ diameter	5.95	10.25	16.20
2440	8′ wide, pipe size 60″ diameter	7.75	13.35	21.10
2460	10′ wide, pipe size 72″ diameter	10.85	18.70	29.55
2480	12′ wide, pipe size 84″ diameter	14.45	25	39.45
2620	Side slope 1 to 1, 1′ wide, pipe size 6″ diameter	.84	1.45	2.29
2640	2′ wide,pipe size 8″ diameter	1.26	2.18	3.44
2660	Pipe size 10″ diameter	1.38	2.39	3.77
2680	Pipe size 12″ diameter	1.52	2.62	4.14

For expanded coverage of these items see *Means Site Work & Landscape Cost Data 2001*

12.3-310	Pipe Bedding	COST PER L.F.		
		MAT.	INST.	TOTAL
2700	3′ wide, pipe size 14″ diameter	2.04	3.52	5.56
2720	Pipe size 15″ diameter	2.11	3.63	5.74
2740	Pipe size 16″ diameter	2.19	3.76	5.95
2760	Pipe size 18″ diameter	2.35	4.06	6.41
2780	4′ wide, pipe size 20″ diameter	2.97	5.10	8.07
2800	Pipe size 21″ diameter	3.06	5.25	8.31
2820	Pipe size 24″ diameter	3.35	5.75	9.10
2840	Pipe size 30″ diameter	3.91	6.75	10.66
2860	6′ wide, pipe size 32″ diameter	5.35	9.20	14.55
2880	Pipe size 36″ diameter	5.80	10.05	15.85
2900	7′ wide, pipe size 48″ diameter	8.15	14.05	22.20
2920	8′ wide, pipe size 60″ diameter	10.80	18.65	29.45
2940	10′ wide, pipe size 72″ diameter	14.90	25.50	40.40
2960	12′ wide, pipe size 84″ diameter	19.65	34	53.65
3000	Side slope 1-1/2 to 1, 1′ wide, pipe size 6″ diameter	1.13	1.94	3.07
3020	2′ wide, pipe size 8″ diameter	1.60	2.77	4.37
3040	Pipe size 10″ diameter	1.78	3.06	4.84
3060	Pipe size 12″ diameter	1.97	3.39	5.36
3080	3′ wide, pipe size 14″ diameter	2.55	4.38	6.93
3100	Pipe size 15″ diameter	2.65	4.57	7.22
3120	Pipe size 16″ diameter	2.76	4.76	7.52
3140	Pipe size 18″ diameter	3.01	5.20	8.21
3160	4′ wide, pipe size 20″ diameter	3.71	6.40	10.11
3180	Pipe size 21″ diameter	3.82	6.60	10.42
3200	Pipe size 24″ diameter	4.23	7.30	11.53
3220	Pipe size 30″ diameter	5.10	8.75	13.85
3240	6′ wide, pipe size 32″ diameter	6.60	11.35	17.95
3260	Pipe size 36″ diameter	7.30	12.55	19.85
3280	7′ wide, pipe size 48″ diameter	10.35	17.80	28.15
3300	8′ wide, pipe size 60″ diameter	13.85	24	37.85
3320	10′ wide, pipe size 72″ diameter	19	33	52
3340	12′ wide, pipe size 84″ diameter	25	43	68
3400	Side slope 2 to 1, 1′ wide, pipe size 6″ diameter	1.45	2.49	3.94
3420	2′ wide, pipe size 8″ diameter	1.94	3.35	5.29
3440	Pipe size 10″ diameter	2.17	3.74	5.91
3460	Pipe size 12″ diameter	2.42	4.17	6.59
3480	3′ wide, pipe size 14″ diameter	3.06	5.25	8.31
3500	Pipe size 15″ diameter	3.19	5.50	8.69
3520	Pipe size 16″ diameter	3.34	5.75	9.09
3540	Pipe size 18″ diameter	3.66	6.30	9.96
3560	4′ wide, pipe size 20″ diameter	4.43	7.65	12.08
3580	Pipe size 21″ diameter	4.59	7.90	12.49
3600	Pipe size 24″ diameter	5.10	8.80	13.90
3620	Pipe size 30″ diameter	6.25	10.75	17
3640	6′ wide, pipe size 32″ diameter	7.85	13.55	21.40
3660	Pipe size 36″ diameter	8.75	15.10	23.85
3680	7′ wide, pipe size 48″ diameter	12.50	21.50	34
3700	8′ wide, pipe size 60″ diameter	16.95	29	45.95
3720	10′ wide, pipe size 72″ diameter	23	39.50	62.50
3740	12′ wide, pipe size 84″ diameter	30	52	82

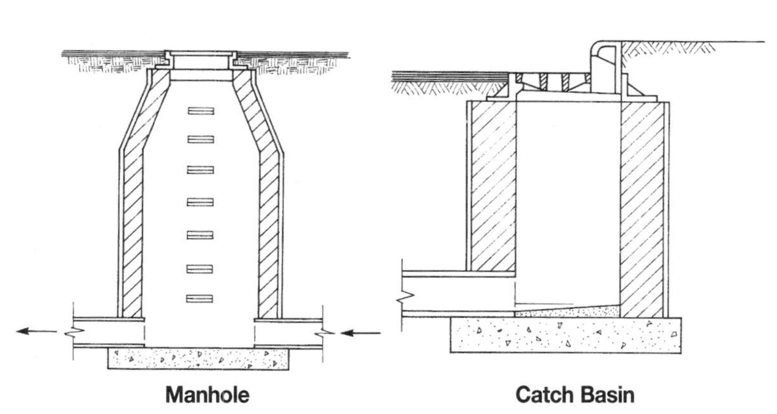

The Manhole and Catch Basin System includes: excavation with a backhoe; a formed concrete footing; frame and cover; cast iron steps and compacted backfill.

The Expanded System Listing shows manholes that have a 4′, 5′ and 6′ inside diameter riser. Depths range from 4′ to 14′. Construction material shown is either concrete, concrete block, precast concrete, or brick.

System Components	QUANTITY	UNIT	COST PER EACH MAT.	COST PER EACH INST.	COST PER EACH TOTAL
SYSTEM 12.3-710-1920					
MANHOLE/CATCH BASIN, BRICK, 4′ I.D. RISER, 4′ DEEP					
Excavation, hydraulic backhoe, 3/8 C.Y. bucket	14.815	C.Y.		75.26	75.26
Trim sides and bottom of excavation	64.000	S.F.		40.32	40.32
Forms in place, manhole base, 4 uses	20.000	SFCA	13.20	66.60	79.80
Reinforcing in place footings, #4 to #7	.019	Ton	11.02	16.06	27.08
Concrete, 3000 psi, incl. place and vibrate under 1 CY, direct chute	.925	C.Y.	65.68	31.47	97.15
Catch basin or MH, brick, 4′ ID, 4′ deep	1.000	Ea.	298	650	948
Catch basin or MH steps; heavy galvanized cast iron	1.000	Ea.	10.40	9.15	19.55
Catch basin or MH frame and cover	1.000	Ea.	174	143.50	317.50
Backfill with wheeled front end loader	12.954	C.Y.	108.17	19.04	127.21
Backfill compaction, 12″ lifts, air tamp	12.954	C.Y.		74.88	74.88
TOTAL			680.47	1,126.28	1,806.75

12.3-710	Manholes & Catch Basins	COST PER EACH MAT.	COST PER EACH INST.	COST PER EACH TOTAL
1920	Manhole/catch basin, brick, 4′ I.D. riser, 4′ deep	680	1,125	1,805
1940	6′ deep	915	1,575	2,490
1960	8′ deep	1,175	2,150	3,325
1980	10′ deep	1,500	2,650	4,150
3000	12′ deep	1,925	2,900	4,825
3020	14′ deep	2,425	4,050	6,475
3200	Block, 4′ I.D. riser, 4′ deep	630	910	1,540
3220	6′ deep	825	1,300	2,125
3240	8′ deep	1,050	1,775	2,825
3260	10′ deep	1,225	2,175	3,400
3280	12′ deep	1,525	2,800	4,325
3300	14′ deep	1,850	3,400	5,250
4620	Concrete, cast-in-place, 4′ I.D. riser, 4′ deep	760	1,600	2,360
4640	6′ deep	1,050	2,125	3,175
4660	8′ deep	1,425	3,075	4,500
4680	10′ deep	1,725	3,800	5,525
4700	12′ deep	2,100	4,725	6,825
4720	14′ deep	2,550	5,650	8,200
5820	Concrete, precast, 4′ I.D. riser, 4′ deep	715	835	1,550
5840	6′ deep	960	1,125	2,085
5860	8′ deep	1,150	1,600	2,750
5880	10′ deep	1,425	1,950	3,375
5900	12′ deep	1,800	2,400	4,200
5920	14′ deep	2,200	3,075	5,275

For expanded coverage of these items see *Means Site Work & Landscape Cost Data 2001*

12.3-710	Manholes & Catch Basins	COST PER EACH		
		MAT.	INST.	TOTAL
6000	5′ I.D. riser, 4′ deep	910	910	1,820
6020	6′ deep	1,200	1,275	2,475
6040	8′ deep	1,525	1,675	3,200
6060	10′ deep	1,925	2,150	4,075
6080	12′ deep	2,375	2,750	5,125
6100	14′ deep	2,875	3,350	6,225
6200	6′ I.D. riser, 4′ deep	1,300	1,200	2,500
6220	6′ deep	1,675	1,600	3,275
6240	8′ deep	2,075	2,200	4,275
6260	10′ deep	2,625	2,825	5,450
6280	12′ deep	3,225	3,525	6,750
6300	14′ deep	3,875	4,275	8,150

Septic Systems

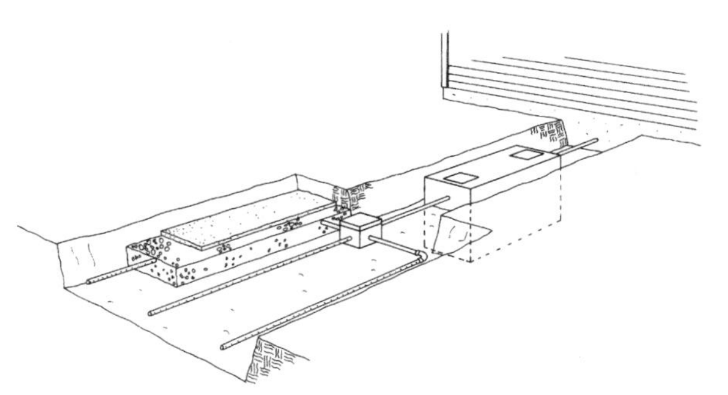

The Septic System includes: a septic tank; leaching field; concrete distribution boxes; plus excavation and gravel backfill.

The Expanded System Listing shows systems with tanks ranging from 1000 gallons to 5000 gallons. Tanks are either concrete or fiberglass. Cost is on a complete unit basis.

System Components	QUANTITY	UNIT	COST EACH MAT.	COST EACH INST.	COST EACH TOTAL
SYSTEM 12.3-900-2010					
1000 GALLON TANK, LEACHING FIELD, 300 S.F., 50′ FROM BLDG.					
Septic tank, precast, 1000 gallon	1.000	Ea.	535	164.55	699.55
Distribution box, precast, 5 outlets	1.000	Ea.	97.50	36	133.50
Sch. 40 sewer pipe, PVC, 4″ diameter	40.000	L.F.	155	259	414
Sch. 40 fitting, wye, & tee, PVC, 4″ diameter	4.000	Ea.	186	43.20	229.20
Excavation, hydraulic backhoe, backfill & load spoil	90.000	C.Y.		323.10	323.10
Stone fill, 3/4″ to 1-1/2″, machine spread	15.000	C.Y.	224.25	368.25	592.50
Haul spoil, 12 C.Y. dump truck, 2 mile round trip	25.000	C.Y.		114.50	114.50
TOTAL			1,197.75	1,308.60	2,506.35

12.3-900	Septic Systems	COST EACH MAT.	COST EACH INST.	COST EACH TOTAL
2010	1000 gal. septic tank, leaching field, 300 S.F., 50′ from bldg.	1,200	1,300	2,500
2110	200′ from bldg.	1,525	1,900	3,425
2210	6′ deep precast pit, one at 6′-6″ diameter, 50′ from bldg.	1,500	1,500	3,000
2310	200′ from bldg.	1,825	2,050	3,875
2410	4′ deep precast galley, 12′ long, 50′ from bldg.	1,550	1,550	3,100
2510	200′ from bldg.	1,875	2,125	4,000
3010	1250 gal. septic tank, leaching field, 400 S.F., 50′ from bldg.	1,475	1,550	3,025
3110	200′ from bldg.	1,800	2,125	3,925
3210	6′ deep precast pit, one at 8′-0″ diameter, 50′ from bldg.	1,950	1,725	3,675
3310	200′ from bldg.	2,275	2,275	4,550
3410	4′ deep precast galley, 16′ long, 50′ from bldg.	1,975	1,800	3,775
3510	200′ from bldg.	2,325	2,400	4,725
4010	1500 gal. septic tank, leaching field, 500 S.F., 50′ from bldg.	1,400	1,700	3,100
4110	200′ from bldg.	1,750	2,250	4,000
4210	6′ deep precast pit, two at 6′-6″ diameter, 50′ from bldg.	2,475	2,525	5,000
4310	200′ from bldg.	2,800	3,075	5,875
4410	4′ deep precast galley, 20′ long, 50′ from bldg.	2,350	2,100	4,450
4510	200′ from bldg.	2,700	2,650	5,350
5010	2000 gal. septic tank, leaching field, 600 S.F., 50′ from bldg.	1,650	2,025	3,675
5110	200′ from bldg.	2,000	2,575	4,575
5210	6′ deep precast pit, two at 8′-0″ diameter, 50′ from bldg.	3,375	3,075	6,450
5310	200′ from bldg.	3,700	3,625	7,325
5410	4′ deep precast galley, 24 ft. long, 50′ from bldg.	2,825	2,525	5,350
5510	200′ from bldg.	3,150	3,100	6,250

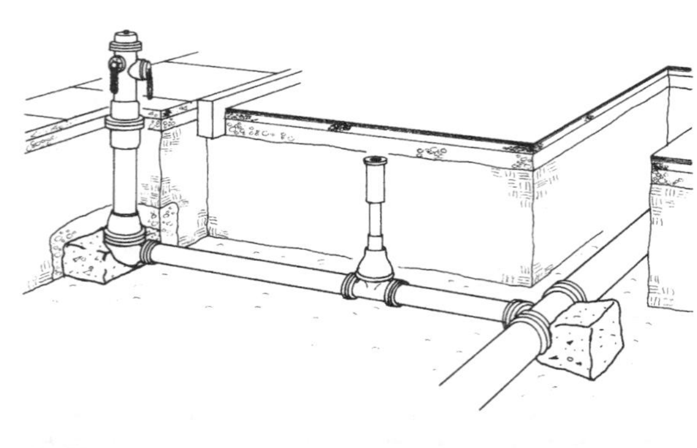

The Fire Hydrant Systems include: four different hydrants with three different lengths of offsets, at several depths of trenching. Excavation and backfill is included with each system as well as thrust blocks as necessary and pipe bedding. Finally spreading of excess material and fine grading complete the components.

System Components	QUANTITY	UNIT	COST PER EACH MAT.	INST.	TOTAL
SYSTEM 12.3-922-1500					
HYDRANT, 4-1/2" VALVE SIZE, TWO WAY, 10′ OFFSET, 2′ DEEP					
Excavation, trench, 1 C.Y. hydraulic backhoe	3.556	C.Y.		18.06	18.06
Sleeve, 12" x 6"	1.000	Ea.	850	73.20	923.20
Offset pipe, 6", CL, 250, ductile iron	8.000	L.F.	80	61.60	141.60
Gate valve, 6"	1.000	Ea.	325	65.80	390.80
Gate box	1.000	Ea.	145	48.50	193.50
Hydrant, incl. lower barrel and shoe	1.000	Ea.	730	131.60	861.60
Thrust blocks, at valve, shoe and sleeve	1.600	C.Y.	166.40	207.36	373.76
Compacted backfill and drainage stone	4.089	C.Y.	58.20	48.83	107.03
Spread excess excavated material	.204	C.Y.		5.30	5.30
Fine grade, hand	400.000	S.F.		168	168
TOTAL			2,354.60	828.25	3,182.85

12.3-922	Fire Hydrants	COST PER EACH MAT.	INST.	TOTAL
1000	Hydrant, 4-1/2" valve size, two way, 0′ offset, 2′ deep	2,275	545	2,820
1100	4′ deep	2,450	545	2,995
1200	6′ deep	2,400	605	3,005
1300	8′ deep	2,575	640	3,215
1400	10′ deep	2,625	685	3,310
1500	10′ offset, 2′ deep	2,350	825	3,175
1600	4′ deep	2,525	960	3,485
1700	6′ deep	2,475	1,250	3,725
1800	8′ deep	2,650	1,650	4,300
1900	10′ deep	2,700	2,475	5,175
2000	20′ offset, 2′ deep	2,450	1,125	3,575
2100	4′ deep	2,625	1,350	3,975
2200	6′ deep	2,575	1,750	4,325
2300	8′ deep	2,750	2,325	5,075
2400	10′ deep	2,800	3,125	5,925
2500	Hydrant, 4-1/2" valve size, three way, 0′ offset, 2′ deep	2,325	555	2,880
2600	4′ deep	2,525	555	3,080
2700	6′ deep	2,475	615	3,090
2800	8′ deep	2,650	655	3,305
2900	10′ deep	2,700	700	3,400
3000	10′ offset, 2′ deep	2,400	840	3,240
3100	4′ deep	2,600	970	3,570
3200	6′ deep	2,550	1,250	3,800
3300	8′ deep	2,725	1,675	4,400
3400	10′ deep	2,775	2,500	5,275
3500	20′ offset, 2′ deep	2,500	1,125	3,625
3600	4′ deep	2,700	1,350	4,050
3700	6′ deep	2,650	1,775	4,425

Important: See the Reference Section for critical supporting data - Reference Nos., Crews, & City Cost Indexes

12.3-922	Fire Hydrants	COST PER EACH		
		MAT.	INST.	TOTAL
3800	8′ deep	2,825	2,375	5,200
3900	10′ deep	2,875	3,150	6,025
5000	5-1/4″ valve size, two way, 0′ offset, 2′ deep	2,450	545	2,995
5100	4′ deep	2,575	560	3,135
5200	6′ deep	2,700	605	3,305
5300	8′ deep	2,825	640	3,465
5400	10′ deep	2,950	685	3,635
5500	10′ offset, 2′ deep	2,525	825	3,350
5600	4′ deep	2,650	975	3,625
5700	6′ deep	2,775	1,250	4,025
5800	8′ deep	2,900	1,650	4,550
5900	10′ deep	3,025	2,475	5,500
6000	20′ offset, 2′ deep	2,625	1,125	3,750
6100	4′ deep	2,750	1,350	4,100
6200	6′ deep	2,875	1,750	4,625
6300	8′ deep	3,000	2,325	5,325
6400	10′ deep	3,125	3,125	6,250
6500	Three way, 0′ offset, 2′ deep	2,525	555	3,080
6600	4′ deep	2,650	570	3,220
6700	6′ deep	2,775	615	3,390
6800	8′ deep	2,900	655	3,555
6900	10′ deep	3,050	700	3,750
7000	10′ offset, 2′ deep	2,600	840	3,440
7100	4′ deep	2,725	980	3,705
7200	6′ deep	2,850	1,250	4,100
7300	8′ deep	3,000	1,675	4,675
7400	10′ deep	3,125	2,500	5,625
7500	20′ offset, 2′ deep	2,700	1,125	3,825
7600	4′ deep	2,825	1,200	4,025
7700	6′ deep	2,950	1,775	4,725

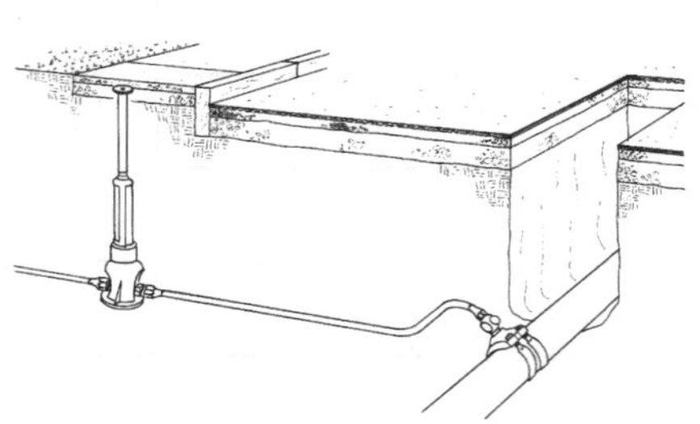

The Water Service Systems are for copper service taps from 1″ to 2″ diameter into pressurized mains from 6″ to 8″ diameter. Costs are given for two offsets with depths varying from 2′ to 10′. Included in system components are excavation and backfill and required curb stops with boxes.

System Components	QUANTITY	UNIT	COST EACH MAT.	COST EACH INST.	COST EACH TOTAL
SYSTEM 12.3-925-1000					
WATER SERVICE, 6″ MAIN, 1″ COPPER SERVICE, 10′ OFFSET, 2′ DEEP					
Trench excavation, 1/2 C.Y. backhoe, 1 laborer	2.220	C.Y.		10.06	10.06
Drill & tap pressurized main, 6″, 1″ to 2″ service	1.000	Ea.		250	250
Corporation, brass, 1″ diameter	1.000	Ea.	28	26	54
Saddle, cast iron, 1″ diameter	1.000	Ea.	25.50		25.50
Copper tubing, type K, 1″ diameter	11.000	L.F.	20.35	25.74	46.09
Curb stop, brass, 1″ diameter	1.000	Ea.	57	26	83
Curb box, cast iron, 1″ diameter	1.000	Ea.	51	34.50	85.50
Backfill by hand, no compaction, heavy soil	2.220	C.Y.		3.27	3.27
Compaction, vibrating plate	2.220	C.Y.		9.16	9.16
TOTAL			181.85	384.73	566.58

12.3-925	Water Service	COST EACH MAT.	COST EACH INST.	COST EACH TOTAL
1000	Water service, 6″ main, 1″ copper service, 10′ offset, 2′ deep	182	385	567
1040	4′ deep	182	425	607
1060	6′ deep	182	475	657
1080	8′ deep	182	545	727
1100	10′ deep	182	575	757
1200	20′ offset, 2′ deep	200	430	630
1240	4′ deep	200	505	705
1260	6′ deep	200	610	810
1280	8′ deep	200	745	945
1300	10′ deep	200	910	1,110
2000	1-1/2″ copper service, 10′ offset, 2′ deep	370	420	790
2040	4′ deep	370	460	830
2060	6′ deep	370	505	875
2080	8′ deep	370	580	950
2100	10′ deep	370	660	1,030
2200	20′ offset, 2′ deep	400	470	870
2240	4′ deep	400	545	945
2260	6′ deep	400	650	1,050
2280	8′ deep	400	785	1,185
2300	10′ deep	400	950	1,350
3000	6″ main, 2″ copper service, 10′ offset, 2′ deep	530	435	965
3040	4′ deep	530	475	1,005
3060	6′ deep	530	525	1,055
3080	8′ deep	530	595	1,125
3100	10′ deep	530	680	1,210
3200	20′ offset, 2′ deep	580	490	1,070
3240	4′ deep	580	570	1,150
3260	6′ deep	580	670	1,250

12.3-925	Water Service	COST EACH		
		MAT.	INST.	TOTAL
3280	8′ deep	580	805	1,385
3300	10′ deep	580	970	1,550
4000	8″ main, 1″ copper service, 10′ offset, 2′ deep	182	405	587
4040	4′ deep	182	445	627
4060	6′ deep	182	495	677
4080	8′ deep	182	565	747
4100	10′ deep	182	650	832
4200	20′ offset, 2′ deep	200	450	650
4240	4′ deep	200	530	730
4260	6′ deep	200	635	835
4280	8′ deep	200	765	965
4300	10′ deep	200	930	1,130
5000	1-1/2″ copper service, 10′ offset, 2′ deep	370	440	810
5040	4′ deep	370	480	850
5060	6′ deep	370	530	900
5080	8′ deep	370	600	970
5100	10′ deep	370	685	1,055
5200	20′ offset, 2′ deep	400	490	890
5240	4′ deep	400	570	970
5260	6′ deep	400	675	1,075
5280	8′ deep	400	805	1,205
5300	10′ deep	400	970	1,370
6000	2″ copper service, 10′ offset, 2′ deep	530	460	990
6040	4′ deep	530	495	1,025
6060	6′ deep	530	545	1,075
6080	8′ deep	530	615	1,145
6100	10′ deep	530	700	1,230
6200	20′ offset, 2′ deep	580	515	1,095
6240	4′ deep	580	590	1,170
6260	6′ deep	580	695	1,275
6280	8′ deep	580	825	1,405
6300	10′ deep	580	990	1,570

A12.7-910 Site Irrigation

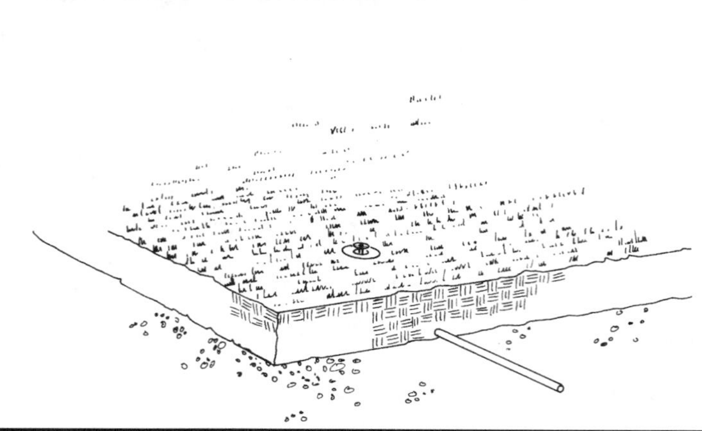

There are three basic types of Site Irrigation Systems: pop-up, riser mounted and quick coupling. Sprinkler heads are spray, impact or gear driven. Each system includes: the hardware for spraying the water; the pipe and fittings needed to deliver the water; and all other accessory equipment such as valves, couplings, nipples, and nozzles. Excavation heads and backfill costs are also included in the system.
The Expanded System Listing shows a wide variety of Site Irrigation Systems.

System Components	QUANTITY	UNIT	COST PER S.F. MAT.	INST.	TOTAL
SYSTEM 12.7-910-1000					
SITE IRRIGATION, POP UP SPRAY, PLASTIC, 10′ RADIUS, 1000 S.F., PVC PIPE					
Excavation, chain trencher	54.000	L.F.		30.78	30.78
Pipe, fittings & nipples, PVC Schedule 40, 1″ dia	5.000	Ea.	30.20	902.50	932.70
Tees, PVC plastic, high pressure schedule, 40, 1″ diameter	5.000	Ea.	15.35	255	270.35
Couplings PVC plastic, high pressure, 1″ diameter	5.000	Ea.	2.25	83.25	85.50
Valves, bronze, globe, 125 lb. rising stem, threaded, 1″ diameter	1.000	Ea.	73.50	22	95.50
Head & nozzle, pop-up spray, PVC plastic	5.000	Ea.	17.35	62.50	79.85
Backfill by hand with compaction	54.000	L.F.		24.30	24.30
TOTAL			138.65	1,380.33	1,518.98
TOTAL			.14	1.38	1.52

12.7-910	Site Irrigation	COST PER S.F. MAT.	INST.	TOTAL
1000	Site irrigation, pop up spray, plastic, 10′ radius, 1000 S.F.,PVC pipe	.14	1.38	1.52
1100	Polyethylene pipe	.13	.49	.62
1200	14′ radius, 8000 S.F., PVC pipe	.05	.26	.31
1300	Polyethylene pipe	.05	.23	.28
1400	18′ square, 1000 S.F. PVC pipe	.12	.39	.51
1500	Polyethylene pipe	.10	.32	.42
1600	24′ square, 8000 S.F., PVC pipe	.04	.16	.20
1700	Polyethylene pipe	.04	.15	.19
1800	4′ x 30′ strip, 200 S.F., PVC pipe	.54	1.67	2.21
1900	Polyethylene pipe	.52	1.50	2.02
2000	6′ x 40′ strip, 800 S.F., PVC pipe	.15	.56	.71
2100	Polyethylene pipe	.14	.50	.64
2200	Economy brass, 11′ radius, 1000 S.F., PVC pipe	.15	.56	.71
2300	Polyethylene pipe	.14	.49	.63
2400	14′ radius, 8000 S.F., PVC pipe	.06	.26	.32
2500	Polyethylene pipe	.06	.23	.29
2600	3′ x 28′ strip, 200 S.F., PVC pipe	.59	1.67	2.26
2700	Polyethylene pipe	.57	1.50	2.07
2800	7′ x 36′ strip, 1000 S.F., PVC pipe	.13	.43	.56
2900	Polyethylene pipe	.12	.38	.50
3000	Hd brass, 11′ radius, 1000 S.F., PVC pipe	.17	.56	.73
3100	Polyethylene pipe	.16	.49	.65
3200	14′ radius, 8000 S.F., PVC pipe	.07	.26	.33
3300	Polyethylene pipe	.07	.23	.30

Important: See the Reference Section for critical supporting data - Reference Nos., Crews, & City Cost Indexes

12.7-910	Site Irrigation	COST PER S.F.		
		MAT.	INST.	TOTAL
3400	Riser mounted spray, plastic,10'radius, 1000S.F., PVC pipe	.13	.56	.69
3500	Polyethylene pipe	.12	.49	.61
3600	12' radius, 5000 S.F., PVC pipe	.06	.32	.38
3700	Polyethylene pipe	.06	.29	.35
3800	19' square, 2000 S.F., PVC pipe	.05	.18	.23
3900	Polyethylene pipe	.04	.16	.20
4000	24' square, 8000 S.F., PVC pipe	.03	.16	.19
4100	Polyethylene pipe	.03	.15	.18
4200	5' x 32' strip, 300 S.F., PVC pipe	.34	1.11	1.45
4300	Polyethylene pipe	.33	1	1.33
4400	6' x 40' strip, 800 S.F., PVC pipe	.14	.56	.70
4500	Polyethylene pipe	.13	.50	.63
4600	Brass, 11' radius, 1000 S.F., PVC pipe	.13	.56	.69
4700	Polyethylene pipe	.12	.49	.61
4800	14' radius, 8000 S.F., PVC pipe	.05	.26	.31
4900	Polythylene pipe	.05	.23	.28
5000	Pop up gear drive stream type,plastic, 30'radius, 10,000 S.F.,PVC pipe	.12	.36	.48
5020	Polyethylene pipe	.05	.11	.16
5040	40,000 S.F., PVC pipe	.10	.36	.46
5060	Polyethylene pipe	.04	.10	.14
5080	Riser mounted gear drive stream type,plas.,30'rad.,10,000S.F.,PVC pipe	.12	.36	.48
5100	Polyethylene pipe	.05	.11	.16
5120	40,000 S.F., PVC pipe	.10	.36	.46
5140	Polyethylene pipe	.04	.10	.14
5200	Q.C. valve thread type w/impact head,brass,75'rad,20,000S.F.,PVC pipe	.06	.12	.18
5250	Polyethylene pipe	.03	.02	.05
5300	100,000 S.F., PVC pipe	.07	.22	.29
5350	Polyethylene pipe	.04	.05	.09
5400	Q.C. valve lug type w/impact head,brass, 75'rad, 20,000 S.F.,PVC pipe	.06	.12	.18
5450	Polyethylene pipe	.03	.02	.05
5500	100,000 S.F. PVC pipe	.08	.22	.30
5550	Polyethylene pipe	.05	.05	.10
6000	Site irrigation,pop up impact type,plastic,40'rad.,10000 S.F.,PVC pipe	.08	.23	.31
6100	Polyethylene pipe	.04	.07	.11
6200	45,000 S.F., PVC pipe	.08	.26	.34
6300	Polyethylene pipe	.04	.07	.11
6400	High medium volume brass, 40' radius, 10000 S.F., PVC pipe	.08	.23	.31
6500	Polyethylene pipe	.04	.07	.11
6600	45,000 S.F. PVC pipe	.08	.26	.34
6700	Polyethylene pipe	.04	.07	.11
6800	High volume brass, 60' radius, 25,000 S.F., PVC pipe	.06	.16	.22
6900	Polyethylene pipe	.03	.03	.06
7000	100,000 S.F., PVC pipe	.06	.17	.23
7100	Polyethylene pipe	.04	.03	.07
7200	Riser mounted part/full impact type,plas.,40'rad.,10,000 S.F.,PVC pipe	.07	.23	.30
7300	Polyethylene pipe	.03	.14	.17
7400	45,000 S.F., PVC pipe	.07	.26	.33
7500	Polyethylene	.03	.07	.10
7600	Low medium volume brass, 40' radius, 10,000 S.F., PVC pipe	.08	.23	.31
7700	Polyethylene pipe	.04	.07	.11
7800	45,000 S.F., PVC pipe	.08	.26	.34
7900	Polyethylene pipe	.04	.07	.11
8000	Medium volume brass, 50' radius, 30,000 S.F., PVC pipe	.07	.19	.26
8100	Polyethylene pipe	.04	.05	.09
8200	70,000 S.F., PVC pipe	.09	.31	.40
8300	Polyethylene pipe	.04	.08	.12
8400	Riser mounted full only impact type,plas.,40'rad.,10000 S.F., PVC pipe	.07	.23	.30
8500	Polyethylene pipe	.03	.07	.10
8600	45,000 S.F., PVC pipe	.07	.26	.33
8700	Polyethylene pipe	.03	.07	.10

12.7-910	Site Irrigation	COST PER S.F.		
		MAT.	INST.	TOTAL
8800	Low medium volume brass, 40' radius, 10,000 S.F., PVC pipe	.07	.23	.30
8900	Polyethylene pipe	.03	.07	.10
9000	45,000 S.F., PVC pipe	.07	.26	.33
9100	Polyethylene pipe	.03	.07	.10
9200	High volume brass, 80' radius, 75,000 S.F., PVC pipe	.04	.12	.16
9300	Polyethylene pipe	.01	.02	.03
9400	150,000 S.F., PVC pipe	.04	.11	.15
9500	Polyethylene pipe	.01	.02	.03
9600	Very high volume brass, 100' radius, 100,000 S.F., PVC pipe	.04	.09	.13
9700	Polyethylene pipe	.01	.02	.03
9800	200,000 S.F., PVC pipe	.04	.09	.13
9900	Polyethylene pipe	.01	.02	.03

A12.7-950 Swimming Pools

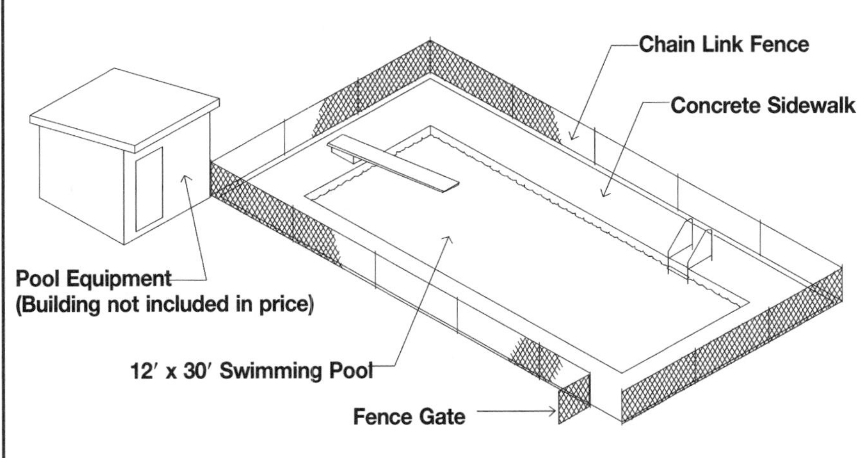

The Swimming Pool System is a complete package. Everything from excavation to deck hardware is included in system costs. Below are three basic types of pool systems: residential, motel, and municipal. Systems elements include: excavation, pool materials, installation, deck hardware, pumps and filters, sidewalk, and fencing.
The Expanded System Listing shows three basic types of pools with a variety of finishes and basic materials. These systems are either vinyl lined with metal sides; gunite shell with a cement plaster finish or tile finish; or concrete sided with vinyl lining. Pool sizes listed here vary from 12′ x 30′ to 60′ x 82.5′. All costs are on a per unit basis.

System Components	QUANTITY	UNIT	COST EACH MAT.	COST EACH INST.	COST EACH TOTAL
SYSTEM 12.7-950-1000					
SWIMMING POOL, RESIDENTIAL, CONC. SIDES, VINYL LINED, 12′X 30′					
Swimming pool, residential, in ground including equipment	360.000	S.F.	4,122	3,333.60	7,455.60
4″ thick reinforced concrete sidewalk, broom finish, no base	400.000	S.F.	476	656	1,132
Chain link fence, residential, 3′ high	124.000	L.F.	343.48	219.48	562.96
Fence gate, chain link	1.000	Ea.	41.50	73.50	115
TOTAL			4,982.98	4,282.58	9,265.56

12.7-950	Swimming Pools	COST EACH MAT.	COST EACH INST.	COST EACH TOTAL
1000	Swimming pool, residential class, concrete sides, vinyl lined, 12′ x 30′	4,975	4,275	9,250
1100	16′ x 32′	5,650	4,850	10,500
1200	20′ x 40′	7,275	6,200	13,475
1500	Tile finish, 12′ x 30′	11,000	18,500	29,500
1600	16′ x 32′	11,500	19,500	31,000
1700	20′ x 40′	17,000	27,400	44,400
2000	Metal sides, vinyl lined, 12′ x 30′	4,300	2,700	7,000
2100	16′ x 32′	4,875	3,050	7,925
2200	20′ x 40′	6,250	3,875	10,125
3000	Gunite shell, cement plaster finish, 12′ x 30′	7,675	7,825	15,500
3100	16′ x 32′	9,675	10,200	19,875
3200	20′ x 40′	13,300	10,100	23,400
4000	Motel class, concrete sides, vinyl lined, 20′ x 40′	10,300	8,650	18,950
4100	28′ x 60′	14,400	12,100	26,500
4500	Tile finish, 20′ x 40′	19,500	31,700	51,200
4600	28′ x 60′	31,800	51,500	83,300
5000	Metal sides, vinyl lined, 20′ x 40′	10,700	6,150	16,850
5100	28′ x 60′	14,400	8,250	22,650
6000	Gunite shell, cement plaster finish, 20′ x 40′	24,600	25,400	50,000
6100	28′ x 60′	36,800	38,000	74,800
7000	Municipal class, gunite shell, cement plaster finish, 42′ x 75′	96,500	89,500	186,000
7100	60′ x 82.5′	124,000	115,000	239,000
7500	Concrete walls, tile finish, 42′ x 75′	116,500	120,500	237,000
7600	60′ x 82.5′	153,000	159,500	312,500
7700	Tile finish and concrete gutter, 42′ x 75′	129,000	120,500	249,500
7800	60′ x 82.5′	167,500	159,500	327,000
7900	Tile finish and stainless gutter, 42′ x 75′	153,000	120,500	273,500
8000	60′x 82.5′	224,000	185,000	409,000

For information about Means Estimating Seminars, see yellow pages 11 and 12 in back of book

For expanded coverage of these items see *Means Site Work & Landscape Cost Data 2001*

Reference Section

All the reference information is in one section making it easy to find what you need to know . . . and easy to use the book on a daily basis. This section is visually identified by a vertical gray bar on the edge of pages.

In the reference number information that follows, you'll see the background that relates to the "reference numbers" that appeared in the Unit Price Section. You'll find reference tables, explanations and estimating information that support how we arrived at the unit price data. Also included are alternate pricing methods, technical data and estimating procedures along with information on design and economy in construction.

Also in this Reference Section, we've included Change Orders, information on pricing changes to contract documents; Crew Listings, a full listing of all the crews, equipment and their costs; Historical Cost Indexes for cost comparisons over time; City Cost Indexes and Location Factors for adjusting costs to the region you are in; and an explanation of all abbreviations used in the book.

Table of Contents

Reference Numbers

Reference Numbers (cont.)

Reference Numbers (cont.)

R01100-005 Tips for Accurate Estimating

1. Use pre-printed or columnar forms for orderly sequence of dimensions and locations and for recording telephone quotations.
2. Use only the front side of each paper or form except for certain pre-printed summary forms.
3. Be consistent in listing dimensions: For example, length x width x height. This helps in rechecking to ensure that, the total length of partitions is appropriate for the building area.
4. Use printed (rather than measured) dimensions where given.
5. Add up multiple printed dimensions for a single entry where possible.
6. Measure all other dimensions carefully.
7. Use each set of dimensions to calculate multiple related quantities.
8. Convert foot and inch measurements to decimal feet when listing. Memorize decimal equivalents to .01 parts of a foot (1/8″ equals approximately .01′).
9. Do not "round off" quantities until the final summary.
10. Mark drawings with different colors as items are taken off.
11. Keep similar items together, different items separate.
12. Identify location and drawing numbers to aid in future checking for completeness.
13. Measure or list everything on the drawings or mentioned in the specifications.
14. It may be necessary to list items not called for to make the job complete.
15. Be alert for: Notes on plans such as N.T.S. (not to scale); changes in scale throughout the drawings; reduced size drawings; discrepancies between the specifications and the drawings.
16. Develop a consistent pattern of performing an estimate. For example:
    a. Start the quantity takeoff at the lower floor and move to the next higher floor.
    b. Proceed from the main section of the building to the wings.
    c. Proceed from south to north or vice versa, clockwise or counterclockwise.
    d. Take off floor plan quantities first, elevations next, then detail drawings.
17. List all gross dimensions that can be either used again for different quantities, or used as a rough check of other quantities for verification (exterior perimeter, gross floor area, individual floor areas, etc.).
18. Utilize design symmetry or repetition (repetitive floors, repetitive wings, symmetrical design around a center line, similar room layouts, etc.). Note: Extreme caution is needed here so as not to omit or duplicate an area.
19. Do not convert units until the final total is obtained. For instance, when estimating concrete work, keep all units to the nearest cubic foot, then summarize and convert to cubic yards.
20. When figuring alternatives, it is best to total all items involved in the basic system, then total all items involved in the alternates. Therefore you work with positive numbers in all cases. When adds and deducts are used, it is often confusing whether to add or subtract a portion of an item; especially on a complicated or involved alternate.

R01100-040 Builder's Risk Insurance

Builder's Risk Insurance is insurance on a building during construction. Premiums are paid by the owner or the contractor. Blasting, collapse and underground insurance would raise total insurance costs above those listed. Floater policy for materials delivered to the job runs $.75 to $1.25 per $100 value. Contractor equipment insurance runs $.50 to $1.50 per $100 value. Insurance for miscellaneous tools to $1,500 value runs from $3.00 to $7.50 per $100 value.

Tabulated below are New England Builder's Risk insurance rates in dollars per $100 value for $1,000 deductible. For $25,000 deductible, rates can be reduced 13% to 34%. On contracts over $1,000,000, rates may be lower than those tabulated. Policies are written annually for the total completed value in place. For "all risk" insurance (excluding flood, earthquake and certain other perils) add $.025 to total rates below.

Coverage	Frame Construction (Class 1)				Brick Construction (Class 4)				Fire Resistive (Class 6)			
	Range			Average	Range			Average	Range			Average
Fire Insurance	$.300	to	$.420	$.394	$.132	to	$.189	$.174	$.052	to	$.080	$.070
Extended Coverage	.115	to	.150	.144	.080	to	.105	.101	.081	to	.105	.100
Vandalism	.012	to	.016	.015	.008	to	.011	.011	.008	to	.011	.010
Total Annual Rate	$.427	to	$.586	$.553	$.220	to	$.305	$.286	$.141	to	$.196	$.180

R01100-050 General Contractor's Overhead

The table below shows a contractor's overhead as a percentage of direct cost in two ways. The figures on the right are for the overhead, markup based on both material and labor. The figures on the left are based on the entire overhead applied only to the labor. This figure would be used if the owner supplied the materials or if a contract is for labor only. Note: Some of these markups are included in the labor rates shown on Reference Table R01100-070.

Items of General Contractor's Indirect Costs	% of Direct Costs	
	As a Markup of Labor Only	As a Markup of Both Material and Labor
Field Supervision	6.0%	2.9%
Main Office Expense (see details below)	16.2	7.7
Tools and Minor Equipment	1.0	0.5
Workers' Compensation & Employers' Liability. See R01100-060	17.5	8.3
Field Office, Sheds, Photos, Etc.	1.5	0.7
Performance and Payment Bond, 0.7% to 1.5%. See R01100-080	2.3	1.1
Unemployment Tax See R01100-100 (Combined Federal and State)	7.0	3.3
Social Security and Medicare, See R01100-100	7.7	3.7
Sales Tax — add if applicable 42/80 x % as markup of total direct costs including both material and labor. See R01100-090		
Sub Total	59.2%	28.2%
*Builder's Risk Insurance ranges from 0.141% to 0.586%. See R01100-040	0.6	0.3
*Public Liability Insurance	3.2	1.5
Grand Total	63.0%	30.0%

*Paid by Owner or Contractor

Main Office Expense

A General Contractor's main office expense consists of many items not detailed in the front portion of the book. The percentage of main office expense declines with increased annual volume of the contractor. Typical main office expense ranges from 2% to 20% with the median about 7.2% of total volume. This equals about 7.7% of direct costs. The following are approximate percentages of total overhead for different items usually included in a General Contractor's main office overhead. With different accounting procedures, these percentages may vary.

Item	Typical Range	Average
Managers', clerical and estimators' salaries	40 % to 55 %	48%
Profit sharing, pension and bonus plans	2 to 20	12
Insurance	5 to 8	6
Estimating and project management (not including salaries)	5 to 9	7
Legal, accounting and data processing	0.5 to 5	3
Automobile and light truck expense	2 to 8	5
Depreciation of overhead capital expenditures	2 to 6	4
Maintenance of office equipment	0.1 to 1.5	1
Office rental	3 to 5	4
Utilities including phone and light	1 to 3	2
Miscellaneous	5 to 15	8
Total		100%

R01100-060 Workers' Compensation Insurance Rates by Trade

The table below tabulates the national averages for Workers' Compensation insurance rates by trade and type of building. The average "Insurance Rate" is multiplied by the "% of Building Cost" for each trade. This produces the "Workers' Compensation Cost" by % of total labor cost, to be added for each trade by building type to determine the weighted average Workers' Compensation rate for the building types analyzed.

Trade	Insurance Rate (% Labor Cost) Range	Insurance Rate Average	% of Building Cost Office Bldgs.	% of Building Cost Schools & Apts.	% of Building Cost Mfg.	Workers' Compensation Office Bldgs.	Workers' Compensation Schools & Apts.	Workers' Compensation Mfg.
Excavation, Grading, etc.	3.8 % to 26.6%	10.9%	4.8%	4.9%	4.5%	.52%	.53%	.49%
Piles & Foundations	5.9 to 80.1	27.7	7.1	5.2	8.7	1.97	1.44	2.41
Concrete	5.9 to 35.5	17.8	5.0	14.8	3.7	.89	2.63	.66
Masonry	5.3 to 48.6	17.0	6.9	7.5	1.9	1.17	1.28	.32
Structural Steel	5.9 to 132.9	40.9	10.7	3.9	17.6	4.38	1.60	7.20
Miscellaneous & Ornamental Metals	4.9 to 34.0	13.2	2.8	4.0	3.6	.37	.53	.48
Carpentry & Millwork	5.9 to 49.9	19.0	3.7	4.0	0.5	.70	.76	.10
Metal or Composition Siding	5.9 to 36.7	17.6	2.3	0.3	4.3	.40	.05	.76
Roofing	5.9 to 88.8	34.3	2.3	2.6	3.1	.79	.89	1.06
Doors & Hardware	4.1 to 28.8	11.4	0.9	1.4	0.4	.10	.16	.05
Sash & Glazing	5.7 to 32.4	14.1	3.5	4.0	1.0	.49	.56	.14
Lath & Plaster	5.2 to 44.6	15.7	3.3	6.9	0.8	.52	1.08	.13
Tile, Marble & Floors	3.7 to 29.5	10.3	2.6	3.0	0.5	.27	.31	.05
Acoustical Ceilings	3.5 to 26.7	11.8	2.4	0.2	0.3	.28	.02	.04
Painting	5.9 to 41.1	14.8	1.5	1.6	1.6	.22	.24	.24
Interior Partitions	5.9 to 49.9	19.0	3.9	4.3	4.4	.74	.82	.84
Miscellaneous Items	2.6 to 128.0	18.5	5.2	3.7	9.7	.96	.68	1.79
Elevators	2.0 to 18.9	8.4	2.1	1.1	2.2	.18	.09	.18
Sprinklers	3.0 to 20.6	8.8	0.5	—	2.0	.04	—	.18
Plumbing	3.0 to 15.9	8.5	4.9	7.2	5.2	.42	.61	.44
Heat., Vent., Air Conditioning	3.5 to 32.5	11.9	13.5	11.0	12.9	1.61	1.31	1.54
Electrical	2.7 to 11.0	6.8	10.1	8.4	11.1	.69	.57	.75
Total	2.0 % to 132.9%	—	100.0%	100.0%	100.0%	17.71%	16.16%	19.85%
		Overall Weighted Average	17.91%					

Workers' Compensation Insurance Rates by States

The table below lists the weighted average Workers' Compensation base rate for each state with a factor comparing this with the national average of 17.5%.

State	Weighted Average	Factor	State	Weighted Average	Factor	State	Weighted Average	Factor
Alabama	30.5%	174	Kentucky	19.8%	113	North Dakota	15.0%	86
Alaska	11.1	63	Louisiana	27.3	156	Ohio	19.1	109
Arizona	12.5	71	Maine	21.8	125	Oklahoma	21.9	125
Arkansas	14.2	81	Maryland	11.9	68	Oregon	19.2	110
California	18.6	106	Massachusetts	17.5	100	Pennsylvania	22.9	131
Colorado	26.5	151	Michigan	20.6	118	Rhode Island	20.9	119
Connecticut	19.4	111	Minnesota	37.7	215	South Carolina	13.8	79
Delaware	12.3	70	Mississippi	22.8	130	South Dakota	14.6	83
District of Columbia	21.4	122	Missouri	15.8	90	Tennessee	13.0	74
Florida	28.0	160	Montana	37.3	213	Texas	16.4	94
Georgia	24.9	142	Nebraska	15.7	90	Utah	12.9	74
Hawaii	16.7	95	Nevada	19.4	111	Vermont	14.9	85
Idaho	10.8	62	New Hampshire	23.2	133	Virginia	10.9	62
Illinois	23.3	133	New Jersey	11.0	63	Washington	10.0	57
Indiana	7.2	41	New Mexico	23.1	132	West Virginia	14.3	82
Iowa	12.1	69	New York	16.4	94	Wisconsin	13.7	78
Kansas	9.6	55	North Carolina	13.3	76	Wyoming	6.5	37
			Weighted Average for U.S. is	17.9% of payroll = 100%				

Rates in the following table are the base or manual costs per $100 of payroll for Workers' Compensation in each state. Rates are usually applied to straight time wages only and not to premium time wages and bonuses.

The weighted average skilled worker rate for 35 trades is 17.5%. For bidding purposes, apply the full value of Workers' Compensation directly to total labor costs, or if labor is 38%, materials 42% and overhead and profit 20% of total cost, carry 38/80 x 17.5% =8.3% of cost (before overhead and profit) into overhead. Rates vary not only from state to state but also with the experience rating of the contractor.

Rates are the most current available at the time of publication.

R01100-060 Workers' Compensation Insurance Rates by Trade and State (cont.)

State	Carpentry — 3 stories or less 5651	Carpentry — interior cab. work 5437	Carpentry — general 5403	Concrete Work — NOC 5213	Concrete Work — flat (flr., sdwk.) 5221	Electrical Wiring — inside 5190	Excavation — earth NOC 6217	Excavation — rock 6217	Glaziers 5462	Insulation Work 5479	Lathing 5443	Masonry 5022	Painting & Decorating 5474	Pile Driving 6003	Plastering 5480	Plumbing 5183	Roofing 5551	Sheet Metal Work (HVAC) 5538	Steel Erection — door & sash 5102	Steel Erection — inter., ornam. 5102	Steel Erection — structure 5040	Steel Erection — NOC 5057	Tile Work — (interior ceramic) 5348	Waterproofing 9014	Wrecking 5701
AL	29.00	14.42	30.31	23.87	18.39	10.36	19.14	19.14	32.36	21.66	14.63	25.12	19.63	49.29	44.55	12.07	75.95	30.15	17.28	17.28	52.35	65.34	20.31	7.12	54.68
AK	8.81	8.42	9.30	9.30	5.59	5.67	8.82	8.82	9.10	10.01	5.64	10.99	8.49	29.93	9.97	5.25	17.33	6.01	9.40	9.40	17.20	17.20	6.23	6.38	25.07
AZ	12.94	7.31	23.11	11.90	7.03	5.46	7.53	7.53	11.21	13.57	6.80	11.90	8.89	17.26	10.66	6.75	17.09	8.80	8.71	8.71	31.81	20.24	6.29	4.29	50.31
AR	16.12	10.60	15.38	16.45	5.97	7.93	9.21	9.21	12.58	15.29	12.66	9.32	11.10	19.14	13.01	6.58	30.74	10.28	8.19	8.19	27.11	25.39	7.80	4.58	27.11
CA	28.28	9.07	28.28	13.18	13.18	9.93	7.88	7.88	16.37	23.34	10.90	14.55	19.45	21.18	18.12	11.59	39.79	15.33	13.55	13.55	23.93	21.91	7.74	19.45	21.91
CO	32.22	13.54	20.95	20.96	15.13	9.20	15.99	15.99	15.98	21.31	14.68	31.06	21.85	41.53	37.40	14.50	59.10	12.54	11.83	11.83	73.46	43.49	15.04	12.26	73.46
CT	22.88	13.79	21.35	15.44	13.85	5.94	8.19	8.19	21.67	26.13	8.60	23.82	16.74	20.45	16.66	9.16	29.83	9.89	11.52	11.52	66.97	27.81	11.14	5.71	50.86
DE	13.83	13.83	11.56	9.28	6.29	5.82	8.54	8.54	11.18	11.56	10.51	9.85	13.41	11.64	10.51	5.41	22.89	10.29	10.21	10.21	26.69	10.21	7.54	9.85	25.69
DC	16.64	7.93	16.96	29.16	16.85	7.03	12.07	12.07	21.57	18.23	11.28	25.62	14.04	32.48	18.37	10.51	23.16	11.62	24.74	24.74	60.38	32.31	17.92	4.39	66.43
FL	36.73	23.28	29.09	29.92	16.15	10.51	15.27	15.27	17.81	25.18	24.27	28.72	28.92	49.35	28.79	13.37	53.20	17.12	18.67	18.67	43.71	50.61	12.56	7.89	43.71
GA	31.04	16.02	32.71	20.67	15.12	9.82	19.55	19.55	21.97	19.15	26.66	23.03	21.20	40.88	19.50	11.14	40.72	18.65	13.74	13.74	36.03	53.09	13.08	10.16	36.03
HI	17.38	11.62	28.20	13.74	12.27	7.10	7.73	7.73	20.55	21.19	10.94	17.87	11.33	20.15	15.61	6.06	33.24	8.02	11.11	11.11	31.71	21.70	9.78	11.54	31.71
ID	8.79	6.95	14.19	10.28	10.45	4.56	6.41	6.41	6.96	10.12	7.12	7.92	7.51	14.30	8.97	3.95	20.53	5.87	8.05	8.05	27.60	20.85	4.78	6.59	26.68
IL	20.76	12.31	21.64	35.48	12.83	8.86	8.69	8.69	23.09	24.90	14.24	20.34	13.15	37.63	13.78	11.68	35.88	14.58	21.53	21.53	58.63	52.47	13.87	6.26	58.63
IN	7.08	4.06	7.07	6.68	3.23	3.00	3.79	3.79	5.72	7.01	3.48	5.32	5.94	9.68	5.16	3.00	11.87	4.29	4.92	4.92	28.84	13.32	3.69	3.26	27.19
IA	10.22	5.19	12.01	9.18	6.42	3.84	5.56	5.56	11.62	16.49	5.95	10.93	7.58	16.30	9.26	5.01	17.38	6.09	9.80	9.80	45.02	27.20	5.60	4.16	45.02
KS	12.10	7.50	9.17	10.12	6.55	4.01	5.02	5.02	7.59	13.37	5.73	9.94	7.28	12.42	8.78	4.96	20.54	5.93	5.93	5.93	19.14	17.05	4.16	3.66	16.49
KY	16.47	15.26	24.34	22.12	9.53	8.95	13.67	13.67	15.96	14.68	10.62	22.00	19.83	30.44	13.94	10.38	32.83	15.92	13.21	13.21	45.21	33.73	10.60	8.85	45.21
LA	29.63	22.95	26.01	20.86	14.80	10.48	24.73	24.73	23.88	20.82	14.00	21.27	27.63	54.39	21.42	13.78	49.11	21.35	15.51	15.51	66.41	36.94	12.82	10.26	66.41
ME	13.43	10.42	43.59	24.29	11.12	10.02	14.75	14.75	13.35	15.39	14.03	17.26	16.07	31.44	17.95	11.12	32.07	17.70	15.08	15.08	37.84	61.81	11.66	9.37	37.84
MD	10.55	5.95	10.55	11.35	5.05	5.15	9.25	9.25	13.20	13.05	6.45	11.35	6.75	27.45	6.65	5.55	22.80	7.00	9.15	9.15	26.80	18.50	6.35	3.50	26.80
MA	11.05	7.61	13.61	22.12	9.81	3.69	5.91	5.91	9.92	18.24	8.84	17.55	9.04	17.59	7.39	5.69	67.48	7.65	11.96	11.96	50.78	46.56	10.48	3.89	34.86
MI	16.08	9.61	17.85	24.50	12.37	5.71	10.99	10.99	9.62	20.95	11.61	19.18	17.86	64.20	18.33	8.19	43.08	10.91	14.52	14.52	36.41	31.20	13.08	10.42	41.58
MN	27.66	28.76	49.94	30.20	22.59	8.28	19.73	19.73	30.01	37.20	23.93	29.89	19.91	80.08	23.93	15.88	88.83	13.91	27.21	27.21	132.92	31.83	29.54	8.71	132.92
MS	24.99	20.10	26.38	17.69	10.13	9.84	15.53	15.53	18.15	21.31	11.97	19.52	16.75	55.90	22.89	10.23	33.43	21.81	12.10	12.10	46.09	36.57	12.26	9.38	37.42
MO	18.72	7.52	11.84	13.28	10.66	5.79	11.05	11.05	8.97	19.99	9.67	13.57	11.77	16.88	14.22	7.10	25.74	9.76	13.82	13.82	47.98	31.96	5.83	5.71	47.98
MT	30.63	15.19	47.53	32.78	24.74	11.01	26.63	26.63	25.04	38.54	19.54	48.55	41.06	53.68	27.00	15.03	85.06	16.93	33.95	33.95	86.98	55.25	13.30	12.10	86.98
NE	16.19	9.23	15.94	17.31	12.73	5.43	10.19	10.19	12.36	16.79	9.34	18.07	11.58	22.01	14.62	9.06	33.91	13.86	10.75	10.75	25.03	26.96	7.88	5.47	25.03
NV	19.86	10.88	14.94	14.89	11.46	9.64	9.26	9.26	13.64	22.78	14.60	14.48	16.01	27.25	17.93	10.76	33.53	18.03	16.53	16.53	45.36	32.56	18.03	7.43	45.36
NH	16.38	9.82	19.68	28.99	9.29	7.10	13.87	13.87	12.87	43.98	10.90	25.97	17.94	28.09	17.33	9.74	58.76	12.48	13.27	13.27	55.90	49.67	12.82	7.37	55.90
NJ	10.40	8.85	10.40	10.83	8.37	4.23	7.95	7.95	7.36	12.78	7.88	10.77	11.49	11.32	7.88	5.48	29.41	6.20	8.03	8.03	22.52	12.96	5.74	5.37	29.92
NM	25.35	13.09	20.45	19.66	13.39	6.34	10.95	10.95	14.58	31.49	16.25	26.97	16.87	33.27	15.64	12.10	43.41	16.46	21.26	21.26	67.82	29.91	9.67	10.65	67.82
NY	15.13	6.82	14.34	22.04	15.14	6.77	8.75	8.75	16.13	11.09	14.58	21.00	13.37	29.90	11.57	8.49	32.78	15.10	14.46	14.46	21.93	20.26	12.11	5.85	32.53
NC	13.08	11.23	15.08	16.19	6.34	7.36	9.18	9.18	8.25	12.80	6.49	11.00	10.42	18.01	15.63	6.06	29.08	10.36	10.19	10.19	29.07	15.79	7.44	3.74	39.03
ND	12.70	12.70	12.70	9.80	9.80	5.57	9.09	9.06	8.83	12.70	11.94	10.99	9.02	25.96	11.94	7.26	28.69	7.26	12.70	12.70	25.96	25.96	8.80	28.69	20.13
OH	17.06	17.06	17.06	15.16	15.16	9.12	15.16	15.16	22.67	17.90	17.90	20.44	22.67	15.16	17.90	10.47	33.45	32.47	15.16	15.16	15.16	15.16	13.22	15.16	15.16
OK	26.16	11.79	19.33	15.51	11.50	7.41	18.45	18.45	13.55	18.00	12.05	16.80	17.10	36.26	18.29	8.72	42.26	12.45	11.17	11.17	69.45	46.76	11.42	8.20	69.45
OR	29.51	9.28	20.03	16.50	10.61	6.63	12.94	12.94	14.95	19.68	12.27	12.43	19.83	22.05	13.85	8.38	38.11	11.85	11.01	11.01	61.69	25.54	9.92	12.10	61.69
PA	15.80	15.80	19.85	28.18	11.84	7.80	12.74	12.74	15.69	19.85	18.94	18.96	22.05	32.12	18.94	11.57	43.30	12.58	24.71	24.71	59.05	24.71	11.83	18.96	81.34
RI	19.53	11.65	18.07	18.23	16.24	4.43	10.38	10.38	12.85	22.78	11.97	25.11	24.13	37.66	17.25	8.52	33.92	10.29	14.07	14.07	59.49	37.50	14.36	7.70	78.79
SC	19.10	13.03	19.24	12.88	6.19	7.02	8.32	8.32	12.22	10.69	7.27	9.55	11.19	16.64	18.89	6.88	28.98	11.42	9.49	9.49	21.80	23.66	6.23	4.65	21.80
SD	13.70	6.67	17.59	16.61	6.07	5.60	9.23	9.23	11.50	15.30	7.77	10.02	12.97	31.49	12.79	8.27	20.70	7.55	9.64	9.64	45.63	21.23	6.90	4.75	46.57
TN	13.22	9.52	11.99	12.39	7.53	5.00	8.68	8.68	8.96	15.04	8.51	12.73	10.88	15.62	12.19	5.91	24.81	8.76	7.36	7.36	45.06	13.95	5.89	4.98	45.06
TX	16.89	12.49	16.89	18.90	10.64	8.69	11.28	11.28	11.91	14.50	12.98	16.51	11.74	22.91	20.99	8.82	27.56	15.13	11.85	11.85	27.79	21.44	9.25	7.91	28.44
UT	11.17	11.17	11.17	17.68	7.99	7.05	6.38	6.38	9.55	10.95	12.32	13.75	15.29	17.24	10.60	6.41	26.27	6.30	8.99	8.99	23.51	23.51	6.06	5.83	26.53
VT	11.93	7.61	14.88	34.92	9.35	4.51	7.91	7.91	11.95	11.32	8.84	13.76	7.82	19.81	10.86	6.71	24.32	10.29	10.29	10.29	32.12	35.79	6.45	8.56	32.12
VA	10.00	6.46	8.65	9.29	5.47	3.45	6.13	6.13	10.17	8.77	7.59	8.07	8.11	24.38	8.20	5.60	17.80	6.74	10.42	10.42	19.28	30.25	6.78	2.57	19.28
WA	7.57	7.57	7.57	6.22	6.22	2.71	7.32	7.32	9.42	7.30	7.57	9.87	8.81	14.33	10.96	4.47	19.48	3.47	12.96	12.96	12.96	12.96	8.93	10.23	12.96
WV	14.98	14.98	14.98	26.36	26.36	5.02	9.10	9.10	6.96	6.96	16.88	14.51	16.81	10.03	16.81	6.65	12.47	6.96	14.98	14.98	11.89	15.59	16.02	4.67	13.27
WI	10.31	8.30	19.64	9.42	8.44	5.53	6.39	6.39	10.37	12.12	13.99	15.11	10.03	18.39	11.59	5.53	32.21	8.32	12.89	12.89	32.11	16.18	10.39	3.63	36.51
WY	5.94	5.94	5.94	5.94	5.94	5.94	5.94	5.94	5.94	5.94	5.94	5.94	5.94	5.94	5.94	5.94	5.94	5.94	5.94	5.94	5.94	5.94	5.94	5.94	5.94
AVG.	17.57	11.40	19.01	17.82	11.14	6.79	10.93	10.93	14.08	17.65	11.76	17.04	14.81	27.68	15.71	8.47	34.33	11.93	13.21	13.21	40.87	29.19	10.30	8.04	42.15

R01100-060 Workers' Compensation (cont.) (Canada in Canadian dollars)

Province		Alberta	British Columbia	Manitoba	Ontario	New Brunswick	Newfndld. & Labrador	Northwest Territories	Nova Scotia	Prince Edward Island	Quebec	Saskat-chewan	Yukon
Carpentry—3 stories or less	Rate	2.53	5.20	5.35	6.66	3.76	6.00	2.50	6.96	5.96	13.01	5.38	2.35
	Code	40400	721028	40102	723	422	403	4-41	4022	401	80110	B12-02	4-042
Carpentry—interior cab. work	Rate	1.70	5.44	5.35	6.66	3.85	6.00	2.50	6.96	3.64	13.01	3.86	2.35
	Code	42133	721021	40102	723	427	403	4-41	4022	402	80110	B11-25	4-042
CARPENTRY—general	Rate	2.53	5.20	5.35	6.66	3.76	6.00	2.50	6.96	5.96	13.01	5.38	2.35
	Code	40400	721028	40102	723	422	403	4-41	4022	401	80110	B12-02	4-042
CONCRETE WORK—NOC	Rate	3.63	7.33	6.72	15.22	3.76	6.00	2.50	6.31	5.96	15.31	7.99	3.25
	Code	42104	721010	40110	745	422	403	4-41	4224	401	80100	B14-04	2-032
CONCRETE WORK—flat	Rate	3.63	7.33	6.72	15.22	3.76	6.00	2.50	6.31	5.96	15.31	7.99	3.25
(flr. sidewalk)	Code	42104	721010	40110	745	422	403	4-41	4224	401	80100	B14-04	2-032
ELECTRICAL Wiring—inside	Rate	1.73	3.93	3.50	3.20	1.49	4.83	2.00	2.70	3.64	6.97	3.86	2.35
	Code	42124	721019	40203	704	426	400	4-46	4261	402	80170	B11-05	4-041
EXCAVATION—earth NOC	Rate	2.19	4.03	5.35	4.54	2.72	6.00	2.50	3.64	3.76	7.40	4.69	3.25
	Code	40604	721031	40706	711	421	403	4-43	4214	404	80030	R11-06	2-016
EXCAVATION—rock	Rate	2.19	4.03	5.35	4.54	2.72	6.00	2.50	3.64	3.76	7.40	4.69	3.25
	Code	40604	721031	40706	711	421	403	4-43	4214	404	80030	R11-06	2-016
GLAZIERS	Rate	2.14	3.37	3.50	8.82	3.85	4.55	2.50	9.30	3.64	16.25	7.99	2.35
	Code	42121	715020	40109	751	423	402	4-41	4233	402	80150	B13-04	4-042
INSULATION WORK	Rate	2.53	6.46	5.35	8.82	3.85	4.55	2.50	9.30	5.96	14.15	5.38	3.25
	Code	42184	721029	40102	751	423	402	4-41	4234	401	80120	B12-07	2-035
LATHING	Rate	4.58	8.12	5.35	6.66	3.85	4.55	2.50	6.34	3.64	14.15	7.99	3.25
	Code	42135	721033	40102	723	427	402	4-41	4271	402	80120	B13-02	2-036
MASONRY	Rate	3.63	8.12	5.35	14.22	3.85	6.00	2.50	9.30	5.96	15.31	7.99	3.25
	Code	42102	721037	40102	741	423	403	4-41	4231	401	80100	B15-01	2-032
PAINTING & DECORATING	Rate	2.93	6.46	6.72	8.87	3.85	4.55	2.50	6.34	3.64	14.15	5.38	3.25
	Code	42111	721041	40105	719	427	402	4-41	4275	402	80120	B12-01	2-036
PILE DRIVING	Rate	3.63	14.60	5.35	5.27	3.76	9.60	2.50	6.31	5.96	7.40	5.38	3.25
	Code	42159	722004	40706	732	422	404	4-43	4221	401	80030	B12-10	2-030
PLASTERING	Rate	4.58	8.12	6.72	8.87	3.85	4.55	2.50	6.34	3.64	14.15	7.99	3.25
	Code	42135	721042	40108	719	427	402	4-41	4271	402	80120	B13-02	2-036
PLUMBING	Rate	1.73	4.30	3.50	4.04	1.88	3.88	2.00	3.64	3.64	7.55	3.86	2.35
	Code	42122	721043	40204	707	424	401	4-46	4241	402	80160	B11-01	4-039
ROOFING	Rate	7.69	10.43	6.72	12.45	7.72	6.00	2.50	9.30	5.96	21.16	7.99	3.25
	Code	42118	721036	40403	728	430	403	4-41	4236	401	80130	B15-02	2-031
SHEET METAL WORK (HVAC)	Rate	1.73	4.30	5.35	4.04	1.88	3.88	2.00	3.64	3.64	7.55	3.86	2.35
	Code	42117	721043	40402	707	424	401	4-46	4244	402	80160	B11-07	4-040
STEEL ERECTION—door & sash	Rate	2.53	14.60	5.35	18.61	3.76	9.60	2.50	9.33	5.96	30.77	7.99	3.25
	Code	42106	722005	40502	748	422	404	4-41	4227	401	80080	B15-03	2-012
STEEL ERECTION—inter., ornam.	Rate	2.53	14.60	5.35	18.61	3.76	6.00	2.50	9.33	5.96	30.77	7.99	3.25
	Code	42106	722005	40502	748	422	403	4-41	4227	401	80080	B15-03	2-012
STEEL ERECTION—structure	Rate	2.53	14.60	5.35	18.61	3.76	9.60	2.50	9.33	5.96	30.77	7.99	3.25
	Code	42106	722005	40502	748	422	404	4-41	4227	401	80080	B15-04	2-012
STEEL ERECTION—NOC	Rate	2.53	14.60	5.35	18.61	3.76	9.60	2.50	9.33	5.96	30.77	7.99	3.25
	Code	42106	722005	40502	748	422	404	4-41	4227	401	80080	B15-03	2-012
TILE WORK—inter. (ceramic)	Rate	2.56	3.57	2.31	8.87	3.85	4.55	2.50	6.34	3.64	14.15	7.99	3.25
	Code	42113	721054	40103	719	427	402	4-41	4276	402	80120	B13-01	2-034
WATERPROOFING	Rate	2.93	6.46	5.35	6.66	3.85	6.00	2.50	9.30	3.64	21.16	3.86	3.25
	Code	42139	721016	40102	723	423	403	4-41	4239	402	80130	B11-17	2-030
WRECKING	Rate	12.86	6.75	6.72	18.61	2.72	6.00	2.50	3.64	5.96	25.35	7.99	3.25
	Code	42108	721005	40106	748	421	403	4-43	4211	401	80220	B14-07	2-030

R01100-070 Contractor's Overhead & Profit

Below are the **average** installing contractor's percentage mark-ups applied to base labor rates to arrive at typical billing rates.

Column A: Labor rates are based on union wages averaged for 30 major U.S. cities. Base rates including fringe benefits are listed hourly and daily. These figures are the sum of the wage rate and employer-paid fringe benefits such as vacation pay, employer-paid health and welfare costs, pension costs, plus appropriate training and industry advancement funds costs.

Column B: Workers' Compensation rates are the national average of state rates established for each trade.

Column C: Column C lists average fixed overhead figures for all trades. Included are Federal and State Unemployment costs set at 7.0%; Social Security Taxes (FICA) set at 7.65%; Builder's Risk Insurance costs set at 0.34%; and Public Liability costs set at 1.55%. All the percentages except those for Social Security Taxes vary from state to state as well as from company to company.

Columns D and E: Percentages in Columns D and E are based on the presumption that the installing contractor has annual billing of $1,500,000 and up. Overhead percentages may increase with smaller annual billing. The overhead percentages for any given contractor may vary greatly and depend on a number of factors, such as the contractor's annual volume, engineering and logistical support costs, and staff requirements. The figures for overhead and profit will also vary depending on the type of job, the job location, and the prevailing economic conditions. All factors should be examined very carefully for each job.

Column F: Column F lists the total of Columns B, C, D, and E.

Column G: Column G is Column A (hourly base labor rate) multiplied by the percentage in Column F (O&P percentage).

Column H: Column H is the total of Column A (hourly base labor rate) plus Column G (Total O&P).

Column I: Column I is Column H multiplied by eight hours.

		A		B	C	D	E	F	G	H	I
		Base Rate Incl. Fringes		Workers' Comp. Ins.	Average Fixed Overhead	Overhead		Total Overhead & Profit		Rate with O & P	
Abbr.	Trade	Hourly	Daily				Profit	%	Amount	Hourly	Daily
Skwk	Skilled Workers Average (35 trades)	$29.85	$238.80	17.5%	16.5%	13.0%	10.0%	57.0%	$17.00	$46.85	$374.80
	Helpers Average (5 trades)	22.15	177.20	19.3		11.0		56.8	12.55	34.70	277.60
	Foreman Average, Inside ($0.50 over trade)	30.35	242.80	17.5		13.0		57.0	17.30	47.65	381.20
	Foreman Average, Outside ($2.00 over trade)	31.85	254.80	17.5		13.0		57.0	18.15	50.00	400.00
Clab	Common Building Laborers	22.85	182.80	19.0		11.0		56.5	12.90	35.75	286.00
Asbe	Asbestos/Insulation Workers/Pipe Coverers	32.15	257.20	17.7		16.0		60.2	19.35	51.50	412.00
Boil	Boilermakers	34.55	276.40	15.9		16.0		58.4	20.20	54.75	438.00
Bric	Bricklayers	29.60	236.80	17.0		11.0		54.5	16.15	45.75	366.00
Brhe	Bricklayer Helpers	22.95	183.60	17.0		11.0		54.5	12.50	35.45	283.60
Carp	Carpenters	29.15	233.20	19.0		11.0		56.5	16.45	45.60	364.80
Cefi	Cement Finishers	27.95	223.60	11.1		11.0		48.6	13.60	41.55	332.40
Elec	Electricians	34.30	274.40	6.8		16.0		49.3	16.90	51.20	409.60
Elev	Elevator Constructors	35.65	285.20	8.4		16.0		50.9	18.15	53.80	430.40
Eqhv	Equipment Operators, Crane or Shovel	31.15	249.20	10.9		14.0		51.4	16.00	47.15	377.20
Eqmd	Equipment Operators, Medium Equipment	30.35	242.80	10.9		14.0		51.4	15.60	45.95	367.60
Eqlt	Equipment Operators, Light Equipment	28.75	230.00	10.9		14.0		51.4	14.80	43.55	348.40
Eqol	Equipment Operators, Oilers	25.75	206.00	10.9		14.0		51.4	13.25	39.00	312.00
Eqmm	Equipment Operators, Master Mechanics	31.80	254.40	10.9		14.0		51.4	16.35	48.15	385.20
Glaz	Glaziers	28.55	228.40	14.1		11.0		51.6	14.75	43.30	346.40
Lath	Lathers	28.20	225.60	11.8		11.0		49.3	13.90	42.10	336.80
Marb	Marble Setters	29.20	233.60	17.0		11.0		54.5	15.90	45.10	360.80
Mill	Millwrights	30.55	244.40	11.2		11.0		48.7	14.90	45.45	363.60
Mstz	Mosaic and Terrazzo Workers	28.30	226.40	10.3		11.0		47.8	13.55	41.85	334.80
Pord	Painters, Ordinary	26.10	208.80	14.8		11.0		52.3	13.65	39.75	318.00
Psst	Painters, Structural Steel	27.15	217.20	50.0		11.0		87.5	23.75	50.90	407.20
Pape	Paper Hangers	26.25	210.00	14.8		11.0		52.3	13.75	40.00	320.00
Pile	Pile Drivers	28.80	230.40	27.7		16.0		70.2	20.20	49.00	392.00
Plas	Plasterers	27.30	218.40	15.7		11.0		53.2	14.50	41.80	334.40
Plah	Plasterer Helpers	23.25	186.00	15.7		11.0		53.2	12.35	35.60	284.80
Plum	Plumbers	34.45	275.60	8.5		16.0		51.0	17.55	52.00	416.00
Rodm	Rodmen (Reinforcing)	32.75	262.00	29.2		14.0		69.7	22.85	55.60	444.80
Rofc	Roofers, Composition	25.70	205.60	34.3		11.0		71.8	18.45	44.15	353.20
Rots	Roofers, Tile and Slate	26.00	208.00	34.3		11.0		71.8	18.65	44.65	357.20
Rohe	Roofer Helpers (Composition)	19.10	152.80	34.3		11.0		71.8	13.70	32.80	262.40
Shee	Sheet Metal Workers	33.55	268.40	11.9		16.0		54.4	18.25	51.80	414.40
Spri	Sprinkler Installers	34.45	275.60	8.8		16.0		51.3	17.65	52.10	416.80
Stpi	Steamfitters or Pipefitters	34.85	278.80	8.5		16.0		51.0	17.75	52.60	420.80
Ston	Stone Masons	29.75	238.00	17.0		11.0		54.5	16.20	45.95	367.60
Sswk	Structural Steel Workers	32.85	262.80	40.9		14.0		81.4	26.75	59.60	476.80
Tilf	Tile Layers	28.15	225.20	10.3		11.0		47.8	13.45	41.60	332.80
Tilh	Tile Layer Helpers	22.70	181.60	10.3		11.0		47.8	10.85	33.55	268.40
Trlt	Truck Drivers, Light	23.40	187.20	15.3		11.0		52.8	12.35	35.75	286.00
Trhv	Truck Drivers, Heavy	23.95	191.60	15.3		11.0		52.8	12.65	36.60	292.80
Sswl	Welders, Structural Steel	32.85	262.80	40.9		14.0		81.4	26.75	59.60	476.80
Wrck	*Wrecking	22.85	182.80	42.2		11.0		79.7	18.20	41.05	328.40

*Not included in Averages.

R01100-080 Performance Bond

This table shows the cost of a Performance Bond for a construction job scheduled to be completed in 12 months. Add 1% of the premium cost per month for jobs requiring more than 12 months to complete. The rates are "standard" rates offered to contractors that the bonding company considers financially sound and capable of doing the work. Preferred rates are offered by some bonding companies based upon financial strength of the contractor. Actual rates vary from contractor to contractor and from bonding company to bonding company. Contractors should prequalify through a bonding agency before submitting a bid on a contract that requires a bond.

Contract Amount	Building Construction Class B Projects	Highways & Bridges	
		Class A New Construction	Class A-1 Highway Resurfacing
First $ 100,000 bid	$25.00 per M	$15.00 per M	$9.40 per M
Next 400,000 bid	$ 2,500 plus $15.00 per M	$ 1,500 plus $10.00 per M	$ 940 plus $7.20 per M
Next 2,000,000 bid	8,500 plus 10.00 per M	5,500 plus 7.00 per M	3,820 plus 5.00 per M
Next 2,500,000 bid	28,500 plus 7.50 per M	19,500 plus 5.50 per M	15,820 plus 4.50 per M
Next 2,500,000 bid	47,250 plus 7.00 per M	33,250 plus 5.00 per M	28,320 plus 4.50 per M
Over 7,500,000 bid	64,750 plus 6.00 per M	45,750 plus 4.50 per M	39,570 plus 4.00 per M

R01100-090 Sales Tax by State

State sales tax on materials is tabulated below (5 states have no sales tax). Many states allow local jurisdictions, such as a county or city, to levy additional sales tax.

Some projects may be sales tax exempt, particularly those constructed with public funds.

State	Tax (%)	State	Tax (%)	State	Tax (%)	State	Tax (%)
Alabama	4	Illinois	6.25	Montana	0	Rhode Island	7
Alaska	0	Indiana	5	Nebraska	4.5	South Carolina	5
Arizona	5	Iowa	5	Nevada	6.875	South Dakota	4
Arkansas	4.625	Kansas	4.9	New Hampshire	0	Tennessee	6
California	7.25	Kentucky	6	New Jersey	6	Texas	6.25
Colorado	3	Louisiana	4	New Mexico	5	Utah	4.75
Connecticut	6	Maine	6	New York	4	Vermont	5
Delaware	0	Maryland	5	North Carolina	4	Virginia	4.5
District of Columbia	5.75	Massachusetts	5	North Dakota	5	Washington	6.5
Florida	6	Michigan	6	Ohio	5	West Virginia	6
Georgia	4	Minnesota	6.5	Oklahoma	4.5	Wisconsin	5
Hawaii	4	Mississippi	7	Oregon	0	Wyoming	4
Idaho	5	Missouri	4.225	Pennsylvania	6	Average	4.71 %

Sales Tax by Province (Canada)

GST - a value-added tax, which the federal government imposes on most goods and services provided in or imported into Canada.
PST - a retail sales tax, which five of the provinces impose on the price of most goods and some services.
QST - a value-added tax, similar to the federal GST, which Quebec imposes.
HST - Three provinces have combined their retail sales tax with the federal GST into one harmonized tax.

Province	PST (%)	QST (%)	GST(%)	HST(%)
Alberta	0	0	7	0
British Columbia	7	0	7	0
Manitoba	7	0	7	0
New Brunswick	0	0	0	15
Newfoundland	0	0	0	15
Northwest Territories	0	0	7	0
Nova Scotia	0	0	0	15
Ontario	8	0	7	0
Prince Edward Island	10.7	0	7	0
Quebec	0	8.02	7	0
Saskatchewan	6	0	7	0
Yukon	0	0	7	0

R01100-100 Unemployment Taxes and Social Security Taxes

Mass. State Unemployment tax ranges from 1.325% to 7.225% plus an experience rating assessment the following year, on the first $10,800 of wages. Federal Unemployment tax is 6.2% of the first $7,000 of wages. This is reduced by a credit for payment to the state. The minimum Federal Unemployment tax is .8% after all credits.

Combined rates in Mass. thus vary from 2.125% to 8.025% of the first $10,800 of wages. Combined average U.S. rate is about 7.0% of the first $7,000. Contractors with permanent workers will pay less since the average annual wages for skilled workers is $29.85 x 2,000 hours or about $59,700 per year. The average combined rate for U.S. would thus be 7.0% x $7,000 ÷ $59,700 = 0.8% of total wages for permanent employees.

Rates vary not only from state to state but also with the experience rating of the contractor.

Social Security (FICA) for 2001 is estimated at time of publication to be 7.65% of wages up to $76,200.

R01100-110 Overtime

One way to improve the completion date of a project or eliminate negative float from a schedule is to compress activity duration times. This can be achieved by increasing the crew size or working overtime with the proposed crew.

To determine the costs of working overtime to compress activity duration times, consider the following examples. Below is an overtime efficiency and cost chart based on a five, six, or seven day week with an eight through twelve hour day. Payroll percentage increases for time and one half and double time are shown for the various working days.

Days per Week	Hours per Day	Production Efficiency					Payroll Cost Factors	
		1 Week	2 Weeks	3 Weeks	4 Weeks	Average 4 Weeks	@ 1-1/2 Times	@ 2 Times
5	8	100%	100%	100%	100%	100 %	100 %	100 %
	9	100	100	95	90	96.25	105.6	111.1
	10	100	95	90	85	91.25	110.0	120.0
	11	95	90	75	65	81.25	113.6	127.3
	12	90	85	70	60	76.25	116.7	133.3
6	8	100	100	95	90	96.25	108.3	116.7
	9	100	95	90	85	92.50	113.0	125.9
	10	95	90	85	80	87.50	116.7	133.3
	11	95	85	70	65	78.75	119.7	139.4
	12	90	80	65	60	73.75	122.2	144.4
7	8	100	95	85	75	88.75	114.3	128.6
	9	95	90	80	70	83.75	118.3	136.5
	10	90	85	75	65	78.75	121.4	142.9
	11	85	80	65	60	72.50	124.0	148.1
	12	85	75	60	55	68.75	126.2	152.4

R01100-710 Unit Gross Area Requirements

The figures in the table below indicate typical ranges in square feet as a function of the "occupant" unit. This table is best used in the preliminary design stages to help determine the probable size requirement for the total project. See R171-100 for the typical total size ranges for various types of buildings.

Building Type	Unit	Gross Area in S.F.		
		1/4	Median	3/4
Apartments	Unit	660	860	1,100
Auditorium & Play Theaters	Seat	18	25	38
Bowling Alleys	Lane		940	
Churches & Synagogues	Seat	20	28	39
Dormitories	Bed	200	230	275
Fraternity & Sorority Houses	Bed	220	315	370
Garages, Parking	Car	325	355	385
Hospitals	Bed	685	850	1,075
Hotels	Rental Unit	475	600	710
Housing for the elderly	Unit	515	635	755
Housing, Public	Unit	700	875	1,030
Ice Skating Rinks	Total	27,000	30,000	36,000
Motels	Rental Unit	360	465	620
Nursing Homes	Bed	290	350	450
Restaurants	Seat	23	29	39
Schools, Elementary	Pupil	65	77	90
Junior High & Middle	↓	85	110	129
Senior High	↓	102	130	145
Vocational	↓	110	135	195
Shooting Ranges	Point		450	
Theaters & Movies	Seat		15	

R01100-720 Floor Area Ratios

Table below lists commonly used gross to net area and net to gross area ratios expressed in % for various building types.

Building Type	Gross to Net Ratio	Net to Gross Ratio	Building Type	Gross to Net Ratio	Net to Gross Ratio
Apartment	156	64	School Buildings (campus type)		
Bank	140	72	Administrative	150	67
Church	142	70	Auditorium	142	70
Courthouse	162	61	Biology	161	62
Department Store	123	81	Chemistry	170	59
Garage	118	85	Classroom	152	66
Hospital	183	55	Dining Hall	138	72
Hotel	158	63	Dormitory	154	65
Laboratory	171	58	Engineering	164	61
Library	132	76	Fraternity	160	63
Office	135	75	Gymnasium	142	70
Restaurant	141	70	Science	167	60
Warehouse	108	93	Service	120	83
			Student Union	172	59

The gross area of a building is the total floor area based on outside dimensions.

The net area of a building is the usable floor area for the function intended and excludes such items as stairways, corridors and mechanical rooms. In the case of a commercial building, it might be considered as the "leasable area."

R01100-730 Occupancy Determinations

Description		S.F. Required per Person		
		BOCA	SBC	UBC
Assembly Areas	Fixed Seats	**	6	7
	Movable Seats		15	15
	Concentrated	7		
	Unconcentrated	15		
	Standing Space	3		
Educational	Unclassified			
	Classrooms	20	40	20
	Shop Areas	50	100	50
Institutional	Unclassified		125	
	In-Patient Areas	240		
	Sleeping Areas	120		
Mercantile	Basement	30	30	20
	Ground Floor	30	30	30
	Upper Floors	60	60	50
Office		100	100	100

BOCA=Building Officials & Code Administrators
SBC=Southern Building Code
UBC=Uniform Building Code

** The occupancy load for assembly area with fixed seats shall be determined by the number of fixed seats installed.

R01100-740 Weather Data and Design Conditions

City	Latitude (1)		Winter Temperatures (1)			Winter Degree Days (2)	Summer (Design Dry Bulb) Temperatures and Relative Humidity		
	0	1′	Med. of Annual Extremes	99%	97½%		1%	2½%	5%
UNITED STATES									
Albuquerque, NM	35	0	5.1	12	16	4,400	96/61	94/61	92/61
Atlanta, GA	33	4	11.9	17	22	3,000	94/74	92/74	90/73
Baltimore, MD	39	2	7	14	17	4,600	94/75	91/75	89/74
Birmingham, AL	33	3	13	17	21	2,600	96/74	94/75	92/74
Bismarck, ND	46	5	-32	-23	-19	8,800	95/68	91/68	88/67
Boise, ID	43	3	1	3	10	5,800	96/65	94/64	91/64
Boston, MA	42	2	-1	6	9	5,600	91/73	88/71	85/70
Burlington, VT	44	3	-17	-12	-7	8,200	88/72	85/70	82/69
Charleston, WV	38	2	3	7	11	4,400	92/74	90/73	87/72
Charlotte, NC	35	1	13	18	22	3,200	95/74	93/74	91/74
Casper, WY	42	5	-21	-11	-5	7,400	92/58	90/57	87/57
Chicago, IL	41	5	-8	-3	2	6,600	94/75	91/74	88/73
Cincinnati, OH	39	1	0	1	6	4,400	92/73	90/72	88/72
Cleveland, OH	41	2	-3	1	5	6,400	91/73	88/72	86/71
Columbia, SC	34	0	16	20	24	2,400	97/76	95/75	93/75
Dallas, TX	32	5	14	18	22	2,400	102/75	100/75	97/75
Denver, CO	39	5	-10	-5	1	6,200	93/59	91/59	89/59
Des Moines, IA	41	3	-14	-10	-5	6,600	94/75	91/74	88/73
Detroit, MI	42	2	-3	3	6	6,200	91/73	88/72	86/71
Great Falls, MT	47	3	-25	-21	-15	7,800	91/60	88/60	85/59
Hartford, CT	41	5	-4	3	7	6,200	91/74	88/73	85/72
Houston, TX	29	5	24	28	33	1,400	97/77	95/77	93/77
Indianapolis, IN	39	4	-7	-2	2	5,600	92/74	90/74	87/73
Jackson, MS	32	2	16	21	25	2,200	97/76	95/76	93/76
Kansas City, MO	39	1	-4	2	6	4,800	99/75	96/74	93/74
Las Vegas, NV	36	1	18	25	28	2,800	108/66	106/65	104/65
Lexington, KY	38	0	-1	3	8	4,600	93/73	91/73	88/72
Little Rock, AR	34	4	11	15	20	3,200	99/76	96/77	94/77
Los Angeles, CA	34	0	36	41	43	2,000	93/70	89/70	86/69
Memphis, TN	35	0	10	13	18	3,200	98/77	95/76	93/76
Miami, FL	25	5	39	44	47	200	91/77	90/77	89/77
Milwaukee, WI	43	0	-11	-8	-4	7,600	90/74	87/73	84/71
Minneapolis, MN	44	5	-22	-16	-12	8,400	92/75	89/73	86/71
New Orleans, LA	30	0	28	29	33	1,400	93/78	92/77	90/77
New York, NY	40	5	6	11	15	5,000	92/74	89/73	87/72
Norfolk, VA	36	5	15	20	22	3,400	93/77	91/76	89/76
Oklahoma City, OK	35	2	4	9	13	3,200	100/74	97/74	95/73
Omaha, NE	41	2	-13	-8	-3	6,600	94/76	91/75	88/74
Philadelphia, PA	39	5	6	10	14	4,400	93/75	90/74	87/72
Phoenix, AZ	33	3	27	31	34	1,800	109/71	107/71	105/71
Pittsburgh, PA	40	3	-1	3	7	6,000	91/72	88/71	86/70
Portland, ME	43	4	-10	-6	-1	7,600	87/72	84/71	81/69
Portland, OR	45	4	18	17	23	4,600	89/68	85/67	81/65
Portsmouth, NH	43	1	-8	-2	2	7,200	89/73	85/71	83/70
Providence, RI	41	4	-1	5	9	6,000	89/73	86/72	83/70
Rochester, NY	43	1	-5	1	5	6,800	91/73	88/71	85/70
Salt Lake City, UT	40	5	0	3	8	6,000	97/62	95/62	92/61
San Francisco, CA	37	5	36	38	40	3,000	74/63	71/62	69/61
Seattle, WA	47	4	22	22	27	5,200	85/68	82/66	78/65
Sioux Falls, SD	43	4	-21	-15	-11	7,800	94/73	91/72	88/71
St. Louis, MO	38	4	-3	3	8	5,000	98/75	94/75	91/75
Tampa, FL	28	0	32	36	40	680	92/77	91/77	90/76
Trenton, NJ	40	1	4	11	14	5,000	91/75	88/74	85/73
Washington, DC	38	5	7	14	17	4,200	93/75	91/74	89/74
Wichita, KS	37	4	-3	3	7	4,600	101/72	98/73	96/73
Wilmington, DE	39	4	5	10	14	5,000	92/74	89/74	87/73
ALASKA									
Anchorage	61	1	-29	-23	-18	10,800	71/59	68/58	66/56
Fairbanks	64	5	-59	-51	-47	14,280	82/62	78/60	75/59
CANADA									
Edmonton, Alta.	53	3	-30	-29	-25	11,000	85/66	82/65	79/63
Halifax, N.S.	44	4	-4	1	5	8,000	79/66	76/65	74/64
Montreal, Que.	45	3	-20	-16	-10	9,000	88/73	85/72	83/71
Saskatoon, Sask.	52	1	-35	-35	-31	11,000	89/68	86/66	83/65
St. John's, N.F.	47	4	1	3	7	8,600	77/66	75/65	73/64
Saint John, N.B.	45	2	-15	-12	-8	8,200	80/67	77/65	75/64
Toronto, Ont.	43	4	-10	-5	-1	7,000	90/73	87/72	85/71
Vancouver, B.C.	49	1	13	15	19	6,000	79/67	77/66	74/65
Winnipeg, Man.	49	5	-31	-30	-27	10,800	89/73	86/71	84/70

(1) Handbook of Fundamentals, ASHRAE, Inc., NY 1989
(2) Local Climatological Annual Survey, USDC Env. Science Services Administration, Asheville, NC

R01100-750 Metric Conversion Factors

Description: This table is primarily for converting customary U.S. units in the left hand column to SI metric units in the right hand column. In addition, conversion factors for some commonly encountered Canadian and non-SI metric units are included.

	If You Know		Multiply By		To Find
Length	Inches	x	25.4[a]	=	Millimeters
	Feet	x	0.3048[a]	=	Meters
	Yards	x	0.9144[a]	=	Meters
	Miles (statute)	x	1.609	=	Kilometers
Area	Square inches	x	645.2	=	Square millimeters
	Square feet	x	0.0929	=	Square meters
	Square yards	x	0.8361	=	Square meters
Volume (Capacity)	Cubic inches	x	16,387	=	Cubic millimeters
	Cubic feet	x	0.02832	=	Cubic meters
	Cubic yards	x	0.7646	=	Cubic meters
	Gallons (U.S. liquids)[b]	x	0.003785	=	Cubic meters[c]
	Gallons (Canadian liquid)[b]	x	0.004546	=	Cubic meters[c]
	Ounces (U.S. liquid)[b]	x	29.57	=	Milliliters[c, d]
	Quarts (U.S. liquid)[b]	x	0.9464	=	Liters[c, d]
	Gallons (U.S. liquid)[b]	x	3.785	=	Liters[c, d]
Force	Kilograms force[d]	x	9.807	=	Newtons
	Pounds force	x	4.448	=	Newtons
	Pounds force	x	0.4536	=	Kilograms force[d]
	Kips	x	4448	=	Newtons
	Kips	x	453.6	=	Kilograms force[d]
Pressure, Stress, Strength (Force per unit area)	Kilograms force per square centimeter[d]	x	0.09807	=	Megapascals
	Pounds force per square inch (psi)	x	0.006895	=	Megapascals
	Kips per square inch	x	6.895	=	Megapascals
	Pounds force per square inch (psi)	x	0.07031	=	Kilograms force per square centimeter[d]
	Pounds force per square foot	x	47.88	=	Pascals
	Pounds force per square foot	x	4.882	=	Kilograms force per square meter[d]
Bending Moment Or Torque	Inch-pounds force	x	0.01152	=	Meter-kilograms force[d]
	Inch-pounds force	x	0.1130	=	Newton-meters
	Foot-pounds force	x	0.1383	=	Meter-kilograms force[d]
	Foot-pounds force	x	1.356	=	Newton-meters
	Meter-kilograms force[d]	x	9.807	=	Newton-meters
Mass	Ounces (avoirdupois)	x	28.35	=	Grams
	Pounds (avoirdupois)	x	0.4536	=	Kilograms
	Tons (metric)	x	1000	=	Kilograms
	Tons, short (2000 pounds)	x	907.2	=	Kilograms
	Tons, short (2000 pounds)	x	0.9072	=	Megagrams[e]
Mass per Unit Volume	Pounds mass per cubic foot	x	16.02	=	Kilograms per cubic meter
	Pounds mass per cubic yard	x	0.5933	=	Kilograms per cubic meter
	Pounds mass per gallon (U.S. liquid)[b]	x	119.8	=	Kilograms per cubic meter
	Pounds mass per gallon (Canadian liquid)[b]	x	99.78	=	Kilograms per cubic meter
Temperature	Degrees Fahrenheit	(F-32)/1.8		=	Degrees Celsius
	Degrees Fahrenheit	(F+459.67)/1.8		=	Degrees Kelvin
	Degrees Celsius	C+273.15		=	Degrees Kelvin

[a]The factor given is exact
[b]One U.S. gallon = 0.8327 Canadian gallon
[c]1 liter = 1000 milliliters = 1000 cubic centimeters
1 cubic decimeter = 0.001 cubic meter
[d]Metric but not SI unit
[e]Called "tonne" in England and "metric ton" in other metric countries

R01100-760 Weights and Measures

Measures of Length
1 Mile = 1760 Yards = 5280 Feet
1 Yard = 3 Feet = 36 inches
1 Foot = 12 Inches
1 Mil = 0.001 Inch
1 Fathom = 2 Yards = 6 Feet
1 Rod = 5.5 Yards = 16.5 Feet
1 Hand = 4 Inches
1 Span = 9 Inches
1 Micro-inch = One Millionth Inch or 0.000001 Inch
1 Micron = One Millionth Meter + 0.00003937 Inch

Surveyor's Measure
1 Mile = 8 Furlongs = 80 Chains
1 Furlong = 10 Chains = 220 Yards
1 Chain = 4 Rods = 22 Yards = 66 Feet = 100 Links
1 Link = 7.92 Inches

Square Measure
1 Square Mile = 640 Acres = 6400 Square Chains
1 Acre = 10 Square Chains = 4840 Square Yards = 43,560 Sq. Ft.
1 Square Chain = 16 Square Rods = 484 Square Yards = 4356 Sq. Ft.
1 Square Rod = 30.25 Square Yards = 272.25 Square Feet = 625 Square Lines
1 Square Yard = 9 Square Feet
1 Square Foot = 144 Square Inches
An Acre equals a Square 208.7 Feet per Side

Cubic Measure
1 Cubic Yard = 27 Cubic Feet
1 Cubic Foot = 1728 Cubic Inches
1 Cord of Wood = 4 x 4 x 8 Feet = 128 Cubic Feet
1 Perch of Masonry = 16½ x 1½ x 1 Foot = 24.75 Cubic Feet

Avoirdupois or Commercial Weight
1 Gross or Long Ton = 2240 Pounds
1 Net or Short ton = 2000 Pounds
1 Pound = 16 Ounces = 7000 Grains
1 Ounce = 16 Drachms = 437.5 Grains
1 Stone = 14 Pounds

Shipping Measure
For Measuring Internal Capacity of a Vessel:
1 Register Ton = 100 Cubic Feet

For Measurement of Cargo:
Approximately 40 Cubic Feet of Merchandise is considered a Shipping Ton, unless that bulk would weigh more than 2000 Pounds, in which case Freight Charge may be based upon weight.

40 Cubic Feet = 32.143 U.S. Bushels = 31.16 Imp. Bushels

Liquid Measure
1 Imperial Gallon = 1.2009 U.S. Gallon = 277.42 Cu. In.
1 Cubic Foot = 7.48 U.S. Gallons

R01107-030 Engineering Fees

Typical **Structural Engineering Fees** based on type of construction and total project size. These fees are included in Architectural Fees.

Type of Construction	Total Project Size (in thousands of dollars)			
	$500	$500-$1,000	$1,000-$5,000	Over $5000
Industrial buildings, factories & warehouses	Technical payroll times 2.0 to 2.5	1.60%	1.25%	1.00%
Hotels, apartments, offices, dormitories, hospitals, public buildings, food stores	↓	2.00%	1.70%	1.20%
Museums, banks, churches and cathedrals	↓	2.00%	1.75%	1.25%
Thin shells, prestressed concrete, earthquake resistive	↓	2.00%	1.75%	1.50%
Parking ramps, auditoriums, stadiums, convention halls, hangars & boiler houses	↓	2.50%	2.00%	1.75%
Special buildings, major alterations, underpinning & future expansion	↓	Add to above 0.5%	Add to above 0.5%	Add to above 0.5%

For complex reinforced concrete or unusually complicated structures, add 20% to 50%.

Typical **Mechanical and Electrical Engineering Fees** are based on the size of the subcontract. The fee structure for Mechanical Engineering is shown below; Electrical Engineering fees range from 2 to 5%. These fees are included in Architectural Fees.

Type of Construction	Subcontract Size							
	$25,000	$50,000	$100,000	$225,000	$350,000	$500,000	$750,000	$1,000,000
Simple structures	6.4%	5.7%	4.8%	4.5%	4.4%	4.3%	4.2%	4.1%
Intermediate structures	8.0	7.3	6.5	5.6	5.1	5.0	4.9	4.8
Complex structures	12.0	9.0	9.0	8.0	7.5	7.5	7.0	7.0

For renovations, add 15% to 25% to applicable fee.

R01250-010 Repair and Remodeling

Cost figures are based on new construction utilizing the most cost-effective combination of labor, equipment and material with the work scheduled in proper sequence to allow the various trades to accomplish their work in an efficient manner.

The costs for repair and remodeling work must be modified due to the following factors that may be present in any given repair and remodeling project.

1. Equipment usage curtailment due to the physical limitations of the project, with only hand-operated equipment being used.
2. Increased requirement for shoring and bracing to hold up the building while structural changes are being made and to allow for temporary storage of construction materials on above-grade floors.
3. Material handling becomes more costly due to having to move within the confines of an enclosed building. For multi-story construction, low capacity elevators and stairwells may be the only access to the upper floors.
4. Large amount of cutting and patching and attempting to match the existing construction is required. It is often more economical to remove entire walls rather than create many new door and window openings. This sort of trade-off has to be carefully analyzed.
5. Cost of protection of completed work is increased since the usual sequence of construction usually cannot be accomplished.
6. Economies of scale usually associated with new construction may not be present. If small quantities of components must be custom fabricated due to job requirements, unit costs will naturally increase. Also, if only small work areas are available at a given time, job scheduling between trades becomes difficult and subcontractor quotations may reflect the excessive start-up and shut-down phases of the job.
7. Work may have to be done on other than normal shifts and may have to be done around an existing production facility which has to stay in production during the course of the repair and remodeling.
8. Dust and noise protection of adjoining non-construction areas can involve substantial special protection and alter usual construction methods.
9. Job may be delayed due to unexpected conditions discovered during demolition or removal. These delays ultimately increase construction costs.
10. Piping and ductwork runs may not be as simple as for new construction. Wiring may have to be snaked through walls and floors.
11. Matching "existing construction" may be impossible because materials may no longer be manufactured. Substitutions may be expensive.
12. Weather protection of existing structure requires additional temporary structures to protect building at openings.
13. On small projects, because of local conditions, it may be necessary to pay a tradesman for a minimum of four hours for a task that is completed in one hour.

All of the above areas can contribute to increased costs for a repair and remodeling project. Each of the above factors should be considered in the planning, bidding and construction stage in order to minimize the increased costs associated with repair and remodeling jobs.

R01540-100 Steel Tubular Scaffolding

On new construction, tubular scaffolding is efficient up to 60′ high or five stories. Above this it is usually better to use a hung scaffolding if construction permits. Swing scaffolding operations may interfere with tenants. In this case, the tubular is more practical at all heights.

In repairing or cleaning the front of an existing building the cost of tubular scaffolding per S.F. of building front increases as the height increases above the first tier. The first tier cost is relatively high due to leveling and alignment.

The minimum efficient crew for erection is three workers. For heights over 50′, a crew of four is more efficient. Use two or more on top and two at the bottom for handing up or hoisting. Four workers can erect and dismantle about nine frames per hour up to five stories. From five to eight stories they will average six frames per hour. With 7′ horizontal spacing this will run about 400 S.F. and 265 S.F. of wall surface, respectively. Time for placing planks must be added to the above. On heights above 50′, five planks can be placed per labor-hour.

The cost per 1,000 S.F. of building front in the table below was developed by pricing the materials required for a typical tubular scaffolding system eleven frames long and two frames high. Planks were figured five wide for standing plus two wide for materials.

Frames are 5′ wide and usually spaced 7′ O.C. horizontally. Sidewalk frames are 6′ wide. Rental rates will be lower for jobs over three months duration.

For jobs under twenty-five frames, add 50% to rental cost. These figures do not include accessories which are listed separately below. Large quantities for long periods can reduce rental rates by 20%.

Item	Unit	Monthly Rent	Per 1,000 S. F. of Building Front	
			No. of Pieces	Rental per Month
5′ Wide Standard Frame, 6′-4″ High	Ea.	$ 3.75	24	$ 90.00
Leveling Jack & Plate		1.50	24	36.00
Cross Brace		.60	44	26.40
Side Arm Bracket, 21″		1.50	12	18.00
Guardrail Post		1.00	12	12.00
Guardrail, 7′ section		.75	22	16.50
Stairway Section		10.00	2	20.00
Stairway Starter Bar		.10	1	.10
Stairway Inside Handrail		5.00	2	10.00
Stairway Outside Handrail		5.00	2	10.00
Walk-Thru Frame Guardrail	▼	2.00	2	4.00
			Total	$243.00
			Per C.S.F., 1 Use/Mo.	$ 24.30

Scaffolding is often used as falsework over 15′ high during construction of cast-in-place concrete beams and slabs. Two foot wide scaffolding is generally used for heavy beam construction. The span between frames depends upon the load to be carried with a maximum span of 5′.

Heavy duty shoring frames with a capacity of 10,000#/leg can be spaced up to 10′ O. C. depending upon form support design and loading.

Scaffolding used as horizontal shoring requires less than half the material required with conventional shoring.

On new construction, erection is done by carpenters.

Rolling towers supporting horizontal shores can reduce labor and speed the job. For maintenance work, catwalks with spans up to 70′ can be supported by the rolling towers.

R01590-100 Contractor Equipment

Rental Rates shown in the front of the book pertain to late model high quality machines in excellent working condition, rented from equipment dealers. Rental rates from contractors may be substantially lower than the rental rates from equipment dealers depending upon economic conditions. For older, less productive machines, reduce rates by a maximum of 15%. Any overtime must be added to the base rates. For shift work, rates are lower. Usual rule of thumb is 150% of one shift rate for two shifts; 200% for three shifts.

For periods of less than one week, operated equipment is usually more economical to rent than renting bare equipment and hiring an operator.

Equipment moving and mobilization costs must be added to rental rates where applicable. A large crane, for instance, may take two days to erect and two days to dismantle.

Rental rates vary throughout the country with larger cities generally having lower rates. Lease plans for new equipment are available for periods in excess of six months with a percentage of payments applying toward purchase.

Monthly rental rates vary from 2% to 5% of the cost of the equipment depending on the anticipated life of the equipment and its wearing parts. Weekly rates are about 1/3 the monthly rates and daily rental rates about 1/3 the weekly rate.

The hourly operating costs for each piece of equipment include costs to the user such as fuel, oil, lubrication, normal expendables for the equipment, and a percentage of mechanic's wages chargeable to maintenance. The hourly operating costs listed do not include the operator's wages.

The daily cost for equipment used in the standard crews is figured by dividing the weekly rate by five, then adding eight times the hourly operating cost to give the total daily equipment cost, not including the operator. This figure is in the right hand column of Division 01590 under Crew Equipment Cost/Day.

Pile Driving rates shown for pile hammer and extractor do not include leads, crane, boiler or compressor. Vibratory pile driving requires an added field specialist during set-up and pile driving operation for the electric model. The hydraulic model requires a field specialist for set-up only. Up to 125 reuses of sheet piling are possible using vibratory drivers. For normal conditions, crane capacity for hammer type and size are as follows.

Crane Capacity	Hammer Type and Size		
	Air or Steam	Diesel	Vibratory
25 ton	to 8,750 ft.-lb.		70 H.P.
40 ton	15,000 ft.-lb.	to 32,000 ft.-lb.	170 H.P.
60 ton	25,000 ft.-lb.		300 H.P.
100 ton		112,000 ft.-lb.	

Cranes should be specified for the job by size, building and site characteristics, availability, performance characteristics, and duration of time required.

Backhoes & Shovels rent for about the same as equivalent size cranes but maintenance and operating expense is higher. Crane operators rate must be adjusted for high boom heights. Average adjustments: for 150′ boom add $.80 per hour; over 185′, add $1.40 per hour; over 210′, add $1.85 per hour; over 250′, add $2.65 per hour and over 295′, add $4.20 per hour.

Tower Cranes

Capacity In Kip-Feet	Typical Jib Length in Feet	Speed at Maximum Reach and Load	Purchase Price (New)		Monthly Rental, to 6 mo.	
			Crane & 80′ Mast	Mast Sections	Crane & 80′ Mast	Mast Sections
725	100	350 FPM	$ 256,000	$ 670 /L.F.	$ 6,400	$16.80 /L.F.
900	100	500	303,000	790	7,600	19.60
*1100	130	1000	418,000	910	10,400	22.70
1450	150	1000	580,000	1,300	14,500	32.40
2150	200	1000	729,000	1,480	18,200	37.00
3000	200	1000	1,018,000	1,510	25,500	37.70

*Most widely used.

Tower Cranes of the climbing or static type have jibs from 50′ to 200′ and capacities at maximum reach range from 4,000 to 14,000 pounds. Lifting capacities increase up to maximum load as the hook radius decreases.

Typical rental rates, based on purchase price are about 2% to 3% per month.

Erection and dismantling runs between $13,000 and $82,000. Climbing operation takes ten labor hours per 20′ climb. Crane dead time is about five hours per 40′ climb. If crane is bolted to side of the building add cost of ties and extra mast sections. Mast sections cost $600 to $1,500 per vertical foot or can be rented at 2% to 3% of purchase price per month. Contractors using climbers claim savings of $1.50 per C.Y. of concrete placed, plus $.21 per S.F. of formwork. Climbing cranes have from 80′ to 180′ of mast while static cranes have 80′ to 800′ of mast.

Truck Cranes can be converted to tower cranes by using tower attachments. Mast heights over 400′ have been used. See Division 01590-600 for rental rates of high boom cranes.

A single 100′ high material **Hoist and Tower** can be erected and dismantled for about $15,600; a double 100′ high hoist and tower for about $20,700. Erection costs for additional heights are $100 and $130 per vertical foot respectively up to 150′ and $100 to $165 per vertical foot over 150′ high. A 40′ high portable Buck hoist costs about $5,300 to erect and dismantle. Additional heights run $80 per vertical foot to 80′ and $100 per vertical foot for the next 100′. Most material hoists do not meet local code requirements for carrying personnel.

A 150′ high **Personnel Hoist** requires about 500 to 800 labor hours to erect and dismantle with costs ranging from $16,400 to $26,300. Budget erection cost is $165 per vertical foot for all trades. Local code requirements or labor scarcity requiring overtime can add up to 50% to any of the above erection costs.

Earthmoving Equipment: The selection of earthmoving equipment depends upon the type and quantity of material, moisture content, haul distance, haul road, time available, and equipment available. Short haul cut and fill operations may require dozers only, while another operation may require excavators, a fleet of trucks, and spreading and compaction equipment. Stockpiled material and granular material are easily excavated with front end loaders. Scrapers are most economically used with hauls between 300′ and 1-1/2 miles if adequate haul roads can be maintained. Shovels are often used for blasted rock and any material where a vertical face of 8′ or more can be excavated. Special conditions may dictate the use of draglines, clamshells, or backhoes. Spreading and compaction equipment must be matched to the soil characteristics, the compaction required and the rate the fill is being supplied.

R02115-200 Underground Storage Tank Removal

Underground Storage Tank Removal can be divided into two categories: Non-Leaking and Leaking. Prior to removing an underground storage tank, tests should be made, with the proper authorities present, to determine whether a tank has been leaking or the surrounding soil has been contaminated.

To safely remove Liquid Underground Storage Tanks:

1. Excavate to the top of the tank.
2. Disconnect all piping.
3. Open all tank vents and access ports.
4. Remove all liquids and/or sludge.
5. Purge the tank with an inert gas.
6. Provide access to the inside of the tank and clean out the interior using proper personal protective equipment (PPE).
7. Excavate soil surrounding the tank using proper PPE for on-site personnel.
8. Pull and properly dispose of the tank.
9. Clean up the site of all contaminated material.
10. Install new tanks or close the excavation.

R02240-900 Wellpoints

A single stage wellpoint system is usually limited to dewatering an average 15′ depth below normal ground water level. Multi-stage systems are employed for greater depth with the pumping equipment installed only at the lowest header level. Ejectors, with unlimited lift capacity, can be economical when two or more stages of wellpoints can be replaced or when horizontal clearance is restricted, such as in deep trenches or tunneling projects, and where low water flows are expected. Wellpoints are usually spaced on 2-1/2′ to 10′ centers along a header pipe. Wellpoint spacing, header size, and pump size are all determined by the expected flow, as dictated by soil conditions.

In almost all soils encountered in wellpoint dewatering, the wellpoints may be jetted into place. Cemented soils and stiff clays may require sand wicks about 12″ in diameter around each wellpoint to increase efficiency and eliminate weeping into the excavation. These sand wicks require 1/2 to 3 C.Y. of washed filter sand and are installed by using a 12″ diameter steel casing and hole puncher jetted into the ground 2′ deeper than the wellpoint. Rock may require predrilled holes.

Labor required for the complete installation and removal of a single stage wellpoint system is in the range of 3/4 to 2 labor-hours per linear foot of header, depending upon jetting conditions, wellpoint spacing, etc.

Continuous pumping is necessary except in some free draining soil where temporary flooding is permissible (as in trenches which are backfilled after each day′s work). Good practice requires provision of a stand-by pump during the continuous pumping operation.

Systems for continuous trenching below the water table should be installed three to four times the length of expected daily progress to insure uninterrupted digging, and header pipe size should not be changed during the job.

For pervious free draining soils, deep wells in place of wellpoints may be economical because of lower installation and maintenance costs. Daily production ranges between two to three wells per day, for 25′ to 40′ depths, to one well per day for depths over 50′.

Detailed analysis and estimating for any dewatering problem is available at no cost from wellpoint manufacturers. Major firms will quote "sufficient equipment" quotes or their affiliates offer lump sum proposals to cover complete dewatering responsibility.

	Description for 200′ System with 8″ Header	Quantities	1st Month Unit	1st Month Total	Thereafter Unit	Thereafter Total
Equipment & Material	Wellpoints 25′ long, 2″ diameter @ 5′ O.C.	40 Each	$ 28.50	$ 1,140.00	$ 21.38	$ 855.00
	Header pipe, 8″ diameter	200 L.F.	5.40	1,080.00	2.70	540.00
	Discharge pipe, 8″ diameter	100 L.F.	3.66	366.00	2.38	238.00
	8″ valves	3 Each	140.00	420.00	91.00	273.00
	Combination Jetting & Wellpoint pump (standby)	1 Each	2,275.00	2,275.00	1,592.50	1,592.50
	Wellpoint pump, 8″ diameter	1 Each	2,225.00	2,225.00	1,557.50	1,557.50
	Transportation to and from site	1 Day	674.40	674.40	—	—
	Fuel 30 days x 60 gal./day	1800 Gallons	1.05	1,890.00	1.05	1,890.00
	Lubricants for 30 days x 16 lbs./day	480 Lbs.	1.10	528.00	1.10	528.00
	Sand for points	40 C.Y.	16.60	664.00	—	—
	Equipment & Materials Subtotal			$11,262.40		$ 7,474.00
Labor	Technician to supervise installation	1 Week	$ 31.80	$ 1,272.00	—	—
	Labor for installation and removal of system	300 Labor-hours	22.85	6,855.00	—	—
	4 Operators straight time 40 hrs./wk. for 4.33 wks.	693 Hrs.	28.75	19,923.75	$ 28.75	$19,923.75
	4 Operators overtime 2 hrs./wk. for 4.33 wks.	35 Hrs.	57.50	2,012.50	57.50	2,012.50
		Labor Subtotal		$30,063.25		$21,936.25
		Total Cost		$41,325.65		$29,410.25
		Monthly Cost/L.F. Header		$ 206.63		$ 147.05

R02250-400 Wood Sheet Piling

Wood sheet piling may be used for depths to 20′ where there is no ground water. If moderate ground water is encountered Tongue & Groove sheeting will help to keep it out. When considerable ground water is present, steel sheeting must be used.

For estimating purposes on trench excavation, sizes are as follows:

Depth	Sheeting	Wales	Braces	B.F. per S.F.
To 8′	3 x 12′s	6 x 8′s, 2 line	6 x 8′s, @ 10′	4.0 @ 8′
8′ x 12′	3 x 12′s	10 x 10′s, 2 line	10 x 10′s, @ 9′	5.0 average
12′ to 20′	3 x 12′s	12 x 12′s, 3 line	12 x 12′s, @ 8′	7.0 average

Sheeting to be toed in at least 2′ depending upon soil conditions. A five person crew with an air compressor and sheeting driver can drive and brace 440 SF/day at 8′ deep, 360 SF/day at 12′ deep, and 320 SF/day at 16′ deep. For normal soils, piling can be pulled in 1/3 the time to install. Pulling difficulty increases with the time in the ground. Production can be increased by high pressure jetting. Figures below assume 50% of lumber is salvaged and includes pulling costs. Some jurisdictions require an equipment operator in addition to Crew B-31.

Sheeting Pulled

Daily Cost Crew B-31	L.H./ Day	Hourly Cost	Daily Cost	8′ Depth, 440 S.F./Day			16′ Depth, 320 S.F./Day		
				To Drive (1 Day)		To Pull (1/3 Day)	To Drive (1 Day)		To Pull (1/3 Day)
1 Foreman	8	$24.85	$ 198.80		$ 198.80	$ 66.20		$ 198.80	$ 66.20
3 Laborers	24	22.85	548.40		548.40	182.62		548.40	182.62
1 Carpenter	8	29.15	233.20		233.20	77.66		233.20	77.66
1 Air Compressor			125.20		125.20	41.69		125.20	41.69
1 Sheeting Driver			12.50		12.50	4.16		12.50	4.16
2 -50 Ft. Air Hoses, 1-1/2″ Diam.			11.80		11.80	3.93		11.80	3.93
Lumber (50% salvage)				1.76 MBF	559.68		2.24 MBF	712.32	
Total			$1,129.90		$1,689.58	$376.26		$1,842.22	$376.26
Total/S.F.					$ 3.84	$ .86		$ 5.76	$ 1.18
Total (Drive and Pull)/S.F.					$ 4.70			$ 6.93	

Sheeting Left in Place

Daily Cost	8′ Depth 440 S.F./Day		10′ Depth 400 S.F./Day		12′ Depth 360 S.F./Day		16′ Depth 320 S.F./Day		18′ Depth 305 S.F./Day		20′ Depth 280 S.F./Day	
Crew B-31		$1,129.90		$1,129.90		$1,129.90		$1,129.90		$1,129.90		$1,129.90
Lumber	1.76 M	1,119.36	1.8 M	1,144.80	1.8 M	1,144.80	2.24 M	1,424.64	2.1 M	1,335.60	1.9 M	1,208.40
Total in Place		$2,249.26		$2,274.70		$2,274.70		$2,554.54		$2,465.50		$2,338.30
Total/S.F.		$ 5.11		$ 5.69		$ 6.32		$ 7.98		$ 8.08		$ 8.35
Total/M.B.F.		$1,277.99		$1,263.72		$1,263.72		$1,140.42		$1,174.05		$1,230.68

R02250-450 Steel Sheet Piling

Limiting weights are 22 to 38#/S.F. of wall surface with 27#/S.F. average for usual types and sizes. (Weights of piles themselves are from 30.7#/L.F. to 57#/L.F. but they are 15″ to 21″ wide.) Lightweight sections 12″ to 28″ wide from 3 ga. to 12 ga. thick are also available for shallow excavations. Piles may be driven two at a time with an impact or vibratory hammer (use vibratory to pull) hung from a crane without leads. A reasonable estimate of the life of steel sheet piling is 10 uses with up to 125 uses possible if a vibratory hammer is used. Used piling costs from 50% to 80% of new piling depending on location and market conditions. Sheet piling and H piles can be rented for about 30% of the delivered mill price for the first month and 5% per month thereafter. Allow 1 labor-hour per pile for cleaning and trimming after driving. These costs increase with depth and hydrostatic head. Vibratory drivers are faster in wet granular soils and are excellent for pile extraction. Pulling difficulty increases with the time in the ground and may cost more than driving. It is often economical to abandon the sheet piling, especially if it can be used as the outer wall form. Allow about 1/3 additional length or more, for toeing into ground. Add bracing, waler and strut costs. Waler costs can equal the cost per ton of sheeting.

Cost of Sheet Piling & Production Rate by Ton & S.F.

Depth of Excavation	15′ Depth		20′ Depth		25′ Depth	
Description of Pile	**22 psf = 90.9 S.F./Ton**		**27 psf = 74 S.F./Ton**		**38 psf = 52.6 S.F./Ton**	
Type of operation: Left in Place or Removed	Drive & Left	Drive & Extract*	Drive & Left	Drive & Extract*	Drive & Left	Drive & Extract*
Labor & Equip. to Drive 1 ton	$ 358.13	$536.20	$ 298.72	$448.08	$ 203.76	$304.12
Piling (75% Salvage)	812.00	203.00	812.00	203.00	812.00	203.00
Cost/Ton in Place	$1,170.13	$739.20	$1,110.72	$651.08	$1,015.76	$507.12
Production Rate, Tons/Day	10.81	7.22	12.96	8.64	19.00	12.73
Cost, S.F. in Place (incl. 33% toe in)	$ 12.87	$ 8.13	$ 15.01	$ 8.80	$ 19.31	$ 9.64
Production Rate, S.F./Day (incl. 33% toe in)	983	656	960	640	1,000	670

Crew B-40	Bare Costs Hr.	Bare Costs Daily	Inc. Subs O&P Hr.	Inc. Subs O&P Daily
1 Pile Driver Foreman	$30.80	$ 246.40	$52.40	$ 419.20
4 Pile Drivers	28.80	921.60	49.00	1,568.00
1 Equip. Oper. Oiler	25.75	206.00	39.00	312.00
2 Equip. Oper. (crane)	31.15	498.40	47.15	754.40
1 Crane, 40 Ton		716.00		787.60
Vibratory Hammer & Gen.		1,283.00		1,411.30
64 L.H., Daily Totals		$3,871.40		$5,252.50

Installation Cost per Ton

for 15′ Deep Excavation plus 33% toe-in

$3,871.40 per day/10.81 tons = $358.13 per ton

Installation Cost per S.F.

$\frac{2000\#}{22\#/S.F.}$ = 90.9 S.F. x 10.81 tons = 983 S.F. per day

$3,871.40 / 983 S.F. per day = $3.94 per S.F.

*For driving & extracting, two mobilizations & demobilizations will be necessary

R02315-260 Trench Excavation and Pipe Bedding

Example: Find the cost per LF for excavating, bedding, and backfilling a 36″ diameter pipe 10′ deep in damp sandy loam.
Method: For trench bottom widths for various size pipe, Table 12.3-310, a 36″ diameter pipe uses a 6′ wide trench bottom.

						6′Wide Bottom of Trench			
						Slope 1/2 to 1	Slope 1 to 1	Slope 1-1/2 to 1	Slope 2 to 1
Excavation & Backfill System	For 1/2 to 1 slope see:	Trench Excavation	System	12.3-110	-2880	$35.05			
	1 to 1 slope		↓	↓	-4050		$47.50		
	For 1/2 to 1 slope see:	Trench Excavation	↓	↓	-5700			$53.50	
	2 to 1 slope		↓	↓	-7300				$60.00
Pipe Bedding System	For 1/2 to 1 slope see:	Trench Pipe Bedding	System	12.3-310	-2400	$11.86			
	1 to 1 slope		↓	↓	-2880		$15.85		
	For 1-1/2 to 1 slope	Trench Pipe Bedding	↓	↓	-3260			$19.85	
	2 to 1 slope		↓	↓	-3660				$23.85
Total Cost per LF (not incl. pipe)						$46.91	$63.35	$73.35	$83.85

R02450-660 Maximum Depth of Frost Penetration in Inches

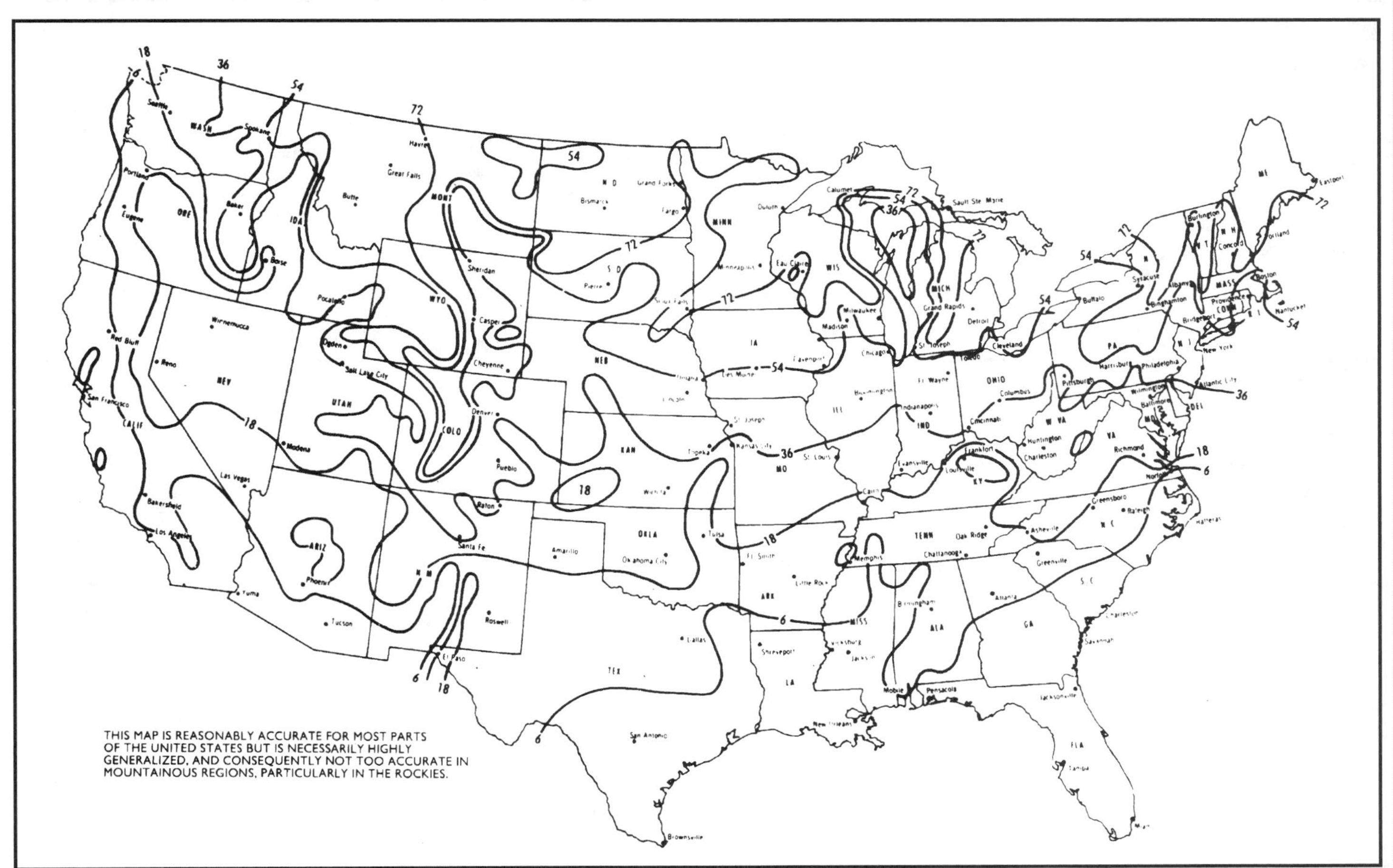
THIS MAP IS REASONABLY ACCURATE FOR MOST PARTS OF THE UNITED STATES BUT IS NECESSARILY HIGHLY GENERALIZED, AND CONSEQUENTLY NOT TOO ACCURATE IN MOUNTAINOUS REGIONS, PARTICULARLY IN THE ROCKIES.
18
36
54
72
6

R02510-800 Piping Designations

There are several systems currently in use to describe pipe and fittings. The following paragraphs will help to identify and clarify classifications of piping systems used for water distribution.

Piping may be classified by schedule. Piping schedules include 5S, 10S, 10, 20, 30, Standard, 40, 60, Extra Strong, 80, 100, 120, 140, 160 and Double Extra Strong. These schedules are dependent upon the pipe wall thickness. The wall thickness of a particular schedule may vary with pipe size.

Ductile iron pipe for water distribution is classified by Pressure Classes such as Class 150, 200, 250, 300 and 350. These classes are actually the rated water working pressure of the pipe in pounds per square inch (psi). The pipe in these pressure classes is designed to withstand the rated water working pressure plus a surge allowance of 100 psi.

The American Water Works Association (AWWA) provides standards for various types of **plastic pipe.** C-900 is the specification for polyvinyl chloride (PVC) piping used for water distribution in sizes ranging from 4″ through 12″. C-901 is the specification for polyethylene (PE) pressure pipe, tubing and fittings used for water distribution in sizes ranging from 1/2″ through 3″. C-905 is the specification for PVC piping sizes 14″ and greater.

PVC pressure-rated pipe is identified using the standard dimensional ratio (SDR) method. This method is defined by the American Society for Testing and Materials (ASTM) Standard D 2241. This pipe is available in SDR numbers 64, 41, 32.5, 26, 21, 17, and 13.5. Pipe with an SDR of 64 will have the thinnest wall while pipe with an SDR of 13.5 will have the thickest wall. When the pressure rating (PR) of a pipe is given in psi, it is based on a line supplying water at 73 degrees F.

The National Sanitation Foundation (NSF) seal of approval is applied to products that can be used with potable water. These products have been tested to ANSI/NSF Standard 14.

Valves and strainers are classified by American National Standards Institute (ANSI) Classes. These Classes are 125, 150, 200, 250, 300, 400, 600, 900, 1500 and 2500. Within each class there is an operating pressure range dependent upon temperature. Design parameters should be compared to the appropriate material dependent, pressure-temperature rating chart for accurate valve selection.

Metals | R050 | Materials, Coatings, & Fastenings

R05090-520 Welded Structural Steel

Usual weight reductions with welded design run 10% to 20% compared with bolted or riveted connections. This amounts to about the same total cost compared with bolted structures since field welding is more expensive than bolts. For normal spans of 18′ to 24′ figure 6 to 7 connections per ton.

Trusses — For welded trusses add 4% to weight of main members for connections. Up to 15% less steel can be expected in a welded truss compared to one that is shop bolted. Cost of erection is the same whether shop bolted or welded.

General — Typical electrodes for structural steel welding are E6010, E6011, E60T and E70T. Typical buildings vary between 2# to 8# of weld rod per ton of steel. Buildings utilizing continuous design require about three times as much welding as conventional welded structures. In estimating field erection by welding, it is best to use the average linear feet of weld per ton to arrive at the welding cost per ton. The type, size and position of the weld will have a direct bearing on the cost per linear foot. A typical field welder will deposit 1.8# to 2# of weld rod per hour manually. Using semiautomatic methods can increase production by as much as 50% to 75%. Below is the cost per hour for manual welding.

Welded Structural Steel in Field, per Hour					
Item			No Operating Engr.	1/2 Operating Engr.	1 Operating Engr.
2 lb. weld rod	$1.34	per lb.	$ 2.68	$ 2.68	$ 2.68
Equipment (for welding only)	$210	/40 hrs.	5.25	5.25	5.25
Operating cost @	$5.25	per hour	5.25	5.25	5.25
Welder Foreman 1 hr. @	$34.85	per hour	34.85	34.85	34.85
Operating engineer @	$28.75	per hour	—	14.38	28.75
Total per Welder Hour (Bare Costs)			$48.03	$62.41	$76.78

The welding costs per ton of structural steel will vary from $48 to $307 per ton depending on Union requirements, crew size, design and inspection required.

R10520-110 Sprinkler Systems (Automatic)

Sprinkler systems may be classified by type as follows:

1. **Wet Pipe System.** A system employing automatic sprinklers attached to a piping system containing water and connected to a water supply so that water discharges immediately from sprinklers opened by a fire.
2. **Dry Pipe System.** A system employing automatic sprinklers attached to a piping system containing air under pressure, the release of which as from the opening of sprinklers permits the water pressure to open a valve known as a "dry pipe valve". The water then flows into the piping system and out the opened sprinklers.
3. **Pre-Action System.** A system employing automatic sprinklers attached to a piping system containing air that may or may not be under pressure, with a supplemental heat responsive system of generally more sensitive characteristics than the automatic sprinklers themselves, installed in the same areas as the sprinklers; actuation of the heat responsive system, as from a fire, opens a valve which permits water to flow into the sprinkler piping system and to be discharged from any sprinklers which may be open.
4. **Deluge System.** A system employing open sprinklers attached to a piping system connected to a water supply through a valve which is opened by the operation of a heat responsive system installed in the same areas as the sprinklers. When this valve opens, water flows into the piping system and discharges from all sprinklers attached thereto.
5. **Combined Dry Pipe and Pre-Action Sprinkler System.** A system employing automatic sprinklers attached to a piping system containing air under pressure with a supplemental heat responsive system of generally more sensitive characteristics than the automatic sprinklers themselves, installed in the same areas as the sprinklers; operation of the heat responsive system, as from a fire, actuates tripping devices which open dry pipe valves simultaneously and without loss of air pressure in the system. Operation of the heat responsive system also opens approved air exhaust valves at the end of the feed main which facilitates the filling of the system with water which usually precedes the opening of sprinklers. The heat responsive system also serves as an automatic fire alarm system.
6. **Limited Water Supply System.** A system employing automatic sprinklers and conforming to these standards but supplied by a pressure tank of limited capacity.
7. **Chemical Systems.** Systems using halon, carbon dioxide, dry chemical or high expansion foam as selected for special requirements. Agent may extinguish flames by chemically inhibiting flame propagation, suffocate flames by excluding oxygen, interrupting chemical action of oxygen uniting with fuel or sealing and cooling the combustion center.
8. **Firecycle System.** Firecycle is a fixed fire protection sprinkler system utilizing water as its extinguishing agent. It is a time delayed, recycling, preaction type which automatically shuts the water off when heat is reduced below the detector operating temperature and turns the water back on when that temperature is exceeded. The system senses a fire condition through a closed circuit electrical detector system which controls water flow to the fire automatically. Batteries supply up to 90 hour emergency power supply for system operation. The piping system is dry (until water is required) and is monitored with pressurized air. Should any leak in the system piping occur, an alarm will sound, but water will not enter the system until heat is sensed by a firecycle detector.

Area coverage sprinkler systems may be laid out and fed from the supply in any one of several patterns as shown below. It is desirable, if possible, to utilize a central feed and achieve a shorter flow path from the riser to the furthest sprinkler. This permits use of the smallest sizes of pipe possible with resulting savings.

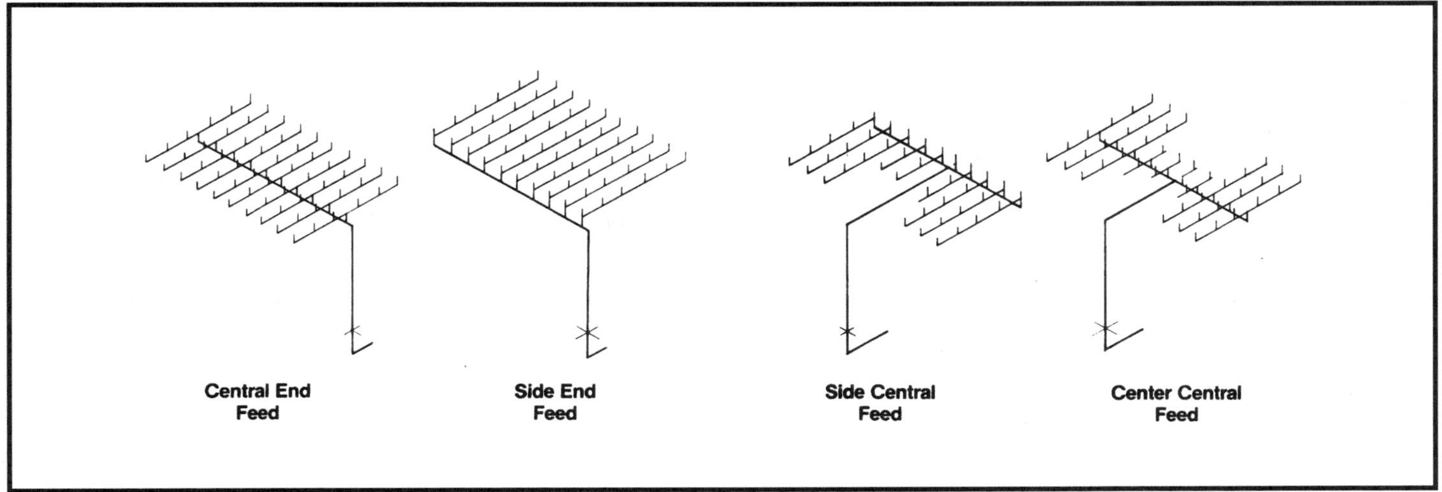

R10520-120 System Classification

System Classification

Rules for installation of sprinkler systems vary depending on the classification of occupancy falling into one of three categories as follows:

Light Hazard Occupancy

The protection area allotted per sprinkler should not exceed 200 S.F. with the maximum distance between lines and sprinklers on lines being 15′. The sprinklers do not need to be staggered. Branch lines should not exceed eight sprinklers on either side of a cross main. Each large area requiring more than 100 sprinklers and without a sub-dividing partition should be supplied by feed mains or risers sized for ordinary hazard occupancy.

Included in this group are:

Auditoriums
Churches
Clubs
Educational
Hospitals
Institutional
Libraries
(except large stack rooms)
Museums
Nursing Homes
Offices
Residential
Restaurants
Schools
Theaters

Ordinary Hazard Occupancy

The protection area allotted per sprinkler shall not exceed 130 S.F. of noncombustible ceiling and 120 S.F. of combustible ceiling. The maximum allowable distance between sprinkler lines and sprinklers on line is 15′. Sprinklers shall be staggered if the distance between heads exceeds 12′. Branch lines should not exceed eight sprinklers on either side of a cross main.

Included in this group are:

Automotive garages
Bakeries
Beverage manufacturing
Bleacheries
Boiler houses
Canneries
Cement plants
Clothing factories
Cold storage warehouses
Dairy products manufacturing
Distilleries
Dry cleaning
Electric generating stations
Feed mills
Grain elevators
Ice manufacturing
Laundries
Machine shops
Mercantiles
Paper mills
Printing and Publishing
Shoe factories
Warehouses
Wood product assembly

Extra Hazard Occupancy

The protection area allotted per sprinkler shall not exceed 90 S.F. of noncombustible ceiling and 80 S.F. of combustible ceiling. The maximum allowable distance between lines and between sprinklers on lines is 12′. Sprinklers on alternate lines shall be staggered if the distance between sprinklers on lines exceeds 8′. Branch lines should not exceed six sprinklers on either side of a cross main.

Included in this group are:

Aircraft hangars
Chemical works
Explosives manufacturing
Linoleum manufacturing
Linseed oil mills
Oil refineries
Paint shops
Shade cloth manufacturing
Solvent extracting
Varnish works
Volatile flammable
liquid manufacturing & use

R10520-130 Sprinkler Quantities for Various Sizes and Types of Pipe

Sprinkler Quantities: The table below lists the usual maximum number of sprinkler heads for each size of copper and steel pipe for both wet and dry systems. These quantities may be adjusted to meet individual structural needs or local code requirements. Maximum area on any one floor for one system is: light hazard and ordinary hazard 52,000 S.F., extra hazardous 25,000 S.F.

Pipe Size	Light Hazard Occupancy		Ordinary Hazard Occupancy		Extra Hazard Occupancy	
Diameter	Steel Pipe	Copper Pipe	Steel Pipe	Copper Pipe	Steel Pipe	Copper Pipe
1"	2 sprinklers	2 sprinklers	2 sprinklers	2 sprinklers	1 sprinklers	1 sprinklers
1-1/4"	3	3	3	3	2	2
1-1/2"	5	5	5	5	5	5
2"	10	12	10	12	8	8
2-1/2"	30	40	20	25	15	20
3"	60	65	40	45	27	30
3-1/2"	100	115	65	75	40	45
4"			100	115	55	65
5"			160	180	90	100
6"			275	300	150	170

Dry Pipe Systems: A dry pipe system should be installed where a wet pipe system is impractical, as in rooms or buildings which cannot be properly heated.

The use of an approved dry pipe system is more desirable than shutting off the water supply during cold weather.

Not more than 750 gallons of system capacity should be controlled by one dry pipe valve. Where two or more dry pipe valves are used, systems should preferably be divided horizontally.

R10520-140 Adjustment for Sprinkler/Standpipe Installations

Quality/Complexity Multiplier (For all installations)

Economy installation, add .... 0 to 5%
Good quality, medium complexity, add .... 5 to 15%
Above average quality and complexity, add .... 15 to 25%

R10520-310 Standpipe Systems

The basis for standpipe system design is National Fire Protection Association NFPA 14, however, the authority having jurisdiction should be consulted for special conditions, local requirements and approval.

Standpipe systems, properly designed and maintained, are an effective and valuable time saving aid for extinquishing fires, especially in the upper stories of tall buildings, the interior of large commercial or industrial malls, or other areas where construction features or access make the laying of temporary hose lines time consuming and/or hazardous. Standpipes are frequently installed with automatic sprinkler systems for maximum protection.

There are three general classes of service for standpipe systems:
Class I for use by fire departments and personnel with special training for heavy streams (2-1/2" hose connections).
Class II for use by building occupants until the arrival of the fire department (1-1/2" hose connector with hose).
Class III for use by either fire departments and trained personnel or by the building occupants (both 2-1/2" and 1-1/2" hose connections or one 2-1/2" hose valve with an easily removable 2-1/2" by 1-1/2" adapter).

Standpipe systems are also classified by the way water is supplied to the system. The four basic types are:
Type 1: Wet standpipe system having supply valve open and water pressure maintained at all times.
Type 2: Standpipe system so arranged through the use of approved devices as to admit water to the system automatically by opening a hose valve.
Type 3: Standpipe system arranged to admit water to the system through manual operation of approved remote control devices located at each hose station.
Type 4: Dry standpipe having no permanent water supply.

R10520-320 NFPA 14 Basic Standpipe Design

Class	Design-Use	Pipe Size Minimums	Water Supply Minimums
Class I	2 1/2″ hose connection on each floor All areas within 150′ of an exit in every exit stairway Fire Department Trained Personnel	Height to 100′, 4″ dia. Heights above 100′, 6″ dia. (275′ max. except with pressure regulators 400′ max.)	For each standpipe riser 500 GPM flow For common supply pipe allow 500 GPM for first standpipe plus 250 GPM for each additional standpipe (2500 GPM max. total) 30 min. duration 65 PSI at 500 GPM
Class II	1 1/2″ hose connection with hose on each floor All areas within 130′ of hose connection measured along path of hose travel Occupant personnel	Height to 50′, 2″ dia. Height above 50′, 2 1/2″ dia.	For each standpipe riser 100 GPM flow For multiple riser common supply pipe 100 GPM 300 min. duration, 65 PSI at 100 GPM
Class III	Both of above. Class I valved connections will meet Class III with additional 2 1/2″ by 1 1/2″ adapter and 1 1/2″ hose.	Same as Class I	Same as Class I

*Note: Where 2 or more standpipes are installed in the same building or section of building they shall be interconnected at the bottom.

Combined Systems

Combined systems are systems where the risers supply both automatic sprinklers and 2-1/2″ hose connection outlets for fire department use. In such a system the sprinkler spacing pattern shall be in accordance with NFPA 13 while the risers and supply piping will be sized in accordance with NFPA 14. When the building is completely sprinklered the risers may be sized by hydraulic calculation.The minimum size riser for buildings not completely sprinklered is 6″.

The minimum water supply of a completely sprinklered, light hazard, high-rise occupancy building will be 500 GPM while the supply required for other types of completely sprinklered high-rise buildings is 1000 GPM.

General System Requirements

1. Approved valves will be provided at the riser for controlling branch lines to hose outlets.
2. A hose valve will be provided at each outlet for attachment of hose.
3. Where pressure at any standpipe outlet exceeds 100 PSI a pressure reducer must be installed to limit the pressure to 100 PSI. Note that the pressure head due to gravity in 100′ of riser is 43.4 PSI. This must be overcome by city pressure, fire pumps, or gravity tanks to provide adequate pressure at the top of the riser.
4. Each hose valve on a wet system having linen hose shall have an automatic drip connection to prevent valve leakage from entering the hose.
5. Each riser will have a valve to isolate it from the rest of the system.
6. One or more fire department connections as an auxiliary supply shall be provided for each Class I or Class III standpipe system. In buildings having two or more zones, a connection will be provided for each zone.
7. There will be no shutoff valve in the fire department connection, but a check valve will be located in the line before it joins the system.
8. All hose connections street side will be identified on a cast plate or fitting as to purpose.

R13128-520 Swimming Pools

Pool prices given per square foot of surface area include pool structure, filter and chlorination equipment where required, pumps, related piping, diving boards, ladders, maintenance kit, skimmer and vacuum system. Decks and electrical service to equipment are not included.

Residential in-ground pool construction can be divided into two categories: vinyl lined and gunite. Vinyl lined pool walls are constructed of different materials including wood, concrete, plastic or metal. The bottom is often graded with sand over which the vinyl liner is installed. Costs are generally in the $16 to $23 per S.F. range. Vermiculite or soil cement bottoms may be substituted for an added cost.

Gunite pool construction is used both in residential and municipal installations. These structures are steel reinforced for strength and finished with a white cement limestone plaster. Residential costs run from $26 to $38 per S.F. surface. Municipal costs vary from $51 to $77 because plumbing codes require more expensive materials, chlorination equipment and higher filtration rates.

Municipal pools greater than 1,800 S.F. require gutter systems to control waves. This gutter may be formed into the concrete wall. Often a vinyl, stainless steel gutter or gutter and wall system is specified, which will raise the pool cost an additional $60 to $150 per L.F. of gutter installed and up to $290 per L.F. if a gutter and wall system is installed.

Competition pools usually require tile bottoms and sides with contrasting lane striping. Add $12 per S.F. of wall or bottom to be tiled.

Special Construction | R132 | Hazardous Material Remediation

R13281-120 Asbestos Removal Process

Asbestos removal is accomplished by a specialty contractor who understands the federal and state regulations regarding the handling and disposal of the material. The process of asbestos removal is divided into many individual steps. An accurate estimate can be calculated only after all the steps have been priced.

The steps are generally as follows:

1. Obtain an asbestos abatement plan from an industrial hygienist.
2. Monitor the air quality in and around the removal area and along the path of travel between the removal area and transport area. This establishes the background contamination.
3. Construct a two part decontamination chamber at entrance to removal area.
4. Install a HEPA filter to create a negative pressure in the removal area.
5. Install wall, floor and ceiling protection as required by the plan, usually 2 layers of fireproof 6 mil polyethylene.
6. Industrial hygienist visually inspects work area to verify compliance with plan.
7. Provide temporary supports for conduit and piping affected by the removal process.
8. Proceed with asbestos removal and bagging process. Monitor air quality as described in Step #2. Discontinue operations when contaminate levels exceed applicable standards.
9. Document the legal disposal of materials in accordance with EPA standards.
10. Thoroughly clean removal area including all ledges, crevices and surfaces.
11. Post abatement inspection by industrial hygienist to verify plan compliance.
12. Provide a certificate from a licensed industrial hygienist attesting that contaminate levels are within acceptable standards before returning area to regular use.

Special Construction | R136 | Solar & Wind Energy Equip.

R13600-610 Solar Heating (Space and Hot Water)

Collectors should face as close to due South as possible, however, variations of up to 20 degrees on either side of true South are acceptable. Local climate and collector type may influence the choice between east or west deviations. Obviously they should be located so they are not shaded from the sun's rays. Incline collectors at a slope of latitude minus 5 degrees for domestic hot water and latitude plus 15 degrees for space heating.

Flat plate collectors consist of a number of components as follows: Insulation to reduce heat loss through the bottom and sides of the collector. The enclosure which contains all the components in this assembly is usually weatherproof and prevents dust, wind and water from coming in contact with the absorber plate. The cover plate usually consists of one or more layers of a variety of glass or plastic and reduces the reradiation by creating an air space which traps the heat between the cover and the absorber plates.

The absorber plate must have a good thermal bond with the fluid passages. The absorber plate is usually metallic and treated with a surface coating which improves absorptivity. Black or dark paints or selective coatings are used for this purpose, and the design of this passage and plate combination helps determine a solar system's effectiveness.

Heat transfer fluid passage tubes are attached above and below or integral with an absorber plate for the purpose of transferring thermal energy from the absorber plate to a heat transfer medium. The heat exchanger is a device for transferring thermal energy from one fluid to another.

Piping and storage tanks should be well insulated to minimize heat losses.

Size domestic water heating storage tanks to hold 20 gallons of water per user, minimum, plus 10 gallons per dishwasher or washing machine. For domestic water heating an optimum collector size is approximately 3/4 square foot of area per gallon of water storage. For space heating of residences and small commercial applications the collector is commonly sized between 30% and 50% of the internal floor area. For space heating of large commercial applications, collector areas less than 30% of the internal floor area can still provide significant heat reductions.

A supplementary heat source is recommended for Northern states for December through February.

The solar energy transmission per square foot of collector surface varies greatly with the material used. Initial cost, heat transmittance and useful life are obviously interrelated.

For information about Means Estimating Seminars, see yellow pages 11 and 12 in back of book

R15050-710 Subcontractors

On the unit cost pages of the R.S. Means Cost Data books, the last column is entitled "Total Incl. O&P". This is normally the cost of the installing contractor. In Division 15, this is the cost of the mechanical contractor. If the particular work being estimated is to be performed by a sub to the mechanical contractor, the mechanical's profit and handling charge (usually 10%) is added to the total of the last column.

R15050-720 Demolition (Selective vs. Removal for Replacement)

Demolition can be divided into two basic categories.

One type of demolition involves the removal of material with no concern for its replacement. The labor-hours to estimate this work are found in Div. 15055 under "Selective Demolition". It is selective in that individual items or all the material installed as a system or trade grouping such as plumbing or heating systems are removed. This may be accomplished by the easiest way possible, such as sawing, torch cutting, or sledge hammer as well as simple unbolting.

The second type of demolition is the removal of some item for repair or replacement. This removal may involve careful draining, opening of unions, disconnecting and tagging of electrical connections, capping of pipes/ducts to prevent entry of debris or leakage of the material contained as well as transport of the item away from its in-place location to a truck/dumpster. An approximation of the time required to accomplish this type of demolition is to use half of the time indicated as necessary to install a new unit. For example; installation of a new pump might be listed as requiring 6 labor-hours so if we had to estimate the removal of the old pump we would allow an additional 3 hours for a total of 9 hours. That is, the complete replacement of a defective pump with a new pump would be estimated to take 9 labor-hours.

R15100-040 Plumbing Approximations for Quick Estimating

Water Control

Water Meter; Backflow Preventer, Shock Absorbers; Vacuum Breakers; Mixer. .... 10 to 15% of Fixtures

Pipe And Fittings .... 30 to 60% of Fixtures

Note: Lower percentage for compact buildings or larger buildings with plumbing in one area.
Larger percentage for large buildings with plumbing spread out.
In extreme cases pipe may be more than 100% of fixtures.
Percentages **do not** include special purpose or process piping.

Plumbing Labor

1 & 2 Story Residential .... Rough-in Labor = 80% of Materials

Apartment Buildings .... Rough-in Labor = 90 to 100% of Materials

Labor for handling and placing fixtures is approximately 25 to 30% of fixtures

Quality/Complexity Multiplier (for all installations)

Economy installation, add.......... 0 to 5%

Good quality, medium complexity, add .... 5 to 15%

Above average quality and complexity, add .... 15 to 25%

R15100-050 Pipe Material Considerations

1. Malleable fittings should be used for gas service.
2. Malleable fittings are used where there are stresses/strains due to expansion and vibration.
3. Cast fittings may be broken as an aid to disassembling of heating lines frozen by long use, temperature and minerals.
4. Cast iron pipe is extensively used for underground and submerged service.
5. Type M (light wall) copper tubing is available in hard temper only and is used for nonpressure and less severe applications than K and L.
6. Type L (medium wall) copper tubing, available hard or soft for interior service.
7. Type K (heavy wall) copper tubing, available in hard or soft temper for use where conditions are severe. For underground and interior service.
8. Hard drawn tubing requires fewer hangers or supports but should not be bent. Silver brazed fittings are recommended, however soft solder is normally used.
9. Type DMV (very light wall) copper tubing designed for drainage, waste and vent plus other non-critical pressure services.

Domestic/Imported Pipe and Fittings Cost

The prices shown in this publication for steel/cast iron pipe and steel, cast iron, malleable iron fittings are based on domestic production sold at the normal trade discounts. The above listed items of foreign manufacture may be available at prices of 1/3 to 1/2 those shown. Some imported items after minor machining or finishing operations are being sold as domestic to further complicate the system.

Caution: Most pipe prices in this book also include a coupling and pipe hangers which for the larger sizes can add significantly to the per meter cost and should be taken into account when comparing "book cost" with quoted supplier's cost.

R15100-070 Piping to 10′ High

When taking off pipe, it is important to identify the different material types and joining procedures, as well as distances between supports and components required for proper support.

During the takeoff, measure through all fittings. Do not subtract the lengths of the fittings, valves, or strainers, etc. This added length plus the final rounding of the totals will compensate for nipples and waste.

When rounding off totals always increase the actual amount to correspond with manufacturer's shipping lengths.

A. Both red brass and yellow brass pipe are normally furnished in 12′ lengths, plain end. Division 15107-220 includes in the linear foot costs two field threads and one coupling per 10′ length. A carbon steel clevis type hanger assembly every 10′ is also prorated into the linear foot costs, including both material and labor.

B. Cast iron soil pipe is furnished in either 5′ or 10′ lengths. For pricing purposes, Division 15107-320 in *Means Plumbing Cost Data* features 10′ lengths with a joint and a carbon steel clevis hanger assembly every 5′ prorated into the per foot costs of both material and labor.

   Three methods of joining are considered in Division 15107-320, lead and oakum poured joints, or push-on gasket type joints for the bell and spigot pipe, and a joint clamp for the no-hub soil pipe. The labor and material costs for each of these individual joining procedures are also prorated into the linear costs per foot.

C. Copper tubing in Division 15107-420 covers types K, L, M, and DWV which are furnished in 20′ lengths. Means pricing data is based on a tubing cut each length and a coupling and two soft soldered joints every 10′. A carbon steel, clevis type hanger assembly every 10′ is also prorated into the per foot costs. The prices for refrigeration tubing are for materials only. Labor for full lengths may be based on the type L labor but short cut measures in tight areas can increase the installation labor-hours from 20 to 40%.

D. Corrosion-resistant piping does not lend itself to one particular standard of hanging or support assembly due to its diversity of application and placement. The several varieties of corrosion-resistant piping do not include any material or labor costs for hanger assemblies (See Division 15060-300 for appropriate selection). The estimator should refer to Division 15107-560 in *Means Plumbing Cost Data* for the compatible corrosion-resistant fittings and piping.

E. Glass pipe, as shown in *Means Plumbing Cost Data* Division 15106-120, is furnished in standard lengths either 5′ or 10′ long, beaded on one end. Special orders for diverse lengths beaded on both ends are also available. For pricing purposes, R.S. Means features 10′ lengths with a coupling and a carbon steel band hanger assembly every 10′ prorated into the per foot linear costs.

   Glass pipe is also available with conical ends and standard lengths ranging from 6″ through 3′ in 6″ increments, then up to 10′ in 12″ increments. Special lengths can be customized for particular installation requirements.

   For pricing purposes, Means has based the labor and material pricing on 10′ lengths. Included in these costs per linear foot are the prorated costs for a flanged assembly every 10′ consisting of two flanges, a gasket, two insertable seals, and the required number of bolts and nuts. A carbon steel band hanger assembly based on 10′ center lines has also been prorated into the costs per foot for labor and materials.

F. Plastic pipe of several compositions and joining methods are considered in Division 15108-520. Fiberglass reinforced pipe (FRP) is priced, based on 10′ lengths (20′ lengths are also available), with coupling and epoxy joints every 10′. FRP is furnished in both "General Service" and "High Strength." A carbon steel clevis hanger assembly, 3 for every 10′, is built into the prorated labor and material costs on a per foot basis.

   The PVC and CPVC pipe schedules 40, 80 and 120, plus SDR ratings are all based on 20′ lengths with a coupling installed every 10′, as well as a carbon steel clevis hanger assembly every 3′. The PVC and ABS type DWV piping is based on 10′ lengths with solvent weld couplings every 10′, and with carbon steel clevis hanger assemblies, 3 for every 10′. The rest of the plastic piping in this section is based on flexible 100′ coils and does not include any coupling or supports.

   This section ends with PVC drain and sewer piping based on 10′ lengths with bell and spigot ends and 0-ring type, push-on joints. Material costs only are given, but the complete installation costs should be similar to those shown in line number 02620-270.

G. Stainless steel piping, Division 15107-900, includes both weld end and threaded piping, both in the type 304 and 316 specification and in the following schedules, 5, 10, 40, 80, and 160. Although this piping is usually furnished in 20′ lengths, this cost grouping has a joint (either heli-arc butt-welded or threads and coupling) every 10′. A carbon steel clevis type hanger assembly is also included at 10′ intervals and prorated into the linear foot costs.

H. Division 15107-620 covers carbon steel pipe, both black and galvanized. This section encompasses schedules 40 (standard) and 80 (extra heavy).

   Several common methods of joining steel pipe — such as thread and coupled, butt welded, and flanged (150 lb. weld neck flanges) are also included.

   For estimating purposes, it is assumed that the piping is purchased in 20′ lengths and that a compatible joint is made up every 10′. These joints are prorated into the labor and material costs per linear foot. The following hanger and support assemblies every 10′ are also included: carbon steel clevis for the T & C pipe, and single rod roll type for both the welded and flanged piping. All of these hangers are oversized to accommodate pipe insulation 3/4″ thick through 5″ pipe size and 1-1/2″ thick from 6″ through 12″ pipe size.

I. Division 15107-690 prices grooved joint steel pipe, both black and galvanized, in schedules 10, 40, and 80, furnished in 20′ lengths. This section describes two joining methods: cut groove and roll groove. The schedule 10 piping is roll-grooved, while the heavier schedules are cut-grooved. The labor and material costs are prorated into per linear foot prices, including a coupled joint every 10′, as well as a carbon steel clevis hanger assembly.

Notes:

The pipe hanger assemblies mentioned in the preceding paragraphs include the described hanger; appropriately sized steel, box-type insert and nut; plus a threaded hanger rod. On average, the distance from the pipe center line to the insert face is 2′.

C clamps are used when the pipe is to be supported from steel shapes rather than anchored in the slab. C clamps are slightly less costly than inserts. However, to save time in estimating, it is advisable to use the given line number cost, rather than substituting a C clamp for the insert.

Add to piping labor for elevated installation:

10′ to 15′ high	10%	30′ to 35′ high	40%
15′ to 20′ high	20%	35′ to 40′ high	50%
20′ to 25′ high	25%	Over 40′ high	55%
25′ to 30′ high	35%		

When using the percentage adds for elevated piping installations as shown above, bear in mind that the given heights are for the pipe supports, even though the insert, anchor, or clamp may be several feet higher than the pipe itself.

An allowance has been included in the piping installation time for testing and minor tightening of leaking joints, fittings, stuffing boxes, packing glands, etc. For extraordinary test requirements such as x-rays, prolonged pressure or demonstration tests, a percentage of the piping labor, based on the estimator's experience, must be added to the labor total. A testing service specializing in weld x-rays should be consulted for pricing if it is an estimate requirement. Equipment installation time includes start-up with associated adjustments.

R15100-080 Valve Materials

VALVE MATERIALS

Bronze:
Bronze is one of the oldest materials used to make valves. It is most commonly used in hot and cold water systems and other non-corrosive services. It is often used as a seating surface in larger iron body valves to ensure tight closure.

Carbon Steel:
Carbon steel is a high strength material. Therefore, valves made from this metal are used in higher pressure services, such as steam lines up to 600 psi at 850°F. Many steel valves are available with butt-weld ends for economy and are generally used in high pressure steam service as well as other higher pressure non-corrosive services.

Forged Steel:
Valves from tough carbon steel are used in service up to 2000 psi and temperatures up to 1000°F in Gate, Globe and Check valves.

Iron:
Valves are normally used in medium to large pipe lines to control non-corrosive fluid and gases, where pressures do not exceed 250 psi at 450° or 500 psi cold water, oil or gas.

Stainless Steel:
Developed steel alloys can be used in over 90% corrosive services.

Plastic PVC:
This is used in a great variety of valves generally in high corrosive service with lower temperatures and pressures.

VALVE SERVICE PRESSURES

Pressure ratings on valves provide an indication of the safe operating pressure for a valve at some elevated temperature. This temperature is dependent upon the materials used and the fabrication of the valve. When specific data is not available, a good "rule-of-thumb" to follow is the temperature of saturated steam on the primary rating indicated on the valve body. Example: The valve has the number 150S printed on the side indicating 150 psi and hence, a maximum operating temperature of 367°F (temperature of saturated steam and 150 psi).

DEFINITIONS

1. "WOG" – Water, oil, gas (cold working pressures).
2. "SWP" – Steam working pressure.
3. 100% area (full port) – means the area through the valve is equal to or greater than the area of standard pipe.
4. "Standard Opening" – means that the area through the valve is less than the area of standard pipe and therefore these valves should be used only where restriction of flow is unimportant.
5. "Round Port" – means the valve has a full round opening through the plug and body, of the same size and area as standard pipe.
6. "Rectangular Port" – valves have rectangular shaped ports through the plug body. The area of the port is either equal to 100% of the area of standard pipe, or restricted (standard opening). In either case it is clearly marked.
7. "ANSI" – American National Standards Institute.

R15100-090 Valve Selection Considerations

INTRODUCTION: In any piping application, valve performance is critical. Valves should be selected to give the best performance at the lowest cost.

The following is a list of performance characteristics generally expected of valves.

1. Stopping flow or starting it.
2. Throttling flow (Modulation).
3. Flow direction changing.
4. Checking backflow (Permitting flow in only one direction).
5. Relieving or regulating pressure.

In order to properly select the right valve, some facts must be determined.

A. What liquid or gas will flow through the valve?
B. Does the fluid contain suspended particles?
C. Does the fluid remain in liquid form at all times?
D. Which metals does fluid corrode?
E. What are the pressure and temperature limits? (As temperature and pressure rise, so will the price of the valve.)
F. Is there constant line pressure?
G. Is the valve merely an on-off valve?
H. Will checking of backflow be required?
I. Will the valve operate frequently or infrequently?

Valves are classified by design type into such classifications as Gate, Globe, Angle, Check, Ball, Butterfly and Plug. They are also classified by end connection, stem, pressure restrictions and material such as bronze, cast iron, etc. Each valve has a specific use. A quality valve used correctly will provide a lifetime of trouble- free service, but a high quality valve installed in the wrong service may require frequent attention.

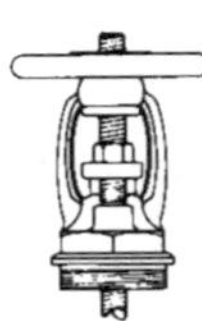

STEM TYPES

(OS & Y)—Rising Stem-Outside Screw and Yoke

Offers a visual indication of whether the valve is open or closed. Recommended where high temperatures, corrosives, and solids in the line might cause damage to inside-valve stem threads. The stem threads are engaged by the yoke bushing so the stem rises through the hand wheel as it is turned.

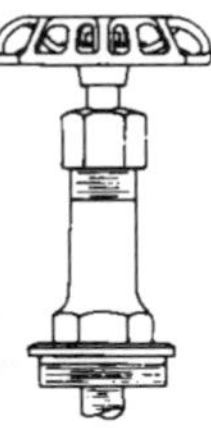

(R.S.)—Rising Stem-Inside Screw

Adequate clearance for operation must be provided because both the hand wheel and the stem rise.
The valve wedge position is indicated by the position of the stem and hand wheel.

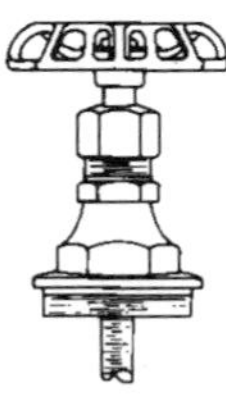

(N.R.S.)—Non-Rising Stem-Inside Screw

A minimum clearance is required for operating this type of valve. Excessive wear or damage to stem threads inside the valve may be caused by heat, corrosion, and solids. Because the hand wheel and stem do not rise, wedge position cannot be visually determined.

VALVE TYPES

Gate Valves

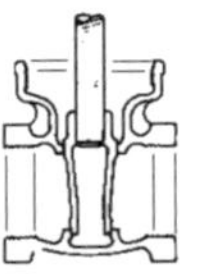

Provide full flow, minute pressure drop, minimum turbulence and minimum fluid trapped in the line.
They are normally used where operation is infrequent.

Globe Valves

Globe valves are designed for throttling and/or frequent operation with positive shut-off. Particular attention must be paid to the several types of seating materials available to avoid unnecessary wear. The seats must be compatible with the fluid in service and may be composition or metal. The configuration of the Globe valve opening causes turbulence which results in increased resistance. Most bronze Globe valves are rising stem-inside screw, but they are also available on O.S. & Y.

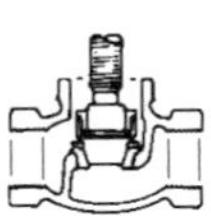

Angle Valves

The fundamental difference between the Angle valve and the Globe valve is the fluid flow through the Angle valve. It makes a 90° turn and offers less resistance to flow than the Globe valve while replacing an elbow. An angle valve thus reduces the number of joints and installation time.

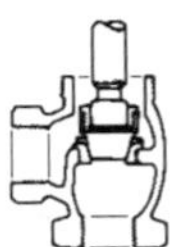

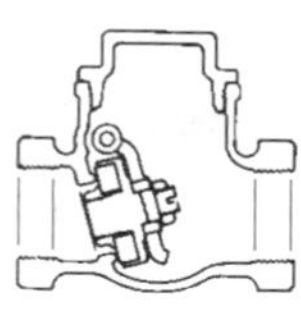

Check Valves

Check valves are designed to prevent backflow by automatically seating when the direction of fluid is reversed.
Swing Check valves are generally installed with Gate-valves, as they provide comparable full flow.
Usually recommended for lines where flow velocities are low and should not be used on lines with pulsating flow. Recommended for horizontal installation, or in vertical lines only where flow is upward.

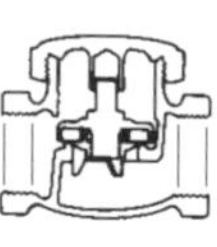

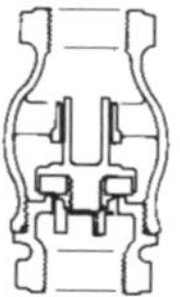

Lift Check Valves

These are commonly used with Globe and Angle valves since they have similar diaphragm seating arrangements and are recommended for preventing backflow of steam, air, gas and water, and on vapor lines with high flow velocities. For horizontal lines, horizontal lift checks should be used and vertical lift checks for vertical lines.

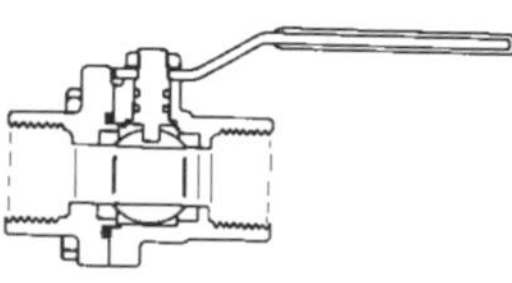

Ball Valves

Ball valves are light and easily installed, yet because of modern elastomeric seats, provide tight closure. Flow is controlled by rotating up to 90° a drilled ball which fits tightly against resilient seals. This ball seats with flow in either direction, and valve handle indicates the degree of opening. Recommended for frequent operation readily adaptable to automation, ideal for installation where space is limited.

Butterfly Valves

Butterfly valves provide bubble-tight closure with excellent throttling characteristics. They can be used for full-open, closed and for throttling applications.

The Butterfly valve consists of a disc within the valve body which is controlled by a shaft. In its closed position, the valve disc seals against a resilient seat. The disc position throughout the full 90° rotation is visually indicated by the position of the operator.

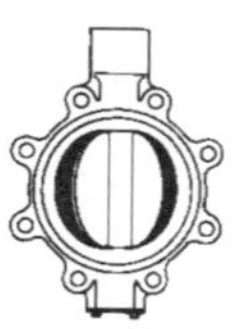

A Butterfly valve is only a fraction of the weight of a Gate valve and requires no gaskets between flanges in most cases. Recommended for frequent operation and adaptable to automation where space is limited.

Wafer and Lug type bodies when installed between two pipe flanges, can be easily removed from the line. The pressure of the bolted flanges holds the valve in place.
Locating lugs makes installation easier.

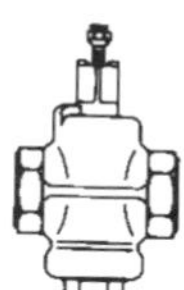

Plug Valves

Lubricated plug valves, because of the wide range of service to which they are adapted, may be classified as all purpose valves. They can be safely used at all pressure and vacuums, and at all temperatures up to the limits of available lubricants. They are the most satisfactory valves for the handling of gritty suspensions and many other destructive, erosive, corrosive and chemical solutions.

R15100-110 Hot Water Consumption Rates

Type of Building	Size Factor	Maximum Hourly Demand	Average Day Demand
Apartment Dwellings	No. of Apartments:		
	Up to 20	12.0 Gal. per apt.	42.0 Gal. per apt.
	21 to 50	10.0 Gal. per apt.	40.0 Gal. per apt.
	51 to 75	8.5 Gal. per apt.	38.0 Gal. per apt.
	76 to 100	7.0 Gal. per apt.	37.0 Gal. per apt.
	101 to 200	6.0 Gal. per apt.	36.0 Gal. per apt.
	201 up	5.0 Gal. per apt.	35.0 Gal. per apt.
Dormitories	Men	3.8 Gal. per man	13.1 Gal. per man
	Women	5.0 Gal. per woman	12.3 Gal. per woman
Hospitals	Per bed	23.0 Gal. per patient	90.0 Gal. per patient
Hotels	Single room with bath	17.0 Gal. per unit	50.0 Gal. per unit
	Double room with bath	27.0 Gal. per unit	80.0 Gal. per unit
Motels	No. of units:		
	Up to 20	6.0 Gal. per unit	20.0 Gal. per unit
	21 to 100	5.0 Gal. per unit	14.0 Gal. per unit
	101 Up	4.0 Gal. per unit	10.0 Gal. per unit
Nursing Homes		4.5 Gal. per bed	18.4 Gal. per bed
Office buildings		0.4 Gal. per person	1.0 Gal. per person
Restaurants	Full meal type	1.5 Gal./max. meals/hr.	2.4 Gal. per meal
	Drive-in snack type	0.7 Gal./max. meals/hr.	0.7 Gal. per meal
Schools	Elementary	0.6 Gal. per student	0.6 Gal. per student
	Secondary & High	1.0 Gal. per student	1.8 Gal. per student

For evaluation purposes, recovery rate and storage capacity are inversely proportional. Water heaters should be sized so that the maximum hourly demand anticipated can be met in addition to allowance for the heat loss from the pipes and storage tank.

R15100-120 Fixture Demands in Gallons Per Fixture Per Hour

Table below is based on 140°F final temperature except for dishwashers in public places (*) where 180°F water is mandatory.

Fixture	Apartment House	Club	Gym	Hospital	Hotel	Indust. Plant	Office	Private Home	School
Bathtubs	20	20	30	20	20			20	
Dishwashers, automatic	15	50-150*		50-150*	50-200*	20-100*		15	20-100*
Kitchen sink	10	20		20	30	20	20	10	20
Laundry, stationary tubs	20	28		28	28			20	
Laundry, automatic wash	75	75		100	150			75	
Private lavatory	2	2	2	2	2	2	2	2	2
Public lavatory	4	6	8	6	8	12	6		15
Showers	30	150	225	75	75	225	30	30	225
Service sink	20	20		20	30	20	20	15	20
Demand factor	0.30	0.30	0.40	0.25	0.25	0.40	0.30	0.30	0.40
Storage capacity factor	1.25	0.90	1.00	0.60	0.80	1.00	2.00	0.70	1.00

To obtain the probable maximum demand multiply the total demands for the fixtures (gal./fixture/hour) by the demand factor. The heater should have a heating capacity in gallons per hour equal to this maximum. The storage tank should have a capacity in gallons equal to the probable maximum demand multiplied by the storage capacity factor.

R15100-410 Minimum Plumbing Fixture Requirements

Type of Building/Use	Water Closets Persons	Water Closets Fixtures	Urinals Persons	Urinals Fixtures	Lavatories Persons	Lavatories Fixtures	Bathtubs or Showers Persons	Bathtubs or Showers Fixtures	Drinking Fountain Fixtures	Other Fixtures
Assembly Halls Auditoriums Theater Public assembly	1-100 101-200 201-400	1 2 3	1-200 201-400 401-600	1 2 3	1-200 201-400 401-750	1 2 3			1 for each 1000 persons	1 service sink
	Over 400 add 1 fixt. for ea. 500 men; 1 fixt. for ea. 300 women		Over 600 add 1 fixture for each 300 men		Over 750 add 1 fixture for each 500 persons					
Assembly Public Worship	300 men 150 women	1 1	300 men	1	men women	1 1			1	
Dormitories	Men: 1 for each 10 persons Women: 1 for each 8 persons		1 for each 25 men; over 150 add 1 fixture for each 50 men		1 for ea. 12 persons 1 separate dental lav. for each 50 persons recom.		1 for ea. 8 persons For women add 1 additional for each 30. Over 150 persons add 1 for each 20.		1 for each 75 persons	Laundry trays 1 for each 50 serv. sink 1 for ea. 100
Dwellings Apartments and homes	1 fixture for each unit				1 fixture for each unit		1 fixture for each unit			
Hospitals Indiv. Room Ward Waiting room	8 persons	1 1 1			10 persons	1 1 1	20 persons	1 1	1 for 100 patients	1 service sink per floor
Industrial Mfg. plants Warehouses	1-10 11-25 26-50 51-75 76-100	1 2 3 4 5	0-30 31-80 81-160 161-240	1 2 3 4	1-100 over 100	1 for ea. 10 1 for ea. 15	1 Shower for each 15 persons subject to excessive heat or occupational hazard		1 for each 75 persons	
	1 fixture for each additional 30 persons									
Public Buildings Businesses Offices	1-15 16-35 36-55 56-80 81-110 111-150	1 2 3 4 5 6	Urinals may be provided in place of water closets but may not replace more than 1/3 required number of men's water closets		1-15 16-35 36-60 61-90 91-125	1 2 3 4 5			1 for each 75 persons	1 service sink per floor
	1 fixture for ea. additional 40 persons				1 fixture for ea. additional 45 persons					
Schools Elementary	1 for ea. 30 boys 1 for ea. 25 girls		1 for ea. 25 boys		1 for ea. 35 boys 1 for ea. 35 girls		For gym or pool shower room 1/5 of a class		1 for each 40 pupils	
Schools Secondary	1 for ea. 40 boys 1 for ea. 30 girls		1 for ea. 25 boys		1 for ea. 40 boys 1 for ea. 40 girls		For gym or pool shower room 1/5 of a class		1 for each 50 pupils	

R15100-420 Plumbing Fixture Installation Time

Item	Rough-In	Set	Total Hours	Item	Rough-In	Set	Total Hours
Bathtub	5	5	10	Shower head only	2	1	3
Bathtub and shower, cast iron	6	6	12	Shower drain	3	1	4
Fire hose reel and cabinet	4	2	6	Shower stall, slate		15	15
Floor drain to 4 inch diameter	3	1	4	Slop sink	5	3	8
Grease trap, single, cast iron	5	3	8	Test 6 fixtures			14
Kitchen gas range		4	4	Urinal, wall	6	2	8
Kitchen sink, single	4	4	8	Urinal, pedestal or floor	6	4	10
Kitchen sink, double	6	6	12	Water closet and tank	4	3	7
Laundry tubs	4	2	6	Water closet and tank, wall hung	5	3	8
Lavatory wall hung	5	3	8	Water heater, 45 gals. gas, automatic	5	2	7
Lavatory pedestal	5	3	8	Water heaters, 65 gals. gas, automatic	5	2	7
Shower and stall	6	4	10	Water heaters, electric, plumbing only	4	2	6

Fixture prices in front of book are based on the cost per fixture set in place. The rough-in cost, which must be added for each fixture, includes carrier, if required, some supply, waste and vent pipe connecting fittings and stops. The lengths of rough-in pipe are nominal runs which would connect to the larger runs and stacks. The supply runs and DWV runs and stacks must be accounted for in separate entries. In the eastern half of the United States it is common for the plumber to carry these to a point 5′ outside the building.

R15100-430 Water Cooler Application

Type of Service	Requirement
Office, School or Hospital	12 persons per gallon per hour
Office, Lobby or Department Store	4 or 5 gallons per hour per fountain
Light manufacturing	7 persons per gallon per hour
Heavy manufacturing	5 persons per gallon per hour
Hot heavy manufacturing	4 persons per gallon per hour
Hotel	.08 gallons per hour per room
Theatre	1 gallon per hour per 100 seats

R15155-210 Drainage Requirements

Drainage lines must have a slope to maintain flow for proper operation. This slope should not be less than 1/4″ per foot for 3″ diameter or smaller pipe and not less than 1/8″ per foot for 4″ diameter or larger pipe. The capacity of building drainage systems is calculated on a basis of "drainage fixture units" (d.f.u.) as per the following chart.

Type of Fixture	d.f.u. Value	Type of Fixture	d.f.u. Value
Automatic clothes washer (2″ standpipe)	3	Service sink (trap standard)	3
Bathroom group (water closet, lavatory and bathtub or shower) tank type closet	6	Service sink (P trap)	2
		Urinal, pedestal, syphon jet blowout	6
Bathtub (with or without overhead shower)	2	Urinal, wall hung	4
Clinic sink	6	Urinal, stall washout	4
Combination sink & tray with food disposal	4	Wash sink (cir. or mult.) per faucet set	2
Dental unit or cuspidor	1	Water closet, tank operated	4
Dental lavatory	1	Water closet, valve operated	6
Drinking fountain	1/2	Fixtures not listed above	
Dishwasher, domestic	2	Trap size 1-1/4″ or smaller	1
Floor drains with 2″ waste	3	Trap size 1-1/2″	2
Kitchen sink, domestic with one 1-1/2″ trap	2	Trap size 2″	3
Kitchen sink, domestic with food disposal	2	Trap size 2-1/2″	4
Lavatory with 1-1/4″ waste	1	Trap size 3″	5
Laundry tray (1 or 2 compartment)	2	Trap size 4″	6
Shower stall, domestic	2		

For continuous or nearly continuous flow into the sytem from a pump, air conditioning equipment or other item, allow 2 fixture units for each gallon per minute of flow.

When the "drainage fixture units" (d.f.u.) for each horizontal branch or vertical stack is computed from the table above, the appropriate pipe size for each branch or stack is determined from the table below.

R15155-220 Allowable Fixture Units (d.f.u.) for Branches and Stacks

Pipe Diam.	Horiz. Branch (not incl. drains)	Stack Size for 3 Stories or 3 Levels	Stack size for Over 3 levels	Maximum for 1 Story building Stack
1-1/2″	3	4	8	2
2″	6	10	24	6
2-1/2″	12	20	42	9
3″	20*	48*	72*	20*
4″	160	240	500	90
5″	360	540	1100	200
6″	620	960	1900	350
8″	1400	2200	3600	600
10″	2500	3800	5600	1000
12″	3900	6000	8400	1500
15″	7000			

*Not more than two water closets or bathroom groups within each branch interval nor more than six water closets or bathroom groups on the stack.

Stacks sized for the total may be reduced as load decreases at each story to a minimum diameter of 1/2 the maximum diameter.

R17100-100 Square Foot Project Size Modifier

One factor that affects the S.F. cost of a particular building is the size. In general, for buildings built to the same specifications in the same locality, the larger building will have the lower S.F. cost. This is due mainly to the decreasing contribution of the exterior walls plus the economy of scale usually achievable in larger buildings. The Area Conversion Scale shown below will give a factor to convert costs for the typical size building to an adjusted cost for the particular project.

The Square Foot Base Size lists the median costs, most typical project size in our accumulated data and the range in size of the projects.

The Size Factor for your project is determined by dividing your project area in S.F. by the typical project size for the particular Building Type. With this factor, enter the Area Conversion Scale at the appropriate Size Factor and determine the appropriate cost multiplier for your building size.

Example: Determine the cost per S.F. for a 100,000 S.F. Mid-rise apartment building.

$$\frac{\text{Proposed building area} = 100{,}000 \text{ S.F.}}{\text{Typical size from below} = 50{,}000 \text{ S.F.}} = 2.00$$

Enter Area Conversion scale at 2.0, intersect curve, read horizontally the appropriate cost multiplier of .94. Size adjusted cost becomes .94 x $69.50 = $65.35 based on national average costs.

Note: For Size Factors less than .50, the Cost Multiplier is 1.1
For Size Factors greater than 3.5, the Cost Multiplier is .90

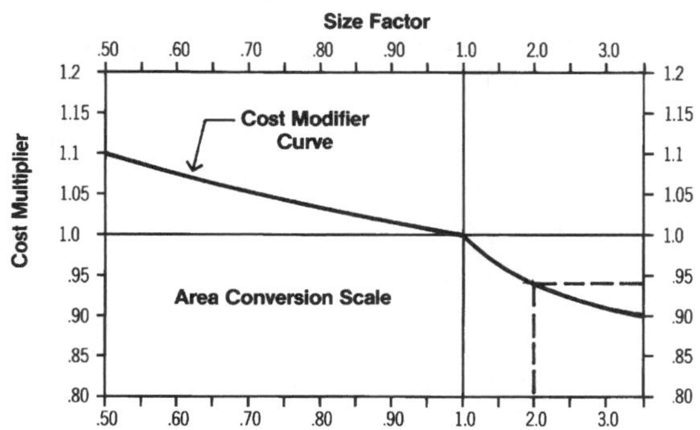

Building Type	Median Cost per S.F.	Typical Size Gross S.F.	Typical Range Gross S.F.	Building Type	Median Cost per S.F.	Typical Size Gross S.F.	Typical Range Gross S.F.
Apartments, Low Rise	$ 55.05	21,000	9,700 - 37,200	Jails	$168.00	40,000	5,500 - 145,000
Apartments, Mid Rise	69.50	50,000	32,000 - 100,000	Libraries	99.15	12,000	7,000 - 31,000
Apartments, High Rise	79.70	145,000	95,000 - 600,000	Medical Clinics	94.80	7,200	4,200 - 15,700
Auditoriums	92.05	25,000	7,600 - 39,000	Medical Offices	89.05	6,000	4,000 - 15,000
Auto Sales	56.95	20,000	10,800 - 28,600	Motels	68.25	40,000	15,800 - 120,000
Banks	123.00	4,200	2,500 - 7,500	Nursing Homes	91.55	23,000	15,000 - 37,000
Churches	83.15	17,000	2,000 - 42,000	Offices, Low Rise	74.30	20,000	5,000 - 80,000
Clubs, Country	82.85	6,500	4,500 - 15,000	Offices, Mid Rise	78.05	120,000	20,000 - 300,000
Clubs, Social	80.65	10,000	6,000 - 13,500	Offices, High Rise	99.85	260,000	120,000 - 800,000
Clubs, YMCA	83.10	28,300	12,800 - 39,400	Police Stations	125.00	10,500	4,000 - 19,000
Colleges (Class)	109.00	50,000	15,000 - 150,000	Post Offices	92.00	12,400	6,800 - 30,000
Colleges (Science Lab)	159.00	45,600	16,600 - 80,000	Power Plants	691.00	7,500	1,000 - 20,000
College (Student Union)	121.00	33,400	16,000 - 85,000	Religious Education	76.25	9,000	6,000 - 12,000
Community Center	86.60	9,400	5,300 - 16,700	Research	130.00	19,000	6,300 - 45,000
Court Houses	118.00	32,400	17,800 - 106,000	Restaurants	112.00	4,400	2,800 - 6,000
Dept. Stores	51.45	90,000	44,000 - 122,000	Retail Stores	54.65	7,200	4,000 - 17,600
Dormitories, Low Rise	88.75	25,000	10,000 - 95,000	Schools, Elementary	79.60	41,000	24,500 - 55,000
Dormitories, Mid Rise	116.00	85,000	20,000 - 200,000	Schools, Jr. High	81.10	92,000	52,000 - 119,000
Factories	49.85	26,400	12,900 - 50,000	Schools, Sr. High	81.10	101,000	50,500 - 175,000
Fire Stations	87.05	5,800	4,000 - 8,700	Schools, Vocational	80.75	37,000	20,500 - 82,000
Fraternity Houses	85.65	12,500	8,200 - 14,800	Sports Arenas	67.70	15,000	5,000 - 40,000
Funeral Homes	95.70	10,000	4,000 - 20,000	Supermarkets	54.85	44,000	12,000 - 60,000
Garages, Commercial	60.80	9,300	5,000 - 13,600	Swimming Pools	127.00	20,000	10,000 - 32,000
Garages, Municipal	77.75	8,300	4,500 - 12,600	Telephone Exchange	147.00	4,500	1,200 - 10,600
Garages, Parking	31.90	163,000	76,400 - 225,300	Theaters	81.20	10,500	8,800 - 17,500
Gymnasiums	80.45	19,200	11,600 - 41,000	Town Halls	89.30	10,800	4,800 - 23,400
Hospitals	152.00	55,000	27,200 - 125,000	Warehouses	36.80	25,000	8,000 - 72,000
House (Elderly)	75.25	37,000	21,000 - 66,000	Warehouse & Office	42.50	25,000	8,000 - 72,000
Housing (Public)	69.70	36,000	14,400 - 74,400				
Ice Rinks	77.40	29,000	27,200 - 33,600				

Change Orders

Change Order Considerations

A Change Order is a written document, usually prepared by the design professional, and signed by the owner, the architect/engineer and the contractor. A change order states the agreement of the parties to: an addition, deletion, or revision in the work; an adjustment in the contract sum, if any; or an adjustment in the contract time, if any. Change orders, or "extras" in the construction process occur after execution of the construction contract and impact architects/engineers, contractors and owners.

Change orders that are properly recognized and managed can ensure orderly, professional and profitable progress for all who are involved in the project. There are many causes for change orders and change order requests. In all cases, change orders or change order requests should be addressed promptly and in a precise and prescribed manner. The following paragraphs include information regarding change order pricing and procedures.

The Causes of Change Orders

Reasons for issuing change orders include:

- Unforeseen field conditions that require a change in the work
- Correction of design discrepancies, errors or omissions in the contract documents
- Owner-requested changes, either by design criteria, scope of work, or project objectives
- Completion date changes for reasons unrelated to the construction process
- Changes in building code interpretations, or other public authority requirements that require a change in the work
- Changes in availability of existing or new materials and products

Procedures

Properly written contract documents must include the correct change order procedures for all parties—owners, design professionals and contractors—to follow in order to avoid costly delays and litigation.

Being "in the right" is not always a sufficient or acceptable defense. The contract provisions requiring notification and documentation must be adhered to within a defined or reasonable time frame.

The appropriate method of handling change orders is by a written proposal and acceptance by all parties involved. Prior to starting work on a project, all parties should identify their authorized agents who may sign and accept change orders, as well as any limits placed on their authority.

Time may be a critical factor when the need for a change arises. For such cases, the contractor might be directed to proceed on a "time and materials" basis, rather than wait for all paperwork to be processed—a delay that could impede progress. In this situation, the contractor must still follow the prescribed change order procedures, including but not limited to, notification and documentation.

All forms used for change orders should be dated and signed by the proper authority. Lack of documentation can be very costly, especially if legal judgments are to be made and if certain field personnel are no longer available. For time and material change orders, the contractor should keep accurate daily records of all labor and material allocated to the change. Forms that can be used to document change order work are available in *Means Forms for Building Construction Professionals.*

Owners or awarding authorities who do considerable and continual building construction (such as the federal government) realize the inevitability of change orders for numerous reasons, both predictable and unpredictable. As a result, the federal government, the American Institute of Architects (AIA), the Engineers Joint Contract Documents Committee (EJCDC) and other contractor, legal and technical organizations have developed standards and procedures to be followed by all parties to achieve contract continuance and timely completion, while being financially fair to all concerned.

In addition to the change order standards put forth by industry associations, there are also many books available on the subject.

Pricing Change Orders

When pricing change orders, regardless of their cause, the most significant factor is *when* the change occurs. The need for a change may be perceived in the field or requested by the architect/engineer *before* any of the actual installation has begun, or may evolve or appear *during* construction when the item of work in question is partially installed. In the latter cases, the original sequence of construction is disrupted, along with all contiguous and supporting systems. Change orders cause the greatest impact when they occur *after* the installation has been completed and must be uncovered, or even replaced. Post-completion changes may be caused by necessary design changes, product failure, or changes in the owner's requirements that are not discovered until the building or the systems begin to function.

Specified procedures of notification and record keeping must be adhered to and enforced regardless of the stage of construction: *before, during,* or *after* installation. Some bidding documents anticipate change orders by requiring that unit prices including overhead and profit percentages—for additional as well as deductible changes—be listed. Generally these unit prices do not fully take into account the ripple effect, or impact on other trades, and should be used for general guidance only.

When pricing change orders, it is important to classify the time frame in which the change occurs. There are two basic time frames for change orders: *pre-installation change orders*, which occur before the start of construction, and *post-installation change orders*, which involve reworking after the original installation. Change orders that occur between these stages may be priced according to the extent of work completed using a combination of techniques developed for pricing *pre-* and *post-installation* changes.

The following factors are the basis for a check list to use when preparing a change order estimate.

Factors To Consider When Pricing Change Orders

As an estimator begins to prepare a change order, the following questions should be reviewed to determine their impact on the final price.

General

- Is the change order work *pre-installation* or *post-installation*?

  Change order work costs vary according to how much of the installation has been completed. Once workers have the project scoped in their mind, even though they have not started, it can be difficult to refocus. Consequently they may spend more than the normal amount of time understanding the change. Also, modifications to work in place such as trimming or refitting usually take more time than was initially estimated. The greater the amount of work in place, the more reluctant workers are to change it. Psychologically they may resent the change and as a result the rework takes longer than normal. Post-installation change order estimates must include demolition of existing work as required to accomplish the change. If the work is performed at a later time, additional obstacles such as building finishes may be present which must be protected. Regardless of whether the change occurs pre-installation or post-installation, attempt to isolate the identifiable factors and price them separately. For example, add shipping costs that may be required pre-installation or any demolition required post-installation. Then analyze the potential impact on productivity of psychological and/or learning curve factors and adjust the output rates accordingly. One approach is to break down the typical workday into segments and quantify the impact on each segment. The following chart may be useful as a guide:

Activities (Productivity) Expressed as Percentages of a Workday			
Task	**Means Mechanical Cost Data (for New Construction)**	**Pre-Installation Change Orders**	**Post-Installation Change Orders**
1. Study plans	3%	6%	6%
2. Material procurement	3%	3%	3%
3. Receiving and storing	3%	3%	3%
4. Mobilization	5%	5%	5%
5. Site movement	5%	5%	8%
6. Layout and marking	8%	10%	12%
7. Actual installation	64%	59%	54%
8. Clean-up	3%	3%	3%
9. Breaks—non-productive	6%	6%	6%
Total	100%	100%	100%

Change Order Installation Efficiency

The labor-hours expressed (for new construction) are based on average installation time, using an efficiency level of approximately 60-65%. For change order situations, adjustments to this efficiency level should reflect the daily labor-hour allocation for that particular occurrence.

If any of the specific percentages expressed in the above chart do not apply to a particular project situation, then those percentage points should be reallocated to the appropriate task(s). Example: Using data for new construction, assume there is no new material being utilized. The percentages for Tasks 2 and 3 would therefore be reallocated to other tasks. If the time required for Tasks 2 and 3 can now be applied to installation, we can add the time allocated for *Material Procurement* and *Receiving and Storing* to the *Actual Installation* time for new construction, thereby increasing the Actual Installation percentage.

This chart shows that, due to reduced productivity, labor costs will be higher than those for new construction by 5% to 15% for pre-installation change orders and by 15% to 25% for post-installation change orders. Each job and change order is unique and must be examined individually. Many factors, covered elsewhere in this section, can each have a significant impact on productivity and change order costs. All such factors should be considered in every case.

- Will the change substantially delay the original completion date?

  A significant change in the project may cause the original completion date to be extended. The extended schedule may subject the contractor to new wage rates dictated by relevant labor contracts. Project supervision and other project overhead must also be extended beyond the original completion date. The schedule extension may also put installation into a new weather season. For example, underground piping scheduled for October installation was delayed until January. As a result, frost penetrated the trench area, thereby changing the degree of difficulty of the task. Changes and delays may have a ripple effect throughout the project. This effect must be analyzed and negotiated with the owner.

- What is the net effect of a deduct change order?

  In most cases, change orders resulting in a deduction or credit reflect only bare costs. The contractor may retain the overhead and profit based on the original bid.

Materials

- Will you have to pay more or less for the new material, required by the change order, than you paid for the original purchase?

  The same material prices or discounts will usually apply to materials purchased for change orders as new construction. In some instances, however, the contractor may forfeit the advantages of competitive pricing for change orders. Consider the following example:

  A contractor purchased over $20,000 worth of fan coil units for an installation, and obtained the maximum discount. Some time later it was determined the project required an additional matching unit. The contractor has to purchase this unit from the original supplier to ensure a match. The supplier at this time may not discount the unit because of the small quantity, and the fact that he is no longer in a competitive situation. The impact of quantity on purchase can add between 0% and 25% to material prices and/or subcontractor quotes.

- If materials have been ordered or delivered to the job site, will they be subject to a cancellation charge or restocking fee?

  Check with the supplier to determine if ordered materials are subject to a cancellation charge. Delivered materials not used as result of a change order may be subject to a restocking fee if returned to the supplier. Common restocking charges run between 20% and 40%. Also, delivery charges to return the goods to the supplier must be added.

Labor

- How efficient is the existing crew at the actual installation?

  Is the same crew that performed the initial work going to do the change order? Possibly the change consists of the installation of a unit identical to one already installed; therefore the change should take less time. Be sure to consider this potential productivity increase and modify the productivity rates accordingly.

- If the crew size is increased, what impact will that have on supervision requirements?

  Under most bargaining agreements or management practices, there is a point at which a working foreman is replaced by a nonworking foreman. This replacement increases project overhead by adding a nonproductive worker. If additional workers are added to accelerate the project or to perform changes while maintaining the schedule, be sure to add additional supervision time if warranted. Calculate the hours involved and the additional cost directly if possible.

- What are the other impacts of increased crew size?

  The larger the crew, the greater the potential for productivity to decrease. Some of the factors that cause this productivity loss are: overcrowding (producing restrictive conditions in the working space), and possibly a shortage of any special tools and equipment required. Such factors affect not only the crew working on the elements directly involved in the change order, but other crews whose movement may also be hampered.

  As the crew increases, check its basic composition for changes by the addition or deletion of apprentices or nonworking foreman and quantify the potential effects of equipment shortages or other logistical factors.

- As new crews, unfamiliar with the project, are brought onto the site, how long will it take them to become oriented to the project requirements?

  The orientation time for a new crew to become 100% effective varies with the site and type of project. Orientation is easiest at a new construction site, and most difficult at existing, very restrictive renovation sites. The type of work also affects orientation time. When all elements of the work are exposed, such as concrete or masonry work, orientation is decreased. When the work is concealed or less visible, such as existing electrical systems, orientation takes longer. Usually orientation can be accomplished in one day or less. Costs for added orientation should be itemized and added to the total estimated cost.

- How much actual production can be gained by working overtime?

  Short term overtime can be used effectively to accomplish more work in a day. However, as overtime is scheduled to run beyond several weeks, studies have shown marked decreases in output. The following chart shows the effect of long term overtime on worker efficiency. If the anticipated change requires extended overtime to keep the job on schedule, these factors can be used as a guide to predict the impact on time and cost. Add project overhead, particularly supervision, that may also be incurred.

Days per Week	Hours per Day	Production Efficiency					Payroll Cost Factors	
		1 Week	2 Weeks	3 Weeks	4 Weeks	Average 4 Weeks	@ 1-1/2 Times	@ 2 Times
5	8	100%	100%	100%	100%	100%	100%	100%
	9	100	100	95	90	96.25	105.6	111.1
	10	100	95	90	85	91.25	110.0	120.0
	11	95	90	75	65	81.25	113.6	127.3
	12	90	85	70	60	76.25	116.7	133.3
6	8	100	100	95	90	96.25	108.3	116.7
	9	100	95	90	85	92.50	113.0	125.9
	10	95	90	85	80	87.50	116.7	133.3
	11	95	85	70	65	78.75	119.7	139.4
	12	90	80	65	60	73.75	122.2	144.4
7	8	100	95	85	75	88.75	114.3	128.6
	9	95	90	80	70	83.75	118.3	136.5
	10	90	85	75	65	78.75	121.4	142.9
	11	85	80	65	60	72.50	124.0	148.1
	12	85	75	60	55	68.75	126.2	152.4

Effects of Overtime

Caution: Under many labor agreements, Sundays and holidays are paid at a higher premium than the normal overtime rate.

The use of long-term overtime is counterproductive on almost any construction job; that is, the longer the period of overtime, the lower the actual production rate. Numerous studies have been conducted, and while they have resulted in slightly different numbers, all reach the same conclusion. The figure above tabulates the effects of overtime work on efficiency.

As illustrated, there can be a difference between the *actual* payroll cost per hour and the *effective* cost per hour for overtime work. This is due to the reduced production efficiency with the increase in weekly hours beyond 40. This difference between actual and effective cost results from overtime work over a prolonged period. Short-term overtime work does not result in as great a reduction in efficiency, and in such cases, effective cost may not vary significantly from the actual payroll cost. As the total hours per week are increased on a regular basis, more time is lost because of fatigue, lowered morale, and an increased accident rate.

As an example, assume a project where workers are working 6 days a week, 10 hours per day. From the figure above (based on productivity studies), the average effective productive hours over a four-week period are:

$$0.875 \times 60 = 52.5$$

Depending upon the locale and day of week, overtime hours may be paid at time and a half or double time. For time and a half, the overall (average) *actual* payroll cost (including regular and overtime hours) is determined as follows:

$$\frac{40 \text{ reg. hrs.} + (20 \text{ overtime hrs.} \times 1.5)}{60 \text{ hrs.}} = 1.167$$

Based on 60 hours, the payroll cost per hour will be 116.7% of the normal rate at 40 hours per week. However, because the effective production (efficiency) for 60 hours is reduced to the equivalent of 52.5 hours, the effective cost of overtime is calculated as follows:

For time and a half:

$$\frac{40 \text{ reg. hrs.} + (20 \text{ overtime hrs.} \times 1.5)}{52.5 \text{ hrs.}} = 1.33$$

Installed cost will be 133% of the normal rate (for labor).

Thus, when figuring overtime, the actual cost per unit of work will be higher than the apparent overtime payroll dollar increase, due to the reduced productivity of the longer workweek. These efficiency calculations are true only for those cost factors determined by hours worked. Costs that are applied weekly or monthly, such as equipment rentals, will not be similarly affected.

Equipment

- What equipment is required to complete the change order?

  Change orders may require extending the rental period of equipment already on the job site, or the addition of special equipment brought in to accomplish the change work. In either case, the additional rental charges and operator labor charges must be added.

Summary

The preceding considerations and others you deem appropriate should be analyzed and applied to a change order estimate. The impact of each should be quantified and listed on the estimate to form an audit trail.

Change orders that are properly identified, documented, and managed help to ensure the orderly, professional and profitable progress of the work. They also minimize potential claims or disputes at the end of the project.

Crew No.	Bare Costs		Incl. Subs O & P		Cost Per Labor-Hour	
Crew A-1	**Hr.**	**Daily**	**Hr.**	**Daily**	**Bare Costs**	**Incl. O&P**
1 Building Laborer	$22.85	$182.80	$35.75	$286.00	$22.85	$35.75
1 Gas Eng. Power Tool		64.80		71.30	8.10	8.91
8 L.H., Daily Totals		$247.60		$357.30	$30.95	$44.66
Crew A-1A	**Hr.**	**Daily**	**Hr.**	**Daily**	**Bare Costs**	**Incl. O&P**
1 Skilled Worker	$29.85	$238.80	$46.85	$374.80	$29.85	$46.85
1 Shot Blaster, 20"		146.40		161.05	18.30	20.13
8 L.H., Daily Totals		$385.20		$535.85	$48.15	$66.98
Crew A-2	**Hr.**	**Daily**	**Hr.**	**Daily**	**Bare Costs**	**Incl. O&P**
2 Laborers	$22.85	$365.60	$35.75	$572.00	$23.03	$35.75
1 Truck Driver (light)	23.40	187.20	35.75	286.00		
1 Light Truck, 1.5 Ton		158.00		173.80	6.58	7.24
24 L.H., Daily Totals		$710.80		$1031.80	$29.61	$42.99
Crew A-2A	**Hr.**	**Daily**	**Hr.**	**Daily**	**Bare Costs**	**Incl. O&P**
2 Laborers	$22.85	$365.60	$35.75	$572.00	$23.03	$35.75
1 Truck Driver (light)	23.40	187.20	35.75	286.00		
1 Light Truck, 1.5 Ton		158.00		173.80		
1 Concrete Saw		98.40		108.25	10.68	11.75
24 L.H., Daily Totals		$809.20		$1140.05	$33.71	$47.50
Crew A-3	**Hr.**	**Daily**	**Hr.**	**Daily**	**Bare Costs**	**Incl. O&P**
1 Truck Driver (heavy)	$23.95	$191.60	$36.60	$292.80	$23.95	$36.60
1 Dump Truck, 12 Ton		370.40		407.45	46.30	50.93
8 L.H., Daily Totals		$562.00		$700.25	$70.25	$87.53
Crew A-3A	**Hr.**	**Daily**	**Hr.**	**Daily**	**Bare Costs**	**Incl. O&P**
1 Truck Driver (light)	$23.40	$187.20	$35.75	$286.00	$23.40	$35.75
1 Pickup Truck (4x4)		89.20		98.10	11.15	12.27
8 L.H., Daily Totals		$276.40		$384.10	$34.55	$48.02
Crew A-3B	**Hr.**	**Daily**	**Hr.**	**Daily**	**Bare Costs**	**Incl. O&P**
1 Equip. Oper. (medium)	$30.35	$242.80	$45.95	$367.60	$27.15	$41.28
1 Truck Driver (heavy)	23.95	191.60	36.60	292.80		
1 Dump Truck, 16 Ton		482.80		531.10		
1 F.E. Loader, 3 C.Y.		391.80		431.00	54.66	60.13
16 L.H., Daily Totals		$1309.00		$1622.50	$81.81	$101.41
Crew A-3C	**Hr.**	**Daily**	**Hr.**	**Daily**	**Bare Costs**	**Incl. O&P**
1 Equip. Oper. (light)	$28.75	$230.00	$43.55	$348.40	$28.75	$43.55
1 Wheeled Skid Steer Loader		220.00		242.00	27.50	30.25
8 L.H., Daily Totals		$450.00		$590.40	$56.25	$73.80
Crew A-4	**Hr.**	**Daily**	**Hr.**	**Daily**	**Bare Costs**	**Incl. O&P**
2 Carpenters	$29.15	$466.40	$45.60	$729.60	$28.13	$43.65
1 Painter, Ordinary	26.10	208.80	39.75	318.00		
24 L.H., Daily Totals		$675.20		$1047.60	$28.13	$43.65
Crew A-5	**Hr.**	**Daily**	**Hr.**	**Daily**	**Bare Costs**	**Incl. O&P**
2 Laborers	$22.85	$365.60	$35.75	$572.00	$22.91	$35.75
.25 Truck Driver (light)	23.40	46.80	35.75	71.50		
.25 Light Truck, 1.5 Ton		39.50		43.45	2.19	2.41
18 L.H., Daily Totals		$451.90		$686.95	$25.10	$38.16
Crew A-6	**Hr.**	**Daily**	**Hr.**	**Daily**	**Bare Costs**	**Incl. O&P**
1 Chief Of Party	$34.55	$276.40	$54.75	$438.00	$32.20	$50.80
1 Instrument Man	29.85	238.80	46.85	374.80		
16 L.H., Daily Totals		$515.20		$812.80	$32.20	$50.80

Crew No.	Bare Costs		Incl. Subs O & P		Cost Per Labor-Hour	
Crew A-7	**Hr.**	**Daily**	**Hr.**	**Daily**	**Bare Costs**	**Incl. O&P**
1 Chief Of Party	$34.55	$276.40	$54.75	$438.00	$30.98	$48.30
1 Instrument Man	29.85	238.80	46.85	374.80		
1 Rodman/Chainman	28.55	228.40	43.30	346.40		
24 L.H., Daily Totals		$743.60		$1159.20	$30.98	$48.30
Crew A-8	**Hr.**	**Daily**	**Hr.**	**Daily**	**Bare Costs**	**Incl. O&P**
1 Chief Of Party	$34.55	$276.40	$54.75	$438.00	$30.38	$47.05
1 Instrument Man	29.85	238.80	46.85	374.80		
2 Rodmen/Chainmen	28.55	456.80	43.30	692.80		
32 L.H., Daily Totals		$972.00		$1505.60	$30.38	$47.05
Crew A-9	**Hr.**	**Daily**	**Hr.**	**Daily**	**Bare Costs**	**Incl. O&P**
1 Asbestos Foreman	$32.65	$261.20	$52.30	$418.40	$32.21	$51.60
7 Asbestos Workers	32.15	1800.40	51.50	2884.00		
64 L.H., Daily Totals		$2061.60		$3302.40	$32.21	$51.60
Crew A-10	**Hr.**	**Daily**	**Hr.**	**Daily**	**Bare Costs**	**Incl. O&P**
1 Asbestos Foreman	$32.65	$261.20	$52.30	$418.40	$32.21	$51.60
7 Asbestos Workers	32.15	1800.40	51.50	2884.00		
64 L.H., Daily Totals		$2061.60		$3302.40	$32.21	$51.60
Crew A-10A	**Hr.**	**Daily**	**Hr.**	**Daily**	**Bare Costs**	**Incl. O&P**
1 Asbestos Foreman	$32.65	$261.20	$52.30	$418.40	$32.32	$51.77
2 Asbestos Workers	32.15	514.40	51.50	824.00		
24 L.H., Daily Totals		$775.60		$1242.40	$32.32	$51.77
Crew A-10B	**Hr.**	**Daily**	**Hr.**	**Daily**	**Bare Costs**	**Incl. O&P**
1 Asbestos Foreman	$32.65	$261.20	$52.30	$418.40	$32.28	$51.70
3 Asbestos Workers	32.15	771.60	51.50	1236.00		
32 L.H., Daily Totals		$1032.80		$1654.40	$32.28	$51.70
Crew A-10C	**Hr.**	**Daily**	**Hr.**	**Daily**	**Bare Costs**	**Incl. O&P**
3 Asbestos Workers	$32.15	$771.60	$51.50	$1236.00	$32.15	$51.50
1 Flatbed Truck		158.00		173.80	6.58	7.24
24 L.H., Daily Totals		$929.60		$1409.80	$38.73	$58.74
Crew A-10D	**Hr.**	**Daily**	**Hr.**	**Daily**	**Bare Costs**	**Incl. O&P**
2 Asbestos Workers	$32.15	$514.40	$51.50	$824.00	$30.30	$47.29
1 Equip. Oper. (crane)	31.15	249.20	47.15	377.20		
1 Equip. Oper. Oiler	25.75	206.00	39.00	312.00		
1 Hydraulic Crane, 33 Ton		804.80		885.30	25.15	27.67
32 L.H., Daily Totals		$1774.40		$2398.50	$55.45	$74.96
Crew A-11	**Hr.**	**Daily**	**Hr.**	**Daily**	**Bare Costs**	**Incl. O&P**
1 Asbestos Foreman	$32.65	$261.20	$52.30	$418.40	$32.21	$51.60
7 Asbestos Workers	32.15	1800.40	51.50	2884.00		
2 Chipping Hammers		35.20		38.70	.55	.61
64 L.H., Daily Totals		$2096.80		$3341.10	$32.76	$52.21
Crew A-12	**Hr.**	**Daily**	**Hr.**	**Daily**	**Bare Costs**	**Incl. O&P**
1 Asbestos Foreman	$32.65	$261.20	$52.30	$418.40	$32.21	$51.60
7 Asbestos Workers	32.15	1800.40	51.50	2884.00		
1 Large Prod. Vac. Loader		520.80		572.90	8.14	8.95
64 L.H., Daily Totals		$2582.40		$3875.30	$40.35	$60.55
Crew A-13	**Hr.**	**Daily**	**Hr.**	**Daily**	**Bare Costs**	**Incl. O&P**
1 Equip. Oper. (light)	$28.75	$230.00	$43.55	$348.40	$28.75	$43.55
1 Large Prod. Vac. Loader		520.80		572.90	65.10	71.61
8 L.H., Daily Totals		$750.80		$921.30	$93.85	$115.16

Crews

Crew B-1

Crew B-1	Bare Costs Hr.	Bare Costs Daily	Incl. Subs O & P Hr.	Incl. Subs O & P Daily	Cost Per Labor-Hour Bare Costs	Cost Per Labor-Hour Incl. O&P
1 Labor Foreman (outside)	$24.85	$198.80	$38.90	$311.20	$23.52	$36.80
2 Laborers	22.85	365.60	35.75	572.00		
24 L.H., Daily Totals		$564.40		$883.20	$23.52	$36.80

Crew B-2

Crew B-2	Hr.	Daily	Hr.	Daily	Bare Costs	Incl. O&P
1 Labor Foreman (outside)	$24.85	$198.80	$38.90	$311.20	$23.25	$36.38
4 Laborers	22.85	731.20	35.75	1144.00		
40 L.H., Daily Totals		$930.00		$1455.20	$23.25	$36.38

Crew B-3

Crew B-3	Hr.	Daily	Hr.	Daily	Bare Costs	Incl. O&P
1 Labor Foreman (outside)	$24.85	$198.80	$38.90	$311.20	$24.80	$38.26
2 Laborers	22.85	365.60	35.75	572.00		
1 Equip. Oper. (med.)	30.35	242.80	45.95	367.60		
2 Truck Drivers (heavy)	23.95	383.20	36.60	585.60		
1 F.E. Loader, T.M., 2.5 C.Y.		770.00		847.00		
2 Dump Trucks, 16 Ton		965.60		1062.15	36.16	39.77
48 L.H., Daily Totals		$2926.00		$3745.55	$60.96	$78.03

Crew B-3A

Crew B-3A	Hr.	Daily	Hr.	Daily	Bare Costs	Incl. O&P
4 Laborers	$22.85	$731.20	$35.75	$1144.00	$24.35	$37.79
1 Equip. Oper. (med.)	30.35	242.80	45.95	367.60		
1 Hyd. Excavator, 1.5 C.Y.		593.20		652.50	14.83	16.31
40 L.H., Daily Totals		$1567.20		$2164.10	$39.18	$54.10

Crew B-3B

Crew B-3B	Hr.	Daily	Hr.	Daily	Bare Costs	Incl. O&P
2 Laborers	$22.85	$365.60	$35.75	$572.00	$25.00	$38.51
1 Equip. Oper. (med.)	30.35	242.80	45.95	367.60		
1 Truck Drivers (heavy)	23.95	191.60	36.60	292.80		
1 Backhoe Loader, 80 H.P.		228.40		251.25		
1 Dump Trucks, 16 Ton		482.80		531.10	22.23	24.45
32 L.H., Daily Totals		$1511.20		$2014.75	$47.23	$62.96

Crew B-3C

Crew B-3C	Hr.	Daily	Hr.	Daily	Bare Costs	Incl. O&P
3 Laborers	$22.85	$548.40	$35.75	$858.00	$24.73	$38.30
1 Equip. Oper. (med.)	30.35	242.80	45.95	367.60		
1 F.E. Loader, 3.75 C.Y.		518.80		570.70	16.21	17.83
32 L.H., Daily Totals		$1310.00		$1796.30	$40.94	$56.13

Crew B-4

Crew B-4	Hr.	Daily	Hr.	Daily	Bare Costs	Incl. O&P
1 Labor Foreman (outside)	$24.85	$198.80	$38.90	$311.20	$23.37	$36.42
4 Laborers	22.85	731.20	35.75	1144.00		
1 Truck Driver (heavy)	23.95	191.60	36.60	292.80		
1 Tractor, 4 x 2, 195 H.P.		320.40		352.45		
1 Platform Trailer		170.40		187.45	10.23	11.25
48 L.H., Daily Totals		$1612.40		$2287.90	$33.60	$47.67

Crew B-5

Crew B-5	Hr.	Daily	Hr.	Daily	Bare Costs	Incl. O&P
1 Labor Foreman (outside)	$24.85	$198.80	$38.90	$311.20	$25.28	$39.11
4 Laborers	22.85	731.20	35.75	1144.00		
2 Equip. Oper. (med.)	30.35	485.60	45.95	735.20		
1 Air Compr., 250 C.F.M.		125.20		137.70		
2 Air Tools & Accessories		39.20		43.10		
2-50 Ft. Air Hoses, 1.5" Dia.		11.80		13.00		
1 F.E. Loader, T.M., 2.5 C.Y.		770.00		847.00	16.90	18.59
56 L.H., Daily Totals		$2361.80		$3231.20	$42.18	$57.70

Crew B-5A

Crew B-5A	Hr.	Daily	Hr.	Daily	Bare Costs	Incl. O&P
1 Foreman	$24.85	$198.80	$38.90	$311.20	$24.94	$38.50
6 Laborers	22.85	1096.80	35.75	1716.00		
2 Equip. Oper. (med.)	30.35	485.60	45.95	735.20		
1 Equip. Oper. (light)	28.75	230.00	43.55	348.40		
2 Truck Drivers (heavy)	23.95	383.20	36.60	585.60		
1 Air Compr. 365 C.F.M.		154.80		170.30		
2 Pavement Breakers		39.20		43.10		
8 Air Hoses w/Coup.,1"		29.20		32.10		
2 Dump Trucks, 12 Ton		740.80		814.90	10.04	11.05
96 L.H., Daily Totals		$3358.40		$4756.80	$34.98	$49.55

Crew B-5B

Crew B-5B	Hr.	Daily	Hr.	Daily	Bare Costs	Incl. O&P
1 Powderman	$29.85	$238.80	$46.85	$374.80	$27.07	$41.42
2 Equip. Oper. (med.)	30.35	485.60	45.95	735.20		
3 Truck Drivers (heavy)	23.95	574.80	36.60	878.40		
1 F.E. Ldr. 2-1/2 CY		391.80		431.00		
3 Dump Trucks, 16 Ton		1448.40		1593.25		
1 Air Compr. 365 C.F.M.		154.80		170.30	41.56	45.72
48 L.H., Daily Totals		$3294.20		$4182.95	$68.63	$87.14

Crew B-5C

Crew B-5C	Hr.	Daily	Hr.	Daily	Bare Costs	Incl. O&P
3 Laborers	$22.85	$548.40	$35.75	$858.00	$25.46	$39.07
1 Equip. Oper. (medium)	30.35	242.80	45.95	367.60		
2 Truck Drivers (heavy)	23.95	383.20	36.60	585.60		
1 Equip. Oper. (crane)	31.15	249.20	47.15	377.20		
1 Equip. Oper. Oiler	25.75	206.00	39.00	312.00		
2 Dump Truck, 16 Ton		965.60		1062.15		
1 F.E. Lder, 3.75 C.Y.		990.80		1089.90		
1 Hyd. Crane, 25 Ton		602.00		662.20	39.98	43.97
64 L.H., Daily Totals		$4188.00		$5314.65	$65.44	$83.04

Crew B-6

Crew B-6	Hr.	Daily	Hr.	Daily	Bare Costs	Incl. O&P
2 Laborers	$22.85	$365.60	$35.75	$572.00	$24.82	$38.35
1 Equip. Oper. (light)	28.75	230.00	43.55	348.40		
1 Backhoe Loader, 48 H.P.		181.00		199.10	7.54	8.30
24 L.H., Daily Totals		$776.60		$1119.50	$32.36	$46.65

Crew B-6A

Crew B-6A	Hr.	Daily	Hr.	Daily	Bare Costs	Incl. O&P
.5 Labor Foreman (outside)	$24.85	$99.40	$38.90	$155.60	$26.25	$40.46
1 Laborers	22.85	182.80	35.75	286.00		
1 Equip. Oper. (med.)	30.35	242.80	45.95	367.60		
1 Vacuum Trk.,5000 Gal.		302.20		332.40	15.11	16.62
20 L.H., Daily Totals		$827.20		$1141.60	$41.36	$57.08

Crew B-7

Crew B-7	Hr.	Daily	Hr.	Daily	Bare Costs	Incl. O&P
1 Labor Foreman (outside)	$24.85	$198.80	$38.90	$311.20	$24.43	$37.98
4 Laborers	22.85	731.20	35.75	1144.00		
1 Equip. Oper. (med.)	30.35	242.80	45.95	367.60		
1 Chipping Machine		190.60		209.65		
1 F.E. Loader, T.M., 2.5 C.Y.		770.00		847.00		
2 Chain Saws, 36"		86.40		95.05	21.81	23.99
48 L.H., Daily Totals		$2219.80		$2974.50	$46.24	$61.97

Crew B-7A

Crew B-7A	Hr.	Daily	Hr.	Daily	Bare Costs	Incl. O&P
2 Laborers	$22.85	$365.60	$35.75	$572.00	$24.82	$38.35
1 Equip. Oper. (light)	28.75	230.00	43.55	348.40		
1 Rake w/Tractor		139.70		153.65		
2 Chain Saws, 18"		52.40		57.65	8.00	8.80
24 L.H., Daily Totals		$787.70		$1131.70	$32.82	$47.15

Crews

Crew No.	Bare Costs		Incl. Subs O & P		Cost Per Labor-Hour	
Crew B-8	**Hr.**	**Daily**	**Hr.**	**Daily**	**Bare Costs**	**Incl. O&P**
1 Labor Foreman (outside)	$24.85	$198.80	$38.90	$311.20	$25.61	$39.31
2 Laborers	22.85	365.60	35.75	572.00		
2 Equip. Oper. (med.)	30.35	485.60	45.95	735.20		
1 Equip. Oper. Oiler	25.75	206.00	39.00	312.00		
2 Truck Drivers (heavy)	23.95	383.20	36.60	585.60		
1 Hyd. Crane, 25 Ton		629.60		692.55		
1 F.E. Loader, T.M., 2.5 C.Y.		770.00		847.00		
2 Dump Trucks, 16 Ton		965.60		1062.15	36.96	40.65
64 L.H., Daily Totals		$4004.40		$5117.70	$62.57	$79.96
Crew B-9	**Hr.**	**Daily**	**Hr.**	**Daily**	**Bare Costs**	**Incl. O&P**
1 Labor Foreman (outside)	$24.85	$198.80	$38.90	$311.20	$23.25	$36.38
4 Laborers	22.85	731.20	35.75	1144.00		
1 Air Compr., 250 C.F.M.		125.20		137.70		
2 Air Tools & Accessories		39.20		43.10		
2-50 Ft. Air Hoses, 1.5" Dia.		11.80		13.00	4.41	4.85
40 L.H., Daily Totals		$1106.20		$1649.00	$27.66	$41.23
Crew B-9A	**Hr.**	**Daily**	**Hr.**	**Daily**	**Bare Costs**	**Incl. O&P**
2 Laborers	$22.85	$365.60	$35.75	$572.00	$23.22	$36.03
1 Truck Driver (heavy)	23.95	191.60	36.60	292.80		
1 Water Tanker		184.00		202.40		
1 Tractor		320.40		352.45		
2-50 Ft. Disch. Hoses		12.80		14.10	21.55	23.71
24 L.H., Daily Totals		$1074.40		$1433.75	$44.77	$59.74
Crew B-9B	**Hr.**	**Daily**	**Hr.**	**Daily**	**Bare Costs**	**Incl. O&P**
2 Laborers	$22.85	$365.60	$35.75	$572.00	$23.22	$36.03
1 Truck Driver (heavy)	23.95	191.60	36.60	292.80		
2-50 Ft. Disch. Hoses		12.80		14.10		
1 Water Tanker		184.00		202.40		
1 Tractor		320.40		352.45		
1 Pressure Washer		57.40		63.15	23.94	26.34
24 L.H., Daily Totals		$1131.80		$1496.90	$47.16	$62.37
Crew B-9C	**Hr.**	**Daily**	**Hr.**	**Daily**	**Bare Costs**	**Incl. O&P**
1 Labor Foreman (outside)	$24.85	$198.80	$38.90	$311.20	$23.25	$36.38
4 Laborers	22.85	731.20	35.75	1144.00		
1 Air Compr., 250 C.F.M.		125.20		137.70		
2-50 Ft. Air Hoses, 1.5" Dia.		11.80		13.00		
2 Breaker, Pavement, 60 lb.		39.20		43.10	4.41	4.85
40 L.H., Daily Totals		$1106.20		$1649.00	$27.66	$41.23
Crew B-10	**Hr.**	**Daily**	**Hr.**	**Daily**	**Bare Costs**	**Incl. O&P**
1 Equip. Oper. (med.)	$30.35	$242.80	$45.95	$367.60	$27.85	$42.55
.5 Laborer	22.85	91.40	35.75	143.00		
12 L.H., Daily Totals		$334.20		$510.60	$27.85	$42.55
Crew B-10A	**Hr.**	**Daily**	**Hr.**	**Daily**	**Bare Costs**	**Incl. O&P**
1 Equip. Oper. (med.)	$30.35	$242.80	$45.95	$367.60	$27.85	$42.55
.5 Laborer	22.85	91.40	35.75	143.00		
1 Roll. Compact., 2K Lbs.		111.00		122.10	9.25	10.18
12 L.H., Daily Totals		$445.20		$632.70	$37.10	$52.73
Crew B-10B	**Hr.**	**Daily**	**Hr.**	**Daily**	**Bare Costs**	**Incl. O&P**
1 Equip. Oper. (med.)	$30.35	$242.80	$45.95	$367.60	[illegible]	$42.55
.5 Laborer	22.85	91.40	35.75	143.00		
1 Dozer, 200 H.P.		735.00		808.50	61.25	67.38
12 L.H., Daily Totals		$1069.20		$1319.10	$89.10	$109.93

Crew No.	Bare Costs		Incl. Subs O & P		Cost Per Labor-Hour	
Crew B-10C	**Hr.**	**Daily**	**Hr.**	**Daily**	**Bare Costs**	**Incl. O&P**
1 Equip. Oper. (med.)	$30.35	$242.80	$45.95	$367.60	$27.85	$42.55
.5 Laborer	22.85	91.40	35.75	143.00		
1 Dozer, 200 H.P.		735.00		808.50		
1 Vibratory Roller, Towed		168.40		185.25	75.28	82.81
12 L.H., Daily Totals		$1237.60		$1504.35	$103.13	$125.36
Crew B-10D	**Hr.**	**Daily**	**Hr.**	**Daily**	**Bare Costs**	**Incl. O&P**
1 Equip. Oper. (med.)	$30.35	$242.80	$45.95	$367.60	$27.85	$42.55
.5 Laborer	22.85	91.40	35.75	143.00		
1 Dozer, 200 H.P.		735.00		808.50		
1 Sheepsft. Roller, Towed		69.80		76.80	67.07	73.77
12 L.H., Daily Totals		$1139.00		$1395.90	$94.92	$116.32
Crew B-10E	**Hr.**	**Daily**	**Hr.**	**Daily**	**Bare Costs**	**Incl. O&P**
1 Equip. Oper. (med.)	$30.35	$242.80	$45.95	$367.60	$27.85	$42.55
.5 Laborer	22.85	91.40	35.75	143.00		
1 Tandem Roller, 5 Ton		133.20		146.50	11.10	12.21
12 L.H., Daily Totals		$467.40		$657.10	$38.95	$54.76
Crew B-10F	**Hr.**	**Daily**	**Hr.**	**Daily**	**Bare Costs**	**Incl. O&P**
1 Equip. Oper. (med.)	$30.35	$242.80	$45.95	$367.60	$27.85	$42.55
.5 Laborer	22.85	91.40	35.75	143.00		
1 Tandem Roller, 10 Ton		210.40		231.45	17.53	19.29
12 L.H., Daily Totals		$544.60		$742.05	$45.38	$61.84
Crew B-10G	**Hr.**	**Daily**	**Hr.**	**Daily**	**Bare Costs**	**Incl. O&P**
1 Equip. Oper. (med.)	$30.35	$242.80	$45.95	$367.60	$27.85	$42.55
.5 Laborer	22.85	91.40	35.75	143.00		
1 Sheepsft. Roll., 130 H.P.		616.00		677.60	51.33	56.47
12 L.H., Daily Totals		$950.20		$1188.20	$79.18	$99.02
Crew B-10H	**Hr.**	**Daily**	**Hr.**	**Daily**	**Bare Costs**	**Incl. O&P**
1 Equip. Oper. (med.)	$30.35	$242.80	$45.95	$367.60	$27.85	$42.55
.5 Laborer	22.85	91.40	35.75	143.00		
1 Diaphr. Water Pump, 2"		30.80		33.90		
1-20 Ft. Suction Hose, 2"		6.50		7.15		
2-50 Ft. Disch. Hoses, 2"		10.80		11.90	4.01	4.41
12 L.H., Daily Totals		$382.30		$563.55	$31.86	$46.96
Crew B-10I	**Hr.**	**Daily**	**Hr.**	**Daily**	**Bare Costs**	**Incl. O&P**
1 Equip. Oper. (med.)	$30.35	$242.80	$45.95	$367.60	$27.85	$42.55
.5 Laborer	22.85	91.40	35.75	143.00		
1 Diaphr. Water Pump, 4"		62.00		68.20		
1-20 Ft. Suction Hose, 4"		12.50		13.75		
2-50 Ft. Disch. Hoses, 4"		17.00		18.70	7.63	8.39
12 L.H., Daily Totals		$425.70		$611.25	$35.48	$50.94
Crew B-10J	**Hr.**	**Daily**	**Hr.**	**Daily**	**Bare Costs**	**Incl. O&P**
1 Equip. Oper. (med.)	$30.35	$242.80	$45.95	$367.60	$27.85	$42.55
.5 Laborer	22.85	91.40	35.75	143.00		
1 Centr. Water Pump, 3"		39.20		43.10		
1-20 Ft. Suction Hose, 3"		9.50		10.45		
2-50 Ft. Disch. Hoses, 3"		12.80		14.10	5.13	5.64
12 L.H., Daily Totals		$395.70		$578.25	$32.98	$48.19

Crew B-10K	Hr.	Daily	Hr.	Daily	Bare Costs	Incl. O&P
1 Equip. Oper. (med.)	$30.35	$242.80	$45.95	$367.60	$27.85	$42.55
.5 Laborer	22.85	91.40	35.75	143.00		
1 Centr. Water Pump, 6"		164.40		180.85		
1-20 Ft. Suction Hose, 6"		22.50		24.75		
2-50 Ft. Disch. Hoses, 6"		41.10		45.20	19.00	20.90
12 L.H., Daily Totals		$562.20		$761.40	$46.85	$63.45

Crew B-10L	Hr.	Daily	Hr.	Daily	Bare Costs	Incl. O&P
1 Equip. Oper. (med.)	$30.35	$242.80	$45.95	$367.60	$27.85	$42.55
.5 Laborer	22.85	91.40	35.75	143.00		
1 Dozer, 75 H.P.		266.00		292.60	22.17	24.38
12 L.H., Daily Totals		$600.20		$803.20	$50.02	$66.93

Crew B-10M	Hr.	Daily	Hr.	Daily	Bare Costs	Incl. O&P
1 Equip. Oper. (med.)	$30.35	$242.80	$45.95	$367.60	$27.85	$42.55
.5 Laborer	22.85	91.40	35.75	143.00		
1 Dozer, 300 H.P.		1027.00		1129.70	85.58	94.14
12 L.H., Daily Totals		$1361.20		$1640.30	$113.43	$136.69

Crew B-10N	Hr.	Daily	Hr.	Daily	Bare Costs	Incl. O&P
1 Equip. Oper. (med.)	$30.35	$242.80	$45.95	$367.60	$27.85	$42.55
.5 Laborer	22.85	91.40	35.75	143.00		
1 F.E. Loader, T.M., 1.5 C.Y		307.20		337.90	25.60	28.16
12 L.H., Daily Totals		$641.40		$848.50	$53.45	$70.71

Crew B-10O	Hr.	Daily	Hr.	Daily	Bare Costs	Incl. O&P
1 Equip. Oper. (med.)	$30.35	$242.80	$45.95	$367.60	$27.85	$42.55
.5 Laborer	22.85	91.40	35.75	143.00		
1 F.E. Loader, T.M., 2.25 C.Y.		457.00		502.70	38.08	41.89
12 L.H., Daily Totals		$791.20		$1013.30	$65.93	$84.44

Crew B-10P	Hr.	Daily	Hr.	Daily	Bare Costs	Incl. O&P
1 Equip. Oper. (med.)	$30.35	$242.80	$45.95	$367.60	$27.85	$42.55
.5 Laborer	22.85	91.40	35.75	143.00		
1 F.E. Loader, T.M., 2.5 C.Y.		770.00		847.00	64.17	70.58
12 L.H., Daily Totals		$1104.20		$1357.60	$92.02	$113.13

Crew B-10Q	Hr.	Daily	Hr.	Daily	Bare Costs	Incl. O&P
1 Equip. Oper. (med.)	$30.35	$242.80	$45.95	$367.60	$27.85	$42.55
.5 Laborer	22.85	91.40	35.75	143.00		
1 F.E. Loader, T.M., 5 C.Y.		990.80		1089.90	82.57	90.82
12 L.H., Daily Totals		$1325.00		$1600.50	$110.42	$133.37

Crew B-10R	Hr.	Daily	Hr.	Daily	Bare Costs	Incl. O&P
1 Equip. Oper. (med.)	$30.35	$242.80	$45.95	$367.60	$27.85	$42.55
.5 Laborer	22.85	91.40	35.75	143.00		
1 F.E. Loader, W.M., 1 C.Y.		227.60		250.35	18.97	20.86
12 L.H., Daily Totals		$561.80		$760.95	$46.82	$63.41

Crew B-10S	Hr.	Daily	Hr.	Daily	Bare Costs	Incl. O&P
1 Equip. Oper. (med.)	$30.35	$242.80	$45.95	$367.60	$27.85	$42.55
.5 Laborer	22.85	91.40	35.75	143.00		
1 F.E. Loader, W.M., 1.5 C.Y.		284.40		312.85	23.70	26.07
12 L.H., Daily Totals		$618.60		$823.45	$51.55	$68.62

Crew B-10T	Hr.	Daily	Hr.	Daily	Bare Costs	Incl. O&P
1 Equip. Oper. (med.)	$30.35	$242.80	$45.95	$367.60	$27.85	$42.55
.5 Laborer	22.85	91.40	35.75	143.00		
1 F.E. Loader, W.M.,2.5 C.Y.		391.80		431.00	32.65	35.92
12 L.H., Daily Totals		$726.00		$941.60	$60.50	$78.47

Crew B-10U	Hr.	Daily	Hr.	Daily	Bare Costs	Incl. O&P
1 Equip. Oper. (med.)	$30.35	$242.80	$45.95	$367.60	$27.85	$42.55
.5 Laborer	22.85	91.40	35.75	143.00		
1 F.E. Loader, W.M., 5.5 C.Y.		814.60		896.05	67.88	74.67
12 L.H., Daily Totals		$1148.80		$1406.65	$95.73	$117.22

Crew B-10V	Hr.	Daily	Hr.	Daily	Bare Costs	Incl. O&P
1 Equip. Oper. (med.)	$30.35	$242.80	$45.95	$367.60	$27.85	$42.55
.5 Laborer	22.85	91.40	35.75	143.00		
1 Dozer, 700 H.P.		2629.00		2891.90	219.08	240.99
12 L.H., Daily Totals		$2963.20		$3402.50	$246.93	$283.54

Crew B-10W	Hr.	Daily	Hr.	Daily	Bare Costs	Incl. O&P
1 Equip. Oper. (med.)	$30.35	$242.80	$45.95	$367.60	$27.85	$42.55
.5 Laborer	22.85	91.40	35.75	143.00		
1 Dozer, 105 H.P.		399.20		439.10	33.27	36.59
12 L.H., Daily Totals		$733.40		$949.70	$61.12	$79.14

Crew B-10X	Hr.	Daily	Hr.	Daily	Bare Costs	Incl. O&P
1 Equip. Oper. (med.)	$30.35	$242.80	$45.95	$367.60	$27.85	$42.55
.5 Laborer	22.85	91.40	35.75	143.00		
1 Dozer, 410 H.P.		1323.00		1455.30	110.25	121.28
12 L.H., Daily Totals		$1657.20		$1965.90	$138.10	$163.83

Crew B-10Y	Hr.	Daily	Hr.	Daily	Bare Costs	Incl. O&P
1 Equip. Oper. (med.)	$30.35	$242.80	$45.95	$367.60	$27.85	$42.55
.5 Laborer	22.85	91.40	35.75	143.00		
1 Vibratory Drum Roller		357.20		392.90	29.77	32.74
12 L.H., Daily Totals		$691.40		$903.50	$57.62	$75.29

Crew B-11	Hr.	Daily	Hr.	Daily	Bare Costs	Incl. O&P
1 Equipment Oper. (med.)	$30.35	$242.80	$45.95	$367.60	$26.60	$40.85
1 Laborer	22.85	182.80	35.75	286.00		
16 L.H., Daily Totals		$425.60		$653.60	$26.60	$40.85

Crew B-11A	Hr.	Daily	Hr.	Daily	Bare Costs	Incl. O&P
1 Equipment Oper. (med.)	$30.35	$242.80	$45.95	$367.60	$26.60	$40.85
1 Laborer	22.85	182.80	35.75	286.00		
1 Dozer, 200 H.P.		735.00		808.50	45.94	50.53
16 L.H., Daily Totals		$1160.60		$1462.10	$72.54	$91.38

Crew B-11B	Hr.	Daily	Hr.	Daily	Bare Costs	Incl. O&P
1 Equipment Oper. (med.)	$30.35	$242.80	$45.95	$367.60	$26.60	$40.85
1 Laborer	22.85	182.80	35.75	286.00		
1 Dozer, 200 H.P.		735.00		808.50		
1 Air Powered Tamper		21.40		23.55		
1 Air Compr. 365 C.F.M.		154.80		170.30		
2-50 Ft. Air Hoses, 1.5" Dia.		11.80		13.00	57.69	63.46
16 L.H., Daily Totals		$1348.60		$1668.95	$84.29	$104.31

Crew B-11C	Hr.	Daily	Hr.	Daily	Bare Costs	Incl. O&P
1 Equipment Oper. (med.)	$30.35	$242.80	$45.95	$367.60	$26.60	$40.85
1 Laborer	22.85	182.80	35.75	286.00		
1 Backhoe Loader, 48 H.P.		181.00		199.10	11.31	12.44
16 L.H., Daily Totals		$606.60		$852.70	$37.91	$53.29

Crew B-11K	Hr.	Daily	Hr.	Daily	Bare Costs	Incl. O&P
1 Equipment Oper. (med.)	$30.35	$242.80	$45.95	$367.60	$26.60	$40.85
1 Laborer	22.85	182.80	35.75	286.00		
1 Trencher, 8' D., 16" W.		711.40		782.55	44.46	48.91
16 L.H., Daily Totals		$1137.00		$1436.15	$71.06	$89.76

Crew No.	Bare Costs		Incl. Subs O & P		Cost Per Labor-Hour	
Crew B-11L	**Hr.**	**Daily**	**Hr.**	**Daily**	**Bare Costs**	**Incl. O&P**
1 Equipment Oper. (med.)	$30.35	$242.80	$45.95	$367.60	$26.60	$40.85
1 Laborer	22.85	182.80	35.75	286.00		
1 Grader, 30,000 Lbs.		459.40		505.35	28.71	31.58
16 L.H., Daily Totals		$885.00		$1158.95	$55.31	$72.43
Crew B-11M	**Hr.**	**Daily**	**Hr.**	**Daily**	**Bare Costs**	**Incl. O&P**
1 Equipment Oper. (med.)	$30.35	$242.80	$45.95	$367.60	$26.60	$40.85
1 Laborer	22.85	182.80	35.75	286.00		
1 Backhoe Loader, 80 H.P.		228.40		251.25	14.28	15.70
16 L.H., Daily Totals		$654.00		$904.85	$40.88	$56.55
Crew B-11N	**Hr.**	**Daily**	**Hr.**	**Daily**	**Bare Costs**	**Incl. O&P**
1 Labor Foreman	$24.85	$198.80	$38.90	$311.20	$25.47	$38.93
2 Equipment Operators (med.)	30.35	485.60	45.95	735.20		
6 Truck Drivers (hvy.)	23.95	1149.60	36.60	1756.80		
1 F.E. Loader, 5.5 C.Y.		814.60		896.05		
1 Dozer, 400 H.P.		1323.00		1455.30		
6 Off Hwy. Tks. 50 Ton		8796.00		9675.60	151.86	167.04
72 L.H., Daily Totals		$12767.60		$14830.15	$177.33	$205.97
Crew B-11Q	**Hr.**	**Daily**	**Hr.**	**Daily**	**Bare Costs**	**Incl. O&P**
1 Equipment Operator (med.)	$30.35	$242.80	$45.95	$367.60	$27.85	$42.55
.5 Laborer	22.85	91.40	35.75	143.00		
1 Dozer, 140 H.P.		493.20		542.50	41.10	45.21
12 L.H., Daily Totals		$827.40		$1053.10	$68.95	$87.76
Crew B-11R	**Hr.**	**Daily**	**Hr.**	**Daily**	**Bare Costs**	**Incl. O&P**
1 Equipment Operator (med.)	$30.35	$242.80	$45.95	$367.60	$27.85	$42.55
.5 Laborer	22.85	91.40	35.75	143.00		
1 Dozer, 215 H.P.		735.00		808.50	61.25	67.38
12 L.H., Daily Totals		$1069.20		$1319.10	$89.10	$109.93
Crew B-11S	**Hr.**	**Daily**	**Hr.**	**Daily**	**Bare Costs**	**Incl. O&P**
1 Equipment Operator	$30.35	$242.80	$45.95	$367.60	$27.85	$42.55
.5 Laborer	22.85	91.40	35.75	143.00		
1 Dozer, 285 H.P.		1027.00		1129.70	85.58	94.14
12 L.H., Daily Totals		$1361.20		$1640.30	$113.43	$136.69
Crew B-11T	**Hr.**	**Daily**	**Hr.**	**Daily**	**Bare Costs**	**Incl. O&P**
1 Equipment Operator (med.)	$30.35	$242.80	$45.95	$367.60	$27.85	$42.55
.5 Laborer	22.85	91.40	35.75	143.00		
1 Dozer, 370 H.P.		1323.00		1455.30	110.25	121.28
12 L.H., Daily Totals		$1657.20		$1965.90	$138.10	$163.83
Crew B-11U	**Hr.**	**Daily**	**Hr.**	**Daily**	**Bare Costs**	**Incl. O&P**
1 Equipment Operator (med.)	$30.35	$242.80	$45.95	$367.60	$27.85	$42.55
.5 Laborer	22.85	91.40	35.75	143.00		
1 Dozer, 520 H.P.		1634.00		1797.40	136.17	149.78
12 L.H., Daily Totals		$1968.20		$2308.00	$164.02	$192.33
Crew B-11V	**Hr.**	**Daily**	**Hr.**	**Daily**	**Bare Costs**	**Incl. O&P**
3 Laborer	$22.85	$548.40	$35.75	$858.00	$22.85	$35.75
1 Roll. Compact., 2K Lbs.		111.00		122.10	4.63	5.09
24 L.H., Daily Totals		$659.40		$980.10	$27.48	$40.84
Crew B-12	**Hr.**	**Daily**	**Hr.**	**Daily**	**Bare Costs**	**Incl. O&P**
1 Equip. Oper. (crane)	$31.15	$249.20	$47.15	$377.20	$28.45	$43.08
1 Equip. Oper. Oiler	25.75	206.00	39.00	312.00		
16 L.H., Daily Totals		$455.20		$689.20	$28.45	$43.08

Crew No.	Bare Costs		Incl. Subs O & P		Cost Per Labor-Hour	
Crew B-12A	**Hr.**	**Daily**	**Hr.**	**Daily**	**Bare Costs**	**Incl. O&P**
1 Equip. Oper. (crane)	$31.15	$249.20	$47.15	$377.20	$28.45	$43.08
1 Equip. Oper. Oiler	25.75	206.00	39.00	312.00		
1 Hyd. Excavator, 1 C.Y.		490.80		539.90	30.68	33.74
16 L.H., Daily Totals		$946.00		$1229.10	$59.13	$76.82
Crew B-12B	**Hr.**	**Daily**	**Hr.**	**Daily**	**Bare Costs**	**Incl. O&P**
1 Equip. Oper. (crane)	$31.15	$249.20	$47.15	$377.20	$28.45	$43.08
1 Equip. Oper. Oiler	25.75	206.00	39.00	312.00		
1 Hyd. Excavator, 1.5 C.Y.		593.20		652.50	37.08	40.78
16 L.H., Daily Totals		$1048.40		$1341.70	$65.53	$83.86
Crew B-12C	**Hr.**	**Daily**	**Hr.**	**Daily**	**Bare Costs**	**Incl. O&P**
1 Equip. Oper. (crane)	$31.15	$249.20	$47.15	$377.20	$28.45	$43.08
1 Equip. Oper. Oiler	25.75	206.00	39.00	312.00		
1 Hyd. Excavator, 2 C.Y.		854.80		940.30	53.43	58.77
16 L.H., Daily Totals		$1310.00		$1629.50	$81.88	$101.85
Crew B-12D	**Hr.**	**Daily**	**Hr.**	**Daily**	**Bare Costs**	**Incl. O&P**
1 Equip. Oper. (crane)	$31.15	$249.20	$47.15	$377.20	$28.45	$43.08
1 Equip. Oper. Oiler	25.75	206.00	39.00	312.00		
1 Hyd. Excavator, 3.5 C.Y.		1974.00		2171.40	123.38	135.71
16 L.H., Daily Totals		$2429.20		$2860.60	$151.83	$178.79
Crew B-12E	**Hr.**	**Daily**	**Hr.**	**Daily**	**Bare Costs**	**Incl. O&P**
1 Equip. Oper. (crane)	$31.15	$249.20	$47.15	$377.20	$28.45	$43.08
1 Equip. Oper. Oiler	25.75	206.00	39.00	312.00		
1 Hyd. Excavator, .5 C.Y.		318.40		350.25	19.90	21.89
16 L.H., Daily Totals		$773.60		$1039.45	$48.35	$64.97
Crew B-12F	**Hr.**	**Daily**	**Hr.**	**Daily**	**Bare Costs**	**Incl. O&P**
1 Equip. Oper. (crane)	$31.15	$249.20	$47.15	$377.20	$28.45	$43.08
1 Equip. Oper. Oiler	25.75	206.00	39.00	312.00		
1 Hyd. Excavator, .75 C.Y.		413.60		454.95	25.85	28.44
16 L.H., Daily Totals		$868.80		$1144.15	$54.30	$71.52
Crew B-12G	**Hr.**	**Daily**	**Hr.**	**Daily**	**Bare Costs**	**Incl. O&P**
1 Equip. Oper. (crane)	$31.15	$249.20	$47.15	$377.20	$28.45	$43.08
1 Equip. Oper. Oiler	25.75	206.00	39.00	312.00		
1 Power Shovel, .5 C.Y.		357.75		393.50		
1 Clamshell Bucket, .5 C.Y.		48.00		52.80	24.23	26.66
16 L.H., Daily Totals		$860.95		$1135.50	$52.68	$69.74
Crew B-12H	**Hr.**	**Daily**	**Hr.**	**Daily**	**Bare Costs**	**Incl. O&P**
1 Equip. Oper. (crane)	$31.15	$249.20	$47.15	$377.20	$28.45	$43.08
1 Equip. Oper. Oiler	25.75	206.00	39.00	312.00		
1 Power Shovel, 1 C.Y.		547.20		601.90		
1 Clamshell Bucket, 1 C.Y.		71.00		78.10	38.64	42.50
16 L.H., Daily Totals		$1073.40		$1369.20	$67.09	$85.58
Crew B-12I	**Hr.**	**Daily**	**Hr.**	**Daily**	**Bare Costs**	**Incl. O&P**
1 Equip. Oper. (crane)	$31.15	$249.20	$47.15	$377.20	$28.45	$43.08
1 Equip. Oper. Oiler	25.75	206.00	39.00	312.00		
1 Power Shovel, .75 C.Y.		473.40		520.75		
1 Dragline Bucket, .75 C.Y.		21.20		23.30	30.91	34.00
16 L.H., Daily Totals		$949.80		$1233.25	$59.36	$77.08

Crews

Crew No.	Bare Costs		Incl. Subs O & P		Cost Per Labor-Hour	
Crew B-12J	**Hr.**	**Daily**	**Hr.**	**Daily**	**Bare Costs**	**Incl. O&P**
1 Equip. Oper. (crane)	$31.15	$249.20	$47.15	$377.20	$28.45	$43.08
1 Equip. Oper. Oiler	25.75	206.00	39.00	312.00		
1 Gradall, 3 Ton, .5 C.Y.		648.40		713.25	40.53	44.58
16 L.H., Daily Totals		$1103.60		$1402.45	$68.98	$87.66
Crew B-12K	**Hr.**	**Daily**	**Hr.**	**Daily**	**Bare Costs**	**Incl. O&P**
1 Equip. Oper. (crane)	$31.15	$249.20	$47.15	$377.20	$28.45	$43.08
1 Equip. Oper. Oiler	25.75	206.00	39.00	312.00		
1 Gradall, 3 Ton, 1 C.Y.		774.80		852.30	48.43	53.27
16 L.H., Daily Totals		$1230.00		$1541.50	$76.88	$96.35
Crew B-12L	**Hr.**	**Daily**	**Hr.**	**Daily**	**Bare Costs**	**Incl. O&P**
1 Equip. Oper. (crane)	$31.15	$249.20	$47.15	$377.20	$28.45	$43.08
1 Equip. Oper. Oiler	25.75	206.00	39.00	312.00		
1 Power Shovel, .5 C.Y.		357.75		393.50		
1 F.E. Attachment, .5 C.Y.		37.00		40.70	23.55	25.90
16 L.H., Daily Totals		$849.95		$1123.40	$52.00	$68.98
Crew B-12M	**Hr.**	**Daily**	**Hr.**	**Daily**	**Bare Costs**	**Incl. O&P**
1 Equip. Oper. (crane)	$31.15	$249.20	$47.15	$377.20	$28.45	$43.08
1 Equip. Oper. Oiler	25.75	206.00	39.00	312.00		
1 Power Shovel, .75 C.Y.		473.40		520.75		
1 F.E. Attachment, .75 C.Y.		40.00		44.00	32.09	35.30
16 L.H., Daily Totals		$968.60		$1253.95	$60.54	$78.38
Crew B-12N	**Hr.**	**Daily**	**Hr.**	**Daily**	**Bare Costs**	**Incl. O&P**
1 Equip. Oper. (crane)	$31.15	$249.20	$47.15	$377.20	$28.45	$43.08
1 Equip. Oper. Oiler	25.75	206.00	39.00	312.00		
1 Power Shovel, 1 C.Y.		547.20		601.90		
1 F.E. Attachment, 1 C.Y.		44.40		48.85	36.98	40.67
16 L.H., Daily Totals		$1046.80		$1339.95	$65.43	$83.75
Crew B-12O	**Hr.**	**Daily**	**Hr.**	**Daily**	**Bare Costs**	**Incl. O&P**
1 Equip. Oper. (crane)	$31.15	$249.20	$47.15	$377.20	$28.45	$43.08
1 Equip. Oper. Oiler	25.75	206.00	39.00	312.00		
1 Power Shovel, 1.5 C.Y.		716.00		787.60		
1 F.E. Attachment, 1.5 C.Y.		134.20		147.60	53.14	58.45
16 L.H., Daily Totals		$1305.40		$1624.40	$81.59	$101.53
Crew B-12P	**Hr.**	**Daily**	**Hr.**	**Daily**	**Bare Costs**	**Incl. O&P**
1 Equip. Oper. (crane)	$31.15	$249.20	$47.15	$377.20	$28.45	$43.08
1 Equip. Oper. Oiler	25.75	206.00	39.00	312.00		
1 Crawler Crane, 40 Ton		716.00		787.60		
1 Dragline Bucket, 1.5 C.Y.		35.20		38.70	46.95	51.65
16 L.H., Daily Totals		$1206.40		$1515.50	$75.40	$94.73
Crew B-12Q	**Hr.**	**Daily**	**Hr.**	**Daily**	**Bare Costs**	**Incl. O&P**
1 Equip. Oper. (crane)	$31.15	$249.20	$47.15	$377.20	$28.45	$43.08
1 Equip. Oper. Oiler	25.75	206.00	39.00	312.00		
1 Hyd. Excavator, 5/8 C.Y.		360.00		396.00	22.50	24.75
16 L.H., Daily Totals		$815.20		$1085.20	$50.95	$67.83
Crew B-12R	**Hr.**	**Daily**	**Hr.**	**Daily**	**Bare Costs**	**Incl. O&P**
1 Equip. Oper. (crane)	$31.15	$249.20	$47.15	$377.20	$28.45	$43.08
1 Equip. Oper. Oiler	25.75	206.00	39.00	312.00		
1 Hyd. Excavator, 1.5 C.Y.		593.20		652.50	37.08	40.78
16 L.H., Daily Totals		$1048.40		$1341.70	$65.53	$83.86

Crew No.	Bare Costs		Incl. Subs O & P		Cost Per Labor-Hour	
Crew B-12S	**Hr.**	**Daily**	**Hr.**	**Daily**	**Bare Costs**	**Incl. O&P**
1 Equip. Oper. (crane)	$31.15	$249.20	$47.15	$377.20	$28.45	$43.08
1 Equip. Oper. Oiler	25.75	206.00	39.00	312.00		
1 Hyd. Excavator, 2.5 C.Y.		1557.00		1712.70	97.31	107.04
16 L.H., Daily Totals		$2012.20		$2401.90	$125.76	$150.12
Crew B-12T	**Hr.**	**Daily**	**Hr.**	**Daily**	**Bare Costs**	**Incl. O&P**
1 Equip. Oper. (crane)	$31.15	$249.20	$47.15	$377.20	$28.45	$43.08
1 Equip. Oper. Oiler	25.75	206.00	39.00	312.00		
1 Crawler Crane, 75 Ton		933.20		1026.50		
1 F.E. Attachment, 3 C.Y.		248.40		273.25	73.85	81.24
16 L.H., Daily Totals		$1636.80		$1988.95	$102.30	$124.32
Crew B-12V	**Hr.**	**Daily**	**Hr.**	**Daily**	**Bare Costs**	**Incl. O&P**
1 Equip. Oper. (crane)	$31.15	$249.20	$47.15	$377.20	$28.45	$43.08
1 Equip. Oper. Oiler	25.75	206.00	39.00	312.00		
1 Crawler Crane, 75 Ton		933.20		1026.50		
1 Dragline Bucket, 3 C.Y.		74.00		81.40	62.95	69.25
16 L.H., Daily Totals		$1462.40		$1797.10	$91.40	$112.33
Crew B-13	**Hr.**	**Daily**	**Hr.**	**Daily**	**Bare Costs**	**Incl. O&P**
1 Labor Foreman (outside)	$24.85	$198.80	$38.90	$311.20	$24.74	$38.29
4 Laborers	22.85	731.20	35.75	1144.00		
1 Equip. Oper. (crane)	31.15	249.20	47.15	377.20		
1 Equip. Oper. Oiler	25.75	206.00	39.00	312.00		
1 Hyd. Crane, 25 Ton		629.60		692.55	11.24	12.37
56 L.H., Daily Totals		$2014.80		$2836.95	$35.98	$50.66
Crew B-13A	**Hr.**	**Daily**	**Hr.**	**Daily**	**Bare Costs**	**Incl. O&P**
1 Foreman	$24.85	$198.80	$38.90	$311.20	$25.59	$39.36
2 Laborers	22.85	365.60	35.75	572.00		
2 Equipment Operator	30.35	485.60	45.95	735.20		
2 Truck Drivers (heavy)	23.95	383.20	36.60	585.60		
1 Crane, 75 Ton		933.20		1026.50		
1 F.E. Lder, 3.75 C.Y.		990.80		1089.90		
2 Dump Trucks, 12 Ton		740.80		814.90	47.59	52.34
56 L.H., Daily Totals		$4098.00		$5135.30	$73.18	$91.70
Crew B-13B	**Hr.**	**Daily**	**Hr.**	**Daily**	**Bare Costs**	**Incl. O&P**
1 Labor Foreman (outside)	$24.85	$198.80	$38.90	$311.20	$24.74	$38.29
4 Laborers	22.85	731.20	35.75	1144.00		
1 Equip. Oper. (crane)	31.15	249.20	47.15	377.20		
1 Equip. Oper. Oiler	25.75	206.00	39.00	312.00		
1 Hyd. Crane, 55 Ton		901.60		991.75	16.10	17.71
56 L.H., Daily Totals		$2286.80		$3136.15	$40.84	$56.00
Crew B-13C	**Hr.**	**Daily**	**Hr.**	**Daily**	**Bare Costs**	**Incl. O&P**
1 Labor Foreman (outside)	$24.85	$198.80	$38.90	$311.20	$24.74	$38.29
4 Laborers	22.85	731.20	35.75	1144.00		
1 Equip. Oper. (crane)	31.15	249.20	47.15	377.20		
1 Equip. Oper. Oiler	25.75	206.00	39.00	312.00		
1 Crawler Crane, 100 Ton		1209.00		1329.90	21.59	23.75
56 L.H., Daily Totals		$2594.20		$3474.30	$46.33	$62.04
Crew B-14	**Hr.**	**Daily**	**Hr.**	**Daily**	**Bare Costs**	**Incl. O&P**
1 Labor Foreman (outside)	$24.85	$198.80	$38.90	$311.20	$24.17	$37.58
4 Laborers	22.85	731.20	35.75	1144.00		
1 Equip. Oper. (light)	28.75	230.00	43.55	348.40		
1 Backhoe Loader, 48 H.P.		181.00		199.10	3.77	4.15
48 L.H., Daily Totals		$1341.00		$2002.70	$27.94	$41.73

Crews

Crew No.	Bare Costs		Incl. Subs O & P		Cost Per Labor-Hour	
Crew B-15	**Hr.**	**Daily**	**Hr.**	**Daily**	**Bare Costs**	**Incl. O&P**
1 Equipment Oper. (med)	$30.35	$242.80	$45.95	$367.60	$25.62	$39.15
.5 Laborer	22.85	91.40	35.75	143.00		
2 Truck Drivers (heavy)	23.95	383.20	36.60	585.60		
2 Dump Trucks, 16 Ton		965.60		1062.15		
1 Dozer, 200 H.P.		735.00		808.50	60.74	66.81
28 L.H., Daily Totals		$2418.00		$2966.85	$86.36	$105.96
Crew B-16	**Hr.**	**Daily**	**Hr.**	**Daily**	**Bare Costs**	**Incl. O&P**
1 Labor Foreman (outside)	$24.85	$198.80	$38.90	$311.20	$23.63	$36.75
2 Laborers	22.85	365.60	35.75	572.00		
1 Truck Driver (heavy)	23.95	191.60	36.60	292.80		
1 Dump Truck, 16 Ton		482.80		531.10	15.09	16.60
32 L.H., Daily Totals		$1238.80		$1707.10	$38.72	$53.35
Crew B-17	**Hr.**	**Daily**	**Hr.**	**Daily**	**Bare Costs**	**Incl. O&P**
2 Laborers	$22.85	$365.60	$35.75	$572.00	$24.60	$37.91
1 Equip. Oper. (light)	28.75	230.00	43.55	348.40		
1 Truck Driver (heavy)	23.95	191.60	36.60	292.80		
1 Backhoe Loader, 48 H.P.		181.00		199.10		
1 Dump Truck, 12 Ton		370.40		407.45	17.23	18.95
32 L.H., Daily Totals		$1338.60		$1819.75	$41.83	$56.86
Crew B-18	**Hr.**	**Daily**	**Hr.**	**Daily**	**Bare Costs**	**Incl. O&P**
1 Labor Foreman (outside)	$24.85	$198.80	$38.90	$311.20	$23.52	$36.80
2 Laborers	22.85	365.60	35.75	572.00		
1 Vibrating Compactor		60.60		66.65	2.53	2.78
24 L.H., Daily Totals		$625.00		$949.85	$26.05	$39.58
Crew B-19	**Hr.**	**Daily**	**Hr.**	**Daily**	**Bare Costs**	**Incl. O&P**
1 Pile Driver Foreman	$30.80	$246.40	$52.40	$419.20	$29.26	$47.71
4 Pile Drivers	28.80	921.60	49.00	1568.00		
2 Equip. Oper. (crane)	31.15	498.40	47.15	754.40		
1 Equip. Oper. Oiler	25.75	206.00	39.00	312.00		
1 Crane, 40 Ton & Access.		716.00		787.60		
60 L.F. Leads, 15K Ft. Lbs.		312.00		343.20		
1 Hammer, 15K Ft. Lbs.		313.40		344.75		
1 Air Compr., 600 C.F.M.		251.00		276.10		
2-50 Ft. Air Hoses, 3" Dia.		34.60		38.05	25.42	27.96
64 L.H., Daily Totals		$3499.40		$4843.30	$54.68	$75.67
Crew B-19A	**Hr.**	**Daily**	**Hr.**	**Daily**	**Bare Costs**	**Incl. O&P**
1 Pile Driver Foreman	$30.80	$246.40	$52.40	$419.20	$29.26	$47.71
4 Pile Drivers	28.80	921.60	49.00	1568.00		
2 Equip. Oper. (crane)	31.15	498.40	47.15	754.40		
1 Equip. Oper. Oiler	25.75	206.00	39.00	312.00		
1 Crawler Crane, 75 Ton		933.20		1026.50		
60 L.F. Leads, 25K Ft. Lbs.		432.00		475.20		
1 Air Compressor, 750 CFM		294.20		323.60		
4-50 Ft. Air Hose, 3" Dia.		69.20		76.10	27.01	29.71
64 L.H., Daily Totals		$3601.00		$4955.00	$56.27	$77.42
Crew B-20	**Hr.**	**Daily**	**Hr.**	**Daily**	**Bare Costs**	**Incl. O&P**
1 Labor Foreman (out)	$24.85	$198.80	$38.90	$311.20	$25.85	$40.50
1 Skilled Worker	29.85	238.80	46.85	374.80		
1 Laborer	22.85	182.80	35.75	286.00		
24 L.H., Daily Totals		$620.40		$972.00	$25.85	$40.50

Crew No.	Bare Costs		Incl. Subs O & P		Cost Per Labor-Hour	
Crew B-20A	**Hr.**	**Daily**	**Hr.**	**Daily**	**Bare Costs**	**Incl. O&P**
1 Labor Foreman	$24.85	$198.80	$38.90	$311.20	$27.43	$42.06
1 Laborer	22.85	182.80	35.75	286.00		
1 Plumber	34.45	275.60	52.00	416.00		
1 Plumber Apprentice	27.55	220.40	41.60	332.80		
32 L.H., Daily Totals		$877.60		$1346.00	$27.43	$42.06
Crew B-21	**Hr.**	**Daily**	**Hr.**	**Daily**	**Bare Costs**	**Incl. O&P**
1 Labor Foreman (out)	$24.85	$198.80	$38.90	$311.20	$26.61	$41.45
1 Skilled Worker	29.85	238.80	46.85	374.80		
1 Laborer	22.85	182.80	35.75	286.00		
.5 Equip. Oper. (crane)	31.15	124.60	47.15	188.60		
.5 S.P. Crane, 5 Ton		142.00		156.20	5.07	5.58
28 L.H., Daily Totals		$887.00		$1316.80	$31.68	$47.03
Crew B-21A	**Hr.**	**Daily**	**Hr.**	**Daily**	**Bare Costs**	**Incl. O&P**
1 Labor Foreman	$24.85	$198.80	$38.90	$311.20	$28.17	$43.08
1 Laborer	22.85	182.80	35.75	286.00		
1 Plumber	34.45	275.60	52.00	416.00		
1 Plumber Apprentice	27.55	220.40	41.60	332.80		
1 Equip. Oper. (crane)	31.15	249.20	47.15	377.20		
1 S.P. Crane, 12 Ton		425.20		467.70	10.63	11.69
40 L.H., Daily Totals		$1552.00		$2190.90	$38.80	$54.77
Crew B-22	**Hr.**	**Daily**	**Hr.**	**Daily**	**Bare Costs**	**Incl. O&P**
1 Labor Foreman (out)	$24.85	$198.80	$38.90	$311.20	$26.91	$41.83
1 Skilled Worker	29.85	238.80	46.85	374.80		
1 Laborer	22.85	182.80	35.75	286.00		
.75 Equip. Oper. (crane)	31.15	186.90	47.15	282.90		
.75 S.P. Crane, 5 Ton		213.00		234.30	7.10	7.81
30 L.H., Daily Totals		$1020.30		$1489.20	$34.01	$49.64
Crew B-22A	**Hr.**	**Daily**	**Hr.**	**Daily**	**Bare Costs**	**Incl. O&P**
1 Labor Foreman (out)	$24.85	$198.80	$38.90	$311.20	$26.06	$40.55
1 Skilled Worker	29.85	238.80	46.85	374.80		
2 Laborers	22.85	365.60	35.75	572.00		
.75 Equipment Oper. (crane)	31.15	186.90	47.15	282.90		
.75 Crane, 5 Ton		213.00		234.30		
1 Generator, 5 KW		46.90		51.60		
1 Butt Fusion Machine		302.20		332.40	14.79	16.27
38 L.H., Daily Totals		$1552.20		$2159.20	$40.85	$56.82
Crew B-22B	**Hr.**	**Daily**	**Hr.**	**Daily**	**Bare Costs**	**Incl. O&P**
1 Skilled Worker	$29.85	$238.80	$46.85	$374.80	$26.35	$41.30
1 Laborer	22.85	182.80	35.75	286.00		
1 Electro Fusion Machine		118.60		130.45	7.41	8.15
16 L.H., Daily Totals		$540.20		$791.25	$33.76	$49.45
Crew B-23	**Hr.**	**Daily**	**Hr.**	**Daily**	**Bare Costs**	**Incl. O&P**
1 Labor Foreman (outside)	$24.85	$198.80	$38.90	$311.20	$23.25	$36.38
4 Laborers	22.85	731.20	35.75	1144.00		
1 Drill Rig, Wells		1870.00		2057.00		
1 Light Truck, 3 Ton		184.20		202.60	51.36	56.49
40 L.H., Daily Totals		$2984.20		$3714.80	$74.61	$92.87
Crew B-23A	**Hr.**	**Daily**	**Hr.**	**Daily**	**Bare Costs**	**Incl. O&P**
1 Labor Foreman (outside)	$24.85	$198.80	$38.90	$311.20	$26.02	$40.20
1 Laborers	22.85	182.80	35.75	286.00		
1 Equip. Operator (medium)	30.35	242.80	45.95	367.60		
1 Drill Rig, Wells		1870.00		2057.00		
1 Pickup Truck, 3/4 Ton		83.00		91.30	81.38	89.51
24 L.H., Daily Totals		$2577.40		$3113.10	$107.40	$129.71

Crews

Crew No.	Bare Costs		Incl. Subs O & P		Cost Per Labor-Hour	
Crew B-23B	**Hr.**	**Daily**	**Hr.**	**Daily**	**Bare Costs**	**Incl. O&P**
1 Labor Foreman (outside)	$24.85	$198.80	$38.90	$311.20	$26.02	$40.20
1 Laborer	22.85	182.80	35.75	286.00		
1 Equip. Operator (medium)	30.35	242.80	45.95	367.60		
1 Drill Rig, Wells		1870.00		2057.00		
1 Pickup Truck, 3/4 Ton		83.00		91.30		
1 Pump, Cntfgl, 6"		164.40		180.85	88.23	97.05
24 L.H., Daily Totals		$2741.80		$3293.95	$114.25	$137.25
Crew B-24	**Hr.**	**Daily**	**Hr.**	**Daily**	**Bare Costs**	**Incl. O&P**
1 Cement Finisher	$27.95	$223.60	$41.55	$332.40	$26.65	$40.97
1 Laborer	22.85	182.80	35.75	286.00		
1 Carpenter	29.15	233.20	45.60	364.80		
24 L.H., Daily Totals		$639.60		$983.20	$26.65	$40.97
Crew B-25	**Hr.**	**Daily**	**Hr.**	**Daily**	**Bare Costs**	**Incl. O&P**
1 Labor Foreman	$24.85	$198.80	$38.90	$311.20	$25.08	$38.82
7 Laborers	22.85	1279.60	35.75	2002.00		
3 Equip. Oper. (med.)	30.35	728.40	45.95	1102.80		
1 Asphalt Paver, 130 H.P		1281.00		1409.10		
1 Tandem Roller, 10 Ton		210.40		231.45		
1 Roller, Pneumatic Wheel		215.60		237.15	19.40	21.34
88 L.H., Daily Totals		$3913.80		$5293.70	$44.48	$60.16
Crew B-25B	**Hr.**	**Daily**	**Hr.**	**Daily**	**Bare Costs**	**Incl. O&P**
1 Labor Foreman	$24.85	$198.80	$38.90	$311.20	$25.52	$39.41
7 Laborers	22.85	1279.60	35.75	2002.00		
4 Equip. Oper. (medium)	30.35	971.20	45.95	1470.40		
1 Asphalt Paver, 130 H.P.		1281.00		1409.10		
2 Rollers, Steel Wheel		420.80		462.90		
1 Roller, Pneumatic Wheel		215.60		237.15	19.97	21.97
96 L.H., Daily Totals		$4367.00		$5892.75	$45.49	$61.38
Crew B-25C	**Hr.**	**Daily**	**Hr.**	**Daily**	**Bare Costs**	**Incl. O&P**
1 Labor Foreman	$24.85	$198.80	$38.90	$311.20	$25.68	$39.67
3 Laborers	22.85	548.40	35.75	858.00		
2 Equip. Oper. (medium)	30.35	485.60	45.95	735.20		
1 Asphalt Paver, 130 H.P.		1281.00		1409.10		
1 Rollers, Steel Wheel		210.40		231.45	31.07	34.18
48 L.H., Daily Totals		$2724.20		$3544.95	$56.75	$73.85
Crew B-26	**Hr.**	**Daily**	**Hr.**	**Daily**	**Bare Costs**	**Incl. O&P**
1 Labor Foreman (outside)	$24.85	$198.80	$38.90	$311.20	$25.76	$40.22
6 Laborers	22.85	1096.80	35.75	1716.00		
2 Equip. Oper. (med.)	30.35	485.60	45.95	735.20		
1 Rodman (reinf.)	32.75	262.00	55.60	444.80		
1 Cement Finisher	27.95	223.60	41.55	332.40		
1 Grader, 30,000 Lbs.		459.40		505.35		
1 Paving Mach. & Equip.		1371.00		1508.10	20.80	22.88
88 L.H., Daily Totals		$4097.20		$5553.05	$46.56	$63.10
Crew B-27	**Hr.**	**Daily**	**Hr.**	**Daily**	**Bare Costs**	**Incl. O&P**
1 Labor Foreman (outside)	$24.85	$198.80	$38.90	$311.20	$23.35	$36.54
3 Laborers	22.85	548.40	35.75	858.00		
1 Berm Machine		144.20		158.60	4.51	4.96
32 L.H., Daily Totals		$891.40		$1327.80	$27.86	$41.50
Crew B-28	**Hr.**	**Daily**	**Hr.**	**Daily**	**Bare Costs**	**Incl. O&P**
2 Carpenters	$29.15	$466.40	$45.60	$729.60	$27.05	$42.32
1 Laborer	22.85	182.80	35.75	286.00		
24 L.H., Daily Totals		$649.20		$1015.60	$27.05	$42.32

Crew No.	Bare Costs		Incl. Subs O & P		Cost Per Labor-Hour	
Crew B-29	**Hr.**	**Daily**	**Hr.**	**Daily**	**Bare Costs**	**Incl. O&P**
1 Labor Foreman (outside)	$24.85	$198.80	$38.90	$311.20	$24.74	$38.29
4 Laborers	22.85	731.20	35.75	1144.00		
1 Equip. Oper. (crane)	31.15	249.20	47.15	377.20		
1 Equip. Oper. Oiler	25.75	206.00	39.00	312.00		
1 Gradall, 3 Ton, 1/2 C.Y.		648.40		713.25	11.58	12.74
56 L.H., Daily Totals		$2033.60		$2857.65	$36.32	$51.03
Crew B-30	**Hr.**	**Daily**	**Hr.**	**Daily**	**Bare Costs**	**Incl. O&P**
1 Equip. Oper. (med.)	$30.35	$242.80	$45.95	$367.60	$26.08	$39.72
2 Truck Drivers (heavy)	23.95	383.20	36.60	585.60		
1 Hyd. Excavator, 1.5 C.Y.		593.20		652.50		
2 Dump Trucks, 16 Ton		965.60		1062.15	64.95	71.45
24 L.H., Daily Totals		$2184.80		$2667.85	$91.03	$111.17
Crew B-31	**Hr.**	**Daily**	**Hr.**	**Daily**	**Bare Costs**	**Incl. O&P**
1 Labor Foreman (outside)	$24.85	$198.80	$38.90	$311.20	$24.51	$38.35
3 Laborers	22.85	548.40	35.75	858.00		
1 Carpenter	29.15	233.20	45.60	364.80		
1 Air Compr., 250 C.F.M.		125.20		137.70		
1 Sheeting Driver		12.50		13.75		
2-50 Ft. Air Hoses, 1.5" Dia.		11.80		13.00	3.74	4.11
40 L.H., Daily Totals		$1129.90		$1698.45	$28.25	$42.46
Crew B-32	**Hr.**	**Daily**	**Hr.**	**Daily**	**Bare Costs**	**Incl. O&P**
1 Laborer	$22.85	$182.80	$35.75	$286.00	$28.48	$43.40
3 Equip. Oper. (med.)	30.35	728.40	45.95	1102.80		
1 Grader, 30,000 Lbs.		459.40		505.35		
1 Tandem Roller, 10 Ton		210.40		231.45		
1 Dozer, 200 H.P.		735.00		808.50	43.90	48.29
32 L.H., Daily Totals		$2316.00		$2934.10	$72.38	$91.69
Crew B-32A	**Hr.**	**Daily**	**Hr.**	**Daily**	**Bare Costs**	**Incl. O&P**
1 Laborer	$22.85	$182.80	$35.75	$286.00	$27.85	$42.55
2 Equip. Oper. (medium)	30.35	485.60	45.95	735.20		
1 Grader, 30,000 Lbs.		459.40		505.35		
1 Roller, Vibratory, 29,000 Lbs.		451.20		496.30	37.94	41.74
24 L.H., Daily Totals		$1579.00		$2022.85	$65.79	$84.29
Crew B-32B	**Hr.**	**Daily**	**Hr.**	**Daily**	**Bare Costs**	**Incl. O&P**
1 Laborer	$22.85	$182.80	$35.75	$286.00	$27.85	$42.55
2 Equip. Oper. (medium)	30.35	485.60	45.95	735.20		
1 Dozer, 200 H.P.		735.00		808.50		
1 Roller, Vibratory, 29,000 Lbs.		451.20		496.30	49.43	54.37
24 L.H., Daily Totals		$1854.60		$2326.00	$77.28	$96.92
Crew B-32C	**Hr.**	**Daily**	**Hr.**	**Daily**	**Bare Costs**	**Incl. O&P**
1 Labor Foreman	$24.85	$198.80	$38.90	$311.20	$26.93	$41.38
2 Laborers	22.85	365.60	35.75	572.00		
3 Equip. Oper. (medium)	30.35	728.40	45.95	1102.80		
1 Grader, 30,000 Lbs.		459.40		505.35		
1 Roller, Steel Wheel		210.40		231.45		
1 Dozer, 200 H.P.		735.00		808.50	29.27	32.19
48 L.H., Daily Totals		$2697.60		$3531.30	$56.20	$73.57
Crew B-33	**Hr.**	**Daily**	**Hr.**	**Daily**	**Bare Costs**	**Incl. O&P**
1 Equip. Oper. (med.)	$30.35	$242.80	$45.95	$367.60	$28.21	$43.04
.5 Laborer	22.85	91.40	35.75	143.00		
.25 Equip. Oper. (med.)	30.35	60.70	45.95	91.90		
14 L.H., Daily Totals		$394.90		$602.50	$28.21	$43.04

Crew No.	Bare Costs		Incl. Subs O & P		Cost Per Labor-Hour	
Crew B-33A	**Hr.**	**Daily**	**Hr.**	**Daily**	**Bare Costs**	**Incl. O&P**
1 Equip. Oper. (med.)	$30.35	$242.80	$45.95	$367.60	$28.21	$43.04
.5 Laborer	22.85	91.40	35.75	143.00		
.25 Equip. Oper. (med.)	30.35	60.70	45.95	91.90		
1 Scraper, Towed, 7 C.Y.		123.35		135.70		
1.25 Dozer, 300 H.P.		1283.75		1412.15	100.51	110.56
14 L.H., Daily Totals		$1802.00		$2150.35	$128.72	$153.60
Crew B-33B	**Hr.**	**Daily**	**Hr.**	**Daily**	**Bare Costs**	**Incl. O&P**
1 Equip. Oper. (med.)	$30.35	$242.80	$45.95	$367.60	$28.21	$43.04
.5 Laborer	22.85	91.40	35.75	143.00		
.25 Equip. Oper. (med.)	30.35	60.70	45.95	91.90		
1 Scraper, Towed, 10 C.Y.		157.25		173.00		
1.25 Dozer, 300 H.P.		1283.75		1412.15	102.93	113.22
14 L.H., Daily Totals		$1835.90		$2187.65	$131.14	$156.26
Crew B-33C	**Hr.**	**Daily**	**Hr.**	**Daily**	**Bare Costs**	**Incl. O&P**
1 Equip. Oper. (med.)	$30.35	$242.80	$45.95	$367.60	$28.21	$43.04
.5 Laborer	22.85	91.40	35.75	143.00		
.25 Equip. Oper. (med.)	30.35	60.70	45.95	91.90		
1 Scraper, Towed, 12 C.Y.		157.25		173.00		
1.25 Dozer, 300 H.P.		1283.75		1412.15	102.93	113.22
14 L.H., Daily Totals		$1835.90		$2187.65	$131.14	$156.26
Crew B-33D	**Hr.**	**Daily**	**Hr.**	**Daily**	**Bare Costs**	**Incl. O&P**
1 Equip. Oper. (med.)	$30.35	$242.80	$45.95	$367.60	$28.21	$43.04
.5 Laborer	22.85	91.40	35.75	143.00		
.25 Equip. Oper. (med.)	30.35	60.70	45.95	91.90		
1 S.P. Scraper, 14 C.Y.		1597.00		1756.70		
.25 Dozer, 300 H.P.		256.75		282.45	132.41	145.65
14 L.H., Daily Totals		$2248.65		$2641.65	$160.62	$188.69
Crew B-33E	**Hr.**	**Daily**	**Hr.**	**Daily**	**Bare Costs**	**Incl. O&P**
1 Equip. Oper. (med.)	$30.35	$242.80	$45.95	$367.60	$28.21	$43.04
.5 Laborer	22.85	91.40	35.75	143.00		
.25 Equip. Oper. (med.)	30.35	60.70	45.95	91.90		
1 S.P. Scraper, 24 C.Y.		1932.00		2125.20		
.25 Dozer, 300 H.P.		256.75		282.45	156.34	171.97
14 L.H., Daily Totals		$2583.65		$3010.15	$184.55	$215.01
Crew B-33F	**Hr.**	**Daily**	**Hr.**	**Daily**	**Bare Costs**	**Incl. O&P**
1 Equip. Oper. (med.)	$30.35	$242.80	$45.95	$367.60	$28.21	$43.04
.5 Laborer	22.85	91.40	35.75	143.00		
.25 Equip. Oper. (med.)	30.35	60.70	45.95	91.90		
1 Elev. Scraper, 11 C.Y.		668.60		735.45		
.25 Dozer, 300 H.P.		256.75		282.45	66.10	72.71
14 L.H., Daily Totals		$1320.25		$1620.40	$94.31	$115.75
Crew B-33G	**Hr.**	**Daily**	**Hr.**	**Daily**	**Bare Costs**	**Incl. O&P**
1 Equip. Oper. (med.)	$30.35	$242.80	$45.95	$367.60	$28.21	$43.04
.5 Laborer	22.85	91.40	35.75	143.00		
.25 Equip. Oper. (med.)	30.35	60.70	45.95	91.90		
1 Elev. Scraper, 20 C.Y.		1178.00		1295.80		
.25 Dozer, 300 H.P.		256.75		282.45	102.48	112.73
14 L.H., Daily Totals		$1829.65		$2180.75	$130.69	$155.77

Crew No.	Bare Costs		Incl. Subs O & P		Cost Per Labor-Hour	
Crew B-33H	**Hr.**	**Daily**	**Hr.**	**Daily**	**Bare Costs**	**Incl. O&P**
.25 Laborer	$22.85	$45.70	$35.75	$71.50	$29.06	$44.19
1 Equipment Operator (med.)	30.35	242.80	45.95	367.60		
.2 Equipment Operator (med.)	30.35	48.56	45.95	73.52		
1 Scraper, 32-44 C.Y.		2385.00		2623.50		
.2 Dozer, 298 kw		264.60		291.05	228.41	251.26
11. L.H., Daily Totals		$2986.66		$3427.17	$257.47	$295.45
Crew B-33J	**Hr.**	**Daily**	**Hr.**	**Daily**	**Bare Costs**	**Incl. O&P**
1 Equipment Operator (med.)	$30.35	$242.80	$45.95	$367.60	$30.35	$45.95
1 Scraper 17 C.Y.		1597.00		1756.70	199.63	219.59
8 L.H., Daily Totals		$1839.80		$2124.30	$229.98	$265.54
Crew B-34A	**Hr.**	**Daily**	**Hr.**	**Daily**	**Bare Costs**	**Incl. O&P**
1 Truck Driver (heavy)	$23.95	$191.60	$36.60	$292.80	$23.95	$36.60
1 Dump Truck, 12 Ton		370.40		407.45	46.30	50.93
8 L.H., Daily Totals		$562.00		$700.25	$70.25	$87.53
Crew B-34B	**Hr.**	**Daily**	**Hr.**	**Daily**	**Bare Costs**	**Incl. O&P**
1 Truck Driver (heavy)	$23.95	$191.60	$36.60	$292.80	$23.95	$36.60
1 Dump Truck, 16 Ton		482.80		531.10	60.35	66.39
8 L.H., Daily Totals		$674.40		$823.90	$84.30	$102.99
Crew B-34C	**Hr.**	**Daily**	**Hr.**	**Daily**	**Bare Costs**	**Incl. O&P**
1 Truck Driver (heavy)	$23.95	$191.60	$36.60	$292.80	$23.95	$36.60
1 Truck Tractor, 40 Ton		413.00		454.30		
1 Dump Trailer, 16.5 C.Y.		140.00		154.00	69.13	76.04
8 L.H., Daily Totals		$744.60		$901.10	$93.08	$112.64
Crew B-34D	**Hr.**	**Daily**	**Hr.**	**Daily**	**Bare Costs**	**Incl. O&P**
1 Truck Driver (heavy)	$23.95	$191.60	$36.60	$292.80	$23.95	$36.60
1 Truck Tractor, 40 Ton		413.00		454.30		
1 Dump Trailer, 20 C.Y.		145.60		160.15	69.83	76.81
8 L.H., Daily Totals		$750.20		$907.25	$93.78	$113.41
Crew B-34E	**Hr.**	**Daily**	**Hr.**	**Daily**	**Bare Costs**	**Incl. O&P**
1 Truck Driver (heavy)	$23.95	$191.60	$36.60	$292.80	$23.95	$36.60
1 Truck, Off Hwy., 25 Ton		901.40		991.55	112.68	123.94
8 L.H., Daily Totals		$1093.00		$1284.35	$136.63	$160.54
Crew B-34F	**Hr.**	**Daily**	**Hr.**	**Daily**	**Bare Costs**	**Incl. O&P**
1 Truck Driver (heavy)	$23.95	$191.60	$36.60	$292.80	$23.95	$36.60
1 Truck, Off Hwy., 22 C.Y.		1100.00		1210.00	137.50	151.25
8 L.H., Daily Totals		$1291.60		$1502.80	$161.45	$187.85
Crew B-34G	**Hr.**	**Daily**	**Hr.**	**Daily**	**Bare Costs**	**Incl. O&P**
1 Truck Driver (heavy)	$23.95	$191.60	$36.60	$292.80	$23.95	$36.60
1 Truck, Off Hwy., 34 C.Y.		1466.00		1612.60	183.25	201.58
8 L.H., Daily Totals		$1657.60		$1905.40	$207.20	$238.18
Crew B-34H	**Hr.**	**Daily**	**Hr.**	**Daily**	**Bare Costs**	**Incl. O&P**
1 Truck Driver (heavy)	$23.95	$191.60	$36.60	$292.80	$23.95	$36.60
1 Truck, Off Hwy., 42 C.Y.		1561.00		1717.10	195.13	214.64
8 L.H., Daily Totals		$1752.60		$2009.90	$219.08	$251.24
Crew B-34J	**Hr.**	**Daily**	**Hr.**	**Daily**	**Bare Costs**	**Incl. O&P**
1 Truck Driver (heavy)	$23.95	$191.60	$36.60	$292.80	$23.95	$36.60
1 Truck, Off Hwy., 60 C.Y.		2259.00		2484.90	282.38	310.61
8 L.H., Daily Totals		$2450.60		$2777.70	$306.33	$347.21

Crew No.	Bare Costs		Incl. Subs O & P		Cost Per Labor-Hour	
Crew B-34K	**Hr.**	**Daily**	**Hr.**	**Daily**	**Bare Costs**	**Incl. O&P**
1 Truck Driver (heavy)	$23.95	$191.60	$36.60	$292.80	$23.95	$36.60
1 Truck Tractor, 240 H.P.		503.60		553.95		
1 Low Bed Trailer		286.60		315.25	98.78	108.65
8 L.H., Daily Totals		$981.80		$1162.00	$122.73	$145.25
Crew B-35	**Hr.**	**Daily**	**Hr.**	**Daily**	**Bare Costs**	**Incl. O&P**
1 Laborer Foreman (out)	$24.85	$198.80	$38.90	$311.20	$28.15	$43.27
1 Skilled Worker	29.85	238.80	46.85	374.80		
1 Welder (plumber)	34.45	275.60	52.00	416.00		
1 Laborer	22.85	182.80	35.75	286.00		
1 Equip. Oper. (crane)	31.15	249.20	47.15	377.20		
1 Equip. Oper. Oiler	25.75	206.00	39.00	312.00		
1 Electric Welding Mach.		100.65		110.70		
1 Hyd. Excavator, .75 C.Y.		413.60		454.95	10.71	11.78
48 L.H., Daily Totals		$1865.45		$2642.85	$38.86	$55.05
Crew B-35A	**Hr.**	**Daily**	**Hr.**	**Daily**	**Bare Costs**	**Incl. O&P**
1 Laborer Foreman (out)	$24.85	$198.80	$38.90	$311.20	$27.39	$42.20
2 Laborers	22.85	365.60	35.75	572.00		
1 Skilled Worker	29.85	238.80	46.85	374.80		
1 Welder (plumber)	34.45	275.60	52.00	416.00		
1 Equip. Oper. (crane)	31.15	249.20	47.15	377.20		
1 Equip. Oper. Oiler	25.75	206.00	39.00	312.00		
1 Welder, 300 amp		84.00		92.40		
1 Crane, 75 Ton		933.20		1026.50	18.16	19.98
56 L.H., Daily Totals		$2551.20		$3482.10	$45.55	$62.18
Crew B-36	**Hr.**	**Daily**	**Hr.**	**Daily**	**Bare Costs**	**Incl. O&P**
1 Labor Foreman (outside)	$24.85	$198.80	$38.90	$311.20	$26.25	$40.46
2 Laborers	22.85	365.60	35.75	572.00		
2 Equip. Oper. (med.)	30.35	485.60	45.95	735.20		
1 Dozer, 200 H.P.		735.00		808.50		
1 Aggregate Spreader		73.00		80.30		
1 Tandem Roller, 10 Ton		210.40		231.45	25.46	28.01
40 L.H., Daily Totals		$2068.40		$2738.65	$51.71	$68.47
Crew B-36A	**Hr.**	**Daily**	**Hr.**	**Daily**	**Bare Costs**	**Incl. O&P**
1 Labor Foreman (outside)	$24.85	$198.80	$38.90	$311.20	$27.42	$42.03
2 Laborers	22.85	365.60	35.75	572.00		
4 Equip. Oper. (med.)	30.35	971.20	45.95	1470.40		
1 Dozer, 200 H.P.		735.00		808.50		
1 Aggregate Spreader		73.00		80.30		
1 Roller, Steel Wheel		210.40		231.45		
1 Roller, Pneumatic Wheel		215.60		237.15	22.04	24.24
56 L.H., Daily Totals		$2769.60		$3711.00	$49.46	$66.27
Crew B-36B	**Hr.**	**Daily**	**Hr.**	**Daily**	**Bare Costs**	**Incl. O&P**
1 Labor Foreman (outside)	$24.85	$198.80	$38.90	$311.20	$26.99	$41.35
2 Laborers	22.85	365.60	35.75	572.00		
4 Equip. Oper. (medium)	30.35	971.20	45.95	1470.40		
1 Truck Driver, Heavy	23.95	191.60	36.60	292.80		
1 Grader, 30,000 Lbs.		459.40		505.35		
1 F.E. Loader, crl, 1.5 C.Y.		358.80		394.70		
1 Dozer, 300 H.P.		1027.00		1129.70		
1 Roller, Vibratory		451.20		496.30		
1 Truck, Tractor, 240 H.P.		503.60		553.95		
1 Water Tanker, 5000 Gal.		184.00		202.40	46.63	51.29
64 L.H., Daily Totals		$4711.20		$5928.80	$73.62	$92.64

Crew No.	Bare Costs		Incl. Subs O & P		Cost Per Labor-Hour	
Crew B-37	**Hr.**	**Daily**	**Hr.**	**Daily**	**Bare Costs**	**Incl. O&P**
1 Labor Foreman (outside)	$24.85	$198.80	$38.90	$311.20	$24.17	$37.58
4 Laborers	22.85	731.20	35.75	1144.00		
1 Equip. Oper. (light)	28.75	230.00	43.55	348.40		
1 Tandem Roller, 5 Ton		133.20		146.50	2.78	3.05
48 L.H., Daily Totals		$1293.20		$1950.10	$26.95	$40.63
Crew B-38	**Hr.**	**Daily**	**Hr.**	**Daily**	**Bare Costs**	**Incl. O&P**
1 Labor Foreman (outside)	$24.85	$198.80	$38.90	$311.20	$25.93	$39.98
2 Laborers	22.85	365.60	35.75	572.00		
1 Equip. Oper. (light)	28.75	230.00	43.55	348.40		
1 Equip. Oper. (medium)	30.35	242.80	45.95	367.60		
1 Backhoe Loader, 48 H.P.		181.00		199.10		
1 Hyd.Hammer, (1200 lb)		190.40		209.45		
1 F.E. Loader (170 H.P.)		518.80		570.70		
1 Pavt. Rem. Bucket		45.60		50.15	23.40	25.73
40 L.H., Daily Totals		$1973.00		$2628.60	$49.33	$65.71
Crew B-39	**Hr.**	**Daily**	**Hr.**	**Daily**	**Bare Costs**	**Incl. O&P**
1 Labor Foreman (outside)	$24.85	$198.80	$38.90	$311.20	$24.17	$37.58
4 Laborers	22.85	731.20	35.75	1144.00		
1 Equip. Oper. (light)	28.75	230.00	43.55	348.40		
1 Air Compr., 250 C.F.M.		125.20		137.70		
2 Air Tools & Accessories		39.20		43.10		
2-50 Ft. Air Hoses, 1.5" Dia.		11.80		13.00	3.67	4.04
48 L.H., Daily Totals		$1336.20		$1997.40	$27.84	$41.62
Crew B-40	**Hr.**	**Daily**	**Hr.**	**Daily**	**Bare Costs**	**Incl. O&P**
1 Pile Driver Foreman (out)	$30.80	$246.40	$52.40	$419.20	$29.26	$47.71
4 Pile Drivers	28.80	921.60	49.00	1568.00		
2 Equip. Oper. (crane)	31.15	498.40	47.15	754.40		
1 Equip. Oper. Oiler	25.75	206.00	39.00	312.00		
1 Crane, 40 Ton		716.00		787.60		
1 Vibratory Hammer & Gen.		1283.00		1411.30	31.23	34.36
64 L.H., Daily Totals		$3871.40		$5252.50	$60.49	$82.07
Crew B-41	**Hr.**	**Daily**	**Hr.**	**Daily**	**Bare Costs**	**Incl. O&P**
1 Labor Foreman (outside)	$24.85	$198.80	$38.90	$311.20	$23.72	$36.99
4 Laborers	22.85	731.20	35.75	1144.00		
.25 Equip. Oper. (crane)	31.15	62.30	47.15	94.30		
.25 Equip. Oper. Oiler	25.75	51.50	39.00	78.00		
.25 Crawler Crane, 40 Ton		179.00		196.90	4.07	4.48
44 L.H., Daily Totals		$1222.80		$1824.40	$27.79	$41.47
Crew B-42	**Hr.**	**Daily**	**Hr.**	**Daily**	**Bare Costs**	**Incl. O&P**
1 Labor Foreman (outside)	$24.85	$198.80	$38.90	$311.20	$25.75	$40.96
4 Laborers	22.85	731.20	35.75	1144.00		
1 Equip. Oper. (crane)	31.15	249.20	47.15	377.20		
1 Equip. Oper. Oiler	25.75	206.00	39.00	312.00		
1 Welder	32.85	262.80	59.60	476.80		
1 Hyd. Crane, 25 Ton		629.60		692.55		
1 Gas Welding Machine		84.00		92.40		
1 Horz. Boring Csg. Mch.		523.20		575.50	19.33	21.26
64 L.H., Daily Totals		$2884.80		$3981.65	$45.08	$62.22
Crew B-43	**Hr.**	**Daily**	**Hr.**	**Daily**	**Bare Costs**	**Incl. O&P**
1 Labor Foreman (outside)	$24.85	$198.80	$38.90	$311.20	$25.05	$38.72
3 Laborers	22.85	548.40	35.75	858.00		
1 Equip. Oper. (crane)	31.15	249.20	47.15	377.20		
1 Equip. Oper. Oiler	25.75	206.00	39.00	312.00		
1 Drill Rig & Augers		1870.00		2057.00	38.96	42.85
48 L.H., Daily Totals		$3072.40		$3915.40	$64.01	$81.57

Crew B-44	Hr.	Daily	Hr.	Daily	Bare Costs	Incl. O&P
1 Pile Driver Foreman	$30.80	$246.40	$52.40	$419.20	$28.89	$47.31
4 Pile Drivers	28.80	921.60	49.00	1568.00		
2 Equip. Oper. (crane)	31.15	498.40	47.15	754.40		
1 Laborer	22.85	182.80	35.75	286.00		
1 Crane, 40 Ton, & Access.		716.00		787.60		
45 L.F. Leads, 15K Ft. Lbs.		234.00		257.40	14.84	16.33
64 L.H., Daily Totals		$2799.20		$4072.60	$43.73	$63.64

Crew B-45	Hr.	Daily	Hr.	Daily	Bare Costs	Incl. O&P
1 Equip. Oper. (med.)	$30.35	$242.80	$45.95	$367.60	$27.15	$41.28
1 Truck Driver (heavy)	23.95	191.60	36.60	292.80		
1 Dist. Tank Truck, 3K Gal.		328.80		361.70		
1 Tractor, 4 x 2, 250 H.P.		397.60		437.35	45.40	49.94
16 L.H., Daily Totals		$1160.80		$1459.45	$72.55	$91.22

Crew B-46	Hr.	Daily	Hr.	Daily	Bare Costs	Incl. O&P
1 Pile Driver Foreman	$30.80	$246.40	$52.40	$419.20	$26.16	$42.94
2 Pile Drivers	28.80	460.80	49.00	784.00		
3 Laborers	22.85	548.40	35.75	858.00		
1 Chain Saw, 36" Long		43.20		47.50	.90	.99
48 L.H., Daily Totals		$1298.80		$2108.70	$27.06	$43.93

Crew B-47	Hr.	Daily	Hr.	Daily	Bare Costs	Incl. O&P
1 Blast Foreman	$24.85	$198.80	$38.90	$311.20	$25.48	$39.40
1 Driller	22.85	182.80	35.75	286.00		
1 Equip. Oper. (light)	28.75	230.00	43.55	348.40		
1 Crawler Type Drill, 4"		455.60		501.15		
1 Air Compr., 600 C.F.M.		251.00		276.10		
2-50 Ft. Air Hoses, 3" Dia.		34.60		38.05	30.88	33.97
24 L.H., Daily Totals		$1352.80		$1760.90	$56.36	$73.37

Crew B-47A	Hr.	Daily	Hr.	Daily	Bare Costs	Incl. O&P
1 Drilling Foreman	$24.85	$198.80	$38.90	$311.20	$27.25	$41.68
1 Equip. Oper. (heavy)	31.15	249.20	47.15	377.20		
1 Oiler	25.75	206.00	39.00	312.00		
1 Quarry Drill		721.20		793.30	30.05	33.06
24 L.H., Daily Totals		$1375.20		$1793.70	$57.30	$74.74

Crew B-47C	Hr.	Daily	Hr.	Daily	Bare Costs	Incl. O&P
1 Laborer	$22.85	$182.80	$35.75	$286.00	$25.80	$39.65
1 Equip. Oper. (light)	28.75	230.00	43.55	348.40		
1 Air Compressor, 750 CFM		294.20		323.60		
2-50' Air Hose, 3"		34.60		38.05		
1 Air Track Drill, 4"		455.60		501.15	49.03	53.93
16 L.H., Daily Totals		$1197.20		$1497.20	$74.83	$93.58

Crew B-47E	Hr.	Daily	Hr.	Daily	Bare Costs	Incl. O&P
1 Laborer Foreman	$24.85	$198.80	$38.90	$311.20	$23.35	$36.54
3 Laborers	22.85	548.40	35.75	858.00		
1 Truck, Flatbed, 3 Ton		184.20		202.60	5.76	6.33
32 L.H., Daily Totals		$931.40		$1371.80	$29.11	$42.87

Crew B-48	Hr.	Daily	Hr.	Daily	Bare Costs	Incl. O&P
1 Labor Foreman (outside)	$24.85	$198.80	$38.90	$311.20	$25.58	$39.41
3 Laborers	22.85	548.40	35.75	858.00		
1 Equip. Oper. (crane)	31.15	249.20	47.15	377.20		
1 Equip. Oper. Oiler	25.75	206.00	39.00	312.00		
1 Equip. Oper. (light)	28.75	230.00	43.55	348.40		
1 Centr. Water Pump, 6"		164.40		180.85		
1-20 Ft. Suction Hose, 6"		22.50		24.75		
1-50 Ft. Disch. Hose, 6"		20.55		22.60		
1 Drill Rig & Augers		1870.00		2057.00	37.10	40.81
56 L.H., Daily Totals		$3509.85		$4492.00	$62.68	$80.22

Crew B-49	Hr.	Daily	Hr.	Daily	Bare Costs	Incl. O&P
1 Labor Foreman (outside)	$24.85	$198.80	$38.90	$311.20	$26.69	$41.82
3 Laborers	22.85	548.40	35.75	858.00		
2 Equip. Oper. (crane)	31.15	498.40	47.15	754.40		
2 Equip. Oper. Oilers	25.75	412.00	39.00	624.00		
1 Equip. Oper. (light)	28.75	230.00	43.55	348.40		
2 Pile Drivers	28.80	460.80	49.00	784.00		
1 Hyd. Crane, 25 Ton		629.60		692.55		
1 Centr. Water Pump, 6"		164.40		180.85		
1-20 Ft. Suction Hose, 6"		22.50		24.75		
1-50 Ft. Disch. Hose, 6"		20.55		22.60		
1 Drill Rig & Augers		1870.00		2057.00	30.76	33.84
88 L.H., Daily Totals		$5055.45		$6657.75	$57.45	$75.66

Crew B-50	Hr.	Daily	Hr.	Daily	Bare Costs	Incl. O&P
2 Pile Driver Foremen	$30.80	$492.80	$52.40	$838.40	$27.93	$45.67
6 Pile Drivers	28.80	1382.40	49.00	2352.00		
2 Equip. Oper. (crane)	31.15	498.40	47.15	754.40		
1 Equip. Oper. Oiler	25.75	206.00	39.00	312.00		
3 Laborers	22.85	548.40	35.75	858.00		
1 Crane, 40 Ton		716.00		787.60		
60 L.F. Leads, 15K Ft. Lbs.		312.00		343.20		
1 Hammer, 15K Ft. Lbs.		313.40		344.75		
1 Air Compr., 600 C.F.M.		251.00		276.10		
2-50 Ft. Air Hoses, 3" Dia.		34.60		38.05		
1 Chain Saw, 36" Long		43.20		47.50	14.91	16.40
112 L.H., Daily Totals		$4798.20		$6952.00	$42.84	$62.07

Crew B-51	Hr.	Daily	Hr.	Daily	Bare Costs	Incl. O&P
1 Labor Foreman (outside)	$24.85	$198.80	$38.90	$311.20	$23.28	$36.28
4 Laborers	22.85	731.20	35.75	1144.00		
1 Truck Driver (light)	23.40	187.20	35.75	286.00		
1 Light Truck, 1.5 Ton		158.00		173.80	3.29	3.62
48 L.H., Daily Totals		$1275.20		$1915.00	$26.57	$39.90

Crew B-52	Hr.	Daily	Hr.	Daily	Bare Costs	Incl. O&P
1 Carpenter Foreman	$31.15	$249.20	$48.75	$390.00	$26.91	$41.99
1 Carpenter	29.15	233.20	45.60	364.80		
3 Laborers	22.85	548.40	35.75	858.00		
1 Cement Finisher	27.95	223.60	41.55	332.40		
.5 Rodman (reinf.)	32.75	131.00	55.60	222.40		
.5 Equip. Oper. (med.)	30.35	121.40	45.95	183.80		
.5 F.E. Ldr., T.M., 2.5 C.Y.		385.00		423.50	6.88	7.56
56 L.H., Daily Totals		$1891.80		$2774.90	$33.79	$49.55

Crew B-53	Hr.	Daily	Hr.	Daily	Bare Costs	Incl. O&P
1 Equip. Oper. (light)	$28.75	$230.00	$43.55	$348.40	$28.75	$43.55
1 Trencher, Chain, 12 H.P.		96.80		106.50	12.10	13.31
8 L.H., Daily Totals		$326.80		$454.90	$40.85	$56.86

Crew B-54	Bare Costs Hr.	Bare Costs Daily	Incl. Subs O & P Hr.	Incl. Subs O & P Daily	Cost Per Labor-Hour Bare Costs	Cost Per Labor-Hour Incl. O&P
1 Equip. Oper. (light)	$28.75	$230.00	$43.55	$348.40	$28.75	$43.55
1 Trencher, Chain, 40 H.P.		196.40		216.05	24.55	27.01
8 L.H., Daily Totals		$426.40		$564.45	$53.30	$70.56

Crew B-54A	Hr.	Daily	Hr.	Daily	Bare Costs	Incl. O&P
.17 Labor Foreman (outside)	$24.85	$33.80	$38.90	$52.90	$29.55	$44.93
1 Equipment Operator (med.)	30.35	242.80	45.95	367.60		
1 Wheel Trencher, 67 H.P.		424.80		467.30	45.38	49.92
9.36 L.H., Daily Totals		$701.40		$887.80	$74.93	$94.85

Crew B-54B	Hr.	Daily	Hr.	Daily	Bare Costs	Incl. O&P
.25 Labor Foreman (outside)	$24.85	$49.70	$38.90	$77.80	$29.25	$44.54
1 Equipment Operator (med.)	30.35	242.80	45.95	367.60		
1 Wheel Trencher, 150 H.P.		636.80		700.50	63.68	70.05
10 L.H., Daily Totals		$929.30		$1145.90	$92.93	$114.59

Crew B-55	Hr.	Daily	Hr.	Daily	Bare Costs	Incl. O&P
2 Laborers	$22.85	$365.60	$35.75	$572.00	$23.03	$35.75
1 Truck Driver (light)	23.40	187.20	35.75	286.00		
1 Auger, 4″ to 36″ Dia		519.40		571.35		
1 Flatbed 3 Ton Truck		184.20		202.60	29.32	32.25
24 L.H., Daily Totals		$1256.40		$1631.95	$52.35	$68.00

Crew B-56	Hr.	Daily	Hr.	Daily	Bare Costs	Incl. O&P
1 Laborer	$22.85	$182.80	$35.75	$286.00	$25.80	$39.65
1 Equip. Oper. (light)	28.75	230.00	43.55	348.40		
1 Crawler Type Drill, 4″		455.60		501.15		
1 Air Compr., 600 C.F.M.		251.00		276.10		
1-50 Ft. Air Hose, 3″ Dia.		17.30		19.05	45.24	49.77
16 L.H., Daily Totals		$1136.70		$1430.70	$71.04	$89.42

Crew B-57	Hr.	Daily	Hr.	Daily	Bare Costs	Incl. O&P
1 Labor Foreman (outside)	$24.85	$198.80	$38.90	$311.20	$26.03	$40.02
2 Laborers	22.85	365.60	35.75	572.00		
1 Equip. Oper. (crane)	31.15	249.20	47.15	377.20		
1 Equip. Oper. (light)	28.75	230.00	43.55	348.40		
1 Equip. Oper. Oiler	25.75	206.00	39.00	312.00		
1 Power Shovel, 1 C.Y.		547.20		601.90		
1 Clamshell Bucket, 1 C.Y.		71.00		78.10		
1 Centr. Water Pump, 6″		164.40		180.85		
1-20 Ft. Suction Hose, 6″		22.50		24.75		
20-50 Ft. Disch. Hoses, 6″		411.00		452.10	25.34	27.87
48 L.H., Daily Totals		$2465.70		$3258.50	$51.37	$67.89

Crew B-58	Hr.	Daily	Hr.	Daily	Bare Costs	Incl. O&P
2 Laborers	$22.85	$365.60	$35.75	$572.00	$24.82	$38.35
1 Equip. Oper. (light)	28.75	230.00	43.55	348.40		
1 Backhoe Loader, 48 H.P.		181.00		199.10		
1 Small Helicopter, w/pilot		1986.00		2184.60	90.29	99.32
24 L.H., Daily Totals		$2762.60		$3304.10	$115.11	$137.67

Crew B-59	Hr.	Daily	Hr.	Daily	Bare Costs	Incl. O&P
1 Truck Driver (heavy)	$23.95	$191.60	$36.60	$292.80	$23.95	$36.60
1 Truck, 30 Ton		320.40		352.45		
1 Water tank, 5000 Gal.		184.00		202.40	63.05	69.36
8 L.H., Daily Totals		$696.00		$847.65	$87.00	$105.96

Crew B-59A	Hr.	Daily	Hr.	Daily	Bare Costs	Incl. O&P
2 Laborers	$22.85	$365.60	$35.75	$572.00	$23.22	$36.03
1 Truck Driver (heavy)	23.95	191.60	36.60	292.80		
1 Water tank, 5K w/pum		184.00		202.40		
1 Truck, 30 Ton		320.40		352.45	21.02	23.12
24 L.H., Daily Totals		$1061.60		$1419.65	$44.24	$59.15

Crew B-60	Hr.	Daily	Hr.	Daily	Bare Costs	Incl. O&P
1 Labor Foreman (outside)	$24.85	$198.80	$38.90	$311.20	$26.42	$40.52
2 Laborers	22.85	365.60	35.75	572.00		
1 Equip. Oper. (crane)	31.15	249.20	47.15	377.20		
2 Equip. Oper. (light)	28.75	460.00	43.55	696.80		
1 Equip. Oper. Oiler	25.75	206.00	39.00	312.00		
1 Crawler Crane, 40 Ton		716.00		787.60		
45 L.F. Leads, 15K Ft. Lbs.		234.00		257.40		
1 Backhoe Loader, 48 H.P.		181.00		199.10	20.20	22.22
56 L.H., Daily Totals		$2610.60		$3513.30	$46.62	$62.74

Crew B-61	Hr.	Daily	Hr.	Daily	Bare Costs	Incl. O&P
1 Labor Foreman (outside)	$24.85	$198.80	$38.90	$311.20	$24.43	$37.94
3 Laborers	22.85	548.40	35.75	858.00		
1 Equip. Oper. (light)	28.75	230.00	43.55	348.40		
1 Cement Mixer, 2 C.Y.		222.00		244.20		
1 Air Compr., 160 C.F.M.		98.40		108.25	8.01	8.81
40 L.H., Daily Totals		$1297.60		$1870.05	$32.44	$46.75

Crew B-62	Hr.	Daily	Hr.	Daily	Bare Costs	Incl. O&P
2 Laborers	$22.85	$365.60	$35.75	$572.00	$24.82	$38.35
1 Equip. Oper. (light)	28.75	230.00	43.55	348.40		
1 Loader, Skid Steer		114.00		125.40	4.75	5.23
24 L.H., Daily Totals		$709.60		$1045.80	$29.57	$43.58

Crew B-63	Hr.	Daily	Hr.	Daily	Bare Costs	Incl. O&P
4 Laborers	$22.85	$731.20	$35.75	$1144.00	$24.03	$37.31
1 Equip. Oper. (light)	28.75	230.00	43.55	348.40		
1 Loader, Skid Steer		114.00		125.40	2.85	3.14
40 L.H., Daily Totals		$1075.20		$1617.80	$26.88	$40.45

Crew B-64	Hr.	Daily	Hr.	Daily	Bare Costs	Incl. O&P
1 Laborer	$22.85	$182.80	$35.75	$286.00	$23.13	$35.75
1 Truck Driver (light)	23.40	187.20	35.75	286.00		
1 Power Mulcher (small)		124.80		137.30		
1 Light Truck, 1.5 Ton		158.00		173.80	17.68	19.44
16 L.H., Daily Totals		$652.80		$883.10	$40.81	$55.19

Crew B-65	Hr.	Daily	Hr.	Daily	Bare Costs	Incl. O&P
1 Laborer	$22.85	$182.80	$35.75	$286.00	$23.13	$35.75
1 Truck Driver (light)	23.40	187.20	35.75	286.00		
1 Power Mulcher (large)		232.80		256.10		
1 Light Truck, 1.5 Ton		158.00		173.80	24.43	26.87
16 L.H., Daily Totals		$760.80		$1001.90	$47.56	$62.62

Crew B-66	Hr.	Daily	Hr.	Daily	Bare Costs	Incl. O&P
1 Equip. Oper. (light)	$28.75	$230.00	$43.55	$348.40	$28.75	$43.55
1 Backhoe Ldr. w/Attchmt.		168.80		185.70	21.10	23.21
8 L.H., Daily Totals		$398.80		$534.10	$49.85	$66.76

Crew No.	Bare Costs		Incl. Subs O & P		Cost Per Labor-Hour	
Crew B-67	**Hr.**	**Daily**	**Hr.**	**Daily**	**Bare Costs**	**Incl. O&P**
1 Millwright	$30.55	$244.40	$45.45	$363.60	$29.65	$44.50
1 Equip. Oper. (light)	28.75	230.00	43.55	348.40		
1 Forklift		198.20		218.00	12.39	13.63
16 L.H., Daily Totals		$672.60		$930.00	$42.04	$58.13
Crew B-68	**Hr.**	**Daily**	**Hr.**	**Daily**	**Bare Costs**	**Incl. O&P**
2 Millwrights	$30.55	$488.80	$45.45	$727.20	$29.95	$44.82
1 Equip. Oper. (light)	28.75	230.00	43.55	348.40		
1 Forklift		198.20		218.00	8.26	9.08
24 L.H., Daily Totals		$917.00		$1293.60	$38.21	$53.90
Crew B-69	**Hr.**	**Daily**	**Hr.**	**Daily**	**Bare Costs**	**Incl. O&P**
1 Labor Foreman (outside)	$24.85	$198.80	$38.90	$311.20	$25.05	$38.72
3 Laborers	22.85	548.40	35.75	858.00		
1 Equip Oper. (crane)	31.15	249.20	47.15	377.20		
1 Equip Oper. Oiler	25.75	206.00	39.00	312.00		
1 Truck Crane, 80 Ton		1325.00		1457.50	27.60	30.36
48 L.H., Daily Totals		$2527.40		$3315.90	$52.65	$69.08
Crew B-69A	**Hr.**	**Daily**	**Hr.**	**Daily**	**Bare Costs**	**Incl. O&P**
1 Labor Foreman	$24.85	$198.80	$38.90	$311.20	$25.28	$38.94
3 Laborers	22.85	548.40	35.75	858.00		
1 Equip. Oper. (medium)	30.35	242.80	45.95	367.60		
1 Concrete Finisher	27.95	223.60	41.55	332.40		
1 Curb Paver		444.60		489.05	9.26	10.19
48 L.H., Daily Totals		$1658.20		$2358.25	$34.54	$49.13
Crew B-69B	**Hr.**	**Daily**	**Hr.**	**Daily**	**Bare Costs**	**Incl. O&P**
1 Labor Foreman	$24.85	$198.80	$38.90	$311.20	$25.28	$38.94
3 Laborers	22.85	548.40	35.75	858.00		
1 Equip. Oper. (medium)	30.35	242.80	45.95	367.60		
1 Cement Finisher	27.95	223.60	41.55	332.40		
1 Curb/Gutter Paver		897.60		987.35	18.70	20.57
48 L.H., Daily Totals		$2111.20		$2856.55	$43.98	$59.51
Crew B-70	**Hr.**	**Daily**	**Hr.**	**Daily**	**Bare Costs**	**Incl. O&P**
1 Labor Foreman (outside)	$24.85	$198.80	$38.90	$311.20	$26.35	$40.57
3 Laborers	22.85	548.40	35.75	858.00		
3 Equip. Oper. (med.)	30.35	728.40	45.95	1102.80		
1 Motor Grader, 30,000 Lb.		459.40		505.35		
1 Grader Attach., Ripper		67.20		73.90		
1 Road Sweeper, S.P.		258.00		283.80		
1 F.E. Loader, 1-3/4 C.Y.		284.40		312.85	19.09	21.00
56 L.H., Daily Totals		$2544.60		$3447.90	$45.44	$61.57
Crew B-71	**Hr.**	**Daily**	**Hr.**	**Daily**	**Bare Costs**	**Incl. O&P**
1 Labor Foreman (outside)	$24.85	$198.80	$38.90	$311.20	$26.35	$40.57
3 Laborers	22.85	548.40	35.75	858.00		
3 Equip. Oper. (med.)	30.35	728.40	45.95	1102.80		
1 Pvmt. Profiler, 450 H.P.		3506.00		3856.60		
1 Road Sweeper, S.P.		258.00		283.80		
1 F.E. Loader, 1-3/4 C.Y.		284.40		312.85	72.29	79.52
56 L.H., Daily Totals		$5524.00		$6725.25	$98.64	$120.09

Crew No.	Bare Costs		Incl. Subs O & P		Cost Per Labor-Hour	
Crew B-72	**Hr.**	**Daily**	**Hr.**	**Daily**	**Bare Costs**	**Incl. O&P**
1 Labor Foreman (outside)	$24.85	$198.80	$38.90	$311.20	$26.85	$41.24
3 Laborers	22.85	548.40	35.75	858.00		
4 Equip. Oper. (med.)	30.35	971.20	45.95	1470.40		
1 Pvmt. Profiler, 450 H.P.		3506.00		3856.60		
1 Hammermill, 250 H.P.		1234.00		1357.40		
1 Windrow Loader		884.00		972.40		
1 Mix Paver 165 H.P.		1428.00		1570.80		
1 Roller, Pneu. Tire, 12 T.		215.60		237.15	113.56	124.91
64 L.H., Daily Totals		$8986.00		$10633.95	$140.41	$166.15
Crew B-73	**Hr.**	**Daily**	**Hr.**	**Daily**	**Bare Costs**	**Incl. O&P**
1 Labor Foreman (outside)	$24.85	$198.80	$38.90	$311.20	$27.79	$42.52
2 Laborers	22.85	365.60	35.75	572.00		
5 Equip. Oper. (med.)	30.35	1214.00	45.95	1838.00		
1 Road Mixer, 310 H.P.		1112.00		1223.20		
1 Roller, Tandem, 12 Ton		210.40		231.45		
1 Hammermill, 250 H.P.		1234.00		1357.40		
1 Motor Grader, 30,000 Lb.		459.40		505.35		
.5 F.E. Loader, 1-3/4 C.Y.		142.20		156.40		
.5 Truck, 30 Ton		160.20		176.20		
.5 Water Tank 5000 Gal.		92.00		101.20	53.28	58.61
64 L.H., Daily Totals		$5188.60		$6472.40	$81.07	$101.13
Crew B-74	**Hr.**	**Daily**	**Hr.**	**Daily**	**Bare Costs**	**Incl. O&P**
1 Labor Foreman (outside)	$24.85	$198.80	$38.90	$311.20	$27.13	$41.46
1 Laborer	22.85	182.80	35.75	286.00		
4 Equip. Oper. (med.)	30.35	971.20	45.95	1470.40		
2 Truck Drivers (heavy)	23.95	383.20	36.60	585.60		
1 Motor Grader, 30,000 Lb.		459.40		505.35		
1 Grader Attach., Ripper		67.20		73.90		
2 Stabilizers, 310 H.P.		1601.60		1761.75		
1 Flatbed Truck, 3 Ton		184.20		202.60		
1 Chem. Spreader, Towed		92.40		101.65		
1 Vibr. Roller, 29,000 Lb.		451.20		496.30		
1 Water Tank 5000 Gal.		184.00		202.40		
1 Truck, 30 Ton		320.40		352.45	52.51	57.76
64 L.H., Daily Totals		$5096.40		$6349.60	$79.64	$99.22
Crew B-75	**Hr.**	**Daily**	**Hr.**	**Daily**	**Bare Costs**	**Incl. O&P**
1 Labor Foreman (outside)	$24.85	$198.80	$38.90	$311.20	$27.58	$42.15
1 Laborer	22.85	182.80	35.75	286.00		
4 Equip. Oper. (med.)	30.35	971.20	45.95	1470.40		
1 Truck Driver (heavy)	23.95	191.60	36.60	292.80		
1 Motor Grader, 30,000 Lb.		459.40		505.35		
1 Grader Attach., Ripper		67.20		73.90		
2 Stabilizers, 310 H.P.		1601.60		1761.75		
1 Dist. Truck, 3000 Gal.		328.80		361.70		
1 Vibr. Roller, 29,000 Lb.		451.20		496.30	51.93	57.13
56 L.H., Daily Totals		$4452.60		$5559.40	$79.51	$99.28
Crew B-76	**Hr.**	**Daily**	**Hr.**	**Daily**	**Bare Costs**	**Incl. O&P**
1 Dock Builder Foreman	$30.80	$246.40	$52.40	$419.20	$29.21	$47.86
5 Dock Builders	28.80	1152.00	49.00	1960.00		
2 Equip. Oper. (crane)	31.15	498.40	47.15	754.40		
1 Equip. Oper. Oiler	25.75	206.00	39.00	312.00		
1 Crawler Crane, 50 Ton		885.80		974.40		
1 Barge, 400 Ton		428.40		471.25		
1 Hammer, 15K Ft. Lbs.		313.40		344.75		
60 L.F. Leads, 15K Ft. Lbs.		312.00		343.20		
1 Air Compr., 600 C.F.M.		251.00		276.10		
2-50 Ft. Air Hoses, 3" Dia.		34.60		38.05	30.91	34.00
72 L.H., Daily Totals		$4328.00		$5893.35	$60.12	$81.86

Crew B-77	Hr.	Daily	Hr.	Daily	Bare Costs	Incl. O&P
1 Labor Foreman	$24.85	$198.80	$38.90	$311.20	$23.36	$36.38
3 Laborers	22.85	548.40	35.75	858.00		
1 Truck Driver (light)	23.40	187.20	35.75	286.00		
1 Crack Cleaner, 25 H.P.		71.20		78.30		
1 Crack Filler, Trailer Mtd.		142.40		156.65		
1 Flatbed Truck, 3 Ton		184.20		202.60	9.95	10.94
40 L.H., Daily Totals		$1332.20		$1892.75	$33.31	$47.32

Crew B-78	Hr.	Daily	Hr.	Daily	Bare Costs	Incl. O&P
1 Labor Foreman	$24.85	$198.80	$38.90	$311.20	$23.28	$36.28
4 Laborers	22.85	731.20	35.75	1144.00		
1 Truck Driver (light)	23.40	187.20	35.75	286.00		
1 Paint Striper, S.P.		201.80		222.00		
1 Flatbed Truck, 3 Ton		184.20		202.60		
1 Pickup Truck, 3/4 Ton		83.00		91.30	9.77	10.75
48 L.H., Daily Totals		$1586.20		$2257.10	$33.05	$47.03

Crew B-79	Hr.	Daily	Hr.	Daily	Bare Costs	Incl. O&P
1 Labor Foreman	$24.85	$198.80	$38.90	$311.20	$23.36	$36.38
3 Laborers	22.85	548.40	35.75	858.00		
1 Truck Driver (light)	23.40	187.20	35.75	286.00		
1 Thermo. Striper, T.M.		390.00		429.00		
1 Flatbed Truck, 3 Ton		184.20		202.60		
2 Pickup Truck, 3/4 Ton		166.00		182.60	18.51	20.36
40 L.H., Daily Totals		$1674.60		$2269.40	$41.87	$56.74

Crew B-80	Hr.	Daily	Hr.	Daily	Bare Costs	Incl. O&P
1 Labor Foreman	$24.85	$198.80	$38.90	$311.20	$24.96	$38.49
1 Laborer	22.85	182.80	35.75	286.00		
1 Truck Driver (light)	23.40	187.20	35.75	286.00		
1 Equip. Oper. (light)	28.75	230.00	43.55	348.40		
1 Flatbed Truck, 3 Ton		184.20		202.60		
1 Fence Post Auger, T.M.		363.00		399.30	17.10	18.81
32 L.H., Daily Totals		$1346.00		$1833.50	$42.06	$57.30

Crew B-80A	Hr.	Daily	Hr.	Daily	Bare Costs	Incl. O&P
3 Laborers	$22.85	$548.40	$35.75	$858.00	$22.85	$35.75
1 Flatbed Truck, 3 Ton		184.20		202.60	7.68	8.44
24 L.H., Daily Totals		$732.60		$1060.60	$30.53	$44.19

Crew B-80B	Hr.	Daily	Hr.	Daily	Bare Costs	Incl. O&P
3 Laborers	$22.85	$548.40	$35.75	$858.00	$24.33	$37.70
1 Equip. Oper. (light)	28.75	230.00	43.55	348.40		
1 Crane, Flatbed Mnt.		222.40		244.65	6.95	7.65
32 L.H., Daily Totals		$1000.80		$1451.05	$31.28	$45.35

Crew B-81	Hr.	Daily	Hr.	Daily	Bare Costs	Incl. O&P
1 Laborer	$22.85	$182.80	$35.75	$286.00	$25.72	$39.43
1 Equip. Oper. (med.)	30.35	242.80	45.95	367.60		
1 Truck Driver (heavy)	23.95	191.60	36.60	292.80		
1 Hydromulcher, T.M.		263.20		289.50		
1 Tractor Truck, 4x2		320.40		352.45	24.32	26.75
24 L.H., Daily Totals		$1200.80		$1588.35	$50.04	$66.18

Crew B-82	Hr.	Daily	Hr.	Daily	Bare Costs	Incl. O&P
1 Laborer	$22.85	$182.80	$35.75	$286.00	$25.80	$39.65
1 Equip. Oper. (light)	28.75	230.00	43.55	348.40		
1 Horiz. Borer, 6 H.P.		46.40		51.05	2.90	3.19
16 L.H., Daily Totals		$459.20		$685.45	$28.70	$42.84

Crew B-83	Hr.	Daily	Hr.	Daily	Bare Costs	Incl. O&P
1 Tugboat Captain	$30.35	$242.80	$45.95	$367.60	$26.60	$40.85
1 Tugboat Hand	22.85	182.80	35.75	286.00		
1 Tugboat, 250 H.P.		466.00		512.60	29.13	32.04
16 L.H., Daily Totals		$891.60		$1166.20	$55.73	$72.89

Crew B-84	Hr.	Daily	Hr.	Daily	Bare Costs	Incl. O&P
1 Equip. Oper. (med.)	$30.35	$242.80	$45.95	$367.60	$30.35	$45.95
1 Rotary Mower/Tractor		179.20		197.10	22.40	24.64
8 L.H., Daily Totals		$422.00		$564.70	$52.75	$70.59

Crew B-85	Hr.	Daily	Hr.	Daily	Bare Costs	Incl. O&P
3 Laborers	$22.85	$548.40	$35.75	$858.00	$24.57	$37.96
1 Equip. Oper. (med.)	30.35	242.80	45.95	367.60		
1 Truck Driver (heavy)	23.95	191.60	36.60	292.80		
1 Aerial Lift Truck, 80'		530.80		583.90		
1 Brush Chipper, 130 H.P.		190.60		209.65		
1 Pruning Saw, Rotary		22.00		24.20	18.59	20.44
40 L.H., Daily Totals		$1726.20		$2336.15	$43.16	$58.40

Crew B-86	Hr.	Daily	Hr.	Daily	Bare Costs	Incl. O&P
1 Equip. Oper. (med.)	$30.35	$242.80	$45.95	$367.60	$30.35	$45.95
1 Stump Chipper, S.P.		166.65		183.30	20.83	22.91
8 L.H., Daily Totals		$409.45		$550.90	$51.18	$68.86

Crew B-86A	Hr.	Daily	Hr.	Daily	Bare Costs	Incl. O&P
1 Equip. Oper. (medium)	$30.35	$242.80	$45.95	$367.60	$30.35	$45.95
1 Grader, 30,000 Lbs.		459.40		505.35	57.43	63.17
8 L.H., Daily Totals		$702.20		$872.95	$87.78	$109.12

Crew B-86B	Hr.	Daily	Hr.	Daily	Bare Costs	Incl. O&P
1 Equip. Oper. (medium)	$30.35	$242.80	$45.95	$367.60	$30.35	$45.95
1 Dozer, 200 H.P.		735.00		808.50	91.88	101.06
8 L.H., Daily Totals		$977.80		$1176.10	$122.23	$147.01

Crew B-87	Hr.	Daily	Hr.	Daily	Bare Costs	Incl. O&P
1 Laborer	$22.85	$182.80	$35.75	$286.00	$28.85	$43.91
4 Equip. Oper. (med.)	30.35	971.20	45.95	1470.40		
2 Feller Bunchers, 50 H.P.		746.00		820.60		
1 Log Chipper, 22" Tree		1789.00		1967.90		
1 Dozer, 105 H.P.		266.00		292.60		
1 Chainsaw, Gas, 36" Long		43.20		47.50	71.11	78.22
40 L.H., Daily Totals		$3998.20		$4885.00	$99.96	$122.13

Crew B-88	Hr.	Daily	Hr.	Daily	Bare Costs	Incl. O&P
1 Laborer	$22.85	$182.80	$35.75	$286.00	$29.28	$44.49
6 Equip. Oper. (med.)	30.35	1456.80	45.95	2205.60		
2 Feller Bunchers, 50 H.P.		746.00		820.60		
1 Log Chipper, 22" Tree		1789.00		1967.90		
2 Log Skidders, 50 H.P.		1026.80		1129.50		
1 Dozer, 105 H.P.		266.00		292.60		
1 Chainsaw, Gas, 36" Long		43.20		47.50	69.13	76.04
56 L.H., Daily Totals		$5510.60		$6749.70	$98.41	$120.53

Crew B-89	Hr.	Daily	Hr.	Daily	Bare Costs	Incl. O&P
1 Equip. Oper. (light)	$28.75	$230.00	$43.55	$348.40	$26.08	$39.65
1 Truck Driver (light)	23.40	187.20	35.75	286.00		
1 Truck, Stake Body, 3 Ton		184.20		202.60		
1 Concrete Saw		98.40		108.25		
1 Water Tank, 65 Gal.		13.50		14.85	18.51	20.36
16 L.H., Daily Totals		$713.30		$960.10	$44.59	$60.01

Crew No.	Bare Costs		Incl. Subs O & P		Cost Per Labor-Hour	
Crew B-89A	**Hr.**	**Daily**	**Hr.**	**Daily**	**Bare Costs**	**Incl. O&P**
1 Skilled Worker	$29.85	$238.80	$46.85	$374.80	$26.35	$41.30
1 Laborer	22.85	182.80	35.75	286.00		
1 Core Drill (large)		101.45		111.60	6.34	6.97
16 L.H., Daily Totals		$523.05		$772.40	$32.69	$48.27
Crew B-89B	**Hr.**	**Daily**	**Hr.**	**Daily**	**Bare Costs**	**Incl. O&P**
1 Equip. Oper. (light)	$28.75	$230.00	$43.55	$348.40	$26.08	$39.65
1 Truck Driver, Light	23.40	187.20	35.75	286.00		
1 Wall Saw, Hydraulic, 10 H.P.		106.05		116.65		
1 Generator, Diesel, 100 KW		211.05		232.15		
1 Water Tank, 65 Gal.		13.50		14.85		
1 Flatbed Truck, 3 Ton		184.20		202.60	32.18	35.39
16 L.H., Daily Totals		$932.00		$1200.65	$58.26	$75.04
Crew B-90	**Hr.**	**Daily**	**Hr.**	**Daily**	**Bare Costs**	**Incl. O&P**
1 Labor Foreman (outside)	$24.85	$198.80	$38.90	$311.20	$24.85	$38.31
3 Laborers	22.85	548.40	35.75	858.00		
2 Equip. Oper. (light)	28.75	460.00	43.55	696.80		
2 Truck Drivers (heavy)	23.95	383.20	36.60	585.60		
1 Road Mixer, 310 H.P.		1112.00		1223.20		
1 Dist. Truck, 2000 Gal.		314.80		346.30	22.29	24.52
64 L.H., Daily Totals		$3017.20		$4021.10	$47.14	$62.83
Crew B-90A	**Hr.**	**Daily**	**Hr.**	**Daily**	**Bare Costs**	**Incl. O&P**
1 Labor Foreman	$24.85	$198.80	$38.90	$311.20	$27.42	$42.03
2 Laborers	22.85	365.60	35.75	572.00		
4 Equip. Oper. (medium)	30.35	971.20	45.95	1470.40		
2 Graders, 30,000 Lbs.		918.80		1010.70		
1 Roller, Steel Wheel		210.40		231.45		
1 Roller, Pneumatic Wheel		215.60		237.15	24.01	26.42
56 L.H., Daily Totals		$2880.40		$3832.90	$51.43	$68.45
Crew B-90B	**Hr.**	**Daily**	**Hr.**	**Daily**	**Bare Costs**	**Incl. O&P**
1 Labor Foreman	$24.85	$198.80	$38.90	$311.20	$26.93	$41.38
2 Laborers	22.85	365.60	35.75	572.00		
3 Equip. Oper. (medium)	30.35	728.40	45.95	1102.80		
1 Roller, Steel Wheel		210.40		231.45		
1 Roller, Pneumatic Wheel		215.60		237.15		
1 Road Mixer, 310 H.P.		1112.00		1223.20	32.04	35.25
48 L.H., Daily Totals		$2830.80		$3677.80	$58.97	$76.63
Crew B-91	**Hr.**	**Daily**	**Hr.**	**Daily**	**Bare Costs**	**Incl. O&P**
1 Labor Foreman (outside)	$24.85	$198.80	$38.90	$311.20	$26.99	$41.35
2 Laborers	22.85	365.60	35.75	572.00		
4 Equip. Oper. (med.)	30.35	971.20	45.95	1470.40		
1 Truck Driver (heavy)	23.95	191.60	36.60	292.80		
1 Dist. Truck, 3000 Gal.		328.80		361.70		
1 Aggreg. Spreader, S.P.		628.60		691.45		
1 Roller, Pneu. Tire, 12 Ton		215.60		237.15		
1 Roller, Steel, 10 Ton		210.40		231.45	21.62	23.78
64 L.H., Daily Totals		$3110.60		$4168.15	$48.61	$65.13
Crew B-92	**Hr.**	**Daily**	**Hr.**	**Daily**	**Bare Costs**	**Incl. O&P**
1 Labor Foreman (outside)	$24.85	$198.80	$38.90	$311.20	$23.35	$36.54
3 Laborers	22.85	548.40	35.75	858.00		
1 Crack Cleaner, 25 H.P.		71.20		78.30		
1 Air Compressor		78.30		86.15		
1 Tar Kettle, T.M.		37.50		41.25		
1 Flatbed Truck, 3 Ton		184.20		202.60	11.60	12.76
32 L.H., Daily Totals		$1118.40		$1577.50	$34.95	$49.30

Crew No.	Bare Costs		Incl. Subs O & P		Cost Per Labor-Hour	
Crew B-93	**Hr.**	**Daily**	**Hr.**	**Daily**	**Bare Costs**	**Incl. O&P**
1 Equip. Oper. (med.)	$30.35	$242.80	$45.95	$367.60	$30.35	$45.95
1 Feller Buncher, 50 H.P.		373.00		410.30	46.63	51.29
8 L.H., Daily Totals		$615.80		$777.90	$76.98	$97.24
Crew B-94A	**Hr.**	**Daily**	**Hr.**	**Daily**	**Bare Costs**	**Incl. O&P**
1 Laborer	$22.85	$182.80	$35.75	$286.00	$22.85	$35.75
1 Diaph. Water Pump, 2"		30.80		33.90		
1-20 Ft. Suction Hose, 2"		6.50		7.15		
2-50 Ft. Disch. Hoses, 2"		10.80		11.90	6.01	6.61
8 L.H., Daily Totals		$230.90		$338.95	$28.86	$42.36
Crew B-94B	**Hr.**	**Daily**	**Hr.**	**Daily**	**Bare Costs**	**Incl. O&P**
1 Laborer	$22.85	$182.80	$35.75	$286.00	$22.85	$35.75
1 Diaph. Water Pump, 4"		62.00		68.20		
1-20 Ft. Suction Hose, 4"		12.50		13.75		
2-50 Ft. Disch. Hoses, 4"		17.00		18.70	11.44	12.58
8 L.H., Daily Totals		$274.30		$386.65	$34.29	$48.33
Crew B-94C	**Hr.**	**Daily**	**Hr.**	**Daily**	**Bare Costs**	**Incl. O&P**
1 Laborer	$22.85	$182.80	$35.75	$286.00	$22.85	$35.75
1 Centr. Water Pump, 3"		39.20		43.10		
1-20 Ft. Suction Hose, 3"		9.50		10.45		
2-50 Ft. Disch. Hoses, 3"		12.80		14.10	7.69	8.46
8 L.H., Daily Totals		$244.30		$353.65	$30.54	$44.21
Crew B-94D	**Hr.**	**Daily**	**Hr.**	**Daily**	**Bare Costs**	**Incl. O&P**
1 Laborer	$22.85	$182.80	$35.75	$286.00	$22.85	$35.75
1 Centr. Water Pump, 6"		164.40		180.85		
1-20 Ft. Suction Hose, 6"		22.50		24.75		
2-50 Ft. Disch. Hoses, 6"		41.10		45.20	28.50	31.35
8 L.H., Daily Totals		$410.80		$536.80	$51.35	$67.10
Crew B-95	**Hr.**	**Daily**	**Hr.**	**Daily**	**Bare Costs**	**Incl. O&P**
1 Equip. Oper. (crane)	$31.15	$249.20	$47.15	$377.20	$27.00	$41.45
1 Laborer	22.85	182.80	35.75	286.00		
16 L.H., Daily Totals		$432.00		$663.20	$27.00	$41.45
Crew B-95A	**Hr.**	**Daily**	**Hr.**	**Daily**	**Bare Costs**	**Incl. O&P**
1 Equip. Oper. (crane)	$31.15	$249.20	$47.15	$377.20	$27.00	$41.45
1 Laborer	22.85	182.80	35.75	286.00		
1 Hyd. Excavator, 5/8 C.Y.		360.00		396.00	22.50	24.75
16 L.H., Daily Totals		$792.00		$1059.20	$49.50	$66.20
Crew B-95B	**Hr.**	**Daily**	**Hr.**	**Daily**	**Bare Costs**	**Incl. O&P**
1 Equip. Oper. (crane)	$31.15	$249.20	$47.15	$377.20	$27.00	$41.45
1 Laborer	22.85	182.80	35.75	286.00		
1 Hyd. Excavator, 1.5 C.Y.		593.20		652.50	37.08	40.78
16 L.H., Daily Totals		$1025.20		$1315.70	$64.08	$82.23
Crew B-95C	**Hr.**	**Daily**	**Hr.**	**Daily**	**Bare Costs**	**Incl. O&P**
1 Equip. Oper. (crane)	$31.15	$249.20	$47.15	$377.20	$27.00	$41.45
1 Laborer	22.85	182.80	35.75	286.00		
1 Hyd. Excavator, 2.5 C.Y.		1557.00		1712.70	97.31	107.04
16 L.H., Daily Totals		$1989.00		$2375.90	$124.31	$148.49
Crew C-1	**Hr.**	**Daily**	**Hr.**	**Daily**	**Bare Costs**	**Incl. O&P**
3 Carpenters	$29.15	$699.60	$45.60	$1094.40	$27.57	$43.14
1 Laborer	22.85	182.80	35.75	286.00		
32 L.H., Daily Totals		$882.40		$1380.40	$27.57	$43.14

Crews

Crew No.	Bare Costs		Incl. Subs O & P		Cost Per Labor-Hour	
Crew C-2	**Hr.**	**Daily**	**Hr.**	**Daily**	**Bare Costs**	**Incl. O&P**
1 Carpenter Foreman (out)	$31.15	$249.20	$48.75	$390.00	$28.43	$44.48
4 Carpenters	29.15	932.80	45.60	1459.20		
1 Laborer	22.85	182.80	35.75	286.00		
48 L.H., Daily Totals		$1364.80		$2135.20	$28.43	$44.48

Crew C-2A	**Hr.**	**Daily**	**Hr.**	**Daily**	**Bare Costs**	**Incl. O&P**
1 Carpenter Foreman (out)	$31.15	$249.20	$48.75	$390.00	$28.23	$43.81
3 Carpenters	29.15	699.60	45.60	1094.40		
1 Cement Finisher	27.95	223.60	41.55	332.40		
1 Laborer	22.85	182.80	35.75	286.00		
48 L.H., Daily Totals		$1355.20		$2102.80	$28.23	$43.81

Crew C-3	**Hr.**	**Daily**	**Hr.**	**Daily**	**Bare Costs**	**Incl. O&P**
1 Rodman Foreman	$34.75	$278.00	$58.95	$471.60	$30.02	$49.55
4 Rodmen (reinf.)	32.75	1048.00	55.60	1779.20		
1 Equip. Oper. (light)	28.75	230.00	43.55	348.40		
2 Laborers	22.85	365.60	35.75	572.00		
3 Stressing Equipment		78.00		85.80		
.5 Grouting Equipment		116.60		128.25	3.04	3.34
64 L.H., Daily Totals		$2116.20		$3385.25	$33.06	$52.89

Crew C-4	**Hr.**	**Daily**	**Hr.**	**Daily**	**Bare Costs**	**Incl. O&P**
1 Rodman Foreman	$34.75	$278.00	$58.95	$471.60	$33.25	$56.44
3 Rodmen (reinf.)	32.75	786.00	55.60	1334.40		
3 Stressing Equipment		78.00		85.80	2.44	2.68
32 L.H., Daily Totals		$1142.00		$1891.80	$35.69	$59.12

Crew C-5	**Hr.**	**Daily**	**Hr.**	**Daily**	**Bare Costs**	**Incl. O&P**
1 Rodman Foreman	$34.75	$278.00	$58.95	$471.60	$31.81	$52.50
4 Rodmen (reinf.)	32.75	1048.00	55.60	1779.20		
1 Equip. Oper. (crane)	31.15	249.20	47.15	377.20		
1 Equip. Oper. Oiler	25.75	206.00	39.00	312.00		
1 Hyd. Crane, 25 Ton		629.60		692.55	11.24	12.37
56 L.H., Daily Totals		$2410.80		$3632.55	$43.05	$64.87

Crew C-6	**Hr.**	**Daily**	**Hr.**	**Daily**	**Bare Costs**	**Incl. O&P**
1 Labor Foreman (outside)	$24.85	$198.80	$38.90	$311.20	$24.03	$37.24
4 Laborers	22.85	731.20	35.75	1144.00		
1 Cement Finisher	27.95	223.60	41.55	332.40		
2 Gas Engine Vibrators		76.00		83.60	1.58	1.74
48 L.H., Daily Totals		$1229.60		$1871.20	$25.61	$38.98

Crew C-7	**Hr.**	**Daily**	**Hr.**	**Daily**	**Bare Costs**	**Incl. O&P**
1 Labor Foreman (outside)	$24.85	$198.80	$38.90	$311.20	$24.79	$38.24
5 Laborers	22.85	914.00	35.75	1430.00		
1 Cement Finisher	27.95	223.60	41.55	332.40		
1 Equip. Oper. (med.)	30.35	242.80	45.95	367.60		
1 Equip. Oper. (oiler)	25.75	206.00	39.00	312.00		
2 Gas Engine Vibrators		76.00		83.60		
1 Concrete Bucket, 1 C.Y.		27.60		30.35		
1 Hyd. Crane, 55 Ton		901.60		991.75	13.96	15.36
72 L.H., Daily Totals		$2790.40		$3858.90	$38.75	$53.60

Crew C-7A	**Hr.**	**Daily**	**Hr.**	**Daily**	**Bare Costs**	**Incl. O&P**
1 Labor Foreman (outside)	$24.85	$198.80	$38.90	$311.20	$23.38	$36.36
5 Laborers	22.85	914.00	35.75	1430.00		
2 Truck Driver (Heavy)	23.95	383.20	36.60	585.60		
2 Conc. Transit Mixers		1684.00		1852.40	26.31	28.94
64 L.H., Daily Totals		$3180.00		$4179.20	$49.69	$65.30

Crew C-7B	**Hr.**	**Daily**	**Hr.**	**Daily**	**Bare Costs**	**Incl. O&P**
1 Labor Foreman (outside)	$24.85	$198.80	$38.90	$311.20	$24.50	$37.98
5 Laborers	22.85	914.00	35.75	1430.00		
1 Equipment Operator (heavy)	31.15	249.20	47.15	377.20		
1 Equipment Oiler	25.75	206.00	39.00	312.00		
1 Conc. Bucket, 2 C.Y.		37.00		40.70		
1 Truck Crane, 165 Ton		1714.00		1885.40	27.36	30.10
64 L.H., Daily Totals		$3319.00		$4356.50	$51.86	$68.08

Crew C-7C	**Hr.**	**Daily**	**Hr.**	**Daily**	**Bare Costs**	**Incl. O&P**
1 Labor Foreman (outside)	$24.85	$198.80	$38.90	$311.20	$24.98	$38.69
5 Laborers	22.85	914.00	35.75	1430.00		
2 Equipment Operator (medium)	30.35	485.60	45.95	735.20		
2 Wheel Loader, 4 C.Y.		1037.60		1141.35	16.21	17.83
64 L.H., Daily Totals		$2636.00		$3617.75	$41.19	$56.52

Crew C-7D	**Hr.**	**Daily**	**Hr.**	**Daily**	**Bare Costs**	**Incl. O&P**
1 Labor Foreman (outside)	$24.85	$198.80	$38.90	$311.20	$24.21	$37.66
5 Laborers	22.85	914.00	35.75	1430.00		
1 Equipment Operator (med.)	30.35	242.80	45.95	367.60		
1 Concrete Conveyer		190.00		209.00	3.39	3.73
56 L.H., Daily Totals		$1545.60		$2317.80	$27.60	$41.39

Crew C-8	**Hr.**	**Daily**	**Hr.**	**Daily**	**Bare Costs**	**Incl. O&P**
1 Labor Foreman (outside)	$24.85	$198.80	$38.90	$311.20	$25.66	$39.31
3 Laborers	22.85	548.40	35.75	858.00		
2 Cement Finishers	27.95	447.20	41.55	664.80		
1 Equip. Oper. (med.)	30.35	242.80	45.95	367.60		
1 Concrete Pump (small)		580.00		638.00	10.36	11.39
56 L.H., Daily Totals		$2017.20		$2839.60	$36.02	$50.70

Crew C-8A	**Hr.**	**Daily**	**Hr.**	**Daily**	**Bare Costs**	**Incl. O&P**
1 Labor Foreman (outside)	$24.85	$198.80	$38.90	$311.20	$24.88	$38.21
3 Laborers	22.85	548.40	35.75	858.00		
2 Cement Finishers	27.95	447.20	41.55	664.80		
48 L.H., Daily Totals		$1194.40		$1834.00	$24.88	$38.21

Crew C-8B	**Hr.**	**Daily**	**Hr.**	**Daily**	**Bare Costs**	**Incl. O&P**
1 Labor Foreman (outside)	$24.85	$198.80	$38.90	$311.20	$24.75	$38.42
3 Laborers	22.85	548.40	35.75	858.00		
1 Equipment Operator	30.35	242.80	45.95	367.60		
1 Vibrating Screed		53.90		59.30		
1 Vibratory Roller		451.20		496.30		
1 Dozer, 200 H.P.		735.00		808.50	31.00	34.10
40 L.H., Daily Totals		$2230.10		$2900.90	$55.75	$72.52

Crew C-8C	**Hr.**	**Daily**	**Hr.**	**Daily**	**Bare Costs**	**Incl. O&P**
1 Labor Foreman (outside)	$24.85	$198.80	$38.90	$311.20	$25.28	$38.94
3 Laborers	22.85	548.40	35.75	858.00		
1 Cement Finisher	27.95	223.60	41.55	332.40		
1 Equipment Operator (med.)	30.35	242.80	45.95	367.60		
1 Shotcrete Rig, 12 CY/hr		341.20		375.30	7.11	7.82
48 L.H., Daily Totals		$1554.80		$2244.50	$32.39	$46.76

Crew C-8D	**Hr.**	**Daily**	**Hr.**	**Daily**	**Bare Costs**	**Incl. O&P**
1 Labor Foreman (outside)	$24.85	$198.80	$38.90	$311.20	$26.10	$39.94
1 Laborers	22.85	182.80	35.75	286.00		
1 Cement Finisher	27.95	223.60	41.55	332.40		
1 Equipment Operator (light)	28.75	230.00	43.55	348.40		
1 Compressor, 250 CFM		125.20		137.70		
2 Hoses, 1", 50'		7.30		8.05	4.14	4.55
32 L.H., Daily Totals		$967.70		$1423.75	$30.24	$44.49

Crew C-8E	Bare Costs Hr.	Bare Costs Daily	Incl. Subs O & P Hr.	Incl. Subs O & P Daily	Cost Per Labor-Hour Bare Costs	Cost Per Labor-Hour Incl. O&P
1 Labor Foreman (outside)	$24.85	$198.80	$38.90	$311.20	$26.10	$39.94
1 Laborers	22.85	182.80	35.75	286.00		
1 Cement Finisher	27.95	223.60	41.55	332.40		
1 Equipment Operator (light)	28.75	230.00	43.55	348.40		
1 Compressor, 250 CFM		125.20		137.70		
2 Hoses, 1", 50'		7.30		8.05		
1 Concrete Pump (small)		580.00		638.00	22.27	24.49
32 L.H., Daily Totals		$1547.70		$2061.75	$48.37	$64.43

Crew C-10	Hr.	Daily	Hr.	Daily	Bare Costs	Incl. O&P
1 Laborer	$22.85	$182.80	$35.75	$286.00	$26.25	$39.62
2 Cement Finishers	27.95	447.20	41.55	664.80		
24 L.H., Daily Totals		$630.00		$950.80	$26.25	$39.62

Crew C-11	Hr.	Daily	Hr.	Daily	Bare Costs	Incl. O&P
1 Struc. Steel Foreman	$34.85	$278.80	$63.20	$505.60	$32.09	$56.33
6 Struc. Steel Workers	32.85	1576.80	59.60	2860.80		
1 Equip. Oper. (crane)	31.15	249.20	47.15	377.20		
1 Equip. Oper. Oiler	25.75	206.00	39.00	312.00		
1 Truck Crane, 150 Ton		1478.00		1625.80	20.53	22.58
72 L.H., Daily Totals		$3788.80		$5681.40	$52.62	$78.91

Crew C-12	Hr.	Daily	Hr.	Daily	Bare Costs	Incl. O&P
1 Carpenter Foreman (out)	$31.15	$249.20	$48.75	$390.00	$28.77	$44.74
3 Carpenters	29.15	699.60	45.60	1094.40		
1 Laborer	22.85	182.80	35.75	286.00		
1 Equip. Oper. (crane)	31.15	249.20	47.15	377.20		
1 Hyd. Crane, 12 Ton		514.80		566.30	10.73	11.80
48 L.H., Daily Totals		$1895.60		$2713.90	$39.50	$56.54

Crew C-13	Hr.	Daily	Hr.	Daily	Bare Costs	Incl. O&P
1 Struc. Steel Worker	$32.85	$262.80	$59.60	$476.80	$31.62	$54.93
1 Welder	32.85	262.80	59.60	476.80		
1 Carpenter	29.15	233.20	45.60	364.80		
1 Gas Welding Machine		84.00		92.40	3.50	3.85
24 L.H., Daily Totals		$842.80		$1410.80	$35.12	$58.78

Crew C-14	Hr.	Daily	Hr.	Daily	Bare Costs	Incl. O&P
1 Carpenter Foreman (out)	$31.15	$249.20	$48.75	$390.00	$28.45	$45.08
5 Carpenters	29.15	1166.00	45.60	1824.00		
4 Laborers	22.85	731.20	35.75	1144.00		
4 Rodmen (reinf.)	32.75	1048.00	55.60	1779.20		
2 Cement Finishers	27.95	447.20	41.55	664.80		
1 Equip. Oper. (crane)	31.15	249.20	47.15	377.20		
1 Equip. Oper. Oiler	25.75	206.00	39.00	312.00		
1 Crane, 80 Ton, & Tools		1325.00		1457.50	9.20	10.12
144 L.H., Daily Totals		$5421.80		$7948.70	$37.65	$55.20

Crew C-14A	Hr.	Daily	Hr.	Daily	Bare Costs	Incl. O&P
1 Carpenter Foreman (out)	$31.15	$249.20	$48.75	$390.00	$29.30	$46.39
16 Carpenters	29.15	3731.20	45.60	5836.80		
4 Rodmen (reinf.)	32.75	1048.00	55.60	1779.20		
2 Laborers	22.85	365.60	35.75	572.00		
1 Cement Finisher	27.95	223.60	41.55	332.40		
1 Equip. Oper. (med.)	30.35	242.80	45.95	367.60		
1 Gas Engine Vibrator		38.00		41.80		
1 Concrete Pump (small)		580.00		638.00	3.09	3.40
200 L.H., Daily Totals		$6478.40		$9957.80	$32.39	$49.79

Crew C-14B	Hr.	Daily	Hr.	Daily	Bare Costs	Incl. O&P
1 Carpenter Foreman (out)	$31.15	$249.20	$48.75	$390.00	$29.25	$46.20
16 Carpenters	29.15	3731.20	45.60	5836.80		
4 Rodmen (reinf.)	32.75	1048.00	55.60	1779.20		
2 Laborers	22.85	365.60	35.75	572.00		
2 Cement Finishers	27.95	447.20	41.55	664.80		
1 Equip. Oper. (med.)	30.35	242.80	45.95	367.60		
1 Gas Engine Vibrator		38.00		41.80		
1 Concrete Pump (small)		580.00		638.00	2.97	3.27
208 L.H., Daily Totals		$6702.00		$10290.20	$32.22	$49.47

Crew C-14C	Hr.	Daily	Hr.	Daily	Bare Costs	Incl. O&P
1 Carpenter Foreman (out)	$31.15	$249.20	$48.75	$390.00	$27.92	$44.15
6 Carpenters	29.15	1399.20	45.60	2188.80		
2 Rodmen (reinf.)	32.75	524.00	55.60	889.60		
4 Laborers	22.85	731.20	35.75	1144.00		
1 Cement Finisher	27.95	223.60	41.55	332.40		
1 Gas Engine Vibrator		38.00		41.80	.34	.37
112 L.H., Daily Totals		$3165.20		$4986.60	$28.26	$44.52

Crew C-14D	Hr.	Daily	Hr.	Daily	Bare Costs	Incl. O&P
1 Carpenter Foreman (out)	$31.15	$249.20	$48.75	$390.00	$29.01	$45.59
18 Carpenters	29.15	4197.60	45.60	6566.40		
2 Rodmen (reinf.)	32.75	524.00	55.60	889.60		
2 Laborers	22.85	365.60	35.75	572.00		
1 Cement Finisher	27.95	223.60	41.55	332.40		
1 Equip. Oper. (med.)	30.35	242.80	45.95	367.60		
1 Gas Engine Vibrator		38.00		41.80		
1 Concrete Pump (small)		580.00		638.00	3.09	3.40
200 L.H., Daily Totals		$6420.80		$9797.80	$32.10	$48.99

Crew C-14E	Hr.	Daily	Hr.	Daily	Bare Costs	Incl. O&P
1 Carpenter Foreman (out)	$31.15	$249.20	$48.75	$390.00	$28.81	$46.47
2 Carpenters	29.15	466.40	45.60	729.60		
4 Rodmen (reinf.)	32.75	1048.00	55.60	1779.20		
3 Laborers	22.85	548.40	35.75	858.00		
1 Cement Finisher	27.95	223.60	41.55	332.40		
1 Gas Engine Vibrator		38.00		41.80	.43	.48
88 L.H., Daily Totals		$2573.60		$4131.00	$29.24	$46.95

Crew C-14F	Hr.	Daily	Hr.	Daily	Bare Costs	Incl. O&P
1 Laborer Foreman (out)	$24.85	$198.80	$38.90	$311.20	$26.47	$39.97
2 Laborers	22.85	365.60	35.75	572.00		
6 Cement Finishers	27.95	1341.60	41.55	1994.40		
1 Gas Engine Vibrator		38.00		41.80	.53	.58
72 L.H., Daily Totals		$1944.00		$2919.40	$27.00	$40.55

Crew C-14G	Hr.	Daily	Hr.	Daily	Bare Costs	Incl. O&P
1 Laborer Foreman (out)	$24.85	$198.80	$38.90	$311.20	$26.05	$39.51
2 Laborers	22.85	365.60	35.75	572.00		
4 Cement Finishers	27.95	894.40	41.55	1329.60		
1 Gas Engine Vibrator		38.00		41.80	.68	.75
56 L.H., Daily Totals		$1496.80		$2254.60	$26.73	$40.26

Crew C-14H	Hr.	Daily	Hr.	Daily	Bare Costs	Incl. O&P
1 Carpenter Foreman (out)	$31.15	$249.20	$48.75	$390.00	$28.83	$45.48
2 Carpenters	29.15	466.40	45.60	729.60		
1 Rodman (reinf.)	32.75	262.00	55.60	444.80		
1 Laborer	22.85	182.80	35.75	286.00		
1 Cement Finisher	27.95	223.60	41.55	332.40		
1 Gas Engine Vibrator		38.00		41.80	.79	.87
48 L.H., Daily Totals		$1422.00		$2224.60	$29.62	$46.35

Crew No.	Bare Costs		Incl. Subs O & P		Cost Per Labor-Hour	

Crew C-15	Hr.	Daily	Hr.	Daily	Bare Costs	Incl. O&P
1 Carpenter Foreman (out)	$31.15	$249.20	$48.75	$390.00	$27.41	$42.88
2 Carpenters	29.15	466.40	45.60	729.60		
3 Laborers	22.85	548.40	35.75	858.00		
2 Cement Finishers	27.95	447.20	41.55	664.80		
1 Rodman (reinf.)	32.75	262.00	55.60	444.80		
72 L.H., Daily Totals		$1973.20		$3087.20	$27.41	$42.88

Crew C-16	Hr.	Daily	Hr.	Daily	Bare Costs	Incl. O&P
1 Labor Foreman (outside)	$24.85	$198.80	$38.90	$311.20	$27.24	$42.93
3 Laborers	22.85	548.40	35.75	858.00		
2 Cement Finishers	27.95	447.20	41.55	664.80		
1 Equip. Oper. (med.)	30.35	242.80	45.95	367.60		
2 Rodmen (reinf.)	32.75	524.00	55.60	889.60		
1 Concrete Pump (small)		580.00		638.00	8.06	8.86
72 L.H., Daily Totals		$2541.20		$3729.20	$35.30	$51.79

Crew C-17	Hr.	Daily	Hr.	Daily	Bare Costs	Incl. O&P
2 Skilled Worker Foremen	$31.85	$509.60	$50.00	$800.00	$30.25	$47.48
8 Skilled Workers	29.85	1910.40	46.85	2998.40		
80 L.H., Daily Totals		$2420.00		$3798.40	$30.25	$47.48

Crew C-17A	Hr.	Daily	Hr.	Daily	Bare Costs	Incl. O&P
2 Skilled Worker Foremen	$31.85	$509.60	$50.00	$800.00	$30.26	$47.48
8 Skilled Workers	29.85	1910.40	46.85	2998.40		
.125 Equip. Oper. (crane)	31.15	31.15	47.15	47.15		
.125 Crane, 80 Ton, & Tools		165.63		182.20	2.04	2.25
81 L.H., Daily Totals		$2616.78		$4027.75	$32.30	$49.73

Crew C-17B	Hr.	Daily	Hr.	Daily	Bare Costs	Incl. O&P
2 Skilled Worker Foremen	$31.85	$509.60	$50.00	$800.00	$30.27	$47.47
8 Skilled Workers	29.85	1910.40	46.85	2998.40		
.25 Equip. Oper. (crane)	31.15	62.30	47.15	94.30		
.25 Crane, 80 Ton, & Tools		331.25		364.40		
.25 Walk Behind Power Tools		13.25		14.60	4.20	4.62
82 L.H., Daily Totals		$2826.80		$4271.70	$34.47	$52.09

Crew C-17C	Hr.	Daily	Hr.	Daily	Bare Costs	Incl. O&P
2 Skilled Worker Foremen	$31.85	$509.60	$50.00	$800.00	$30.28	$47.47
8 Skilled Workers	29.85	1910.40	46.85	2998.40		
.375 Equip. Oper. (crane)	31.15	93.45	47.15	141.45		
.375 Crane, 80 Ton & Tools		496.88		546.55	5.99	6.59
83 L.H., Daily Totals		$3010.33		$4486.40	$36.27	$54.06

Crew C-17D	Hr.	Daily	Hr.	Daily	Bare Costs	Incl. O&P
2 Skilled Worker Foremen	$31.85	$509.60	$50.00	$800.00	$30.29	$47.46
8 Skilled Workers	29.85	1910.40	46.85	2998.40		
.5 Equip. Oper. (crane)	31.15	124.60	47.15	188.60		
.5 Crane, 80 Ton & Tools		662.50		728.75	7.89	8.68
84 L.H., Daily Totals		$3207.10		$4715.75	$38.18	$56.14

Crew C-17E	Hr.	Daily	Hr.	Daily	Bare Costs	Incl. O&P
2 Skilled Worker Foremen	$31.85	$509.60	$50.00	$800.00	$30.25	$47.48
8 Skilled Workers	29.85	1910.40	46.85	2998.40		
1 Hyd. Jack with Rods		73.25		80.60	.92	1.01
80 L.H., Daily Totals		$2493.25		$3879.00	$31.17	$48.49

Crew No.	Bare Costs		Incl. Subs O & P		Cost Per Labor-Hour	

Crew C-18	Hr.	Daily	Hr.	Daily	Bare Costs	Incl. O&P
.125 Labor Foreman (out)	$24.85	$24.85	$38.90	$38.90	$23.07	$36.10
1 Laborer	22.85	182.80	35.75	286.00		
1 Concrete Cart, 10 C.F.		50.60		55.65	5.62	6.18
9 L.H., Daily Totals		$258.25		$380.55	$28.69	$42.28

Crew C-19	Hr.	Daily	Hr.	Daily	Bare Costs	Incl. O&P
.125 Labor Foreman (out)	$24.85	$24.85	$38.90	$38.90	$23.07	$36.10
1 Laborer	22.85	182.80	35.75	286.00		
1 Concrete Cart, 18 C.F.		76.00		83.60	8.44	9.29
9 L.H., Daily Totals		$283.65		$408.50	$31.51	$45.39

Crew C-20	Hr.	Daily	Hr.	Daily	Bare Costs	Incl. O&P
1 Labor Foreman (outside)	$24.85	$198.80	$38.90	$311.20	$24.68	$38.14
5 Laborers	22.85	914.00	35.75	1430.00		
1 Cement Finisher	27.95	223.60	41.55	332.40		
1 Equip. Oper. (med.)	30.35	242.80	45.95	367.60		
2 Gas Engine Vibrators		76.00		83.60		
1 Concrete Pump (small)		580.00		638.00	10.25	11.28
64 L.H., Daily Totals		$2235.20		$3162.80	$34.93	$49.42

Crew C-21	Hr.	Daily	Hr.	Daily	Bare Costs	Incl. O&P
1 Labor Foreman (outside)	$24.85	$198.80	$38.90	$311.20	$24.68	$38.14
5 Laborers	22.85	914.00	35.75	1430.00		
1 Cement Finisher	27.95	223.60	41.55	332.40		
1 Equip. Oper. (med.)	30.35	242.80	45.95	367.60		
2 Gas Engine Vibrators		76.00		83.60		
1 Concrete Conveyer		190.00		209.00	4.16	4.57
64 L.H., Daily Totals		$1845.20		$2733.80	$28.84	$42.71

Crew C-22	Hr.	Daily	Hr.	Daily	Bare Costs	Incl. O&P
1 Rodman Foreman	$34.75	$278.00	$58.95	$471.60	$32.93	$55.64
4 Rodmen (reinf.)	32.75	1048.00	55.60	1779.20		
.125 Equip. Oper. (crane)	31.15	31.15	47.15	47.15		
.125 Equip. Oper. Oiler	25.75	25.75	39.00	39.00		
.125 Hyd. Crane, 25 Ton		78.70		86.55	1.87	2.06
42 L.H., Daily Totals		$1461.60		$2423.50	$34.80	$57.70

Crew C-23	Hr.	Daily	Hr.	Daily	Bare Costs	Incl. O&P
2 Skilled Worker Foremen	$31.85	$509.60	$50.00	$800.00	$29.97	$46.72
6 Skilled Workers	29.85	1432.80	46.85	2248.80		
1 Equip. Oper. (crane)	31.15	249.20	47.15	377.20		
1 Equip. Oper. Oiler	25.75	206.00	39.00	312.00		
1 Crane, 90 Ton		1290.00		1419.00	16.13	17.74
80 L.H., Daily Totals		$3687.60		$5157.00	$46.10	$64.46

Crew C-24	Hr.	Daily	Hr.	Daily	Bare Costs	Incl. O&P
2 Skilled Worker Foremen	$31.85	$509.60	$50.00	$800.00	$29.97	$46.72
6 Skilled Workers	29.85	1432.80	46.85	2248.80		
1 Equip. Oper. (crane)	31.15	249.20	47.15	377.20		
1 Equip. Oper. Oiler	25.75	206.00	39.00	312.00		
1 Truck Crane, 150 Ton		1478.00		1625.80	18.48	20.32
80 L.H., Daily Totals		$3875.60		$5363.80	$48.45	$67.04

Crew C-25	Hr.	Daily	Hr.	Daily	Bare Costs	Incl. O&P
2 Rodmen (reinf.)	$32.75	$524.00	$55.60	$889.60	$25.93	$44.20
2 Rodmen Helpers	19.10	305.60	32.80	524.80		
32 L.H., Daily Totals		$829.60		$1414.40	$25.93	$44.20

Crews

Crew No.	Bare Costs		Incl. Subs O & P		Cost Per Labor-Hour	
Crew D-1	**Hr.**	**Daily**	**Hr.**	**Daily**	**Bare Costs**	**Incl. O&P**
1 Bricklayer	$29.60	$236.80	$45.75	$366.00	$26.28	$40.60
1 Bricklayer Helper	22.95	183.60	35.45	283.60		
16 L.H., Daily Totals		$420.40		$649.60	$26.28	$40.60
Crew D-2	**Hr.**	**Daily**	**Hr.**	**Daily**	**Bare Costs**	**Incl. O&P**
3 Bricklayers	$29.60	$710.40	$45.75	$1098.00	$27.14	$41.99
2 Bricklayer Helpers	22.95	367.20	35.45	567.20		
.5 Carpenter	29.15	116.60	45.60	182.40		
44 L.H., Daily Totals		$1194.20		$1847.60	$27.14	$41.99
Crew D-3	**Hr.**	**Daily**	**Hr.**	**Daily**	**Bare Costs**	**Incl. O&P**
3 Bricklayers	$29.60	$710.40	$45.75	$1098.00	$27.05	$41.82
2 Bricklayer Helpers	22.95	367.20	35.45	567.20		
.25 Carpenter	29.15	58.30	45.60	91.20		
42 L.H., Daily Totals		$1135.90		$1756.40	$27.05	$41.82
Crew D-4	**Hr.**	**Daily**	**Hr.**	**Daily**	**Bare Costs**	**Incl. O&P**
1 Bricklayer	$29.60	$236.80	$45.75	$366.00	$26.06	$40.05
2 Bricklayer Helpers	22.95	367.20	35.45	567.20		
1 Equip. Oper. (light)	28.75	230.00	43.55	348.40		
1 Grout Pump, 50 C.F./hr		83.70		92.05		
1 Hoses & Hopper		33.20		36.50		
1 Accessories		11.00		12.10	4.00	4.40
32 L.H., Daily Totals		$961.90		$1422.25	$30.06	$44.45
Crew D-5	**Hr.**	**Daily**	**Hr.**	**Daily**	**Bare Costs**	**Incl. O&P**
1 Bricklayer	$29.60	$236.80	$45.75	$366.00	$29.60	$45.75
8 L.H., Daily Totals		$236.80		$366.00	$29.60	$45.75
Crew D-6	**Hr.**	**Daily**	**Hr.**	**Daily**	**Bare Costs**	**Incl. O&P**
3 Bricklayers	$29.60	$710.40	$45.75	$1098.00	$26.39	$40.80
3 Bricklayer Helpers	22.95	550.80	35.45	850.80		
.25 Carpenter	29.15	58.30	45.60	91.20		
50 L.H., Daily Totals		$1319.50		$2040.00	$26.39	$40.80
Crew D-7	**Hr.**	**Daily**	**Hr.**	**Daily**	**Bare Costs**	**Incl. O&P**
1 Tile Layer	$28.15	$225.20	$41.60	$332.80	$25.43	$37.58
1 Tile Layer Helper	22.70	181.60	33.55	268.40		
16 L.H., Daily Totals		$406.80		$601.20	$25.43	$37.58
Crew D-8	**Hr.**	**Daily**	**Hr.**	**Daily**	**Bare Costs**	**Incl. O&P**
3 Bricklayers	$29.60	$710.40	$45.75	$1098.00	$26.94	$41.63
2 Bricklayer Helpers	22.95	367.20	35.45	567.20		
40 L.H., Daily Totals		$1077.60		$1665.20	$26.94	$41.63
Crew D-9	**Hr.**	**Daily**	**Hr.**	**Daily**	**Bare Costs**	**Incl. O&P**
3 Bricklayers	$29.60	$710.40	$45.75	$1098.00	$26.28	$40.60
3 Bricklayer Helpers	22.95	550.80	35.45	850.80		
48 L.H., Daily Totals		$1261.20		$1948.80	$26.28	$40.60
Crew D-10	**Hr.**	**Daily**	**Hr.**	**Daily**	**Bare Costs**	**Incl. O&P**
1 Bricklayer Foreman	$31.60	$252.80	$48.80	$390.40	$27.65	$42.52
1 Bricklayer	29.60	236.80	45.75	366.00		
2 Bricklayer Helpers	22.95	367.20	35.45	567.20		
1 Equip. Oper. (crane)	31.15	249.20	47.15	377.20		
1 Truck Crane, 12.5 Ton		425.20		467.70	10.63	11.69
40 L.H., Daily Totals		$1531.20		$2168.50	$38.28	$54.21

Crew No.	Bare Costs		Incl. Subs O & P		Cost Per Labor-Hour	
Crew D-11	**Hr.**	**Daily**	**Hr.**	**Daily**	**Bare Costs**	**Incl. O&P**
1 Bricklayer Foreman	$31.60	$252.80	$48.80	$390.40	$28.05	$43.33
1 Bricklayer	29.60	236.80	45.75	366.00		
1 Bricklayer Helper	22.95	183.60	35.45	283.60		
24 L.H., Daily Totals		$673.20		$1040.00	$28.05	$43.33
Crew D-12	**Hr.**	**Daily**	**Hr.**	**Daily**	**Bare Costs**	**Incl. O&P**
1 Bricklayer Foreman	$31.60	$252.80	$48.80	$390.40	$26.78	$41.36
1 Bricklayer	29.60	236.80	45.75	366.00		
2 Bricklayer Helpers	22.95	367.20	35.45	567.20		
32 L.H., Daily Totals		$856.80		$1323.60	$26.78	$41.36
Crew D-13	**Hr.**	**Daily**	**Hr.**	**Daily**	**Bare Costs**	**Incl. O&P**
1 Bricklayer Foreman	$31.60	$252.80	$48.80	$390.40	$27.90	$43.03
1 Bricklayer	29.60	236.80	45.75	366.00		
2 Bricklayer Helpers	22.95	367.20	35.45	567.20		
1 Carpenter	29.15	233.20	45.60	364.80		
1 Equip. Oper. (crane)	31.15	249.20	47.15	377.20		
1 Truck Crane, 12.5 Ton		425.20		467.70	8.86	9.74
48 L.H., Daily Totals		$1764.40		$2533.30	$36.76	$52.77
Crew E-1	**Hr.**	**Daily**	**Hr.**	**Daily**	**Bare Costs**	**Incl. O&P**
1 Welder Foreman	$34.85	$278.80	$63.20	$505.60	$32.15	$55.45
1 Welder	32.85	262.80	59.60	476.80		
1 Equip. Oper. (light)	28.75	230.00	43.55	348.40		
1 Gas Welding Machine		84.00		92.40	3.50	3.85
24 L.H., Daily Totals		$855.60		$1423.20	$35.65	$59.30
Crew E-2	**Hr.**	**Daily**	**Hr.**	**Daily**	**Bare Costs**	**Incl. O&P**
1 Struc. Steel Foreman	$34.85	$278.80	$63.20	$505.60	$31.88	$55.39
4 Struc. Steel Workers	32.85	1051.20	59.60	1907.20		
1 Equip. Oper. (crane)	31.15	249.20	47.15	377.20		
1 Equip. Oper. Oiler	25.75	206.00	39.00	312.00		
1 Crane, 90 Ton		1290.00		1419.00	23.04	25.34
56 L.H., Daily Totals		$3075.20		$4521.00	$54.92	$80.73
Crew E-3	**Hr.**	**Daily**	**Hr.**	**Daily**	**Bare Costs**	**Incl. O&P**
1 Struc. Steel Foreman	$34.85	$278.80	$63.20	$505.60	$33.52	$60.80
1 Struc. Steel Worker	32.85	262.80	59.60	476.80		
1 Welder	32.85	262.80	59.60	476.80		
1 Gas Welding Machine		84.00		92.40	3.50	3.85
24 L.H., Daily Totals		$888.40		$1551.60	$37.02	$64.65
Crew E-4	**Hr.**	**Daily**	**Hr.**	**Daily**	**Bare Costs**	**Incl. O&P**
1 Struc. Steel Foreman	$34.85	$278.80	$63.20	$505.60	$33.35	$60.50
3 Struc. Steel Workers	32.85	788.40	59.60	1430.40		
1 Gas Welding Machine		84.00		92.40	2.63	2.89
32 L.H., Daily Totals		$1151.20		$2028.40	$35.98	$63.39
Crew E-5	**Hr.**	**Daily**	**Hr.**	**Daily**	**Bare Costs**	**Incl. O&P**
2 Struc. Steel Foremen	$34.85	$557.60	$63.20	$1011.20	$32.37	$57.02
5 Struc. Steel Workers	32.85	1314.00	59.60	2384.00		
1 Equip. Oper. (crane)	31.15	249.20	47.15	377.20		
1 Welder	32.85	262.80	59.60	476.80		
1 Equip. Oper. Oiler	25.75	206.00	39.00	312.00		
1 Crane, 90 Ton		1290.00		1419.00		
1 Gas Welding Machine		84.00		92.40	17.18	18.89
80 L.H., Daily Totals		$3963.60		$6072.60	$49.55	$75.91

Crew E-6	Bare Costs Hr.	Bare Costs Daily	Incl. Subs O & P Hr.	Incl. Subs O & P Daily	Cost Per Labor-Hour Bare Costs	Cost Per Labor-Hour Incl. O&P
3 Struc. Steel Foremen	$34.85	$836.40	$63.20	$1516.80	$32.42	$57.21
9 Struc. Steel Workers	32.85	2365.20	59.60	4291.20		
1 Equip. Oper. (crane)	31.15	249.20	47.15	377.20		
1 Welder	32.85	262.80	59.60	476.80		
1 Equip. Oper. Oiler	25.75	206.00	39.00	312.00		
1 Equip. Oper. (light)	28.75	230.00	43.55	348.40		
1 Crane, 90 Ton		1290.00		1419.00		
1 Gas Welding Machine		84.00		92.40		
1 Air Compr., 160 C.F.M.		98.40		108.25		
2 Impact Wrenches		50.40		55.45	11.90	13.09
128 L.H., Daily Totals		$5672.40		$8997.50	$44.32	$70.30

Crew E-7	Hr.	Daily	Hr.	Daily	Bare Costs	Incl. O&P
1 Struc. Steel Foreman	$34.85	$278.80	$63.20	$505.60	$32.37	$57.02
4 Struc. Steel Workers	32.85	1051.20	59.60	1907.20		
1 Equip. Oper. (crane)	31.15	249.20	47.15	377.20		
1 Equip. Oper. Oiler	25.75	206.00	39.00	312.00		
1 Welder Foreman	34.85	278.80	63.20	505.60		
2 Welders	32.85	525.60	59.60	953.60		
1 Crane, 90 Ton		1290.00		1419.00		
2 Gas Welding Machines		168.00		184.80	18.23	20.05
80 L.H., Daily Totals		$4047.60		$6165.00	$50.60	$77.07

Crew E-8	Hr.	Daily	Hr.	Daily	Bare Costs	Incl. O&P
1 Struc. Steel Foreman	$34.85	$278.80	$63.20	$505.60	$32.17	$56.38
4 Struc. Steel Workers	32.85	1051.20	59.60	1907.20		
1 Welder Foreman	34.85	278.80	63.20	505.60		
4 Welders	32.85	1051.20	59.60	1907.20		
1 Equip. Oper. (crane)	31.15	249.20	47.15	377.20		
1 Equip. Oper. Oiler	25.75	206.00	39.00	312.00		
1 Equip. Oper. (light)	28.75	230.00	43.55	348.40		
1 Crane, 90 Ton		1290.00		1419.00		
4 Gas Welding Machines		336.00		369.60	15.63	17.20
104 L.H., Daily Totals		$4971.20		$7651.80	$47.80	$73.58

Crew E-9	Hr.	Daily	Hr.	Daily	Bare Costs	Incl. O&P
2 Struc. Steel Foremen	$34.85	$557.60	$63.20	$1011.20	$32.42	$57.21
5 Struc. Steel Workers	32.85	1314.00	59.60	2384.00		
1 Welder Foreman	34.85	278.80	63.20	505.60		
5 Welders	32.85	1314.00	59.60	2384.00		
1 Equip. Oper. (crane)	31.15	249.20	47.15	377.20		
1 Equip. Oper. Oiler	25.75	206.00	39.00	312.00		
1 Equip. Oper. (light)	28.75	230.00	43.55	348.40		
1 Crane, 90 Ton		1290.00		1419.00		
5 Gas Welding Machines		420.00		462.00	13.36	14.70
128 L.H., Daily Totals		$5859.60		$9203.40	$45.78	$71.91

Crew E-10	Hr.	Daily	Hr.	Daily	Bare Costs	Incl. O&P
1 Welder Foreman	$34.85	$278.80	$63.20	$505.60	$33.85	$61.40
1 Welder	32.85	262.80	59.60	476.80		
4 Gas Welding Machines		336.00		369.60		
1 Truck, 3 Ton		184.20		202.60	32.51	35.76
16 L.H., Daily Totals		$1061.80		$1554.60	$66.36	$97.16

Crew E-11	Hr.	Daily	Hr.	Daily	Bare Costs	Incl. O&P
2 Painters, Struc. Steel	$27.15	$434.40	$50.90	$814.40	$26.48	$45.28
1 Building Laborer	22.85	182.80	35.75	286.00		
1 Equip. Oper. (light)	28.75	230.00	43.55	348.40		
1 Air Compressor 250 C.F.M.		125.20		137.70		
1 Sand Blaster		33.20		36.50		
1 Sand Blasting Accessories		11.00		12.10	5.29	5.82
32 L.H., Daily Totals		$1016.60		$1635.10	$31.77	$51.10

Crew E-12	Hr.	Daily	Hr.	Daily	Bare Costs	Incl. O&P
1 Welder Foreman	$34.85	$278.80	$63.20	$505.60	$31.80	$53.38
1 Equip. Oper. (light)	28.75	230.00	43.55	348.40		
1 Gas Welding Machine		84.00		92.40	5.25	5.78
16 L.H., Daily Totals		$592.80		$946.40	$37.05	$59.16

Crew E-13	Hr.	Daily	Hr.	Daily	Bare Costs	Incl. O&P
1 Welder Foreman	$34.85	$278.80	$63.20	$505.60	$32.82	$56.65
.5 Equip. Oper. (light)	28.75	115.00	43.55	174.20		
1 Gas Welding Machine		84.00		92.40	7.00	7.70
12 L.H., Daily Totals		$477.80		$772.20	$39.82	$64.35

Crew E-14	Hr.	Daily	Hr.	Daily	Bare Costs	Incl. O&P
1 Welder Foreman	$34.85	$278.80	$63.20	$505.60	$34.85	$63.20
1 Gas Welding Machine		84.00		92.40	10.50	11.55
8 L.H., Daily Totals		$362.80		$598.00	$45.35	$74.75

Crew E-16	Hr.	Daily	Hr.	Daily	Bare Costs	Incl. O&P
1 Welder Foreman	$34.85	$278.80	$63.20	$505.60	$33.85	$61.40
1 Welder	32.85	262.80	59.60	476.80		
1 Gas Welding Machine		84.00		92.40	5.25	5.78
16 L.H., Daily Totals		$625.60		$1074.80	$39.10	$67.18

Crew E-17	Hr.	Daily	Hr.	Daily	Bare Costs	Incl. O&P
1 Structural Steel Foreman	$34.85	$278.80	$63.20	$505.60	$33.85	$61.40
1 Structural Steel Worker	32.85	262.80	59.60	476.80		
1 Power Tool		2.90		3.20	.18	.20
16 L.H., Daily Totals		$544.50		$985.60	$34.03	$61.60

Crew E-18	Hr.	Daily	Hr.	Daily	Bare Costs	Incl. O&P
1 Structural Steel Foreman	$34.85	$278.80	$63.20	$505.60	$32.75	$57.59
3 Structural Steel Workers	32.85	788.40	59.60	1430.40		
1 Equipment Operator (med.)	30.35	242.80	45.95	367.60		
1 Crane, 20 Ton		554.80		610.30	13.87	15.26
40 L.H., Daily Totals		$1864.80		$2913.90	$46.62	$72.85

Crew E-19	Hr.	Daily	Hr.	Daily	Bare Costs	Incl. O&P
1 Structural Steel Worker	$32.85	$262.80	$59.60	$476.80	$32.15	$55.45
1 Structural Steel Foreman	34.85	278.80	63.20	505.60		
1 Equip. Oper. (light)	28.75	230.00	43.55	348.40		
1 Power Tool		2.90		3.20		
1 Crane, 20 Ton		554.80		610.30	23.24	25.56
24 L.H., Daily Totals		$1329.30		$1944.30	$55.39	$81.01

Crew E-20	Hr.	Daily	Hr.	Daily	Bare Costs	Incl. O&P
1 Structural Steel Foreman	$34.85	$278.80	$63.20	$505.60	$32.00	$55.92
5 Structural Steel Workers	32.85	1314.00	59.60	2384.00		
1 Equip. Oper. (crane)	31.15	249.20	47.15	377.20		
1 Oiler	25.75	206.00	39.00	312.00		
1 Power Tool		2.90		3.20		
1 Crane, 40 Ton		751.60		826.75	11.79	12.97
64 L.H., Daily Totals		$2802.50		$4408.75	$43.79	$68.89

Crew E-22	Hr.	Daily	Hr.	Daily	Bare Costs	Incl. O&P
1 Skilled Worker Foreman	$31.85	$254.80	$50.00	$400.00	$30.52	$47.90
2 Skilled Worker	29.85	477.60	46.85	749.60		
24 L.H., Daily Totals		$732.40		$1149.60	$30.52	$47.90

Crew No.	Bare Costs		Incl. Subs O & P		Cost Per Labor-Hour	
Crew E-24	**Hr.**	**Daily**	**Hr.**	**Daily**	**Bare Costs**	**Incl. O&P**
3 Structural Steel Worker	$32.85	$788.40	$59.60	$1430.40	$32.22	$56.19
1 Equipment Operator (medium)	30.35	242.80	45.95	367.60		
1-25 Ton Crane		629.60		692.55	19.68	21.64
32 L.H., Daily Totals		$1660.80		$2490.55	$51.90	$77.83

Crew F-3	Hr.	Daily	Hr.	Daily	Bare Costs	Incl. O&P
4 Carpenters	$29.15	$932.80	$45.60	$1459.20	$29.55	$45.91
1 Equip. Oper. (crane)	31.15	249.20	47.15	377.20		
1 Hyd. Crane, 12 Ton		514.80		566.30	12.87	14.16
40 L.H., Daily Totals		$1696.80		$2402.70	$42.42	$60.07

Crew F-4	Hr.	Daily	Hr.	Daily	Bare Costs	Incl. O&P
4 Carpenters	$29.15	$932.80	$45.60	$1459.20	$28.92	$44.76
1 Equip. Oper. (crane)	31.15	249.20	47.15	377.20		
1 Equip. Oper. Oiler	25.75	206.00	39.00	312.00		
1 Hyd. Crane, 55 Ton		901.60		991.75	18.78	20.66
48 L.H., Daily Totals		$2289.60		$3140.15	$47.70	$65.42

Crew F-5	Hr.	Daily	Hr.	Daily	Bare Costs	Incl. O&P
1 Carpenter Foreman	$31.15	$249.20	$48.75	$390.00	$29.65	$46.39
3 Carpenters	29.15	699.60	45.60	1094.40		
32 L.H., Daily Totals		$948.80		$1484.40	$29.65	$46.39

Crew F-6	Hr.	Daily	Hr.	Daily	Bare Costs	Incl. O&P
2 Carpenters	$29.15	$466.40	$45.60	$729.60	$27.03	$41.97
2 Building Laborers	22.85	365.60	35.75	572.00		
1 Equip. Oper. (crane)	31.15	249.20	47.15	377.20		
1 Hyd. Crane, 12 Ton		514.80		566.30	12.87	14.16
40 L.H., Daily Totals		$1596.00		$2245.10	$39.90	$56.13

Crew F-7	Hr.	Daily	Hr.	Daily	Bare Costs	Incl. O&P
2 Carpenters	$29.15	$466.40	$45.60	$729.60	$26.00	$40.67
2 Building Laborers	22.85	365.60	35.75	572.00		
32 L.H., Daily Totals		$832.00		$1301.60	$26.00	$40.67

Crew G-1	Hr.	Daily	Hr.	Daily	Bare Costs	Incl. O&P
1 Roofer Foreman	$27.70	$221.60	$47.60	$380.80	$24.10	$41.40
4 Roofers, Composition	25.70	822.40	44.15	1412.80		
2 Roofer Helpers	19.10	305.60	32.80	524.80		
1 Application Equipment		181.60		199.75		
1 Tar Kettle/Pot		60.95		67.05		
1 Crew Truck		90.40		99.45	5.95	6.54
56 L.H., Daily Totals		$1682.55		$2684.65	$30.05	$47.94

Crew G-2	Hr.	Daily	Hr.	Daily	Bare Costs	Incl. O&P
1 Plasterer	$27.30	$218.40	$41.80	$334.40	$24.47	$37.72
1 Plasterer Helper	23.25	186.00	35.60	284.80		
1 Building Laborer	22.85	182.80	35.75	286.00		
1 Grouting Equipment		233.20		256.50	9.72	10.69
24 L.H., Daily Totals		$820.40		$1161.70	$34.19	$48.41

Crew G-3	Hr.	Daily	Hr.	Daily	Bare Costs	Incl. O&P
2 Sheet Metal Workers	$33.55	$536.80	$51.80	$828.80	$28.20	$43.78
2 Building Laborers	22.85	365.60	35.75	572.00		
32 L.H., Daily Totals		$902.40		$1400.80	$28.20	$43.78

Crew No.	Bare Costs		Incl. Subs O & P		Cost Per Labor-Hour	
Crew G-4	**Hr.**	**Daily**	**Hr.**	**Daily**	**Bare Costs**	**Incl. O&P**
1 Labor Foreman (outside)	$24.85	$198.80	$38.90	$311.20	$23.52	$36.80
2 Building Laborers	22.85	365.60	35.75	572.00		
1 Light Truck, 1.5 Ton		158.00		173.80		
1 Air Compr., 160 C.F.M.		98.40		108.25	10.68	11.75
24 L.H., Daily Totals		$820.80		$1165.25	$34.20	$48.55

Crew G-5	Hr.	Daily	Hr.	Daily	Bare Costs	Incl. O&P
1 Roofer Foreman	$27.70	$221.60	$47.60	$380.80	$23.46	$40.30
2 Roofers, Composition	25.70	411.20	44.15	706.40		
2 Roofer Helpers	19.10	305.60	32.80	524.80		
1 Application Equipment		181.60		199.75	4.54	4.99
40 L.H., Daily Totals		$1120.00		$1811.75	$28.00	$45.29

Crew G-6A	Hr.	Daily	Hr.	Daily	Bare Costs	Incl. O&P
2 Roofers Composition	$25.70	$411.20	$44.15	$706.40	$25.70	$44.15
1 Small Compressor		25.60		28.15		
2 Pneumatic Nailers		40.30		44.35	4.12	4.53
16 L.H., Daily Totals		$477.10		$778.90	$29.82	$48.68

Crew G-7	Hr.	Daily	Hr.	Daily	Bare Costs	Incl. O&P
1 Carpenter	$29.15	$233.20	$45.60	$364.80	$29.15	$45.60
1 Small Compressor		25.60		28.15		
1 Pneumatic Nailer		20.15		22.15	5.72	6.29
8 L.H., Daily Totals		$278.95		$415.10	$34.87	$51.89

Crew H-1	Hr.	Daily	Hr.	Daily	Bare Costs	Incl. O&P
2 Glaziers	$28.55	$456.80	$43.30	$692.80	$30.70	$51.45
2 Struc. Steel Workers	32.85	525.60	59.60	953.60		
32 L.H., Daily Totals		$982.40		$1646.40	$30.70	$51.45

Crew H-2	Hr.	Daily	Hr.	Daily	Bare Costs	Incl. O&P
2 Glaziers	$28.55	$456.80	$43.30	$692.80	$26.65	$40.78
1 Building Laborer	22.85	182.80	35.75	286.00		
24 L.H., Daily Totals		$639.60		$978.80	$26.65	$40.78

Crew H-3	Hr.	Daily	Hr.	Daily	Bare Costs	Incl. O&P
1 Glazier	$28.55	$228.40	$43.30	$346.40	$25.35	$39.00
1 Helper	22.15	177.20	34.70	277.60		
16 L.H., Daily Totals		$405.60		$624.00	$25.35	$39.00

Crew J-1	Hr.	Daily	Hr.	Daily	Bare Costs	Incl. O&P
3 Plasterers	$27.30	$655.20	$41.80	$1003.20	$25.68	$39.32
2 Plasterer Helpers	23.25	372.00	35.60	569.60		
1 Mixing Machine, 6 C.F.		59.00		64.90	1.48	1.62
40 L.H., Daily Totals		$1086.20		$1637.70	$27.16	$40.94

Crew J-2	Hr.	Daily	Hr.	Daily	Bare Costs	Incl. O&P
3 Plasterers	$27.30	$655.20	$41.80	$1003.20	$26.10	$39.78
2 Plasterer Helpers	23.25	372.00	35.60	569.60		
1 Lather	28.20	225.60	42.10	336.80		
1 Mixing Machine, 6 C.F.		59.00		64.90	1.23	1.35
48 L.H., Daily Totals		$1311.80		$1974.50	$27.33	$41.13

Crew J-3	Hr.	Daily	Hr.	Daily	Bare Costs	Incl. O&P
1 Terrazzo Worker	$28.30	$226.40	$41.85	$334.80	$25.70	$38.00
1 Terrazzo Helper	23.10	184.80	34.15	273.20		
1 Terrazzo Grinder, Electric		60.50		66.55		
1 Terrazzo Mixer		80.20		88.20	8.79	9.67
16 L.H., Daily Totals		$551.90		$762.75	$34.49	$47.67

Crew No.	Bare Costs		Incl. Subs O & P		Cost Per Labor-Hour	
Crew J-4	**Hr.**	**Daily**	**Hr.**	**Daily**	**Bare Costs**	**Incl. O&P**
1 Tile Layer	$28.15	$225.20	$41.60	$332.80	$25.43	$37.58
1 Tile Layer Helper	22.70	181.60	33.55	268.40		
16 L.H., Daily Totals		$406.80		$601.20	$25.43	$37.58
Crew K-1	**Hr.**	**Daily**	**Hr.**	**Daily**	**Bare Costs**	**Incl. O&P**
1 Carpenter	$29.15	$233.20	$45.60	$364.80	$26.27	$40.67
1 Truck Driver (light)	23.40	187.20	35.75	286.00		
1 Truck w/Power Equip.		184.20		202.60	11.51	12.66
16 L.H., Daily Totals		$604.60		$853.40	$37.78	$53.33
Crew K-2	**Hr.**	**Daily**	**Hr.**	**Daily**	**Bare Costs**	**Incl. O&P**
1 Struc. Steel Foreman	$34.85	$278.80	$63.20	$505.60	$30.37	$52.85
1 Struc. Steel Worker	32.85	262.80	59.60	476.80		
1 Truck Driver (light)	23.40	187.20	35.75	286.00		
1 Truck w/Power Equip.		184.20		202.60	7.68	8.44
24 L.H., Daily Totals		$913.00		$1471.00	$38.05	$61.29
Crew L-1	**Hr.**	**Daily**	**Hr.**	**Daily**	**Bare Costs**	**Incl. O&P**
1 Electrician	$34.30	$274.40	$51.20	$409.60	$34.38	$51.60
1 Plumber	34.45	275.60	52.00	416.00		
16 L.H., Daily Totals		$550.00		$825.60	$34.38	$51.60
Crew L-2	**Hr.**	**Daily**	**Hr.**	**Daily**	**Bare Costs**	**Incl. O&P**
1 Carpenter	$29.15	$233.20	$45.60	$364.80	$25.65	$40.15
1 Carpenter Helper	22.15	177.20	34.70	277.60		
16 L.H., Daily Totals		$410.40		$642.40	$25.65	$40.15
Crew L-3	**Hr.**	**Daily**	**Hr.**	**Daily**	**Bare Costs**	**Incl. O&P**
1 Carpenter	$29.15	$233.20	$45.60	$364.80	$31.54	$48.55
.5 Electrician	34.30	137.20	51.20	204.80		
.5 Sheet Metal Worker	33.55	134.20	51.80	207.20		
16 L.H., Daily Totals		$504.60		$776.80	$31.54	$48.55
Crew L-3A	**Hr.**	**Daily**	**Hr.**	**Daily**	**Bare Costs**	**Incl. O&P**
1 Carpenter Foreman (outside)	$31.15	$249.20	$48.75	$390.00	$31.95	$49.77
.5 Sheet Metal Worker	33.55	134.20	51.80	207.20		
12 L.H., Daily Totals		$383.40		$597.20	$31.95	$49.77
Crew L-4	**Hr.**	**Daily**	**Hr.**	**Daily**	**Bare Costs**	**Incl. O&P**
2 Skilled Workers	$29.85	$477.60	$46.85	$749.60	$27.28	$42.80
1 Helper	22.15	177.20	34.70	277.60		
24 L.H., Daily Totals		$654.80		$1027.20	$27.28	$42.80
Crew L-5	**Hr.**	**Daily**	**Hr.**	**Daily**	**Bare Costs**	**Incl. O&P**
1 Struc. Steel Foreman	$34.85	$278.80	$63.20	$505.60	$32.89	$58.34
5 Struc. Steel Workers	32.85	1314.00	59.60	2384.00		
1 Equip. Oper. (crane)	31.15	249.20	47.15	377.20		
1 Hyd. Crane, 25 Ton		629.60		692.55	11.24	12.37
56 L.H., Daily Totals		$2471.60		$3959.35	$44.13	$70.71
Crew L-5A	**Hr.**	**Daily**	**Hr.**	**Daily**	**Bare Costs**	**Incl. O&P**
1 Structural Steel Foreman	$34.85	$278.80	$63.20	$505.60	$32.92	$57.39
2 Structural Steel Workers	32.85	525.60	59.60	953.60		
1 Equip. Oper. (crane)	31.15	249.20	47.15	377.20		
1 Crane,SP, 25 Ton		602.00		662.20	18.81	20.69
32 L.H., Daily Totals		$1655.60		$2498.60	$51.73	$78.08

Crew No.	Bare Costs		Incl. Subs O & P		Cost Per Labor-Hour	
Crew L-6	**Hr.**	**Daily**	**Hr.**	**Daily**	**Bare Costs**	**Incl. O&P**
1 Plumber	$34.45	$275.60	$52.00	$416.00	$34.40	$51.73
.5 Electrician	34.30	137.20	51.20	204.80		
12 L.H., Daily Totals		$412.80		$620.80	$34.40	$51.73
Crew L-7	**Hr.**	**Daily**	**Hr.**	**Daily**	**Bare Costs**	**Incl. O&P**
2 Carpenters	$29.15	$466.40	$45.60	$729.60	$28.09	$43.59
1 Building Laborer	22.85	182.80	35.75	286.00		
.5 Electrician	34.30	137.20	51.20	204.80		
28 L.H., Daily Totals		$786.40		$1220.40	$28.09	$43.59
Crew L-8	**Hr.**	**Daily**	**Hr.**	**Daily**	**Bare Costs**	**Incl. O&P**
2 Carpenters	$29.15	$466.40	$45.60	$729.60	$30.21	$46.88
.5 Plumber	34.45	137.80	52.00	208.00		
20 L.H., Daily Totals		$604.20		$937.60	$30.21	$46.88
Crew L-9	**Hr.**	**Daily**	**Hr.**	**Daily**	**Bare Costs**	**Incl. O&P**
1 Labor Foreman (inside)	$23.35	$186.80	$36.55	$292.40	$26.46	$42.94
2 Building Laborers	22.85	365.60	35.75	572.00		
1 Struc. Steel Worker	32.85	262.80	59.60	476.80		
.5 Electrician	34.30	137.20	51.20	204.80		
36 L.H., Daily Totals		$952.40		$1546.00	$26.46	$42.94
Crew L-10	**Hr.**	**Daily**	**Hr.**	**Daily**	**Bare Costs**	**Incl. O&P**
1 Structural Steel Foreman	$34.85	$278.80	$63.20	$505.60	$32.95	$56.65
1 Structural Steel Worker	32.85	262.80	59.60	476.80		
1 Equip. Oper. (crane)	31.15	249.20	47.15	377.20		
1 Hyd. Crane, 12 Ton		514.80		566.30	21.45	23.60
24 L.H., Daily Totals		$1305.60		$1925.90	$54.40	$80.25
Crew M-1	**Hr.**	**Daily**	**Hr.**	**Daily**	**Bare Costs**	**Incl. O&P**
3 Elevator Constructors	$35.65	$855.60	$53.80	$1291.20	$33.86	$51.10
1 Elevator Apprentice	28.50	228.00	43.00	344.00		
5 Hand Tools		95.00		104.50	2.97	3.27
32 L.H., Daily Totals		$1178.60		$1739.70	$36.83	$54.37
Crew M-3	**Hr.**	**Daily**	**Hr.**	**Daily**	**Bare Costs**	**Incl. O&P**
1 Electrician Foreman (out)	$36.30	$290.40	$54.20	$433.60	$30.80	$46.64
1 Common Laborer	22.85	182.80	35.75	286.00		
.25 Equipment Operator, Medium	30.35	60.70	45.95	91.90		
1 Elevator Constructor	35.65	285.20	53.80	430.40		
1 Elevator Apprentice	28.50	228.00	43.00	344.00		
.25 Crane, SP, 4 x 4, 20 ton		134.35		147.80	3.95	4.35
34 L.H., Daily Totals		$1181.45		$1733.70	$34.75	$50.99
Crew M-4	**Hr.**	**Daily**	**Hr.**	**Daily**	**Bare Costs**	**Incl. O&P**
1 Electrician Foreman (out)	$36.30	$290.40	$54.20	$433.60	$30.56	$46.29
1 Common Laborer	22.85	182.80	35.75	286.00		
.25 Equipment Operator, Crane	31.15	62.30	47.15	94.30		
.25 Equipment Operator, Oiler	25.75	51.50	39.00	78.00		
1 Elevator Constructor	35.65	285.20	53.80	430.40		
1 Elevator Apprentice	28.50	228.00	43.00	344.00		
.25 Crane, Hyd, SP, 4WD, 40 Ton		218.00		239.80	6.06	6.66
36 L.H., Daily Totals		$1318.20		$1906.10	$36.62	$52.95
Crew Q-1	**Hr.**	**Daily**	**Hr.**	**Daily**	**Bare Costs**	**Incl. O&P**
1 Plumber	$34.45	$275.60	$52.00	$416.00	$31.00	$46.80
1 Plumber Apprentice	27.55	220.40	41.60	332.80		
16 L.H., Daily Totals		$496.00		$748.80	$31.00	$46.80

Crew No.	Bare Costs		Incl. Subs O & P		Cost Per Labor-Hour	
Crew Q-1C	**Hr.**	**Daily**	**Hr.**	**Daily**	**Bare Costs**	**Incl. O&P**
1 Plumber	$34.45	$275.60	$52.00	$416.00	$30.78	$46.52
1 Plumber Apprentice	27.55	220.40	41.60	332.80		
1 Equip. Oper. (medium)	30.35	242.80	45.95	367.60		
1 Trencher, Chain		711.40		782.55	29.64	32.61
24 L.H., Daily Totals		$1450.20		$1898.95	$60.42	$79.13
Crew Q-2	**Hr.**	**Daily**	**Hr.**	**Daily**	**Bare Costs**	**Incl. O&P**
2 Plumbers	$34.45	$551.20	$52.00	$832.00	$32.15	$48.53
1 Plumber Apprentice	27.55	220.40	41.60	332.80		
24 L.H., Daily Totals		$771.60		$1164.80	$32.15	$48.53
Crew Q-3	**Hr.**	**Daily**	**Hr.**	**Daily**	**Bare Costs**	**Incl. O&P**
1 Plumber Foreman (inside)	$34.95	$279.60	$52.75	$422.00	$32.85	$49.59
2 Plumbers	34.45	551.20	52.00	832.00		
1 Plumber Apprentice	27.55	220.40	41.60	332.80		
32 L.H., Daily Totals		$1051.20		$1586.80	$32.85	$49.59
Crew Q-4	**Hr.**	**Daily**	**Hr.**	**Daily**	**Bare Costs**	**Incl. O&P**
1 Plumber Foreman (inside)	$34.95	$279.60	$52.75	$422.00	$32.85	$49.59
1 Plumber	34.45	275.60	52.00	416.00		
1 Welder (plumber)	34.45	275.60	52.00	416.00		
1 Plumber Apprentice	27.55	220.40	41.60	332.80		
1 Electric Welding Mach.		100.65		110.70	3.15	3.46
32 L.H., Daily Totals		$1151.85		$1697.50	$36.00	$53.05
Crew Q-5	**Hr.**	**Daily**	**Hr.**	**Daily**	**Bare Costs**	**Incl. O&P**
1 Steamfitter	$34.85	$278.80	$52.60	$420.80	$31.38	$47.38
1 Steamfitter Apprentice	27.90	223.20	42.15	337.20		
16 L.H., Daily Totals		$502.00		$758.00	$31.38	$47.38
Crew Q-6	**Hr.**	**Daily**	**Hr.**	**Daily**	**Bare Costs**	**Incl. O&P**
2 Steamfitters	$34.85	$557.60	$52.60	$841.60	$32.53	$49.12
1 Steamfitter Apprentice	27.90	223.20	42.15	337.20		
24 L.H., Daily Totals		$780.80		$1178.80	$32.53	$49.12
Crew Q-7	**Hr.**	**Daily**	**Hr.**	**Daily**	**Bare Costs**	**Incl. O&P**
1 Steamfitter Foreman (inside)	$35.35	$282.80	$53.40	$427.20	$33.24	$50.19
2 Steamfitters	34.85	557.60	52.60	841.60		
1 Steamfitter Apprentice	27.90	223.20	42.15	337.20		
32 L.H., Daily Totals		$1063.60		$1606.00	$33.24	$50.19
Crew Q-8	**Hr.**	**Daily**	**Hr.**	**Daily**	**Bare Costs**	**Incl. O&P**
1 Steamfitter Foreman (inside)	$35.35	$282.80	$53.40	$427.20	$33.24	$50.19
1 Steamfitter	34.85	278.80	52.60	420.80		
1 Welder (steamfitter)	34.85	278.80	52.60	420.80		
1 Steamfitter Apprentice	27.90	223.20	42.15	337.20		
1 Electric Welding Mach.		100.65		110.70	3.15	3.46
32 L.H., Daily Totals		$1164.25		$1716.70	$36.39	$53.65
Crew Q-9	**Hr.**	**Daily**	**Hr.**	**Daily**	**Bare Costs**	**Incl. O&P**
1 Sheet Metal Worker	$33.55	$268.40	$51.80	$414.40	$30.20	$46.63
1 Sheet Metal Apprentice	26.85	214.80	41.45	331.60		
16 L.H., Daily Totals		$483.20		$746.00	$30.20	$46.63
Crew Q-10	**Hr.**	**Daily**	**Hr.**	**Daily**	**Bare Costs**	**Incl. O&P**
2 Sheet Metal Workers	$33.55	$536.80	$51.80	$828.80	$31.32	$48.35
1 Sheet Metal Apprentice	26.85	214.80	41.45	331.60		
24 L.H., Daily Totals		$751.60		$1160.40	$31.32	$48.35

Crew No.	Bare Costs		Incl. Subs O & P		Cost Per Labor-Hour	
Crew Q-11	**Hr.**	**Daily**	**Hr.**	**Daily**	**Bare Costs**	**Incl. O&P**
1 Sheet Metal Foreman (inside)	$34.05	$272.40	$52.55	$420.40	$32.00	$49.40
2 Sheet Metal Workers	33.55	536.80	51.80	828.80		
1 Sheet Metal Apprentice	26.85	214.80	41.45	331.60		
32 L.H., Daily Totals		$1024.00		$1580.80	$32.00	$49.40
Crew Q-12	**Hr.**	**Daily**	**Hr.**	**Daily**	**Bare Costs**	**Incl. O&P**
1 Sprinkler Installer	$34.45	$275.60	$52.10	$416.80	$31.00	$46.90
1 Sprinkler Apprentice	27.55	220.40	41.70	333.60		
16 L.H., Daily Totals		$496.00		$750.40	$31.00	$46.90
Crew Q-13	**Hr.**	**Daily**	**Hr.**	**Daily**	**Bare Costs**	**Incl. O&P**
1 Sprinkler Foreman (inside)	$34.95	$279.60	$52.90	$423.20	$32.85	$49.70
2 Sprinkler Installers	34.45	551.20	52.10	833.60		
1 Sprinkler Apprentice	27.55	220.40	41.70	333.60		
32 L.H., Daily Totals		$1051.20		$1590.40	$32.85	$49.70
Crew Q-14	**Hr.**	**Daily**	**Hr.**	**Daily**	**Bare Costs**	**Incl. O&P**
1 Asbestos Worker	$32.15	$257.20	$51.50	$412.00	$28.93	$46.33
1 Asbestos Apprentice	25.70	205.60	41.15	329.20		
16 L.H., Daily Totals		$462.80		$741.20	$28.93	$46.33
Crew Q-15	**Hr.**	**Daily**	**Hr.**	**Daily**	**Bare Costs**	**Incl. O&P**
1 Plumber	$34.45	$275.60	$52.00	$416.00	$31.00	$46.80
1 Plumber Apprentice	27.55	220.40	41.60	332.80		
1 Electric Welding Mach.		100.65		110.70	6.29	6.92
16 L.H., Daily Totals		$596.65		$859.50	$37.29	$53.72
Crew Q-16	**Hr.**	**Daily**	**Hr.**	**Daily**	**Bare Costs**	**Incl. O&P**
2 Plumbers	$34.45	$551.20	$52.00	$832.00	$32.15	$48.53
1 Plumber Apprentice	27.55	220.40	41.60	332.80		
1 Electric Welding Mach.		100.65		110.70	4.19	4.61
24 L.H., Daily Totals		$872.25		$1275.50	$36.34	$53.14
Crew Q-17	**Hr.**	**Daily**	**Hr.**	**Daily**	**Bare Costs**	**Incl. O&P**
1 Steamfitter	$34.85	$278.80	$52.60	$420.80	$31.38	$47.38
1 Steamfitter Apprentice	27.90	223.20	42.15	337.20		
1 Electric Welding Mach.		100.65		110.70	6.29	6.92
16 L.H., Daily Totals		$602.65		$868.70	$37.67	$54.30
Crew Q-17A	**Hr.**	**Daily**	**Hr.**	**Daily**	**Bare Costs**	**Incl. O&P**
1 Steamfitter	$34.85	$278.80	$52.60	$420.80	$31.30	$47.30
1 Steamfitter Apprentice	27.90	223.20	42.15	337.20		
1 Equip. Oper. (crane)	31.15	249.20	47.15	377.20		
1 Truck Crane, 12 Ton		514.80		566.30		
1 Electric Welding Mach.		100.65		110.70	25.64	28.21
24 L.H., Daily Totals		$1366.65		$1812.20	$56.94	$75.51
Crew Q-18	**Hr.**	**Daily**	**Hr.**	**Daily**	**Bare Costs**	**Incl. O&P**
2 Steamfitters	$34.85	$557.60	$52.60	$841.60	$32.53	$49.12
1 Steamfitter Apprentice	27.90	223.20	42.15	337.20		
1 Electric Welding Mach.		100.65		110.70	4.19	4.61
24 L.H., Daily Totals		$881.45		$1289.50	$36.72	$53.73
Crew Q-19	**Hr.**	**Daily**	**Hr.**	**Daily**	**Bare Costs**	**Incl. O&P**
1 Steamfitter	$34.85	$278.80	$52.60	$420.80	$32.35	$48.65
1 Steamfitter Apprentice	27.90	223.20	42.15	337.20		
1 Electrician	34.30	274.40	51.20	409.60		
24 L.H., Daily Totals		$776.40		$1167.60	$32.35	$48.65

Crew No.	Bare Costs		Incl. Subs O & P		Cost Per Labor-Hour	
Crew Q-20	**Hr.**	**Daily**	**Hr.**	**Daily**	**Bare Costs**	**Incl. O&P**
1 Sheet Metal Worker	$33.55	$268.40	$51.80	$414.40	$31.02	$47.54
1 Sheet Metal Apprentice	26.85	214.80	41.45	331.60		
.5 Electrician	34.30	137.20	51.20	204.80		
20 L.H., Daily Totals		$620.40		$950.80	$31.02	$47.54
Crew Q-21	**Hr.**	**Daily**	**Hr.**	**Daily**	**Bare Costs**	**Incl. O&P**
2 Steamfitters	$34.85	$557.60	$52.60	$841.60	$32.97	$49.64
1 Steamfitter Apprentice	27.90	223.20	42.15	337.20		
1 Electrician	34.30	274.40	51.20	409.60		
32 L.H., Daily Totals		$1055.20		$1588.40	$32.97	$49.64
Crew Q-22	**Hr.**	**Daily**	**Hr.**	**Daily**	**Bare Costs**	**Incl. O&P**
1 Plumber	$34.45	$275.60	$52.00	$416.00	$31.00	$46.80
1 Plumber Apprentice	27.55	220.40	41.60	332.80		
1 Truck Crane, 12 Ton		514.80		566.30	32.18	35.39
16 L.H., Daily Totals		$1010.80		$1315.10	$63.18	$82.19
Crew Q-22A	**Hr.**	**Daily**	**Hr.**	**Daily**	**Bare Costs**	**Incl. O&P**
1 Plumber	$34.45	$275.60	$52.00	$416.00	$29.00	$44.13
1 Plumber Apprentice	27.55	220.40	41.60	332.80		
1 Laborer	22.85	182.80	35.75	286.00		
1 Equip. Oper. (crane)	31.15	249.20	47.15	377.20		
1 Truck Crane, 12 Ton		514.80		566.30	16.09	17.70
32 L.H., Daily Totals		$1442.80		$1978.30	$45.09	$61.83
Crew Q-23	**Hr.**	**Daily**	**Hr.**	**Daily**	**Bare Costs**	**Incl. O&P**
1 Plumber Foreman	$36.45	$291.60	$55.05	$440.40	$33.75	$51.00
1 Plumber	34.45	275.60	52.00	416.00		
1 Equip. Oper. (medium)	30.35	242.80	45.95	367.60		
1 Power Tools		2.90		3.20		
1 Crane, 20 Ton		554.80		610.30	23.24	25.56
24 L.H., Daily Totals		$1367.70		$1837.50	$56.99	$76.56
Crew R-1	**Hr.**	**Daily**	**Hr.**	**Daily**	**Bare Costs**	**Incl. O&P**
1 Electrician Foreman	$34.80	$278.40	$51.95	$415.60	$30.33	$45.83
3 Electricians	34.30	823.20	51.20	1228.80		
2 Helpers	22.15	354.40	34.70	555.20		
48 L.H., Daily Totals		$1456.00		$2199.60	$30.33	$45.83
Crew R-1A	**Hr.**	**Daily**	**Hr.**	**Daily**	**Bare Costs**	**Incl. O&P**
1 Electrician	$34.30	$274.40	$51.20	$409.60	$28.23	$42.95
1 Helper	22.15	177.20	34.70	277.60		
16 L.H., Daily Totals		$451.60		$687.20	$28.23	$42.95
Crew R-2	**Hr.**	**Daily**	**Hr.**	**Daily**	**Bare Costs**	**Incl. O&P**
1 Electrician Foreman	$34.80	$278.40	$51.95	$415.60	$30.45	$46.01
3 Electricians	34.30	823.20	51.20	1228.80		
2 Helpers	22.15	354.40	34.70	555.20		
1 Equip. Oper. (crane)	31.15	249.20	47.15	377.20		
1 S.P. Crane, 5 Ton		284.00		312.40	5.07	5.58
56 L.H., Daily Totals		$1989.20		$2889.20	$35.52	$51.59
Crew R-3	**Hr.**	**Daily**	**Hr.**	**Daily**	**Bare Costs**	**Incl. O&P**
1 Electrician Foreman	$34.80	$278.40	$51.95	$415.60	$33.87	$50.69
1 Electrician	34.30	274.40	51.20	409.60		
.5 Equip. Oper. (crane)	31.15	124.60	47.15	188.60		
.5 S.P. Crane, 5 Ton		142.00		156.20	7.10	7.81
20 L.H., Daily Totals		$819.40		$1170.00	$40.97	$58.50

Crew No.	Bare Costs		Incl. Subs O & P		Cost Per Labor-Hour	
Crew R-4	**Hr.**	**Daily**	**Hr.**	**Daily**	**Bare Costs**	**Incl. O&P**
1 Struc. Steel Foreman	$34.85	$278.80	$63.20	$505.60	$33.54	$58.64
3 Struc. Steel Workers	32.85	788.40	59.60	1430.40		
1 Electrician	34.30	274.40	51.20	409.60		
1 Gas Welding Machine		84.00		92.40	2.10	2.31
40 L.H., Daily Totals		$1425.60		$2438.00	$35.64	$60.95
Crew R-5	**Hr.**	**Daily**	**Hr.**	**Daily**	**Bare Costs**	**Incl. O&P**
1 Electrician Foreman	$34.80	$278.40	$51.95	$415.60	$29.93	$45.27
4 Electrician Linemen	34.30	1097.60	51.20	1638.40		
2 Electrician Operators	34.30	548.80	51.20	819.20		
4 Electrician Groundmen	22.15	708.80	34.70	1110.40		
1 Crew Truck		90.40		99.45		
1 Tool Van		113.90		125.30		
1 Pickup Truck, 3/4 Ton		83.00		91.30		
.2 Crane, 55 Ton		180.32		198.35		
.2 Crane, 12 Ton		102.96		113.25		
.2 Auger, Truck Mtd.		374.00		411.40		
1 Tractor w/Winch		263.00		289.30	13.72	15.09
88 L.H., Daily Totals		$3841.18		$5311.95	$43.65	$60.36
Crew R-6	**Hr.**	**Daily**	**Hr.**	**Daily**	**Bare Costs**	**Incl. O&P**
1 Electrician Foreman	$34.80	$278.40	$51.95	$415.60	$29.93	$45.27
4 Electrician Linemen	34.30	1097.60	51.20	1638.40		
2 Electrician Operators	34.30	548.80	51.20	819.20		
4 Electrician Groundmen	22.15	708.80	34.70	1110.40		
1 Crew Truck		90.40		99.45		
1 Tool Van		113.90		125.30		
1 Pickup Truck, 3/4 Ton		83.00		91.30		
.2 Crane, 55 Ton		180.32		198.35		
.2 Crane, 12 Ton		102.96		113.25		
.2 Auger, Truck Mtd.		374.00		411.40		
1 Tractor w/Winch		263.00		289.30		
3 Cable Trailers		453.15		498.45		
.5 Tensioning Rig		149.75		164.75		
.5 Cable Pulling Rig		882.00		970.20	30.60	33.66
88 L.H., Daily Totals		$5326.08		$6945.35	$60.53	$78.93
Crew R-7	**Hr.**	**Daily**	**Hr.**	**Daily**	**Bare Costs**	**Incl. O&P**
1 Electrician Foreman	$34.80	$278.40	$51.95	$415.60	$24.26	$37.58
5 Electrician Groundmen	22.15	886.00	34.70	1388.00		
1 Crew Truck		90.40		99.45	1.88	2.07
48 L.H., Daily Totals		$1254.80		$1903.05	$26.14	$39.65
Crew R-8	**Hr.**	**Daily**	**Hr.**	**Daily**	**Bare Costs**	**Incl. O&P**
1 Electrician Foreman	$34.80	$278.40	$51.95	$415.60	$30.33	$45.83
3 Electrician Linemen	34.30	823.20	51.20	1228.80		
2 Electrician Groundmen	22.15	354.40	34.70	555.20		
1 Pickup Truck, 3/4 Ton		83.00		91.30		
1 Crew Truck		90.40		99.45	3.61	3.97
48 L.H., Daily Totals		$1629.40		$2390.35	$33.94	$49.80
Crew R-9	**Hr.**	**Daily**	**Hr.**	**Daily**	**Bare Costs**	**Incl. O&P**
1 Electrician Foreman	$34.80	$278.40	$51.95	$415.60	$28.29	$43.04
1 Electrician Lineman	34.30	274.40	51.20	409.60		
2 Electrician Operators	34.30	548.80	51.20	819.20		
4 Electrician Groundmen	22.15	708.80	34.70	1110.40		
1 Pickup Truck, 3/4 Ton		83.00		91.30		
1 Crew Truck		90.40		99.45	2.71	2.98
64 L.H., Daily Totals		$1983.80		$2945.55	$31.00	$46.02

Crew No.	Bare Costs		Incl. Subs O & P		Cost Per Labor-Hour	
Crew R-10	**Hr.**	**Daily**	**Hr.**	**Daily**	**Bare Costs**	**Incl. O&P**
1 Electrician Foreman	$34.80	$278.40	$51.95	$415.60	$32.36	$48.58
4 Electrician Linemen	34.30	1097.60	51.20	1638.40		
1 Electrician Groundman	22.15	177.20	34.70	277.60		
1 Crew Truck		90.40		99.45		
3 Tram Cars		485.70		534.25	12.00	13.20
48 L.H., Daily Totals		$2129.30		$2965.30	$44.36	$61.78
Crew R-11	**Hr.**	**Daily**	**Hr.**	**Daily**	**Bare Costs**	**Incl. O&P**
1 Electrician Foreman	$34.80	$278.40	$51.95	$415.60	$31.00	$46.74
4 Electricians	34.30	1097.60	51.20	1638.40		
1 Helper	22.15	177.20	34.70	277.60		
1 Common Laborer	22.85	182.80	35.75	286.00		
1 Crew Truck		90.40		99.45		
1 Crane, 12 Ton		514.80		566.30	10.81	11.89
56 L.H., Daily Totals		$2341.20		$3283.35	$41.81	$58.63
Crew R-12	**Hr.**	**Daily**	**Hr.**	**Daily**	**Bare Costs**	**Incl. O&P**
1 Carpenter Foreman	$29.65	$237.20	$46.40	$371.20	$27.35	$43.40
4 Carpenters	29.15	932.80	45.60	1459.20		
4 Common Laborers	22.85	731.20	35.75	1144.00		
1 Equip. Oper. (med.)	30.35	242.80	45.95	367.60		
1 Steel Worker	32.85	262.80	59.60	476.80		
1 Dozer, 200 H.P.		735.00		808.50		
1 Pickup Truck, 3/4 Ton		83.00		91.30	9.30	10.23
88 L.H., Daily Totals		$3224.80		$4718.60	$36.65	$53.63
Crew R-13	**Hr.**	**Daily**	**Hr.**	**Daily**	**Bare Costs**	**Incl. O&P**
1 Electrician Foreman	$34.80	$278.40	$51.95	$415.60	$32.43	$48.62
3 Electricians	34.30	823.20	51.20	1228.80		
1 Equipment Operator	31.15	249.20	47.15	377.20		
1 Equipment Oiler	25.75	206.00	39.00	312.00		
.25-1 Hyd. Crane, 33 Ton		201.20		221.30	4.19	4.61
48 L.H., Daily Totals		$1758.00		$2554.90	$36.62	$53.23
Crew R-15	**Hr.**	**Daily**	**Hr.**	**Daily**	**Bare Costs**	**Incl. O&P**
1 Electrician Foreman	$34.80	$278.40	$51.95	$415.60	$33.46	$50.05
4 Electricians	34.30	1097.60	51.20	1638.40		
1 Equipment Operator	28.75	230.00	43.55	348.40		
1 Aerial Lift Truck		211.00		232.10	4.40	4.84
48 L.H., Daily Totals		$1817.00		$2634.50	$37.86	$54.89
Crew R-18	**Hr.**	**Daily**	**Hr.**	**Daily**	**Bare Costs**	**Incl. O&P**
.25 Electrician Foreman	$34.80	$69.60	$51.95	$103.90	$26.86	$41.10
1 Electrician	34.30	274.40	51.20	409.60		
2 Helpers	22.15	354.40	34.70	555.20		
26 L.H., Daily Totals		$698.40		$1068.70	$26.86	$41.10
Crew R-19	**Hr.**	**Daily**	**Hr.**	**Daily**	**Bare Costs**	**Incl. O&P**
.5 Electrician Foreman	$34.80	$139.20	$51.95	$207.80	$34.40	$51.35
2 Electricians	34.30	548.80	51.20	819.20		
20 L.H., Daily Totals		$688.00		$1027.00	$34.40	$51.35
Crew R-21	**Hr.**	**Daily**	**Hr.**	**Daily**	**Bare Costs**	**Incl. O&P**
1 Electrician Foreman	$34.80	$278.40	$51.95	$415.60	$34.33	$51.25
3 Electricians	34.30	823.20	51.20	1228.80		
.1 Equip. Oper. (med.)	30.35	24.28	45.95	36.76		
.1 Hyd. Crane 25 Ton		60.20		66.20	1.84	2.02
32. L.H., Daily Totals		$1186.08		$1747.36	$36.17	$53.27

Crew No.	Bare Costs		Incl. Subs O & P		Cost Per Labor-Hour	
Crew R-22	**Hr.**	**Daily**	**Hr.**	**Daily**	**Bare Costs**	**Incl. O&P**
.66 Electrician Foreman	$34.80	$183.74	$51.95	$274.30	$29.16	$44.22
2 Helpers	22.15	354.40	34.70	555.20		
2 Electricians	34.30	548.80	51.20	819.20		
37.28 L.H., Daily Totals		$1086.94		$1648.70	$29.16	$44.22
Crew R-30	**Hr.**	**Daily**	**Hr.**	**Daily**	**Bare Costs**	**Incl. O&P**
.25 Electricians	$36.30	$72.60	$54.20	$108.40	$27.41	$41.92
1 Electricians	34.30	274.40	51.20	409.60		
2 Laborers, (Semi-Skilled)	22.85	365.60	35.75	572.00		
26 L.H., Daily Totals		$712.60		$1090.00	$27.41	$41.92

Historical Cost Indexes

The table below lists both the Means Historical Cost Index based on Jan. 1, 1993 = 100 as well as the computed value of an index based on Jan. 1, 2001 costs. Since the Jan. 1, 2001 figure is estimated, space is left to write in the actual index figures as they become available through either the quarterly "Means Construction Cost Indexes" or as printed in the "Engineering News-Record." To compute the actual index based on Jan. 1, 2001 = 100, divide the Historical Cost Index for a particular year by the actual Jan. 1, 2001 Construction Cost Index. Space has been left to advance the index figures as the year progresses.

Year	Historical Cost Index Jan. 1, 1993 = 100		Current Index Based on Jan. 1, 2001 = 100	
	Est.	Actual	Est.	Actual
Oct 2001				
July 2001				
April 2001				
Jan 2001	121.8		100.0	100.0
July 2000		120.9	99.3	
1999		117.6	96.6	
1998		115.1	94.5	
1997		112.8	92.6	
1996		110.2	90.5	
1995		107.6	88.3	
1994		104.4	85.7	
1993		101.7	83.5	
1992		99.4	81.6	
1991		96.8	79.5	
1990		94.3	77.4	
1989		92.1	75.6	
1988		89.9	73.8	
1987		87.7	72.0	

Year	Historical Cost Index Jan. 1, 1993 = 100	Current Index Based on Jan. 1, 2001 = 100	
	Actual	Est.	Actual
July 1986	84.2	69.2	
1985	82.6	67.8	
1984	82.0	67.3	
1983	80.2	65.8	
1982	76.1	62.5	
1981	70.0	57.5	
1980	62.9	51.6	
1979	57.8	47.5	
1978	53.5	43.9	
1977	49.5	40.6	
1976	46.9	38.5	
1975	44.8	36.8	
1974	41.4	34.0	
1973	37.7	31.0	
1972	34.8	28.6	
1971	32.1	26.4	
1970	28.7	23.6	
1969	26.9	22.1	

Year	Historical Cost Index Jan. 1, 1993 = 100	Current Index Based on Jan. 1, 2001 = 100	
	Actual	Est.	Actual
July 1968	24.9	20.4	
1967	23.5	19.3	
1966	22.7	18.6	
1965	21.7	17.8	
1964	21.2	17.4	
1963	20.7	17.0	
1962	20.2	16.6	
1961	19.8	16.3	
1960	19.7	16.2	
1959	19.3	15.8	
1958	18.8	15.4	
1957	18.4	15.1	
1956	17.6	14.4	
1955	16.6	13.6	
1954	16.0	13.1	
1953	15.8	13.0	
1952	15.4	12.6	
1951	15.0	12.3	

Adjustments to Costs

The Historical Cost Index can be used to convert National Average building costs at a particular time to the approximate building costs for some other time.

Example:

Estimate and compare construction costs for different years in the same city.

To estimate the National Average construction cost of a building in 1970, knowing that it cost $900,000 in 2001:

INDEX in 1970 = 28.7

INDEX in 2001 = 121.8

Note: The City Cost Indexes for Canada can be used to convert U.S. National averages to local costs in Canadian dollars.

Time Adjustment using the Historical Cost Indexes:

$$\frac{\text{Index for Year A}}{\text{Index for Year B}} \times \text{Cost in Year B} = \text{Cost in Year A}$$

$$\frac{\text{INDEX 1970}}{\text{INDEX 2001}} \times \text{Cost 2001} = \text{Cost 1970}$$

$$\frac{28.7}{121.8} \times \$900{,}000 = .236 \times \$900{,}000 = \$212{,}400$$

The construction cost of the building in 1970 is $212,400.

How to Use the City Cost Indexes

What you should know before you begin

Means City Cost Indexes (CCI) are an extremely useful tool to use when you want to compare costs from city to city and region to region.

This publication contains average construction cost indexes for 305 major U.S. and Canadian cities and Location Factors covering over 930 three-digit zip code locations.

Keep in mind that a City Cost Index number is a *percentage ratio* of a specific city's cost to the national average cost of the same item at a stated time period.

In other words, these index figures represent relative construction *factors* (or, if you prefer, multipliers) for Material and Installation costs, as well as the weighted average for Total In Place costs for each CSI MasterFormat division. Installation costs include both labor and equipment rental costs.

The 30 City Average Index is the average of 30 major U.S. cities and serves as a National Average.

Index figures for both material and installation are based on the 30 major city average of 100 and represent the cost relationship as of July 1, 2000. The index for each division is computed from representative material and labor quantities for that division. The weighted average for each city is a weighted total of the components listed above it, but does not include relative productivity between trades or cities.

As changes occur in local material prices, labor rates and equipment rental rates, the impact of these changes should be accurately measured by the change in the City Cost Index for each particular city (as compared to the 30 City Average).

Therefore, if you know (or have estimated) building costs in one city today, you can easily convert those costs to expected building costs in another city.

In addition, by using the Historical Cost Index, you can easily convert National Average building costs at a particular time to the approximate building costs for some other time. The City Cost Indexes can then be applied to calculate the costs for a particular city.

Quick Calculations

Location Adjustment Using the City Cost Indexes:

$$\frac{\text{Index for City A}}{\text{Index for City B}} \text{ x Cost in City B = Cost in City A}$$

Time Adjustment for the National Average Using the Historical Cost Index:

$$\frac{\text{Index for Year A}}{\text{Index for Year B}} \text{ x Cost in Year B = Cost in Year A}$$

Adjustment from the National Average:

$$\frac{\text{Index for City A}}{100} \text{ x National Average Cost = Cost in City A}$$

Since each of the other R.S. Means publications contains many different items, any *one* item multiplied by the particular city index may give incorrect results. However, the larger the number of items compiled, the closer the results should be to actual costs for that particular city.

The City Cost Indexes for Canadian cities are calculated using Canadian material and equipment prices and labor rates, in Canadian dollars. Therefore, indexes for Canadian cities can be used to convert U.S. National Average prices to local costs in Canadian dollars.

How to use this section

1. Compare costs from city to city.

In using the Means Indexes, remember that an index number is not a fixed number but a *ratio:* It's a percentage ratio of a building component's cost at any stated time to the National Average cost of that same component at the same time period. Put in the form of an equation:

$$\frac{\text{Specific City Cost}}{\text{National Average Cost}} \text{ x 100 = City Index Number}$$

Therefore, when making cost comparisons between cities, do not subtract one city's index number from the index number of another city and read the result as a percentage difference. Instead, divide one city's index number by that of the other city. The resulting number may then be used as a multiplier to calculate cost differences from city to city.

The formula used to find cost differences between cities for the purpose of comparison is as follows:

$$\frac{\text{City A Index}}{\text{City B Index}} \text{ x City B Cost (Known) = City A Cost (Unknown)}$$

In addition, you can use *Means CCI* to calculate and compare costs division by division between cities using the same basic formula. (Just be sure that you're comparing similar divisions.)

2. Compare a specific city's construction costs with the National Average.

When you're studying construction location feasibility, it's advisable to compare a prospective project's cost index with an index of the National Average cost.

For example, divide the weighted average index of construction costs of a specific city by that of the 30 City Average, which = 100.

$$\frac{\text{City Index}}{100} = \text{\% of National Average}$$

As a result, you get a ratio that indicates the relative cost of construction in that city in comparison with the National Average.

3. Convert U.S. National Average to actual costs in Canadian City

$$\frac{\text{Index for Canadian City}}{100} \times \text{National Average Cost = Cost in Canadian City in \$ CAN}$$

COST INDEXES

4. Adjust construction cost data based on a National Average.

When you use a source of construction cost data which is based on a National Average (such as *Means cost data publications*), it is necessary to adjust those costs to a specific location.

$$\frac{\text{City Index}}{100} \times \begin{matrix}\text{"Book" Cost Based on}\\ \text{National Average Costs}\end{matrix} = \begin{matrix}\text{City Cost}\\ \text{(Unknown)}\end{matrix}$$

5. When applying the City Cost Indexes to demolition projects, use the appropriate division index. For example, for removal of existing doors and windows, use the Division 8 index.

What you might like to know about how we developed the Indexes

To create a reliable index, R.S. Means researched the building type most often constructed in the United States and Canada. Because it was concluded that no one type of building completely represented the building construction industry, nine different types of buildings were combined to create a composite model.

The exact material, labor and equipment quantities are based on detailed analysis of these nine building types, then each quantity is weighted in proportion to expected usage. These various material items, labor hours, and equipment rental rates are thus combined to form a composite building representing as closely as possible the actual usage of materials, labor and equipment used in the North American Building Construction Industry.

The following structures were chosen to make up that composite model:

1. Factory, 1 story
2. Office, 2–4 story
3. Store, Retail
4. Town Hall, 2–3 story
5. High School, 2–3 story
6. Hospital, 4–8 story
7. Garage, Parking
8. Apartment, 1–3 story
9. Hotel/Motel, 2–3 story

For the purposes of ensuring the timeliness of the data, the components of the index for the composite model have been streamlined. They currently consist of:

- specific quantities of 66 commonly used construction materials;
- specific labor-hours for 21 building construction trades; and
- specific days of equipment rental for 6 types of construction equipment (normally used to install the 66 material items by the 21 trades.)

A sophisticated computer program handles the updating of all costs for each city on a quarterly basis. Material and equipment price quotations are gathered quarterly from over 305 cities in the United States and Canada. These prices and the latest negotiated labor wage rates for 21 different building trades are used to compile the quarterly update of the City Cost Index.

The 30 major U.S. cities used to calculate the National Average are:

Atlanta, GA	Memphis, TN
Baltimore, MD	Milwaukee, WI
Boston, MA	Minneapolis, MN
Buffalo, NY	Nashville, TN
Chicago, IL	New Orleans, LA
Cincinnati, OH	New York, NY
Cleveland, OH	Philadelphia, PA
Columbus, OH	Phoenix, AZ
Dallas, TX	Pittsburgh, PA
Denver, CO	St. Louis, MO
Detroit, MI	San Antonio, TX
Houston, TX	San Diego, CA
Indianapolis, IN	San Francisco, CA
Kansas City, MO	Seattle, WA
Los Angeles, CA	Washington, DC

F.Y.I.: The CSI MasterFormat Divisions

1. General Requirements
2. Site Work
3. Concrete
4. Masonry
5. Metals
6. Wood & Plastics
7. Thermal & Moisture Protection
8. Doors & Windows
9. Finishes
10. Specialties
11. Equipment
12. Furnishings
13. Special Construction
14. Conveying Systems
15. Mechanical
16. Electrical

The information presented in the CCI is organized according to the Construction Specifications Institute (CSI) MasterFormat.

What the CCI does not *indicate*

The weighted average for each city is a total of the components listed above weighted to reflect typical usage, but it does *not* include the productivity variations between trades or cities.

In addition, the CCI does not take into consideration factors such as the following:

- managerial efficiency
- competitive conditions
- automation
- restrictive union practices
- Unique local requirements
- regional variations due to specific building codes

DIVISION		UNITED STATES 30 CITY AVERAGE			ALABAMA BIRMINGHAM			HUNTSVILLE			MOBILE			MONTGOMERY			TUSCALOOSA		
		MAT.	INST.	TOTAL	MAT.	INST.	TOTAL	MAT.	INST.	TOTAL	MAT.	INST.	TOTAL	MAT.	INST.	TOTAL	MAT.	INST.	TOTAL
02	**SITE CONSTRUCTION**	100.0	100.0	100.0	88.6	93.8	92.6	86.6	93.0	91.5	97.9	86.8	89.4	98.4	87.3	89.9	87.1	92.7	91.4
03100	CONCRETE FORMS & ACCESSORIES	100.0	100.0	100.0	94.4	77.4	79.7	96.0	57.6	62.9	96.0	60.9	65.7	94.2	55.0	60.4	96.0	45.2	52.2
03200	CONCRETE REINFORCEMENT	100.0	100.0	100.0	92.8	91.0	91.8	92.8	89.2	90.8	95.8	67.2	79.5	95.8	90.0	92.5	92.8	90.3	91.4
03300	CAST-IN-PLACE CONCRETE	100.0	100.0	100.0	93.9	68.1	83.3	88.8	65.3	79.1	93.8	63.3	81.2	95.5	57.0	79.6	92.5	54.5	76.8
03	**CONCRETE**	100.0	100.0	100.0	89.8	77.7	83.8	87.4	67.7	77.6	90.2	64.4	77.3	90.9	63.9	77.4	89.2	58.8	74.0
04	**MASONRY**	100.0	100.0	100.0	85.8	72.4	77.5	85.7	58.1	68.7	86.3	58.5	69.2	86.8	40.7	58.5	86.0	43.0	59.5
05	**METALS**	100.0	100.0	100.0	98.6	95.9	97.6	98.2	93.9	96.7	96.8	86.2	93.0	96.8	93.7	95.7	97.3	94.3	96.2
06	**WOOD & PLASTICS**	100.0	100.0	100.0	95.3	77.6	86.2	94.4	56.1	74.6	94.4	61.2	77.2	92.3	56.1	73.5	94.4	44.5	68.6
07	**THERMAL & MOISTURE PROTECTION**	100.0	100.0	100.0	96.8	80.1	88.9	96.6	72.7	85.3	96.6	73.9	85.8	96.3	66.9	82.3	96.5	65.4	81.8
08	**DOORS & WINDOWS**	100.0	100.0	100.0	98.3	77.9	93.4	98.3	60.4	89.1	98.3	60.3	89.1	98.3	61.9	89.5	98.3	59.9	89.0
09200	PLASTER & GYPSUM BOARD	100.0	100.0	100.0	110.0	77.5	90.1	107.2	55.2	75.4	107.2	60.5	78.6	107.2	55.2	75.4	107.2	43.3	68.1
095,098	CEILINGS & ACOUSTICAL TREATMENT	100.0	100.0	100.0	95.1	77.5	83.5	95.1	55.2	68.8	95.1	60.5	72.3	95.1	55.2	68.8	95.1	43.3	60.9
09600	FLOORING	100.0	100.0	100.0	101.6	60.9	91.7	101.6	52.7	89.8	110.2	66.0	99.5	110.2	34.1	91.8	101.6	50.1	89.1
097,099	WALL FINISHES, PAINTS & COATINGS	100.0	100.0	100.0	92.9	54.9	70.8	92.9	53.8	70.2	92.9	62.3	75.1	92.9	61.2	74.5	92.9	52.7	69.5
09	**FINISHES**	100.0	100.0	100.0	98.8	71.7	85.0	98.2	55.2	76.3	102.1	61.6	81.5	102.2	50.7	75.9	98.2	45.5	71.3
10 - 14	**TOTAL DIV. 10000 - 14000**	100.0	100.0	100.0	100.0	84.7	96.8	100.0	78.4	95.4	100.0	74.5	94.6	100.0	76.0	94.9	100.0	73.0	94.3
15	**MECHANICAL**	100.0	100.0	100.0	99.9	72.2	87.1	99.9	54.2	78.8	99.9	67.0	84.7	99.9	48.8	76.3	99.9	43.0	73.6
16	**ELECTRICAL**	100.0	100.0	100.0	96.3	67.1	76.2	96.7	73.9	81.0	96.7	61.5	72.5	96.3	72.4	79.9	96.7	67.1	76.3
01 - 16	**WEIGHTED AVERAGE**	100.0	100.0	100.0	96.6	77.2	87.2	96.2	68.6	82.8	97.1	68.0	83.0	97.2	63.9	81.0	96.3	61.1	79.1

DIVISION		ALASKA ANCHORAGE			FAIRBANKS			JUNEAU			ARIZONA FLAGSTAFF			MESA/TEMPE			PHOENIX		
		MAT.	INST.	TOTAL	MAT.	INST.	TOTAL	MAT.	INST.	TOTAL	MAT.	INST.	TOTAL	MAT.	INST.	TOTAL	MAT.	INST.	TOTAL
02	**SITE CONSTRUCTION**	132.6	135.4	134.8	117.3	135.4	131.2	128.6	135.4	133.8	86.9	99.6	96.7	81.6	99.3	95.1	82.0	100.0	95.8
03100	CONCRETE FORMS & ACCESSORIES	132.5	115.8	118.1	134.1	119.6	121.6	133.8	115.8	118.3	108.9	74.3	79.0	103.6	72.2	76.5	105.0	79.5	83.0
03200	CONCRETE REINFORCEMENT	141.4	106.5	121.5	118.8	106.6	111.8	105.2	106.5	105.9	103.6	80.9	90.7	103.9	77.9	89.0	102.0	81.4	90.3
03300	CAST-IN-PLACE CONCRETE	186.6	115.3	157.2	158.0	115.9	140.6	187.4	115.4	157.7	96.8	84.9	91.9	96.8	76.6	88.4	96.9	84.9	91.9
03	**CONCRETE**	152.2	113.2	132.7	130.0	115.0	122.5	148.1	113.2	130.6	122.5	78.9	100.7	99.6	74.6	87.1	99.2	81.4	90.3
04	**MASONRY**	195.5	123.6	151.3	182.8	123.6	146.4	184.8	123.6	147.1	103.1	65.7	80.2	110.7	57.8	78.2	97.5	73.6	82.8
05	**METALS**	130.0	99.3	119.0	130.1	99.7	119.1	130.3	99.4	119.2	98.1	72.7	88.9	98.3	72.5	89.0	99.6	75.1	90.8
06	**WOOD & PLASTICS**	123.8	113.8	118.6	124.1	118.5	121.2	123.8	113.8	118.6	109.7	74.1	91.3	101.4	79.1	89.9	102.7	80.9	91.4
07	**THERMAL & MOISTURE PROTECTION**	199.7	114.1	159.1	195.7	116.4	158.1	196.4	114.1	157.3	111.5	74.9	94.1	106.5	70.1	89.2	106.4	76.8	92.4
08	**DOORS & WINDOWS**	131.3	107.3	125.6	128.4	110.2	124.0	128.4	107.3	123.3	103.0	74.0	96.0	99.8	74.3	93.6	98.9	77.7	93.8
09200	PLASTER & GYPSUM BOARD	129.9	114.2	120.3	129.9	119.0	123.3	129.9	114.2	120.3	93.2	73.4	81.1	93.8	78.4	84.4	94.3	80.2	85.7
095,098	CEILINGS & ACOUSTICAL TREATMENT	131.7	114.2	120.1	131.7	119.0	123.3	131.7	114.2	120.1	99.2	73.4	82.2	103.3	78.4	86.9	103.3	80.2	88.1
09600	FLOORING	126.3	128.2	126.8	126.4	128.2	126.9	126.3	128.2	126.8	99.0	61.6	89.9	101.4	69.9	93.8	101.8	69.9	94.1
097,099	WALL FINISHES, PAINTS & COATINGS	117.6	110.6	113.5	117.6	116.0	116.7	117.6	110.6	113.5	99.1	58.5	75.5	109.9	62.8	82.5	109.9	70.5	87.0
09	**FINISHES**	140.1	117.8	128.7	138.0	121.2	129.4	138.7	117.8	128.0	96.8	69.9	83.1	98.2	70.9	84.3	98.4	76.9	87.4
10 - 14	**TOTAL DIV. 10000 - 14000**	100.0	114.8	103.1	100.0	115.5	103.3	100.0	114.8	103.1	100.0	83.9	96.6	100.0	78.7	95.5	100.0	85.0	96.8
15	**MECHANICAL**	106.4	108.0	107.1	106.4	118.5	112.0	106.4	108.6	107.4	100.3	82.2	91.9	100.1	73.8	87.9	100.1	82.3	91.8
16	**ELECTRICAL**	161.7	110.5	126.5	164.6	110.5	127.5	164.6	110.5	127.5	100.4	57.4	70.9	98.1	38.8	57.3	107.2	63.5	77.2
01 - 16	**WEIGHTED AVERAGE**	134.5	114.1	124.5	130.1	117.0	123.7	132.9	114.2	123.8	102.8	74.8	89.2	99.6	68.7	84.6	99.6	78.4	89.3

DIVISION		ARIZONA PRESCOTT			TUCSON			ARKANSAS FORT SMITH			JONESBORO			LITTLE ROCK			PINE BLUFF		
		MAT.	INST.	TOTAL	MAT.	INST.	TOTAL	MAT.	INST.	TOTAL	MAT.	INST.	TOTAL	MAT.	INST.	TOTAL	MAT.	INST.	TOTAL
02	**SITE CONSTRUCTION**	73.7	99.2	93.3	77.9	99.8	94.7	76.7	83.9	82.2	101.7	99.0	99.6	76.7	83.9	82.2	79.9	83.9	83.0
03100	CONCRETE FORMS & ACCESSORIES	103.4	68.2	73.0	104.4	79.2	82.6	98.7	48.2	55.1	83.1	55.6	59.4	93.0	52.9	58.4	75.0	52.8	55.9
03200	CONCRETE REINFORCEMENT	103.6	82.3	91.5	100.9	80.9	89.5	95.5	72.8	82.6	91.4	54.5	70.4	95.8	71.9	82.2	95.7	71.9	82.2
03300	CAST-IN-PLACE CONCRETE	96.8	73.7	87.2	89.7	84.7	87.6	84.7	52.0	71.2	89.2	63.7	78.7	85.7	52.0	71.8	85.9	52.0	71.9
03	**CONCRETE**	106.9	72.5	89.7	95.5	81.0	88.3	84.1	55.0	69.5	85.0	60.0	72.5	84.2	56.9	70.6	85.6	56.8	71.2
04	**MASONRY**	103.6	71.6	83.9	97.5	65.7	77.9	97.0	58.4	73.3	91.3	54.9	68.9	95.2	58.4	72.6	115.0	58.4	80.2
05	**METALS**	98.1	72.0	88.7	98.9	73.4	89.7	96.6	72.8	88.0	90.9	82.3	87.8	96.2	72.6	87.7	95.4	72.5	87.1
06	**WOOD & PLASTICS**	104.2	67.3	85.1	102.8	80.9	91.5	100.4	47.3	73.0	83.3	57.0	69.7	96.9	53.6	74.5	75.7	53.6	64.3
07	**THERMAL & MOISTURE PROTECTION**	109.7	72.3	91.9	107.1	71.3	90.1	98.3	54.5	77.5	108.0	59.3	84.9	96.9	55.1	77.1	96.8	55.1	77.0
08	**DOORS & WINDOWS**	103.0	69.3	94.9	97.6	77.7	92.8	98.1	50.0	86.5	99.5	54.9	88.8	98.1	54.2	87.5	93.5	54.2	84.0
09200	PLASTER & GYPSUM BOARD	91.5	66.3	76.1	94.5	80.2	85.8	97.6	46.6	66.4	102.1	56.1	74.0	97.6	53.0	70.4	91.2	53.0	67.9
095,098	CEILINGS & ACOUSTICAL TREATMENT	99.2	66.3	77.5	104.7	80.2	88.5	101.1	46.6	65.1	101.7	56.1	71.6	101.1	53.0	69.4	97.0	53.0	68.0
09600	FLOORING	97.3	61.3	88.6	100.9	61.6	91.4	118.7	74.4	107.9	79.3	53.2	73.0	120.0	74.4	108.9	107.2	74.4	99.3
097,099	WALL FINISHES, PAINTS & COATINGS	99.1	58.5	75.5	107.4	58.5	78.9	95.6	67.7	79.4	84.3	60.9	70.7	95.6	51.8	70.1	95.6	51.8	70.1
09	**FINISHES**	94.9	65.6	80.0	98.1	73.8	85.7	102.0	54.9	78.0	92.6	55.8	73.8	102.4	56.7	79.1	96.8	56.7	76.3
10 - 14	**TOTAL DIV. 10000 - 14000**	100.0	82.6	96.3	100.0	85.1	96.8	100.0	70.5	93.8	100.0	65.4	92.7	100.0	71.4	94.0	100.0	71.4	94.0
15	**MECHANICAL**	100.3	71.7	87.1	100.0	76.2	89.0	100.1	50.6	77.2	100.3	50.7	77.3	100.1	57.3	80.3	100.1	53.5	78.5
16	**ELECTRICAL**	100.0	55.3	69.3	101.6	64.1	75.8	95.6	70.0	78.0	100.7	57.1	70.8	96.7	72.7	80.2	94.5	72.7	79.5
01 - 16	**WEIGHTED AVERAGE**	100.2	71.2	86.1	98.4	75.8	87.4	96.3	61.6	79.4	95.6	62.5	79.5	96.2	64.0	80.6	95.9	63.3	80.1

	DIVISION	ARKANSAS			CALIFORNIA															
		TEXARKANA			ANAHEIM			BAKERSFIELD			FRESNO			LOS ANGELES			OAKLAND			
		MAT.	INST.	TOTAL	MAT.	INST.	TOTAL	MAT.	INST.	TOTAL	MAT.	INST.	TOTAL	MAT.	INST.	TOTAL	MAT.	INST.	TOTAL	
02	**SITE CONSTRUCTION**	96.0	84.6	87.3	94.1	110.2	106.5	98.7	107.1	105.2	100.2	107.2	105.5	89.5	109.0	104.5	128.3	104.0	109.7	
03100	CONCRETE FORMS & ACCESSORIES	84.1	46.1	51.3	105.2	119.2	117.2	96.3	118.9	115.8	101.8	121.7	118.9	108.8	118.7	117.3	107.7	136.7	132.7	
03200	CONCRETE REINFORCEMENT	95.2	52.8	71.0	106.8	117.3	112.8	106.4	117.1	112.5	106.9	117.1	112.7	108.3	117.2	113.4	98.8	118.0	109.8	
03300	CAST-IN-PLACE CONCRETE	85.8	49.7	70.9	102.3	116.3	108.1	97.7	115.1	104.9	106.9	115.7	110.5	89.4	113.5	99.3	127.9	118.4	124.0	
03	**CONCRETE**	81.1	49.6	65.4	112.1	116.9	114.5	109.2	116.2	112.7	114.2	117.7	115.9	108.2	115.6	111.9	125.9	125.6	125.7	
04	**MASONRY**	97.1	38.5	61.1	87.8	116.5	105.4	107.2	113.3	111.0	109.7	112.4	111.4	97.8	116.5	109.3	145.5	125.5	133.2	
05	**METALS**	87.7	64.2	79.2	110.5	101.4	107.2	105.0	100.5	103.4	106.9	101.0	104.8	111.2	99.3	106.9	101.7	106.4	103.4	
06	**WOOD & PLASTICS**	85.7	49.4	66.9	99.5	117.7	108.9	89.1	117.9	104.0	101.4	120.6	111.4	99.6	117.3	108.7	106.4	139.1	123.3	
07	**THERMAL & MOISTURE PROTECTION**	97.7	48.5	74.4	121.8	116.7	119.4	105.9	110.1	107.9	102.3	111.2	106.5	114.2	114.6	114.4	111.7	130.0	120.4	
08	**DOORS & WINDOWS**	98.5	47.0	86.1	105.8	115.1	108.1	104.6	113.4	106.7	104.6	115.3	107.2	99.1	114.8	102.9	108.6	129.7	113.7	
09200	PLASTER & GYPSUM BOARD	94.2	48.7	66.4	94.8	118.4	109.3	94.8	118.4	109.3	94.0	121.3	110.7	88.9	118.4	106.9	89.1	140.0	120.2	
095,098	CEILINGS & ACOUSTICAL TREATMENT	103.8	48.7	67.5	118.1	118.4	118.3	118.1	118.4	118.3	118.1	121.3	120.2	115.2	118.4	117.3	116.9	140.0	132.2	
09600	FLOORING	110.1	50.8	95.7	123.1	110.5	120.0	118.6	73.3	107.6	119.3	143.2	125.1	115.7	110.5	114.4	108.8	120.7	111.7	
097,099	WALL FINISHES, PAINTS & COATINGS	95.6	42.7	64.8	111.3	116.0	114.0	114.4	106.0	109.5	113.9	101.7	106.8	108.6	114.8	112.2	118.3	133.2	126.9	
09	**FINISHES**	99.8	46.7	72.7	112.9	117.1	115.1	114.1	109.6	111.8	114.2	123.9	119.2	108.5	116.6	112.7	112.6	134.4	123.7	
10 - 14	**TOTAL DIV. 10000 - 14000**	100.0	46.6	88.7	100.0	115.8	103.3	100.0	130.8	106.5	100.0	133.3	107.0	100.0	114.5	103.1	100.0	136.6	107.7	
15	**MECHANICAL**	100.1	44.8	74.5	100.3	114.1	106.7	100.3	112.7	106.0	100.3	115.1	107.1	100.2	114.0	106.6	100.1	136.4	116.9	
16	**ELECTRICAL**	97.7	47.6	63.2	95.0	109.9	105.2	95.4	94.2	94.6	94.1	99.6	97.9	109.8	113.6	112.4	113.6	130.5	125.2	
01 - 16	**WEIGHTED AVERAGE**	94.9	51.5	73.8	104.7	113.0	108.7	104.1	108.5	106.3	105.2	112.0	108.5	104.3	112.9	108.5	110.7	126.3	118.3	

	DIVISION	CALIFORNIA																	
		OXNARD			REDDING			RIVERSIDE			SACRAMENTO			SAN DIEGO			SAN FRANCISCO		
		MAT.	INST.	TOTAL	MAT.	INST.	TOTAL	MAT.	INST.	TOTAL	MAT.	INST.	TOTAL	MAT.	INST.	TOTAL	MAT.	INST.	TOTAL
02	**SITE CONSTRUCTION**	99.8	104.8	103.6	105.0	106.0	105.8	92.0	107.9	104.1	95.9	110.5	107.1	100.6	102.1	101.8	130.3	110.5	115.2
03100	CONCRETE FORMS & ACCESSORIES	103.1	119.3	117.1	102.7	120.8	118.3	106.3	119.1	117.3	106.2	121.2	119.2	106.6	113.8	112.8	108.0	137.5	133.4
03200	CONCRETE REINFORCEMENT	106.4	117.1	112.5	103.1	117.2	111.1	105.4	117.1	112.1	99.9	117.2	109.8	110.5	116.9	114.1	112.6	118.4	115.9
03300	CAST-IN-PLACE CONCRETE	103.7	115.4	108.5	117.4	114.5	116.2	101.3	116.3	107.5	111.0	114.8	112.6	105.2	105.0	105.1	127.8	120.1	124.6
03	**CONCRETE**	112.6	116.5	114.6	123.0	116.8	119.9	111.5	116.8	114.2	117.6	117.1	117.4	116.1	110.5	113.3	127.6	126.6	127.1
04	**MASONRY**	112.4	112.4	112.4	113.8	112.1	112.7	85.3	113.7	102.8	117.2	112.1	114.1	105.4	110.9	108.7	145.8	130.6	136.4
05	**METALS**	104.4	100.8	103.1	108.9	100.3	105.8	110.5	101.1	107.1	97.7	100.3	98.6	110.6	100.2	106.8	106.8	107.9	107.2
06	**WOOD & PLASTICS**	98.1	117.9	108.3	99.2	120.6	110.3	99.5	117.7	108.9	101.1	120.7	111.2	100.1	111.1	105.8	106.4	139.3	123.4
07	**THERMAL & MOISTURE PROTECTION**	111.3	114.1	112.6	112.0	112.8	112.4	120.8	115.4	118.3	121.3	113.4	117.5	114.7	107.8	111.4	111.7	132.0	121.3
08	**DOORS & WINDOWS**	103.4	115.1	106.3	106.3	112.5	107.8	105.8	115.1	108.1	120.6	116.5	119.6	104.6	110.9	106.1	111.1	129.3	115.5
09200	PLASTER & GYPSUM BOARD	94.8	118.4	109.3	95.3	121.3	111.2	94.6	118.4	109.2	87.8	121.3	108.2	88.1	111.4	102.4	90.9	140.0	120.9
095,098	CEILINGS & ACOUSTICAL TREATMENT	118.1	118.4	118.3	127.6	121.3	123.4	116.7	118.4	117.8	125.0	121.3	122.5	117.9	111.4	113.6	123.4	140.0	134.4
09600	FLOORING	118.6	110.5	116.6	117.8	111.4	116.3	123.0	110.5	120.0	111.2	111.4	111.3	119.1	109.8	116.8	108.8	120.7	111.7
097,099	WALL FINISHES, PAINTS & COATINGS	113.9	109.3	111.2	113.9	114.3	114.1	111.3	116.0	114.0	116.7	114.3	115.3	111.9	116.0	114.3	118.3	134.1	127.5
09	**FINISHES**	113.9	116.4	115.2	115.9	119.4	117.7	112.4	117.1	114.8	113.1	119.5	116.4	111.5	113.2	112.4	114.1	134.6	124.6
10 - 14	**TOTAL DIV. 10000 - 14000**	100.0	116.1	103.4	100.0	132.7	106.9	100.0	115.8	103.3	100.0	132.8	106.9	100.0	115.2	103.2	100.0	137.3	107.9
15	**MECHANICAL**	100.3	114.1	106.7	100.3	113.1	106.2	100.2	114.1	106.6	100.0	118.2	108.4	100.1	114.1	106.6	100.1	159.3	127.5
16	**ELECTRICAL**	99.2	113.1	108.8	104.4	99.5	101.0	95.1	105.5	102.2	104.0	100.4	101.5	92.8	93.9	93.6	109.3	140.3	130.6
01 - 16	**WEIGHTED AVERAGE**	105.2	112.4	108.7	108.3	110.6	109.4	104.3	111.7	107.9	107.5	112.4	109.9	105.6	107.0	106.3	111.9	133.9	122.6

	DIVISION	CALIFORNIA												COLORADO					
		SAN JOSE			SANTA BARBARA			STOCKTON			VALLEJO			COLORADO SPRINGS			DENVER		
		MAT.	INST.	TOTAL	MAT.	INST.	TOTAL	MAT.	INST.	TOTAL	MAT.	INST.	TOTAL	MAT.	INST.	TOTAL	MAT.	INST.	TOTAL
02	**SITE CONSTRUCTION**	136.3	100.4	108.8	100.2	107.2	105.5	98.1	106.4	104.5	92.5	110.7	106.4	110.1	98.3	101.1	108.4	107.8	107.9
03100	CONCRETE FORMS & ACCESSORIES	105.5	136.9	132.6	103.8	119.1	117.0	97.5	121.6	118.3	108.1	136.2	132.3	93.4	76.2	78.6	103.8	81.3	84.4
03200	CONCRETE REINFORCEMENT	103.4	118.3	111.9	106.4	116.9	112.4	106.9	117.3	112.8	98.5	118.2	109.7	98.9	80.8	88.6	98.4	81.5	88.7
03300	CAST-IN-PLACE CONCRETE	123.2	118.8	121.4	103.3	115.2	108.2	103.2	115.6	108.3	114.8	117.1	115.7	110.8	85.6	100.4	103.7	85.9	96.3
03	**CONCRETE**	122.0	126.0	124.0	112.4	116.3	114.4	112.1	117.7	114.9	119.4	124.7	122.0	103.8	80.7	92.2	108.3	83.1	95.7
04	**MASONRY**	149.4	129.1	136.9	107.6	109.8	109.0	112.1	115.6	114.2	83.3	129.4	111.6	108.3	69.4	84.4	107.6	83.3	92.7
05	**METALS**	111.1	108.4	110.2	104.3	100.4	102.9	106.5	101.4	104.7	101.0	102.5	101.5	101.6	85.9	96.0	105.2	86.8	98.5
06	**WOOD & PLASTICS**	106.0	139.0	123.1	98.1	117.9	108.3	93.4	120.6	107.5	99.6	138.8	119.8	91.4	76.5	83.7	101.5	82.2	91.5
07	**THERMAL & MOISTURE PROTECTION**	107.9	130.8	118.7	107.0	112.0	109.4	111.4	113.4	112.4	125.6	129.9	127.7	107.5	81.4	95.1	106.8	84.8	96.4
08	**DOORS & WINDOWS**	95.4	129.7	103.7	104.6	115.1	107.2	103.9	116.5	106.9	122.0	129.0	123.7	99.2	82.1	95.1	98.6	85.2	95.3
09200	PLASTER & GYPSUM BOARD	95.3	140.0	122.6	94.8	118.4	109.3	95.7	121.3	111.3	88.2	140.0	119.9	89.2	75.7	81.0	97.4	81.9	87.9
095,098	CEILINGS & ACOUSTICAL TREATMENT	109.9	140.0	129.8	118.1	118.4	118.3	118.1	121.3	120.2	125.0	140.0	134.9	97.5	75.7	83.1	96.2	81.9	86.7
09600	FLOORING	115.5	120.7	116.8	118.6	99.1	113.8	118.6	111.4	116.8	115.5	120.7	116.8	111.8	94.1	107.5	112.8	86.6	106.4
097,099	WALL FINISHES, PAINTS & COATINGS	115.6	131.4	124.8	113.9	109.3	111.2	113.9	98.9	105.2	114.6	133.2	125.4	111.9	62.4	83.1	111.9	76.8	91.5
09	**FINISHES**	113.1	134.2	123.9	114.1	114.5	114.3	114.1	118.1	116.1	111.7	134.1	123.1	100.5	76.9	88.5	101.6	81.5	91.4
10 - 14	**TOTAL DIV. 10000 - 14000**	100.0	136.4	107.7	100.0	116.1	103.4	100.0	133.4	107.0	100.0	135.6	107.5	100.0	85.0	96.8	100.0	85.8	97.0
15	**MECHANICAL**	100.3	146.3	121.5	100.3	114.1	106.7	100.3	114.0	106.6	100.0	130.9	114.3	100.2	80.9	91.3	100.1	84.7	93.0
16	**ELECTRICAL**	109.9	137.7	129.0	91.9	106.0	101.6	103.4	109.1	107.4	99.8	117.9	112.2	95.2	85.7	88.7	97.3	90.4	92.6
01 - 16	**WEIGHTED AVERAGE**	110.2	129.7	119.7	104.4	110.8	107.5	105.7	113.0	109.3	106.0	123.6	114.6	101.3	82.4	92.1	102.5	87.4	95.2

	DIVISION	COLORADO												CONNECTICUT					
		FORT COLLINS			GRAND JUNCTION			GREELEY			PUEBLO			BRIDGEPORT			BRISTOL		
		MAT.	INST.	TOTAL	MAT.	INST.	TOTAL	MAT.	INST.	TOTAL	MAT.	INST.	TOTAL	MAT.	INST.	TOTAL	MAT.	INST.	TOTAL
02	**SITE CONSTRUCTION**	121.3	101.5	106.2	131.4	100.1	107.5	106.3	100.3	101.7	122.7	94.9	101.4	107.0	105.4	105.8	106.1	105.4	105.6
03100	CONCRETE FORMS & ACCESSORIES	103.2	75.1	79.0	113.0	51.6	60.0	100.2	55.9	62.0	107.0	76.5	80.7	99.3	102.7	102.3	99.3	108.0	106.8
03200	CONCRETE REINFORCEMENT	103.1	80.4	90.2	108.9	80.2	92.5	102.7	58.1	77.2	101.5	80.7	89.7	111.9	119.7	116.3	111.9	119.6	116.3
03300	CAST-IN-PLACE CONCRETE	120.4	82.1	104.6	119.4	51.2	91.3	100.6	68.7	87.4	107.1	86.3	98.5	105.2	114.5	109.0	98.5	114.4	105.1
03	**CONCRETE**	123.0	78.9	101.0	120.0	57.8	88.9	105.7	61.3	83.5	108.6	81.1	94.9	110.0	109.9	109.9	106.7	112.1	109.4
04	**MASONRY**	121.3	69.1	89.3	148.4	56.5	91.9	114.9	47.7	73.6	107.9	67.3	82.9	107.4	115.1	112.2	96.8	115.1	108.1
05	**METALS**	101.0	83.6	94.7	101.5	81.5	94.3	101.0	66.8	88.6	103.0	86.9	97.2	99.0	112.9	104.0	99.0	112.7	103.9
06	**WOOD & PLASTICS**	101.2	76.2	88.3	108.6	53.0	79.8	98.1	56.8	76.7	103.0	76.8	89.4	100.9	98.6	99.7	100.9	106.1	103.6
07	**THERMAL & MOISTURE PROTECTION**	107.4	76.3	92.6	105.5	62.4	85.1	106.5	66.1	87.4	104.9	80.1	93.1	99.4	112.4	105.5	99.5	107.5	103.3
08	**DOORS & WINDOWS**	96.6	74.4	91.2	101.4	58.2	91.0	96.6	49.6	85.2	95.8	82.2	92.6	110.0	111.7	110.4	110.0	111.0	110.3
09200	PLASTER & GYPSUM BOARD	96.4	75.7	83.8	102.3	51.6	71.3	95.6	55.6	71.1	89.5	75.7	81.1	97.8	97.4	97.5	97.8	105.0	102.2
095,098	CEILINGS & ACOUSTICAL TREATMENT	90.7	75.7	80.8	93.8	51.6	66.0	90.7	55.6	67.6	104.7	75.7	85.6	102.6	97.4	99.1	102.6	105.0	104.2
09600	FLOORING	112.7	83.0	105.5	120.7	83.0	111.5	111.2	83.0	104.4	117.0	86.6	109.6	99.9	109.3	102.2	99.9	109.3	102.2
097,099	WALL FINISHES, PAINTS & COATINGS	111.9	61.9	82.8	122.6	36.3	72.5	111.9	38.5	69.2	122.6	57.8	84.9	95.6	111.5	104.8	95.6	110.7	104.4
09	**FINISHES**	101.1	75.2	87.9	106.8	55.7	80.7	99.8	58.7	78.8	105.3	75.9	90.3	101.2	103.9	102.6	101.2	108.1	104.8
10 - 14	**TOTAL DIV. 10000 - 14000**	100.0	71.6	94.0	100.0	63.9	92.4	100.0	65.1	92.6	100.0	85.9	97.0	100.0	112.9	102.7	100.0	113.9	102.9
15	**MECHANICAL**	100.1	83.3	92.3	100.1	40.2	72.4	100.1	77.2	89.5	100.1	79.2	90.4	99.8	104.6	102.0	99.8	104.5	102.0
16	**ELECTRICAL**	90.1	70.4	76.6	94.0	75.0	80.9	90.1	70.4	76.6	95.3	81.9	86.1	104.0	103.9	104.0	104.0	105.0	104.7
01 - 16	**WEIGHTED AVERAGE**	104.1	79.2	91.9	107.1	63.6	85.9	100.9	68.4	85.1	102.6	80.9	92.0	103.3	107.6	105.4	102.3	108.5	105.3

	DIVISION	CONNECTICUT																	
		HARTFORD			NEW BRITAIN			NEW HAVEN			NORWALK			STAMFORD			WATERBURY		
		MAT.	INST.	TOTAL	MAT.	INST.	TOTAL	MAT.	INST.	TOTAL	MAT.	INST.	TOTAL	MAT.	INST.	TOTAL	MAT.	INST.	TOTAL
02	**SITE CONSTRUCTION**	106.3	105.4	105.6	106.3	105.4	105.6	106.1	106.3	106.3	106.8	105.4	105.8	107.5	105.5	105.9	106.5	105.4	105.7
03100	CONCRETE FORMS & ACCESSORIES	98.4	108.0	106.6	99.7	108.0	106.9	99.1	108.2	106.9	99.3	102.9	102.4	99.3	103.1	102.6	99.3	108.1	106.9
03200	CONCRETE REINFORCEMENT	111.9	119.6	116.3	111.9	119.6	116.3	111.9	119.7	116.3	111.9	119.8	116.4	111.9	119.8	116.4	111.9	119.7	116.3
03300	CAST-IN-PLACE CONCRETE	98.2	114.4	104.9	100.1	114.4	106.0	101.8	118.7	108.8	103.5	119.7	110.2	105.2	119.8	111.2	105.2	114.4	109.0
03	**CONCRETE**	106.5	112.1	109.3	107.6	112.1	109.8	109.3	113.7	111.5	109.2	111.8	110.5	110.0	111.9	111.0	110.0	112.2	111.1
04	**MASONRY**	96.8	115.1	108.1	96.9	115.1	108.1	97.0	115.1	108.2	97.2	117.4	109.6	97.3	117.4	109.6	97.3	115.1	108.3
05	**METALS**	99.6	112.7	104.3	95.8	112.7	101.9	96.0	112.9	102.1	99.0	113.2	104.1	99.0	113.4	104.2	99.0	112.8	104.0
06	**WOOD & PLASTICS**	100.9	106.1	103.6	100.9	106.1	103.6	100.9	106.1	103.6	100.9	98.6	99.7	100.9	98.6	99.7	100.9	106.1	103.6
07	**THERMAL & MOISTURE PROTECTION**	98.3	107.5	102.7	99.5	109.5	104.3	99.6	110.1	104.6	99.5	113.7	106.2	99.5	113.7	106.2	99.5	109.5	104.3
08	**DOORS & WINDOWS**	110.0	111.0	110.3	110.0	111.0	110.3	110.0	115.7	111.4	110.0	111.7	110.4	110.0	111.7	110.4	110.0	115.7	111.4
09200	PLASTER & GYPSUM BOARD	97.8	105.0	102.2	97.8	105.0	102.2	97.8	105.0	102.2	97.8	97.4	97.5	97.8	97.4	97.5	97.8	105.0	102.2
095,098	CEILINGS & ACOUSTICAL TREATMENT	102.6	105.0	104.2	102.6	105.0	104.2	102.6	105.0	104.2	102.6	97.4	99.1	102.6	97.4	99.1	102.6	105.0	104.2
09600	FLOORING	99.9	109.3	102.2	99.9	109.3	102.2	99.9	109.3	102.2	99.9	109.3	102.2	99.9	109.3	102.2	99.9	109.3	102.2
097,099	WALL FINISHES, PAINTS & COATINGS	95.6	110.7	104.4	95.6	110.7	104.4	95.6	104.7	100.8	95.6	111.5	104.8	95.6	111.5	104.8	95.6	109.6	103.7
09	**FINISHES**	101.2	108.1	104.7	101.2	108.1	104.8	101.3	107.5	104.4	101.2	103.9	102.6	101.3	103.9	102.6	101.1	108.0	104.6
10 - 14	**TOTAL DIV. 10000 - 14000**	100.0	113.9	102.9	100.0	113.9	102.9	100.0	113.9	102.9	100.0	112.9	102.7	100.0	113.1	102.8	100.0	113.9	102.9
15	**MECHANICAL**	99.8	104.5	102.0	99.8	104.5	102.0	99.8	104.6	102.0	99.8	104.6	102.0	99.8	104.6	102.0	99.8	104.6	102.0
16	**ELECTRICAL**	103.2	105.3	104.7	104.0	105.0	104.7	104.0	105.0	104.7	104.0	101.4	102.2	104.0	133.9	124.5	103.2	103.9	103.7
01 - 16	**WEIGHTED AVERAGE**	102.2	108.5	105.3	101.9	108.5	105.1	102.1	109.0	105.5	102.6	107.7	105.1	102.7	113.2	107.8	102.6	108.5	105.5

	DIVISION	D.C.			DELAWARE			FLORIDA											
		WASHINGTON			WILMINGTON			DAYTONA BEACH			FORT LAUDERDALE			JACKSONVILLE			MELBOURNE		
		MAT.	INST.	TOTAL	MAT.	INST.	TOTAL	MAT.	INST.	TOTAL	MAT.	INST.	TOTAL	MAT.	INST.	TOTAL	MAT.	INST.	TOTAL
02	**SITE CONSTRUCTION**	103.0	90.4	93.4	87.0	113.7	107.5	120.4	87.1	94.9	105.6	73.6	81.1	120.5	87.5	95.2	128.5	87.3	96.9
03100	CONCRETE FORMS & ACCESSORIES	97.0	79.3	81.7	100.4	102.0	101.8	96.6	70.6	74.2	93.9	65.1	69.0	96.3	59.2	64.3	90.3	72.2	74.7
03200	CONCRETE REINFORCEMENT	99.2	93.3	95.8	98.5	96.2	97.2	95.8	89.0	91.9	95.8	80.6	87.1	95.8	62.4	76.8	96.8	89.1	92.4
03300	CAST-IN-PLACE CONCRETE	112.6	85.5	101.4	76.8	91.7	82.9	91.0	74.0	84.0	95.5	69.7	84.8	91.9	58.6	78.2	105.7	75.9	93.4
03	**CONCRETE**	105.1	85.6	95.4	97.5	98.5	98.0	88.9	76.4	82.7	90.9	71.0	81.0	89.3	61.3	75.3	98.3	77.8	88.1
04	**MASONRY**	90.2	82.0	85.1	100.7	88.0	92.9	86.7	70.2	76.6	86.9	63.4	72.5	86.1	56.7	68.0	83.0	73.6	77.2
05	**METALS**	96.9	110.2	101.7	97.8	116.3	104.5	99.1	95.5	97.8	98.5	93.3	96.6	98.7	85.1	93.8	107.1	95.8	103.0
06	**WOOD & PLASTICS**	94.8	79.8	87.0	104.3	104.1	104.2	95.1	71.9	83.1	90.1	66.8	78.1	95.1	59.8	76.8	88.6	71.9	80.0
07	**THERMAL & MOISTURE PROTECTION**	94.1	83.6	89.1	99.5	103.7	101.5	96.5	73.6	85.7	96.5	68.6	83.3	96.8	63.9	81.2	96.8	76.0	86.9
08	**DOORS & WINDOWS**	101.9	85.9	98.0	97.4	104.4	99.1	100.7	69.5	93.2	98.3	65.1	90.3	100.7	56.3	90.0	99.9	72.9	93.4
09200	PLASTER & GYPSUM BOARD	102.3	79.0	88.1	105.9	104.1	104.8	107.2	71.5	85.4	106.7	66.3	82.0	107.2	59.1	77.8	104.2	71.5	84.2
095,098	CEILINGS & ACOUSTICAL TREATMENT	103.8	79.0	87.5	101.1	104.1	103.1	95.1	71.5	79.5	95.1	66.3	76.1	95.1	59.1	71.3	91.0	71.5	78.2
09600	FLOORING	96.3	88.3	94.4	80.3	93.1	83.4	115.9	74.2	105.8	115.9	70.8	105.0	115.9	58.1	101.9	112.6	74.2	103.3
097,099	WALL FINISHES, PAINTS & COATINGS	105.7	92.4	97.9	87.1	99.7	94.4	109.2	87.5	96.6	105.6	60.6	79.4	109.2	58.4	79.6	109.2	102.4	105.2
09	**FINISHES**	97.0	81.9	89.3	97.6	100.1	98.9	106.9	73.1	89.7	104.9	65.2	84.7	106.9	58.6	82.3	105.1	75.7	90.1
10 - 14	**TOTAL DIV. 10000 - 14000**	100.0	94.1	98.8	100.0	109.6	102.0	100.0	82.2	96.2	100.0	82.3	96.3	100.0	73.8	94.5	100.0	83.6	96.5
15	**MECHANICAL**	100.1	89.2	95.0	100.2	106.9	103.3	99.9	70.5	86.3	99.9	68.5	85.4	99.9	59.3	81.1	99.9	72.5	87.2
16	**ELECTRICAL**	97.4	96.4	96.7	99.8	96.2	97.3	96.7	65.6	75.3	96.7	83.6	87.7	96.2	70.3	78.4	96.4	72.8	80.2
01 - 16	**WEIGHTED AVERAGE**	99.2	90.1	94.8	98.5	102.8	100.6	98.7	75.0	87.2	97.9	73.5	86.0	98.7	66.6	83.1	100.8	77.7	89.5

	DIVISION	FLORIDA																	
		MIAMI			ORLANDO			PANAMA CITY			PENSACOLA			ST. PETERSBURG			TALLAHASSEE		
		MAT.	INST.	TOTAL	MAT.	INST.	TOTAL	MAT.	INST.	TOTAL	MAT.	INST.	TOTAL	MAT.	INST.	TOTAL	MAT.	INST.	TOTAL
02	**SITE CONSTRUCTION**	105.0	73.5	80.9	121.1	86.8	94.8	135.6	84.6	96.6	133.2	86.8	97.7	121.4	86.6	94.7	121.9	86.2	94.6
03100	CONCRETE FORMS & ACCESSORIES	93.7	65.1	69.1	96.4	61.9	66.6	95.2	32.6	41.2	84.7	59.8	63.2	93.7	55.6	60.8	96.3	45.7	52.6
03200	CONCRETE REINFORCEMENT	95.8	80.6	87.2	95.8	84.5	89.4	100.0	61.2	77.9	102.5	61.5	79.1	99.2	70.4	82.8	95.8	61.8	76.4
03300	CAST-IN-PLACE CONCRETE	93.0	70.1	83.5	98.9	73.8	88.6	96.6	39.7	73.1	96.6	62.5	82.5	102.9	63.5	86.7	95.2	55.1	78.7
03	**CONCRETE**	89.7	71.2	80.4	92.8	71.7	82.3	97.5	42.3	69.9	96.3	62.7	79.5	95.1	62.8	78.9	91.0	54.0	72.5
04	**MASONRY**	84.4	63.9	71.8	87.3	70.2	76.8	90.9	33.0	55.3	88.1	59.5	70.5	132.4	58.9	87.2	87.6	46.2	62.2
05	**METALS**	99.0	93.1	96.9	108.4	93.7	103.1	97.4	71.3	88.0	97.3	84.7	92.8	101.1	87.3	96.1	99.0	83.5	93.4
06	**WOOD & PLASTICS**	90.1	66.8	78.1	95.1	60.6	77.3	93.8	32.8	62.2	82.8	60.6	71.3	92.1	55.6	73.3	95.1	44.1	68.7
07	**THERMAL & MOISTURE PROTECTION**	99.5	70.1	85.6	96.9	72.5	85.3	97.1	36.1	68.2	96.8	60.5	79.6	96.5	57.0	77.7	97.0	53.4	76.3
08	**DOORS & WINDOWS**	98.3	65.7	90.4	100.7	61.5	91.2	98.4	30.7	82.0	98.3	57.7	88.5	99.4	53.4	88.3	100.7	47.0	87.7
09200	PLASTER & GYPSUM BOARD	106.7	66.3	82.0	110.0	59.9	79.4	105.7	31.2	60.2	101.8	59.9	76.2	105.3	54.8	74.4	107.2	42.8	67.8
095,098	CEILINGS & ACOUSTICAL TREATMENT	95.1	66.3	76.1	95.1	59.9	71.9	89.7	31.2	51.1	89.7	59.9	70.0	89.7	54.8	66.6	95.1	42.8	60.6
09600	FLOORING	123.5	71.7	110.9	115.9	74.2	105.8	115.3	22.3	92.8	109.5	61.8	97.9	114.3	61.6	101.5	115.9	45.2	98.8
097,099	WALL FINISHES, PAINTS & COATINGS	105.6	60.6	79.4	109.2	67.0	84.7	109.2	29.8	63.1	109.2	67.8	85.1	109.2	56.5	78.5	109.2	48.1	73.6
09	**FINISHES**	107.3	65.4	86.0	107.3	64.2	85.3	106.6	29.9	67.5	103.9	61.0	82.0	105.1	56.6	80.3	107.0	45.0	75.4
10 - 14	**TOTAL DIV. 10000 - 14000**	100.0	82.2	96.2	100.0	80.6	95.9	100.0	55.7	90.7	100.0	62.3	92.0	100.0	65.2	92.7	100.0	72.9	94.3
15	**MECHANICAL**	99.9	72.2	87.1	99.9	65.2	83.9	99.9	30.3	67.7	99.9	60.3	81.6	99.9	60.1	81.5	99.9	48.0	75.9
16	**ELECTRICAL**	97.2	78.5	84.4	97.3	53.1	66.9	94.9	39.7	57.0	100.4	64.3	75.6	97.1	57.4	69.8	97.3	49.2	64.2
01 - 16	**WEIGHTED AVERAGE**	98.1	73.4	86.1	100.8	69.5	85.6	99.8	43.6	72.5	99.4	66.1	83.2	102.1	64.3	83.7	99.2	56.2	78.2

	DIVISION	FLORIDA			GEORGIA														
		TAMPA			ALBANY			ATLANTA			AUGUSTA			COLUMBUS			MACON		
		MAT.	INST.	TOTAL	MAT.	INST.	TOTAL	MAT.	INST.	TOTAL	MAT.	INST.	TOTAL	MAT.	INST.	TOTAL	MAT.	INST.	TOTAL
02	**SITE CONSTRUCTION**	121.7	86.6	94.8	105.9	76.4	83.3	110.3	95.2	98.8	106.5	92.8	96.0	105.8	76.5	83.4	106.6	94.4	97.3
03100	CONCRETE FORMS & ACCESSORIES	97.9	55.8	61.6	95.9	49.9	56.2	97.7	79.8	82.2	94.4	60.2	64.9	97.5	63.8	68.4	94.9	64.5	68.6
03200	CONCRETE REINFORCEMENT	95.8	70.4	81.4	95.8	92.2	93.7	102.9	93.8	97.7	108.7	85.2	95.3	95.8	92.4	93.9	98.2	92.4	94.9
03300	CAST-IN-PLACE CONCRETE	100.7	63.6	85.4	97.2	47.4	76.7	99.0	75.2	89.2	93.6	55.8	78.0	96.8	48.0	76.7	95.5	51.7	77.5
03	**CONCRETE**	93.8	62.9	78.4	91.9	58.7	75.3	94.6	80.6	87.6	90.9	63.7	77.3	91.8	65.0	78.4	91.3	66.6	79.0
04	**MASONRY**	87.7	58.9	70.0	89.0	37.6	57.4	90.9	72.1	79.3	91.2	47.4	64.3	88.9	41.4	59.7	103.7	45.2	67.7
05	**METALS**	102.0	87.5	96.8	96.4	92.5	95.0	92.3	81.2	88.3	91.1	74.8	85.2	96.8	93.1	95.5	91.7	93.5	92.4
06	**WOOD & PLASTICS**	96.7	55.6	75.5	94.4	50.1	71.5	98.6	82.4	90.3	95.4	62.8	78.5	96.2	69.0	82.1	100.7	67.7	83.6
07	**THERMAL & MOISTURE PROTECTION**	96.8	58.0	78.4	96.6	56.2	77.5	92.4	77.0	85.1	91.8	57.2	75.4	96.3	59.1	78.7	95.2	63.3	80.1
08	**DOORS & WINDOWS**	100.7	53.7	89.3	98.3	55.5	88.0	98.3	78.9	93.6	94.5	61.2	86.5	98.3	66.0	90.5	96.7	66.3	89.4
09200	PLASTER & GYPSUM BOARD	107.2	54.8	75.1	107.2	49.0	71.6	110.1	82.3	93.1	109.2	62.0	80.3	107.2	68.5	83.6	114.6	67.2	85.7
095,098	CEILINGS & ACOUSTICAL TREATMENT	95.1	54.8	68.5	95.1	49.0	64.7	100.9	82.3	88.6	102.4	62.0	75.7	95.1	68.5	77.6	87.5	67.2	74.1
09600	FLOORING	115.9	61.6	102.7	115.9	37.7	96.9	87.2	72.3	83.6	86.0	48.0	76.8	115.9	43.9	98.5	90.5	44.3	79.3
097,099	WALL FINISHES, PAINTS & COATINGS	109.2	56.5	78.5	105.6	48.1	72.2	94.1	78.2	84.9	94.1	45.7	66.0	105.6	46.1	71.0	107.2	56.4	77.7
09	**FINISHES**	106.9	56.6	81.3	105.0	46.3	75.0	93.5	78.3	85.7	93.0	56.5	74.4	104.9	58.4	81.1	91.6	59.7	75.3
10 - 14	**TOTAL DIV. 10000 - 14000**	100.0	75.9	94.9	100.0	73.1	94.3	100.0	82.0	96.2	100.0	69.8	93.6	100.0	75.4	94.8	100.0	76.9	95.1
15	**MECHANICAL**	99.9	60.3	81.6	99.9	53.4	78.4	100.1	80.3	91.0	100.1	50.6	77.2	99.9	43.7	73.9	99.9	49.2	76.4
16	**ELECTRICAL**	96.2	57.4	69.5	90.7	66.0	73.8	94.0	84.0	87.1	96.1	54.1	67.2	92.0	46.7	60.8	90.1	66.2	73.7
01 - 16	**WEIGHTED AVERAGE**	99.9	64.7	82.8	97.5	60.3	79.4	96.4	81.4	89.1	95.2	60.9	78.5	97.6	58.7	78.7	95.9	65.7	81.2

	DIVISION	GEORGIA						HAWAII			IDAHO								
		SAVANNAH			VALDOSTA			HONOLULU			BOISE			LEWISTON			POCATELLO		
		MAT.	INST.	TOTAL	MAT.	INST.	TOTAL	MAT.	INST.	TOTAL	MAT.	INST.	TOTAL	MAT.	INST.	TOTAL	MAT.	INST.	TOTAL
02	**SITE CONSTRUCTION**	106.4	77.6	84.3	116.6	76.8	86.1	127.9	108.2	112.8	82.7	101.5	97.1	88.0	95.2	93.5	84.3	101.5	97.5
03100	CONCRETE FORMS & ACCESSORIES	95.7	60.1	65.0	81.4	51.8	55.9	106.9	150.1	144.2	101.5	87.3	89.3	112.9	76.6	81.6	101.5	87.0	89.0
03200	CONCRETE REINFORCEMENT	101.5	85.5	92.4	101.7	51.0	72.8	106.4	126.7	118.0	100.3	81.6	89.6	113.3	99.0	105.2	100.6	81.1	89.5
03300	CAST-IN-PLACE CONCRETE	93.8	56.4	78.4	95.1	57.1	79.5	190.4	128.9	165.0	101.2	95.3	98.8	109.7	93.0	102.8	100.4	95.1	98.2
03	**CONCRETE**	90.9	65.0	78.0	95.9	55.4	75.6	155.7	136.7	146.2	104.2	88.8	96.5	116.1	86.6	101.4	103.9	88.6	96.2
04	**MASONRY**	92.3	59.0	71.8	95.3	53.3	69.5	132.9	133.4	133.2	138.3	83.1	104.4	138.7	91.3	109.5	133.1	74.9	97.4
05	**METALS**	97.1	90.9	94.9	96.6	79.3	90.4	116.4	110.3	114.2	113.2	81.3	101.7	96.7	89.9	94.3	113.5	80.4	101.6
06	**WOOD & PLASTICS**	94.5	59.1	76.2	79.3	48.8	63.5	102.2	154.3	129.2	96.3	86.6	91.3	102.0	71.0	86.0	96.3	86.6	91.3
07	**THERMAL & MOISTURE PROTECTION**	96.7	59.0	78.9	96.4	62.2	80.2	113.7	132.2	122.5	98.1	84.2	91.5	168.2	84.1	128.3	98.3	77.7	88.6
08	**DOORS & WINDOWS**	98.3	58.5	88.7	93.8	45.2	82.1	109.0	143.5	117.4	95.5	81.2	92.0	115.8	75.3	106.0	95.5	75.6	90.7
09200	PLASTER & GYPSUM BOARD	107.2	58.4	77.3	101.0	47.7	68.4	111.9	156.0	138.9	88.2	86.0	86.9	152.1	70.1	102.0	88.2	86.0	86.9
095,098	CEILINGS & ACOUSTICAL TREATMENT	95.1	58.4	70.9	89.7	47.7	62.0	118.1	156.0	143.1	104.7	86.0	92.3	151.2	70.1	97.7	104.7	86.0	92.3
09600	FLOORING	115.9	56.6	101.5	107.7	45.2	92.5	119.4	131.9	122.5	103.8	65.3	94.5	143.7	95.9	132.1	104.1	65.3	94.7
097,099	WALL FINISHES, PAINTS & COATINGS	105.6	57.2	77.4	105.6	41.7	68.5	113.9	150.0	134.9	108.7	60.3	80.6	136.5	75.5	101.0	108.7	63.2	82.2
09	**FINISHES**	105.1	59.0	81.6	101.0	48.9	74.4	118.6	148.1	133.6	98.6	80.6	89.4	167.2	79.1	122.3	98.7	80.9	89.6
10 - 14	**TOTAL DIV. 10000 - 14000**	100.0	71.5	94.0	100.0	70.1	93.7	100.0	129.1	106.2	100.0	84.1	96.7	100.0	92.9	98.5	100.0	84.1	96.7
15	**MECHANICAL**	99.9	54.2	78.8	99.9	47.5	75.7	100.3	123.6	111.1	100.1	84.8	93.0	101.5	90.7	96.5	100.1	84.7	93.0
16	**ELECTRICAL**	93.4	62.7	72.3	89.0	36.0	52.5	114.3	129.9	125.1	81.4	80.3	80.6	83.6	89.6	87.7	84.5	79.9	81.3
01 - 16	**WEIGHTED AVERAGE**	97.9	64.5	81.7	97.4	54.0	76.3	116.6	129.0	122.7	102.1	85.1	93.8	113.7	88.1	101.2	102.1	83.7	93.2

City Cost Indexes

	DIVISION	IDAHO			ILLINOIS														
		TWIN FALLS			CHICAGO			DECATUR			EAST ST. LOUIS			JOLIET			PEORIA		
		MAT.	INST.	TOTAL	MAT.	INST.	TOTAL	MAT.	INST.	TOTAL	MAT.	INST.	TOTAL	MAT.	INST.	TOTAL	MAT.	INST.	TOTAL
02	**SITE CONSTRUCTION**	91.6	99.6	97.7	86.0	91.0	89.8	82.2	95.3	92.2	100.1	95.8	96.8	86.0	90.1	89.1	90.2	94.4	93.4
03100	CONCRETE FORMS & ACCESSORIES	104.9	41.2	49.9	105.9	130.5	127.1	100.3	104.5	103.9	92.6	109.5	107.2	106.9	130.7	127.4	98.4	108.2	106.9
03200	CONCRETE REINFORCEMENT	109.3	57.0	79.5	94.9	158.7	131.3	96.5	110.4	104.4	102.1	118.2	111.2	94.9	144.0	122.9	96.5	114.5	106.7
03300	CAST-IN-PLACE CONCRETE	103.0	51.0	81.6	103.3	130.2	114.4	96.2	93.2	95.0	90.1	112.3	99.2	103.2	114.9	108.0	94.1	105.9	98.9
03	**CONCRETE**	113.2	48.7	80.9	100.6	134.6	117.6	93.8	101.9	97.8	85.2	112.9	99.0	100.6	126.5	113.5	92.5	108.7	100.6
04	**MASONRY**	136.7	46.0	80.9	93.9	131.5	117.0	69.0	101.3	88.9	73.1	110.4	96.0	95.5	126.4	114.5	113.0	109.7	111.0
05	**METALS**	113.4	68.4	97.2	96.4	123.7	106.3	95.5	108.1	100.0	93.0	123.9	104.1	94.5	115.2	102.0	95.5	112.0	101.4
06	**WOOD & PLASTICS**	99.7	41.2	69.5	103.3	128.9	116.5	102.6	103.5	103.1	96.8	108.9	103.1	105.3	131.3	118.7	102.6	106.0	104.4
07	**THERMAL & MOISTURE PROTECTION**	99.4	53.6	77.7	99.3	128.7	113.3	97.0	97.5	97.2	92.0	106.0	98.6	99.1	125.6	111.7	96.8	105.2	100.8
08	**DOORS & WINDOWS**	98.8	42.0	85.1	104.1	136.4	111.9	99.0	102.4	99.8	88.2	116.1	95.0	103.9	134.2	111.2	99.0	106.4	100.8
09200	PLASTER & GYPSUM BOARD	87.8	39.1	58.1	99.4	129.4	117.7	106.2	103.3	104.4	103.5	108.8	106.7	97.7	132.0	118.7	106.2	105.8	106.0
095,098	CEILINGS & ACOUSTICAL TREATMENT	97.9	39.1	59.1	99.2	129.4	119.1	99.2	103.3	101.9	93.8	108.8	103.7	99.2	132.0	120.8	99.2	105.8	103.6
09600	FLOORING	105.4	65.3	95.7	85.1	126.3	95.1	101.3	96.6	100.2	111.0	61.5	99.0	84.7	122.0	93.7	101.3	102.0	101.5
097,099	WALL FINISHES, PAINTS & COATINGS	108.7	37.2	67.1	80.7	134.4	111.9	91.0	102.0	97.4	99.2	101.7	100.6	78.4	123.0	104.3	91.0	96.2	94.0
09	**FINISHES**	98.4	44.9	71.1	89.4	129.9	110.1	97.9	103.3	100.6	97.8	99.4	98.6	88.9	128.7	109.2	97.8	105.6	101.8
10 - 14	**TOTAL DIV. 10000 - 14000**	100.0	57.7	91.1	100.0	123.7	105.0	100.0	101.0	100.2	100.0	105.5	101.2	100.0	122.7	104.8	100.0	102.8	100.6
15	**MECHANICAL**	100.1	43.0	73.7	100.0	124.5	111.3	100.2	99.6	99.9	100.1	99.8	99.9	100.2	119.7	109.2	100.2	105.1	102.4
16	**ELECTRICAL**	87.0	37.5	53.0	101.1	130.7	121.4	100.6	83.6	88.9	96.9	102.1	100.5	100.7	91.9	94.7	99.5	93.1	95.1
01 - 16	**WEIGHTED AVERAGE**	104.2	51.9	78.8	98.2	125.4	111.4	95.9	98.3	97.1	93.5	105.7	99.4	98.0	115.2	106.4	98.4	103.6	100.9

	DIVISION	ILLINOIS						INDIANA											
		ROCKFORD			SPRINGFIELD			ANDERSON			BLOOMINGTON			EVANSVILLE			FORT WAYNE		
		MAT.	INST.	TOTAL	MAT.	INST.	TOTAL	MAT.	INST.	TOTAL	MAT.	INST.	TOTAL	MAT.	INST.	TOTAL	MAT.	INST.	TOTAL
02	**SITE CONSTRUCTION**	90.4	94.7	93.7	88.7	95.3	93.8	88.0	94.7	93.1	78.5	95.3	91.4	84.0	130.5	119.6	88.8	94.9	93.5
03100	CONCRETE FORMS & ACCESSORIES	103.2	112.5	111.3	102.6	104.6	104.3	91.5	82.2	83.5	103.1	83.4	86.1	91.0	84.1	85.0	91.1	77.1	79.0
03200	CONCRETE REINFORCEMENT	96.5	141.5	122.2	96.5	110.4	104.4	96.8	86.9	91.1	88.8	86.7	87.6	97.3	85.8	90.7	96.8	83.2	89.0
03300	CAST-IN-PLACE CONCRETE	96.2	104.5	99.6	89.4	100.7	94.1	101.7	83.3	94.1	102.7	81.8	94.1	98.1	93.5	96.2	108.0	81.4	97.0
03	**CONCRETE**	94.0	115.2	104.6	90.6	104.6	97.6	96.7	84.0	90.3	104.2	83.0	93.6	107.0	87.8	97.4	99.7	80.4	90.1
04	**MASONRY**	83.9	114.5	102.7	67.7	100.4	87.8	92.4	83.6	87.0	92.2	81.9	85.9	88.3	85.8	86.8	89.7	83.2	85.7
05	**METALS**	95.5	123.7	105.7	95.5	108.9	100.3	102.1	92.7	98.8	97.4	77.4	90.2	90.7	89.6	90.3	102.1	91.0	98.1
06	**WOOD & PLASTICS**	102.6	110.1	106.4	102.4	103.5	103.0	101.9	82.0	91.6	116.9	82.7	99.2	92.7	82.5	87.4	101.5	75.2	87.9
07	**THERMAL & MOISTURE PROTECTION**	96.7	112.8	104.3	96.6	103.5	99.9	100.0	82.3	91.6	92.1	86.7	89.5	95.3	88.5	92.1	99.3	86.4	93.2
08	**DOORS & WINDOWS**	99.0	116.5	103.2	99.0	102.4	99.8	101.9	84.8	97.8	105.5	83.4	100.2	98.0	83.0	94.4	101.9	77.9	96.1
09200	PLASTER & GYPSUM BOARD	106.2	110.0	108.5	106.2	103.3	104.4	106.5	82.0	91.5	105.0	82.9	91.5	101.1	80.8	88.7	102.2	74.9	85.5
095,098	CEILINGS & ACOUSTICAL TREATMENT	99.2	110.0	106.3	99.2	103.3	101.9	93.8	82.0	86.0	83.7	82.9	83.2	95.1	80.8	85.7	93.8	74.9	81.3
09600	FLOORING	101.3	103.0	101.7	101.5	96.8	100.4	87.0	77.4	84.7	102.2	93.7	100.2	95.7	86.6	93.5	87.0	82.6	85.9
097,099	WALL FINISHES, PAINTS & COATINGS	91.0	111.8	103.1	91.0	101.8	97.3	90.9	77.5	83.1	90.2	90.3	90.3	96.5	89.0	92.2	90.9	77.5	83.1
09	**FINISHES**	97.8	110.2	104.1	97.9	103.0	100.5	91.7	81.0	86.3	96.9	86.1	91.4	96.2	84.9	90.4	90.9	77.7	84.1
10 - 14	**TOTAL DIV. 10000 - 14000**	100.0	112.0	102.5	100.0	100.6	100.1	100.0	89.2	97.7	100.0	89.3	97.7	100.0	92.3	98.4	100.0	86.5	97.2
15	**MECHANICAL**	100.2	106.8	103.2	100.2	99.4	99.8	99.8	85.1	93.0	99.8	86.8	93.8	100.0	87.7	94.3	99.8	84.0	92.5
16	**ELECTRICAL**	99.5	113.1	108.8	100.5	95.1	96.8	81.3	92.8	89.2	101.0	88.4	92.3	95.7	89.6	91.5	81.8	86.7	85.2
01 - 16	**WEIGHTED AVERAGE**	97.0	111.2	103.9	95.6	100.6	98.1	97.1	87.2	92.3	99.3	85.9	92.8	97.1	91.8	94.6	97.3	84.7	91.1

	DIVISION	INDIANA															IOWA		
		GARY			INDIANAPOLIS			MUNCIE			SOUTH BEND			TERRE HAUTE			CEDAR RAPIDS		
		MAT.	INST.	TOTAL	MAT.	INST.	TOTAL	MAT.	INST.	TOTAL	MAT.	INST.	TOTAL	MAT.	INST.	TOTAL	MAT.	INST.	TOTAL
02	**SITE CONSTRUCTION**	88.6	96.8	94.9	88.0	98.7	96.2	78.3	95.5	91.5	88.1	95.1	93.4	85.1	130.7	120.0	84.8	94.3	92.1
03100	CONCRETE FORMS & ACCESSORIES	92.0	108.3	106.1	92.0	87.4	88.0	88.5	82.0	82.9	96.8	78.4	80.9	92.6	84.9	86.0	100.9	78.3	81.4
03200	CONCRETE REINFORCEMENT	96.8	102.8	100.2	96.1	86.6	90.7	98.2	86.8	91.7	96.8	79.7	87.0	97.3	84.2	89.8	100.1	82.2	89.9
03300	CAST-IN-PLACE CONCRETE	106.2	107.5	106.7	97.8	87.5	93.6	107.9	82.7	97.5	98.9	86.6	93.8	94.9	88.3	92.2	105.3	83.4	96.3
03	**CONCRETE**	98.9	107.0	102.9	99.1	86.8	92.9	102.2	83.8	93.0	91.2	83.2	87.2	106.8	86.1	96.4	97.7	81.2	89.5
04	**MASONRY**	93.8	101.9	98.8	99.6	85.1	90.7	94.3	83.6	87.7	87.1	83.3	84.7	95.2	83.8	88.2	107.3	79.9	90.4
05	**METALS**	102.1	104.1	102.9	104.1	80.3	95.5	99.0	92.8	96.8	102.1	105.7	103.4	91.4	89.2	90.6	89.6	86.4	88.4
06	**WOOD & PLASTICS**	100.9	109.8	105.5	99.2	87.5	93.1	103.6	81.8	92.3	102.1	76.8	89.0	95.2	84.5	89.6	105.2	76.6	90.4
07	**THERMAL & MOISTURE PROTECTION**	99.2	104.0	101.4	98.7	88.1	93.7	93.6	82.6	88.4	98.9	86.9	93.2	95.4	86.6	91.3	98.2	81.0	90.0
08	**DOORS & WINDOWS**	101.9	105.7	102.8	108.1	87.7	103.1	100.1	84.7	96.4	94.9	78.7	91.0	98.6	85.4	95.4	103.0	79.7	97.4
09200	PLASTER & GYPSUM BOARD	103.0	110.7	107.7	103.0	87.4	93.4	100.0	82.0	89.0	106.1	76.6	88.0	101.1	82.9	89.9	100.4	76.4	85.7
095,098	CEILINGS & ACOUSTICAL TREATMENT	93.8	110.7	104.9	92.4	87.4	89.1	85.1	82.0	83.0	93.8	76.6	82.4	95.1	82.9	87.0	112.6	76.4	88.7
09600	FLOORING	87.0	108.4	92.2	86.8	97.7	89.4	95.1	77.4	90.8	87.0	78.7	85.0	95.7	96.4	95.9	126.4	67.1	112.0
097,099	WALL FINISHES, PAINTS & COATINGS	90.9	91.0	91.0	90.9	90.4	90.6	90.2	77.5	82.8	90.9	83.3	86.5	96.5	84.9	89.8	111.4	77.0	91.4
09	**FINISHES**	91.2	106.7	99.1	91.2	89.7	90.5	93.8	81.0	87.3	91.7	78.5	84.9	96.2	87.1	91.5	112.6	75.1	93.5
10 - 14	**TOTAL DIV. 10000 - 14000**	100.0	108.4	101.8	100.0	91.1	98.1	100.0	88.6	97.6	100.0	85.5	96.9	100.0	93.8	98.7	100.0	86.6	97.2
15	**MECHANICAL**	99.8	100.9	100.3	99.8	91.0	95.7	99.7	85.0	92.9	99.8	81.3	91.3	100.0	85.9	93.5	100.4	86.2	93.8
16	**ELECTRICAL**	88.3	105.6	100.2	99.0	92.7	94.7	89.4	87.0	87.8	94.2	87.6	89.7	92.9	88.4	89.8	96.2	83.5	87.4
01 - 16	**WEIGHTED AVERAGE**	97.9	103.7	100.7	100.0	89.6	94.9	97.5	86.3	92.1	96.2	86.2	91.3	97.6	91.2	94.5	99.6	83.5	91.8

	DIVISION	IOWA																	
		COUNCIL BLUFFS			DAVENPORT			DES MOINES			DUBUQUE			SIOUX CITY			WATERLOO		
		MAT.	INST.	TOTAL	MAT.	INST.	TOTAL	MAT.	INST.	TOTAL	MAT.	INST.	TOTAL	MAT.	INST.	TOTAL	MAT.	INST.	TOTAL
02	**SITE CONSTRUCTION**	88.1	91.9	91.0	83.6	99.4	95.7	77.1	101.3	95.6	83.0	92.0	89.9	92.5	96.6	95.7	84.1	95.9	93.1
03100	CONCRETE FORMS & ACCESSORIES	76.5	64.7	66.3	100.5	93.6	94.5	102.6	80.7	83.7	78.5	69.5	70.8	100.9	70.2	74.4	101.5	56.2	62.4
03200	CONCRETE REINFORCEMENT	102.2	80.2	89.7	100.1	96.1	97.8	100.1	78.5	87.8	98.7	82.0	89.2	100.1	68.5	82.1	100.1	81.9	89.7
03300	CAST-IN-PLACE CONCRETE	109.6	72.8	94.4	101.5	94.2	98.5	105.7	85.0	97.2	103.1	68.6	88.9	104.4	62.3	87.1	105.3	53.6	84.0
03	**CONCRETE**	99.4	71.2	85.3	95.8	94.3	95.0	98.0	82.2	90.1	94.0	72.2	83.1	97.3	67.8	82.5	97.7	61.1	79.4
04	**MASONRY**	107.6	75.0	87.6	106.9	84.7	93.3	101.4	81.8	89.3	108.3	73.5	86.9	101.5	68.1	81.0	101.6	62.6	77.6
05	**METALS**	94.9	84.9	91.3	89.6	94.9	91.5	89.5	86.8	88.5	88.1	85.5	87.2	89.6	80.1	86.2	89.6	85.0	87.9
06	**WOOD & PLASTICS**	79.5	62.9	70.9	105.2	92.4	98.6	106.8	79.0	92.4	81.8	67.4	74.3	105.2	70.0	87.0	105.9	56.9	80.6
07	**THERMAL & MOISTURE PROTECTION**	97.3	68.9	83.8	97.5	88.2	93.1	98.3	78.1	88.7	97.6	71.2	85.1	97.5	64.0	81.6	97.1	58.5	78.8
08	**DOORS & WINDOWS**	102.0	66.6	93.5	103.0	92.0	100.3	103.0	78.8	97.2	102.0	74.8	95.5	103.0	67.8	94.5	98.3	65.8	90.5
09200	PLASTER & GYPSUM BOARD	92.7	62.2	74.1	100.4	92.4	95.5	97.3	78.6	85.8	93.1	66.9	77.1	100.4	69.3	81.4	100.4	55.8	73.1
095,098	CEILINGS & ACOUSTICAL TREATMENT	108.5	62.2	78.0	112.6	92.4	99.3	111.3	78.6	89.7	108.5	66.9	81.1	112.6	69.3	84.0	112.6	55.8	75.1
09600	FLOORING	101.0	54.8	89.8	112.1	86.4	105.8	111.8	50.0	96.9	115.6	46.7	98.9	112.6	70.7	102.4	113.7	71.0	103.4
097,099	WALL FINISHES, PAINTS & COATINGS	102.5	58.1	76.7	106.9	101.9	104.0	106.9	83.7	93.4	110.1	61.1	81.6	108.2	71.1	86.6	108.2	42.2	69.8
09	**FINISHES**	101.8	61.1	81.0	107.6	93.0	100.2	106.1	74.5	90.0	107.1	63.3	84.7	108.8	70.4	89.2	108.3	57.0	82.1
10 - 14	**TOTAL DIV. 10000 - 14000**	100.0	81.5	96.1	100.0	91.8	98.3	100.0	88.3	97.5	100.0	84.0	96.6	100.0	84.2	96.7	100.0	78.4	95.4
15	**MECHANICAL**	100.4	79.7	90.8	100.4	100.0	100.2	100.4	89.5	95.3	100.4	87.0	94.2	100.4	85.7	93.6	100.4	54.8	79.3
16	**ELECTRICAL**	104.5	84.9	91.0	93.6	88.4	90.0	96.9	90.1	92.2	102.2	81.9	88.3	96.2	77.9	83.6	96.2	64.3	74.2
01 - 16	**WEIGHTED AVERAGE**	99.8	77.5	88.9	98.6	93.6	96.1	98.4	86.2	92.5	98.4	79.2	89.1	99.0	77.6	88.6	98.2	66.2	82.7

	DIVISION	KANSAS															KENTUCKY		
		DODGE CITY			KANSAS CITY			SALINA			TOPEKA			WICHITA			BOWLING GREEN		
		MAT.	INST.	TOTAL	MAT.	INST.	TOTAL	MAT.	INST.	TOTAL	MAT.	INST.	TOTAL	MAT.	INST.	TOTAL	MAT.	INST.	TOTAL
02	**SITE CONSTRUCTION**	105.8	93.2	96.1	86.1	90.4	89.4	95.6	93.5	93.9	88.4	90.8	90.2	89.7	93.7	92.8	70.8	100.9	93.9
03100	CONCRETE FORMS & ACCESSORIES	95.8	50.7	56.9	102.4	90.3	92.0	90.8	53.4	58.5	101.3	51.6	58.5	98.0	63.6	68.3	82.2	85.3	84.9
03200	CONCRETE REINFORCEMENT	104.4	69.7	84.7	99.3	92.9	95.6	103.8	86.9	94.2	96.5	92.6	94.3	96.5	87.8	91.5	88.1	73.8	79.9
03300	CAST-IN-PLACE CONCRETE	111.7	65.8	92.8	87.6	93.1	89.9	96.9	56.0	80.0	91.3	60.8	78.7	85.8	64.3	76.9	89.6	85.9	88.0
03	**CONCRETE**	110.1	61.1	85.6	92.0	92.4	92.2	97.5	62.2	79.8	91.4	64.1	77.8	88.4	69.7	79.1	95.3	83.4	89.4
04	**MASONRY**	103.1	50.3	70.7	103.1	88.7	94.3	117.7	48.4	75.1	96.7	60.6	74.5	91.3	68.9	77.6	95.2	86.2	89.7
05	**METALS**	92.8	86.3	90.4	97.6	100.5	98.6	92.6	93.4	92.9	95.5	97.4	96.2	95.5	95.2	95.4	94.8	80.8	89.8
06	**WOOD & PLASTICS**	94.2	50.8	71.7	102.4	91.4	96.7	89.4	54.6	71.4	98.8	48.3	72.7	96.3	62.8	79.0	88.5	82.8	85.6
07	**THERMAL & MOISTURE PROTECTION**	97.9	56.7	78.4	95.6	91.3	93.6	97.3	61.7	80.4	97.2	72.8	85.6	96.8	71.4	84.8	86.2	84.6	85.4
08	**DOORS & WINDOWS**	98.9	54.7	88.2	97.7	88.5	95.5	98.8	59.0	89.2	99.0	64.0	90.5	99.0	68.3	91.6	97.7	77.7	92.9
09200	PLASTER & GYPSUM BOARD	104.1	48.8	70.3	100.5	90.8	94.6	102.8	52.8	72.3	106.2	46.3	69.6	106.2	61.3	78.7	97.3	82.5	88.3
095,098	CEILINGS & ACOUSTICAL TREATMENT	92.4	48.8	63.7	99.2	90.8	93.7	92.4	52.8	66.3	99.2	46.3	64.3	99.2	61.3	74.2	91.0	82.5	85.4
09600	FLOORING	100.6	59.3	90.6	90.7	63.3	84.0	98.6	43.3	85.2	102.2	55.1	90.8	101.3	79.4	96.0	91.7	73.1	87.2
097,099	WALL FINISHES, PAINTS & COATINGS	91.0	58.6	72.2	99.2	98.5	98.8	91.0	46.2	64.9	91.0	68.4	77.9	91.0	68.7	78.0	96.5	78.8	86.2
09	**FINISHES**	97.0	52.6	74.4	96.2	86.1	91.0	95.6	50.0	72.3	98.1	52.2	74.7	97.9	66.6	82.0	93.7	82.2	87.8
10 - 14	**TOTAL DIV. 10000 - 14000**	100.0	60.1	91.6	100.0	90.7	98.0	100.0	72.7	94.2	100.0	79.5	95.7	100.0	76.8	95.1	100.0	74.5	94.6
15	**MECHANICAL**	100.2	56.8	80.1	100.0	97.1	98.7	100.2	42.9	73.7	100.2	72.0	87.2	100.2	73.8	88.0	100.2	87.5	94.3
16	**ELECTRICAL**	95.2	60.0	71.0	104.5	78.8	86.8	94.7	74.2	80.6	101.4	72.2	81.4	99.3	74.2	82.1	93.2	90.4	91.3
01 - 16	**WEIGHTED AVERAGE**	99.6	63.0	81.8	98.0	90.2	94.2	98.3	63.7	81.5	97.4	71.4	84.8	96.6	75.6	86.4	95.6	86.6	91.2

	DIVISION	KENTUCKY									LOUISIANA								
		LEXINGTON			LOUISVILLE			OWENSBORO			ALEXANDRIA			BATON ROUGE			LAKE CHARLES		
		MAT.	INST.	TOTAL	MAT.	INST.	TOTAL	MAT.	INST.	TOTAL	MAT.	INST.	TOTAL	MAT.	INST.	TOTAL	MAT.	INST.	TOTAL
02	**SITE CONSTRUCTION**	78.3	102.9	97.1	67.9	100.4	92.8	83.2	130.4	119.3	104.2	85.4	89.8	114.7	87.0	93.4	116.4	86.7	93.7
03100	CONCRETE FORMS & ACCESSORIES	93.0	73.1	75.8	91.6	82.2	83.5	86.7	75.4	77.0	78.3	50.9	54.7	98.7	56.2	62.0	98.8	57.1	62.8
03200	CONCRETE REINFORCEMENT	97.3	98.1	97.7	97.3	99.1	98.3	88.2	99.4	94.6	97.0	68.2	80.6	98.1	59.2	75.9	98.1	59.0	75.8
03300	CAST-IN-PLACE CONCRETE	96.7	76.0	88.2	93.6	80.6	88.2	92.4	87.8	90.5	97.9	48.9	77.7	88.6	58.8	76.3	93.8	60.5	80.1
03	**CONCRETE**	97.3	79.1	88.2	95.7	84.9	90.3	104.9	84.5	94.7	90.0	54.6	72.3	93.3	58.5	75.9	95.8	59.5	77.7
04	**MASONRY**	92.3	56.6	70.3	92.7	81.8	86.0	92.3	61.3	73.3	114.7	50.8	75.4	102.3	60.7	76.7	101.1	61.0	76.5
05	**METALS**	96.3	90.7	94.3	96.3	91.4	94.5	86.2	92.1	88.3	86.6	76.6	83.0	101.0	72.2	90.6	100.6	72.3	90.4
06	**WOOD & PLASTICS**	95.2	75.3	84.9	98.2	82.8	90.3	88.4	74.9	81.4	80.0	51.4	65.2	105.8	56.6	80.4	103.7	57.5	79.8
07	**THERMAL & MOISTURE PROTECTION**	95.8	68.9	83.0	86.4	83.0	84.8	95.3	76.1	86.2	98.3	55.5	78.0	99.3	62.3	81.7	101.9	62.3	83.1
08	**DOORS & WINDOWS**	98.6	82.9	94.8	98.6	87.3	95.9	95.9	83.3	92.9	100.2	56.9	89.7	104.6	56.7	93.1	104.6	59.0	93.6
09200	PLASTER & GYPSUM BOARD	101.1	73.4	84.1	103.9	82.5	90.8	96.8	73.0	82.3	92.2	50.8	66.9	105.1	56.1	75.1	105.1	56.9	75.7
095,098	CEILINGS & ACOUSTICAL TREATMENT	95.1	73.4	80.8	95.1	82.5	86.8	81.5	73.0	75.9	99.7	50.8	67.4	100.7	56.1	71.3	100.7	56.9	71.8
09600	FLOORING	96.6	50.8	85.5	95.1	77.8	90.9	94.0	73.1	88.9	107.4	58.2	95.5	110.7	70.2	100.9	110.9	61.3	98.9
097,099	WALL FINISHES, PAINTS & COATINGS	96.5	69.5	80.8	96.5	83.8	89.1	96.5	100.7	98.9	95.6	49.0	68.5	97.1	56.2	73.3	97.1	51.8	70.8
09	**FINISHES**	96.5	68.1	82.1	96.2	81.2	88.6	92.6	77.2	84.7	98.3	51.6	74.5	104.0	58.7	80.9	104.0	57.2	80.1
10 - 14	**TOTAL DIV. 10000 - 14000**	100.0	92.0	98.3	100.0	92.4	98.4	100.0	94.0	98.7	100.0	84.9	96.8	100.0	73.5	94.4	100.0	73.7	94.5
15	**MECHANICAL**	100.0	57.3	80.3	100.0	83.5	92.4	100.2	63.0	83.0	100.1	45.3	74.7	100.2	51.6	77.7	100.2	63.6	83.2
16	**ELECTRICAL**	94.0	60.5	70.9	94.0	90.4	91.5	92.4	89.0	90.0	95.5	55.1	67.7	96.4	60.1	71.4	96.4	63.1	73.5
01 - 16	**WEIGHTED AVERAGE**	96.9	71.9	84.7	96.1	87.2	91.8	95.5	83.1	89.5	96.9	58.1	78.0	100.7	62.1	81.9	101.0	65.0	83.5

	DIVISION	LOUISIANA									MAINE								
		MONROE			NEW ORLEANS			SHREVEPORT			AUGUSTA			BANGOR			LEWISTON		
		MAT.	INST.	TOTAL	MAT.	INST.	TOTAL	MAT.	INST.	TOTAL	MAT.	INST.	TOTAL	MAT.	INST.	TOTAL	MAT.	INST.	TOTAL
02	**SITE CONSTRUCTION**	104.2	85.2	89.7	118.5	89.1	96.0	102.8	85.1	89.2	87.2	100.9	97.7	87.1	103.2	99.5	85.3	103.2	99.0
03100	CONCRETE FORMS & ACCESSORIES	77.6	51.7	55.3	97.7	64.2	68.8	100.1	54.8	61.0	97.8	39.0	47.0	91.4	90.6	90.7	97.9	90.3	91.4
03200	CONCRETE REINFORCEMENT	96.0	68.3	80.2	98.1	69.6	81.8	95.5	68.2	80.0	91.1	40.9	62.5	91.1	117.9	106.4	111.9	117.9	115.3
03300	CAST-IN-PLACE CONCRETE	97.9	53.8	79.7	92.9	62.5	80.4	96.5	57.9	80.6	81.6	48.9	68.1	81.6	74.4	78.6	90.8	74.3	84.0
03	**CONCRETE**	89.8	56.6	73.2	95.3	65.3	80.3	90.0	59.3	74.7	100.5	44.5	72.5	99.5	89.3	94.4	102.8	89.2	96.0
04	**MASONRY**	109.6	52.6	74.6	103.2	61.4	77.5	102.6	47.4	68.7	93.6	43.4	62.7	111.1	63.7	82.0	94.7	63.7	75.6
05	**METALS**	86.6	76.1	82.8	103.3	78.1	94.2	87.4	75.6	83.2	95.1	75.1	87.9	94.9	86.9	92.0	98.1	86.6	94.0
06	**WOOD & PLASTICS**	79.1	51.9	65.0	99.5	65.2	81.8	102.6	56.5	78.7	99.8	37.3	67.5	93.1	93.2	93.1	99.8	93.2	96.4
07	**THERMAL & MOISTURE PROTECTION**	98.2	57.9	79.1	102.7	65.9	85.3	97.4	57.1	78.3	99.8	42.3	72.5	99.6	64.2	82.8	99.5	64.2	82.8
08	**DOORS & WINDOWS**	100.2	62.0	91.0	103.4	68.7	95.0	98.1	59.7	88.8	106.8	45.3	92.0	106.8	82.1	100.8	110.0	82.1	103.3
09200	PLASTER & GYPSUM BOARD	91.7	51.3	67.0	104.2	64.9	80.2	97.6	56.0	72.2	96.7	34.1	58.4	93.9	91.7	92.6	98.2	91.7	94.3
095,098	CEILINGS & ACOUSTICAL TREATMENT	99.7	51.3	67.8	99.2	64.9	76.6	101.1	56.0	71.4	94.4	34.1	54.6	93.1	91.7	92.2	102.6	91.7	95.4
09600	FLOORING	107.0	50.6	93.3	111.4	60.1	98.9	118.5	53.1	102.6	99.9	50.9	88.0	97.5	50.9	86.2	99.9	50.9	88.0
097,099	WALL FINISHES, PAINTS & COATINGS	95.6	50.6	69.4	99.0	65.9	79.8	95.6	49.0	68.5	95.6	28.2	56.4	95.6	39.0	62.7	95.6	39.0	62.7
09	**FINISHES**	98.1	50.7	73.9	104.2	63.4	83.4	102.9	53.5	77.7	98.4	39.1	68.2	97.0	78.1	87.3	99.8	78.1	88.7
10 - 14	**TOTAL DIV. 10000 - 14000**	100.0	70.9	93.9	100.0	75.7	94.9	100.0	71.2	93.9	100.0	64.5	92.5	100.0	87.7	97.4	100.0	87.5	97.4
15	**MECHANICAL**	100.1	54.6	79.0	100.2	65.4	84.1	100.1	58.2	80.7	99.8	30.8	67.9	99.8	85.0	93.0	99.8	85.0	93.0
16	**ELECTRICAL**	98.6	59.0	71.4	96.7	67.7	76.8	96.6	69.6	78.1	104.0	86.1	91.7	100.6	86.2	90.7	104.0	53.1	69.0
01 - 16	**WEIGHTED AVERAGE**	96.8	60.7	79.2	101.4	69.1	85.7	96.9	63.3	80.5	99.5	56.9	78.8	99.8	84.4	92.3	100.8	78.7	90.0

	DIVISION	MAINE			MARYLAND						MASSACHUSETTS								
		PORTLAND			BALTIMORE			HAGERSTOWN			BOSTON			BROCKTON			FALL RIVER		
		MAT.	INST.	TOTAL	MAT.	INST.	TOTAL	MAT.	INST.	TOTAL	MAT.	INST.	TOTAL	MAT.	INST.	TOTAL	MAT.	INST.	TOTAL
02	**SITE CONSTRUCTION**	84.7	103.2	98.9	94.5	92.3	92.8	84.7	88.1	87.3	92.3	110.2	106.0	89.9	106.0	102.3	88.9	106.2	102.1
03100	CONCRETE FORMS & ACCESSORIES	97.1	90.3	91.3	101.1	76.2	79.6	88.4	77.9	79.3	103.9	131.3	127.6	103.5	114.5	113.0	103.5	110.9	109.9
03200	CONCRETE REINFORCEMENT	111.9	117.9	115.3	97.2	93.1	94.9	85.9	79.2	82.1	110.7	141.3	128.1	111.9	141.0	128.5	111.9	128.0	121.1
03300	CAST-IN-PLACE CONCRETE	84.5	74.3	80.3	99.0	80.7	91.5	84.2	66.4	76.9	105.2	142.0	120.4	100.2	135.7	114.8	96.8	138.9	114.2
03	**CONCRETE**	99.7	89.2	94.4	100.4	82.2	91.3	85.7	75.4	80.6	113.3	136.0	124.6	110.4	126.2	118.3	108.8	123.3	116.0
04	**MASONRY**	92.6	63.7	74.8	88.1	74.6	79.8	92.3	77.7	83.3	113.9	142.6	131.5	109.5	135.5	125.5	109.2	138.3	127.1
05	**METALS**	98.2	86.6	94.0	95.9	102.6	98.3	95.0	95.4	95.1	100.9	123.7	109.1	98.0	120.4	106.1	98.0	115.8	104.4
06	**WOOD & PLASTICS**	99.8	93.2	96.4	101.2	77.3	88.8	88.5	77.1	82.6	104.0	130.6	117.8	103.0	110.8	107.1	103.0	106.2	104.7
07	**THERMAL & MOISTURE PROTECTION**	99.4	64.2	82.7	96.9	80.7	89.2	96.1	77.0	87.1	100.1	138.1	118.1	99.9	130.9	114.6	99.8	125.1	111.8
08	**DOORS & WINDOWS**	110.0	82.1	103.3	92.3	86.7	91.0	91.6	78.0	88.3	103.7	131.2	110.3	103.6	120.5	107.7	103.6	112.9	105.9
09200	PLASTER & GYPSUM BOARD	98.2	91.7	94.3	100.9	77.2	86.4	97.5	77.2	85.1	98.2	130.4	117.9	98.2	109.9	105.4	98.2	104.8	102.2
095,098	CEILINGS & ACOUSTICAL TREATMENT	102.6	91.7	95.4	89.7	77.2	81.4	91.2	77.2	81.9	99.5	130.4	119.9	102.6	109.9	107.4	102.6	104.8	104.0
09600	FLOORING	99.9	50.9	88.0	91.6	81.7	89.2	86.7	78.3	84.7	99.7	152.1	112.4	100.2	152.1	112.8	99.9	152.1	112.6
097,099	WALL FINISHES, PAINTS & COATINGS	95.6	39.0	62.7	98.2	80.2	87.7	98.2	47.8	68.9	96.3	139.3	121.2	95.6	125.7	113.1	95.6	125.7	113.1
09	**FINISHES**	99.9	78.1	88.7	91.1	77.1	84.0	88.7	74.6	81.5	99.0	136.3	118.0	99.7	122.2	111.2	99.6	119.6	109.8
10 - 14	**TOTAL DIV. 10000 - 14000**	100.0	87.5	97.4	100.0	85.5	96.9	100.0	88.1	97.5	100.0	126.4	105.6	100.0	122.3	104.7	100.0	122.8	104.8
15	**MECHANICAL**	99.8	85.0	93.0	99.9	86.2	93.6	99.9	85.3	93.2	99.8	122.0	110.1	99.8	101.9	100.8	99.8	102.2	100.9
16	**ELECTRICAL**	104.2	53.1	69.1	101.9	94.5	96.8	98.0	85.6	89.5	99.1	125.1	116.9	98.7	101.8	100.8	98.1	101.8	100.6
01 - 16	**WEIGHTED AVERAGE**	100.3	78.7	89.8	96.8	86.7	91.9	94.0	82.7	88.5	102.6	128.0	115.0	101.5	115.0	108.1	101.2	113.6	107.3

	DIVISION	MASSACHUSETTS																	
		HYANNIS			LAWRENCE			LOWELL			NEW BEDFORD			PITTSFIELD			SPRINGFIELD		
		MAT.	INST.	TOTAL	MAT.	INST.	TOTAL	MAT.	INST.	TOTAL	MAT.	INST.	TOTAL	MAT.	INST.	TOTAL	MAT.	INST.	TOTAL
02	**SITE CONSTRUCTION**	86.6	106.0	101.5	90.7	106.0	102.5	89.8	106.0	102.2	88.6	106.2	102.0	90.9	104.9	101.6	90.1	105.0	101.5
03100	CONCRETE FORMS & ACCESSORIES	93.2	110.4	108.0	103.4	114.5	113.0	99.7	114.5	112.5	103.5	110.9	109.9	99.7	95.1	95.7	99.8	103.0	102.5
03200	CONCRETE REINFORCEMENT	89.7	124.9	109.8	111.0	126.3	119.7	111.9	125.8	119.8	111.9	128.0	121.1	92.9	106.2	100.5	111.9	119.4	116.2
03300	CAST-IN-PLACE CONCRETE	91.3	138.0	110.6	100.4	138.2	116.0	91.8	138.2	110.9	94.0	138.9	112.5	100.1	105.2	102.2	94.0	110.9	101.0
03	**CONCRETE**	100.4	122.2	111.3	110.4	124.4	117.4	100.9	124.1	112.5	107.4	123.3	115.3	102.6	100.4	101.5	102.0	108.3	105.1
04	**MASONRY**	108.5	138.4	126.9	108.5	138.4	126.9	94.7	129.5	116.1	109.0	138.3	127.0	95.4	106.4	102.1	94.9	111.4	105.0
05	**METALS**	94.7	113.9	101.6	95.7	114.8	102.6	95.6	110.9	101.2	98.0	115.8	104.4	95.5	98.3	96.5	98.0	104.1	100.2
06	**WOOD & PLASTICS**	92.7	105.8	99.5	103.0	110.8	107.1	101.9	110.8	106.5	103.0	106.2	104.7	101.9	92.5	97.0	101.9	101.3	101.6
07	**THERMAL & MOISTURE PROTECTION**	99.3	125.0	111.5	99.8	132.0	115.1	99.6	128.3	113.2	99.8	125.1	111.8	99.6	104.9	102.1	99.6	108.0	103.6
08	**DOORS & WINDOWS**	99.7	111.9	102.6	103.6	116.6	106.7	110.0	116.6	111.6	103.6	112.9	105.9	110.0	97.5	107.0	110.0	105.7	109.0
09200	PLASTER & GYPSUM BOARD	93.0	104.8	100.2	98.2	109.9	105.4	98.2	109.9	105.4	98.2	104.8	102.2	98.2	91.0	93.8	98.2	100.1	99.4
095,098	CEILINGS & ACOUSTICAL TREATMENT	93.1	104.8	100.8	102.6	109.9	107.4	102.6	109.9	107.4	102.6	104.8	104.0	102.6	91.0	94.9	102.6	100.1	101.0
09600	FLOORING	96.3	152.1	109.8	99.9	152.1	112.6	99.9	152.1	112.6	99.9	152.1	112.6	100.2	112.2	103.1	99.7	117.3	104.0
097,099	WALL FINISHES, PAINTS & COATINGS	95.6	125.7	113.1	95.6	125.7	113.1	95.6	125.7	113.1	95.6	125.7	113.1	95.6	94.1	94.7	96.8	94.1	95.3
09	**FINISHES**	95.7	119.3	107.7	99.6	122.2	111.1	99.6	122.2	111.1	99.6	119.6	109.8	99.7	97.8	98.7	99.6	104.5	102.1
10 - 14	**TOTAL DIV. 10000 - 14000**	100.0	121.7	104.6	100.0	122.4	104.7	100.0	122.4	104.7	100.0	122.8	104.8	100.0	107.7	101.6	100.0	109.7	102.0
15	**MECHANICAL**	99.8	102.1	100.9	99.8	105.2	102.3	99.8	117.6	108.0	99.8	102.2	100.9	99.8	88.3	94.5	99.8	98.4	99.2
16	**ELECTRICAL**	93.9	101.8	99.3	103.2	105.9	105.0	104.0	105.8	105.3	99.1	101.8	100.9	104.0	84.0	90.2	104.0	91.6	95.4
01 - 16	**WEIGHTED AVERAGE**	98.2	113.2	105.5	101.4	115.7	108.4	100.3	116.7	108.2	101.1	113.6	107.2	100.5	95.9	98.3	100.8	102.6	101.6

City Cost Indexes

	DIVISION	MASSACHUSETTS			MICHIGAN														
		WORCESTER			ANN ARBOR			DEARBORN			DETROIT			FLINT			GRAND RAPIDS		
		MAT.	INST.	TOTAL	MAT.	INST.	TOTAL	MAT.	INST.	TOTAL	MAT.	INST.	TOTAL	MAT.	INST.	TOTAL	MAT.	INST.	TOTAL
02	SITE CONSTRUCTION	89.8	105.9	102.2	81.2	93.9	90.9	80.9	94.0	91.0	98.2	95.8	96.4	69.6	93.2	87.7	83.8	88.0	87.0
03100	CONCRETE FORMS & ACCESSORIES	100.3	113.7	111.8	97.1	116.9	114.2	97.0	120.6	117.4	98.8	120.6	117.6	98.3	98.9	98.8	95.9	79.0	81.3
03200	CONCRETE REINFORCEMENT	111.9	137.0	126.2	93.8	119.2	108.3	93.8	119.6	108.5	93.2	119.5	108.2	93.8	118.7	108.0	97.3	91.5	94.0
03300	CAST-IN-PLACE CONCRETE	91.8	132.5	108.6	92.5	122.3	104.8	90.4	124.2	104.3	99.2	124.2	109.5	93.1	105.4	98.2	96.9	89.5	93.8
03	CONCRETE	100.9	123.8	112.4	89.9	119.7	104.8	88.8	122.0	105.4	93.3	120.4	106.8	90.3	105.8	98.0	97.6	84.5	91.0
04	MASONRY	94.7	132.3	117.8	99.2	115.8	109.4	99.0	123.1	113.8	97.9	123.1	113.4	99.2	103.7	102.0	96.7	58.9	73.5
05	METALS	98.0	114.7	104.0	100.2	126.4	109.6	100.3	127.2	110.0	100.8	103.1	101.6	100.3	123.7	108.7	97.4	84.9	92.9
06	WOOD & PLASTICS	102.5	110.8	106.8	93.5	117.1	105.7	93.5	120.3	107.4	95.4	120.3	108.3	93.5	97.8	95.7	96.6	79.6	87.8
07	THERMAL & MOISTURE PROTECTION	99.6	119.3	108.9	94.0	116.4	104.6	93.1	124.8	108.2	91.5	124.8	107.3	92.3	99.5	95.7	91.7	67.1	80.1
08	DOORS & WINDOWS	110.0	119.6	112.3	97.8	113.9	101.7	97.8	115.6	102.1	97.4	115.7	101.8	97.8	101.1	98.6	93.3	73.8	88.6
09200	PLASTER & GYPSUM BOARD	98.2	109.9	105.4	107.9	115.8	112.7	107.9	119.2	114.8	107.9	119.2	114.8	107.9	95.9	100.6	101.1	72.9	83.9
095,098	CEILINGS & ACOUSTICAL TREATMENT	102.6	109.9	107.4	101.1	115.8	110.8	101.1	119.2	113.0	102.6	119.2	113.5	101.1	95.9	97.7	95.1	72.9	80.5
09600	FLOORING	99.9	143.8	110.6	90.4	121.8	98.0	89.9	123.5	98.1	90.1	123.5	98.2	90.1	88.3	89.7	95.3	52.1	84.8
097,099	WALL FINISHES, PAINTS & COATINGS	95.6	116.9	108.0	93.2	117.6	107.4	93.2	119.6	108.5	95.1	119.6	109.3	93.2	94.2	93.8	96.5	51.8	70.5
09	FINISHES	99.6	119.5	109.7	97.3	117.7	107.7	97.2	121.2	109.4	98.3	121.2	110.0	96.7	96.0	96.3	96.1	70.6	83.1
10 - 14	TOTAL DIV. 10000 - 14000	100.0	113.2	102.8	100.0	112.4	102.6	100.0	113.7	102.9	100.0	113.7	102.9	100.0	106.4	101.3	100.0	109.8	102.1
15	MECHANICAL	99.8	100.7	100.2	100.0	104.9	102.3	100.0	115.6	107.2	100.0	115.9	107.4	100.0	103.1	101.4	100.0	63.7	83.2
16	ELECTRICAL	104.0	97.2	99.3	95.2	104.1	101.3	95.2	115.4	109.1	96.5	115.4	109.5	95.2	105.3	102.1	93.5	72.6	79.1
01 - 16	WEIGHTED AVERAGE	100.6	111.9	106.1	97.0	111.1	103.9	96.9	117.0	106.7	98.1	114.8	106.2	96.7	103.8	100.1	96.7	74.4	85.9

	DIVISION	MICHIGAN												MINNESOTA					
		KALAMAZOO			LANSING			MUSKEGON			SAGINAW			DULUTH			MINNEAPOLIS		
		MAT.	INST.	TOTAL	MAT.	INST.	TOTAL	MAT.	INST.	TOTAL	MAT.	INST.	TOTAL	MAT.	INST.	TOTAL	MAT.	INST.	TOTAL
02	SITE CONSTRUCTION	84.4	88.4	87.4	88.6	93.0	92.0	82.5	88.3	87.0	71.9	92.9	88.0	82.3	103.3	98.4	82.0	109.3	102.9
03100	CONCRETE FORMS & ACCESSORIES	95.0	92.2	92.5	100.7	96.9	97.4	95.9	89.4	90.3	97.1	95.6	95.8	97.5	121.4	118.1	98.2	139.9	134.2
03200	CONCRETE REINFORCEMENT	97.3	87.9	91.9	93.8	118.5	107.9	97.9	92.8	95.0	93.8	118.2	107.7	95.6	108.1	102.7	95.8	119.4	109.2
03300	CAST-IN-PLACE CONCRETE	98.6	110.7	103.6	92.5	102.5	96.6	96.3	108.2	101.2	91.3	102.7	96.0	109.5	115.1	111.8	107.9	127.3	115.9
03	CONCRETE	100.8	97.0	98.9	90.2	103.8	97.0	95.6	95.8	95.7	89.3	103.3	96.3	96.7	117.2	106.9	97.5	131.7	114.6
04	MASONRY	99.5	84.7	90.4	92.7	98.4	96.2	96.5	83.0	88.2	100.6	92.3	95.5	108.0	127.3	119.9	108.1	137.0	125.8
05	METALS	97.5	86.0	93.4	99.5	123.1	108.0	95.6	87.9	92.8	100.3	122.4	108.3	96.8	125.8	107.2	98.8	133.7	111.4
06	WOOD & PLASTICS	95.2	92.0	93.5	96.9	96.8	96.9	92.9	88.7	90.7	93.5	95.8	94.7	104.2	119.6	112.2	104.7	139.8	122.9
07	THERMAL & MOISTURE PROTECTION	91.4	88.5	90.1	93.6	96.8	95.1	90.6	81.5	86.3	92.8	93.8	93.3	99.0	126.7	112.2	98.9	135.1	116.1
08	DOORS & WINDOWS	93.3	80.4	90.2	97.8	100.6	98.4	92.6	87.9	91.4	97.8	100.0	98.3	101.0	127.7	107.5	102.2	149.4	113.6
09200	PLASTER & GYPSUM BOARD	101.1	85.7	91.7	110.7	95.0	101.1	92.3	82.3	86.2	107.9	93.9	99.3	100.6	121.0	113.1	100.6	141.8	125.7
095,098	CEILINGS & ACOUSTICAL TREATMENT	95.1	85.7	88.9	101.1	95.0	97.0	100.6	82.3	88.5	101.1	93.9	96.3	90.9	121.0	110.8	90.9	141.8	124.5
09600	FLOORING	95.3	60.4	86.8	90.4	77.2	87.2	94.4	75.2	89.8	90.4	70.2	85.5	112.1	129.2	116.2	109.2	118.1	111.4
097,099	WALL FINISHES, PAINTS & COATINGS	96.5	89.5	92.5	93.2	89.5	91.1	95.7	75.1	83.8	93.2	90.0	91.3	99.2	111.2	106.2	106.0	122.4	115.5
09	FINISHES	96.0	85.5	90.7	98.2	91.9	95.0	94.9	84.7	89.7	97.1	89.7	93.3	102.3	122.1	112.4	101.8	135.1	118.8
10 - 14	TOTAL DIV. 10000 - 14000	100.0	114.1	103.0	100.0	105.3	101.1	100.0	113.2	102.8	100.0	104.8	101.0	100.0	110.1	102.1	100.0	116.1	103.4
15	MECHANICAL	100.0	91.8	96.2	100.0	94.0	97.2	99.7	94.9	97.5	100.0	95.5	97.9	99.8	109.1	104.1	99.7	119.5	108.9
16	ELECTRICAL	93.1	85.0	87.5	93.9	100.7	98.6	93.4	79.8	84.0	97.6	88.1	91.0	102.3	104.4	103.8	103.4	101.4	102.0
01 - 16	WEIGHTED AVERAGE	97.2	89.0	93.2	96.8	99.8	98.3	95.7	88.5	92.2	96.9	96.8	96.9	99.6	115.0	107.1	100.1	123.6	111.6

	DIVISION	MINNESOTA									MISSISSIPPI								
		ROCHESTER			SAINT PAUL			ST. CLOUD			BILOXI			GREENVILLE			JACKSON		
		MAT.	INST.	TOTAL	MAT.	INST.	TOTAL	MAT.	INST.	TOTAL	MAT.	INST.	TOTAL	MAT.	INST.	TOTAL	MAT.	INST.	TOTAL
02	SITE CONSTRUCTION	81.3	102.5	97.6	84.2	103.7	99.1	77.3	105.3	98.7	107.8	87.1	91.9	112.2	86.8	92.8	104.2	86.8	90.9
03100	CONCRETE FORMS & ACCESSORIES	98.1	109.9	108.3	88.3	132.1	126.1	79.1	108.8	104.7	99.0	52.6	58.9	77.2	43.6	48.2	93.5	49.8	55.8
03200	CONCRETE REINFORCEMENT	95.6	118.2	108.5	92.2	119.3	107.7	103.4	118.7	112.2	95.8	68.7	80.4	103.8	52.9	74.8	95.8	57.4	73.9
03300	CAST-IN-PLACE CONCRETE	105.8	105.0	105.5	108.8	123.7	115.0	95.7	117.9	104.8	105.7	54.0	84.4	106.2	47.3	81.9	103.7	50.2	81.7
03	CONCRETE	94.9	110.6	102.8	98.9	127.1	113.0	88.1	114.8	101.4	96.3	57.9	77.1	98.7	48.7	73.7	94.9	53.3	74.1
04	MASONRY	107.8	105.4	106.3	119.5	137.0	130.2	110.4	116.0	113.9	89.5	47.0	63.4	132.4	44.6	78.4	92.4	44.6	63.0
05	METALS	96.7	130.9	109.0	95.4	133.4	109.1	96.5	133.0	109.7	98.1	87.3	94.2	96.4	80.7	90.8	98.0	82.5	92.4
06	WOOD & PLASTICS	104.7	108.1	106.5	94.5	129.5	112.6	85.2	103.4	94.6	99.0	53.1	75.3	75.6	43.1	58.8	92.8	51.2	71.3
07	THERMAL & MOISTURE PROTECTION	98.8	107.9	103.1	98.7	133.9	115.4	96.8	124.2	109.8	96.5	54.5	76.6	96.5	49.5	74.2	96.2	51.1	74.8
08	DOORS & WINDOWS	101.0	132.3	108.6	98.1	143.8	109.1	94.7	129.7	103.2	98.3	56.2	88.1	97.7	45.8	85.2	98.7	51.3	87.3
09200	PLASTER & GYPSUM BOARD	100.6	109.2	105.9	97.0	131.2	117.9	87.7	104.4	97.9	108.0	52.2	73.9	101.0	41.8	64.8	108.0	50.2	72.7
095,098	CEILINGS & ACOUSTICAL TREATMENT	90.9	109.2	103.0	88.2	131.2	116.6	69.0	104.4	92.3	95.1	52.2	66.8	89.7	41.8	58.1	95.1	50.2	65.5
09600	FLOORING	111.8	92.7	107.2	104.7	118.1	107.9	100.2	129.2	107.3	115.9	49.4	99.8	106.3	42.2	90.8	115.9	48.8	99.7
097,099	WALL FINISHES, PAINTS & COATINGS	101.8	107.0	104.8	106.0	116.8	112.3	106.9	122.4	115.9	105.6	43.7	69.6	105.6	43.1	69.2	105.6	43.1	69.2
09	FINISHES	102.4	106.9	104.7	99.6	128.4	114.3	93.5	113.1	103.5	105.0	50.7	77.3	100.2	43.0	71.0	105.0	48.9	76.4
10 - 14	TOTAL DIV. 10000 - 14000	100.0	105.9	101.2	100.0	114.3	103.0	100.0	106.6	101.4	100.0	68.1	93.3	100.0	65.9	92.8	100.0	67.0	93.0
15	MECHANICAL	99.7	98.8	99.3	99.7	119.4	108.9	99.7	110.7	104.8	99.9	63.5	83.1	99.9	43.8	73.9	99.9	43.4	73.8
16	ELECTRICAL	102.2	94.5	96.9	101.3	104.2	103.3	98.9	104.2	102.6	96.7	56.1	68.8	95.6	46.7	62.0	97.2	46.7	62.5
01 - 16	WEIGHTED AVERAGE	99.3	106.4	102.7	99.5	121.7	110.3	96.2	113.4	104.6	98.9	62.4	81.1	100.5	53.3	77.6	98.7	55.1	77.5

	DIVISION	MISSISSIPPI			MISSOURI														
		MERIDIAN			CAPE GIRARDEAU			COLUMBIA			JOPLIN			KANSAS CITY			SPRINGFIELD		
		MAT.	INST.	TOTAL	MAT.	INST.	TOTAL	MAT.	INST.	TOTAL	MAT.	INST.	TOTAL	MAT.	INST.	TOTAL	MAT.	INST.	TOTAL
02	**SITE CONSTRUCTION**	103.1	86.8	90.6	93.9	93.3	93.4	98.1	94.5	95.4	99.2	95.9	96.7	92.0	94.4	93.9	90.6	93.6	92.9
03100	CONCRETE FORMS & ACCESSORIES	75.5	42.3	46.9	83.8	84.5	84.4	85.5	76.0	77.3	105.5	69.5	74.4	105.0	99.7	100.5	103.6	58.6	64.8
03200	CONCRETE REINFORCEMENT	102.5	56.9	76.5	106.0	91.6	97.8	100.4	104.4	102.7	104.0	76.6	88.3	98.8	105.4	102.6	96.5	104.4	101.0
03300	CAST-IN-PLACE CONCRETE	100.2	50.2	79.6	85.1	92.7	88.2	86.1	75.1	81.6	102.9	77.6	92.4	95.6	99.4	97.2	97.4	71.8	86.8
03	**CONCRETE**	91.4	49.9	70.7	85.9	90.5	88.2	81.7	83.0	82.3	97.8	75.0	86.4	94.0	101.2	97.6	94.6	73.3	83.9
04	**MASONRY**	89.2	36.4	56.7	114.3	83.0	95.1	130.9	75.5	96.9	94.7	61.8	74.5	100.1	99.7	99.9	86.5	73.8	78.7
05	**METALS**	96.4	81.7	91.1	93.7	116.1	101.8	93.0	120.1	102.8	95.8	94.8	95.4	99.2	110.5	103.2	96.3	105.0	99.4
06	**WOOD & PLASTICS**	73.9	41.3	57.0	85.3	84.2	84.7	89.4	72.7	80.8	102.7	69.9	85.7	102.8	99.4	101.1	102.6	54.9	78.0
07	**THERMAL & MOISTURE PROTECTION**	95.9	48.0	73.2	95.1	86.8	91.2	92.2	79.6	86.2	98.5	68.1	84.0	97.3	99.8	98.5	96.4	73.2	85.4
08	**DOORS & WINDOWS**	97.7	45.1	85.0	96.6	85.0	93.8	93.2	92.9	93.1	94.4	77.6	90.4	98.8	100.8	99.3	99.0	70.2	92.0
09200	PLASTER & GYPSUM BOARD	100.1	40.0	63.4	98.0	83.3	89.0	100.9	71.5	82.9	108.6	68.6	84.2	104.0	99.1	101.0	106.2	53.1	73.8
095,098	CEILINGS & ACOUSTICAL TREATMENT	89.7	40.0	56.9	92.6	83.3	86.4	93.8	71.5	79.1	98.5	68.6	78.8	104.0	99.1	100.7	99.2	53.1	68.8
09600	FLOORING	105.2	34.3	88.0	99.0	90.1	96.8	107.9	75.0	99.9	118.9	60.6	104.8	96.8	98.1	97.2	110.4	60.6	98.3
097,099	WALL FINISHES, PAINTS & COATINGS	105.6	42.5	68.9	100.8	90.2	94.6	99.2	71.5	83.1	92.6	53.3	69.7	97.4	104.8	101.7	93.5	76.3	83.5
09	**FINISHES**	99.1	40.2	69.1	94.0	85.5	89.6	96.3	73.1	84.5	105.8	65.7	85.3	100.6	99.7	100.1	101.1	59.3	79.8
10 - 14	**TOTAL DIV. 10000 - 14000**	100.0	65.8	92.8	100.0	81.5	96.1	100.0	95.8	99.1	100.0	86.3	97.1	100.0	95.9	99.1	100.0	84.1	96.6
15	**MECHANICAL**	99.9	43.4	73.8	100.1	96.9	98.6	100.1	97.8	99.0	100.3	63.2	83.1	100.1	99.5	99.8	100.2	72.7	87.5
16	**ELECTRICAL**	95.0	61.5	72.0	95.3	113.3	107.7	93.8	85.9	88.4	89.2	69.2	75.4	103.4	104.4	104.1	100.2	75.0	82.9
01 - 16	**WEIGHTED AVERAGE**	96.7	54.8	76.3	96.3	96.2	96.2	96.4	89.4	93.0	98.0	73.4	86.0	99.0	101.1	100.0	97.7	76.7	87.5

	DIVISION	MISSOURI						MONTANA											
		ST. JOSEPH			ST. LOUIS			BILLINGS			BUTTE			GREAT FALLS			HELENA		
		MAT.	INST.	TOTAL	MAT.	INST.	TOTAL	MAT.	INST.	TOTAL	MAT.	INST.	TOTAL	MAT.	INST.	TOTAL	MAT.	INST.	TOTAL
02	**SITE CONSTRUCTION**	93.4	91.5	91.9	93.3	96.6	95.8	82.7	98.2	94.6	88.6	96.9	95.0	92.1	97.7	96.4	93.8	97.7	96.8
03100	CONCRETE FORMS & ACCESSORIES	104.9	75.2	79.3	101.4	109.2	108.1	94.4	85.7	86.9	80.4	88.5	87.4	100.8	86.3	88.3	100.9	86.8	88.7
03200	CONCRETE REINFORCEMENT	97.6	99.0	98.4	97.0	108.8	103.7	100.1	104.9	102.9	108.6	104.9	106.5	100.1	104.9	102.9	103.7	99.9	101.5
03300	CAST-IN-PLACE CONCRETE	95.6	94.3	95.0	85.1	112.3	96.3	116.3	84.1	103.0	119.7	85.3	105.5	126.5	71.2	103.7	128.9	83.7	110.3
03	**CONCRETE**	93.8	87.5	90.6	86.0	111.3	98.6	102.6	89.1	95.9	102.3	90.7	96.5	108.1	84.8	96.5	109.7	88.4	99.1
04	**MASONRY**	99.4	88.2	92.5	96.4	110.6	105.1	122.6	96.2	106.4	121.0	96.9	106.2	124.7	99.5	109.2	121.9	96.5	106.3
05	**METALS**	97.8	107.1	101.2	96.1	125.3	106.7	98.4	99.4	98.8	97.7	99.1	98.2	98.8	99.2	98.9	98.0	96.7	97.5
06	**WOOD & PLASTICS**	103.4	71.7	87.0	103.4	107.8	105.7	96.6	86.0	91.1	83.8	91.1	87.6	105.0	86.0	95.2	105.2	86.0	95.2
07	**THERMAL & MOISTURE PROTECTION**	98.0	88.9	93.7	95.1	105.5	100.1	98.1	87.8	93.2	97.3	86.2	92.0	98.3	86.4	92.7	98.3	87.7	93.3
08	**DOORS & WINDOWS**	99.5	84.2	95.8	93.6	112.7	98.2	102.7	85.8	98.7	99.3	88.5	96.7	103.0	85.8	98.9	102.4	84.7	98.2
09200	PLASTER & GYPSUM BOARD	109.4	70.5	85.6	104.8	107.7	106.5	100.4	86.1	91.6	93.8	91.4	92.3	100.4	86.1	91.6	99.9	86.1	91.4
095,098	CEILINGS & ACOUSTICAL TREATMENT	102.6	70.5	81.4	96.7	107.7	103.9	112.6	86.1	95.1	109.9	91.4	97.7	112.6	86.1	95.1	109.9	86.1	94.2
09600	FLOORING	99.5	97.1	98.9	106.7	101.6	105.4	110.5	63.1	99.0	102.1	50.2	89.5	110.5	68.1	100.2	110.5	66.0	99.7
097,099	WALL FINISHES, PAINTS & COATINGS	93.2	88.4	90.4	100.8	109.6	105.9	102.5	78.4	88.5	102.5	60.5	78.1	102.5	68.1	82.5	102.5	75.6	86.8
09	**FINISHES**	101.8	79.5	90.4	98.3	107.5	103.0	106.7	80.4	93.3	102.4	78.3	90.1	106.8	80.6	93.5	106.3	81.4	93.6
10 - 14	**TOTAL DIV. 10000 - 14000**	100.0	88.6	97.6	100.0	106.3	101.3	100.0	84.7	96.8	100.0	84.8	96.8	100.0	85.6	97.0	100.0	76.5	95.0
15	**MECHANICAL**	100.3	88.2	94.7	100.1	108.1	103.8	100.4	90.4	95.7	100.4	87.8	94.5	100.4	91.6	96.3	100.4	90.5	95.8
16	**ELECTRICAL**	102.8	81.5	88.1	99.0	116.5	111.0	95.0	84.6	87.9	106.5	80.7	88.8	94.9	82.3	86.2	94.9	82.3	86.2
01 - 16	**WEIGHTED AVERAGE**	99.0	87.7	93.5	96.2	110.6	103.2	101.5	89.8	95.8	101.3	88.6	95.1	102.8	89.3	96.2	102.6	88.9	96.0

	DIVISION	MONTANA			NEBRASKA												NEVADA		
		MISSOULA			GRAND ISLAND			LINCOLN			NORTH PLATTE			OMAHA			CARSON CITY		
		MAT.	INST.	TOTAL	MAT.	INST.	TOTAL	MAT.	INST.	TOTAL	MAT.	INST.	TOTAL	MAT.	INST.	TOTAL	MAT.	INST.	TOTAL
02	**SITE CONSTRUCTION**	72.8	96.9	91.3	93.1	90.7	91.3	84.2	90.7	89.2	93.9	90.4	91.2	76.0	89.4	86.2	64.2	102.3	93.4
03100	CONCRETE FORMS & ACCESSORIES	84.8	83.2	83.4	99.6	57.3	63.1	105.1	50.2	57.8	99.5	55.4	61.4	100.1	74.5	78.0	101.1	96.9	97.5
03200	CONCRETE REINFORCEMENT	110.5	110.3	110.4	105.0	78.9	90.2	96.5	78.7	86.4	106.4	78.7	90.6	101.0	80.3	89.2	108.4	117.3	113.5
03300	CAST-IN-PLACE CONCRETE	88.2	81.7	85.5	112.4	64.4	92.6	101.5	67.0	87.3	112.4	60.1	90.9	105.4	75.8	93.2	117.8	94.1	108.0
03	**CONCRETE**	81.1	88.1	84.6	104.2	65.2	84.7	96.7	63.0	79.8	104.4	63.0	83.7	98.0	76.2	87.1	113.4	99.5	106.5
04	**MASONRY**	147.4	91.2	112.9	104.7	54.5	73.8	94.6	51.8	68.3	89.9	47.7	64.0	100.4	78.9	87.2	136.3	84.2	104.3
05	**METALS**	97.8	101.0	99.0	92.9	88.0	91.1	95.8	87.6	92.9	93.1	91.8	92.6	98.4	79.4	91.6	103.9	101.6	103.1
06	**WOOD & PLASTICS**	88.6	85.7	87.1	98.9	53.6	75.5	104.2	44.3	73.2	98.7	53.6	75.4	99.6	75.6	87.2	95.4	96.7	96.1
07	**THERMAL & MOISTURE PROTECTION**	96.4	92.8	94.7	96.7	63.0	80.7	97.1	55.5	77.4	96.6	54.5	76.6	92.5	75.2	84.3	104.8	91.6	98.6
08	**DOORS & WINDOWS**	99.4	87.9	96.6	92.5	56.7	83.9	98.1	53.8	87.4	91.8	62.1	84.6	101.8	70.8	94.3	95.1	105.4	97.6
09200	PLASTER & GYPSUM BOARD	95.5	85.8	89.6	103.2	51.7	71.7	106.2	42.1	67.0	103.5	51.7	71.8	111.9	75.3	89.5	86.9	96.4	92.7
095,098	CEILINGS & ACOUSTICAL TREATMENT	109.9	85.8	94.0	92.4	51.7	65.6	99.2	42.1	61.6	93.8	51.7	66.0	138.8	75.3	96.9	97.9	96.4	96.9
09600	FLOORING	104.0	78.5	97.8	99.5	42.1	85.6	101.3	43.8	87.4	99.5	43.8	86.0	130.2	52.7	111.4	105.2	83.4	99.9
097,099	WALL FINISHES, PAINTS & COATINGS	102.5	68.1	82.5	91.0	49.9	67.1	91.0	50.3	67.3	91.0	60.3	73.1	146.4	60.3	96.3	108.7	93.7	100.0
09	**FINISHES**	102.5	80.5	91.3	96.1	52.4	73.8	98.2	47.4	72.3	96.4	53.0	74.2	123.9	68.6	95.7	97.0	93.5	95.2
10 - 14	**TOTAL DIV. 10000 - 14000**	100.0	72.4	94.2	100.0	82.3	96.3	100.0	81.1	96.0	100.0	66.1	92.8	100.0	83.3	96.5	100.0	120.9	104.4
15	**MECHANICAL**	100.4	82.9	92.3	100.2	84.4	92.9	100.2	84.4	92.9	100.2	82.1	91.8	99.8	83.2	92.2	100.2	96.9	98.7
16	**ELECTRICAL**	103.1	85.4	90.9	91.0	52.5	64.5	100.5	52.5	67.5	94.3	52.5	65.6	88.3	86.2	86.9	92.7	91.9	92.1
01 - 16	**WEIGHTED AVERAGE**	99.5	87.8	93.8	97.5	68.7	83.5	97.8	67.0	82.8	96.9	67.3	82.5	100.5	80.1	90.6	102.0	96.5	99.3

City Cost Indexes

	DIVISION	NEVADA						NEW HAMPSHIRE									NEW JERSEY		
		LAS VEGAS			RENO			MANCHESTER			NASHUA			PORTSMOUTH			CAMDEN		
		MAT.	INST.	TOTAL	MAT.	INST.	TOTAL	MAT.	INST.	TOTAL	MAT.	INST.	TOTAL	MAT.	INST.	TOTAL	MAT.	INST.	TOTAL
02	**SITE CONSTRUCTION**	64.3	103.9	94.7	64.6	102.3	93.5	90.0	102.2	99.3	92.1	102.2	99.8	86.2	101.7	98.0	95.5	106.3	103.8
03100	CONCRETE FORMS & ACCESSORIES	97.2	110.0	108.3	101.6	97.0	97.6	99.4	72.6	76.2	99.8	72.6	76.3	85.7	69.5	71.8	99.8	123.5	120.2
03200	CONCRETE REINFORCEMENT	100.1	126.3	115.0	100.1	125.8	114.8	111.9	99.3	104.7	111.9	99.3	104.7	89.5	99.2	95.0	111.9	113.5	112.8
03300	CAST-IN-PLACE CONCRETE	112.1	112.0	112.1	120.0	94.1	109.3	95.1	89.5	92.8	92.8	89.5	91.4	88.0	85.6	87.0	80.7	122.0	97.7
03	**CONCRETE**	109.3	113.2	111.2	113.5	101.1	107.3	105.1	83.5	94.3	103.9	83.5	93.7	95.1	80.8	88.0	98.0	119.9	108.9
04	**MASONRY**	130.3	98.5	110.8	136.4	84.2	104.3	96.5	94.9	95.6	96.6	94.9	95.6	91.0	88.4	89.4	88.8	123.0	109.8
05	**METALS**	104.6	107.5	105.6	104.4	104.8	104.6	98.0	89.3	94.8	98.0	89.3	94.8	94.7	88.0	92.2	97.8	102.0	99.3
06	**WOOD & PLASTICS**	90.4	108.5	99.8	95.9	96.7	96.3	102.0	67.1	83.9	101.9	67.1	83.9	87.3	67.1	76.8	101.9	123.4	113.0
07	**THERMAL & MOISTURE PROTECTION**	104.4	102.2	103.3	104.8	91.6	98.6	99.4	96.5	98.0	99.8	96.5	98.2	99.3	97.0	98.2	99.2	118.6	108.4
08	**DOORS & WINDOWS**	95.5	114.0	99.9	95.5	105.4	97.9	110.0	74.3	101.4	110.0	74.3	101.4	111.0	67.4	100.5	110.0	116.6	111.6
09200	PLASTER & GYPSUM BOARD	87.4	108.6	100.3	88.2	96.4	93.2	98.2	64.8	77.8	98.2	64.8	77.8	91.3	64.8	75.1	98.2	122.9	113.3
095,098	CEILINGS & ACOUSTICAL TREATMENT	104.7	108.6	107.2	104.7	96.4	99.2	102.6	64.8	77.6	102.6	64.8	77.6	93.1	64.8	74.4	102.6	122.9	116.0
09600	FLOORING	105.2	99.3	103.8	105.2	83.4	99.9	100.2	101.9	100.6	99.9	101.9	100.4	94.7	101.9	96.5	99.9	123.2	105.6
097,099	WALL FINISHES, PAINTS & COATINGS	108.7	113.2	111.3	108.7	93.7	100.0	95.6	96.4	96.1	95.6	96.4	96.1	95.6	47.3	67.5	95.6	144.1	123.8
09	**FINISHES**	98.2	108.0	103.2	98.4	93.5	95.9	99.8	79.5	89.4	100.0	79.5	89.5	95.2	72.2	83.5	100.3	126.0	113.4
10 - 14	**TOTAL DIV. 10000 - 14000**	100.0	112.6	102.7	100.0	120.9	104.4	100.0	93.6	98.7	100.0	93.6	98.7	100.0	91.0	98.1	100.0	118.1	103.8
15	**MECHANICAL**	100.2	111.9	105.6	100.2	97.0	98.7	99.8	86.1	93.5	99.8	86.1	93.5	99.8	82.5	91.8	99.8	118.3	108.4
16	**ELECTRICAL**	94.8	110.2	105.3	92.7	91.9	92.1	104.2	82.3	89.1	104.0	82.3	89.1	100.7	82.3	88.0	104.0	120.3	115.2
01 - 16	**WEIGHTED AVERAGE**	101.5	108.6	105.0	102.3	97.0	99.7	101.3	86.8	94.2	101.2	86.8	94.2	98.3	83.8	91.3	100.1	117.5	108.6

	DIVISION	NEW JERSEY															NEW MEXICO		
		ELIZABETH			JERSEY CITY			NEWARK			PATERSON			TRENTON			ALBUQUERQUE		
		MAT.	INST.	TOTAL	MAT.	INST.	TOTAL	MAT.	INST.	TOTAL	MAT.	INST.	TOTAL	MAT.	INST.	TOTAL	MAT.	INST.	TOTAL
02	**SITE CONSTRUCTION**	110.3	107.0	107.8	95.5	106.8	104.1	114.1	106.7	108.4	108.5	106.7	107.1	96.6	106.9	104.5	83.9	110.1	103.9
03100	CONCRETE FORMS & ACCESSORIES	111.1	125.6	123.6	99.8	126.5	122.9	98.1	126.6	122.7	98.6	126.6	122.8	98.9	124.0	120.6	99.5	73.3	76.9
03200	CONCRETE REINFORCEMENT	86.2	131.2	111.9	111.9	131.4	123.0	111.9	131.4	123.0	111.9	131.4	123.0	111.9	125.1	119.4	100.8	72.3	84.6
03300	CAST-IN-PLACE CONCRETE	88.9	128.7	105.3	80.7	130.8	101.4	90.9	130.8	107.3	102.8	130.7	114.3	90.8	124.8	104.8	110.2	79.5	97.6
03	**CONCRETE**	101.4	126.7	114.0	98.0	127.6	112.8	102.9	127.9	115.4	108.8	127.9	118.3	102.9	123.3	113.1	108.6	76.4	92.5
04	**MASONRY**	110.2	125.3	119.5	88.8	125.3	111.3	98.2	125.3	114.9	93.4	125.3	113.0	89.7	122.8	110.1	111.4	74.2	88.5
05	**METALS**	94.4	111.9	100.7	97.9	109.7	102.2	97.9	112.9	103.3	97.9	112.9	103.3	98.7	107.0	101.7	105.6	92.2	100.8
06	**WOOD & PLASTICS**	117.6	126.5	122.2	101.9	126.2	114.5	103.8	126.5	115.5	103.8	126.5	115.5	102.0	123.1	112.9	96.3	74.8	85.2
07	**THERMAL & MOISTURE PROTECTION**	100.1	121.9	110.4	99.2	122.9	110.4	99.5	121.9	110.1	99.9	122.6	110.7	98.1	121.7	109.3	103.6	79.5	92.2
08	**DOORS & WINDOWS**	111.4	126.6	115.0	110.0	126.8	114.1	116.5	126.6	118.9	116.5	126.6	118.9	110.0	123.6	113.3	95.4	76.1	90.8
09200	PLASTER & GYPSUM BOARD	101.2	126.1	116.4	98.2	125.8	115.1	98.2	126.1	115.3	98.2	126.1	115.3	98.2	122.7	113.2	91.1	73.4	80.3
095,098	CEILINGS & ACOUSTICAL TREATMENT	93.1	126.1	114.9	102.6	125.8	117.9	102.6	126.1	118.1	102.6	126.1	118.1	102.6	122.7	115.8	104.7	73.4	84.1
09600	FLOORING	104.5	66.4	95.3	99.9	133.2	108.0	100.2	133.2	108.2	99.9	133.2	108.0	100.2	123.3	105.8	105.2	72.6	97.3
097,099	WALL FINISHES, PAINTS & COATINGS	95.5	144.2	123.8	95.6	144.2	123.8	95.5	144.2	123.8	95.5	144.2	123.8	95.6	144.2	123.8	108.7	67.2	84.6
09	**FINISHES**	101.2	116.0	108.7	100.3	129.7	115.3	100.8	129.9	115.7	100.4	129.9	115.5	100.4	125.8	113.3	99.2	72.5	85.6
10 - 14	**TOTAL DIV. 10000 - 14000**	100.0	99.5	99.9	100.0	100.4	100.1	100.0	100.4	100.1	100.0	100.4	100.1	100.0	117.8	103.8	100.0	82.3	96.3
15	**MECHANICAL**	99.8	116.5	107.6	99.8	127.8	112.8	99.8	129.1	113.4	99.8	127.9	112.8	99.8	125.4	111.7	100.1	77.5	89.6
16	**ELECTRICAL**	98.8	115.7	110.4	105.7	133.7	124.9	105.4	121.9	116.8	105.7	118.9	114.8	104.2	127.6	120.3	87.2	77.8	80.8
01 - 16	**WEIGHTED AVERAGE**	101.7	117.4	109.3	100.3	124.2	111.9	102.8	122.7	112.5	103.1	122.0	112.3	100.9	121.4	110.9	100.6	81.2	91.2

	DIVISION	NEW MEXICO												NEW YORK					
		FARMINGTON			LAS CRUCES			ROSWELL			SANTA FE			ALBANY			BINGHAMTON		
		MAT.	INST.	TOTAL	MAT.	INST.	TOTAL	MAT.	INST.	TOTAL	MAT.	INST.	TOTAL	MAT.	INST.	TOTAL	MAT.	INST.	TOTAL
02	**SITE CONSTRUCTION**	89.3	110.1	105.2	96.0	85.5	88.0	94.2	110.1	106.4	83.1	110.1	103.8	72.8	107.7	99.5	96.6	90.4	91.8
03100	CONCRETE FORMS & ACCESSORIES	99.5	73.4	76.9	96.9	71.3	74.8	99.5	73.3	76.8	99.5	73.3	76.9	98.4	88.4	89.8	104.3	79.5	82.9
03200	CONCRETE REINFORCEMENT	110.1	72.3	88.5	104.0	66.2	82.5	109.3	67.0	85.2	108.1	72.3	87.7	100.8	87.8	93.4	99.8	87.5	92.8
03300	CAST-IN-PLACE CONCRETE	110.9	79.6	97.9	96.4	68.7	85.0	102.1	79.5	92.8	104.2	79.5	94.0	83.7	104.0	92.1	106.0	92.6	100.5
03	**CONCRETE**	112.2	76.4	94.3	88.5	70.1	79.3	112.7	75.3	94.0	106.5	76.4	91.5	100.1	95.0	97.6	101.8	88.0	94.9
04	**MASONRY**	127.2	74.2	94.6	111.8	63.8	82.3	127.7	74.2	94.8	116.3	74.2	90.4	90.6	100.0	96.3	106.5	84.8	93.1
05	**METALS**	104.8	92.2	100.3	98.9	79.7	92.0	104.8	89.8	99.4	104.8	92.2	100.2	100.1	109.0	103.3	95.2	121.2	104.6
06	**WOOD & PLASTICS**	96.3	74.8	85.2	89.3	73.2	81.0	96.3	74.8	85.2	96.3	74.8	85.2	100.6	85.8	92.9	110.4	77.0	93.1
07	**THERMAL & MOISTURE PROTECTION**	104.6	79.5	92.7	89.9	70.5	80.7	104.9	79.5	92.8	103.9	79.5	92.3	91.1	95.1	93.0	100.8	86.8	94.2
08	**DOORS & WINDOWS**	98.4	76.1	93.0	87.6	73.6	84.2	94.4	74.5	89.6	94.5	76.1	90.1	98.1	82.8	94.4	93.1	77.0	89.2
09200	PLASTER & GYPSUM BOARD	86.4	73.4	78.5	88.5	73.4	79.3	86.4	73.4	78.5	86.4	73.4	78.5	102.4	85.0	91.8	108.1	75.7	88.3
095,098	CEILINGS & ACOUSTICAL TREATMENT	95.1	73.4	80.8	99.9	73.4	82.5	95.1	73.4	80.8	95.1	73.4	80.8	97.9	85.0	89.4	97.9	75.7	83.2
09600	FLOORING	105.2	72.6	97.3	138.1	62.6	119.8	105.2	72.6	97.3	105.2	72.6	97.3	84.8	93.7	87.0	94.6	92.2	94.0
097,099	WALL FINISHES, PAINTS & COATINGS	108.7	74.8	89.0	102.1	67.2	81.8	108.7	67.2	84.6	108.7	67.2	84.6	81.8	82.2	82.0	84.1	79.2	81.3
09	**FINISHES**	97.1	73.3	85.0	114.7	69.2	91.5	97.5	72.5	84.8	96.9	72.5	84.4	95.6	88.2	91.8	96.1	81.2	88.5
10 - 14	**TOTAL DIV. 10000 - 14000**	100.0	82.3	96.3	100.0	77.6	95.3	100.0	82.3	96.3	100.0	82.3	96.3	100.0	95.0	98.9	100.0	93.4	98.6
15	**MECHANICAL**	100.1	77.5	89.6	100.4	76.9	89.5	100.1	77.4	89.6	100.1	77.5	89.6	100.3	92.5	96.7	100.5	87.2	94.4
16	**ELECTRICAL**	84.5	77.8	79.9	85.7	65.7	71.9	85.7	77.8	80.3	87.2	77.8	80.8	101.9	96.1	97.9	102.8	80.1	87.2
01 - 16	**WEIGHTED AVERAGE**	102.0	81.3	91.9	97.7	72.7	85.6	101.8	80.7	91.6	100.2	81.2	90.9	97.9	96.3	97.1	99.0	88.2	93.7

DIVISION		NEW YORK																	
		BUFFALO			HICKSVILLE			NEW YORK			RIVERHEAD			ROCHESTER			SCHENECTADY		
		MAT.	INST.	TOTAL	MAT.	INST.	TOTAL	MAT.	INST.	TOTAL	MAT.	INST.	TOTAL	MAT.	INST.	TOTAL	MAT.	INST.	TOTAL
02	**SITE CONSTRUCTION**	101.6	93.3	95.2	116.3	131.1	127.7	139.5	128.6	131.1	116.6	131.1	127.7	77.3	107.6	100.5	73.0	107.8	99.6
03100	CONCRETE FORMS & ACCESSORIES	101.7	115.0	113.2	89.0	156.2	147.0	109.7	171.5	163.1	93.6	156.3	147.7	101.4	98.8	99.2	103.8	89.2	91.2
03200	CONCRETE REINFORCEMENT	101.6	103.9	102.9	101.1	177.9	144.9	105.1	181.2	148.5	103.0	177.9	145.7	102.5	89.8	95.2	99.5	87.8	92.8
03300	CAST-IN-PLACE CONCRETE	115.5	120.9	117.8	99.1	156.7	122.8	129.6	163.4	143.5	97.3	156.7	121.8	109.9	105.1	107.9	97.3	105.2	100.6
03	**CONCRETE**	106.2	114.1	110.1	105.8	158.9	132.3	127.6	168.6	148.1	104.7	159.0	131.8	112.0	100.4	106.2	107.0	95.8	101.4
04	**MASONRY**	107.2	119.0	114.4	108.8	159.6	140.0	110.9	167.5	145.7	115.7	159.6	142.7	103.1	101.2	101.9	91.2	102.1	97.9
05	**METALS**	100.0	95.7	98.4	109.8	137.9	119.9	109.5	141.0	120.8	109.8	138.0	120.0	103.3	110.5	105.9	100.1	109.0	103.3
06	**WOOD & PLASTICS**	103.4	114.9	109.4	90.2	157.9	125.2	112.4	172.6	143.6	94.9	157.9	127.5	99.9	100.0	99.9	107.0	85.8	96.0
07	**THERMAL & MOISTURE PROTECTION**	100.8	108.6	104.5	104.7	150.4	126.4	107.0	160.6	132.4	104.7	150.4	126.4	97.6	99.0	98.3	91.2	96.0	93.5
08	**DOORS & WINDOWS**	93.9	106.6	96.9	90.6	159.5	107.2	97.8	168.4	114.8	90.6	159.5	107.2	99.5	93.5	98.0	98.1	82.8	94.4
09200	PLASTER & GYPSUM BOARD	96.7	114.9	107.8	97.5	159.7	135.5	105.8	174.7	147.9	98.8	159.7	136.0	98.0	99.8	99.1	102.4	85.0	91.8
095,098	CEILINGS & ACOUSTICAL TREATMENT	92.2	114.9	107.2	81.5	159.7	133.1	106.5	174.7	151.5	81.5	159.7	133.1	96.5	99.8	98.7	97.9	85.0	89.4
09600	FLOORING	89.9	119.5	97.1	94.3	106.8	97.3	99.0	166.3	115.3	95.6	106.8	98.3	79.7	101.7	85.1	84.8	93.7	87.0
097,099	WALL FINISHES, PAINTS & COATINGS	84.6	118.1	104.1	106.7	145.9	129.5	97.5	151.8	129.1	106.7	150.7	132.3	81.8	105.1	95.4	81.8	82.2	82.0
09	**FINISHES**	92.7	116.9	105.0	105.6	145.2	125.8	108.8	168.8	139.4	106.1	145.7	126.3	93.9	100.4	97.2	95.3	88.7	91.9
10 - 14	**TOTAL DIV. 10000 - 14000**	100.0	106.4	101.3	100.0	138.0	108.0	100.0	144.8	109.5	100.0	138.1	108.0	100.0	96.1	99.2	100.0	95.9	99.1
15	**MECHANICAL**	99.8	97.1	98.6	99.7	145.2	120.8	100.4	163.8	129.7	99.7	145.2	120.8	99.8	92.8	96.6	100.3	93.7	97.2
16	**ELECTRICAL**	99.5	99.5	99.5	102.4	157.8	140.5	111.1	177.9	157.0	103.6	170.2	149.4	104.5	95.4	98.2	102.3	96.1	98.0
01 - 16	**WEIGHTED AVERAGE**	99.7	104.8	102.2	102.8	149.1	125.3	108.5	161.9	134.5	103.3	151.3	126.6	101.0	99.4	100.2	98.9	97.0	98.0

DIVISION		NEW YORK															NORTH CAROLINA		
		SYRACUSE			UTICA			WATERTOWN			WHITE PLAINS			YONKERS			ASHEVILLE		
		MAT.	INST.	TOTAL	MAT.	INST.	TOTAL	MAT.	INST.	TOTAL	MAT.	INST.	TOTAL	MAT.	INST.	TOTAL	MAT.	INST.	TOTAL
02	**SITE CONSTRUCTION**	95.2	107.6	104.7	71.0	106.8	98.4	79.5	108.7	101.8	128.5	126.1	126.7	138.9	125.9	129.0	101.3	72.9	79.5
03100	CONCRETE FORMS & ACCESSORIES	103.1	86.1	88.5	104.3	76.2	80.1	83.8	84.0	84.0	110.5	125.7	123.6	109.5	135.7	132.1	93.0	45.4	52.0
03200	CONCRETE REINFORCEMENT	100.8	86.0	92.3	100.8	83.0	90.6	101.4	67.2	81.9	98.4	176.6	143.0	102.3	176.6	144.7	92.7	43.5	64.6
03300	CAST-IN-PLACE CONCRETE	97.3	95.6	96.6	89.9	95.0	92.0	104.7	97.1	101.6	114.1	128.3	120.0	127.8	129.3	128.4	95.7	53.1	78.1
03	**CONCRETE**	105.1	90.7	97.9	103.5	85.6	94.6	117.1	87.2	102.2	114.6	135.3	125.0	126.6	140.1	133.3	96.9	49.9	73.4
04	**MASONRY**	96.6	95.8	96.1	89.9	91.4	90.8	91.0	101.1	97.2	105.0	127.8	119.0	110.6	129.8	122.4	77.5	38.9	53.8
05	**METALS**	99.9	107.2	102.6	98.2	105.9	101.0	98.3	106.7	101.3	99.6	134.1	112.0	106.4	134.2	116.4	93.9	80.4	89.1
06	**WOOD & PLASTICS**	107.0	84.0	95.1	107.0	72.2	89.0	85.8	79.7	82.6	114.6	123.2	119.0	113.2	135.5	124.7	94.3	46.2	69.4
07	**THERMAL & MOISTURE PROTECTION**	99.0	94.2	96.8	91.2	91.5	91.3	91.5	95.4	93.3	107.9	134.2	120.4	108.3	138.1	122.4	96.1	46.7	72.7
08	**DOORS & WINDOWS**	95.9	79.1	91.9	98.1	73.9	92.3	98.1	79.1	93.5	94.4	140.8	105.6	97.9	147.6	109.9	93.1	44.1	81.3
09200	PLASTER & GYPSUM BOARD	102.4	83.1	90.6	102.4	71.0	83.2	96.4	78.7	85.6	102.4	123.7	115.4	105.5	136.4	124.4	100.3	44.3	66.1
095,098	CEILINGS & ACOUSTICAL TREATMENT	97.9	83.1	88.1	97.9	71.0	80.1	97.9	78.7	85.2	86.1	123.7	110.9	105.2	136.4	125.8	88.2	44.3	59.2
09600	FLOORING	86.4	85.6	86.2	84.8	85.7	85.1	77.4	80.9	78.2	95.1	165.3	112.1	94.7	165.3	111.8	92.4	48.1	81.6
097,099	WALL FINISHES, PAINTS & COATINGS	87.1	86.2	86.6	81.8	86.2	84.4	81.8	85.8	84.1	95.0	150.7	127.4	95.0	150.7	127.4	106.5	45.7	71.2
09	**FINISHES**	96.7	85.5	91.0	95.3	78.0	86.5	92.6	82.8	87.6	102.4	135.2	119.1	106.9	142.9	125.2	92.6	45.8	68.7
10 - 14	**TOTAL DIV. 10000 - 14000**	100.0	95.1	99.0	100.0	91.5	98.2	100.0	95.8	99.1	100.0	129.7	106.3	100.0	133.3	107.0	100.0	66.7	93.0
15	**MECHANICAL**	100.3	89.1	95.1	100.3	88.6	94.9	100.3	86.1	93.7	100.5	119.5	109.3	100.5	121.5	110.2	100.1	50.1	77.0
16	**ELECTRICAL**	102.3	85.3	90.6	102.2	84.7	90.2	102.2	87.1	91.8	95.5	130.9	119.9	108.6	141.0	130.9	99.6	47.7	63.9
01 - 16	**WEIGHTED AVERAGE**	99.7	92.2	96.1	98.1	89.3	93.8	99.5	91.7	95.7	102.6	129.7	115.8	107.6	134.1	120.5	95.6	53.1	75.0

DIVISION		NORTH CAROLINA																	
		CHARLOTTE			DURHAM			FAYETTEVILLE			GREENSBORO			RALEIGH			WILMINGTON		
		MAT.	INST.	TOTAL	MAT.	INST.	TOTAL	MAT.	INST.	TOTAL	MAT.	INST.	TOTAL	MAT.	INST.	TOTAL	MAT.	INST.	TOTAL
02	**SITE CONSTRUCTION**	101.6	72.9	79.6	101.3	82.9	87.2	99.4	82.9	86.8	101.2	82.9	87.2	102.4	82.9	87.5	102.3	72.9	79.8
03100	CONCRETE FORMS & ACCESSORIES	101.5	45.5	53.2	95.2	45.7	52.5	90.3	45.7	51.9	95.2	45.7	52.5	98.3	45.7	52.9	94.8	45.7	52.5
03200	CONCRETE REINFORCEMENT	93.1	43.5	64.8	93.1	57.7	72.9	92.2	57.7	72.5	93.1	57.6	72.9	93.1	57.6	72.9	93.4	57.7	73.0
03300	CAST-IN-PLACE CONCRETE	98.1	53.1	79.6	96.0	53.2	78.3	92.6	53.2	76.4	95.1	53.2	77.8	101.5	53.2	81.6	95.3	53.2	77.9
03	**CONCRETE**	98.1	50.0	74.1	96.6	52.8	74.7	94.2	52.8	73.5	96.2	52.7	74.5	99.6	52.7	76.2	96.8	52.8	74.8
04	**MASONRY**	79.2	38.9	54.5	77.7	38.9	53.9	81.5	38.9	55.4	77.5	38.9	53.8	84.5	38.9	56.5	67.8	38.9	50.1
05	**METALS**	95.2	80.4	89.9	95.3	86.1	92.0	94.2	86.1	91.3	96.0	86.1	92.4	95.3	86.1	92.0	94.3	86.1	91.4
06	**WOOD & PLASTICS**	104.4	46.2	74.3	96.8	46.2	70.6	91.1	46.2	67.9	96.8	46.2	70.6	100.5	46.2	72.4	96.2	46.2	70.3
07	**THERMAL & MOISTURE PROTECTION**	96.1	47.8	73.2	96.8	47.4	73.3	96.4	47.4	73.1	96.8	47.4	73.3	96.6	46.8	72.9	96.1	47.4	73.0
08	**DOORS & WINDOWS**	97.1	44.1	84.3	97.1	48.3	85.3	93.2	48.3	82.4	97.1	48.3	85.3	93.9	48.3	82.9	93.3	48.3	82.4
09200	PLASTER & GYPSUM BOARD	105.8	44.3	68.2	105.8	44.3	68.2	100.5	44.3	66.2	105.8	44.3	68.2	108.6	44.3	69.3	101.0	44.3	66.3
095,098	CEILINGS & ACOUSTICAL TREATMENT	92.2	44.3	60.6	92.2	44.3	60.6	89.5	44.3	59.7	92.2	44.3	60.6	92.2	44.3	60.6	89.5	44.3	59.7
09600	FLOORING	95.8	48.1	84.2	96.0	48.1	84.4	92.5	48.1	81.7	96.0	48.1	84.4	96.0	48.1	84.4	93.3	48.1	82.3
097,099	WALL FINISHES, PAINTS & COATINGS	106.5	45.7	71.2	106.5	45.7	71.2	106.5	45.7	71.2	106.5	45.7	71.2	106.5	45.7	71.2	106.5	45.7	71.2
09	**FINISHES**	95.2	45.8	70.0	95.4	45.8	70.1	92.9	45.8	68.8	95.4	45.8	70.1	95.8	45.8	70.3	93.2	45.8	69.0
10 - 14	**TOTAL DIV. 10000 - 14000**	100.0	66.7	93.0	100.0	73.5	94.4	100.0	73.5	94.4	100.0	66.7	93.0	100.0	73.5	94.4	100.0	73.5	94.4
15	**MECHANICAL**	100.1	50.2	77.0	100.1	50.4	77.1	100.1	50.4	77.1	100.1	50.4	77.1	100.1	50.4	77.1	100.1	50.4	77.1
16	**ELECTRICAL**	99.1	42.8	60.4	97.5	45.6	61.8	92.7	45.6	60.3	98.6	45.6	62.1	98.2	45.6	62.0	99.3	45.6	62.4
01 - 16	**WEIGHTED AVERAGE**	96.9	52.4	75.2	96.5	55.1	76.4	95.0	55.1	75.6	96.6	54.9	76.3	97.0	55.1	76.6	95.2	54.1	75.2

	DIVISION	NORTH CAROLINA			NORTH DAKOTA												OHIO		
		WINSTON-SALEM			BISMARCK			FARGO			GRAND FORKS			MINOT			AKRON		
		MAT.	INST.	TOTAL	MAT.	INST.	TOTAL	MAT.	INST.	TOTAL	MAT.	INST.	TOTAL	MAT.	INST.	TOTAL	MAT.	INST.	TOTAL
02	**SITE CONSTRUCTION**	101.6	82.9	87.3	85.1	96.4	93.8	85.1	96.4	93.8	93.0	95.8	95.2	90.7	96.4	95.1	102.4	109.4	107.8
03100	CONCRETE FORMS & ACCESSORIES	96.7	45.5	52.5	93.2	58.2	63.0	93.8	58.9	63.7	89.6	51.9	57.1	84.1	58.0	61.6	99.9	102.3	102.0
03200	CONCRETE REINFORCEMENT	93.1	43.5	64.8	108.6	67.6	85.2	100.1	67.6	81.6	107.0	61.2	80.9	110.7	67.6	86.1	99.1	99.7	99.4
03300	CAST-IN-PLACE CONCRETE	98.2	53.1	79.6	104.9	63.3	87.8	107.4	66.0	90.3	104.9	61.7	87.1	104.9	61.6	87.1	101.2	109.8	104.7
03	**CONCRETE**	97.8	50.0	73.9	98.0	62.6	80.3	98.2	63.8	81.0	102.8	58.0	80.4	101.3	61.9	81.6	100.3	103.2	101.8
04	**MASONRY**	77.8	38.9	53.9	107.3	60.5	78.6	107.4	60.1	78.3	107.3	52.9	73.9	107.7	60.6	78.8	93.4	103.6	99.7
05	**METALS**	95.2	80.5	89.9	95.6	74.5	88.0	95.5	74.5	87.9	95.6	71.0	86.7	95.9	74.3	88.1	91.8	83.5	88.8
06	**WOOD & PLASTICS**	96.8	46.2	70.6	85.3	54.5	69.4	85.3	55.4	69.8	81.9	50.1	65.5	76.4	54.5	65.1	95.2	101.5	98.4
07	**THERMAL & MOISTURE PROTECTION**	96.8	47.4	73.3	98.8	59.3	80.1	99.3	60.5	80.9	99.6	55.9	78.9	99.2	59.1	80.2	106.4	102.9	104.7
08	**DOORS & WINDOWS**	97.1	44.5	84.4	103.4	51.5	90.9	103.4	52.0	91.0	103.4	47.6	90.0	103.6	51.5	91.0	107.2	101.1	105.7
09200	PLASTER & GYPSUM BOARD	105.8	44.3	68.2	117.4	53.6	78.4	117.4	54.5	78.9	116.5	49.1	75.3	114.8	53.6	77.4	105.0	100.8	102.4
095,098	CEILINGS & ACOUSTICAL TREATMENT	92.2	44.3	60.6	143.1	53.6	84.1	143.1	54.5	84.6	143.1	49.1	81.1	143.1	53.6	84.1	99.5	100.8	100.4
09600	FLOORING	96.0	48.1	84.4	111.0	55.3	97.5	110.8	54.5	97.1	109.3	54.5	96.0	106.6	55.5	94.2	107.9	102.7	106.6
097,099	WALL FINISHES, PAINTS & COATINGS	106.5	45.7	71.2	102.5	45.9	69.6	102.5	42.2	67.5	102.5	35.1	63.3	102.5	38.4	65.3	103.9	122.1	114.5
09	**FINISHES**	95.4	45.8	70.1	115.2	55.8	84.9	115.0	55.7	84.8	115.0	50.1	81.9	113.7	55.0	83.8	104.4	104.5	104.5
10 - 14	**TOTAL DIV. 10000 - 14000**	100.0	66.7	93.0	100.0	61.2	91.8	100.0	61.2	91.8	100.0	57.8	91.1	100.0	61.2	91.8	100.0	102.4	100.5
15	**MECHANICAL**	100.1	50.2	77.0	100.6	69.5	86.2	100.6	76.1	89.3	100.6	55.5	79.7	100.6	59.4	81.5	100.1	105.3	102.5
16	**ELECTRICAL**	98.6	45.6	62.1	96.9	70.8	79.0	96.8	50.9	65.3	101.7	69.4	79.5	106.8	70.8	82.0	98.0	93.8	95.1
01 - 16	**WEIGHTED AVERAGE**	96.7	53.9	75.9	100.8	68.1	84.9	100.8	66.1	84.0	102.0	62.4	82.7	102.0	65.9	84.4	99.8	100.9	100.4

	DIVISION	OHIO																	
		CANTON			CINCINNATI			CLEVELAND			COLUMBUS			DAYTON			LORAIN		
		MAT.	INST.	TOTAL	MAT.	INST.	TOTAL	MAT.	INST.	TOTAL	MAT.	INST.	TOTAL	MAT.	INST.	TOTAL	MAT.	INST.	TOTAL
02	**SITE CONSTRUCTION**	102.6	108.4	107.0	78.6	111.0	103.4	102.5	109.0	107.5	84.4	100.4	96.6	77.4	111.0	103.1	101.8	108.7	107.1
03100	CONCRETE FORMS & ACCESSORIES	99.9	90.7	91.9	96.8	90.5	91.3	100.0	109.3	108.0	94.4	90.6	91.1	96.7	87.2	88.5	99.9	91.5	92.7
03200	CONCRETE REINFORCEMENT	99.1	84.3	90.7	93.4	88.9	90.8	99.5	100.0	99.8	103.2	88.2	94.7	93.4	85.1	88.6	99.1	99.7	99.5
03300	CAST-IN-PLACE CONCRETE	102.2	106.8	104.1	90.0	92.5	91.1	99.3	118.0	107.0	90.0	98.3	93.4	82.0	92.7	86.4	96.3	98.6	97.3
03	**CONCRETE**	100.8	94.2	97.5	93.2	90.9	92.0	99.4	109.3	104.4	95.0	92.3	93.6	89.3	88.2	88.7	97.9	94.7	96.3
04	**MASONRY**	94.3	93.9	94.0	87.6	98.4	94.2	98.0	114.2	108.0	92.7	97.6	95.7	87.0	93.3	90.9	90.2	103.4	98.3
05	**METALS**	91.8	76.7	86.4	94.8	89.0	92.7	93.2	86.9	90.9	96.3	82.4	91.3	94.1	78.8	88.6	92.3	84.1	89.4
06	**WOOD & PLASTICS**	95.2	89.8	92.4	97.7	88.2	92.8	94.4	106.7	100.7	95.5	89.2	92.2	99.9	85.8	92.6	95.2	87.9	91.4
07	**THERMAL & MOISTURE PROTECTION**	106.9	98.5	102.9	89.9	98.0	93.7	105.1	115.9	110.2	101.7	97.9	99.9	95.3	94.0	94.7	106.7	105.4	106.1
08	**DOORS & WINDOWS**	101.4	83.1	97.0	97.2	85.8	94.5	95.6	104.7	97.8	99.8	87.5	96.8	97.5	84.1	94.2	101.4	94.5	99.7
09200	PLASTER & GYPSUM BOARD	105.0	88.7	95.0	102.5	88.2	93.7	104.3	106.1	105.4	102.9	88.6	94.1	102.5	85.7	92.2	105.0	86.7	93.8
095,098	CEILINGS & ACOUSTICAL TREATMENT	99.5	88.7	92.4	90.9	88.2	89.1	98.2	106.1	103.4	96.5	88.6	91.3	92.4	85.7	88.0	99.5	86.7	91.1
09600	FLOORING	108.1	88.9	103.4	92.9	103.7	95.5	107.6	112.6	108.9	97.1	95.8	96.8	95.4	91.1	94.4	108.1	110.2	108.6
097,099	WALL FINISHES, PAINTS & COATINGS	103.9	95.0	98.7	101.5	97.5	99.2	103.9	122.5	114.7	97.7	100.7	99.4	101.5	95.0	97.7	103.9	122.5	114.7
09	**FINISHES**	104.5	90.2	97.2	93.7	93.4	93.6	104.1	111.2	107.7	97.9	92.4	95.0	94.7	88.4	91.5	104.5	97.7	101.0
10 - 14	**TOTAL DIV. 10000 - 14000**	100.0	92.4	98.4	100.0	91.0	98.1	100.0	107.3	101.5	100.0	94.0	98.7	100.0	89.5	97.8	100.0	101.6	100.3
15	**MECHANICAL**	100.1	90.4	95.6	99.7	94.6	97.4	100.1	109.5	104.5	100.1	96.8	98.6	100.9	90.4	96.0	100.1	95.9	98.1
16	**ELECTRICAL**	96.5	97.5	97.2	95.8	87.9	90.4	97.1	105.8	103.1	102.1	89.4	93.4	94.5	90.1	91.5	96.5	96.4	96.4
01 - 16	**WEIGHTED AVERAGE**	99.2	92.9	96.1	95.3	93.9	94.6	98.7	107.3	102.9	97.9	93.1	95.5	95.2	90.9	93.1	98.6	97.2	97.9

	DIVISION	OHIO									OKLAHOMA								
		SPRINGFIELD			TOLEDO			YOUNGSTOWN			ENID			LAWTON			MUSKOGEE		
		MAT.	INST.	TOTAL	MAT.	INST.	TOTAL	MAT.	INST.	TOTAL	MAT.	INST.	TOTAL	MAT.	INST.	TOTAL	MAT.	INST.	TOTAL
02	**SITE CONSTRUCTION**	78.0	110.0	102.5	83.7	101.4	97.2	102.3	109.4	107.8	110.2	89.2	94.1	105.5	90.8	94.3	95.9	85.2	87.7
03100	CONCRETE FORMS & ACCESSORIES	96.7	88.7	89.8	94.4	103.9	102.6	99.9	92.8	93.7	94.3	46.5	53.1	98.6	64.1	68.8	100.4	42.2	50.2
03200	CONCRETE REINFORCEMENT	93.4	85.0	88.6	103.2	88.4	94.8	99.1	93.3	95.8	95.3	84.9	89.4	95.5	84.9	89.5	95.3	45.5	66.9
03300	CAST-IN-PLACE CONCRETE	86.0	92.2	88.5	90.0	108.0	97.4	100.2	108.6	103.7	96.5	57.9	80.6	93.4	57.9	78.8	86.9	44.4	69.3
03	**CONCRETE**	91.2	88.7	89.9	94.9	101.8	98.3	99.9	97.5	98.7	91.8	57.8	74.8	88.4	65.5	76.9	83.8	44.8	64.3
04	**MASONRY**	87.3	92.5	90.5	103.0	101.6	102.1	93.5	99.8	97.4	102.0	60.1	76.2	97.6	60.1	74.6	110.6	51.0	73.9
05	**METALS**	94.1	78.6	88.5	96.1	86.3	92.6	91.8	81.3	88.0	93.4	68.4	84.4	95.4	68.5	85.7	93.3	64.4	82.9
06	**WOOD & PLASTICS**	99.9	88.6	94.0	95.5	105.1	100.4	95.2	90.8	92.9	96.5	43.9	69.3	100.4	67.6	83.4	102.6	43.4	72.0
07	**THERMAL & MOISTURE PROTECTION**	95.3	94.0	94.7	103.6	106.4	105.0	107.0	100.3	103.8	98.6	64.6	82.5	98.4	67.0	83.5	98.2	51.8	76.2
08	**DOORS & WINDOWS**	97.5	80.2	93.3	99.7	97.6	99.2	101.4	91.7	99.0	96.6	57.2	87.1	98.1	70.0	91.3	96.5	42.7	83.5
09200	PLASTER & GYPSUM BOARD	102.5	88.6	94.0	102.9	105.0	104.1	105.0	89.7	95.6	95.2	43.3	63.5	97.6	67.6	79.3	96.9	42.6	63.7
095,098	CEILINGS & ACOUSTICAL TREATMENT	92.4	88.6	89.9	96.5	105.0	102.1	99.5	89.7	93.0	92.9	43.3	60.2	101.1	67.6	79.0	92.9	42.6	59.7
09600	FLOORING	95.4	91.1	94.4	96.3	74.1	90.9	108.1	95.9	105.1	115.6	62.3	102.7	118.7	62.3	105.0	119.9	50.2	103.0
097,099	WALL FINISHES, PAINTS & COATINGS	101.5	95.0	97.7	97.7	109.4	104.5	103.9	103.0	103.4	95.6	66.5	78.7	95.6	66.5	78.7	95.6	42.3	64.6
09	**FINISHES**	94.7	89.8	92.2	97.6	98.9	98.3	104.6	93.9	99.1	100.9	49.9	74.9	103.3	63.7	83.1	101.5	43.9	72.1
10 - 14	**TOTAL DIV. 10000 - 14000**	100.0	89.5	97.8	100.0	98.9	99.8	100.0	99.7	99.9	100.0	66.7	93.0	100.0	69.8	93.6	100.0	39.2	87.2
15	**MECHANICAL**	100.9	89.9	95.8	100.1	104.2	102.0	100.1	96.3	98.4	100.1	69.1	85.8	100.1	69.1	85.8	100.1	35.6	70.3
16	**ELECTRICAL**	94.5	89.4	91.0	102.8	98.1	99.6	96.5	92.0	93.4	95.0	70.8	78.4	97.7	65.1	75.3	94.4	41.9	58.3
01 - 16	**WEIGHTED AVERAGE**	95.4	90.6	93.1	98.5	99.7	99.1	99.0	95.7	97.4	97.6	65.7	82.1	97.8	68.5	83.6	96.8	48.8	73.4

	DIVISION	OKLAHOMA						OREGON											
		OKLAHOMA CITY			TULSA			EUGENE			MEDFORD			PORTLAND			SALEM		
		MAT.	INST.	TOTAL	MAT.	INST.	TOTAL	MAT.	INST.	TOTAL	MAT.	INST.	TOTAL	MAT.	INST.	TOTAL	MAT.	INST.	TOTAL
02	**SITE CONSTRUCTION**	105.8	91.4	94.8	102.6	86.2	90.0	99.4	106.4	104.8	108.5	106.4	106.9	100.5	106.4	105.0	99.6	106.4	104.8
03100	CONCRETE FORMS & ACCESSORIES	100.1	55.0	61.2	99.9	55.8	61.8	103.2	110.8	109.8	97.5	110.7	108.9	104.8	111.2	110.3	104.7	111.1	110.2
03200	CONCRETE REINFORCEMENT	95.5	84.9	89.5	95.5	84.9	89.5	109.3	105.5	107.1	100.9	105.5	103.5	104.5	105.8	105.2	110.4	105.8	107.8
03300	CAST-IN-PLACE CONCRETE	93.4	57.0	78.4	94.6	56.5	78.9	101.2	109.8	104.7	104.5	109.7	106.6	108.6	109.9	109.2	104.9	109.9	107.0
03	**CONCRETE**	88.5	61.2	74.8	89.1	62.5	75.8	111.7	109.1	110.4	119.0	109.0	114.0	114.9	109.4	112.1	113.8	109.3	111.6
04	**MASONRY**	99.5	62.7	76.8	96.8	62.8	75.9	112.3	106.4	108.7	109.1	106.4	107.4	113.6	111.9	112.6	114.3	111.9	112.8
05	**METALS**	95.6	68.4	85.8	96.0	84.6	91.9	95.1	103.7	98.2	94.7	103.6	97.9	93.8	104.4	97.7	95.7	104.4	98.8
06	**WOOD & PLASTICS**	102.6	54.1	77.5	101.2	55.7	77.7	98.2	109.8	104.2	92.2	109.8	101.3	100.1	109.8	105.1	100.1	109.8	105.1
07	**THERMAL & MOISTURE PROTECTION**	97.6	66.7	82.9	98.1	64.9	82.4	111.0	100.7	106.1	111.9	99.7	106.1	110.8	108.3	109.6	110.8	108.3	109.6
08	**DOORS & WINDOWS**	98.1	62.7	89.5	98.1	63.0	89.6	101.8	112.6	104.4	104.5	112.6	106.5	99.2	112.6	102.4	101.1	112.6	103.9
09200	PLASTER & GYPSUM BOARD	97.6	53.7	70.8	97.6	55.3	71.7	94.8	110.0	104.1	92.4	110.0	103.2	94.0	110.0	103.8	96.0	110.0	104.6
095,098	CEILINGS & ACOUSTICAL TREATMENT	101.1	53.7	69.8	101.1	55.3	70.9	118.1	110.0	112.8	116.7	110.0	112.3	118.1	110.0	112.8	129.0	110.0	116.5
09600	FLOORING	118.7	62.3	105.0	118.5	64.8	105.4	117.1	104.5	114.1	114.7	104.5	112.2	117.1	102.0	113.5	117.1	104.5	114.1
097,099	WALL FINISHES, PAINTS & COATINGS	95.6	66.5	78.7	95.6	62.3	76.3	119.6	82.9	98.3	119.6	70.1	90.9	119.6	82.9	98.3	119.6	82.9	98.3
09	**FINISHES**	103.4	56.4	79.4	102.8	57.5	79.7	113.9	106.6	110.2	113.1	105.2	109.0	113.8	106.2	109.9	116.0	106.6	111.2
10 - 14	**TOTAL DIV. 10000 - 14000**	100.0	69.1	93.5	100.0	71.9	94.1	100.0	110.2	102.2	100.0	110.1	102.1	100.0	110.2	102.2	100.0	110.2	102.2
15	**MECHANICAL**	100.1	70.5	86.4	100.1	62.8	82.8	100.3	112.5	105.9	100.3	112.5	105.9	100.3	112.5	105.9	100.3	112.5	105.9
16	**ELECTRICAL**	96.5	70.8	78.8	97.7	54.6	68.1	104.2	102.9	103.3	109.8	93.6	98.6	104.0	112.4	109.8	103.8	102.9	103.2
01 - 16	**WEIGHTED AVERAGE**	97.9	68.2	83.4	97.8	65.3	82.0	103.8	107.3	105.5	105.3	105.5	105.4	103.8	109.7	106.7	104.4	108.2	106.2

	DIVISION	PENNSYLVANIA																	
		ALLENTOWN			ALTOONA			ERIE			HARRISBURG			PHILADELPHIA			PITTSBURGH		
		MAT.	INST.	TOTAL	MAT.	INST.	TOTAL	MAT.	INST.	TOTAL	MAT.	INST.	TOTAL	MAT.	INST.	TOTAL	MAT.	INST.	TOTAL
02	**SITE CONSTRUCTION**	94.1	107.7	104.6	98.5	107.7	105.6	94.9	108.4	105.3	82.8	106.3	100.8	101.9	95.5	97.0	101.3	110.3	108.2
03100	CONCRETE FORMS & ACCESSORIES	102.9	107.8	107.1	81.8	91.8	90.4	101.3	101.9	101.8	94.3	87.0	88.0	104.0	131.2	127.5	101.2	103.5	103.2
03200	CONCRETE REINFORCEMENT	100.8	108.5	105.2	97.8	98.3	98.1	99.8	98.2	98.9	100.8	93.2	96.4	100.2	123.3	113.3	102.6	107.3	105.2
03300	CAST-IN-PLACE CONCRETE	89.1	99.5	93.4	99.3	92.2	96.4	97.7	95.2	96.7	87.4	92.5	89.5	94.6	124.4	106.9	97.5	101.4	99.1
03	**CONCRETE**	99.8	106.2	103.0	94.2	94.7	94.4	93.2	100.2	96.7	98.2	91.7	95.0	106.8	126.9	116.8	95.1	104.8	99.9
04	**MASONRY**	95.0	95.7	95.4	96.8	82.2	87.8	88.1	98.4	94.4	92.4	88.5	90.0	89.9	125.5	111.8	89.2	102.3	97.2
05	**METALS**	98.1	125.0	107.8	91.8	118.9	101.6	91.9	120.8	102.3	99.9	117.6	106.3	100.1	122.1	108.0	98.6	126.1	108.5
06	**WOOD & PLASTICS**	107.0	110.5	108.8	81.7	93.1	87.6	102.5	102.1	102.3	99.0	85.8	92.2	107.7	131.3	119.9	103.0	103.7	103.3
07	**THERMAL & MOISTURE PROTECTION**	99.0	112.1	105.3	97.9	93.2	95.7	98.0	100.2	99.1	102.3	105.1	103.6	100.1	131.1	114.8	95.0	103.5	99.0
08	**DOORS & WINDOWS**	95.9	108.0	98.8	90.3	101.4	93.0	90.4	99.3	92.6	95.9	95.9	95.9	97.4	134.0	106.2	92.4	111.5	97.0
09200	PLASTER & GYPSUM BOARD	102.4	110.5	107.4	88.8	92.6	91.1	96.8	101.9	99.9	102.4	85.0	91.8	98.8	132.1	119.2	94.3	103.4	99.8
095,098	CEILINGS & ACOUSTICAL TREATMENT	97.9	110.5	106.2	92.4	92.6	92.5	97.9	101.9	100.5	97.9	85.0	89.4	99.9	132.1	121.1	96.8	103.4	101.1
09600	FLOORING	86.4	87.1	86.5	80.7	69.1	77.9	87.5	90.4	88.2	86.6	83.2	85.8	82.3	136.6	95.5	96.5	103.9	98.3
097,099	WALL FINISHES, PAINTS & COATINGS	87.1	102.2	95.9	82.7	104.0	95.1	90.9	88.0	89.2	87.1	88.9	88.1	89.0	138.8	117.9	92.0	116.0	105.9
09	**FINISHES**	96.7	103.7	100.3	92.1	88.4	90.2	96.9	98.3	97.6	95.7	85.8	90.6	96.9	133.2	115.4	98.6	104.6	101.7
10 - 14	**TOTAL DIV. 10000 - 14000**	100.0	103.9	100.8	100.0	100.9	100.2	100.0	105.5	101.2	100.0	93.6	98.6	100.0	124.9	105.3	100.0	105.2	101.1
15	**MECHANICAL**	100.3	102.7	101.4	99.7	88.3	94.4	99.7	98.6	99.2	100.3	93.4	97.1	99.9	124.7	111.4	99.8	100.4	100.1
16	**ELECTRICAL**	102.2	89.1	93.2	87.6	102.2	97.6	90.0	88.4	88.9	101.3	81.1	87.4	97.5	126.7	117.5	95.0	102.2	100.0
01 - 16	**WEIGHTED AVERAGE**	98.7	103.4	101.0	94.6	96.6	95.6	94.8	100.3	97.5	98.1	93.6	95.9	99.6	123.9	111.4	97.0	106.0	101.4

	DIVISION	PENNSYLVANIA									PUERTO RICO			RHODE ISLAND			SOUTH CAROLINA		
		READING			SCRANTON			YORK			SAN JUAN			PROVIDENCE			CHARLESTON		
		MAT.	INST.	TOTAL	MAT.	INST.	TOTAL	MAT.	INST.	TOTAL	MAT.	INST.	TOTAL	MAT.	INST.	TOTAL	MAT.	INST.	TOTAL
02	**SITE CONSTRUCTION**	99.4	113.5	110.2	94.6	107.7	104.7	83.5	106.4	101.0	125.2	94.5	101.7	84.5	105.6	100.6	93.6	82.0	84.7
03100	CONCRETE FORMS & ACCESSORIES	103.5	93.1	94.6	103.1	93.9	95.1	81.6	87.5	86.7	98.9	21.4	32.1	102.5	111.2	110.0	95.2	44.6	51.6
03200	CONCRETE REINFORCEMENT	98.5	108.2	104.0	100.8	106.5	104.0	99.8	93.1	96.0	201.2	14.5	94.7	111.9	119.3	116.1	93.1	65.8	77.6
03300	CAST-IN-PLACE CONCRETE	69.2	98.8	81.4	93.1	95.3	94.0	87.7	93.6	90.1	109.8	36.2	79.4	86.1	118.9	99.6	81.5	55.6	70.8
03	**CONCRETE**	94.0	99.5	96.7	101.8	98.2	100.0	100.0	92.3	96.2	114.4	26.1	70.2	103.4	114.9	109.2	89.5	54.4	72.0
04	**MASONRY**	91.9	91.1	91.4	95.3	100.9	98.8	93.7	88.9	90.8	221.3	19.4	97.2	104.3	119.2	113.4	84.7	42.6	58.8
05	**METALS**	97.1	124.8	107.1	99.9	124.3	108.7	97.3	117.5	104.6	129.6	34.4	95.3	98.0	110.2	102.4	95.2	85.7	91.8
06	**WOOD & PLASTICS**	107.7	91.0	99.1	107.0	91.9	99.2	89.0	85.5	87.2	94.2	21.1	56.4	102.9	109.1	106.1	96.8	44.1	69.6
07	**THERMAL & MOISTURE PROTECTION**	100.1	108.8	104.2	98.9	108.0	103.2	97.4	106.2	101.6	168.9	27.1	101.7	98.5	112.4	105.1	96.3	50.5	74.6
08	**DOORS & WINDOWS**	97.4	97.4	97.4	95.9	103.3	97.7	92.8	95.7	93.5	160.0	18.4	125.8	103.6	111.1	105.4	97.1	46.9	85.0
09200	PLASTER & GYPSUM BOARD	105.9	90.5	96.5	102.4	91.4	95.7	97.9	84.7	89.8	225.4	18.2	98.8	98.2	108.2	104.3	105.8	42.2	66.9
095,098	CEILINGS & ACOUSTICAL TREATMENT	101.1	90.5	94.1	97.9	91.4	93.6	89.7	84.7	86.4	364.2	18.2	135.9	102.6	108.2	106.3	92.2	42.2	59.2
09600	FLOORING	80.5	85.9	81.8	86.4	93.8	88.2	82.4	83.2	82.6	201.6	19.9	157.6	99.9	126.2	106.3	96.0	47.6	84.3
097,099	WALL FINISHES, PAINTS & COATINGS	87.1	95.1	91.8	87.1	98.2	93.5	87.1	88.9	88.1	209.3	18.4	98.3	95.6	121.2	110.5	106.5	46.0	71.3
09	**FINISHES**	97.5	91.2	94.3	96.7	93.7	95.2	92.3	86.2	89.2	242.3	20.9	129.4	99.5	115.1	107.5	95.6	44.8	69.7
10 - 14	**TOTAL DIV. 10000 - 14000**	100.0	101.0	100.2	100.0	101.5	100.3	100.0	94.3	98.8	100.0	24.9	84.1	100.0	112.7	102.7	100.0	66.1	92.8
15	**MECHANICAL**	100.2	102.6	101.3	100.3	93.2	97.0	100.3	94.4	97.6	106.9	18.8	66.1	99.8	102.0	100.9	100.1	52.6	78.1
16	**ELECTRICAL**	100.1	81.0	87.0	102.3	86.6	91.5	91.9	81.1	84.5	144.8	16.1	56.3	98.6	97.0	97.5	98.5	39.6	58.0
01 - 16	**WEIGHTED AVERAGE**	97.9	99.0	98.4	99.2	98.8	99.0	96.3	94.0	95.2	143.2	29.0	87.6	100.1	108.1	104.0	95.9	54.6	75.8

DIVISION		SOUTH CAROLINA												SOUTH DAKOTA					
		COLUMBIA			FLORENCE			GREENVILLE			SPARTANBURG			ABERDEEN			PIERRE		
		MAT.	INST.	TOTAL	MAT.	INST.	TOTAL	MAT.	INST.	TOTAL	MAT.	INST.	TOTAL	MAT.	INST.	TOTAL	MAT.	INST.	TOTAL
02	SITE CONSTRUCTION	93.3	82.0	84.6	104.4	82.0	87.2	99.4	81.6	85.8	99.2	81.6	85.7	81.5	95.2	92.0	79.9	95.2	91.6
03100	CONCRETE FORMS & ACCESSORIES	100.6	47.2	54.5	78.9	47.2	51.6	95.1	47.0	53.6	99.3	47.0	54.2	93.5	44.4	51.2	91.7	46.7	52.9
03200	CONCRETE REINFORCEMENT	93.1	65.5	77.4	92.8	65.8	77.4	92.7	46.4	66.3	92.7	46.4	66.3	107.2	57.5	78.9	106.7	64.5	82.6
03300	CAST-IN-PLACE CONCRETE	79.9	51.5	68.2	72.1	55.5	65.3	72.1	55.4	65.2	72.1	55.4	65.2	100.0	55.9	81.9	97.2	51.8	78.4
03	CONCRETE	89.0	54.1	71.6	91.3	55.5	73.4	90.0	51.8	70.9	90.3	51.8	71.1	96.1	52.2	74.2	93.8	53.1	73.5
04	MASONRY	83.7	37.3	55.2	71.2	42.6	53.6	69.3	42.6	52.9	71.2	42.6	53.6	108.5	57.9	77.4	105.8	48.8	70.8
05	METALS	95.2	85.0	91.5	94.2	85.5	91.0	94.0	78.7	88.5	94.0	78.7	88.5	107.4	72.7	94.9	107.3	75.0	95.7
06	WOOD & PLASTICS	103.3	48.0	74.7	79.3	48.0	63.1	96.3	48.0	71.3	101.0	48.0	73.6	96.4	42.1	68.3	94.5	45.5	69.1
07	THERMAL & MOISTURE PROTECTION	96.3	48.4	73.5	96.5	50.9	74.9	96.6	50.9	74.9	96.7	50.9	74.9	97.1	52.4	75.9	97.2	50.3	75.0
08	DOORS & WINDOWS	97.1	49.0	85.5	93.2	49.0	82.5	93.2	44.4	81.4	93.2	44.4	81.4	98.5	45.8	85.8	102.0	49.8	89.4
09200	PLASTER & GYPSUM BOARD	105.8	46.2	69.4	96.7	46.2	65.8	101.2	46.2	67.6	102.4	46.2	68.1	101.7	40.8	64.5	100.6	44.3	66.2
095,098	CEILINGS & ACOUSTICAL TREATMENT	92.2	46.2	61.9	89.5	46.2	60.9	88.2	46.2	60.5	88.2	46.2	60.5	110.8	40.8	64.6	109.4	44.3	66.5
09600	FLOORING	95.8	47.6	84.1	87.0	47.6	77.5	93.6	48.5	82.7	95.3	48.5	83.9	111.4	68.4	100.9	110.5	48.7	95.5
097,099	WALL FINISHES, PAINTS & COATINGS	106.5	46.0	71.3	106.5	46.0	71.3	106.5	46.0	71.3	106.5	46.0	71.3	102.5	40.9	66.7	102.5	40.9	66.7
09	FINISHES	95.5	47.1	70.8	91.7	47.1	68.9	93.9	47.2	70.1	94.7	47.2	70.5	106.7	48.2	76.8	105.9	46.2	75.4
10 - 14	TOTAL DIV. 10000 - 14000	100.0	66.6	93.0	100.0	66.5	92.9	100.0	66.6	92.9	100.0	66.6	92.9	100.0	67.9	93.2	100.0	68.4	93.3
15	MECHANICAL	100.1	44.3	74.3	100.1	44.3	74.3	100.1	44.3	74.3	100.1	44.3	74.3	100.2	41.6	73.1	100.2	41.6	73.1
16	ELECTRICAL	99.6	41.7	59.8	95.4	23.8	46.2	98.9	37.9	57.0	98.9	37.9	57.0	102.5	48.1	65.1	96.1	56.4	68.8
01 - 16	WEIGHTED AVERAGE	95.9	53.2	75.1	94.2	51.0	73.1	94.4	52.0	73.8	94.7	52.0	73.9	101.2	55.7	79.1	100.6	56.5	79.1

DIVISION		SOUTH DAKOTA						TENNESSEE											
		RAPID CITY			SIOUX FALLS			CHATTANOOGA			JACKSON			JOHNSON CITY			KNOXVILLE		
		MAT.	INST.	TOTAL	MAT.	INST.	TOTAL	MAT.	INST.	TOTAL	MAT.	INST.	TOTAL	MAT.	INST.	TOTAL	MAT.	INST.	TOTAL
02	SITE CONSTRUCTION	80.1	94.9	91.5	81.1	97.2	93.4	104.9	97.6	99.3	105.0	95.6	97.8	114.5	86.3	92.9	92.0	86.3	87.6
03100	CONCRETE FORMS & ACCESSORIES	104.8	42.6	51.1	91.7	44.9	51.4	95.4	52.5	58.4	86.2	38.0	44.6	79.2	54.7	58.1	94.3	54.8	60.2
03200	CONCRETE REINFORCEMENT	100.1	64.6	79.9	100.1	65.4	80.3	94.0	49.4	68.5	99.7	48.8	70.7	98.0	49.6	70.4	94.0	49.6	68.7
03300	CAST-IN-PLACE CONCRETE	96.5	48.7	76.8	100.5	56.1	82.2	103.0	54.6	83.0	101.7	38.8	75.8	82.8	61.0	73.8	96.7	55.9	79.9
03	CONCRETE	93.6	50.2	71.9	94.7	54.0	74.3	93.9	54.6	74.3	94.0	42.6	68.3	99.2	57.8	78.5	91.0	56.1	73.5
04	MASONRY	105.9	46.0	69.1	103.0	64.1	79.1	96.9	51.8	69.2	110.8	31.1	61.8	107.5	51.6	73.1	74.9	51.6	60.6
05	METALS	109.4	75.3	97.1	109.9	76.8	98.0	97.1	83.1	92.0	95.1	79.7	89.6	94.7	82.8	90.4	97.3	82.9	92.1
06	WOOD & PLASTICS	105.0	42.6	72.7	94.5	42.1	67.4	95.8	53.5	73.9	83.8	40.2	61.3	71.4	56.8	63.8	86.7	56.8	71.2
07	THERMAL & MOISTURE PROTECTION	97.7	49.2	74.7	97.2	56.0	77.6	103.0	54.5	80.0	100.7	40.3	72.0	97.3	55.3	77.4	94.9	54.8	75.8
08	DOORS & WINDOWS	103.0	48.2	89.8	103.0	47.9	89.7	100.7	52.8	89.2	101.7	44.1	87.8	97.0	56.2	87.1	93.7	56.2	84.6
09200	PLASTER & GYPSUM BOARD	101.6	41.3	64.7	101.6	40.8	64.5	95.1	52.7	69.2	98.3	38.9	62.0	104.7	56.0	74.9	110.4	56.0	77.2
095,098	CEILINGS & ACOUSTICAL TREATMENT	114.9	41.3	66.3	114.9	40.8	66.0	93.8	52.7	66.6	95.8	38.9	58.2	85.3	56.0	66.0	90.7	56.0	67.8
09600	FLOORING	110.5	57.3	97.6	110.5	79.3	102.9	99.4	59.1	89.7	90.7	28.6	75.7	92.7	60.8	84.9	98.2	60.8	89.1
097,099	WALL FINISHES, PAINTS & COATINGS	102.5	40.9	66.7	102.5	40.9	66.7	104.9	49.9	72.9	94.4	32.6	58.5	102.8	63.9	80.2	102.8	63.9	80.2
09	FINISHES	107.1	44.7	75.3	107.1	50.4	78.2	97.5	53.5	75.0	95.9	36.3	65.5	98.1	57.0	77.1	92.8	57.0	74.5
10 - 14	TOTAL DIV. 10000 - 14000	100.0	65.7	92.8	100.0	68.0	93.2	100.0	65.3	92.7	100.0	58.9	91.3	100.0	68.8	93.4	100.0	68.9	93.4
15	MECHANICAL	100.2	38.6	71.7	100.2	41.7	73.1	100.2	50.7	77.3	100.1	44.8	74.5	99.8	56.1	79.6	99.8	59.8	81.3
16	ELECTRICAL	96.9	56.4	69.1	95.8	72.1	79.5	103.5	63.9	76.3	99.3	36.6	56.2	90.9	39.1	55.3	100.4	62.8	74.5
01 - 16	WEIGHTED AVERAGE	101.3	54.9	78.7	101.2	61.6	81.9	98.9	62.1	81.0	98.8	49.2	74.7	98.3	58.9	79.1	95.0	63.3	79.6

DIVISION		TENNESSEE						TEXAS											
		MEMPHIS			NASHVILLE			ABILENE			AMARILLO			AUSTIN			BEAUMONT		
		MAT.	INST.	TOTAL	MAT.	INST.	TOTAL	MAT.	INST.	TOTAL	MAT.	INST.	TOTAL	MAT.	INST.	TOTAL	MAT.	INST.	TOTAL
02	SITE CONSTRUCTION	98.7	92.1	93.7	95.6	100.3	99.2	103.0	85.1	89.3	103.1	86.1	90.1	88.1	85.9	86.4	96.7	87.4	89.6
03100	CONCRETE FORMS & ACCESSORIES	94.7	62.7	67.1	95.6	68.3	72.0	96.6	48.8	55.3	100.2	58.0	63.8	98.6	62.8	67.7	105.9	64.3	70.0
03200	CONCRETE REINFORCEMENT	92.3	64.2	76.3	95.0	63.1	76.8	95.5	59.9	75.2	95.5	59.5	75.0	94.0	59.9	74.6	94.8	54.0	71.6
03300	CAST-IN-PLACE CONCRETE	95.9	66.5	83.8	91.4	65.5	80.7	98.2	47.6	77.3	101.8	56.0	82.9	91.9	57.9	77.9	91.6	64.1	80.2
03	CONCRETE	89.9	66.2	78.1	89.4	68.0	78.7	90.6	51.5	71.0	92.6	58.4	75.5	82.6	61.1	71.8	88.7	63.0	75.9
04	MASONRY	84.2	62.5	70.8	85.5	64.9	72.9	100.2	53.0	71.2	104.0	47.8	69.4	97.2	53.9	70.6	103.3	65.3	80.0
05	METALS	96.5	93.9	95.6	99.7	91.8	96.9	96.6	70.3	87.1	96.6	70.0	87.0	95.7	69.6	86.3	95.9	70.7	86.8
06	WOOD & PLASTICS	93.4	64.5	78.5	90.9	70.5	80.4	97.2	49.5	72.5	100.4	60.6	79.8	98.2	66.2	81.6	107.6	66.1	86.1
07	THERMAL & MOISTURE PROTECTION	99.1	64.8	82.8	98.4	66.2	83.1	98.3	55.7	78.1	100.4	52.1	77.5	95.2	59.3	78.2	100.3	65.9	84.0
08	DOORS & WINDOWS	98.8	67.2	91.2	96.0	68.9	89.5	93.6	52.2	83.6	93.6	55.4	84.4	95.9	63.7	88.1	97.4	59.9	88.3
09200	PLASTER & GYPSUM BOARD	103.2	63.6	79.0	102.4	70.1	82.7	97.6	48.8	67.8	97.6	60.3	74.8	98.7	66.0	78.7	99.9	65.9	79.1
095,098	CEILINGS & ACOUSTICAL TREATMENT	89.2	63.6	72.3	87.7	70.1	76.1	101.1	48.8	66.6	101.1	60.3	74.2	90.6	66.0	74.3	108.4	65.9	80.3
09600	FLOORING	92.0	54.0	82.8	103.4	73.4	96.1	118.7	64.7	105.6	118.5	52.4	102.4	98.3	60.3	89.0	117.3	76.4	107.4
097,099	WALL FINISHES, PAINTS & COATINGS	93.5	64.8	76.8	109.1	60.2	80.7	94.4	64.6	77.1	94.4	46.1	66.3	95.0	51.8	69.9	91.0	62.2	74.3
09	FINISHES	93.2	60.9	76.7	101.7	68.6	84.8	102.8	53.3	77.6	102.8	55.6	78.8	93.9	61.3	77.3	98.7	66.5	82.3
10 - 14	TOTAL DIV. 10000 - 14000	100.0	71.7	94.0	100.0	73.1	94.3	100.0	68.3	93.3	100.0	70.7	93.8	100.0	71.3	93.9	100.0	74.9	94.7
15	MECHANICAL	100.0	66.7	84.6	100.0	66.2	84.4	100.1	43.1	73.7	100.1	59.3	81.2	99.7	67.7	84.9	100.0	65.7	84.1
16	ELECTRICAL	99.1	93.8	95.4	101.6	70.3	80.1	97.7	46.2	62.3	98.7	65.6	75.9	97.1	72.5	80.2	96.6	78.3	84.0
01 - 16	WEIGHTED AVERAGE	96.3	75.2	86.0	97.4	73.3	85.7	97.7	54.9	76.9	98.4	62.3	80.8	95.1	67.2	81.5	97.5	70.1	84.2

City Cost Indexes

	DIVISION	TEXAS																	
		CORPUS CHRISTI			DALLAS			EL PASO			FORT WORTH			HOUSTON			LAREDO		
		MAT.	INST.	TOTAL	MAT.	INST.	TOTAL	MAT.	INST.	TOTAL	MAT.	INST.	TOTAL	MAT.	INST.	TOTAL	MAT.	INST.	TOTAL
02	SITE CONSTRUCTION	123.0	81.1	90.9	131.9	82.5	94.1	103.5	85.1	89.4	102.8	85.4	89.5	124.5	85.1	94.4	88.2	85.7	86.2
03100	CONCRETE FORMS & ACCESSORIES	102.2	47.0	54.6	95.4	62.6	67.1	97.3	57.5	63.0	99.3	62.2	67.3	92.8	71.6	74.5	90.2	47.1	53.0
03200	CONCRETE REINFORCEMENT	92.2	57.0	72.1	94.4	61.7	75.7	95.5	59.9	75.2	95.5	61.4	76.1	97.5	62.6	77.6	94.0	57.1	73.0
03300	CAST-IN-PLACE CONCRETE	108.9	53.6	86.1	102.1	62.4	85.7	100.5	43.1	76.8	96.5	57.6	80.5	94.4	72.4	85.3	83.9	60.4	74.2
03	CONCRETE	93.0	53.2	73.1	89.7	64.0	76.8	91.8	53.6	72.7	89.9	61.1	75.5	88.0	71.9	80.0	82.2	54.4	68.3
04	MASONRY	87.9	52.8	66.4	95.9	64.1	76.3	99.0	52.3	70.3	95.1	60.3	73.7	98.3	68.8	80.2	96.9	55.3	71.3
05	METALS	95.2	84.1	91.2	96.5	86.6	92.9	96.5	68.7	86.4	96.4	71.4	87.4	100.2	92.3	97.4	96.5	65.2	85.2
06	WOOD & PLASTICS	109.5	47.1	77.2	99.9	63.9	81.2	98.5	62.4	79.8	105.3	63.9	83.9	91.8	72.7	81.9	87.4	46.9	66.5
07	THERMAL & MOISTURE PROTECTION	97.9	54.1	77.1	93.8	64.6	79.9	97.9	58.9	79.4	98.8	60.0	80.4	92.7	69.6	81.7	94.1	57.0	76.5
08	DOORS & WINDOWS	102.9	47.2	89.5	106.6	60.2	95.4	93.6	55.7	84.5	87.1	60.2	80.6	105.7	68.6	96.8	96.5	45.9	84.3
09200	PLASTER & GYPSUM BOARD	99.7	46.1	66.9	100.8	63.7	78.1	97.6	62.1	75.9	97.6	63.7	76.9	98.2	72.5	82.5	96.5	46.1	65.7
095,098	CEILINGS & ACOUSTICAL TREATMENT	96.0	46.1	63.1	102.1	63.7	76.8	101.1	62.1	75.4	101.1	63.7	76.4	108.0	72.5	84.6	90.6	46.1	61.2
09600	FLOORING	110.4	57.0	97.4	110.5	73.0	101.4	118.7	66.7	106.1	153.9	58.4	130.7	104.7	65.5	95.2	95.1	57.0	85.9
097,099	WALL FINISHES, PAINTS & COATINGS	109.1	46.2	72.6	101.8	60.4	77.7	94.4	44.3	65.3	95.6	65.8	78.3	99.6	70.3	82.6	95.0	53.4	70.8
09	FINISHES	100.3	48.2	73.8	105.7	64.4	84.6	102.8	58.5	80.2	114.5	62.0	87.8	103.9	70.4	86.8	92.7	48.9	70.4
10 - 14	TOTAL DIV. 10000 - 14000	100.0	70.6	93.8	100.0	72.7	94.2	100.0	71.3	93.9	100.0	72.7	94.2	100.0	78.3	95.4	100.0	69.5	93.6
15	MECHANICAL	99.7	56.9	79.9	99.9	65.8	84.1	100.1	42.4	73.4	100.1	65.2	84.0	99.9	74.3	88.1	99.7	51.1	77.2
16	ELECTRICAL	94.0	58.8	69.8	98.9	68.6	78.0	96.5	61.7	72.6	96.5	66.3	75.7	98.5	71.6	80.0	97.1	69.0	77.8
01 - 16	WEIGHTED AVERAGE	98.3	60.0	79.7	100.0	69.1	85.0	97.7	58.6	78.7	97.8	66.4	82.6	99.8	75.0	87.7	95.0	59.8	77.8

	DIVISION	TEXAS															UTAH		
		LUBBOCK			ODESSA			SAN ANTONIO			WACO			WICHITA FALLS			LOGAN		
		MAT.	INST.	TOTAL	MAT.	INST.	TOTAL	MAT.	INST.	TOTAL	MAT.	INST.	TOTAL	MAT.	INST.	TOTAL	MAT.	INST.	TOTAL
02	SITE CONSTRUCTION	133.7	83.3	95.1	103.0	85.6	89.7	87.7	90.1	89.6	101.7	85.4	89.2	102.5	85.1	89.2	96.0	99.8	98.9
03100	CONCRETE FORMS & ACCESSORIES	98.5	50.6	57.2	96.6	48.5	55.1	90.3	64.6	68.1	98.7	48.3	55.2	98.7	50.8	57.4	104.3	68.9	73.8
03200	CONCRETE REINFORCEMENT	94.8	59.5	74.7	97.7	59.3	75.8	94.0	58.2	73.6	95.5	59.7	75.1	95.5	51.6	70.5	100.6	75.7	86.4
03300	CAST-IN-PLACE CONCRETE	98.4	56.2	81.0	98.2	51.5	78.9	82.3	65.8	75.5	88.6	59.6	76.6	94.7	54.4	78.1	93.1	78.1	86.9
03	CONCRETE	89.4	56.3	72.8	90.8	52.6	71.7	81.4	64.4	72.9	86.0	55.4	70.7	89.0	53.4	71.2	112.3	73.7	93.0
04	MASONRY	99.7	44.1	65.5	100.2	44.2	65.7	96.7	62.3	75.6	97.3	56.3	72.1	97.7	54.2	71.0	122.1	72.5	91.6
05	METALS	100.7	86.3	95.5	96.8	69.3	86.9	97.1	71.0	87.7	96.6	70.0	87.0	96.5	70.7	87.2	105.3	78.3	95.6
06	WOOD & PLASTICS	100.5	52.0	75.4	97.2	49.3	72.4	87.4	65.7	76.2	104.1	45.8	74.0	104.1	51.8	77.1	90.1	67.4	78.4
07	THERMAL & MOISTURE PROTECTION	90.8	54.9	73.8	98.3	50.7	75.7	94.1	69.1	82.2	99.0	55.5	78.4	99.0	56.2	78.7	104.5	76.7	91.3
08	DOORS & WINDOWS	105.7	51.3	92.6	93.6	49.3	82.9	96.5	62.6	88.4	87.1	46.0	77.2	87.1	53.3	79.0	90.2	66.5	84.5
09200	PLASTER & GYPSUM BOARD	98.1	51.2	69.5	97.6	48.6	67.7	96.5	65.4	77.5	97.6	45.0	65.5	97.6	51.2	69.3	88.2	66.2	74.8
095,098	CEILINGS & ACOUSTICAL TREATMENT	103.8	51.2	69.1	101.1	48.6	66.5	90.6	65.4	73.9	101.1	45.0	64.1	101.1	51.2	68.2	104.7	66.2	79.3
09600	FLOORING	109.7	49.6	95.1	118.7	48.9	101.8	95.1	61.3	86.9	154.1	45.4	127.8	154.6	66.7	133.3	105.2	72.3	97.2
097,099	WALL FINISHES, PAINTS & COATINGS	106.4	41.4	68.6	94.4	41.4	63.6	95.0	53.4	70.8	95.6	42.7	64.8	97.5	67.7	80.1	108.7	57.7	79.1
09	FINISHES	104.4	49.0	76.1	102.8	47.4	74.5	92.7	62.6	77.4	114.5	46.2	79.7	114.8	55.4	84.5	100.0	67.8	83.6
10 - 14	TOTAL DIV. 10000 - 14000	100.0	69.1	93.5	100.0	70.3	93.7	100.0	74.9	94.7	100.0	70.3	93.7	100.0	68.7	93.4	100.0	78.5	95.5
15	MECHANICAL	99.7	54.2	78.7	100.1	46.4	75.3	99.7	73.3	87.5	100.1	62.9	82.9	100.1	52.0	77.9	100.1	74.4	88.2
16	ELECTRICAL	96.2	56.3	68.8	97.7	52.4	66.6	97.1	69.0	77.8	97.7	66.6	76.4	102.6	65.2	76.9	92.7	82.2	85.5
01 - 16	WEIGHTED AVERAGE	100.3	59.3	80.4	97.8	55.0	77.0	94.9	70.0	82.8	97.6	61.9	80.2	98.3	60.6	80.0	101.8	77.3	89.9

	DIVISION	UTAH									VERMONT						VIRGINIA		
		OGDEN			PROVO			SALT LAKE CITY			BURLINGTON			RUTLAND			ALEXANDRIA		
		MAT.	INST.	TOTAL	MAT.	INST.	TOTAL	MAT.	INST.	TOTAL	MAT.	INST.	TOTAL	MAT.	INST.	TOTAL	MAT.	INST.	TOTAL
02	SITE CONSTRUCTION	83.3	99.8	95.9	92.3	97.8	96.5	82.9	99.7	95.8	83.3	101.5	97.3	83.3	101.5	97.3	112.6	83.8	90.5
03100	CONCRETE FORMS & ACCESSORIES	104.3	68.9	73.8	106.1	69.0	74.1	102.9	69.0	73.6	88.5	58.3	62.4	100.2	58.3	64.1	91.6	73.7	76.1
03200	CONCRETE REINFORCEMENT	100.6	75.7	86.4	108.8	75.7	89.9	98.5	75.7	85.5	111.9	66.6	86.1	111.9	66.6	86.1	82.7	91.3	87.6
03300	CAST-IN-PLACE CONCRETE	94.5	78.1	87.8	93.2	78.1	87.0	102.2	78.1	92.3	97.2	64.5	83.7	92.6	64.5	81.0	99.1	80.9	91.6
03	CONCRETE	101.2	73.7	87.4	111.7	73.7	92.7	104.6	73.7	89.2	105.3	62.5	83.9	103.9	62.5	83.2	96.6	80.6	88.6
04	MASONRY	115.3	72.5	89.0	127.6	72.5	93.7	133.9	72.5	96.2	96.3	70.1	80.2	84.0	70.1	75.4	83.9	70.7	75.7
05	METALS	105.7	78.3	95.8	102.7	78.3	93.9	104.5	78.3	95.1	98.6	72.6	89.2	98.0	72.6	88.8	95.2	97.4	96.0
06	WOOD & PLASTICS	90.1	67.4	78.4	92.0	67.4	79.3	90.2	67.4	78.4	88.6	56.9	72.2	102.5	56.9	78.9	95.7	74.4	84.7
07	THERMAL & MOISTURE PROTECTION	102.9	76.7	90.5	107.0	76.7	92.6	106.0	76.7	92.1	99.2	64.0	82.5	99.0	64.0	82.4	96.5	79.8	88.6
08	DOORS & WINDOWS	90.2	66.5	84.5	94.5	66.5	87.8	90.2	66.5	84.5	110.0	52.4	96.1	110.0	52.4	96.1	97.1	74.9	91.8
09200	PLASTER & GYPSUM BOARD	88.2	66.2	74.8	87.1	66.2	74.3	88.2	66.2	74.8	98.2	54.2	71.3	98.2	54.2	71.3	105.8	73.4	86.0
095,098	CEILINGS & ACOUSTICAL TREATMENT	104.7	66.2	79.3	96.5	66.2	76.5	104.7	66.2	79.3	102.6	54.2	70.7	102.6	54.2	70.7	92.2	73.4	79.8
09600	FLOORING	105.2	72.3	97.2	106.0	72.3	97.8	104.8	72.3	96.9	99.9	81.9	95.6	99.9	81.9	95.6	96.0	80.0	92.1
097,099	WALL FINISHES, PAINTS & COATINGS	108.7	57.7	79.1	108.7	67.0	84.5	108.7	67.0	84.5	95.6	46.9	67.3	95.6	46.9	67.3	117.5	77.8	94.4
09	FINISHES	98.9	67.8	83.0	99.0	68.9	83.6	99.2	68.9	83.7	99.1	61.2	79.7	99.0	61.2	79.7	96.1	74.9	85.3
10 - 14	TOTAL DIV. 10000 - 14000	100.0	78.5	95.5	100.0	78.5	95.5	100.0	78.5	95.5	100.0	96.7	99.3	100.0	96.7	99.3	100.0	86.0	97.0
15	MECHANICAL	100.1	74.4	88.2	100.1	74.4	88.2	100.3	74.4	88.3	99.8	60.1	81.4	99.8	60.1	81.4	100.1	84.6	92.9
16	ELECTRICAL	92.7	82.2	85.5	93.2	82.2	85.6	93.6	82.2	85.7	105.2	70.6	81.4	104.0	70.6	81.0	98.7	93.2	94.9
01 - 16	WEIGHTED AVERAGE	99.6	77.3	88.7	102.1	77.2	90.0	101.1	77.4	89.6	101.0	69.2	85.5	100.1	69.2	85.1	97.3	83.5	90.6

City Cost Indexes

DIVISION		VIRGINIA																	
		ARLINGTON			NEWPORT NEWS			NORFOLK			PORTSMOUTH			RICHMOND			ROANOKE		
		MAT.	INST.	TOTAL	MAT.	INST.	TOTAL	MAT.	INST.	TOTAL	MAT.	INST.	TOTAL	MAT.	INST.	TOTAL	MAT.	INST.	TOTAL
02	**SITE CONSTRUCTION**	123.0	83.4	92.7	105.2	85.0	89.7	104.4	86.1	90.4	103.6	84.6	89.1	105.8	85.7	90.4	101.9	81.7	86.4
03100	CONCRETE FORMS & ACCESSORIES	88.5	72.5	74.7	95.2	53.7	59.4	100.2	53.8	60.2	81.6	53.7	57.5	96.1	63.0	67.5	95.0	41.8	49.1
03200	CONCRETE REINFORCEMENT	93.3	77.9	84.5	93.1	76.4	83.6	93.1	76.4	83.6	92.8	78.7	84.8	93.1	76.6	83.7	93.1	77.8	84.4
03300	CAST-IN-PLACE CONCRETE	96.4	74.9	87.5	97.0	59.9	81.7	99.8	59.9	83.4	96.0	61.8	81.9	103.1	62.1	86.2	100.5	53.3	81.0
03	**CONCRETE**	104.2	75.5	89.8	97.1	61.8	79.5	98.9	61.9	80.4	95.7	62.9	79.3	100.2	66.7	83.5	98.8	54.5	76.7
04	**MASONRY**	95.0	68.2	78.6	86.0	57.9	68.7	93.7	57.9	71.7	90.7	56.2	69.5	82.5	55.8	66.1	85.6	39.8	57.5
05	**METALS**	94.2	92.4	93.5	95.3	91.6	93.9	94.5	91.7	93.5	94.5	91.5	93.4	95.3	92.2	94.1	95.1	87.1	92.2
06	**WOOD & PLASTICS**	91.3	74.4	82.6	96.8	54.1	74.7	102.9	54.1	77.7	82.1	54.1	67.6	98.1	65.8	81.4	96.8	41.5	68.2
07	**THERMAL & MOISTURE PROTECTION**	97.2	76.0	87.2	96.3	53.6	76.0	96.1	53.6	76.0	96.2	53.2	75.8	95.8	56.6	77.2	96.2	47.5	73.1
08	**DOORS & WINDOWS**	95.1	71.9	89.5	97.1	55.0	87.0	97.1	58.2	87.7	97.2	58.7	87.9	97.1	60.3	88.2	97.1	47.4	85.1
09200	PLASTER & GYPSUM BOARD	101.8	73.4	84.5	105.8	51.6	72.6	105.8	51.6	72.6	98.0	51.6	69.6	105.8	63.6	80.0	105.8	39.5	65.3
095,098	CEILINGS & ACOUSTICAL TREATMENT	89.5	73.4	78.9	92.2	51.6	65.4	92.2	51.6	65.4	92.2	51.6	65.4	92.2	63.6	73.3	92.2	39.5	57.4
09600	FLOORING	94.5	80.0	91.0	96.0	46.6	84.0	95.8	46.6	83.9	88.1	46.6	78.1	95.8	69.3	89.4	96.0	41.8	82.9
097,099	WALL FINISHES, PAINTS & COATINGS	117.5	93.3	103.4	106.5	49.7	73.5	106.5	64.3	82.0	106.5	43.6	69.9	106.5	65.0	82.4	106.5	38.8	67.2
09	**FINISHES**	95.4	76.2	85.6	95.4	51.4	72.9	95.3	53.1	73.8	91.7	50.7	70.8	95.2	64.6	79.6	95.2	40.9	67.5
10 - 14	**TOTAL DIV. 10000 - 14000**	100.0	85.0	96.8	100.0	76.7	95.1	100.0	76.8	95.1	100.0	76.8	95.1	100.0	84.2	96.7	100.0	70.7	93.8
15	**MECHANICAL**	100.1	83.2	92.3	100.1	68.1	85.3	100.1	65.2	83.9	100.1	65.2	83.9	100.1	67.7	85.1	100.1	47.6	75.8
16	**ELECTRICAL**	95.1	90.9	92.2	98.6	62.7	73.9	98.6	64.7	75.3	95.7	64.7	74.4	99.6	69.3	78.8	98.6	41.3	59.2
01 - 16	**WEIGHTED AVERAGE**	98.4	81.4	90.1	97.2	66.4	82.2	97.8	66.6	82.6	96.4	66.1	81.6	97.5	70.1	84.2	97.3	53.4	75.9

DIVISION		WASHINGTON																	
		EVERETT			RICHLAND			SEATTLE			SPOKANE			TACOMA			VANCOUVER		
		MAT.	INST.	TOTAL	MAT.	INST.	TOTAL	MAT.	INST.	TOTAL	MAT.	INST.	TOTAL	MAT.	INST.	TOTAL	MAT.	INST.	TOTAL
02	**SITE CONSTRUCTION**	93.3	118.0	112.2	97.2	90.7	92.2	97.3	115.7	111.4	98.0	90.7	92.4	96.1	118.0	112.9	108.8	103.4	104.7
03100	CONCRETE FORMS & ACCESSORIES	109.7	100.7	102.0	113.5	85.6	89.4	92.6	101.2	100.1	122.0	85.7	90.7	92.6	101.0	99.9	97.2	99.9	99.5
03200	CONCRETE REINFORCEMENT	107.1	95.6	100.6	101.1	92.7	96.4	101.1	95.7	98.0	101.8	92.9	96.7	101.1	95.6	98.0	104.0	95.5	99.2
03300	CAST-IN-PLACE CONCRETE	93.1	105.8	98.3	112.1	89.6	102.8	97.7	108.2	102.0	116.1	89.7	105.2	95.7	108.0	100.8	106.9	104.4	105.9
03	**CONCRETE**	100.3	100.9	100.6	109.2	88.2	98.7	102.3	102.0	102.2	111.7	88.2	99.9	101.3	101.8	101.6	112.9	100.3	106.6
04	**MASONRY**	129.6	104.9	114.5	113.8	87.2	97.4	123.1	105.6	112.3	115.5	90.2	99.9	123.0	105.6	112.3	124.7	105.7	113.0
05	**METALS**	99.3	88.8	95.5	93.1	85.4	90.4	100.8	90.9	97.2	92.9	85.7	90.3	100.3	88.8	96.2	98.8	91.5	96.2
06	**WOOD & PLASTICS**	111.3	99.0	105.0	103.2	84.6	93.6	93.9	99.0	96.5	113.5	84.6	98.5	92.6	99.0	95.9	90.7	98.6	94.7
07	**THERMAL & MOISTURE PROTECTION**	107.4	100.5	104.2	171.3	84.3	130.0	107.2	101.5	104.5	169.0	84.9	129.1	107.0	101.3	104.3	110.5	97.1	104.1
08	**DOORS & WINDOWS**	101.5	97.2	100.5	115.7	81.1	107.3	101.7	97.7	100.7	115.4	80.2	106.9	102.1	97.7	101.0	98.5	96.2	98.0
09200	PLASTER & GYPSUM BOARD	96.8	98.6	97.9	123.8	84.1	99.5	91.6	98.6	95.9	126.8	84.1	100.7	94.2	98.6	96.9	95.4	98.6	97.4
095,098	CEILINGS & ACOUSTICAL TREATMENT	110.9	98.6	102.8	122.2	84.1	97.0	112.0	98.6	103.2	122.2	84.1	97.0	113.7	98.6	103.8	122.0	98.6	106.6
09600	FLOORING	123.6	107.9	119.8	112.5	53.0	98.1	115.3	107.9	113.5	116.0	79.7	107.2	116.0	107.9	114.0	121.4	97.6	115.6
097,099	WALL FINISHES, PAINTS & COATINGS	115.0	91.5	101.3	117.6	76.4	93.6	115.0	91.5	101.3	117.6	80.9	96.3	115.0	91.5	101.3	123.1	75.3	95.3
09	**FINISHES**	115.9	100.9	108.3	129.2	77.8	103.0	112.8	101.0	106.8	130.8	83.8	106.8	113.6	101.0	107.2	114.0	96.6	105.1
10 - 14	**TOTAL DIV. 10000 - 14000**	100.0	104.2	100.9	100.0	92.4	98.4	100.0	104.5	100.9	100.0	92.4	98.4	100.0	104.4	100.9	100.0	99.2	99.8
15	**MECHANICAL**	100.1	105.2	102.5	100.7	99.9	100.4	100.1	111.8	105.5	100.7	89.8	95.7	100.2	102.3	101.2	100.4	101.4	100.8
16	**ELECTRICAL**	106.3	93.6	97.6	104.2	96.3	98.8	106.0	105.6	105.7	94.7	88.0	90.1	106.0	99.0	101.2	119.1	108.3	111.6
01 - 16	**WEIGHTED AVERAGE**	104.1	101.4	102.8	108.4	90.1	99.5	103.8	104.9	104.4	108.3	87.8	98.3	103.7	102.0	102.9	106.1	101.1	103.7

DIVISION		WASHINGTON			WEST VIRGINIA												WISCONSIN		
		YAKIMA			CHARLESTON			HUNTINGTON			PARKERSBURG			WHEELING			EAU CLAIRE		
		MAT.	INST.	TOTAL	MAT.	INST.	TOTAL	MAT.	INST.	TOTAL	MAT.	INST.	TOTAL	MAT.	INST.	TOTAL	MAT.	INST.	TOTAL
02	**SITE CONSTRUCTION**	99.3	117.3	113.1	100.6	86.0	89.4	102.1	86.4	90.1	106.6	86.0	90.9	107.3	85.1	90.3	82.2	102.3	97.6
03100	CONCRETE FORMS & ACCESSORIES	95.1	96.9	96.6	103.5	95.5	96.6	95.9	97.8	97.5	84.4	94.2	92.8	86.6	90.6	90.1	95.9	95.5	95.6
03200	CONCRETE REINFORCEMENT	103.6	94.8	98.6	93.1	85.7	88.8	93.1	92.1	92.5	91.8	97.6	95.1	91.1	93.1	92.2	103.7	99.3	101.2
03300	CAST-IN-PLACE CONCRETE	102.2	90.9	97.6	94.9	108.7	100.6	102.3	109.6	105.3	95.6	97.9	96.6	95.6	101.9	98.2	97.8	92.3	95.5
03	**CONCRETE**	106.8	93.8	100.3	96.6	99.0	97.8	99.8	101.5	100.6	99.8	96.8	98.3	99.8	95.9	97.8	95.0	95.1	95.1
04	**MASONRY**	114.9	87.1	97.8	84.4	98.1	92.8	85.8	102.1	95.8	74.1	93.7	86.1	96.1	92.2	93.7	93.6	97.6	96.1
05	**METALS**	99.2	85.4	94.2	95.3	101.8	97.6	95.3	104.5	98.6	94.1	105.7	98.3	94.2	104.2	97.8	94.0	96.8	95.0
06	**WOOD & PLASTICS**	95.3	99.0	97.2	106.2	94.9	100.4	96.8	96.8	96.8	84.6	92.6	88.8	86.7	89.5	88.2	106.9	94.5	100.5
07	**THERMAL & MOISTURE PROTECTION**	107.3	85.5	96.9	96.3	93.0	94.7	96.5	94.7	95.6	96.6	90.7	93.8	96.8	89.9	93.5	96.7	89.7	93.4
08	**DOORS & WINDOWS**	101.7	85.9	97.9	98.3	85.8	95.3	97.1	88.2	95.0	97.7	86.3	95.0	98.6	90.3	96.6	104.3	91.3	101.2
09200	PLASTER & GYPSUM BOARD	94.3	98.6	96.9	105.8	94.6	99.0	105.8	96.5	100.1	99.7	92.2	95.1	100.1	89.0	93.3	112.5	94.8	101.7
095,098	CEILINGS & ACOUSTICAL TREATMENT	109.6	98.6	102.4	92.2	94.6	93.8	92.2	96.5	95.1	89.5	92.2	91.3	89.5	89.0	89.2	108.5	94.8	99.5
09600	FLOORING	117.3	79.7	108.2	95.8	104.6	97.9	95.8	114.2	100.3	91.1	97.9	92.7	92.0	103.1	94.7	92.6	72.3	87.7
097,099	WALL FINISHES, PAINTS & COATINGS	115.0	80.9	95.2	106.5	100.2	102.8	106.5	95.1	99.9	106.5	104.8	105.5	106.5	93.3	98.8	91.7	80.7	85.3
09	**FINISHES**	113.5	92.2	102.7	95.4	98.1	96.7	95.2	100.9	98.1	92.7	96.0	94.4	93.0	93.2	93.1	101.1	89.4	95.1
10 - 14	**TOTAL DIV. 10000 - 14000**	100.0	101.5	100.3	100.0	99.4	99.9	100.0	100.2	100.1	100.0	99.5	99.9	100.0	95.6	99.1	100.0	96.1	99.2
15	**MECHANICAL**	100.2	100.4	100.3	100.1	89.1	95.0	100.1	93.0	96.8	100.1	95.8	98.1	100.1	93.3	97.0	100.2	90.8	95.9
16	**ELECTRICAL**	110.8	96.3	100.8	98.6	77.9	84.3	98.6	90.8	93.2	99.4	97.2	97.8	94.6	91.0	92.1	105.0	88.2	93.5
01 - 16	**WEIGHTED AVERAGE**	104.2	95.9	100.2	97.2	91.6	94.4	97.4	96.0	96.7	96.4	95.4	96.0	97.6	93.1	95.4	98.6	93.3	96.0

DIVISION		WISCONSIN																	
		GREEN BAY			KENOSHA			LA CROSSE			MADISON			MILWAUKEE			RACINE		
		MAT.	INST.	TOTAL	MAT.	INST.	TOTAL	MAT.	INST.	TOTAL	MAT.	INST.	TOTAL	MAT.	INST.	TOTAL	MAT.	INST.	TOTAL
02	**SITE CONSTRUCTION**	84.4	98.1	94.9	89.2	101.7	98.8	76.6	102.3	96.2	84.4	105.0	100.2	85.7	93.5	91.7	84.8	105.5	100.7
03100	CONCRETE FORMS & ACCESSORIES	115.7	95.2	98.1	109.9	103.8	104.7	78.6	95.6	93.3	97.1	100.6	100.1	100.0	108.3	107.2	96.8	103.8	102.9
03200	CONCRETE REINFORCEMENT	100.1	91.2	95.0	111.7	96.8	103.2	102.2	89.8	95.1	104.6	90.1	96.4	104.6	97.0	100.3	104.6	96.8	100.1
03300	CAST-IN-PLACE CONCRETE	101.2	97.9	99.8	109.9	100.4	106.0	88.0	91.9	89.6	99.1	98.4	98.8	99.1	104.2	101.2	99.1	100.0	99.4
03	**CONCRETE**	96.7	95.5	96.1	104.0	101.2	102.6	86.1	93.3	89.7	96.9	97.8	97.4	97.3	103.8	100.6	96.9	101.1	99.0
04	**MASONRY**	114.5	98.7	104.8	100.4	108.8	105.6	92.8	99.9	97.1	104.5	103.3	103.8	104.4	112.9	109.6	104.5	108.8	107.1
05	**METALS**	96.6	93.8	95.6	96.7	98.3	97.3	94.0	93.2	93.7	97.0	94.9	96.2	98.6	88.6	95.0	97.0	98.3	97.5
06	**WOOD & PLASTICS**	122.6	94.5	108.1	117.3	102.6	109.7	88.3	94.5	91.5	108.1	100.2	104.0	111.4	107.3	109.3	108.4	102.6	105.4
07	**THERMAL & MOISTURE PROTECTION**	99.4	90.1	95.0	96.0	101.9	98.8	95.9	89.5	92.9	95.2	98.0	96.6	94.6	105.1	99.5	95.8	101.4	98.4
08	**DOORS & WINDOWS**	102.8	87.9	99.2	99.7	101.6	100.1	104.3	85.3	99.7	104.6	97.7	102.9	104.5	104.2	104.4	104.6	101.6	103.9
09200	PLASTER & GYPSUM BOARD	96.4	94.8	95.4	97.4	103.2	100.9	106.3	94.8	99.3	107.3	100.7	103.3	107.3	107.9	107.7	107.3	103.2	104.8
095,098	CEILINGS & ACOUSTICAL TREATMENT	103.5	94.8	97.8	85.8	103.2	97.3	103.1	94.8	97.6	94.8	100.7	98.7	94.8	107.9	103.4	94.8	103.2	100.3
09600	FLOORING	111.5	102.2	109.2	108.5	92.0	104.5	85.3	102.2	89.4	91.9	97.1	93.1	94.6	109.4	98.2	91.9	92.0	91.9
097,099	WALL FINISHES, PAINTS & COATINGS	102.5	72.8	85.3	103.4	96.3	99.3	91.7	80.7	85.3	92.6	92.2	92.4	95.2	104.9	100.8	92.6	97.0	95.2
09	**FINISHES**	105.0	94.4	99.6	100.9	101.4	101.1	96.5	95.6	96.0	97.8	99.4	98.6	98.9	108.5	103.8	97.9	101.4	99.7
10 - 14	**TOTAL DIV. 10000 - 14000**	100.0	95.0	98.9	100.0	101.1	100.2	100.0	96.1	99.2	100.0	96.9	99.4	100.0	102.9	100.6	100.0	101.1	100.2
15	**MECHANICAL**	100.6	90.7	96.0	100.4	98.4	99.5	100.2	91.6	96.2	100.1	94.8	97.6	100.1	99.0	99.6	100.1	98.4	99.3
16	**ELECTRICAL**	96.1	89.3	91.5	93.3	99.4	97.5	105.4	88.2	93.6	92.1	95.4	94.4	92.9	101.5	98.8	91.0	99.2	96.7
01 - 16	**WEIGHTED AVERAGE**	100.4	93.3	97.0	99.4	101.0	100.2	96.6	93.5	95.1	98.7	98.0	98.3	99.2	101.6	100.3	98.6	101.3	99.9

DIVISION		WYOMING									CANADA								
		CASPER			CHEYENNE			ROCK SPRINGS			CALGARY, ALBERTA			EDMONTON, ALBERTA			HAMILTON, ONTARIO		
		MAT.	INST.	TOTAL	MAT.	INST.	TOTAL	MAT.	INST.	TOTAL	MAT.	INST.	TOTAL	MAT.	INST.	TOTAL	MAT.	INST.	TOTAL
02	**SITE CONSTRUCTION**	84.1	99.2	95.7	84.0	99.2	95.7	87.3	98.8	96.1	107.1	97.4	99.7	108.6	97.4	100.0	110.7	97.6	100.7
03100	CONCRETE FORMS & ACCESSORIES	102.0	50.2	57.3	102.0	50.2	57.3	108.0	35.5	45.4	119.2	80.4	85.7	116.5	80.4	85.3	118.6	104.4	106.3
03200	CONCRETE REINFORCEMENT	107.2	54.2	77.0	100.6	54.4	74.2	109.3	54.0	77.8	161.1	64.3	105.9	161.1	64.3	105.9	174.4	97.3	130.4
03300	CAST-IN-PLACE CONCRETE	104.9	65.0	88.4	104.9	65.0	88.4	105.3	48.9	82.1	162.8	82.8	129.8	163.6	82.8	130.3	157.2	106.8	136.4
03	**CONCRETE**	106.9	56.9	81.9	106.1	56.9	81.5	107.9	44.8	76.4	146.0	78.8	112.4	146.3	78.8	112.5	144.9	103.6	124.3
04	**MASONRY**	109.0	45.9	70.2	108.0	48.5	71.4	175.8	37.5	90.8	146.7	81.4	106.5	146.7	81.4	106.6	148.4	111.6	125.8
05	**METALS**	105.1	66.7	91.2	106.3	66.9	92.1	105.1	65.4	90.8	100.8	83.0	94.4	100.8	83.0	94.4	100.9	95.9	99.1
06	**WOOD & PLASTICS**	96.8	48.7	71.9	95.8	48.7	71.4	103.7	34.5	67.9	116.2	79.7	97.3	111.9	79.7	95.3	117.0	103.8	110.1
07	**THERMAL & MOISTURE PROTECTION**	104.5	61.9	84.3	104.5	62.6	84.6	105.2	47.1	77.7	104.3	80.8	93.1	104.2	80.7	93.1	104.3	101.6	103.0
08	**DOORS & WINDOWS**	94.3	50.0	83.6	95.0	50.0	84.2	101.3	42.4	87.1	92.3	75.6	88.3	92.3	75.6	88.3	92.3	99.2	94.0
09200	PLASTER & GYPSUM BOARD	86.7	46.9	62.3	88.2	46.9	62.9	90.1	32.2	54.7	147.0	79.1	105.5	144.4	79.1	104.5	209.2	104.0	144.9
095,098	CEILINGS & ACOUSTICAL TREATMENT	96.5	46.9	63.7	104.7	46.9	66.5	96.5	32.2	54.1	102.4	79.1	87.0	102.4	79.1	87.0	102.4	104.0	103.4
09600	FLOORING	105.2	49.6	91.7	105.2	72.9	97.4	107.5	48.6	93.2	132.6	83.1	120.6	132.6	83.1	120.6	132.6	108.8	126.8
097,099	WALL FINISHES, PAINTS & COATINGS	108.7	65.7	83.7	108.5	65.7	83.7	108.7	39.9	68.7	109.2	78.1	91.1	109.3	78.1	91.2	109.2	110.3	109.8
09	**FINISHES**	97.4	50.9	73.7	99.0	55.6	76.9	98.6	37.5	67.4	120.4	81.0	100.3	120.1	81.0	100.2	129.5	106.3	117.6
10 - 14	**TOTAL DIV. 10000 - 14000**	100.0	77.6	95.3	100.0	77.6	95.3	100.0	59.8	91.5	100.0	89.0	97.7	100.0	84.6	96.8	100.0	110.0	102.1
15	**MECHANICAL**	100.1	69.3	85.9	100.1	52.5	78.1	100.1	37.1	70.9	102.5	73.3	89.0	102.5	73.3	89.0	102.5	102.2	102.4
16	**ELECTRICAL**	92.6	63.1	72.4	92.6	62.2	71.7	91.6	53.6	65.5	113.7	80.3	90.8	113.7	80.3	90.8	127.0	102.0	109.8
01 - 16	**WEIGHTED AVERAGE**	100.3	63.9	82.6	100.6	61.4	81.5	105.3	50.7	78.7	111.8	80.9	96.8	111.8	80.8	96.7	113.8	102.8	108.4

DIVISION		CANADA																	
		LONDON, ONTARIO			MONTREAL, QUEBEC			OTTAWA, ONTARIO			QUEBEC, QUEBEC			TORONTO, ONTARIO			VANCOUVER, B C		
		MAT.	INST.	TOTAL	MAT.	INST.	TOTAL	MAT.	INST.	TOTAL	MAT.	INST.	TOTAL	MAT.	INST.	TOTAL	MAT.	INST.	TOTAL
02	**SITE CONSTRUCTION**	110.7	97.4	100.5	91.1	97.8	96.2	111.0	97.4	100.6	92.6	97.8	96.6	111.7	98.1	101.2	109.3	98.9	101.4
03100	CONCRETE FORMS & ACCESSORIES	118.5	99.0	101.7	128.8	96.1	100.6	116.8	100.1	102.4	128.8	96.3	100.7	120.3	112.3	113.4	111.0	100.7	102.1
03200	CONCRETE REINFORCEMENT	122.9	95.4	107.2	159.2	97.5	124.0	174.4	95.3	129.3	147.8	97.5	119.1	170.7	101.9	131.5	161.1	97.8	125.0
03300	CAST-IN-PLACE CONCRETE	157.2	104.0	135.3	140.1	102.7	124.7	159.6	104.0	136.6	151.5	102.8	131.4	164.5	114.9	144.0	159.6	104.6	136.9
03	**CONCRETE**	138.5	100.0	119.2	135.3	98.5	116.9	146.0	100.4	123.2	139.5	98.6	119.1	148.2	110.8	129.5	150.2	101.3	125.7
04	**MASONRY**	147.6	108.3	123.4	143.9	97.6	115.5	148.6	107.8	123.5	144.0	97.6	115.5	149.0	121.4	132.0	146.4	104.6	120.7
05	**METALS**	99.2	95.4	97.8	101.0	93.8	98.4	100.9	95.4	98.9	101.0	94.0	98.5	100.9	99.3	100.3	101.1	95.2	99.0
06	**WOOD & PLASTICS**	117.0	97.7	107.0	131.0	96.3	113.1	116.0	99.7	107.5	131.0	96.3	113.1	118.2	111.2	114.6	103.8	98.6	101.1
07	**THERMAL & MOISTURE PROTECTION**	104.3	99.4	102.0	104.5	99.3	102.1	104.2	98.5	101.5	104.5	99.3	102.1	104.6	109.9	107.1	104.1	100.9	102.5
08	**DOORS & WINDOWS**	93.2	94.8	93.6	92.3	85.2	90.6	92.3	95.8	93.1	92.3	90.4	91.9	91.4	106.1	95.0	92.3	96.1	93.2
09200	PLASTER & GYPSUM BOARD	209.2	97.7	141.1	143.8	96.3	114.8	264.5	99.7	163.8	214.8	96.3	142.4	143.0	111.6	123.8	154.3	98.6	120.3
095,098	CEILINGS & ACOUSTICAL TREATMENT	102.4	97.7	99.3	102.4	96.3	98.4	102.4	99.7	100.6	102.4	96.3	98.4	102.4	111.6	108.5	102.4	98.6	99.9
09600	FLOORING	132.6	108.8	126.8	132.6	109.6	127.0	132.6	107.2	126.4	132.6	109.6	127.0	132.6	115.3	128.4	132.6	107.1	126.4
097,099	WALL FINISHES, PAINTS & COATINGS	109.2	110.3	109.8	109.2	102.6	105.4	109.2	102.7	105.5	109.2	102.6	105.4	112.3	120.3	116.9	109.2	116.0	113.2
09	**FINISHES**	129.5	102.3	115.6	119.0	99.8	109.2	137.9	102.1	119.6	129.8	99.8	114.5	119.7	114.2	116.9	122.1	103.1	112.4
10 - 14	**TOTAL DIV. 10000 - 14000**	100.0	108.4	101.8	100.0	94.4	98.8	100.0	106.1	101.3	100.0	94.4	98.8	100.0	112.8	102.7	100.0	105.3	101.1
15	**MECHANICAL**	102.5	98.1	100.5	102.5	89.0	96.3	102.5	98.4	100.6	102.5	89.0	96.3	102.5	108.6	105.3	102.5	100.2	101.5
16	**ELECTRICAL**	127.0	98.2	107.2	123.6	88.9	99.7	123.6	99.2	106.8	123.6	88.9	99.7	127.0	108.1	114.0	131.0	100.3	109.9
01 - 16	**WEIGHTED AVERAGE**	112.8	99.7	106.4	110.6	94.0	102.5	114.6	99.9	107.4	112.3	94.2	103.5	113.1	108.9	111.1	113.7	100.5	107.3

City Cost Indexes

DIVISION		CANADA																
		WINNIPEG, MANITOBA																
		MAT.	INST.	TOTAL	MAT.	INST.	TOTAL	MAT.	INST.	TOTAL	MAT.	INST.	TOTAL	MAT.	INST.	TOTAL	MAT.	INST.	TOTAL
02	**SITE CONSTRUCTION**	108.4	96.3	99.2															
03100	CONCRETE FORMS & ACCESSORIES	119.3	79.2	84.7															
03200	CONCRETE REINFORCEMENT	161.1	64.8	106.2															
03300	CAST-IN-PLACE CONCRETE	146.7	79.3	118.9															
03	**CONCRETE**	131.1	77.1	104.1															
04	**MASONRY**	145.5	75.7	102.6															
05	**METALS**	100.8	82.0	94.0															
06	**WOOD & PLASTICS**	116.0	80.4	97.6															
07	**THERMAL & MOISTURE PROTECTION**	104.3	79.7	92.6															
08	**DOORS & WINDOWS**	92.3	71.7	87.3															
09200	PLASTER & GYPSUM BOARD	135.2	79.9	101.4															
095,098	CEILINGS & ACOUSTICAL TREATMENT	102.4	79.9	87.6															
09600	FLOORING	132.6	76.1	118.9															
097,099	WALL FINISHES, PAINTS & COATINGS	109.3	65.9	84.1															
09	**FINISHES**	118.6	77.6	97.7															
10 - 14	**TOTAL DIV. 10000 - 14000**	100.0	74.6	94.6															
15	**MECHANICAL**	102.5	81.8	92.9															
16	**ELECTRICAL**	128.8	85.1	98.8															
01 - 16	**WEIGHTED AVERAGE**	110.8	81.5	96.5															

Location Factors

Costs shown in *Means cost data publications* are based on National Averages for materials and installation. To adjust these costs to a specific location, simply multiply the base cost by the factor and divide by 100 for that city. The data is arranged alphabetically by state and postal zip code numbers. For a city not listed, use the factor for a nearby city with similar economic characteristics.

STATE/ZIP	CITY	MAT.	INST.	TOTAL
ALABAMA				
350-352	Birmingham	96.6	77.2	87.2
354	Tuscaloosa	96.3	61.1	79.1
355	Jasper	97.4	57.5	77.9
356	Decatur	96.3	65.6	81.3
357-358	Huntsville	96.2	68.6	82.8
359	Gadsden	96.9	65.0	81.4
360-361	Montgomery	97.2	63.9	81.0
362	Anniston	95.9	52.5	74.8
363	Dothan	96.9	57.2	77.6
364	Evergreen	96.1	58.7	77.9
365-366	Mobile	97.1	68.0	83.0
367	Selma	96.4	58.7	78.1
368	Phenix City	97.2	62.9	80.5
369	Butler	96.6	57.2	77.4
ALASKA				
995-996	Anchorage	134.5	114.1	124.5
997	Fairbanks	130.1	117.0	123.7
998	Juneau	132.9	114.2	123.8
999	Ketchikan	144.2	114.2	129.6
ARIZONA				
850,853	Phoenix	99.6	78.4	89.3
852	Mesa/Tempe	99.6	68.7	84.6
855	Globe	100.5	69.4	85.4
856-857	Tucson	98.4	75.8	87.4
859	Show Low	100.7	71.3	86.4
860	Flagstaff	102.8	74.8	89.2
863	Prescott	100.2	71.2	86.1
864	Kingman	98.7	72.7	86.1
865	Chambers	98.8	70.8	85.2
ARKANSAS				
716	Pine Bluff	95.9	63.3	80.1
717	Camden	94.3	45.9	70.7
718	Texarkana	94.9	51.5	73.8
719	Hot Springs	93.4	45.3	70.0
720-722	Little Rock	96.2	64.0	80.6
723	West Memphis	95.6	62.5	79.5
724	Jonesboro	95.6	62.5	79.5
725	Batesville	94.3	56.9	76.1
726	Harrison	95.8	56.9	76.9
727	Fayetteville	92.8	39.5	66.9
728	Russellville	94.4	54.9	75.2
729	Fort Smith	96.3	61.6	79.4
CALIFORNIA				
900-902	Los Angeles	104.3	112.9	108.5
903-905	Inglewood	100.6	111.1	105.7
906-908	Long Beach	102.4	111.1	106.6
910-912	Pasadena	101.3	111.1	106.1
913-916	Van Nuys	105.3	110.9	108.1
917-918	Alhambra	104.0	111.2	107.5
919-921	San Diego	105.6	107.0	106.3
922	Palm Springs	102.3	110.3	106.2
923-924	San Bernardino	99.8	110.6	105.0
925	Riverside	104.3	111.7	107.9
926-927	Santa Ana	102.1	111.0	106.4
928	Anaheim	104.7	113.0	108.7
930	Oxnard	105.2	112.4	108.7
931	Santa Barbara	104.4	110.8	107.5
932-933	Bakersfield	104.1	108.5	106.3
934	San Luis Obispo	106.2	110.7	108.4
935	Mojave	102.7	108.2	105.4
936-938	Fresno	105.2	112.0	108.5
939	Salinas	107.1	116.2	111.6
940-941	San Francisco	111.9	133.9	122.6
942,956-958	Sacramento	107.5	112.4	109.9
943	Palo Alto	105.5	129.2	117.0
944	San Mateo	108.7	126.1	117.2
945	Vallejo	106.0	123.6	114.6
946	Oakland	110.7	126.3	118.3
947	Berkeley	110.2	125.7	117.7
948	Richmond	110.0	124.3	116.9
949	San Rafael	111.9	127.2	119.3
950	Santa Cruz	110.9	116.3	113.6

STATE/ZIP	CITY	MAT.	INST.	TOTAL
CALIFORNIA (CONT'D)				
951	San Jose	110.2	129.7	119.7
952	Stockton	105.7	113.0	109.3
953	Modesto	105.8	113.8	109.7
954	Santa Rosa	107.2	127.1	116.9
955	Eureka	108.5	111.7	110.0
959	Marysville	107.3	110.8	109.0
960	Redding	108.3	110.6	109.4
961	Susanville	108.4	110.7	109.5
COLORADO				
800-802	Denver	102.5	87.4	95.2
803	Boulder	100.7	68.4	85.0
804	Golden	103.3	82.0	92.9
805	Fort Collins	104.1	79.2	91.9
806	Greeley	100.9	68.4	85.1
807	Fort Morgan	101.5	81.0	91.5
808-809	Colorado Springs	101.3	82.4	92.1
810	Pueblo	102.6	80.9	92.0
811	Alamosa	105.1	70.0	88.0
812	Salida	105.1	70.1	88.0
813	Durango	105.7	65.9	86.3
814	Montrose	103.9	63.2	84.1
815	Grand Junction	107.1	63.6	85.9
816	Glenwood Springs	104.9	76.1	90.9
CONNECTICUT				
060	New Britain	101.9	108.5	105.1
061	Hartford	102.2	108.5	105.3
062	Willimantic	102.6	107.1	104.8
063	New London	99.2	108.7	103.9
064	Meriden	101.9	106.9	104.3
065	New Haven	102.1	109.0	105.5
066	Bridgeport	103.3	107.6	105.4
067	Waterbury	102.6	108.5	105.5
068	Norwalk	102.6	107.7	105.1
069	Stamford	102.7	113.2	107.8
D.C.				
200-205	Washington	99.2	90.1	94.8
DELAWARE				
197	Newark	99.2	102.7	100.9
198	Wilmington	98.5	102.8	100.6
199	Dover	99.1	102.7	100.9
FLORIDA				
320,322	Jacksonville	98.7	66.6	83.1
321	Daytona Beach	98.7	75.0	87.2
323	Tallahassee	99.2	56.2	78.2
324	Panama City	99.8	43.6	72.5
325	Pensacola	99.4	66.1	83.2
326,344	Gainesville	100.2	62.0	81.6
327-328,347	Orlando	100.8	69.5	85.6
329	Melbourne	100.8	77.7	89.5
330-332,340	Miami	98.1	73.4	86.1
333	Fort Lauderdale	97.9	73.5	86.0
334,349	West Palm Beach	96.8	70.2	83.9
335-336,346	Tampa	99.9	64.7	82.8
337	St. Petersburg	102.1	64.3	83.7
338	Lakeland	98.8	64.4	82.1
339,341	Fort Myers	98.2	61.4	80.3
342	Sarasota	100.0	61.8	81.4
GEORGIA				
300-303,399	Atlanta	96.4	81.4	89.1
304	Statesboro	96.7	35.7	67.0
305	Gainesville	95.5	53.5	75.1
306	Athens	94.6	69.0	82.2
307	Dalton	96.6	36.8	67.5
308-309	Augusta	95.2	60.9	78.5
310-312	Macon	95.9	65.7	81.2
313-314	Savannah	97.9	64.5	81.7
315	Waycross	97.9	49.7	74.4
316	Valdosta	97.4	54.0	76.3
317	Albany	97.5	60.3	79.4
318-319	Columbus	97.6	58.7	78.7

Location Factors

STATE/ZIP	CITY	MAT.	INST.	TOTAL
HAWAII				
967	Hilo	115.9	129.0	122.3
968	Honolulu	116.6	129.0	122.7
STATES & POSS.				
969	Guam	211.9	60.7	138.4
IDAHO				
832	Pocatello	102.1	83.7	93.2
833	Twin Falls	104.2	51.9	78.8
834	Idaho Falls	101.3	62.9	82.6
835	Lewiston	113.7	88.1	101.2
836-837	Boise	102.1	85.1	93.8
838	Coeur d'Alene	113.4	63.2	89.0
ILLINOIS				
600-603	North Suburban	98.0	120.9	109.1
604	Joliet	98.0	115.2	106.4
605	South Suburban	98.0	119.6	108.5
606	Chicago	98.2	125.4	111.4
609	Kankakee	94.4	104.2	99.2
610-611	Rockford	97.0	111.2	103.9
612	Rock Island	94.9	100.2	97.5
613	La Salle	96.4	99.7	98.0
614	Galesburg	96.0	101.4	98.6
615-616	Peoria	98.4	103.6	100.9
617	Bloomington	95.2	103.4	99.2
618-619	Champaign	99.1	101.9	100.4
620-622	East St. Louis	93.5	105.7	99.4
623	Quincy	95.3	95.4	95.4
624	Effingham	94.6	99.3	96.9
625	Decatur	95.9	98.3	97.1
626-627	Springfield	95.6	100.6	98.1
628	Centralia	92.8	104.4	98.4
629	Carbondale	92.4	100.7	96.4
INDIANA				
460	Anderson	97.1	87.2	92.3
461-462	Indianapolis	100.0	89.6	94.9
463-464	Gary	97.9	103.7	100.7
465-466	South Bend	96.2	86.2	91.3
467-468	Fort Wayne	97.3	84.7	91.1
469	Kokomo	95.5	86.0	90.9
470	Lawrenceburg	93.4	84.3	89.0
471	New Albany	95.2	80.1	87.9
472	Columbus	97.4	86.2	91.9
473	Muncie	97.5	86.3	92.1
474	Bloomington	99.3	85.9	92.8
475	Washington	95.6	89.9	92.8
476-477	Evansville	97.1	91.8	94.6
478	Terre Haute	97.6	91.2	94.5
479	Lafayette	97.2	84.6	91.1
IOWA				
500-503,509	Des Moines	98.4	86.2	92.5
504	Mason City	96.9	65.2	81.4
505	Fort Dodge	97.2	59.5	78.8
506-507	Waterloo	98.2	66.2	82.7
508	Creston	97.7	71.6	85.0
510-511	Sioux City	99.0	77.6	88.6
512	Sibley	97.9	57.1	78.0
513	Spencer	100.2	56.1	78.7
514	Carroll	96.7	63.0	80.3
515	Council Bluffs	99.8	77.5	88.9
516	Shenandoah	96.7	57.3	77.5
520	Dubuque	98.4	79.2	89.1
521	Decorah	97.8	60.9	79.9
522-524	Cedar Rapids	99.6	83.5	91.8
525	Ottumwa	97.9	75.3	86.9
526	Burlington	96.7	76.2	86.7
527-528	Davenport	98.6	93.6	96.1
KANSAS				
660-662	Kansas City	98.0	90.2	94.2
664-666	Topeka	97.4	71.4	84.8
667	Fort Scott	97.1	70.2	84.0
668	Emporia	96.8	64.4	81.0
669	Belleville	98.8	63.0	81.4
670-672	Wichita	96.6	75.6	86.4
673	Independence	98.6	59.1	79.4
674	Salina	98.3	63.7	81.5
675	Hutchinson	93.7	58.0	76.3
676	Hays	98.1	63.0	81.1
677	Colby	98.9	63.0	81.4

STATE/ZIP	CITY	MAT.	INST.	TOTAL
KANSAS (CONT'D)				
678	Dodge City	99.6	63.0	81.8
679	Liberal	97.9	52.4	75.8
KENTUCKY				
400-402	Louisville	96.1	87.2	91.8
403-405	Lexington	96.9	71.9	84.7
406	Frankfort	96.1	76.4	86.5
407-409	Corbin	94.7	50.7	73.3
410	Covington	95.6	93.9	94.8
411-412	Ashland	93.8	101.9	97.7
413-414	Campton	95.7	50.5	73.7
415-416	Pikeville	96.7	70.9	84.1
417-418	Hazard	94.9	50.7	73.4
420	Paducah	93.6	91.1	92.4
421-422	Bowling Green	95.6	86.6	91.2
423	Owensboro	95.5	83.1	89.5
424	Henderson	93.1	94.1	93.6
425-426	Somerset	92.6	50.5	72.1
427	Elizabethtown	92.2	87.0	89.7
LOUISIANA				
700-701	New Orleans	101.4	69.1	85.7
703	Thibodaux	101.9	68.3	85.6
704	Hammond	98.7	67.0	83.3
705	Lafayette	100.8	61.6	81.7
706	Lake Charles	101.0	65.0	83.5
707-708	Baton Rouge	100.7	62.1	81.9
710-711	Shreveport	96.9	63.3	80.5
712	Monroe	96.8	60.7	79.2
713-714	Alexandria	96.9	58.1	78.0
MAINE				
039	Kittery	97.4	59.7	79.1
040-041	Portland	100.3	78.7	89.8
042	Lewiston	100.8	78.7	90.0
043	Augusta	99.5	56.9	78.8
044	Bangor	99.8	84.4	92.3
045	Bath	98.8	54.7	77.4
046	Machias	98.3	66.8	83.0
047	Houlton	98.5	65.3	82.3
048	Rockland	97.3	67.6	82.9
049	Waterville	98.9	56.9	78.5
MARYLAND				
206	Waldorf	97.3	78.3	88.1
207-208	College Park	97.2	82.0	89.8
209	Silver Spring	96.6	82.1	89.5
210-212	Baltimore	96.8	86.7	91.9
214	Annapolis	96.5	83.8	90.3
215	Cumberland	93.7	82.8	88.4
216	Easton	95.4	51.2	73.9
217	Hagerstown	94.0	82.7	88.5
218	Salisbury	95.9	58.5	77.7
219	Elkton	92.8	74.5	83.9
MASSACHUSETTS				
010-011	Springfield	100.8	102.6	101.6
012	Pittsfield	100.5	95.9	98.3
013	Greenfield	98.7	100.1	99.4
014	Fitchburg	97.1	111.5	104.1
015-016	Worcester	100.6	111.9	106.1
017	Framingham	96.7	117.2	106.7
018	Lowell	100.3	116.7	108.2
019	Lawrence	101.4	115.7	108.4
020-022, 024	Boston	102.6	128.0	115.0
023	Brockton	101.5	115.0	108.1
025	Buzzards Bay	95.9	113.2	104.3
026	Hyannis	98.2	113.2	105.5
027	New Bedford	101.1	113.6	107.2
MICHIGAN				
480,483	Royal Oak	94.7	108.5	101.4
481	Ann Arbor	97.0	111.1	103.9
482	Detroit	98.1	114.8	106.2
484-485	Flint	96.7	103.8	100.1
486	Saginaw	96.9	96.8	96.9
487	Bay City	96.5	96.7	96.6
488-489	Lansing	96.8	99.8	98.3
490	Battle Creek	96.8	91.8	94.4
491	Kalamazoo	97.2	89.0	93.2
492	Jackson	95.5	97.7	96.6
493,495	Grand Rapids	96.7	74.4	85.9
494	Muskegon	95.7	88.5	92.2

STATE/ZIP	CITY	MAT.	INST.	TOTAL
MICHIGAN (CONT'D)				
496	Traverse City	94.9	75.2	85.3
497	Gaylord	96.4	81.3	89.1
498-499	Iron Mountain	98.4	91.7	95.2
MINNESOTA				
550-551	Saint Paul	99.5	121.7	110.3
553-555	Minneapolis	100.1	123.6	111.6
556-558	Duluth	99.6	115.0	107.1
559	Rochester	99.3	106.4	102.7
560	Mankato	97.3	105.4	101.3
561	Windom	96.1	85.4	90.9
562	Willmar	95.3	93.1	94.3
563	St. Cloud	96.2	113.4	104.6
564	Brainerd	97.1	104.4	100.7
565	Detroit Lakes	99.7	93.4	96.6
566	Bemidji	98.9	101.5	100.2
567	Thief River Falls	98.0	94.1	96.1
MISSISSIPPI				
386	Clarksdale	96.9	35.0	66.8
387	Greenville	100.5	53.3	77.6
388	Tupelo	98.4	44.5	72.2
389	Greenwood	98.4	37.5	68.8
390-392	Jackson	98.7	55.1	77.5
393	Meridian	96.7	54.8	76.3
394	Laurel	98.3	38.8	69.4
395	Biloxi	98.9	62.4	81.1
396	McComb	96.7	36.1	67.2
397	Columbus	98.3	44.0	71.9
MISSOURI				
630-631	St. Louis	96.2	110.6	103.2
633	Bowling Green	96.2	93.7	95.0
634	Hannibal	94.9	92.3	93.6
635	Kirksville	97.3	83.6	90.6
636	Flat River	97.3	98.2	97.7
637	Cape Girardeau	96.3	96.2	96.2
638	Sikeston	94.9	90.0	92.5
639	Poplar Bluff	94.4	88.8	91.7
640-641	Kansas City	99.0	101.1	100.0
644-645	St. Joseph	99.0	87.7	93.5
646	Chillicothe	96.3	75.0	85.9
647	Harrisonville	95.8	96.8	96.3
648	Joplin	98.0	73.4	86.0
650-651	Jefferson City	95.0	88.9	92.0
652	Columbia	96.4	89.4	93.0
653	Sedalia	95.9	87.6	91.9
654-655	Rolla	94.7	85.2	90.0
656-658	Springfield	97.7	76.7	87.5
MONTANA				
590-591	Billings	101.5	89.8	95.8
592	Wolf Point	102.1	88.8	95.7
593	Miles City	100.0	89.7	95.0
594	Great Falls	102.8	89.3	96.2
595	Havre	100.8	89.2	95.1
596	Helena	102.6	88.9	96.0
597	Butte	101.3	88.6	95.1
598	Missoula	99.5	87.8	93.8
599	Kalispell	98.8	86.8	93.0
NEBRASKA				
680-681	Omaha	100.5	80.1	90.6
683-685	Lincoln	97.8	67.0	82.8
686	Columbus	96.6	48.5	73.2
687	Norfolk	99.1	66.9	83.4
688	Grand Island	97.5	68.7	83.5
689	Hastings	97.2	58.5	78.4
690	Mccook	96.9	53.5	75.8
691	North Platte	96.9	67.3	82.5
692	Valentine	100.9	46.0	74.2
693	Alliance	100.0	41.8	71.7
NEVADA				
889-891	Las Vegas	101.5	108.6	105.0
893	Ely	102.3	88.1	95.4
894-895	Reno	102.3	97.0	99.7
897	Carson City	102.0	96.5	99.3
898	Elko	100.9	87.7	94.5
NEW HAMPSHIRE				
030	Nashua	101.2	86.8	94.2
031	Manchester	101.3	86.8	94.2

STATE/ZIP	CITY	MAT.	INST.	TOTAL
NEW HAMPSHIRE (CONT'D)				
032-033	Concord	99.0	86.8	93.1
034	Keene	98.3	60.2	79.7
035	Littleton	98.3	66.6	82.9
036	Charleston	97.7	56.7	77.8
037	Claremont	96.8	56.7	77.3
038	Portsmouth	98.3	83.8	91.3
NEW JERSEY				
070-071	Newark	102.8	122.7	112.5
072	Elizabeth	101.7	117.4	109.3
073	Jersey City	100.3	124.2	111.9
074-075	Paterson	103.1	122.0	112.3
076	Hackensack	100.1	122.7	111.1
077	Long Branch	99.7	119.0	109.1
078	Dover	100.5	119.7	109.8
079	Summit	100.6	116.0	108.1
080,083	Vineland	98.0	118.2	107.8
081	Camden	100.1	117.5	108.6
082,084	Atlantic City	99.0	118.4	108.5
085-086	Trenton	100.9	121.4	110.9
087	Point Pleasant	100.4	119.6	109.7
088-089	New Brunswick	100.8	119.7	110.0
NEW MEXICO				
870-872	Albuquerque	100.6	81.2	91.2
873	Gallup	102.0	81.2	91.9
874	Farmington	102.0	81.3	91.9
875	Santa Fe	100.2	81.2	90.9
877	Las Vegas	100.1	81.2	90.9
878	Socorro	99.6	81.2	90.6
879	Truth/Consequences	99.6	75.4	87.8
880	Las Cruces	97.7	72.7	85.6
881	Clovis	100.3	80.7	90.8
882	Roswell	101.8	80.7	91.6
883	Carrizozo	102.7	81.2	92.2
884	Tucumcari	101.3	80.7	91.3
NEW YORK				
100-102	New York	108.5	161.9	134.5
103	Staten Island	104.8	153.5	128.5
104	Bronx	102.0	156.0	128.3
105	Mount Vernon	102.9	131.5	116.8
106	White Plains	102.6	129.7	115.8
107	Yonkers	107.6	134.1	120.5
108	New Rochelle	103.6	131.4	117.1
109	Suffern	103.5	126.7	114.8
110	Queens	103.0	156.9	129.2
111	Long Island City	104.9	156.9	130.2
112	Brooklyn	105.2	156.9	130.3
113	Flushing	105.8	156.9	130.6
114	Jamaica	103.7	157.9	130.1
115,117,118	Hicksville	102.8	149.1	125.3
116	Far Rockaway	105.9	156.9	130.7
119	Riverhead	103.3	151.3	126.6
120-122	Albany	97.9	96.3	97.1
123	Schenectady	98.9	97.0	98.0
124	Kingston	103.4	115.7	109.3
125-126	Poughkeepsie	102.5	122.2	112.1
127	Monticello	101.7	113.2	107.3
128	Glens Falls	93.3	93.0	93.1
129	Plattsburgh	99.1	86.9	93.2
130-132	Syracuse	99.7	92.2	96.1
133-135	Utica	98.1	89.3	93.8
136	Watertown	99.5	91.7	95.7
137-139	Binghamton	99.0	88.2	93.7
140-142	Buffalo	99.7	104.8	102.2
143	Niagara Falls	98.2	107.0	102.5
144-146	Rochester	101.0	99.4	100.2
147	Jamestown	97.3	89.7	93.6
148-149	Elmira	96.6	90.3	93.6
NORTH CAROLINA				
270,272-274	Greensboro	96.6	54.9	76.3
271	Winston-Salem	96.7	53.9	75.9
275-276	Raleigh	97.0	55.1	76.6
277	Durham	96.5	55.1	76.4
278	Rocky Mount	95.2	40.8	68.7
279	Elizabeth City	95.8	43.4	70.3
280	Gastonia	96.7	52.0	75.0
281-282	Charlotte	96.9	52.4	75.2
283	Fayetteville	95.0	55.1	75.6
284	Wilmington	95.2	54.1	75.2
285	Kinston	93.7	40.4	67.8

Location Factors

STATE/ZIP	CITY	MAT.	INST.	TOTAL
NORTH CAROLINA (CONT'D)				
286	Hickory	93.9	39.4	67.4
287-288	Asheville	95.6	53.1	75.0
289	Murphy	95.4	38.4	67.6
NORTH DAKOTA				
580-581	Fargo	100.8	66.1	84.0
582	Grand Forks	102.0	62.4	82.7
583	Devils Lake	102.1	60.3	81.7
584	Jamestown	101.9	61.3	82.1
585	Bismarck	100.8	68.1	84.9
586	Dickinson	102.7	60.3	82.1
587	Minot	102.0	65.9	84.4
588	Williston	101.2	60.3	81.3
OHIO				
430-432	Columbus	97.9	93.1	95.5
433	Marion	95.3	89.0	92.2
434-436	Toledo	98.5	99.7	99.1
437-438	Zanesville	95.4	87.2	91.4
439	Steubenville	97.2	98.4	97.8
440	Lorain	98.6	97.2	97.9
441	Cleveland	98.7	107.3	102.9
442-443	Akron	99.8	100.9	100.4
444-445	Youngstown	99.0	95.7	97.4
446-447	Canton	99.2	92.9	96.1
448-449	Mansfield	96.7	92.7	94.7
450	Hamilton	95.3	92.3	93.8
451-452	Cincinnati	95.3	93.9	94.6
453-454	Dayton	95.2	90.9	93.1
455	Springfield	95.4	90.6	93.1
456	Chillicothe	94.9	97.2	96.0
457	Athens	98.4	75.3	87.1
458	Lima	99.0	89.7	94.5
OKLAHOMA				
730-731	Oklahoma City	97.9	68.2	83.4
734	Ardmore	96.1	67.6	82.2
735	Lawton	97.8	68.5	83.6
736	Clinton	97.6	65.7	82.1
737	Enid	97.6	65.7	82.1
738	Woodward	96.2	65.8	81.4
739	Guymon	97.5	38.0	68.6
740-741	Tulsa	97.8	65.3	82.0
743	Miami	94.9	71.1	83.3
744	Muskogee	96.8	48.8	73.4
745	Mcalester	94.4	61.1	78.2
746	Ponca City	95.0	66.7	81.2
747	Durant	95.0	66.7	81.2
748	Shawnee	96.7	64.9	81.2
749	Poteau	94.0	68.0	81.4
OREGON				
970-972	Portland	103.8	109.7	106.7
973	Salem	104.4	108.2	106.2
974	Eugene	103.8	107.3	105.5
975	Medford	105.3	105.5	105.4
976	Klamath Falls	106.3	104.9	105.6
977	Bend	105.1	107.3	106.2
978	Pendleton	98.5	106.2	102.2
979	Vale	96.0	98.2	97.1
PENNSYLVANIA				
150-152	Pittsburgh	97.0	106.0	101.4
153	Washington	94.6	105.5	99.9
154	Uniontown	94.7	103.9	99.2
155	Bedford	95.8	97.3	96.5
156	Greensburg	96.0	104.3	100.0
157	Indiana	94.6	102.2	98.3
158	Dubois	96.2	97.9	97.0
159	Johnstown	95.8	99.4	97.6
160	Butler	92.8	104.0	98.2
161	New Castle	92.7	104.0	98.2
162	Kittanning	93.3	105.7	99.4
163	Oil City	92.7	99.0	95.8
164-165	Erie	94.8	100.3	97.5
166	Altoona	94.6	96.6	95.6
167	Bradford	96.4	99.0	97.7
168	State College	96.1	98.0	97.0
169	Wellsboro	97.2	92.9	95.1
170-171	Harrisburg	98.1	93.6	95.9
172	Chambersburg	96.7	92.9	94.9
173-174	York	96.3	94.0	95.2
175-176	Lancaster	95.3	91.7	93.6
PENNSYLVANIA (CONT'D)				
177	Williamsport	93.6	88.0	90.9
178	Sunbury	96.0	93.5	94.8
179	Pottsville	95.1	94.5	94.8
180	Lehigh Valley	96.9	105.4	101.0
181	Allentown	98.7	103.4	101.0
182	Hazleton	96.0	96.0	96.0
183	Stroudsburg	96.0	107.4	101.6
184-185	Scranton	99.2	98.8	99.0
186-187	Wilkes-Barre	95.7	96.4	96.0
188	Montrose	95.2	98.1	96.6
189	Doylestown	95.4	117.6	106.2
190-191	Philadelphia	99.6	123.9	111.4
193	Westchester	96.8	113.9	105.1
194	Norristown	95.5	119.5	107.2
195-196	Reading	97.9	99.0	98.4
PUERTO RICO				
009	San Juan	143.2	29.0	87.6
RHODE ISLAND				
028	Newport	99.8	108.1	103.8
029	Providence	100.1	108.1	104.0
SOUTH CAROLINA				
290-292	Columbia	95.9	53.2	75.1
293	Spartanburg	94.7	52.0	73.9
294	Charleston	95.9	54.6	75.8
295	Florence	94.2	51.0	73.1
296	Greenville	94.4	52.0	73.8
297	Rock Hill	94.5	38.0	67.0
298	Aiken	95.6	38.6	67.8
299	Beaufort	96.2	42.5	70.1
SOUTH DAKOTA				
570-571	Sioux Falls	101.2	61.6	81.9
572	Watertown	100.1	55.7	78.5
573	Mitchell	98.9	55.5	77.8
574	Aberdeen	101.2	55.7	79.1
575	Pierre	100.6	56.5	79.1
576	Mobridge	99.7	55.7	78.3
577	Rapid City	101.3	54.9	78.7
TENNESSEE				
370-372	Nashville	97.4	73.3	85.7
373-374	Chattanooga	98.9	62.1	81.0
375,380-381	Memphis	96.3	75.2	86.0
376	Johnson City	98.3	58.9	79.1
377-379	Knoxville	95.0	63.3	79.6
382	Mckenzie	97.3	38.9	68.9
383	Jackson	98.8	49.2	74.7
384	Columbia	95.7	54.9	75.9
385	Cookeville	97.0	37.6	68.1
TEXAS				
750	Mckinney	99.5	63.6	82.1
751	Waxahackie	99.5	63.9	82.2
752-753	Dallas	100.0	69.1	85.0
754	Greenville	99.7	45.0	73.1
755	Texarkana	98.8	54.7	77.4
756	Longview	99.3	47.5	74.1
757	Tyler	99.9	58.5	79.8
758	Palestine	95.8	48.8	72.9
759	Lufkin	96.9	55.0	76.5
760-761	Fort Worth	97.8	66.4	82.6
762	Denton	98.1	59.2	79.2
763	Wichita Falls	98.3	60.6	80.0
764	Eastland	97.4	47.7	73.2
765	Temple	95.9	56.8	76.9
766-767	Waco	97.6	61.9	80.2
768	Brownwood	98.3	44.3	72.1
769	San Angelo	97.9	51.1	75.2
770-772	Houston	99.8	75.0	87.7
773	Huntsville	98.7	47.8	73.9
774	Wharton	100.3	52.4	77.0
775	Galveston	98.1	74.4	86.5
776-777	Beaumont	97.5	70.1	84.2
778	Bryan	95.0	68.7	82.2
779	Victoria	100.4	56.9	79.2
780	Laredo	95.0	59.8	77.8
781-782	San Antonio	94.9	70.0	82.8
783-784	Corpus Christi	98.3	60.0	79.7
785	Mc Allen	98.8	53.3	76.7
786-787	Austin	95.1	67.2	81.5

Location Factors

STATE/ZIP	CITY	MAT.	INST.	TOTAL
TEXAS (CONT'D)				
788	Del Rio	98.2	37.0	68.4
789	Giddings	95.2	47.7	72.1
790-791	Amarillo	98.4	62.3	80.8
792	Childress	97.9	57.9	78.4
793-794	Lubbock	100.3	59.3	80.4
795-796	Abilene	97.7	54.9	76.9
797	Midland	101.1	56.6	79.4
798-799,885	El Paso	97.7	58.6	78.7
UTAH				
840-841	Salt Lake City	101.1	77.4	89.6
842,844	Ogden	99.6	77.3	88.7
843	Logan	101.8	77.3	89.9
845	Price	102.4	57.6	80.6
846-847	Provo	102.1	77.2	90.0
VERMONT				
050	White River Jct.	99.8	42.9	72.1
051	Bellows Falls	98.2	46.6	73.1
052	Bennington	98.5	42.6	71.3
053	Brattleboro	99.0	46.9	73.6
054	Burlington	101.0	69.2	85.5
056	Montpelier	98.2	69.1	84.0
057	Rutland	100.1	69.2	85.1
058	St. Johnsbury	100.0	49.5	75.4
059	Guildhall	98.4	48.9	74.3
VIRGINIA				
220-221	Fairfax	97.7	81.6	89.9
222	Arlington	98.4	81.4	90.1
223	Alexandria	97.3	83.5	90.6
224-225	Fredericksburg	96.2	73.1	84.9
226	Winchester	97.0	60.9	79.5
227	Culpeper	96.7	62.0	79.8
228	Harrisonburg	97.1	54.1	76.2
229	Charlottesville	97.3	65.2	81.7
230-232	Richmond	97.5	70.1	84.2
233-235	Norfolk	97.8	66.6	82.6
236	Newport News	97.2	66.4	82.2
237	Portsmouth	96.4	66.1	81.6
238	Petersburg	97.1	70.1	84.0
239	Farmville	96.5	48.3	73.1
240-241	Roanoke	97.3	53.4	75.9
242	Bristol	96.2	52.5	74.9
243	Pulaski	95.9	47.7	72.4
244	Staunton	96.7	52.4	75.1
245	Lynchburg	96.7	55.8	76.8
246	Grundy	96.2	47.7	72.6
WASHINGTON				
980-981,987	Seattle	103.8	104.9	104.4
982	Everett	104.1	101.4	102.8
983-984	Tacoma	103.7	102.0	102.9
985	Olympia	103.6	102.0	102.8
986	Vancouver	106.1	101.1	103.7
988	Wenatchee	104.9	91.2	98.3
989	Yakima	104.2	95.9	100.2
990-992	Spokane	108.3	87.8	98.3
993	Richland	108.4	90.1	99.5
994	Clarkston	107.6	89.6	98.8
WEST VIRGINIA				
247-248	Bluefield	95.0	85.1	90.2
249	Lewisburg	96.5	86.2	91.5
250-253	Charleston	97.2	91.6	94.4
254	Martinsburg	96.5	56.1	76.8
255-257	Huntington	97.4	96.0	96.7
258-259	Beckley	94.7	89.2	92.0
260	Wheeling	97.6	93.1	95.4
261	Parkersburg	96.4	95.4	96.0
262	Buckhannon	96.4	91.7	94.1
263-264	Clarksburg	96.8	91.7	94.3
265	Morgantown	96.9	91.7	94.4
266	Gassaway	96.2	92.2	94.2
267	Romney	96.2	87.2	91.8
268	Petersburg	96.2	90.1	93.2
WISCONSIN				
530,532	Milwaukee	99.2	101.6	100.3
531	Kenosha	99.4	101.0	100.2
534	Racine	98.6	101.3	99.9
535	Beloit	98.5	97.5	98.0
537	Madison	98.7	98.0	98.3

STATE/ZIP	CITY	MAT.	INST.	TOTAL
WISCONSIN (CONT'D)				
538	Lancaster	96.9	82.5	89.9
539	Portage	95.2	97.8	96.5
540	New Richmond	96.9	93.6	95.3
541-543	Green Bay	100.4	93.3	97.0
544	Wausau	96.1	93.1	94.6
545	Rhinelander	99.5	89.3	94.5
546	La Crosse	96.6	93.5	95.1
547	Eau Claire	98.6	93.3	96.0
548	Superior	96.8	98.9	97.8
549	Oshkosh	96.7	90.7	93.8
WYOMING				
820	Cheyenne	100.6	61.4	81.5
821	Yellowstone Nat'l Park	100.0	55.1	78.2
822	Wheatland	101.7	54.3	78.7
823	Rawlins	103.3	50.3	77.5
824	Worland	100.8	50.7	76.4
825	Riverton	102.0	53.9	78.6
826	Casper	100.3	63.9	82.6
827	Newcastle	100.5	50.3	76.1
828	Sheridan	101.6	59.2	81.0
829-831	Rock Springs	105.3	50.7	78.7
CANADIAN FACTORS (reflect Canadian currency)				
ALBERTA				
	Calgary	111.8	80.9	96.8
	Edmonton	111.8	80.8	96.7
	Fort McMurray	110.4	80.8	96.0
	Lethbridge	110.5	80.9	96.1
	Lloydminster	110.4	80.8	96.0
	Medicine Hat	110.5	80.9	96.1
	Red Deer	110.2	80.9	96.0
BRITISH COLUMBIA				
	Kamloops	110.6	97.3	104.2
	Prince George	114.1	97.3	106.0
	Vancouver	113.7	100.5	107.3
	Victoria	114.1	97.3	106.0
MANITOBA				
	Brandon	111.9	81.5	97.1
	Portage la Prairie	111.1	81.5	96.7
	Winnipeg	110.8	81.5	96.5
NEW BRUNSWICK				
	Bathurst	109.9	72.0	91.5
	Dalhousie	109.9	72.0	91.5
	Fredericton	110.6	77.2	94.4
	Moncton	109.5	72.0	91.3
	Newcastle	109.9	72.0	91.5
	Saint John	110.9	77.2	94.5
NEWFOUNDLAND				
	Corner Brook	115.9	71.5	94.3
	St. John's	115.9	71.5	94.3
NORTHWEST TERRITORIES				
	Yellowknife	108.1	73.3	91.1
NOVA SCOTIA				
	Dartmouth	111.0	79.5	95.7
	Halifax	111.0	81.0	96.4
	New Glasgow	111.0	79.5	95.7
	Sydney	108.2	79.5	94.3
	Yarmouth	111.0	79.5	95.7
ONTARIO				
	Barrie	115.7	98.1	107.2
	Brantford	113.5	103.9	108.8
	Cornwall	114.1	99.2	106.8
	Hamilton	113.8	102.8	108.4
	Kingston	114.1	98.9	106.7
	Kitchener	108.4	97.4	103.0
	London	112.8	99.7	106.4
	North Bay	113.5	97.0	105.5
	Oshawa	113.6	100.3	107.1
	Ottawa	114.6	99.9	107.4
	Owen Sound	115.8	96.6	106.5
	Peterborough	113.5	99.0	106.4
	Sarnia	112.7	104.6	108.8
	St. Catharines	107.3	97.3	102.4
	Sudbury	107.2	97.0	102.3

STATE/ZIP	CITY	MAT.	INST.	TOTAL
ONTARIO (CONT'D)				
	Thunder Bay	109.0	97.2	103.3
	Toronto	113.1	108.9	111.1
	Windsor	107.6	100.2	104.0
PRINCE EDWARD ISLAND				
	Charlottetown	113.1	67.0	90.7
	Summerside	113.3	67.0	90.8
QUEBEC				
	Cap-de-la-Madeleine	111.3	94.2	103.0
	Charlesbourg	111.3	94.2	103.0
	Chicoutimi	110.0	93.9	102.2
	Gatineau	108.3	94.0	101.4
	Laval	110.5	94.0	102.5
	Montreal	110.6	94.0	102.5
	Quebec	112.3	94.2	103.5
	Sherbrooke	110.5	94.0	102.5
	Trois Rivieres	111.3	94.2	103.0
SASKATCHEWAN				
	Moose Jaw	108.2	75.1	92.1
	Prince Albert	107.6	74.9	91.7
	Regina	108.9	75.1	92.5
	Saskatoon	107.8	74.9	91.8
YUKON				
	Whitehorse	108.1	73.3	91.2

Abbreviations

A	Area Square Feet; Ampere
ABS	Acrylonitrile Butadiene Stryrene; Asbestos Bonded Steel
A.C.	Alternating Current; Air-Conditioning; Asbestos Cement; Plywood Grade A & C
A.C.I.	American Concrete Institute
AD	Plywood, Grade A & D
Addit.	Additional
Adj.	Adjustable
af	Audio-frequency
A.G.A.	American Gas Association
Agg.	Aggregate
A.H.	Ampere Hours
A hr.	Ampere-hour
A.H.U.	Air Handling Unit
A.I.A.	American Institute of Architects
AIC	Ampere Interrupting Capacity
Allow.	Allowance
alt.	Altitude
Alum.	Aluminum
a.m.	Ante Meridiem
Amp.	Ampere
Anod.	Anodized
Approx.	Approximate
Apt.	Apartment
Asb.	Asbestos
A.S.B.C.	American Standard Building Code
Asbe.	Asbestos Worker
A.S.H.R.A.E.	American Society of Heating, Refrig. & AC Engineers
A.S.M.E.	American Society of Mechanical Engineers
A.S.T.M.	American Society for Testing and Materials
Attchmt.	Attachment
Avg.	Average
A.W.G.	American Wire Gauge
AWWA	American Water Works Assoc.
Bbl.	Barrel
B. & B.	Grade B and Better; Balled & Burlapped
B. & S.	Bell and Spigot
B. & W.	Black and White
b.c.c.	Body-centered Cubic
B.C.Y.	Bank Cubic Yards
BE	Bevel End
B.F.	Board Feet
Bg. cem.	Bag of Cement
BHP	Boiler Horsepower; Brake Horsepower
B.I.	Black Iron
Bit.; Bitum.	Bituminous
Bk.	Backed
Bkrs.	Breakers
Bldg.	Building
Blk.	Block
Bm.	Beam
Boil.	Boilermaker
B.P.M.	Blows per Minute
BR	Bedroom
Brg.	Bearing
Brhe.	Bricklayer Helper
Bric.	Bricklayer
Brk.	Brick
Brng.	Bearing
Brs.	Brass
Brz.	Bronze
Bsn.	Basin
Btr.	Better
BTU	British Thermal Unit
BTUH	BTU per Hour
B.U.R.	Built-up Roofing
BX	Interlocked Armored Cable
c	Conductivity, Copper Sweat
C	Hundred; Centigrade
C/C	Center to Center, Cedar on Cedar
Cab.	Cabinet
Cair.	Air Tool Laborer
Calc	Calculated
Cap.	Capacity
Carp.	Carpenter
C.B.	Circuit Breaker
C.C.A.	Chromate Copper Arsenate
C.C.F.	Hundred Cubic Feet
cd	Candela
cd/sf	Candela per Square Foot
CD	Grade of Plywood Face & Back
CDX	Plywood, Grade C & D, exterior glue
Cefi.	Cement Finisher
Cem.	Cement
CF	Hundred Feet
C.F.	Cubic Feet
CFM	Cubic Feet per Minute
c.g.	Center of Gravity
CHW	Chilled Water; Commercial Hot Water
C.I.	Cast Iron
C.I.P.	Cast in Place
Circ.	Circuit
C.L.	Carload Lot
Clab.	Common Laborer
C.L.F.	Hundred Linear Feet
CLF	Current Limiting Fuse
CLP	Cross Linked Polyethylene
cm	Centimeter
CMP	Corr. Metal Pipe
C.M.U.	Concrete Masonry Unit
CN	Change Notice
Col.	Column
CO_2	Carbon Dioxide
Comb.	Combination
Compr.	Compressor
Conc.	Concrete
Cont.	Continuous; Continued
Corr.	Corrugated
Cos	Cosine
Cot	Cotangent
Cov.	Cover
C/P	Cedar on Paneling
CPA	Control Point Adjustment
Cplg.	Coupling
C.P.M.	Critical Path Method
CPVC	Chlorinated Polyvinyl Chloride
C.Pr.	Hundred Pair
CRC	Cold Rolled Channel
Creos.	Creosote
Crpt.	Carpet & Linoleum Layer
CRT	Cathode-ray Tube
CS	Carbon Steel, Constant Shear Bar Joist
Csc	Cosecant
C.S.F.	Hundred Square Feet
CSI	Construction Specifications Institute
C.T.	Current Transformer
CTS	Copper Tube Size
Cu	Copper, Cubic
Cu. Ft.	Cubic Foot
cw	Continuous Wave
C.W.	Cool White; Cold Water
Cwt.	100 Pounds
C.W.X.	Cool White Deluxe
C.Y.	Cubic Yard (27 cubic feet)
C.Y./Hr.	Cubic Yard per Hour
Cyl.	Cylinder
d	Penny (nail size)
D	Deep; Depth; Discharge
Dis.;Disch.	Discharge
Db.	Decibel
Dbl.	Double
DC	Direct Current
DDC	Direct Digital Control
Demob.	Demobilization
d.f.u.	Drainage Fixture Units
D.H.	Double Hung
DHW	Domestic Hot Water
Diag.	Diagonal
Diam.	Diameter
Distrib.	Distribution
Dk.	Deck
D.L.	Dead Load; Diesel
DLH	Deep Long Span Bar Joist
Do.	Ditto
Dp.	Depth
D.P.S.T.	Double Pole, Single Throw
Dr.	Driver
Drink.	Drinking
D.S.	Double Strength
D.S.A.	Double Strength A Grade
D.S.B.	Double Strength B Grade
Dty.	Duty
DWV	Drain Waste Vent
DX	Deluxe White, Direct Expansion
dyn	Dyne
e	Eccentricity
E	Equipment Only; East
Ea.	Each
E.B.	Encased Burial
Econ.	Economy
EDP	Electronic Data Processing
EIFS	Exterior Insulation Finish System
E.D.R.	Equiv. Direct Radiation
Eq.	Equation
Elec.	Electrician; Electrical
Elev.	Elevator; Elevating
EMT	Electrical Metallic Conduit; Thin Wall Conduit
Eng.	Engine, Engineered
EPDM	Ethylene Propylene Diene Monomer
EPS	Expanded Polystyrene
Eqhv.	Equip. Oper., Heavy
Eqlt.	Equip. Oper., Light
Eqmd.	Equip. Oper., Medium
Eqmm.	Equip. Oper., Master Mechanic
Eqol.	Equip. Oper., Oilers
Equip.	Equipment
ERW	Electric Resistance Welded
E.S.	Energy Saver
Est.	Estimated
esu	Electrostatic Units
E.W.	Each Way
EWT	Entering Water Temperature
Excav.	Excavation
Exp.	Expansion, Exposure
Ext.	Exterior
Extru.	Extrusion
f.	Fiber stress
F	Fahrenheit; Female; Fill
Fab.	Fabricated
FBGS	Fiberglass
F.C.	Footcandles
f.c.c.	Face-centered Cubic
f'c.	Compressive Stress in Concrete; Extreme Compressive Stress
F.E.	Front End
FEP	Fluorinated Ethylene Propylene (Teflon)
F.G.	Flat Grain
F.H.A.	Federal Housing Administration
Fig.	Figure
Fin.	Finished
Fixt.	Fixture
Fl. Oz.	Fluid Ounces
Flr.	Floor
F.M.	Frequency Modulation; Factory Mutual
Fmg.	Framing
Fndtn.	Foundation
Fori.	Foreman, Inside
Foro.	Foreman, Outside

Abbreviations

Fount.	Fountain
FPM	Feet per Minute
FPT	Female Pipe Thread
Fr.	Frame
F.R.	Fire Rating
FRK	Foil Reinforced Kraft
FRP	Fiberglass Reinforced Plastic
FS	Forged Steel
FSC	Cast Body; Cast Switch Box
Ft.	Foot; Feet
Ftng.	Fitting
Ftg.	Footing
Ft. Lb.	Foot Pound
Furn.	Furniture
FVNR	Full Voltage Non-Reversing
FXM	Female by Male
Fy.	Minimum Yield Stress of Steel
g	Gram
G	Gauss
Ga.	Gauge
Gal.	Gallon
Gal./Min.	Gallon per Minute
Galv.	Galvanized
Gen.	General
G.F.I.	Ground Fault Interrupter
Glaz.	Glazier
GPD	Gallons per Day
GPH	Gallons per Hour
GPM	Gallons per Minute
GR	Grade
Gran.	Granular
Grnd.	Ground
H	High; High Strength Bar Joist; Henry
H.C.	High Capacity
H.D.	Heavy Duty; High Density
H.D.O.	High Density Overlaid
Hdr.	Header
Hdwe.	Hardware
Help.	Helpers Average
HEPA	High Efficiency Particulate Air Filter
Hg	Mercury
HIC	High Interrupting Capacity
HM	Hollow Metal
H.O.	High Output
Horiz.	Horizontal
H.P.	Horsepower; High Pressure
H.P.F.	High Power Factor
Hr.	Hour
Hrs./Day	Hours per Day
HSC	High Short Circuit
Ht.	Height
Htg.	Heating
Htrs.	Heaters
HVAC	Heating, Ventilation & Air-Conditioning
Hvy.	Heavy
HW	Hot Water
Hyd.;Hydr.	Hydraulic
Hz.	Hertz (cycles)
I.	Moment of Inertia
I.C.	Interrupting Capacity
ID	Inside Diameter
I.D.	Inside Dimension; Identification
I.F.	Inside Frosted
I.M.C.	Intermediate Metal Conduit
In.	Inch
Incan.	Incandescent
Incl.	Included; Including
Int.	Interior
Inst.	Installation
Insul.	Insulation/Insulated
I.P.	Iron Pipe
I.P.S.	Iron Pipe Size
I.P.T.	Iron Pipe Threaded
I.W.	Indirect Waste
J	Joule
J.I.C.	Joint Industrial Council
K	Thousand; Thousand Pounds; Heavy Wall Copper Tubing, Kelvin
K.A.H.	Thousand Amp. Hours
KCMIL	Thousand Circular Mils
KD	Knock Down
K.D.A.T.	Kiln Dried After Treatment
kg	Kilogram
kG	Kilogauss
kgf	Kilogram Force
kHz	Kilohertz
Kip.	1000 Pounds
KJ	Kiljoule
K.L.	Effective Length Factor
K.L.F.	Kips per Linear Foot
Km	Kilometer
K.S.F.	Kips per Square Foot
K.S.I.	Kips per Square Inch
kV	Kilovolt
kVA	Kilovolt Ampere
K.V.A.R.	Kilovar (Reactance)
KW	Kilowatt
KWh	Kilowatt-hour
L	Labor Only; Length; Long; Medium Wall Copper Tubing
Lab.	Labor
lat	Latitude
Lath.	Lather
Lav.	Lavatory
lb.; #	Pound
L.B.	Load Bearing; L Conduit Body
L. & E.	Labor & Equipment
lb./hr.	Pounds per Hour
lb./L.F.	Pounds per Linear Foot
lbf/sq.in.	Pound-force per Square Inch
L.C.L.	Less than Carload Lot
Ld.	Load
LE	Lead Equivalent
LED	Light Emitting Diode
L.F.	Linear Foot
Lg.	Long; Length; Large
L & H	Light and Heat
LH	Long Span Bar Joist
L.H.	Labor Hours
L.L.	Live Load
L.L.D.	Lamp Lumen Depreciation
L-O-L	Lateralolet
lm	Lumen
lm/sf	Lumen per Square Foot
lm/W	Lumen per Watt
L.O.A.	Length Over All
log	Logarithm
L.P.	Liquefied Petroleum; Low Pressure
L.P.F.	Low Power Factor
LR	Long Radius
L.S.	Lump Sum
Lt.	Light
Lt. Ga.	Light Gauge
L.T.L.	Less than Truckload Lot
Lt. Wt.	Lightweight
L.V.	Low Voltage
M	Thousand; Material; Male; Light Wall Copper Tubing
M^2CA	Meters Squared Contact Area
m/hr; M.H.	Man-hour
mA	Milliampere
Mach.	Machine
Mag. Str.	Magnetic Starter
Maint.	Maintenance
Marb.	Marble Setter
Mat; Mat'l.	Material
Max.	Maximum
MBF	Thousand Board Feet
MBH	Thousand BTU's per hr.
MC	Metal Clad Cable
M.C.F.	Thousand Cubic Feet
M.C.F.M.	Thousand Cubic Feet per Minute
M.C.M.	Thousand Circular Mils
M.C.P.	Motor Circuit Protector
MD	Medium Duty
M.D.O.	Medium Density Overlaid
Med.	Medium
MF	Thousand Feet
M.F.B.M.	Thousand Feet Board Measure
Mfg.	Manufacturing
Mfrs.	Manufacturers
mg	Milligram
MGD	Million Gallons per Day
MGPH	Thousand Gallons per Hour
MH, M.H.	Manhole; Metal Halide; Man-Hour
MHz	Megahertz
Mi.	Mile
MI	Malleable Iron; Mineral Insulated
mm	Millimeter
Mill.	Millwright
Min., min.	Minimum, minute
Misc.	Miscellaneous
ml	Milliliter, Mainline
M.L.F.	Thousand Linear Feet
Mo.	Month
Mobil.	Mobilization
Mog.	Mogul Base
MPH	Miles per Hour
MPT	Male Pipe Thread
MRT	Mile Round Trip
ms	Millisecond
M.S.F.	Thousand Square Feet
Mstz.	Mosaic & Terrazzo Worker
M.S.Y.	Thousand Square Yards
Mtd.	Mounted
Mthe.	Mosaic & Terrazzo Helper
Mtng.	Mounting
Mult.	Multi; Multiply
M.V.A.	Million Volt Amperes
M.V.A.R.	Million Volt Amperes Reactance
MV	Megavolt
MW	Megawatt
MXM	Male by Male
MYD	Thousand Yards
N	Natural; North
nA	Nanoampere
NA	Not Available; Not Applicable
N.B.C.	National Building Code
NC	Normally Closed
N.E.M.A.	National Electrical Manufacturers Assoc.
NEHB	Bolted Circuit Breaker to 600V.
N.L.B.	Non-Load-Bearing
NM	Non-Metallic Cable
nm	Nanometer
No.	Number
NO	Normally Open
N.O.C.	Not Otherwise Classified
Nose.	Nosing
N.P.T.	National Pipe Thread
NQOD	Combination Plug-on/Bolt on Circuit Breaker to 240V.
N.R.C.	Noise Reduction Coefficient
N.R.S.	Non Rising Stem
ns	Nanosecond
nW	Nanowatt
OB	Opposing Blade
OC	On Center
OD	Outside Diameter
O.D.	Outside Dimension
ODS	Overhead Distribution System
O.G.	Ogee
O.H.	Overhead
O & P	Overhead and Profit
Oper.	Operator
Opng.	Opening
Orna.	Ornamental
OSB	Oriented Strand Board
O. S. & Y.	Outside Screw and Yoke
Ovhd.	Overhead
OWG	Oil, Water or Gas

Abbreviations

Oz.	Ounce
P.	Pole; Applied Load; Projection
p.	Page
Pape.	Paperhanger
P.A.P.R.	Powered Air Purifying Respirator
PAR	Weatherproof Reflector
Pc., Pcs.	Piece, Pieces
P.C.	Portland Cement; Power Connector
P.C.F.	Pounds per Cubic Foot
P.C.M.	Phase Contract Microscopy
P.E.	Professional Engineer; Porcelain Enamel; Polyethylene; Plain End
Perf.	Perforated
Ph.	Phase
P.I.	Pressure Injected
Pile.	Pile Driver
Pkg.	Package
Pl.	Plate
Plah.	Plasterer Helper
Plas.	Plasterer
Pluh.	Plumbers Helper
Plum.	Plumber
Ply.	Plywood
p.m.	Post Meridiem
Pntd.	Painted
Pord.	Painter, Ordinary
pp	Pages
PP; PPL	Polypropylene
P.P.M.	Parts per Million
Pr.	Pair
P.E.S.B.	Pre-engineered Steel Building
Prefab.	Prefabricated
Prefin.	Prefinished
Prop.	Propelled
PSF; psf	Pounds per Square Foot
PSI; psi	Pounds per Square Inch
PSIG	Pounds per Square Inch Gauge
PSP	Plastic Sewer Pipe
Pspr.	Painter, Spray
Psst.	Painter, Structural Steel
P.T.	Potential Transformer
P. & T.	Pressure & Temperature
Ptd.	Painted
Ptns.	Partitions
Pu	Ultimate Load
PVC	Polyvinyl Chloride
Pvmt.	Pavement
Pwr.	Power
Q	Quantity Heat Flow
Quan.;Qty.	Quantity
Q.C.	Quick Coupling
r	Radius of Gyration
R	Resistance
R.C.P.	Reinforced Concrete Pipe
Rect.	Rectangle
Reg.	Regular
Reinf.	Reinforced
Req'd.	Required
Res.	Resistant
Resi.	Residential
Rgh.	Rough
RGS	Rigid Galvanized Steel
R.H.W.	Rubber, Heat & Water Resistant; Residential Hot Water
rms	Root Mean Square
Rnd.	Round
Rodm.	Rodman
Rofc.	Roofer, Composition
Rofp.	Roofer, Precast
Rohe.	Roofer Helpers (Composition)
Rots.	Roofer, Tile & Slate
R.O.W.	Right of Way
RPM	Revolutions per Minute
R.R.	Direct Burial Feeder Conduit
R.S.	Rapid Start
Rsr	Riser
RT	Round Trip
S.	Suction; Single Entrance; South
SCFM	Standard Cubic Feet per Minute
Scaf.	Scaffold
Sch.; Sched.	Schedule
S.C.R.	Modular Brick
S.D.	Sound Deadening
S.D.R.	Standard Dimension Ratio
S.E.	Surfaced Edge
Sel.	Select
S.E.R.;	Service Entrance Cable
S.E.U.	Service Entrance Cable
S.F.	Square Foot
S.F.C.A.	Square Foot Contact Area
S.F.G.	Square Foot of Ground
S.F. Hor.	Square Foot Horizontal
S.F.R.	Square Feet of Radiation
S.F. Shlf.	Square Foot of Shelf
S4S	Surface 4 Sides
Shee.	Sheet Metal Worker
Sin.	Sine
Skwk.	Skilled Worker
SL	Saran Lined
S.L.	Slimline
Sldr.	Solder
SLH	Super Long Span Bar Joist
S.N.	Solid Neutral
S-O-L	Socketolet
sp	Standpipe
S.P.	Static Pressure; Single Pole; Self-Propelled
Spri.	Sprinkler Installer
spwg	Static Pressure Water Gauge
Sq.	Square; 100 Square Feet
S.P.D.T.	Single Pole, Double Throw
SPF	Spruce Pine Fir
S.P.S.T.	Single Pole, Single Throw
SPT	Standard Pipe Thread
Sq. Hd.	Square Head
Sq. In.	Square Inch
S.S.	Single Strength; Stainless Steel
S.S.B.	Single Strength B Grade
sst	Stainless Steel
Sswk.	Structural Steel Worker
Sswl.	Structural Steel Welder
St.;Stl.	Steel
S.T.C.	Sound Transmission Coefficient
Std.	Standard
STK	Select Tight Knot
STP	Standard Temperature & Pressure
Stpi.	Steamfitter, Pipefitter
Str.	Strength; Starter; Straight
Strd.	Stranded
Struct.	Structural
Sty.	Story
Subj.	Subject
Subs.	Subcontractors
Surf.	Surface
Sw.	Switch
Swbd.	Switchboard
S.Y.	Square Yard
Syn.	Synthetic
S.Y.P.	Southern Yellow Pine
Sys.	System
t.	Thickness
T	Temperature; Ton
Tan	Tangent
T.C.	Terra Cotta
T & C	Threaded and Coupled
T.D.	Temperature Difference
T.E.M.	Transmission Electron Microscopy
TFE	Tetrafluoroethylene (Teflon)
T. & G.	Tongue & Groove; Tar & Gravel
Th.;Thk.	Thick
Thn.	Thin
Thrded	Threaded
Tilf.	Tile Layer, Floor
Tilh.	Tile Layer, Helper
THHN	Nylon Jacketed Wire
THW.	Insulated Strand Wire
THWN;	Nylon Jacketed Wire
T.L.	Truckload
T.M.	Track Mounted
Tot.	Total
T-O-L	Threadolet
T.S.	Trigger Start
Tr.	Trade
Transf.	Transformer
Trhv.	Truck Driver, Heavy
Trlr	Trailer
Trlt.	Truck Driver, Light
TV	Television
T.W.	Thermoplastic Water Resistant Wire
UCI	Uniform Construction Index
UF	Underground Feeder
UGND	Underground Feeder
U.H.F.	Ultra High Frequency
U.L.	Underwriters Laboratory
Unfin.	Unfinished
URD	Underground Residential Distribution
US	United States
USP	United States Primed
UTP	Unshielded Twisted Pair
V	Volt
V.A.	Volt Amperes
V.C.T.	Vinyl Composition Tile
VAV	Variable Air Volume
VC	Veneer Core
Vent.	Ventilation
Vert.	Vertical
V.F.	Vinyl Faced
V.G.	Vertical Grain
V.H.F.	Very High Frequency
VHO	Very High Output
Vib.	Vibrating
V.L.F.	Vertical Linear Foot
Vol.	Volume
VRP	Vinyl Reinforced Polyester
W	Wire; Watt; Wide; West
w/	With
W.C.	Water Column; Water Closet
W.F.	Wide Flange
W.G.	Water Gauge
Wldg.	Welding
W. Mile	Wire Mile
W-O-L	Weldolet
W.R.	Water Resistant
Wrck.	Wrecker
W.S.P.	Water, Steam, Petroleum
WT., Wt.	Weight
WWF	Welded Wire Fabric
XFER	Transfer
XFMR	Transformer
XHD	Extra Heavy Duty
XHHW; XLPE	Cross-Linked Polyethylene Wire Insulation
XLP	Cross-linked Polyethylene
Y	Wye
yd	Yard
yr	Year
Δ	Delta
%	Percent
~	Approximately
Ø	Phase
@	At
#	Pound; Number
<	Less Than
>	Greater Than

Index

D

E

F

INDEX

G

Q

R

S

T

U

V

INDEX

W

X

Y

Notes

CMD Group...

R.S. Means Company, Inc., a CMD Group company, which is a division of Cahners Business Information, the leading provider of construction cost data in North America, supplies comprehensive construction cost guides, related technical publications and educational services.

CMD Group, a leading worldwide provider of total construction information solutions, is comprised of three synergistic product groups crafted to be the complete resource for reliable, timely and actionable project, product and cost data. In North America, CMD Group encompasses:

- Architects' First Source
- Construction Market Data (CMD)
- Manufacturer's Survey Associates (MSA)
- R.S. Means
- CMD Canada
- BIMSA/Mexico
- Worldwide, CMD Group includes Byggfakta Scandinavia (Denmark, Estonia, Finland, Norway and Sweden) and Cordell Building Information Services (Australia).

First Source for Products, available in print, on the Means CostWorks CD, and on the Internet, is a comprehensive product information source. In alliance with The Construction Specifications Institute (CSI) and Thomas Register, Architects' First Source also produces CSI's SPEC-DATA® and MANU-SPEC® as well as CADBlocks.℠ Together, these products offer commercial building product information for the building team at each stage of the construction process.

CMD Exchange, a single electronic community where all members of the construction industry can communicate, collaborate, and conduct business, better, faster, easier.

Construction Market Data provides complete, accurate and timely project information through all stages of construction. Construction Market Data supplies industry data through productive leads, project reports, contact lists, market penetration analysis and sales evaluation reports. Any of these products can pinpoint a county, look at a state, or cover the country. Data is delivered via paper, e-mail or the Internet.

CMD Group Canada serves the Canadian construction market with reliable and comprehensive information services that cover all facets of construction. Core services include: Buildcore, product selection and specification tools available in print and on the Internet; CMD Building Reports, a national construction project lead service; CanaData, statistical and forecasting information; Daily Commercial News, a construction newspaper reporting on news and projects in Ontario; and Journal of Commerce, reporting news in British Columbia and Alberta.

Manufacturers' Survey Associates is a quantity survey and specification service whose experienced estimating staff examines project documents throughout the bidding process and distributes edited information to its clients. Material estimates, edited plans and specifications and distribution of addenda form the heart of the Manufacturers' Survey Associates product line, which is available in print and on CD-ROM.

Clark Reports is the premier provider of industrial construction project data, noted for providing earlier, more complete project information in the planning and pre-planning stages for all segments of industrial and institutional construction.

Byggfakta Scandinavia AB, founded in 1936, is the parent company for the leaders of customized construction market data for Denmark, Estonia, Finland, Norway and Sweden. Each company fully covers the local construction market and provides information across several platforms including subscription, ad-hoc basis, electronically and on paper.

Cordell Building Information Services, with its complete range of project and cost and estimating services, is Australia's specialist in the construction information industry. Cordell provides in-depth and historical information on all aspects of construction projects and estimation, including several customized reports, construction and sales leads, and detailed cost information among others.

For more information, please visit our web site at www.cmdg.com.

CMD Group Corporate Offices
30 Technology Parkway South #100
Norcross, GA 30092-2912
(800) 793-0304
(770) 417-4002 (fax)
info@cmdg.com
www.cmdg.com

Means Project Cost Report

By filling out and returning the Project Description, you can receive a discount of $20.00 off any one of the Means products advertised in the following pages. The cost information required includes all items marked (✓) except those where no costs occurred. The sum of all major items should equal the Total Project Cost.

$20.00 Discount per product for each report you submit.

DISCOUNT PRODUCTS AVAILABLE—FOR U.S. CUSTOMERS ONLY—STRICTLY CONFIDENTIAL

Project Description (No remodeling projects, please.)

✓ Type Building ____________________

✓ Location ____________________

Capacity ____________________

✓ Frame ____________________

✓ Exterior ____________________

✓ Basement: full ☐ partial ☐ none ☐ crawl ☐

✓ Height in Stories ____________________

✓ Total Floor Area ____________________

Ground Floor Area ____________________

✓ Volume in C.F. ____________________

% Air Conditioned ____________ Tons ________

Comments ____________________

	✓	**Total Project Cost**		$
A	✓	**General Conditions**		$
B	✓	**Site Work**		$
BS		**Site Clearing & Improvement**		
BE		*Excavation*	(C.Y.)	
BF		*Caissons & Piling*	(L.F.)	
BU		*Site Utilities*		
BP		*Roads & Walks Exterior Paving*	(S.Y.)	
C	✓	**Concrete**		$
C		*Cast in Place*	(C.Y.)	
CP		*Precast*	(S.F.)	
D	✓	**Masonry**		$
DB		*Brick*	(M)	
DC		*Block*	(M)	
DT		*Tile*	(S.F.)	
DS		*Stone*	(S.F.)	
E	✓	**Metals**		$
ES		*Structural Steel*	(Tons)	
EM		*Misc. & Ornamental Metals*		
F	✓	**Wood & Plastics**		$
FR		*Rough Carpentry*	(MBF)	
FF		*Finish Carpentry*		
FM		*Architectural Millwork*		
G	✓	**Thermal & Moisture Protection**		$
GW		*Waterproofing-Dampproofing*	(S.F.)	
GN		*Insulation*	(S.F.)	
GR		*Roofing & Flashing*	(S.F.)	
GM		*Metal Siding/Curtain Wall*	(S.F.)	
H	✓	**Doors and Windows**		$
HD		*Doors*	(Ea.)	
HW		*Windows*	(S.F.)	
HH		*Finish Hardware*		
HG		*Glass & Glazing*	(S.F.)	
HS		*Storefronts*	(S.F.)	

Product Name ____________________

Product Number ____________________

Your Name ____________________

Title ____________________

Company ____________________

☐ Company

☐ Home Street Address ____________________

City, State, Zip ____________________

☐ Please send ________ forms.

Owner ____________________

Architect ____________________

General Contractor ____________________

✓ Bid Date ____________________

Typical Bay Size ____________________

✓ Labor Force: _____ % Union _____ % Non-Union

✓ Project Description (Circle one number in each line)

1. Economy 2. Average 3. Custom 4. Luxury

1. Square 2. Rectangular 3. Irregular 4. Very Irregular

J	✓	**Finishes**		$
JL		*Lath & Plaster*	(S.Y.)	
JD		*Drywall*	(S.F.)	
JM		*Tile & Marble*	(S.F.)	
JT		*Terrazzo*	(S.F.)	
JA		*Acoustical Treatment*	(S.F.)	
JC		*Carpet*	(S.Y.)	
JF		*Hard Surface Flooring*	(S.F.)	
JP		*Painting & Wall Covering*	(S.F.)	
K	✓	**Specialties**		$
KB		*Bathroom Partitions & Access.*	(S.F.)	
KF		*Other Partitions*	(S.F.)	
KL		*Lockers*	(Ea.)	
L	✓	**Equipment**		$
LK		*Kitchen*		
LS		*School*		
LO		*Other*		
M	✓	**Furnishings**		$
MW		*Window Treatment*		
MS		*Seating*	(Ea.)	
N	✓	**Special Construction**		$
NA		*Acoustical*	(S.F.)	
NB		*Prefab. Bldgs.*	(S.F.)	
NO		*Other*		
P	✓	**Conveying Systems**		$
PE		*Elevators*	(Ea.)	
PS		*Escalators*	(Ea.)	
PM		*Material Handling*		
Q	✓	**Mechanical**		$
QP		*Plumbing*	*(No. of fixtures)*	
QS		*Fire Protection (Sprinklers)*		
QF		*Fire Protection (Hose Standpipes)*		
QB		*Heating, Ventilating & A.C.*		
QH		*Heating & Ventilating*	*(BTU Output)*	
QA		*Air Conditioning*	(Tons)	
R	✓	**Electrical**		$
RL		*Lighting*	(S.F.)	
RP		*Power Service*		
RD		*Power Distribution*		
RA		*Alarms*		
RG		*Special Systems*		
S	✓	**Mech./Elec. Combined**		$

Please specify the Means product you wish to receive. Complete the address information as requested and return this form with your check (product cost less $20.00) to address below.

R.S. Means Company, Inc.,
Square Foot Costs Department
P.O. Box 800
Kingston, MA 02364-9988

For more information
visit Means Web Site
at www.rsmeans.com

R.S. Means Company, Inc. . . a tradition of excellence in Construction Cost Information and Services since 1942.

Table of Contents

Book Selection Guide

The following table provides definitive information on the content of each cost data publication. The number of lines of data provided in each unit price or assemblies division, as well as the number of reference tables and crews is listed for each book. The presence of other elements such as an historical cost index, city cost indexes, square foot models or cross-referenced index is also indicated. You can use the table to help select the Means' book that has the quantity and type of information you most need in your work.

Unit Cost Divisions	Building Construction Costs	Mechanical	Electrical	Repair & Remodel.	Square Foot	Site Work Landsc.	Assemblies	Interior	Concrete Masonry	Open Shop	Heavy Construc.	Resi-dential	Light Commercial	Facil. Construc.	Plumbing	Western Construction Costs
1	1098	586	667	862		1089		491	987	1096	1132	416	621	1612	659	1091
2	3141	1467	509	1881		9149		852	1655	3090	5621	1064	984	5015	1826	3114
3	1516	91	71	768		1321		198	1937	1510	1494	268	233	1378	43	1514
4	871	27	0	688		751		636	1174	853	651	313	396	1124	0	849
5	1857	177	174	955		803		985	667	1824	1056	779	809	1849	95	1841
6	1388	97	84	1314		493		1336	339	1379	612	1532	1473	1411	59	1772
7	1365	180	91	1329		490		546	451	1365	353	805	1036	1357	188	1364
8	1899	60	0	1904		337		1821	726	1880	9	1133	1184	2032	0	1900
9	1650	50	0	1473		188		1746	327	1598	113	1251	1390	1845	50	1644
10	991	58	31	565		217		822	196	995	0	250	467	996	249	991
11	1158	435	213	622		151		917	34	1052	95	116	237	1221	371	1051
12	352	0	0	48		228		1536	30	344	0	72	68	1538	0	343
13	1384	1221	498	587		443		1075	106	1358	302	237	562	2094	1083	1339
14	364	43	0	261		36		330	0	364	40	9	18	362	17	362
15	2510	14742	763	2243		1809		1477	80	2515	2028	1087	1553	12312	10902	2545
16	1517	541	10191	1200		879		1292	62	1535	916	706	1290	9872	481	1459
17	421	0	0	421		0		421	421	460	0	421	421	421	0	421
Totals	**23482**	**19775**	**13292**	**17121**		**18384**		**16508**	**9192**	**23218**	**14422**	**10459**	**12742**	**46439**	**16023**	**23600**

Assembly Divisions	Building Construction Costs	Mechanical	Electrical	Repair & Remodel.	Square Foot	Site Work Landsc.	Assemblies	Interior	Concrete Masonry	Open Shop	Heavy Construc.	Resi-dential	Light Commercial	Facil. Construc.	Plumbing	Western Construction Costs
1		0	0	202	131	738	775	0	689		752	841	131	89	0	
2		0	0	40	33	35	48	0	49		0	584	32	0	0	
3		0	0	443	1114	0	3064	228	1245		0	1546	805	174	0	
4		0	0	714	1382	0	3201	255	1273		0	2050	1200	26	0	
5		0	0	288	227	0	459	0	0		0	1081	223	25	0	
6		0	0	1006	843	0	1295	1723	152		0	1077	738	330	0	
7		0	0	42	89	0	179	159	0		0	709	33	71	0	
8		2257	149	998	1689	0	2718	926	0		0	1686	1189	1114	2084	
9		0	1383	355	367	0	1320	306	0		0	238	358	319	0	
10		0	0	0	0	0	0	0	0		0	0	0	0	0	
11		0	0	365	465	0	724	153	0		0	0	466	201	0	
12		501	160	539	84	2392	699	0	698		541	0	84	119	663	
Totals		**2758**	**1692**	**4992**	**6424**	**3165**	**14482**	**3750**	**4106**		**1293**	**9812**	**5259**	**2468**	**2747**	

Reference Section	Building Construction Costs	Mechanical	Electrical	Repair & Remodel.	Square Foot	Site Work Landsc.	Assemblies	Interior	Concrete Masonry	Open Shop	Heavy Construc.	Resi-dential	Light Commercial	Facil. Construc.	Plumbing	Western Construction Costs
Tables	150	46	84	66	4	84	223	59	83	146	55	49	70	86	49	148
Models					102							32	43			
Crews	408	408	408	389		408		408	408	391	408	391	391	389	408	408
City Cost Indexes	yes	yes	yes	yes	yes	yes	yes	yes	yes	yes	yes	yes	yes	yes	yes	yes
Historical Cost Indexes	yes	yes	yes	yes	yes	yes	yes	yes	yes	yes	yes	no	yes	yes	yes	yes
Index	yes	yes	yes	yes	no	yes	yes	yes	yes	yes	yes	yes	yes	yes	yes	yes

Means Building Construction Cost Data 2001

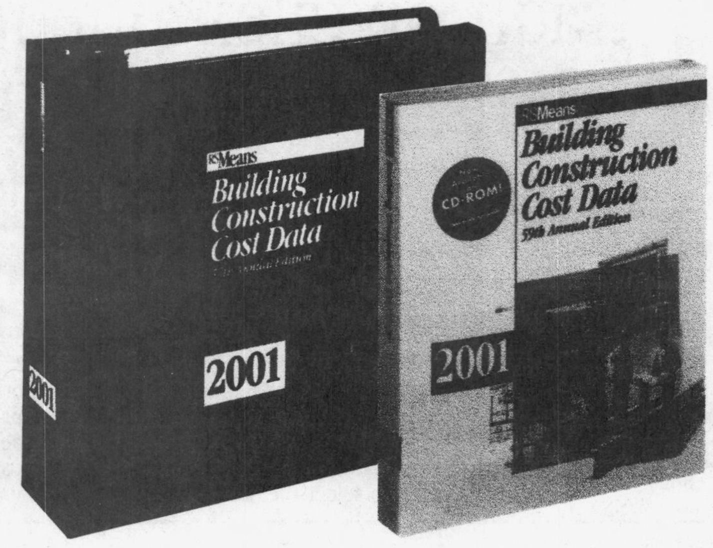

Available in Both Softbound and Looseleaf Editions

The "Bible" of the industry comes in the standard softcover edition or the looseleaf edition.

Many customers enjoy the convenience and flexibility of the looseleaf binder, which increases the usefulness of *Means Building Construction Cost Data 2001* by making it easy to add and remove pages. You can insert your own cost information pages, so everything is in one place. Copying pages for faxing is easier also. Whichever edition you prefer, softbound or the convenient looseleaf edition, you'll get the *DesignBuildIntelligence* newsletter at no extra cost.

$89.95 per copy, Softbound
Catalog No. 60011

$116.95 per copy, Looseleaf
Catalog No. 61011

Means Building Construction Cost Data 2001

Offers you unchallenged unit price reliability in an easy-to-use arrangement. Whether used for complete, finished estimates or for periodic checks, it supplies more cost facts better and faster than any comparable source. Over 23,000 unit prices for 2001. The City Cost Indexes cover over 930 areas, for indexing to any project location in North America. Order and get the *DesignBuildIntelligence* newsletter sent to you FREE. You'll have year-long access to the Means Estimating **Hotline** FREE with your subscription. Expert assistance when using Means data is just a phone call away.

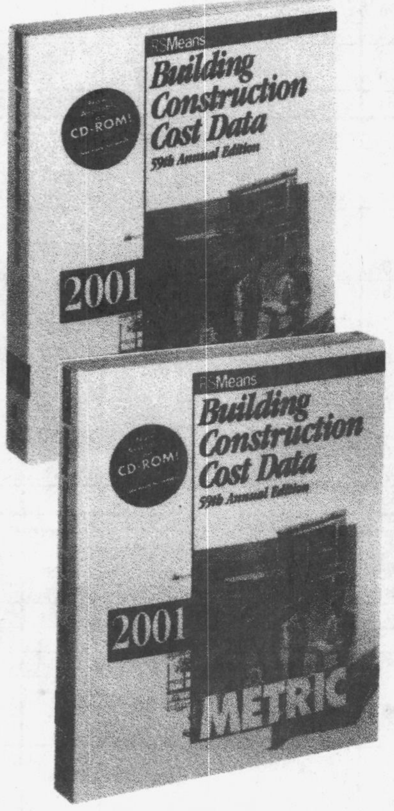

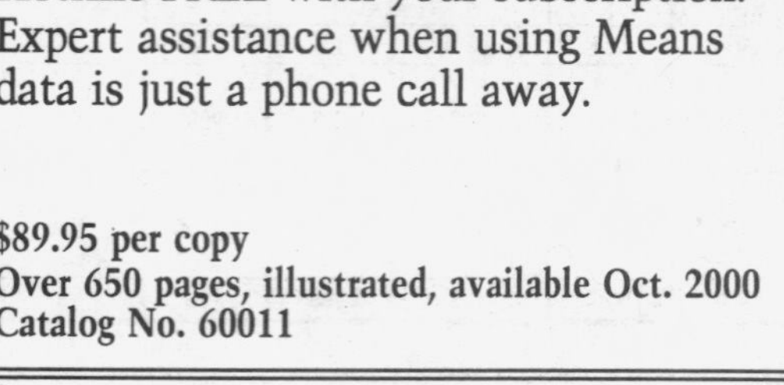

$89.95 per copy
Over 650 pages, illustrated, available Oct. 2000
Catalog No. 60011

Means Building Construction Cost Data 2001

Metric Version

The Federal Government has stated that all federal construction projects must now use metric documentation. The *Metric Version* of *Means Building Construction Cost Data 2001* is presented in metric measurements covering all construction areas. Don't miss out on these billion dollar opportunities. Make the switch to metric today.

$89.95 per copy
Over 650 pages, illustrated, available Nov. 2000
Catalog No. 63011

Annual Cost Guides

Means Mechanical Cost Data 2001

• HVAC • Controls

Total unit and systems price guidance for mechanical construction . . . materials, parts, fittings, and complete labor cost information. Includes prices for piping, heating, air conditioning, ventilation, and all related construction.

Plus new 2001 unit costs for:

- Over 2500 installed HVAC/controls assemblies
- "On Site" Location Factors for over 930 cities and towns in the U.S. and Canada
- Crews, labor and equipment

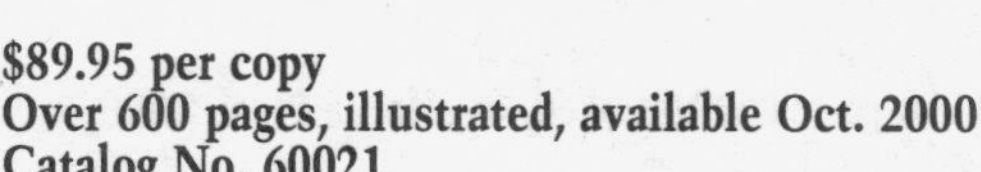

$89.95 per copy
Over 600 pages, illustrated, available Oct. 2000
Catalog No. 60021

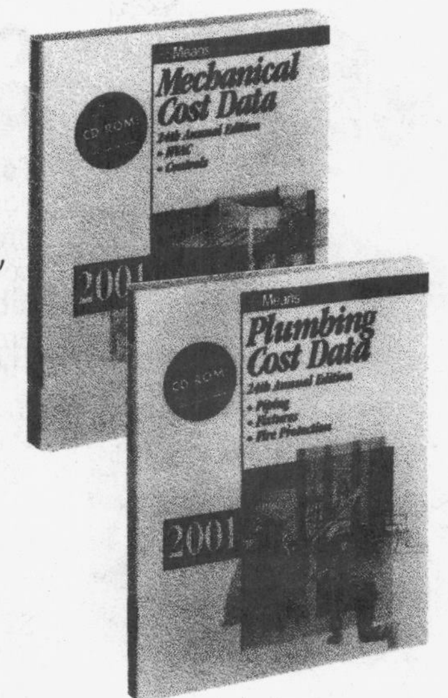

Means Plumbing Cost Data 2001

Comprehensive unit prices and assemblies for plumbing, irrigation systems, commercial and residential fire protection, point-of-use water heaters, and the latest approved materials. This publication and its companion, *Means Mechanical Cost Data,* provide full-range cost estimating coverage for all the mechanical trades.

$89.95 per copy
Over 500 pages, illustrated, available Nov. 2000
Catalog No. 60211

Means Electrical Cost Data 2001

Pricing information for every part of electrical cost planning: More than 15,000 unit and systems costs with design tables; clear specifications and drawings; engineering guides and illustrated estimating procedures; complete labor-hour and materials costs for better scheduling and procurement; the latest electrical products and construction methods.

- A Variety of Special Electrical Systems including Cathodic Protection
- Costs for maintenance, demolition, HVAC/ mechanical, specialties, equipment, and more

$89.95 per copy
Over 450 pages, illustrated, available Oct. 2000
Catalog No. 60031

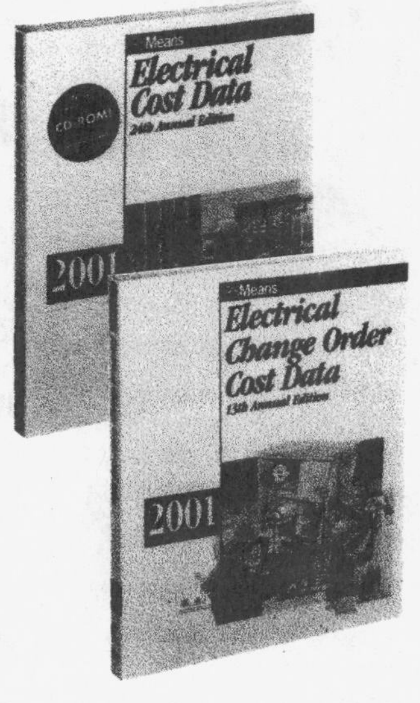

Means Electrical Change Order Cost Data 2001

You are provided with electrical unit prices exclusively for pricing change orders—based on the recent, direct experience of contractors and suppliers. Analyze and check your own change order estimates against the experience others have had doing the same work. It also covers productivity analysis and change order cost justifications. With useful information for calculating the effects of change orders and dealing with their administration.

$89.95 per copy
Over 450 pages, available Oct. 2000
Catalog No. 60231

Means Facilities Maintenance & Repair Cost Data 2001

Published in a looseleaf format, *Means Facilities Maintenance & Repair Cost Data* gives you a complete system to manage and plan your facility repair and maintenance costs and budget efficiently. Guidelines for auditing a facility and developing an annual maintenance plan. Budgeting is included, along with reference tables on cost and management and information on frequency and productivity of maintenance operations.

The only nationally recognized source of maintenance and repair costs. Developed in cooperation with the Army Corps of Engineers.

$199.95 per copy
Over 600 pages, illustrated, available Dec. 2000
Catalog No. 60301

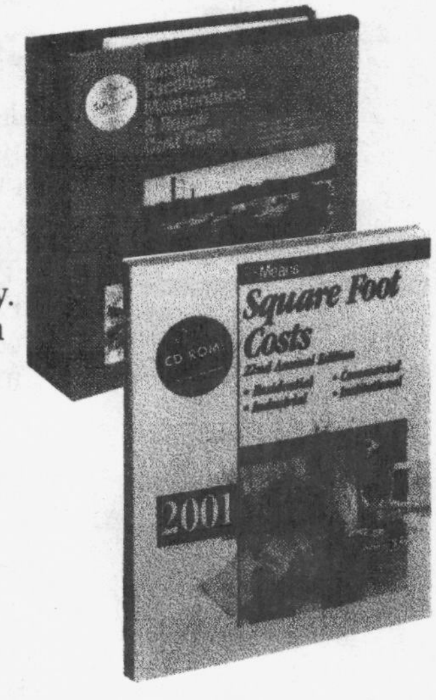

Means Square Foot Costs 2001

It's Accurate and Easy To Use!

- **Updated 2001 price information,** based on nationwide figures from suppliers, estimators, labor experts and contractors.
- "How-to-Use" Sections, with **clear examples** of commercial, residential, industrial, and institutional structures.
- Realistic graphics, offering true-to-life illustrations of building projects.
- Extensive information on using square foot cost data, including **sample estimates** and **alternate pricing methods.**

$99.95 per copy
Over 450 pages, illustrated, available Oct. 2000
Catalog No. 60051

Annual Cost Guides

For more information
visit Means Web Site
at www.rsmeans.com

Means Repair & Remodeling Cost Data 2001

Commercial/Residential

You can use this valuable tool to estimate commercial and residential renovation and remodeling.

Includes: New costs for hundreds of unique methods, materials and conditions that only come up in repair and remodeling. PLUS:

- Unit costs for over 16,000 construction components
- Installed costs for over 90 assemblies
- Costs for 300+ construction crews
- Over 930 "On Site" localization factors for the U.S. and Canada.

$79.95 per copy
Over 600 pages, illustrated, available Nov. 2000
Catalog No. 60041

Means Facilities Construction Cost Data 2001

For the maintenance and construction of commercial, industrial, municipal, and institutional properties. Costs are shown for new and remodeling construction and are broken down into materials, labor, equipment, overhead, and profit. Special emphasis is given to sections on mechanical, electrical, furnishings, site work, building maintenance, finish work, and demolition. More than 45,000 unit costs plus assemblies and reference sections are included.

$219.95 per copy
Over 1150 pages, illustrated, available Nov. 2000
Catalog No. 60201

Means Residential Cost Data 2001

Contains square foot costs for 30 basic home models with the look of today—plus hundreds of custom additions and modifications you can quote right off the page. With costs for the 100 residential systems you're most likely to use in the year ahead. Complete with blank estimating forms, sample estimates and step-by-step instructions.

$79.95 per copy
Over 550 pages, illustrated, available Dec. 2000
Catalog No. 60171

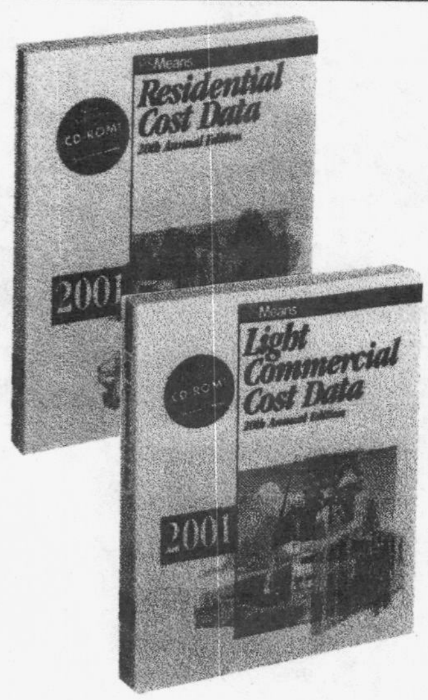

Means Light Commercial Cost Data 2001

Specifically addresses the light commercial market, which is an increasingly specialized niche in the industry. Aids you, the owner/designer/contractor, in preparing all types of estimates, from budgets to detailed bids. Includes new advances in methods and materials. Assemblies section allows you to evaluate alternatives in the early stages of design/planning.

Over 13,000 unit costs for 2001 ensure you have the prices you need... when you need them.

$79.95 per copy
Over 600 pages, illustrated, available Dec. 2000
Catalog No. 60181

Means Assemblies Cost Data 2001

Means Assemblies Cost Data 2001 takes the guesswork out of preliminary or conceptual estimates. Now you don't have to try to calculate the assembled cost by working up individual components costs. We've done all the work for you.

Presents detailed illustrations, descriptions, specifications and costs for every conceivable building assembly—240 types in all—arranged in the easy-to-use UniFormat system. Each illustrated "assembled" cost includes a complete grouping of materials and associated installation costs including the installing contractor's overhead and profit.

$149.95 per copy
Over 550 pages, illustrated, available Oct. 2000
Catalog No. 60061

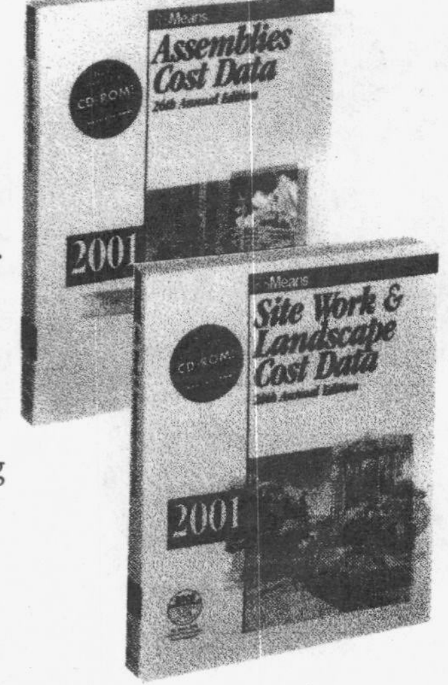

Means Site Work & Landscape Cost Data 2001

Means Site Work & Landscape Cost Data 2001 is organized to assist you in all your estimating needs. Hundreds of fact-filled pages help you make accurate cost estimates efficiently.

Updated for 2001!

- Demolition features—including ceilings, doors, electrical, flooring, HVAC, millwork, plumbing, roofing, walls and windows
- State-of-the-art segmental retaining walls
- Flywheel trenching costs and details
- Updated Wells section
- Thousands of landscape materials, flowers, shrubs and trees

$89.95 per copy
Over 600 pages, illustrated, available Oct. 2000
Catalog No. 60281

Annual Cost Guides

Means Open Shop Building Construction Cost Data 2001

The latest costs for accurate budgeting and estimating of new commercial and residential construction... renovation work... change orders... cost engineering. *Means Open Shop BCCD* will assist you to...

- Develop benchmark prices for change orders
- Plug gaps in preliminary estimates, budgets
- Estimate complex projects
- Substantiate invoices on contracts
- Price ADA-related renovations

$89.95 per copy
Over 650 pages, illustrated, available Nov. 2000
Catalog No. 60151

Means Heavy Construction Cost Data 2001

A comprehensive guide to heavy construction costs. Includes costs for highly specialized projects such as tunnels, dams, highways, airports, and waterways. Information on different labor rates, equipment, and material costs is included. Has unit price costs, systems costs, and numerous reference tables for costs and design. Valuable not only to contractors and civil engineers, but also to government agencies and city/town engineers.

$89.95 per copy
Over 450 pages, illustrated, available Nov. 2000
Catalog No. 60161

Means Building Construction Cost Data 2001

Western Edition

This regional edition provides more precise cost information for western North America. Labor rates are based on union rates from 13 western states and western Canada. Included are western practices and materials not found in our national edition: tilt-up concrete walls, glu-lam structural systems, specialized timber construction, seismic restraints, landscape and irrigation systems.

$89.95 per copy
Over 600 pages, illustrated, available Dec. 2000
Catalog No. 60221

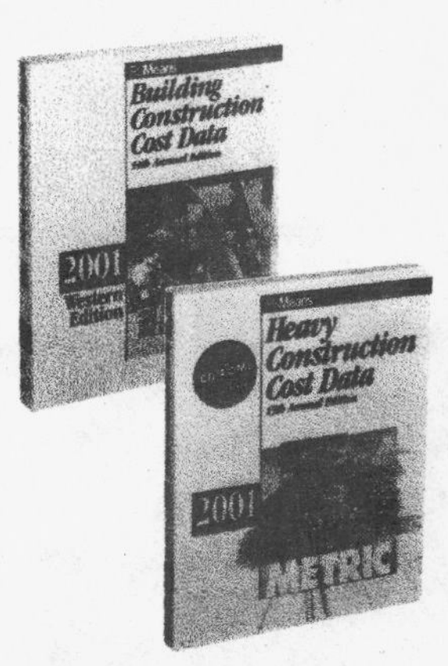

Means Heavy Construction Cost Data 2001

Metric Version

Make sure you have the Means industry standard metric costs for the federal, state, municipal and private marketplace. With thousands of up-to-date metric unit prices in tables by CSI standard divisions. Supplies you with assemblies costs using the metric standard for reliable cost projections in the design stage of your project. Helps you determine sizes, material amounts, and has tips for handling metric estimates.

$89.95 per copy
Over 450 pages, illustrated, available Nov. 2000
Catalog No. 63161

Means Construction Cost Indexes 2001

Who knows what 2001 holds? What materials and labor costs will change unexpectedly? By how much?

- Breakdowns for 305 major cities.
- National averages for 30 key cities.
- Expanded five major city indexes.
- Historical construction cost indexes.

$198.00 per year/$49.50 individual quarters
Catalog No. 60141

Means Interior Cost Data 2001

Provides you with prices and guidance needed to make accurate interior work estimates. Contains costs on materials, equipment, hardware, custom installations, furnishings, labor costs . . . every cost factor for new and remodel commercial and industrial interior construction, including updated information on office furnishings, plus more than 50 reference tables. For contractors, facility managers, owners.

$89.95 per copy
Over 550 pages, illustrated, available Oct. 2000
Catalog No. 60091

Means Concrete & Masonry Cost Data 2001

Provides you with cost facts for virtually all concrete/masonry estimating needs, from complicated formwork to various sizes and face finishes of brick and block, all in great detail. The comprehensive unit cost section contains more than 8,500 selected entries. Also contains an assemblies cost section, and a detailed reference section which supplements the cost data.

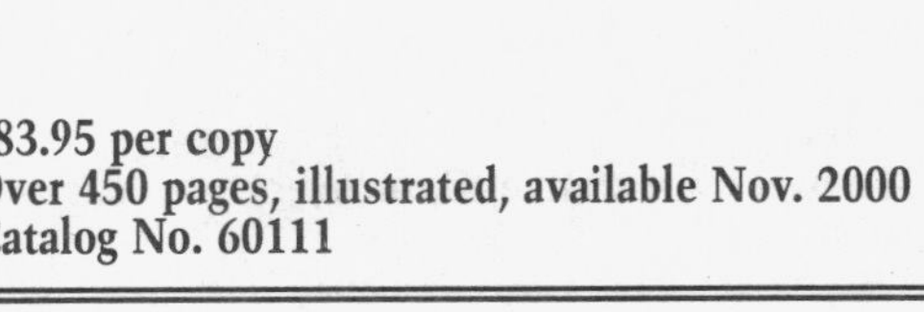

$83.95 per copy
Over 450 pages, illustrated, available Nov. 2000
Catalog No. 60111

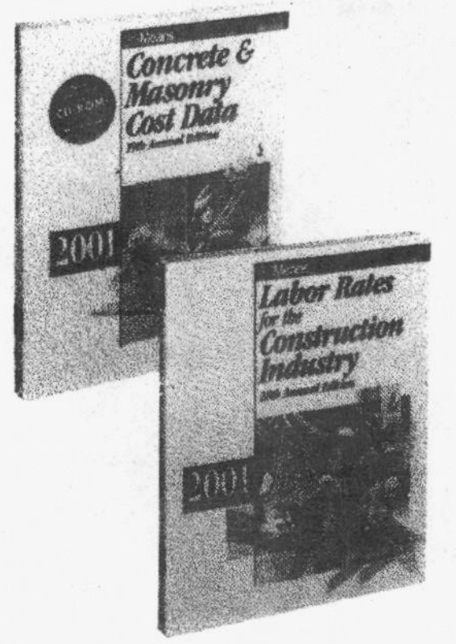

Means Labor Rates for the Construction Industry 2001

Complete information for estimating labor costs, making comparisons and negotiating wage rates by trade for over 300 cities (United States and Canada). With 46 construction trades listed by local union number in each city, and historical wage rates included for comparison. No similar book is available through the trade.

Each city chart lists the county and is alphabetically arranged with handy visual flip tabs for quick reference.

$199.95 per copy
Over 300 pages, available Dec. 2000
Catalog No. 60121

For more information visit Means Web Site at www.rsmeans.com

Project Scheduling & Management for Construction

New 2nd Edition

By David R. Pierce, Jr.

A comprehensive, yet easy-to-follow guide to construction project scheduling and control—from vital project management principles through the latest scheduling, tracking, and controlling techniques. The author is a leading authority on scheduling with years of field and teaching experience at leading academic institutions. Spend a few hours with this book and come away with a solid understanding of this essential management topic.

$64.95 per copy
Over 250 pages; illustrated, Hardcover
Catalog No. 67247A

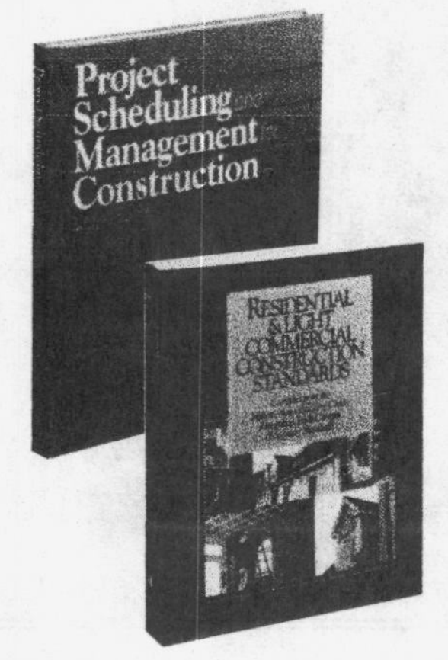

Residential & Light Commercial Construction Standards

A Unique Collection of Industry Standards That Define Quality in Construction

For Contractors & Subcontractors, Owners, Developers, Architects & Engineers, Attorneys & Insurance Personnel

Compiled from the nation's major building codes, professional associations, and product manufacturer institutes, this unique resource enables you to:

- Set a standard for subcontractors and employees
- Protect yourself against defect claims
- Resolve disputes
- Answer client questions with authority

$59.95 per copy
Over 500 pages, illustrated, Softcover
Catalog No. 67322

The Building Professional's Guide to Contract Documents

New 3rd Edition
By Waller S. Poage, AIA, CSI, SAVE

This comprehensive treatment of Contract Documents is an important reference for owners, design professionals, contractors, and students.

- Structure your Documents for Maximum Efficiency
- Understand the Roles and Responsibilities of Construction Professionals
- Improve Methods of Project Delivery

$64.95 per copy, 400 pages
Diagrams and construction forms, Hardcover
Catalog No. 67261A

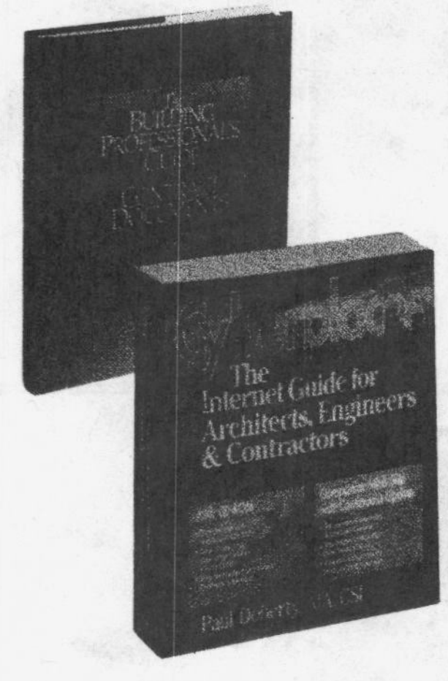

Cyberplaces II: The Internet Guide for AECs & Facility Managers 2nd Edition

By Paul Doherty

With an added section on the New Economy and e-commerce—such as AEC arbitrage, aggregate purchasing, auctions, e-supply chains, and more. Plus the best Knowledge Management and other Internet tools, including PDAs and the newest Web software for scheduling and project management.

The **Web site** keeps you current on the best Internet tools and applications for building design, construction and management—with workshops to put it all into practice.

$59.95 per copy
Over 700 pages, illustrated, Softcover
Catalog No. 67317A

Value Engineering: Practical Applications

... For Design, Construction, Maintenance & Operations

By Alphonse Dell'Isola, PE, leading authority on VE in construction

A tool for immediate application—for engineers, architects, facility managers, owners and contractors. Includes: Making the Case for VE—The Management Briefing, Integrating VE into Planning and Budgeting, Conducting Life Cycle Costing, Integrating VE into the Design Process, Using VE Methodology in Design Review and Consultant Selection, Case Studies, VE Workbook, and a Life Cycle Costing program on disk.

$79.95 per copy
Over 450 pages, illustrated, Hardcover
Catalog No. 67319

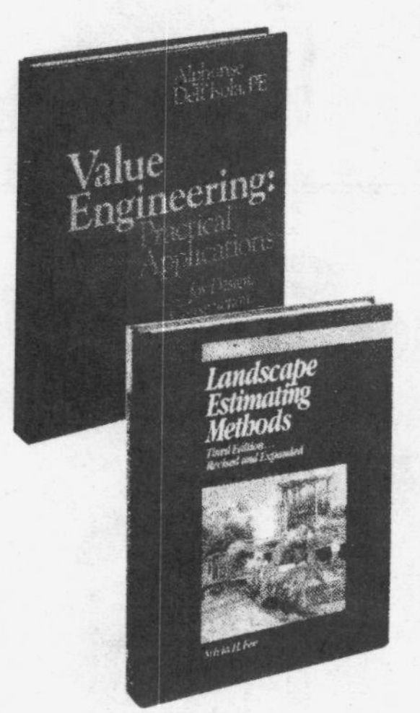

Means Landscape Estimating Methods

3rd Edition

By Sylvia H. Fee

This revised edition offers expert guidance for preparing accurate estimates for new landscape construction and grounds maintenance. Includes a complete project estimate featuring the latest equipment and methods, and **two totally new chapters—Life Cycle Costing and Landscape Maintenance Estimating.**

$62.95 per copy
Over 300 pages, illustrated, Hardcover
Catalog No. 67295A

Reference Books

Preventive Maintenance Guidelines for School Facilities

NEW! By John C. Maciha

This new publication gives you a complete program for school maintenance that ensures: sustained security, safety, property integrity, facility acceptance, user satisfaction, and reasonable ongoing expenditures.

Includes schedules for weekly, monthly, half-yearly and annual maintenance. 3-ring binder with hard copy forms, electronic forms on disk, and laminated wall chart.

$149.95 per copy
Over 150 pages, Hardcover
Catalog No. 67326

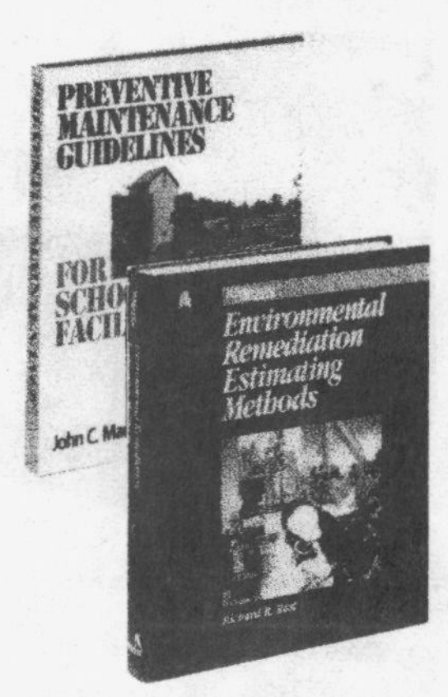

Means Environmental Remediation Estimating Methods

By Richard R. Rast

Guidelines for estimating 50 standard remediation technologies. Use it to: prepare preliminary budgets, develop detailed estimates, compare costs and solutions, estimate liability, review quotes, negotiate settlements. A valuable support tool for ***Means Envitonmental Remediation Unit Price*** and ***Assemblies books.***

$99.95 per copy
Over 600 pages, illustrated, Hardcover
Catalog No. 64777

Means Facilities Maintenance Standards

By Roger W. Liska, PE, AIC

Unique features of this one-of-a-kind working guide for facilities maintenance

A working encyclopedia that points the way to solutions to every kind of maintenance and repair dilemma. With a labor-hours section to provide productivity figures for over 180 maintenance tasks. Included are ready-to-use forms, checklists, worksheets and comparisons, as well as analysis of materials systems and remedies for deterioration and wear.

$159.95 per copy, 600 pages, 205 tables, checklists and diagrams, Hardcover
Catalog No. 67246

HVAC: Design Criteria, Options, Selection

Expanded Second Edition

By William H. Rowe III, AIA, PE

Includes Indoor Air Quality, CFC Removal, Energy Efficient Systems and Special Systems by Building Type. Helps you solve a wide range of HVAC system design and selection problems effectively and economically. Gives you clear explanations of the latest ASHRAE standards.

$84.95 per copy
Over 600 pages, illustrated, Hardcover
Catalog No. 67306

The Facilities Manager's Reference

By Harvey H. Kaiser, PhD

The tasks and tools the facility manager needs to accomplish the organization's objectives, and develop individual and staff skills. Includes Facilities and Property Management, Administrative Control, Planning and Operations, Support Services, and a complete building audit with forms and instructions, widely used by facilities managers nationwide.

$86.95 per copy, over 250 pages, with prototype forms and graphics, Hardcover
Catalog No. 67264

Facilities Maintenance Management

By Gregory H. Magee, PE

Now you can get successful management methods and techniques for all aspects of facilities maintenance. This comprehensive reference explains and demonstrates successful management techniques for all aspects of maintenance, repair and improvements for buildings, machinery, equipment and grounds. Plus, guidance for outsourcing and managing internal staffs.

$86.95 per copy
Over 280 pages with illustrations, Hardcover
Catalog No. 67249

Cost Planning & Estimating for Facilities Maintenance

In this unique book, a team of facilities management authorities shares their expertise at:

- Evaluating and budgeting maintenance operations
- Maintaining & repairing key building components
- Applying *Means Facilities Maintenance & Repair Cost Data* to your estimating

With the special maintenance requirements of the 10 major building types.

$82.95 per copy
Over 475 pages, Hardcover
Catalog No. 67314

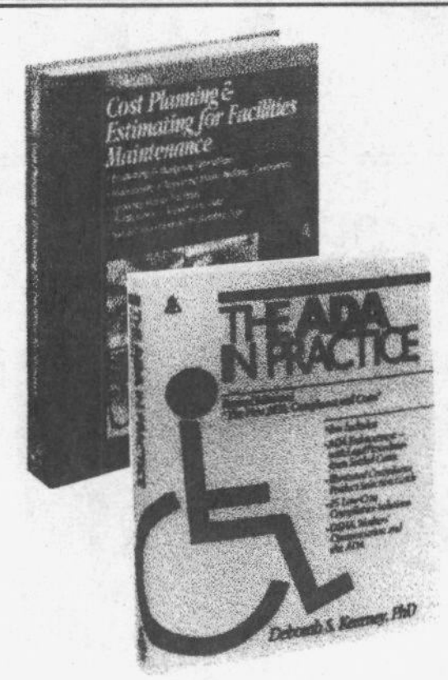

The ADA in Practice

By Deborah S. Kearney, PhD

Helps you meet and budget for the requirements of the Americans with Disabilities Act. Shows how to do the job right by understanding what the law requires, allows, and enforces. Includes a "buyers guide" for 70 ADA-compliant products.

*Winner of IFMA's "Distinguished Author of the Year" award.

See page 9 for another ADA resource.

$72.95 per copy
Over 600 pages, illustrated, Softcover
Catalog No. 67147A

Reference Books

For more information visit Means Web Site at www.rsmeans.com

Basics for Builders: Plan Reading & Material Takeoff

By Wayne J. DelPico

For Residential and Light Commercial Construction

A valuable tool for understanding plans and specs, and accurately calculating material quantities. Step-by-step instructions and takeoff procedures based on a full set of working drawings.

$35.95 per copy
Over 420 pages, Softcover
Catalog No. 67307

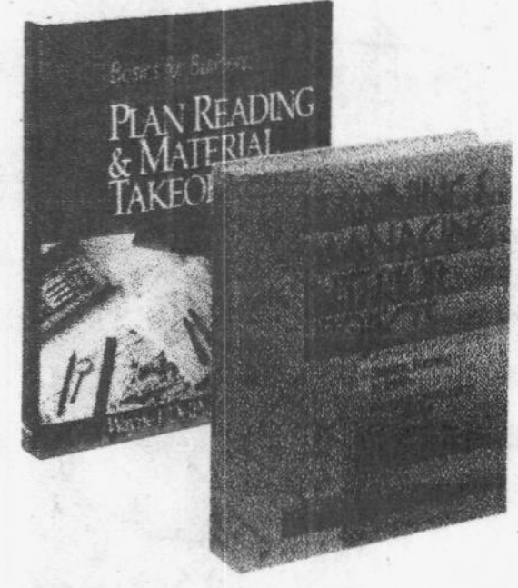

Planning & Managing Interior Projects New 2nd Edition, By Carol E. Farren, CFM

Addresses changes in technology and business, guiding you through commercial design and construction from initial client meetings to post-project administration. Includes: evaluating space requirements, alternative work models, telecommunications and data management, and environmental issues.

$69.95 per copy
Over 400 pages, illustrated, Hardcover
Catalog No. 67245A

Superintending for Contractors: How to Bring Jobs in On-time, On-Budget

By Paul J. Cook

This book examines the complex role of the superintendent/field project manager, and provides guidelines for the efficient organization of this job. Includes administration of contracts, change orders, purchase orders, and more.

$35.95 per copy
Over 220 pages, illustrated, Softcover
Catalog No. 67233

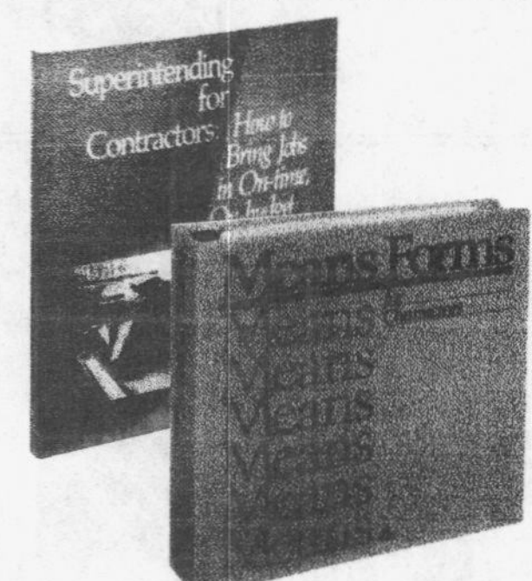

Means Forms for Contractors

For general and specialty contractors

Means editors have created and collected the most needed forms as requested by contractors of various-sized firms and specialties. This book covers all project phases. Includes sample correspondence and personnel administration.
Full-size forms on durable paper for photocopying or reprinting.

$79.95 per copy
Over 400 pages, three-ring binder, more than 80 forms
Catalog No. 67288

Total Productive Facilities Management

By Richard W. Sievert, Jr.

The best of today's cost and project management, quality, and value engineering principles. Includes Benchmarking, Value Engineering, and Scheduling.

$79.95 per copy
Over 270 pages, illustrated, Hardcover
Catalog No. 67321

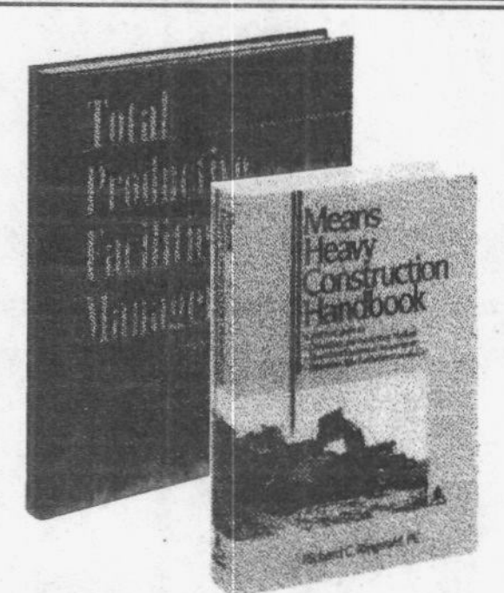

Means Heavy Construction Handbook

Informed guidance for planning, estimating and performing today's heavy construction projects. Provides expert advice on every aspect of heavy construction work including hazardous waste remediation and estimating. To assist planning, estimating, performing, or overseeing work.

$74.95 per copy
Over 430 pages, illustrated, Hardcover
Catalog No. 67148

Estimating for Contractors: How to Make Estimates that Win Jobs

By Paul J. Cook

Estimating for Contractors is a reference that will be used over and over, whether to check a specific estimating procedure, or to take a complete course in estimating.

$35.95 per copy
Over 225 pages, illustrated, Softcover
Catalog No. 67160

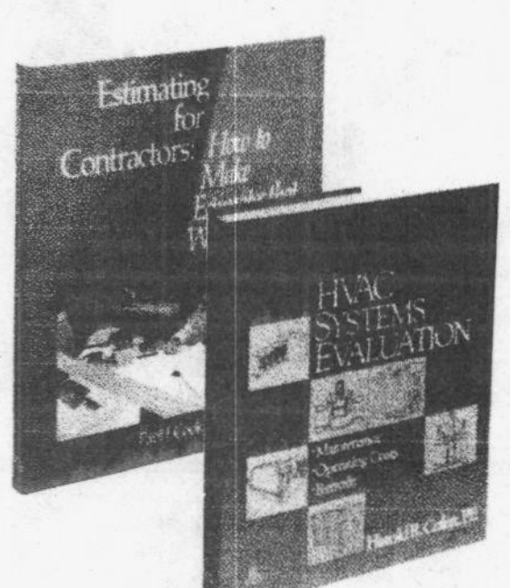

HVAC Systems Evaluation

By Harold R. Colen, PE

You get direct comparisons of how each type of system works, with the relative costs of installation, operation, maintenance and applications by type of building. With requirements for hooking up electrical power to HVAC components. Contains experienced advice for repairing operational problems in existing HVAC systems, ductwork, fans, cooling coils, and much more!

$84.95 per copy
Over 500 pages, illustrated, Hardcover
Catalog No. 67281

Quantity Takeoff for Contractors: How to Get Accurate Material Counts

By Paul J. Cook

Contractors who are new to material takeoffs or want to be sure they are using the best techniques will find helpful information in this book, organized by CSI MasterFormat division.

Now $17.98 per copy, limited quantity
Over 250 pages, illustrated, Softcover
Catalog No. 67262

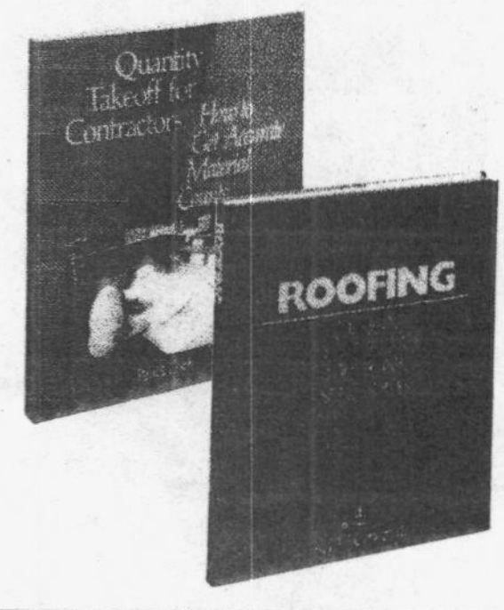

Roofing: Design Criteria, Options, Selection

By R.D. Herbert, III

This book is required reading for those who specify, install or have to maintain roofing systems. It covers all types of roofing technology and systems. You'll get the facts needed to intelligently evaluate and select both traditional and new roofing systems.

Now $31.48 per copy
Over 225 pages with illustrations, Hardcover
Catalog No. 67253

Reference Books

Means Estimating Handbook

This comprehensive reference covers a full spectrum of technical data for estimating, with information on sizing, productivity, equipment requirements, codes, design standards and engineering factors.
Means Estimating Handbook will help you: evaluate architectural plans and specifications, prepare accurate quantity takeoffs, prepare estimates from conceptual to detailed, and evaluate change orders.

$99.95 per copy
Over 900 pages, Hardcover
Catalog No. 67276

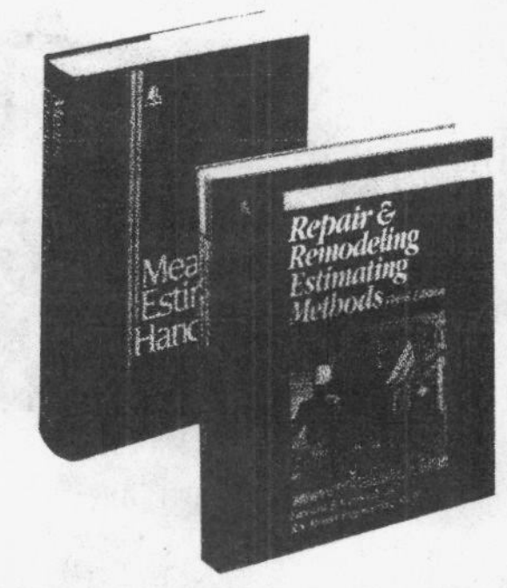

Means Repair and Remodeling Estimating Third Edition

By Edward B. Wetherill & R.S. Means

Focuses on the unique problems of estimating renovations of existing structures. It helps you determine the true costs of remodeling through careful evaluation of architectural details and a site visit.
New section on disaster restoration costs.

$69.95 per copy
Over 450 pages, illustrated, Hardcover
Catalog No. 67265A

Successful Estimating Methods: From Concept to Bid

By John D. Bledsoe, PhD, PE

A highly practical, all-in-one guide to the tips and practices of today's successful estimator. Presents techniques for all types of estimates, *and* advanced topics such as life cycle cost analysis, value engineering, and automated estimating.
Estimate spreadsheets available at Means Web site.

$64.95 per copy
Over 300 pages, illustrated, Hardcover
Catalog No. 67287

Means Productivity Standards for Construction

Expanded Edition *(Formerly* Man-Hour Standards)

Here is the working encyclopedia of labor productivity information for construction professionals, with labor requirements for thousands of construction functions in CSI MasterFormat.
Completely updated, with over 3,000 new work items.

$159.95 per copy
Over 800 pages, Hardcover
Catalog No. 67236A

Means Electrical Estimating Methods Second Edition

Expanded version includes sample estimates and cost information in keeping with the latest version of the CSI MasterFormat. Contains new coverage of Fiber Optic and Uninterruptible Power Supply electrical systems, broken down by components and explained in detail. A practical companion to *Means Electrical Cost Data.*

$64.95 per copy
Over 325 pages, Hardcover
Catalog No. 67230A

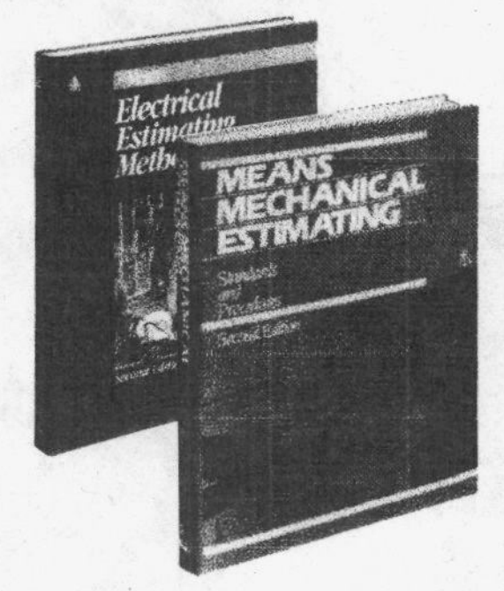

Means Mechanical Estimating Methods Second Edition

This guide assists you in making a review of plans, specs and bid packages with suggestions for takeoff procedures, listings, substitutions and pre-bid scheduling. Includes suggestions for budgeting labor and equipment usage. Compares materials and construction methods to allow you to select the best option.

$64.95 per copy
Over 350 pages, illustrated, Hardcover
Catalog No. 67294

Means Scheduling Manual

Third Edition

By F. William Horsley

Fast, convenient expertise for keeping your scheduling skills right in step with today's cost-conscious times. Covers bar charts, PERT, precedence and CPM scheduling methods. Now updated to include computer applications.

$64.95 per copy
Over 200 pages, spiral-bound, Softcover
Catalog No. 67291

Maintenance Management Audit

This annual audit program is essential for every organization in need of a proper assessment of its maintenance operation. Downloadable electronic forms from Means Web Site allow managers to identify and correct problems and enhance productivity.

Now $32.48 per copy, limited quantity
125 pages, spiral bound, illustrated, Hardcover
Catalog No. 67299

Means Square Foot Estimating Methods Second Edition

By Billy J. Cox and F. William Horsley

Proven techniques for conceptual and design-stage cost planning. Steps you through the square foot cost process, demonstrating faster, better ways to relate the design to the budget. Now updated to the latest version of UniFormat.

$69.95 per copy
Over 300 pages, illustrated, Hardcover
Catalog No. 67145A

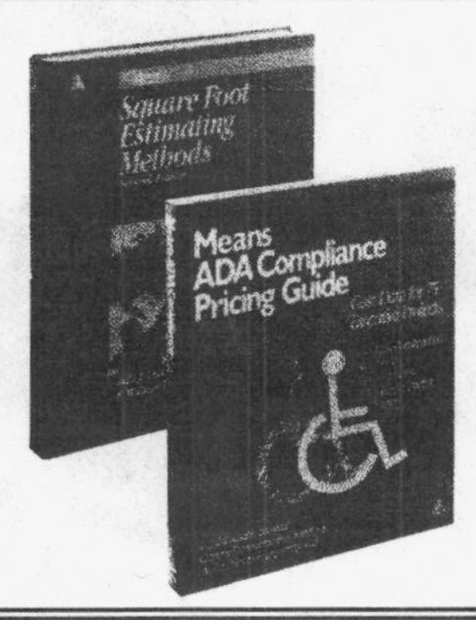

Means ADA Compliance Pricing Guide

Gives you detailed cost estimates for budgeting modification projects, including estimates for each of 260 alternates. You get the 75 most commonly needed modifications for ADA compliance, each an assembly estimate with detailed cost breakdown.

$72.95 per copy
Over 350 pages, illustrated, Softcover
Catalog No. 67310

Reference Books

For more information visit Means Web Site at www.rsmeans.com

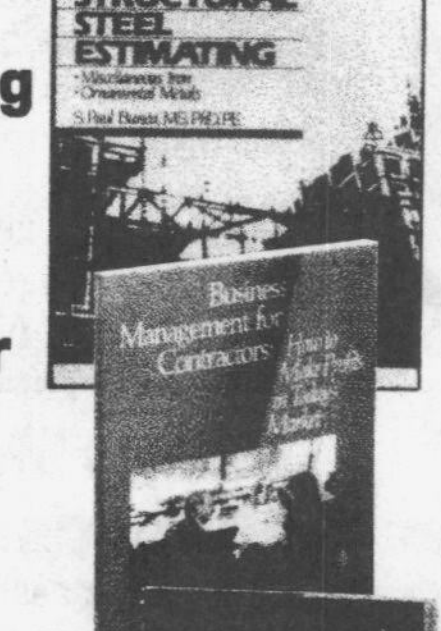

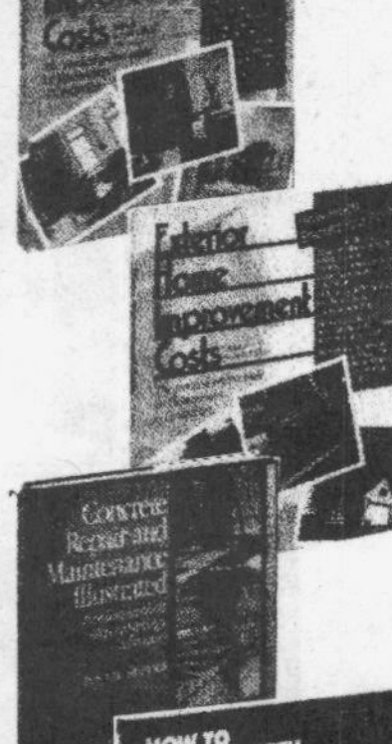
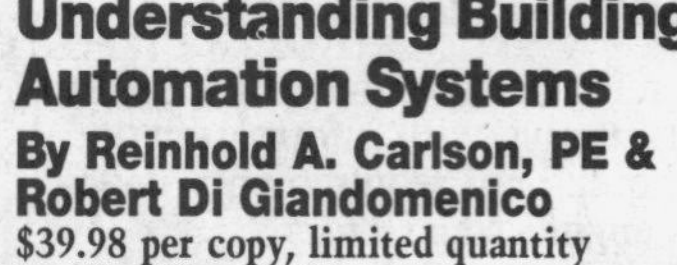

Means Unit Price Estimating Methods
2nd Edition
$59.95 per copy
Catalog No. 67303

Legal Reference for Design & Construction
By Charles R. Heuer, Esq., AIA
Now $54.98 per copy, limited quantity
Catalog No. 67266

Means Plumbing Estimating Methods
New 2nd Edition
By Joseph J. Galeno & Sheldon T. Greene
$59.95 per copy
Catalog No. 67283A

Structural Steel Estimating
By S. Paul Bunea, PhD
Now $39.98 per copy, limited quantity
Catalog No. 67241

Business Management for Contractors
How to Make Profits in Today's Market
By Paul J. Cook
Now $17.98 per copy, limited quantity
Catalog No. 67250

Understanding Legal Aspects of Design/Build
By Timothy R. Twomey, Esq., AIA
$79.95 per copy
Catalog No. 67259

Risk Management for Building Professionals
By Thomas E. Papageorge, RA
Now $29.98 per copy, limited quantity
Catalog No. 67254

Contractor's Business Handbook
By Michael S. Milliner
Now $21.48 per copy, limited quantity
Catalog No. 67255

Basics for Builders: How to Survive and Prosper in Construction
By Thomas N. Frisby
$34.95 per copy
Catalog No. 67273

Successful Interior Projects Through Effective Contract Documents
By Joel Downey & Patricia K. Gilbert
Now $34.98 per copy, limited quantity
Catalog No. 67313

Understanding Building Automation Systems
By Reinhold A. Carlson, PE & Robert Di Giandomenico
$39.98 per copy, limited quantity
Catalog No. 67284

Illustrated Construction Dictionary, Condensed
$59.95 per copy
Catalog No. 67282

Managing Construction Purchasing
By John G. McConville, CCC, CPE
Now $31.48 per copy, limited quantity
Catalog No. 67302

Hazardous Material & Hazardous Waste
By Francis J. Hopcroft, PE, David L. Vitale, M. Ed., & Donald L. Anglehart, Esq.
Now $44.98 per copy, limited quantity
Catalog No. 67258

Fundamentals of the Construction Process
By Kweku K. Bentil, AIC
Now $34.98 per copy, limited quantity
Catalog No. 67260

Construction Delays
By Theodore J. Trauner, Jr., PE, PP
$59.95 per copy
Catalog No. 67278

Basics for Builders: Framing & Rough Carpentry
By Scot Simpson
$24.95 per copy
Catalog No. 67298

Interior Home Improvement Costs
New 7th Edition
$19.95 per copy
Catalog No. 67308B

Exterior Home Improvement Costs
New 7th Edition
$19.95 per copy
Catalog No. 67309B

Concrete Repair and Maintenance Illustrated
By Peter H. Emmons
$69.95 per copy
Catalog No. 67146

How to Estimate with Metric Units
Now $24.98 per copy, limited quantity
Catalog No. 67304

Developing Facility Assessment Programs

This two-day program concentrates on the management process required for planning, conducting, and documenting the physical condition and functional adequacy of buildings and other facilities. Regular facilities condition inspections are one of the facility department's most important duties. However, gathering reliable data hinges on the design of the overall program used to identify and gauge deferred maintenance requirements. Knowing where to look . . . and reporting results effectively are the keys. This seminar is designed to give the facility executive essential steps for conducting facilities inspection programs.

Inspection Program Requirements • Where is the deficiency? • What is the nature of the problem? • How can it be remedied? • How much will it cost in labor, equipment and materials? • When should it be accomplished? • Who is best suited to do the work?

Note: Because of its management focus, this course will not address trade practices and procedures.

Repair and Remodeling Estimating

Repair and remodeling work is becoming increasingly competitive as more professionals enter the market. Recycling existing buildings can pose difficult estimating problems. Labor costs, energy use concerns, building codes, and the limitations of working with an existing structure place enormous importance on the development of accurate estimates. Using the exclusive techniques associated with Means' widely acclaimed **Repair & Remodeling Cost Data**, this seminar sorts out and discusses solutions to the problems of building alteration estimating. Attendees will receive two intensive days of eye-opening methods for handling virtually every kind of repair and remodeling situation . . . from demolition and removal to final restoration.

Mechanical and Electrical Estimating

This seminar is tailored to fit the needs of those seeking to develop or improve their skills and to have a better understanding of how mechanical and electrical estimates are prepared during the conceptual, planning, budgeting and bidding stages. Learn how to avoid costly omissions and overlaps between these two interrelated specialties by preparing complete and thorough cost estimates for both trades. Featured are order of magnitude, assemblies, and unit price estimating. In combination with the use of **Means Mechanical Cost Data**, **Means Plumbing Cost Data** and **Means Electrical Cost Data**, this seminar will ensure more accurate and complete Mechanical/Electrical estimates for both unit price and preliminary estimating procedures.

Unit Price Estimating

This seminar shows how today's advanced estimating techniques and cost information sources can be used to develop more reliable unit price estimates for projects of any size. It demonstrates how to organize data, use plans efficiently, and avoid embarrassing errors by using better methods of checking.

You'll get down-to-earth help and easy-to-apply guidance for:

- making maximum use of construction cost information sources
- organizing estimating procedures in order to save time and reduce mistakes
- sorting out and identifying unusual job requirements to improve estimating accuracy.

Square Foot Cost Estimating

Learn how to make better preliminary estimates with a limited amount of budget and design information. You will benefit from examples of a wide range of systems estimates with specifications limited to building use requirements, budget, building codes, and type of building. And yet, with minimal information, you will obtain a remarkable degree of accuracy.

Workshop sessions will provide you with model square foot estimating problems and other skill-building exercises. The exclusive Means building assemblies square foot cost approach shows how to make very reliable estimates using preliminary budget and design information.

Scheduling and Project Management

This seminar helps you successfully establish project priorities, develop realistic schedules, and apply today's advanced management techniques to your construction projects. Hands-on exercises familiarize participants with network approaches such as the Critical Path Method. Special emphasis is placed on cost control, including use of computer-based systems. Through this seminar you'll perfect your scheduling and management skills, ensuring completion of your projects *on time* and *within budget*. Includes hands-on application of **Means Building Construction Cost Data.**

Facilities Maintenance and Repair Estimating

With our Facilities Maintenance and Repair Estimating seminar, you'll learn how to plan, budget, and estimate the cost of ongoing and preventive maintenance and repair for all your buildings and grounds. Based on R.S. Means' groundbreaking cost estimating book, this two-day seminar will show you how to decide to either contract out or retain maintenance and repair work in-house. In addition, you'll learn how to prepare budgets and schedules that help cut down on unplanned and costly emergency repair projects. Facilities Maintenance and Repair Estimating crystallizes what facilities professionals have learned over the years, but never had time to organize or document. This program covers a variety of maintenance and repair projects, from underground storage tank removal, roof repair and maintenance, exterior wall renovations, and energy source conversions, to service upgrades and estimating energy-saving alternatives.

Managing Facilities Construction and Maintenance

In you're involved in new facility construction, renovation or maintenance projects and are concerned about getting quality work done on time and on or below budget, in Means' seminar **Managing Facilities Construction and Maintenance** you'll learn management techniques needed to effectively plan, organize, control and get the most out of your limited facilities resources.

Learn how to develop budgets, reduce expenditures and check productivity. With the knowledge gained in this course, you'll be better prepared to successfully sell accurate project budgets, timing and manpower needs to senior management . . . plus understand how to evaluate the impact of today's facility decisions on tomorrow's budget.

Call 1-800-448-8182 for more information

Seminars

For more information
visit Means Web Site
at www.rsmeans.com

2001 Means Seminar Schedule

Location	Dates
Las Vegas, NV	March 12 - 15
Washington, DC	April 23 - 26
Denver, CO	May 21-24
San Francisco, CA	June 11 - 14
Cape Cod, MA	September TBD
Washington, DC	October 1 - 4
San Diego, CA	October TBD
Atlantic City, NJ	November TBD
Orlando, FL	November 5-8

Note: Call for exact dates and detail

Registration Information

Register Early... Save up to $150! Register 30 days before the start date of a seminar and save $150 off your total fee. ***Note: This discount can be applied only once per order.***

How to Register Register by phone today! Means toll-free number for making reservations is: **1-800-448-8182.**

Individual Seminar Registration Fee $875 To register by mail, complete the registration form and return with your full fee to: Seminar Division, R.S. Means Company, Inc., 63 Smiths Lane, Kingston, MA 02364.

Federal Government Pricing All Federal Government employees save 25% off regular seminar price. Other promotional discounts cannot be combined with Federal Government discount.

Team Discount Program Two to four seminar registrations: $760 per person—Five or more seminar registrations: $710 per person—Ten or more seminar registrations: Call for pricing.

Consecutive Seminar Offer One individual signing up for two separate courses at the same location during the designated time period pays only $1,400. You get the second course for only $525 (**a 40% discount**). Payment must be received at least ten days prior to seminar dates to confirm attendance.

Refunds Cancellations will be accepted up to ten days prior to the seminar start. There are no refunds for cancellations postmarked later than ten working days prior to the first day of the seminar. A $150 processing fee will be charged for all cancellations. Written notice or telegram is required for all cancellations. Substitutions can be made at any time before the session starts. **No-shows are subject to the full seminar fee.**

AACE Approved Courses The R.S. Means Construction Estimating and Management Seminars described and offered to you here have each been approved for 14 hours (1.4 recertification credits) of credit by the AACE International Certification Board toward meeting the continuing education requirements for re-certification as a Certified Cost Engineer/Certified Cost Consultant.

AIA Continuing Education R.S. Means is registered with the AIA Continuing Education System (AIA/CES) and is committed to developing quality learning activities in accordance with the CES criteria. R.S. Means seminars meet the AIA/CES criteria for Quality Level 2. AIA members will receive (28) learning units (LUs) for each two day R.S. Means Course.

Daily Course Schedule The first day of each seminar session begins at 8:30 A.M. and ends at 4:30 P.M. The second day is 8:00 A.M.–4:00 P.M. Participants are urged to bring a hand-held calculator since many actual problems will be worked out in each session.

Continental Breakfast Your registration includes the cost of a continental breakfast, a morning coffee break, and an afternoon break. These informal segments will allow you to discuss topics of mutual interest with other members of the seminar. (You are free to make your own lunch and dinner arrangements.)

Hotel/Transportation Arrangements R.S. Means has arranged to hold a block of rooms at each hotel hosting a seminar. To take advantage of special group rates when making your reservation be sure to mention that you are attending the Means Seminar. You are of course free to stay at the lodging place of your choice. (**Hotel reservations and transportation arrangements should be made directly by seminar attendees.)**

Important Class sizes are limited, so please register as soon as possible.

Registration Form

Call 1-800-448-8182 x5115/to register or FAX 1-800-632-6732. Visit our web site www.rsmeans.com

Please register the following people for the Means Construction Seminars as shown here. Full payment or deposit is enclosed, and we understand that we must make our own hotel reservations if overnight stays are necessary.

☐ Full payment of $ ______________________ enclosed.

☐ Bill me

Name of Registrant(s)
(To appear on certificate of completion)

____

____

____

P.O. #: ______________________________

GOVERNMENT AGENCIES MUST SUPPLY PURCHASE ORDER NUMBER

Firm Name ______________________________

Address ______________________________

City/State/Zip ______________________________

Telephone No. ______________ Fax No. ______________

E-Mail Address ______________________________

Charge our registration(s) to: ☐ MasterCard ☐ VISA ☐ American Express ☐ Discover

Account No. ______________ Exp. Date ______________

Cardholder's Signature ______________________________

Seminar Name City Dates

____

____

Please mail check to: R.S. MEANS COMPANY, INC., 63 Smiths Lane, P.O. Box 800, Kingston, MA 02364 USA

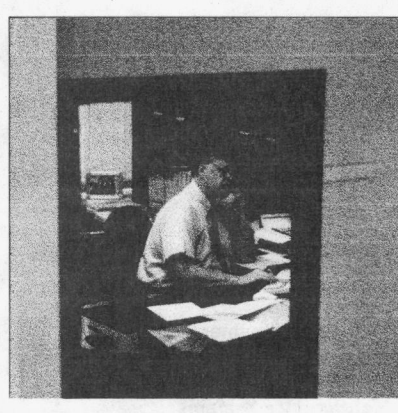

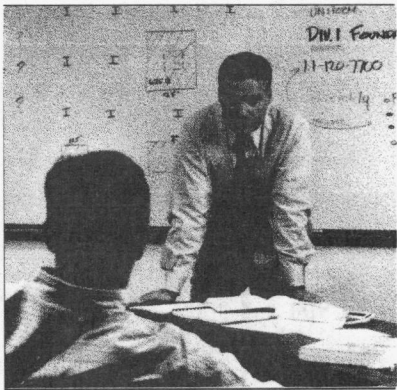

MeansData™

CONSTRUCTION COSTS FOR SOFTWARE APPLICATIONS

Your construction estimating software is only as good as your cost data.

Software Integration

A proven construction cost database is a mandatory part of any estimating package. We have linked MeansData™ directly into the industry's leading software applications. The following list of software providers can offer you MeansData™ as an added feature for their estimating systems. Visit them on-line at ***www.rsmeans.com/demo/*** for more information and free demos. Or call their numbers listed below.

3D International
713-871-7000 budwit@3di.com

ACT
Applied Computer Technologies
Facility Management Software
919-859-1335 patw@srs.net

AEPCO, Inc.
301-670-4642 blueworks@aepco.com

ArenaSoft ESTIMATING
888-370-8806 info@arenasoft.com

Ares Corporation
650-401-7100
sales@arescorporation.com

ASSETWORKS, Inc.
Facility Management Software
800-659-9001 info@assetworks.com

Benchmark, Inc.
800-393-9193 sales@benchmark-inc.com

BSD
Building Systems Design, Inc.
888-273-7638 bsd@bsdsoftlink.com

cManagement
800-945-7093 sales@cmanagement.com

CDCI
Construction Data Controls, Inc.
800-285-3929 sales@cdci.com

CMS
Computerized Micro Solutions
800-255-7407 cms@proest.com

CONAC GROUP
800-663-2338 sales@concac.com

ESTIMATING SYSTEMS, Inc.
800-967-8572 pulsar@capecod.net

G2 Estimator
A Div. of Valli Info. Syst., Inc.
800-657-6312 info@g2estimator.com

GEAC COMMERCIAL SYSTEMS, Inc.
800-554-9865 info@geac.com

GRANTLUN CORPORATION
480-897-7750 support@grantlun.com

HCI SYSTEMS, Inc.
800-750-4424 info@hcisystems.com

IQ BENECO
801-565-1122 mbrown@beneco.com

Lugs International
888-682-5573 info@lugs.com

MC²
Management Computer Controls
800-225-5622 vkeys@mc2-ice.com

PRISM COMPUTER CORPORATION
Facility Management Software
800-774-7622 nalbadri@prismcc.com

PYXIS TECHNOLOGIES
888-841-0004 ldenigris@ebid98.com

QUEST SOLUTIONS, Inc.
800-452-2342 info@questsolutions.com

RICHARDSON ENGINEERING SERVICES, Inc.
480-497-2062 info@resi.net

SANDERS SOFTWARE, Inc.
800-280-9760 hsander@vallnet.com

STN, Inc.
Workline Maintenance Systems
800-321-1969 ask@stn.com

TIMBERLINE SOFTWARE CORP.
800-628-6583
product.info@timberline.com

TMA SYSTEMS, Inc.
Facility Management Software
800-862-1130 sales@tmasys.com

US COST, Inc.
800-372-4003
usc-software@compuserve.com

VERTIGRAPH, Inc.
800-989-4243
info-request@vertigraph.com

WENDLWARE
714-895-7222 sales@corecon.com

WINESTIMATOR, Inc.
800-950-2374 sszabo@winest.com

DemoSource™

One-stop shopping for the latest cost estimating software for just $19.95. This evaluation tool includes product literature and demo diskettes for ten or more estimating systems, all of which link to MeansData™. **Call 1-800-334-3509 to order.**

FOR MORE INFORMATION ON ELECTRONIC PRODUCTS CALL 1-800-448-8182 OR FAX 1-800-632-6732.

For more information
visit Means Web Site
at www.rsmeans.com

New Titles

Means Illustrated Construction Dictionary, 3rd Edition

New Updated Edition with interactive CD-ROM

Long regarded as the Industry's finest, the *Means Illustrated Construction Dictionary* is now even better. With the addition of over 1,000 new terms and hundreds of new illustrations, it is the clear choice for the most comprehensive and current information.

The companion CD-ROM that comes with this new edition adds many extra features: larger graphics, expanded definitions, and links to both CSI MasterFormat numbers and product information.

- 19,000 construction words, terms, phrases, symbols, weights, measures, and equivalents
- 1,000 new entries
- 1,200 helpful illustrations
- Contemporary coverage of professional & trade terms
- Easy-to-use format, with thumbtabs

$99.95 per copy
Over 790 pages, illustrated, Hardcover
Catalog No. 67292A

Historic Preservation: Project Planning & Estimating

- ***Managing Historic Restoration, Rehabilitation and Preservation Building Projects and***
- ***Determining and Controlling Their Costs***

By Swanke Hayden Connell Architects

- Applicable standards and financing
- Site survey and documentation methods
- Protecting finishes and features
- Hazardous materials
- Identifying and qualifying specialty contractors
- Machanical/Electrical systems upgrades
- Restoring old building materials
- How to assemble a reliable, detailed estimate and budget for future maintenance

$99.95 per copy
Over 450 pages, Hardcover
Catalog No. 67323

Facilities Operations & Engineering Reference

An all-in-one technical reference for planning & managing facility projects & solving day-to-day operations problems

Ten major sections address:

- **Management Skills**
- **Economics** (budgeting/cost control, financial analysis, VE, etc.)
- **Civil Engineering & Construction Practices**
- **Maintenance** (detailed staffing guidance and job descriptions, CMMS, planning/scheduling, training, work orders, preventive/predictive maintenance)
- **HVAC & Energy Efficiencies** (optimizing energy use: heating, cooling, lighting, water)
- **Mechanical and Electrical Engineering**
- **Instrumentation & Controls**
- **Environment, Health & Safety**

Produced jointly by R.S. Means and the Association for Facilities Engineering.

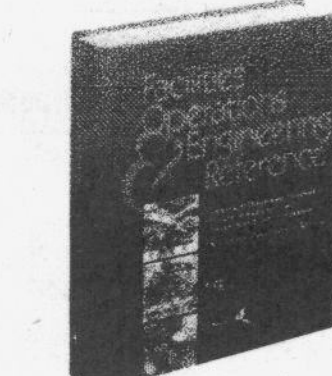

$99.95 per copy
Over 700 pages, illustrated, Hardcover
Catalog No. 67318

ORDER TOLL FREE 1-800-334-3509
OR FAX 1-800-632-6732.

2001 Order Form

Qty.	Book No.	COST ESTIMATING BOOKS	Unit Price	Total
	60061	Assemblies Cost Data 2001	$149.95	
	60011	Building Construction Cost Data 2001	89.95	
	61011	Building Const. Cost Data–Looseleaf Ed. 2001	116.95	
	63011	Building Const. Cost Data–Metric Version 2001	89.95	
	60221	Building Const. Cost Data–Western Ed. 2001	89.95	
	60111	Concrete & Masonry Cost Data 2001	83.95	
	50141	Construction Cost Indexes 2001	198.00	
	60141A	Construction Cost Index–January 2001	49.50	
	60141B	Construction Cost Index–April 2001	49.50	
	60141C	Construction Cost Index–July 2001	49.50	
	60141D	Construction Cost Index–October 2001	49.50	
	60311	Contr. Pricing Guide: Framing/Carpentry 2001	36.95	
	60331	Contr. Pricing Guide: Resid. Detailed 2001	36.95	
	60321	Contr. Pricing Guide: Resid. Sq. Ft. 2001	39.95	
	64021	ECHOS Assemblies Cost Book 2001	149.95	
	64011	ECHOS Unit Cost Book 2001	99.95	
	54001	ECHOS (Combo set of both books)	214.95	
	60231	Electrical Change Order Cost Data 2001	89.95	
	60031	Electrical Cost Data 2001	89.95	
	60201	Facilities Construction Cost Data 2001	219.95	
	60301	Facilities Maintenance & Repair Cost Data 2001	199.95	
	60161	Heavy Construction Cost Data 2001	89.95	
	63161	Heavy Const. Cost Data–Metric Version 2001	89.95	
	60091	Interior Cost Data 2001	89.95	
	60121	Labor Rates for the Const. Industry 2001	199.95	
	60181	Light Commercial Cost Data 2001	79.95	
	60021	Mechanical Cost Data 2001	89.95	
	60151	Open Shop Building Const. Cost Data 2001	89.95	
	60211	Plumbing Cost Data 2001	89.95	
	60041	Repair and Remodeling Cost Data 2001	79.95	
	60171	Residential Cost Data 2001	79.95	
	60281	Site Work & Landscape Cost Data 2001	89.95	
	60051	Square Foot Costs 2001	99.95	
		REFERENCE BOOKS		
	67147A	ADA in Practice	72.95	
	67310	ADA Pricing Guide	72.95	
	67298	Basics for Builders: Framing & Rough Carpentry	24.95	
	67273	Basics for Builders: How to Survive and Prosper	34.95	
	67307	Basics for Builders: Plan Reading & Takeoff	35.95	
	67261	Bldg. Prof. Guide to Contract Documents-3rd Ed.	64.95	
	67312	Building Spec Homes Profitably	29.95	
	67250	Business Management for Contractors	17.98	
	67146	Concrete Repair & Maintenance Illustrated	69.95	
	67278	Construction Delays	59.95	
	67255	Contractor's Business Handbook	21.48	
	67314	Cost Planning & Est. for Facil. Maint.	82.95	
	67317A	Cyberplaces: The Internet Guide-2nd Ed.	59.95	
	67230A	Electrical Estimating Methods–2nd Ed.	64.95	
	64777	Environmental Remediation Est. Methods	99.95	
	67160	Estimating for Contractors	35.95	
	67276	Estimating Handbook	99.95	
	67249	Facilities Maintenance Management	86.95	

Qty.	Book No.	REFERENCE BOOKS (Cont.)	Unit Price	Total
	67246	Facilities Maintenance Standards	$159.95	
	67264	Facilities Manager's Reference	86.95	
	67318	Facilities Operations & Engineering Reference	99.95	
	67301	Facilities Planning & Relocation	109.95	
	67231	Forms for Building Const. Professional	94.95	
	67288	Forms for Contractors	79.95	
	67260	Fundamentals of the Construction Process	34.98	
	67258	Hazardous Material & Hazardous Waste	44.98	
	67148	Heavy Construction Handbook	74.95	
	67323	Historic Preservation: Proj. Planning & Est.	99.95	
	67308C	Home Improvement Costs–Interior Projects	19.95	
	67309C	Home Improvement Costs–Exterior Projects	19.95	
	67304	How to Estimate with Metric Units	24.98	
	67306	HVAC: Design Criteria, Options, Select.–2nd Ed.	84.95	
	67281	HVAC Systems Evaluation	84.95	
	67282	Illustrated Construction Dictionary, Condensed	59.95	
	67292A	Illustrated Construction Dictionary, w/CD-ROM	99.95	
	67295A	Landscape Estimating–3rd Ed.	62.95	
	67266	Legal Reference for Design & Construction	54.98	
	67299	Maintenance Management Audit	32.48	
	67302	Managing Construction Purchasing	31.48	
	67294	Mechanical Estimating–2nd Ed.	64.95	
	67245A	Planning and Managing Interior Projects-2nd Ed.	69.95	
	67283A	Plumbing Estimating Methods-2nd Ed.	59.95	
	67326	Preventive Maint. Guidelines for School Facil.	149.95	
	67236A	Productivity Standards for Constr.–3rd Ed.	159.95	
	67247A	Project Scheduling & Management for Constr.	64.95	
	67262	Quantity Takeoff for Contractors	17.98	
	67265A	Repair & Remodeling Estimating-3rd Ed.	69.95	
	67322	Residential & Light Commercial Const. Stds.	59.95	
	67254	Risk Management for Building Professionals	29.98	
	67291	Scheduling Manual–3rd Ed.	64.95	
	67145A	Square Foot Estimating Methods–2nd Ed.	69.95	
	67287	Successful Estimating Methods	64.95	
	67313	Successful Interior Projects	34.98	
	67233	Superintending for Contractors	35.95	
	67321	Total Productive Facilities Management	79.95	
	67284	Understanding Building Automation Systems	39.98	
	67259	Understanding Legal Aspects of Design/Build	79.95	
	67303	Unit Price Estimating Methods–2nd Ed.	59.95	
	67319	Value Engineering: Practical Applications	79.95	
		MA residents add 5% state sales tax		
		Shipping & Handling**		
		Total (U.S. Funds)*		

Prices are subject to change and are for U.S. delivery only. *Canadian customers may call for current prices. **Shipping & handling charges: Add 7% of total order for check and credit card payments. Add 9% of total order for invoiced orders.

Send Order To: **ADDV-1001**

Name (Please Print) ______________________

Company ______________________

☐ **Company**
☐ **Home** Address ______________________

____

City/State/Zip ______________________

Phone # ______________ P.O. # ______________

Mail To: **R.S. Means Company, Inc.,** P.O. Box 800, Kingston, MA 02364-0800

(Must accompany all orders being billed)